AUTOMOTIVE ENGINE PERFORMANCE

TUNEUP, TESTING, AND SERVICE

VOLUME 2: PRACTICE MANUAL

KEN LAYNE CMAT, SAE

AUTOMOTIVE ENGINE PERFORMANCE

TUNEUP, TESTING, AND SERVICE

VOLUME 2: PRACTICE MANUAL

Second Edition

PRENTICE HALL CAREER & TECHNOLOGY
Englewood Cliffs, New Jersey 07632

Library of Congress Cataloging-in-Publication Data

Layne, Ken.
Automotive engine performance.
CONTENTS: v. 1. Text—v. 2. Practice Manual.
Includes index.
1. Automobiles—Motors. 2. Automobiles—Motors—Maintenance and repair. I. Title.
TD210.L38 1993 629.25'04 92-24357
ISBN 0-13-061177-8 (v. 2)

Acquisitions Editor: Robert Koehler
Cover Designer: Marianne Frasco
Production: Ruth Cottrell
Manufacturing Buyer: Ed O'Dougherty
Prepress Buyer: Ilene Levy

Prentice-Hall, Inc.
A Paramount Communications Company
Englewood Cliffs, New Jersey 07632

Printed in the United States of America.

10 9 8 7 6 5 4 3

ISBN 0-13-061177-8

Prentice-Hall International (UK) Limited, *London*
Prentice-Hall of Australia Pty. Limited, *Sydney*
Prentice-Hall Canada, Inc., *Toronto*
Prentice-Hall Hispanoamericana, S.A., *Mexico*
Prentice-Hall of India Private Limited, *New Delhi*
Prentice-Hall of Japan, Inc., *Tokyo*
Simon & Schuster Asia Pte. Ltd., *Singapore*
Editora Prentice-Hall do Brasil, Ltda., *Rio de Janeiro*

DEDICATION

A lot of writers dedicate books to their wives, husbands, or other family members whose forbearance has helped them through the trials and tribulations of getting a book into print. I think Roseann would find that patronizing. In fact, my wife is a far better teacher than I. She is also a skilled librarian. Working in both capacities, she contributed many of the chapter review questions to this text and catalogued several thousand illustrations. If you find value in these areas, you may thank the lady.

The acknowledgments page of this book lists the contributions of many people and companies who helped make this work possible. I want to extend special thanks to two people. Bob Mahaffay, former engineering director of the Gousha/Chek-Chart company, has done more in his career to aid the automobile service profession than any other individual that I can think of. Don Nilson, retired head of the automotive technology department at Las Positas College in Livermore, California, is a "teacher's teacher." His career defines "professionalism" in automobile service. Both of these gentlemen have been looking over my shoulder and keeping me straight in automobile service publications for more years than I care to recount. Without their assistance, this book would not exist.

Additionally, I must acknowledge and thank one of the outstanding college automotive technology programs in the country: the Auto Tech Department at DeAnza College in Cupertino, California. This school has been a leader in the automotive service education field for well over 20 years even though it provides me a classroom forum to hold forth on the subject of automotive electronics.

Any textbook, any educational program, is less than half complete without skilled teachers. The best information in print and the best training plans are enhanced twofold by a teacher who can bring a subject to life and demonstrate its importance. Without teachers, there is no learning; so this book is dedicated to:

Bill Abbott
Paul "X" Derry
Paul Hebert
Hank Nesel
Karl Pape
and especially,
Jerry Talbert . . . teachers.

The information in this book is based on the best available sources from the automobile industry. Persons using this book are cautioned to refer to vehicle and equipment manufacturers' procedures and specifications for specific service information. Neither the publisher nor the author can be responsible for changes in procedures and specifications made by carmakers or parts and equipment suppliers.

Persons using this book also are advised to observe all safety precautions in this text and refer to manufacturers' procedures for specific vehicle and personnel safety direction. The publisher and author are not responsible for anyone's failure to understand and follow *vehicle manufacturers' complete and specific service instructions*.

BRIEF CONTENTS

EXPANDED CONTENTS

PART ONE

PART SIX

PREFACE

No technical service profession demands more knowledge and higher skills from its members than the automobile service profession. In the early days of the automobile, a person could do a lot of service jobs with a good measure of horse sense and a box full of blacksmith's tools. Those days are long gone.

We cannot underestimate the knowledge and skills of an old-time mechanic, however. Professional mechanics of past generations laid the foundation for modern automobile service. A professional service technician today, though, must have equal measures of:

1. Knowledge of automobile system operation

2. The service skills necessary to keep those systems operating correctly.

TWO VOLUMES—SYSTEM DESCRIPTION AND SERVICE

Because a professional technician must know how systems operate and how to service them, we have presented this body of information in two volumes: AUTOMOTIVE ENGINE PERFORMANCE: TUNEUP, TESTING, AND SERVICE, the first volume, will help you learn how engine combustion control systems operate. The second volume, AUTOMOTIVE ENGINE PERFORMANCE: TUNEUP, TESTING, AND SERVICE PRACTICE MANUAL, will help you learn the service skills that are the true measure of a master mechanic.

AUTOMOTIVE ENGINE PERFORMANCE explains the operation of fuel, ignition, emission control, and electrical and electronic systems on late-model vehicles. It progresses from the fundamentals of combustion control through specific examples of these systems from major carmakers. These descriptions emphasize the similarities among various systems and the basic combustion control principles used by all manufacturers.

The Practice Manual provides the testing and repair instructions that will be the foundation of your professional service skills. This volume contains step-by-step procedures similar to those you will find in service manuals from carmakers and service guide publishers. These instructions will help you to understand professional reference manuals and use them as basic tools in your service work. Moreover, the Practice Manual explains the *reasons* for different service procedures and the common features of all carmakers' service methods.

SERVICE INSTRUCTIONS COMPLEMENT INDUSTRY REFERENCE MANUALS

We do not have the room in this book to include every manufacturer's specific test and service procedures. No textbook can do that. The service procedures in the Practice Manual are basic guidelines for engine performance testing and service. They will familiarize you with the principles and common points of engine maintenance. These service instructions will also acquaint you with the ways in which carmakers and independent manual publishers produce service procedures and specifications.

Reference manuals are essential tools for service of late-model cars. The second chapter of the Practice Manual explains how to use various service publications. For every job that you do, you will need the following information from factory shop manuals or independent service guides:

- Specifications for parts and adjustments
- Step-by-step test and service procedures
- Carmakers' preventive maintenance schedules
- Test and service equipment operating instructions
- Flat-rate labor time standards.

The basic service skills that you learn as you study engine performance service will lay the foundation for your career as a professional automobile service technician.

SYSTEM ENGINEERING AND SYSTEM SERVICE

This text applies a system-engineering and system-service viewpoint to engine performance maintenance. The first chapter of AUTOMOTIVE ENGINE PERFORMANCE lays this foundation. On today's cars, fuel delivery, ignition, and emission control are fully integrated combustion control systems, regulated by on-board computers. You can't think of emission controls as items that are separate from fuel and ignition components. Also, when you work on fuel and ignition components, you must know how a change in one part or system will affect other systems. In addition, you must know how the operation of any one part will affect overall performance, economy, and emission control.

Don't get the idea that a good mechanic ever ignored any fuel, ignition, or emission system component when servicing an engine for maximum efficiency. The point is that on today's cars, you can't think about them separately for a moment. You have to understand total engine operation. Engine performance service, or "tune-up" has always been an interesting and profitable automotive service field. Whether you specialize in this area or in general automobile service, this book will help you understand modern engine service.

Each volume is divided into six major parts that contain chapters on related subjects and engine systems:

- An introduction to the principles of engine performance and to common shop practices, tools, and materials
- Basic engine lubrication, ventilation, cooling, and exhaust system operation and maintenance
- Engine electrical system operation and maintenance
- Ignition system operation and service
- Fuel systems and emission controls
- Electronic engine control systems

These divisions follow the traditional viewpoint toward comprehensive engine performance service. This arrangement also highlights the fact that all engine systems on today's cars are fully integrated through electronic controls.

INDUSTRY TESTING AND TRAINING STANDARDS

The arrangement of the subjects in these volumes covers the skill requirements outlined in the specifications for the Automotive Service Excellence (ASE) engine performance certification test. Review questions at the end of the chapters are written in the style of ASE test questions. These will not only help you review the chapter contents, they will help to prepare you for the actual certification tests.

Additionally, the organization and contents of these volumes are based on automotive industry training standards contained in American National Standards Institute (ANSI) publication D18.1-1980. This is your assurance that this text covers the knowledge and skills that the service industry requires of its members.

FULLY ILLUSTRATED TEXT

The two volumes contain almost 2,000 illustrations. Many are original drawings and photographs, prepared expressly for these volumes. Others have been selected from carmakers' late-model service manuals and adapted to illustrate the contents of these books. The result is a fully integrated combination of text and illustrations that *shows and tells* you how vehicle systems operate and how to service them.

The Practice Manual contains several step-by-step photographic overhaul procedures for alternators, starter motors, carburetors, and ignition distributors. These procedures provide detailed illustrations and instructions for typical methods required to service these vital engine performance components.

Major sections of the chapters in the Practice Manual also explain the use of electrical diagrams, troubleshooting charts, and carmakers' illustrated service procedures.

OTHER FEATURES OF THESE VOLUMES

In addition to the fully illustrated text, these volumes contain the following features that will help you learn and practice professional engine performance service:

- *Safety*—Safety is good business, and following all necessary precautions will ensure the safety of yourself, your shop, and your customers. The first chapter of the Practical Manual summarizes the general safety practices for engine performance service. Specific CAUTIONS and WARNINGS appear in the procedures of other chapters to help you do various repair jobs safely.
- *Learning Objectives*—Each chapter begins with a list of basic knowledge or skills that you should acquire from studying the chapter. Again, these objectives follow the general requirements of ASE certification testing and industry training standards.
- *Technical Terms*—Key words and phrases important to a service technician are printed in boldface type. These terms are explained in the text and collected in the glossary at the end of each volume. Studying and understanding these will help you build the vocabulary of a professional mechanic.
- *Essays on Related Topics*—Short essays on topics related to engine performance service are included throughout the first volume. These short subjects deal with specific service tips, system designs, and the historical development of automobile engine systems. They will add depth to your understanding of the automobile industry and help you in the service profession.
- *Measurement Data*—An appendix at the end of each volume contains important information on threaded fasteners and on customary and metric measurement standards that you will use in your service work.

NEW FEATURES FOR THE SECOND EDITION

Additions and changes to these volumes for the second edition consist basically of bringing the technology up to date. Engine control systems and engine performance service have evolved rapidly during the 1980's. Fuel injection became the "conventional" fuel delivery system for most automobiles, and direct ignition systems (DIS) are showing the same trend in the 1990's. The self-diagnostic capabilities of on-board computer systems also have improved dramatically. The new on-board diagnostic II (OBD II) systems of 1994 models promise standardization of basic test methods among all carmakers.

The second edition of these volumes offers:

- Increased coverage of DIS installations on domestic engines.

- Improved explanations of computer principles shared by all carmakers.
- Expanded overviews of fuel injection systems and basic service procedures for domestic and imported vehicles.
- Comprehensive insight into the self-diagnostic methods used on GM, Ford, and Chrysler vehicles.

ACKNOWLEDGMENTS

The author and the publisher wish to thank the vehicle manufacturers and other companies in the automotive service industry that contributed information and illustrations for this book. Automobile service is a highly technical, rapidly advancing profession. This book would not have been possible without the assistance of the following companies:

American Isuzu Motors Inc.
Atlas Supply Company
Automotive Technician Associates, Inc. (An Organization Dedicated to Better Mechanics)
Balco, Inc.
B&M Automotive Products
Champion Spark Plug Company
Chilton Book Company
Chrysler Corporation
Clayton Industries
Federal-Mogul Corporation
FEL-PRO Incorporated
Ford Motor Company (Ford Parts and Service Division)
Garrett Air-Research Industrial Division
The Gates Rubber Company
General Motors Corporation:
  AC-Delco Division
  Buick Motor Division
  Cadillac Motor Car Division
  Chevrolet Motor Division
  Delco-Remy Division
  Oldsmobile Division
  Pontiac Motor Division
  Rochester Products Division
  Saginaw Steering Gear Division
Hennessy Industries, Inc.: Coats Diagnostic Division
Industrial Indemnity Company
K-D Manufacturing Company
Kent-Moore Tool Division
Mazda Motor Corporation
McCord Gasket Corporation
Mitchell Manuals, Inc.
Mitsubishi Motor Sales of America, Inc.
Motor Publications (The Hearst Corporation)
Nissan Motor Corporation in U.S.A.
National Institute for Occupational Safety and Health
Porsche + Audi
Prestolite Division of Bunker Ramo-Eltra Corporation (a subsidiary of Allied Corporation)
Snap-on Tools Corporation
Stanley Proto Industrial Tools
Society of Automotive Engineers (SAE)
The L. S. Starrett Company
Sun Electric Corporation
Robert Bosch Sales Corporation (Robert Bosch GmbH)
Volkswagen of America, Inc.
Volvo of America Corporation
Weatherhead Division, Dana Corporation

Additionally, the following individuals contributed information and research materials or reviewed selected parts of the manuscript. The author wishes to thank:

Robert "Charlie" Chapman—Software Engineering, Tujunga, CA
Michael DeMiniconi
Peter Felice—Red Line Synthetic Oil Corporation, Martinez, CA
James Geddes
William J. Hanley
Carlton Hardy—Ford Motor Company Training Center, Milpitas, CA
Bob Kruze—Saturn Corp.
Russ Ostler—Chrysler Corporation, Livermore, CA
Ben Sims—California State Automobile Association
Thomas M. Terrell—Garrett Air-Research Industrial Division, Torrance, CA
Chris Wood—Grant & El Camino Chevron, Mountain View, CA

Photographs for the component overhaul sequences and many other photos are by *Kalton C. Lahue.*

Finally, this text wouldn't exist at all if Ruth Cottrell hadn't put it all together.

PART ONE COMMON SHOP PRACTICES, TOOLS, MATERIALS, AND REPAIR METHODS

INTRODUCTION

This volume contains practical instructions and general procedures for testing and servicing all automobile systems. We can't include every detailed repair procedure for every car on the road, but we can provide basic information to guide you through the important points of most service tasks.

Many testing and repair methods and tools are common to all vehicle systems. The first five chapters of this volume introduce these common shop practices, tools, materials, and service methods. As you continue your study of automotive technology, you will find that the correct methods for cleaning and inspecting parts; handling gaskets, seals, and sealants; and working with tubing, pipe, and hose fittings apply to all vehicle systems. These skills are fundamental parts of your job.

Safety is the most fundamental shop practice for all automobile service. Chapter 1 summarizes basic safety practices for all shop work.

No tools are more important to a skilled technician than reference manuals. Late-model automobiles are too complex to be serviced by guesswork when it comes to specifications and repair procedures. Chapter 2 introduces you to the kinds of specifications and procedures you will use in the service profession, and to the sources of this information in reference manuals.

Chapter 3 outlines the use of common repair materials, such as hoses, tubing, gaskets, seals, adhesives, and sealants. Too often, these items are taken for granted, but understanding and working with them properly can make the difference between doing a job right the first time and doing the job a second time to get it right.

Chapter 4 explains the operation of basic test equipment that you will use for engine performance service. You will use these principles in all service jobs covered in the rest of this book. Chapter 5 explains the principles of logical troubleshooting, as well as the basic procedures for specific engine tests and adjustments.

Everything you learn in the next five chapters will be the foundation for the specific system service instructions in the other parts of this volume.

1 WORKING SAFELY

INTRODUCTION

There are many tasks and many places in an auto repair shop where accidents can occur, but a professional mechanic—a skilled technician—can work an entire lifetime without suffering personal injury or damage to equipment. A professional can do this because a professional *thinks safety*. A professional knows where potential hazards exist and knows the precautions to avoid them.

Safety is smart business. The safe way to do a job is the fastest and most efficient way. Any time lost because of injury to personnel or damage to vehicles and equipment is time in which you can't do your job. Working with unsafe equipment will slow you down and keep you from doing a job correctly; so will using the wrong tool for a job. A professional in any field knows that the right way to do a job is the safe way. He or she makes safety an instinctive part of every job.

This manual contains service and repair instructions for engine combustion control systems. These are guidelines that you will follow as you learn to be a professional mechanic. The procedures in this manual contain specific *CAUTIONS* and *WARNINGS*. *CAUTIONS* alert you to hazards that could damage a vehicle or equipment. *WARNINGS* alert you to hazards that could injure you or other people.

Although certain procedures require specific precautions, there are also many general safety practices that you must follow from the first moment that you walk into a shop. This chapter summarizes the basic information that you need for equipment and personnel safety. Whenever you work with engine combustion control systems, you work with highly flammable fuel and with electrical devices. By following proper precautions, you can work with these materials and parts safely and avoid fire, injury, and equipment damage. Make these safety precautions an instinctive part of every job. Never guess about safe shop practices. If in doubt, ask your instructor before proceeding with a job. Safety shortcuts will cost you in the long run.

GOALS

After studying this chapter, you should be able to:

1. Understand, list, and follow the basic safety principles for the equipment and practices listed below:

- **a.** Personal actions
- **b.** Fire prevention
- **c.** Hoisting and jacking equipment
- **d.** Electric power tools and equipment
- **e.** Compressed air and pneumatic tools
- **g.** Cleaning equipment, solvents, and other chemicals

2. Understand, list, and follow general safety precautions needed when working with engines and with cooling, exhaust, fuel, electrical, and emission systems.

PERSONAL AND GENERAL SHOP SAFETY

The following precautions apply to all activities in the shop.

1. Flammable liquids and combustible materials are present in all auto repair shops. You can minimize fire hazards by *not smoking at any time in the shop area.*

2. Before starting to work, remove all jewelry, such as rings and watches. Remove sweaters and tuck in loose clothing, figure 1-1. If you have long hair, secure it behind your head or wear a suitable hat.

3. Always wear safety glasses in any area or for any job where an eye hazard could exist.

4. If in doubt about the use of any tool or machine, ask your instructor about

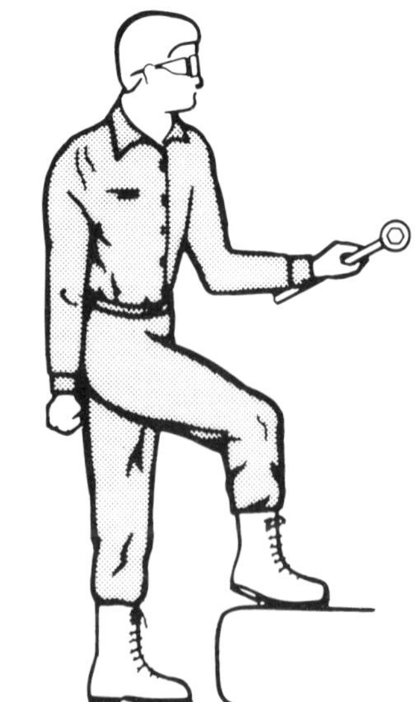

Figure 1-1. Remove jewelry and wear proper clothing and safety glasses when required for any shop work.

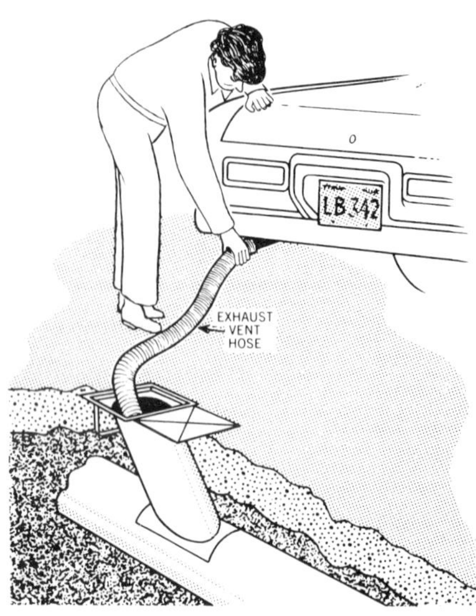

Figure 1-3. Connect vehicle exhausts to the ventilation system before operating the engine in a closed area. (NIOSH)

Figure 1-4. Lift a heavy object by bending at the knees and raising it straight up. Do not lift with your back.

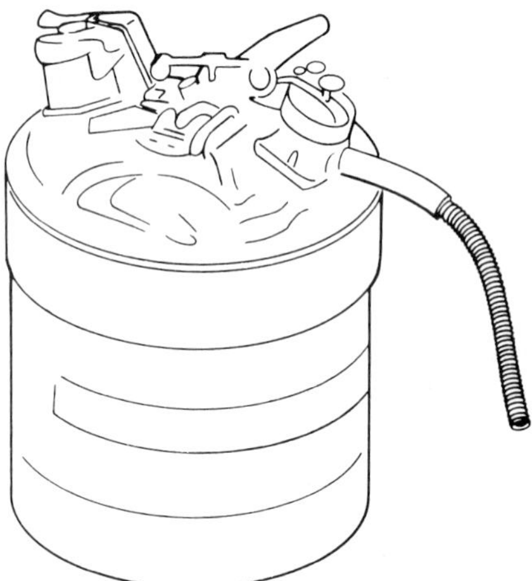

Figure 1-5. Gasoline and other flammable liquids must be stored in closed, unvented, clearly labeled containers.

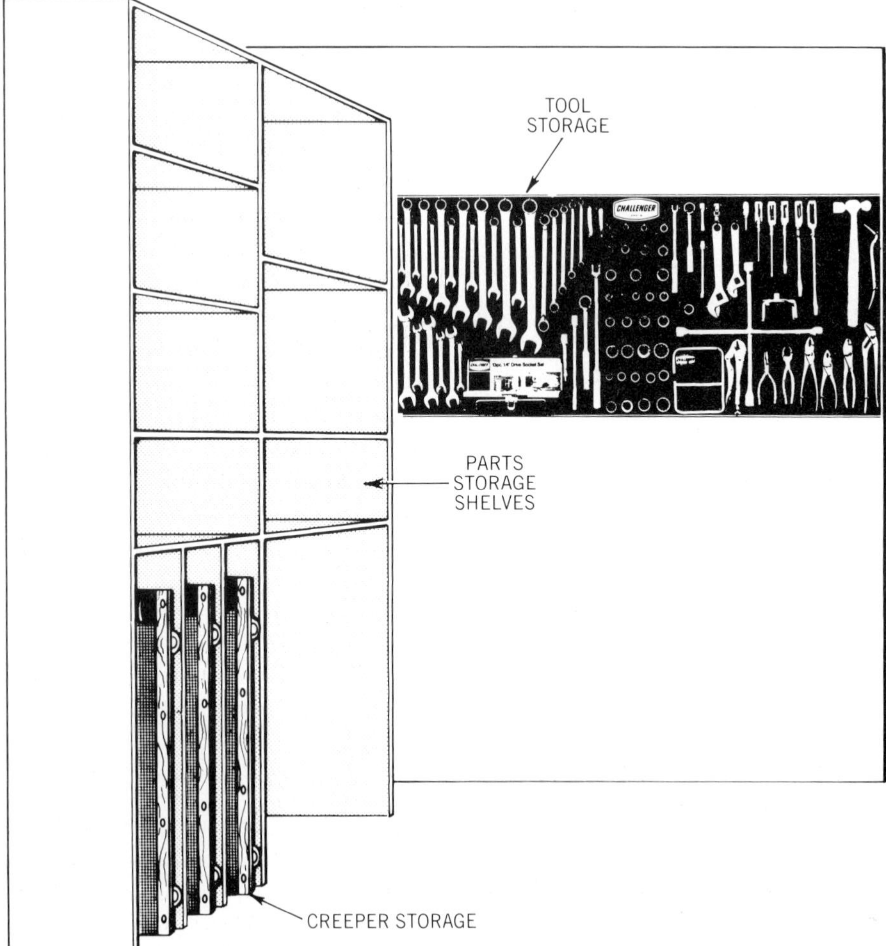

Figure 1-2. Keep tools, parts, and supplies in assigned storage areas. Stand creepers on end when not in use.

safe operation before using the equipment.

5. Before turning on any machine, be sure all persons are clear and that all machine setups and adjustments are correct. Be sure that all safety equipment is installed and functioning on any machine you use.

6. Always stay with a running machine until you have turned it off and it has stopped completely.

7. Keep floors and aisles clear of tools, parts, and materials, figure 1-2.

8. Be sure that all creepers are put away by standing them on end in assigned locations, figure 1-2.

9. Before starting an engine, set the parking brake and put a manual transmission in neutral or an automatic transmission in park.

10. Never start an engine when someone is under an automobile or leaning

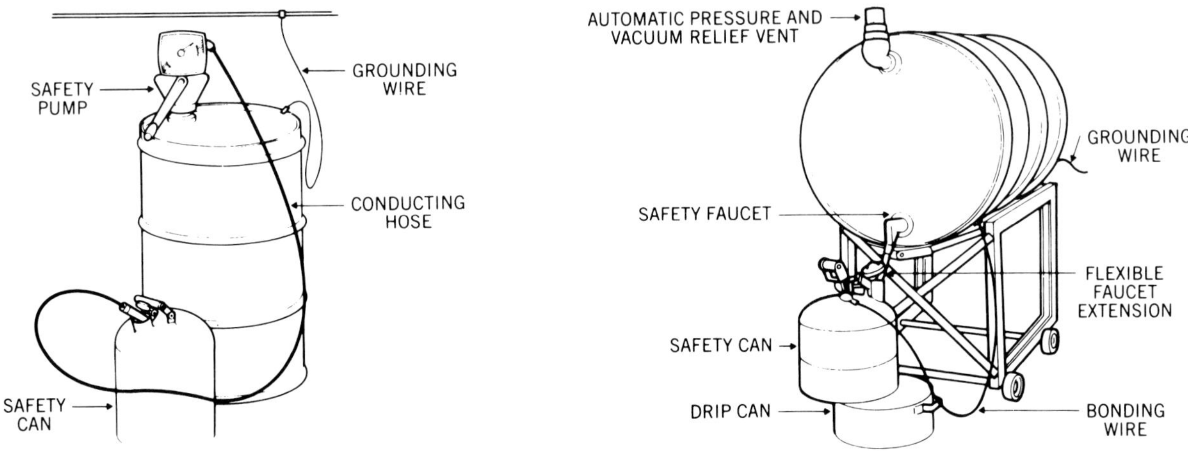

Figure 1-6. When flammable liquids are transferred from one container to another, the containers must be electrically bonded and grounded. (Industrial Indemnity Company)

into an engine compartment. Do not start an engine with tools or parts lying on the air cleaner.

11. Do not work under an automobile when the engine is running.

12. Never choke an engine with your hand over the carburetor. Never pour gasoline into a carburetor when starting an engine or when it is running.

13. Never run an engine in a closed room. Connect the vehicle exhaust to shop exhaust ducts, figure 1-3, or otherwise ensure good ventilation.

14. Keep gasoline and diesel fuel away from sparks, flames, or hot objects.

15. Do not splash cleaning solvents when placing parts into, or removing them from, a cleaning tank.

16. Keep vehicle doors closed whenever possible, even when working inside the vehicle.

17. Do not carry sharp tools, such as chisels and scrapers, in your pockets. Carry them in your hand with the cutting edge down.

18. When handing a tool to another person, extend the handle first.

19. Be sure that any item held in a vise is clamped securely.

20. Do not leave tools or parts protruding from a vise or hanging off a workbench.

21. When working under a vehicle, be sure no loose objects are on the fenders or bumpers.

22. Keep your hands and other parts of your body away from hot exhaust manifolds, exhaust pipes, and catalytic converters. A catalytic converter stays very hot long after an engine is shut off.

23. Do not drive a vehicle faster than five miles per hour within, or when entering or leaving, the shop.

24. Pick up and set down heavy objects by bending at your knees and lifting with your back straight, figure 1-4. Do not bend at the waist and lift with your back.

25. Know the location of shop first-aid supplies and equipment. Know where to get emergency medical service if needed.

26. Wear hearing protection when working near noisy equipment.

FIRE PREVENTION SAFETY

For complete fire prevention safety, you must be aware of two major subjects:

- Safe handling and storage of flammable and combustible materials
- Correct use of fire extinguishers

Flammable and Combustible Material Safety

Gasoline, diesel fuel, oil, and grease are flammable, as are some gases, solvents, and thinners. Additionally, paper, wood, and plastic materials used in a shop may be combustible, particularly if they are covered with oil or grease. Follow these basic precautions for using and storing these materials in any shop.

1. Keep sparks, open flame, hot metal, and any smoking material away from combustible and flammable materials.

2. Always keep fuels, lubricants, and other flammable materials in approved storage containers and locations. Keep all containers closed when not in use, figure 1-5.

3. Clean up any spilled fuel, lubricant, or thinner immediately. Keep work areas clear of oily rags and paper.

4. Keep combustible waste materials in approved metal containers and dispose of them daily.

5. When you transfer a flammable liquid from one container to another, be sure the containers are electrically bonded together and to ground, figure 1-6, to prevent sparks from static electricity.

Fire Extinguisher Safety

Everyone working in a shop must know the locations of all fire extinguishers and which kind of extinguisher to use for different kinds of fires.

Fire extinguishers are grouped in four classes—A, B, C, and D—according to the kind of fire on which they can be used.

Class A can be used on fires of ordi-

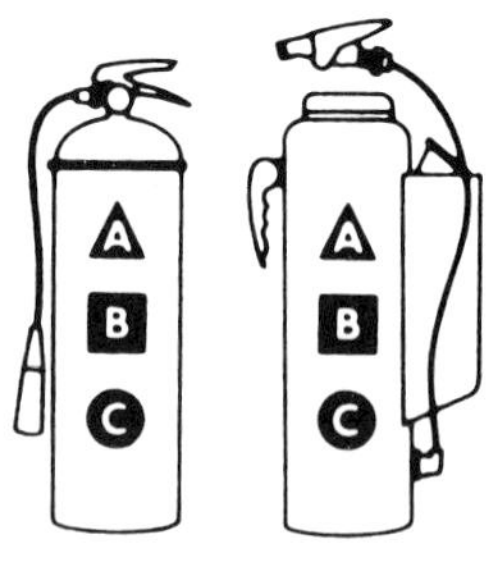

Figure 1-7. Every fire extinguisher is marked with its class rating. Multipurpose extinguishers are most common in automobile shops. (Industrial Indemnity Company)

Fire Extinguishers			
Type of Fire → / Extinguisher to Use ↓	CLASS A • Ordinary Combustibles • Wood • Paper • Cloth	CLASS B • Flammable liquids, grease • Gasoline • Paints • Oils	CLASS C • Energized electrical equipment • Motors • Switches • Fuse boxes
Multi-Purpose Dry Chemical • Stored Pressure • Cartridge Operated			
Ordinary Dry Chemical • Stored Pressure • Cartridge Operated	No		
Carbon Dioxide • Self-Propelling	No		
Water • Stored Pressure • Pump Tank		No	No
AFFF Foam • Stored Pressure			No
Halon 1211 • Stored Pressure			
CLASS D Involve Combustible Metals	Combustible Metals Include: • Magnesium • Titanium • Zirconium • Sodium • Potassium • Lithium • Dry Powder Aluminum • Calcium • Thorium • Sodium-Potassium	Met-L-X Chemical	Cartridge operated

Figure 1-8. This chart summarizes the classes of fire extinguishers and their uses. (Industrial Indemnity Company)

nary combustibles such as wood, paper, and cloth. They are not safe for use on fires of flammable liquids or electrical fires.

Class B can be used on fires of flammable greases and liquids such as gasoline, oil, paints, and thinners.

Class C can be used on electrical fires.

Class D can be used only on fires of combustible metals such as magnesium, sodium, or powdered aluminum.

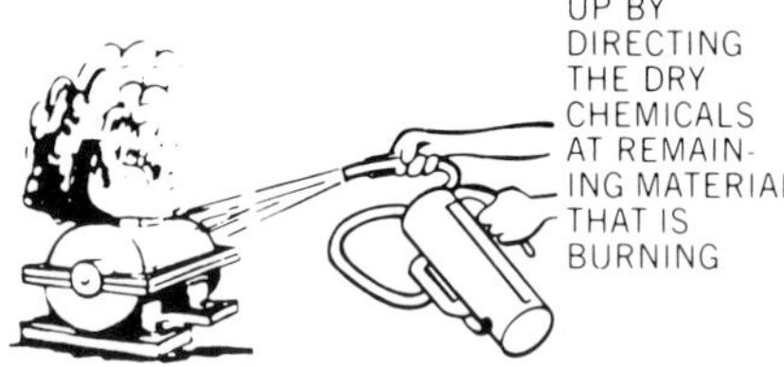

Figure 1-9. Here are the general ways to use different kinds of extinguishers on different fires.

Extinguishers that contain plain water or bicarbonate of soda solutions are suitable for Class A use but must not be used on electrical fires or oil or other liquid fires. Mixing water with an electrical fire or a petroleum fire can be as dangerous as the fire itself. Similarly, carbon dioxide (CO_2) extinguishers will work on oil and electrical fires, but they are not suitable for a Class A wood or paper fire.

Every fire extinguisher has its class rating marked clearly on its side, figure 1-7. Many extinguishers are rated for two or three classes of use. Most multipurpose dry-chemical extinguishers have Class A, B, and C ratings. These are the kind most often found in automobile shops. Some shops may have Class D extinguishers for combustible metal fires. Figure 1-8 summarizes the classes of fire extinguishers, the kinds of materials they contain, and the types of fires on which they can be used. Figure 1-9 illustrates some important points about using different kinds of extinguishers on different kinds of fires.

JACK, HOIST, AND LIFTING EQUIPMENT SAFETY

Various kinds of lifting equipment are used throughout the shop. Observe these guidelines for safe use.

1. Bumper jacks and other jacks provided with vehicles are for emergency road use only. Do not use them to lift or support a vehicle in the shop.

2. Never guess about correct jacking or lifting points for any vehicle. Consult a service manual, figure 1-10, or ask your instructor. All carmakers' manuals and most independent service guides have diagrams that show correct lift points for:

- Single-post, frame-contact hoists
- Twin-post, suspension-contact hoists
- Drive-on hoists
- Jacks and support stands

3. Before raising or lowering a vehicle, be sure all personnel and equipment are clear.

4. Before lifting one end of a vehicle, be sure the wheels at the other end are securely blocked to keep the vehicle from rolling off the jack, figure 1-11.

5. After raising a vehicle on a jack, place safety stands under the axle, sus-

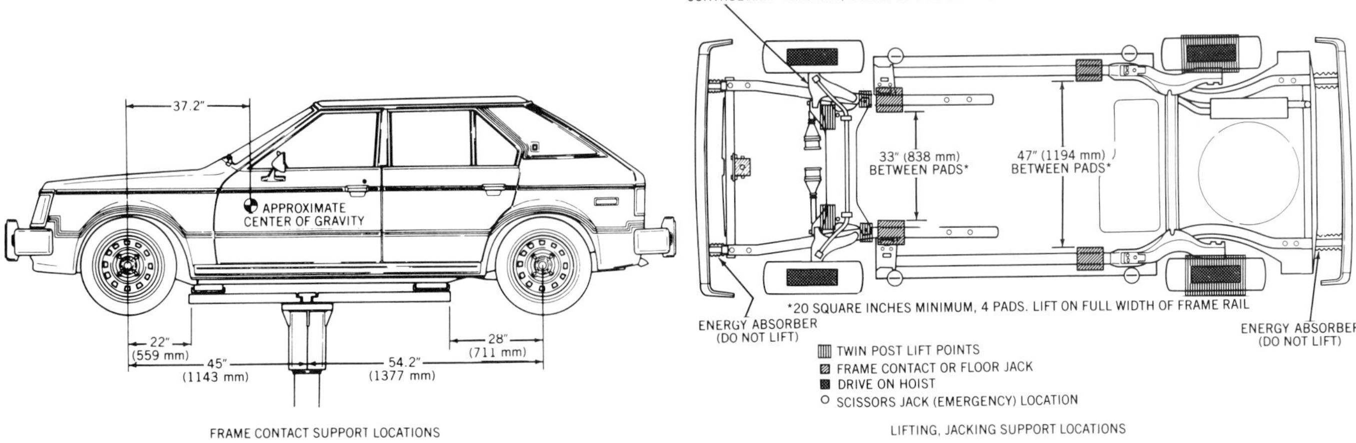

Figure 1-10. This drawing from a carmaker's manual shows the correct lifting points for different jacks and hoists. (Chrysler)

Figure 1-11. Before raising a vehicle at one end, block the wheels at the other. Support the vehicle with safety stands, not the jack alone.

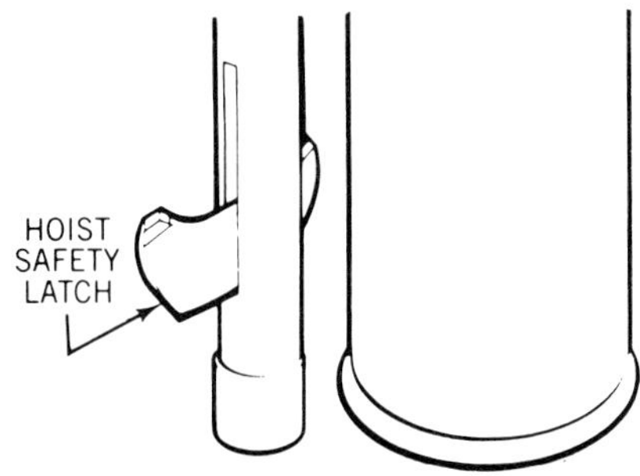

Figure 1-12. After raising a vehicle on a hoist, be sure the safety latch is engaged. (NIOSH)

Figure 1-13. Most shops use a single-post, frame-contact hoist to raise a vehicle for general under-chassis service.

pension, or frame according to the carmaker's lifting instructions, figure 1-11. Do not guess about safe support points. If in doubt, check a service manual or ask your instructor.

6. Do not stand in front of a vehicle when directing it onto a hoist.

7. After raising a vehicle on a hydraulic hoist, be sure the hoist safety latch is securely locked, figure 1-12. Some shops require that you place tall safety stands under the frame or suspension before working under a vehicle on a hoist.

Raising and Lowering a Vehicle on a Single-Post Hoist

Most shops have single-post, frame-contact hoists to raise vehicles for general under-chassis service, figure 1-13. These hoists have arms that are adjustable for length and width, and contact pads that are adjustable for the height of the lift points. You probably will use this kind of hoist to raise a vehicle for an oil change or for any general under-vehicle service that involves the engine and combustion control systems. Refer to the following instructons whenever you raise a vehicle on a single-post, frame-contact hoist. Follow your instructor's directions and the equipment maker's instructions to use other lifting and hoisting equipment safely. Correct lifting and support are important because:

- Support must be placed securely under the frame or specified suspension members. Supporting a vehicle on its torsion bars, springs, suspension links, or front-wheel-drive axles may damage the vehicle or cause it to slip off the hoist.
- If the vehicle is raised on a single-post hoist, the center of gravity must be near the center of the post. If the

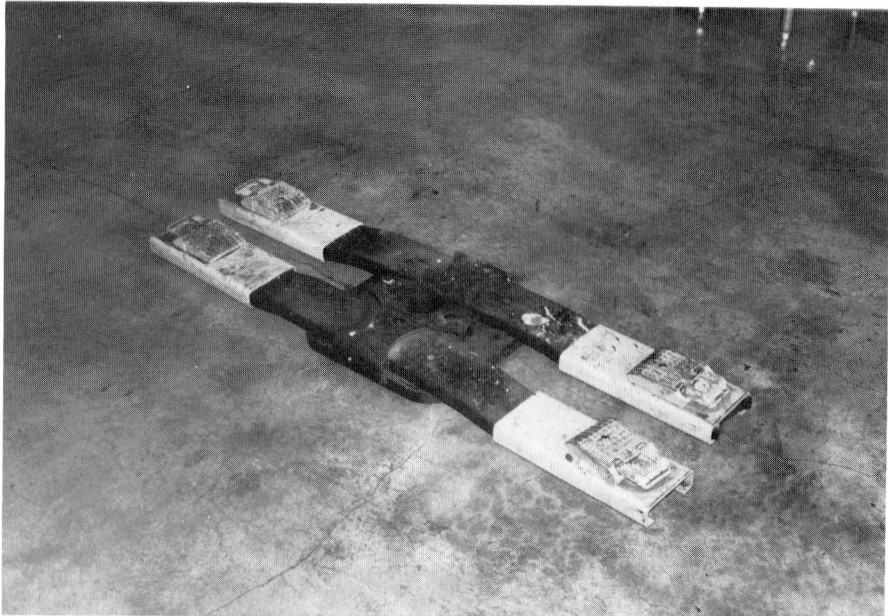

Figure 1-14. Be sure the hoist is lowered completely and that all pads and arms are retracted before driving the vehicle onto the hoist.

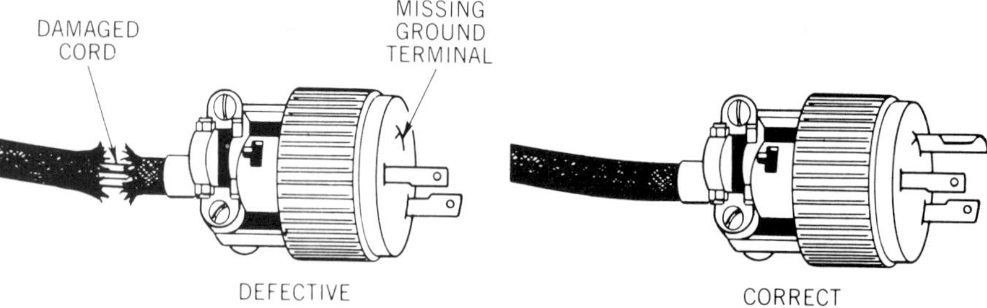

Figure 1-15. Cords and plugs on power tools must not have damaged or frayed cables or broken terminals.

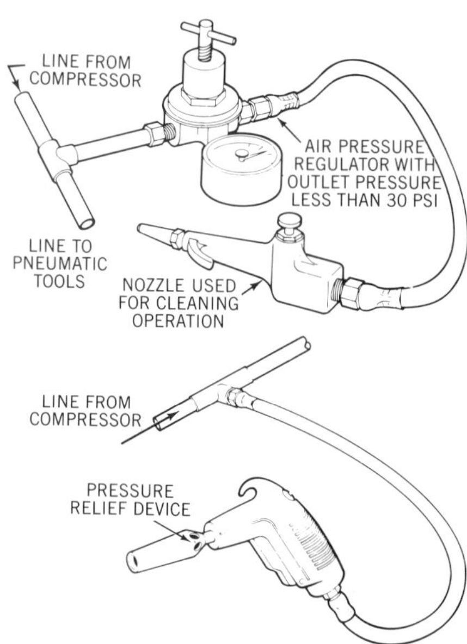

Figure 1-16. These safety nozzles limit air pressure to 30 psi when the nozzle is dead ended. (NIOSH)

center of gravity is offset from the post, the vehicle may slip and fall.

Follow these basic steps to raise a vehicle safely on a single-post hoist:

1. Before driving the vehicle onto the hoist, be sure the hoist is all the way down and that all arms and pads are retracted, figure 1-14.

2. Slowly drive the vehicle onto the hoist so that the center of gravity is over the post. Have another person guide you onto the hoist if possible.

3. Turn the engine off, place the transmission in neutral, and release the parking brake.

4. Place the hoist arms and lifting pads under the specified lift points, figure 1-10. Raise the lifting pads, if necessary, so that they contact the chassis to keep the vehicle level during lifting.

5. Raise the hoist so that the lift pads just contact the chassis. Stop. Check to be sure that the pads or other parts of the hoist are not hitting fuel or brake lines, brake cables, suspension links, or other parts that could be damaged by vehicle weight.

6. Raise the hoist so that the tires are one or two inches off the floor. Push the vehicle forward, backward, and sideways to be sure the vehicle is supported securely.

7. If the vehicle is secure on the hoist and the pads are contacting the specified chassis points, raise the hoist to a convenient working height.

8. Set and lock any safety device, figure 1-12 to prevent accidental lowering.

After you have finished any under-chassis service, lower the vehicle as follows:

1. Remove tools, parts, and equipment from under the vehicle.

2. Release the hoist safety lock.

3. Slowly lower the vehicle to the floor.

4. Retract the lifting pads and hoist arms so that you can drive the vehicle off the hoist.

ELECTRIC POWER TOOL SAFETY

Electric drills, grinders, and buffers are used for many jobs. Observe these general precautions when using any electric hand tools.

1. Always wear safety glasses when using any electric tool.

2. Be sure your hands, the floor, and the entire work area are dry before touching electrical switches or plugs.

3. Be sure that cords and plugs on all electric tools are not damaged and that power cords are properly grounded, figure 1-15.

4. Lay power cords so that no one will trip over them.

5. Hold the tool with both hands. If grinding or drilling a small part or piece of material off a vehicle, clamp it in a vise.

6. Be sure drill bits, grinding wheels,

and buffer pads are secure and that the chuck key is not in the chuck before starting the motor.

7. Be sure the tool motor cannot be started accidentally when a drill or grinding wheel is being tightened in the chuck.

8. Always use a sharp drill, ground properly for the material being drilled. Improperly ground drills will dig and gouge the work.

9. Ease up on the feed pressure as the drill breaks through the work to decrease the chance of the drill catching in the work.

10. Stop the drill immediately if the drill catches in the work.

11. Stop the drill before removing chips. Use a brush to remove chips.

12. Keep your head away from the tool handle and from cutting or grinding edges.

COMPRESSED AIR AND PNEUMATIC TOOL SAFETY

A compressed air supply and air-operated (pneumatic) tools are essential in a well-equipped shop, but pneumatic equipment must be used safely.

1. High-pressure compressed air can be dangerous. Never point an air nozzle toward your body, particularly your face, or toward another person.

2. Wear safety glasses when working with pneumatic equipment.

3. Always check air hoses, fittings, and air chucks for looseness and damage before using. Be sure chucks and fittings are secure. Route air hoses during use so that no one will trip on them.

4. Be sure that an air nozzle, or "blow gun," used for cleaning has a safety tip that limits air outlet pressure to 30 psi when the nozzle outlet is blocked (dead ended), figure 1-16.

5. When using an air nozzle to clean parts, direct the airstream away from yourself and other people.

6. Be sure that belt guards and other safety devices on air compressors are securely in place before operating the compressor, figure 1-17.

7. Periodically turn off the compressor and drain the tank to remove moisture and prevent tank corrosion. Check compressor oil level periodically as instructed by the manufacturer.

CLEANING EQUIPMENT AND CHEMICAL SAFETY

You won't go through a day in the shop without having to clean some part before you do something else to it. Some cleaning methods are as simple as wiping dirt off a part with a clean cloth. Others involve using special equipment and solvents to remove grease, varnish, carbon, and other dirt. As with any job, there is a right way and a wrong way to clean various automobile parts. Fuel system and electrical parts especially can be damaged by improper cleaning.

Cleaning with Brushes and Scrapers

Various wire brushes and sharp scrapers are made to remove heavy dirt and hardened gum, varnish, and carbon deposits. The wire brushes that you use will be simple hand-held brushes or rotary brushes powered by electric motors, figure 1-18. You will use various rotary wire brushes either in portable electric drill motors or in bench grinders. Follow these guidelines to use these tools.

1. Always wear eye protection when working with brushes and scrapers.

2. Most wire brushes have hardened steel bristles and can gouge parts made of plastic and soft metals such as alu-

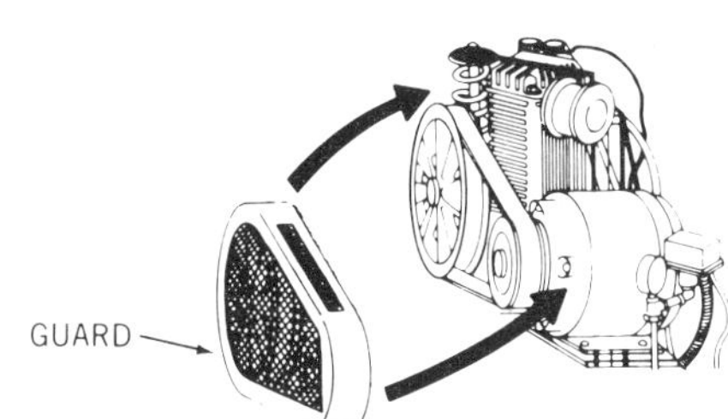

Figure 1-17. Air compressors must have belt guards installed whenever they are operating. (NIOSH)

Figure 1-18. These are some of the kinds of wire brushes you will use to clean parts. (K-D Tools)

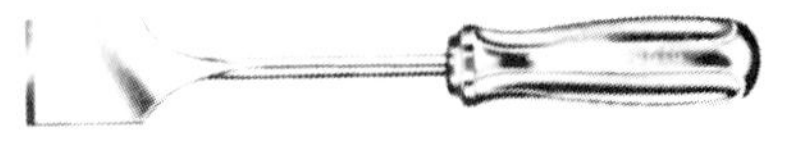

RIGID SCRAPERS

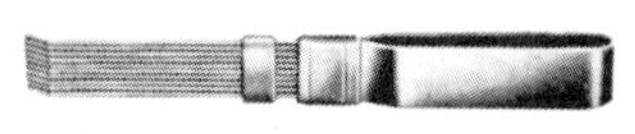

FLEXIBLE SCRAPER

Figure 1-19. Use rigid and flexible scrapers to remove carbon and other hard deposits from parts. (Proto)

minum and zinc. Be careful using wire brushes on these kinds of materials, particularly on machined surfaces, such as gasket flanges.

3. Do not use hand-held or electric-powered brushes to clean precision aluminum or other soft metal parts such as carburetor bodies or bearing and seal bores in aluminum castings.

4. When using a rotary brush in a drill motor, hold it so that brush rotation throws debris away from you.

5. When using a wire brush in a bench grinder, be sure the tool rest and safety shields are in place correctly. Hold the part being cleaned securely against the tool rest, above the centerline of the brush rotation.

6. Use special thin, "rat-tail" brushes to clean oil, coolant, and other fluid passages and tubing.

For many cleaning jobs, you will use hand-held scrapers with both rigid and flexible blades, figure 1-19. These are cutting tools, so observe the same safety precautions that you follow when using chisels and punches. Be careful not to gouge the metal when using a scraper on aluminum or zinc parts.

Cleaning with Solvents and Special Cleaning Equipment

Fuels, lubricants, cleaning solvents, and adhesives are a few of the common chemicals that you work with daily. While all of these chemicals have essential uses, many can be toxic or otherwise harmful to people, vehicles, and equipment if used incorrectly. Follow these precautions for safe handling of solvents and other chemicals:

1. Never use gasoline as a cleaning solvent.

2. Don't wash your hands with gasoline, cleaning solvent, or paint thinner. These solvents will dry the natural oil in your skin and may cause rashes or other illness.

3. After using any chemicals, especially caustic or toxic solvents, wash your hands before eating, drinking, or smoking.

4. Keep gasoline away from your mouth and do not breathe the vapors. As well as being flammable, gasoline is toxic if you swallow it or breathe the vapor in large amounts.

5. Gasoline and other solvents will harm vehicle painted surfaces and rubber parts. Unless proven otherwise, assume that any solvent will harm vehicle paint and rubber parts. If any of these chemicals spill on the vehicle finish, wash the area immediately with mild soap and water.

6. Wear safety glasses when working with solvents and steam cleaners.

7. Place parts into cleaning tanks carefully; do not splash solvent.

8. Use all solvents and other chemicals in well-ventilated areas and avoid breathing vapors. Toxic vapors from some chemicals have no odor; don't rely on your nose to tell you if ventilation is safe.

9. Keep chemical containers closed when not in use.

10. Wear rubber gloves and arm protection when working with cleaning solvents, particularly caustic solvents used to clean carburetors and fuel system parts. Keep solvents off your skin, or if they do contact your skin, wash immediately. Some toxic chemicals can be absorbed through the skin and cause nausea and other illness.

11. Don't guess about the correct use for any solvent. Some caustic cleaners used for iron and steel parts will harm aluminum, brass, and pot metal. Some cleaners will remove anodized coatings and other surface treatments from some metal parts. Always read the instructions for any cleaning material and ask your instructor for guidance on its safe use.

12. Do not pour fuel, motor oil, or any used chemicals into lavatories or waste water drains. Dispose of them only in approved containers and according to local health and safety regulations.

Solvent and Material Compatability

Most, but not all, general-purpose cleaning solvents are noncaustic and nonflammable. They are also safe for use with most metal, plastic, rubber, and synthetic parts of an automobile. Don't assume, however, that any solvent is safe for use with *all* materials and for *all* purposes. Check the manufacturer's instructions and ask your instructor for guidance before using any solvent.

General-purpose solvent will remove grease, oil, and most dirt from metal and plastic parts. It won't, however, dissolve hardened carbon deposits or gum and varnish. Special solvents are made to clean fuel system deposits from carburetors and carbon from engine parts. The caustic qualities of these solvents can damage plastics and soft metals, such as aluminum, zinc, and brass, and can remove protective coatings from some parts.

Some solvents will destroy the rubber materials of seals and gaskets. Even a slight residue of these solvents on a metal part that contacts a seal or gasket can harm the rubber or plastic material. Use special solvents (usually alcohol based) to clean hydraulic parts. Petroleum-based solvents may leave a residue on metal parts that will contaminate the system fluid and may damage rubber seals.

Parts Washing

A simple parts-washing tank contains a general-purpose solvent to remove grease, oil, and dirt from most parts. Never fill a parts washer with gasoline or another flammable solvent. You can clean parts in any, or all, of three ways in a parts washer.

1. Submerge the parts and let the solvent dissolve the dirt.

2. Brush the parts with a soft-bristle brush.

3. Use a parts washer with an agitator that circulates the solvent or with a nozzle that sprays a solvent stream onto the parts.

Parts washers have safety covers—even for use with nonflammable solvents—that are held open by a chain and fusible link. In case of fire or other high temperature, the link melts and lets the lid close. Don't prop a cover open with a stick and defeat this safety feature.

Special Chemical Dip Tanks

Carburetors usually are cleaned by submerging them in a tank with special solvent. These solvents are quite harmful to your skin and may be toxic. Wear gloves and put the parts in perforated metal baskets, figure 1-20, or attach wires to them so that you don't have to reach into the solvent to remove them from the tank.

Parts are usually left in these solvents for one-half to three hours, depending on the kind of solvent and the parts being cleaned. If you leave zinc or aluminum parts in some solvents too long, the chemicals will attack the metal or protective coatings on the metal

Some of these solvents can be used in tanks with agitators; others cannot. Some solvents will get overheated if agitated, so do not agitate a solvent unless the manufacturer's directions allow it.

After cleaning any parts in a special solvent, rinse them with general-purpose solvent or warm water, or both, and dry them thoroughly with low-pressure compressed air.

VEHICLE SYSTEM SAFETY

The service procedures of this manual contain safety *CAUTIONS* and *WARNINGS* for specific vehicle systems and parts. However, when you are working on one system of a vehicle, you usually are working with or near other systems and parts as well. Every vehicle system has its own basic safety precautions that you must know and follow whenever you do any work on an automobile.

Engine and Fuel System Safety

All of the fire prevention and flammable liquid precautions listed previously apply to any job that involves the engine or fuel system. Also, follow these general practices.

1. Keep your hands, clothing, face, hair, and tools away from the fan and other moving engine parts whenever you work under the hood with the engine running, figure 1-21.

2. Keep your body and clothes away from hot exhaust manifolds and other exhaust parts. Keep gasoline, other flammable liquids, and oily cloths away from running engines and hot exhaust parts. Catalytic converters stay very hot long after an engine is shut off.

3. Cap or plug all fuel lines when disconnected for service to prevent fuel leakage, figure 1-22.

4. Keep the engine air cleaner installed whenever possible to avoid a backfire or the possibility of dropping something into the engine. Keep small loose parts and tools away from the engine air intake.

Ignition, Electrical System, and Accessory Safety

1. Disconnect the battery ground cable (negative cable), figure 1-23, before removing an engine or removing and replacing electrical parts other than lamp bulbs, fuses, and spark plugs.

2. Do not short circuit or ground any solid-state electronic parts or electrical terminals or apply battery voltage directly to electronic parts unless specifically instructed to do so in a service procedure.

3. Keep metal tools away from exposed electrical terminals, particularly the alternator output terminal, unless servicing a specific connection.

4. A battery gives off explosive hydrogen gas, and some hydrogen may always be present near a battery's vents. Keep open flame, sparks, smoking materials, and hot objects away from a battery to prevent explosion.

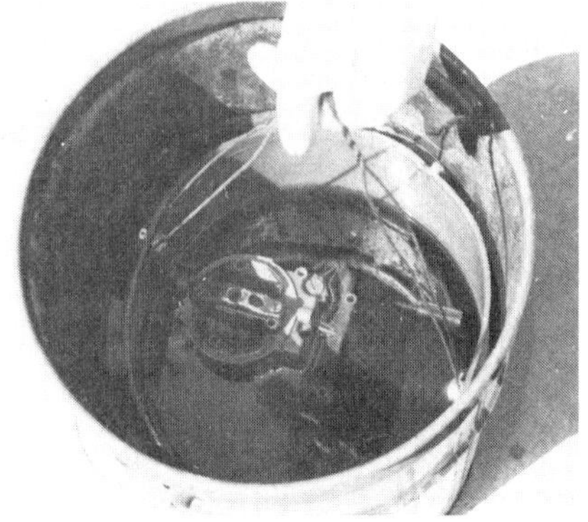

Figure 1-20. Put carburetor parts in a perforated basket for solvent cleaning. Wear rubber gloves and safety glasses when working with strong solvents.

Figure 1-21. Keep your hands, face, hair, and tools away from moving fans and drive belts.

Figure 1-22. Cap or plug all disconnected fuel lines.

Figure 1-23. Disconnect the battery ground cable before removing electrical parts other than bulbs, fuses, and spark plugs.

5. Never short-circuit the two terminals of a battery and always keep metal tools off the top of a battery to avoid sparks and possible explosion.

6. Battery electrolyte contains sulfuric acid (H_2SO_4). If electrolyte spills on a vehicle, wash it off immediately with a mixture of baking soda and water. Use baking soda and water to clean electrolyte condensation and corrosion from the outside of a battery. Wash your hands immediately if electrolyte gets on them.

REVIEW QUESTIONS

Multiple Choice

1. When working around an engine, you should remove rings, watches, and other jewelry *primarily* because:

- **a.** the objects might distract you from what you are doing and cause a mistake.
- **b.** the objects might conduct heat from hot manifold and cause a burn.
- **c.** the objects might cause a short between a "hot" electrical terminal and ground, causing a burn and possible vehicle damage.
- **d.** the objects might get caught in equipment test leads.

2. Mechanic A says that you don't have to set the parking brake before starting an engine if the automatic transmission is in park. Mechanic B says that before starting an engine on a car with a vacuum-operated parking brake, you should set the brake and disconnect and plug the vacuum lines. Who is right?

- **a.** A only
- **b.** B only
- **c.** Both A and B
- **d.** neither A nor B

3. All of the following are combustibles EXCEPT:

- **a.** ungrounded metal parts of an electrical system
- **b.** plastics
- **c.** grease and oil
- **d.** diesel fuel

4. To extinguish an electrical fire, you should use a fire extinguisher that is rate as:

- **a.** class A
- **b.** class B
- **c.** class C
- **d.** class D

5. To extinguish a grease or oil fire, a fire extinguisher must be rated as:

- **a.** class A
- **b.** class B
- **c.** class C
- **d.** class D

6. To extinguish a combustible-metal fire, a fire extinguisher must be rated as:

- **a.** class A
- **b.** class B
- **c.** class C
- **d.** class D

7. If you spray water on an electrical fire, you will:

- **a.** put out the fire
- **b.** retard, but not put out, the fire
- **c.** make the fire worse
- **d.** none of the above

8. When working with cleaning solvents:

- **a.** you can smell any toxic fumes immediately if they are released
- **b.** used solvent should be poured down a waste-water drain as soon as possible after use
- **c.** safety glasses should be worn at all times
- **d.** baking soda can be used to neutralize gasoline spilled on painted surfaces

9. Mechanic A says that battery electrolyte contains corrosive sulfuric acid. Mechanic B says that a battery can release explosive hydrogen gas. Who is right?

- **a.** A only
- **b.** B only
- **c.** Both A and B
- **d.** neither A nor B

10. Mechanic A says that safety glasses are necessary when working around shop equipment with movable parts. Mechanic B says that safety glasses are needed only when working with batteries and flammable liquids. Who is right?

a. A only
b. B only
c. Both A and B
d. neither A nor B

11. Mechanic A says that safety stands should be used to support the front wheels of a front-wheel-drive (fwd) car when the car is raised on a jack. Mechanic B says that safety stands should be used to support the front wheels of a rear-wheel-drive (rwd) car when the car is raised on a jack. Who is right?

a. A only
b. B only
c. Both A and B
d. neither A nor B

12. All of the following are true about using compressed-air equipment EXCEPT:

a. Check compressor oil level periodically.
b. Turn off the compressor and drain the tank periodically to remove moisture.
c. Wear safety glasses whenver you work with pneumatic equipment.
d. Use a high-pressure air nozzle to clean or dry parts with compressed air.

13. Mechanic A says you can use a rotary wire brush to clean gasket flanges on valve covers and oil pans. Mechanic B says you can use a rotary wire brush to remove carbon from piston ring grooves. Who is right?

a. A only
b. B only
c. Both A and B
d. neither A nor B

14. Mechanic A says that you should always place hoist lifting pads evenly under the drive axles. Mechanic B says that you should check the carmaker's hoisting instructions to be sure the vehicle center of gravity is properly balanced. Who is right?

a. A only
b. B only
c. Both A and B
d. neither A nor B

15. Mechanic A says that you can use gasoline as a solvent to remove hardened gum and varnish from engine parts. Mechanic B says that you can use gasoline only to clean carburetor parts, and then only in a well-ventilated area. Who is right?

a. A only
b. B only
c. Both A and B
d. neither A nor B

16. When you remove a carburetor from an engine for service, you should:

a. plug or cap all disconnected fuel lines
b. disconnect the battery ground cable
c. place a protective cover across the intake manifold opening
d. all of the above

17. Mechanic A says that safety CAUTIONS in the book warn about vehicle or equipment damage. Mechanic B says that safety WARNINGS warn about personal injury. Who is right?

a. A only
b. B only
c. Both A and B
d. neither A nor B

2 WORKING WITH SPECIFICATIONS AND REFERENCE MANUALS

INTRODUCTION

Back in the good old days, a mechanic could do most repair jobs with a portable box of hand tools. He didn't have, and didn't need, a lot of sophisticated test and service equipment. Cars were simpler 40 or 50 years ago, and a mechanic could remember most of the service specifications and repair procedures that he needed because they were similar for most cars.

Today, as a professional mechanic—a skilled service technician—you work with special tools and electronic test equipment worth tens of thousands of dollars. Automobiles are more sophisticated today. Different models have unique specifications and need special test and repair procedures. You can't possibly remember *all* of the specifications and procedures for *all* of the cars that you work on. Moreover, this manual can't provide you with all of the special information that you need for your job. This book will give you basic guidelines and procedures for most common service and repair jobs. This book also explains the principles used by all carmakers in their service procedures.

You must understand one thing clearly, however, to get the most benefit from the information in this book. For specifications and procedures to service a particular car correctly, *you must be able to use reference manuals*. This book and professional reference manuals complement each other. This book will help you understand the similarities among many manufacturers' procedures. But, you must follow the specific procedures and specifications for a particular vehicle.

Professional manuals are essential tools for a professional mechanic. As with any tool, you must be able to use a reference manual accurately and efficiently. You must know where to look for particular information, and you must know how to interpret instructions and test results. This chapter introduces you to the use of reference manuals as necessary tools for your profession.

GOALS

After studying this chapter, you should be able to:

1. Using vehicle identification numbers, correctly identify the manufacturer, model, year, engine, transmission, and other key features of several vehicles.

2. Identify the part or model number on various components such as distributors, carburetors, and electronic modules.

3. Explain the information on engine decals.

4. Using information from vehicle identification numbers and component part and model numbers, locate tuneup specifications and engine test procedures in reference manuals.

5. Using information from vehicle identification numbers and part numbers, identify correct replacement parts from various parts catalogues.

6. Using various flat-rate manuals, determine time standards for various repair jobs.

SPECIFICATIONS AND PROCEDURES

Basically, specifications tell you "what"; procedures tell you "how."

Specifications

For easy understanding, we can divide specifications into three categories:

1. Descriptive specifications

2. Service and operating specifications

3. Parts specifications

Descriptive specifications give you general information about a vehicle or

Chevrolet—1984

Engine Code	Type	Displacement Liter (cid)	Carburetor	Spark Plug AC Part No.	Gap
T	V-6	2.8-L (193)	2-barrel	R43CTS	0.045
H	V-8	5.0-L (305)	4-barrel	R45TS	0.045
S	V-8	5.0-L (305)	CFI	R45TS	0.045

DESCRIPTIVE SPECIFICATIONS (Engine Code through Displacement) — PART NUMBER (Spark Plug AC Part No.) — SERVICE SPECIFICATION (Gap)

Figure 2-1. These spark plug specifications contain descriptive, service, and part number specifications.

Chevrolet 5.0 Liter V8 Passenger Car:

Part #	Eng.	RPO	Type	Applications
17084201		LG4	E4ME	NAT, A/T, A/C, NON A/C, (G, B, & F)
17084205	HO	L69	E4ME	NAT, M/T, A/C, NON A/C, (F)
17084208	HO	L69	E4ME	NAT, A/T, A/C, NON A/C, (G)
17084209		LG4	E4ME	NAT, M/T, A/C, NON A/C, (F)

Figure 2-2. Different engine and transmission calibrations require four different part numbers for the basic Rochester model E4ME carburetor for 1984 Chevrolet 5.0-liter V-8's. (Rochester)

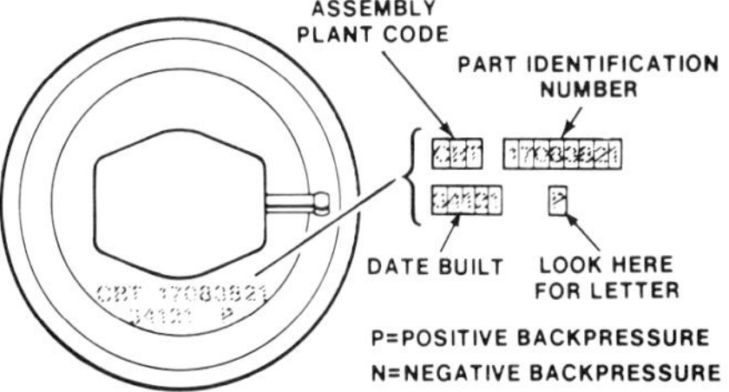

Figure 2-3. The assembly plant code and the date-built code work like serial numbers to give more information about this exhaust gas recirculation (EGR) valve than the part number alone provides. (Rochester)

a component. When we speak of a 3.8-liter (232-cid) engine that develops 150 horsepower, we are using a descriptive specification. Similarly, when we speak of a THM 700-R4 transmission, we are citing a descriptive specification.

Service and operating specifications are information about the adjustment and installation of parts. A valve clearance specification of 0.012 inch is a service specification. So is a spark plug gap specification of 0.045 inch. Service and operating specifications also include carmakers' recommendations for preventive maintenance intervals, as well as fuels and lubricants.

Parts specifications are descriptive, but they describe, or identify, a particular part by number. R45TS is a part number specification for a specific spark plug. Distributor number 1103370 is a part number specification for a particular distributor.

When you set the gap of an R45TS spark plug to 0.045 inch for a 1984 5.0-liter (305-cid) engine with a 4-barrel carburetor, figure 2-1, you have used all three kinds of specifications: descriptive, service, and part number. If you ignore any one, you won't be doing the job correctly.

Model, Part, and Serial Numbers. All carmakers identify major components with model numbers, part numbers, serial numbers, or a combination of all three. A model number is a general identification, similar to the general name of a carline, such as Ford Escort or Toyota Corolla. References to a Rochester M4MC or E4MC carburetor or a Delco-Remy 10SI alternator are all references to model numbers.

The same basic model of a carburetor or an alternator can be used on several different vehicles with differences in internal parts or calibrations. Therefore, part numbers distinguish among variations of a basic component, figure 2-2. A part number identifies a specific part or assembly. Every water pump, carburetor, alternator, and other assembly on a car has a part number. Moreover, each individual piece of an assembly has a part number. Two parts with the same, or equivalent, part numbers will interchange with each other. Two assemblies with the same model number may or may not interchange.

For more precise identification, manufacturers use serial numbers on major assemblies. Serial numbers do not necessarily indicate the interchangeability of one part with another, but they are useful to identify the time, place, and sequence in which a part or assembly was built, figure 2-3.

Procedures

Procedures complement specifications. Procedures are the step-by-step instructions for specific jobs. They tell you in logical order how to test a vehicle system or how to disassemble, repair, and reassemble various parts. When you follow a procedure, you also must refer to specifications. This is because you can use a single procedure to disassemble and reassemble a basic component, but the adjustment specifications for that component as it is used on different cars will vary. For example, you can follow the same repair procedure for a Rochester M4MC carburetor as used on a Chev-

1984 Adjustment Specifications—Quadrajet M4ME—M4MC

Carb. Number	Float Setting ±1/16″	Pump Setting	Air Valve Spring	Choke Stat Level	Choke Link Cam ±2.5°	Vac. Break Front ±2.5°	Vac. Break Rear ±3.5°	Air Valve Link	Unloader ±4°
17083298 CHV 5.7 V8	3/8	INNER 9/32	1	.120G	37°	23°	30°	.025	40°
17084200 CHV 5.7 V8	13/32	INNER 9/32	7/8	.120G	46°	—	26°	.025	39°

Figure 2-4. Two different part numbers of the same basic carburetor have different adjustment specifications for use on two different 5.7-liter Chevrolet V-8's. (Rochester)

rolet, Buick, or Oldsmobile. However, the service specifications for float level, choke settings, idle mixture, and idle speed will vary for that basic carburetor used on different engines, figure 2-4.

The procedures in this manual give you general instructions for basic service and repair jobs. To do a particular job on a particular car, however, you should locate and follow the carmaker's procedure in a reference manual. Similarly, we have used examples of typical specifications, but you will have to use other reference manuals to find exact specifications for individual cars.

Flat-Rate Time Standards

All carmakers and many independent manual publishers compile standards of the average time required for various jobs and publish these standards in flat-rate manuals. After you determine exactly what work is to be done on a car, you can look up the flat-rate time for that job on that particular year and model. You will find time standards listed in hours and tenths of hours, such as these examples for a 1977 full-size Buick:

Carburetor—R & R, rebuild, adjust: 1.5 hrs. (for 4-barrel carburetor, add 0.4 hr.)

Spark plugs—clean or R & R, adjust: 0.3 hr. (with air conditioning, add 0.2 hr.)

These examples show how flat-rate times are given for very specific jobs. A complete job often consists of several flat-rate tasks, so you must know the flat-rate time for each and add them together. In this example, flat-rate time to remove, overhaul, reinstall, and adjust the carburetor is 1.5 hours. Some flat-rate standards may separate the removal and replacement (R & R) time and the overhaul time.

Most shops charge customers for labor on the basis of flat-rate times multiplied by the shop's hourly labor rate. Also, many shops pay mechanics a percentage of the flat-rate labor charge for each job that he or she does. If you can do a 1.5-hour job in 1.5 hours, you earn flat-rate wages. If you can do it in 1.2 hours, you make more money per hour. If you take 2 hours for a 1.5-hour job, you earn less money.

Flat-rate times are average times that an experienced mechanic should take to do a job. Someone starting in the profession probably will take longer. As you gain experience, you will find that you do most jobs in flat-rate time or a little faster. A mechanic being paid on a flat-rate basis might be tempted to rush a job to make more money, but jobs done in haste often result in having to do the job over. If a job comes back to be redone, the shop doesn't charge the customer for a mechanic's errors. And, if a mechanic is working for flat-rate wages, he or she doesn't get paid for the comeback rework.

Flat-rate times are not only the basis for a shop's labor charges, they are the basis for labor allowances that manufacturers pay to car dealers for warranty repairs. A mechanic who takes longer than flat-rate times to do various jobs is not doing his or her fair share of the labor in a shop. Similarly, a mechanic who rushes jobs to beat flat-rate time and creates too many comebacks is losing money for the shop and himself or herself.

VEHICLE AND COMPONENT IDENTIFICATION

Before you can find the right specifications, procedures, and flat-rate times for a vehicle, you must know the year and model and other facts about it. This may sound obvious, but it's not as simple as looking at the name on the hood and saying, "It's a Chevrolet" or "It's a Toyota" or "It's a Volkswagen" and then looking under the hood and seeing that it has a 4-cylinder or a V-8 engine. All carmakers build several models with various engine, transmission, and accessory combinations. Moreover, engine calibrations for emission controls vary depending on the transmission and accessories of a particular car and where it was sold. Specifications seldom remain the same from year to year, even for the same make and model.

Vehicle Identification Numbers

All cars and light trucks built since the 1968 model year have a vehicle identification number (VIN) plate visible through the windshield on the driver's side. Most VIN plates are on the top left of the instrument panel, inside the base of the windshield figure 2-5. This is where you first look to identify a specific car or truck. You can think of a VIN as a combination model, part, and serial number for a specific vehicle. On 1967 and earlier models, you can find similar identification plates on a door post or inside the engine compartment.

Vehicle identification numbers follow a form standardized by the SAE. For 1968-80 vehicles, the VIN standard allowed up to 15 digits of information, but most carmakers used 11 or 13 letters or numbers, figure 2-6. The first 5 to 8 digits (letters or numbers) indicate the carline, the series, the body type, the engine type, the model year, and the manufacturer's assembly plant. The last 6 digits are the sequence number, or serial number, of that vehicle built in the specified assembly plant. For the 1981 model year, the SAE expanded the VIN to 17 digits to include codes for the company, country of origin, restraint system, and a check digit that prevents altering the VIN, figure 2-7. The VIN contains standard information about all vehicles. But, as figures 2-6 and 2-7 show, manufacturers are free to use different letters and numbers in various positions for various items. About the only thing that is uniform is that the last six digits are the assembly sequence number.

Don't try to memorize all of the letter and number code combinations for each carmaker's VIN plates. That's where your reference manuals come in. Whenever a manual or parts catalogue has different specifications or procedures for different engines, years, carlines, or model series from one manufacturer, that manual will tell you the codes and digit positions to check. Remember, though, that the VIN is the starting point for accurate vehicle identification for any service or repair job.

Component Identificaton

Some of the jobs that you do will require you to know the code or identification numbers for engines, transmissions, and accessories of a specific vehicle. You won't find all of this information on the VIN plate, but you will find it on other identification plates. Ford Motor Company, for example, puts a certification label on the door post of its vehicles that repeats the VIN and has codes for transmission type, drive axle ratio, body series and trim, paint, air conditioning, springs, sun-

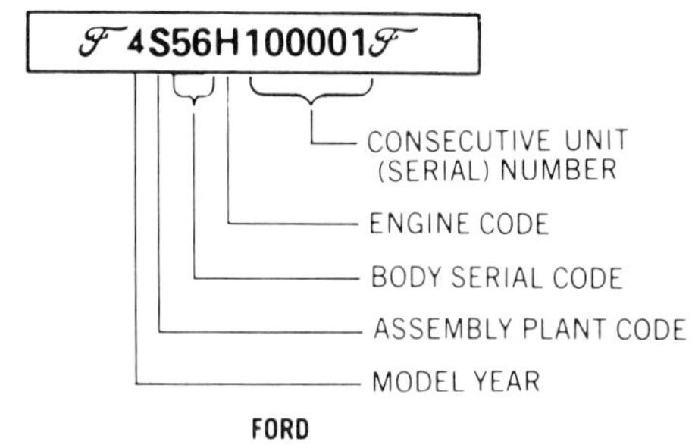

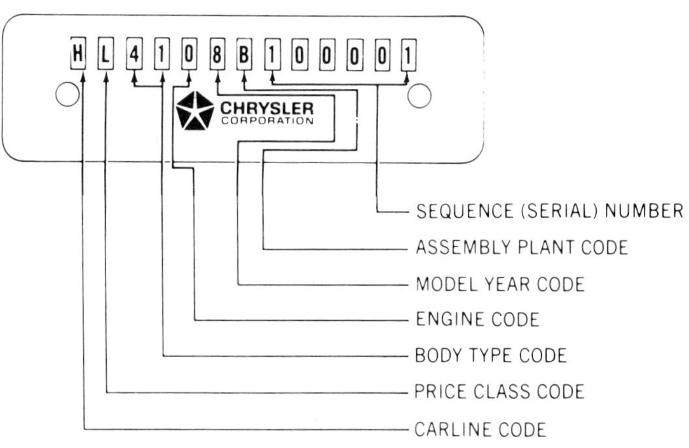

CHRYSLER

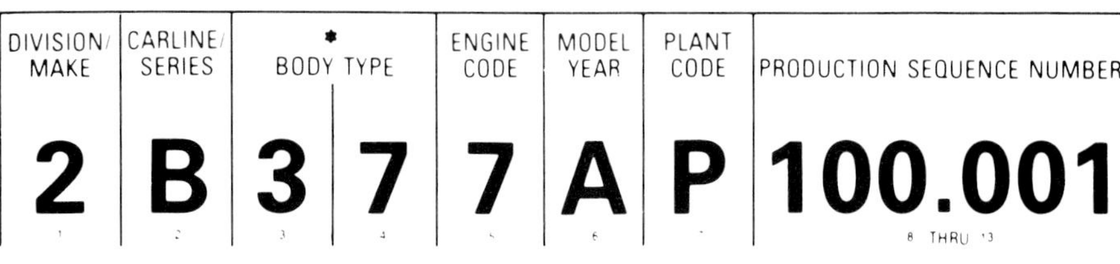

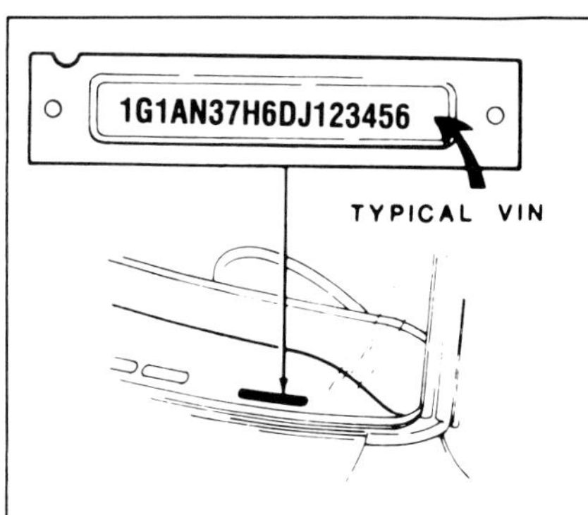

Figure 2-5. The vehicle identification number (VIN) is a combination model, part, and serial number for a specific car or truck. (Chevrolet)

Figure 2-6. These examples of 1968–80 VIN's show the codes and digit positions used by different carmakers. (Ford) (Chrysler) (General Motors)

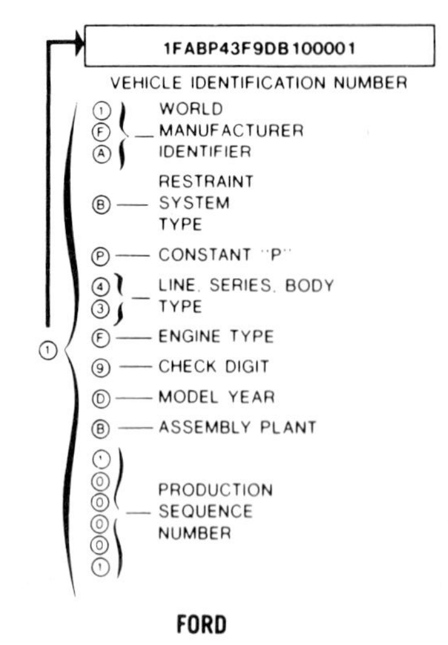

FORD

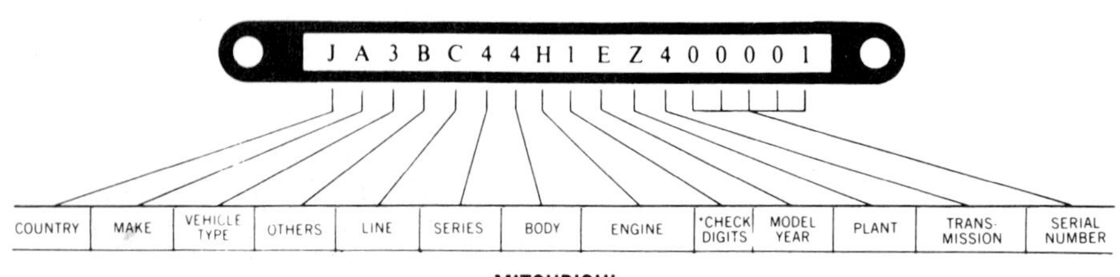

MITSUBISHI

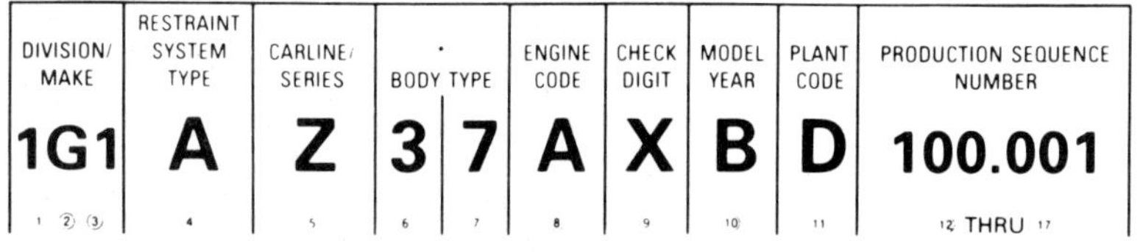

DIVISION/ MAKE	RESTRAINT SYSTEM TYPE	CARLINE/ SERIES	BODY TYPE		ENGINE CODE	CHECK DIGIT	MODEL YEAR	PLANT CODE	PRODUCTION SEQUENCE NUMBER
1G1	A	Z	3	7	A	X	B	D	100.001
1 2 3	4	5	6	7	8	9	10	11	12 THRU 17

GENERAL MOTORS

Figure 2-7. These examples of 1981 and later VIN's show the expanded code fields and the digit positions used by several carmakers. (Ford) (General Motors) (Mitsubishi)

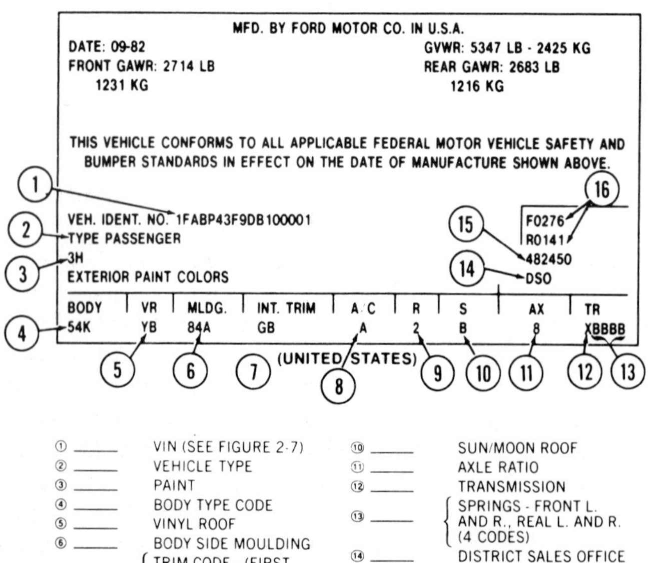

MFD. BY FORD MOTOR CO. IN U.S.A.
DATE: 09-82 GVWR: 5347 LB - 2425 KG
FRONT GAWR: 2714 LB REAR GAWR: 2683 LB
1231 KG 1216 KG

THIS VEHICLE CONFORMS TO ALL APPLICABLE FEDERAL MOTOR VEHICLE SAFETY AND BUMPER STANDARDS IN EFFECT ON THE DATE OF MANUFACTURE SHOWN ABOVE.

(1) VEH. IDENT. NO. 1FABP43F9DB100001
(2) TYPE PASSENGER
(3) 3H
EXTERIOR PAINT COLORS
(16) F0276
(15) R0141
(14) 482450
DSO

BODY	VR	MLDG.	INT. TRIM	A/C	R	S	AX	TR
(4) 54K	YB	84A	GB	A	2	B	8	XBBBB

(5) (6) (7) (8) (UNITED STATES) (9) (10) (11) (12) (13)

① ____ VIN (SEE FIGURE 2-7)
② ____ VEHICLE TYPE
③ ____ PAINT
④ ____ BODY TYPE CODE
⑤ ____ VINYL ROOF
⑥ ____ BODY SIDE MOULDING
⑦ ____ TRIM CODE - (FIRST CODE LETTER - FABRIC AND SEAT TYPE, SECOND CODE - COLOR)
⑧ ____ AIR CONDITIONING
⑨ ____ RADIO
⑩ ____ SUN/MOON ROOF
⑪ ____ AXLE RATIO
⑫ ____ TRANSMISSION
⑬ ____ SPRINGS - FRONT L. AND R., REAL L. AND R. (4 CODES)
⑭ ____ DISTRICT SALES OFFICE
⑮ ____ PTO/SPL ORDER NUMBER
⑯ ____ ACCESSORY RESERVE LOAD

Figure 2-8. This Ford vehicle certification label is an example of other decals or plates that carmakers use to give you information about vehicle components and options. (Ford)

LINE NO.	
6	1 2 3 4 5 6 7 8 9 10 11 12 13 14 15 16 17 18 19 20 21 22 23
5	3 4 5 6 7 8 9 10 11 12 13 14 15 16 17 18 19 20 21
4	1 2 3 4 5 6 7 8 9 10 11 12 13 14 15 16 17 18 19 20 21 22 23
3	1 2 3 5 6 7 9 10 11 14 17 18 19 20 21 22 23
2	1 2 3 5 6 7 8 10 11 12 14 15 16 18 19 20 21 22 23
1	1 2 3 4 6 7 8 9 10 11 12 13 14 15 16 18 19 20 21 22 23

FOR FACTORY USE ONLY

BODY CODE PLATE INTERPRETATION

Line No.	Code Position	Code Interpretation
1	1 thru 4	Model
	5	(blank)
	6 thru 16	V.I.N. (see V.I.N. code chart)
	17	(blank)
	18 thru 23	V.I.N. (see V.I.N. code chart)
2	1 thru 3	1st. Body Paint Color
	4	(blank)
	5 thru 8	Trim
	9	(blank)
	10 thru 12	Engine (sales code)
	13	(blank)
	14	Month (Vehicle Sales Order Data)
	15 & 16	Day of Month (Vehicle Sales Order Data)
	17	(blank)
	18 thru 23	Order Number (Vehicle Sales Order Data)
3	1 thru 3	2nd. Body Paint Color
	4	(blank)
	5 thru 7	Vinyl Roof
	8	(blank)
	9 thru 11	Transmission (sales code)
	12 & 13	(blank)
	14	U = U.S. Order C = Canadian Order I = International Order
	15 & 16	(blank)

Figure 2-9. This Chrysler body code plate has codes for engine, transmission, paint, trim, and other accessories. (Chrysler)

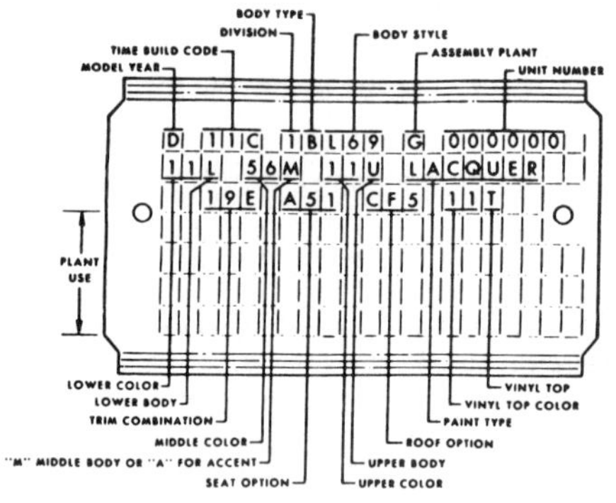

Figure 2-10. This is a General Motors body code plate that lists the body style, paint, trim, and other information about a specific vehicle. (Chevrolet)

roof, and other accessories, figure 2-8. Chrysler products have a body code plate that repeats the VIN and lists codes for the engine, transmission, body series, paint, trim, and accessories, figure 2-9. GM uses similar body identification plates on its vehicles, figure 2-10.

The engine, transmission, and other major assemblies of a car have identification numbers that are combina-

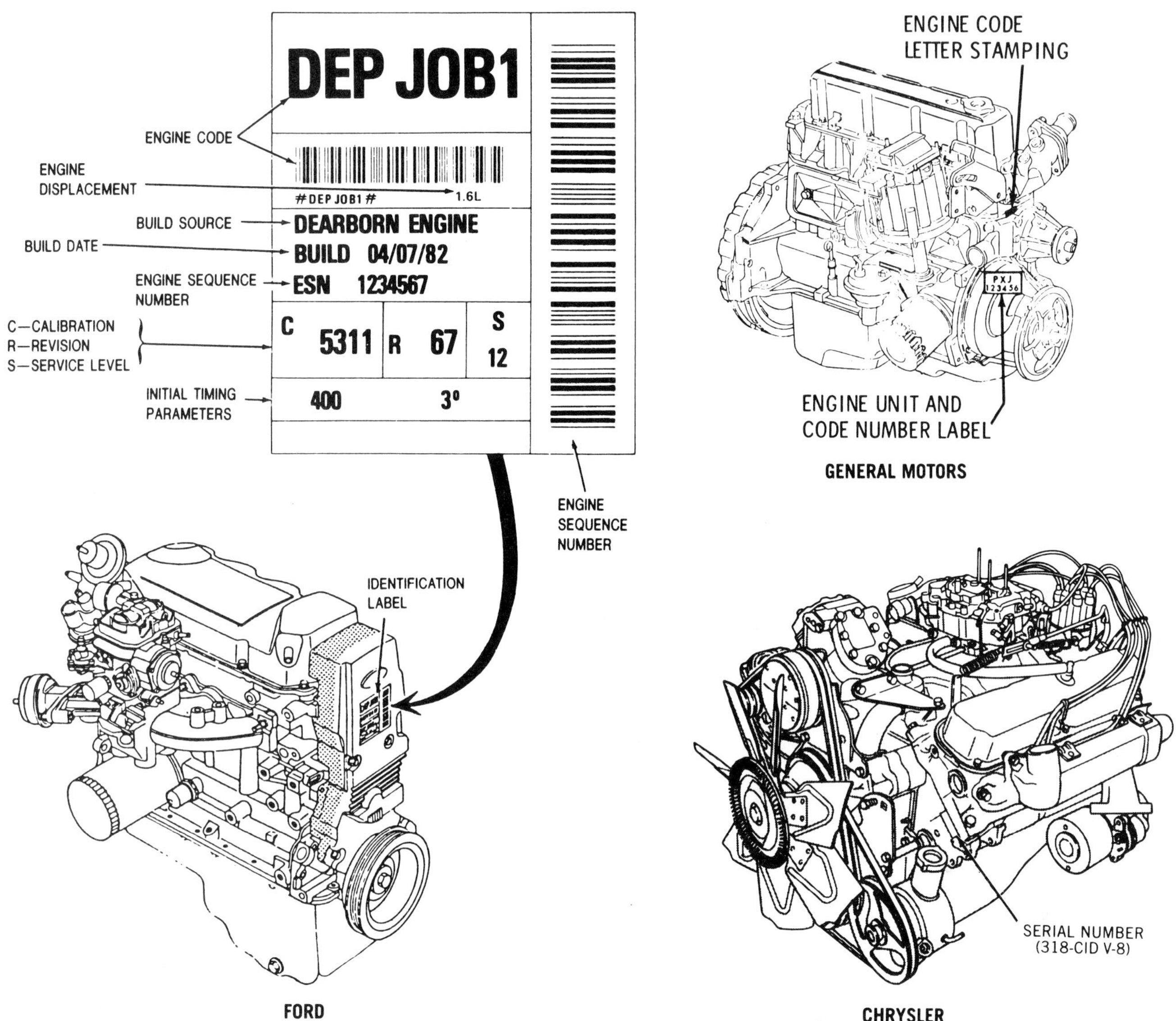

Figure 2-11. These are typical locations of engine and transmission numbers used by several carmakers. Reference manuals will show you exact locations for specific manufacturers. (Ford) (Chrysler) (Oldsmobile)

tions of model, part, and serial numbers. Engine and transmission numbers correlate to vehicle identification numbers. They are usually stamped on a machined surface on the engine block or transmission case or on a decal or plate, figure 2-11.

Many service specifications differ for different part numbers of a basic model. This is why you will see distributor spark advance specifications listed by distributor part number, figure 2-12, for example. Carburetor float, idle, and choke adjustments usually are specified by carburetor model number or part number and by the year and model of the engine on which the carburetor is used.

These examples are just a few of the ways that carmakers use combinations of descriptive, service, and parts specifications to give you the information you need to do your job. But there are other kinds of service information besides identification numbers that appear on each vehicle you service.

Engine Decals

Federal law requires every carmaker to place an "emission control decal" in the engine compartment of every vehicle that lists basic engine and tuneup specifications and instructions, figure 2-13. They give you basic data for tuneup and emission control adjustments, and they supplement printed specifications and procedures. Because carmakers often change engine calibrations or introduce new model and engine combinations in the middle of a year, data printed in manuals may not always be up to date. If data on an emission control decal conflicts with data in a manual, follow the information on the decal.

Some carmakers, particularly Ford, also put decals on their engines to in-

Dist. Part No.	Rotation	Centrigual Advance—Dist° and Dist RPM							
		Start		Intermediate		Intermediate		Maximum	
		RPM	DEG	RPM	DEG	RPM	DEG	RPM	DEG
1112168	CC	600	0–2	900	4–6	1200	6–9	2100	10–12

Figure 2-12. Spark advance specifications are listed by distributor part number. The part number is stamped on the distributor body or on a tag attached to the distributor.

dicate specific calibrations. Tuneup and engine service specifications often refer to these codes, decal numbers, or calibrations. GM, Chrysler, and other manufacturers list the engine codes on the emission control decal. Ford engine code labels are on the engine valve cover or cam cover. Figure 2-14 shows how to determine a Ford engine code and calibration from the emission control decal and the engine code label.

Along with the VIN plate, the engine decal is your most important starting point to determine engine performance service specifications for a particular vehicle. Engine decal styles vary from manufacturer to manufacturer, but all contain the same kinds of information. Every carmaker's shop manual has a section toward the front that tells you how to interpret the information on decals and VIN plates for specific years and models. Also, most independent service manuals that require you to check a calibration or code number will tell you how to find and interpret that number. The specifications on the vehicle itself are your starting point for understanding and using reference manuals.

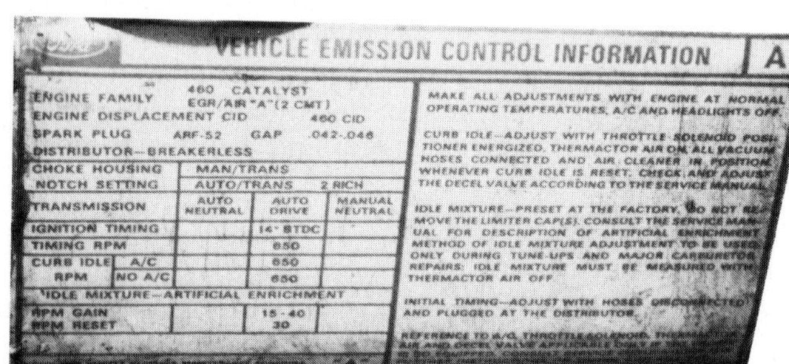

VEHICLE EMISSION CONTROL INFORMATION A

ENGINE FAMILY 460 CATALYST EGR/AIR "A" (2 CMT)
ENGINE DISPLACEMENT CID 460 CID
SPARK PLUG ARF-52 GAP .042-.046
DISTRIBUTOR—BREAKERLESS

CHOKE HOUSING NOTCH SETTING	MAN/TRANS		
	AUTO/TRANS 2 RICH		
TRANSMISSION	AUTO NEUTRAL	AUTO DRIVE	MANUAL NEUTRAL
IGNITION TIMING		14° BTDC	
TIMING RPM		650	
CURB IDLE RPM A/C		650	
CURB IDLE RPM NO A/C		650	
IDLE MIXTURE—ARTIFICIAL ENRICHMENT			
RPM GAIN / RPM RESET		15-40 / 30	

MAKE ALL ADJUSTMENTS WITH ENGINE AT NORMAL OPERATING TEMPERATURES, A/C AND HEADLIGHTS OFF.

CURB IDLE—ADJUST WITH THROTTLE SOLENOID POSITIONER ENERGIZED, THERMACTOR AIR ON, ALL VACUUM HOSES CONNECTED AND AIR CLEANER IN POSITION. WHENEVER CURB IDLE IS RESET, CHECK AND ADJUST THE DECEL VALVE ACCORDING TO THE SERVICE MANUAL.

IDLE MIXTURE—PRESET AT THE FACTORY. DO NOT REMOVE THE LIMITER CAP(S). CONSULT THE SERVICE MANUAL FOR DESCRIPTION OF ARTIFICIAL ENRICHMENT METHOD OF IDLE MIXTURE ADJUSTMENT TO BE USED ONLY DURING TUNE-UPS AND MAJOR CARBURETOR REPAIRS. IDLE MIXTURE MUST BE MEASURED WITH THERMACTOR AIR OFF.

INITIAL TIMING—ADJUST WITH HOSES DISCONNECTED AND PLUGGED AT THE DISTRIBUTOR.

REFERENCE TO A/C, THROTTLE SOLENOID, THERMACTOR AIR AND DECEL VALVE APPLICABLE ONLY IF ...

VEHICLE EMISSION CONTROL INFORMATION · HONDA ACCORD

ENGINE FAMILY IDENTIFICATION ... DISPLACEMENT 112 CID
EVAPORATIVE FAMILY IDENTIFICATION

CATALYST

TUNE UP SPECIFICATIONS
TUNE UP CONDITIONS: ENGINE AT NORMAL OPERATING TEMPERATURE, HEADLIGHTS OFF, COOLING FAN OFF, HEATER FAN OFF, AIR CONDITIONER OFF

IDLE SPEED	5 SPEED TRANSMISSION	750	± 50 rpm
	AUTOMATIC (IN GEAR)	700	
FAST IDLE SPEED	5 SPEED TRANSMISSION	2500	± 500 rpm
	AUTOMATIC (IN NEUTRAL)		
VALVE LASH	IN (MAIN & AUX.)	0.15 mm	COLD
	EX	0.28 mm	COLD
IGNITION TIMING AT IDLE	5 SPEED TRANSMISSION	18 ± 2°	BTDC
	AUTOMATIC (IN GEAR)	18 ± 2°	
PLUG TYPE	NGK: BUR4EB-11, BUR5EB-11; ND: W14EKR-S11, W16EKR-S11 OR EQUIVALENT	SPARK PLUG GAP	1.1 +0 −0.1 mm

IDLE MIXTURE ADJUSTMENT
ADJUSTMENT DURING TUNE UP IS NOT RECOMMENDED. FOR MAJOR REPAIR, ADJUSTING MIXTURE SETTING BY OTHER THAN APPROVED SERVICE PROCEDURE MAY VIOLATE FEDERAL AND STATE LAWS.

THIS VEHICLE CONFORMS TO U.S. EPA AND STATE OF CALIFORNIA REGULATIONS APPLICABLE TO 1985 MODEL YEAR NEW MOTOR VEHICLES PROVIDED THAT THIS VEHICLE IS ONLY INTRODUCED INTO COMMERCE FOR SALE IN THE STATE OF CALIFORNIA.

HONDA MOTOR CO., LTD. JAPAN

TOYOTA — VEHICLE EMISSION CONTROL INFORMATION — TOYOTA MOTOR CORPORATION

ENGINE FAMILY: ETY2.0V5FBB2 EVAP. FAMILY: EV-E
ENGINE DISPLACEMENT: 121.7 CID
EXHAUST EMISSION CONTROL SYSTEM: EFI/EGR/O₂S/TWC

MAKE ALL ADJUSTMENTS WITH ENGINE AT NORMAL OPERATING TEMPERATURE, AIR CLEANER INSTALLED, ALL ACCESSORIES TURNED OFF, COOLING FAN OFF AND TRANSMISSION IN NEUTRAL.
ENGINE TUNE-UP SPECIFICATIONS FOR ALL ALTITUDES

IDLE SPEED (RPM)	700 FOR MANUAL TRANSMISSION 750 FOR AUTOMATIC TRANSMISSION
IGNITION TIMING (°BTDC)	5° @ 950 RPM MAX. WITH VACUUM HOSES DISCONNECTED FROM DISTRIBUTOR AND SEALED.
IDLE MIXTURE SETTING	IDLE MIXTURE SCREW IS PRESET AND SEALED AT FACTORY. ADJUSTMENT DURING TUNE-UP IS NOT RECOMMENDED.
FAST IDLE SPEED	N/A
VALVE CLEARANCE	N/A

THIS VEHICLE CONFORMS TO U.S. EPA AND STATE OF CALIFORNIA REGULATIONS APPLICABLE TO 1985 MODEL YEAR NEW LIGHT-DUTY VEHICLES.

2S-E CAL — CATALYST

DTY 5.0 LITER * D1G5.0W4NDA7 384-TC

VEHICLE EMISSION CONTROL INFORMATION
GENERAL MOTORS CORPORATION

CATALYST AIR/BPEGR/ORC/OC

SET PARKING BRAKE AND BLOCK DRIVE WHEELS.
MAKE ALL ADJUSTMENTS WITH ENGINE AT NORMAL OPERATING TEMPERATURE, EMISSION CONTROL SYSTEM IN CLOSED LOOP MODE, CHOKE FULL OPEN, AIR CLEANER INSTALLED, AND AIR CONDITIONING (A/C) OFF, EXCEPT WHERE NOTED.

1. DISTRIBUTOR: DISCONNECT FOUR WIRE CONNECTOR. SET IGNITION TIMING SPECIFICATION WITH ENGINE RUNNING AT SPEED INDICATED. RECONNECT FOUR WIRE CONNECTOR. CLEAR ECM TROUBLE CODE.
2. IDLE SPEED SCREW: DISCONNECT ELECTRICAL LEAD FROM IDLE SOLENOID, IF EQUIPPED. ADJUST CARBURETOR IDLE SPEED SCREW TO OBTAIN SPECIFIED SPEED. RECONNECT LEAD TO SOLENOID.
3. IDLE SPEED SOLENOID: (WITH A/C) DISCONNECT ELECTRICAL LEAD FROM A/C COMPRESSOR AND TURN A/C ON. (WITHOUT A/C) DISCONNECT ELECTRICAL LEAD FROM SOLENOID AND CONNECT A JUMPER WIRE FROM A +12 VOLT SUPPLY. (ALL) OPEN THROTTLE MOMENTARILY TO ASSURE SOLENOID IS FULLY EXTENDED. ADJUST SOLENOID TO OBTAIN SPECIFIED SPEED. DISCONNECT JUMPER WIRE (IF USED). RECONNECT ELECTRICAL LEAD AND (WHERE EQUIPPED) TURN A/C OFF.
4. FAST IDLE SPEED: DISCONNECT AND PLUG VACUUM HOSE AT EGR VALVE. WITH THROTTLE ON HIGH STEP OF FAST IDLE CAM, ADJUST FAST IDLE SCREW TO OBTAIN SPECIFIED SPEED. NOTE: THIS ADJUSTMENT MUST BE CHECKED WITHIN 15 SECONDS AFTER INCREASING SPEED ABOVE 700 RPM (AUTO) OR 900 RPM (MAN). OPEN THROTTLE TO RELEASE FAST IDLE CAM AND STOP ENGINE. UNPLUG AND RECONNECT VACUUM HOSE TO EGR VALVE.

	LOW ALTITUDE SPECIFICATIONS TRANSMISSION AUTOMATIC	LOW ALTITUDE SPECIFICATIONS TRANSMISSION MANUAL
TIMING (°BTC @ RPM)	6° @ 500 (DR)	6° @ 700 (N)
SPARK PLUG GAP (IN.)	0.045	0.045
IDLE SPEED SCREW (RPM) (SOLENOID INACTIVE)	500 (DR)	700 (N)
IDLE SPEED SOLENOID (RPM) (SOLENOID ACTIVE)	650 (DR)	800 (N)
FAST IDLE SPEED (RPM)	2200 (P) OR (N)	1800 (N)

NOTE: IDLE MIXTURE SCREWS ARE PRESET AND SEALED AT FACTORY. PROVISION FOR ADJUSTMENT DURING TUNE UP IS NOT PROVIDED.

SEE SERVICE MANUAL, MAINTENANCE SCHEDULE AND EMISSION HOSE ROUTING DIAGRAM FOR ADDITIONAL INFORMATION.

THIS VEHICLE CONFORMS TO CALIFORNIA REGULATIONS APPLICABLE TO 1983 MODEL YEAR NEW PASSENGER CARS AND SIMILARLY TO U.S. EPA REGULATIONS APPLICABLE TO THE STATE OF CALIFORNIA. PT. NO. 14065571

Figure 2-13. These engine decals, or "emission control decals," give you basic tuneup and emission specifications and service instructions. Note that the Toyota engine code number, "2S-E," appears on the decal. Toyota does not include an engine code in the VIN.

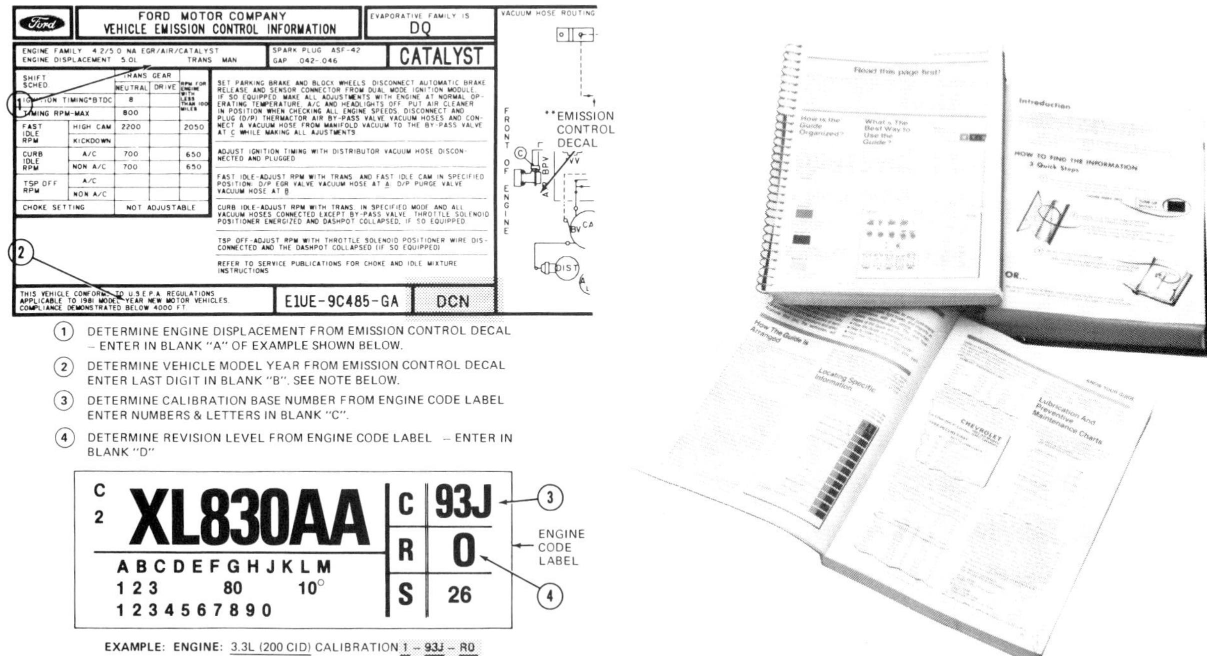

Figure 2-14. Many Ford service procedures refer to calibrations such as 1-93J-R0 or to decal numbers such as DCN. (Ford)

Figure 2-15. The introduction to every reference manual will help you quickly find the information for a specific job.

KINDS OF SPECIFICATIONS AND PROCEDURES

Reference manuals are published by vehicle makers, independent publishers, and suppliers of tools and parts. Before we look at some of these sources, however, we will look at the kinds of information that you will use as tools for your job.

Every reference manual has an introduction, a table of contents, and an index to help you find the information you need fast. These sections tell you how the manual is organized, figure 2-15, and where to find various specifications and procedures. Before using any reference manual, take a few moments to read the introduction and understand how the book is organized. This will save you time in the long run and ensure that you find all of the information for a specific job.

Manuals are usually organized by vehicle systems and major assemblies and by the kind of job to be done. This book is organized in a similar way and for a reason. As you study and use this book, you will become familiar with the general organization of most service and repair manuals. The examples in the following sections illustrate some of the kinds of information you will find in automobile manuals and use on the job.

Preventive Maintenance Schedules

All carmakers publish maintenance schedules for their products. These are the recommended time and mileage intervals for preventive maintenance (PM), and they are the most basic kinds of specifications you will use. The idea of preventive maintenance is simple. It refers to scheduled services that extend vehicle life and help prevent major problems or breakdowns.

Carmakers specify maintenance schedules for three reasons:

1. To help customers get the best service from their vehicles.

2. To protect the manufacturer's new vehicle warranty by listing the necessary maintenance that will keep components in proper condition.

3. To specify necessary services for emission control systems that will keep them operating correctly for 5 years or 50,000 miles.

Federal law requires that the emission control systems of all new cars and light trucks sold in the U.S. since 1972 be warranted for 5 years or 50,000 miles. This means that if any component of the system fails in that time, the manufacturer will replace it. But, the manufacturer can require basic PM services to keep all parts of the emission system working properly. Therefore, maintenance schedules for cars and light trucks cover specified intervals for a 5-year or 50,000-mile period, figure 2-16.

When you think of PM, you probably think of changing the engine oil and filter, lubricating the chassis, and replacing spark plugs and air filters. These are typical jobs that all maintenance schedules include, but they also include other inspections and adjustments, such as:

- Checking all fluid and lubricant levels
- Draining and flushing the cooling system
- Checking tire wear and inflation pressure
- Inspecting consumable or limited-life parts such as brake linings, hoses, and drive belts
- Testing engine exhaust emissions

LUBRICATION AND MAINTENANCE SCHEDULES

SCHEDULED MAINTENANCE SERVICES FOR EMISSION CONTROL AND PROPER VEHICLE PERFORMANCE

Inspection and Service should be performed anytime a malfunction is observed or suspected.

Emission Control System Maintenance	Service Intervals	Mileage In Thousands 7.5	15	22.5	30	37.5	45
		Kilometers In Thousands 12	24	36	48	60	72
Engine Oil Change—Every 12 Months	OR	X	X	X	X	X	X
Engine Oil Filter-Replace At Every Second Oil Change (1)	OR		X		X		X
Apply Solvent To And Check For Freedom Of Operation Of Choke Shaft, Fast Idle Cam, And Pivot Pin (Except E.F.I. Equipped Vehicles)	AT				X		
Replace Spark Plugs (With Cat. Converter)	AT				X		
Replace Spark Plugs (Without Cat. Converter)	AT		X		X		X
Inspect And Adjust Tension On Drive Belts; Replace As Necessary	AT		X(2)		X		X(2)

(1) NOTE: If mileage is less than 7,500 miles each 12 months, replace oil filter at each oil change.
(2) For California vehicles, this maintenance is recommended by Chryser but is not required to maintain the warranty on the air pump drive belt.
NOTE CANADA ONLY: Leaded Fuel Vehicles . . . Check EGR system every 15,000 miles (24 000 km).

GENERAL MAINTENANCE SERVICES FOR PROPER VEHICLE PERFORMANCE

General Maintenance		Service Intervals	Mileage In Thousands 7.5	15	22.5	30	37.5	45
			Kilometers In Thousands 12	24	36	48	60	72
Cooling System	Check & Service As Required Every 12 Months							
	Drain, Flush And Refill At 36 Months Or 52,500 Miles (84 000 Kilometers) And Every 24 Months Or 30,000 Miles (48 000 Kilometers) Thereafter							
Brake Hoses	Inspect For Deterioration And Leaks Whenever Brake System Is Serviced And Every 7,500 Miles Or 12 Months, Whichever Occurs First (Every Engine Oil Change). Replace If Necessary.	OR	•	•	•	•	•	•
Brake Linings & Drums	Inspect	AT				•		
Front Wheel Bearings	Inspect	AT				•		
Ball Joints & Tie Rod Ends	Lubricate	AT				•		

SEVERE SERVICE. . . Severe service is defined as: Stop and go driving, driving in dusty conditions, extensive idling, frequent short trips, operating at sustained high speeds during hot weather (above +90°F, +32°C). The following maintenance intervals apply for this type of operation.

Engine Oil	Change Every 3 Months Or 3,000 Miles (4 800 Kilometers)
Engine Oil Filter	Replace At Every 2nd Oil Change
Transmission Fluid	Change At 15,000 Miles (24 000 Kilometers) Change Filter, Adjust Bands
Axle Oil	Change At 36,000 Miles (58 000 Kilometers)
Front Wheel Bearings	Inspect & Lubricate Whenever The Drums Or Rotors Are Removed To Inspect Or Service The Brake System, Or At Least Every 9,000 Miles (14 000 Kilometers)
Brake Linings	Inspect Every 9,000 Miles (14 000 Kilometers)
Ball Joints & Tie Rod Ends	Lubricate Every 18 Months Or 15,000 Miles (24 000 Kilometers)
Universal Joints	Inspect At Every Oil Change

Figure 2-16. This maintenance schedule for a new car lists recommended services and service intervals for 5 years or 50,000 miles. Notice the different "normal" and "severe-service" recommendations for oil and filter change intervals. (Chrysler)

You can't remember all of the PM jobs and recommended intervals for every car or truck, but your reference manuals will tell you what they are and when to do them.

Normal and Severe Service. Carmakers' PM schedules recommend minimum services and service intervals for vehicles used in "normal" operation. They do this for good reasons. If one manufacturer's car requires half as many oil changes or tuneups as another manufacturer's during five years, the first manufacturer has a marketing advantage. His car is cheaper to operate and maintain. Also, Federal emission warranty regulations encourage carmakers to require minimum service and long life for system components.

Most PM schedules, however, list services and service intervals for "normal" use and for "severe service," figure 2-16. A vehicle in normal use is one that is driven at moderate speed, in moderate temperatures, carrying moderate loads, and on trips long enough to thoroughly warm up the engine and drive train. Normal use is really ideal use and doesn't include carrying heavy loads, continual stop-and-go driving, or operation in extreme temperature and climate conditions. Severe service includes these conditions, which are "normal" for many drivers. When a vehicle is operated under severe-service conditions, it should be serviced according to the carmaker's severe-service schedule.

Tuneup and Emission Service Specifications

Engine tuneup is the collection of engine services needed to maintain the best performance, economy, and emission control. Carmakers include all necessary tuneup specifications and procedures in their shop manuals. But, tuneup specifications can be so extensive and varied that many independent service manuals contain only tuneup data.

On older vehicles with simple emission controls and without electronic engine control systems, tuneup specifications include such things as:

- Battery size and ampere ratings
- Compression pressure
- Spark plug part number, gap, and torque
- Ignition dwell and breaker point gap
- Coil, condenser, and ignition resistor data
- Timing and spark advance specifications
- Fuel pump pressure and volume data
- Carburetor choke adjustment specifications
- Idle speed and mixture adjustment instructions
- Alternator output current and field current draw data
- Starter cranking voltage and current draw
- Valve clearance specifications

Tuneup specifications for late-model vehicles include many of these same items. But, as electronic engine controls have become more sophisticated, many traditional tuneup services are no longer required.

You also may have to test and adjust a vehicle to exhaust emission specifications for a particular city, state, or region. If you work in an area that has such emission regulations, the specifications will be provided by the governing state or local organization and take precedence over the carmaker's specifications for a particular year and model of vehicle, figure 2-17.

Repair and Test Procedures

Many of the instructions you will follow for specific jobs are numbered, step-by-step procedures that take you through the job in an efficient, logical, and safe sequence. Most of the pro-

Vehicle Emission Limits

Vehicle Type	Number Of Cylinders	Year	Conditioning Mode		Pass–Fail	
			HC ppm	CO %	HC ppm	CO %
Passenger Cars With 4-Stroke Engines	4 or fewer	1981 & newer	100	0.50	250	1.50
		1979–80	120	1.00	250	2.50
		1975–78	120	1.00	250	2.50
		1972–74	380	3.50	450	6.00
		1968–71	450	4.25	800	6.50
	6 or 8	1981 & newer	100	0.50	250	1.50
		1979–80	120	1.00	250	2.20
		1975–78	120	1.00	250	2.20
		1972–74	300	3.00	400	5.50
		1968–71	380	3.50	750	6.50
Trucks And Vans	All	1979 & newer (more than 8500 lbs)	300	3.00	350	5.00
		1975–78 (more than 6000 lbs)	300	3.00	350	5.00
		1968–74	Same as passenger cars			

Figure 2-17. If your state, county, or city has a vehicle emission inspection program, you will test vehicles to specifications such as these.

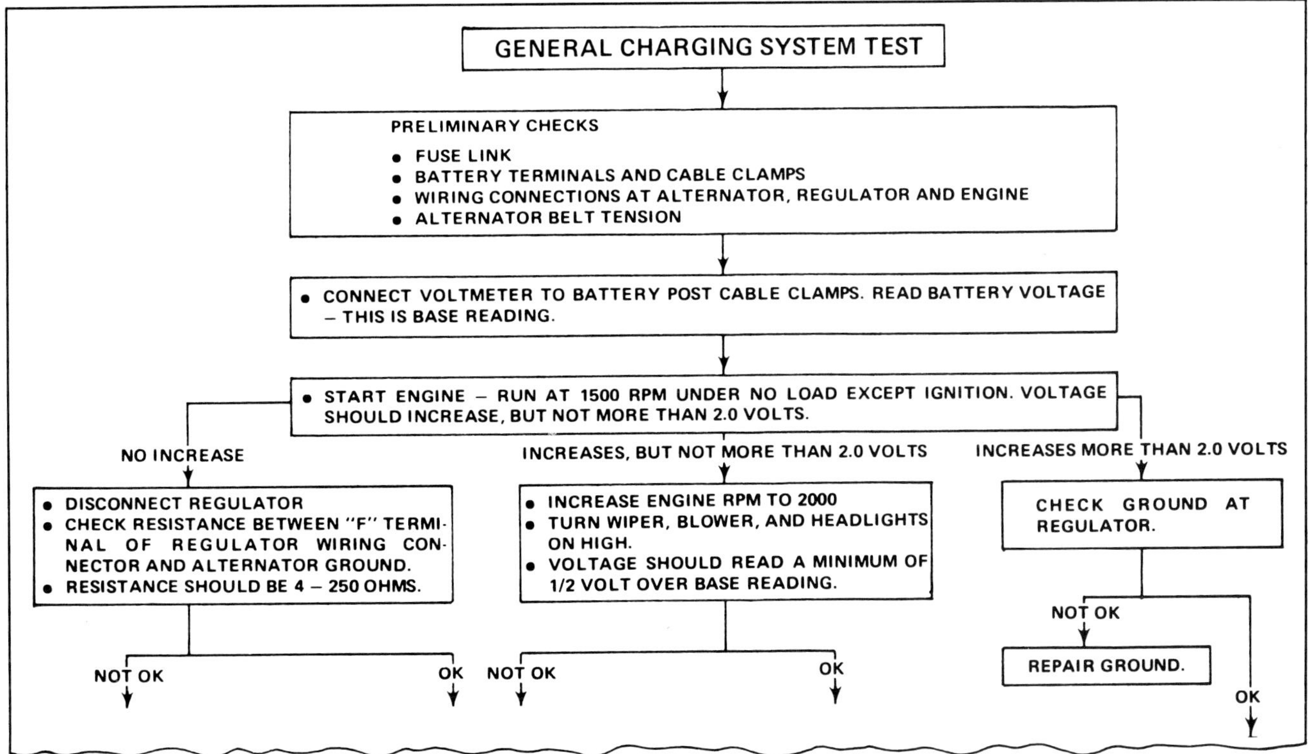

Figure 2-18. This troubleshooting flowchart has "yes/no" or "pass/fail" decisions at key test points. (Ford)

cedures in this manual are written in that style, even though they are more general than a carmaker's instructions.

Most service procedures are written for specific jobs that make up a logical service operation. These tasks usually have corresponding flat-rate times. Thus, you will find a procedure for removing and replacing a carburetor and a matching flat-rate time for that job. Then you will find another procedure for overhauling the carburetor and a flat-rate time for that job. This organization of job procedures and flat-rate times helps shop managers and mechanics organize their time, understand exactly what work is to be done, and charge the customer a fair price.

Some of the reference manuals you use also will have troubleshooting charts and diagrams. These are test procedures in chart or picture form that will lead you through a troubleshooting or diagnostic job that has many variables. When you troubleshoot a hard-starting problem on an engine, for example, there may be several possible causes. You must check each possibility in a logical and efficient sequence until you identify the cause, or several causes, of the problem.

Troubleshooting charts come in many styles. Some are flowcharts, or "Christmas tree" charts, figure 2-18, with "yes/no" or "pass/fail" decisions at certain test points. Others are pictorial charts, figure 2-19. Still others are 3-column tables that list a symptom, possible causes, and cures, figure 2-20. Some charts are combinations of step-by-step instructions, tables, and diagrams. No one style is any better than the other. Each is designed by a particular publisher to lead you through a specific job, and you will work with all kinds.

Introduction to Diagrams

Electrical, vacuum, and hydraulic diagrams are reference information in pictorial form. These diagrams show you, on paper, what the systems contain and how the parts work together. Diagrams are necessary tools for troubleshooting.

All diagrams illustrate a circuit, or several circuits. We usually think of circuits as part of an electrical system, but vacuum and hydraulic systems also operate with circuits. A diagram shows the source of electric current, vacuum, or hydraulic pressure. It then shows the current, vacuum, or pressure passing through wires or lines to different devices in the circuit to do various jobs. When you use any kind

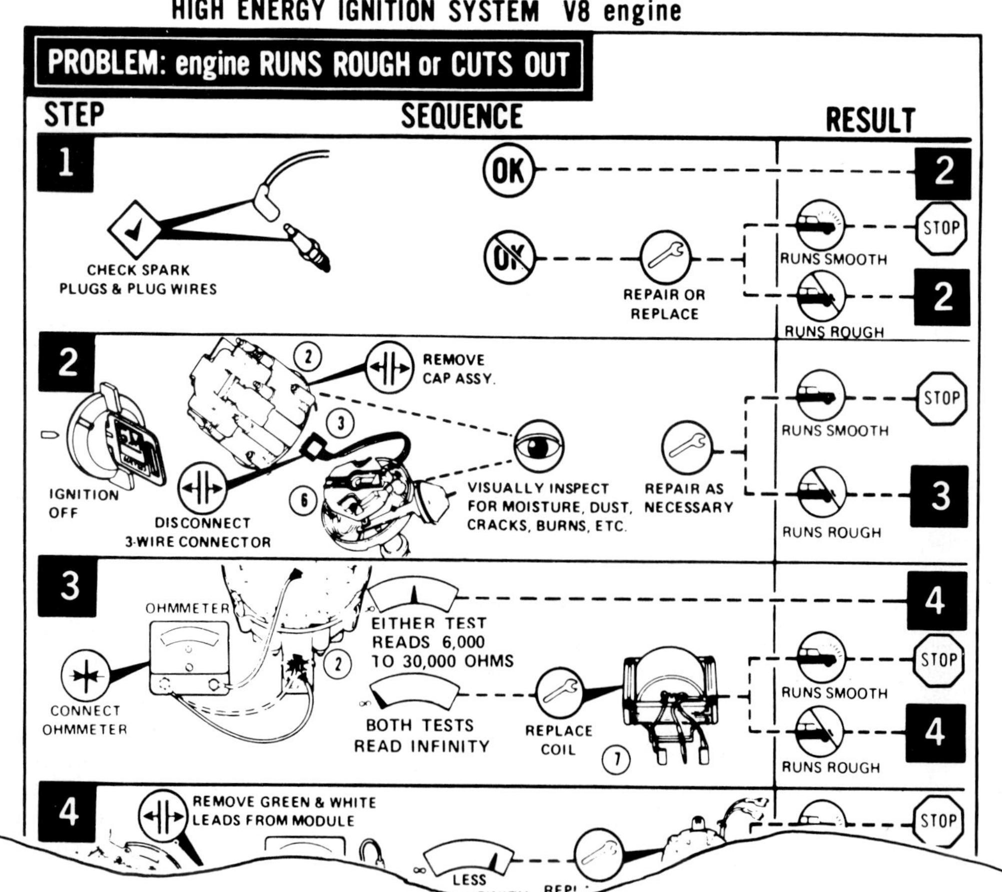

Figure 2-19. This kind of pictorial troubleshooting chart was developed by General Motors for many test and repair procedures. (Chevrolet)

ENGINES

THROTTLE BODY FUEL INJECTION SYSTEM

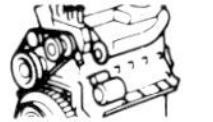

CODE INTERPRETATION AND DIAGNOSIS

The chart below contains a list of troubles, possible causes and suggested corrections.

TROUBLE CODE/CONDITION	POSSIBLE CAUSE	CORRECTION
CODE 1 (unsatisfactory low air temp. engine performance).	Manifold air/fuel temperature (MAT) sensor resistance is not less than 1000 ohms (HOT) or more than 100 kohms (VERY COLD).	Replace the MAT sensor if not within specifications. Refer to the MAT sensor test procedure.
CODE 2 (unsatisfactory warm temp. engine performance-engine lacks power).	Coolant temperature sensor resistance is less than 300 ohms or more than 300 kohms (10 kohms at room temp.).	Replace the coolant temperature sensor. Test the MAT sensor. Refer to the coolant temp. sensor test and MAT sensor test procedures.
CODE 3 (unsatisfactory fuel economy, hard cold engine starting, stalling, and rough idle).	Defective wide open throttle (WOT) switch or closed (idle) throttle switch or both, and/or associated wire harness.	Test WOT switch operation and the associated circuit. Refer to the WOT switch test procedure. Test closed throttle switch operation and the associated circuit. Refer to the closed throttle switch test procedure.
CODE 4 (unsatisfactory engine acceleration, sluggish per-formance ...	Simultaneous closed throttle switch and manifold absolute ...ure (MAP) sensor failure.	Test the closed throttle switch and repair/replace as necessary. Refer to the clos...

Figure 2-20. Three-column tables are another form of troubleshooting chart that you will use.

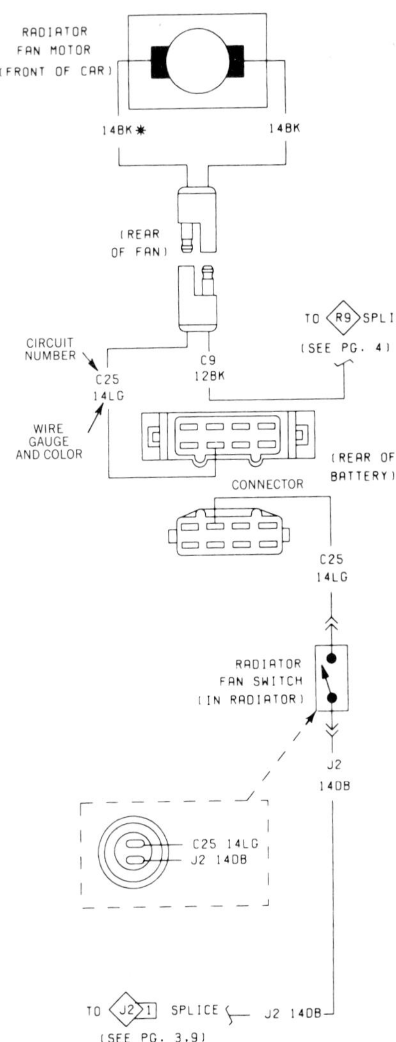

Figure 2-21. This electrical diagram of a simple motor circuit shows you how the motor is connected and how it works, but it doesn't show you what the motor looks like physically. (Chrysler)

ELECTRICAL DEVICE	SYMBOL: UNITED STATES	SYMBOL: EUROPE
BATTERY		
SWITCH, SINGLE POLE, NORMALLY OPEN		
SWITCH, SINGLE POLE, NORMALLY CLOSED		
SWITCH, GANGED		
RESISTOR		
VARIABLE RESISTOR		
DIODE		
ZENER DIODE		
TRANSISTOR	(PNP TYPE)	(NPN TYPE)
THERMISTOR		
CIRCUIT BREAKER		
FUSE		
CAPACITOR (CONDENSER)		
LAMP, SINGLE FILAMENT		
LAMP, MULTIFILAMENT		
CONNECTOR	MALE FEMALE	
MULTIPLE CONNECTOR		
IGNITION COIL		
RELAY		
GROUND		
SOLENOID		
MOTOR	M	M
GENERATOR (ALTERNATOR)	G	G
WIRING SPLICE		

of diagram, you will be tracing a circuit.

Electrical, vacuum, and hydraulic diagrams use symbols to represent parts of the systems. The symbols don't show exactly what the parts look like physically. They show how the parts function and how they are connected to each other, figure 2-21. Many symbols for switches, connectors, fuses, motors, valves, and pumps are standardized and used by all carmakers, figure 2-22. Other symbols are not, but

Figure 2-22. These are some of the standard symbols that you will find in most electrical diagrams.

most diagrams have legends, or callouts, that tell you what the symbols represent.

Electrical Diagrams. Electrical diagrams are either schematic diagrams, wiring diagrams, or installation diagrams. Schematic diagrams show the connectors, switches, loads, and other devices. Schematic symbols show how the loads and switches work. Wiring diagrams show the wires, connectors, and connections to loads and switches but not necessarily how the loads and switches work. Installation diagrams show where and how the wiring and electrical devices are installed on a vehicle. They include bulbs, fasteners, and mounting hardware, but they don't show the individual wires, circuits, and load functions.

Most carmakers combine the functional features of schematics and wiring diagrams and provide installation diagrams to show you how parts and wiring are installed. Figure 2-21 is a diagram that combines the schematic symbols for a motor, a relay, and a switch with the circuit numbers, color coding, and connection points of a wiring diagram. Figure 2-23 is an installation diagram that shows how those parts are installed and connected on a car.

You may work with electrical diagrams that show a complete vehicle electrical system or just one circuit and its parts.

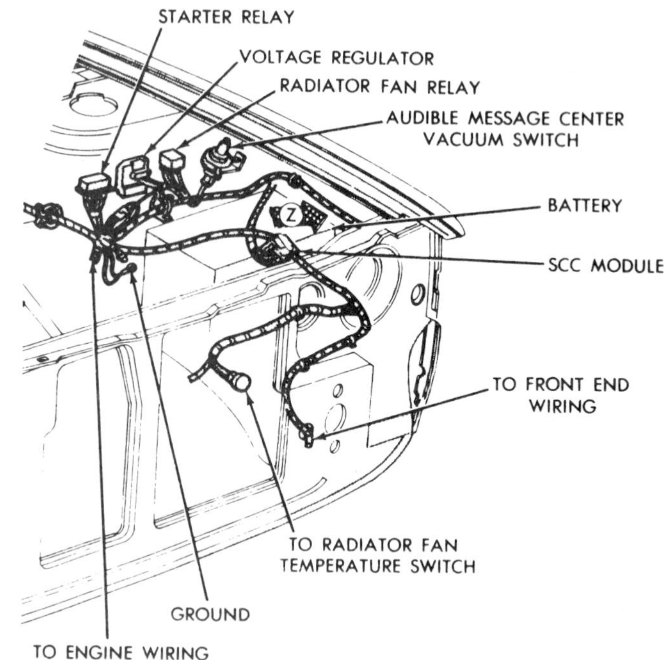

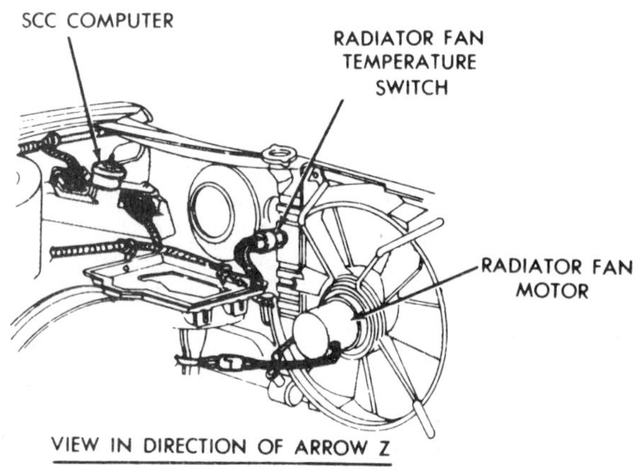

Figure 2-23. This installation diagram shows how the motor from figure 2-21 is installed. (Chrysler)

Vacuum Diagrams. Vacuum control systems are part of all vehicle emission and engine control systems. You will find several vacuum systems on every vehicle, and they often are interconnected. Vacuum diagrams show the routing and connections of vacuum hoses, figure 2-24, as electrical diagrams show the same features of wiring. Vacuum diagrams, however, do not always show the internal functions of vacuum valves and actuators. Late-model vehicles have at least one vacuum diagram on a decal in the engine compartment that shows the basic vacuum system for the emission controls on that vehicle.

The wires and hoses in most electrical and vacuum systems are colored, and diagrams use color coding to show the routing and connections of hoses and wires. Many carmakers print their diagrams in color. Others list the color coding on the diagram.

Hydraulic Diagrams. When most people think of hydraulic diagrams, they think of diagrams for brake systems and automatic transmissions. But, fuel injection systems—gasoline and diesel—are complex hydraulic systems that work on the same principles of controlled force and pressure as other hydraulic systems.

Hydraulic diagrams for fuel injection systems are complex drawings that show fluid circuits and pressures in different operating conditions, figure 2-25. Because several fuel system circuits operate at the same time and with different pressures, carmakers often color code their fuel injection diagrams.

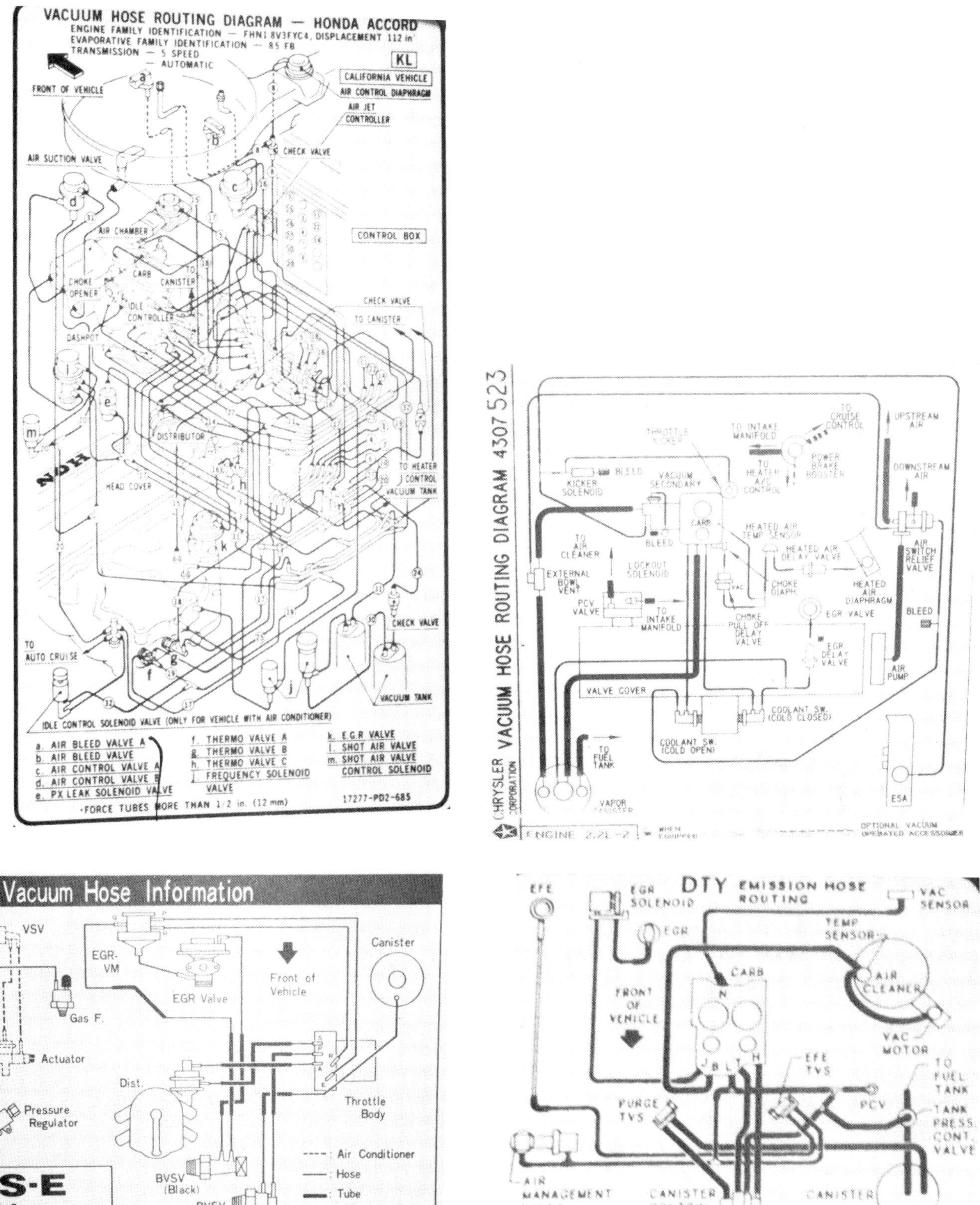

Figure 2-24. Vacuum diagrams show you the routing and connections of vacuum hoses. They are often color coded like electrical diagrams.

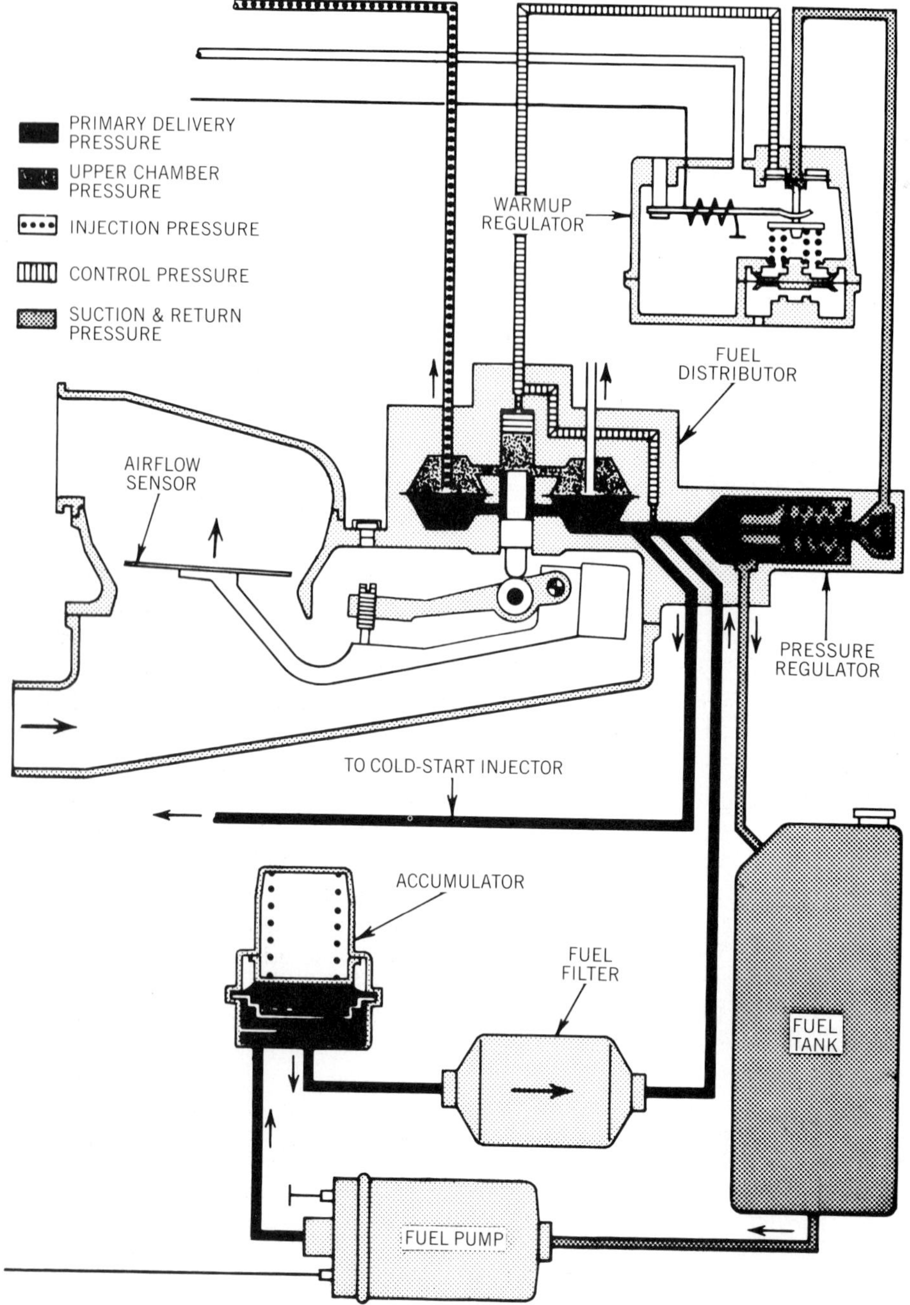

Figure 2-25. This fuel injection diagram illustrates different hydraulic pressures in the system.

REFERENCE MANUAL SOURCES

You will find specifications, procedures, flat-rate times, diagrams, and other information in manuals from three basic sources:

- Vehicle manufacturers' shop and service manuals and parts catalogues
- Service and repair manuals from independent publishers
- Service and repair manuals and parts catalogues from tool, equipment, and parts manufacturers

Carmakers' Manuals

Vehicle manufacturers usually publish car and truck shop manuals for one model year only. However, many of the repair procedures are carried over from one year to the next for unchanged components. Because specifications can change greatly from one year to another, even though repair procedures may not, always be sure that you use a shop manual for the specific year of the vehicle you are servicing.

Carmakers' manuals contain comprehensive specifications and test and repair procedures for all vehicle systems. Many are published as several volumes for each model year. Some manufacturers print separate manuals for different models; others combine several carlines in one manual or set of manuals. These manuals are intended primarily for dealers' service departments, but they are available to vehicle owners and independent mechanics.

Manufacturers also supply their dealers with parts catalogues, flat-rate manuals, and periodic service bulletins for the parts and service departments. These usually are not available directly to independent mechanics, but some of the information is republished by independent publishers.

You can buy factory manuals directly from the vehicle manufacturers or from car dealers. Most imported car dealers sell the factory manuals through their parts departments, or the dealer can provide an address so that you can order manuals from the manufacturer. Most dealers for domestic vehicles do not sell shop manuals, but you can order them by mail from the manufacturers. The addresses for ordering shop manuals from the major domestic carmakers are listed below. Most of these sources will send you a catalogue of available manuals upon request.

Chrysler Corporation
(including AMC, Renault, Jeep, and Eagle)
Dyment Distribution
20770 Westwood Road
Strongsville, OH 44132
(216) 572-0725

Ford Motor Company
(including training materials)
Helm Incorporated
P. O. Box 07150
Detroit, MI 48207
(313) 883-1430

General Motors Corporation

Buick Division
Tuar Co.
P. O. Box 1910
Flint, MI 48501
(313) 239-5552

Cadillac, Chevrolet, Pontiac, and Hydra-matic Divisions
Helm Incorporated
P. O. Box 07130
Detroit, MI 48207
(313) 883-1430

Oldsmobile Division
Lansing Lithographer
P. O. Box 23188
Lansing, MI 48909
(517) 482-0697

GMC Truck & Coach Division
Adista Corporation
171 Hamilton Street
Plymouth, MI 48107
(313) 455-0055

Service Technology Group (training materials)
Kent-Moore Tool & Equipment Division
SPX Corporation
28635 Mound Road
Warren, MI 48092-9923
(800) 468-6657

Honda
American Honda Service Department
100 West Alondra Blvd.
Gardena, CA 90257
(213) 327-8280

American Isuzu Motors
Helm Incorporated
P. O. Box 07280
Detroit, MI 48207
(313) 883-1430

Mazda Motors
Helm Incorporated
P. O. Box 07280
Detroit, MI 48207
(313) 883-1430

Mitsubishi Motor Sales
6400 W. Katella Ave.
Cypress, CA 90630
(714) 963-7677

Nissan
Dyment Distribution
20770 Westwood Road
Strongsville, OH 44132
(216) 572-0725

Toyota Motor Sales
Service Department
19001 S. Western Avenue
P. O. Box 2991
Torrance, CA 90509
(800) 443-7656—Calif.
(800) 622-2033—49 states

Volkswagen
Dyment Distribution
20770 Westwood Road
Strongsville, OH 44132
(216) 572-0725

Volvo of America Corporation
P. O. Box 627
Madison Heights, MI 48071

Robert Bosch Sales Corporation
UA/AMA Library
2800 South 25th Avenue
Broadview, IL 60153
(800) 937-2672

Tool, Equipment, and Parts Manufacturers' Manuals

Every piece of test and repair equipment in your shop has an instruction manual that tells you how to operate the equipment and how to use it on different vehicles, figure 2-26. Many tool and equipment manufacturers also provide mechanics with specification manuals to help use their tools and equipment. Figure 2-27 shows tuneup specifications and test procedures provided by equipment companies.

These specification books usually cover all popular cars and trucks. Sometimes they have information for one model year, sometimes for 5, 7, or 10 years. All of the specifications are the original carmakers' specifications. The tool and equipment companies simply gather them from the

Figure 2-26. These equipment manuals have instructions for equipment operation and test or service specifications for various cars.

Figure 2-27. These service specification books from parts and equipment manufacturers contain the carmakers' original engine tuneup data.

carmakers' manuals and put them in one place to make your job easier.

Parts manufacturers and tire, battery, and accessory (TBA) suppliers also publish specification manuals, as well as parts catalogues. The specification manuals are similar to those from tool and equipment companies. The parts catalogues have replacement part descriptions and numbers that are equivalent to the original equipment manufacturer's (OEM) part numbers. Many parts catalogues also have cross reference listings of equivalent part numbers from several suppliers, figure 2-28.

Figure 2-28. These part number cross reference lists from TBA parts books show equivalent part numbers from several suppliers.

Independent Manual Publishers

It's often worthwhile to have the complete factory shop manuals for some cars you service, particularly your own or those in which you specialize. But if you built a reference library of every shop manual for every year, you would soon fill the shop with books and spend a lot of money. Therefore, independent publishers have been serving professional mechanics for years by compiling the carmakers' specifications, procedures, and flat-rate standards into independent service and repair manuals.

The following sections summarize the kinds of manuals available from some of these publishers. Some companies sell their manuals through local sales representatives throughout the country. Others sell them by mail or through automobile book and parts stores. We have listed the mailing address for each publisher so that you can write for more information about their manuals.

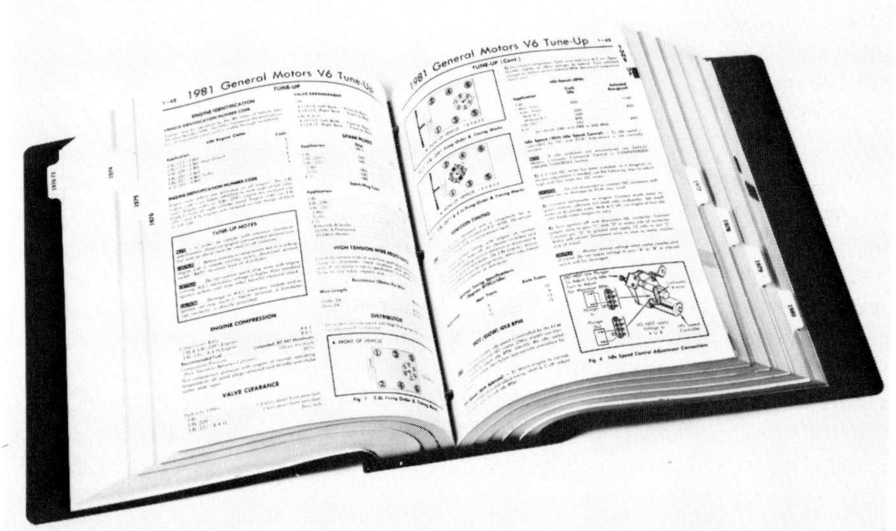

Figure 2-29. Mitchell publishes tuneup procedures and specifications by model year. (Mitchell)

Mitchell Manuals

Address:

Mitchell Manuals, Inc
P. O. Box 26260
San Diego, CA 92126
(800) 854-7030

Mitchell Manuals publishes over 25 separate service and repair manuals. These include tuneup and engine diagnosis, figure 2-29, general mechanical repair, transmission repair, heat-

Figure 2-30. This is the basic 4-volume Mitchell Manual tuneup and electrical service set for domestic cars. (Mitchell)

ing and air conditioning service, collision repair, labor and parts estimating guides, and vacuum and electrical diagram manuals. Mitchell gathers much of the information in its manuals from vehicle makers' shop manuals, but Mitchell also does much independent research and development of repair procedures, particularly for collision repair manuals and flat-rate labor guides.

The basic Mitchell manuals are for domestic cars, imported cars, and light trucks. The domestic car set of manuals is 12 volumes, organized by vehicle system. It contains four volumes of tuneup and electrical data, figure 2-30; two volumes of emission specifications, procedures, and diagrams; two volumes of transmission specifications and service procedures; two volumes of all other mechanical specifications and procedures; and two volumes of air conditioning data.

The master Mitchell Manuals sets for domestic cars, domestic trucks, and imported cars and trucks cover 10 model years. Mitchell manuals include carmakers' test procedures for electronic engine control systems, as well as fuel, ignition, and emission control service information. Mitchell also publishes specialized volumes on heating and air conditioning, transmission repair, and electrical component locations.

Most of the complete Mitchell repair manual library is available on a computerized reference library system, or database, called *Mitchell On-Demand*. This system stores information on compact discs and displays it on the screen of an IBM-type personal computer (PC). Data also can be printed for use on the job.

Mitchell also operates a telephone assistance "hotline" service, *Mitchell On-Call*, which allows technicians to pay a subscription fee and then telephone for assistance with difficult test and repair problems. Additionally, Mitchell provides training seminars on engine performance service and diagnosis.

Mitchell's parts and labor guides supplement the basic repair manuals and contain flat-rate times, OEM part numbers and prices, and mechanic skill codes. The Mitchell interchange guides list parts that are interchangeable among different vehicles, along with part numbers, prices, and modification instructions.

Motor's Manuals

Address

Motor Publications
The Hearst Corporation
5600 Crooks Road
Troy, MI 48098
(800) 426-6867

Motor publishes separate 1-volume repair manuals for domestic cars, imported cars, and light and heavy trucks annually, figure 2-31. They contain specifications and repair procedures for current model-year vehicles, plus the six preceding years (seven years total coverage). Motor's manuals have two major sections. The general repair section contains test and repair procedures and specifications for common assemblies such as alternators, transmissions, brakes, and other parts used by several manufacturers. The car information section is organized alphabetically by carmaker and has detailed specifications and procedures for various years and models built by each manufacturer, figure 2-32.

Motor compiles information for its manuals from carmakers' shop manuals and factory sources. The information represents OEM specifications and procedures. Besides the basic car and truck repair manuals, Motor also publishes detailed special manuals on:

- Flat-rate time and parts estimating
- Heavy truck and diesel repair
- Automatic transmission repair
- Engine and electrical system repair
- Air conditioning service
- Tuneup and emission control service
- Vacuum and electrical system diagrams

The Hearst Corporation also publishes the monthly *Motor* magazine for professional mechanics, which contains articles on new service equipment and procedures, update information for Motor's manuals, and selected information from carmaker's service bulletins.

Chilton Manuals

Address:

The Chilton Book Company
Chilton Way
Radnor, PA 19089
(800) 695-1214

Chilton publishes three annual 1-volume service manuals for domestic cars, imported cars, and trucks and vans. Figure 2-33 shows a few of Chilton's professional manuals. Each manual provides seven model years of coverage, and all specifications and procedures are compiled from carmakers' shop manuals and factory sources.

Figure 2-31. Motor publishes domestic and imported car and light truck manuals annually with seven years of specifications and procedures. (Motor)

Chilton's manuals contain a manufacturer section and a unit repair section. The manufacturer section is organized by corporation (AMC, Chrysler, Ford, and GM, for example) and then by division or carline. This section contains specifications and detailed procedures for all models built by each carmaker. The unit repair section has common test and repair procedures for major assemblies used on several carlines by various manufacturers, figure 2-34.

Additionally, Chilton publishes several special manuals, including:

- Heating and air conditioning manual
- Labor guide (flat-rate) and parts manual
- Emission control service manual
- Wiring diagram manuals

For professional mechanics Chilton also publishes the monthly *Motor Age* magazine that has articles on new service procedures and equipment, engineering trends, and selected information from factory service bulletins.

Gousha/Chek-Chart

Address:

Gousha/Chek-Chart
P. O. Box 49006
San Jose, CA 95161-9006
(408) 296-1060

Gousha/Chek-Chart publishes preventive maintenance, lubrication, and tuneup specifications. The annual *Car Care Guide*, figure 2-35, contains all PM schedules, lubrication data, and tuneup specifications for seven years of domestic cars and light trucks and major imports. The *Car Care Guide* does not contain the extensive repair procedures and mechanical specifications found in other independent manuals, but it has maintenance schedules and lubrication data not readily available from other publishers. Moreover, Gousha/Chek-Chart compiles all information from OEM sources in a uniform, simple format for fast reference. The *Car Care Guide* is distributed principally through oil companies to their service station dealers, but it also can be ordered directly from the publisher.

Haynes Publications

Address

Haynes Publications
861 Lawrence Drive
P. O. Box 978
Newbury Park, CA 91230
(800) 533-4124—Calif.
(800) 442-9637—49 states

Haynes publishes maintenance and repair manuals for individual models of domestic and imported cars and light trucks. Haynes manuals are intended primarily for the do-it-yourself market, but all information comes from factory service manuals or independent research by the Haynes staff.

Haynes manuals are alternatives to factory shop manuals, particularly for

CHRYSLER CORP.—Front Wheel Drive

TUNE UP SPECIFICATIONS—Continued

The following specifications are published from the latest information available. This data should be used only in the absence of a decal affixed in the engine compartment.

▲Before removing wires from distributor cap, determine location of the No. 1 wire in cap, as distributor position may have been altered from that shown at the end of this chart.

● When checking compression, lowest cylinder must be within 25 PSI of the highest.

Spark plug types shown in this chart are recommendations of the original vehicle manufacturer and not MOTOR. Check local sources for other spark plug manufacturers listings.

Year & Engine/VIN Code	Spark Plug		Ignition Timing BTDC①★				Curb Idle Speed		Fast Idle Speed		Fuel Pump Pressure
	Type	Gap	Firing Order Fig. ▲	Man. Trans.	Auto. Trans.	Mark Fig.	Man. Trans.	Auto. Trans.	Man. Trans.	Auto. Trans.	
1983											
4-97/A	⑬	.035	E	12°	—	H	⑭	—	1400	—	4½–6
4-105/B	⑬	.035	A	20°⑫	12°	⑮	850	900N	1400	1350	4½–6
4-135/C⑯	⑬	.035	D	10°	10°	C	775	900N	1400	1500	4½–6
4-135/C⑰	⑬	.035	D	6°	6°	C	900	850N	1350	1375	4½–6
4-135/F	②⑬	.035	D	15°	—	C	850	—	1500	—	4½–6
4-135/D	⑬	.035	D	—	—	⑮	—	⑱	—	—	—
4-156/G	⑬	.040	F	—	7°	G	—	800N	—	—	4½–6
1984											
4-97/A	⑬	.035	E	12°	12°	H	850	1000N	—	—	4½–6
4-135/C	⑬	.035	D	10°	10°	⑮	800	900N	1500	1600	4½–6
4-135/F	②⑬	.035	D	15°	—	C	850	—	1500	—	4½–6
4-135/D	⑬	.035	D	6°	6°	⑮	⑲	⑳	—	—	—
4-135/E	⑬	.035	D	12°	12°	C	㉑	㉑	—	—	—
4-156/G	⑬	.040	F	—	7°	G	—	800	—	950	4½–6

Figure 2-32. This is an example of tuneup data from the Motor domestic car manual. (Motor)

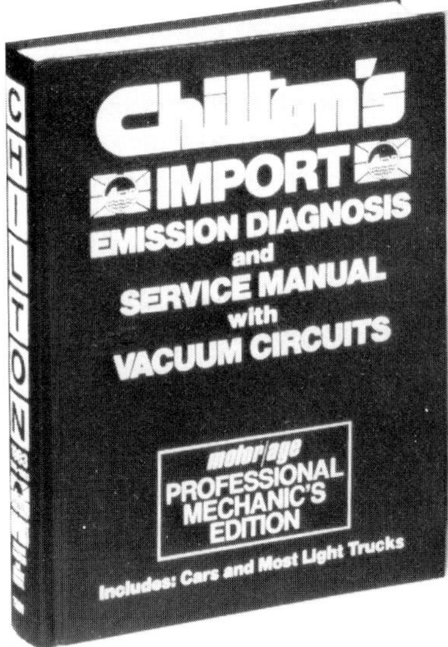

Figure 2-33. Besides its annual domestic and imported car and truck manuals, Chilton also publishes manuals on air conditioning, emission control, and other special services. (Chilton)

older obscure imports. They are available in many auto parts stores, or they can be ordered from the publisher.

Robert Bentley, Inc.

Address

Robert Bentley, Inc.
1000 Massachusetts Avenue
Cambridge, MA 02138
(800) 423-4595

Robert Bentley is the factory-authorized publisher of Volkswagen and Audi shop manuals for North America. Bentley also publishes comprehensive service manuals for other selected domestic and imported vehicles, particularly older British models. Bentley manuals are available through many book dealers or directly from the publisher.

Computerized Reference Libraries and Telephone Hotline Services

The last several years have seen an increase in the production of repair and service manuals on computer systems. These systems basically package traditional manuals on computer discs or compact disc read-only memory (CD-ROM), and they display the information on the screen of a personal computer. When the user enters the

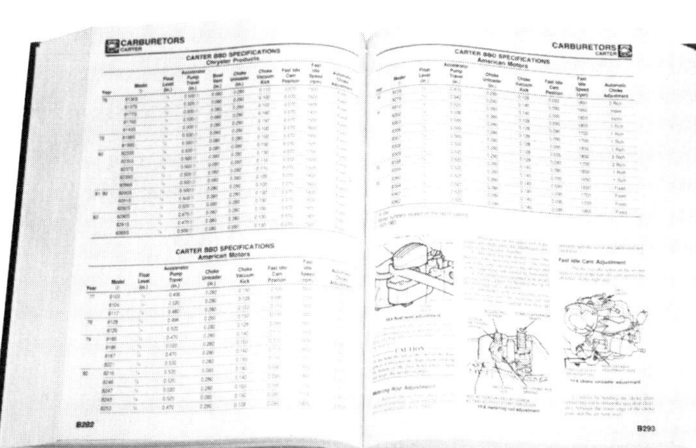

Figure 2-34. This is an example of overhaul instructions from Chilton's unit repair section.

Figure 2-35. The Gousha/Chek-Chart *Car Care Guide* contains PM schedules, lubrication data, and tuneup specifications in a uniform, quick-reference format.

vehicle identification into the computer by following on-screen instructions, the computer sorts the data and displays the information requested by the user. In this way, the computer "turns the pages" of the manuals for you.

In the mid-1980's General Motors was an innovator in this field with its *Expertec* database system, which is now produced by Mitchell International. Other systems of this kind are Mitchell's own *On-Demand* reference library and the system by AllData of Sacramento, California. In the fast-moving field of computerized information technology, it is still too early to judge the ultimate effectiveness of these systems.

Also in the mid-1980's several companies began offering telephone "hot-line" diagnostic assistance services. These programs allow a technician or garage owner to pay a subscription fee and then call a toll-free number for help in repairing or troubleshooting vehicles. Such services rely on the experience of expert technicians who answer the phone calls, supported by comprehensive reference manual libraries. Although these services have been popular and somewhat successful, it is often difficult for even the most experienced technician to fix a car several thousand miles away when he cannot see and touch it.

SUMMARY

Reference manuals are essential tools for modern automobile service. A professional serviceperson knows how to use reference manuals and how to understand specifications and procedures.

Descriptive specifications identify a vehicle, an assembly or system, or an individual part by type and by name. Service specifications indicate an adjustment or a measurement. Model, part, and serial numbers are specifications that identify specific items.

The most important specification number for any vehicle is the vehicle identification number (VIN), located on a plate inside the windshield. Engine decals contain specifications and instructions for engine testing and adjustment. These decals are your starting point for information on engine tuneup.

Carmakers' service schedules are preventive maintenance (PM) specifications that list specific vehicle services and time or mileage intervals. These schedules distinguish between vehicle use in normal service and severe service.

Specifications tell you *what*. Procedures tell you *how* to do a job, step by step. Procedures may be a numbered sequence of steps for a job. Procedures also may be in chart form. Flat-rate time standards tell you how long a job should take. When you do any job on an automobile, you will use all three kinds of reference information: specifications, procedures, and flat-rate standards.

You will find essential specifications and procedures in carmaker's service manuals and in reference manuals from a number of independent publishers.

REVIEW QUESTIONS

Multiple Choice

1. Flat-rate time standards are:

a. the maximum amount of time it takes to do a job
b. the carmaker's recommended hourly wage for specific jobs
c. the carmaker's suggested average time for a job
d. the carmaker's recommended price for a job

2. Descriptive specifications give you:

a. step-by-step instructions for a service job
b. general information about a vehicle or a component
c. information about the adjustment and operation of specific parts
d. part number interchangeability information

3. An engine decal tells you the:

a. engine serial number
b. vehicle identification number (VIN)
c. ignition and fuel system major part numbers
d. the basic tuneup and emission service specifications

4. Service and operating specifications give you:

a. step-by-step instructions for a service job
b. general information about a vehicle or a component
c. information about the adjustment and operation of specific parts
d. part number interchangeability information

5. To determine the basic specifications for any vehicle, you should start by checking the:

a. engine decal
b. vehicle identifcation number (VIN)
c. vehicle registration number
d. factory shop manual or equivalent repair manual

6. Procedures give you:

a. step-by-step instructions for a service job
b. general information about a vehicle or a component
c. information about the adjustment and operation of specific parts
d. part number interchangeability information

7. Mechanic A says that two components with the same model number will interchange with each other. Mechanic B says that two components with the same part number will interchange with each other. Who is right?

a. A only
b. B only
c. both A and B
d. neither A nor B

8. When you adjust the gap of an R46TS spark plug to 0.045 inch for a 1986 302-cid (5-liter) V-8 engine with fuel injection, you are using:

a. service specifications
b. descriptive specifications
c. part numbers
d. all of the above

9. Emission control systems on new cars are warranted for at least 5 years or 50,000 miles because:

a. Mr. Ferrari wants it that way
b. most car owners keep their vehicles for 5 years or less
c. the Federal government requires it
d. that is the useful life of most emission control systems

10. Mechanic A says that "severe service" maintenance schedules are for taxicabs and commercial vehicles. Mechanic B says that "severe service" maintenance schedules apply to many privately owned cars. Who is right?

a. A only
b. B only
c. both A and B
d. neither A nor B

11. The primary reason for troubleshooting charts and diagrams is to:

a. replace words with pictures
b. help a service person deal with many variables
c. help estimate the time needed for a job
d. determine warranty labor cost allowances

12. Mechanic A says that the engine code is always part of the vehicle identification number (VIN). Mechanic B says that the engine code is the same as the engine serial number. Who is right?

a. A only
b. B only
c. both A and B
d. neither A nor B

13. If you need to know how a particular switch works internally, you should check:

a. an installation drawing
b. a wiring diagram
c. a schematic diagram
d. any of the above

14. Distributor spark advance specifications are usually listed by:

a. distributor model
b. engine code
c. vehicle identification number (VIN)
d. distributor part number

15. Preventive maintenance schedules are for:

a. lubrication service
b. emission control and tuneup service
c. vehicle chassis, consumable-parts, and fluid services
d. all of the above

16. Mechanic A says that most engine vacuum hoses are color coded for testing and service. Mechanic B says that most cars have a vacuum diagram decal in the engine compartment to aid testing and service. Who is right?

a. A only
b. B only
c. both A and B
d. neither A nor B

17. Mechanic A says that you can always check a factory shop manual for accurate specifications and procedures. Mechanic B says that independent service manuals contain carmaker's OEM specifications and procedures. Who is right?

a. A only
b. B only
c. both A and B
d. neither A nor B

18. Hydraulic diagrams are used to test and service:

a. brake systems
b. diesel fuel injection systems
c. gasoline fuel injection systems
d. all of the above

19. On a late-model car, you can find the vehicle identification number:

a. on the VIN plate, usually behind the windshield
b. on a body code plate or an engine certification label
c. on the engine or transmission housing
d. all of the above

3 WORKING WITH COMMON REPAIR MATERIALS

INTRODUCTION

When you service fuel, electrical, ignition, and emission control systems, you will work with several kinds of common repair materials. These include tubing, fittings, hoses, gaskets, seals, sealants, and adhesives. Some of these items, such as metal tubing and fittings, have been used unchanged on automobiles for decades. Many other materials, such as modern adhesives and sealants, have benefited from technological development and have features unknown in automobile service a generation ago. This chapter summarizes the features of these materials and the correct ways to work with them.

GOALS

After studying this chapter, you should be able to:

1. Identify and explain the uses of various kinds of metallic tubing, plastic tubing, and hoses.

2. Select and install various kinds of tube fittings as required for different applications.

3. Identify and demonstrate the correct uses and installations of various gaskets, seals, adhesives, and sealants.

TUBING, PIPE, AND HOSE

As a professional mechanic, you will work with several kinds of metal and plastic tubing, pipe, hoses, and pipe and tube fittings. The following sections summarize the important facts about these materials.

When you work with these parts, you won't have to deal with different U.S. customary and metric sizes. There are no separate metric standards for tubing, pipe, and hose sizes. U.S. inch sizes also are metric sizes. Similarly, U.S. inch-size pipe threads are also metric standard pipe threads. Some tube fittings, however, have metric threads.

Metal and plastic tubing and tube fitting sizes are specified by the nominal outside diameter (OD) of the tubing. Pipe, pipe fitting, and hose sizes are specified by the nominal inside diameter (ID) of the pipe or hose, figure 3-1.

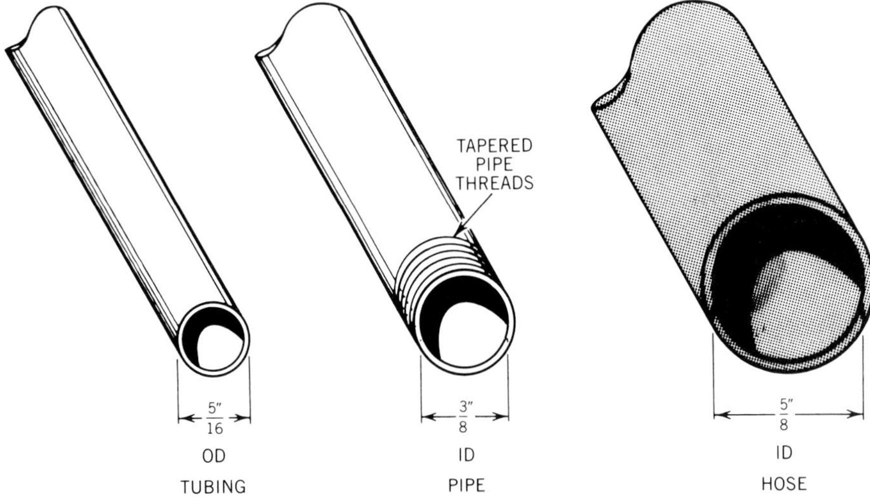

Figure 3-1. Tubing sizes are specified by the outside diameter (OD), pipe and hose sizes by the inside diameter (ID).

Tubing

Copper, steel, and plastic are the most common tubing materials used on automobiles. Copper is easy to bend, forms good joints at fittings, and is rustproof. But copper also is subject to **work hardening**, which means that continual vibration and high fluid pressures make it brittle and cause it to fail. Copper tubing is suitable for vacuum lines, coolant lines, and oil lines that don't receive a lot of vibration and high pressure. *Never use copper tubing for high-pressure hydraulic lines or fuel system lines.*

Several kinds of steel tubing are used widely on automobiles. The double-wrapped, brazed, and tin-plated kind is the most common for fuel and hydraulic lines. This kind of steel tubing handles fluid pressures of several thousand pounds per square inch, resists vibration and work hardening, and withstands corrosion. All brake, power steering, fuel injection, and automatic transmission lines must be steel. Even though copper can withstand the pres-

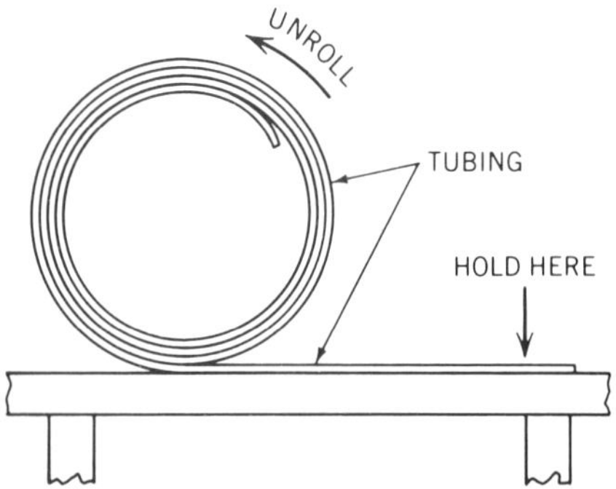

Figure 3-2. Unroll the desired length of tubing from the bulk roll.

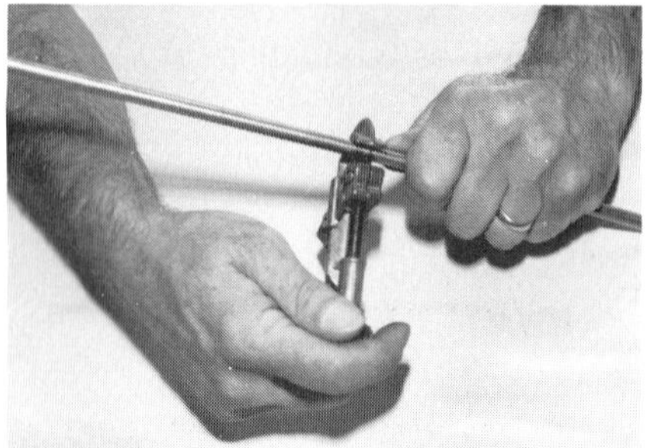

Figure 3-3. Rotate the tubing cutter toward the jaw opening. Tighten the cutter lightly after each revolution.

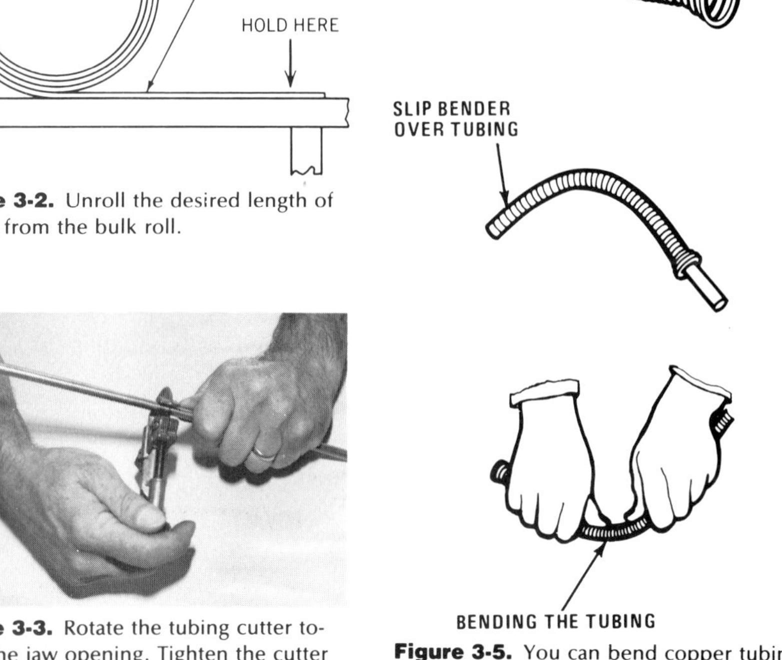

Figure 3-5. You can bend copper tubing with a bending spring. (Chrysler)

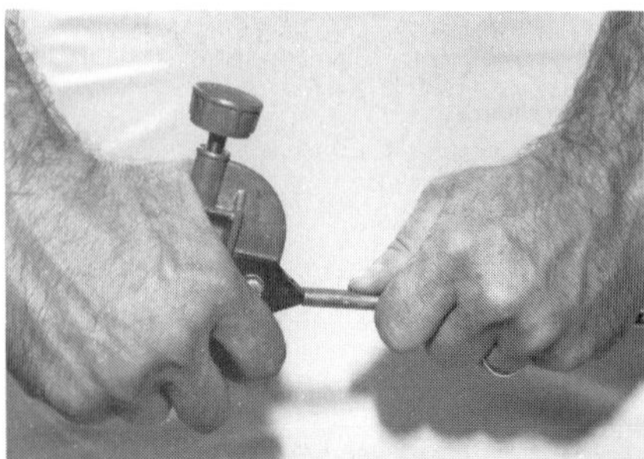

Figure 3-4. Ream the end of the cut tubing to remove burrs.

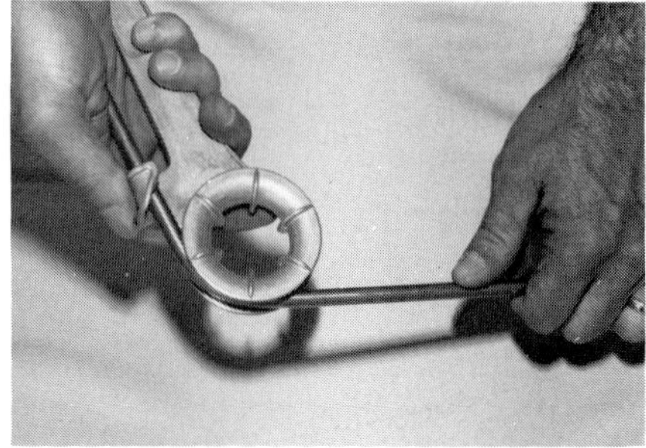

Figure 3-6. Use a lever-type bending tool on steel tubing.

sures in fuel lines from tank to pump, the length and location of these lines subjects them to more vibration than copper can safely withstand. *Never substitute copper tubing for steel tubing used by the carmaker.*

Polyethylene and nylon tubing are two kinds of plastic tubing used for some fuel, vacuum, and oil lines. These materials withstand corrosion and work hardening, but they will not stand high pressures and temperatures. Special fittings are used to attach plastic tubing to standard tube fittings. Any plastic tubing material must be compatible with the fluid that passes through it. Some plastic tubing can react with fuel or oil, soften, and burst. Or the plastic may dissolve in fuel or oil and clog filters or small orifices. Whenever you install plastic tubing, be sure it is a kind that is approved for use with the fluid that it carries.

Cutting and Bending Metal Tubing. Bulk steel and copper tubing comes in large rolls. To form a replacement length of tubing, you must cut it from a roll as follows:

1. Hold the free outer end of the tubing against a bench top with one hand and unroll the roll with the other hand, figure 3-2. Don't lay the roll flat and pull one end toward you. This may twist and kink the tubing.

2. To create an even cut without deforming the tubing, cut it with a tubing cutter, not a hacksaw.

3. Tighten the cutter so that the cutting wheel contacts the tubing firmly but not too tightly.

4. Turn the cutter around the tubing toward the open side of the cutter jaws, figure 3-3. After each revolution, tighten the cutter slightly until the cut is made.

5. After the tubing is cut, ream the end with a reaming tool to remove burrs and sharp edges. Hold the end downward so metal chips fall out, figure 3-4. Ream only enough to remove burrs.

6. You can make slight bends in most tubing by hand, but to avoid kinking or flattening, use a tube bender for most bending.

7. Bend soft steel and copper tubing with a bending spring as shown in figure 3-5.

8. Bend hard, double-wall steel tubing and make any tight bends with a lever-type bender, figure 3-6.

9. If the bend is near the end of a tube where you will install a fitting, leave about 1 1/2 inches or two fitting lengths of straight tubing before the bend.

You usually must bend tubing in several directions for proper installation. Here are a few more guidelines for correct tubing fabrication.

1. Avoid straight lengths, or runs, of tubing from fitting to fitting. They are hard to install and subject to damage from vibration.

2. Support long runs of tubing with clamps and brackets.

3. Bend tubing around hot spots, such as exhaust pipes, and route it away from moving suspension parts.

4. Be sure tubing ends align with fittings easily before installation.

5. When installing tubing, connect the longest section first. Connect the fit-

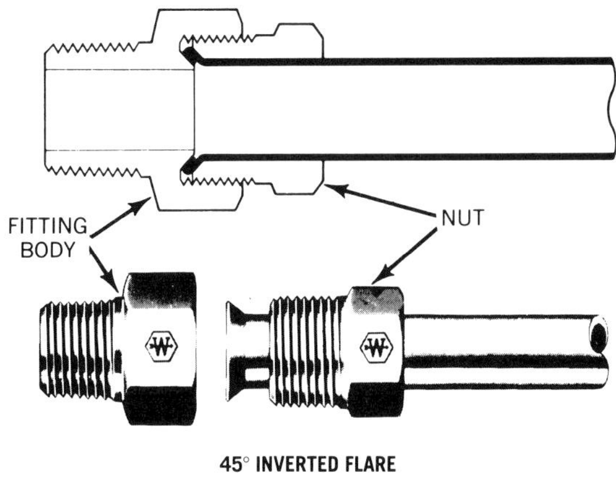

45° INVERTED FLARE

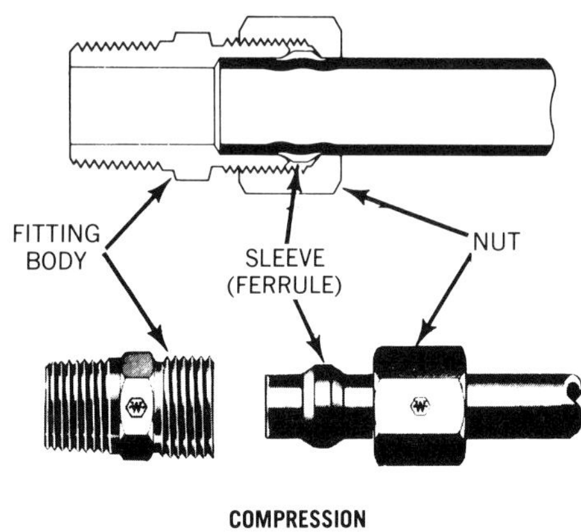

COMPRESSION

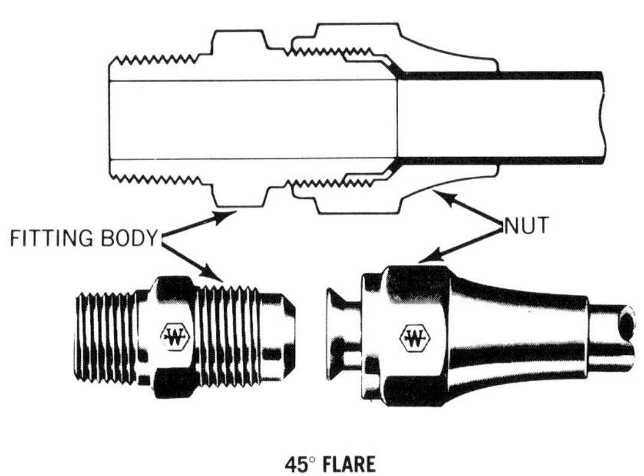

45° FLARE

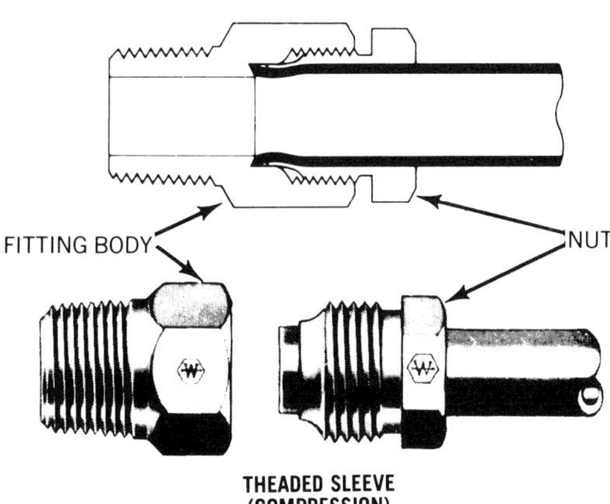

THEADED SLEEVE (COMPRESSION)

Figure 3-7. These are some of the flare and compression fittings used for automotive tubing. (Weatherhead)

tings at both ends loosely by hand before final tightening with a wrench.

Tube Fittings

Most tube fittings on automobiles are either the compression type or the flare type. They have straight threads and create a seal by squeezing the tubing into the body of the fitting. When you assemble a tube fitting, be sure the threads are clean and dry. Do not use a lubricant or thread sealant. Figure 3-7 shows some of the common compression and flare fittings used on automobiles.

Compression Fittings. Compression fittings use a separate sleeve, or ferrule, or the nose of the fitting nut to compress the tubing into the fitting body for a leakproof seal. Simple compression fittings are used for low-pressure vacuum, fuel, and lubrication lines. They are not used on high-pressure hydraulic systems. Special high-pressure compression fittings are used on some hydraulic system hose, but these are usually assembled by the hose manufacturer.

To install a simple compression fitting, cut and bend the tubing as explained previously, then proceed as follows:

1. Slide the fitting nut and sleeve on the end of the tubing.

2. Insert the tubing into the fitting body and hold it firmly.

3. Align the fitting nut and body and tighten the nut by hand. Do not cross thread the fitting.

4. Use a tubing wrench to tighten the nut until the sleeve grasps the tubing and it won't move in the fitting, figure 3-8. Then tighten the nut another 1 1/4 to 1 3/4 turns.

Flare Fittings. Most high-pressure and fuel tubing connections on an automobile are the 45-degree flare type. Some flare fittings, however, have a 37-degree flare, so the first thing to determine when replacing a flare fitting is the angle of the flare.

A tubing flare can be a single or double flare. Single flares are suitable for soft tubing in low-pressure systems, but double flares are preferable because they are stronger. Double-

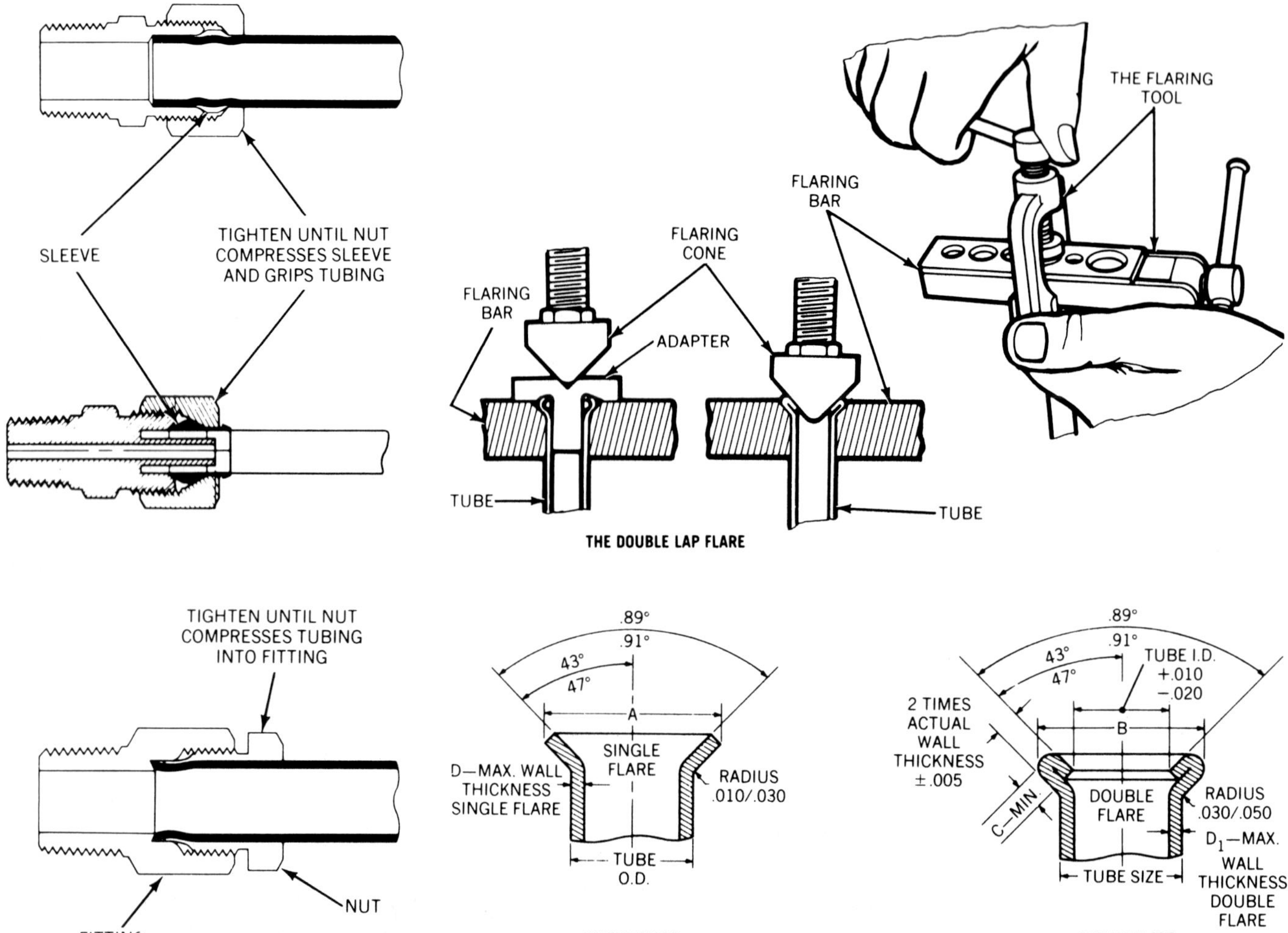

Figure 3-8. Tighten the compression fitting carefully until the sleeve grips the tubing. (Weatherhead)

Figure 3-9. Double-wrapped hard steel tubing must have a double flare, formed like this.

wrapped hard steel tubing must *always* have a double flare because a single flare will split in this material.

To form a double flare, cut and ream the tubing as explained previously. Determine the right flare angle and proceed as follows, figure 3-9:

1. Place the flare nut on the end of the tubing and place the tubing in the right-sized opening of the flaring tool.

2. Place the correct adapter over the end of the tubing and adjust the tubing height against the adapter. Tighten the tool clamp securely so the tubing doesn't slip.

3. Tighten the flaring cone against the adapter to form a bell end on the tubing. Then loosen the cone and remove the adapter.

4. With the tubing still clamped securely in the tool, tighten the flaring cone into the bell end to complete the double flare.

Pipe Fittings

You won't find a lot of steel or iron pipe on an automobile, but you will find fittings with pipe threads. Unlike the threads of tube fittings, pipe threads are tapered and use the compression of the mating threads to form a seal. Most pipe threads on automobiles are the dryseal type, indicated by the abbreviation NPTF, which stands for "national pipe, tapered, fine." Most pipe thread sizes used on automobiles range from 1/8 inch to 3/4 inch. The sizes are specified by the ID of the pipe, which is why a 1/4-inch pipe fitting has an outside diameter of about 1/2 inch. Figure 3-10 shows some common pipe fittings used on automobiles.

When you assemble a pipe fitting, you can often use a thread sealant or Teflon tape to ease assembly and ensure a leakproof seal. Be sure, however, that the sealant is compatible with the fluid or gas that is in the system. Keep the sealant off the first two threads of male fittings. Don't use plumber's "pipe dope" on fuel or oil lines.

Hoses and Hose Fittings

Many different kinds of hoses are used on the cooling, lubrication, fuel, and emission systems of automobiles. To

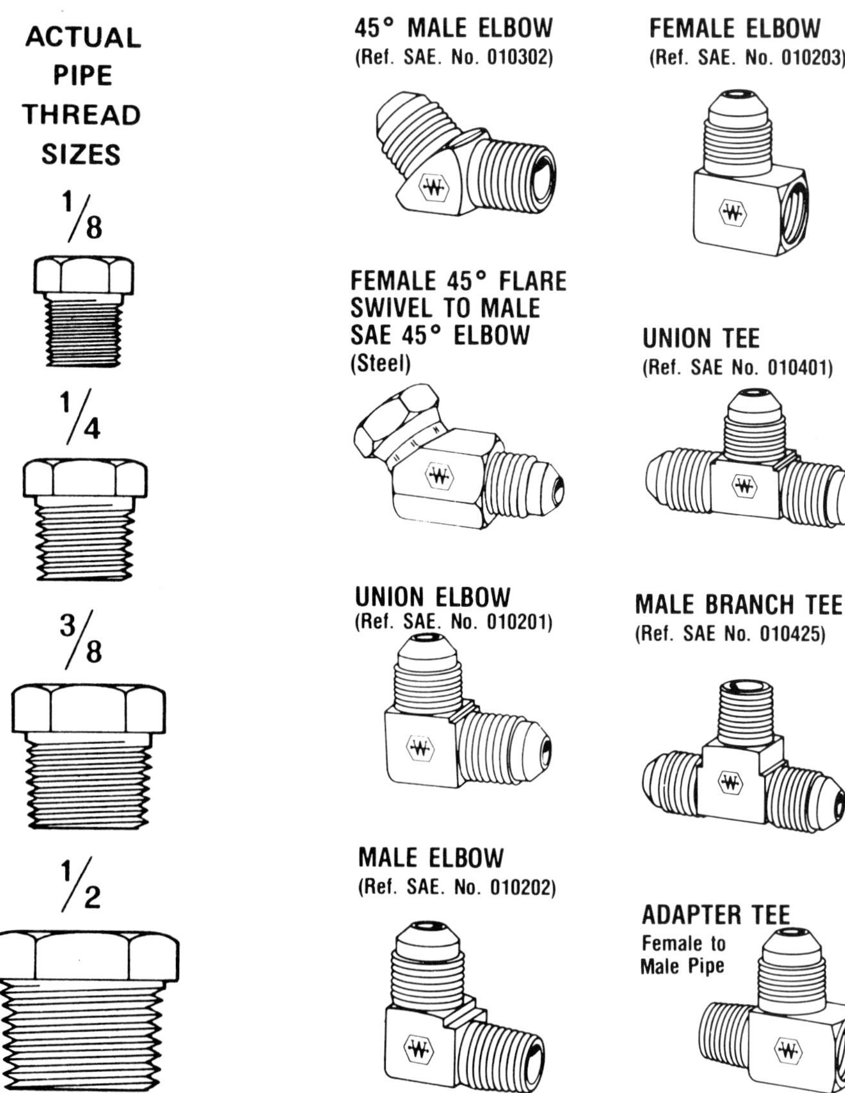

Figure 3-10. These are some pipe fittings used on an automobile. Many are pipe-to-tube adapters. (Weatherhead)

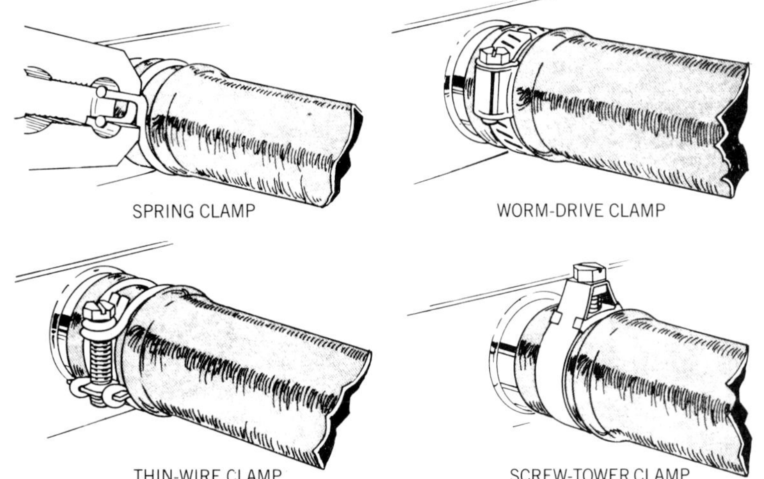

Figure 3-11. Spring and screw-type clamps are used to attach most hoses on an automobile. (Gates)

select correct replacement hoses, you must know the size, type of construction, pressure capacity, and materials the hose can be used with.

Cooling system and heater hoses carry liquid at low pressure but high temperature. These hoses are either single- or double-ply construction. Radiator hoses usually have wire reinforcing to prevent collapse.

Hoses used on fuel, lubrication, and emission systems also operate at low pressure, but they are made of neoprene or Buna N rubber materials that are compatible with fuel, oil, and emission vapors. Many fuel and emission hoses and heater hoses are of the same size and similar construction, but heater hose may not be compatible with petroleum materials. Do not use hose designed only for coolant service on a fuel, oil, or emission system.

Low-pressure hoses in cooling, heating, emission, and some fuel systems usually are mounted with screw or spring-type clamps, figure 3-11. High-pressure hoses and hose connections that must be absolutely leakproof use a variety of tubing and pipe fittings. The same installation guidelines you have learned for these fittings on solid tubing apply to their use on hoses as well.

Hoses with permanent fittings usually have one swivel fitting and one nonswivel fitting. Install and tighten the end without the swivel fitting first. When you install a hose, avoid kinking and twisting it and always allow some slack. Route hoses away from hot exhaust components and moving engine or suspension parts. Any bend should have a radius at least five times the outside diameter of the hose.

GASKETS

Gaskets are used to prevent leaks between two parts in a nonmoving (stationary) joint. All gaskets have some flexibility and compressibility so that they will conform to the shape of the joint, but they are generally classified in two categories: hard and soft. Hard gaskets are made of metal or a com-

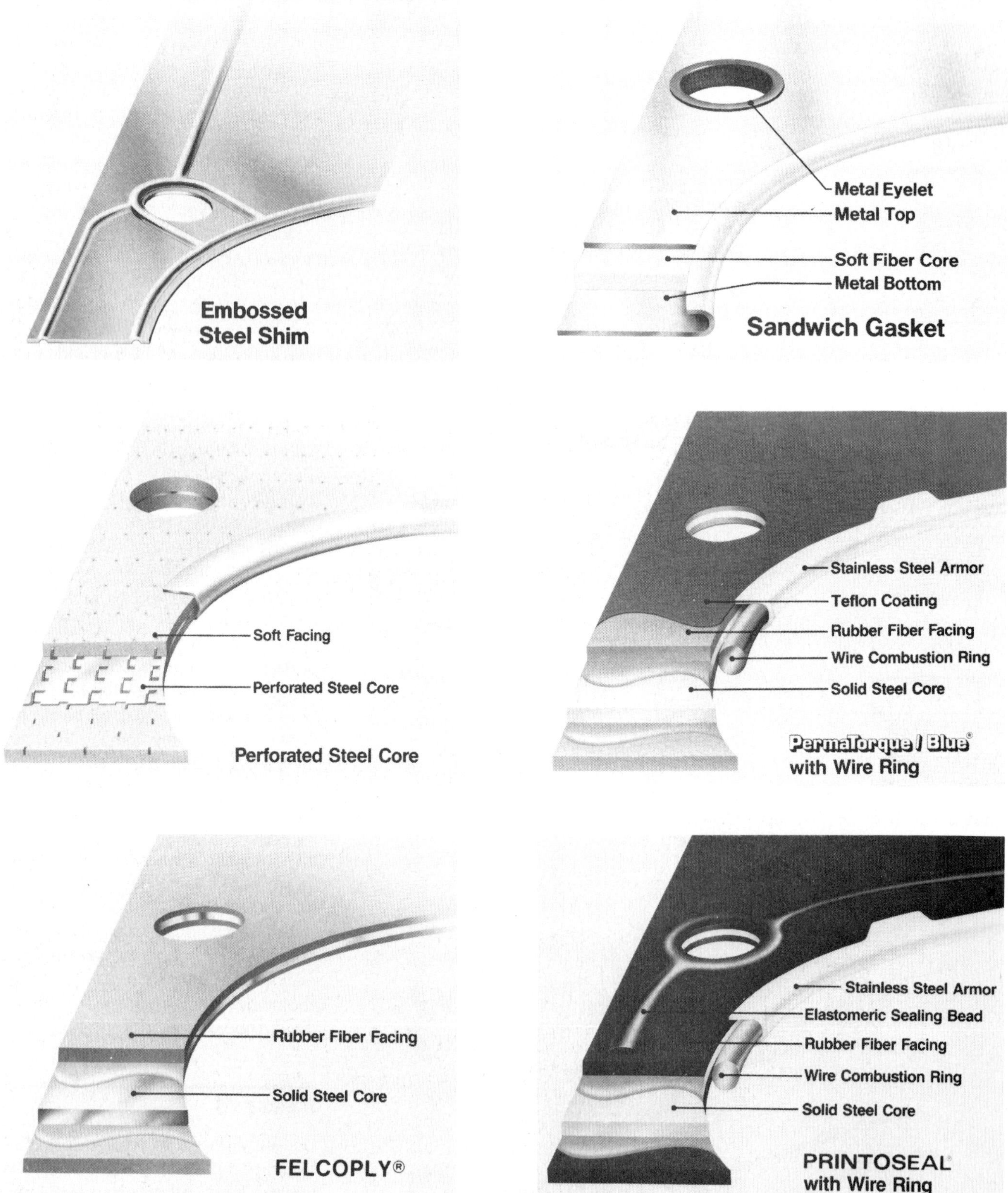

Figure 3-12. Hard gaskets may be embossed steel or a composite construction of metal and asbestos or other materials. (Fel-Pro)

bination of metal and another material, figure 3-12. They are used in high-temperature locations, such as on engine cylinder heads, manifolds, and exhaust systems. Thin metal gaskets are usually embossed steel. The gasket has raised ridges or beads embossed in a pattern around the openings it seals. When the gasket is installed and tightened, the embossed pattern compresses to form a seal.

Hard gaskets made of metal and asbestos or other high-temperature materials are often called sandwich gaskets. The metal may be copper, aluminum, or soft steel, and the gaskets are made in many styles and construction methods, figure 3-12. Many of these gaskets are marked for installation in a specific direction or with one surface up or down. Gasket manufacturers usually include instructions for installation.

Soft gaskets are used almost every place two parts form a leak-free joint. One of the simplest and oldest soft gasket materials is cork. Other soft gaskets are made of special paper, soft plastic foam, rubber, or cork-and-rubber combination material, figure 3-13.

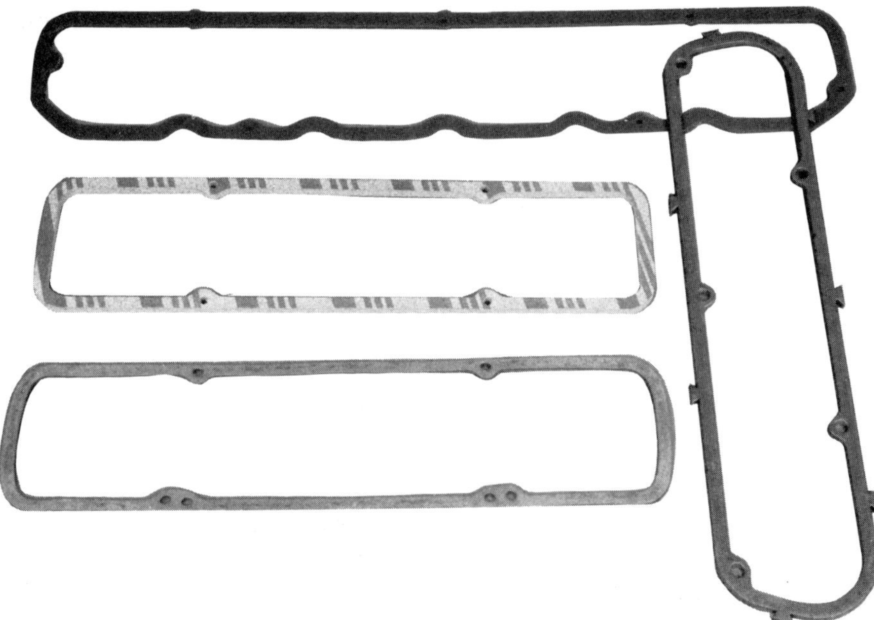

Figure 3-13. Soft gaskets are made of several different materials.

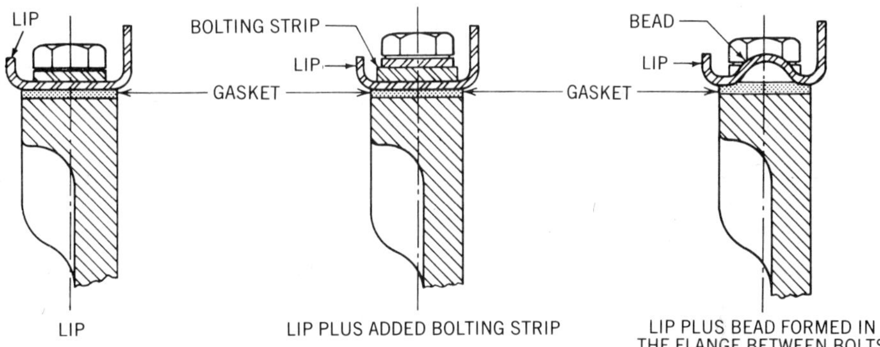

Figure 3-14. Many thin flanges have bolting strips or reinforcing beads to spread clamping force evenly over the gasket.

Gasket Retention and Gasket Creep

Gaskets usually are held by two or more bolts or studs between two flanges or mating surfaces. Some flanges have a lip or protrusion to help hold the gasket, figure 3-14. Some light sheet metal flanges have reinforcing beads or bolting strips to spread the clamping force evenly over the joint. If a flange is bent or the bolt holes are distorted, the flange should be straightened before installation or the gasket will be tight in some places and loose in others.

Many gaskets are used between large parts with 10 or 20 or more bolts or studs. Even the heaviest cast iron cylinder head or intake manifold will flex as the fasteners are tightened and the gasket compresses. Carmakers provide instructions for specific fastener torque and tightening sequences, figure 3-15. When these instructions are available, follow them to reassemble any parts. If specific instructions aren't available for a large gasketed joint with many fasteners, start the tightening sequence toward the center of the part and work outward toward the edges. Use standard torque values for the size and type of fasteners. Tighten them in a sequence of about 25, 50, and 75 percent until you reach the full torque value. Don't overtighten the fasteners or you may break or distort the gasket.

All gaskets compress somewhat when tightened between mating surfaces. This compression is called **gasket creep**, figure 3-16, and it ranges from about 10 percent of the original thickness for some metal gaskets to 60 percent for some soft gaskets. The amount of creep depends on gasket

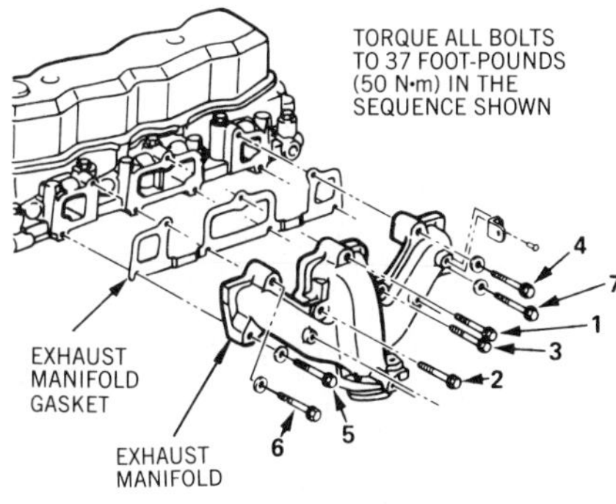

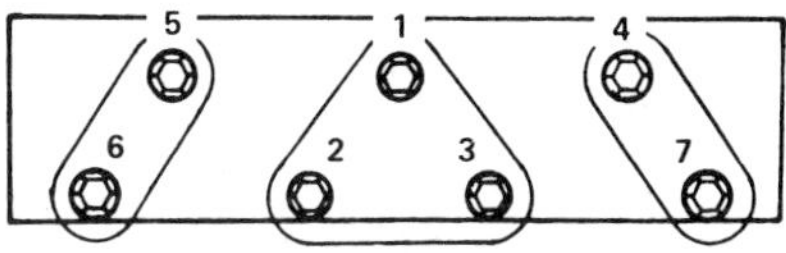

BOLT LOCATIONS

Figure 3-15. Follow the carmaker's tightening sequence and torque instructions for any gasketed joint. (Chevrolet)

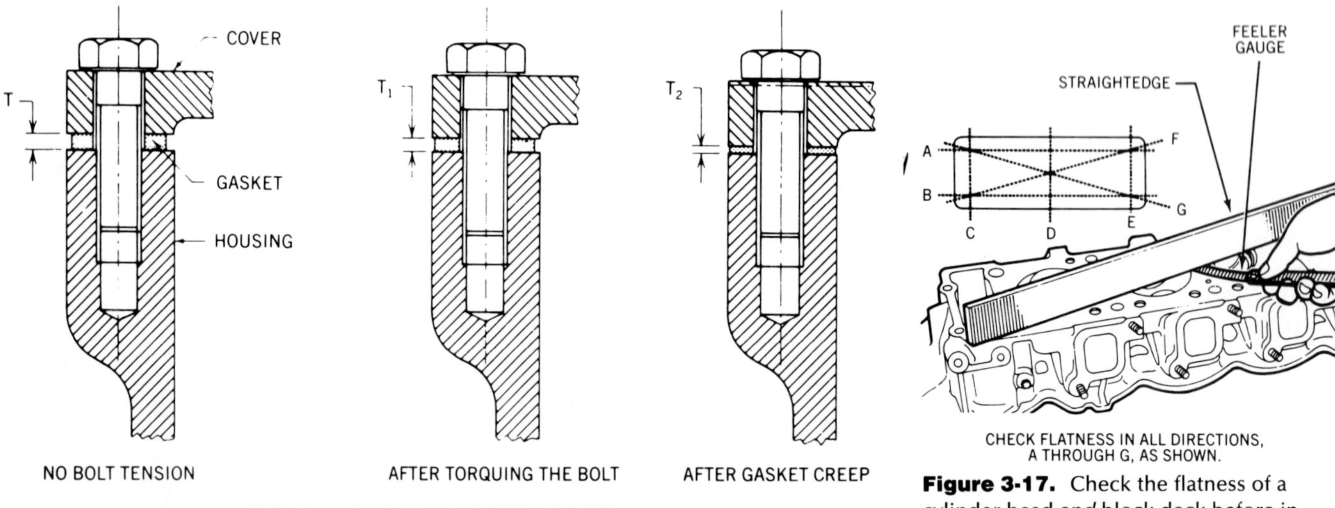

Figure 3-16. Gasket creep is the compression that occurs after the bolts are tightened.

Figure 3-17. Check the flatness of a cylinder head *and* block deck before installing a new gasket. Use the same technique to check for a warped manifold. (Mitsubishi)

thickness as well as material. Gasket creep combined with vibration and changing temperature can cause a joint to loosen after a period of time. Gasket creep can occur gradually after a joint is tightened, and many carmakers require that some gasketed joints be retorqued at some time after assembly. The period of time varies from several minutes to several months. For example, new cylinder head gaskets may need retorquing after the engine is run for 15 to 30 minutes. Retorquing may be required again after 500 to 1,000 miles of driving. Some carmakers call for retorquing manifolds or engine and transmission pans after several months or several thousand miles of operation.

Gasket Installation Guidelines

Soft gaskets will shrink and distort if not stored properly. Always store gaskets flat. Don't hang them vertically. Store gaskets away from extreme heat or cold to keep them from shrinking, drying out, or getting brittle. Handle them carefully and avoid bending them unnecessarily during installation. If a gasket is cracked or torn or doesn't fit, discard it.

Don't reuse an old gasket. Once a gasket is compressed, it loses flexibility and will not return to its original thickness. Embossed metal gaskets can be compressed only once. They won't seal on a second use.

Always allow engine parts to cool before disassembling them. If you loosen bolts while a part is hot, it can warp as it cools. This is particularly critical for aluminum cylinder heads and manifolds, but even a timing cover can warp if it is removed while hot. This warpage can prevent a proper seal and create a leak when the parts are reassembled.

Be sure that the gasket mating surfaces of parts are clean and dry before assembly. Remove dirt with suitable solvents, scrapers, and brushes. Be careful not to gouge or scratch the sealing surfaces of aluminum parts. Straighten bent sheet metal surfaces. If tapped holes in cast metal parts are distorted, remove sharp edges with a file and clean the threads with a tap.

Threads of bolt holes for cylinder heads, manifolds, and engine covers often distort from bolt tension. Also, if a block or head is resurfaced, the top threads of a hole may project into the sealing surface. Before installing a gasket, inspect the threads of tapped holes. Chamfer the holes with a rotary file or grinder to be sure that the threads do not pull above the clamping surface when the bolts are tightened. Distorted threads will create a false torque reading and uneven clamping force, which can lead to gasket failure.

Before installing a cylinder head gasket, use a feeler gauge and machinist's straightedge to check the flatness of the head and block, figure 3-17. Check both the length and the width of the head and the block. Across the width, sealing surfaces should be flat within 0.003 to 0.004 inch. Across the length, surfaces should be flat within 0.003 inch for V-6 engines, 0.004 inch for 4-cylinder and V-8 engines, and 0.006 inch for inline 6-cylinder engines. You can use the same methods to check for warpage on manifold surfaces. If a head, block, or manifold surface is warped, it must be resurfaced to ensure a proper gasket seal.

Many replacement gaskets are made differently and require different installation methods than original equipment gaskets. After several thousand miles of operation and hundreds of heating and cooling cycles, the surfaces of an engine are harder to seal than when they are new. Replacement gaskets often are made with special coatings such as Teflon or graphite to improve sealing. Additionally, a gasket may have a printed bead of silicone rubber around critical openings for improved sealing. Many replacement gaskets for cylinder heads and manifolds do not require retorquing as original equipment gaskets do.

Manufacturers provide specific instructions for installation of different kinds of gaskets. These include directions on whether or not to use adhesives and on whether or not to retorque after installation. Follow the gasket maker's instructions exactly to ensure a leakproof installation.

When you use adhesive or sealant on a gasket, be careful to keep the sealing material out of bolt holes, fluid passages, and cylinders. Sealant or adhesive on bolt threads will cause an inaccurate torque reading, at least, and may crack a casting as a bolt is tightened.

ADHESIVES, CEMENTS, AND SEALANTS

Various adhesives, or cements, and sealants are used in automobile service, and many of them are used with, or in place of, gaskets. These materials can be applied from a tube, with a brush, or from a spray can. It's impossible to describe every brand and variety of adhesive or sealant in this text, but here are some general features.

Adhesives and Cements

Gasket cements are either hardening or nonhardening. Hardening cements are used where the joint isn't likely to be taken apart for some time and where joint movement is minimal. Nonhardening cements are used where the joint may be disassembled for later service and where there may be some movement or vibration. Nonhardening cement is used more often for automotive service.

To use either hardening or nonhardening gasket cement, apply a thin coat to one or both sides of the gasket and to the mating metal surfaces. Allow the cement to dry until it is tacky and then assemble the joint. Don't use cement on gaskets in carburetors or other assemblies where fluid passes through small holes in the gasket. Any cement that gets into fluid passages can plug the passages and damage the system.

Some gasket adhesives are packaged in spray cans and are easier to use than those applied from a tube or with a brush. Follow the directions on the spray can and use them only for applications where they are recommended. Rubber gaskets, such as those on valve covers and oil pans, may need special adhesives. Some adhesives may lubricate the rubber and cause it to slide out of the joint as it is tightened. Adhesives for use with rubber gaskets are designed to prevent this slip. Check the instructions of the gasket or adhesive manufacturer before installing rubber gaskets.

Sealants

Most of the sealants used on automobiles are used in place of gaskets. They are classified as **aerobic** or **anaerobic**. Aerobic sealants cure in the presence of oxygen and humidity in the air. Anaerobic sealants cure in the absence of oxygen.

Aerobic sealants are room-temperature-vulcanizing (RTV) silicone rubber compounds. Many carmakers use them in place of soft gaskets on valve covers, oil pans, and other sheet metal covers because the material forms a thick seal that fills large gaps. When you assemble parts with aerobic, or RTV, sealant, you must join them within 10 minutes of applying the sealant or it won't seal properly. When assembled, the sealant cures from the outer edges inward and will withstand temperatures up to 450° F (232° C).

Anaerobic sealants cure only after the mating pieces are bolted together and air is forced out of the joint. This means that you can apply the sealant and leave the parts unassembled longer than you can with aerobic sealants. When assembled, the sealant cures from the inside outward and withstands temperatures up to 400° F (204° C). Anaerobic sealants form a thinner seal than aerobic materials and usually are used between machined flat surfaces.

Other sealants used on automobiles are various thread-sealing compounds. We talked briefly about some of these in relation to pipe and tube fittings. Other thread sealants are available for use on nut and bolt threads to help ensure against loosening. These are particularly useful on high-torque fasteners. Thread sealants usually come in liquid in small tubes and are anaerobic materials.

To use a thread sealant, be sure the nut and bolt threads are clean and dry. You may have to clean them with a special solvent supplied by the sealant manufacturer. Apply a small drop of sealant to the threads and torque the fasteners to the required tension.

SEALS

Seals, like gaskets, are used between two parts to prevent grease, fluid, or gas leakage. Most seals on automobiles are lip seals, O-rings, wick seals, and square-cut seals. They are used between a rotating shaft and a stationary housing or in high-pressure hydraulic and pneumatic systems. Most of the seals you will work with during engine performance service are lip seals and O-rings.

Lip Seals

Lip seals are low-pressure seals used on a rotating shaft or on a hub that rotates on a shaft. They are often used near bearings to retain lubricant. They are pressed into a bore in a housing, and the lip of the seal contacts the shaft. They are always installed with the lip toward the lubricant pressure, figure 3-18. You will find lip seals in engine timing covers and engine accessories. Because sealing occurs only at the thin edge of the lip, these seals will stand pressures of 3 to 7 psi (20 to 45 kPa). If a rotating hub or shaft must be sealed against higher fluid or gas pressures, other kinds of seals are used.

Wick Seals and Packings

Wick seals and packings are used on rotating shafts to seal higher pressures

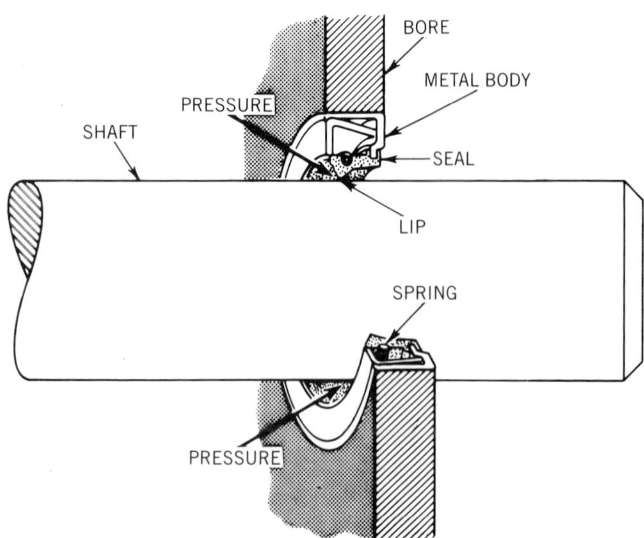

Figure 3-18. A lip seal is always installed with the lip toward the lubricant. Many lip seals have garter springs to improve the seal action.

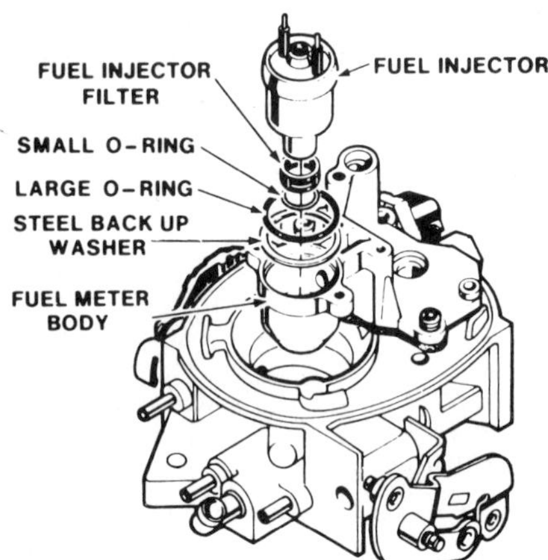

Figure 3-19. Typical O-ring installations.

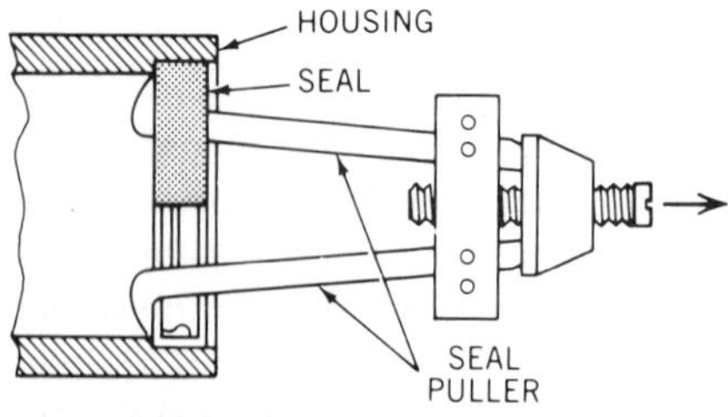

Figure 3-20. Be careful not to knick the seal bore when removing a seal.

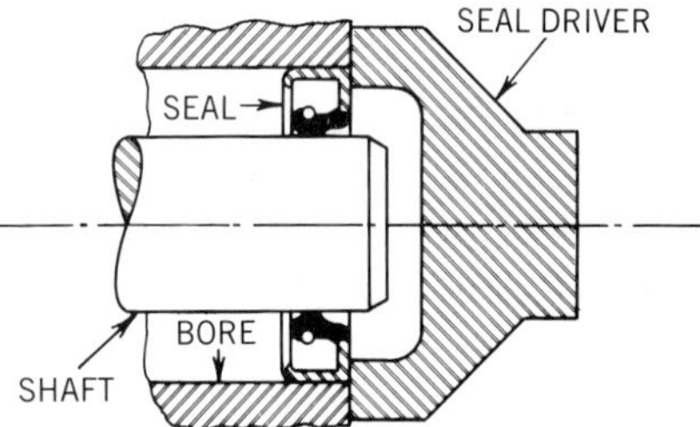

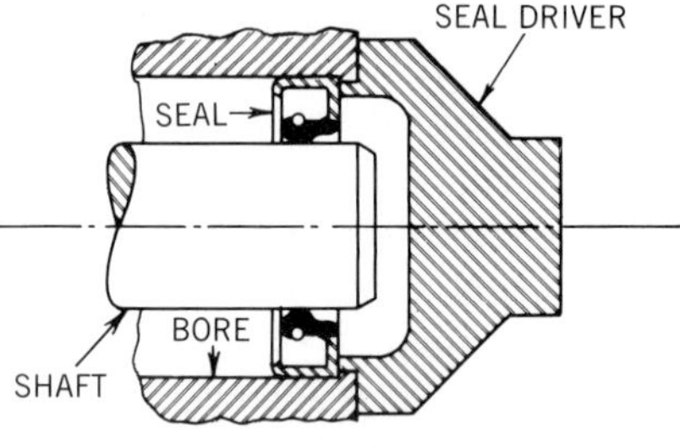

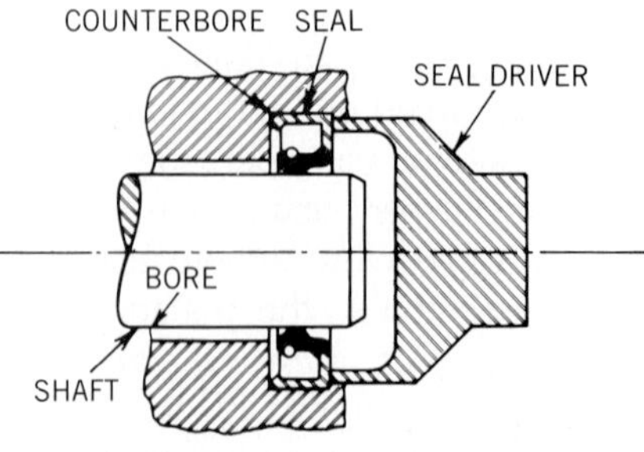

Figure 3-21. Drive the seal squarely to seat against a counterbore or until the seal body is flush with the housing opening.

than lip seals can handle. Wick seals also have a longer life and are used in many installations where replacement is difficult. The rear main bearing seals on most engines are wick seals made of a ropelike material held in a groove in the bearing cap and block.

O-Rings

An O-ring is a soft ring usually made of neoprene or Buna N rubber. O-rings are installed in grooves in one or both mating pieces and create a seal when compressed as the pieces are fastened together, figure 3-19. O-rings are most often used for high-pressure hydraulic or pneumatic seals where there is little or no rotating or reciprocating motion. They work best for such static connections, and you will find them in hydraulic pumps and valves, fuel injection pumps, and other fuel injection components.

Seal Installation Guidelines

Most seals used on an automobile are made of rubber or similar compounds. Like gaskets, seals must be compatible with the fluids or gases that they seal. Also like gaskets, seals should not be reused. If you select a replacement seal by the correct part number, you will get one designed to work in the system. Don't guess about seal materials and don't substitute seals of the same size but the wrong part number.

Many seals need lubrication when you install them. The general rule is to lubricate a seal with the fluid or lubricant used in the system. That is, if a seal is used in a hydraulic system, lubricate it with that system's fluid. If a seal is used in a transmission, lubricate it with transmission fluid. If it is used in an air conditioning system, lubricate it with refrigeration oil. Using the wrong lubricant can destroy the seal and contaminate the system.

Most of the seals that you remove and replace on an automobile are lip seals made with a rubber sealing ring in a metal case, or body. Here are a few general rules for seal installation and removal:

1. Remove a seal with a seal puller, a seal driver, or a suitable prybar, figure 3-20. Be careful not to damage the housing or shaft on which the seal is installed.

2. Before installing a new seal, clean the counterbore or shaft on which the seal is installed. Remove nicks or burrs with a fine file or fine emery cloth.

3. If required, coat the inside of the counterbore or housing that holds the metal part of the seal with a nonhardening sealer or a special sealant used for seal installation.

4. Lubricate the seal lip with system fluid or lubricant.

5. Use a seal driver and hammer to drive the seal into its housing, figure 3-21. If a seal driver is unavailable, use a block of wood or soft metal. Don't drive a seal with a sharp punch or the face of a hammer or you may distort its case.

6. If the seal seats on a shoulder at the bottom of a counterbore, drive it in squarely until seated. If the seal does not seat on the bottom of a counterbore, drive it to the specified depth or until flush with the opening of the housing.

SUMMARY

Copper and steel tubing materials are used in various engine and combustion control system applications. Copper is suitable for low-pressure lubricant, coolant, and vacuum lines. Steel must be used for rigid fuel system lines. Do not substitute copper tubing for an OEM steel tubing installation. Steel tubing must always be double flared for use with flare-type fittings.

Plastic tubing and rubber hoses are also used as fuel, lubricant, emission control, and coolant lines. Rubber and plastic materials must always be compatible with the fluid or gas that passes through them and the pressures and temperatures of the system in which they are used.

Gaskets are classified as either hard or soft gaskets. Hard gaskets are made of metal, or metal and a temperature-resistant material. Soft gaskets are made of plastic, rubber, cork, or paper materials. For a secure gasketed joint, follow the carmaker's instructions for fastener tightening sequence and torque values. Some gasket installations require an adhesive; others do not.

Sealants are often used in place of gaskets. Aerobic sealants cure in the presence of oxygen and humidity in the air. Anaerobic sealants cure in the absence of oxygen. Aerobic and anaerobic sealants are not interchangeable.

Seals are used between many parts to prevent liquid, gas, or lubricant leakage. Lip seals and O-rings are the most common kinds used on engine and combustion control systems.

REVIEW QUESTIONS

Multiple Choice

1. Copper tubing should not be used for fuel lines because:

- **a.** it cannot withstand high temperatures in the engine compartment
- **b.** it is unnecessarily expensive
- **c.** it can work harden and become brittle
- **d.** none of the above

2. Mechanic A says the metal tubing sizes are specified by outside diameter (OD). Mechanic B says that pipe and hose sizes are specified by inside diameter (ID). Who is right?

- **a.** A only
- **b.** B only
- **c.** both A and B
- **d.** neither A nor B

3. Mechanic A has installed a length of ½-inch heater hose to replace a broken PCV valve hose. Mechanic B has installed a steel fuel line with compression fittings at the fuel injection throttle body inlet. Who has done the job correctly?

- **a.** A only
- **b.** B only
- **c.** both A and B
- **d.** neither A nor B

4. On most automotive flare fittings, the flare angle is:

- **a.** 30 degrees
- **b.** 37 degrees
- **c.** 45 degrees
- **d.** 54 degrees

5. Mechanic A has installed a pipe fitting with teflon tape on the threads to ensure a leakproof seal. Mechanic B has installed a flared tube fitting with teflon tape on the threads to ensure a leakproof seal. Who has done the job correctly?

- **a.** A only
- **b.** B only
- **c.** both A and B
- **d.** neither A nor B

6. Soft gaskets are sometimes made of:

- **a.** cork
- **b.** thin, embossed metal
- **c.** metal and asbestos
- **d.** soft metals only, such as copper and aluminum

7. The compression of a gasket after it is tightened is called:

- **a.** gasket retention
- **b.** gasket creep
- **c.** gasket compression
- **d.** gasket distortion

8. Mechanic A says that intake and exhaust manifold bolts should be checked for tightness and retorqued periodically. Mechanic B says that cylinder head fasteners should be retorqued at 20,000- to 30,000-mile, or 3-year intervals. Who is right?

- **a.** A only
- **b.** B only
- **c.** both A and B
- **d.** neither A nor B

9. When installing a new manifold gasket, mechanic B applies gasket cement to bolt threads to ensure a tight seal. Mechanic A chamfers the distorted top threads of a bolt hole with a rotary file. Who is doing the job correctly?

a. A only
b. B only
c. both A and B
d. neither A nor B

10. Mechanic A says that you should disassemble engine parts when they are as warm, or hot, as possible to relieve tension on bolt threads. Mechanic B says that you should apply motor oil to gasket flange and tapped bolt holes before reassembly to ensure a proper seal and torque application. Who is right?

a. A only
b. B only
c. both A and B
d. neither A nor B

11. Sealants that cure in the absence of oxygen are called:

a. aerobic sealants
b. RTV rubber sealants
c. anaerobic sealants
d. doped-silicon sealants

12. Instead of soft gaskets on valve covers, timing covers, oil pans, and other sheet metal parts, many carmakers use:

a. aerobic sealants
b. hardenable gasket sealants
c. anaerobic sealants
d. doped-silicon sealants

13. The most common kind of seal used on a rotating shaft to withstand high fluid pressure is:

a. an O-ring
b. a lip seal
c. an anaerobic seal
d. a wick seal

14. A common high-pressure hydraulic or pneumatic seal used where there is little rotary motion is:

a. an O-ring
b. a lip seal
c. an anaerobic seal
d. a wick seal

15. Mechanic A says that a lip seal should always be installed with the lip away from the fluid pressure. Mechanic B says the lip seals are the most common seals in high-pressure fuel injection throttle bodies. Who is right?

a. A only
b. B only
c. both A and B
d. neither A nor B

16. Mechanic A has installed an O-ring seal in a fuel injection assembly with a thin coating of brake lubricant, which is waterproof and eases installation. Mechanic B has installed a lip seal in a timing cover with the sealing lip clean and completely dry to ensure positive seal contact with the crankshaft or timing gear. Who has done the job correctly?

a. A only
b. B only
c. both A and B
d. neither A nor B

17. Mechanic A says that an anaerobic sealant cures only after the mating parts are assembled. Mechanic B says that parts sealed with an aerobic sealant must be assembled within 10 minutes after applying the sealant. Who is right?

a. A only
b. B only
c. both A and B
d. neither A nor B

18. For proper sealing, a cylinder head should be flat across its surface (side to side) within:

a. 0.010 to 0.020 inch
b. 0.100 to 0.150 inch
c. 0.005 to 0.010 inch
d. 0.003 to 0.004 inch

19. When you install a fuel or lubricant hose that has permanent fittings, one swivel and one nonswivel fitting, you should:

a. be sure that the hose is stretched tightly, with minimum length between both ends
b. install the swivel fitting first and tighten it securely
c. install the nonswivel fitting first and tighten it.
d. double the hose back upon itself three times to protect it from vibration

20. Low-pressure hoses are often mounted with:

a. spring clamps
b. pipe clamps
c. flare fittings
d. compression fittings

4 BASIC TEST EQUIPMENT AND TROUBLESHOOTING

INTRODUCTION

Efficient engine service requires accurate diagnosis. Accurate diagnosis depends on:

- Knowing how engine systems work, separately and together
- Using the right test equipment
- Following the right test procedure and testing in a logical, organized manner
- Checking and testing to the carmaker's specifications
- Interpreting test results accurately

This chapter introduces several kinds of general-purpose test equipment and explains the basic use of this equipment. You will learn about electrical meters, vacuum and pressure gauges, and other common engine test equipment. The chapter concludes with an outline of basic troubleshooting principles that apply to all vehicle systems.

GOALS

After studying this chapter, you should be able to:

1. Explain the uses and demonstrate the basic operation of the following test equipment:

a. An ammeter, a voltmeter, an ohmmeter, and other test meters

b. Test lamps and a jumper wire

c. A timing light

d. Pressure and vacuum gauges and a hand-operated vacuum pump

e. An oscilloscope, or engine analyzer

f. An exhaust gas analyzer

g. A chassis dynamometer

2. Identify and use diagnostic connectors on vehicles that have them.

3. Perform organized troubleshooting, based on the driver's complaint and apparent vehicle symptoms.

ELECTRICAL TESTING

Much engine testing involves troubleshooting electrical systems to isolate faults. There are two basic kinds of electrical faults:

1. High-resistance faults

2. Low-resistance faults

High-Resistance Faults

High resistance in a circuit will decrease or stop current. High resistance can be caused by a loose, damaged, or corroded connection or by a defective part, figure 4-1. This resistance may not stop current completely, but it decreases current enough to keep the circuit from working properly.

Infinitely high resistance is caused by an **open circuit**, which stops current completely, figure 4-2. We use

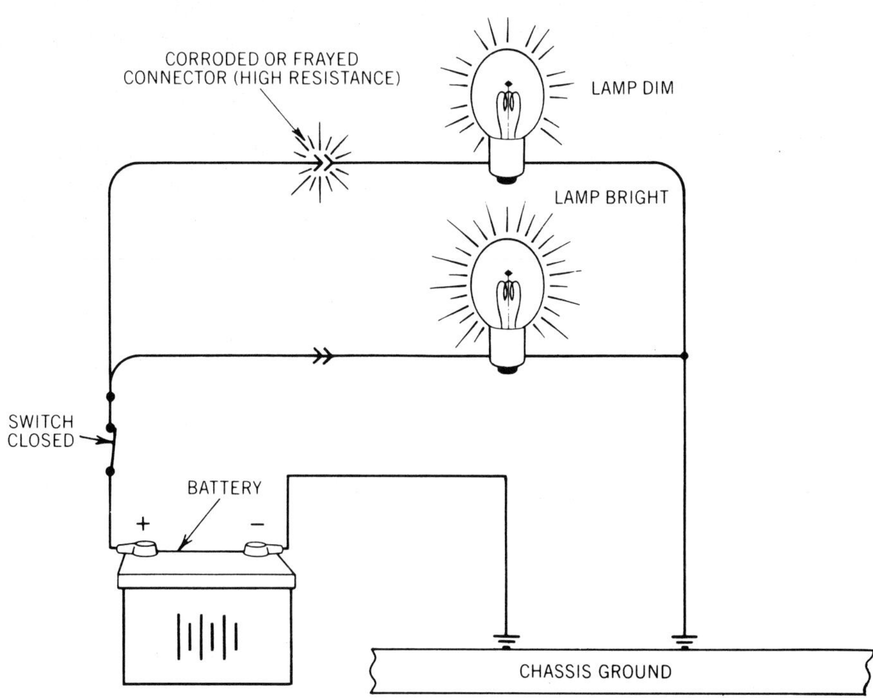

Figure 4-1. This damaged connection creates high resistance and decreases current.

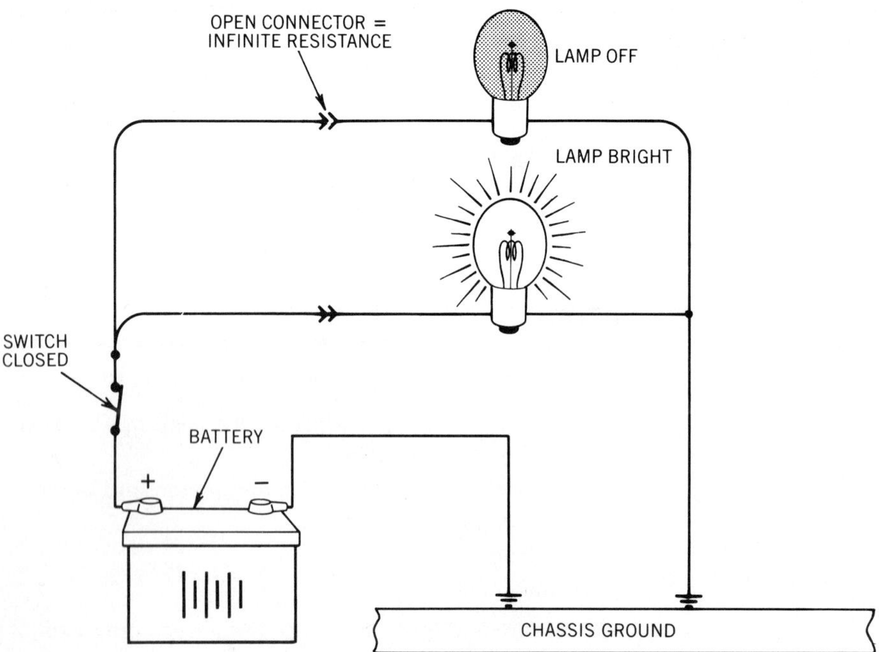

Figure 4-2. An open circuit causes infinite resistance and keeps the circuit from working.

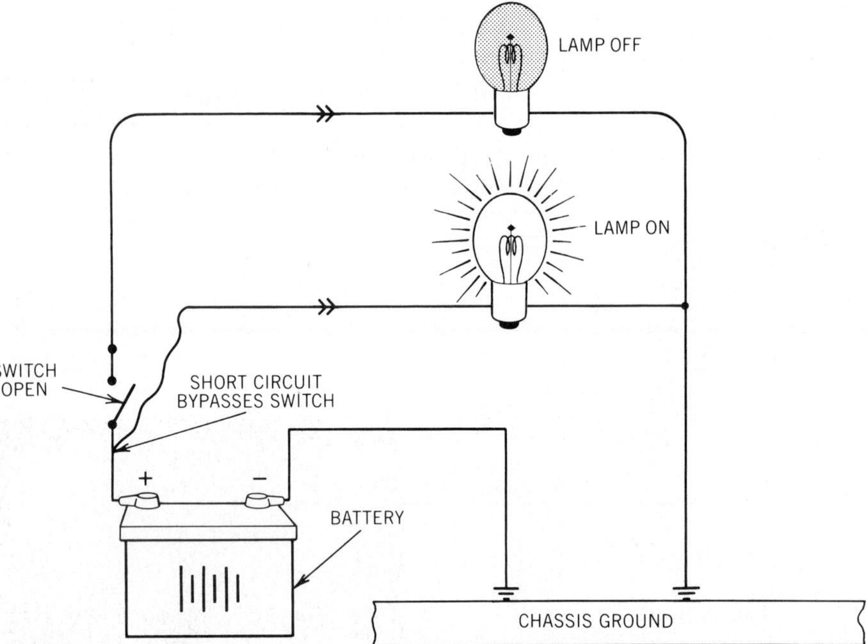

Figure 4-3. A short circuit allows current to bypass part of the circuit.

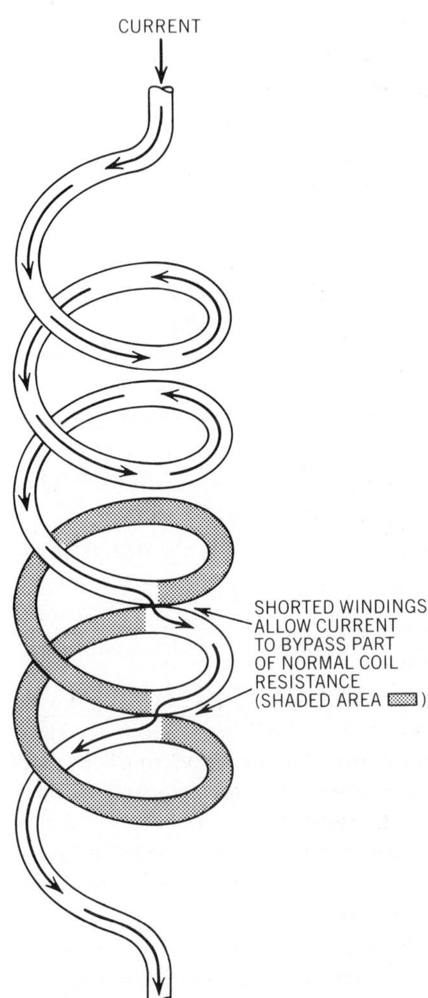

Figure 4-4. If the turns of a coil touch each other, they can cause a short circuit. This can happen in an ignition coil, a relay, a solenoid, a motor, or an alternator.

switches to open circuits and control their operation. An unwanted open circuit can be caused by broken wiring, a damaged part, or no connection between connectors.

Low-Resistance Faults

A low-resistance fault is an unwanted current path that bypasses all or part of a circuit. We generally speak of two kinds of low-resistance faults in automotive electrical systems: a **short circuit** and a **grounded circuit**. We often call them simply "shorts and grounds."

A short circuit is one that is completed in the wrong way. A short usually occurs when two wires of a circuit touch each other where they should not. This allows current to bypass other parts of the circuit, figure 4-3. The current finds a shorter circuit of lower resistance. A short also can occur inside a component, such as when the turns of a coil touch each other, figure 4-4.

A grounded circuit is a short circuit to ground, figure 4-5. In an automobile, a ground usually occurs through the unwanted connection of a wire or component to the chassis or engine. From the point of the unwanted connection, current bypasses all other parts of the circuit and flows to ground.

Automobiles use single-wire electrical systems with the vehicle chassis as a common ground. The battery ground terminal is connected to the engine or frame, and all other parts are similarly grounded. Therefore, most short-circuit faults usually happen as a short to ground.

To troubleshoot electrical faults in the engine and elsewhere, you will use test meters and other simple tools.

BASIC ELECTRICAL METERS

The three basic electrical meters are the:

- Ammeter, which measures amperage
- Voltmeter, which measures voltage
- Ohmmeter, which measures resistance

Tachometers and dwell meters also operate on the same principles as these basic meters.

All dial-indicating meters (not digital display meters) use a mechanism called a **D'Arsonval movement**. This is a small coil of fine wire mounted on bearings within the field of a permanent magnet. The meter needle is attached to the coil, figure 4-6. Small wire springs are connected to the coil. They act as conductors and return the coil to the zero position when the meter is off.

When you connect a D'Arsonval meter to a circuit or a part to be tested, current flows through the coil and creates a magnetic field that interacts with the permanent magnet. This rotates the coil and meter needle in one direction or the other, depending on the direction of current flow. Meters are calibrated so that amperage, voltage, and resistance measurements are proportional to current through the meter coil. Tachometers and dwell meters that measure engine speed and ignition dwell display these values based on current and voltage pulses from the ignition.

Meters that use a D'Arsonval movement and a movable needle, or pointer, are called analog meters because the pointer indicates a continuous range of values as it moves across the scale. Analog means that measurements are given in continuously variable quantities. Electronic digital meters display measurements in discrete digits (numbers). Digital meters are explained later in this section.

Many of the meters that you use are called **multimeters**, figure 4-7, which means that they can measure several combinations of amperage, voltage, resistance, and other values. They have

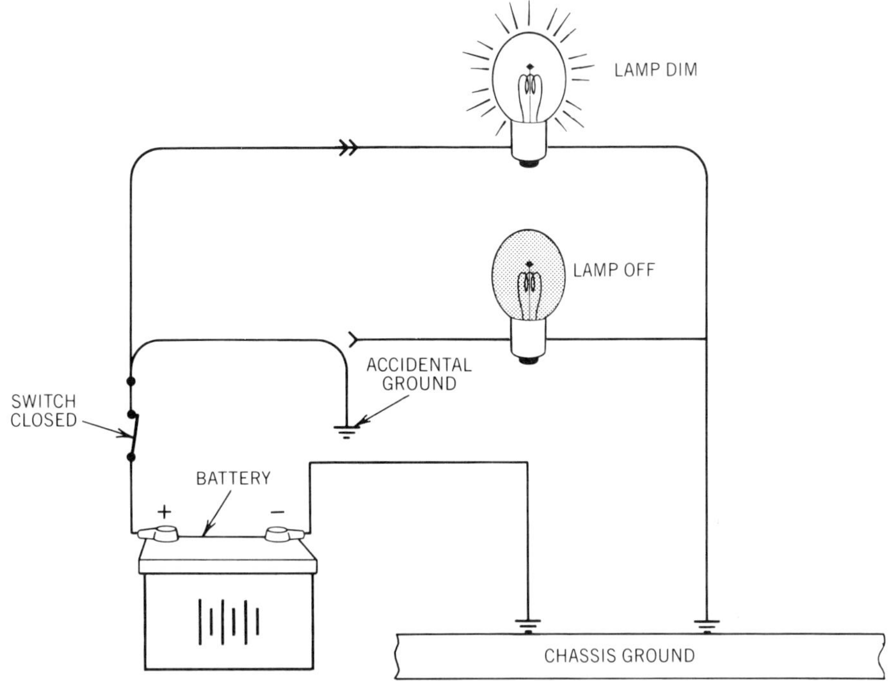

Figure 4-5. A grounded circuit is a type of short circuit.

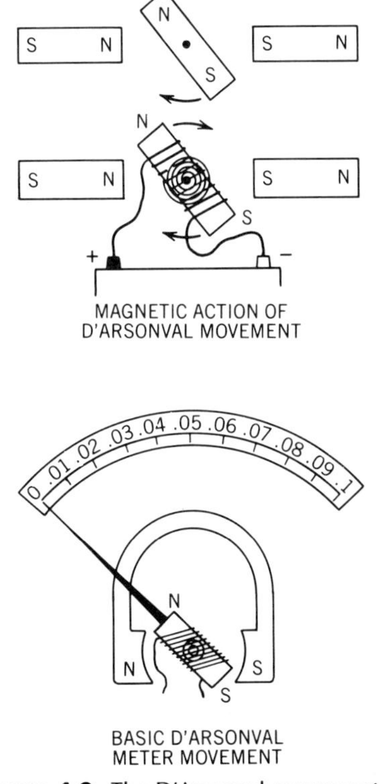

Figure 4-6. The D'Arsonval movement mechanism is the basis of all dial-indicating meters. (Chevrolet)

Figure 4-7. This is a typical analog automotive multimeter that you can use for electrical tests.

internal switching circuits that connect the basic meter movement in different ways to measure different values and various ranges of these values. Multimeters can be either analog or digital instruments. All meters are calibrated for accuracy by the manufacturer, and most have adjustment screws that allow the meter to be recalibrated.

All meters have test leads to contact the circuit or part being tested. The leads may be permanently attached to the meter or they may plug into various sockets for different uses. When you measure amperage or voltage in a circuit, you must be sure that the polarity of the meter and the leads matches the polarity of the circuit. One lead is usually red for positive (+) and is connected to the positive side of the circuit. The other lead is usually black for negative (−) and is connected to the negative side of the circuit. For example, when you measure battery voltage with a voltmeter, connect the red (+) meter lead to the battery positive terminal and the black (−) lead to the negative terminal. The following sections explain the construction and use of basic meters.

Ammeter

You use an ammeter to measure current in a circuit, and you can use it to test circuit **continuity**. To do these things, current must flow through the meter coil. However, the coil is made of very fine wire with little current capacity. Therefore, a low-resistance resistor, called a **shunt** resistor, is in parallel with the coil inside the meter, figure 4-8. Most of the current flows through the shunt resistor. Only a small portion flows through the meter coil. The current through the coil is always proportional to the total current flow. The meter is calibrated to indicate total current based on the percentage that flows through the coil.

> **Caution.** Always connect an ammeter in series with the circuit or object being tested. Because the shunt resistor in parallel with the meter coil has low resistance, the total resistance of the meter is very low and does not affect current measurements. If you connect an ammeter in parallel with a circuit, it will allow too much current to flow and be damaged.

When using an ammeter, connect the positive lead to the positive side of the circuit (nearest the battery + terminal) and the negative lead to the negative side of the circuit (usually toward ground or the battery − terminal), figure 4-9. Some ammeters have scales that read in both directions from zero to measure reverse current flow, but you still must observe system polarity when connecting the meter.

Some ammeters have inductive test leads so the meter can measure current without being connected into the circuit. An inductive test lead is a magnetic coil in a large clamp that fits around the wire being tested, figure 4-10. The meter measures current developed in the inductive lead, which is proportional to the current in the wire being tested. Because the direction of current flow and the direction of magnetic flux relate to each other, you must connect an inductive lead in the direction shown by the arrow on the clamp. Follow the equipment maker's directions.

Ammeter Tests.

Before connecting an ammeter into a circuit, set the range selector to the range above the maximum expected current draw. Then, if the current reading is below the full-scale indication of the next lowest range, switch to that range. If you set an ammeter on a range below the current draw of a circuit, the high current can damage the meter. Here are three general rules about the readings you may get from an ammeter in series with a circuit, figure 4-11:

1. If the meter shows *no current,* the circuit is *open* at some point. There is no circuit continuity, figure 4-11A.

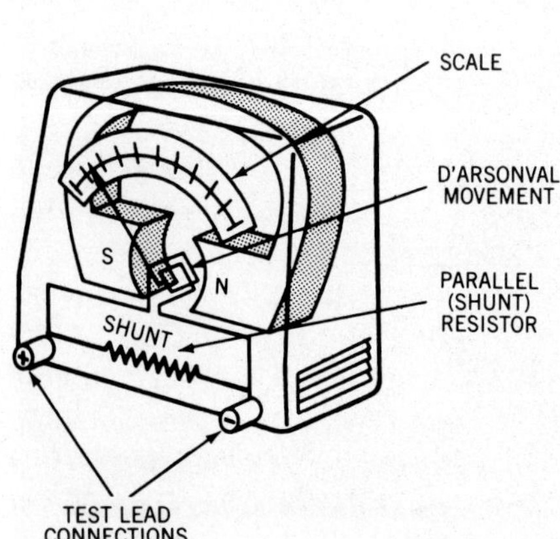

Figure 4-8. Typical ammeter construction. (Delco-Remy)

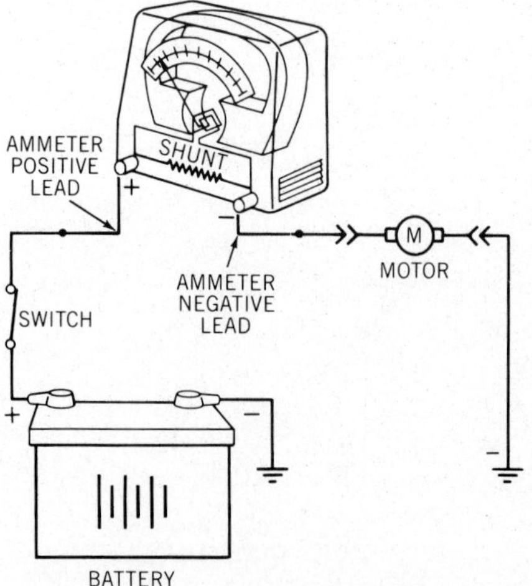

Figure 4-9. Connect an ammeter in series, + to + and − to −.

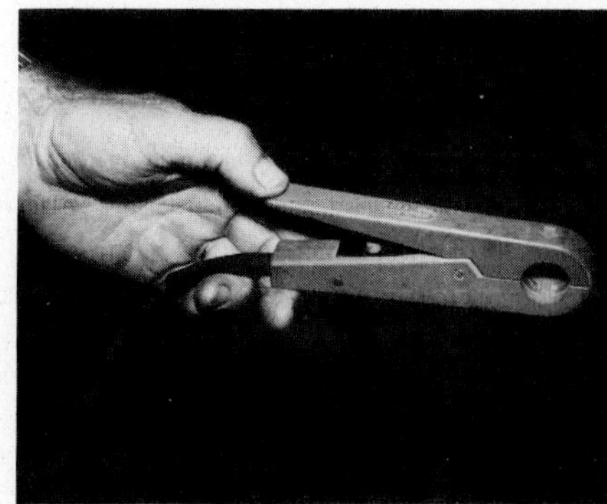

Figure 4-10. Follow the equipment maker's instructions for connection polarity when using an inductive ammeter.

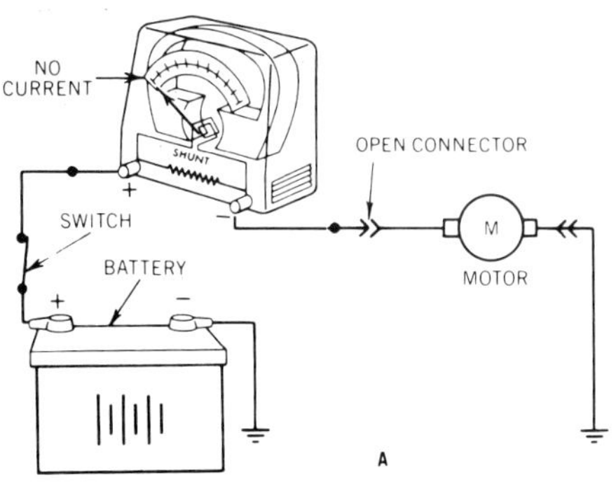

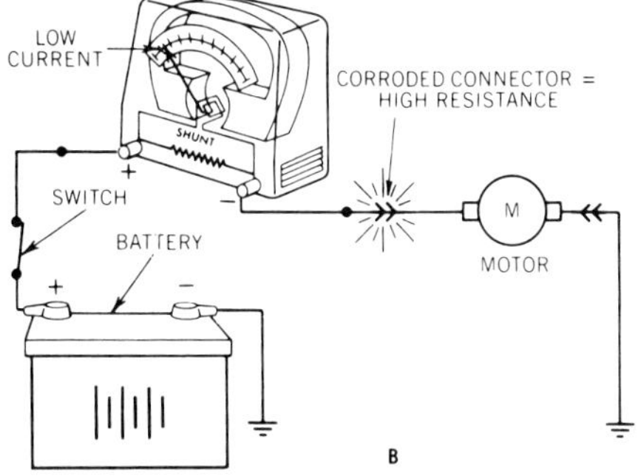

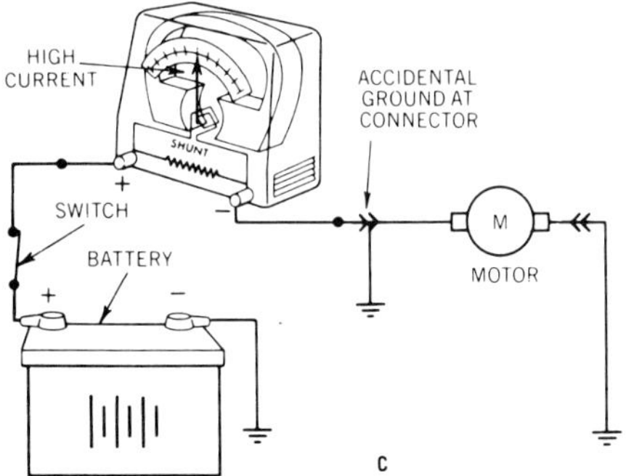

Figure 4-11. Ammeter readings will indicate these basic circuit problems.

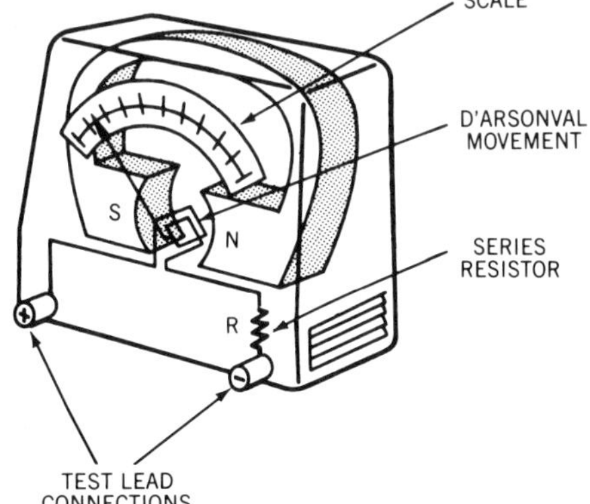

Figure 4-12. Typical voltmeter construction. (Delco-Remy)

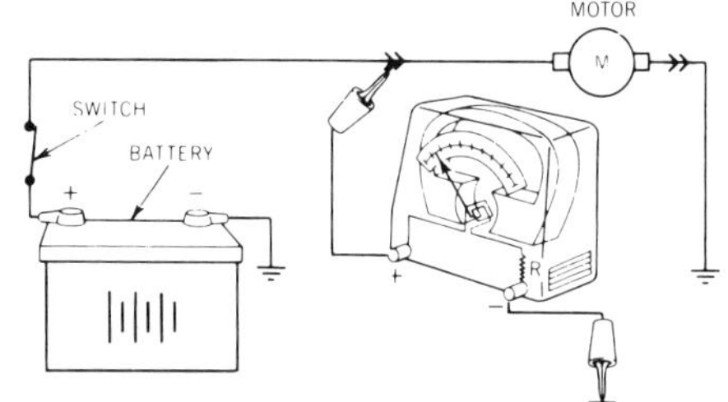

Figure 4-13. Usually, you will connect a voltmeter in parallel, or across a circuit.

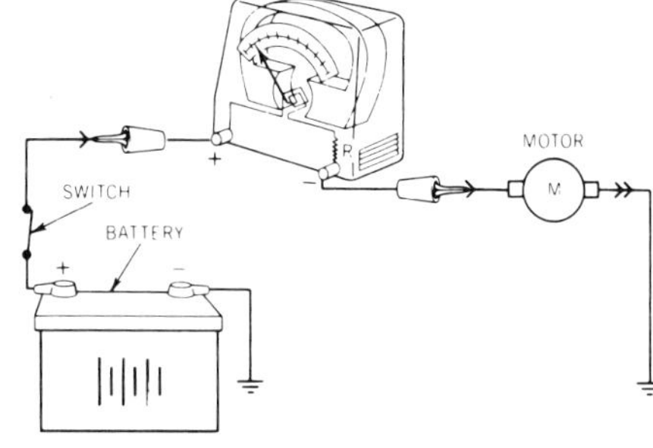

Figure 4-14. You can connect a voltmeter in series to check continuity.

2. If the meter shows *low current*, the circuit is complete but has *high resistance*, figure 4-11B.

3. If the meter shows *current that is too high*, some of the normal resistance has been bypassed to *ground* or through a *short* circuit, figure 4-11C.

You will learn specific applications of these ammeter test principles in the electrical and ignition service chapters of this book.

Current flow is equal everywhere in a series circuit or in one branch of a parallel circuit. A high or low ammeter reading doesn't tell you immediately where a problem is. To pinpoint the high- or low-resistance problem, you must relocate your ammeter or use a voltmeter or an ohmmeter.

Voltmeter

You can use a voltmeter to:

- Measure the source voltage of a circuit
- Measure the voltage drop across a load
- Measure the voltage at any point in a circuit
- Test circuit continuity.

You normally connect a voltmeter in parallel with a circuit or a load or across a voltage source. A voltmeter has a high-resistance resistor in series with the meter coil to protect the movement from high current but allow full voltage across the coil, figure 4-12. The small current that a voltmeter draws does not significantly affect the circuit voltage drop. Most voltmeters have range selectors that switch resistors of different values into series with the coil to measure different voltage ranges. A typical automotive voltmeter might have scales of 0 to 5 volts, 0 to 10 volts, and 0 to 20 or 25 volts.

As with an ammeter, always connect the voltmeter positive (+) lead to the positive side of the circuit or load and the negative (−) lead to the negative side, figure 4-13. Before connecting a voltmeter, set the range selector to the range *above* the maximum circuit voltage. Then if the reading is below the full-scale indication of the next lowest range, switch to that range. High voltage applied across the low range of a voltmeter can damage the meter.

Even though a voltmeter usually is connected in parallel with a circuit or load, it can be connected in series to check continuity, figure 4-14. If the meter shows normal system voltage at that point, the circuit is complete between the test point and ground. If it shows no voltage, the circuit is open (no continuty) between the test point and ground.

You can measure voltage in a circuit with or without current flowing. When current is flowing, the voltmeter shows the voltage drop caused by circuit loads because voltage decreases wherever a load is between the test point and the battery positive terminal. With no current, voltage should be approximately battery voltage (12 to 12.6 volts) at

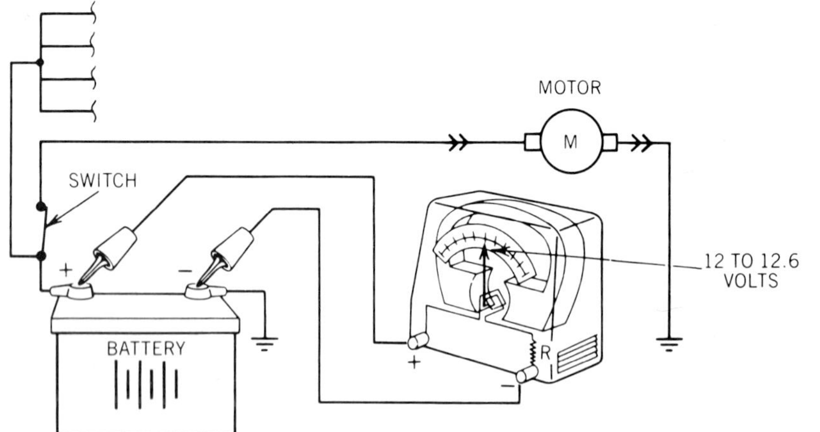

Figure 4-15. Connect your voltmeter across the battery to measure battery voltage.

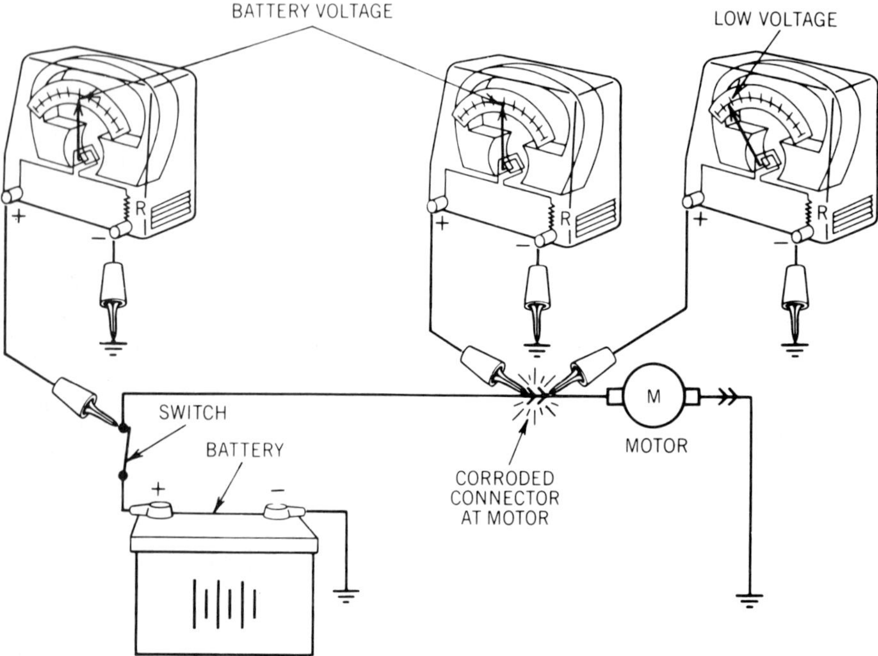

Figure 4-16. Available voltage tests will locate high resistance and an unwanted voltage drop.

every point above ground in a circuit. This is **open-circuit (no-load) voltage.**

There are many ways that you can use a voltmeter to measure available voltage and voltage drop at any point in a circuit and to test circuit continuity. You will learn specific test procedures in the electrical and ignition service chapters of this manual, but all are based on the following principles.

Available Voltage Voltmeter Tests. You can measure voltage available in a circuit with or without current flowing. Voltage without current flow is open-circuit voltage and should equal battery, or source, voltage. When current flows through a circuit, the circuit devices use, or drop, some voltage as they operate. Voltage on one side of a load will be different from the voltage on the other side when the circuit is closed and operating. Battery voltage also will drop when a circuit is operating and will continue to drop until recharged by an alternator or battery charger.

As an example, measure available source voltage at the battery by connecting your voltmeter + lead to the battery + terminal and the − lead to the − terminal, figure 4-15. Be sure all electrical circuits are open (off). The meter reading should be about 12 to 12.6 volts. Now turn on the headlamps to complete a circuit and read the meter again. The available voltage will be lower than open-circuit voltage, depending on the condition of the battery and the circuit current draw.

You also can use your voltmeter to locate high-resistance problems in a circuit with current flowing. Connect the voltmeter − lead to ground and use the + lead to test for available voltage at several points in the circuit, figure 4-16. This drawing shows that poor motor operation has been caused by high resistance and an unwanted voltage drop at a corroded connection.

Voltage-Drop Voltmeter Tests. When an electrical device operates, it uses, or drops, a specific amount of voltage that depends on the resistance of the device and the current in the curcuit. Unwanted voltage drop can result from a high-resistance connection or a defective device. An important rule for voltage-drop testing is:

> *The sum (total) of the voltage drops around a circuit equals the source voltage.*

A voltage-drop test can tell you if:

- An electrical device is using too much voltage because of high resistance in the device
- An electrical device is using too little voltage because of a short or grounded circuit in the device
- High resistance from a loose or corroded connection is causing an unwanted voltage drop.

A circuit must be closed and operating for voltage-drop testing. You can calculate voltage drop indirectly or measure it directly for any part of the circuit.

To calculate voltage drop indirectly, connect your voltmeter as you would for available voltage tests at different points in the circuit, figure 4-17. Compare the voltage on one side of a device to the voltage on the other. In figure 4-17:

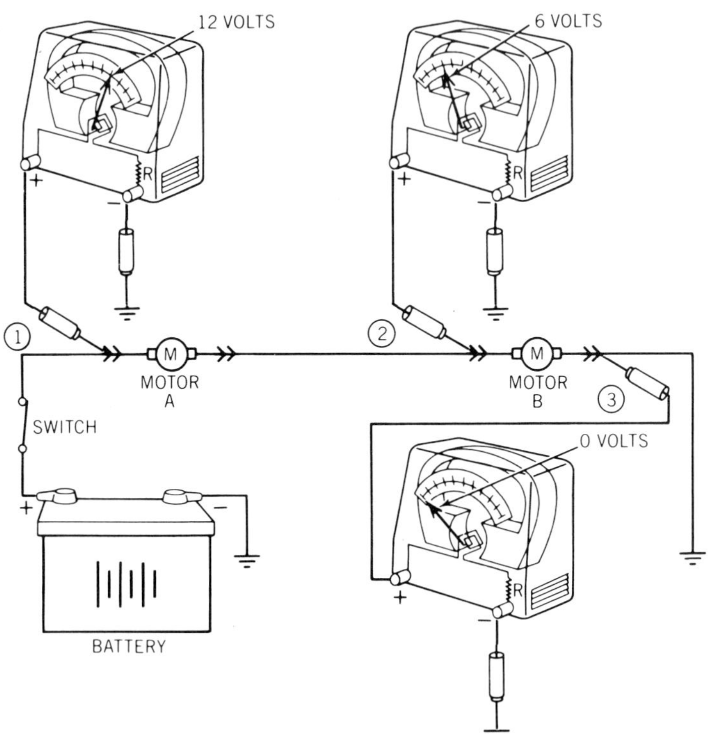

Figure 4-17. You can calculate voltage drops from a series of available-voltage tests, like this.

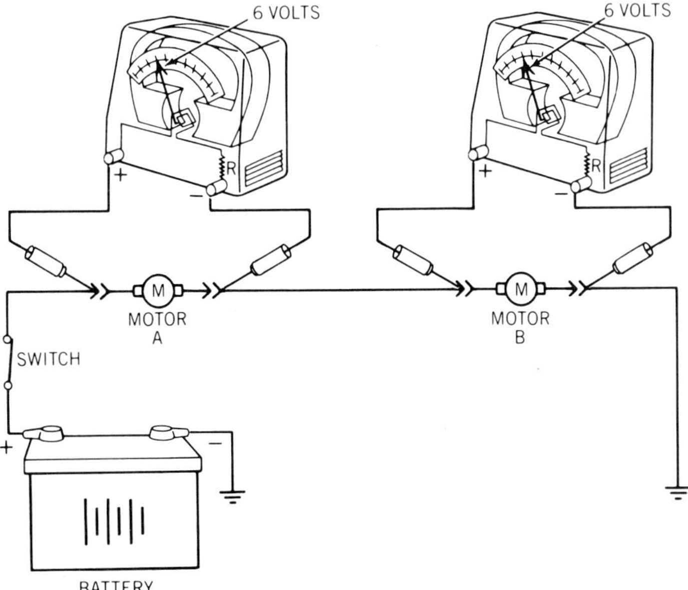

Figure 4-18. You also can measure voltage drops directly across individual loads.

1. The meter at test point 1 reads 12 volts or battery voltage because there are no voltage-dropping loads between the battery and test point 1.

2. The meter at test point 2 reads 6 volts. The voltage drop across motor A is, therefore, 6 volts (12 − 6 = 6).

3. The meter at test point 3 reads 0 volts. The voltage drop across motor B is, therefore, 6 volts (6 − 6 = 0).

The sum of the voltage drops (6 + 6) equals the source voltage, 12 volts.

To measure voltage drop directly, connect the meter + lead to the side of the device nearer the battery + terminal, figure 4-18, and the − lead to the side nearer the battery − terminal, or ground. With the circuit closed and operating, the meter reading indicates the voltage dropped by that one device.

Whether you measure voltage drop directly, figure 4-18, or calculate voltage drop by measuring between ground and both sides of a device, figure 4-17, depends on which way is easier to connect your meter. For available voltage and voltage-drop tests, remember that a voltmeter is connected in parallel with a circuit or load. An ammeter is *always* connected in series.

Ohmmeter

An ohmmeter, too, is based on a D'Arsonval movement, but an ohmmeter has its own voltage source. It measures resistance of a circuit or component and always is connected to an *open* circuit or a part removed from a circuit.

Caution. Never connect an ohmmeter to a circuit in which current is flowing or which has voltage applied to it. Current from an outside source will damage an ohmmeter.

An ohmmeter has a low-voltage dc power supply (usually dry-cell batteries) and a resistor in series with the meter coil to limit current flow, figure 4-19. Because an ohmmeter does not use system voltage, it is not affected by system polarity. You can connect the test leads to either side of a part to be tested. The ohmmeter reacts to current flow from its power source to move the needle.

One exception to this rule applies to diodes. A diode conducts current in one direction but blocks it in the other. Therefore, an ohmmeter will indicate continuity (a closed circuit) when connected to a diode in one direction, figure 4-20, but it will indicate an open circuit (infinite resistance) when the leads are reversed.

Ammeters and voltmeters read from left to right to show increasing current and voltage. An ohmmeter shows low resistance at the right of its scale, which is equivalent to high current flow. It shows infinite resistance at the left of

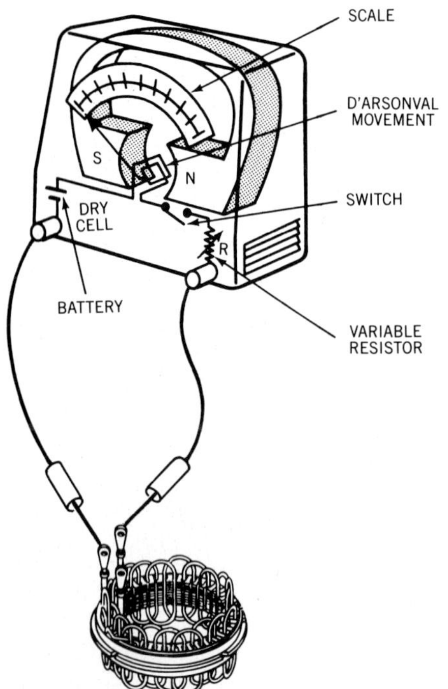

Figure 4-19. Typical ohmmeter construction and connection. (Delco-Remy)

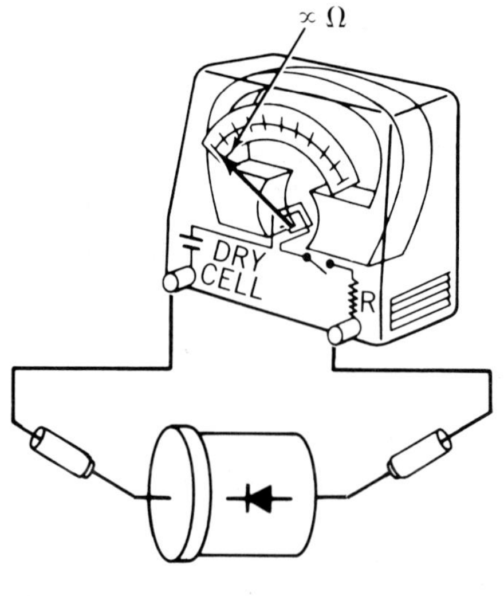

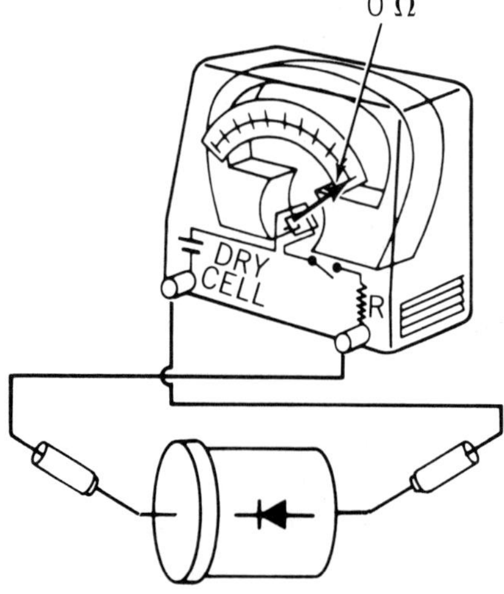

Figure 4-20. When connected to a diode, an ohmmeter will show infinite resistance in one direction and no resistance (continuity) in the other.

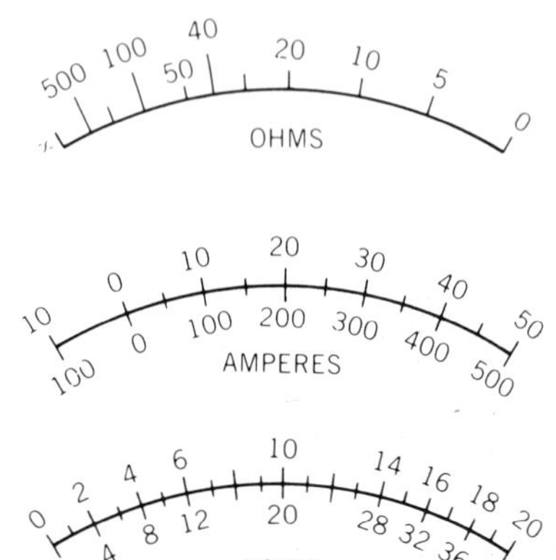

Figure 4-21. An ohmmeter scale is the opposite of ammeter and voltmeter scales.

the scale, which is equivalent to no current flow, figure 4-21. Ohmmeters have range selectors to measure resistance of different values. When you use an ohmmeter, begin testing on the lowest range and then switch to a higher range that gives you a more precise reading. Because voltage and current in an ohmmeter are limited by its power supply and internal resistors, you can't damage the meter by setting it on a low or high scale.

Temperature and the condition of the batteries will affect an ohmmeter's accuracy, so you must adjust an ohmmeter each time that you use it. Do this by touching the two test leads together and turning the adjustment knob until the needle indicates 0 ohms, or continuity, through the meter on the lowest meter scale.

You can measure resistance of a circuit or any part directly with an ohmmeter, or you can compute the resistance from voltage drop measurements made with a voltmeter on an operating circuit. Heat and vibration often increase resistance at points in a circuit. Such high-resistance faults may not show up when a circuit is open or a part is removed for testing. In these cases, voltage drop tests may locate a problem that you might miss with an ohmmeter. Voltage drop tests also have the following advantages:

• You can test a complete circuit quickly because you can move the meter quickly from point to point while the circuit is operating.

• Reading a voltmeter is often easier because small increases in wiring resistance appear as sharp increases in voltage drop.

An ohmmeter, however, has advantages for other test conditions:

• Testing parts that have specific resistance values under specified conditions, such as the windings of ignition coils, solenoids, and relays.

• Testing high-resistance parts such as spark plug cables.

• Testing internal parts of alternators and motors that require disassembly for test access.

You will learn more about ohmmeter use in later chapters of this book.

Multimeters

Many automotive test meters are multimeters that have ammeter, voltmeter, and ohmmeter functions in a single unit. A battery-starter tester is a multimeter that has an ammeter and a voltmeter along with a variable resistance load device. A volt-amp tester is a similar instrument. Tachometers and dwell meters (or a combined tach-dwell meter) are voltmeters that are calibrated to measure engine speed and ignition dwell in response to the frequency of voltage pulses.

Digital Meters

Many service operations on late-model automobile electronic systems require digital test meters. In a digital meter, the D'Arsonval movement is replaced by electronic circuitry that senses current, voltage, or resistance and converts analog signals to a digital display, figure 4-22. Digital meters are required for testing many electronic systems because of their high input **impedance**. Impedance is the com-

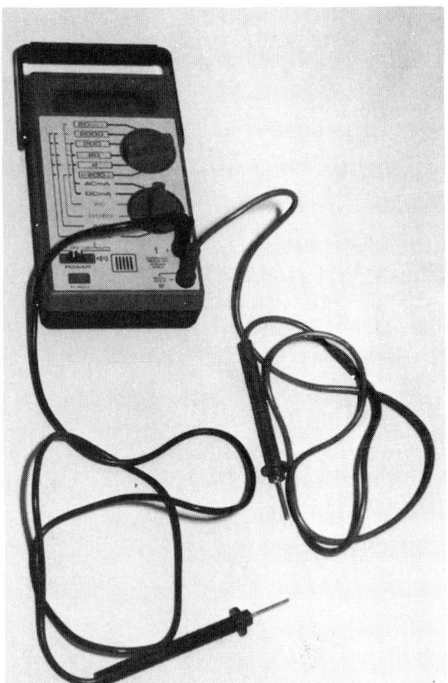

Figure 4-22. High-impedance digital voltmeters are required to test late-model electronic systems.

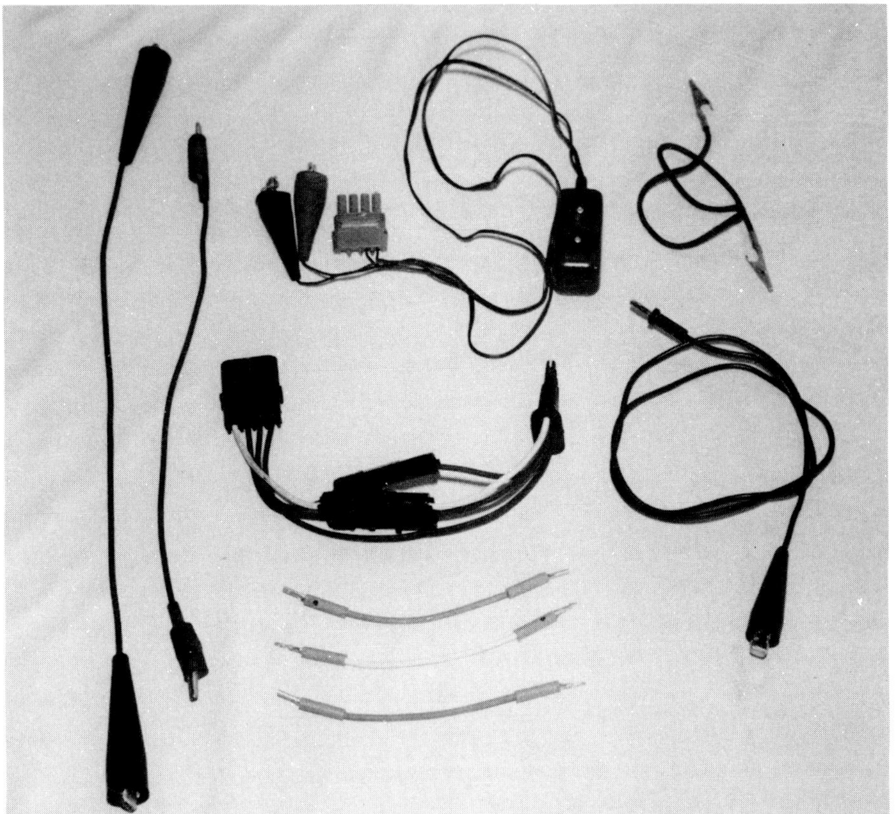

Figure 4-23. Jumper wires with different connectors are among your simplest troubleshooting tools.

bined opposition to current created by the resistance, capacitance, and inductance of the meter. The input impedance of a voltmeter is measured in ohms per volt. This is the total impedance of the meter on any voltage range divided by the full-scale voltage of that range. For example, a meter with a total impedance of 100,000 ohms (100 kilohms) and a 0- to 10-volt scale has an input impedance of 100,000/10, or 10,000 ohms per volt.

Meter input impedance is important for testing accuracy and safety of electronic parts. When you connect a voltmeter across a circuit you add parallel resistance (another conductor) to the circuit. This increases current through the circuit and changes the voltage drop across the load resistance being measured. A meter with low input impedance can draw enough current to produce an inaccurate voltage drop measurement. An ideal voltmeter would have infinite resistance, or impedance, and draw no current from the circuit. This is not possible, however, but a high-impedance meter will draw little current and produce more accurate readings.

Many electronic parts operate with very low current, only a few milliamperes. A low-impedance meter connected across such a device may draw excessive current and damage the electronic circuit. Exhaust gas oxygen (EGO) sensors are an example of this. An EGO sensor is a galvanic battery that generates a few hundred milliamperes and voltage of 0 to about 1 volt. A low-impedance meter connected to an EGO sensor may draw more current than the sensor can provide. This creates a short circuit that can destroy the sensor.

For component safety and testing accuracy, carmakers specify that EGO sensors and other electronic devices must be tested with voltmeters that have at least 10 megohms per volt (10 million ohms) of input impedance. Digital meters have at least 10 megohms per volt impedance. Analog voltmeters have input impedance from about 10 to several hundred kilohms per volt, depending on meter quality.

Test Lamps and Jumper Wires

You can do many electrical tests with a jumper wire and simple test lamps. A jumper wire is a length of wire with a test probe or clip at each end, figure 4-23. You can use a jumper wire to bypass switches, connectors, wires, and any *nonresistive* part of a circuit, figure 4-24.

Caution. Never bypass a bulb, motor, coil, or any resistive device (load) in a circuit. High current will damage the circuit.

A 12-volt test lamp is a 12-volt bulb with two leads, figure 4-25. They are used to test circuits or components while current is flowing. You can use a 12-volt test lamp to check for voltage or for ground.

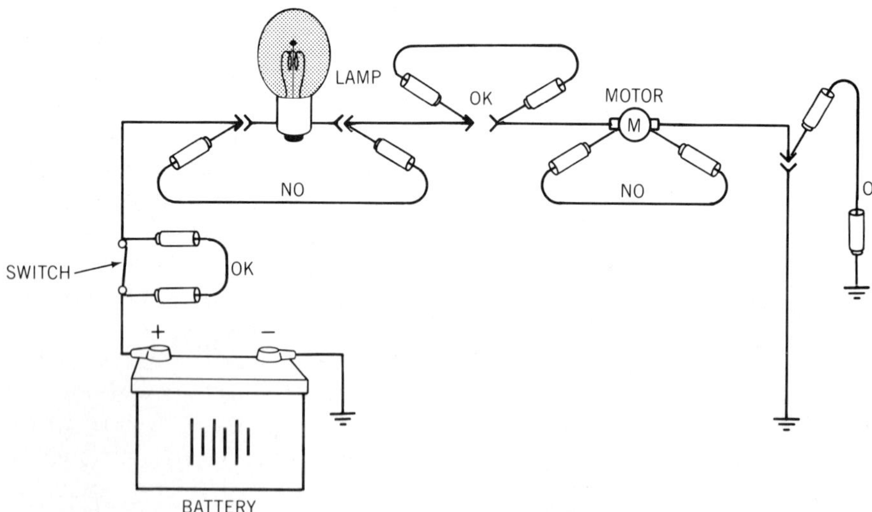

Figure 4-24. Use a jumper wire only on nonresistive parts of a circuit. Never bypass a load.

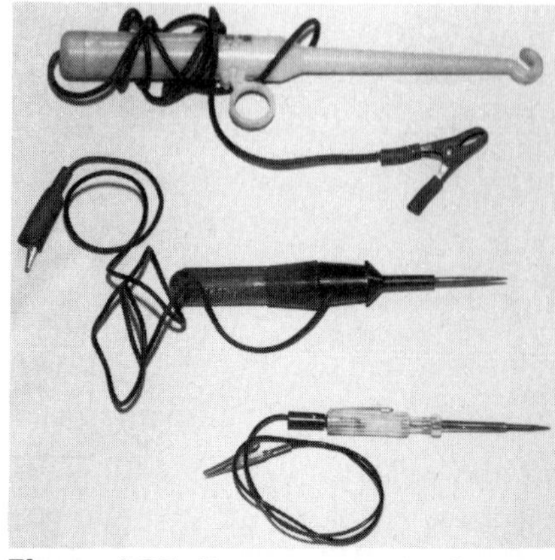

Figure 4-25. You can use a 12-volt test lamp to check a circuit while current is flowing.

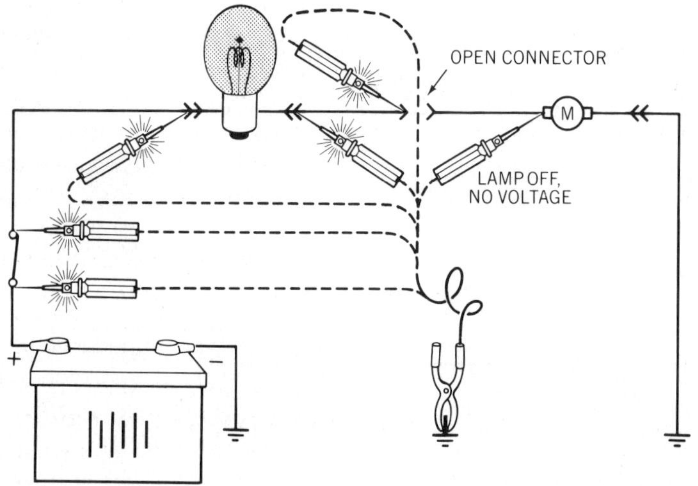

Figure 4-26. Connect the 12-volt test lamp to ground and various circuit points to check for available voltage.

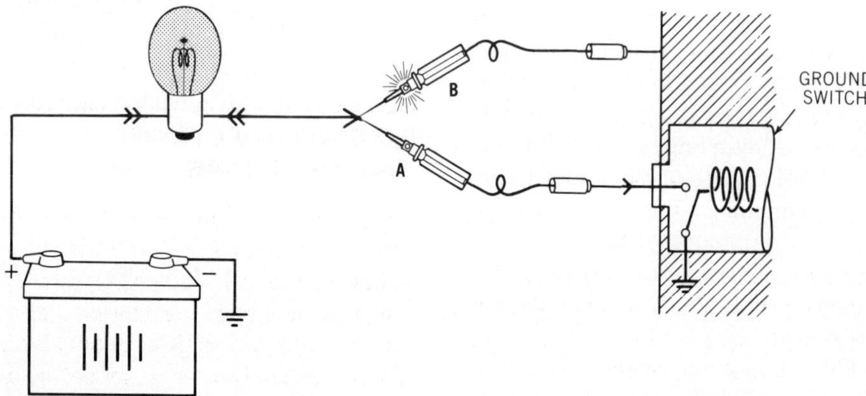

Figure 4-27. You can use a 12-volt test lamp to check for proper circuit ground or continuity.

Caution. Many carmakers advise against using a test lamp on electronic circuits. Like a low-impedance meter, the lamp may draw excessive current and damage electronic parts.

Voltage Check Check for voltage as follows, figure 4-26:

1. Connect one lead to a known chassis ground.

2. Touch the other lead to a source of battery voltage, such as the battery + terminal, to verify the ground connection.

3. Then touch the test lead to various parts of the circuit.

If the lamp lights at one point, but not at the next, the circuit is open between the two points.

Ground Check Check for ground as follows:

1. Disconnect the feed wire from the circuit test point.

2. Turn the circuit switch on.

3. Connect one lamp test lead to the wire and the other to the disconnected connector, figure 4-27A.

4. Move the second lead from the connector to a known ground point, figure 4-27B.

If the lamp lights in both steps, the circuit is complete at that point. If the lamp lights in step 4 but not step 3, the circuit is open (not properly grounded).

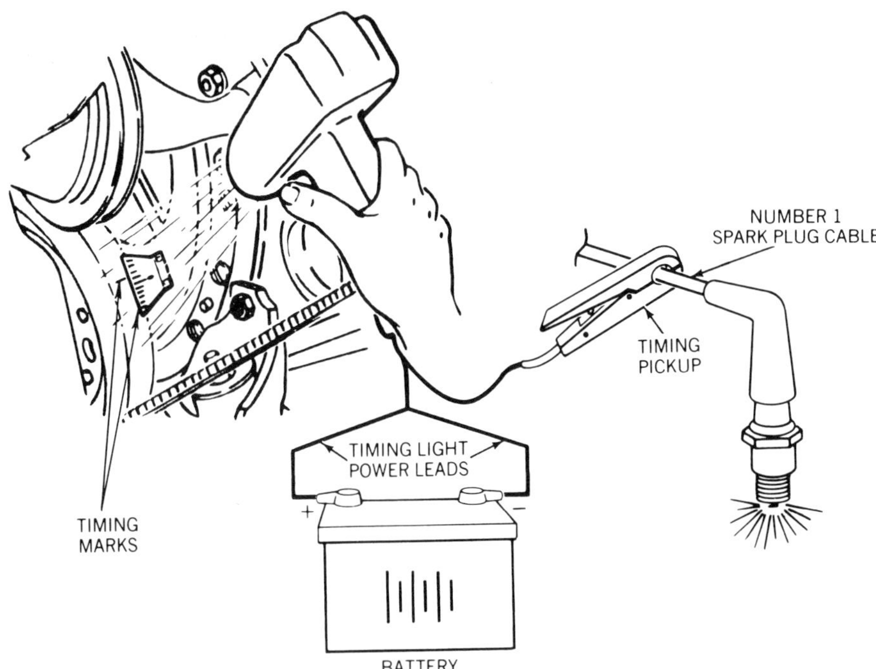

Figure 4-28. Connect a timing light to number 1 spark plug and the battery and aim it at the timing marks to check ignition timing.

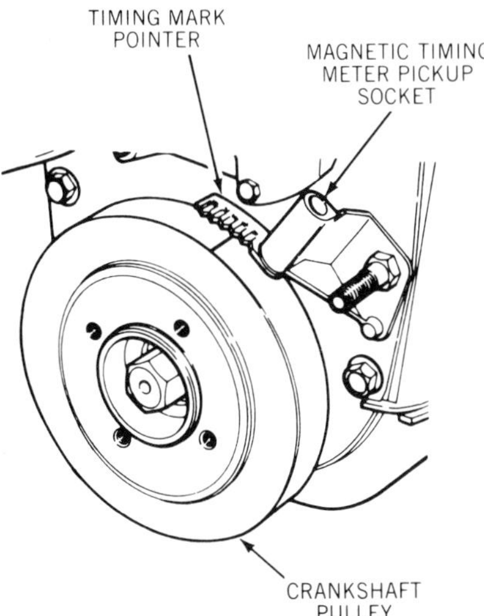

Figure 4-29. Most late-model engines have sockets for magnetic timing meter pickups. (Oldsmobile)

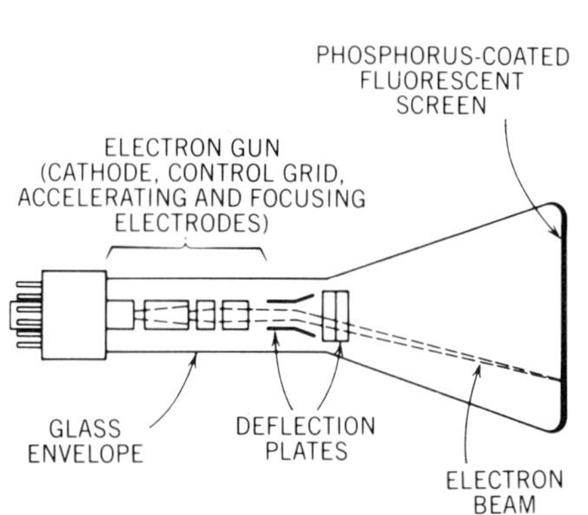

Figure 4-30. The electron gun in a cathode ray tube (CRT) fires electrons at a phosphorus-coated screen. Deflection plates move the electron trace in relation to voltage and time being measured.

Test Lamps and Electronic Control Systems

Incandescent probe lamps can draw excessive current and should not be used to test electronic control systems. An ordinary test lamp may draw 200 to 400 milliamps (0.2 to 0.4 amperes) of current, which may overload an electronic circuit. Some carmakers, such as BMW, forbid use of traditional test lamps on electronic systems. Most electronic systems, however, can be tested with an LED test lamp, which draws 15 to 20 milliamps (0.015 to 0.020 ampere) of current.

IGNITION TIMING LIGHTS AND MAGNETIC TIMING METERS

A timing light is a **stroboscopic** lamp that flashes in parallel with a spark plug when connected to the ignition system. When connected to the specified cylinder (usually number 1 in the firing order) and aimed at the engine timing marks, the timing light indicates the point at which the spark plug fires. The stroboscopic flash makes the marks seem to stand still so you can check and adjust timing to the specified number of degrees before or after top dead center, figure 4-28.

Some timing lights are powered by the car battery. Others are powered by alternating current through an engine analyzer. Many lights have inductive pickups (similar to inductive ammeter pickups) that clamp around the number 1 plug cable. Others need an adapter between the plug or distributor tower and the light.

A simple timing light lets you adjust timing and make basic checks of advance mechanisms. For precise spark advance tests, you need an adjustable timing light, or power timing light. An adjustable light can be set to flash before or after the plug fires.

Most late-model engines have pickup sockets for magnetic timing meters, figure 4-29. Magnetic timing measurement provides a more accurate timing signal than timing marks and a light can. The probe from the meter is inserted in the engine socket, and the meter records a timing signal directly from crankshaft position. The chapter on ignition timing explains the operation of timing lights and magnetic timing meters in detail.

ENGINE ANALYZER OSCILLOSCOPE

The central instrument of most comprehensive engine analyzers is an **oscilloscope**. It has a screen that displays changing voltage levels over a period of time. It is a 2-dimensional voltmeter. You will use a "scope" principally to test and service ignition systems, but it is equally useful for testing alternators, fuel injection solenoids, and other electrical devices with varying voltage.

The scope screen is a **cathode ray tube (CRT)**, figure 4-30, similar to a

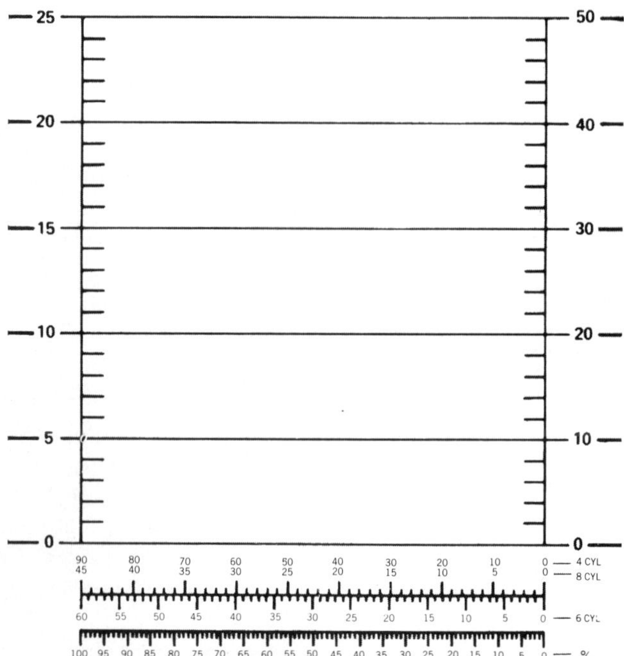

Figure 4-31. The oscilloscope screen has vertical voltage scales and horizontal scales that indicate ignition degrees, time in milliseconds, or percentage. (Coats Diagnostic)

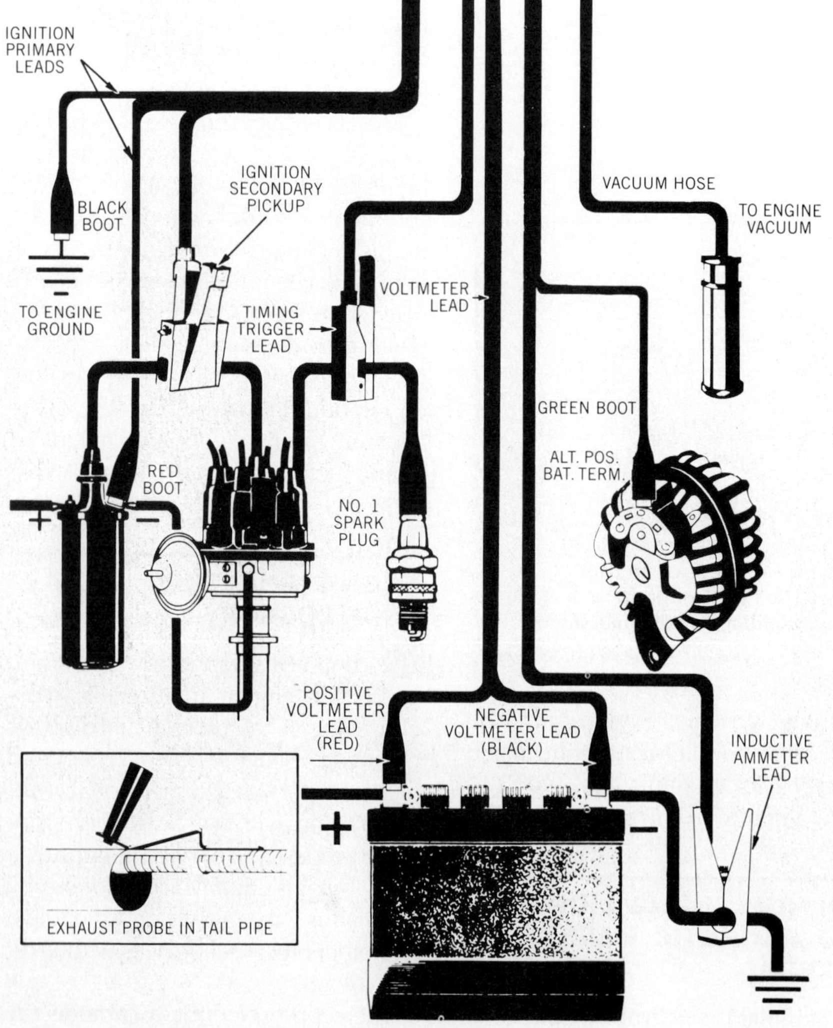

Figure 4-32. These are typical test lead connections for an oscilloscope engine analyzer. (Sun)

television screen or a computer monitor. An electron gun fires electrons at the inside of the phosphorus-coated screen. The electrons create a voltage trace on the screen that is moved by deflection plates in the CRT. The vertical deflection plates move the electron trace up (positive) or down (negative) in relation to the voltage being measured. The horizontal deflection plates move the trace from left to right in relation to time. Horizontal deflection is controlled by the frequency of the circuit being tested. For ignition testing, this is engine speed.

The oscilloscope screen has vertical scales to measure different voltage levels, figure 4-31. Most automotive scopes have a low-voltage scale from 0 to 20 or 40 volts and two high-voltage scales for ignition testing. The ignition voltage scales may be from 0 to 20 or 25 kV and from 0 to 40 or 50 kV. The horizontal scales at the bottom of the screen indicate ignition dwell in degrees, usually for 4-, 6-, and 8-cylinder engines. The horizontal scale also may show time in milliseconds and a percentage scale for testing duty cycle.

An engine analyzer with an oscilloscope has multiple test leads that are connected to various points on an engine for voltage and current measurements. Exact connections will vary from one analyzer to another, but most are similar to the example in figure 4-32.

Digital Engine Analyzers and Dual-Trace Scopes

In the early 1980's several test equipment manufacturers introduced engine analyzers with programmed, automatic test sequences. These analyzers display engine operating conditions in words and numbers on a cathode ray tube (CRT), figure 4-33. Programmed analyzers show readings of voltage, engine speed (rpm), vacuum, and other conditions in a comparison display with the carmaker's specifications for the vehicle being tested. To do this, the analyzer employs its own digital computer.

The analyzer takes analog readings from various test leads and processes

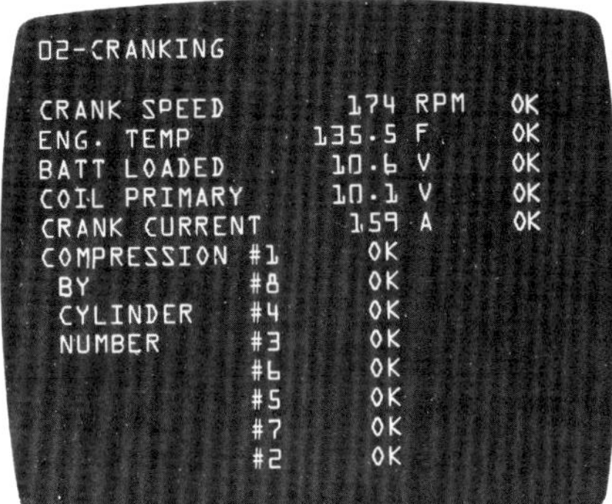

Figure 4-33. A digital engine analyzer displays the results of programmed tests on a cathode ray tube (CRT).

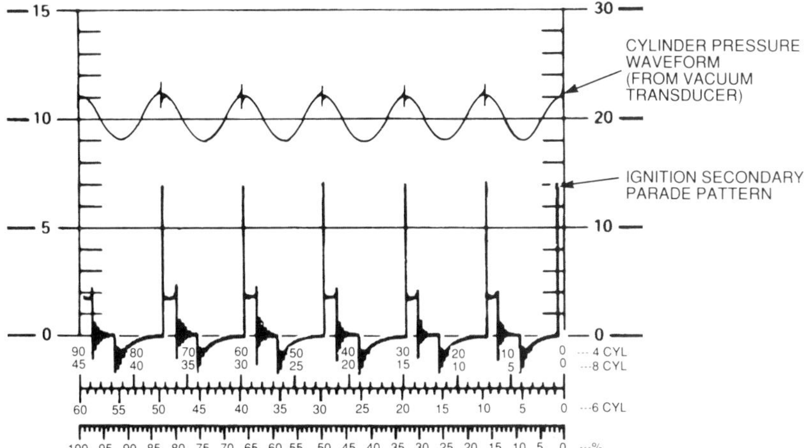

Figure 4-34. A dual-trace scope can display voltage signals at both ends of a circuit or it can display two waveforms for comparison, such as this ignition secondary pattern (bottom) and a cylinder pressure waveform (top).

the signals through analog-to-digital (AD) converter circuits. The analog signals are converted to digital values and displayed as numbers on the CRT. The carmaker's specifications are loaded in the analyzer computer program in programmable read-only memory (PROM).

All programmed analyzer test sequences require the technician to enter certain vehicle identification numbers into the analyzer to begin the tests. This is usually done through a keyboard or a light pen on the analyzer. The analyzer then "looks up" the specifications for the vehicle in its ROM and compares them to the actual values measured during a test.

Programmed test sequences usually include general tests of engine cranking conditions, idle operation, and operation at speeds above 2,000 rpm. The analyzer measures ignition and charging system voltage, engine vacuum, cranking current, rpm, and other conditions. Most modern analyzers also include programs that follow the test sequences of carmakers' automatic testers for onboard computer systems. For example, a comprehensive engine analyzer contains programs that follow Ford's prescribed tests for the self-test automatic readout (STAR) tester, Chrysler's diagnostic readout (DCR) tester, and GM's test routines for the computer command control (CCC) system.

Dual-Trace Oscilloscopes. The standard oscilloscope engine analyzer that has been used in automobile service for decades is a single-trace scope. It measures one voltage waveform at a time. Multiple-trace oscilloscopes that measure two, three, four, or more voltage traces simultaneously have been common in electronics laboratories even longer. In the mid-1980's, dual-trace scopes were introduced to automobile service.

Dual-trace scopes basically have multiple test leads and multiple input channels that allow two waveforms to appear on the CRT simultaneously. For example, you can connect one set of leads to a computer system sensor and another set of leads to the sensor harness at the computer connector. This allows you to compare the voltage signal sent by the sensor to the voltage signal received by the computer. Similarly, you can connect one set of leads to the computer output terminals for an actuator, such as a fuel injector driver circuit, then connect the other set of leads to the injector and compare the voltage signal sent by the computer to the voltage received by the injector.

A dual-trace scope allows you to compare voltage at both ends of a circuit and to isolate faults in the sensor or actuator, the wiring, or the computer.

Allen Test Products was the first test equipment maker to market a dual-trace scope specifically for automotive service. Other manufacturers soon followed. Many modern dual-trace automotive oscilloscopes have special transducers that allow you to connect a vacuum line to the engine and generate an analog voltage signal that indicates vacuum, or manifold pressure. The scope displays the vacuum waveform simultaneously with the ignition waveform, figure 4-34. This allows you to distinguish between abnormal conditions caused by bad plug wires or fouled plugs in the ignition secondary and low compression or manifold air leaks in the engine.

DIAGNOSTIC CONNECTORS

Most automobiles manufactured since 1980 have some kind of diagnostic, or test, connector. These are generally found in the engine compartment and are used to connect special testers or various types of standard test equipment.

Early Diagnostic Connectors

Diagnostic connectors, figure 4-35A, first appeared in the engine compartments of 1976 GM vehicles and were used through 1982. Chrysler vehicles also had similar connectors from 1978 to 1981. These connectors had terminals connected in parallel with test points in the electrical system. They allowed you to connect test equipment easily, at a single location for many standard engine tests, and were designed to let you run a series of prescribed tests quickly. From 1976 to 1982 some GM vehicles with air conditioning also had a second connector to check the air conditioning electrical circuits. Some engine analyzers had special test leads with connectors that matched the vehicle connectors. Several equipment companies made special testers that plugged into the connectors and ran a programmed series of tests, Figure 4-35B. Figure 4-36 is a wiring diagram for a diagnostic connector on a late 1970's full-size GM car. These early test connectors were not installed on cars with onboard computers, but they started the trend toward diagnostic connectors for the computer systems on late-model vehicles.

A

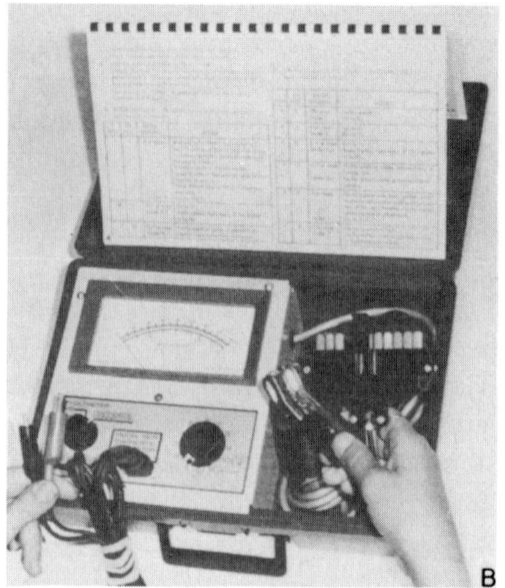

B

Figure 4-35. (A) This vehicle diagnostic connector provides a single connection point for several engine tests. (B) One of the diagnostic testers developed for use with the Chevette diagnostic connector.

GM ALCL or ALDL Connector

A test connector is located in the passenger compartment on all GM vehicles built since 1980 and equipped with either a C-4 or CCC engine control system. The connector is known as an assembly line communications link (ALCL) or assembly line diagnostic link (ALDL) connector. Wired to the electronic control module (ECM), it contains terminals used to diagnose the vehicle systems. The design and complexity of the connector have changed with the evolving engine control system, figure 4-37. When the diagnostic or test terminal is connected to the ground terminal with a jumper wire, the CHECK ENGINE or SERVICE ENGINE SOON lamp in the instrument cluster will flash any stored trouble codes. The remaining terminals in the connector are used to diagnose other vehicle systems by jumpering them to the ground terminal. One terminal in the ALDL connector is a serial data stream connector, which allows you to read sensor and actuator signals from the computer on a hand-held scan tool or an engine analyzer. The number of terminals in the connector and the exact systems that can be diagnosed differ according to vehicle and model year.

GM vehicles equipped with a body computer module (BCM) have a cover over the ALDL connector. The cover should always be in place when the connector is not being used for testing. The BCM system uses a serial data link (SDL) in the form of a loop. This provides two redundant communication paths in the serial data link; one routed clockwise and the other routed counterclockwise around the vehicle. A break in one path will not affect vehicle operation since the other path is still complete. The ALDL connector cover has a built-in shorting plug that completes one part of the SDL loop. Opening the cover to connect a tester breaks one path; connecting the tester restores the path. Leaving the cover off when the tester is removed will reopen that path and can result in an unnecessary vehicle shutdown if the other path should develop a break.

Ford STAR Connector

Late-model Ford vehicles with an MCU engine control system and all Ford vehicles with EEC-IV have a diagnostic connector in the engine compartment through which trouble codes can be retrieved. Although designated as a self-test input, self-test output connector, it is sometimes called the STAR connector because it is the connection point for the self-test automatic readout (STAR) tester used to diagnose problems with MCU and EEC-IV systems. Early systems used just a 6-terminal connector, figure 4-38; later systems also have a seventh terminal on a pigtail, figure 4-39. A standard analog voltmeter also can be used to retrieve trouble codes from the connector, figure 4-40.

The diagnostic connector is located near the module on MCU-equipped vehicles. On EEC-IV vehicles, it is generally near the vacuum tree on the driver's side of the cowl. (Some vehicles, especially trucks, locate it on the passenger's side of the cowl.) Late-model vehicles may have a push-in cover to protect the connector terminals from engine compartment contamination and corrosion.

Chrysler Diagnostic Connector

Chrysler added self-diagnostic capability to its front-wheel-drive cars in

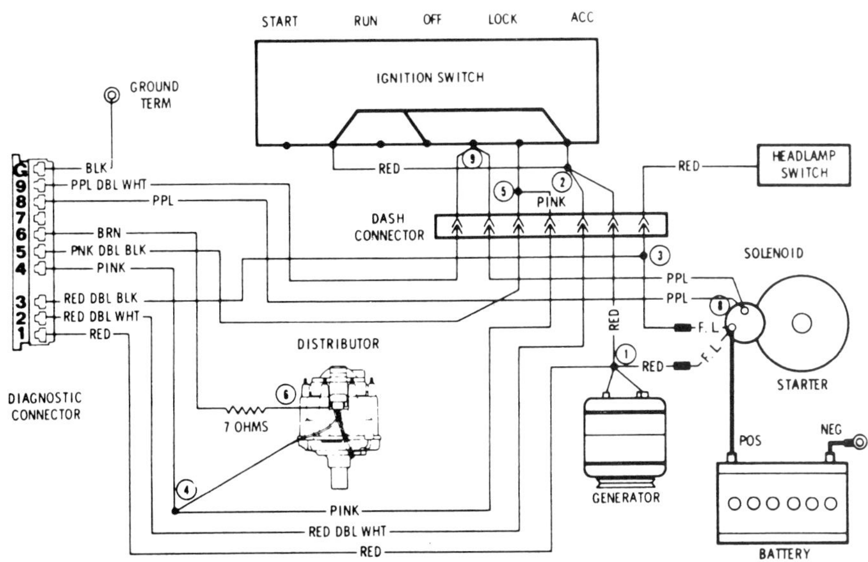

Figure 4-36. Diagnostic connector wiring diagram for full-size GM cars. (Oldsmobile)

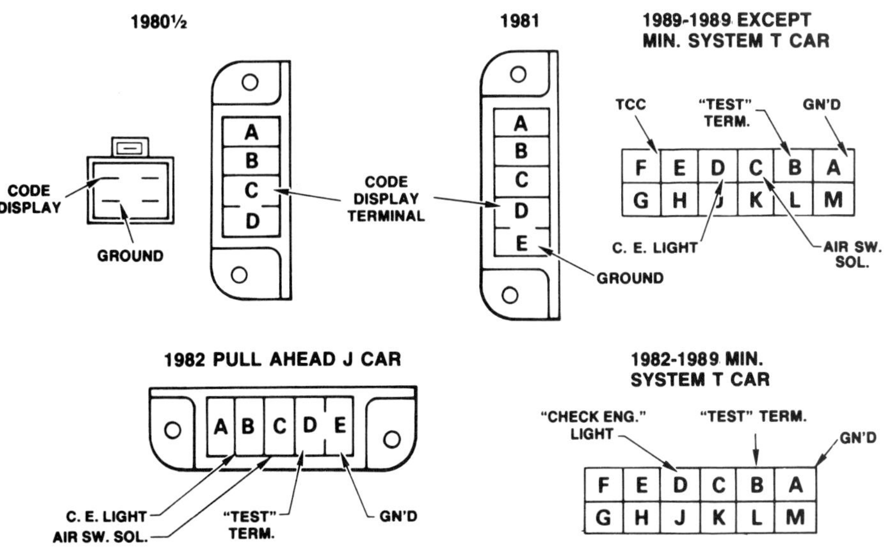

Figure 4-37. GM ALCL and ALDL connectors take varying forms. (GM)

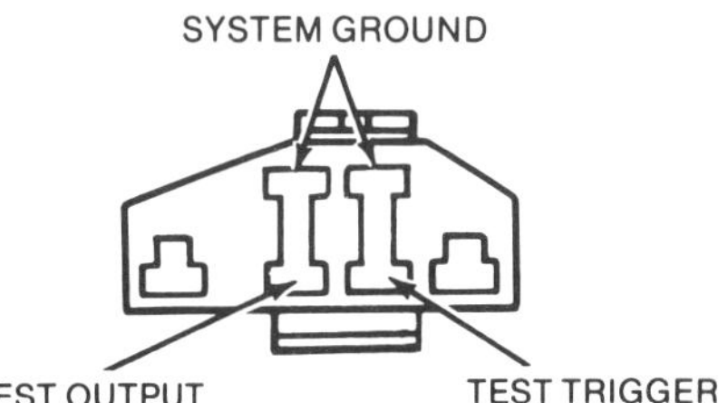

Figure 4-38. The 6-terminal diagnostic connector used with early Ford systems. (Ford)

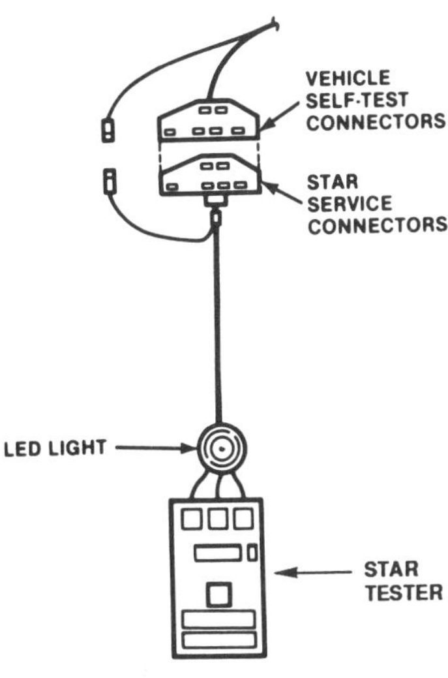

Figure 4-39. The Ford STAR tester, adapter cable, and 7-terminal diagnostic connector. (Ford)

1984. A self-test connector in the engine compartment permits connection of the special tester necessary to check the system. On most Chrysler models, the self-test connector is located in the wiring harness beside a front shock tower, figure 4-41. On models using the single-module engine controller (SMEC), the connector is generally located in the wiring harness near the relay bank, figure 4-42.

On Chrysler vehicles with a body computer module (BCM), a separate BCM diagnostic connector is located under the instrument panel to the right of the steering column. The connector is clipped in an upright position to a bracket and should be unclipped and removed before attempting to attach the tester connector.

Toyota Diagnostic Connector

The Toyota computer-controlled system (TCCS) diagnostic connector is a small box with a snap-top cover located near the cowl on the driver's side of most Toyota models so equipped. Unsnapping the cover exposes

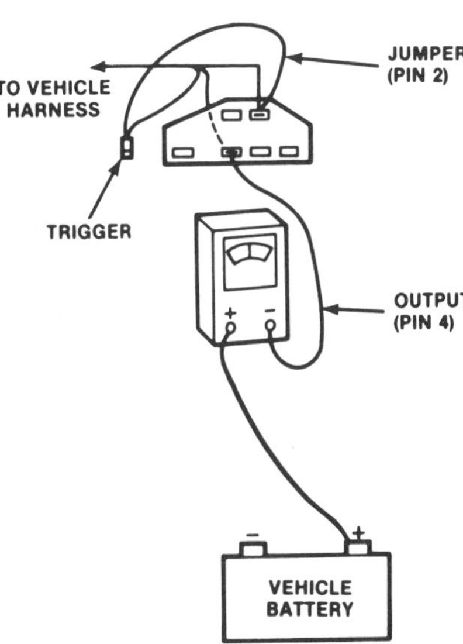

Figure 4-40. Trouble codes can be retrieved from the Ford diagnostic connector with a voltmeter and jumper wire. (Ford)

the test terminals, which are marked according to function inside the cover, figure 4-43. To retrieve stored trouble codes, the ignition switch is turned on and two terminals inside the connector connected with a jumper wire, figure 4-44. The terminals to be connected may differ according to vehicle application and model year, but if they differ from those shown in figure 4-44, the cover markings will indicate the correct pair. The terminals are filled with a silicone grease to inhibit corrosion and improve electrical conductivity. This grease should not be removed.

Special Test Connectors for ABS, Electronic Suspension, and Other Vehicle Systems

Individual electronic systems not integrated with the engine management system, such as antilock braking or electronic suspension, may have their own diagnostic pigtail connectors. Figure 4-45 shows the connector for the 1988 Lincoln Continental air suspension system. Such connectors allow the service technician to run a separate diagnostic check of the system operation.

Other non-integrated electronic systems have a self-diagnostic mode in their control module. By performing a specified procedure, the techni-

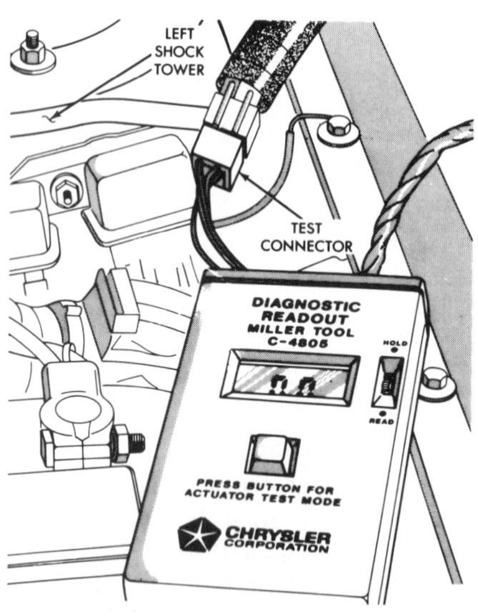

Figure 4-41. The Chrysler diagnostic readout or DRB I attaches to a diagnostic connector near the shock tower. (Chrysler)

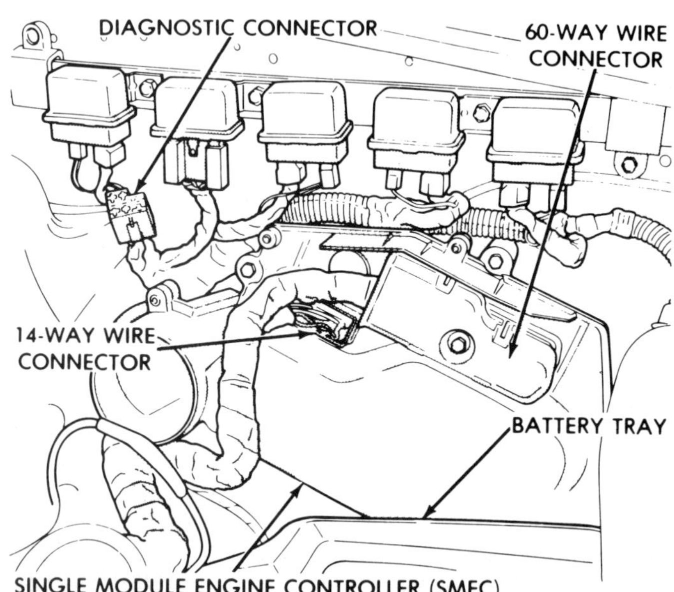

Figure 4-42. Typical diagnostic connector location on vehicles with a SMEC. (Chrysler)

Figure 4-43. The cover on the Toyota diagnostic connector box is coded to indicate test terminal function.

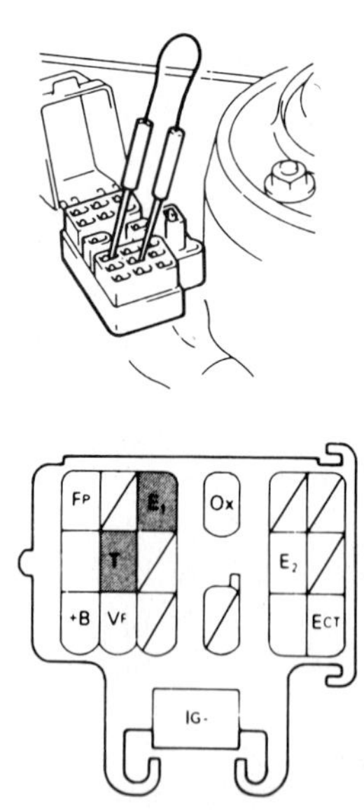

Figure 4-44. Trouble codes are retrieved from the Toyota diagnostic connector box by inserting a jumper wire between the appropriate terminals. (Toyota)

cian can put the system into its self-diagnostic mode. This allows him to retrieve any trouble codes stored in the module. As an example, putting the 1988 Chrysler ABS system module into self-diagnostics causes the module to check its memory and flash the brake warning indicator lamp on the instrument cluster to indicate the stored code. The method used to enter self-diagnostics differs according to the system and its manufacturer.

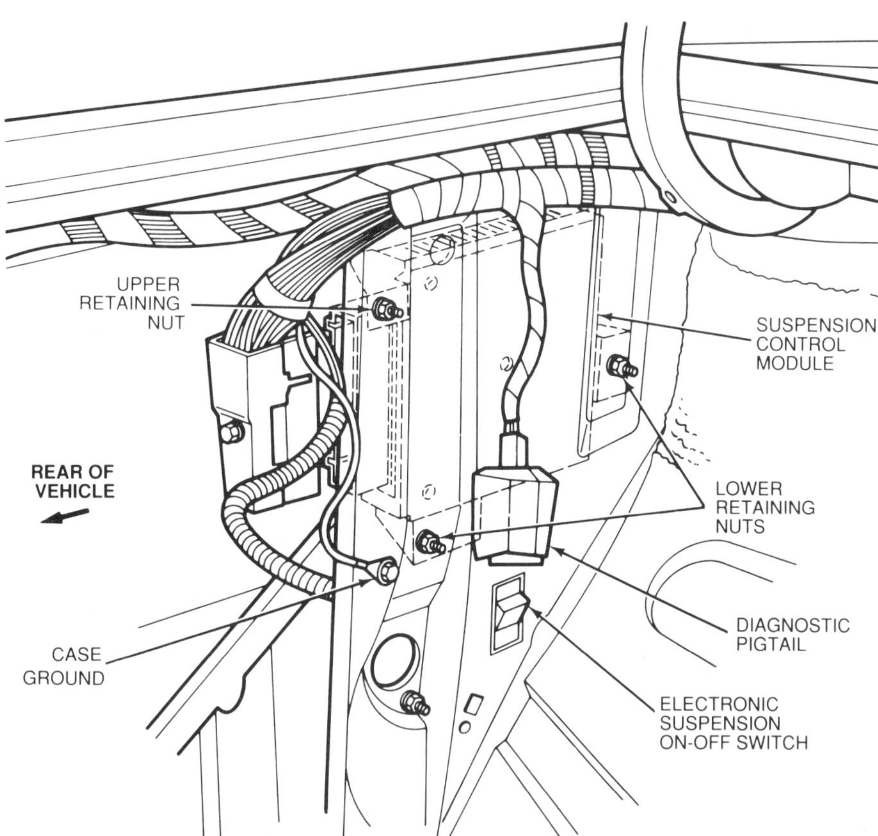

Figure 4-45. The diagnostic pigtail for the 1988 Lincoln Continental air suspension system is located in the trunk behind a trim panel. (Ford)

SCAN TOOLS AND BREAKOUT BOXES

The speed at which the automotive electronics revolution has taken place has made it necessary to develop a great variety of special test instruments to simplify troubleshooting procedures on electronic systems. Some of these test instruments are limited in their capability; once the system for which they were designed has been improved or modified, they must be replaced with new instruments. However, the industry has attempted to design test instruments that can evolve with the electronics they are designed to check, and many current testers fall into this category.

Such test equipment can be used to check individual circuits and will display any trouble codes relating to system operation. In many cases, however, they act only as pass-fail devices. This means that they will tell you if the circuit is functioning correctly or if it has a problem—they will not pinpoint the exact cause and location of the problem. They make troubleshooting quicker and easier for the technician, but they will not think for him. To understand what the tester is telling you, you must understand the operation of the system in question, as well as the test procedure.

Figure 4-46. The OTC Monitor 2000, GM's Tech 1, and the MPSI Pro-Link are typical of modern scan tools for diagnosing computer system problems.

Scan Tools

A special tester offered by most independent test equipment manufacturers is called a scan tool or scanner. Figure 4-46 shows three typical scan tools that can be used on late-model GM, Ford, and Chrysler vehicles. The scan tool is powered by the car's battery, either by direct connection, through the cigarette lighter or an adapter, or connection to the diagnostic connector. On GM and late-model Chrysler vehicles, the scan tool can monitor the continuous datastream transmitted over the serial data link between the engine computer and other computers or between the computer and its sensors and actuators. This datastream is actually information about the operation of all the parts of the system. By putting the scan tool into its proper operational mode, the technician can read the information he needs about a given part of the circuit through a digital display on the face of the tool. In a sense, the scan tool gives the technician the ability to eavesdrop or listen in on the communication between selected parts of the system to determine if there are problems and, if so, where they are located.

There is a wide variation in the capabilities of scan tools. This variation depends upon the design and cost of a particular unit. It also depends upon the year of the system it is designed to monitor. Ford products before 1992, for example, do not make the computer datastream available to a scan tool. The scan tool can only read trouble codes or trigger self-test in the vehicle computer and record a pass or a fail condition. Early scan tools had built-in memory and were restricted to the systems for which they were originally designed. Later scan tools use interchangeable memory cartridges, figure 4-46, and can thus be adapted to work with more than one system.

Some scan tools designed for 1986 and later GM vehicles let the technician manually override the computer and actually control the operating conditions of the engine. This allows him to see if the required control factors are working correctly. It also lets him see how any change in the control factors will affect overall engine operation.

The operating software cartridge and instructions for such scan tools are designed to duplicate the carmakers' test procedures. With each change or upgrade in system functioning, the software and its accompanying instructions are also changed. This allows the scan tool to be used with the vehicle system for which it was originally designed, as well as newer systems that have been modified.

Scan tools are generally designed to perform the following tests:

- *Trouble or fault code display*—The scan tool retrieves any codes stored in the computer memory for display on its built-in screen.
- *Open and closed loop*—The scan tool reads EGO and temperature sensor data sent to the computer. From this data, it can tell the technician whether the system is operating in open- or closed-loop mode.
- *EGO sensor signal voltage*—Scan tools can typically display the EGO sensor signal voltage either in millivolts or as an index number that the technician converts.
- *Air injection switching*—Some scan tools will let the technician trigger the air switching solenoid to momentarily create a lean exhaust condition. This function also serves as a doublecheck on EGO sensor operation.
- *Mixture-control solenoid dwell*—Some scan tools can display the duty cycle dwell readings of a mixture control solenoid.
- *EGR solenoid*—This function allows a check of EGR vacuum solenoid operation, if the vehicle is so equipped.

Some scan tools can perform numerous other functions, according to their particular design. Among the specific sensor and actuator circuits that can be checked for proper operation are:

- MAP and BMAP pressure sensors
- Throttle position switch or sensor
- Coolant and air temperature sensors
- Vehicle speed sensor
- Vapor canister purge solenoid
- Torque converter lockup solenoid
- Air conditioner switch
- Idle speed control (ISC) or idle air control (IAC) actuators

A useful accessory sold by some manufacturers is a data recorder. This device connects to the serial data terminal of the system and lets the technician record up to 60 minutes of actual driving conditions. Once this is done, the data recorder is connected to a suitable engine analyzer, which can then identify any operational problems that were recorded.

Breakout Boxes and Pin Testers

Carmakers make a wide variety of electronic testers available for use with their electronic control systems. Ford provides its dealers with numerous breakout boxes or pin testers for use with its EEC systems, figure 4-47. These testers are connected between the vehicle wiring harness and the electronic engine control assembly (ECA)—the vehicle computer—figure 4-48. When used with a digital volt-ohmmeter (DVOM), a breakout box monitors and displays all EEC signals. A diagnostic manual provided with the breakout box guides the technician through a series of tests, allowing him to spot problems and verify system operation.

The breakout box differs in one very important way from a scan tool. When a scan tool reads a sensor output, it may be displaying the sensor output as determined by the onboard computer, not the actual sensor output. Because the breakout box taps directly into the sensor circuit, it displays the same voltage signal that is being received by the onboard computer, without translation or interpretation.

Ford also provides single-system breakout boxes, such as its electronic message center (EMC) tester. These operate in essentially the same way as the Chrysler ABS tester described above.

The primary Ford tester, however, is called the self-test and automatic readout (STAR) tester, figure 4-49. This test unit is used with all late-model MCU and all EEC-IV systems to retrieve system trouble codes for display on its digital panel. The STAR tester requires a special adapter cable to connect into the self-test input, self-test output (STAR) connector and is

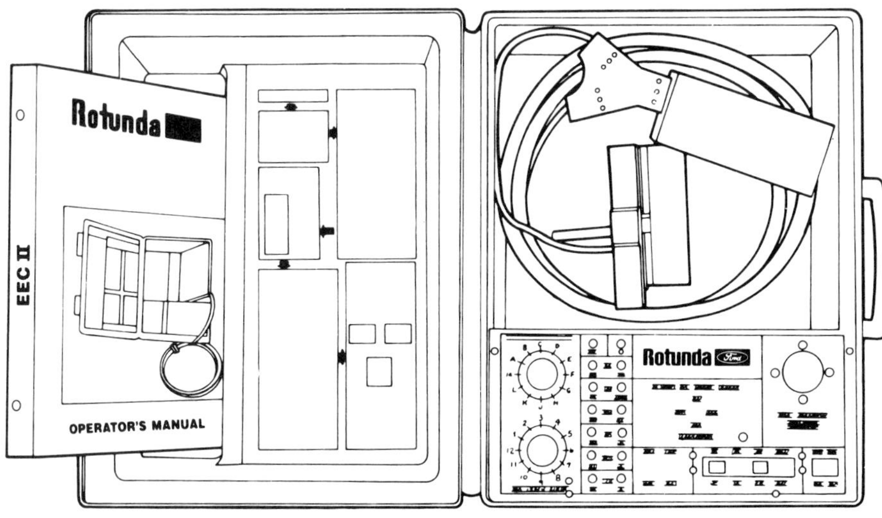

Figure 4-47. A typical Ford breakout box. (Ford)

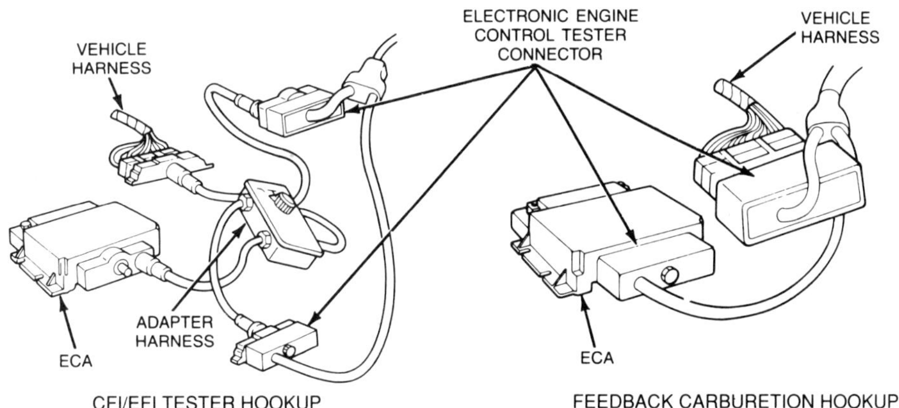

Figure 4-48. The breakout box is connected between the ECA and the vehicle wiring harness. Some systems require an adapter cable. (Ford)

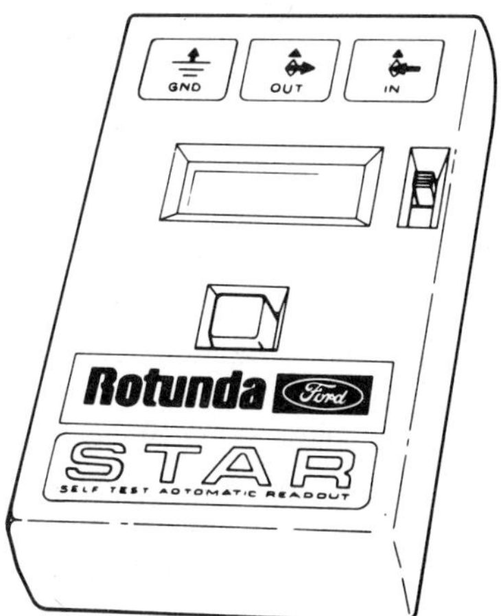

Figure 4-49. The Ford STAR tester. (Ford)

powered by an internal 9-volt battery.

Before 1988, Chrysler sold its dealers a diagnostic readout box (DRB-I), figure 4-41, for use with vehicles that had an onboard diagnostic system. It also provided driveability test procedure booklets for use with the tester. The combined use of the tester and booklet allows a technician to pinpoint a system problem quickly.

With the introduction of its single-module engine controller (SMEC) on some 1987 models, Chrysler marketed the DRB II, figure 4-50, for use with 1988 vehicles. The DRB II is really a scan tool and is specifically designed to put the SMEC into the various test modes required to diagnose the control system and its components. The DRB II is far more sophisticated in its capabilities than the DRB I, but both testers deal with essentially the same five test modes on vehicles without a body computer module (BCM):

- The diagnostic test mode, which retrieves and displays any fault codes stored in the diagnostic system memory.
- The circuit actuator test mode (ATM test), which checks a specific circuit by turning it on and off.
- The switch test mode, which checks to see if specific switch inputs are received by the logic module or SMEC.
- The sensor test mode, which checks to see if specific sensor output signals are received by the logic module or

Figure 4-50. Chrysler's DRB II uses interchangeable software cartridges.

SMEC. The DRB II checks the output signals with the engine off.

- Engine running test mode, which determines if the EGO sensor and idle speed motor are operating properly and establishes specific engine operational conditions. In this mode, the DRB II also can check sensor output signals with the engine running.

On Chrysler vehicles with a BCM, the DRB II deals with four test modes:

- The diagnostic test mode, which retrieves and displays any fault codes that are stored in the diagnostic system memory
- The switch test mode, which checks to see if specific switch inputs are received by the SMEC
- The sensor test mode, which checks to see if specific sensor output signals are received by the SMEC
- The circuit actuator test mode (ATM test), which checks a specific circuit by turning it on and off

Each module in the BCM system is able to perform switch and actuator tests.

Chrysler has also adopted the use of breakout boxes, or pin testers, for specific individual circuits not incorporated in its engine control system, such as the 1988 ABS system. The name "pin tester" derives from the fact that the breakout box can be set to check the signals in a given circuit at the module pin terminals used for that particular circuit.

General Motors relies on the use of scan tools, along with a series of individual component testers. Components such as HEI ignition modules, EGO sensors, and idle speed control (ISC) motors can be checked with the proper test units. These testers let the technician monitor the operation and even activate a particular component without having to remove it from the vehicle system. Some tool companies also sell breakout boxes for GM vehicles.

PRESSURE GAUGES

Testing fuel pump pressure, cooling system pressure, engine vacuum and compression, and hydraulic pressure are some of the jobs for which you will use a pressure gauge. Pressure gauges are based on a **bourdon tube** mechanism, which is a flexible metal tube that is slightly flattened and wound into a coil. One end is sealed; the other is attached to a fitting that is connected to the pressure source. When gas or fluid pressure enters the tube, the pressure tries to uncoil the tube. As the tube tries to straighten out, it turns a gear that drives the gauge pointer, figure 4-51. The gauge is calibrated so that the movement of the tube, gear, and pointer corresponds to the pressure applied to the gauge. Vacuum gauges work exactly like pressure gauges, except that atmospheric pressure outside the tube tries to coil the tube tighter.

Pressure gauges are graduated in pounds per square inch (psi) and in kilopascals (kPa), figure 4-52. Vacuum gauges are graduated in inches of mercury (Hg) or millimeters of mercury. Compound gauges measure both pressure and vacuum and are graduated in pressure and vacuum units. Compound gauges indicate pressure clockwise from the zero point and vacuum counterclockwise from zero.

Vacuum gauges usually are graduated from 0 to 30 inches Hg (−100 kPa). Pressure gauges range from low-pressure gauges graduated from 0 to 10 psi (70 kPa) to high-pressure gauges that measure several thousand psi or kPa. All bourdon tube gauges are most accurate in the midrange of their scales. That is, a gauge that reads from 0 to

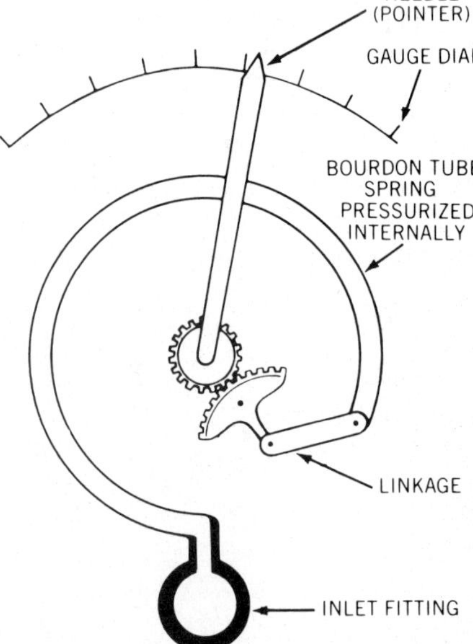

Figure 4-51. Pressure and vacuum gauges use a bourdon tube mechanism, like this.

Figure 4-52. This pressure gauge is graduated in pounds per square inch (psi) and in bars for gasoline fuel injection testing.

100 psi will be most accurate from about 30 to 70 psi. Whenever possible, use a gauge with a midrange that corresponds to the pressure that you expect to measure. Don't try to measure 30 to 60 psi of oil pressure on a 0- to 1,000-psi gauge. It may not even register.

Warning. Do not use a gauge with a maximum pressure indication equal to or less than the pressure you expect to measure. High pressure applied to a low-pressure gauge can cause it to explode.

HAND-OPERATED VACUUM PUMP AND VACUUM CONTROLS

A hand-operated vacuum pump, figure 4-53, is a useful tool to test vacuum devices. Often, you can test an item without removing it from the car. Connect the pump hose to the device being tested and operate the pump handle to apply the desired vacuum, as indicated on the gauge. For some tests, you may need a second vacuum gauge, attached to another port on the device being tested.

Vacuum Devices

Vacuum devices fall into one of these three categories:

1. *Vacuum actuators*—Most actuators are diaphragms that operate mechanical linkage, figure 4-54. This can be an EGR valve, a distributor advance unit, an air conditioning control mechanism, or some other device. Test them by applying vacuum, closing the pump shutoff valve, and watching the gauge for a vacuum drop that indicates leakage.

Some vacuum diaphragm devices, such as some carburetor vacuum breaks, have air bleeds or calibrated leakage holes, figure 4-55. These devices are designed to release vacuum

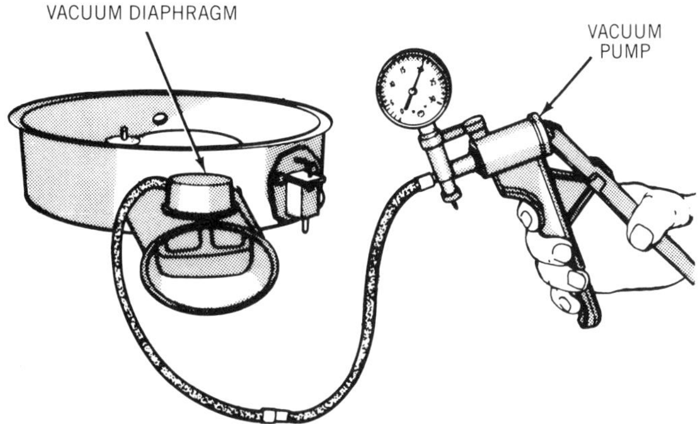

Figure 4-53. A hand-operated vacuum pump is useful for testing vacuum actuators and valves. (Chrysler)

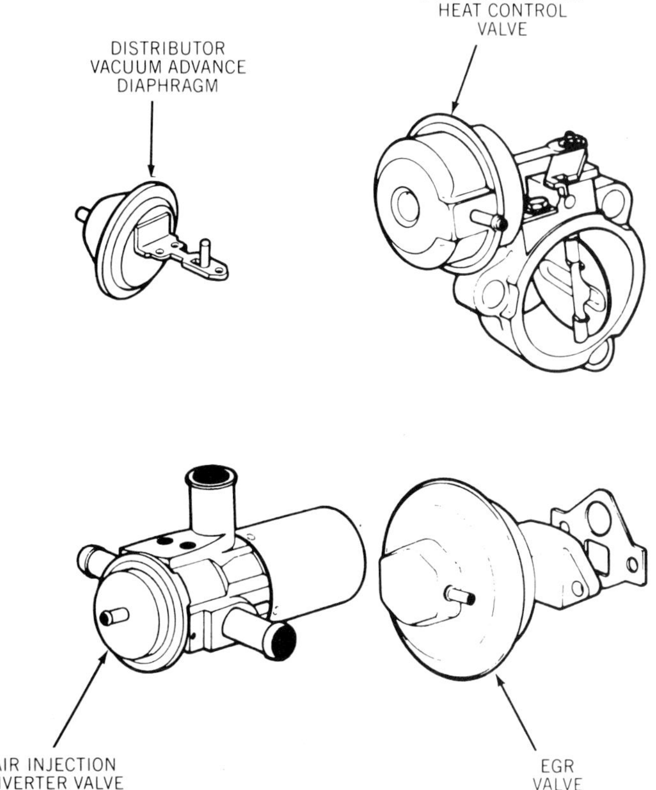

Figure 4-54. Most vacuum actuators are diaphragm-operated devices that move mechanical linkage. (GMWD)

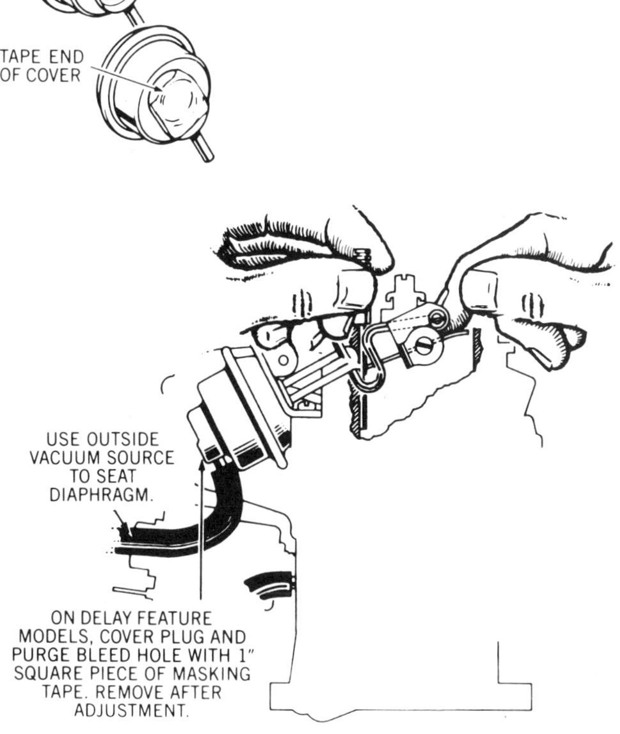

Figure 4-55. The air bleed hole on this vacuum diaphragm must be plugged before testing or adjustment. (Rochester)

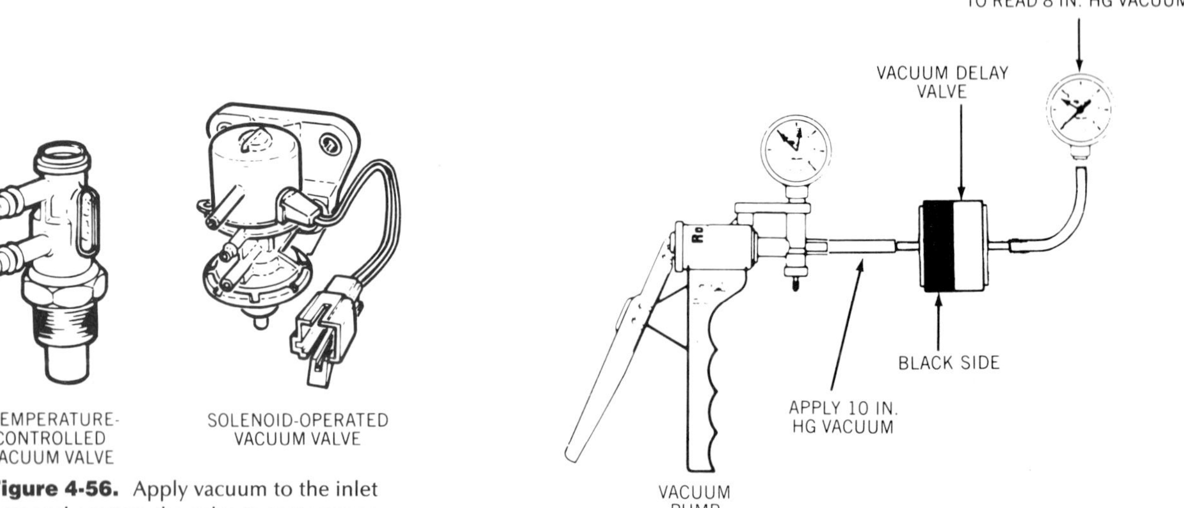

Figure 4-56. Apply vacuum to the inlet port and operate the valve to test vacuum switching operation. (GMWD)

Figure 4-57. Use a vacuum pump and a second gauge to test a vacuum-delay valve. (Ford)

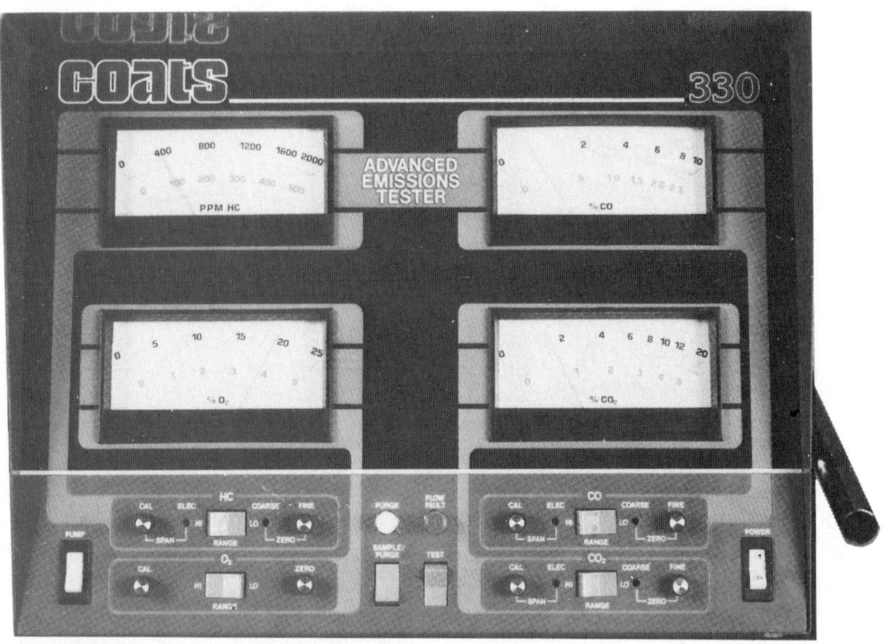

Figure 4-58. Exhaust analyzers, like this 4-gas unit, are essential tools for engine testing and service.

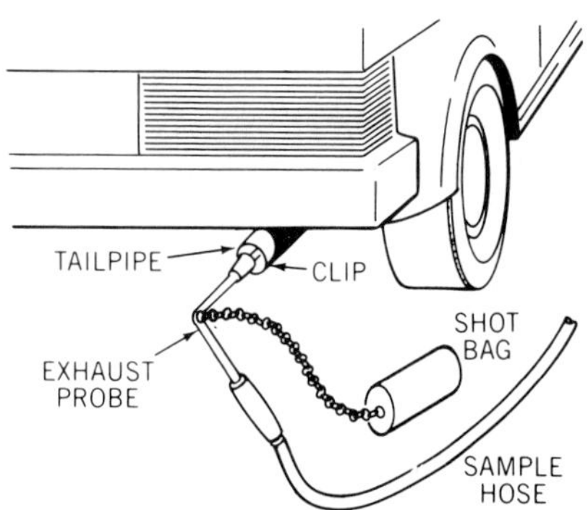

Figure 4-59. Insert the probe securely in the tailpipe.

at a controlled rate. If you test them without blocking the bleed opening, they will not hold vacuum. Check the carmaker's specifications for vacuum diaphragms to avoid inaccurate test results.

2. *Vacuum-switching valves*—These are solenoid-operated valves, or temperature-operated valves that open and close a vacuum line in response to voltage or a temperature change, figure 4-56. To test a vacuum-switching valve, apply vacuum to the inlet, or other specified port. Then operate the valve electrically or by a temperature change and measure the vacuum flow through the device.

3. *Vacuum-delay valves*—These valves contain a restriction that slows vacuum application through a line. They are used to delay distributor vacuum advance or carburetor choke opening. Test a delay valve by applying vacuum to the inlet with a second gauge attached to the outlet, figure 4-57. Compare the carmaker's specifications for vacuum delay to the time required for the gauge readings to equalize.

Test procedures in later chapters will help you apply these test principles to specific vacuum devices.

EXHAUST GAS ANALYZER

Infrared analyzers, figure 4-58, that measure carbon monoxide (CO) and hydrocarbon (HC) in exhaust have been common test equipment since the early 1970's. In recent years, 3- and 4-gas analyzers that also measure carbon dioxide (CO_2) and oxygen (O_2) have become popular.

An infrared analyzer measures exhaust HC, CO, and CO_2 by shining infrared light beams through an exhaust gas sample and through a sample of ambient air. Impurities in the

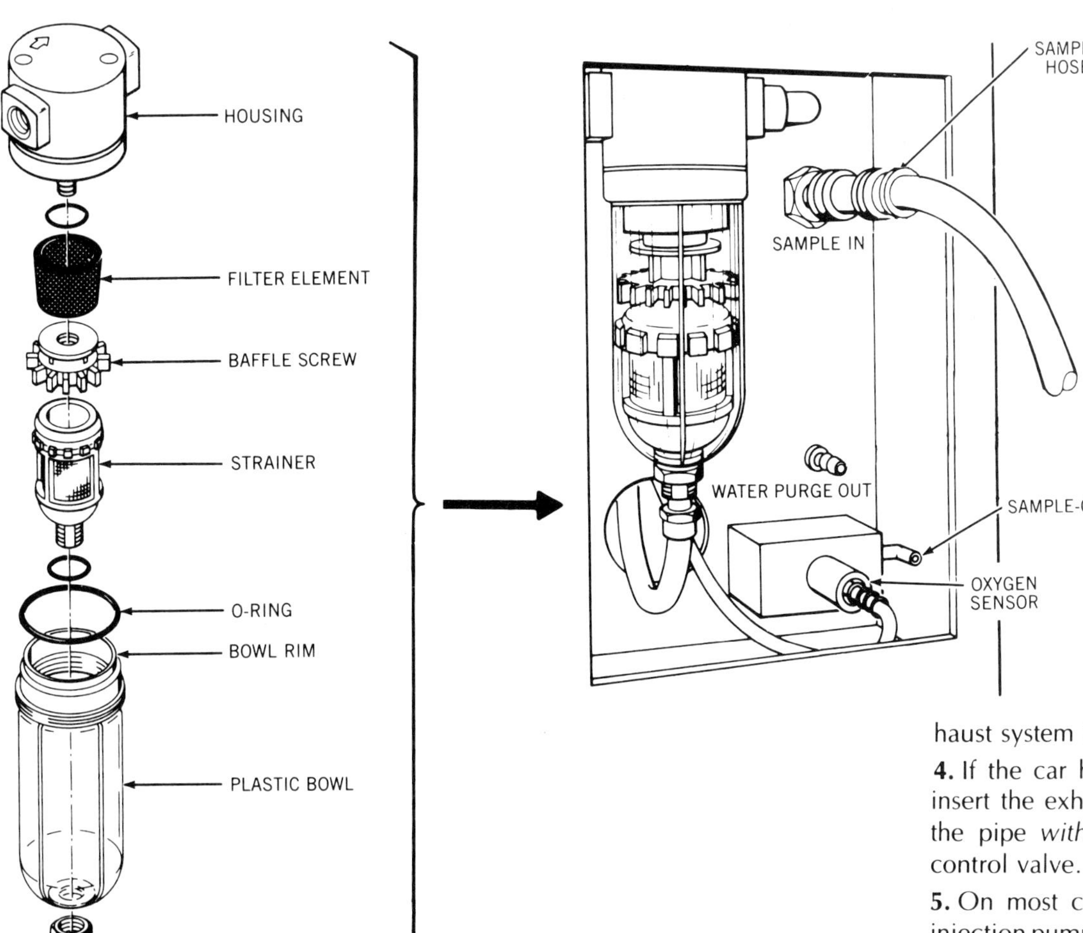

Figure 4-60. Replace the analyzer filter at intervals specified by the manufacturer or whenever the machine indicates an exhaust flow restriction or dirty filter. (Coats Diagnostic)

exhaust bend, or refract, the light beam. The analyzer measures the amount of refraction and converts it to electrical signals on the analyzer meters. A 4-gas analyzer measures exhaust oxygen content with an EGO sensor similar to the kind used in feedback fuel systems. The analyzer takes the exhaust sample through a probe inserted in the vehicle tailpipe.

The analyzer indicates CO in the exhaust by percentage and HC in parts per million (ppm). (One ppm = 0.0001 percent.) The analyzer may display measurements with a digital readout or on an analog meter. Meter-type analyzers usually have dials graduated with high and low ranges. Typical CO meters range from 0 to 2.5 or 3 percent on the low scale and 0 to 10 or 12 percent on the high scale. HC meters range from 0 to about 500 ppm (low range) and 0 to about 2,000 ppm (high range). Four-gas analyzers measure CO_2 from 0 to 20 percent and O_2 from 0 to 25 percent.

Analyzer Operation

Follow the equipment manufacturer's exact instructions and these guidelines when using an exhaust analyzer:

1. Do all testing in a well-ventilated area. Exhaust in the air can affect test accuracy by adding impurities to the reference air sample.

2. Locate the analyzer away from drafts. Airflow across analyzer circuit boards can cause inaccurate test results with some instruments.

3. Be sure the manifold heat control valve moves freely and that the exhaust system is free from leaks.

4. If the car has dual exhaust pipes, insert the exhaust sample probe into the pipe *without* the manifold heat control valve.

5. On most cars, disconnect the air injection pump during exhaust testing. Check the carmaker's instructions for converter-equipped cars.

6. Insert the exhaust probe securely into the exhaust pipe. Use an adapter if the tailpipe has a screen baffle, figure 4-59.

7. After turning on the analyzer, allow it to warm up for the specified time (about 15 minutes) before testing.

Exhaust analyzers are affected by altitude (barometric pressure) and ambient air quality. Many analyzers have ZERO and SPAN adjustment points on the dials or displays. After warming up the analyzer, follow the manufacturer's directions for ZERO and SPAN adjustments before any testing. Recheck the adjustments at least hourly during use. Some mechanics prefer to readjust the analyzer each time they test a vehicle.

Every analyzer draws the exhaust sample through a filter to remove dirt and water before the exhaust enters the analyzer chamber. Filters, figure 4-60, must be changed at the intervals specified by the manufacturer.

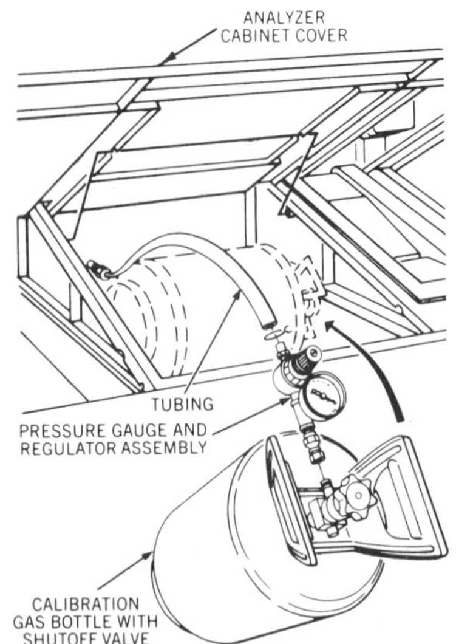

Figure 4-61. Many analyzers have an onboard bottle of calibration gas. (Coats Diagnostic)

Analyzers also require periodic calibration to maintain accuracy. Some states require that analyzers be calibrated regularly with a reference gas sample, particularly if the machine is used for mandatory emission testing. Many analyzers contain a pressurized, replaceable cylinder of calibration gas, figure 4-61. The cylinder contains known, precise quantities of HC and CO. The analyzer draws a sample from the bottle, and the operator compares that measurement with the known gas concentration. Calibration adjustments must be made if the test sample is out of limits.

CHASSIS DYNAMOMETER

A chassis dynamometer, or dyno, is a device that allows you to test vehicle performance under simulated driving conditions. A dynamometer has two large rollers installed in the shop floor. The vehicle to be tested is placed with its drive wheels between the rollers so that it can be "driven" while remaining stationary, figure 4-62. During testing, one of the rollers is adjusted to place a load on the drive wheels. Because the loaded roller is harder to turn, the engine and drive train must develop more power. The load can be varied to simulate driving load conditions such as acceleration, cruising, and hill climbing. A power meter on the dynamometer indicates power applied by the drive wheels to the dyno rollers. The dynamometer load can be created hydraulically, electrically, or mechanically.

The drive roller of a hydraulic dyno turns a large turbine or rotor in a tank of water, figure 4-63. The turning rotor forces water against the stationary blades of a stator, which resists water movement. This develops a load on the dyno drive roller. An electrically controlled valve varies the amount of water in the tank and thus changes the load on the dyno.

The drive roller of an electric dynamometer turns a large armature between the magnetic fields of electromagnet pole shoes. The magnetic fields of the armature and the electromagnetic poles oppose shaft rotation and create the dyno load. Varying the current through the poles varies the magnetic fields and changes the dyno load.

The drive roller of a mechanical dynamometer turns a drum or rotor between large friction brakes, similar to a car's brakes. Applying the friction brakes with varying amounts of pressure changes the dyno load.

Dynamometer Testing

A basic dynamometer will measure vehicle power output at the drive wheels under different load conditions. A dynamometer also allows you to test ignition, fuel, and emission control operation with engine and exhaust analyzers under simulated driving conditions. A dyno is the only piece of test equipment that allows accurate performance testing under loaded conditions at speeds other than idle. Government emission control certification procedures use dynamometers to simulate city and highway driving cycles.

Figure 4-62. A dynamometer produces simulated driving loads for engine performance testing. (Clayton)

TROUBLESHOOTING PRINCIPLES

The following sections are a 10-step method, or checklist, to help you troubleshoot engine performance problems.

1. Ask the Driver

Even if an owner brings you a car for routine maintenance (change oil, filters, spark plugs, and so on), he or she may have a general complaint about performance or a vague idea that, "It's not running the way it should." Ask the driver about specific problems. Also ask about overall performance. Even a general answer such as, "Well, sometimes it's hard to start," or "Mileage isn't too good," can be a helpful clue for your testing.

If the car has a specific problem, it's helpful to talk to the person who was driving when it occurred. Ask questions such as:

• Does the problem occur at specific times or temperatures—idle, cruising, acceleration, at night with heavy elec-

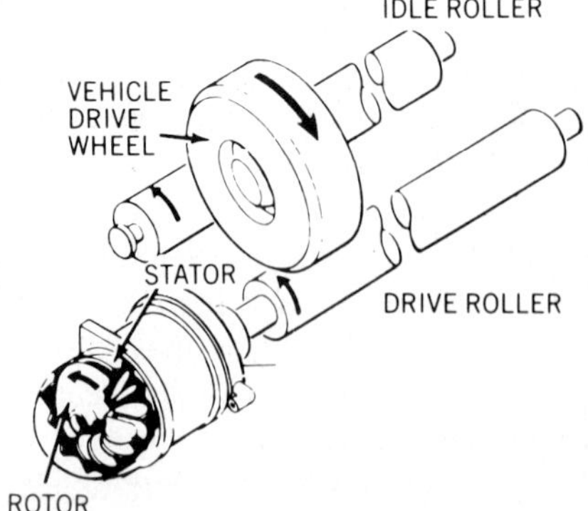

Figure 4-63. Most dynamometers are hydraulically loaded. (Clayton)

trical loads, cold starting, hot starting, or any combination?

- Does the problem occur regularly or at random?
- What are *all* the symptoms—noises, smells, vibrations, or combination of signs?
- Has the problem occurred before and what was done to fix it?
- When was the car last serviced, and what was done then?

Many large shops have service writers who prepare work orders and get information from customers. If you can't talk to the owner directly, read the work order carefully and ask the service writer for details.

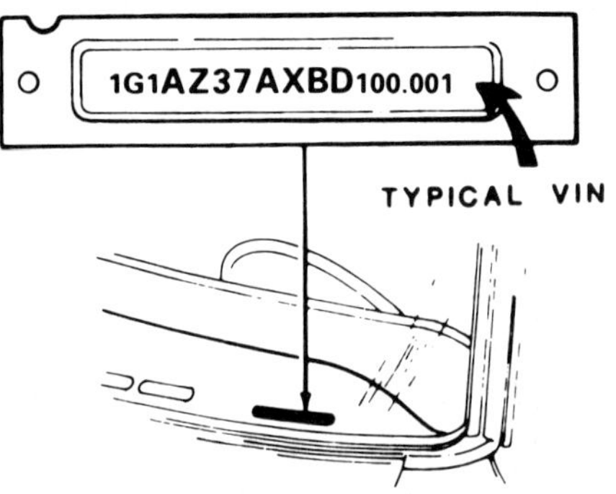

Figure 4-64. The vehicle identification number (VIN) plate is your starting point for correct vehicle identification. (Chevrolet)

2. Ask the Car

Equally, important, drive the car and check performance for yourself. The owner may complain about a rough idle and never notice the mushy automatic transmission shifting and slight high-speed surge caused by the same vacuum leak. Combined clues that lead to a single cause can save you time. Also, by driving the car to analyze the owner's comments, you will get an idea of what the owner expects in the way of driveability. A "hard-starting" problem may occur because the driver does not follow specific starting instructions for the vehicle.

You may operate the car immediately or later during your testing. You may have to let it cool down to check a cold-engine problem. When you do drive it, observe the symptom carefully to be sure when it does occur and when it *doesn't*. If the problem is a "rough idle," does it happen with the engine hot or cold, or both? Does it happen with the air conditioner on or off, or both. Does the car overheat? Does it smoke? Are there other related symptoms?

3. Know the Car, Know the Specs

Before you start troubleshooting, be sure you know what you are working on. That sounds simple, but a 1984 Chevrolet Camaro can have one of four different engines. It can have one of three different transmissions. It may or may not have air conditioning. It may be a California car, a 49-state car, a 50-state car, or a Canadian car.

Start to identify the car by checking the vehicle identification number on the VIN plate, figure 4-64. Then check the engine "emission control" decal for basic specs, such as timing, idle speed, and spark plug gap, figure 4-65. The engine decal also has summary instructions for idle and timing adjustments and, along with other calibration labels, will identify a car with

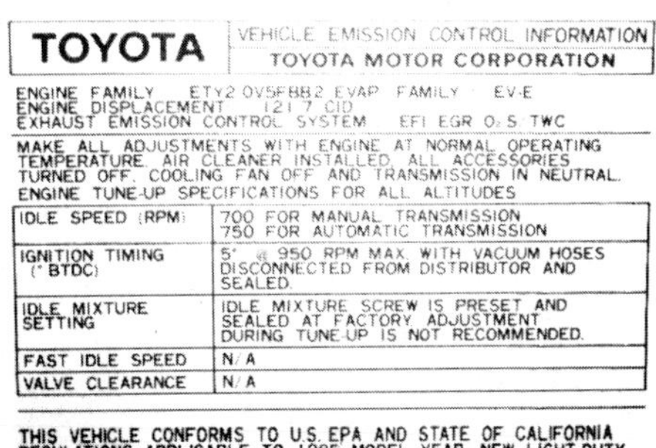

TOYOTA — VEHICLE EMISSION CONTROL INFORMATION — TOYOTA MOTOR CORPORATION

ENGINE FAMILY ETY2.0V5FBB2 EVAP FAMILY EV-E
ENGINE DISPLACEMENT 121.7 CID
EXHAUST EMISSION CONTROL SYSTEM EFI EGR O2S TWC

MAKE ALL ADJUSTMENTS WITH ENGINE AT NORMAL OPERATING TEMPERATURE. AIR CLEANER INSTALLED. ALL ACCESSORIES TURNED OFF. COOLING FAN OFF AND TRANSMISSION IN NEUTRAL.
ENGINE TUNE-UP SPECIFICATIONS FOR ALL ALTITUDES

IDLE SPEED (RPM)	700 FOR MANUAL TRANSMISSION 750 FOR AUTOMATIC TRANSMISSION
IGNITION TIMING (°BTDC)	5° @ 950 RPM MAX. WITH VACUUM HOSES DISCONNECTED FROM DISTRIBUTOR AND SEALED.
IDLE MIXTURE SETTING	IDLE MIXTURE SCREW IS PRESET AND SEALED AT FACTORY. ADJUSTMENT DURING TUNE-UP IS NOT RECOMMENDED.
FAST IDLE SPEED	N/A
VALVE CLEARANCE	N/A

THIS VEHICLE CONFORMS TO U.S. EPA AND STATE OF CALIFORNIA REGULATIONS APPLICABLE TO 1985 MODEL YEAR NEW LIGHT-DUTY VEHICLES.

2S-E CAL CATALYST

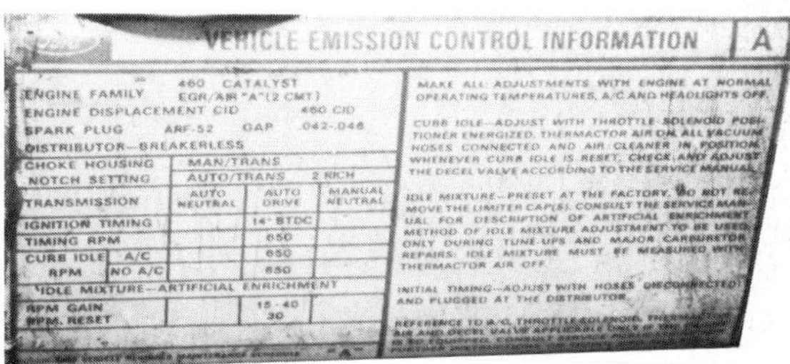

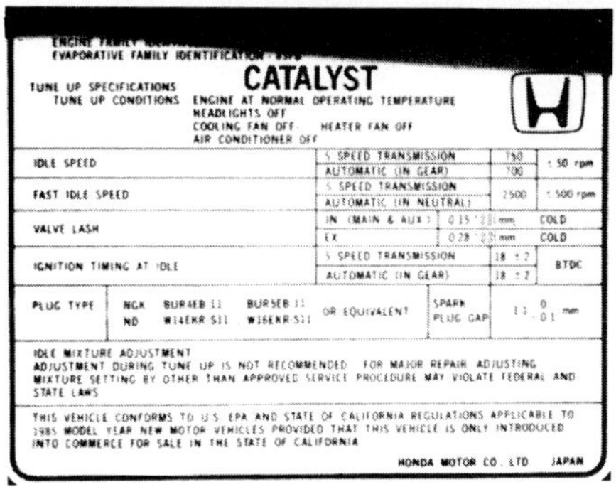

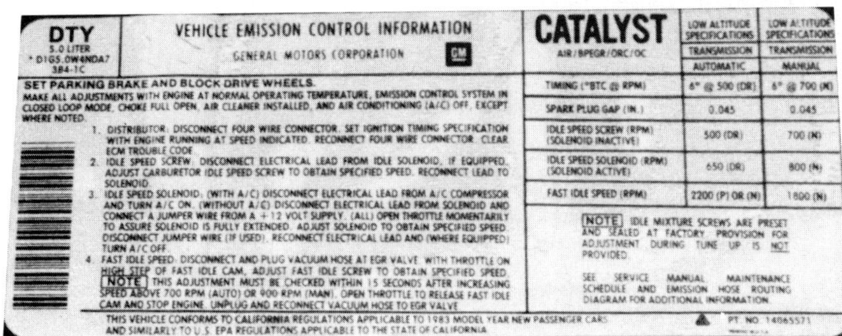

DTY 5.0 LITER *D1G5.0W4NDA7 384-1C — VEHICLE EMISSION CONTROL INFORMATION — GENERAL MOTORS CORPORATION — CATALYST AIR/BPEGR/ORC/OC

SET PARKING BRAKE AND BLOCK DRIVE WHEELS.
MAKE ALL ADJUSTMENTS WITH ENGINE AT NORMAL OPERATING TEMPERATURE, EMISSION CONTROL SYSTEM IN CLOSED LOOP MODE, CHOKE FULL OPEN, AIR CLEANER INSTALLED, AND AIR CONDITIONING (A/C) OFF, EXCEPT WHERE NOTED.

1. DISTRIBUTOR: DISCONNECT FOUR WIRE CONNECTOR. SET IGNITION TIMING SPECIFICATION WITH ENGINE RUNNING AT SPEED INDICATED. RECONNECT FOUR WIRE CONNECTOR. CLEAR ECM TROUBLE CODE.
2. IDLE SPEED SCREW: DISCONNECT ELECTRICAL LEAD FROM IDLE SOLENOID, IF EQUIPPED. ADJUST CARBURETOR IDLE SPEED SCREW TO OBTAIN SPECIFIED SPEED. RECONNECT LEAD TO SOLENOID.
3. IDLE SPEED SOLENOID: (WITH A/C) DISCONNECT ELECTRICAL LEAD FROM A/C COMPRESSOR AND TURN A/C ON. (WITHOUT A/C) DISCONNECT ELECTRICAL LEAD FROM SOLENOID AND CONNECT A JUMPER WIRE FROM A +12 VOLT SUPPLY. (ALL) OPEN THROTTLE MOMENTARILY TO ASSURE SOLENOID IS FULLY EXTENDED. ADJUST SOLENOID TO OBTAIN SPECIFIED SPEED. DISCONNECT JUMPER WIRE (IF USED). RECONNECT ELECTRICAL LEAD AND (WHERE EQUIPPED) TURN A/C OFF.
4. FAST IDLE SPEED: DISCONNECT AND PLUG VACUUM HOSE AT EGR VALVE. WITH THROTTLE ON HIGH STEP OF FAST IDLE CAM, ADJUST FAST IDLE SCREW TO OBTAIN SPECIFIED SPEED. NOTE: THIS ADJUSTMENT MUST BE CHECKED WITHIN 15 SECONDS AFTER INCREASING SPEED ABOVE 700 RPM (AUTO) OR 900 RPM (MAN). OPEN THROTTLE TO RELEASE FAST IDLE CAM AND STOP ENGINE. UNPLUG AND RECONNECT VACUUM HOSE TO EGR VALVE.

	LOW ALTITUDE SPECIFICATIONS TRANSMISSION AUTOMATIC	LOW ALTITUDE SPECIFICATIONS TRANSMISSION MANUAL
TIMING (°BTC @ RPM)	6° @ 500 (DR)	6° @ 700 (N)
SPARK PLUG GAP (IN.)	0.045	0.045
IDLE SPEED SCREW (RPM) (SOLENOID INACTIVE)	500 (DR)	700 (N)
IDLE SPEED SOLENOID (RPM) (SOLENOID ACTIVE)	650 (DR)	800 (N)
FAST IDLE SPEED (RPM)	2200 (P) OR (N)	1800 (N)

NOTE: IDLE MIXTURE SCREWS ARE PRESET AND SEALED AT FACTORY. PROVISION FOR ADJUSTMENT DURING TUNE UP IS NOT PROVIDED.

SEE SERVICE MANUAL MAINTENANCE SCHEDULE AND EMISSION HOSE ROUTING DIAGRAM FOR ADDITIONAL INFORMATION.

THIS VEHICLE CONFORMS TO CALIFORNIA REGULATIONS APPLICABLE TO 1983 MODEL YEAR NEW PASSENGER CARS AND SIMILARLY TO U.S. EPA REGULATIONS APPLICABLE TO THE STATE OF CALIFORNIA. PT. NO. 14065571

Figure 4-65. The engine "emission control" decal contains basic specs and adjustment instructions.

running changes made during the model year.

The VIN and engine decals will lead you to detailed specifications and procedures in reference manuals. You also may need electrical and vacuum diagrams. Look them up, using the VIN and engine decal numbers. Most cars have a decal under the hood that shows engine vacuum hose connections, figure 4-66.

4. Identify the Possible Causes and Cures

You actually start this step when you learn the owner's complaint, and you continue identifying and eliminating causes until you finish the job. Make a list, in your head or on paper, of the symptoms and possible causes. Then check the possibilities one at a time. Begin with the simplest. Don't condemn an alternator for a low-voltage problem without checking for a loose drive belt first.

5. Test from the General to the Specific

Testing is a process of elimination. If you immediately look for the cause of a rough idle in the ignition system, you eliminate other possible causes in the fuel and emission systems without checking them. To isolate a problem accurately, you must begin by checking general engine condition and performance. Then, test each system—fuel, electrical, emission—for overall operation. Finally, check individual components.

Sun Electric Corporation, a leading maker of test equipment, defines general tests of overall system operation as **area tests**. These tests allow you to narrow down the cause of a problem. Detailed, or **pinpoint tests**, then allow you to isolate a bad component. Carmakers follow the same principle with their test procedures. Figure 4-67 is a Ford diagnostic index chart that lists possible system area tests and component tests for one symptom, along with references to detailed test procedures. Late-model GM manuals for computer command control (CCC) systems start diagnostic routines with a circuit check, followed by evaluating driver complaints and testing system performance, figure 4-68.

6. Know the System, Isolate the Problem

Suppose the driver's complaint is a rough acceleration when cold, and your testing shows that the exhaust gas recirculation (EGR) valve is opening when it should not. You then must ask yourself, "Why?" Is vacuum to the valve controlled by a coolant temperature vacuum valve or by a solenoid, figure 4-69? If a solenoid is used, how is it switched, by a coolant switch or an engine computer? Is the EGR valve broken and stuck open? To isolate the problem, you must know what parts are in the system and how they work. This is another way to identify possible causes. The EGR valve may open at the wrong time because of:

- A mechanical problem—broken valve
- A wrong electric signal to a solenoid
- A broken coolant temperature sensor
- A cooling system problem—overheating
- Incorrect vacuum hose connections

7. Test Logically and Systematically

Now that you have isolated the problem to one system, you can use vacuum diagrams to determine hose con-

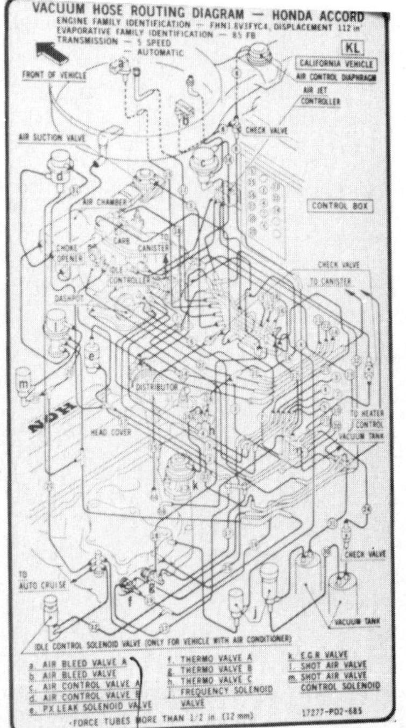

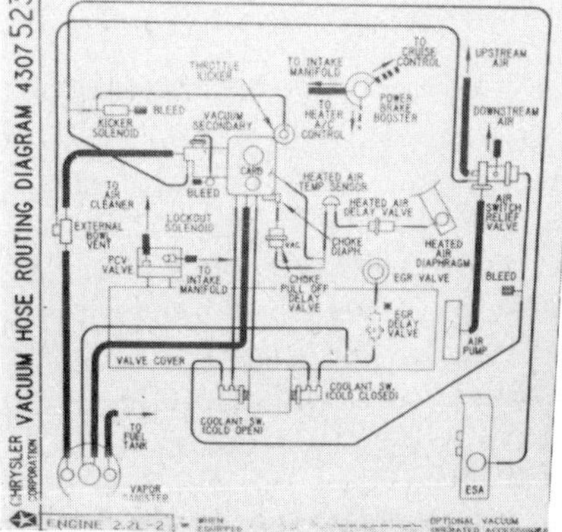

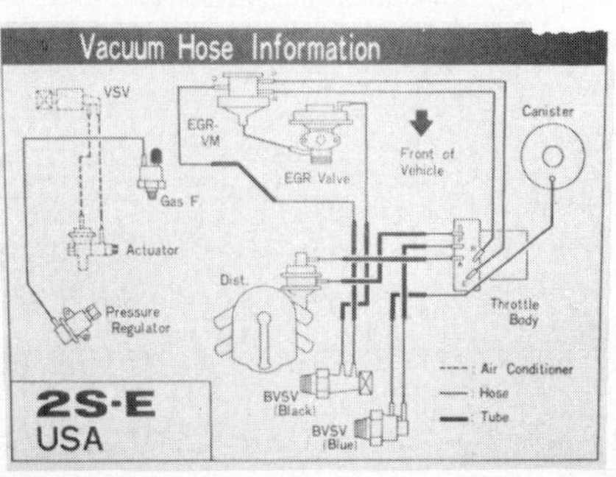

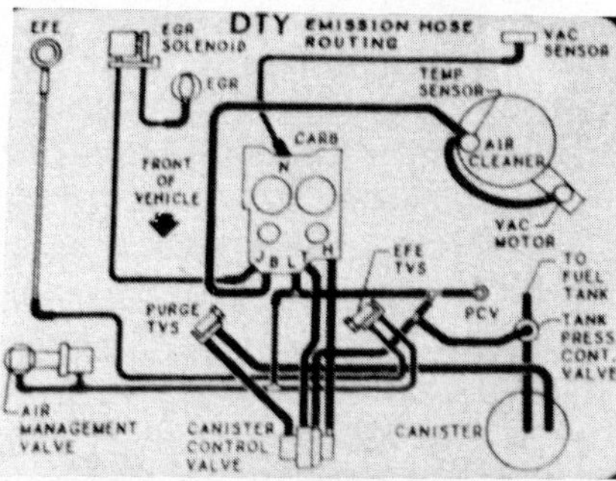

Figure 4-66. Most cars have vacuum diagrams, like this, in the engine compartment.

205	MISSES UNDER LOAD	
System	**Component**	**Reference**
Ignition	Scope Engine For: Spark Plug, Coil, Secondary Wires, Distributor Cap, Adapter and Rotor	Section 15
EEC/MCU	Component Diagnostics	Sections 16-24
Fuel Delivery	Filter Pump Lines Sender Filter	Visual Section 11 (for mechanical pumps) Sections 16-28 (for electrical pumps) and Group 24
External Carburetor/Fuel Charging Assy/Throttle Body	Electrical and Vacuum Connections Choke and Linkage Cold Enrichment Rod and Linkage (7200) Venturi Valves (7200)	Visual Section 4 and Group 24
Internal Carburetor	Basic: Idle, Main, and Accelerator Pump Float/Inlet Needle and Seat Main Metering Fuel Enrichment	Visual Section 4 and Group 24
Ignition Timing	Base plus Advance and Retard Functions	Section 15

Figure 4-67. This Ford index chart identifies systems, components, and pinpoint test procedures to diagnose a performance problem. (Ford)

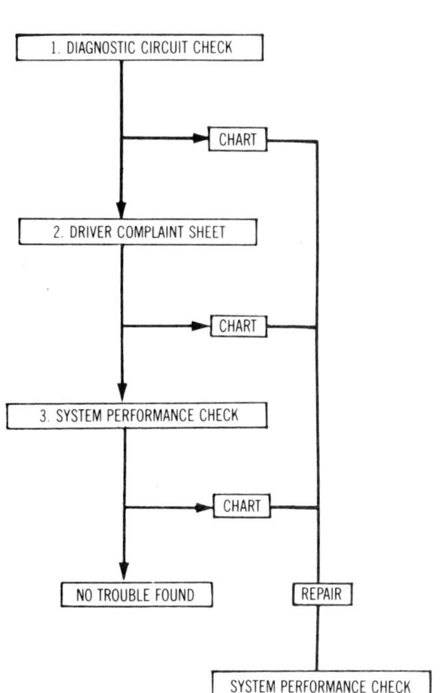

Figure 4-68. This GM flowchart summarizes general system test procedures. Notice that it ends with testing the repair. (Chevrolet)

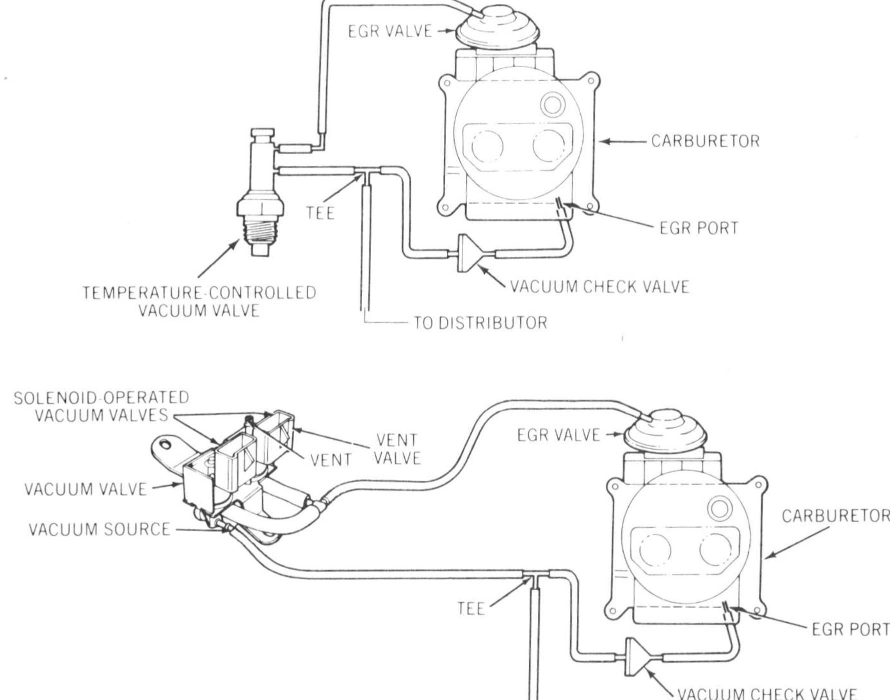

Figure 4-69. To isolate the cause of a problem, you must know what parts are in the system and how they work. (Ford)

nections and vacuum sources. Use electrical diagrams to determine the power source, the ground connection, and how a solenoid is energized. You can determine when vacuum should and should not be present at the valve and when a solenoid should be energized and deenergized. If engine roughness occurs only on cold acceleration, the problem probably is *not* a bad EGR valve. If the valve were stuck open, performance also would suffer at hot and cold idle and at full throttle.

If you find vacuum at the valve as you accelerate a cold engine, you must check the vacuum control devices. If a solenoid is energized to open a vacuum valve, check for voltage at the solenoid, figure 4-70. If voltage is not present, but the solenoid valve is open, you probably have a bad solenoid. If voltage is present when it should not be, you must check the electrical circuit. Check the electrical diagram and manufacturer's test procedures to trace voltage to the source.

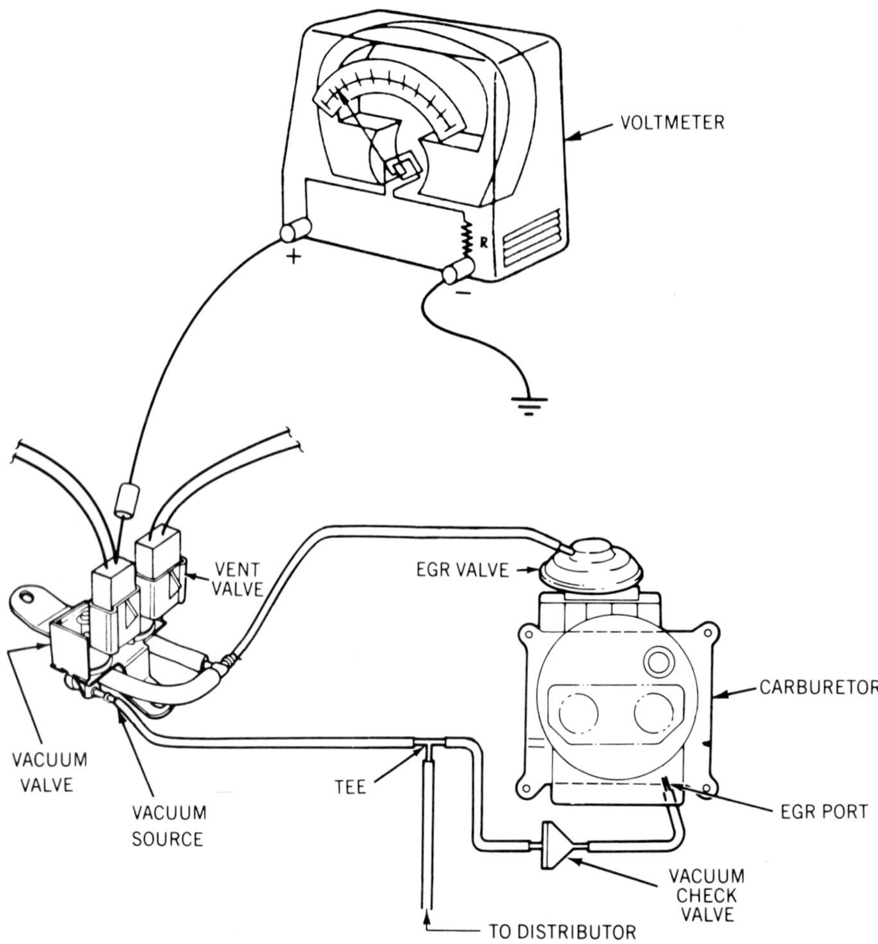

Figure 4-70. Test for voltage at the vacuum control solenoid. (Ford)

8. Doublecheck Your Test Results

Suppose your tests in step 7 seem to show that a vacuum solenoid is permanently grounded and energized. The specific cause could be:

- A ground inside the solenoid
- A grounded wire between the switch and the solenoid
- A defective switch
- An overheating engine that closes the switch too soon

Disconnect the switch wire, first at the solenoid and then at the switch, to see if the solenoid stays energized. Check for voltage and short circuits to ground at the solenoid, in the wire, and at the switch, figure 4-71. Check the switch itself to see if it is closing when it should not.

We have skimmed over several test methods here, but you will learn exactly how to do them as you study and practice various test procedures. The important principle, whether you are testing electrical, vacuum, hydraulic, or mechanical systems, is to test systematically from one point to another and doublecheck your results before making repairs. You often work with several systems—vacuum and electrical, for example—to isolate a problem. The symptom may come from a vacuum component, but the final cause may be an electrical fault.

9. Repair and Retest

Suppose your results in steps 7 and 8 show that the coolant switch is bad. Replace the switch and reconnect and check all vacuum and electrical connections in the system. Then, retest the *complete* system. Be sure the EGR valve opens only when it should and that the switch and solenoid work properly. Most importantly, be sure you have cured the original "roughness on cold acceleration" problem. You may have replaced one bad part and eliminated a problem in one system. But if you haven't corrected the owner's original complaint, you haven't finished the job. If the roughness still exists, there is another problem, or problems, in another system that you must find and fix, using these same steps.

SUMMARY

Engine performance testing requires equipment that ranges from simple jumper wires and test lamps to complex engine analyzers. Troubleshooting electrical problems requires checking for a high-resistance fault (usually an open circuit) or a low-resistance fault (short circuit or ground). Ammeters, voltmeters, and ohmmeters are universal test instruments, all based on the same kind of meter movement. These meters often are combined into multimeters. Others have special functions, such as dwell meters and tachometers.

Pressure and vacuum gauges are based on bourdon tube gauge movements and are used to test engine vacuum, oil pressure, coolant pressure, and other fluid and air systems. Oscilloscopes, timing lights, and exhaust analyzers are other test instruments used widely for engine service.

To test engine performance, you must listen to what the driver and the car tell you. You also must be able to read and understand diagrams for electrical and vacuum systems, as introduced in chapter 6. You should begin by testing the general area, or complete system; then go to specific tests to pinpoint a problem. Begin with the simplest tests and adjustments; then go to the more complex tasks. Beware of multiple causes for one general complaint. You must understand specifications and be able to interpret results in relation to specifications. Make adjustments, repairs, or replacements one at a time. Then, check the results of one change before making another. Finally, test the results of your complete repairs and adjustments. Be sure you have cured the problem and satisfied the owner's complaint.

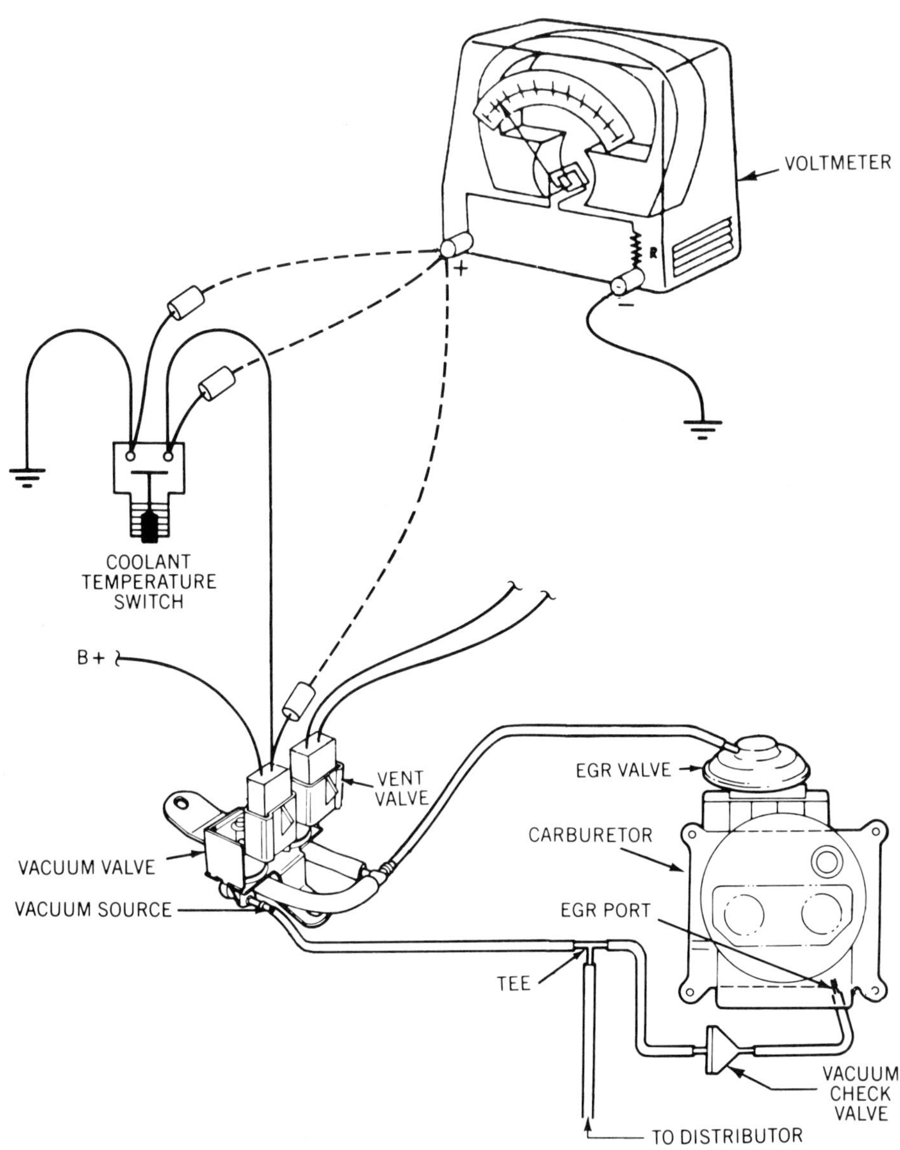

Figure 4-71. Check all points between the solenoid and the switch for an improper ground. (Ford)

REVIEW QUESTIONS

Multiple Choice

1. All of the following can cause a high-resistance electrical fault EXCEPT:
   - **a.** an open circuit
   - **b.** a grounded circuit
   - **c.** a loose connection
   - **d.** a burned out lamp bulb

2. All of the following can cause a low-resistance electrical fault EXCEPT:
   - **a.** an open circuit
   - **b.** a grounded circuit
   - **c.** a short circuit
   - **d.** two bare wires touching in a connector

3. The common ground side of all vehicle electrical circuits is the:
   - **a.** fuse block
   - **b.** bulkhead connector
   - **c.** engine and chassis main wiring harnesses
   - **d.** vehicle chassis

4. Electrical meters that display readings with a needle that moves across a graduated dial are called:
   - **a.** digital meters
   - **b.** multimeters
   - **c.** analog meters
   - **d.** impedance meters

5. An ammeter must always be connected in series with a circuit because it contains:
   - **a.** a high-resistance series resistor
   - **b.** its own power supply that must match the test circuit polarity
   - **c.** a low-resistance series resistor
   - **d.** a low-resistance parallel (shunt) resistor

6. Mechanic A has connected an inductive ammeter clamp around the battery ground cable to measure starter current draw. Mechanic B has connected a series ammeter across the battery terminals to measure starter current draw. Who is doing the job correctly?

a. A only
b. B only
c. both A and B
d. neither A nor B

7. A test ammeter connected to a circuit shows lower than normal current. This means that the circuit:

a. is open
b. has an accidental ground
c. is operating but has high resistance
d. has an accidental short

8. Mechanic A connects a voltmeter between the battery positive terminal and ground to measure battery voltage during cranking. Mechanic B connects a voltmeter in series with a lamp circuit to test continuity. Who is doing the job correctly?

a. A only
b. B only
c. both A and B
d. neither A nor B

9. Voltage at the battery (supply) side of a motor equals 10.5 volts. Voltage at the ground side of the motor equals 9 volts. Voltage drop across the motor is:

a. 9 volts
b. 10.5 volts
c. 1.5 volts
d. 12 volts

10. Mechanic A says that the sum of the voltage drops around a circuit must always equal source voltage. Mechanic B says that the sum of the voltage drops depends on current flow. Who is right?

a. A only
b. B only
c. both A and B
d. neither A nor B

11. Mechanic A has connected an ohmmeter across an ignition ballast resistor to measure resistance with the ignition on, breaker points closed and engine not running. Mechanic B has connected an ohmmeter to an electronic distributor pickup coil connector to test pickup coil continuity. Who is doing the job correctly?

a. A only
b. B only
c. both A and B
d. neither A nor B

12. Mechanic A says that you should always check and "zero" an ohmmeter before using it. Mechanic B says that you should begin ohmmeter testing with the range selector on the lowest setting and then switch to higher settings for accurate measurements. Who is right?

a. A only
b. B only
c. both A and B
d. neither A nor B

13. Many carmakers specify that voltage tests of electronic components should only be made with a digital voltmeter that an input impedance of at least:

a. 10 kilohms per volt
b. 10 megohms per volt
c. 10,000 ohms per volt
d. 100,000 ohms per volt

14. With the circuit closed, Mechanic A has connected a 12-volt probe light to a motor switch connector to check available voltage. Mechanic B has connected a jumper wire across the motor to verify continuity. Who is doing the job correctly?

a. A only
b. B only
c. both A and B
d. neither A nor B

15. An oscilloscope screen displays changing levels of:

a. current
b. resistance
c. voltage
d. all of the above

16. Most vacuum actuators in engine systems are:

a. diaphragms
b. solenoids
c. delay valves
d. potentiometers

17. An infrared exhaust analyzer can measure all of the following exhaust emissions EXCEPT:

a. CO
b. CO_2
c. HC
d. NO_x

18. An engine decal often identifies a car with:

a. disconnected emission controls
b. a void warranty
c. running changes made during the model year
d. none of the above

19. An overall test of an engine system is often called:

a. a certification test
b. an area test
c. a pinpoint test
d. an emission validation test

5 COMPREHENSIVE ENGINE TESTING

INTRODUCTION

Whether you call them comprehensive, area, or general tests, accurate diagnosis begins with checking overall engine performance and condition. This chapter explains the basic points of engine inspection and then outlines general tests for:

- Battery, starting system, and charging system performance
- Engine mechanical condition
- Fuel system operation
- Ignition system operation
- Emission control operation

Some tests focus on a single area. A compression test, for example, principally indicates an engine's mechanical condition. Other tests, such as a power balance test, will check several systems (fuel, ignition, emission) and mechanical condition at the same time. All tests, whether concentrating on one system or covering several, will lead you to specific pinpoint tests and adjustments that you will learn in later chapters.

This chapter has basic instructions for using the test equipment you learned about in chapter 4. Specialized testing, such as with an ignition oscilloscope, and specific service procedures will be covered in later chapters. The information in this chapter will help you recognize the common features of carmakers' test procedures. This chapter also will help you follow specific test equipment instructions quickly and accurately.

GOALS

After studying this chapter, you should be able to:

1. Inspect and operate the vehicle to determine the general condition of engine parts and systems, fluid leaks, exhaust smoke, oil pressure, and engine noises and performance.

2. Using the correct equipment, perform the following tests in twice the allowable flat-rate time, or less, and interpret the results.

a. Perform cranking current draw, battery load, and voltage-drop tests.

b. Perform cranking vacuum and speed tests.

c. Perform various engine vacuum tests.

d. Using both a simple tachometer and an engine analyzer, perform engine power balance tests.

e. Test engine compression and cylinder leakage.

f. Compare the results of power balance, compression, and leakage tests to determine engine mechanical condition.

g. Using an infrared analyzer, measure HC and CO emissions at different engine speeds on vehicles with and without catalytic converters.

3. Adjust the valves on engines with mechanical valve lifters or cam followers.

TESTING SAFETY

As with any automobile service work, safety is paramount during engine testing. Review the safety guidelines in chapter 1 and remember these key points:

1. You will be working with operating engines. Keep your hands, your hair, test equipment leads, and tools away from drive belts, fans, and other moving parts and from hot exhaust manifolds.

2. You will be working with the fuel system and flammable gasoline and diesel fuel. Keep sparks and open flame away from fuel system components.

3. Keep sparks and open flame away from batteries. Do not smoke.

4. Disconnect cables before charging the battery. Disconnect the ground cable before removing or replacing electrical components.

5. Turn the ignition switch off before disconnecting or reconnecting electric terminals.

6. Operate the engine only in a well-ventilated area or use an exhaust collection system to remove fumes from the shop.

7. Avoid any condition that allows fuel vapor to enter a catalytic converter. Follow precautions for converter-equipped cars.

VISUAL INSPECTION AND OBSERVATION

You can find and fix many basic problems with a thorough inspection and operating checkout during the early steps of your 9-step procedure (chapter 4) before you start testing with special equipment:

1. Check for loose or damaged drive belts, figure 5-1, and loose or cracked fan blades.

2. Check the battery, figure 5-2, for
- Electrolyte level
- Damaged case and caps
- Dirty or corroded terminals

Figure 5-1. Defective drive belts can affect engine cooling, electrical system operation, and emission controls such as air injection.

Figure 5-2. An engine won't crank properly and the ignition won't get full voltage if the battery—the electric power source—is in bad condition. (Atlas)

Figure 5-3. Check the electrical system for bad wiring and connections.

Figure 5-4. A loose or missing nut or capscrew on a carburetor or manifold can cause a vacuum leak.

Figure 5-5. Damaged fuel system components create a safety hazard, as well as performance problems.

- Loose, frayed, or worn cables and cable terminals

Also check the cable connections to the starter solenoid or relay and to the engine or chassis ground.

3. Using your vacuum diagram, if necessary, check all vacuum hoses for looseness, damage, and wrong connections.

4. Check electrical harnesses and connectors for broken or frayed wiring, loose terminals, and corrosion. Look for cracked or brittle insulation on wiring, figure 5-3.

5. Look for loose or missing bolts and nuts on carburetors, manifolds, and engine accessories, figure 5-4.

6. Check the carburetor, the injection system, the fuel pump, the filter, and other fuel system parts, figure 5-5, for

- Signs of leakage or loose fittings
- Worn or brittle hoses
- Kinked or otherwise damaged lines
- Disconnected or damaged linkage

7. Remove the air cleaner cover and check the filter for dirt or blockage, figure 5-6. Also check the PCV intake filter and the linkage and ducts on the thermostatic air cleaner.

8. Examine the coil cable and spark plug cables for damaged insulation, loose connections, and corrosion, figure 5-7. Also inspect the coil, the dis-

Figure 5-6. This dirty air filter will upset engine performance and emission control.

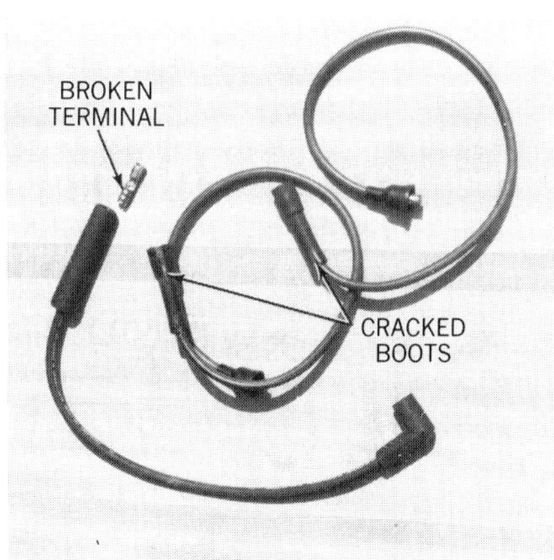

Figure 5-7. Inspect ignition secondary cables for wear and damage.

Figure 5-8. Check for coolant leaks as part of any engine inspection.

tributor cap, the cables, and the spark plugs for carbon tracks or dirt that could short circuit secondary ignition voltage.

9. Pull the engine dipstick and check the oil. Is it full and clean? Is it thick and dirty? Does it show signs of coolant contamination? Does it smell of gasoline, indicating a fuel pump leak?

10. Examine the engine for oil leakage around the pan, the valve covers, or any exposed oil lines.

11. Check the radiator for dirt or rust in the coolant and any other damage.

12. Check the radiator and heater hoses, the water pump, the thermostat housing, head gasket areas, and core plugs for leakage figure 5-8. Check all hoses for looseness, brittleness, and deterioration.

13. Look for burned air injection hoses, which indicate a leaking check valve.

14. Check the general condition of the manifold heat control valve or carburetor heat system.

15. Look for any other obvious damage to the engine and exhaust systems.

Noises, smells, and smoke will tell you a lot when you drive the car or simply run the engine in the shop. Listen and look for:

1. Squealing from loose or glazed drive belts.

2. Rattling of loose accessory brackets.

3. Rumbling from bad bearings in a water pump, an alternator, or other accessory.

4. Rumbling or knocking from engine bearings.

5. Hissing, caused by a vacuum leak.

6. Uneven or random hesitation, caused by misfiring cylinders.

7. Uneven cranking rhythm, caused by uneven compression or starting system problems.

8. Black exhaust smoke, indicating a rich mixture or stuck choke.

9. Blue smoke and an oil smell, indicating excessive oil burning.

10. Crankcase vapors from the oil filler cap, indicating excessive blowby or a clogged PCV system.

PRETEST PROCEDURES

Several engine tests—cranking current draw, cranking vacuum, and compression, for example—require that the ignition be disabled and the engine cranked on the starter motor. Because these tests require standard pretest procedures, we will give you the instructions here so you can refer to them for later tests.

Disabling the Ignition

Some engine analyzers automatically disable the ignition by grounding it through the analyzer for certain tests. If your equipment does not do this, you can disable the ignition in several ways:

1. Remove the coil lead from the distributor tower and ground it, figure 5-9.

2. On a breaker-point distributor, ground the points, either at the distributor connector or at the coil primary terminal, figure 5-10.

Caution. Do not allow an ignition *primary* feed wire to touch ground during cranking. High current will damage the circuit.

3. Disconnect the coil primary feed wire (usually +), figure 5-11. Do not ground it.

Figure 5-9. You can disable many ignitions by grounding the coil lead.

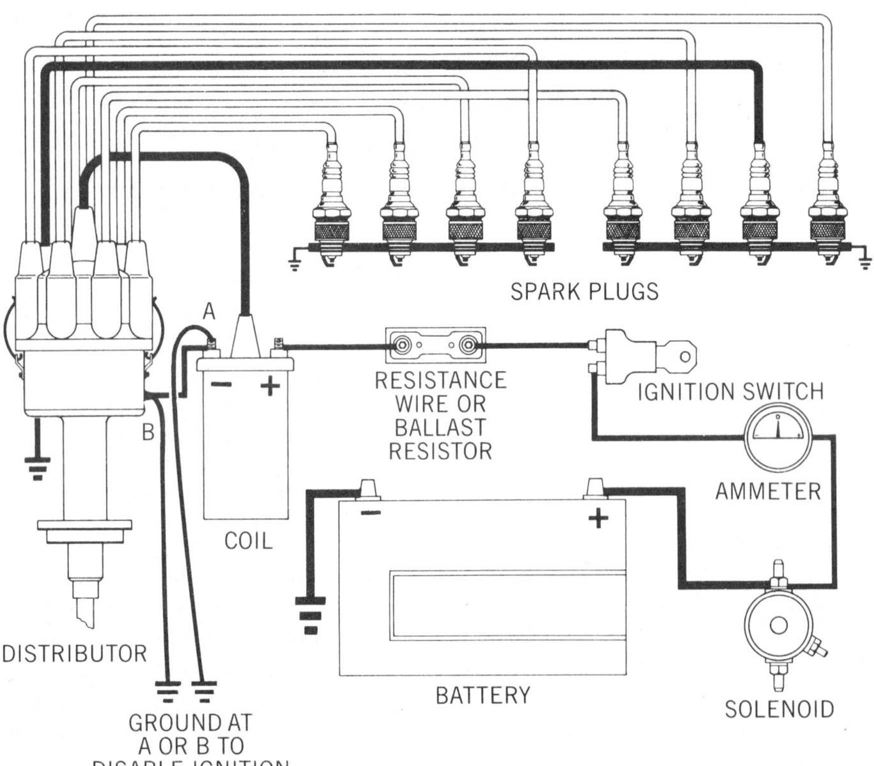

Figure 5-10. You can ground the ignition primary circuit at the coil–terminal or at the distributor to disable a breaker-point ignition. (Prestolite)

Figure 5-11. You also can disable many ignitions by disconnecting the coil primary lead (battery side). Do not ground it.

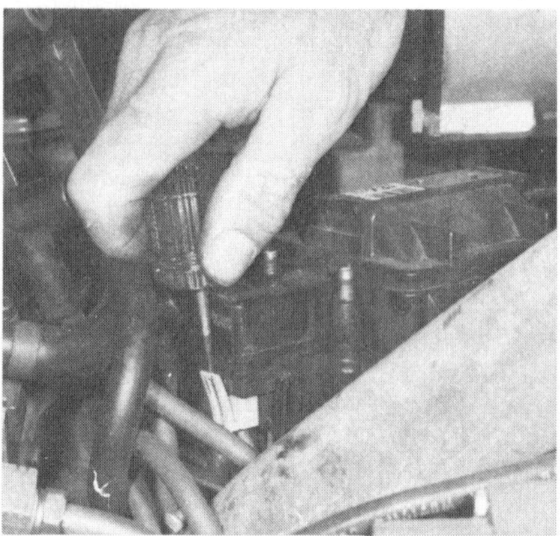

Figure 5-12. Disable Delco HEI systems by disconnecting the outer harness connector from the distributor cap.

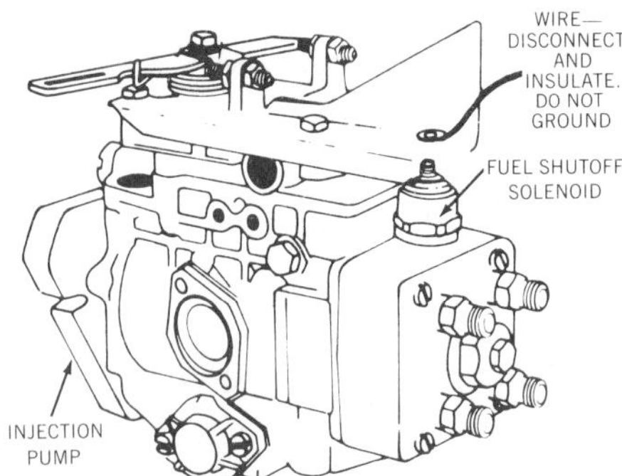

Figure 5-13. Many diesel injection pumps can be disabled by disconnecting the fuel shutoff solenoid. (Volkswagen)

4. On Delco HEI systems, unplug the outer harness connector from the distributor cap, figure 5-12. Do not ground it.

5. On a diesel engine, disconnect the injection pump fuel shutoff solenoid, figure 5-13, or close a shutoff valve.

Many manufacturers have special instructions for disabling ignition. Check

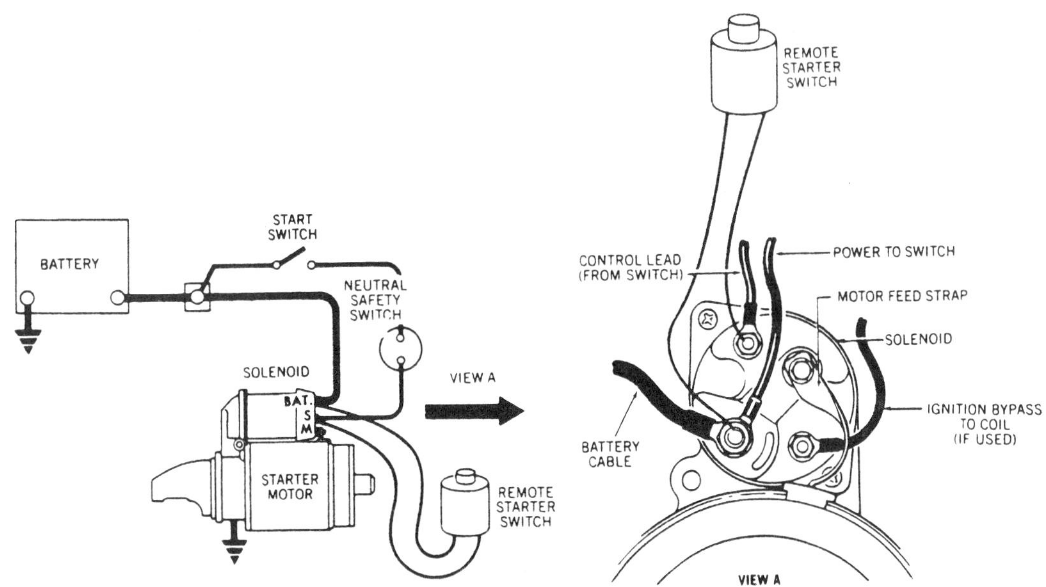

Figure 5-14. On GM cars and others with solenoid-operated starters, connect the remote switch to the starter solenoid. (Delco-Remy) (Ford)

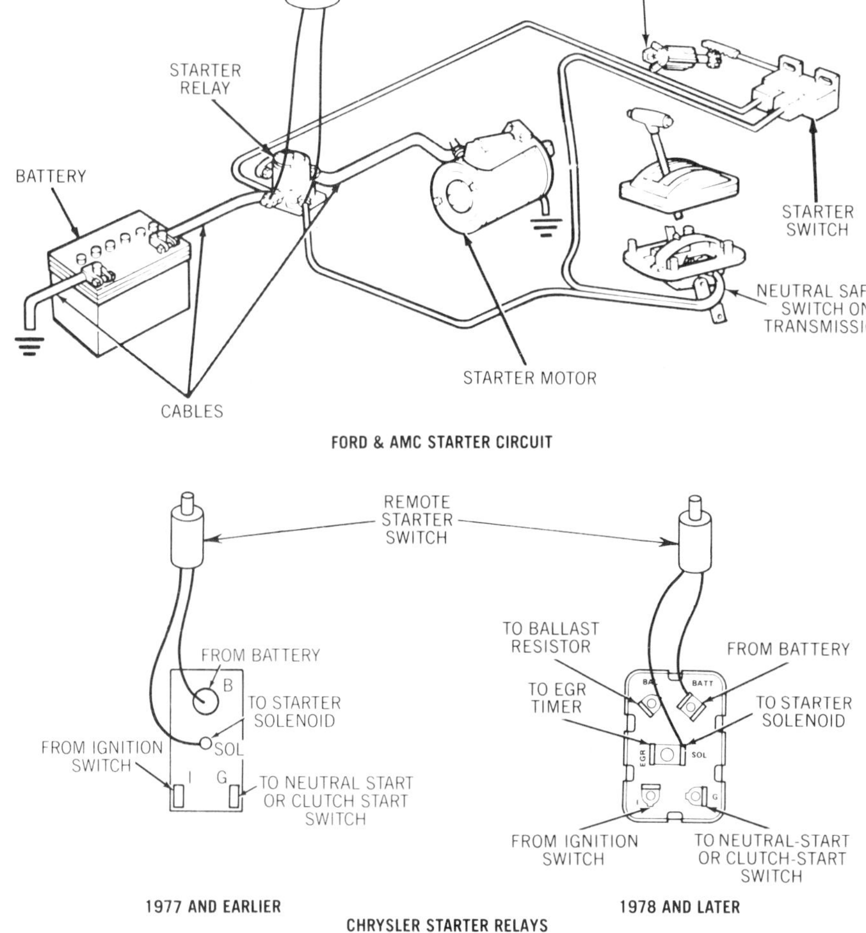

Figure 5-15. On Ford, Chrysler, AMC, and some other cars, connect the switch to the starter relay. (Chrysler)

the carmaker's instructions and ask your instructor for directions.

Remote Starter Switch Connections

A remote starter switch allows you to bypass the starter control circuit and crank the engine from outside the car. The remote switch also bypasses any starting safety switches. Connect a remote starter switch as follows:

Caution. Place the transmission in neutral or park, set the parking brake, and block the drive wheels before using a remote starter switch.

1. On most GM cars, connect the switch between the ignition terminal and the safety switch terminal on the starter solenoid, figure 5-14. If the car has a diagnostic connector, connect the switch between terminals 1 and 8.

2. On most Ford, Chrysler, and AMC cars, connect the remote switch between the battery terminal and the switch terminal of the starter relay, figure 5-15.

Figure 5-16. A battery-starter tester or a volt-amp tester can be used for general electrical tests.

Figure 5-17. This engine analyzer combines a voltmeter and an ammeter with other test instruments.

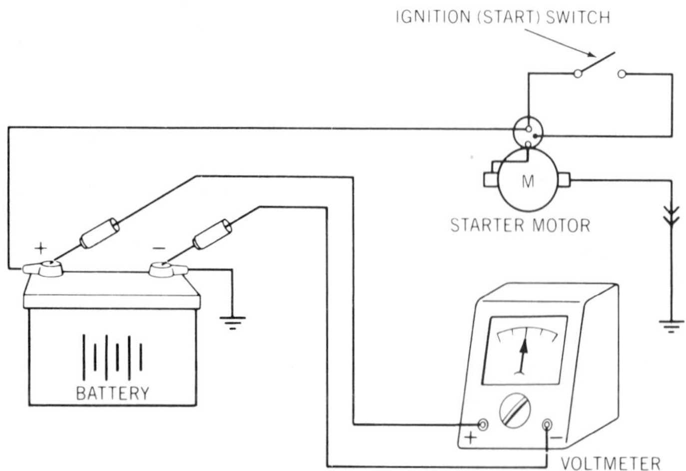

Figure 5-18. Connect the voltmeter across the battery, + to + and − to −.

BATTERY, STARTING SYSTEM, AND CHARGING SYSTEM PERFORMANCE

An engine must crank and start properly before it can run properly. You can check basic battery, starting, and charging performance as you begin testing because the engine is off and you must start it for further tests. The basic checks you can make are the:

- Battery load test
- Cranking current draw and voltage-drop test
- Alternator output test

For these tests, you will use a voltmeter to measure battery and alternator voltage, and an ammeter to measure starter current draw and alternator output current. The voltmeter and the ammeter may be separate instruments, but most shops use multimeters, such as battery-starter testers and volt-ampere testers, figure 5-16. Large engine analyzers combine voltmeters and ammeters in a console with other instruments such as a tachometer, an oscilloscope, and an exhaust analyzer, figure 5-17. Follow the manufacturer's instructions for test lead connections. In all cases, however, you will be taking signals for these tests from the voltmeter and ammeter leads, which you will connect according to these guidelines:

1. Adjust the meters to zero, if required, and turn the load control off. Set the voltage selector to the scale higher than vehicle system voltage (usually 18-volt scale).

2. Connect voltmeter leads in parallel, across the battery, figure 5-18, as follows:

  a. Positive, or red (+), lead to battery positive (+) terminal.

  b. Negative, or black (−), lead to battery negative (−) terminal.

Caution. Never connect an ammeter in parallel with a circuit. Never ground an ammeter lead when connected to a circuit. Low internal resistance of the meter will allow high current and destroy the meter.

Some carmakers warn against using series-connected ammeters with shunt resistors and knife switches on vehicles with solid-state ignition. Voltage surges may damage electronic parts.

3. If your ammeter must be connected in series with the battery, disconnect the negative cable and connect the ammeter leads as shown in figure 5-19 or as instructed.

4. If your ammeter has an inductive pickup, place the pickup clamp around either battery cable or around one of the voltmeter leads, according to the manufacturer's instructions. Inductive ammeter leads have an arrow to show current direction. Usually, point the arrow toward the battery on a positive cable and away from the battery on a negative cable, figure 5-20. Some testers do not require that the arrow point in a specific direction for a load test.

Battery Load Test

A battery load, or capacity, test places a specific current load on the battery to show how it performs under heavy demands, such as cranking. A good battery should deliver a specified high current while maintaining voltage of 9.6 volts or more for 15 seconds.

The resistance load of the battery-starter tester or engine analyzer simulates the load put on the battery by the starting system. You can do a load test with the battery in or out of the car. Battery temperature should be about 70° F (21° C). If the battery has removable cell caps, check the temperature with a thermometer. Many testers provide compensation to test batteries at higher or lower temperatures. A cold battery delivers less current than a warm battery.

Connect the test equipment as described in the previous section. Be sure test leads do not touch each other and proceed as follows:

1. Observe the voltmeter reading before placing a load on the battery. It should be about 12 to 12.6 volts. This is open-circuit, no-load voltage and does *not* indicate battery capacity. But, if voltage is less than 12 volts, check the battery and tester connections to be sure they are secure. If voltage remains less than about 12 volts, the battery may need charging before testing.

2. Determine the battery test load as follows:

a. Some batteries have the load-test current specification printed on the battery top. If a battery does not have this printed specification, use method b or c.

b. Multiply the ampere-hour rating by 3. If the ampere-hour rating is 70, apply a 210-ampere load (3 × 70 = 210).

c. Divide the cold-cranking ampere rating by 2. If the cold-cranking rating is 400 amperes, apply a 200-ampere load (400/2 = 200).

3. If specifications are unavailable, estimate the test load as follows:

a. Test load for batteries used with 4-cylinder or 6-cylinder gasoline engines—170 to 190 amperes.

b. Test load for batteries used with small V-8 gasoline engines (300 cid or 5 liters)—175 to 250 amperes.

c. Test load for batteries used with large V-8 gasoline engines (more than 300 cid or 5 liters) or any diesel engine—225 to 300 amperes.

4. If the battery has been charged recently (including charging by the vehicle alternator), apply a 300-ampere load for 15 seconds to remove the surface charge. Then wait 15 seconds more before proceeding with the actual load test.

5. To begin the load test, turn the tester control knob to draw the current determined in step 2 or 3.

6. Maintain the current load for 15 seconds and note the voltmeter reading.

7. Turn the tester control knob to off.

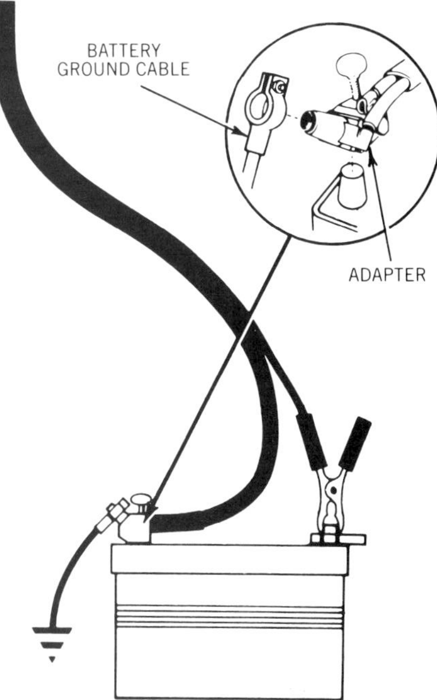

Figure 5-19. Ammeters connected in series with the battery require a battery post adapter. (Sun)

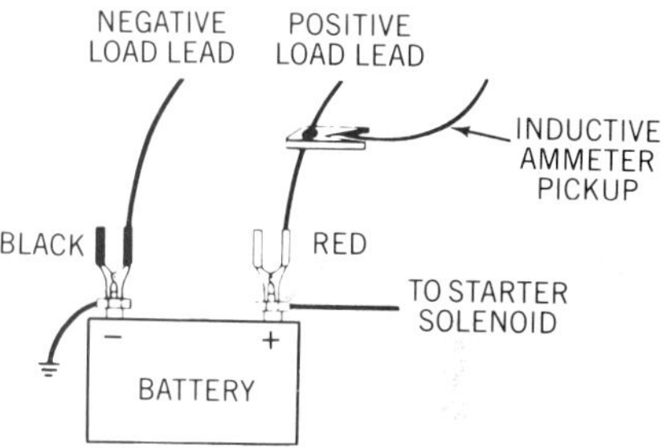

TEST LEAD CONNECTIONS FOR TESTING BATTERY PERFORMANCE

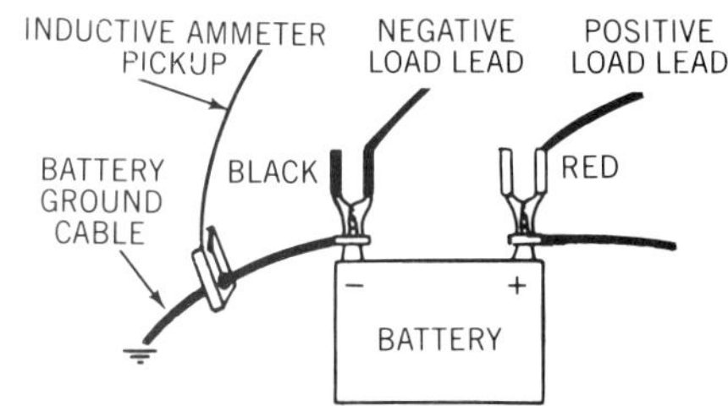

TEST LEAD CONNECTIONS FOR TESTING STARTER SYSTEM, CHARGING SYSTEM, AND VOLTAGE REGULATOR.

Figure 5-20. Place the inductive ammeter pickup around a battery cable or voltmeter lead, as directed. (Sun)

Electrolyte Temperature	*Minimum Voltage (12-Volt Battery)*
70° F (21° C)	9.6
60° F (16° C)	9.5
50° F (10° C)	9.4
40° F (5° C)	9.3
30° F (−1° C)	9.1
20° F (−7° C)	8.9
10° F (−12° C)	8.7
0° F (−18° C)	8.5

Figure 5-21. The minimum acceptable load-test voltage is lower at low temperature.

During the test, voltage should remain above 10 volts (9.6 volts, minimum) for a 12-volt battery at 80° F (27° C). If voltage was below 9.6 volts or if the specified current could not be reached, the battery needs further testing (chapter 9). If voltage was 10.0 volts or more with the full load for 15 seconds, the battery is good, and you can make further tests. If voltage was between 9.6 and 10.0 volts, charge and retest the battery before more engine testing.

If battery temperature is lower than 70° F (21° C), the minimum acceptable voltage level will be lower. For an open battery, check electrolyte temperature with a thermometer. For a maintenance-free battery, estimate temperature from the air temperature the battery was exposed to for one or two hours before testing. Figure 5-21 lists minimum load test voltages for batteries colder than 70° F (21° C). Refer to chapter 9 for further battery test procedures.

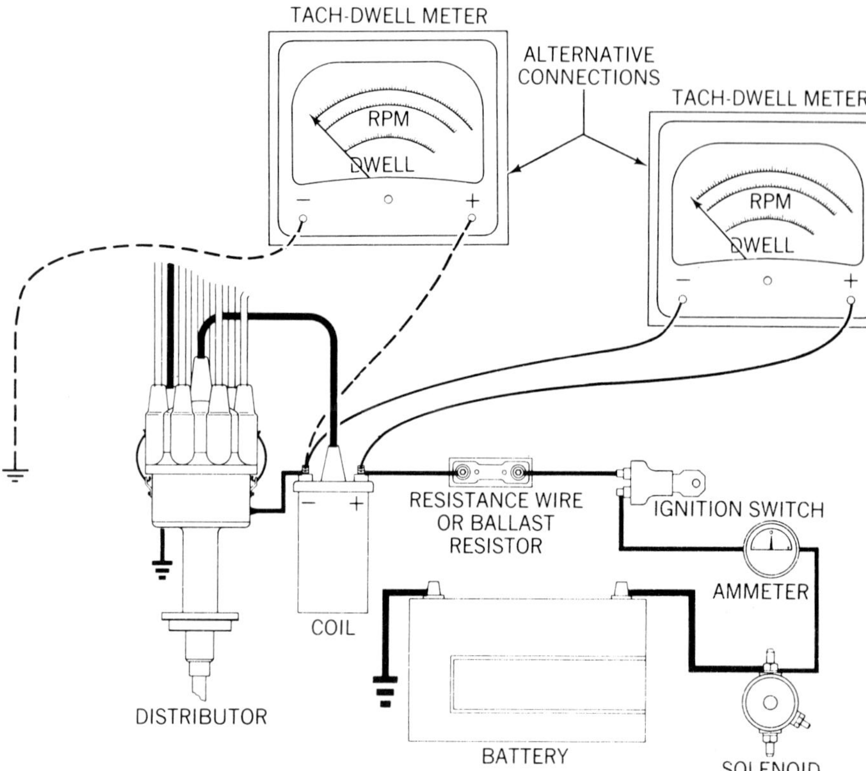

Figure 5-22. Tachometer, or tach-dwell, leads are usually connected across the ignition coil or between the coil and ground. Follow the equipment instructions. (Prestolite)

Cranking Current Draw, Voltage Drop, and Speed Tests

If the battery is good, you can now test the engine and starting system for cranking performance. Use a voltmeter and an ammeter for these tests, connected as described previously. If you want to check cranking speed, connect a tachometer to the engine. On a gasoline engine, connect the tach leads across the ignition coil or between the coil distributor terminal and ground, figure 5-22. Many electronic ignitions require special test adapters. Follow the equipment maker's instructions. To check cranking speed of a diesel engine, you will need a magnetic, or diesel, tachometer, figure 5-23.

This test measures battery capacity during actual cranking, and it tests:

• Starting system current draw
• Primary electrical system (ignition) voltage
• Engine cranking speed

Many engine analyzers include other measurements, such as engine temperature and voltage drop to the ignition coil. Perform the basic test as follows:

1. Connect test leads to the engine and disable the ignition.

2. Turn off all lights and accessories; close all doors.

3. Set the tester controls to the proper positions to remove any tester resistance from the circuit.

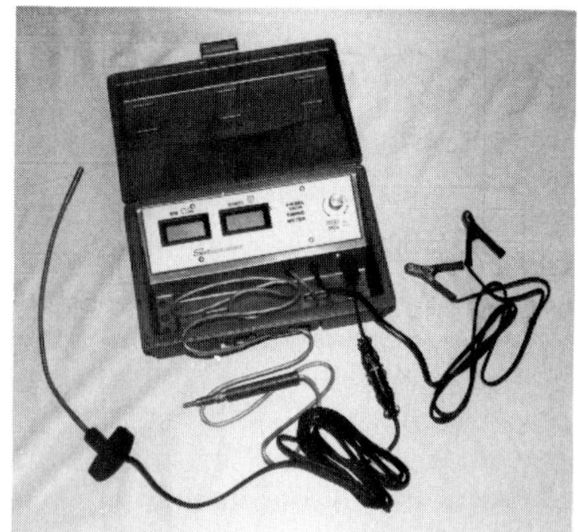

Figure 5-23. Diesel engines require a special tachometer to check engine speed.

Caution. Do not crank a converter-equipped car for more than 15 seconds to avoid drawing fuel into the catalyst.

4. Using the ignition switch or a remote starter switch, crank the engine for 10 to 15 seconds. Note the ammeter and voltmeter readings. Note the cranking speed if using a tachometer.

5. Compare the current, the cranking

voltage, and the rpm readings to the carmaker's specifications.

Cranking voltage should not drop below 9.6 to 10.0 volts.

Cranking current should be within the manufacturer's specifications. If current is *above* specifications, you can suspect:

- A short circuit in the starter
- The starter or engine is binding or tight

If current is *below* specifications, you can suspect:

- High resistance in the starting system
- The battery is undercharged (cranking voltage is lower than 9.6 volts)
- A worn engine (accompanied by good cranking speed and high voltage).

Cranking speeds for most engines are approximately 200 ± 20 rpm. Check the carmaker's specifications. If any specifications are out of limits, you should perform detailed battery and starting system tests (chapters 9 and 10) and engine mechanical tests.

Charging System Output Tests

After cranking the engine, the battery is slightly discharged. The alternator will deliver high current and voltage as soon as the engine is started. You can check general charging system operation, as follows:

1. Connect a voltmeter and ammeter to the engine as described previously.

2. Turn the ignition switch to the run position and note the discharge current on the ammeter. This is the ignition current and, on some cars, the alternator field current and blower motor current.

3. Turn off all accessories and close the doors. Start and run the engine at about 2,000 rpm.

4. Note the voltage and current readings on the voltmeter and ammeter.

5. Hold engine speed at about 2,000 rpm until current drops below 10 amperes.

6. Note the voltmeter reading again and return the engine to idle.

7. Add the current readings from steps 2 and 4 to determine total output current.

The regulated voltage for most alternators is 12.6 to 15.5 volts. This test should produce voltage in the upper half of this range. When current drops below 10 amperes, voltage should be at the regulated maximum specification. If voltage is below 12.6 volts and current is low or negative, the charging system has high resistance or an open circuit.

Current should be above 10 amperes when the engine starts. It may be within 10 or 20 percent of the rated alternator output. Normal, or low, voltage and high current indicate a discharged or defective battery.

- Low current may indicate a defective regulator or alternator.

If charging voltage and current are out of the broad limits for this test, perform detail tests on the charging system (chapter 11).

CRANKING VACUUM AND SPEED TESTS

Cranking vacuum and speed tests are basic mechanical tests of gasoline engine condition. If the engine is in good shape, all air entering the engine is drawn through the carburetor or injection system, past the valves, and compressed in the cylinders. If the engine is worn, air can leak into the cylinders past valve guides, piston rings, valve seats, or bad gaskets, figure 5-24. By checking cranking vacuum, you can tell if all cylinders are drawing air through the induction system. If they are not, air leaks will cause a low or uneven vacuum reading and keep you from tuning the engine for best performance.

Different engines produce cranking vacuum readings from 3 to 15 inches of mercury. Some carmakers publish cranking vacuum specifications; others do not. The important things to look for are a steady vacuum and cranking speed. If the battery and starting system are in good condition, do the test as follows:

1. Warm the engine to normal temperature.

2. Connect a vacuum gauge to a manifold vacuum source, figure 5-25. Do not connect the gauge to ported vacuum. Check a vacuum diagram to ensure correct connection.

3. Connect a tachometer, figure 5-22.

4. Close the throttle and disable the ignition.

Caution. Do not crank a converter-equipped car for more than 15 seconds to avoid drawing fuel into the catalyst.

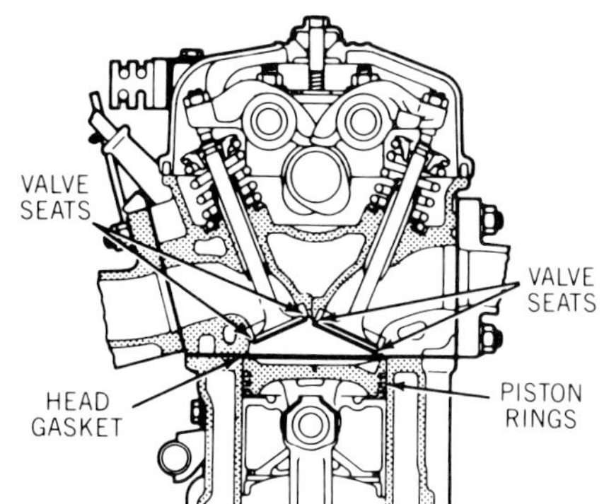

Figure 5-24. Leakage past the valves, rings, and other points will cause uneven cranking vacuum readings. (Chrysler)

Figure 5-25. Connect a vacuum gauge to a manifold vacuum port for cranking vacuum readings.

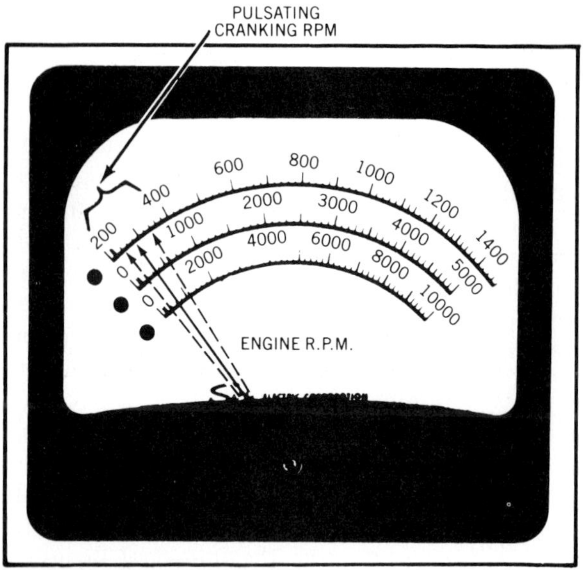

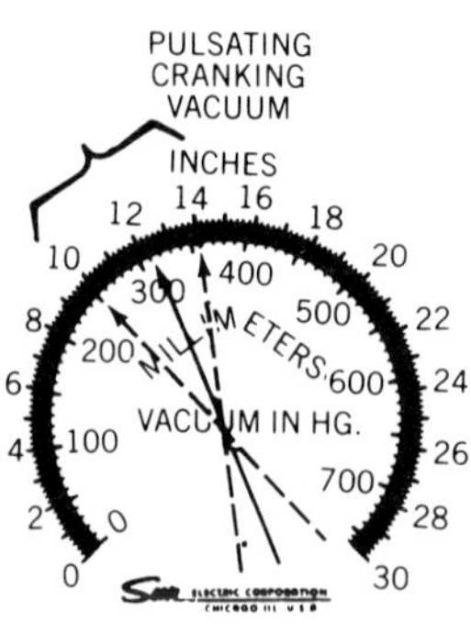

Figure 5-26. Uneven cranking vacuum and speed readings indicate leakage past valves, pistons, or gaskets. (Sun)

Figure 5-27. Disconnect and plug the vapor canister purge hose before a cranking vacuum test.

5. Crank the engine for 10 to 15 seconds and note the vacuum gauge and tachometer readings.

- *Vacuum and cranking speed (approximately 200 rpm) are steady*—The engine probably is mechanically sound.
- *Vacuum and cranking speed are uneven* figure 5-26—The engine probably has leakage past valves, rings, or the head gasket.
- *Speed is uneven, but vacuum is steady*—You may have a bad starter or worn flywheel ring gear.
- *Cranking speed is normal or high and vacuum is low and slightly uneven*—The engine probably has low compression or retarded valve timing.

Further testing will pinpoint possible problems indicated by this cranking test.

You can do a quick check of the PCV system while cranking the engine. Test cranking vacuum as explained above. Then pinch the PCV hose to the manifold closed with a pair of pliers and repeat the test. Vacuum with the PCV hose closed should be higher than with it open. If there is no change in vacuum, test the PCV system for blockage (chapter 6).

MANIFOLD VACUUM TESTS

Manifold vacuum tests with the engine running at idle, low cruising speed, and high cruising speed supplement the cranking vacuum test for gasoline engines. Vacuum tests can indicate problems in manifold and combustion chamber sealing, valve action and condition, ignition timing, air-fuel mixture, PCV operation, and exhaust restrictions.

1. Connect a tachometer, figure 5-22, and a vacuum gauge, figure 5-25, to the engine.

2. Disconnect the purge hose from the fuel vapor canister and plug it, figure 5-27.

> **Caution.** Put the transmission in neutral or park, set the parking brake, and block the drive wheels.

3. Run the engine to normal operating temperature.

4. Run the engine at idle, low cruise (1,800 to 2,200 rpm), and high cruise (2,500 to 3,000 rpm).

5. Note the vacuum reading and any fluctuations at each speed.

6. Run the engine at about 2,500 rpm for 15 seconds and observe the gauge. Release the throttle and observe the reading as the speed drops to idle. Vacuum should jump to about 24 in. Hg or more as you release the throttle.

Normal vacuum for most engines is 15 to 22 inches of mercury. Low vacuum at idle is common on some cars with high-lift cams, long valve overlap, and some emission controls. Vacuum gauge readings indicate the following general conditions, figure 5-28:

- *Steady vacuum, 15 to 22 inches, at all speeds with little fluctuation*—Engine, fuel, and ignition systems are normal.
- *Steady vacuum but lower than normal*—late ignition or valve timing, low compression (worn cylinders, rings or valves), intake manifold leak, incorrect valve adjustment, excessive drag in the engine.
- *Vacuum reading fluctuates within normal range*—Uneven compression, incorrect idle air-fuel mixture, uneven ignition timing and dwell, misfiring spark plugs, uneven valve adjustment, intake manifold leak near one or two cylinders.

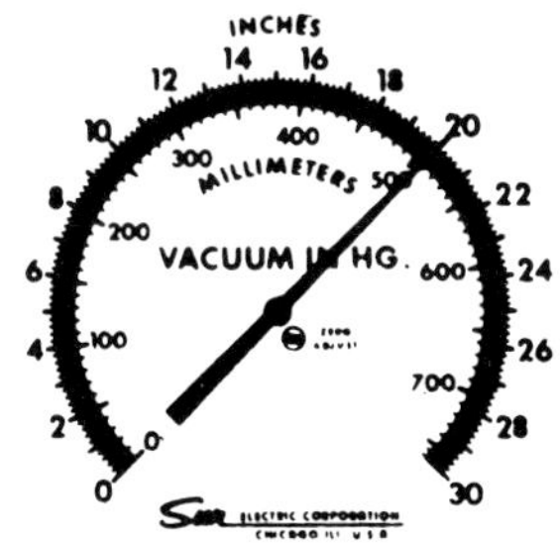

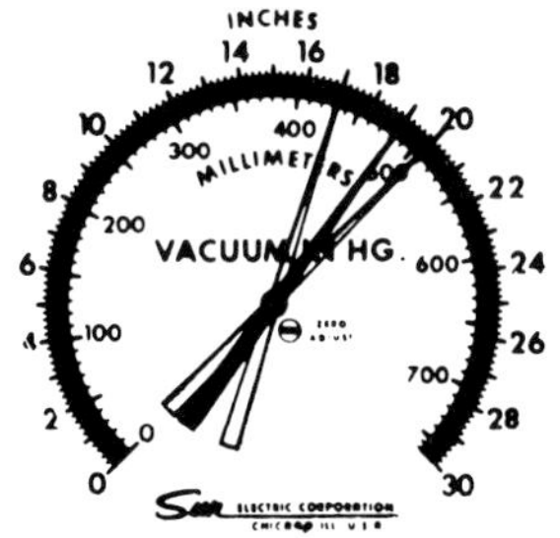

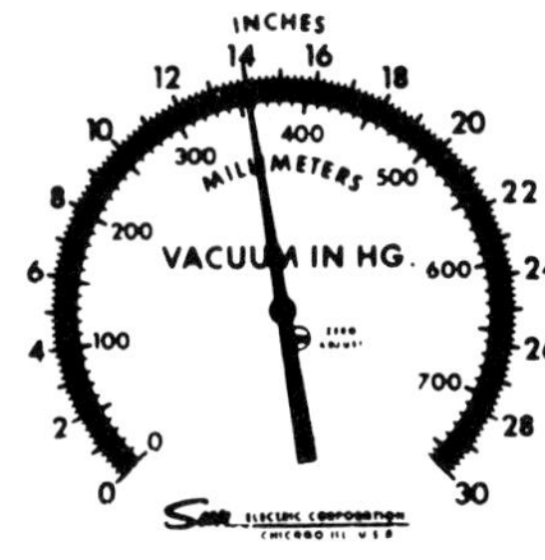

WITH MOTOR AT IDLING SPEED VACCUUM POINTER SHOULD HOLD STEADY.

WITH MOTOR AT IDLING SPEED DROPPING BACK OF VACUUM POINTER INDICATES STICKY VALVES.

WITH MOTOR AT IDLING SPEED FLOATING MOTION RIGHT AND LEFT OF VACUUM POINTER INDICATES CARBURETOR TOO RICH OR TOO LEAN.

WITH MOTOR AT IDLING SPEED LOW READING OF VACUUM POINTER INDICATES LATE TIMING OR INTAKE MANIFOLD AIR LEAK.

Figure 5-28. These are some of the indications you can get with a manifold vacuum test at different engine speeds. (Sun)

- *Vacuum drops irregularly at idle*—Sticking valves.
- *Higher than normal vacuum at idle*—Overly advanced ignition timing.
- *Vacuum does not increase when throttle closes from high cruise*—Worn rings, cylinders, or valves.
- *Vacuum drops during 15 seconds at high cruise, does not increase when throttle closes*—Restricted exhaust system.

These vacuum tests indicate general engine condition and possible problems in one or more systems or cylinders. They do not isolate a worn cylinder, and they do not distinguish between retarded ignition timing and valve timing, for example. Vacuum testing will lead you to detailed system tests and to further area testing, such as a power balance test.

POWER BALANCE TESTING

A power balance test indicates whether or not all cylinders are contributing equally to engine power. The principle of power balance testing is to short circuit the ignition for each cylinder, or groups of cylinders, and observe changes in engine speed and manifold vacuum. Using an infrared exhaust analyzer, you also can observe changes in exhaust HC emissions during power balance testing.

You can do a power balance test with just a tachometer and a vacuum gauge or with an engine analyzer. An engine analyzer automatically short circuits individual spark plugs and provides faster and more accurate testing. To use a tachometer and vacuum gauge, you must connect the meter and gauge to the engine as explained in previous sections. Then, you must remove the plug cables, one at a time, from each spark plug as the engine is running and ground the cables while noting the changes in engine speed and manifold vacuum. This method works well enough on older cars with breaker-point ignitions but not on late-model cars with electronic ignition, catalytic converters, and electronic engine controls. The procedures in the following sections are based on using an engine analyzer. But, before doing a power balance test, you must understand these precautions.

Power Balance Precautions

Observe these precautions to avoid inaccurate test results or possible vehicle damage. Check carmakers' and equipment manufacturers' instructions.

- On vehicles with EGR valves, disconnect and plug the EGR vacuum line before testing. Cycling of the EGR system with fluctuating manifold vacuum can give inaccurate power balance indications.
- Catalytic converters must be protected from unburned fuel. On converter-equipped cars do not short circuit any cylinder for more than 15 seconds. Let the engine run for 30 seconds between cylinder tests to purge unburned fuel from the converter. Some engine analyzers automatically cycle the test intervals to prevent overloading the converter.
- Engines with exhaust gas oxygen (EGO) sensors and feedback carburetors or injection systems may not yield accurate test results. Oxygen in the exhaust when a cylinder is shorted may drive the system to a fully rich condition and increase engine speed. Follow carmakers' instructions to place feedback systems in open-loop operating modes for power balance testing.
- Engines with electronic idle speed control cannot be tested at idle. The speed control system will allow a momentary stumble in idle speed and then return to normal, preventing accurate power balance measurement.

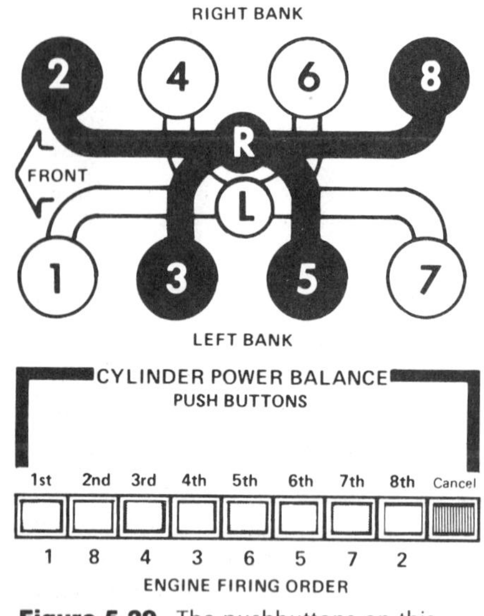

Figure 5-29. The pushbuttons on this panel allow you to "kill" selected cylinders. The numbers are not cylinder numbers, but firing order (first, second, third cylinder in the sequence, etc.). (Sun)

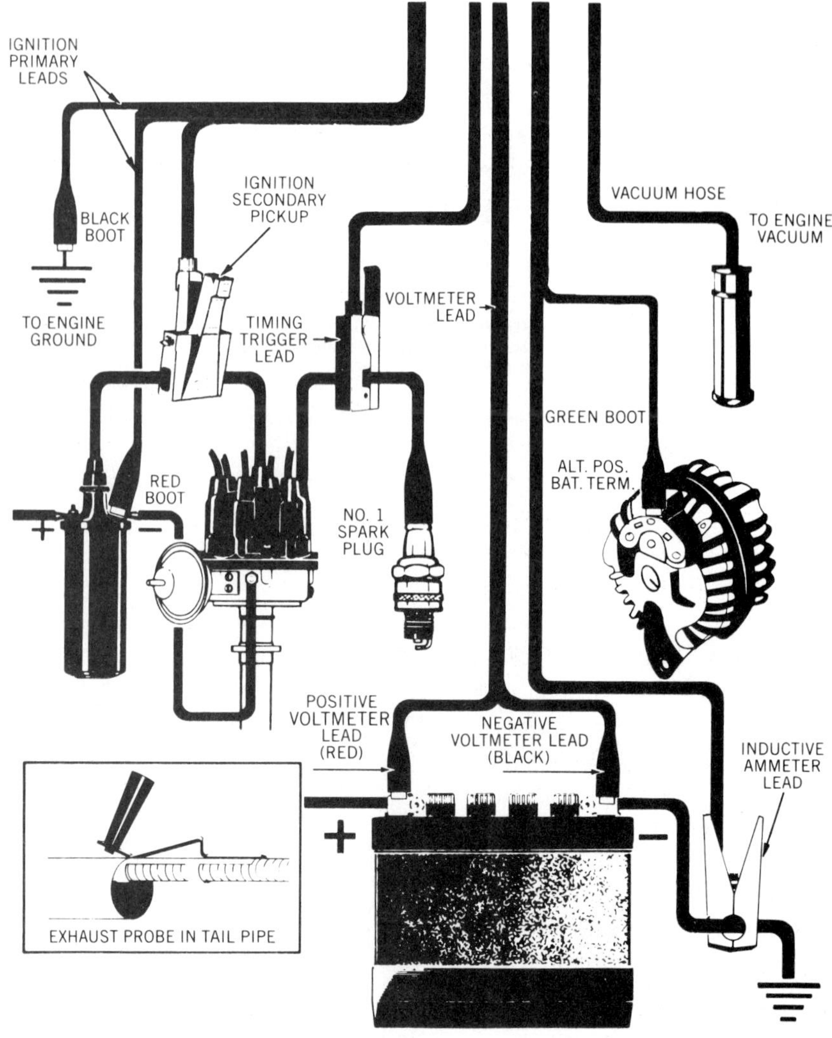

Figure 5-30. These are typical test lead connections for an engine analyzer. (Sun)

Power Balance Test Procedure

Engine analyzers have pushbuttons that let you control the short circuiting of separate cylinders or groups of cylinders, figure 5-29. The numbers on the buttons are *not* cylinder numbers, but firing order. Button 1 is usually cylinder 1, but button 2 may be cylinder 8 or 5 or 3 or some other, depending on firing order. Do a power balance test as follows:

Caution. Put the transmission in neutral or park, set the parking brake, and block the drive wheels.

1. Connect the analyzer to the engine, according to the manufacturer's instructions. Figure 5-30 shows typical test connections.

2. Disconnect and plug the vacuum hose to the EGR valve.

3. Run the engine to normal operating temperature. Be sure the choke is open.

4. Run the engine at a steady speed of 1,200 to 1,500 rpm.

5. Press the button to kill one cylinder and note the decrease in engine speed and vacuum. If you are using an infrared exhaust analyzer, also note the change in HC emissions.

6. Repeat step 5 for each cylinder. Allow engine speed to stabilize between cylinder tests.

7. Evaluate the results after all cylinders are tested.

If speed and vacuum decrease equally as each cylinder is shorted, each cylinder is contributing an equal amount to engine power. If speed and vacuum do not decrease, or decrease less, for any cylinder, it is not producing full power. The exact problem can be in the fuel or ignition systems or in the cylinder condition (low compression), but you have isolated it to one cylinder or a group of cylinders.

Exhaust HC emissions should increase when you kill a good cylinder. If HC emissions are higher than normal overall but don't increase when one cylinder is killed, you can look for a misfire or incomplete combustion in that cylinder. It could be a fouled plug, an open secondary circuit to the cylinder, a burned valve, or a leaking head gasket. Further tests will pinpoint the problem.

You can also use power balance testing to check air-fuel distribution from each side of a 2- or 4-barrel carburetor on V-type engines. You can use the same test methods to balance idle mixture adjustments on each side of the carburetor for a V-type engine. You will learn about this kind of power balancing in chapter 17. Also, during power balance testing with an engine analyzer, you can use the ignition oscilloscope to check overall and individual cylinder ignition. You will learn about oscilloscope use in the chapters on ignition service (chapters 12 and 13).

Diesel Power Balance Testing

Obviously, you can't do a power balance test on a diesel by short circuiting spark plugs. You can, however, measure the relative power contribution of individual diesel cylinders in two ways:

1. Loosen the fitting at an injection nozzle to relieve pressure and stop ignition.

2. Use an ohmmeter to measure glow plug resistance when the engine is running.

You will learn how to make these tests in chapter 18 on diesel performance diagnosis and service.

COMPRESSION TESTING

Power balance testing can lead you to an engine compression test to verify mechanical condition of one or all cylinders. An engine needs good and equal compression in all cylinders for maximum power. If compression leaks past worn rings or valves, or around the head gasket or spark plug, power is lost. Compression testing is not always necessary if power balance and vacuum test results are good. But, if power balance and vacuum tests indicate one or more weak cylinders, a compression test will help to determine if those cylinders are mechanically sound.

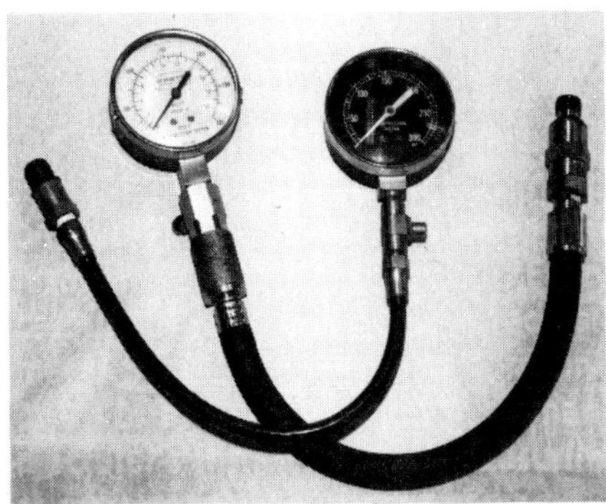

Figure 5-31. A typical compression gauge measures compression pressure in psi or metric pressure units and has adapters for connection to the cylinders.

For a compression test on a gasoline engine, you will need a compression gauge that reads from 0 to 250 or 300 psi (1,722 to 2,067 kPa), figure 5-31. For a diesel engine, you will need a gauge that reads to 600 or 800 psi (4,134 to 5,512 kPa). Carmakers publish compression specifications in one of two ways:

1. Specifications may list a minimum pressure and an allowable variation between cylinders. For example, minimum compression may be 120 psi with a 20-psi difference between cylinders. Compression, then, would be okay if each cylinder is at least 120 to 140 psi.

2. Specifications may say that the lowest cylinder must be within 75 percent of the highest cylinder. If the highest cylinder is 140 psi, the lowest should be 105 psi (75 percent of 140) or higher.

In either case, less compression variation among cylinders indicates an engine in better condition.

Even if a carmaker does not list a minimum compression, you can suspect a worn engine if compression is below 85 to 100 psi.

Compression Test Procedure

Perform a compression test as follows:

1. Run the engine to normal operating temperature; then stop the engine.

2. Use compressed air to blow dirt away from the spark plugs or glow plugs.

3. Remove all spark plugs (gaskets and plug tubes, if installed) or glow plugs and keep them in cylinder number order for reinstallation.

4. Disable the ignition or diesel injection system.

5. On a gasoline engine, block the throttle and choke linkage fully open to allow air to enter the engine. This also keeps fuel from being drawn through the idle and low-speed circuits of the carburetor, which could flood the engine and reduce cylinder lubrication.

6. Connect a remote starter switch to the engine, if desired.

7. Connect the gauge, figure 5-32, to the plug opening of cylinder number 1 by:

a. screwing a gauge adapter into the threaded opening or

b. inserting a tapered rubber gauge adapter into the plug opening and holding it firmly during testing.

8. Using the ignition switch or a remote starter switch, crank the engine for *four complete compression strokes* on that cylinder. Note the gauge reading on the *first* and *fourth* strokes.

9. Disconnect the gauge from the first cylinder and repeat steps 7 and 8 on all other cylinders. Compare the gauge readings to the carmaker's specifications and the following test guidelines.

- *Compression increases steadily on all four strokes and is within specifications*—the cylinder compression is okay.
- *Compression is low on the first stroke and increases on following strokes but does not reach specifications*—the piston rings or cylinder is probably worn.
- *Compression is low on the first stroke and increases only slightly on following strokes*—the valves may be sticking or burned, or rings may be broken.
- *Compression in two side-by-side cylinders is low*—the head gasket probably is leaking between the cylinders.
- *The gauge reading is higher than normal*—that cylinder may have excessive carbon deposits in the combustion chamber.

If compression is low in any cylinder, you can continue with a wet compression test as follows:

Caution. Do not perform a wet compression test on a diesel engine. Oil in the cylinders may cause premature ignition and engine damage.

Wet compression testing is not valid on a horizontal engine (Subaru or air-

Figure 5-32. Connect the compression gauge to the cylinder and crank the engine with a remote starter switch.

Figure 5-33. Typical cylinder leakage test equipment. (Sun)

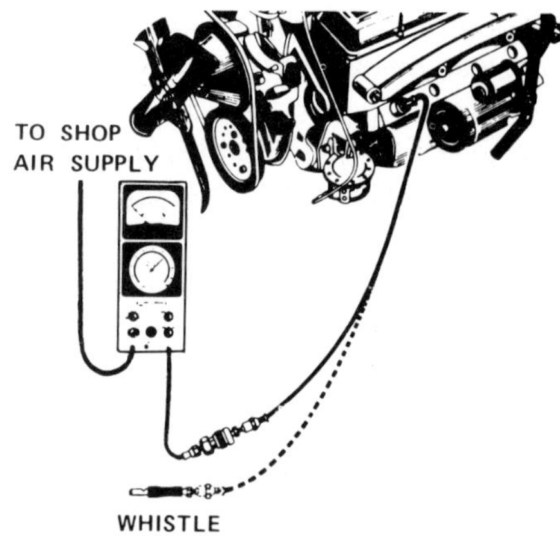

Figure 5-34. Some cylinder leakage test equipment has a whistle indicator to signal when the piston is at tdc. (Sun)

cooled Volkswagen or Porsche) because oil will not flow evenly around a horizontal piston.

1. Pour a small amount of oil (about 1 tablespoon or 8 squirts from an oil can) into the cylinder through the spark plug opening.

2. Allow about 30 seconds for the oil to flow around the top of the piston.

3. Repeat steps 7 and 8 of the basic compression test.

4. Compare the wet test results to the basic test results for that cylinder.

If compression increases by 5 percent or 12 to 14 psi, or more, that cylinder probably has worn rings or cylinder walls. If compression does not increase, that cylinder probably has leakage past the valves or head gasket.

CYLINDER LEAKAGE TESTING

A cylinder leakage test supplements basic power balance and compression testing. This test gives precise indications of cylinder leakage, such as:

- The exact location of a leak
- The size of the leak as a percentage of total cylinder compression

A cylinder leakage test is made by pressurizing an engine cylinder with compressed air. The special leakage test equipment may be built into an engine analyzer, or it may be a separate piece of test equipment, figure 5-33. The equipment consists of an adapter that screws into the spark plug hole for a cylinder, an air pressure regulator, a test gauge, and connecting hoses.

To do a cylinder leakage test, the piston of the test cylinder must be at or near tdc on the compression. This ensures that both valves are closed and the combustion chamber is airtight except for a small amount of leakage through the piston ring gaps. The test adapter is installed in the spark plug hole, and regulated compressed air is applied to the cylinder.

The pressure gauge is graduated from 0 to 100 percent to indicate any cylinder leakage as a percent of total compression pressure. A perfectly sealed combustion chamber would produce 0-percent leakage, but this is not possible because of the small amount of leakage through the ring gaps. A test reading of 100 percent indicates that no pressure is held in the cylinder. A well-sealed cylinder will produce a lower percentage of leakage than a poorly sealed cylinder. You can locate the leakage source as explained in the following procedure.

Leakage Test Procedure

Some leakage test equipment may include a whistle indicator, figure 5-34. When the whistle is connected by a hose to the cylinder adapter, it will sound as the piston is cranked to tdc on the compression stroke. If your equipment includes a whistle indicator, follow the manufacturer's instructions for its use. You also can use the following general procedure to test cylinder leakage:

1. Run the engine to normal operating temperature. Then shut it off and disable the ignition.
2. Remove all spark plugs and gaskets, along with plug tubes if used.
3. Remove the air cleaner and block the choke and throttle valves fully open.
4. Disconnect the PCV valve inlet hose from the crankcase.
5. Remove the radiator cap. Be sure the coolant is at the specified level. Add water if necessary.
6. Adjust the leakage test equipment according to the manufacturer's instructions.
7. Remove the distributor cap and mark the location of the number 1 cylinder tower on the distributor body.
8. Using a wrench on the crankshaft pulley, rotate the engine until the distributor rotor points to the number 1 cylinder location on the distributor body.
9. Install the test adapter in the number 1 spark plug hole (if not already installed) and connect the tester and the compressed air source.
10. Follow the manufacturer's instructions to pressurize the cylinder.
11. Read the leakage percentage shown on the tester gauge. Interpret the reading as follows:

Less than 10%	=	good compression
10% to 20%	=	fair compression
20% to 30%	=	poor compression
More than 30%	=	a leakage problem

12. If cylinder leakage is more than 20 percent, listen for air leakage at the following sources to isolate the problem:

- Air escaping through the crankcase indicates worn or broken rings, a cracked or burned piston, or worn cylinder walls.
- Air escaping through the exhaust pipe indicates a leaking exhaust valve.
- Air escaping through the carburetor indicates a leaking intake valve
- Air bubbles in the radiator indicate a leaking head gasket or a cracked head or block.
- Two cylinders next to each other with high leakage indicate a leaking head gasket or an engine crack between the cylinders.

13. Turn off the compressed air and disconnect the air source from the cylinder adapter.
14. Using a wrench on the crankshaft pulley, rotate the engine until the distributor rotor points at the next cylinder tower location in firing order sequence. This places that cylinder at tdc on the compression stroke.
15. Install the test adapter in the spark plug hole of the next cylinder.
16. Repeat steps 10 through 12 for all remaining cylinders.

Comparing Power Balance, Compression, and Cylinder Leakage Test Results

Compare the results of cylinder leakage tests, compression tests, and power balance tests as explained in the following sections to isolate possible engine problems.

High Leakage With Good Power Balance and Compression. If cylinder leakage is above 20 percent but the engine has equal power balance and compression among cylinders, it is probably a worn, high-mileage engine. Other symptoms include lack of power, poor fuel economy, and excessive blowby. Combustion chamber deposits can raise compression in a worn engine.

Good Compression and Leakage but Poor Power Balance Among Cylinders. An engine with good compression and leakage but poor power balance can have several problems. The oscilloscope tests in chapters 12 and 13 will help you isolate ignition and combustion problems. If oscilloscope testing indicates good ignition and combustion, the problem may be:

- A broken valve spring
- A broken rocker arm
- A broken or bent pushrod
- A collapsed valve lifter
- Worn camshaft lobes
- Leaking valve guides
- Leaking intake manifold gaskets

Good Leakage but Poor Power Balance and Compression. If a valve does not open, opens incompletely, or does not open at the right time, a cylinder will have good (low) leakage but poor compression and a poor power balance rating. If all cylinders have good leakage but poor compression, the camshaft may be out of time because of a slipped timing chain or gears.

INFRARED EXHAUST ANALYSIS

The final step in general engine testing is exhaust analysis. We have already touched on the use of an infrared exhaust analyzer to measure HC emissions during power balance testing. However, you should also check the exhaust, itself, for an overall indication of engine condition. For thorough testing, check the exhaust at idle and at 2,500 rpm with the engine warmed to normal operating temperature. Checking the exhaust at idle and at cruising-speed rpm allows you to test the idle and main-metering fuel circuits and emissions at basic and advanced timing. Chapter 4 has general instructions for exhaust analyzer operaton. Review those instructions and follow the equipment maker's procedures for analyzer use.

HC and CO Specifications

Whether you use a 2-, 3-, or 4-gas analyzer, you will be most interested in HC and CO emission levels. CO_2 and O_2 analysis with a 4-gas analyzer will supplement HC and CO readings and help you diagnose combustion problems on converter-equipped cars that do not emit measurable HC and CO.

HC and CO emissions should be no higher than the maximum allowed by law for a particular model-year vehicle. Regulations vary according to state

HC and CO Limits at Idle
Cars and Light Trucks

Vehicle Year	*HC (PPM)*	*CO (%)*
1967 & earlier	300–500	2.5–3.0
1968–1969	200–300	2.0–2.5
1970–1972	150–250	1.5–2.0
1973–1974	100–200	1.0–1.5
1975–1978	50–100	0.5–1.0
1979 & later	50	0.0–0.5

Figure 5-35. These are general guidelines that you can use to test HC and CO emissions at idle, but manufacturer's specifications and local regulations take precedence over these limits. Generally, emissions will decrease after a tuneup.

and local laws. Figure 5-35 lists guidelines for HC and CO emissions at idle for cars and light trucks of various years. You can use these as general guides if you do not have specific state or local standards.

Exhaust Test Procedure

Sample the exhaust at idle and at 2,500 rpm as follows:

1. Using the engine inspection procedure from the beginning of this chapter, be sure that:

- **a.** The air filter is installed properly and not clogged
- **b.** The choke is fully open and is not stuck
- **c.** All vacuum line connections are secure and not leaking
- **d.** The exhaust system is not leaking and the manifold heat control valve will open correctly

2. Be sure the engine is at normal operating temperature and the analyzer probe is installed properly in the tailpipe.

3. Disconnect the outlet line from the air injection pump or pulse air valve.

4. Check and adjust the ZERO and SPAN settings on the analyzer, if required.

5. Run the engine at normal slow idle and note the HC and CO readings (also CO_2 and O_2 if using a 4-gas analyzer).

6. Increase engine to a steady 2,500 rpm and again note all meter readings.

7. Return the engine to idle and note any change in meter readings as the engine decelerates. Note the meter readings again as the engine runs at steady idle speed.

HC and CO emissions should be within legal limits or within the guidelines of figure 5-35. Emissions will increase during closed-throttle deceleration, but the readings in step 7 should be as low as, or lower than, the readings in step 5.

Exhaust Analysis

CO emissions are directly related to air-fuel mixture. High CO results from a rich mixture, too much gasoline, or not enough air. Common causes of high CO emissions are:

1. A restricted air filter

2. Restricted air passages in the carburetor or injection system

3. A rich carburetor or injection fuel adjustment

4. High float level in the carburetor

5. A stuck or improperly adjusted choke

6. Leaking power valve or accelerator pump in the carburetor

7. Wrong idle speed

8. Engine oil contaminated by fuel due to excessive blowby or a leaking fuel pump. You can isolate this problem by disconnecting the PCV valve from the crankcase and letting it draw fresh air while monitoring exhaust CO. If CO drops by 0.5 percent or more, the oil is probably contaminated with fuel.

High HC emissions indicate unburned fuel in the exhaust. Incomplete combustion due to a lack of ignition (misfire) or a lean mixture will cause high HC. Although ignition problems often cause high HC, engine mechanical problems and vacuum leaks can also increase HC emissions. Common causes of high HC emissions are:

1. Wrong ignition timing

2. Fouled or worn spark plugs, defective secondary cables, worn breaker points, and other ignition problems that cause a misfire

3. An overly rich or lean air-fuel ratio

4. Low compression (incomplete combustion)

5. Vacuum leaks at the carburetor, the manifold, the injection system, or at engine accessories

6. Worn valve train parts

7. Worn cylinders and piston rings

Combined high readings for HC and CO can be caused by:

1. A defective catalytic converter or PCV system

2. A defective thermostatic air cleaner or restricted air filter

3. A defective manifold heat control valve.

A couple rules of thumb for basic HC and CO testing are:

- High HC emissions indicate a lean mixture or misfire—unburned fuel
- High CO emissions indicate a rich mixture—not enough oxygen for complete combustion.

Figure 5-36 lists more abnormal HC and CO readings, at idle and at 2,500 rpm, along with possible causes and cures.

O_2 and CO_2 Analysis

Late-model cars with catalytic converters can reduce HC and CO at the tailpipe to levels where they are almost not measurable. CO_2 and O_2, however, can be measured. On a noncatalyst car, high HC indicates unburned gasoline in the exhaust. This is usually due to a lean misfire. On a converter-equipped car, O_2 indicates a lean condition. If O_2 is above 2.0 percent, the air-fuel mixture is probably lean. If O_2 is above 4 percent, the mixture is definitely too lean.

At the stoichiometric 14.7:1 air-fuel ratio, HC and CO emissions should be low. Similarly, O_2 should be low, but CO_2 should be high. Figure 5-37 shows these relationships. At lean air-fuel ratios, O_2 increases, and CO_2 decreases. Also at a 14.7:1 air-fuel ratio, the total percentage of CO and CO_2 is about 14.7 percent. Figure 5-38 shows the relationship of combined CO and CO_2 percentage to air-fuel ratio. Acceptable combustion in a converter-equipped car will produce the following 4-gas analyzer readings:

- HC and CO—within specificatons
- O_2—1.0 to 2.0 percent

CONDITION	*CAUSE*
Abnormal readings at idle:	
High HC, Low CO	1. Open spark plug cable (misfire) 2. Fouled spark plug (misfire) 3. Lean air-fuel ratio (misfire)
High HC, Normal CO	1. Arcing ignition breaker points 2. Vacuum leak 3. Low compression
High HC, High CO	1. Rich air-fuel ratio (possibly a high fuel level in the carburetor fuel bowl)
Normal HC, High CO	1. Restricted PCV system 2. Restricted intake air filter 3. Choke stuck partially closed 4. Wrong adjustments for carburetor or fuel injection
Abnormal readings at 2,500 rpm:	
High HC, Low CO	1. Open spark plug cable (misfire) 2. Fouled spark plug (misfire) 3. High-speed misfire 4. Floating exhaust valve
High HC, Normal CO	1. Arcing or misaligned breaker points; point bounce 2. Vacuum leak
High HC, High CO	3. High fuel level in carburetor fuel bowl or otherwise rich air-fuel ratio
Normal HC, High CO	1. Restricted intake air filter 2. Choke stuck partially closed 3. Defective high-speed or power circuit in the carburetor; accelerator pump pullover

Figure 5-36. HC and CO troubleshooting guide.

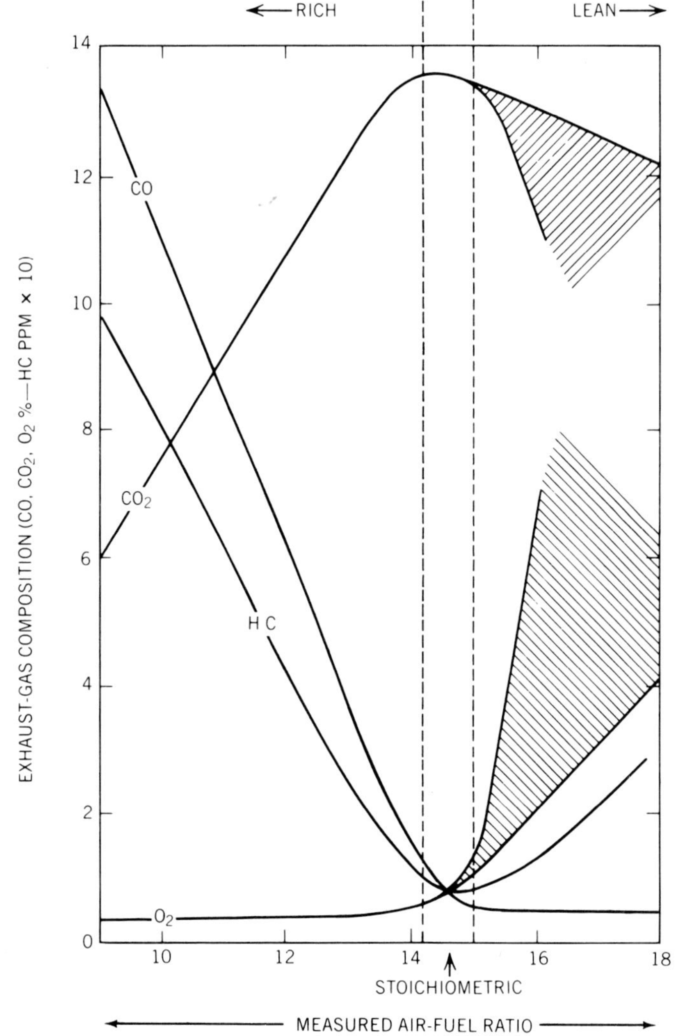

Figure 5-37. This graph shows the relationships of HC, CO, CO_2, and O_2 to air-fuel ratios.

$CO + CO_2$ Percentage		*Approximate Air-Fuel Ratio*
13.5	↑	16.0:1
14.0	Lean	15.5:1
14.5	↓	15.0:1
14.7	—	**14.7:1**
15.0	↑	14.2:1
15.5		13.7:1
16.0		13.0:1
16.5	Rich	12.5:1
17.0	↓	11.7:1

Figure 5-38. CO plus CO_2 percentages indicate the approximate air-fuel ratio.

• CO_2—above 10 percent (ideally 13 to 15 percent)

If O_2 exceeds CO and CO is above 0.5 percent, the catalytic converter may be defective. If CO is above 0.5 percent and also higher than O_2, the converter probably is okay, but the air-fuel mixture is rich. Figure 5-39 summarizes other relationships between HC, CO, CO_2, and O_2 readings on a 4-gas analyzer.

MECHANICAL VALVE ADJUSTMENT

Mechanical valve lifter adjustment is an engine service rather than a test. However, incorrect valve clearance (valve lash) can cause combustion problems that you will detect during engine testing.

CONDITION	*EMISSION LEVELS* *HC*	*CO*	CO_2	O_2
Normal	Within Specifications		13–15%	1.0–2.0%
HC Problem	High	High or Low	Below 10%	Above 2%
CO Problem	High	High	Below 10%	Below 1.0%
HC & CO Problem	High	High or Low	Below 10%	Above 2.0% or Below 1.0%

Figure 5-39. Four-gas relationships for exhaust analysis.

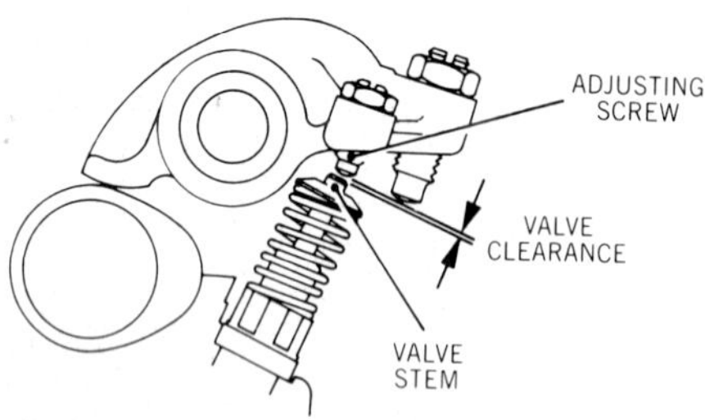

Figure 5-40. Valve clearance is the gap between the tip of the valve stem and the rocker arm or cam follower. (Chrysler)

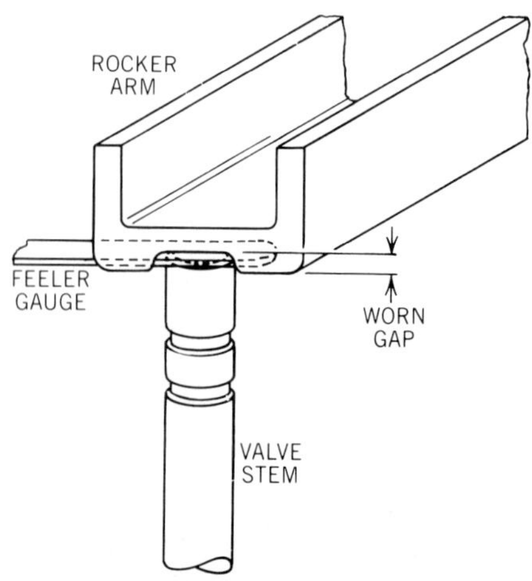

Figure 5-41. A worn rocker arm can create a false clearance measurement.

Compression problems caused by incorrect valve clearance will upset cylinder power balance, ignition operation, and fuel system adjustments. Improper adjustment, either too loose or too tight, will change the valve timing. Uneven adjustment also will unbalance compression pressures among cylinders. Tight valve clearances can cause a rough idle, or worse, burned valves. Loose clearances cause noisy valves and rocker arm or valve stem wear. You often can cure a rough idle or an unbalanced carburetor by adjusting the valves for equal clearance.

Accurate engine tuning is impossible if valve clearances are too tight or too loose.

Hydraulic valve lifters, which require no periodic adjustment, have been used on many passenger car engines for decades. Many other engines, however, have mechanical lifters, or cam followers, that do require adjustment. You will find adjustable lifters, or followers, on several domestic high-performance engines and on many European and Japanese imports.

Valve Clearance Specifications

Valve clearance, or valve lash, is the gap between the rocker arm or the cam follower and the tip of the valve stem when the valve is closed, figure 5-40. Clearance specifications are measured when the heel or base of the cam lobe is against the tappet or follower and the gap is at its maximum.

Valve clearance specifications are given either as hot or cold measurements. Cold clearances usually are greater than hot clearances for the same engine. This allows for metal expansion in a hot engine. If you adjust the valves with the engine cold, coolant temperature should be close to ambient air temperature.

If you adjust the valves hot, the engine must be fully warmed to its normal operating temperature. You can adjust the valves with the engine running or with it off, as explained below. If you adjust the valves with the engine off and adjustment takes more than 5 to 10 minutes, reinstall the valve covers and run the engine to keep it hot.

Intake valve clearances range from 0.004 to 0.025 inch (0.10 to 0.64 mm). Exhaust valve clearances range from 0.004 to 0.030 inch (0.10 to 0.76 mm). Exhaust valve clearances usually are greater than intake valve clearances because exhaust valves run hotter and expand more.

Checking Valve Clearance

To check the clearance of any valve accurately, hot or cold, you must rotate the engine so that the valve is fully closed and the heel, or base circle, of the cam lobe is on the tappet. This provides maximum clearance, figure 5-40.

Insert a flat feeler gauge of the specified thickness between the valve stem and the rocker arm or cam follower. Special feeler gauges for valve adjustment may be bent for easier adjustment of valves near manifolds or at the rear of an engine. Some gauges are the go/no-go type. These have two thicknesses on one blade. The outer, thinner, part of the gauge will fit into the gap when clearance is correct. The inner, thicker, part will not.

On a high-mileage engine, or one that has run for a long time with incorrectly adjusted valves, the rocker arms or followers may be worn as shown in figure 5-41. In this case, a feeler gauge may give an inaccurate measurement, and clearance will be too large. The best solution to this problem is to reface the rockers or followers to remove the wear. However, you may be able to adjust such worn valve train parts by using a narrow feeler gauge that fits into the wear area.

Valve Adjustment Methods

For ohv engines with the cam in the block, the most common valve adjustment methods are as follows:

- An adjusting nut on the center pivot of the rocker arm, figure 5-42, which can be adjusted with a single wrench.
- An interference-fit capscrew or jamscrew on the end of the rocker arm, figure 5-43, which also can be adjusted with a single wrench.
- An adjusting screw and locknut, figure 5- 44, which require a screwdriver

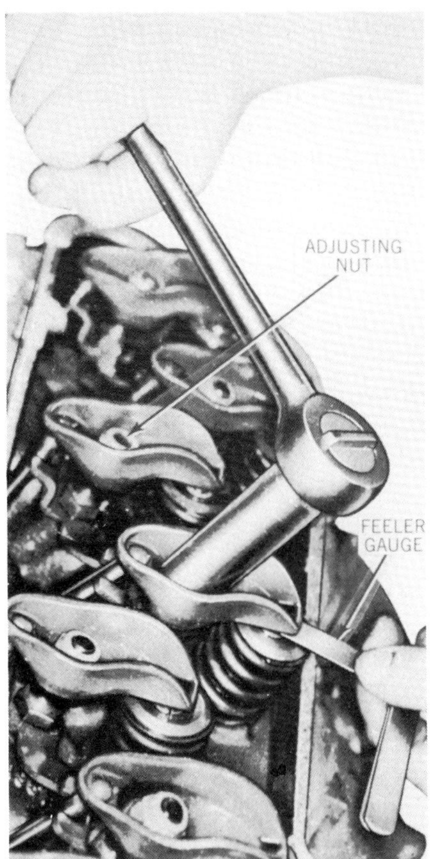

Figure 5-42. Turn the adjusting nut at the center of the rocker arm to adjust valve clearance on some engines. (Chevrolet)

Figure 5-43. Some engines have interference-fit jamscrews in the rocker arms for valve adjustment.

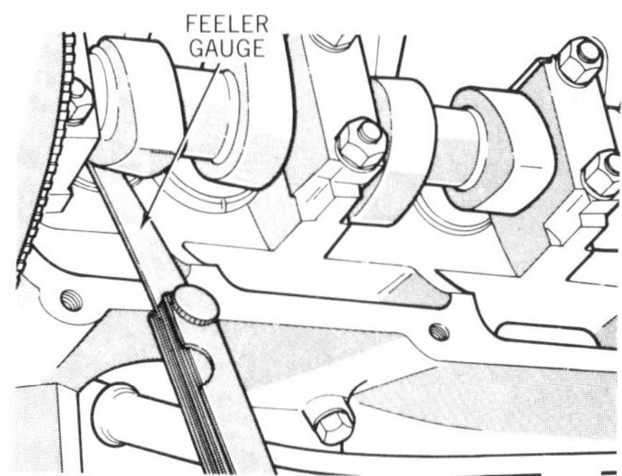

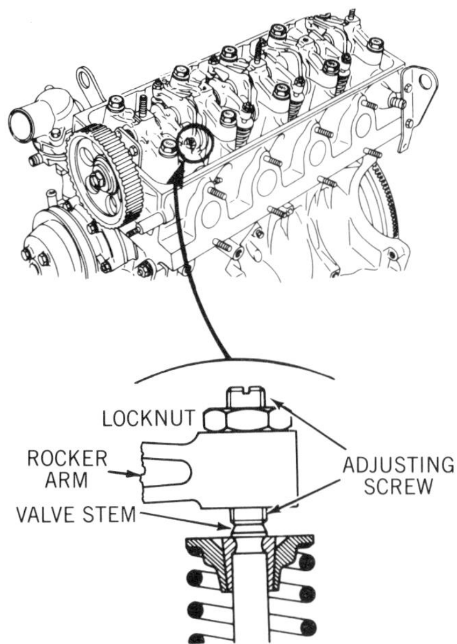

Figure 5-44. Some engines have locknuts and adjusting screws in the rocker arms.

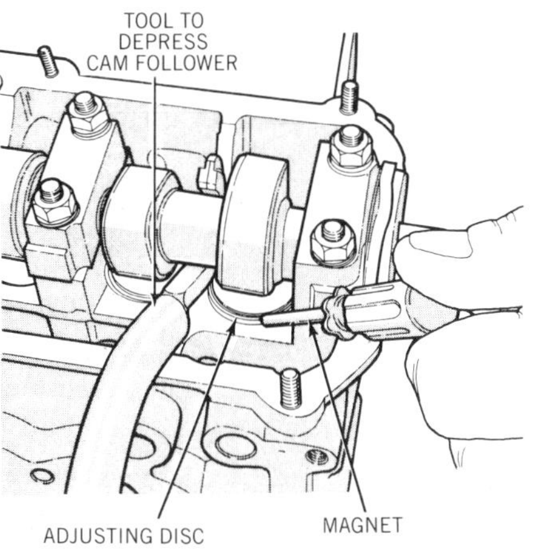

Figure 5-45. Valve clearance measurement and adjustment on an ohc engine with removable discs in the cam followers. (Chrysler)

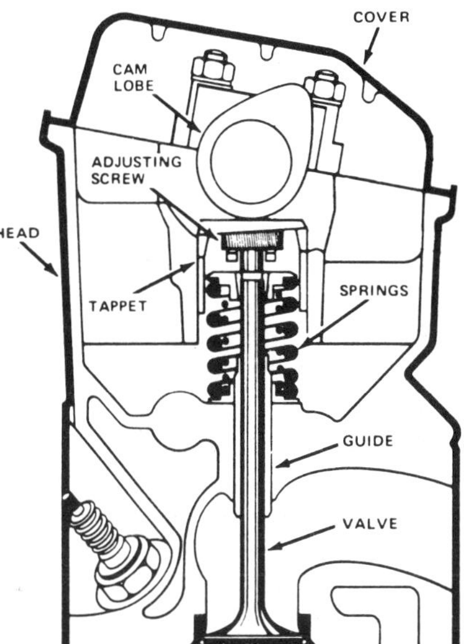

Figure 5-46. Some ohc engines have tapered adjusting screws in the cam followers.

and wrench or a special adjusting tool for adjustment.

For ohc engines, the most common valve adjustment methods are as follows:

- An interference-fit jamscrew or an adjusting screw and locknut in the end of the rocker arm. These are similar to the ohv adjustment methods shown in figures 5-43 and 5-44.
- Removable disc-type shims in the cam follower, figure 5-45. To adjust the clearance, you must depress the cam follower with a special tool and remove the disc with a small screwdriver or a magnet. Installing a disc of a different thickness changes the valve clearance.
- A tapered adjustment screw in the cam follower, figure 5-46. The screw is a wedge that changes the height of the cam follower and thus the valve clearance. The screw has one flat side and must be turned one complete revolution for each adjustment step.

Adjusting Valves with the Engine Off. You can adjust the valves of any engine with the engine off. Overhead-cam engines with removable cam follower discs or tapered adjusting screws *must* be adjusted with the engine off. To adjust valves with the engine off, either hot or cold, you must rotate the engine so that the valves for a particular cylinder are fully closed. You can do this accurately and quickly by using the cylinder running mate principle:

1. Write down the engine firing order. A typical 4-cylinder engine fires 1-3-4-2.

2. Write the first half of the firing order above the second half:

1-3
4-2

The cylinders written above and below each other are running mates. When one cylinder is at tdc on the exhaust stroke, the other is at tdc on the compression stroke with both valves closed.

3. With the valve cover removed, rotate the engine until the exhaust valve of number 4 cylinder just closes.

4. Adjust the intake and exhaust valves of number 1 cylinder. Both valves are fully closed because the piston is at tdc on the compression stroke.

5. Rotate the engine until the exhaust valve of number 2 cylinder just closes and adjust the valves for number 3 cylinder.

6. Continue to rotate the engine and adjust the valves for cylinders 4 and 3 in the same way.

Carmakers often provide instructions that are a variation on this method, but you can use the above procedure on any engine.

Adjusting Valves with the Engine Running. If an engine has adjustment screws in the rocker arms (either in the ends or in the center pivot), you usually can adjust the valves with the engine running. Be sure the engine is completely warmed up and use hot valve clearance specificatons. Follow the carmaker's procedures and these guidelines for easy and accurate adjustment.

- Work in a logical pattern from one end of the engine to the other.
- Run the engine at its slowest idle speed.
- Adjust all intake or all exhaust valves first.
- Some engines, such as Honda CVCC engines and Chrysler-Mitsubishi engines, have auxiliary intake valves. Adjust these before adjusting the main intake valves, figure 5-47.
- Insert the specified feeler gauge between the rocker arm an the valve stem.
  * If the gauge slides freely with a slow, steady drag and engine speed does not change, clearance is correct.
  * If you have to force the gauge and the engine misfires or stumbles, clearance is too tight.
  * If the gauge slides easily with a jerky feel, clearance is too loose.
- Turn the adjusting screw as needed to change the clearance, figures 5-42 through 5- 44. Check your adjustment by inserting a gauge that is 0.002 inch thicker or thinner or use a go/no-go feeler gauge.

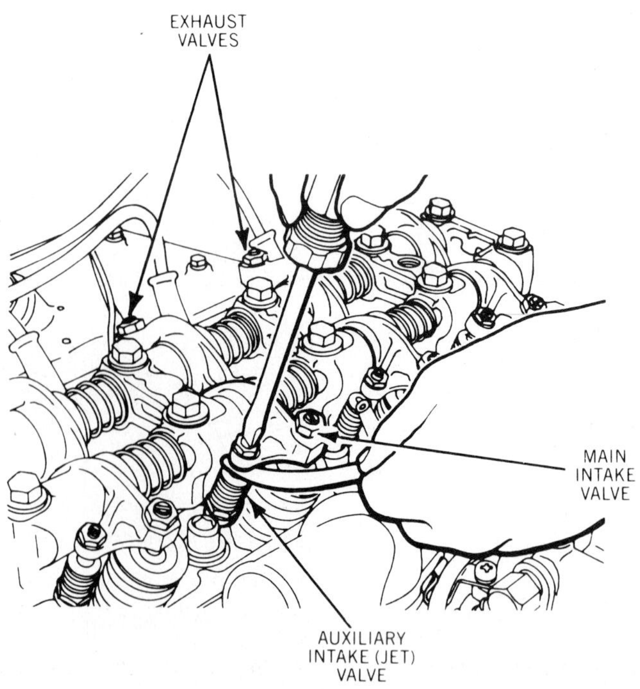

Figure 5-47. Adjust the auxiliary intake (jet) valves before adjusting the main intake valves. (Chrysler)

SUMMARY

General engine testing begins with inspecting the engine and its related systems for obvious defects. Area tests of the electrical system (battery, starter, and alternator) indicate whether the engine will crank and start normally and has enough electrical power for proper ignition. Other area tests of vacuum, power balance, compression, cylinder leakage, and exhaust emissions indicate overall engine condition and show whether the fuel, ignition, and emission systems are operating within generally normal limits. Some area tests concentrate on a single system; others check several systems and system relationships. The results of comprehensive engine tests will lead you to specific pinpoint tests and adjustments or repairs for individual components.

Mechanical valve lifter adjustment is an engine service, rather than a test. However, power balance, compression, and leakage tests can detect improperly adjusted valves. Proper adjustments of ignition, fuel, and emission systems cannot be made if valve clearances are incorrect.

REVIEW QUESTIONS

Multiple Choice

1. Mechanic A says that uneven cranking speed can be caused by unequal compression among cylinders. Mechanic B says that uneven cranking speed can be caused by a worn flywheel ring gear or starting system problems. Who is right?

- **a.** A only
- **b.** B only
- **c.** both A and B
- **d.** neither A nor B

2. You can disable a breaker-point ignition system for testing in any of the following ways EXCEPT:

- **a.** removing the coil high-voltage lead from the distributor and grounding it.
- **b.** removing the primary feed wire from the coil + or BAT terminal and grounding it.
- **c.** grounding the breaker points at the distributor.
- **d.** grounding the breaker points at the coil—or DIST terminal.

3. Mechanic A says that you can connect a remote starter switch on most Ford products between the battery terminal and the switch terminal of the starter relay. Mechanic B says that you can connect a remote starter switch on a GM car in the same way. Who is right?

- **a.** A only
- **b.** B only
- **c.** both A and B
- **d.** neither A nor B

4. The absolute minimum cranking voltage specified by most carmakers is:

- **a.** 9.0 volts
- **b.** 9.6 volts
- **c.** 10.0 volts
- **d.** 10.6 volts

5. If specifications are not available, Mechanic A says that you can determine a battery's load-test current by multiplying the ampere-hour rating by 3. Mechanic B says that you can determine load-test current by dividing the cold-cranking current rating by 2. Who is right?

- **a.** A only
- **b.** B only
- **c.** both A and B
- **d.** neither A nor B

6. All of the following can cause *low* current readings during a cranking current draw test EXCEPT:

- **a.** a worn engine
- **b.** high resistance in the starting system
- **c.** a short circuit in the starter motor
- **d.** an undercharged battery

7. Mechanic A says that the ignition discharge current (measured with the engine off) must be added to alternator output current (measured with the engine running) to accurately test alternator output. Mechanic B says that regulated voltage should not be above 12.6 volts during an alternator current output test. Who is right?

- **a.** A only
- **b.** B only
- **c.** both A and B
- **d.** neither A nor B

8. During a cranking vacuum test, an engine has normal, steady cranking speed with low and slightly uneven vacuum readings. A probable cause is that:

- **a.** the engine has leakage past valves, rings, or the head gasket
- **b.** the engine has low compression or retarded valve timing
- **c.** the flywheel ring gear is worn or broken
- **d.** any of the above

9. During a vacuum test with the engine running, manifold vacuum is irregular (uneven) at idle. This probably indicates:

- **a.** worn rings, cylinders, or valves
- **b.** retarded ignition timing or low compression
- **c.** a restricted exhaust system
- **d.** sticking valves

10. During a vacuum test with the engine running, manifold vacuum drops during 15 seconds at high cruise and does not increase when the throttle closes. This probably indicates:

- **a.** worn rings, cylinders, or valves
- **b.** retarded ignition timing or low compression
- **c.** a restricted exhaust system
- **d.** sticking valves

11. Mechanic A says that you should disconnect and plug the EGR valve vacuum line before doing a power balance test. Mechanic B says that a power balance test can only be done at idle on an engine with electronic idle speed control. Who is right?

- **a.** A only
- **b.** B only
- **c.** both A and B
- **d.** neither A and B

12. During a power balance test, HC emissions are higher than normal overall and do not increase when one cylinder is killed. A probable cause is:

- **a.** a misfire in the killed cylinder

b. a rich air-fuel mixture

c. low compression in the killed cylinder

d. the EGR valve is stuck open

13. Mechanic A says that you should do a compression test on a gasoline engine with the throttle fully closed. Mechanic B says that you should crank an engine for at least four full strokes per cylinder during a compression test. Who is right?

a. A only

b. B only

c. both A and B

d. neither A nor B

14. During a compression test, compression in one cylinder is low on the first stroke and increases on the following strokes but does not reach specifications. A probable cause is:

a. a valve is burned

b. a valve is sticking

c. the rings or cylinder walls are worn

d. the combustion chamber has excessive carbon deposits

15. During a cylinder leakage test on one cylinder, air escapes from the oil filler cap. This probably indicates:

a. a burned exhaust valve

b. a cracked head

c. a worn cylinder or broken rings

d. a burned intake valve

16. During a cylinder leakage test on one cylinder, air escapes from the carburetor. This probably indicates:

a. a burned exhaust valve

b. a cracked head

c. a worn cylinder or broken rings

d. a burned intake valve

17. If an engine has poor compression test results but good leakage test results, it may have:

a. high-mileage cylinder or piston ring wear

b. ignition problems

c. fuel system problems

d. valve problems

18. All of the following are common causes of high CO emissions EXCEPT:

a. a rich air-fuel mixture

b. a stuck or misadjusted choke

c. an ignition misfire

d. a high float level in the carburetor

19. Mechanic A says that high HC emissions often indicate a lean air-fuel mixture or a misfire. Mechanic B says that high CO emissions often indicate a rich mixture. Who is right?

a. A only

b. B only

c. both A and B

d. neither A nor B

20. An engine tested with a 4-gas exhaust analyzer shows a CO_2 reading of 15% and an O_2 reading of 2% at idle. This indicates:

a. normal operation

b. a CO control problem

c. an HC control problem

d. a NO_x control problem

21. An engine tested at 2,500 rpm shows normal HC and high CO readings. A probable cause is:

a. an open spark plug cable

b. a choke stuck partly closed

c. a floating exhaust valve

d. a fouled spark plug

22. With a firing order of 1-5-3-6-2-4, number 1 cylinder is at tdc with both valves closed when which cylinder is at tdc with the exhaust valve fully open?

a. 4

b. 5

c. 3

d. 6

PART TWO

ENGINE LUBRICATION, VENTILATION, COOLING AND EXHAUST SYSTEMS

INTRODUCTION

When most motorists think of an oil and filter change, they think of "lube service;" they don't think of engine performance service. It's true, of course, that engine oil and filter changes are fundamental to complete vehicle lubrication. They also are fundamental to complete preventive maintenance and engine performance service. Complete car care begins on the lube rack.

Engine oiling affects engine wear and thus *all* factors of engine performance. All-new ignition and fuel system parts won't cure a problem caused by engine wear due to improper lubrication.

Changing the oil and filter may seem like the simplest of all service jobs. Selecting the correct oil, however, is essential for proper engine performance. Close attention to the details of oil changing and filter replacement can add thousands of miles to the operating life of an engine. Moreover, your ability to diagnose lubrication problems can solve many performance problems. Chapter 6 outlines these service skills.

The positive crankcase ventilation (PCV) system is an integral part of the engine lubrication system *and* the emission control system. Carmakers specify PCV services as regularly scheduled preventive maintenance. Chapter 6 also covers these tasks.

Many motorists take the cooling and exhaust systems of their cars for granted. As long as the engine doesn't overheat and the muffler doesn't make noise, everything must be okay. This is a tribute to the reliability of the engineering that has gone into these systems. As an automotive service professional, you know that these systems are not that simple. Incorrect engine temperature can upset the entire operation of an electronic engine control system. Improper exhaust flow and exhaust backpressure can affect air-fuel ratios and emission control in all engine-operating modes. Chapter 7 outlines the basic procedures for cooling system and exhaust system service.

6 ENGINE LUBRICATION AND VENTILATION SERVICE

INTRODUCTION

Engine oil and filter changes and crankcase ventilation service are important parts of engine maintenance. Some car owners don't think of these as "tuneup-related" services, but they are. Dirty, diluted, or contaminated oil or oil of the wrong viscosity or service classification can cause engine wear, poor performance, and increased emissions. The positive crankcase ventilation (PCV) system provides fresh air to the crankcase and removes blowby vapors. If the PCV system does not work correctly, emissions can increase and fuel economy, performance, and oil consumption can suffer.

This chapter explains how to change the engine oil and filter and outlines basic lubrication system testing and inspection. It also explains PCV system testing and service for gasoline and diesel engines.

GOALS

After studying this chapter, you should be able to do the following jobs in twice the flat-rate labor time, or less:

1. Drain and refill engine oil and change the filter.
2. Inspect engine oil, analyze symptoms of contamination, and determine possible causes.
3. Using a pressure gauge, test engine oil pressure.
4. Remove and replace valve cover gaskets and oil pan gaskets.
5. Inspect a PCV system for defective parts and improper operation; replace parts as required.
6. Perform vacuum and speed-drop tests on a closed PCV system and interpret the results.
7. Service a diesel PCV system.

OIL AND FILTER CHANGE

The major steps of an oil and filter change are:

- Raising and supporting the vehicle safely
- Draining the oil and removing the old filter
- Reinstalling the drain plug and installing a new filter
- Filling the crankcase with the proper grade and quantity of oil
- Checking for pressure and leakage

Raising and Lowering the Vehicle

Most shops have single-post, frame-contact hoists to raise vehicles for general under-chassis service, figure 6-1. Chapter 1 of this volume has instructions for raising a vehicle on a single-post hoist. If you use a jack and stands or a hoist other than a single-post type, check a service guide for instructions and ask your instructor for directions. If you use a single-post hoist, follow the instructions in chapter 1.

Figure 6-1. Following required safety procedures, raise the vehicle on a hoist.

Changing the Oil and Filter

Carmakers specify oil change mileage intervals that range from 2,000 to 7,500 miles. Time intervals range from 3 to 12 months. Generally, diesel and turbocharged engines require more frequent oil changes than normally aspirated gasoline engines do. Carmakers usually recommend that the filter be changed at every other oil change, but some specify that diesel and turbocharged engine filters should be changed at every oil change.

Figure 6-2. Use a socket or a box wrench to remove the drain plug, not an open-end or adjustable wrench.

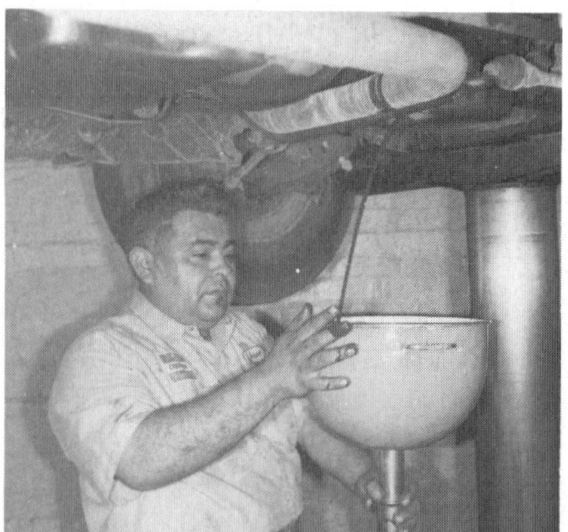

Figure 6-3. Do not splash hot oil onto yourself while draining it from the engine.

Figure 6-4. Sometimes, you can save stripped threads in an oil pan by installing a replacement plug with self-tapping oversize threads.

Carmakers also provide "severe-service" recommendations that are about half the "normal" service intervals. Most severe-service recommendations call for changing the filter at each oil change. Many owners don't realize it, but their cars qualify for severe-service maintenance if they are used for:

- Prolonged stop-and-go driving
- Frequent short trips of 10 miles or less
- Trailer towing
- Regular operation in climatic extremes of hot or cold weather or dusty conditions

Check the carmaker's recommendations and the owner's use of the vehicle to determine when to change oil. Many owners prefer to change the oil and filter more often than recommended. The oil and filter definitely should be changed if dirty or contaminated with fuel or coolant. Before changing the oil, warm the engine to operating temperature so that sludge and dirt are suspended in the hot oil and will be carried out of the engine with it. Change the oil and filter as follows:

1. Raise the vehicle properly and place support stands as required.

2. Place the drain oil container under the engine so that drain oil will flow into it.

> **Warning.** Hot oil may burn or irritate your skin. Do not let it splash while draining.

3. Using a socket or box wrench, loosen and remove the drain plug from the oil pan, figure 6-2.

4. While the oil is draining, examine the drain plug and gasket for stripped threads, cracks, and other damage. Replace a defective plug or gasket.

5. Also, inspect the oil pan gasket area, the front of the crankshaft behind the harmonic balancer, the rear of the oil pan, and the transmission or bellhousing for signs of motor oil or transmission oil leaks.

6. Allow 5 or 10 minutes for oil to drain from the engine, figure 6-3. Then wipe the drain hole area of the pan clean and check the pan for damaged threads. Check the plug fit by screwing it in about halfway by hand. It should turn freely without binding, but it should not wobble or slip when you pull it.

7. If the oil pan threads are stripped, you often can save them by installing an oversize replacement plug with hardened, self-tapping threads, figure 6-4. Install such a plug with a wrench until the gasket seats on the pan. Then remove it and clean any metal slivers from the plug and pan.

8. After all oil has drained from the engine, install the drain plug by hand until the gasket seats on the pan.

9. Move the drain oil container under the oil filter and remove the filter as follows:

a. Most engines have spin-on disposable filters, accessible from under the vehicle. Use a filter wrench, figure 6-5, to loosen the filter. Then unscrew it by hand and let oil drain into the container.

b. Some filters are more accessible from above the engine. To change one of these, lower the vehicle after draining the oil.

c. If the filter is installed vertically, with the closed end upward (Mazda rotary and Chrysler 6-cylinder engines), drive a hole into the end of the filter with a punch before draining oil from the engine. This allows old oil to drain easily from the filter before you remove it.

d. If the engine has an auxiliary bypass oil filter (some diesels and turbo engines), replace the cartridge and clean the filter housing, figure 6-6.

10. Lubricate the rubber gasket on the new filter with oil or light grease, figure 6-7, and clean the filter mounting base on the engine.

11. Screw the filter onto the mounting nipple by hand until the gasket contacts the mounting base, figure 6-8. Continue tightening the filter by hand or with a filter wrench another 2/3 to 1 1/4 turns. Follow the filter maker's directions. Do not overtighten.

12. Move the drain oil container back under the oil pan and remove the drain plug. Let any remaining oil in the pan drain into the container. Install the drain plug by hand until the gasket seats on the pan. Then use a socket or box wrench to tighten the plug until the gasket is just slightly compressed. Do not overtighten.

13. Remove the drain oil container from under the vehicle and lower the vehicle.

14. Fill the crankcase with the quantity, service classification, and viscosity of oil specified by the manufacturer. Check oil level on the dipstick after filling the engine. The dipstick will show a higher than normal level because the filter and oil galleries are not yet filled.

15. Start the engine, run it at idle, and check for oil pressure. If pressure does not rise on the gauge or if the warning lamp does not go out in 10 seconds or less, stop the engine and check for leaks.

16. If the engine has normal oil pressure, let it run for 2 or 3 minutes and check around the filter and drain plug for leaks.

17. Stop the engine and retract the hoist arms and lifting pads.

18. After 2 or 3 minutes, check the oil level on the dipstick, figure 6-9. Add more oil if needed to bring the level

Figure 6-5. This is one kind of filter wrench, or pliers, you can use to remove a filter cartridge.

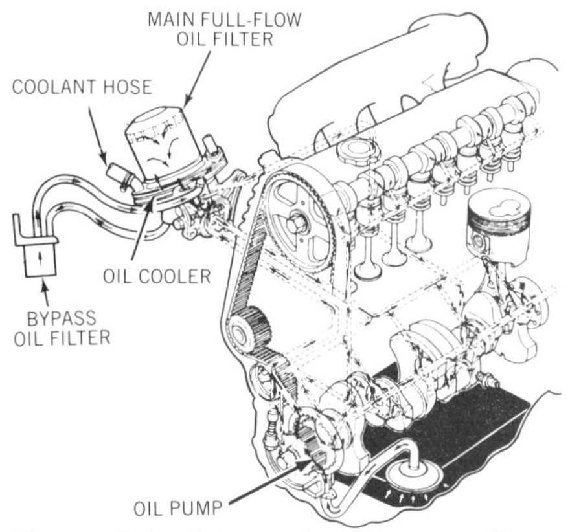

Figure 6-6. If the engine has an auxiliary bypass filter, in addition to the main full-flow filter, change it as well. (Ford)

Figure 6-8. Screw the filter onto the mounting nipple by hand. After the gasket contacts the base, tighten it another 2/3 to 1 1/4 turns.

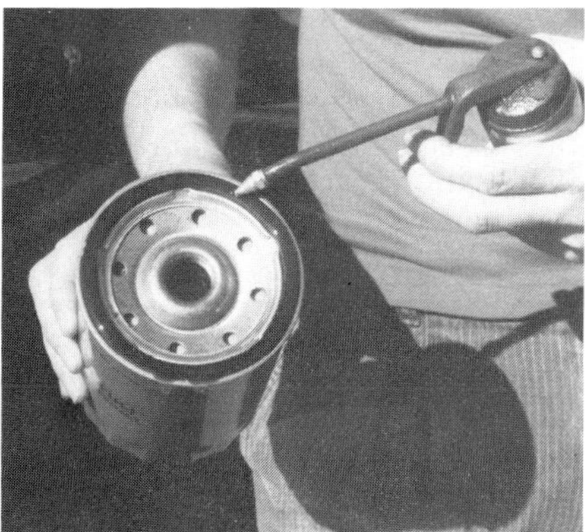

Figure 6-7. Lubricate the filter gasket with oil or light grease before installation.

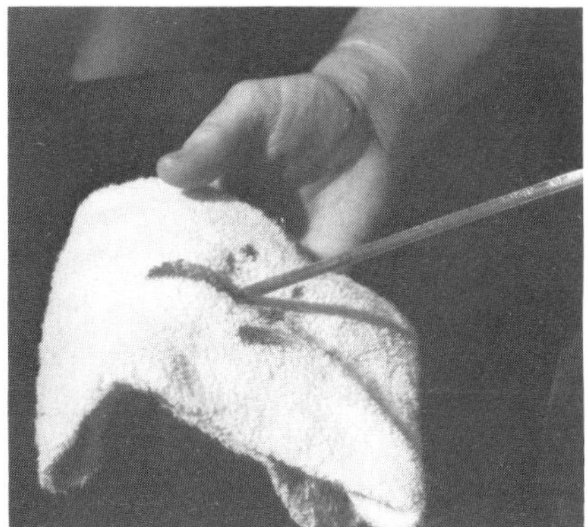

Figure 6-9. After stopping the engine, allow 2 or 3 minutes for oil to drain into the pan. Then check the level on the dipstick.

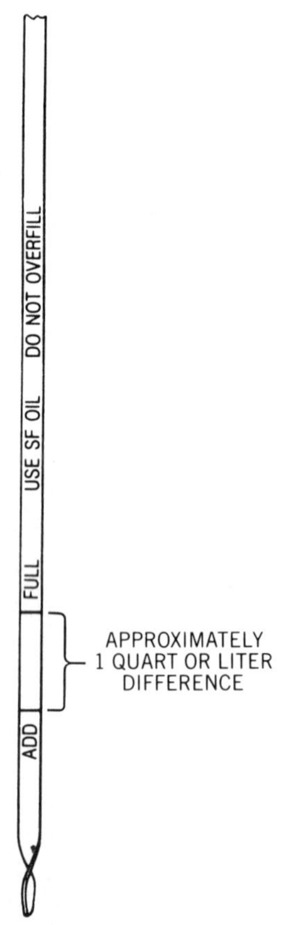

Figure 6-10. Typical dipstick markings.

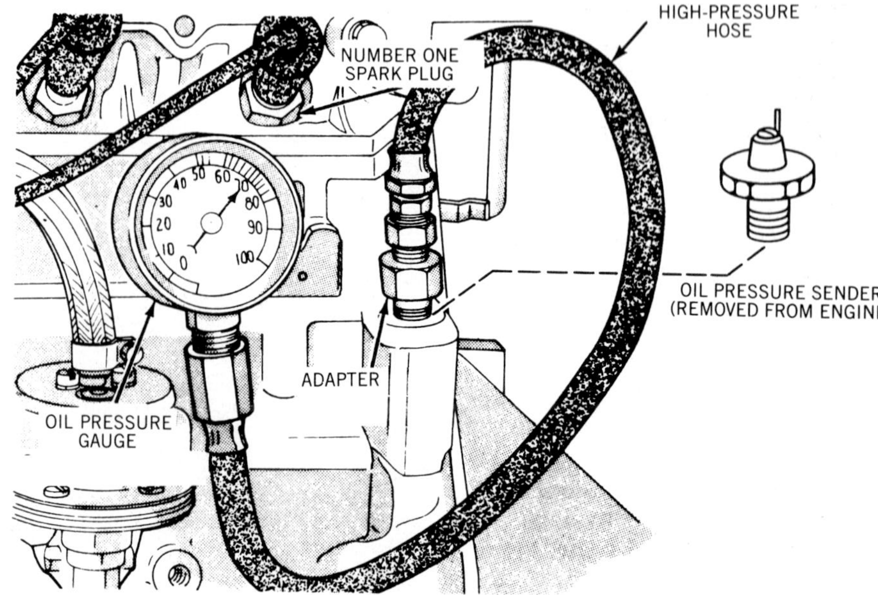

Figure 6-11. Remove the pressure sender and connect the gauge with a length of high-pressure hose or copper tubing and the necessary adapter fittings. (Chrysler)

to the full or safe mark, figure 6-10. Then drive the car off the hoist.

Drain Oil Analysis

Used motor oil can be analyzed in a laboratory for exact kinds and amounts of contamination, viscosity breakdown, and additive depletion. You probably won't need to do this very often for routine oil changes, but many truck, bus, taxicab, and construction equipment companies do it regularly. You can, however, look at the drain oil for telltale signs of engine problems:

• Shiny metallic particles indicate excessive bearing wear and possible failure.

• Heavy, thick, black oil can indicate excessive carbon or soot deposits. This can indicate that the oil simply has not been changed often enough. It also can indicate engine overheating.

• Ethylene glycol coolant leaking into the oil also will cause thickening. This may be accompanied by traces of water, foam, or sludge.

• Foam, traces of water, and excessive sludge combine to indicate a coolant leak. Test for water in the oil by putting 2 or 3 drops on a piece of aluminum foil and heating it with a match. If the oil snaps or crackles, it indicates water boiling as the oil heats.

• Oil that appears sludgy and gummy or that feels lumpy often indicates a sludge-filled engine. This can be caused by infrequent oil changes, continual low-temperature operation, or using oil of the wrong classification. A clogged PCV system also increases sludge formation

• Thin oil accompanied by a gasoline smell indicates fuel dilution. This can be caused by a continually rich mixture, cold operation, a continual misfire, or a fuel pump leaking into the crankcase.

OIL PRESSURE TESTING

Oil pressure that is too high or too low can increase engine wear and lead to major damage. It may not seem that high oil pressure can damage an engine, but it can. High pressure can erode soft bearing metal and lead to crankshaft or camshaft failure. It also can increase leakage around seals and even force plugs out of oil galleries and rocker arm shafts. High pressure can cause leakage around the filter or cause the filter to burst.

Low pressure can be indicated by:

• Knocking or rumbling noises from the crankshaft bearings

• A low-pressure reading on the instrument panel gauge or a warning lamp that flickers or stays on

• Excessive oil consumption

High pressure can be indicated by:

• Leakage around seals, external oil gallery plugs, and valve cover or pan gaskets

• A badly expanded, cracked, or leaking oil filter

• Excessive oil consumption

If you suspect oil pressure that is out of limits, test it as follows:

1. Locate the engine sending unit for the instrument panel gauge or warning lamp. It is screwed into the main oil gallery from the outside of the engine block, figure 6-11. It also may be on or near the filter adapter.

2. Start and run the engine to normal operating temperature. Carmakers' oil pressure specifications are for a com-

Low, or No, Oil Pressure

Cause	*Correction*
Low oil level	Add oil.
Low-viscosity or diluted oil	Change oil. Check drain oil for coolant or fuel dilution.
Disconnected oil pressure sender on engine	Reconnect sender wire.
Defective sender, warning lamp, or vehicle gauge	Test sender, lamp, or gauge electrically. Repair as required.
Oil pump problems Worn pump Broken or worn pump drive gear Clogged oil pickup Weak or broken relief valve spring; valve stuck open	Remove oil pan. Remove, inspect, and service oil pump.
Worn crankshaft or camshaft bearings	Remove oil pan. Check bearing clearances. Disassemble, inspect, and repair as required.
Internal oil leakage Broken oil line Loose or cracked oil gallery plug	Disassemble engine for further testing and service.

Figure 6-12. These are some general causes and cures for low oil pressure.

High Oil Pressure

Cause	*Correction*
Thick oil Wrong viscosity Contaminated	Change oil and filter. Check oil for contamination.
Oil pump relief valve broken, clogged, or stuck closed	Remove oil pan. Remove, inspect, and service oil pump.
Main oil gallery clogged	Disassemble engine for more testing and service.

Figure 6-13. These are the common causes and cures for high oil pressure.

Excessive Oil Consumption (Oil Burning)

Cause	*Correction*
Low-viscosity or diluted oil	Change oil. Check drain oil for coolant or fuel dilution.
High oil level	Lower the oil level.
PCV system clogged or PCV valve stuck wide open	Inspect and service PCV system.
Worn crankshaft or camshaft bearings	Remove oil pan. Check bearing clearances. Disassemble, inspect, and repair as required.
Internal oil leakage Broken oil line Loose or cracked oil gallery plug	Disassemble engine for further testing and service.
Worn or broken piston rings, worn cylinders	Disassemble engine for further testing and service.
Worn valve stems, guides, or valve oil seals	Disassemble engine for further testing and service.
Sludge buildup in oil drain holes of the cylinder heads	Remove valve covers, inspect, and clean as required.

Figure 6-14. Table of general causes and cures for high oil consumption.

pletely warm engine. Pressure will be high when the oil is cold.

Caution. Most sending units require special sockets for removal and installation. Do not use a 12-point socket or you may damage the sender.

3. Disconnect the wire from the sending unit and use a suitable wrench to remove the unit.

4. Look up the carmaker's pressure specifications and select a pressure gauge with a midscale range that includes the highest specification. Usually, a zero to 75- or 100-psi (500- or 700-kPa) gauge is satisfactory. Manufacturers give pressure specifications at a specified rpm or at idle and medium cruising speed.

5. Using a length of high-pressure hose or copper tubing and suitable fittings, connect the gauge to the sending unit opening in the engine, figure 6-11.

6. Start the engine and run it at slow idle for about 1 minute. Compare the gauge reading to specifications.

7. Slowly accelerate the engine to 1,500 to 2,000 rpm and let the speed and pressure stabilize. Compare the gauge reading to specifications.

8. Stop the engine, remove the pressure gauge and reinstall and reconnect the oil pressure sending unit.

Oil Pressure and Consumption Diagnosis

The most obvious cause of low oil pressure is low oil level. Before looking for serious engine problems, check the oil level on the dipstick and be sure the engine is properly full. Oil of the wrong viscosity (too light) or oil that is diluted with fuel or coolant also can cause low pressure. Refer to figure 6-12 for further diagnosis. Internal defects and engine wear require that you tear down the engine for a complete diagnosis and repair.

The most obvious cause of high oil pressure is oil that is too heavy. The viscosity may be too high; the owner may have added an improper "oil conditioner"; the oil may be dirty and thickened. If you know that the oil is clean and of the right viscosity, the oil pump relief valve probably is stuck fully or partly closed or restricted. Figure 6-13 lists these and other possible causes.

An engine can consume oil in only two ways: by leaking it or by burning it. You will learn general ways to correct some common oil leaks in the next section. There is no firm definition for "normal" or "high" oil consumption. A worn engine can burn and consume more oil than an engine in good condition, but that can still be "normal" for a high-mileage engine. Some engines may not visibly consume any oil between changes. On the other hand, some manufacturers say that normal consumption can be 1 quart every 1,000 miles. Some even say that 1 quart every 500 to 700 miles may be normal for a high-mileage but otherwise sound engine. Figure 6-14 lists some common engine problems that can cause oil burning. Engine disassembly or pinpoint engine testing may be needed to isolate and verify these causes.

REPAIRING OIL LEAKS FROM VALVE COVER AND PAN GASKETS

An engine can leak oil at any place there is an opening into the engine. This includes front and rear crankshaft seals, camshaft seals on overhead-cam engines, oil filters, external oil lines, oil pressure sending units and switches, pushrod covers on inline engines, and front timing covers. Two common places for oil leakage are at valve covers and oil pan gaskets. That's why carmakers often include the jobs of checking and tightening valve cover and oil pan screws in their maintenance schedules. Occasionally, a valve cover or pan gasket may break and leak badly enough that it must be replaced. In fact, a gasket can be damaged by overtightening the cover or pan screws. These gaskets are usually soft rubber or cork. If the screws are overtightened, the gaskets can be compressed too much and broken.

If you have to remove and replace a valve cover or oil pan to repair a gasket leak or for any other reason, follow the carmaker's instructions and the guidelines in the following sections.

Figure 6-15. Most valve covers are made of sheet metal and secured with machine screws.

Figure 6-16. Some valve covers and many ohc cam covers are made of cast aluminum.

Valve Cover Removal and Replacement

The valve covers of most ohv engines are stamped from sheet metal and fastened to the head by machine screws around the flange, figure 6-15. The cam covers of ohc engines also can be made of sheet metal, but many are made of cast aluminum, figure 6-16. Cam covers on ohc engines can be secured with screws around the flange or by nuts on studs that pass through the top. Either kind of cover is sealed to the head by a soft rubber or cork gasket or by aerobic sealant (RTV rubber). Remove and replace a valve cover as follows:

1. Remove any accessories, brackets, or hoses that interfere with valve cover removal, figure 6-17. If you remove the air cleaner, cover the carburetor or manifold inlet to keep anything from falling into it.

2. If spark plug cables cross the valve cover, disconnect them from the plugs and lay them out of the way, figure 6-18.

3. Remove the screws or nuts that secure the cover to the head.

4. Gently tap the cover with a rubber hammer to loosen it, figure 6-19. Do not pry with a screwdriver, which can nick or bend the cover and create a leak.

5. Remove the cover and use a scraper or wire brush to carefully clean old gasket material or sealant from the cover flange, figure 6-20. Be careful not to nick the flange.

6. Using a scraper or wire brush, clean old gasket material or sealant from the head. Do not let dirt fall into the engine.

Figure 6-17. Remove any brackets, clamps, hoses, and other accessories that interfere with valve cover removal.

Figure 6-18. If spark plug cables cross the valve cover, disconnect them from the plugs.

Figure 6-19. Gently tap the cover with a rubber hammer to loosen it.

Figure 6-20. Remove old gasket material from the valve cover with a scraper or a wire brush. (Fel-Pro)

Figure 6-21. Straighten the cover flange and screw holes with a ball peen hammer. (Fel-Pro)

Figure 6-22. Apply gasket cement evenly to the cover flange, let it dry, and then press the gasket into place. (Fel-Pro)

7. Inspect the oil return (drainback) holes in each end of the head for clogging. Carefully remove any deposits with a small scraper and round wire brush. Do not let dirt particles fall into the engine.

8. Wash the cover and gasket surface of the head with solvent and wipe clean with a clean cloth.

9. Inspect the cover for nicks, cracks, and distorted screw holes. Remove burrs and nicks with a small file. If the flange of a sheetmetal cover is distorted around the screw holes, straighten it with a ball peen hammer before reinstallation, figure 6-21.

10. If you are installing a rubber or cork gasket, follow the manufacturer's directions and these guidelines, figure 6-22:

- **a.** Apply a thin coat of the specified gasket cement to the cover flange and let it dry until it is tacky.
- **b.** Place the gasket onto the flange and press it into place.

11. If you are installing the cover with aerobic (RTV) sealant, follow the maker's directions and these guidelines:

- **a.** Clean and dry the cover flange and the head surface.
- **b.** Apply a uniform bead of sealant to either the cover flange or head surface, figure 6-23, or both, as directed.

12. Reinstall the cover on the head and tighten the screws or nuts to the carmaker's specified torque, in the specified sequence. Do not overtighten.

13. Reinstall all accessories, brackets, and hoses removed in step 1 and reconnect spark plug cables disconnected in step 2.

14. Start the engine and check for leakage. Add oil if necessary.

Oil Pan Removal and Replacement

Remove and replace an oil pan as follows:

1. Raise the vehicle on a hoist as you would for an oil change.

2. Drain the oil. If the oil is fresh, you can save it for reuse.

3. Following the carmaker's procedures, loosen or remove any of the following parts to provide clearance for oil pan removal:

- **a.** Motor mounts
- **b.** Exhaust crossover pipe
- **c.** Steering linkage
- **d.** Front-wheel-drive axles or transaxle parts
- **e.** Fuel, brake fluid, and oil cooler lines

4. Remove the oil pan screws. The gasket will hold the pan to the engine as you remove the screws, figure 6-24.

5. Tap the pan gently with a rubber hammer to loosen it. You may have to lower the pan 2 or 3 inches and reach inside to remove the oil pump pickup before removing the pan completely. Check the carmaker's procedures.

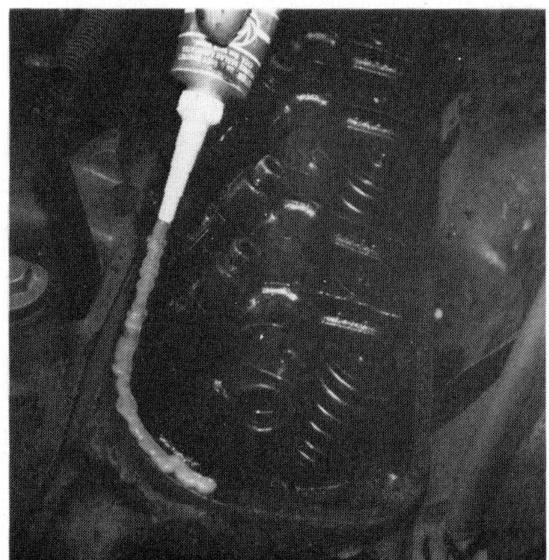

Figure 6-23. Apply a uniform bead of RTV sealant to the head and let it dry a few minutes before installing the valve cover.

6. Remove the oil pan.

7. Using a scraper or wire brush, remove old gasket material and sealant from the pan flange and lower surface of the engine, figure 6-25.

8. Wash the pan and lower surface of the engine with solvent. Wipe them dry with a clean cloth.

9. Inspect the pan for cracks, dents, and other damage. If necessary, remove dents and straighten the pan flange around the screw holes with a ball peen hammer. Remove nicks with a small file.

10. Remove the oil pump pickup screen, clean it in solvent, figure 6-26, blow it dry with compressed air, and reinstall it.

Figure 6-24. Remove the screws; then tap the pan with a rubber hammer to loosen it.

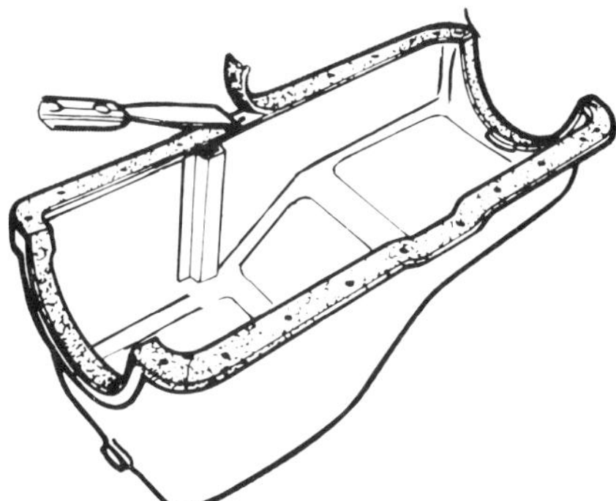

Figure 6-25. Clean the oil pan flange and the bottom of the block with a scraper or a wire brush.

Figure 6-26. While the pan is off the engine, clean the oil pump pickup screen.

11. If you are installing a rubber or cork gasket, follow the maker's directions and these guidelines:

a. Apply a thin, uniform coat of gasket cement to the pan flange and end surfaces. Allow it to dry until tacky.

b. Place the gasket on the flange and press it into place.

c. Install end seals on the pan or on the lower surface of the engine (main bearing caps), as directed, figure 6-27.

12. If you are installing the pan with aerobic (RTV) sealant, follow the maker's directions and these guidelines:

a. Clean and dry the pan flange and the lower engine surface.

b. Apply a uniform bead of sealant to either the pan flange or engine surface, or both, as directed.

13. Install the pan (and oil pickup, if removed) on the engine and tighten the screws evenly in the specified sequence to the specified torque. Do not overtighten.

14. Reinstall parts removed in step 3 and lower the vehicle.

15. Refill the engine with oil and check the oil level.

16. Start the engine and check for oil pressure and leaks.

Rear Main Bearing Oil Seal Replacement

Oil leaks often occur at the rear main bearing seal. You may have to replace the rear main seal, as well as the pan gasket, to cure a leak. To change most rear seals, you will have to remove the pan as explained above. You can replace some seals with the engine in the car and the crankshaft installed. Some engines, however, require that the crankshaft be removed. The manufacturer's procedures are necessary for seal replacement because of variations in seal retainers and engine installations. The following sections, however, outine the most common seal installations and service methods.

Most rear main seals are one of the following kinds:

• A 2-piece wick, or "rope," seal. One-half is installed in a groove at the rear of the block. The other half is installed

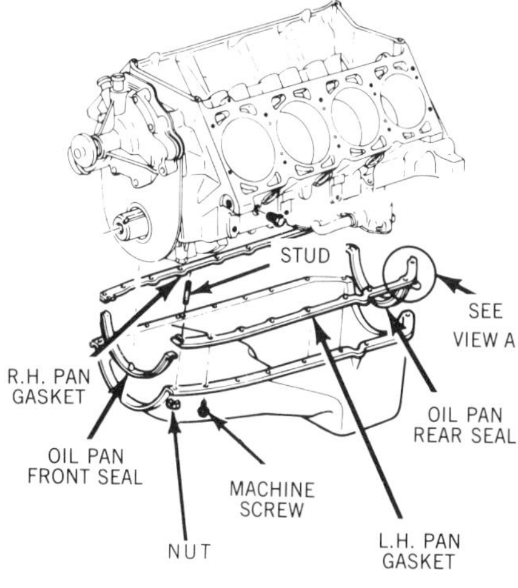

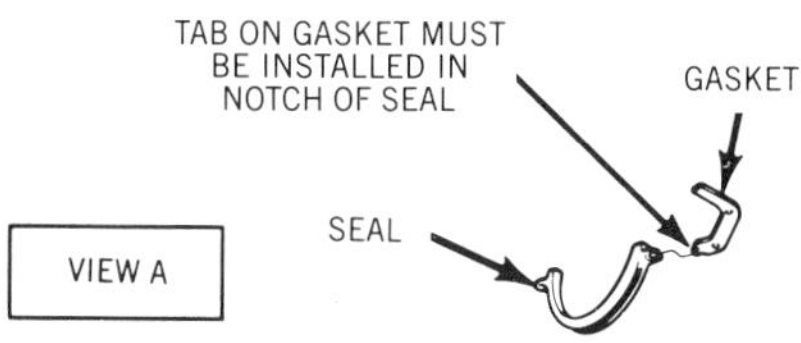

Figure 6-27. Install the pan end seals carefully, according to the gasket manufacturer's directions. (Cadillac)

in a matching groove in the rear main bearing cap.

- A 1-piece lip seal, held in a retainer at the rear of the crankshaft.
- A 2-piece lip seal, installed similarly to a rope seal.

Wick Seal Service. One common way to correct a leak at a wick, or rope-type, rear main seal includes these basic steps:

1. With the oil pan and rear main bearing cap removed, use a seal packing tool to compress the upper half of the seal more tightly into the block groove, figure 6-28. Do this at both sides of the seal.

2. Measure the depth that you have compressed the seal from the parting line surface of the block on both sides. This is usually 1/4 to 3/4 inch, depending on the condition of the seal.

3. Add 1/16 inch to the measurements taken in step 2 and cut sections of these lengths from the old lower seal. Remove the bearing insert from the bearing cap and use the cap as a cutting fixture, figure 6-29.

4. Use two small screwdrivers to work the cut seal pieces into both ends of the cylinder block groove. Compress them with the packing tool and cut off any excess length with a razor blade.

5. Install a new lower seal half in the main bearing cap with a seal installer, figure 6-30. Rotate the installer and cut off any excess seal length with a razor blade.

6. Following the carmaker's instructions place the specified sealer on the ends of the seal and the bearing cap parting line surfaces, figure 6-30.

7. Reinstall the insert in the bearing cap and install the cap on the engine. Torque to specifications.

One-Piece Lip Seal Replacement. To replace a 1-piece lip seal, you must remove the transmission or transaxle and the flywheel or flexplate. The seal is either pressed into the rear of the block or held in a retainer that bolts to the rear of the block, figure 6-31. If the seal is pressed into the block, remove it in one of two ways:

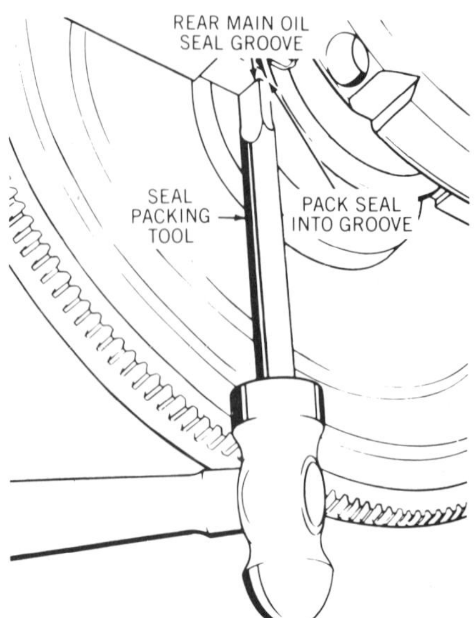

Figure 6-28. Pack the old upper rope seal tightly into the block groove. (Cadillac)

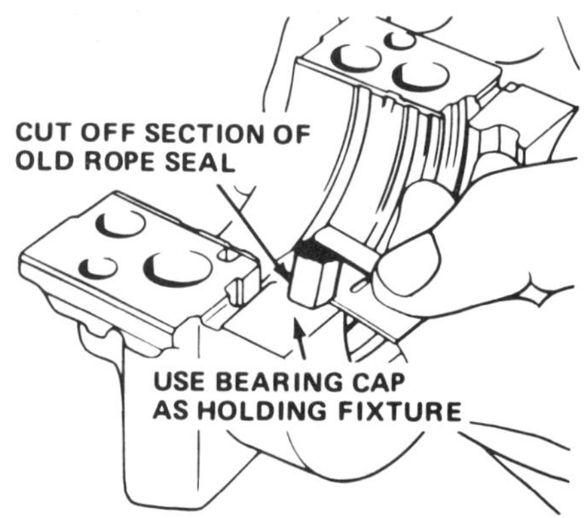

Figure 6-29. Cut sections of the old lower seal to pack into the ends of the upper block groove. (Cadillac)

1. Pry it from its bore with a suitable prying tool, using the crankshaft flange for leverage.

2. Drill a small hole in the steel body of the seal and remove it with a slide hammer.

In either case, do not scratch the seal bore in the block.

If the seal is held in a retainer, remove the retainer from the block and remove the seal with a seal driver, figure 6-31.

> **Caution.** Be sure that the lip of the seal faces inward, toward the front of the engine. Installing a seal backward will create a severe oil leak.

Install a new seal into the block or the seal retainer with a seal driver.

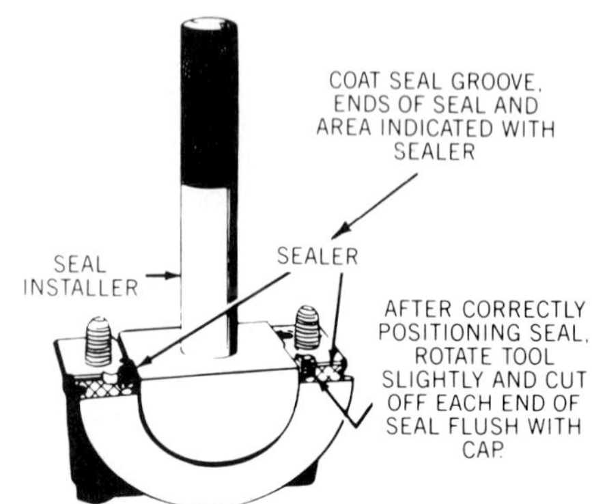

Figure 6-30. Pack the new lower seal into the main bearing cap with a seal installer. (Cadillac)

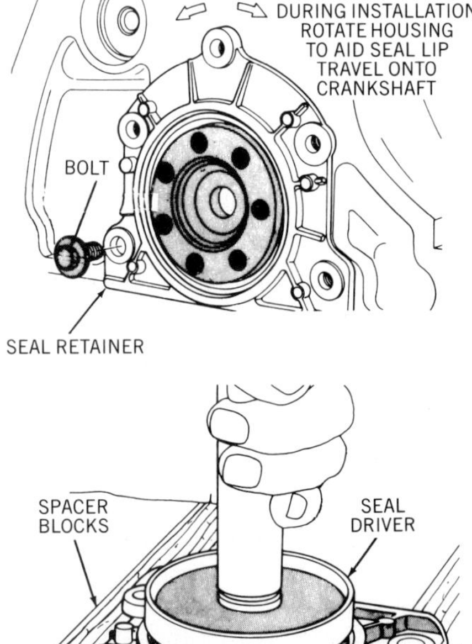

Figure 6-31. Remove the seal retainer from the rear of the block and drive out the old seal. (Chrysler)

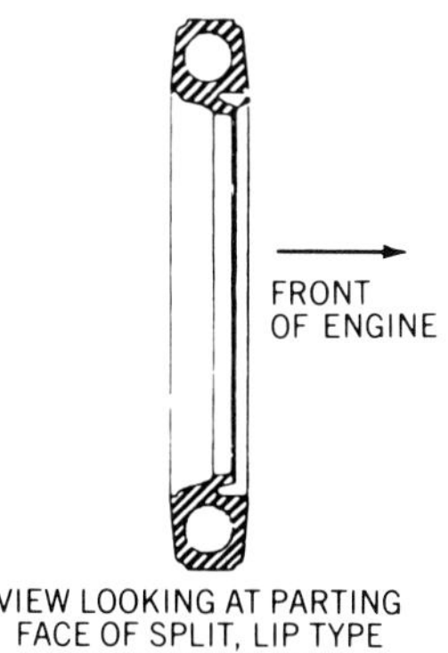

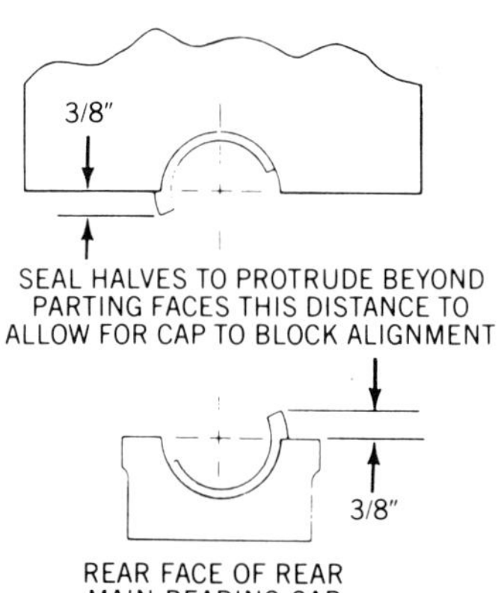

Figure 6-32. Typical split lip seal installation. (Ford)

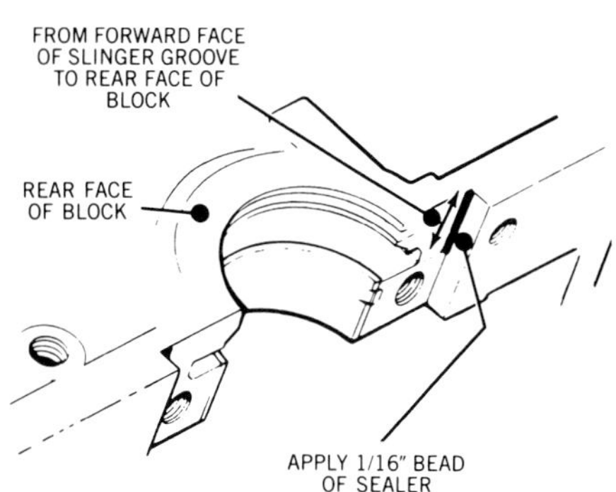

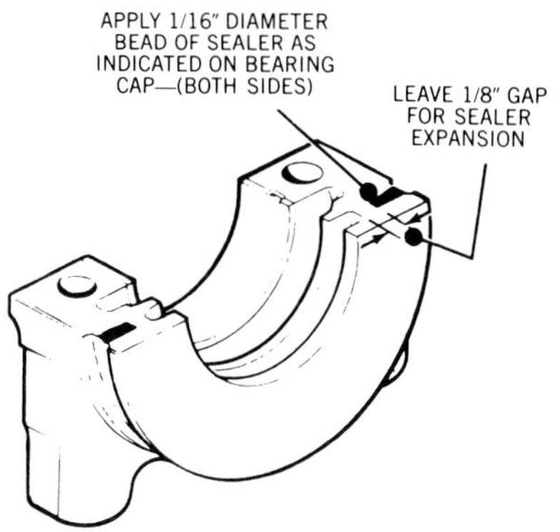

Figure 6-33. Sealant use varies for different engines. Check the manufacturer's instructions. (Ford)

Two-Piece Lip Seal Replacement. Some carmakers recommend that you loosen all main bearing caps to let the crankshaft drop slightly—no more than 1/32 or 0.030 of an inch. Other carmakers do not specify loosening the crankshaft. Follow these guidelines and the manufacturer's directions to replace 2-piece lip seal:

1. With the oil pan and rear main bearing cap removed, carefully force the upper half of the seal into the block with a small punch or screwdriver.

2. When the other end of the seal protrudes from the block, grasp it with pliers and carefully pull it from its groove.

> **Caution.** During installation, do not scrape the back of the upper seal against the cylinder block groove. An oil leak will result.
>
> Be sure that the lip of the seal faces inward, toward the front of the engine. Installing a seal backward will create a severe oil leak.

3. Lubricate the new upper seal as specified and install it in the block. Rotate the crankshaft to ease installation. On some engines, the seal should extend below one side of the block so that the parting lines of the upper and lower seals do not align with the parting line of the main bearing, figure 6-32.

4. Install the lower half of the seal in the main bearing cap. Apply sealant to specified areas of the seal and the bearing cap, figure 6-33.

5. Install the main bearing cap and torque to specifications.

PCV SYSTEM INSPECTION AND SERVICE

Correct PCV operation is essential for proper engine lubrication and emission control. A clogged PCV system will increase oil contamination, sludge formation, and emissions. If the PCV valve is broken or stuck fully open, manifold vacuum can draw excessive oil vapor into the engine and increase oil consumption. On a car with feedback-controlled fuel metering, this can upset overall engine operation. If the PCV system is clogged, it also can increase oil consumption. High crankcase pressure caused by a clogged system will force blowby vapors out of the oil filler, the dipstick tube, valve cover gaskets, and other crankcase openings. Blowby also can back up into the air cleaner, clog the air filter, and cause continually rich air-fuel mixtures.

Vehicle maintenance schedules call for regular inspection and maintenance of the PCV system, including PCV valve replacement. PCV service begins with basic inspection.

PCV System Inspection

1. Inspect all hoses for cracks, brittleness, loose connections, and clogging, figure 6-34.

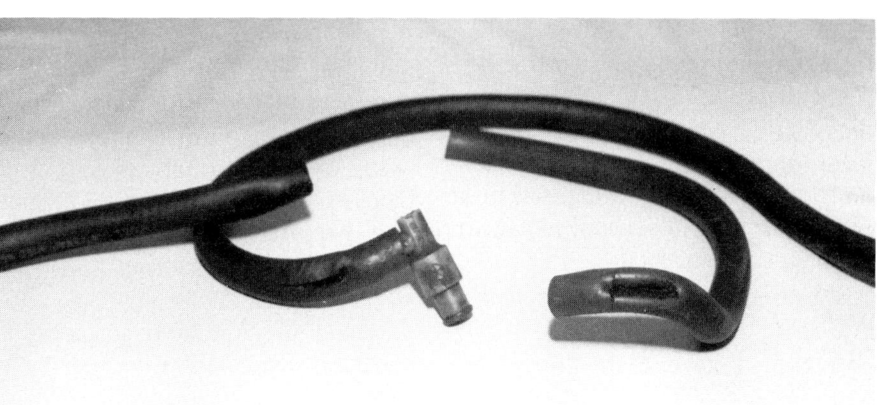

Figure 6-34. Inspect PCV hoses for cracks, brittleness, loose connections, and leakage.

Figure 6-35. Inspect the carburetor air filter for dirt from excessive blowby.

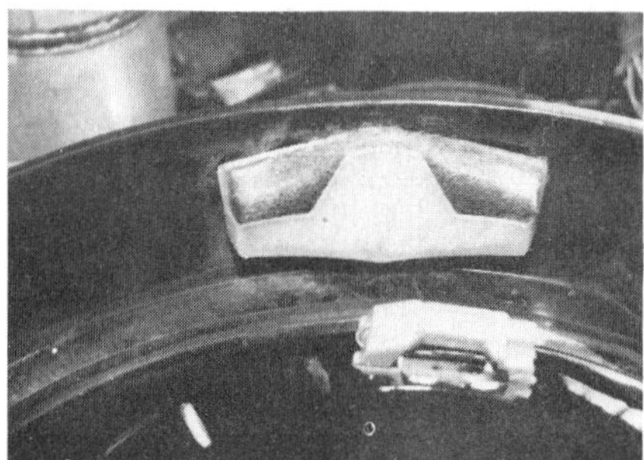

Figure 6-36. Many PCV systems have inlet air filters inside the engine air cleaner housing.

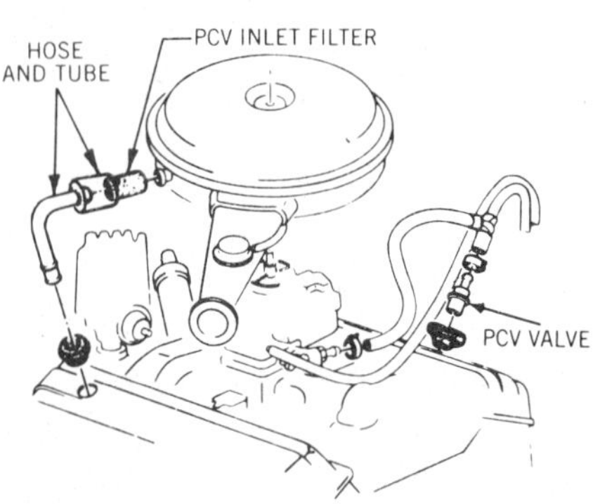

Figure 6-37. Some systems have the inlet filter inside the inlet hose.

Figure 6-38. Chrysler V-8's and a few other engines have a wire-mesh PCV filter in the oil filler cap.

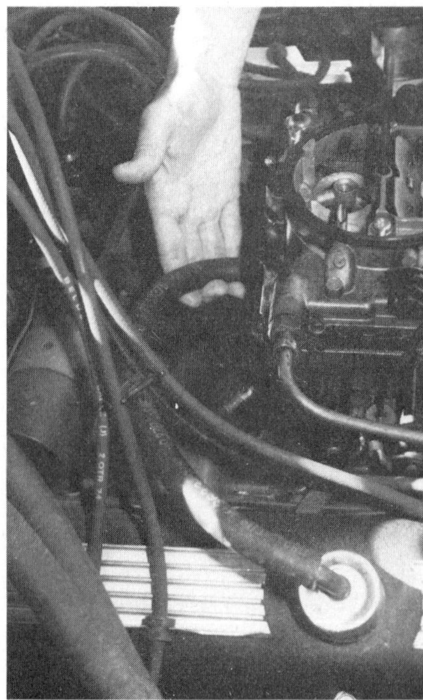

Figure 6-39. If the manifold inlet port for the PCV system is clogged, clean it by hand with solvent and a wire brush or a drill bit.

2. Inspect the engine intake air filter for excessive crankcase blowby deposits, figure 6-35.

3. Locate the crankcase ventilation filter in one of the following locations and inspect it for clogging. Clean a wire-mesh filter in solvent; replace a foam-type filter:

a. In the intake air cleaner, figure 6-36.

b. In the crankcase inlet hose, figure 6-37.

c. In the valve cover oil filler cap, figure 6-38.

4. Inspect PCV inlet port at the carburetor base or intake manifold, figure 6-39, for deposits that can restrict or block PCV flow to the manifold. If this port is restricted, the system will not work right, even if the hoses, filter, and valve are in good shape.

5. Inspect the PCV valve for deposits and clogging. Shake it to see if the valve plunger moves freely. If the valve does not rattle, it probably is clogged, but just because it moves freely does not guarantee that it is working correctly. To verify correct valve and system operation, do one of the following operating tests.

PCV Vacuum Test

Several testers are available to measure crankcase vacuum with the engine running at idle. This indicates PCV system airflow. To use one of these testers, remove the oil filler cap from the valve cover, install the tester in the opening, figure 6-40, and note the reading.

If you don't have a tester available, you can make a quick check of the system by placing a small card (such as a business card or a 3 × 5 index card) over the oil filler opening with the engine running, figure 6-41. Crankcase vacuum should be just strong enough to hold the card onto the opening. If the card does not stick to the opening, or if it is blown away, it indicates pressure in the crankcase. This means that the system is restricted.

If either the PCV tester or the card test indicates pressure instead of vacuum in the crankcase, look for a restricted PCV valve, hose, or port. Crankcase pressure also can be caused by excessive blowby on a worn engine, even with the PCV system unrestricted.

Figure 6-40. This PCV system tester indicates pressure or vacuum in the crankcase. (Atlas)

Figure 6-41. If crankcase vacuum holds a small card against the oil filler opening, the PCV system probably is working correctly.

Figure 6-42. Pull the PCV valve out of the valve cover grommet.

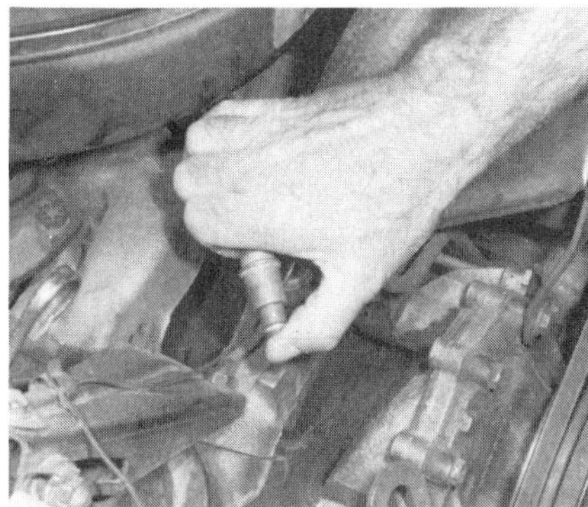

Figure 6-43. You should feel strong vacuum when you put your finger over the PCV valve.

PCV Vacuum and Speed-Drop Test

Use the following method to verify PCV system airflow and valve operation:

1. Connect a tachometer and run the engine at idle.

2. Disconnect the PCV valve and line from the crankcase opening, usually by pulling the valve out of the engine valve cover, figure 6-42.

3. Listen for a hissing sound as air passes through the valve and note a slight increase in idle speed (50 to 75 rpm).

4. Put your finger over the valve, figure 6-43. You should feel a strong vacuum and note a speed drop of 50 to 75 rpm below normal idle speed.

If speed does not increase about 50 rpm when the valve is open to the atmosphere and decrease about 50 rpm from normal idle when you block the valve, the valve, the hose to the manifold, or the manifold port is probably restricted. A speed change of less than 40 or 50 rpm also can indicate that the wrong PCV valve is installed in the engine. PCV valves are made with different flow rates for different engines. Check the part number of the valve against the carmaker's or valve manufacturer's part number for that engine.

On many Ford products and some other cars, the evaporative emission control (EEC) vapor canister is purged through the PCV line. If the PCV system on one of these engines fails the vacuum and speed-drop test, disconnect and plug the purge hose from the canister, figure 6-44, and repeat the test. If the PCV system now responds

Figure 6-44. If the vapor canister purge hose connects to the PCV line, disconnect it and plug the opening to doublecheck PCV valve operation.

Figure 6-45. Some PCV valves are installed through the intake manifold to the lifter valley.

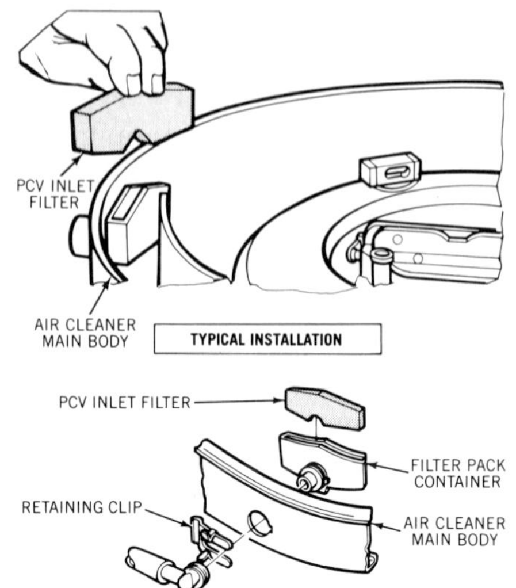

Figure 6-46. Disconnect the hose, remove the retaining clip, and remove the old filter.

correctly, look for a leak in the canister purge hose.

PCV Cranking Vacuum Test

You can verify airflow or lack of airflow through a PCV valve with a cranking vacuum test:

1. Connect a vacuum gauge to the intake manifold and disable the ignition as instructed in chapter 5.

2. Crank the engine and note the vacuum gauge reading.

3. Disconnect the PCV valve from the engine valve cover or other crankcase inlet and plug it with your finger, figure 6-43. Crank the engine and again note the vacuum reading.

4. Compare the vacuum readings in steps 2 and 3. If the reading in step 3 is not higher, the valve, hose, or manifold port is restricted.

PCV Valve and Hose Replacement

Most PCV valves are pressed into a grommet in the engine valve cover, figure 6-42, or installed inline in the crankcase vent hose. On some V-type engines, the valve may be installed through the intake manifold or valley cover to draw vapors from the lifter valley, figure 6-45. On some 4-cylinder engines, the valve may be in a hose from the valve cover to the EEC vapor storage canister, figure 6-44. If the valve location is not easily apparent, find the crankcase vent hose connection at the carburetor or intake manifold and trace it backward to locate the valve.

To replace a valve, simply remove the old valve from the hose or grommet and press the new one into place. If the valve is marked with a flow arrow, it should point toward the intake manifold. Check the manufacturer's part numbers to be sure you install the correct valve.

Most PCV hoses simply press onto nipples of PCV valves and fittings. Some hoses are secured with clamps.

Caution. When replacing a PCV hose, use only hose material that is marked for PCV system use. Vacuum and heater hose material may not be compatible with oil vapors and will deteriorate and block the PCV system.

To replace a PCV hose, remove the old one, cut a new hose to length, and install it. Install new hose clamps, if required.

PCV Inlet Filter Service

Some older PCV systems relied on the engine air filter to clean the crankcase ventilation air. In these systems, the crankcase inlet hose is connected to the "clean side" of the engine air cleaner, and has a wire-mesh flame arrester to protect the crankcase from an intake backfire. These systems do not require special filter service, but you should clean the flame arrester periodically in solvent.

Other systems have a wire-mesh filter in the crankcase air inlet. This often is in the oil breather cap, figure 6-38. Clean this filter in solvent at specified intervals and allow it to dry in the air. Do not use compressed air because the air pressure may damage the wire mesh. If you can't clean the filter satisfactorily, replace it.

Most late-model PCV systems have polyurethane foam or plastic gauze inlet air filters that must be replaced at specified intervals. Most are in the air cleaner housing. To replace these filters:

1. Remove the air cleaner cover.

2. Disconnect the PCV inlet hose from the filter nipple.

3. Remove the retaining clip from the

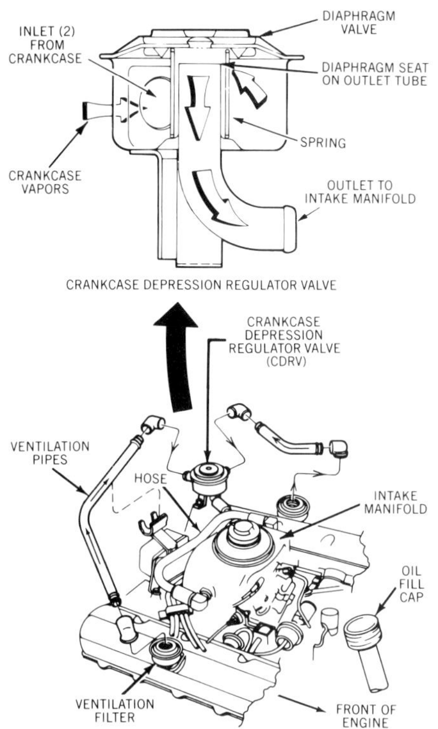

Figure 6-47. GM diesels use this valve to control PCV flow rate from the crankcase. (Cadillac)

filter and remove the filter from the air cleaner, figure 6-46.

4. Install the new filter with a new retaining clip and reconnect the hose.

Diesel Crankcase Ventilation Service

A diesel engine crankcase must be ventilated just as a gasoline engine crankcase is. Diesel PCV systems,

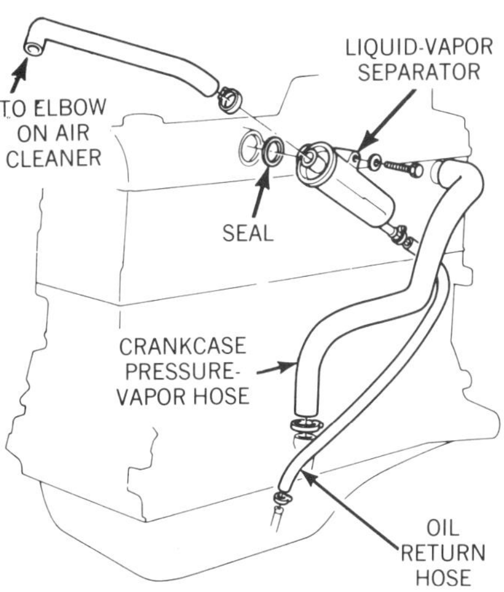

Figure 6-48. Do not remove or bypass a liquid-vapor separator in a diesel PCV system. (Ford)

however, do not have the same kinds of PCV valves as gasoline engines because intake manifold depression (vacuum) is less than on a gasoline engine. Many diesels simply have a tube or hose that connects the crankcase to the intake manifold or air cleaner. GM V-type diesels have a crankcase depression regulator valve (CDRV). It is a diaphragm-operated valve that regulates crankcase ventilation airflow to the intake manifold, figure 6-47. GM diesels also have inlet air filters and check valves in the engine valve covers, and hoses or pipes to connect the ventilation lines to the intake manifold, figure 6-47.

Inspect and clean diesel PCV components according to the carmaker's service schedule. Do not use solvent on the diaphragm of a GM CDRV; it will damage the diaphragm. If the engine has a liquid-vapor separator in the PCV system, figure 6-48, clean it and allow it to dry before reinstalling it.

Caution. Do not remove or bypass a liquid-vapor separator in a diesel PCV system. If excessive blowby or oil enters the intake manifold, the engine will run on its own crankcase oil, and extensive damage may result.

SUMMARY

Oil and filter changes are the most basic engine services. Regular oil changes reduce engine wear, contamination, and sludge formation. Clean motor oil also minimizes HC emissions, both from crankcase blowby to the atmosphere and from blowby routed to the intake manifold by the PCV system.

The PCV system eliminates crankcase HC emissions and provides continuous fresh air ventilation for the engine. PCV systems require periodic inspection and replacement of worn or damaged parts. The systems are trouble free and require no adjustment. Most carmakers call for PCV valve replacement at regular intervals.

Diesel PCV systems do not have the same kinds of valves as gasoline engine systems do, but they require periodic inspection and cleaning.

REVIEW QUESTIONS

1. Which of the following vehicles would require the most frequent oil changes?

- **a.** A naturally aspirated gasoline engine vehicle used mostly for long trips.
- **b.** A turbocharged diesel engine vehicle used for stop-and-go driving.
- **c.** A naturally aspirated diesel engine vehicle used under normal service conditions.
- **d.** A turbocharged gasoline engine used under normal service conditions.

2. Mechanic A says the engine and oil should be warmed to operating temperature before the oil is drained. Mechanic B says the oil should be drained after the vehicle has sat for several hours so that all of the oil is out of the upper engine areas. Who is right?

- **a.** A only
- **b.** B only
- **c.** Both A and B
- **d.** Neither A nor B

3. Mechanic A says an oil filter should be tightened at least 4 full turns after the gasket has contacted the sealing surface. Mechanic B says the gasket on a new oil filter must be left dry to seal properly. Who is right?

- **a.** A only
- **b.** B only
- **c.** Both A and B
- **d.** Neither A nor B

4. Fuel dilution of the crankcase oil can be caused by any of the following problems EXCEPT:

a. excessive bearing wear
b. leaking fuel pump
c. cooling system thermostat stuck open
d. defective (misfiring) spark plug.

5. Too-low oil pressure can be indicated by:

a. leaking oil filter
b. poor fuel economy
c. excessive oil consumption
d. engine running too cold.

6. Mechanic A says an oil pressure check should be made when the engine is cold. Mechanic B says oil pressure should be checked at maximum engine speed. Who is right?

a. A only
b. B only
c. Both A and B
d. Neither A nor B

7. A vehicle is brought in for service because the oil pressure warning light flickers during normal operation. Mechanic A says the engine oil may be of the wrong viscosity. Mechanic B says the engine oil level should be checked. Who is right?

a. A only
b. B only
c. Both A and B
d. Neither A nor B

8. Which of the following would NOT be considered normal oil consumption?

a. No visible consumption between changes
b. 1 quart every 200 miles
c. 2 quarts every 1,000 miles
d. 1 quart every 700 miles

9. Mechanic A says oil leakage at a valve cover gasket can often be stopped by tightening the cover screws more than specified. Mechanic B says valve cover leakage may be caused by not cleaning off old gasket material before installing a new gasket. Who is right?

a. A only
b. B only
c. Both A and B
d. Neither A nor B

10. Mechanic A says a damaged sheet-metal valve cover flange can sometimes be repaired. Mechanic B says such damage may be caused by not letting RTV sealant cure before installation. Who is right?

a. A only
b. B only
c. Both A and B
d. Neither A nor B

11. Which of the following is NOT a part of oil pan gasket replacement?

a. Draining the engine oil
b. Washing out the oil pan
c. Applying gasket sealant
d. Prying the pan from the engine

12. Mechanic A says wick-type rear main seals are usually made in one piece. Mechanic B says lip-type seals can be installed facing either direction. Who is right?

a. A only
b. B only
c. Both A and B
d. Neither A nor B

13. A poorly maintained PCV system could cause any of the following problems EXCEPT:

a. Increased oil consumption
b. Rich air-fuel mixtures
c. Excessive backpressure
d. Clogged air filter

14. A car with a road draft crankcase ventilation tube has fuel-diluted oil. Mehanic A says the PCV valve may be plugged. Mechanic B says the vehicle may be operating mostly at low speeds. Who is right?

a. A only
b. B only
c. Both A and B
d. Neither A nor B

15. Which of the following is NOT a typical location for the crankcase ventilation filter?

a. PCV valve outlet
b. Valve cover oil filler cap
c. Intake air cleaner
d. Crankcase inlet hose

16. Which of the following indicates incorrect PCV system operation?

a. Increase in idle speed when PCV valve is pulled from the valve cover
b. Business card pushed away from the oil filler opening at idle
c. Decrease idle speed when PCV valve is blocked by your finger
d. Intake manifold vacuum increases when PCV valve is blocked by your finger

17. Mechanic A says the arrow in a PCV valve should point toward the crankcase. Mechanic B says PCV systems use regular vacuum hose material. Who is right?

a. A only
b. B only
c. Both A and B
d. Neither A nor B

18. Which of the following is NOT a recommended part of PCV system inlet air filter service?

a. Drying wire-mesh filter with compressed air
b. Cleaning flame arrester in solvent
c. Replacing plastic gauze inlet filter
d. Cleaning wire-mesh filter in solvent

19. Mechnic A says solvent can damage a diesel's crankcase depression regulator valve. Mechanic B says a diesel's liquid-vapor separator can be cleaned in solvent. Who is right?

a. A only
b. B only
c. Both A and B
d. Neither A nor B

20. Which of the following is NOT considered a type of "severe service"?

a. Frequent short trips
b. Regular operation in dusty conditions
c. Regular operation on superhighways
d. Trailer towing

21. After an oil change, the oil pressure warning lamp stays on. What should you do?

a. Check the oil level with the engine still running.
b. Increase engine speed to help prime the oil pump.
c. Shut the engine off and drain the oil a second time.
d. Shut the engine off and check the oil level.

22. The oil drained from an engine is sludgy and lumpy. Which of the following is NOT a likely cause?

a. Cooling system thermostat stuck open
b. A defective (misfiring) spark plug
c. Too long between engine oil changes
d. Wrong type of oil in use

7 COOLING SYSTEM, DRIVE BELT, AND EXHAUST SYSTEM SERVICE

INTRODUCTION

Besides doing the obvious jobs of cooling the engine and removing exhaust, the cooling and exhaust systems control engine temperature and affect overall performance. Precise adjustments of the fuel, ignition, and emission systems will not allow an engine to run right if it continually overheats or runs cold, or if the exhaust system is restricted or leaking. The water pump, the alternator, the air injection pump, and other accessories cannot work properly if drive belts are loose or slipping. If a drive belt breaks, a car can be disabled.

This chapter contains basic procedures for the maintenance of cooling and exhaust systems and engine drive belts.

GOALS

After studying this chapter, you should be able to do the following jobs in twice the flat-rate labor time, or less:

1. Inspect the cooling system for damage, deterioration, and leakage; replace components as required.

2. Perform the following cooling system and component tests, analyze the results, and repair or replace parts as required: system and component pressure, coolant flow and circulation, coolant specific gravity and contamination, temperature and thermostat operation, and fan operation.

3. Flush, clean, and refill the cooling system.

4. Inspect drive belts for damage or deterioration, test and adjust belt tension; replace drive belts.

5. Inspect the exhaust system for damage, deterioration and leakage; replace parts as required.

COOLING AND EXHAUST SYSTEM SAFETY

Before servicing a cooling or exhaust system, review the shop and vehicle safety information in chapter 1. Observe these precautions:

1. Before running an engine with the hood up, inspect the fan for cracked or damaged blades.

2. Keep yourself and your tools away from the fan, drive belts, and other moving engine parts when working under the hood.

3. Do not remove a radiator cap when the system is pressurized unless specifically directed to do so. Release system pressure with the cap vent valve, figure 7-1, or by turning the cap one-half turn and releasing pressure before removal.

4. Hot coolant can burn you, and ethylene glycol can make you ill or irritate your skin. Do not allow hot coolant to splash onto you; keep coolant away from your face; do not inhale vapors.

5. Do not spill coolant on vehicle paint. It may remove wax or damage the finish.

6. Do not pour cold water into an overheated engine. Let it cool; then run it at idle while adding coolant.

7. Do not run an engine with a pressure tester attached to the radiator unless specifically directed. A pressure tester does not have an automatic relief valve, and the system may be overpressurized and damaged.

8. Keep your hands, arms, tools, and flammable liquids away from hot exhaust manifolds and pipes. Catalytic converters stay hot long after an engine is shut off.

COOLING SYSTEM AND DRIVE BELT INSPECTION

As with any maintenance job, cooling system service begins with inspection. Inspect the cooling system and all drive belts as follows:

1. Inspect drive belts for:

Figure 7-1. Release system pressure by carefully opening the cap vent valve or turning the cap one-half turn. Remove the cap with a shop cloth.

a. Proper tension. They should not be too loose or too tight. Adjust belts as required.

b. Fraying, glazing, oil, grease, cuts, breaks, and other damage, figure 7-2. Replace defective belts.

2. Inspect radiator hoses, heater hoses, and coolant bypass hoses for:

a. Loose or broken clamps, figure 7-3. Tighten screw-type clamps if needed; replace defective clamps.

b. Cracks or brittleness. Heat and ozone in the air cause hoses to

Figure 7-2. Replace a drive belt that is frayed, cut, cracked, glazed, oil soaked, or otherwise defective.

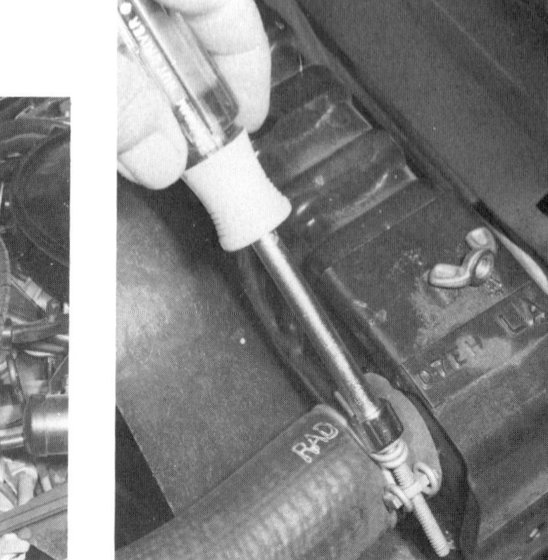

Figure 7-3. Tighten or replace loose or broken hose clamps.

Figure 7-4. A hose can look good on the outside but be deteriorated on the inside.

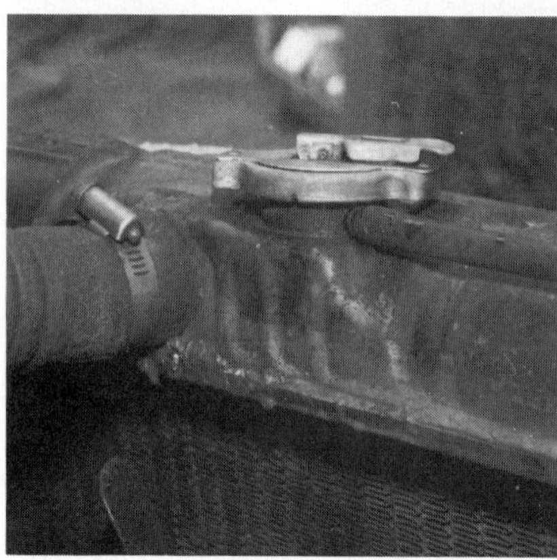

Figure 7-5. Inspect the radiator core and tank seams for leakage.

harden. Replace cracked or brittle hoses.

c. Swelling and softness. Grease, oil, and hot coolant can soften hoses, figure 7-4. Squeeze them with your hand and replace any that are spongy.

3. Inspect the radiator for:

a. Oil, rust, and scale inside the filler neck and in the coolant. These indicate that the radiator and cooling system need to be flushed and cleaned.

b. Leakage around the inlet and outlet tank seams and core fins, figure 7-5. This often appears as rust or mineral deposits.

c. Broken or bent overflow tube from the filler neck.

d. Nicked or bent surfaces on the cap seat at the filler neck. Clean these surfaces with a wire brush, a small file, or a resurfacing tool.

e. Oil or coolant leakage around fittings and lines on an automatic transmission cooler or engine oil cooler at the radiator.

f. Dirt, leaves, dead bugs, and other debris in the radiator core fins or air conditioning condenser. Remove debris from the radiator or condenser with low-pressure compressed air, blown from the engine side of the radiator.

4. Inspect the radiator cap for:

a. Looseness on the radiator when installed. If loose, replace the cap.

b. Broken or hardened rubber gaskets, figure 7-6. If gaskets are not soft and pliable, replace the cap.

c. A pressure rating that matches system specifications.

d. A broken or sticking spring or vacuum relief valve. If the spring or valve is broken or stuck, replace the cap.

5. Inspect the radiator fan shroud for cracks and missing fasteners.

6. Inspect the water pump and thermostat gaskets for leakage.

7. Inspect the water pump for leakage around shaft seals and from the ven-

Figure 7-6. A radiator cap with broken gaskets will not maintain system pressure.

tilation hole on the bottom of the pump housing, figure 7-7.

8. Inspect core plugs for rust and leakage.

9. Inspect the heater enclosure, heater hose connections, and coolant control valves for leakage.

10. Inspect the fan for loose, cracked, or broken blades.

11. Inspect a fluid-drive (viscous-drive) fan for:

a. Fluid leakage

b. Binding or sticking when you turn the fan blades by hand

c. Excessive endplay in the fan drive

12. Inspect an electric fan, figure 7-8, for loose or broken wiring connections, damaged blades, and loose mounting of the fan and motor.

13. Inspect the coolant recovery reservoir, figure 7-9, for leakage, defective hoses, loose mounting, and a missing or damaged cap.

In addition to this visual inspection, start the engine and listen to the cooling system for these noises:

1. Screeching noise of a loose or glazed drive belt.

2. A thumping noise at normal temperature that may indicate a restricted water jacket or head gasket.

3. Buzzing or whistling from a loose pressure cap.

4. Rumbling or grinding from a bad water pump bearing or bent pulley or shaft.

5. Gurgling from the radiator that indicates air or an exhaust leak into the cooling system.

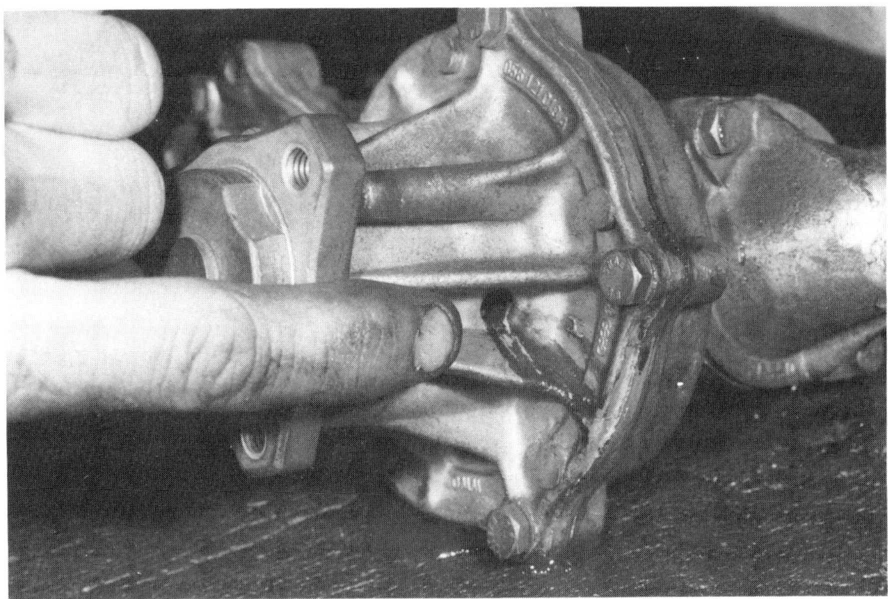

Figure 7-7. Inspect the pump ventilation hole for signs of leakage.

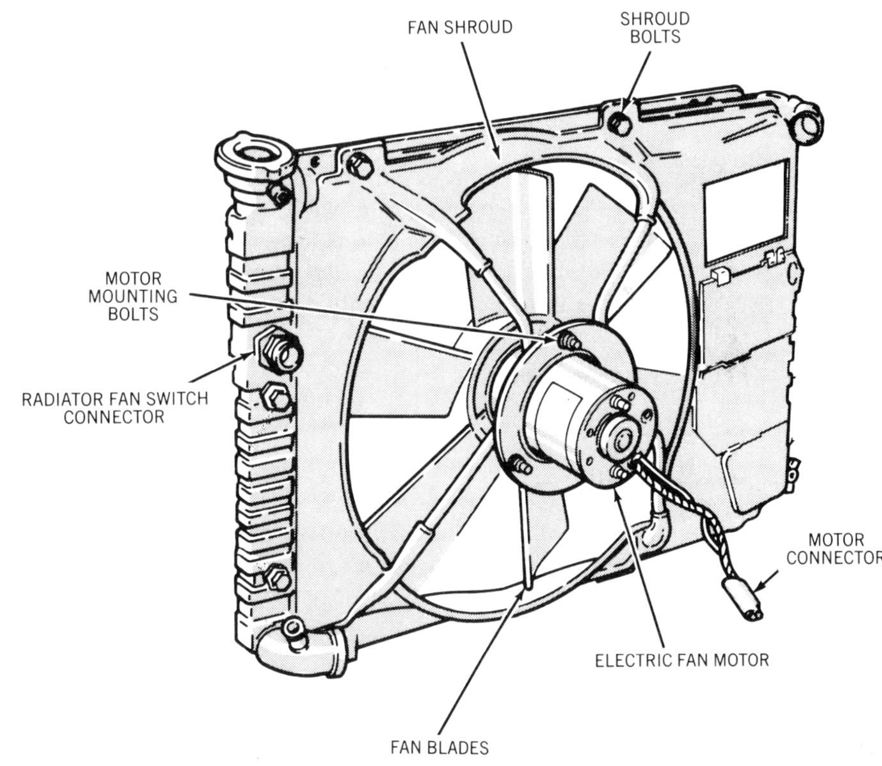

Figure 7-8. Inspect these points of an electric fan installation. (Chrysler)

COOLING SYSTEM TESTING

If your inspection indicates cooling system problems, perform the following tests.

Coolant Circulation Check

For an engine to run at proper temperature, coolant must circulate evenly through the radiator. Check the coolant circulation by running the engine to normal temperature; then shut it off. Run your hand from the top of the

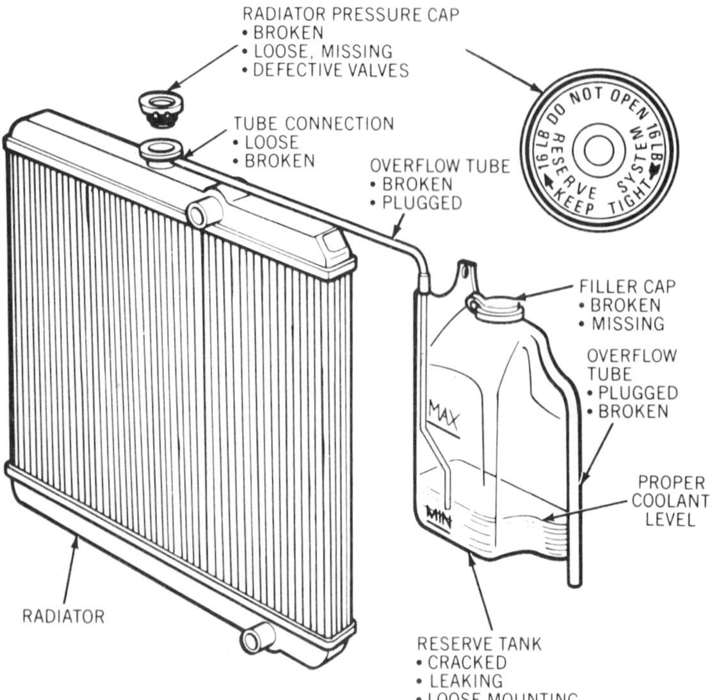

Figure 7-9. Inspect the coolant recovery system for leakage, loose bolts, and a damaged hose or cap. (Chrysler)

Ethylene Glycol Percentage (Approximate)	*Protects to °F (Approximate)*	*(°C)*
25	10	(−12)
30	4	(−16)
33	0	(−18)
40	−12	(−24)
50	−34	(−37)
60	−62	(−52)
65	−75	(−60)

Figure 7-10. This table lists the freeze protection temperature in relation to coolant specific gravity.

radiator inlet tank to the bottom of the outlet tank. Do this across the entire radiator. The radiator should be hot near the inlet and warm near the outlet. The temperature of the core should be uniform. Cold spots indicate clogged sections of the core.

Coolant Inspection and Specific Gravity Test

If your inspection shows that the coolant contains dirt, rust, oil, or scale, the system probably needs a thorough flushing and cleaning and a fresh mixture of water and ethylene glycol. The coolant can look clean, however, and still not provide the antifreeze and boiling protection that the engine needs. Over a period of time, coolant gets diluted as water is added to the system, and additives lose effectiveness against rust and corrosion.

You can test coolant freeze protection by measuring its **specific gravity**. Specific gravity is the weight of a liquid in relation to the weight of water. Water has a specific gravity of 1.00. Heavier liquids have specific gravities greater than 1.00; lighter liquids have specific gravities less than 1.00. Coolant should have a specific gravity greater than 1, and the higher the specific gravity, the more freeze protection the coolant provides. Figure 7-10 lists the freeze protection provided by coolant in relation to its specific gravity.

To measure coolant specific gravity, you will need a **hydrometer**, figure 7-11. This is similar to the instrument you will use to test the state of charge of a battery by measuring the specific gravity of the electrolyte. A cooling system hydrometer has a built-in thermometer, and the coolant should be hot when tested. Test coolant specific gravity as follows:

> **Caution.** Release system pressure slowly. Do not allow hot coolant to splash when removing the cap.

1. Warm the engine to normal temperature and shut it off. Release cooling system pressure and carefully remove the cap.

2. Insert the hydrometer tube into the radiator and draw a coolant sample into the hydrometer. Return the sample to the radiator and repeat this step 3 or 4 times until the thermometer reading stabilizes.

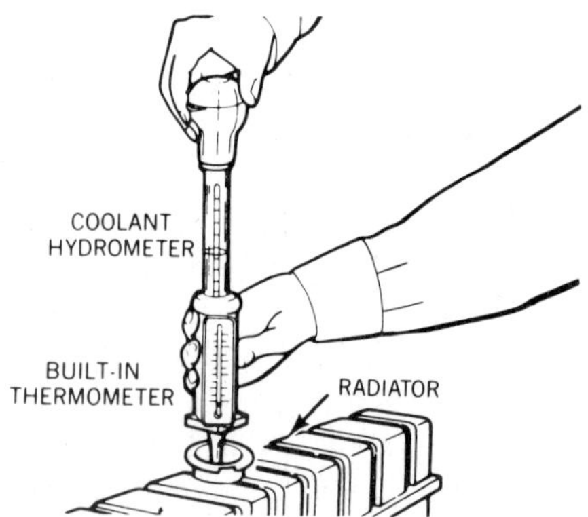

Figure 7-11. Typical cooling system hydrometer.

3. Following the hydrometer manufacturer's instructions, adjust the thermometer index to the coolant temperature. Note the thermometer reading.

4. Hold the hydrometer straight upward, figure 7-12, and draw in enough

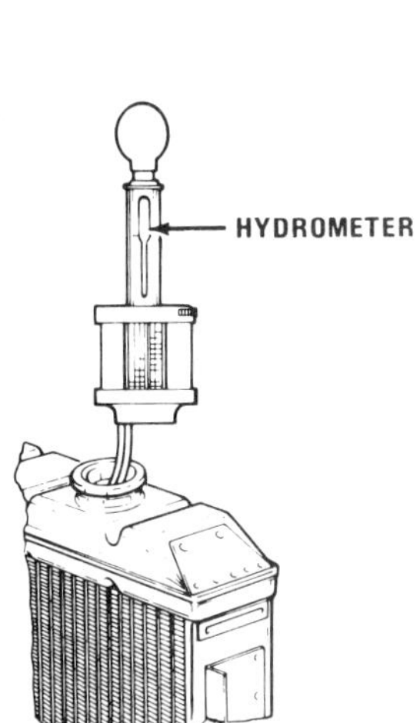

Figure 7-12. Hold the hydrometer vertically and draw in a sample. Don't let the float touch the sides of the tube. (Chrysler)

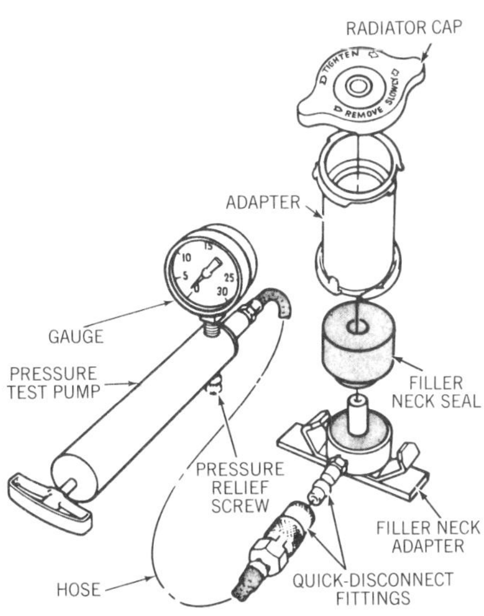

Figure 7-13. Use a pressure tester to test radiator cap and system pressure. (Ford)

Figure 7-14. Install the cap on the tester adapter and slowly apply pressure.

coolant to raise the float. Do not let the float touch the sides of the hydrometer tube.

5. Observing the float at eye level, read the top marking where the float touches the coolant sample level. Return the sample to the radiator.

6. Refer to the hydrometer reading, the manufacturer's instructions, and figure 7-10 to determine the coolant specific gravity and the degree of freeze protection it provides.

Cooling System Pressure Tests

Small leaks or internal leaks may not show up under visual inspection, but they will show up under pressure. Pressure tests also verify the ability of the system to maintain specified pressure, which is critical for proper engine temperature control and combustion. To do these tests, you will need a cooling system pressure tester, figure 7-13, which is a small air pump with a built-in pressure gauge.

Radiator Cap Test. Most pressure losses occur through a radiator cap that does not maintain a proper seal. Test the cap as follows:

1. With the engine cold, remove the cap from the radiator and attach it to the tester adapter, figure 7-14. Dampen the cap gaskets with water to help them seal.

2. If required, place a filler neck sealing ring in the cap adapter and attach the tester to the cap adapter.

3. Slowly operate the pump until the reading on the pressure gauge stops increasing. This is the pressure at which the cap pressure relief valve opens. Fast pumping causes an inaccurate reading.

4. Release the tester pressure and repeat the test twice more.

5. Compare the test results to the system pressure specifications. If the cap relief valve opens below the minimum specified pressure or 2 to 4 psi above the maximum pressure, replace the cap.

Radiator and System Test. If the radiator cap is okay but the system still appears to lose pressure, test the system as follows:

1. With the engine cold, remove the cap from the radiator and add water, if necessary, to bring the coolant level to 3/4 inch below the neck of a downflow radiator or 2 1/2 inches below the neck of a crossflow radiator.

2. Attach the pressure tester to the radiator filler neck. If required, use a filler neck sealing ring to seal the overflow tube opening, figure 7-15.

> **Caution.** Do not overpressurize the system, which may damage the radiator.

3. With the tester securely attached to the radiator, operate the pump until the pressure gauge indicates the specified system pressure.

Figure 7-15. Install the tester on the radiator filler neck with a sealing ring adapter, if required.

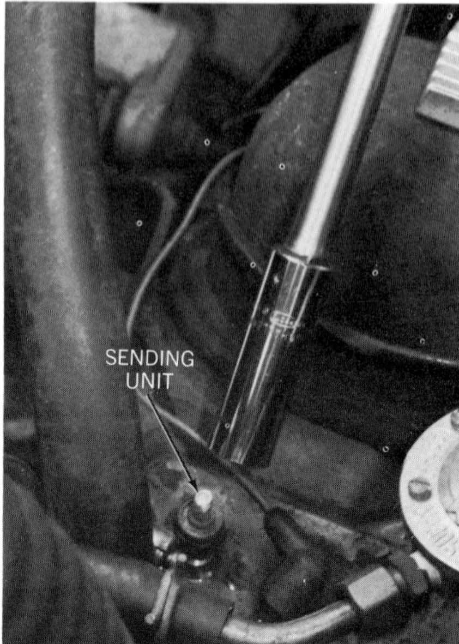

Figure 7-16. Remove the temperature sending unit and install the pressure tester to check the entire system.

4. Watch the gauge for at least 2 minutes. It should hold steady. If pressure drops, pump the tester to maintain pressure and inspect for leaks. You can apply a soap-and-water solution to radiator seams, hose connections, and gaskets to check for leakage.

If you do not find an external leak, the system may be leaking inside the engine. Test for air in the coolant or a cylinder compression leak as described later. Also, examine the oil on the engine dipstick and oil filler cap for signs of coolant (chapter 6).

Combined Cap and System Test. Using the same cooling system pressure tester, you can check the system pressure and the sealing action of the cap when it is installed on the radiator:

1. Locate the engine temperature gauge sending unit and remove it from the engine, figure 7-16.

2. Install a pipe fitting adapter to which you can connect the pressure tester.

3. Check the coolant level in the radiator and add water if necessary. Clean the cap sealing gaskets and reinstall the cap.

4. Disconnect the overflow hose from the radiator neck and attach a length of hose. Put the other end of the hose in a container of water.

5. With the tester connected to the adapter, apply pressure until the gauge shows the lower specified cap pressure.

6. Observe the hose in the water container. No bubbles should appear.

7. Increase pressure until bubbles appear and note the gauge reading. This is the upper limit of the cap pressure. Replace the cap if it is out of limits.

Cooling System Aeration Tests

Air or gases can enter the cooling system from an air leak at the water pump inlet or from an exhaust leak at a cylinder head gasket or a cracked cylinder. Special chemical test equipment is available to check for exhaust gas in the coolant, but there are several other checks you can make by observation and with test equipment you have already used.

Radiator Overflow Test

1. Warm the engine to normal operating temperature and shut it off.

2. Observing necessary safety precautions, use a shop towel or other protection to release system pressure and remove the radiator cap.

3. Add water to bring the coolant level to the top of the radiator neck.

4. Install a test cap that seals only the top of the filler neck and leaves the overflow tube open. You can make one by removing the lower sealing ring from a spare pressure cap.

5. Disconnect the overflow hose from the radiator neck and attach a length of hose. Put the other end of the hose in a container of water.

6. Start the engine and run it at a steady 1,500 to 2,000 rpm for 30 seconds. Watch the hose in the container.

7. If you see bubbles, air is entering the system from an exhaust leak or at the water pump. Perform any of the following tests for an exhaust leak. If these tests do not indicate exhaust in the coolant, the leak is probably at the water pump.

Coolant Exhaust Test with an Exhaust Analyzer. You can test for exhaust in the coolant with an infrared exhaust analyzer. Be sure the analyzer is warmed up and calibrated. Then proceed as follows:

1. Run the engine to normal operating

temperature and, observing necessary precautions, carefully remove the radiator cap.

Caution. Do not immerse the analyzer probe in the radiator. Do not let coolant splash onto the probe. Coolant will damage the analyzer.

3. Hold the analyzer probe at the top of the filler neck.

4. Accelerate the engine to about 2,000 rpm and let it return to idle. Repeat this two or three times while watching the analyzer meters.

If HC and CO readings increase as you accelerate the engine, exhaust is leaking into the cooling system.

Exhaust Leak Pressure Test. You can test for exhaust leakage into the cooling system with the same pressure tester you used to check system pressure, as follows:

1. Run the engine to normal operating temperature. Do not turn it off.

2. Observing necessary precautions, carefully remove the radiator cap.

3. Install the sealing ring in the filler neck and attach the tester to the radiator with the relief valve open, figure 7-17.

Caution. Do not allow pressure to rise above the maximum specification. System damage may result.

4. Close the tester relief valve and watch the gauge.

5. If the gauge reading rises quickly, there is a large exhaust leak into the system. If the gauge reading does not rise, operate the pump until the gauge reads within the pressure range of the cap but below the maximum specification. If the needle vibrates, there is a small exhaust leak. If the needle is steady, there is no leak.

6. Isolate the source of a leak by shorting each spark plug as you would for a power balance test. Gauge needle vibration will decrease when you short the leaking cylinder.

Thermostat Tests

An engine must have the proper thermostat for proper temperature control. If a thermostat is removed or opens at a lower temperature than specified, the engine may not reach normal temperature and continually run rich. More importantly, on an engine with electronic combustion controls, coolant temperature sensors may never signal the computer that the engine is at normal temperature. This can keep the fuel system in a continuous open-loop, cold-operating mode. On the other hand, if the thermostat sticks closed or opens at a temperature that is too high, the engine may overheat.

You can check thermostat operation with it installed in, or removed from, the system.

Testing a Thermostat on the Vehicle. The fastest basic check is with the thermostat installed. You can measure system temperature with a thermometer or with a temperature-indicating label, figure 7-18, applied to the radiator inlet tank.

1. With the engine cold, remove the radiator cap and add coolant if necessary to bring it to the proper level.

2. Place a thermometer in the filler neck or apply a temperature label to the radiator tank.

3. Start the engine and watch the thermometer and coolant circulation in the radiator tank as the engine warms up. You can speed the warmup by blocking the front of the radiator or the grille with a fender cover until the engine reaches normal temperature.

4. You will see coolant start to circulate in the top of the tank when the thermostat opens. Note the thermometer or label indication at this point. Interpret the indications as follows:

a. If temperature increases only slightly as the engine warms, and then rises quickly to the specified range as the thermostat opens, the system is working properly.

b. If temperature is above or below specifications when the thermostat opens and coolant starts to circulate, the thermostat is bad.

c. If coolant starts to circulate almost immediately as you run the engine and temperature rises steadily to 120° to 150° F (50° to 66° C), the thermostat is stuck open or missing from the system.

Testing a Thermostat out of the System. You can test the exact opening and closing temperatures of a thermostat when it is out of the system:

1. Open the radiator drain or loosen the lower radiator hose to drain coolant from the system until it is below the height of the thermostat. Save the

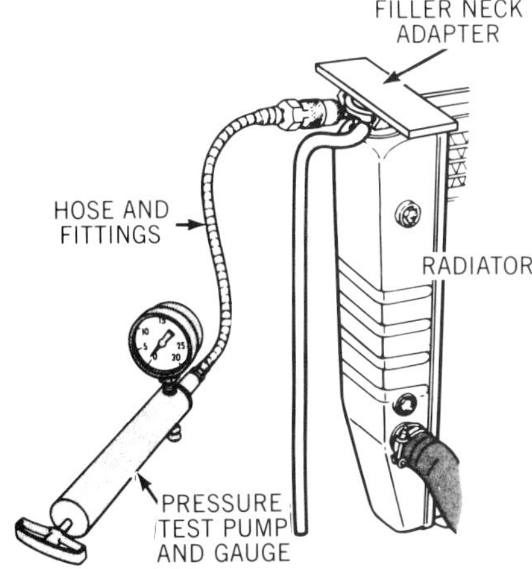

Figure 7-17. Attach the tester to the radiator and run the engine to check for an exhaust leak. (Ford)

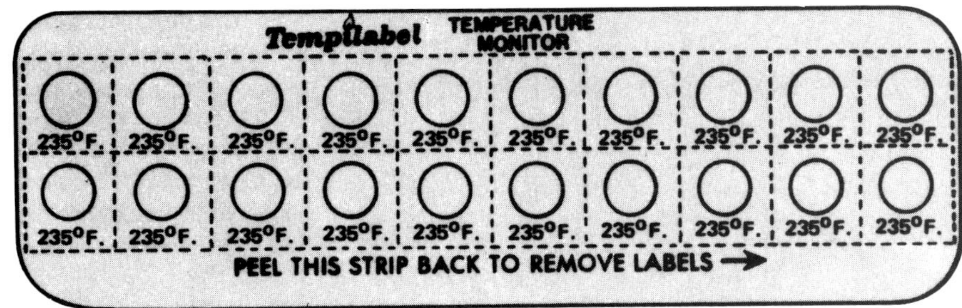

Figure 7-18. These labels change color to indicate cooling system temperature. (Ford)

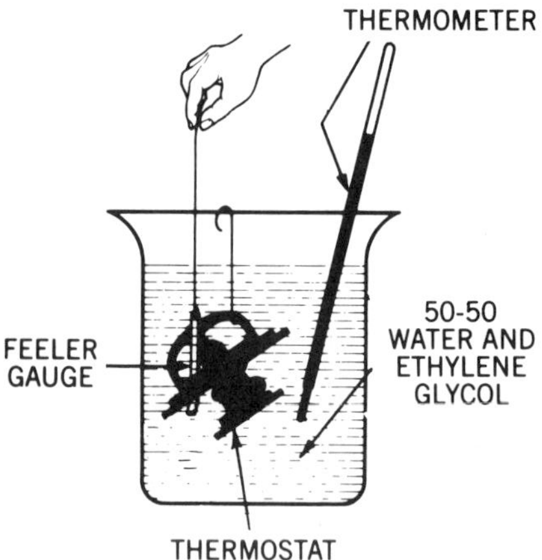

Figure 7-19. Heat the thermostat in a glycol-and-water mixture to check the temperature at which it opens.

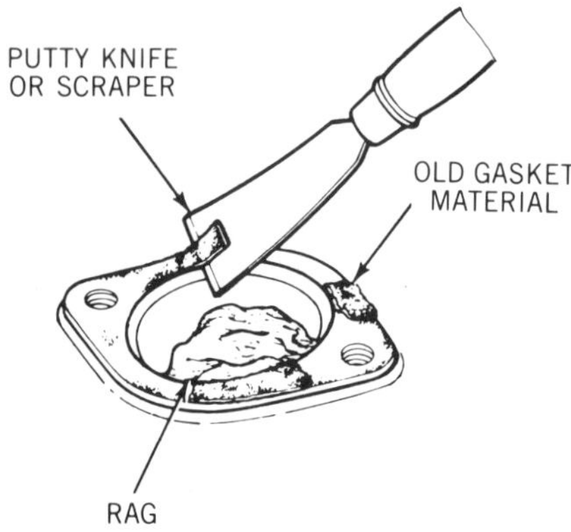

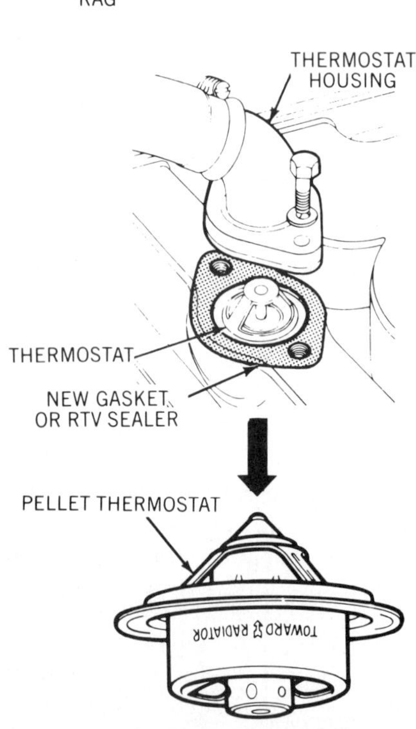

Figure 7-20. Install the thermostat with the sensing bulb toward the engine. Use a new gasket or RTV sealant. (Chrysler)

coolant for reuse if it is in good condition.

2. Remove the thermostat from its housing. On most vehicles, you must disconnect the upper radiator hose to do this.

3. Place the thermostat in a heat-proof container with a thermometer or in a special thermostat tester. Do not let the thermostat touch the sides or bottom of the container, figure 7-19. Hang it on a wire if necessary.

4. Fill the container with a 50–50 mixture of water and ethylene glycol coolant. Heat the container on a stove or with the tester's electric heater and watch the thermometer.

5. Note the temperature at which the thermostat starts to open and the temperature at which it is fully open.

6. Turn off the heat, allow the thermostat to cool, and note the temperature at which it closes.

7. For a more accurate temperature measurement, insert a 0.002- or 0.003-inch feeler gauge in the thermostat valve, figure 7-19. When you can pull the gauge easily from the thermostat, the valve has just opened.

8. Replace the thermostat if any of the following temperatures are out of limits:

- **a.** The temperature at which the thermostat is fully open should match its rated temperature.
- **b.** The temperature at which it starts to open should be 5 to 10 degrees lower than its rated temperature.
- **c.** The temperature at which it closes should be equal to, or slightly less than, its opening temperature.

9. Install the correct thermostat in the housing with the sensing bulb toward the engine, figure 7-20.

10. Clean the housing surfaces and install a new gasket or seal the housing with RTV sealant.

11. Reconnect the radiator hose and refill the system with coolant.

Thermostatic Fan Tests

Test the temperature switch of an electric fan or the viscous-drive clutch of a clutch fan with the following procedures.

Electric Fan Switch Test. The motor for an electric cooling fan is switched on and off by a coolant temperature switch. The switch is usually installed in a radiator tank or in the thermostat housing, figure 7-21. Test the switch with an ohmmeter or self-powered test lamp as follows:

1. With the engine cold, disconnect the electrical connector from the switch and connect the ohmmeter or test lamp across the terminals of the switch (not the connector), figure 7-21. The lamp should be off or the ohmmeter should show infinite resistance, indicating an open switch.

2. Reconnect the switch connector and run the engine to normal, or slightly above normal, temperature. Check coolant temperature with a thermometer in the radiator, if you wish. If the fan turns on, the switch is okay.

3. If the fan does not turn on, stop the engine, disconnect the switch connector, and test the switch as in step 1. If the lamp does not light or the ohmmeter does not show continuity, the switch is not closing when it should. Replace it.

Clutch Fan Test. You can check the operation of a clutch fan with a tachometer, a thermometer, and a timing light. You also need to know the temperature engagement point for the fan clutch, which you can find in the carmaker's specifications.

1. Using wire or duct tape, securely attach a thermometer to the engine side of the radiator where you can see it during testing. You may have to get access from under the vehicle or remove the fan shroud to do this. Be sure the thermometer is clear of the fan blades.

2. Connect a timing light and tachometer to the engine (chapter 4).

3. Run the engine at about 2,000 rpm.

4. Note the thermometer reading when the engine is cold and aim the timing light at the fan blades. They should appear to turn slowly.

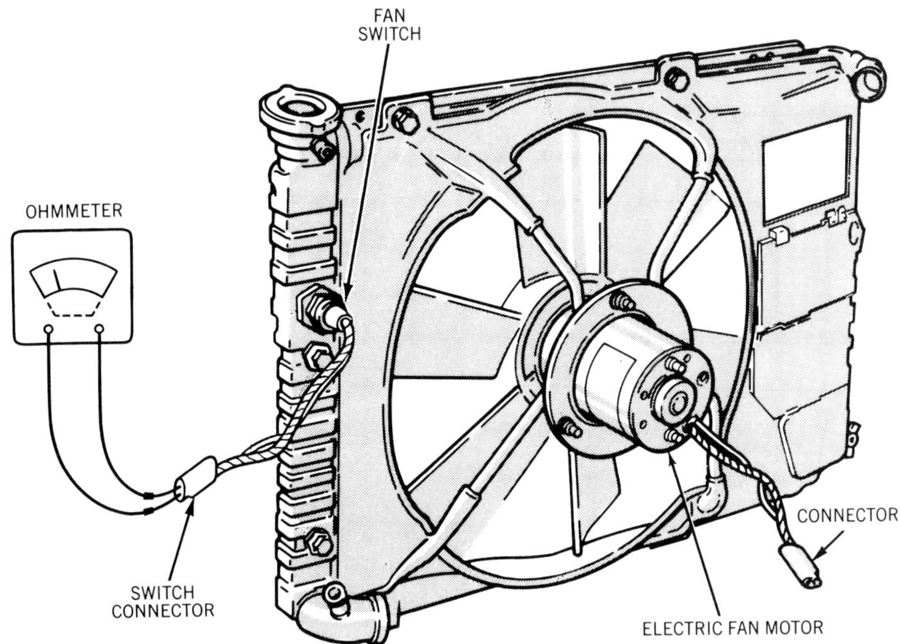

Figure 7-21. Check an electric fan coolant temperature switch with a self-powered test lamp or an ohmmeter. (Chrysler)

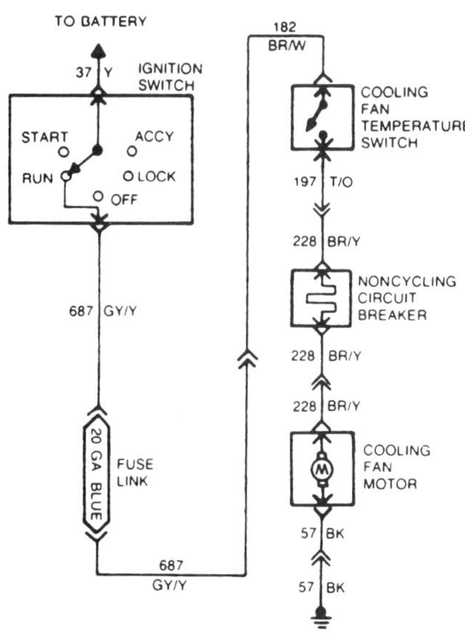

Figure 7-22. A basic cooling fan circuit used with non-air conditioned vehicles. (Ford)

Caution. Do not allow the engine to overheat during the rest of this test.

5. Block the radiator with a fender cover or cardboard and continue watching the thermometer and timing light aimed at the fan.

6. When the thermometer indicates the fan clutch engagement point, remove the radiator cover, if used. Fan speed should increase. The blades will appear to move faster in the timing light beam, and noise and airflow will increase.

7. With the radiator unblocked, watch the thermometer and fan blades as temperature drops. When temperature is below the engagement point, fan speed should decrease.

If the fan clutch does not operate within these general guidelines, it should be repaired or replaced.

ELECTRIC COOLING FAN TESTING

Carmakers use a wide variety of electric cooling fan circuits ranging from basic systems used on vehicles without air conditioning to complex, multiple fan systems that change fan speed according to vehicle speed. This section covers basic circuit and component testing. To test more complex systems, you will need the manufacturer's specifications and circuit diagrams. The procedures given for component testing can be used with any system that is not controlled by a computer.

Circuit Testing

The circuit shown in figure 7-22 is an example of a basic circuit without a relay used with vehicles that are not air conditioned. When coolant temperature reaches a specified point, the temperature switch in the fan circuit closes, completing the circuit to the fan motor. To troubleshoot this simple circuit:

1. Unplug the electrical connector at the fan motor. Connect the (+) terminal of the disconnected connector to the battery (+) terminal with a jumper wire; connect the (−) terminal of the connector to a good engine ground with a jumper wire.

- **a.** If the fan motor operates, continue testing.
- **b.** If the fan motor does not operate, replace it.

2. Remove the jumper wires and reattach the electrical connector to the fan motor. Disconnect the electrical connector at the temperature switch. Check circuit 37, figure 7-22, for battery voltage:

- **a.** If voltage is shown, continue testing.
- **b.** If voltage is not shown, check the fusible link and test the circuit breaker. If both are satisfactory, there is an open or short in circuit 37.

3. Install a jumper wire across the temperature switch connector pins:

- **a.** If the fan motor operates, replace the temperature switch.
- **b.** If the fan motor still does not operate, continue testing with the jumper wire connected across the switch connector pins.

4. Unplug the electrical connector at the fan motor. Check circuit 228, figure 7-22, for battery voltage:

- **a.** If voltage is shown, continue testing.

b. If voltage is not shown, there is an open in circuit 228 or in circuit 182.

5. Check ground circuit 57 for continuity:

a. If continuity is shown, replace the fan motor.

b. If there is no continuity shown, there is an open in the ground circuit.

Cooling Fan Controllers

Solid-state fan motor controllers are used on 1983 and later Ford front-wheel-drive vehicles with air conditioning. A large number of different fan controllers have been used because of the variety of powertrain combinations and model years. When a problem develops in a system that uses a controller, Ford recommends that you check the controller before troubleshooting the system. You cannot troubleshoot a circuit with a fan controller unless you know exactly which controller is used and have the correct circuit diagram and test procedures for the controller.

Temperature Switches

A temperature switch used in an electric cooling fan circuit can be tested to determine whether it works without removal from the engine. To determine whether the switch works properly within manufacturer's specifications, you will have to remove it from the engine for testing. The procedures for both methods of testing are provided below.

Testing in the Engine

1. Start the engine and run at fast idle until normal operating temperature is reached; then shut the engine off.

2. Check for continuity between the switch body and the component in which it is installed (cylinder head, radiator, or thermostat-water housing outlet) with an ohmmeter to make sure the switch is properly grounded. If it is not, tighten the switch to see if continuity can be restored. If it cannot, replace the switch.

3. Disconnect the electrical connector at the switch and connect it to ground with a jumper wire.

4. Turn the ignition switch to the Run position. If the fan motor runs with the connector grounded, but does not run when the connector is reattached to the switch, replace the switch.

Testing Out of the Engine. You will need the manufacturer's specifications for the temperature switch for this procedure.

1. Remove the temperature switch.

2. Place a container of engine coolant on a heating device, such as a hot plate, and put a thermometer in the container.

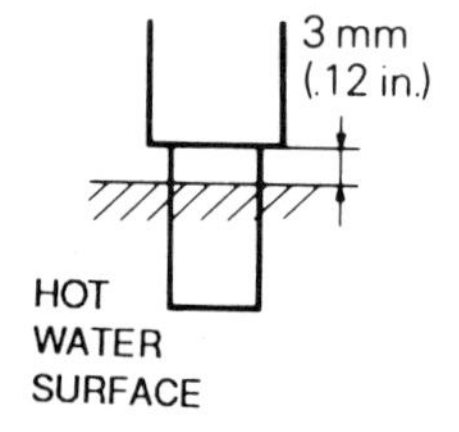

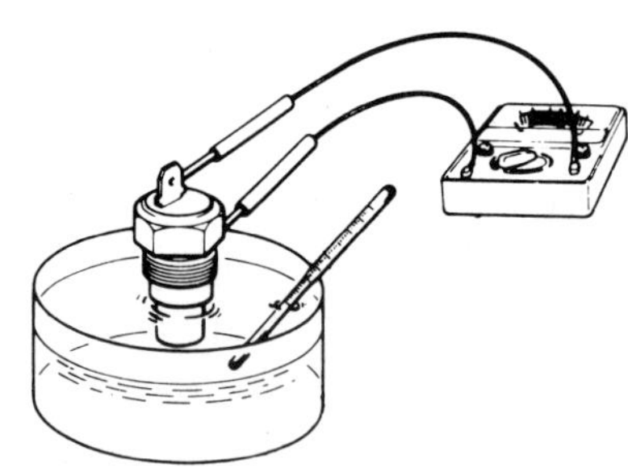

Figure 7-23. Temperature switch test. (Mitsubishi)

3. Suspend the temperature switch element in the coolant and connect the ohmmeter leads as shown in figure 7-23.

4. Heat the container of coolant while watching the ohmmeter scale. Compare the readings obtained at each specified temperature with the manufacturer's specifications. If switch resistance does not match that specified at specific temperatures, replace the switch.

Some vehicles may use a dual-function temperature switch. This type of switch looks exactly like a single function switch, but has two terminals instead of one. A dual function switch is tested in the same way as the single function switch, but you must know which terminal serves the cooling fan circuit.

Relays

You can test the relays used in electric cooling fan systems either with a voltmeter or an ohmmeter.

Fan Motors

Fan motor operation can be checked in the vehicle by unplugging the wiring harness electrical connector. Connect the fan motor + terminal to a 12-volt battery with a suitable jumper wire. Replace the fan motor if it does not run.

The fan motor is generally attached to a radiator shroud on air conditioned vehicles, and to a 3-legged support bracket with a finger guard on vehicles without air conditioning, figure 7-24. The shroud or support

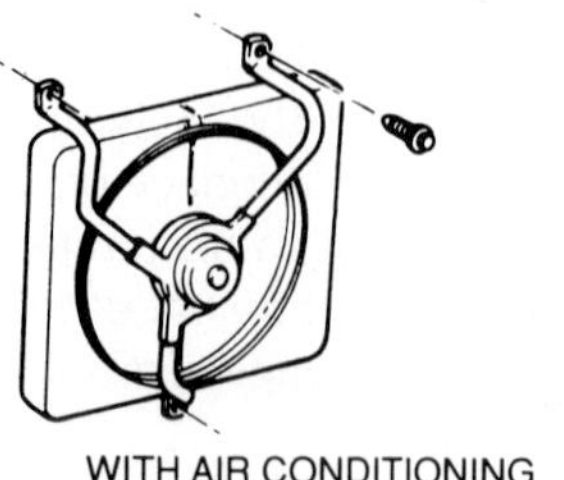

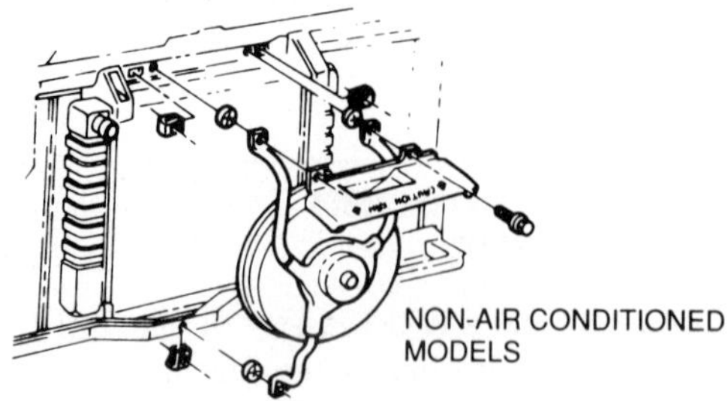

Figure 7-24. Electric cooling fan motor mounting. (Chevrolet)

bracket must be removed with the fan motor as a unit; then the motor assembly can be separated from its mounting for motor or fan blade replacement.

ELECTRONICALLY CONTROLLED ENGINE FAN SYSTEM TESTING

Electric cooling fan circuits used on late-model vehicles may have some type of computer control. An electronically controlled cooling fan system contains the same three components as one that is not computer controlled:

- A coolant temperature switch installed in the radiator, cylinder head, intake manifold, or thermostat housing
- A small electric motor and fan installed on a shroud or fan support located behind (puller fans), or in front of (pusher fans), the radiator
- A relay

On vehicles with air conditioning (A/C), fan operation is controlled by the compressor head pressure or the A/C relay. For this reason, the fan circuit is interconnected with the A/C. If the circuit has some type of computer control, a trouble code will be set if a problem develops in the fan circuit (or A/C circuit on air conditioned vehicles).

Circuits, Sensors, and Control Modules

You have already seen that there are many variations of electric cooling fan circuits used with late-model vehicles. As computer control of the fan circuit increases, there does not seem to be any relief in sight. The number and variations will continue to expand. As we cover the computer control of some late-model GM systems, you will see how necessary it is to use the automaker's circuit diagrams and diagnostic charts to troubleshoot computer-controlled cooling fan circuits.

The GM continuously variable dual cooling fan system was introduced on 1986 Cadillac Eldorado and Seville, Buick Riviera, and Oldsmobile Toronado models. Figure 7-25 is the circuit diagram for this ECM-controlled system. One fan draws or pulls air through the radiator; the other one pushes it through.

Power for the cooling fan is provided through fusible link A; relay coil power is supplied by the 5-ampere relay fuse. The low fan speed is obtained by grounding through ECM terminal D2 or terminal B of the A/C pressure switch. Ground for the high fan speed is provided through the single-wire temperature switch or terminal A of the A/C pressure switch.

The ECM turns the standard (puller) fan on its low speed under the following conditions:

- Coolant temperature exceeds 208° F (98° C).
- Vehicle speed is less than 45 mph.
- The A/C pressure switch senses a pressure of 260 psi (1,792 kPa).

The ECM also turns the fan on and off whenever the system is in its self-diagnostic mode.

The ECM turns the standard (puller) fan on its high speed under the following conditions:

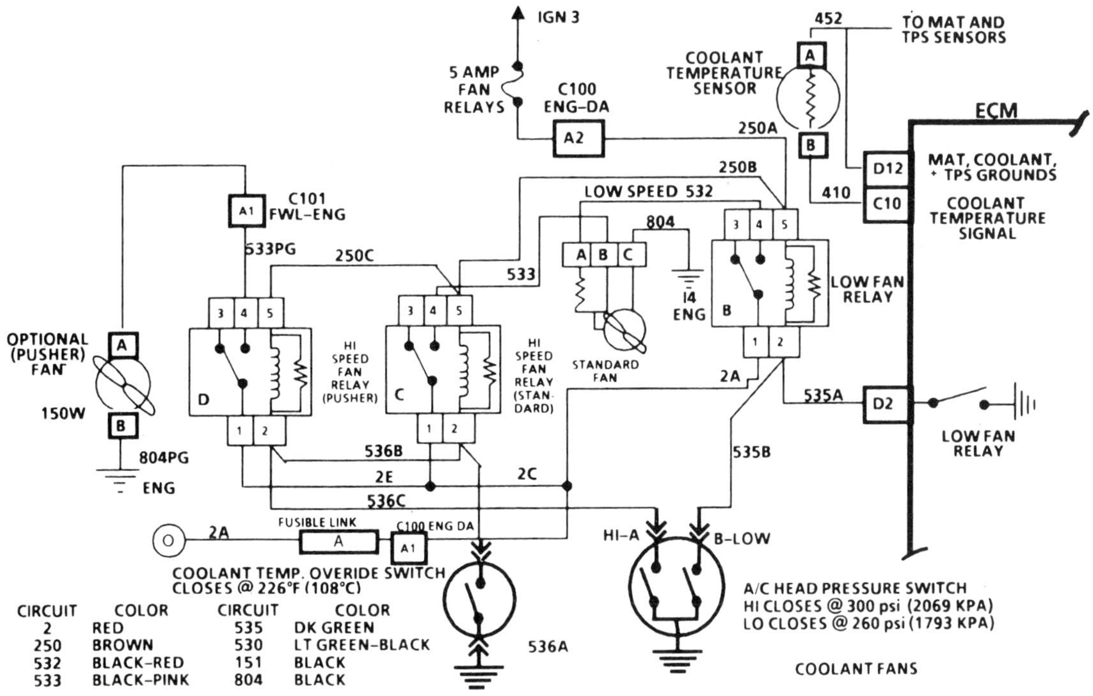

Figure 7-25. Circuit diagram for the ECM-controlled cooling fan system used on some 1986 GM vehicles. (Buick)

• Coolant temperature exceeds 227° F (108° C).

• The A/C pressure switch senses a pressure of 300 psi (2,069 kPa).

The optional (pusher) fan comes on any time the standard fan is operating at its high speed.

The fans are energized by one low-speed and two high-speed relays. Power from fusible link A is sent to terminal 1 of each relay. Voltage from the fan relay fuse energizes the low fan relay coil between terminals 5 and 2; the relay switch closes, sending current through a resistor to the 2-speed fan. Because the resistor limits current flow, the fan runs at low speed.

Low-speed operation of the standard fan uses voltage from terminal 4 of the low-speed relay to fan connector terminal A. Terminal 4 of the high-speed relay provides voltage to terminal B of the connector for high-speed operation of the standard fan. Since there is no resistor in the relay, the fan runs at high speed. Connector terminal C provides the ground.

The optional pusher fan uses voltage from terminal 4 of the relay to fan connector terminal A. Connector terminal B provides the ground.

Now look at the circuit diagram for the same fan system used on 1988 models, figure 7-26. At first glance there doesn't seem to be much difference except the way in which the diagram is arranged. But when you compare the system as described below with the 1986 system, you will find considerable differences that will affect your troubleshooting.

Power for the cooling fan is provided through the fusible link; relay coil power is supplied by the 20-ampere cooling fan fuse. The low fan speed is obtained by grounding through ECM terminal 2B5. Ground for the high fan speed is provided through ECM terminal 3C2 or terminal B of the A/C pressure switch.

The ECM turns the standard (puller) fan on its low speed under the following conditions:

• Coolant temperature exceeds 208° F (98° C).

• The A/C pressure switch senses a pressure of 150 psi (1,034 kPa).

The ECM also turns the fan on and off whenever the system is in its self-diagnostic mode.

The ECM turns the standard (puller) fan on its high speed under the following conditions:

• Coolant temperature exceeds 227° F (108° C).

• The A/C pressure switch senses a pressure of 275 psi (1,896 kPa).

The optional (pusher) fan comes on any time the standard fan is operating at its high speed.

The fans are energized by one low-speed and two high-speed relays. Power from fusible link 2 is sent to terminal 1 of each relay. Voltage from the fan relay fuse energizes the low fan relay coil between terminals 3 and 2; the relay switch closes, turning the fan on.

Low-speed operation of the standard fan uses voltage from terminal 4 of the low-speed relay to fan connector terminal A. Terminal 4 of the

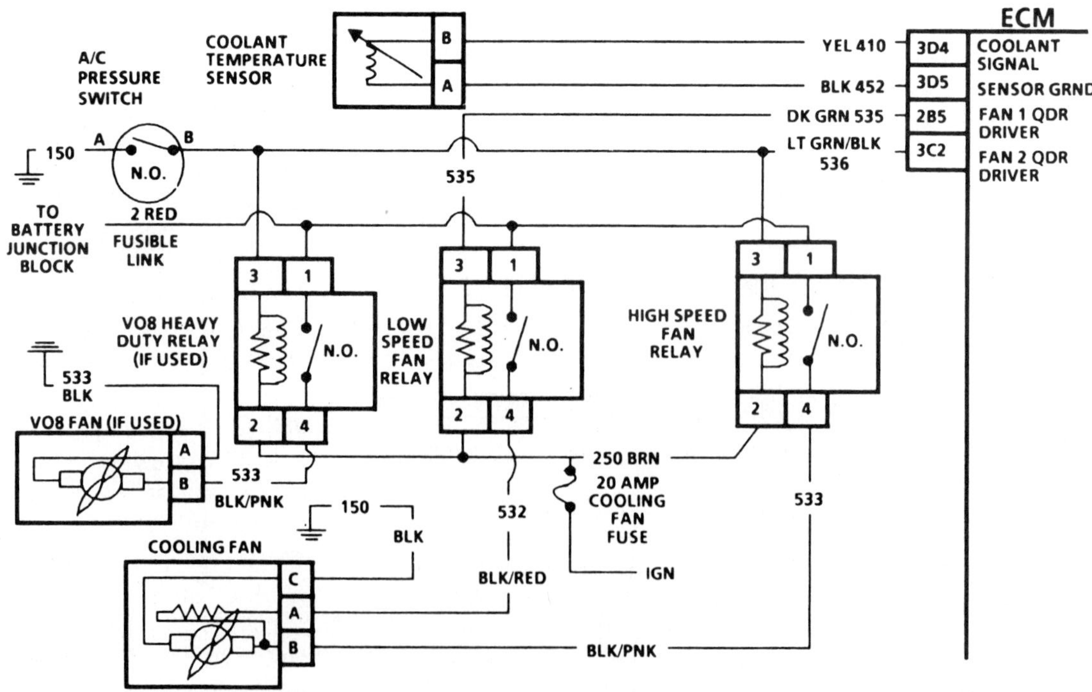

Figure 7-26. Circuit diagram for the ECM-controlled cooling fan system used on the same GM cars for 1988. Note the differences in circuit components, wiring, and operation. (Oldsmobile)

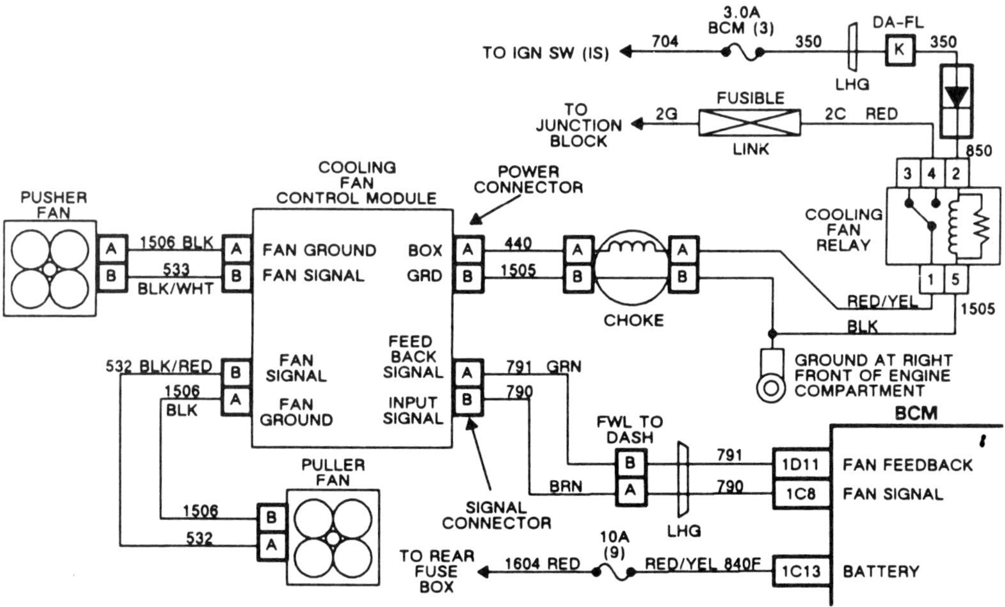

Figure 7-27. Circuit diagram of the BCM-controlled dual fan system used on the Cadillac Allanté.

high-speed relay provides voltage to terminal B of the connector for high-speed operation of the standard fan. Connector terminal C provides the ground.

The optional pusher fan uses voltage from terminal 4 of the relay to fan connector terminal B. Connector terminal A provides the ground.

The dual cooling fan circuit used with the Cadillac Allanté, figure 7-27, is controlled by the body computer module (BCM) and differs considerably in circuit design and operation. Power for cooling fan operation is provided through the fusible link between 2C and 2G. Relay coil power is supplied by the 3-ampere BCM fuse. System control is based on engine coolant temperature and A/C system high-side temperature using a single relay and a cooling fan control module. The BCM constantly calculates two fan speeds, using the lower speed when the A/C is off and the higher speed when the A/C is on.

The BCM turns control line voltage on and off to send pulse width modulation to the fan control module, which controls fan speed by switching the ground. The longer the ground is retained during a pulse period, the longer the fan runs. If either temperature sensor fails, the BCM sends the fan control module a duty cycle that runs the fans at maximum speed. If the BCM fails, the fan module also will run the fans at their higher speed. A feedback generator, or current sensor, in the fan control module sends the BCM a 12-volt signal when the fans are off. This signal drops to zero when the fans are operating. The pulsing voltage from the current sensor informs the BCM whether or not fan operation is normal. If system operation is not normal, the BCM will set a code B441 to help you diagnose the malfunction.

COOLING SYSTEM CLEANING AND FLUSHING

Carmakers specify draining and refilling the cooling system at intervals that range from 12 to 48 months. Regardless of the recommended service interval, the system should be drained and refilled if the coolant is contaminated or has lost its effectiveness, if the system is clogged, or if the engine continues to overheat after passing all other system tests. When you drain the cooling system, you can flush it with water or a chemical cleaner, according to the following procedures.

Flushing with Water

1. Following the carmaker's lifting instructions and appropriate safety procedures, raise the vehicle to a convenient working height.
2. Open the radiator drain and open or remove the block drain plugs to drain old coolant into a suitable container, figure 7-28. Some vehicles do not have radiator or block drains. Drain these by disconnecting the lower hose from the radiator. If the system is badly contaminated, disconnect the lower radiator hose because flushing can loosen large deposits that may clog block drains.
3. Remove the thermostat and reinstall the thermostat housing and hose on the engine.
4. With all drains open or the lower hose disconnected, put a water hose in the radiator. Turn on the water and adjust the flow to keep the radiator full as water flows from the drains.
5. Flush the engine and radiator with running water for about 10 minutes. Run the engine at idle and turn on the

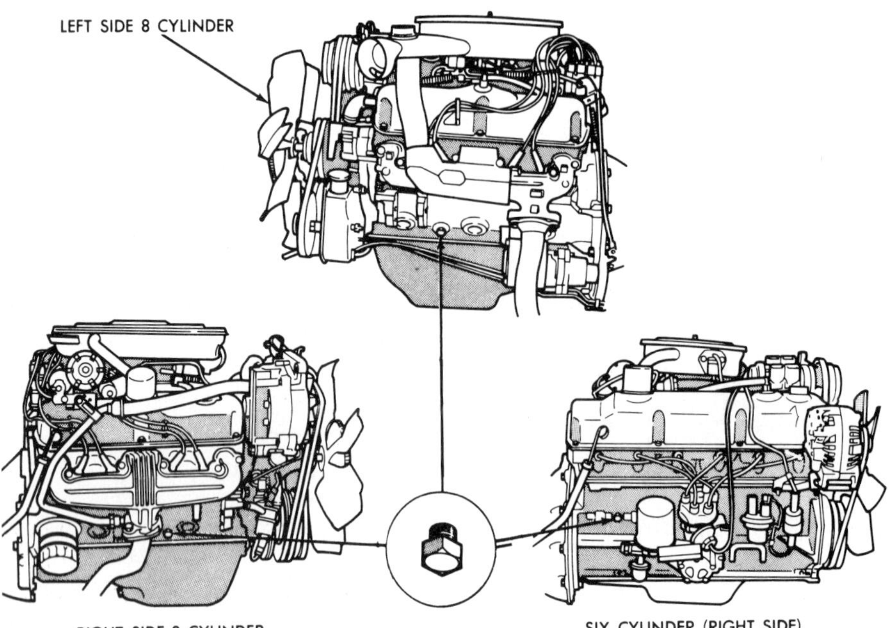

Figure 7-28. Check a service manual to find the block drain locations. (Chrysler)

heater to the maximum heat position to improve circulation. Do not accelerate the engine or you will splash water all over the engine compartment.

6. After 10 minutes, stop the engine. Turn off the water and close all drains after water has drained from the system.

7. Remove the coolant recovery reservoir from the vehicle, clean it thoroughly with fresh water, and reinstall it.

8. Reinstall the thermostat and refill the system with the kind and amount of coolant specified by the carmaker. Add water to bring the coolant level to the full mark and run the engine to operating temperature until the thermostat opens.

9. With the engine off, check the coolant level, and add water and ethylene glycol if necessary.

Cooling System Chemical Cleaning

If the cooling system is badly contaminated, you may want to use a chemical cleaner on the system as follows:

1. Drain the old coolant and remove the thermostat as directed in the preceding section.

2. Close the drains and reinstall the thermostat housing.

3. Be sure the chemical cleaner is safe for use on the vehicle. Some cleaners are not compatible with aluminum engine and radiator parts.

4. Fill the radiator with the specified amount of cleaner and fresh water.

Caution. Do not allow the cleaning solution to boil when running the engine.

5. Run the engine for about 30 minutes at a fast idle. Turn on the heater to the maximum heat position for complete circulation. Cover the car grille or lower part of the radiator to raise the temperature if necessary.

Warning. Do not splash hot cleaning solution onto yourself. It may burn or irritate your skin.

6. Stop the engine and drain the cleaning solution from the system while it is still warm.

7. If the cleaner requires a neutralizer, close the drains and add it to the system along with fresh water. This is important because acids in some cleaners will attack metal parts if left in the system.

8. Run the engine for 10 to 15 minutes at fast idle. Then drain the neutralizer.

9. If desired, flush the system with fresh water as directed in the preceding section.

10. Remove the coolant recovery reservoir from the vehicle, clean it thoroughly with fresh water, and reinstall it.

11. Reinstall the thermostat, close the drains, and refill the system with the kind and amount of specified coolant. Add water to bring the coolant level to the full mark and run the engine to operating temperature until the thermostat opens. Check the coolant level and add water and ethylene glycol if needed.

If the system was badly contaminated, deposits may continue to break loose from water jackets and the radiator after they are refilled. You may want to inspect the cooling system again in a week or two and drain it, flush it, and refill it with fresh coolant. This extra effort is cheaper and simpler than expensive engine repairs caused by a badly contaminated cooling system.

COOLING SYSTEM REPAIR

Most repairs that you will do to a cooling system will be replacing damaged or worn components. Radiators, heater cores, and water pumps can be rebuilt, but these are usually jobs for specialty shops. Radiators on some late-model Ford and GM products have plastic inlet and outlet tanks crimped and sealed to the core with O-rings, figure 7-29. You can disassemble these radiators and replace tanks and cores according to the manufacturer's procedures.

The following sections provide guidelines for removing and replacing hoses, thermostats, engine core plugs, water pumps, and radiators.

Hose and Thermostat Replacement

We touched on hose and thermostat installation in the procedures for cleaning and flushing the cooling sys-

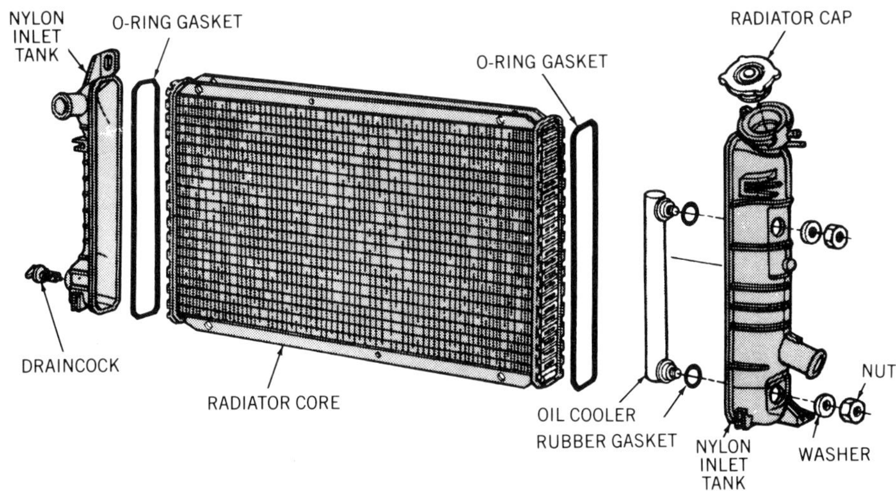

Figure 7-29. These Ford radiators have removable, reinforced nylon tanks. (Ford)

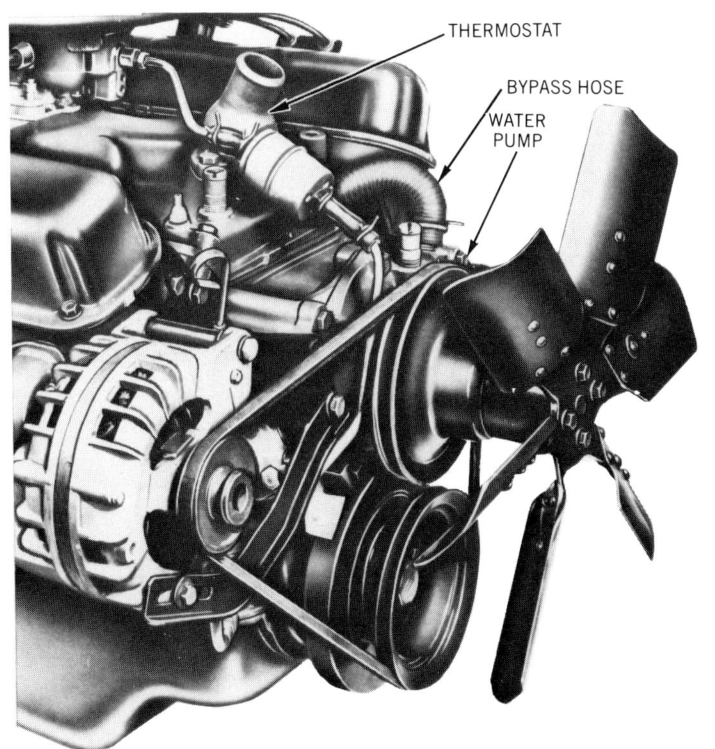

Figure 7-30. Don't overlook the bypass hose on many cooling systems. (Chrysler)

tem, but here are a few more guidelines for these jobs.

Don't overlook heater hoses and the small thermostat bypass hoses, figure 7-30, used on some engines. They are subject to the same deterioration from temperature and age that main radiator hoses are. If the radiator hoses need replacement, the other hoses in the system may also.

Original equipment radiator hoses are molded to fit the engine and radiator installations in different vehicles. Carmakers usually recommend replacing worn hoses with similar molded parts, but flexible replacement hoses work well on many engines. When installing any hose, particularly a flexible one, be sure it is not kinked or stretched and it does not rub against any engine or chassis parts.

Thermostat housings are often hard to remove because of a sticking gasket or mineral deposits. Do not hammer on a housing to loosen it. After removing the hose, put a hammer handle or large screwdriver handle in the housing outlet and *carefully* pry the housing from the engine. Do not use a metal bar or you may crack the housing.

Thermostat housings are made of cast iron or aluminum and can suffer corrosion from rust or **electrolysis**. Before reinstalling a housing, clean off all old gasket or sealant material and remove corrosion with a scraper or brush. Replace the housing if it is cracked or badly corroded.

Engine Core Plug Replacement

Another important reason for using ethylene glycol coolant is to keep the engine core plugs from rusting. Core plugs are steel plugs installed in casting openings in cylinder heads and blocks. If the cooling system contains only water, the core plugs can rust and leak.

Removing and replacing a core plug requires cutting out the old plug and driving in a new one. However, the job can be time consuming because plugs often are obstructed by manifolds, engine accessories, and body parts. You may have to remove accessories for access to drive in the replacement.

It is usually easier to cut out the old plug than to drive in the new one. Often you can save time by installing a copper expansion plug, figure 7-31. You do not have to drive the expansion plug into place. Rather, you insert it into the core hole and turn the hex nut 1 to 2 turns with a wrench to expand the plug and lock it into the hole. The soft copper conforms to the hole opening and provides a leakproof seal.

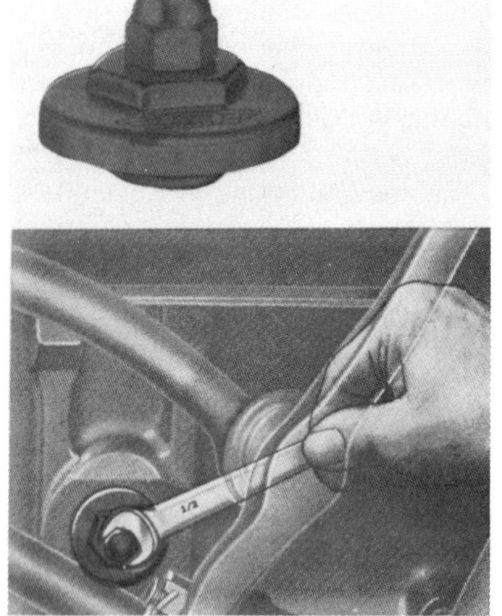

Figure 7-31. These copper expansion plugs make core plug replacement easier on many engines. (Dorman)

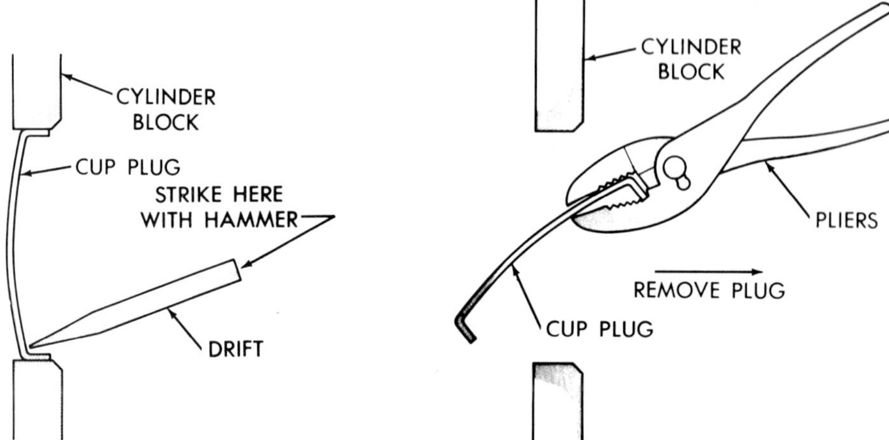

Figure 7-32. Loosen the core plug with a drift punch. Remove it with pliers. (Chrysler)

Use the following methods to remove an old core plug. If you choose to install a drive-in replacement, continue with the procedure to install the new plug.

1. Drain the cooling system completely. If the coolant is in good condition, save it for reuse.

2. Remove any accessories, if necessary, for access to the old plug.

3. If access permits, punch or drill a small hole (about 1/16 inch) through the old plug and install a slide hammer puller.

4. Use the slide hammer to extract the old plug from the block or head.

5. If you don't have access for a slide hammer, tap one side of the plug with a drift punch and hammer to turn it in the opening, figure 7-32.

Caution. Do not leave the old plug in the water jacket. It will obstruct coolant flow and promote rust deposits in the system.

6. Use a pair of pliers to reach into the water jacket and extract the old plug.

7. Using a wire brush or scraper, remove rust and corrosion from the core hole.

8. Select a replacement plug of the same diameter, depth, and type as the original. Do not install a deep plug where a shallow one is required. A deep plug can obstruct coolant circulation.

9. Coat the sealing edge of the core hole and the plug with nonhardening sealer.

10. If the new plug is a flat type (actually slightly curved), place it against the shoulder of the core hole with the curved side outward. Tap the plug into place with a 1/2-inch drift punch until it is almost flat to seal it in the hole.

11. If the new plug is a cup plug or a cup-expansion plug, place it squarely into the core hole as shown in figure 7-33. Drive it into place with a driver or large drift punch until the rim is flush with the surface of the block. Do not cock the plug in the hole, or it will leak.

12. Refill the cooling system, start the engine, and warm it to normal temperature.

13. Carefully remove the radiator cap, install a pressure tester, pressurize the system, and check the new core plug for leakage.

14. If there is any sign of leakage, remove the plug and install another. Do not reuse a core plug.

Radiator Removal and Replacement

We can't show you the exact steps for replacing a radiator on all cars, but

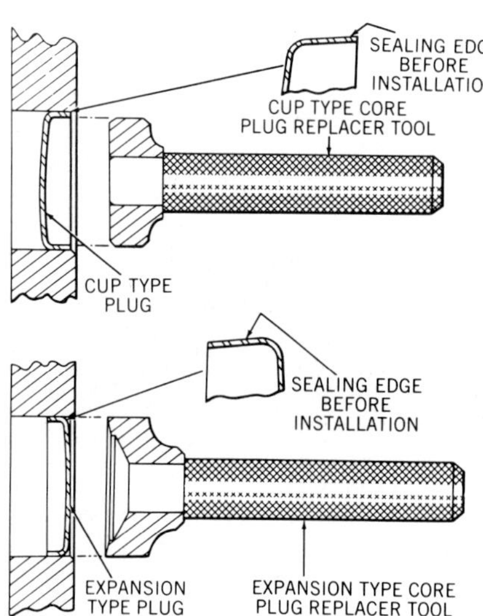

Figure 7-33. Typical core plug installations. (Ford)

the following photo procedure illustrates the principles. Before doing this job on a specific car, consult a repair manual for the carmaker's instructions. Use the photo procedure and the following guidelines to help with the job:

1. Before draining the cooling system, inspect the coolant and test the specific gravity. If it is in good condition, save it for reuse.

2. Before loosening or removing any parts, inspect the job to determine what parts you must remove for radiator clearance. These include:

- Fan shrouds and air shields
- Transmission or oil cooler lines at the radiator outlet tank

RADIATOR REMOVAL

1. The radiator in this Chevrolet is typical of many large cars. After draining the coolant, remove the upper radiator hose.

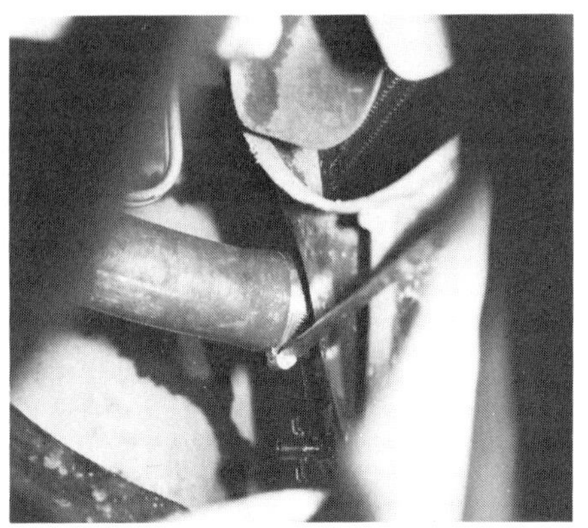

2. On many cars, you will have to go underneath for access to the lower hose. Remove it from the radiator.

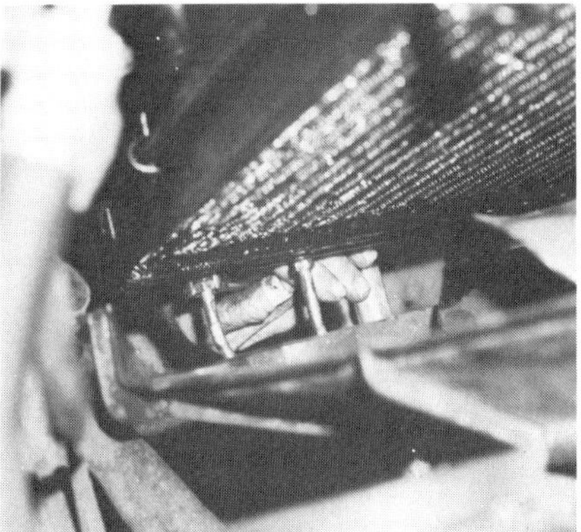

3. Disconnect the transmission cooler lines from the radiator. Remember to check and add transmission fluid after replacing the radiator.

4. Loosen the fan shroud bolts from the radiator support brackets. You can remove it now or just before you lift out the radiator.

5. Remove the capscrews that hold the radiator to the front bulkhead of the engine compartment.

6. Carefully lift the radiator straight up and out. You don't have to remove the fan on this car, but don't nick the radiator as you remove it.

WATER PUMP REMOVAL AND REPLACEMENT

1. Begin by draining the system and removing the hoses and radiator. Remove bypass and heater hoses, if necessary.

2. Remove accessories from the front of the engine, such as the alternator, air pump, A/C compressor, and power steering pump.

3. This Toyota pump is typical of the tight clearance on some water pump bolts. Some pump bolts go through the timing cover, into the block. Remove them carefully.

WATER PUMP REMOVAL AND REPLACEMENT (*continued*)

4. Hold the pump and fan assembly securely and lift it off the engine. On some cars, you may want to remove the fan first.

5. Clamp the old pump in a vise and remove the fan bolts. If the fan sticks to the pulley, tap it with a rubber mallet to loosen it.

6. Remove the fan. When you install the fan on a new pump, clamp it in the vise with a shop cloth to avoid damage.

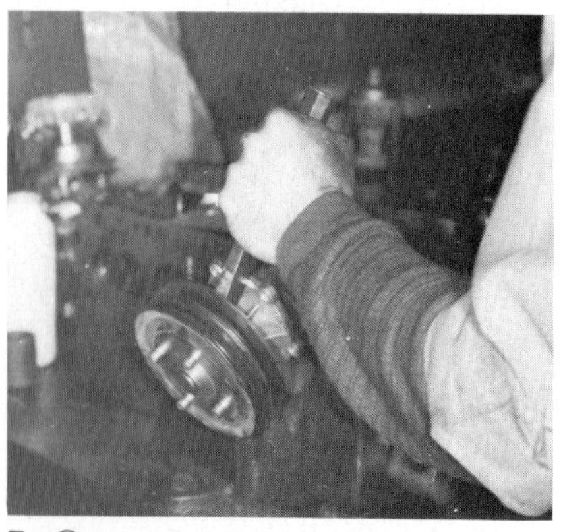

7. On some cars, you must remove the pulley from the old pump and install it on the new one. Remove it with a puller or drive it off carefully with a blunt chisel and hammer.

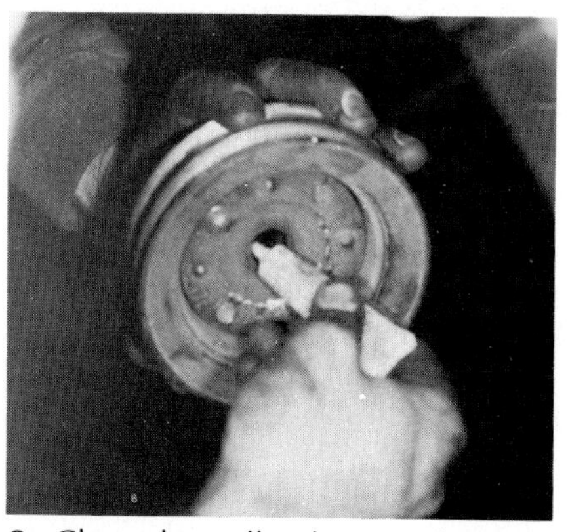

8. Clean the pulley bore and coat it with stud-mount sealant. Use a press to install it on the new pump.

9. Here is the new pump with pulley and fan installed, ready for installation on the engine. Coat the flange with sealant or gasket cement.

10. This Toyota has studs and nuts for pump mounting. Coat the new gasket with sealant or cement, align it properly, and put it in place.

11. If the pump is secured by capscrews, place the gasket on the pump, align it with two capscrews, and install the pump.

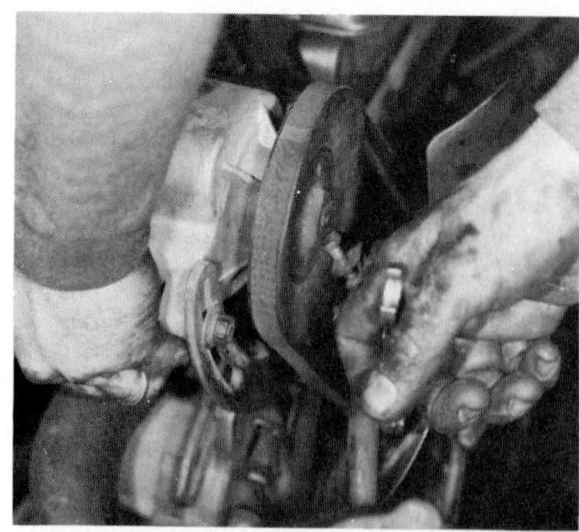

12. Reinstall engine accessories. Tighten all belts to the proper tension, reinstall the radiator and hoses, and refill the system.

- The air conditioning condenser
- The fan (belt driven or electric)
- Coolant temperature switches and connections in the radiator inlet tank
- The coolant recovery reservoir

3. Inspect the hoses before removing them. If they are in good condition, remove them carefully. If the hoses are to be replaced, you can cut them off the radiator and engine to speed removal.

Water Pump Removal and Replacement

The preceding photo procedure for a Toyota water pump replacement outlines the principles of this job. As with radiator replacement, refer to a repair manual for the carmaker's instructions and observe the following guidelines:

1. When removing accessories such as air conditioning compressors and power steering pumps, do not put strain on wires, hoses, and fittings. Remove the mounting bolts and support the components with wire while removing the water pump.

2. Inspect all hoses and drive belts and replace any that are defective.

3. Note the locations of bolts of different lengths and reinstall them in their original locations. Many water pump bolts go through timing covers and into the water jackets. Use thread sealant when reinstalling bolts.

4. Align water pump gaskets correctly so that coolant bypass holes are not blocked.

5. Be sure all mating surfaces are clean. Install gaskets correctly.

DRIVE BELT SERVICE

Drive belt service consists of inspecting belts for wear, damage, and deterioration; adjusting belt tension; and replacing belts. You learned about belt inspection earlier in this chapter. The following sections outline the principles of belt tension adjustment and belt replacement.

Belt Tension Testing and Adjustment

Proper belt tension is important because:

- A belt that is too tight puts a constant high load on rotating shaft bearings. Excessive tension also strains the belt and can cause tension failure.
- A belt that is too loose may slip on the pulleys and become glazed. Slippage also reduces the efficiency of components such as alternators and water pumps. Also, a loose belt puts high impact loads on shaft bearings because of the whipping action of the belt.

You can check the tension on a V-belt quickly by pressing on the belt midway between the pulleys on the longest belt section. It should deflect 1/8 to 3/8 inch (3 to 9 mm) with the amount of pressure needed to replace a crimp-type bottle cap (about 20 pounds). Place a straightedge along the belt and use a short ruler to measure deflection, figure 7-34.

Beware that this deflection test gives you a *general* idea of the tension of a V-belt. It is no substitute for accurate tension measurement with a tension gauge. Also, a deflection test won't work on a V-ribbed, or serpentine, belt. A V-ribbed belt will bend too much for accurate deflection measurement, and belts with spring-loaded idler pulleys will deflect too far to indicate tension.

Use a tension gauge for accurate tension measurements on V-belts and V-ribbed serpentine belts. Place the tension gauge in the middle of the belt length between two pulleys with the feet on one side of the belt and the hook on the other, figure 7-35. Pull or push the handle according to the manufacturer's instructions and read the tension in pounds or newtons on the gauge dial. When testing a matched pair of belts, measure the tension of each separately.

Carmakers provide tension specifications for both new and used belts. New-belt tension specifications are almost twice as high as used-belt specifications. For example, a new-belt tension may be 120 pounds (530 N), while the same belt when used may require tension of only 70 pounds (310 N).

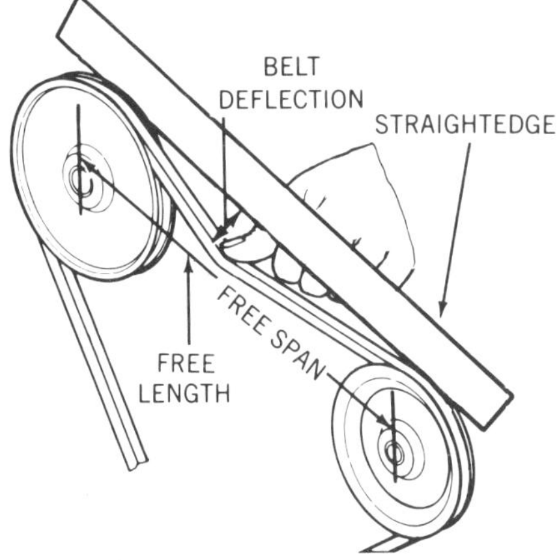

Figure 7-34. Use a straightedge and ruler to measure belt deflection for a quick check of belt tension. (Chrysler)

Figure 7-35. Use a belt tension gauge for accurate belt adjustment.

A new belt is exactly that, *brand new, never installed and tensioned.* Carmakers' definitions of used belts vary, however. Some say that a belt is "used" after it is installed, tensioned, and run just one revolution at engine speed. Others say that a belt is "used" after it has been run for 15 minutes. In any case, you can adjust a new drive belt properly if you:

1. Install the belt and adjust it to new-belt tension.

2. Run the engine at idle for 10 to 15 minutes.

3. Stop the engine, measure belt tension with a gauge, and readjust it to used-belt specifications.

Drive Belt Installation

Drive belt installation varies in complexity with the design of an engine compartment and the arrangement of driven accessories. No two car models are exactly alike, but you can use these guidelines to install most belts properly.

1. To replace an inner belt, you must remove all outer belts. Doublecheck the condition of all belts and replace any that are worn. A car owner will be money and time ahead in the long run to change all worn belts at one time.

2. Always replace both belts of a matched pair at the same time. If one belt needs replacement, the other will soon. And, you can't adjust tension equally on mismatched belts.

3. Adjust V-belt tension by loosening the pivot and mounting bolts of the driven accessories and moving the accessories on their brackets. Carmakers provide three basic ways to do this:

Caution. Do not pry against the soft aluminum casting of an air pump, an alternator, or other accessory. Prying force may break the casting.

a. Use a large screwdriver or bar to pry against the driven accessory and set belt tension, figure 7-36. Refer to the manufacturer's installation diagram and pry only in specified locations. Check tension with a tension gauge.

b. Use a breaker bar or ratchet handle to apply force at a specified hole or slot, figure 7-37. If a manufacturer provides torque specifications, follow the instructions for belt installation.

c. Turn an adjusting screw on the accessory bracket, figure 7-38, while measuring belt tension with a gauge.

4. To install a V-ribbed serpentine belt, loosen the idler pulley and remove the old belt. Place the new belt on the pulleys of all accessories, making sure

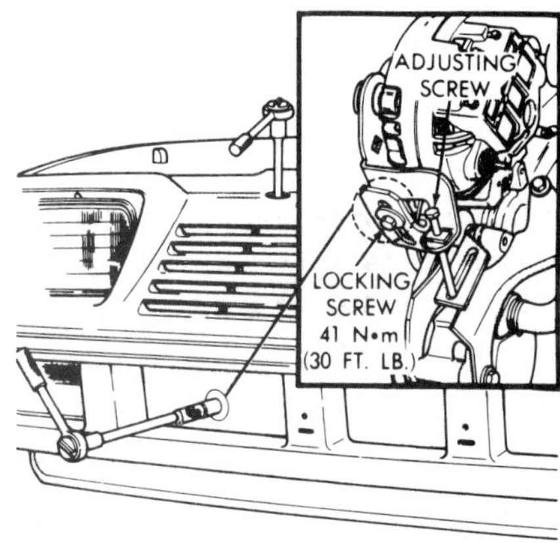

Figure 7-38. Turn the adjusting screw to position the accessory for belt adjustment. (Chrysler)

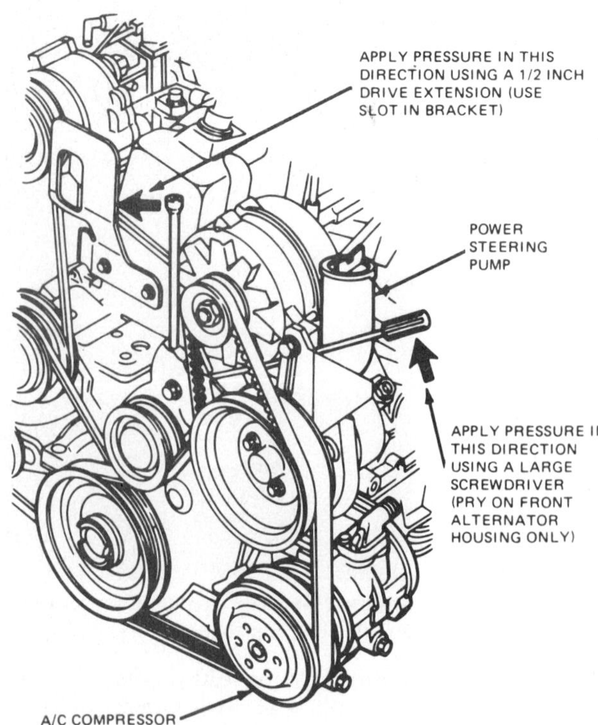

Figure 7-36. Pry only at specified locations to adjust belt tension. (Ford)

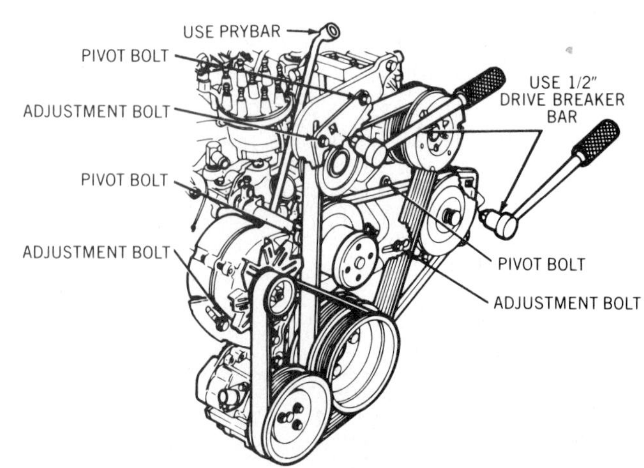

Figure 7-37. Use a breaker bar or ratchet handle at the specified locations to move accessories for belt adjustment.

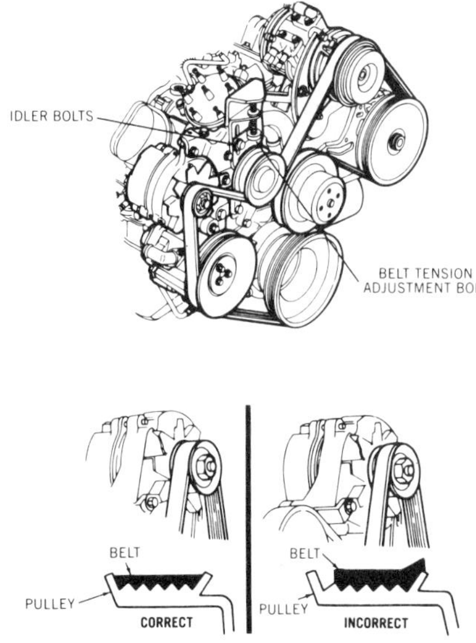

Figure 7-39. Install a V-ribbed belt so that all V-grooves make proper contact. Adjust the idler for proper tension. (Ford)

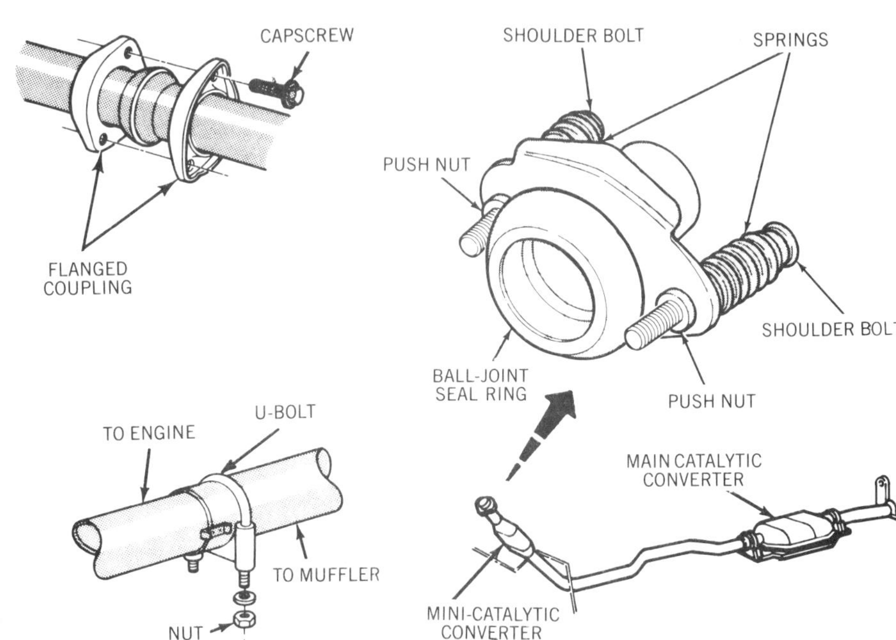

Figure 7-40. Typical exhaust system connections. (Chrysler)

that all V-grooves make proper contact with each pulley, figure 7-39. Adjust the belt idler pulley tension according to the maker's instructions.

EXHAUST SYSTEM SERVICE

Most exhaust system service consists of replacing damaged pipes, brackets, mufflers, and catalytic converters to repair leaks, restrictions, loose mountings, and emission control problems. Servicing manifold heat control valves, and changing the catalyst in a pellet-type converter are covered in later chapters. The following procedure outlines pipe, muffler, and converter replacement.

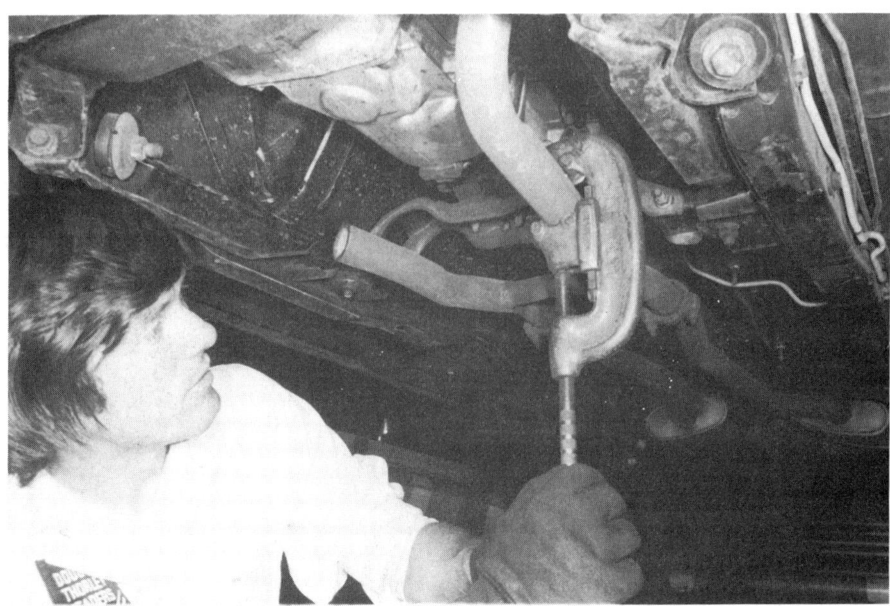

Figure 7-41. Use a pipe cutter to cut off an old exhaust pipe.

Pipe, Muffler, and Converter Replacement

Warning. Although most muffler shops use torches to remove and replace mufflers and pipes, heating and welding equipment must be used with extreme caution under a vehicle. Use the following methods to replace exhaust system parts without a torch.

1. Following the carmaker's lifting instructions and safety precautions, raise the car on a drive-on hoist or on jacks and frame stands for unobstructed access to the exhaust system.

2. Apply penetrating oil to nuts, bolts, and U-bolt clamps to loosen rust and aid disassembly.

3. For easiest disassembly, work from the rear of the system forward. The outlet of a pipe or muffler usually fits inside the inlet of the following part, figure 7-40. Loosen or remove brackets and clamps necessary to remove damaged parts.

4. Work the pipe, muffler, or converter up and down until the joint is free. If the joint is rusted or welded, separate it as follows:

a. Cut the pipe with a pipe cutter, figure 7-41, or a hacksaw.

b. Cut the outer pipe with a chisel and spread it to separate the joint.

5. To remove a catalytic converter, remove any heat shields or air injection tubes, figure 7-42, if installed. Then unbolt and remove the converter.

6. There are three basic kinds of exhaust system joints (connections), figures 7-40 and 7-42. Install new parts as follows:

a. For a slip-joint connection, install the outlet end of one pipe into the inlet of the other about 2 inches and secure the joint with a U-bolt.

b. For a flange joint with a flat gasket, clean the mating surfaces of the flanges with a wire brush or scraper. Install a new gasket and the nuts and bolts. Do not use cement or sealer.

c. For a ball-joint connection, install a new spherical gasket or sealing ring, if required. Assemble the clamping flanges and tighten the bolts evenly for a leak-free seal.

7. As a final step for exhaust component replacement, install all nuts, bolts, and brackets loosely. Check the alignment of all components and be sure that pipes, mufflers, and converters do not hit chassis parts or fuel and brake lines. Then tighten all fasteners securely.

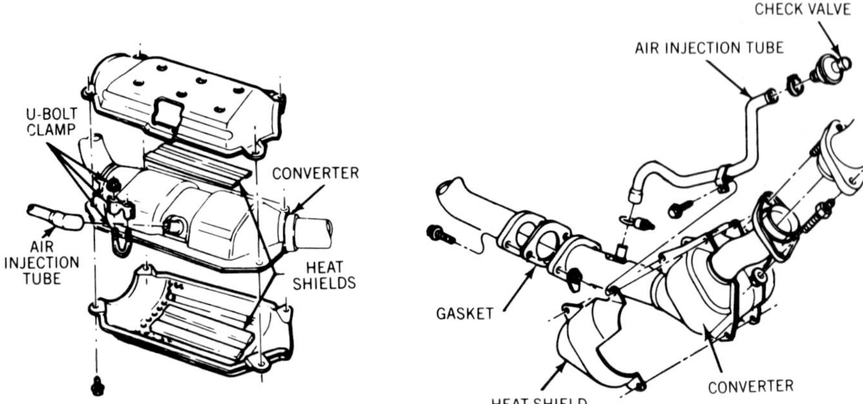

Figure 7-42. Remove heat shields and air injection tubes before removing a converter. Reinstall all parts securely during reassembly. (Ford)

SUMMARY

The cooling and exhaust systems are important combustion control systems because they regulate engine temperature and remove combustion byproducts. Cooling system service includes maintaining the proper quality of ethylene glycol and water coolant for engine protection and temperature control. Preventive maintenance consists of regular inspection, flushing, and refilling. The cooling system also must be free of leaks and maintain the required system pressure. Water pumps, thermostats, hoses, radiators, and other parts must be replaced if defective. Rebuilding radiators and other parts are jobs for specialty shops and not part of general system service.

Proper drive belt condition and tension are essential for correct operation of the water pump and other engine accessories. Check belt tension with a tension gauge and adjust it to the manufacturer's specifications.

Exhaust system service consists of inspecting for leaking or restricted pipes, mufflers, and catalytic converters and replacing defective components.

REVIEW QUESTIONS

1. Mechanic A says that debris in the radiator core fins should be removed with a fine wire brush. Mechanic B says such debris should be blown away with compressed air from the front of the radiator. Who is right?

a. A only
b. B only
c. Both A and B
d. Neither A nor B

2. Mechanic A says the specific gravity of coolant should be greater than 1.00. Mechanic B says you don't have to check the specific gravity if the coolant looks clean. Who is right?

a. A only
b. B only
c. Both A and B
d. Neither A nor B

3. Coolant specific gravity can be tested with a:

a. thermometer
b. exhaust gas analyzer
c. hydrometer
d. pressure tester.

4. A cooling system pressure test can indicate any of the following problems EXCEPT:

a. water jacket leaks
b. faulty radiator cap
c. stuck thermostat
d. deteriorated radiator hose.

5. Mechanic A says cooling system aeration can only be caused by an exhaust leak into the coolant. Mechanic B says you can check for an exhaust leak using a pressure tester. Who is right?

a. A only
b. B only
c. Both A and B
d. Neither A nor B

6. When a cold engine is started, coolant starts to circulate almost immediately in the top tank and temperature rises steadily to about 150° F (66° C). What is the most likely conclusion?

a. The thermostat has been removed from the system.

b. The system is operating normally.
c. An exhaust leak is heating the coolant.
d. The thermostat is stuck closed.

7. Mechanic A says the coolant temperature switch controlling an electric fan should open (show no continuity) when the coolant reaches normal operating temperature. Mechanic B says the switch can be tested with a test light or ohmmeter. Who is right?
a. A only
b. B only
c. Both A and B
d. Neither A nor B

8. While testing a clutch fan using a timing light, fan speed increased at the clutch's engagement temperature. What is the most likely conclusion?
a. The clutch fluid is leaking.
b. The bimetal sensor is faulty.
c. The fan is out of balance.
d. The fan is operating normally.

9. Which of the following is NOT part of flushing the cooling system with plain water?
a. Turn the heater to its maximum heat position.
b. Leave the thermostat housing off while flushing.
c. Open the radiator and block drain plugs.
d. Clean the coolant recovery reservoir.

10. Mechanic A says some chemical cleaners cannot be used in aluminum engines. Mechanic B says chemical cleaners should be allowed to boil in the cooling system. Who is right?
a. A only
b. B only
c. Both A and B
d. Neither A nor B

11. After replacing a core plug, you notice coolant leakage at the new plug. You should:
a. apply more nonhardening sealer at the leak
b. tap the plug farther into the core hole
c. remove the plug and install a new one
d. install a copper expansion plug on top of the leaking plug.

12. Which of the following parts should NOT have to be removed for radiator removal?
a. Thermostat housing
b. Cooling fan
c. Transmission cooler lines
d. Fan shrouds

13. Mechanic A says the water pump gasket must be aligned correctly for proper coolant flow. Mechanic B says water pump bolts should be coated with thread sealant. Who is right?
a. A only
b. B only
c. Both A and B
d. Neither A nor B

14. Mechanic A says that during a belt deflection test, the belt should move 1/8 to 3/8 in. Mechanic B says the belt deflection test can be used on V-ribbed serpentine belts. Who is right?
a. A only
b. B only
c. Both A and B
d. Neither A nor B

15. Which of the following statements about the use of a belt tension gauge is NOT true?
a. Most gauges read in either pounds or newtons.
b. New-belt tension specifications are much higher than those for used belts.
c. Any belt is considered "used" after it has been run for 15 minutes.
d. Tension should be checked on both belts of a matched set at the same time.

16. Which of the following is NOT an accepted method of applying tension to a drive belt?
a. Turning an adjustment screw on the accessory bracket.
b. Prying on the aluminum body of an accessory unit.
c. Prying against an accessory unit in a specified location.
d. Using a ratchet handle to apply force to a hole or slot.

17. Mechanic A says an exhaust system should be disassembled from the rear of the system forward. Mechanic B says the joints will be easier to separate if the system is at operating temperature. Who is right?
a. A only
b. B only
c. Both A and B
d. Neither A nor B

18. Which of the following is NOT a common exhaust system joint?
a. Flange joint
b. Ball-joint
c. U-joint
d. Slip-joint

19. During a coolant circulation test, which of the following describes normal system operation?
a. Radiator is cool near the inlet and warm near the outlet.
b. Radiator has large cold spots in the center of the core.
c. Radiator is hot at both the inlet and the outlet.
d. Radiator is hot at the inlet and warm at the outlet.

20. Mechanic A says that pressure should be pumped up rapidly when testing the radiator cap pressure relief valve. Mechanic B says the pressure will stop increasing when the valve has opened. Who is right?
a. A only
b. B only
c. Both A and B
d. Neither A nor B

21. During an exhaust leak pressure test, the pressure tester needle vibrates within the pressure range of the cap. Which of the following is true?
a. The system is operating normally.
b. Shorting each spark plug in turn will reveal the leaking cylinder.
c. The radiator cap pressure relief valve is faulty.
d. Increasing pressure beyond system specifications will reveal the source of the leak.

22. Mechanic A says that after a badly contaminated cooling system is chemically flushed, it should be drained and flushed again within a week or two. Mechanic B says some types of chemical flushing agents must be followed by a neutralizer. Who is right?
a. A only
b. B only
c. Both A and B
d. Neither A nor B

PART THREE
ELECTRICAL SYSTEM SERVICE

INTRODUCTION

You can't service a late-model engine system if you can't trace an electrical circuit. The skills of an electrical or electronic service technician are essential for a professional automobile service technician. Electrical service has always been a challenging field of automotive service. There is no mystery to it, however, if you master the principles of electricity and electronics that you have studied already.

Chapter 8 outlines how to use wiring diagrams and explains the symbols and different kinds of drawings used by various carmakers. This chapter also covers circuit protection devices and power distribution, as well as the basic procedures for repairing wiring and connectors. You will use the information in this chapter throughout the rest of this volume.

Because the battery is the heart of the complete electrical system on any vehicle, all electrical service begins at the battery. Chapter 9 has general instructions for battery inspection, cleaning, testing, charging, and replacement.

The starting and charging systems directly affect engine performance. You can't service an ignition system or an engine electronic module or computer if the engine doesn't start properly or if charging system voltage and current are out of limits. Chapters 10 and 11 contain the basic procedures for testing and servicing the starting and charging systems. These chapters also present photo sequences as examples of typical overhaul methods for starter motors and alternators.

8 ELECTRICAL CIRCUIT TRACING AND WIRING REPAIR

INTRODUCTION

Chapter 2 of this volume introduced reference manuals as essential tools for engine performance service. Among the most important items in reference manuals are electrical diagrams. Modern engine performance service requires that you understand and use circuit diagrams for engine electrical and electronic systems. They are essential day-to-day tools.

Many engine performance problems can be traced to defects in wiring and connectors. Often, faulty operation of an electronic sensor or actuator or a system computer is not caused by a defective part but by a problem in the wiring or a connector. Engine electronic systems operate on low voltage and current. Any excess resistance due to a poor connection or a worn wire can upset system operation. Correct diagnosis and repair of a $3.00 connector can avoid unnecessary replacement of a $300.00 computer.

This chapter explains the principles of electrical circuit diagrams and summarizes the basics of wiring and connector service. You will use these skills in all areas of engine performance service.

GOALS

After studying this chapter, you should be able to:

1. Trace a wiring or current flow diagram for an engine electrical circuit and identify the current path, circuit components, circuit numbers, color coding, and connections.

2. Select replacement wiring or cables of the correct type and gauge and repair defective wiring by crimping or soldering.

3. Repair various connectors used on late-model engine wiring harnesses.

ELECTRICAL DIAGRAMS

Chapter 2 introduced three kinds of diagrams: schematic, wiring, and installation. Installation diagrams are mechanical drawings that show wire harness routing, connectors, and the location of system parts. These drawings also often include fasteners and other mounting hardware to show you how to install the harnesses and system parts. Installation drawings sometimes show connector and pin numbers. They do not show circuit numbers, wire color codes, the operation of system parts, current paths, or other electrical data.

Installation drawings are some of the simplest electrical diagrams and are essential for easy identification and service of connectors and other parts. Figures 8-1 and 8-2 are typical installation drawings for engine compartment electrical systems.

Schematic diagrams use symbols to represent circuit devices, figure 8-3. The symbols also show how the devices operate. Wiring diagrams show the wires, connectors, circuit numbers, and color coding. Most carmakers combine the features of schematic and wiring diagrams into overall electrical system diagrams. Whether you call them "wiring diagrams," "electrical diagrams," "circuit diagrams," or "current flow diagrams," they contain the information you need to trace a circuit and test or repair a part.

To trace a circuit with any kind of diagram, you will use the basic troubleshooting methods explained in chapter 3. Be sure you understand the basic uses of electrical meters, test lamps, and jumper wires before using a diagram to test and repair a circuit.

Wiring Diagrams

Many complete wiring diagrams are drawn to follow the general arrangement of parts and wiring on the vehicle. Notice in figure 8-4 that the front lamps are at the left edge of the page, with the right front at the top. This establishes the orientation of the entire diagram in relation to the vehicle. The battery is shown behind the right front lamps, which indicates its location on the car. Similarly, the alternator is shown at the bottom of the page, which indicates its location at the left front of the engine compartment.

Some diagrams have reference co-

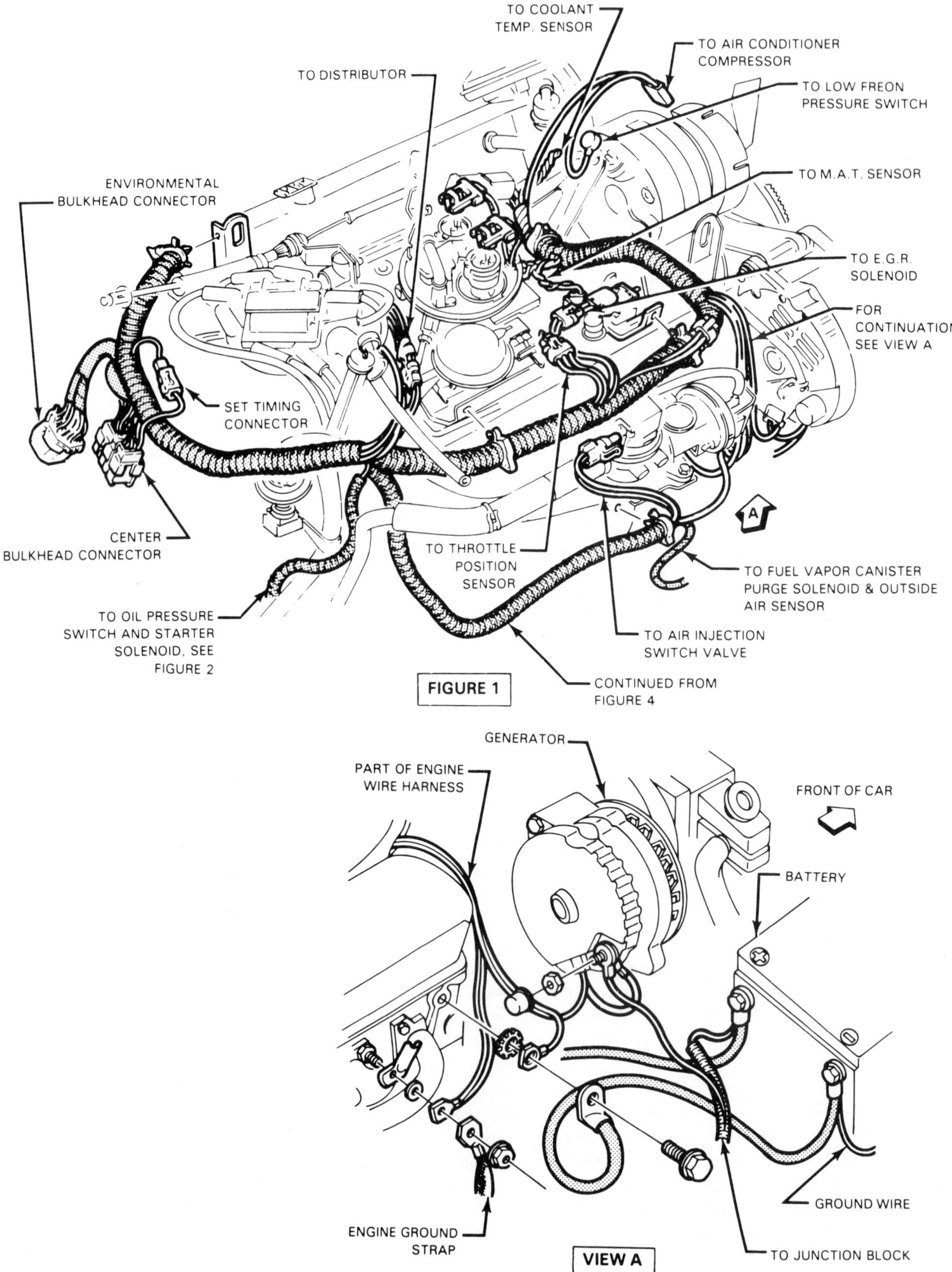

Figure 8-1. This is the main engine harness installation drawing for a late-model car. (Cadillac)

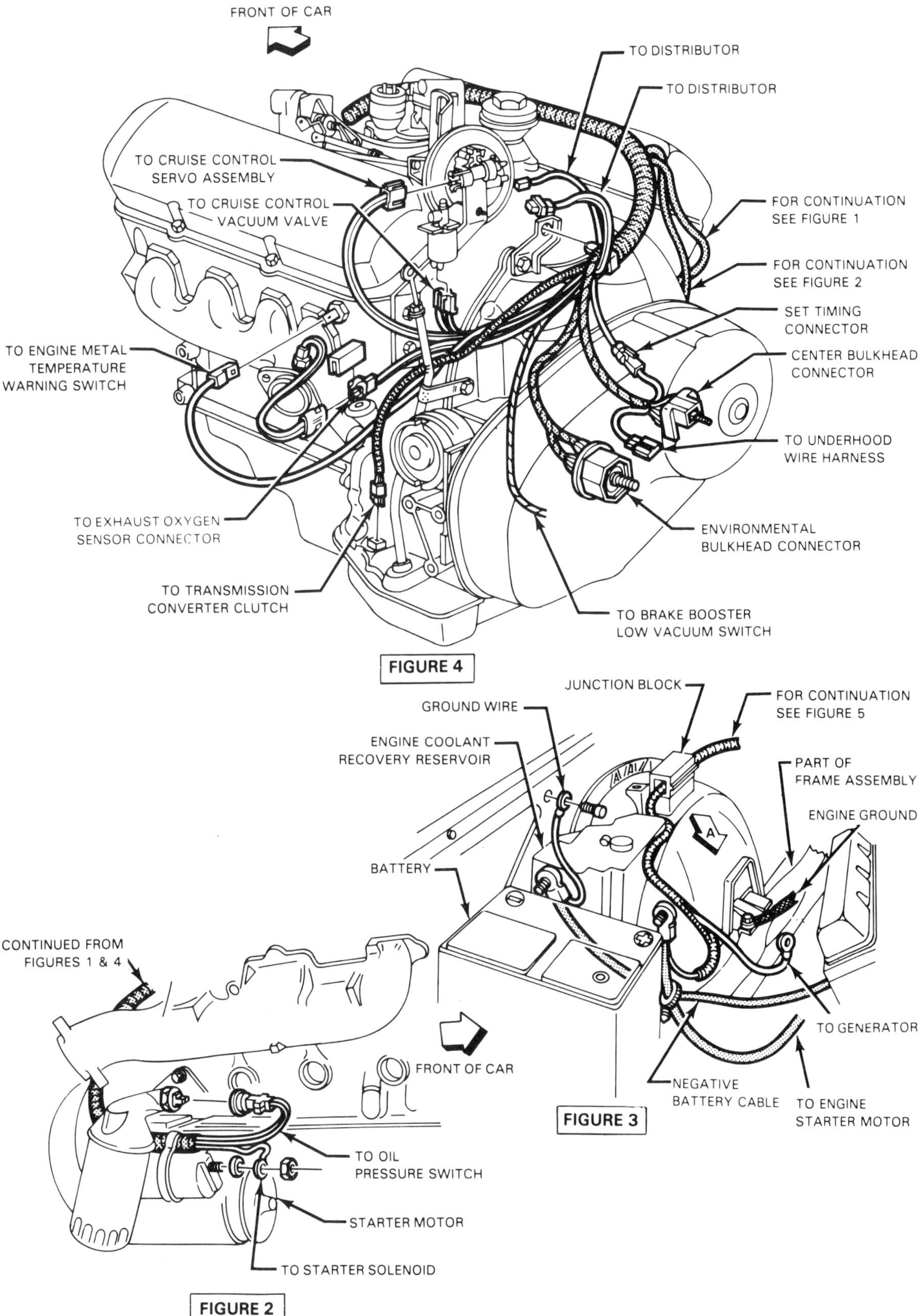

Figure 8-2. These drawings are continuations of the installation diagrams in figure 8-1. (Cadillac)

ELECTRICAL DEVICE	SYMBOL	
	UNITED STATES	EUROPE
BATTERY		
SWITCH, SINGLE POLE, NORMALLY OPEN		
SWITCH, SINGLE POLE, NORMALLY CLOSED		
SWITCH, GANGED		
RESISTOR		
VARIABLE RESISTOR		
DIODE		
ZENER DIODE		
TRANSISTOR	(PNP TYPE)	(NPN TYPE)
THERMISTOR		
CIRCUIT BREAKER		
FUSE		
CAPACITOR (CONDENSER)		
LAMP, SINGLE FILAMENT		
LAMP, MULTIFILAMENT		
CONNECTOR	MALE FEMALE	
MULTIPLE CONNECTOR		
IGNITION COIL		
RELAY		
GROUND		
SOLENOID		
MOTOR		
GENERATOR (ALTERNATOR)		
WIRING SPLICE		

Figure 8-3. These are the basic electrical symbols used in most automobile electrical diagrams.

ordinates along the edges to divide the drawing into a series of grids to locate parts, figure 8-5. An index accompanies such drawings that lists system parts alphabetically, along with the grid coordinates to find them on the diagram. For example, the index in figure 8-6 indicates that the electric choke is located at coordinates C-3.

Some carmakers arrange wiring diagrams to show related circuits or subsystems on sequential pages. Figure 8-7 is a Chrysler index that lists the electronic spark advance system on sheet 11 of the complete wiring diagrams for one carline. The diagram for that spark advance system, figure 8-8, contains references to other sheets for circuit continuations.

Circuit Numbers and Color Coding. The diagrams in figures 8-4, 8-5, and 8-8 list the circuit numbers for each wire and the color of the wire insulation. The Ford electric choke in figure 8-5 is part of circuit 4, and the wire is white with a black stripe. Circuit 4 is a simple circuit that normally receives power directly from the alternator and is grounded through the electric choke. Circuit 4 also provides power to the regulator field relay.

The General Motors diagram, figure 8-4, shows circuit numbers as part of each connector symbol. The choke heater is part of circuit 78, and the wire is light blue. GM also lists the wire size in metric specifications of square millimeters, cross-sectional area. The choke heater wire is 0.8-mm^2 wire.

The Chrysler diagram in figure 8-8 also lists circuit numbers and color codes for each wire. The Chrysler choke heater is circuit J20. The wire is 14 gauge with dark blue insulation.

Wire color code abbreviations are fairly standard throughout the industry, but reference manuals have tables that list the color codes and abbreviations, figure 8-9. Each manufacturer has its own circuit numbering system, but again, diagram manuals have tables that explain the numbers. Figure 8-10 is an example.

Wire Size. We mentioned that the GM diagram, figure 8-4, showed wire size in metric specifications and that the Chrysler diagram, figure 8-8, listed wire size by gauge number. Two standard systems for wire sizes are the **American Wire Gauge (AWG)** and the metric systems. AWG wire sizes are indicated by number. The higher the number, the smaller the wire cross-sectional area. Number 18 wire is smaller than number 14 wire, for example. Engineers determine wire sizes by the current loads of each circuit. High-current circuits require heavier wire.

Automobile electrical system wiring typically uses 14-, 16-, and 18-gauge wire. High-current power distribution circuits may use 10- or 12-gauge wire. Low-current electronic circuits may use 20-gauge wire. All of these wire sizes often are called **primary wiring** because it is the kind of wiring used for

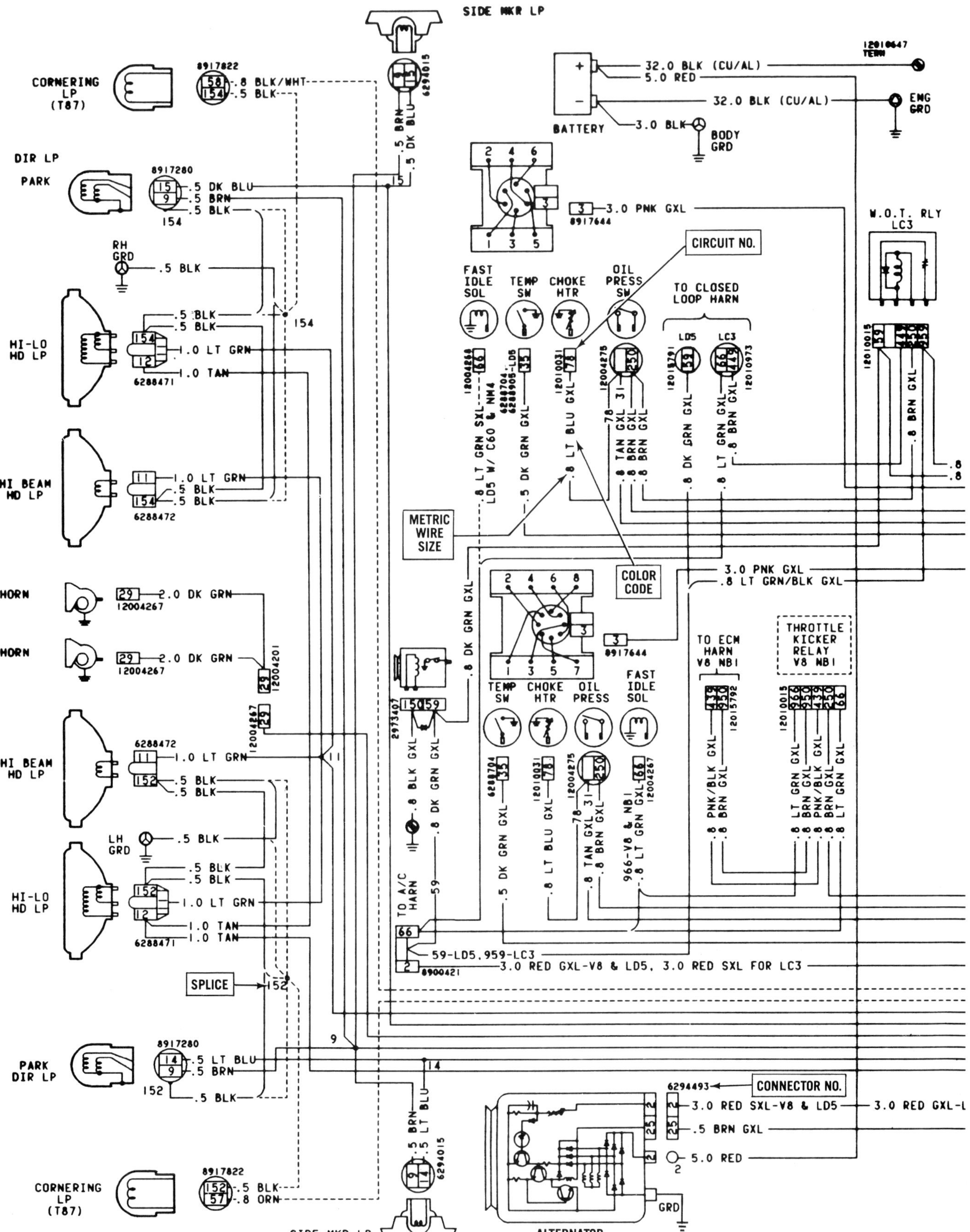

Figure 8-4. This is the first part of a typical chassis wiring diagram for a late-model GM car. (Chevrolet)

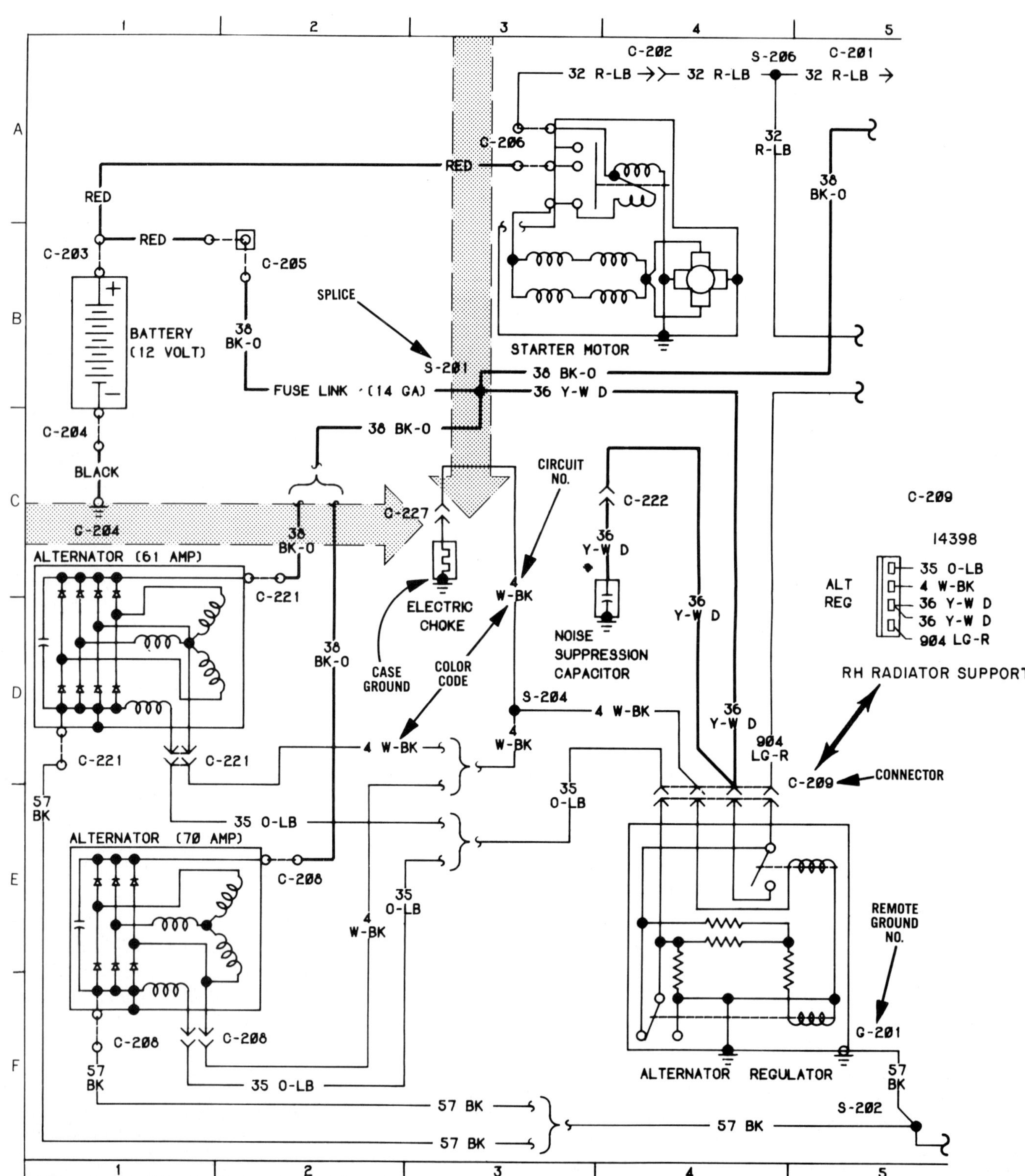

Figure 8-5. This is part of a chassis wiring diagram for a late-model Ford car. (Ford)

C	DISTRIBUTOR	D-6.
	ELECTRIC CHOKE	C-3.
	EMERGENCY WARNING FLASHER	D-82.
	FADER CONTROL	C-99.
	FUEL GAUGE	C-20.
	FUEL GAUGE SENDER	F-20.

Figure 8-6. The diagram index shows that the electric choke is in area C-3. (Ford)

ALPHABETICAL INDEX

Name	Wiring Diagram Sheet Number
Air Conditioning System 1.7L	19
Air Conditioning System 2.2L	21
Air Conditioning Low Pressure Cutout Switch-ALL	17
Air Conditioning Push Button Switch-ALL	17
Electric Choke Heater 1.6L	9
Electric Choke Heater 1.7L-2.2L	11
Electrically Heated Rear Window	63
Electronic Spark Advance System 1.6L	9
Electronic Spark Advance System 1.7L-2.2L	11
Fast Idle Stop Solenoid 1.7L	16
Fast Idle Stop Solenoid 2.2L	13-16
Front End Lighting MZ-24-28	25
Front End Lighting MZ-44	27

Figure 8-7. This Chrysler diagram index lists the electronic spark advance system on sheet 11. (Chrysler)

low-voltage, low-current ignition primary circuits. For comparison, battery cables are typically 2-, 4-, or 6-gauge wire.

Metric wire sizes are listed by the cross-sectional area of the wire in square millimeters. The smaller the number, the smaller the wire. Figure 8-11 lists equivalent metric and AWG wire sizes.

Other Diagram Information. Wiring diagrams also show connector and splice symbols and identify them by number. The Ford diagram, figure 8-5, indicates splices with the letter "S" and a number: S-201, S-202, and S-204, for example. This diagram indicates connectors with the letter "C" and a number. C-209 is the 4-terminal connector for the alternator regulator. A separate detail drawing shows the terminal arrangement in this connector and its location on the right-hand fender support. The drawing also shows that connector C-209 is attached to harness number 14398.

Connectors on the GM diagram, figure 8-4, are identified by 7-digit part numbers. The connector on the rear of the alternator, for example, is number 6294493.

The Chrysler diagram, figure 8-8, indicates splices with the circuit number inside a diamond symbol. If there is more than one splice per circuit, the splice number is in a small box attached to the diamond symbol. This diagram shows the physical shape of the connectors with detail drawings that indicate terminal location, circuit number, and wire identification for each connector.

Circuit Grounds. Automobile electrical systems use the vehicle chassis and engine as the ground side of all circuits, figure 8-12. This completes the low-voltage return side of each circuit to the battery. Diagrams indicate the ground locations of all circuits and loads.

Circuit loads can be grounded directly through the case of the device or remotely by system wiring. The electric chokes in figures 8-4, 8-5, and 8-8 all have direct **case grounds**. Case grounds are used where a device is installed with electrical continuity to the engine or chassis. **Remote grounds** are used where continuity does not exist. Because of the nonmetallic materials used in late-model instrument panels and exterior lamp housings, gauge and lamp circuits usually have remote grounds.

Circuit 57 in the Ford diagram, figure 8-5, is a remote ground connection for the alternator, the regulator, and other devices. Circuits with remote grounds usually have ground connection numbers and locations listed on the diagrams. Remote ground wiring is usually black.

Switches and Relays

The electrical symbols in figure 8-3 are the basic, simple symbols used by most carmakers. Some diagrams, however, have symbols that appear more complex. Most of these really are combinations of simple symbols, or symbols with more detail. Switches are good examples. The simple switch in figure 8-3 is a single-pole, single-throw (spst) switch that can be shown in open (off) or closed (on) positions. The term "pole" indicates the number of power (input) connections on the switch. The term "throw" indicates the number of output connections on the switch. An spst switch can open and close one circuit. A single-pole, double-throw (spdt) switch, figure 8-13, can alternately open and close two output circuits.

A switch can have as many poles and throws as it needs to do its job and as the designer can fit into the switch body. Some switches are "ganged" switches that have multiple poles, throws, and wipers. The wiper

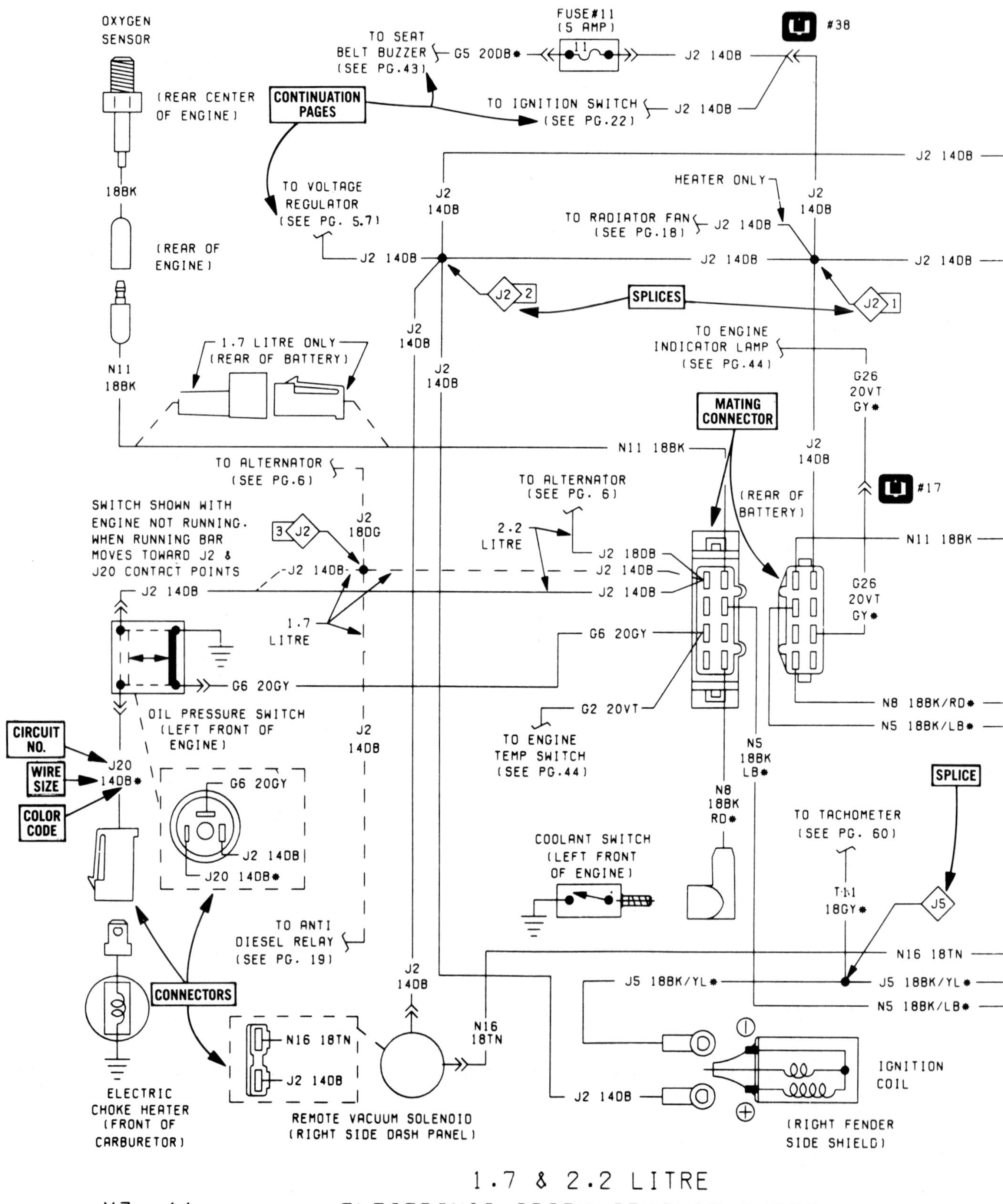

Figure 8-8. This is a typical sheet from Chrysler wiring diagrams. (Chrysler)

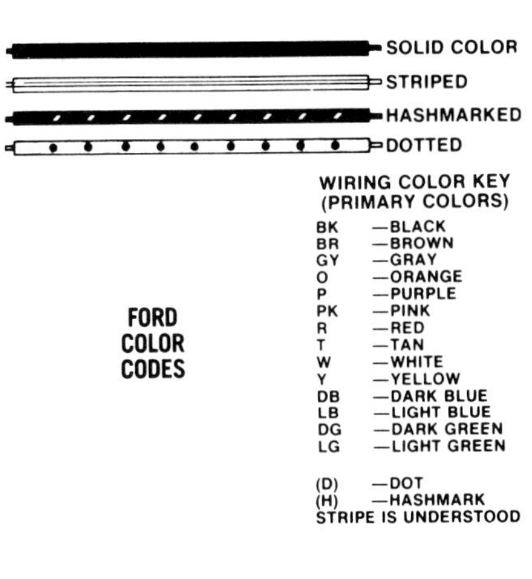

WIRING COLOR CODE CHART					
COLOR CODE	COLOR	STANDARD TRACER COLOR	COLOR CODE	COLOR	STANDARD TRACER CODE
BK	BLACK	WH	PK	PINK	BK OR WH
BR	BROWN	WH	RD	RED	WH
DB	DARK BLUE	WH	TN	TAN	BK
DG	DARK GREEN	WH	VT	VIOLET	WH
GY	GRAY	BK	WT	WHITE	BK
LB	LIGHT BLUE	BK	YL	YELLOW	BK
LG	LIGHT GREEN	BK	•	WITH TRACER	
OR	ORANGE	BK			

CHRYSLER COLOR CODES

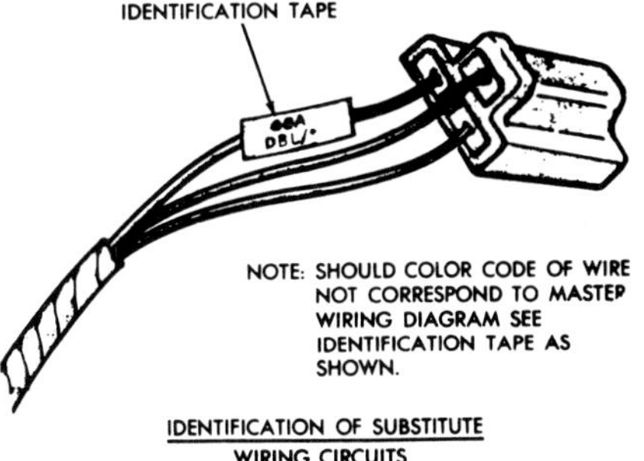

Figure 8-9. Typical wire color code tables. (Ford) (Chrysler)

Metric Size	AWG Size
.22	24
.35	22
.5	20
.8	18
1.0	16
2.0	14
3.0	12
5.0	10
8.0	8
13.0	6
19.0	4
32.0	2

Figure 8-11. These are approximately equivalent AWG and metric wire sizes.

MAIN CIRCUIT IDENTIFICATION CODES

A1 Battery Circuit to Ammeter.
A2 Battery Circuit to Ammeter.
B Back Up Lamp Circuit.
C Air Conditioning and Heater Circuits.
D Emergency, Stop Lamp and Turn Signal Circuits.
E Instrument Panel Cluster, Switches and Illumination Circuits.
F Radio Speakers and Power Seat Circuits.
G Gauges and Warning Lamp Circuits.
H Horn Circuit.
J Ignition System Run Circuit.
J1 Ignition Switch Feed Circuit.
J3 Ignition Switch Start Circuit.
K Trailer Tow.
L Lighting Circuit (Exterior Lights).
M Lighting Circuit (Interior Lights).
P Brake Checking Circuit.
Q2 Accessory Buss Bar Feed (Fuse Block).
Q3 Battery Buss Bar Feed (Feed).
R3 Alternator Circuit to Electronic Voltage Regulator (Field).
R6 Alternator Circuit to Ammeter (Feed).
S Starter Motor and Starter Relay Circuit.
T Trunk Lamp Circuit.
V Windshield Wiper and Washer Circuit.
W Power Window Circuit.
X Radio, Cigar Lighter, Lamp Grounds, Clock, Speed Control, Power Antenna, Deck Lid and Door Locks.

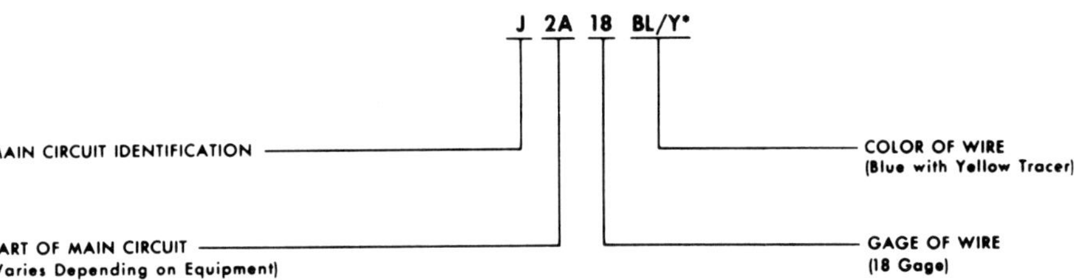

Figure 8-10. Typical circuit identification table. (Chrysler)

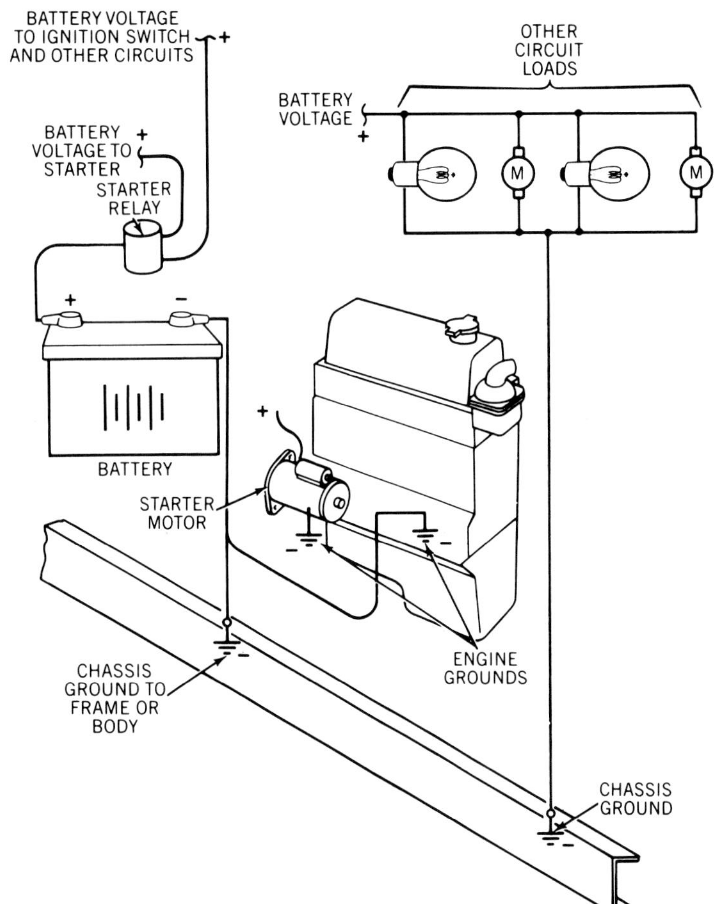

Figure 8-12. All vehicle electrical circuits have a ground connection to the chassis or the engine.

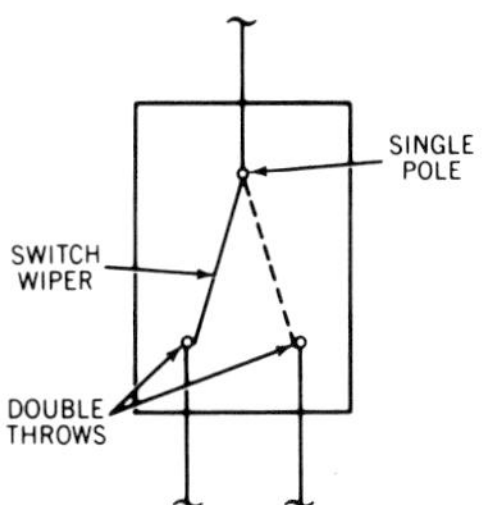

Figure 8-13. Typical single-pole, double-throw (spdt) switch symbol.

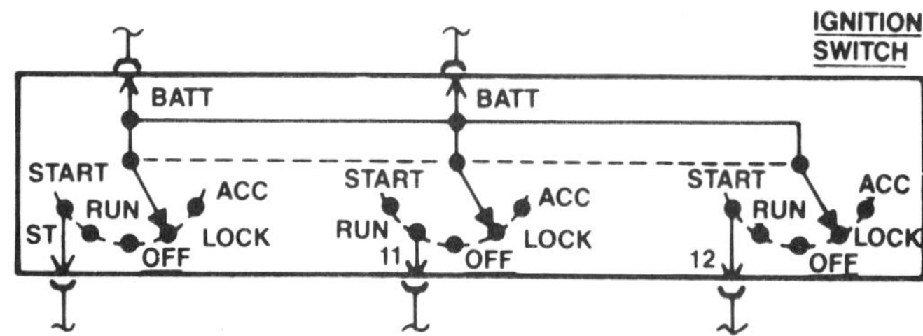

Figure 8-14. This typical ignition switch symbol illustrates a switch with multiple poles and throws and ganged wipers. (Ford)

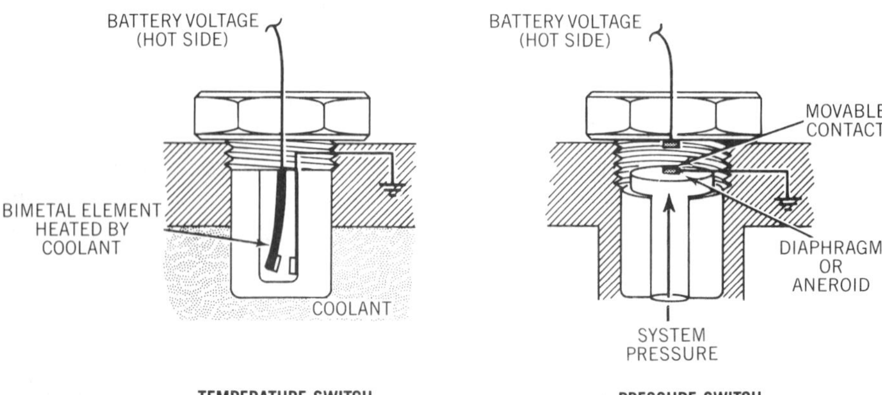

Figure 8-15. Temperature and pressure switches can be normally open or normally closed.

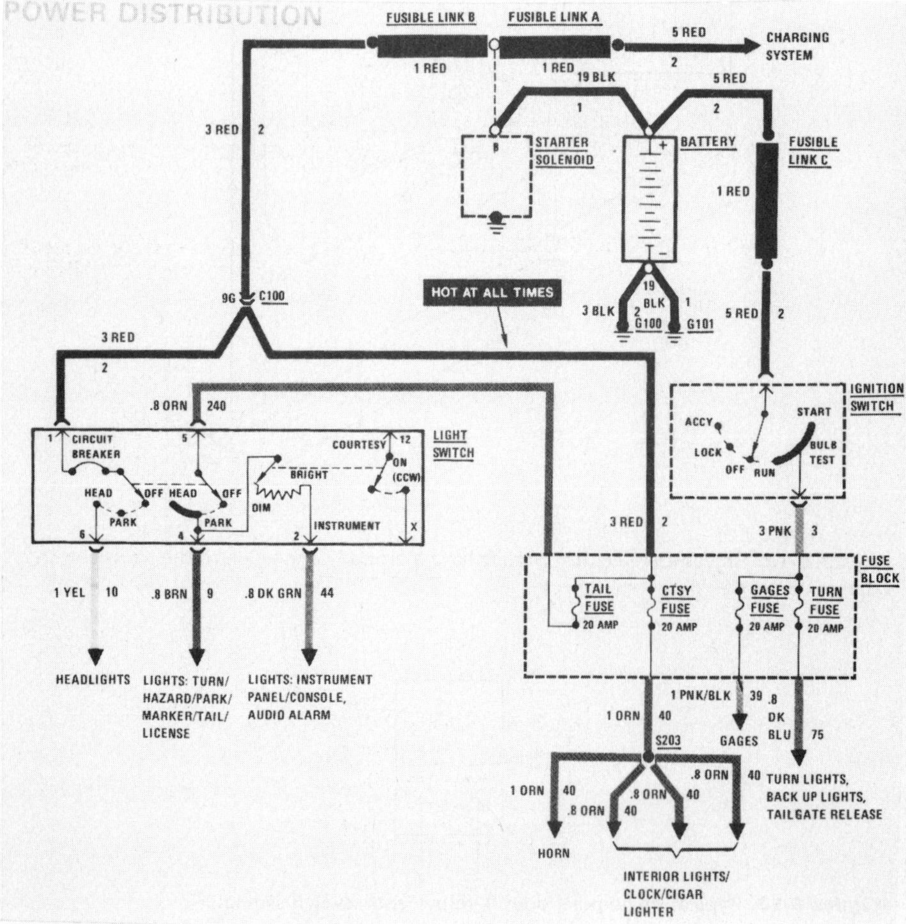

Figure 8-16. This is a typical GM power distribution schematic. (Chevrolet)

is the movable part of the switch that opens and closes circuits. An ignition switch is a ganged switch with several poles, throws, and wipers. Figure 8-14 is a typical ignition switch symbol used by most carmakers. The dashed line between the wipers indicates that they are ganged and move together.

Engine electrical systems have other switches besides simple and complex manual switches. Many switches are operated by temperature or pressure. A simple temperature switch used to sense coolant temperature has a bimetal element that bends as it heats and cools. The arm opens and closes the switch contacts as temperature changes.

An engine oil pressure switch can open and close circuits for other devices besides warning lamps. Many engines have oil pressure switches to control electric fuel pumps or electric chokes. The switch opens the pump circuit when the engine is off or oil pressure is too low. A vacuum switch is another pressure switch that can control engine circuits or send a voltage signal to a computer in response to engine manifold pressure. Figure 8-15 shows examples of typical temperature and pressure switches.

Engine electronic control systems also use relays as remote switches. Every gasoline fuel injection system has an electric fuel pump relay to control pump operation and other related system devices. Similarly, most electronic engine control systems have a main power relay, energized by the ignition switch, to provide power to the computer.

Switch and Relay Locations. Knowing the locations and operating positions of switches and relays is important for troubleshooting. Switch and relay positions determine which circuit devices are hot (have voltage applied to them) under various conditions.

Switches are described as normally open or normally closed. "Normal" refers to the switch position when it is *not* actuated, or has no outside force applied to it. A circuit with a normally open switch is open until the switch

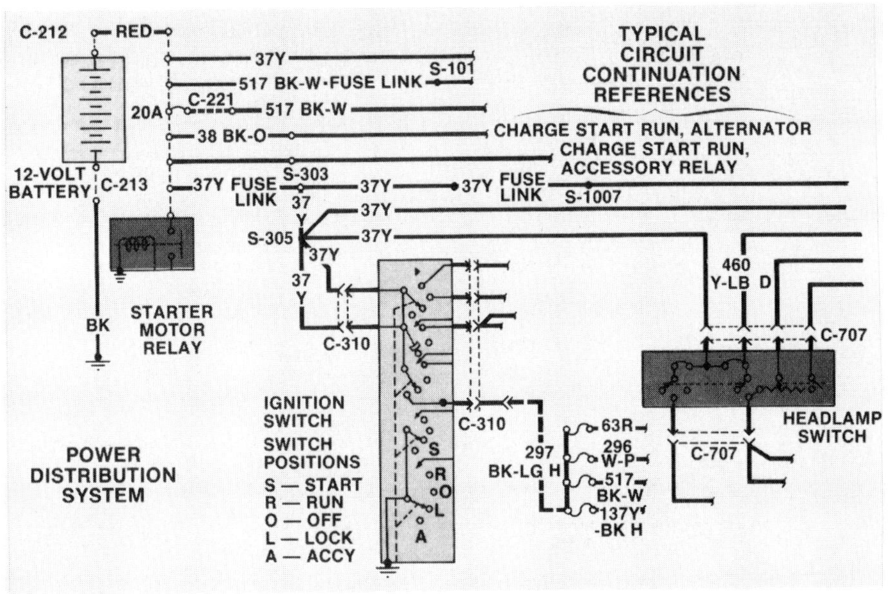

Figure 8-17. This is a typical power distribution drawing from a Ford wiring diagram. (Ford)

is operated. A normally closed switch provides the opposite kind of control. These descriptions apply to temperature and pressure switches, as well as manual switches. The control and power circuits of a relay also may be normally open or normally closed, depending on circuit requirements. Most relays, however, have normally open circuits and need outside force to operate them.

Switches can be placed on the hot side or the ground side of circuit devices. The hot side of a circuit has voltage applied to it; the ground side is connected to ground. Switch locations determine the conditions when circuit devices are hot, which is important for accurate testing. The ignition switch is always on the hot side

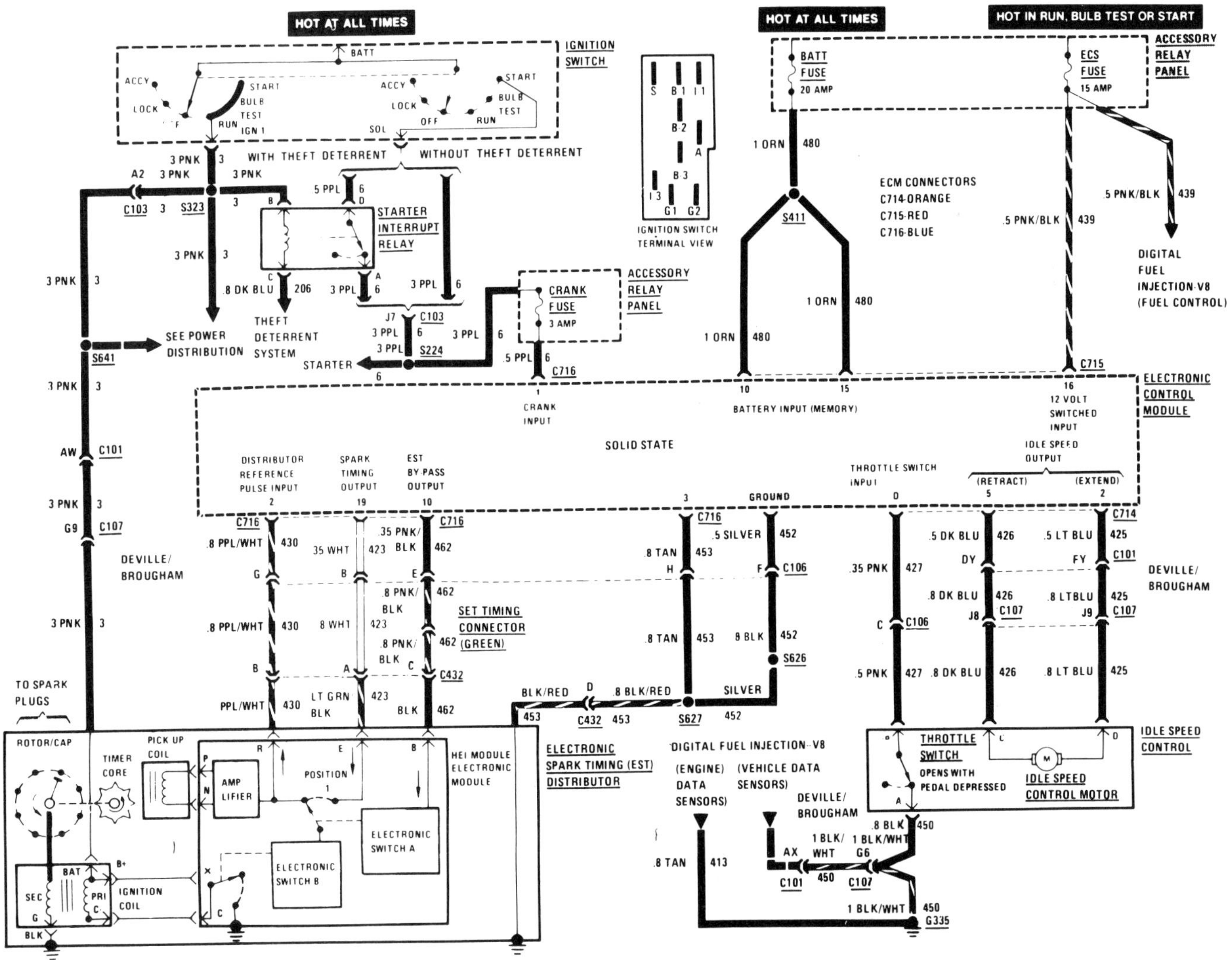

Figure 8-18. This circuit schematic shows the electronic control module (ECM) power, idle speed control, and electronic spark timing parts of an electronic engine control system. (Cadillac)

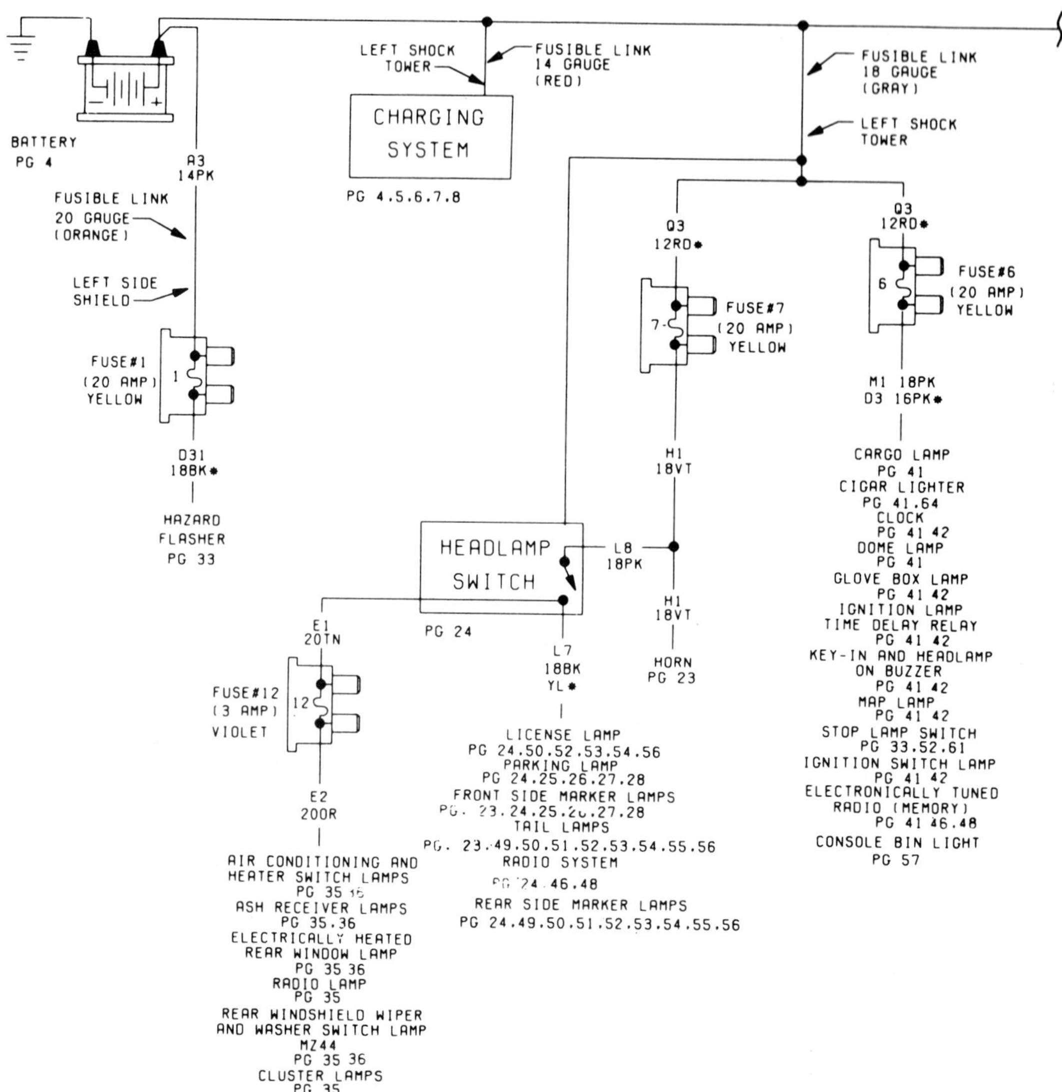

Figure 8-19. This Chrysler fuse application chart has references to other pages for various circuits. (Chrysler)

of the circuits it controls. Relays may be on the hot side or the ground side of a circuit.

Power Distribution and Current Flow Diagrams

In addition to complete vehicle wiring diagrams, many carmakers provide power distribution and current flow diagrams to aid troubleshooting. These diagrams usually do not show wiring and part locations in relation to the vehicle. They are schematic illustrations of circuit functions. Domestic carmakers call these diagrams "power distribution diagrams" or "circuit schematics." European carmakers usually call them "current flow diagrams." All have similar basic features to show circuit operation.

Figure 8-16 is a typical GM power distribution schematic that shows battery connections to the complete electrical system through portions of the ignition switch, the light switch, and the fuse block. Figure 8-17 is part of a similar Ford power distribution diagram. Notice that power enters the system at the top or the upper left part of the drawing. Grounds are shown toward the bottom or lower right. Power distribution diagrams or current flow drawings are drawn so that voltage drops from top to bottom on the page and current flows downward.

Although the battery is the source of electrical power for all systems, most engine electrical circuits receive battery voltage from the ignition switch or the fuse block. Therefore, individual circuit schematics are usually drawn with power originating at these points. Figure 8-18 shows power distribution for the electronic control module (ECM), electronic ignition, and idle speed control circuits of an electronic engine system. The legends, HOT AT ALL TIMES, indicate where battery voltage is applied to the circuits. In this diagram, you can identify switches and relays that are on the hot and the ground sides of different load devices. Circuit grounds are at the bottom of the drawing.

Other carmakers include fuse application charts in their wiring diagrams. These are similar to power distribution schematics. Figure 8-19 is a Chrysler example.

Figure 8-20 is an example of current

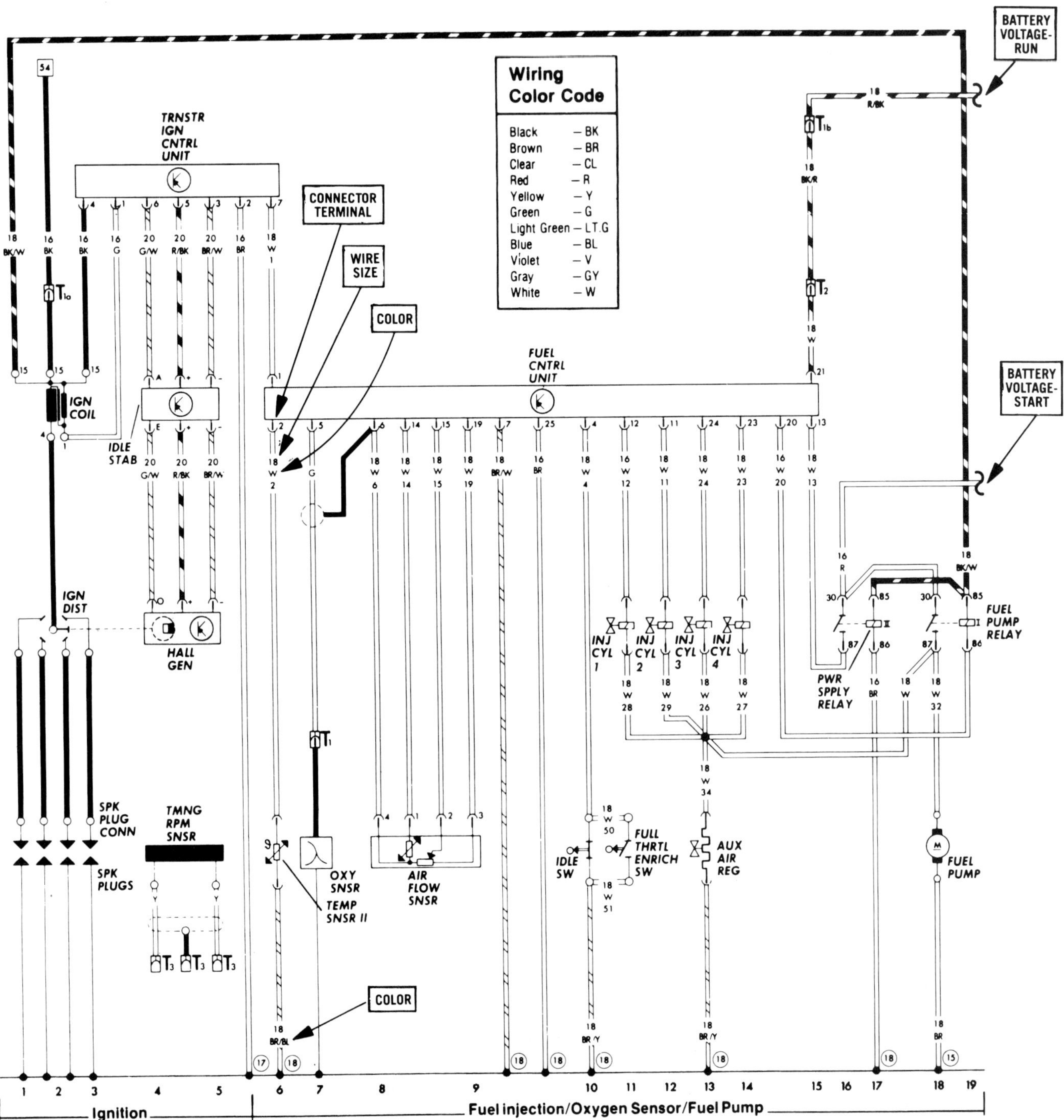

Note: All wire sizes American Wire Gauge

Figure 8-20. This Volkswagen current flow diagram is typical of schematics provided by several European carmakers. (Volkswagen)

flow diagrams used by many European carmakers. Again, hot sides of circuits are in the upper areas of the diagram. Voltage drops down the page, and current flows downward. Ground connections are shown by circled numbers at the bottom of the diagram. Notice that wires that do not end with ground connections at the bottom of the diagram are grounded through other components, such as the ignition control unit and fuel control unit. Numbers 1 through 19 at the bottom are "current paths." The index for this diagram lists current paths by number and circuits by name, figure 8-21. The index also lists the locations of ground connections and circuit connectors.

These current flow diagrams also show connector terminal numbers, wire size, and color coding. In figure 8-20, for example, current path 6 is the engine temperature sensor. It has 18-gauge white wire and is connected to terminal 2 on the control unit. It ends at ground connection 18 (current path 17). The wire color changes to brown with a blue tracer between the sensor and ground. This diagram does not assign current path numbers to

Description	Current Track
Air flow sensor	8, 9
Alternator	24-26
Alternator warning light	26
Auxiliary air regulator	13
Battery	20
Back-up light, left	55
Back-up light, right	56

Wire connectors

T1	-single, in engine compart. left
T1a	-single, in connector housing
T1b	-single, in connector housing
T1c	-single, near alternator in engine compart.
T1d	-single, behind dash

Ground connectors

①	-from battery to body
②	-from transmission to body

⑮	-near fuel pump at floor board/crossmember
⑰	-near ignition distributor
⑱	-left at cylinder head
⑲	-near ignition coil, in engine compart., left
㉑	-plus connection, in connector housing

Figure 8-21. The VW current flow diagram index lists circuit names, current paths, connector locations, and ground locations. (Volkswagen)

control unit ground connections that have no loads.

BASIC ELECTRICAL SYSTEM PARTS

Other basic electrical parts are common to most circuits. These include switches, relays, solenoids, motors, and other electromagnetic devices.

Switches

A switch controls circuit operation by opening and closing the current path. Like a connector, a switch is an extension of the circuit conductors. It should have little or no resistance and little or no measurable voltage drop across it.

The simple switch in figure 8-22A is a single-pole, single-throw (spst) switch. The symbol for a switch can

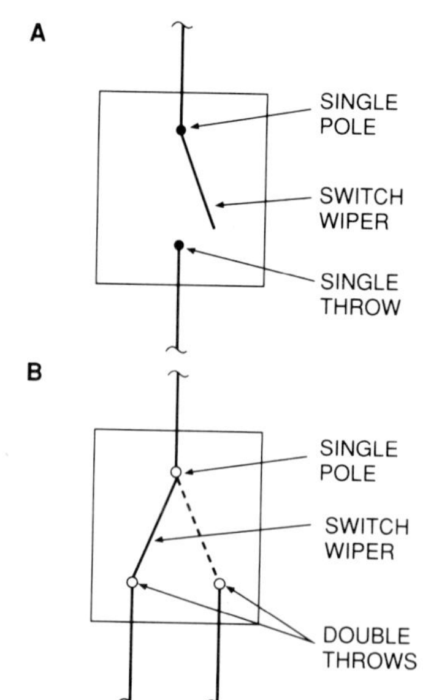

Figure 8-22. The symbol shown in **A** is used for a typical single-pole, single-throw (spst) switch; the symbol shown in **B** is a typical single-pole, double-throw (spdt) switch.

be drawn to show the open (off) or the closed (on) positions. The term "pole" indicates the number of power (input) connections on the switch. The term "throw" indicates the number of output connections on the switch. An spst switch can open and close one circuit. A single-pole, double-throw (spdt) switch, figure 8-22B, can alternately open and close the two output circuits.

A switch can have as many poles and throws as it needs to do its job and as the designer can fit into the switch body. Some switches are "ganged" switches that have multiple poles, throws, and wipers. The wiper is the movable part of the switch that opens and closes circuits. An ignition switch is a ganged switch with several poles, throws, and wipers. Figure 8-14 is a typical ignition switch symbol used by most carmakers. The dashed line between the wipers indicates that they are ganged and move together.

Besides identifying switches by the number of poles and throws, we also identify them by their normal contact positions. "Normal" refers to the switch position when it is not actuated or has no outside force applied to it.

- In a **normally open (NO) switch**, the contacts are open until an outside force closes them to complete the circuit.
- In a **normally closed (NC) switch**, the contacts are closed to complete the circuit until an outside force opens them.

You will find both kinds of switches, NO and NC, in vehicle electrical systems.

Vehicle electrical systems have other switches besides simple and complex manual switches. Many switches are operated by temperature or pressure. A simple temperature switch used to sense coolant temperature has a bimetal element that bends as it heats and cools. The arm opens and closes the switch contacts as the temperature changes. The contacts of oil pressure switches and vacuum switches open and close in response to pressure. Figure 8-15 shows examples of typical temperature and pressure switches.

Mercury switches and inertia switches operate automatically to open and close circuits in response to movement. A mercury switch has a closed capsule partly filled with mercury, which is a good conductor. In a normally open position, the mercury is at the end of the capsule away from the switch contacts. When the capsule rotates or moves in an arc, the mercury flows to the other end of the capsule and closes the switch. Mercury switches are often used as automatic switches to turn on trunk and engine compartment lamps. They also can be used as rollover switches to open electric fuel pump circuits in a vehicle accident.

An inertia switch is usually a normally closed switch with its contacts held together by a calculated amount of friction or spring tension. When sudden physical movement overcomes the friction or tension, the contacts open. They respond to a sudden change in inertia. Inertia switches are often used as safety switches to open electric fuel pump circuits upon the impact of a vehicle accident. The switch must be reset to its normally closed position by hand.

Switches can be placed on the hot side or the ground side of circuit devices. The hot side of a circuit has voltage applied to it; the ground side is connected to ground. Switch locations determine the conditions when circuit devices are hot, which is important for accurate testing.

Relays

A relay is an electromagnetic switch that works on the principles you learned in Volume 1. Current through a coil creates an electromagnetic field. Relays have two important features that make them common in automotive systems.

1. A relay provides remote control of one or more other circuits by opening and closing a switch in a control circuit.
2. A relay allows a low-current control circuit to switch high current on and off in a power circuit.

Every relay has two circuits in parallel, figure 8-23. The control circuit of the relay has an electromagnetic coil wound around an iron core. The coil has resistance and is the load on the control circuit. The power circuit, or output circuit, of the relay has two switch contacts, one of which is on a movable armature. In this example, the armature is made of spring steel, and its tension holds the contacts open. The contacts are the switch for the power circuit and contain little or no resistance. Here is how the relay control and power circuits in figure 8-23 work:

1. When the switch in the control circuit closes, current flows through the coil and creates an electromagnetic field in the core.
2. The core attracts the movable armature downward against spring tension to close the contacts.
3. The armature contacts close the circuit for the motor so that current flows through the motor.
4. When the control switch opens, current stops through the relay coil, and the electromagnetic field collapses.
5. Spring tension opens the armature contacts, and current to the motor stops.

Like switches, relays can be normally open or normally closed. Both the control circuit and the power circuit of the relay in figure 8-23 are normally open. This is the kind of relay you will find most often in automotive systems. Normally open relays control horns, starter motors, and electric fuel pumps on many cars.

A relay with a single control winding is generally used only for short-term operation, such as in a horn circuit. Relays for continuous operation usually have two control windings. One winding creates the magnetic field and moves the armature. A second, lighter winding maintains the field and holds the armature. A stronger field is needed to move the armature than to hold it. When the armature moves, it breaks the contacts for the first or pull-in winding. The second or hold-in winding keeps the relay energized with less current.

Using the same electromagnetic principles, engineers can design relays with other operating features.

• The relay in figure 8-24 has two sets of contacts to operate two power circuits with one control circuit.

• The relay in figure 8-25 has normally closed power circuit contacts. The coil field opens the power circuit when the control circuit closes.

• The relay in figure 8-26 has opposing normally open and normally closed contacts. Control circuit operation opens one power circuit and closes the other.

Knowing the operating positions of switches and relays is important for

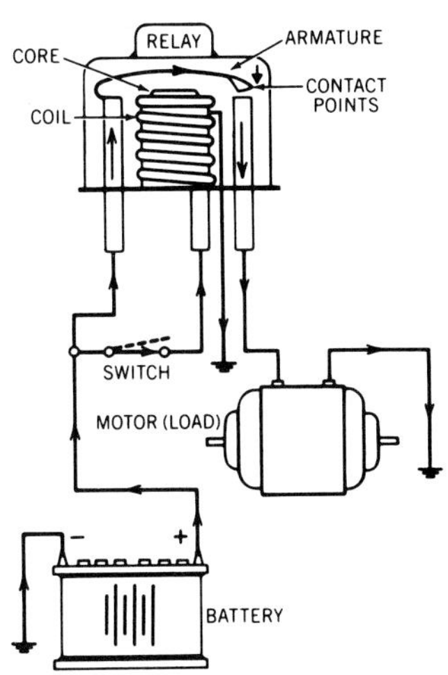

Figure 8-23. This relay is a remote control switch that opens and closes a motor circuit.

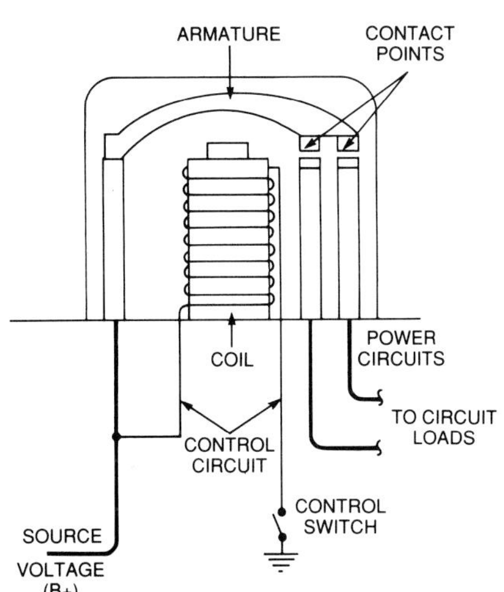

Figure 8-24. A double-contact relay.

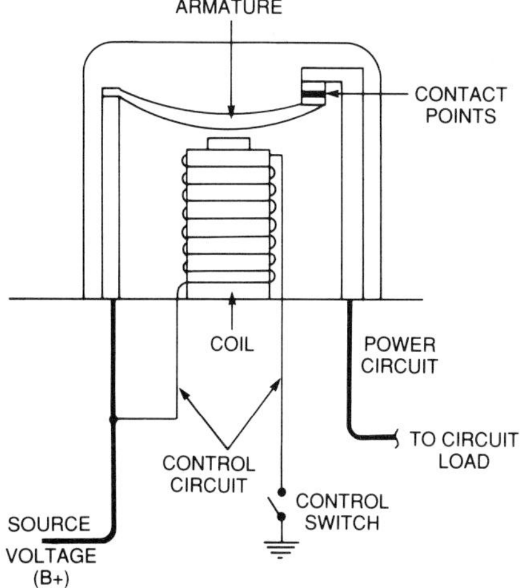

Figure 8-25. A normally closed relay.

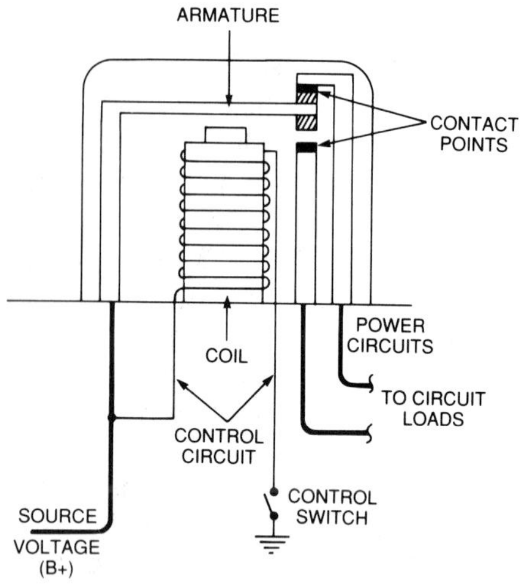

Figure 8-26. A double-acting relay.

troubleshooting. Switch and relay positions determine which circuit devices are hot (have voltage applied to them) under various conditions. Relay operation is also the basis for computer logic and control operations.

Solenoids

A solenoid is an electromagnetic device that works like a relay. A solenoid, however, uses electromagnetism and a movable plunger or armature to do mechanical work. Like a relay, a solenoid has an electromagnetic coil and a core. The solenoid also has a spring-loaded, movable plunger. Electromagnetism in the coil and core attracts the plunger, which moves mechanical linkage. The solenoid is then energized. When current stops in the coil, the magnetic field collapses, and the spring returns the plunger to the deenergized position. A solenoid is a circuit load device. The resistance of its coil changes electrical energy to magnetic and mechanical energy.

The solenoid plunger can open and close a vacuum valve or a fuel valve. It can also operate linkage for door and trunk locks. The most common solenoids on many cars engage the drive gears for the starter motor. Many starter solenoids also function as relays, or electromagnetic switches. The solenoid plunger moves linkage to engage the starter drive gear with the engine flywheel. The plunger also closes the contacts for the high-current circuit from the battery to the starter motor. Chapter 10 explains starter solenoid design and operation in detail.

High-current starter solenoids and solenoids for continous operation usually have separate pull-in and hold-in windings, similar to those in some relays. Plunger movement opens the contacts on the pull-in winding. The hold-in winding maintains the magnetic field and holds the plunger with less current.

Relays and solenoids are direct current devices. They can operate as described here only if current flows continuously in one direction through their coils. Remember that current direction establishes the direction of a magnetic field. If current reversed continually as it does in ac devices, the field would continually reverse. A relay armature or a solenoid plunger would not stay energized. It would vibrate back and forth. Because automotive electrical systems are dc systems, however, relays and solenoids are ideal devices to use for circuit control and electromechanical operations.

Many solenoids in computer-controlled systems operate with intermittent, on-off direct current. You will learn how these solenoids operate as electromechanical actuators in chapter 21. On-off electromagnetic fields are also used in other common electrical parts, such as warning buzzers.

Buzzers

Many circuits contain buzzers to warn the driver of various conditions, such as seatbelts unfastened, keys left in the ignition switch, and doors ajar. A buzzer has an electromagnetic coil and core, as well as a movable armature and a set of contact points. The coil and the armature contacts are in series, however, figure 8-27. Current flows through the coil and the normally closed contacts to ground. Electromagnetism in the core moves the armature to open the contacts and break the circuit. The magnetic field

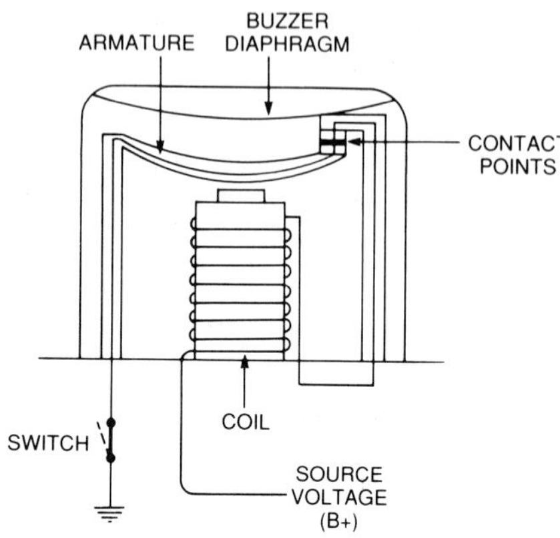

Figure 8-27. In a buzzer, the coil is in series with the armature and contact points. Opening and closing the circuit causes the armature to vibrate against the buzzer diaphragm.

then collapses, and the contacts reclose. The circuit opens and closes many times per second. The resulting vibrations create the buzzing sound.

CIRCUIT PROTECTION

All electrical circuits are protected from excess current by circuit breakers, fuses, and fusible links (fuse links). High current, usually caused by short or grounded circuits, can overheat and damage wiring and components. Circuit protection devices will open a circuit before high current can cause permanent damage. All circuit protection devices are sensitive to current, not voltage, and are rated by current-carrying capacity. A 10-ampere fuse or circuit breaker will conduct up to 10 amperes of current in a 6-, 12-, or 24-volt electrical system. Circuit protection devices are located at, or near, the power source for the circuits they protect. This makes them an important starting point for many troubleshooting sequences.

Circuit Breakers

Circuit breakers are used in circuits where it is important to restore power quickly or where temporary overloads may occur. A circuit breaker has a bimetal element, figure 8-28, similar to the kind in a temperature switch. The two strips of metal expand at different rates when heated, which causes the

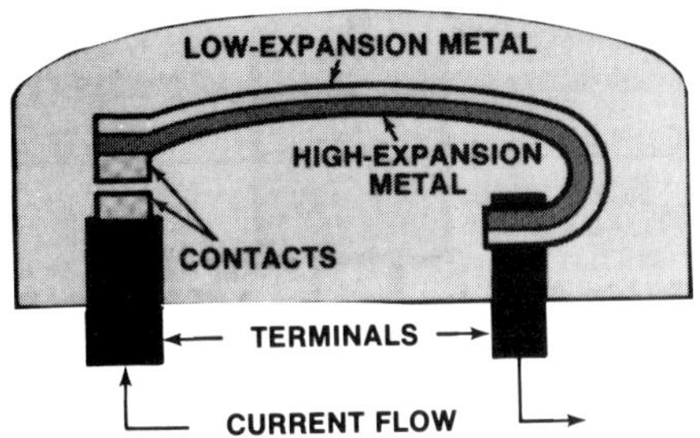

Figure 8-28. Typical self-resetting circuit breaker construction. (Ford)

element to bend. The bimetal element is part of the circuit being protected, and the contacts close the circuit through the breaker. Low or normal current through the breaker does not overheat the bimetal element, and the contacts stay closed. Heat from excess current causes the element to bend and open the circuit.

All headlamp circuits have self-resetting, type I breakers. When current stops in an open type I breaker, the bimetal element cools and closes the contacts. Current flow resumes, but if it remains too high, the breaker reopens. Continuously high current in such a circuit causes the breaker to open and close with a clicking sound. Self-setting type I breakers may be used in a few other circuits of some cars.

Type II breakers remain open when current is too high. They do not reset. A small wire is coiled around the bimetal element to act as a heater and keep the circuit open as long as high current is present. Type II breakers are used in a few electronic system power supply circuits.

Fuses

Most fuses are installed in a central fuse block, under the instrument panel or on the firewall. Most circuits other than the headlamps, the starter, and the ignition system receive power through the fuse block. Battery voltage is applied to a main **bus bar** in the fuse block, which is connected to one end of each fuse. The other end of each fuse is connected to the circuit, or circuits, that it protects.

Fuses are shown and numbered on wiring diagrams and individual circuit schematics. Because fuse and circuit number identification is important for troubleshooting, however, carmakers also provide detailed drawings of fuse blocks and fuse locations, figure 8-29.

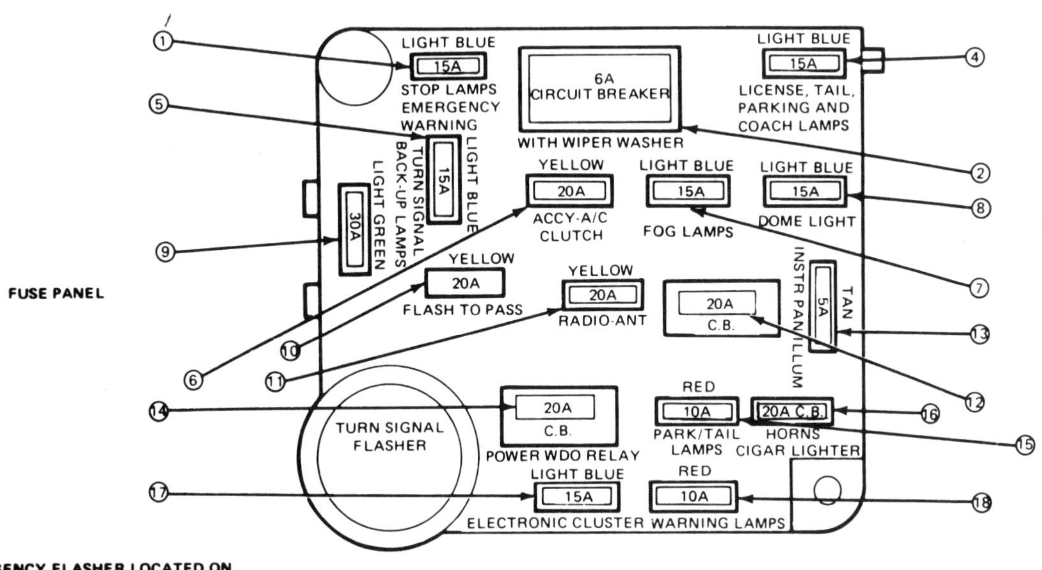

CAVITY NUMBER	SYSTEMS
1	Shop Lamps, Hazard Warning Lamps, Speed Control Module, Cornering Lamp Relay
2	Windshield Wiper, Windshield Washer Pump, Interval Wiper Module
4	Taillamps, Parking Lamps, Side Marker/Coach Lamps, Cluster Illumination Lamps, License Lamps, Digital Clock Illuminator, Autolamp Relay, Chime Isolation Relay
5	Turn Signal Lamps, Back-Up Lamps, Illuminate/Keyless Entry
6	A/C Clutch, Heated Backlite Control, Deck Lid Release, Electronic Chimes Module, Lamp Outage Module, Throttle Solenoid Positioner, and Cornering Lamps
7	Fog Lamps (2.3L EFI Turbo Only)
8	Courtesy Lamps, Key Warning Chime, Illuminated Entry Module, Keyless Entry Module, Trunk Lamp, Visor Mirror Lights*, Remote Electric Mirrors, Electronic Radio Memory, Clock, Tripminder Module

CAVITY NUMBER	SYSTEMS
9	A/C-Heater Blower Motor
10	Flash-to-Pass
11	Radio, Tape Player, Power Antenna, Premium Sound Amplifier
12	Power Seat, Door Locks, Fuel Filler Door Lock Solenoid
13	Instrument Cluster, Radio, Climate Control, Clock (Analog), Cigar Lighters, Ash Tray Illumination Lamps, Tripminder Module
14	Power Window
15	Tail/Park Lamps, Coach Lamps, License Lamps
16	Horn, Cigar Lighter
17	Electronic Cluster, Tripminder Module
18	Warning Indicator Lamps, Auto Lamp System, Electronic Chimes; Fuel Vapor Solenoid, Low Washer Fluid Level Lamp, Low Fuel Module

* The visor mirror uses two fuses: a 15-amp fuse located in the fuse panel and a 2-amp fuse located in the visor assembly. The garage door opener is not fused within the visor.

Figure 8-29. This typical late-model fuse panel holds circuit breakers and flashers as well as fuses. (Ford)

Fuse blocks also may hold warning buzzers and flashers for some lamp circuits.

Some circuits have inline fuses. These are installed in plastic fuse holders, spliced into the circuit wiring, figure 8-30.

Automobile fuse ratings range from 0.5 to 35 amperes, but 4-ampere to 20-ampere fuses are most common. Fuse sizes and ratings are established by the Society of Fuse Engineers (SFE). All SFE fuses are the same diameter, but the length varies with the current rating, figure 8-31. Some cars have AGA, AGC, and AGW series fuses. The lengths of all fuses in one series are the same, even though the current ratings vary, figure 8-32.

Caution. Never install a higher capacity fuse in a circuit than specified. A high-amperage fuse may allow excessive current, which will damage circuit parts.

Some European cars have cartridge fuses with pointed ends that fit into special fuse holders, figure 8-33. Many late-model vehicles have blade-type fuses, figure 8-34. These were first used on some 1977 GM cars. All fuses have the current rating marked on the body. Blade fuses also are color coded to indicate current ratings, figure 8-35.

Fusible Links

Besides fuses and circuit breakers, some circuits have fusible links, or fuse links. The diagrams in figures 8-5, 8-16, and 8-19 show fusible link locations. Fusible links are used in circuits for three general reasons:

1. The circuit is not otherwise fused.

2. Maximum current control is not as critical as in other fused circuits.

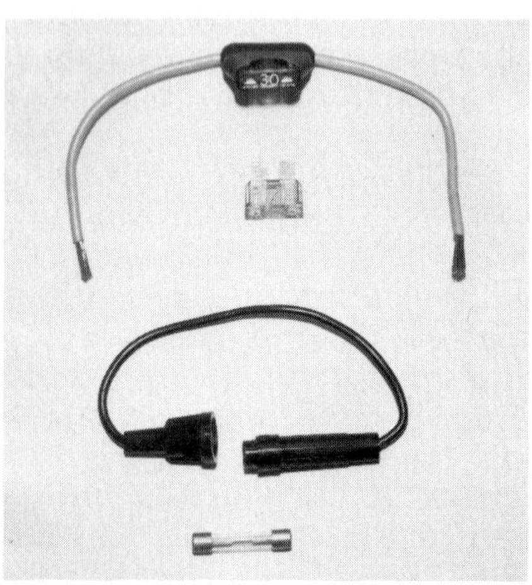

Figure 8-30. Some circuits are protected by inline fuses in the main power feed for the circuit.

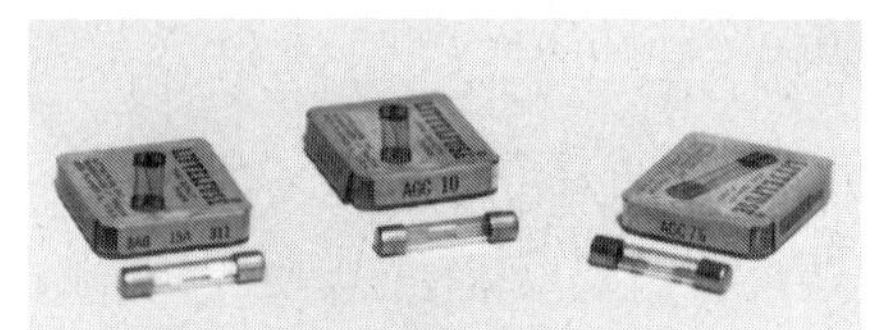

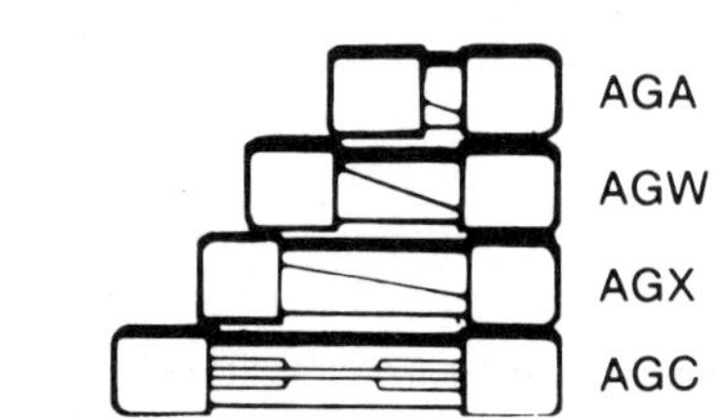

Figure 8-32. Typical AGA, AGC, and AGW fuses.

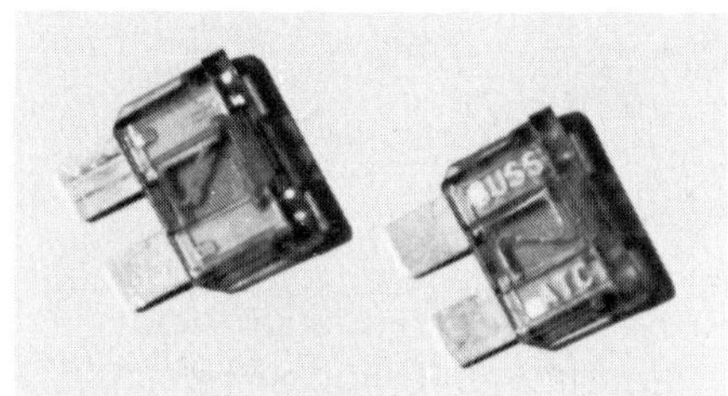

Figure 8-34. First used in 1977 GM cars, blade fuses are now in many late-model cars.

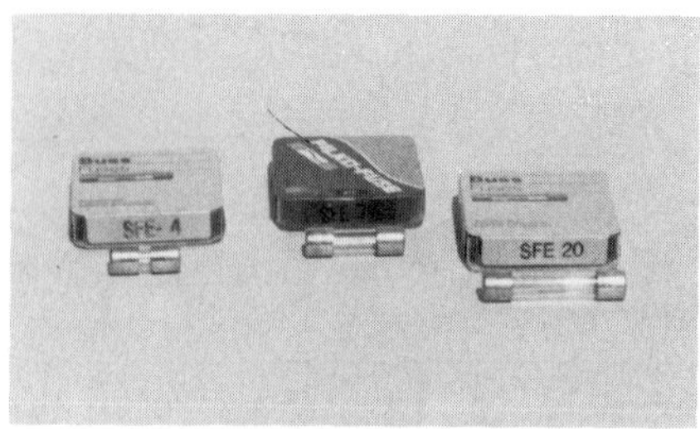

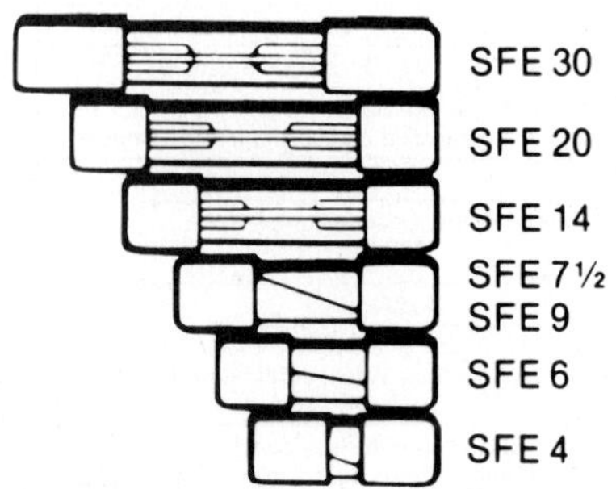

Figure 8-31. Typical SFE series fuses.

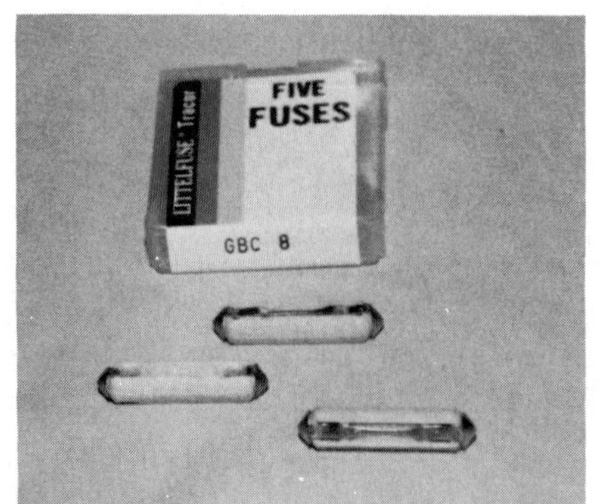

Figure 8-33. These fuses with pointed ends are used in many European cars. Some have exposed elements.

FUSE (AMPERAGE)	COLOR STRIPE/BODY
3	VIOLET
5	TAN
7.5	BROWN
10	RED
15	BLUE
20	YELLOW
25	WHITE

Figure 8-35. Color codes for blade-type fuses. (Cadillac)

3. Circuit protection is needed immediately near a circuit splice or junction connection.

A fusible link is a short length of smaller gauge wire installed in a larger conductor, figure 8-36. Excessive current will melt the link before it harms the rest of the circuit. Links are usually four gauges smaller than the circuit wiring. Thus, a circuit with 16-gauge wiring would have a 20-gauge fusible link. Fusible links have special insulation that blisters, or bubbles, when excessive current melts the link wire. This shows that the link is blown and must be replaced.

When you replace a blown fusible link, install a new link of the specified gauge and length. Do not substitute ordinary smaller gauge wire because it will not have the special insulation. Install most replacement links with crimp-type connectors and insulate the connectors with electrical tape. When installing a 16-, 18-, or 20-gauge fusible link into a crimp connector for heavier wire, bend over the stripped end of the link wire to double its diameter for a secure connection. You can install some replacement links by soldering, but be careful not to burn the link wiring or insulation with the soldering iron.

The first real advance in fusible link design made its appearance on domestic automobiles with the 1988 Lincoln Continental and 1989 Ford Probe. Instead of a length of wire installed in the circuit, this fuse link, as its manufacturer calls it, is a plug-in type that looks much like a long, fat blade-type fuse when installed. The housing contains a short length of wire suitable for the rated current load. As figure 8-37 shows, the top of the fuse link housing is transparent and allows you to visually determine the condition of the link inside. The fuse link has no exposed blades, but plugs into a combination fuse and fuse link box that contains blade terminals. Color coded according to amperage value, fuse links are used in high-amperage circuits (from 30 to 80 amperes).

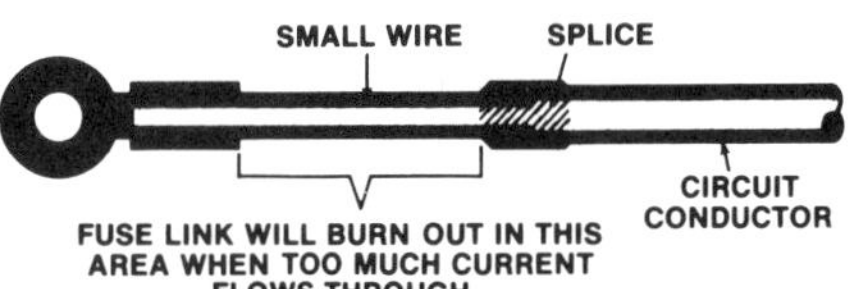

Figure 8-36. A fusible link, or fuse link, is light-gauge wire that protects heavier circuit wire from excessive current. (Ford)

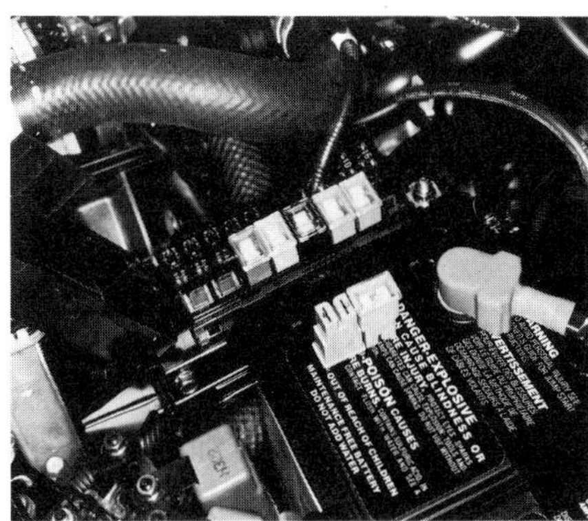

Figure 8-37. These color-coded fuse links are used for high-current, 30- to 80-ampere circuits on some 1988 and later vehicles.

Factory-installed fusible links are generally color coded to indicate current-carrying capacity, although replacement links may not be color coded. As with fuses, a burned-out fusible link should not be replaced with ordinary wire or a link of greater capacity. The resulting overload could result in a fire or even complete loss of the vehicle.

A final important point about circuit protection devices is that they "blow," or open a circuit, because of a problem elsewhere in the circuit. Do not expect to repair an electrical problem simply by replacing a fuse or fusible link. The fault that caused the first device to blow will soon blow the replacement. Always locate and repair the *cause* of the blown fuse or link.

To test for a short or grounded circuit that repeatedly blows a fuse or a fusible link, you can attach a resetting circuit breaker to a pair of jumper wire leads and connect it in the circuit in place of the fuse or link. Apply power to the circuit and test for the short or ground. The circuit breaker may open, but it will reset itself so that you can continue to test the circuit until you locate the problem.

WIRING REPAIRS

Most automobile primary wiring is multistrand copper wire, covered with plastic insulation. Solid, single-strand wire is used inside components such as alternators and starter motors, but repair of such wiring is not part of normal engine performance service.

Some GM cars use aluminum wiring in forward body harnesses. Engine control harnesses often contain shielded cable or wiring with twisted leads for low-current circuits. These kinds of wiring require special repair methods.

Whenever you replace damaged wiring, install new wire of the same, or larger, size. As explained earlier, wire sizes are specified by AWG or metric specifications. Whenever possible, select replacement wire with the same color insulation as the damaged wire. This maintains the color coding for future troubleshooting and service.

The following sections explain the basic methods for repairing damaged wiring with soldered or crimped connections. These sections also explain insulation repair and special repair methods for aluminum, twisted, and shielded wiring.

Insulation Repair

All wiring repair requires insulation repair. However, worn insulation can expose conductors to short circuits without damaging the wire. There are two basic ways to repair insulation:

1. Wrap the repaired or damaged area with plastic electrical tape.

2. Cover a repaired wiring splice with heat-shrink tubing insulation.

When you use electrical tape to cover a splice or damaged insulation, put three turns of tape over the repair area. Overlap the tape about 1/2 inch on the undamaged insulation at either end of the repair.

Heat-shrink tubing is plastic tubing that shrinks in diameter when heat is applied to it. Heat-shrink tubing is available in small diameters to cover wires of various gauges and in large diameters to cover several wires in a harness. It is available at electronic supply stores in assorted lengths and colors.

To insulate a wire with heat-shrink tubing, you must slide the tubing over the wire before it is spliced or attached to a connector. Tubing is ideal for insulating replacement wire splices, but cannot be used for simple insulation repairs without cutting the wire. Slide a length of tubing over the wire before splicing or attaching a connector terminal. After the wire is repaired, slide the tubing over the splice or connection.

Warning. Do not use open flame to shrink tubing over wire installed on an engine. Open flame near a battery or engine fuel components may cause an explosion.

Heat the tubing with a hot air gun or—if the wiring is off the vehicle—a match, figure 8-38. It will shrink to the diameter of the wire and insulate the repair area.

Soldering

You can make a wire splice by soldering two wire ends into a splice clip, figure 8-39, or by soldering the wire ends directly to each other.

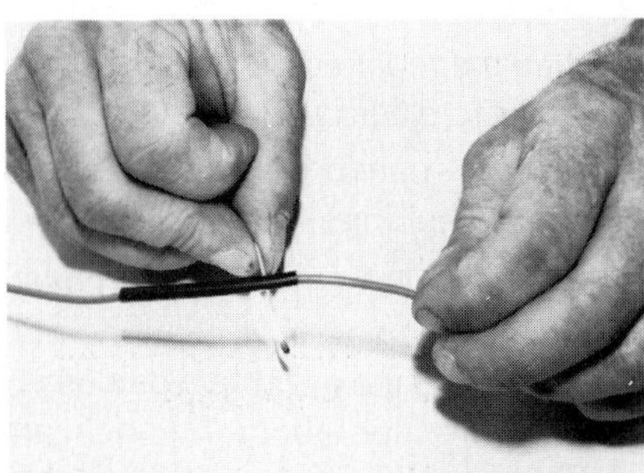

Figure 8-38. Heat the tubing with a heat gun or a match to shrink it.

Caution. Use only rosin-core solder for electrical repairs. Acid-core solder will corrode wire strands and create high-resistance connections.

To solder two wires together:

1. Use a stripping tool to remove about 1 inch of insulation from the end of each wire, figure 8-40. Use the correct opening in the stripping tool for the wire gauge. Be careful not to nick the wire strands.

2. Braid or lace the strands of each end together and twist them securely, figure 8-41.

3. Heat the twisted connection with the broad part of a soldering iron tip, figure 8-42. A soldering iron or gun that supplies 80 to 120 watts of heat is satisfactory for most wire soldering. You can use up to 150 watts for 12- or 10-gauge wire.

4. Touch the rosin-core solder to the strands of the heated splice, *not* to the iron. Solder must melt from the heat of the splice and flow evenly among all the strands, figure 8-43. Use only enough solder to cover the splice securely. Too much solder creates a weak splice and may trap dirt that can cause high resistance.

5. Insulate the soldered splice with electrical tape or heat-shrink tubing.

You can solder wire to some terminal connectors with the same methods used for splicing.

STRAIGHT LEADS

1. LOCATE DAMAGED WIRE.
2. REMOVE INSULATION AS REQUIRED.

3. SPLICE TWO WIRES TOGETHER USING SPLICE CLIP. SOLDER WITH ROSIN CORE SOLDER.

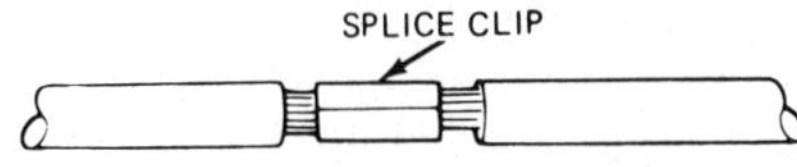

4. COVER SPLICE WITH ELECTRICAL TAPE TO INSULATE FROM OTHER WIRES.

Figure 8-39. You can use a splice clip to make a soldered splice.

To solder electronic components that cannot stand high temperature, clamp the component in an alligator clip attached to a vise or other heat sink. This will draw the heat away from the component while you solder the leads.

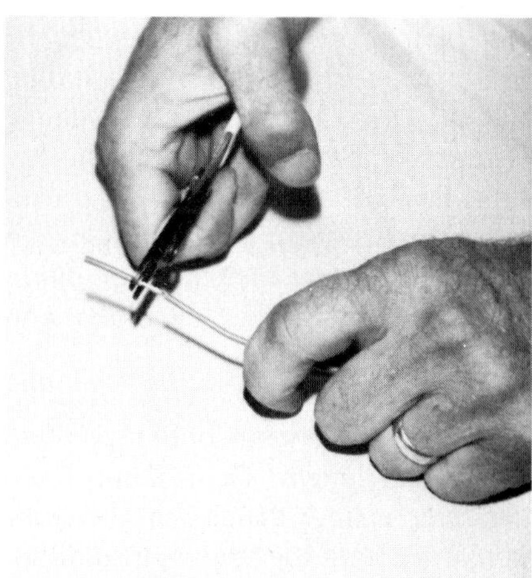

Figure 8-40. Strip the insulation from the wire without nicking or cutting the strands.

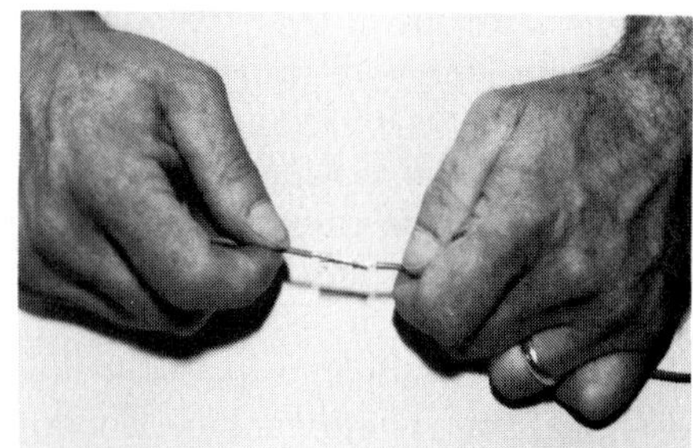

Figure 8-41. Twist or braid the strands together before soldering.

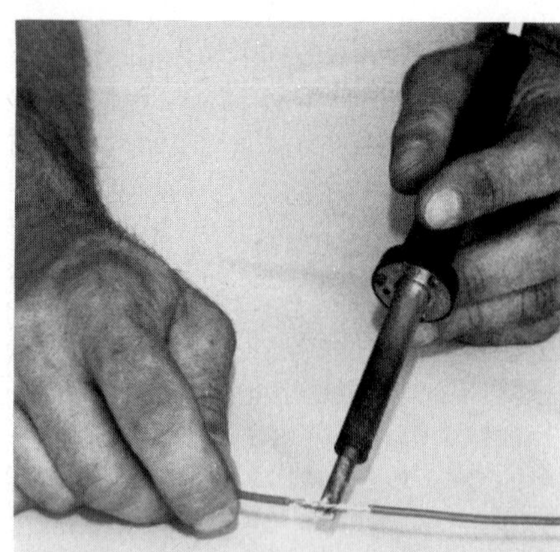

Figure 8-42. Heat the splice with the broad part of the soldering iron tip.

Crimping

Some carmakers recommend crimp-type butt connectors for wire splices and crimp connections for connector terminal replacement. A standard crimping tool, figure 8-44, has separate areas for cutting, stripping, and crimping.

1. Using the proper stripping opening on the tool, remove enough insulation so that the wire will penetrate the connector opening completely. Do not nick the wire strands. Do not remove more insulation than necessary.

2. Insert the wire into the proper size connector and crimp the connector to the stripped wire, figure 8-44.

3. Also crimp the connector in the area of the unstripped insulation.

4. Be sure the wire is compressed under the crimped areas of the connector.

5. If the butt connector or terminal is not insulated, cover the connection with electrical tape or heat-shrink tubing. You also can use tape or tubing for extra protection on insulated splice connectors.

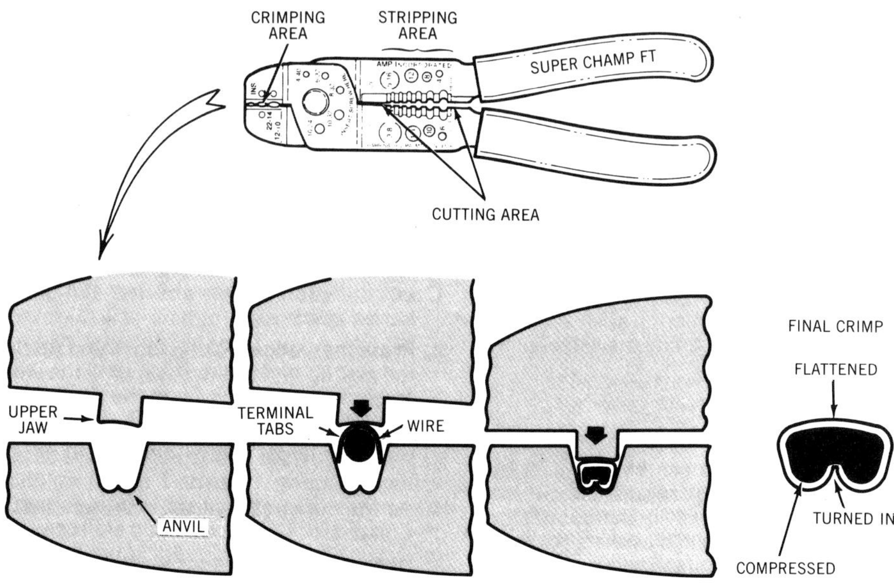

Figure 8-44. Typical crimping tool and a properly formed crimp splice. (Ford)

Splicing Wires into Wiring Harnesses

Most vehicle wiring is contained in wiring harnesses as shown in the installation drawings at the beginning of this chapter. A single harness may contain a dozen or more wires in a plastic jacket. If one or two wires are damaged, you don't need to replace the complete harness.

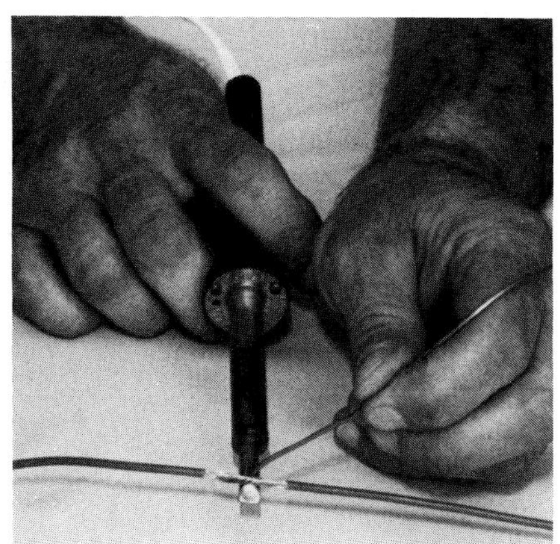

Figure 8-43. Melt the solder with the heat of the splice, not the iron.

First, isolate the open or shorted wire using standard troubleshooting procedures. Follow the color coding and circuit numbers on your diagrams. Then, open the harness if you can do so to check for wire damage. If the harness is inaccessible in the damaged area, cut the defective wire where it enters the harness at both ends. Install a length of replacement wire along the outside of the harness and solder or crimp it to the undamaged wire ends where you cut them, figure 8-45. Tape the replacement wire to the outside of the harness jacket.

Figure 8-45. You can bypass a damaged wire inside a harness by splicing in a replacement length. (Ford)

Aluminum Wire Repair

The aluminum wiring in some GM front body and engine compartment harnesses requires special repair methods, figure 8-46. To splice aluminum wiring, strip about 1/4 inch of insulation from the wire ends and insert them into an aluminum butt connector. Crimp the connection with adjustable pliers. Do not solder aluminum wiring. Cover the splice with petroleum jelly to prevent oxidation and corrosion. Then wrap the splice with electrical tape. Use similar methods to splice an aluminum pigtail with a connector terminal to a harness wire.

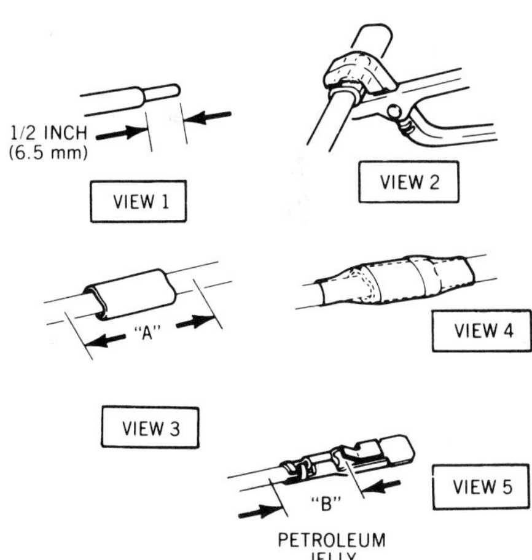

Figure 8-46. Aluminum wiring repair for GM cars. (Cadillac)

Low-Current Wiring Repairs

Many harnesses for engine electronic systems contain wire that carries very low current, sometimes only 0.1 or 0.2 ampere (100 or 200 mA). Figure 8-47 outlines the repair methods for twisted, shielded cable and for small, twisted lead wiring.

CONNECTOR REPAIRS

Electrical connectors join any number of wires from 1 pair to as many as 40. Connector and terminal identification

TWISTED/SHIELDED CABLE

1. REMOVE OUTER JACKET.

OUTER JACKET

2. UNWRAP ALUMINUM/MYLAR TAPE. DO NOT REMOVE MYLAR.

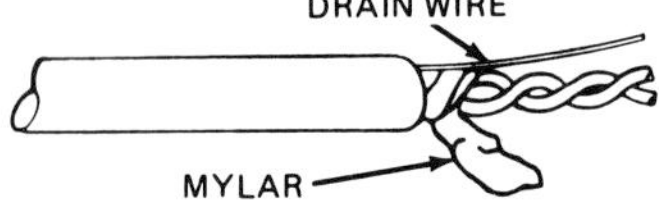

3. UNTWIST CONDUCTORS. STRIP INSULATION AS NECESSARY.

4. SPLICE WIRES AS NECESSARY USING SPLICE CLIPS (5297428) AND ROSIN CORE SOLDER. WRAP EACH WITH TAPE AND SPLICE TO INSULATE. SPLICES SHOULD BE STAGGERED.

5. WRAP WITH MYLAR AND DRAIN (INSULATED) WIRE. SPLICE DRAIN WIRE AS PER STEP 4.

6. TAPE OVER WHOLE BUNDLE TO SECURE AS BEFORE.

TWISTED LEADS

1. LOCATE DAMAGED WIRE.
2. REMOVE INSULATION AS REQUIRED.

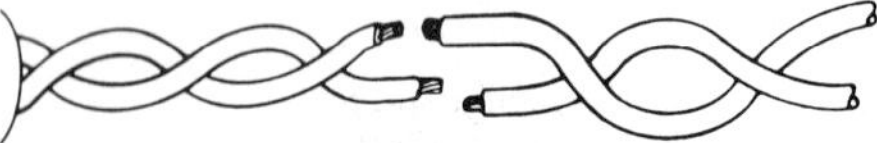

3. SPLICE TWO WIRES TOGETHER USING SPLICE CLIPS 5297428 AND ROSIN CORE SOLDER. SPLICES SHOULD BE STAGGERED.

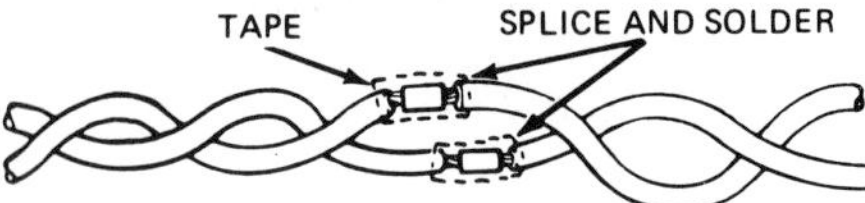

4. COVER EACH SPLICE WITH ELECTRICAL TAPE TO INSULATE FROM OTHER WIRES.

5. RETWIST AS BEFORE AND TAPE WITH ELECTRICAL TAPE TO HOLD IN PLACE.

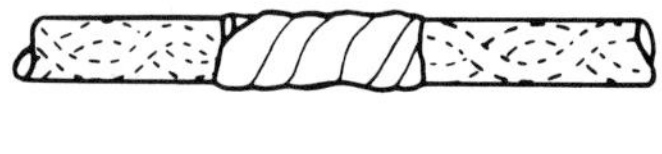

Figure 8-47. Typical repair methods for low-current engine wiring. (Cadillac)

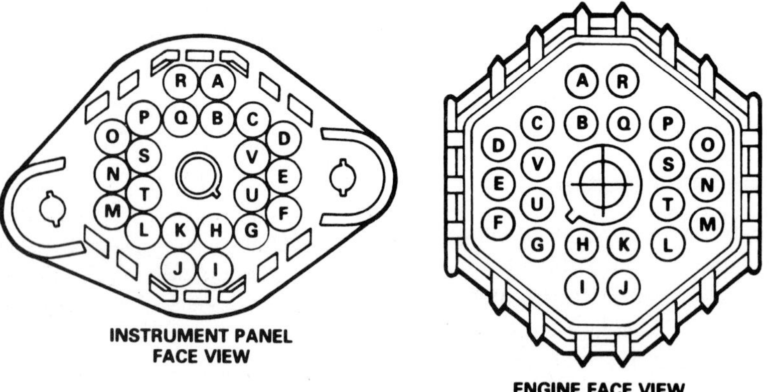

CAVITY	CIRCUIT #	WIRE COLOR	CIRCUIT
A	428	DARK GREEN/ YELLOW	ECM TO CANISTER PURGE SOLENOID
B•	423	WHITE	ELECTRONIC SPARK TIMING SIGNAL FROM ECM TO DISTRIBUTOR
C	427	PINK	ECM TO ISC MOTOR
D•	410	YELLOW	ECM TO COOLANT TEMPERATURE SENSOR
E	462	PINK/BLACK	HEI BYPASS SIGNAL FROM ECM TO DISTRIBUTOR
F•	452	DRAIN WIRE	LOW LEVEL GROUND
G•	430	PURPLE/WHITE	DIST. REF. SIGNAL—ECM TO DISTRIBUTOR
H•	453	BLACK/RED	DIST. REF. GROUND—ECM TO DISTRIBUTOR
I	910	LIGHT GREEN/ BLACK	ECM TO OUTSIDE TEMPERATURE SENSOR

Figure 8-48. This connector diagram identifies connector sockets and pins and circuit numbers. (Cadillac)

is important for troubleshooting. The wiring diagrams and circuit schematics earlier in this chapter show connectors several different ways. Most diagrams list a connector by number or show its location and shape. Many diagrams show connectors with multiple terminals by dashed lines between the individual terminals. This is a common way to show connectors for ignition switches and electronic modules. Most carmakers also provide drawings of each connector to identify the terminals, or cavities, and the circuit numbers, figure 8-48.

Some connectors (usually with 1 to 4 wires) are 1-piece molded parts, figure 8-49. You can't separate individual wires and terminals for repair. To repair a molded connector, you must cut the damaged wire, or wires, out of both halves of the connector and splice in a new connector with pigtail leads. If only one wire of a molded multiple connector is damaged, you can splice in a new single-wire bullet connector to bypass the damaged area, figure 8-50.

Hard-Shell Connectors

Many multiple-wire connectors have hard plastic shells that hold the mating pins and sockets (male and female terminals) of individual connectors. Figure 8-51 shows several common types. These connectors also allow you to probe the rear of the individual connections to test circuit operation without separating the connector.

To replace individual terminals in these connectors, you must separate the connector halves or unlatch the connector from the part to which it is attached. Connectors that pass through bulkheads (the firewall, for example) usually are secured with a screw through the center of both halves.

After separating the connector halves, use a small screwdriver or a special connector tool to release the damaged terminal from the connector shell. Figure 8-52 shows several examples. Cut the damaged terminal from the wire and install a new one by crimping or soldering. Reinstall the terminal in the connector shell.

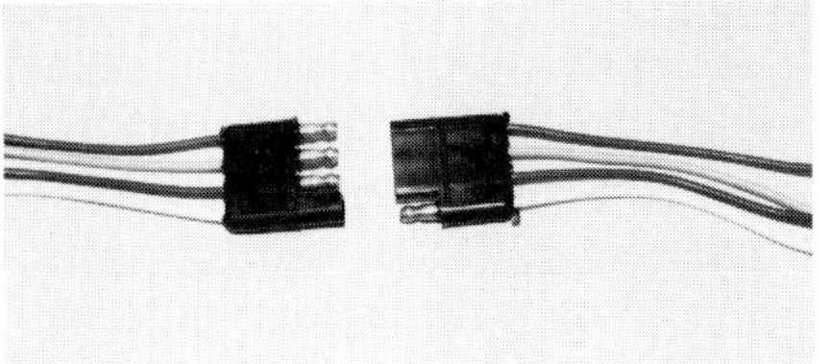

Figure 8-49. Typical molded connector halves with multiple circuit wires.

Figure 8-50. Double the end of the fusible link wire for a secure splice to a larger wire. (Ford)

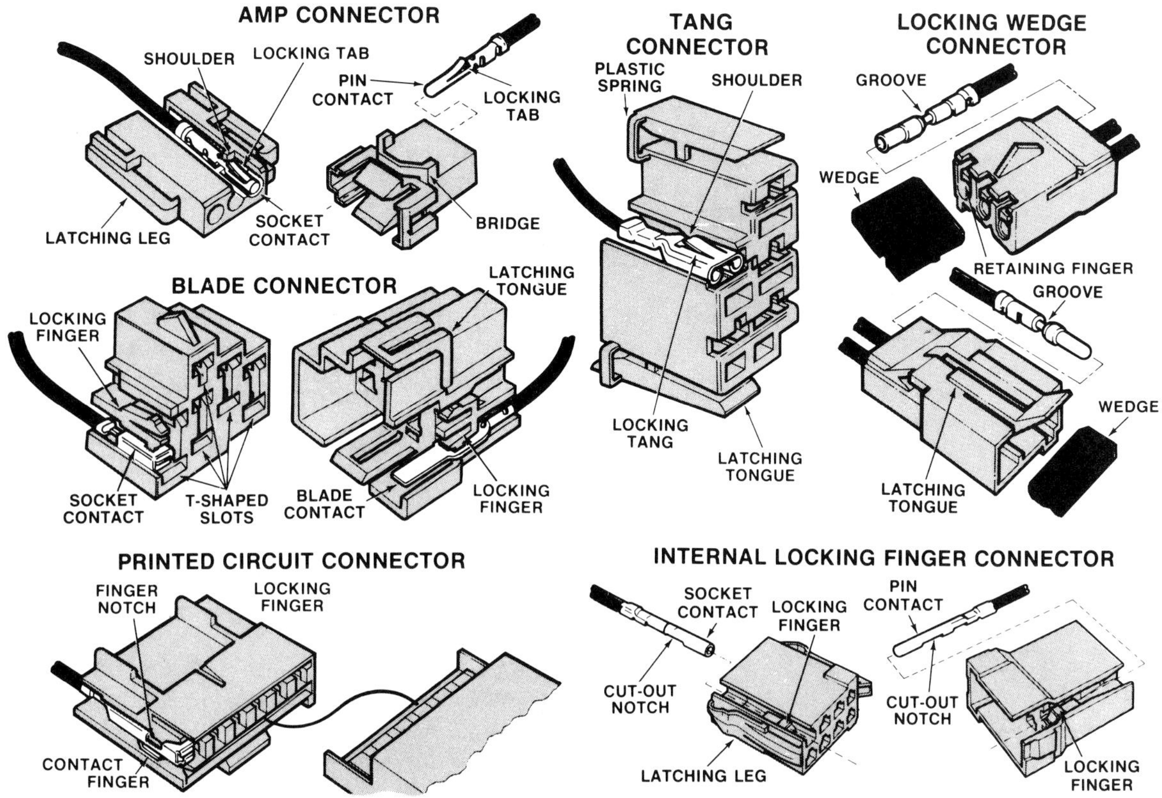

Figure 8-51. These are six typical kinds of hard-shell connectors. (Ford)

Weather-Pack Connector Repair

Late-model GM cars have special weather-proof connectors throughout the engine and body harnesses. These points where temperature and humidity are a high concern. These environmental connectors have rubber seals on the wire ends of the terminals and secondary sealing covers on the rear of each connector half.

Weather-Pack connectors should not be checked by probing through the seals; you should always check the circuit with jumper wires. These connectors should not be replaced with standard connectors. Figure 8-53 shows how to open a Weather-Pack connector and replace the individual terminals. Special tools are recommended for removal and installation of the pin and sleeve terminals.

Metri-Pack Connector Repair

These environmental connectors also are used on late-model GM cars. They are sometimes called "pull-to-seat" connectors because of the method used to install terminals or wires. A terminal or wire is replaced by sliding the seal back on the wire and using a suitable tool to release the locking tab on the terminal. With the tab released, the wire and terminal can be pushed out through the connector. If the terminal is to be reused, check and reshape the locking tab, if necessary.

If a new wire is to be installed, insert it through the connector seal and connector. Crimp the terminal onto the wire end securely and pull the terminal back into the connector housing until it seats in place. Figure 8-54 shows Metri-Pack terminal replacement.

Micro-Pack Connector Repair

These GM connectors are serviced much like other hard-shell connectors, figure 8-55.

INTERFERENCE SUPPRESSION

When you studied ignition cables, you learned something about the subject of radiofrequency interference (RFI). RFI is a form of electromagnetic inter-

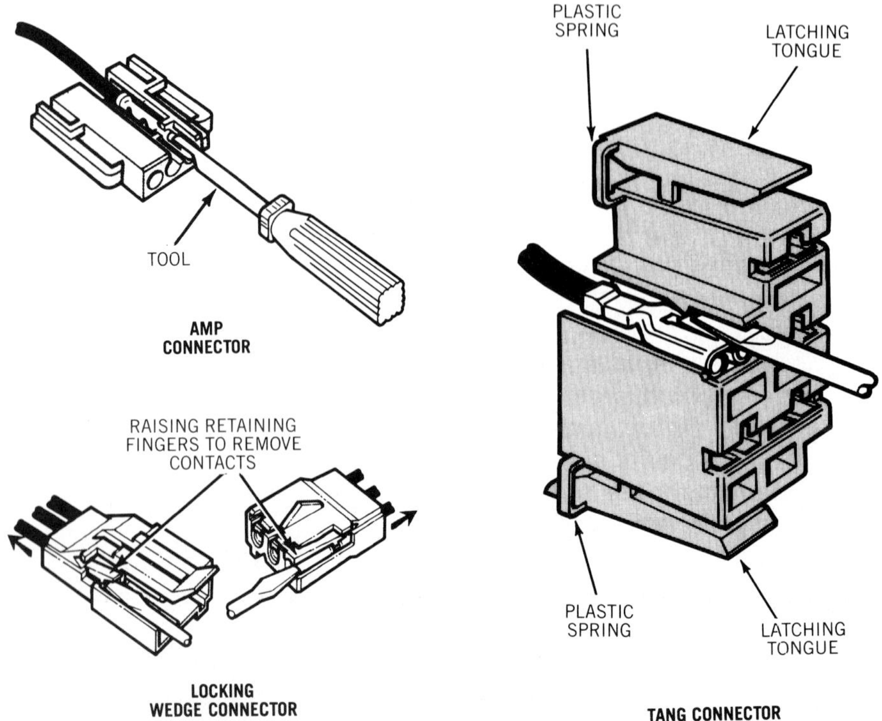

Figure 8-52. Use a screwdriver or a terminal removal tool to release the terminals from the connector body. (Ford)

ference (EMI). RFI suppression has been an important part of automotive electrical system design for decades. The purpose is to protect radio and television broadcast signals against interference from vehicle systems, principally the ignition. As the use of electronic systems increases on late-model cars, the area of interference suppression has grown to include the broader field of EMI suppression. Low-power digital integrated circuits are vulnerable to EMI signals that were unimportant a decade ago.

Radio transmission is the generation of high-frequency electromagnetic waves. An electromagnetic field is formed and its strength is varied many thousand or million times per second. As field strength and polarity change, the field passes through a cycle similar to an alternating current cycle. We measure cycles per second with the unit called the hertz (Hz). High-frequency signals of thousands of cycles per second are measured in

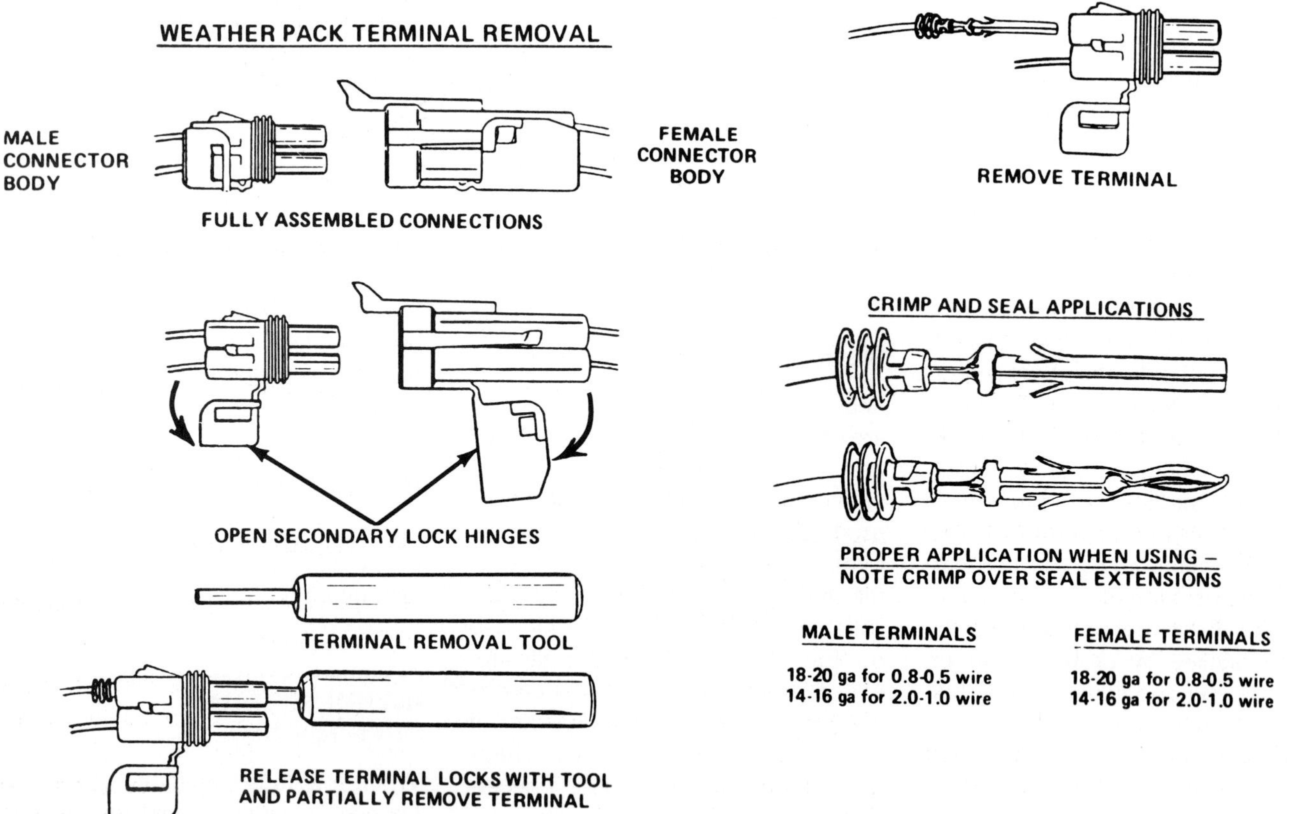

Figure 8-53. These are the basic repair methods for GM weather-pack connectors.

kilohertz (kHz). Millions of cycles per second are measured in megahertz (MHz). These are the transmission frequencies of AM and FM radio and television broadcasting, figure 8-56. AM transmitters send signals by varying the strength, or amplitude, of the transmission frequencies. FM and TV transmitters send signals by varying the frequency itself. Radio and television receivers change the high-frequency electromagnetic waves back to voltage signals that create sound and pictures.

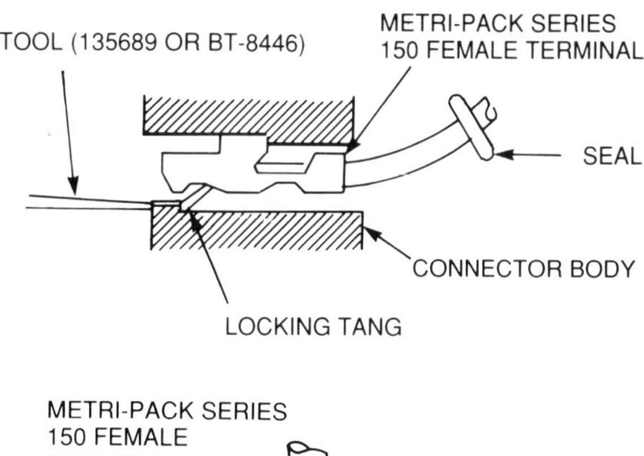

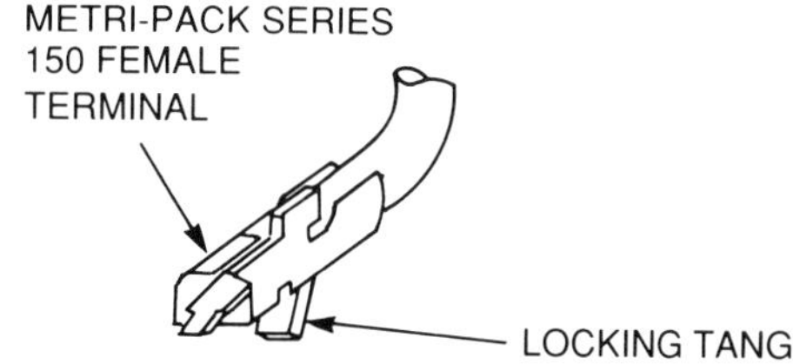

Figure 8-54. The basic repair methods for GM Metri-Pack connectors. (Chevrolet)

Interference Generation and Transmission

An electromagnetic field is created whenever current flows in a conductor, and field strength changes whenever current stops and starts. Each change in field strength creates an electromagnetic signal wave. If current changes fast enough, it creates high-frequency electromagnetic waves that interfere with radio transmission or with other electronic systems. EMI is an undesirable form of electromagnetism.

Electromagnetic waves are created whenever current starts and stops in a spark plug cable. The length of the cables, the operating frequency of the system, and the high voltage of the spark plug circuits make the ignition system an excellent EMI transmitter.

An electromagnetic wave is also created whenever a switch opens or closes in a circuit. High-frequency waves are created by a revolving commutator in a motor. (A commutator is a rotating switch.) EMI waves are created by the slight ac fluctuations in the alternator output circuit. Relays, solenoids, electromechanical voltage regulators, and vibrating horn contacts all produce electromagnetic waves. Figure 8-57 shows common sources of EMI on a typical automobile.

Additionally, EMI can be generated by static electric charges that develop due to friction at the following locations:

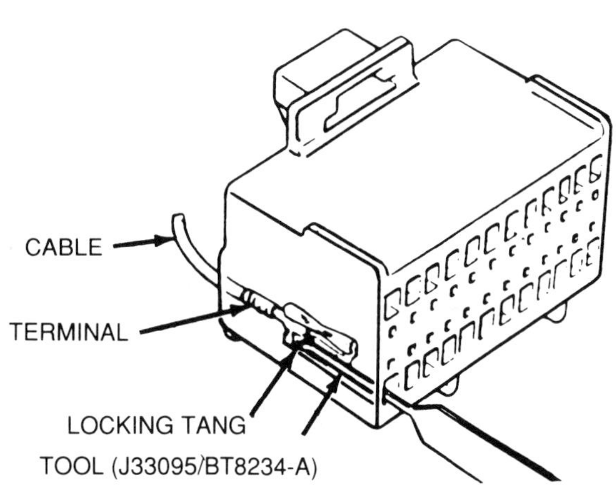

Figure 8-55. The basic repair method for GM Micro-Pack connectors. (Chevrolet)

- Tire contact with the road surface
- Engine drive belts
- Drive axles and shafts
- Clutch and brake lining surfaces

EMI is transmitted in four ways:

1. By coupling through circuit conductors, figure 8-58

2. By capacitive coupling in an electrostatic field between two devices or conductors with voltage applied to them, figure 8-59

3. By inductive coupling as magnetic fields form and collapse between two conductors, figure 8-60

4. By electromagnetic radiation, figure 8-61

All of these interference transmission sources exist in an automobile.

EMI Suppression

There are four general approaches used to reduce EMI:

1. The addition of resistance to conductors to suppress conductive transmission and radiation

2. The installation of capacitors and **choke coil** combinations to reduce an inductive capacitive coupling

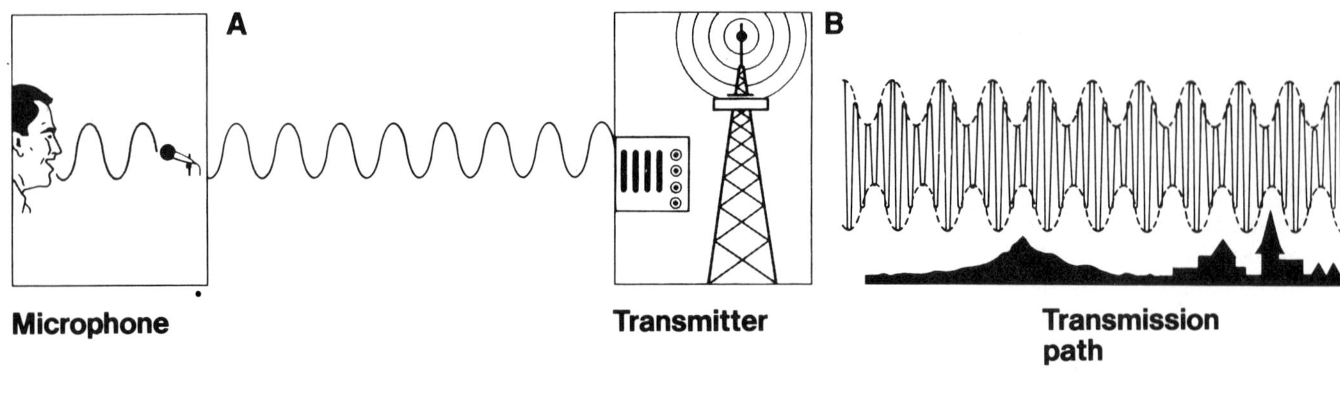

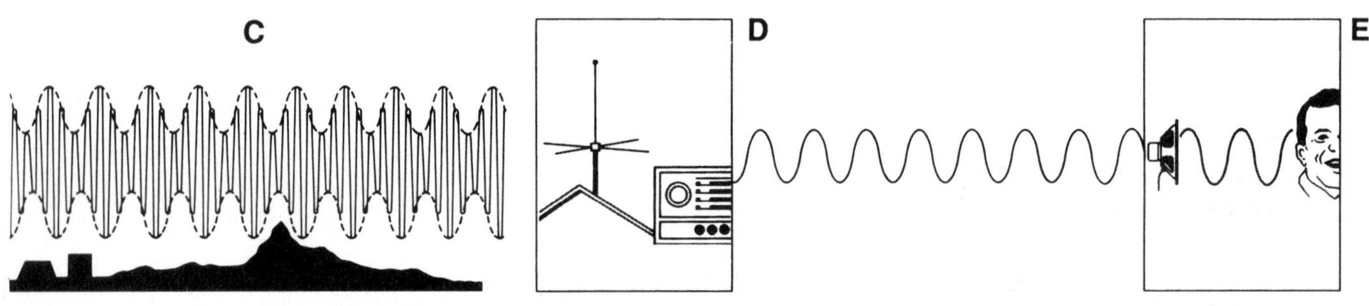

Figure 8-56. Wireless communication begins with acoustic oscillations converted into alternating voltage of the same frequency by a microphone (**A**). The low-frequency signal created in **A** is superimposed on an amplitude-modulated carrier wave (**B**). The high-frequency wave created in **B** is transmitted and received by an antenna (**C**). The receiver separates the low-frequency signal from the high-frequency carrier wave (**D**). The speaker translates the signal back into acoustic oscillations (**E**). These are the principles of radio and tv communication. (Bosch)

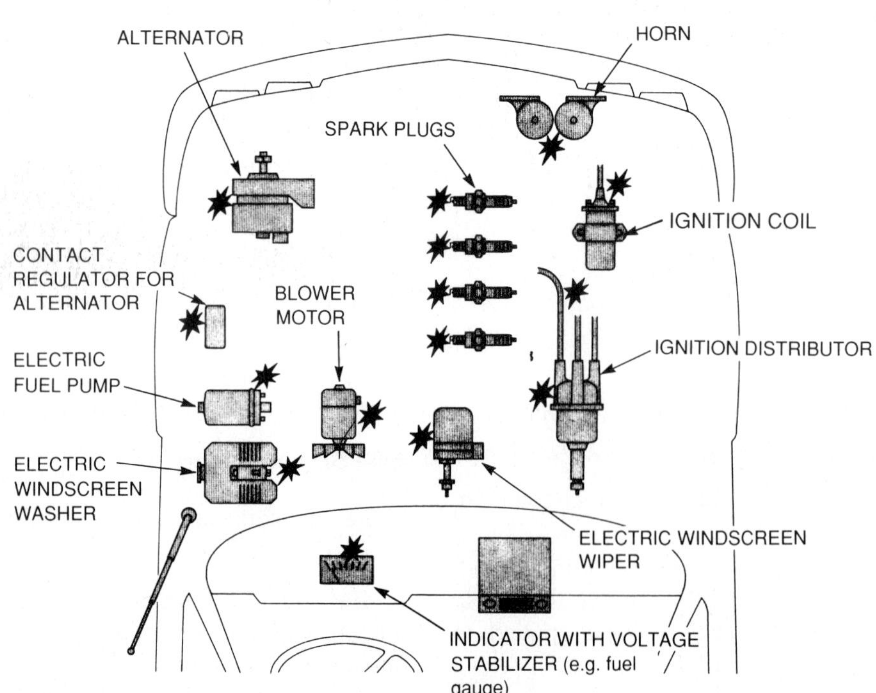

Figure 8-57. The major sources of electromagnetic interference (EMI) found in a typical automobile. (Bosch)

3. The installation of ground straps that will bypass signals to ground and thus reduce conductive transmission and radiation

4. The use of various forms of metal shielding to reduce EMI radiation, as well as capacitive and inductive coupling

Suppressing with Resistance. Circuit resistance works only in a high-voltage system to suppress RFI. Interference-suppression resistors are not practical in low-voltage circuits because they create too much voltage drop and power loss.

The ignition secondary circuit is the only high-voltage system on most vehicles and can be the largest source of EMI. The distributor switches thousands of volts across separate spark plug circuits and turns current on and off thousands of times per second. An 8-cylinder engine running at 2,000 rpm produces electromagnetic waves at a frequency of 8 kHz. Current in metallic high-voltage ignition cables

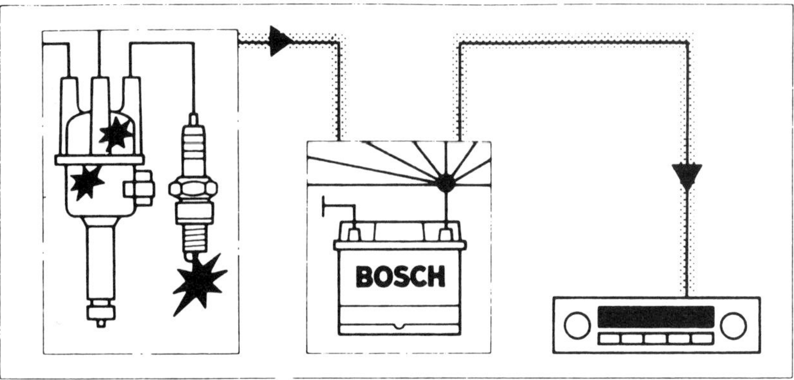

Figure 8-58. In conductive coupling interference, the wiring transmits the interference directly from the source to the receiver. (Bosch)

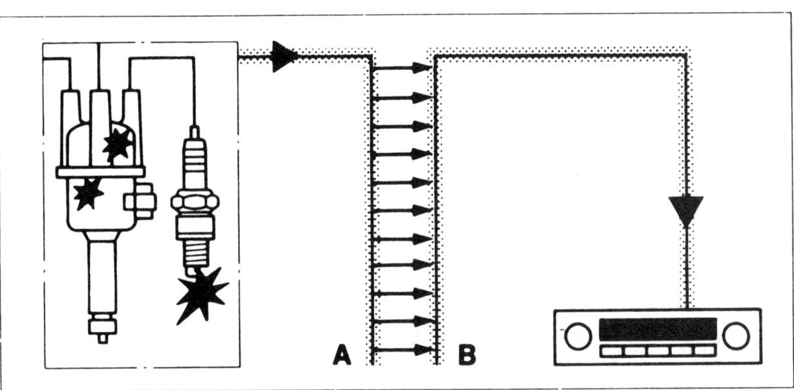

Figure 8-59. A capacitive field between adjacent wiring results in capacitive coupling interference. (Bosch)

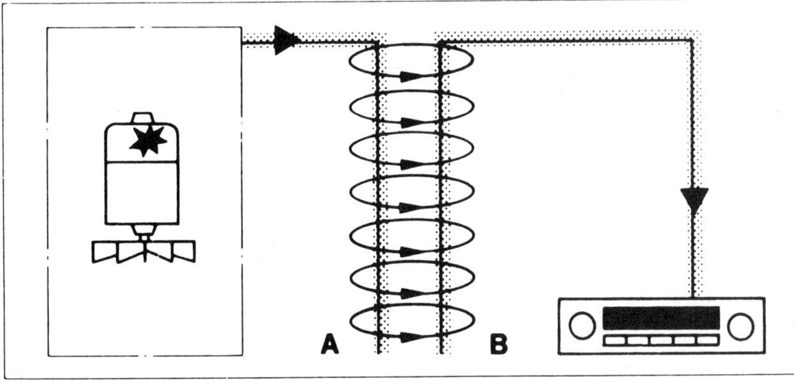

Figure 8-60. An electromagnetic field between adjacent wiring results in inductive coupling interference. (Bosch)

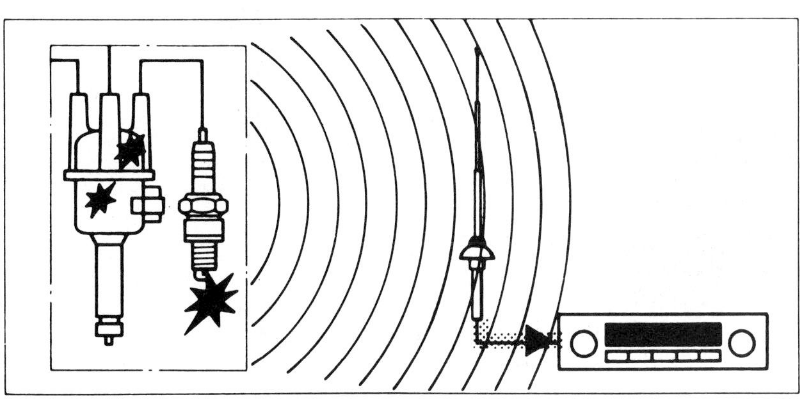

Figure 8-61. In radiation interference, the EMI waves are picked up by wiring that serves as a receiving antenna. (Bosch)

can create electromagnetic waves strong enough to interfere with radio and TV signals and other electronic systems. This interference can upset radio reception on a vehicle. It can also disturb public radio and TV transmission, as well as emergency radio service for police and other public service agencies.

Because the ignition system is the largest EMI source on a vehicle, engineers attacked that problem more than 30 years ago. Nonmetallic conductors in ignition cables add resistance and reduce the current in the cables, thus reducing the strength of the magnetic fields. High resistance also reduces voltage fluctuations in the circuits. Resistors added to spark plug electrodes and to distributor rotors also help to reduce RFI, figure 8-62. Some electronic ignitions require silicone grease on distributor cap and rotor terminals to suppress RFI radiation.

Suppressing with Coils and Capacitors. Remember from volume 1 that a capacitor can be used as a voltage "shock absorber." Capacitors are installed in parallel with (across) many circuits and switching points to absorb changing voltage. Electronic ignition systems do not require ignition condensers, or capacitors, as do breaker-point ignitions. Many, however, have a capacitor across the primary circuit at the ignition module, figure 8-63, to absorb voltage changes as the primary current switches on and off. Breaker-point ignitions often

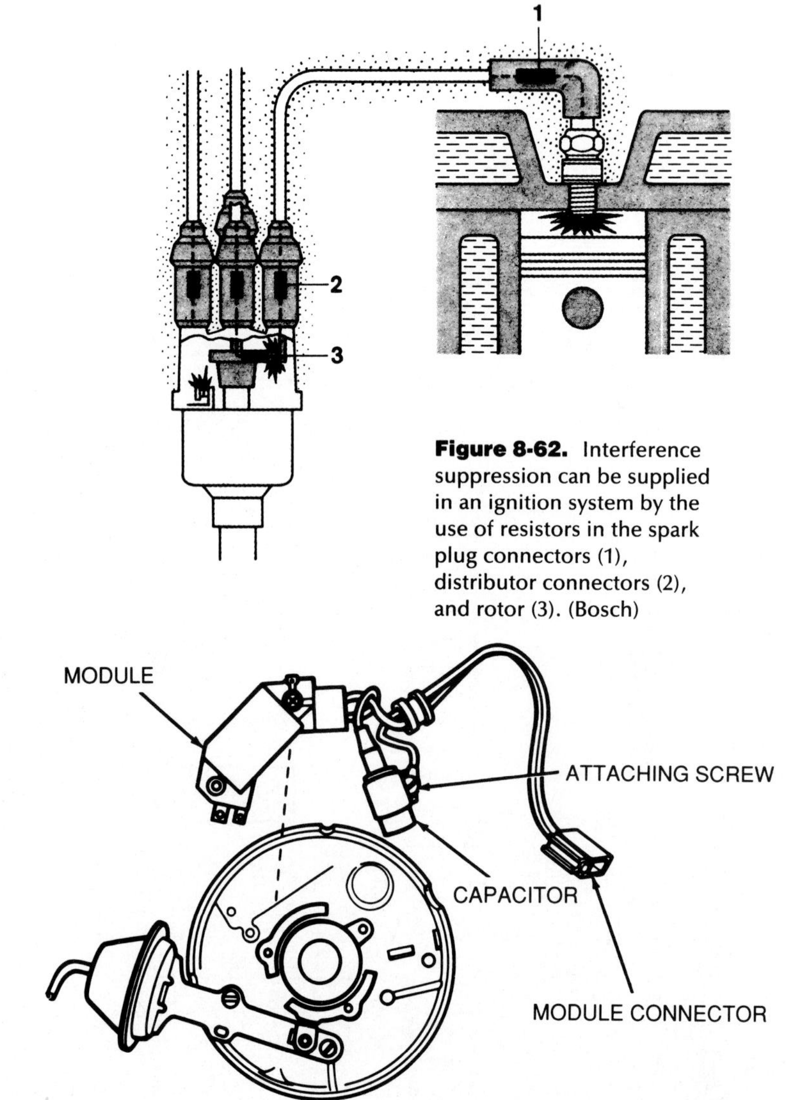

Figure 8-62. Interference suppression can be supplied in an ignition system by the use of resistors in the spark plug connectors (1), distributor connectors (2), and rotor (3). (Bosch)

Figure 8-63. The capacitor in this HEI distributor is not an ignition condenser. It suppresses radiofrequency interference.

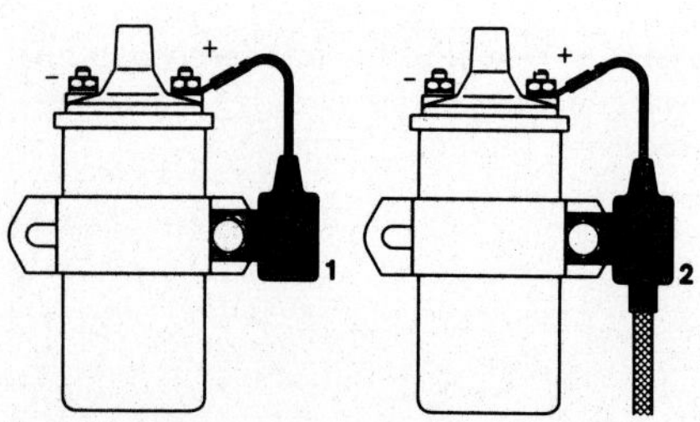

Figure 8-64. Installation of an ignition coil suppression capacitor when the coil is connected directly to the chassis (1). If the coil is not connected directly to the chassis, a separate ground strap is used (2). (Bosch)

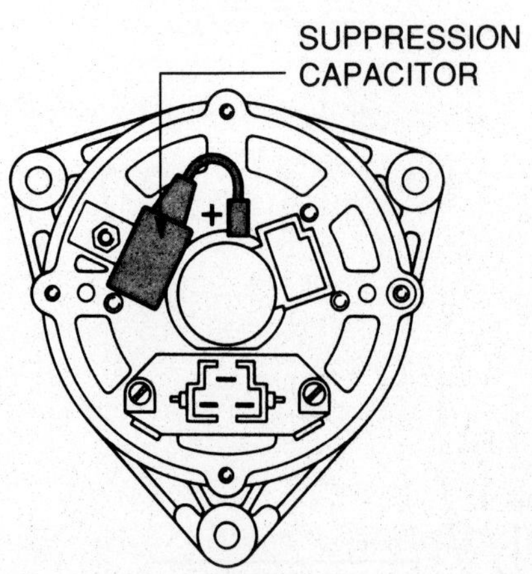

Figure 8-65. Installation of a suppression capacitor on an alternator. (Bosch)

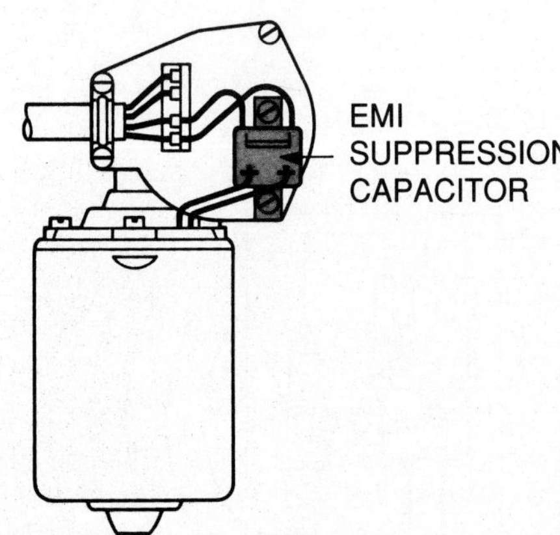

Figure 8-66. Choke coils and suppression capacitors are often used with automotive electric motors. (Bosch)

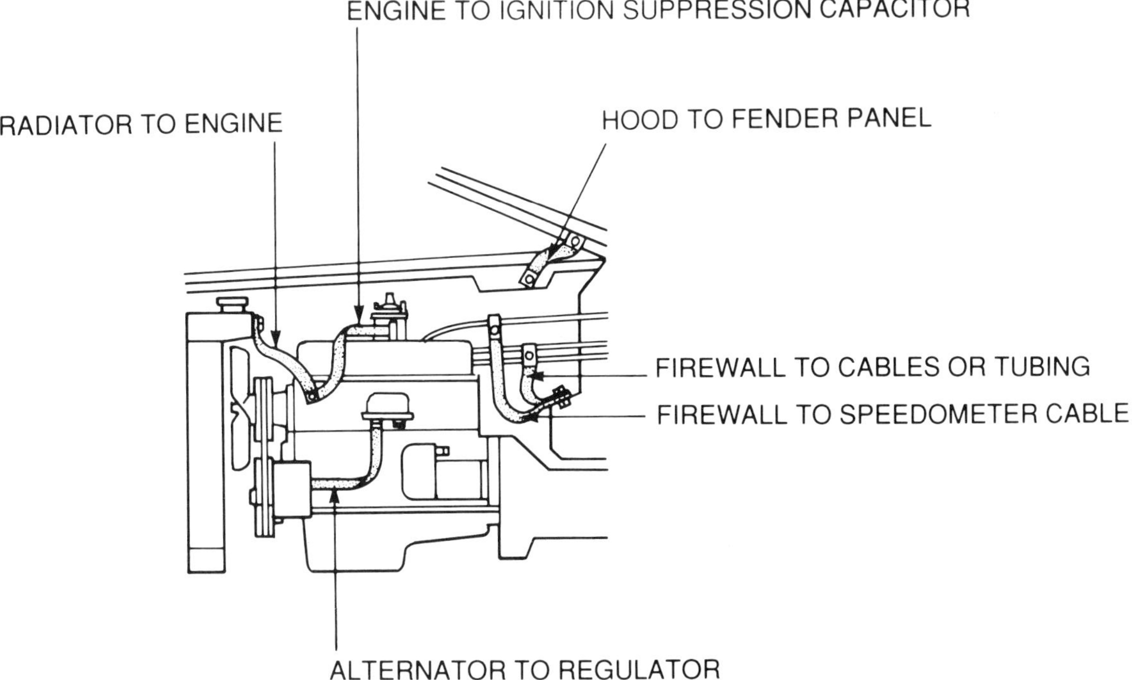

Figure 8-67. EMI suppression is accomplished in many late-model vehicles by the use of separate ground straps. (Bosch)

Figure 8-68. This bonding strap reduces the electrostatic field between the hood and the engine compartment.

have interference suppression capacitors at the coil + terminal, figure 8-64. Most alternators have a capacitor across the output terminal to smooth out the ac voltage fluctuations, figure 8-65. This not only reduces EMI, but it also smooths the voltage applied to the vehicle electrical system. Electric motors, too, often have capacitors across the armature circuit to reduce voltage pulses at the commutator. Many horns have capacitors to suppress interference from their vibrating contacts.

Choke coils are installed in some motor circuits and radio power circuits to reduce current fluctuations created by self-induction. On automobiles, choke coils are generally used in combination with capacitors as EMI filter circuits. Wiper motors and electric fuel pump motors often have such filters, figure 8-66.

Suppressing with Ground Straps. You learned earlier that many automobiles have ground or bonding straps between the chassis and the engine to ensure a low-resistance circuit ground path. These straps also help to suppress EMI conduction and radiation. Resistance in a ground path creates an unwanted voltage drop. It also creates a voltage difference across the resistance, which can be an EMI source. Many vehicles have suppression ground straps between body parts and rubber-mounted components to ensure good conductivity and "short circuit" any interference, figure 8-67.

Some late-model cars have ground straps between body parts where no electrical circuit exists. Figure 8-68 shows such a strap between the hood and a fender panel. This strap is solely an EMI suppression device. Without it, the hood and the car body would form a large capacitor. The opening between the hood and fender would form an electrostatic field that would couple with computer circuits in the wiring harness routed along the fender panel.

Suppressing with Shielding. Metal shields around components that generate RFI signals block the waves. The distributors on a few engines are completely covered by metal shields to reduce RFI radiation from breaker points, condensers, and rotors. The metal housings of most vehicle computers help to shield the circuits from external electromagnetic waves.

The Need for EMI Suppression

Interference suppression has become more important as electronic systems have increased on late-model vehicles. The greater use of citizens-band (CB) and other 2-way radios, as well as mobile telephones, has increased the need for EMI suppression. Electromagnetic interference can also damage the operation of vehicle computer systems. These systems operate on voltage signals of a few millivolts (thousandths of a volt) and milliamperes (thousandths of an ampere) of current. Any of the four kinds of interference transmission listed earlier can generate false voltage signals and excessive current in computer systems. False voltage signals can upset computer operation. Excessive current can destroy microelectronic circuits.

As the use of increasingly sophisticated automotive electronics becomes more widespread with each model year, interference suppression has become a critical engineering field. Much electrical service work requires working with suppression devices and ensuring that they are installed and operating properly. Engineers are currently studying the use of fiberoptics as a means of reducing EMI signals. Since fiberoptics transmit light instead of electrical signals, they will neither radiate nor be affected by EMI transmissions.

SUMMARY

Electrical diagrams are essential tools for an engine performance technician, and the ability to read and understand a wiring diagram or circuit schematic is a necessary skill. Diagrams show circuit connections, wire colors and sizes, and the functions of circuit parts.

Some wiring diagrams are drawn to show component and wire locations in reference to their positions on the car. Other circuit schematics show current flow and power distribution. Illustrations of connectors, fuse blocks, and other electrical parts supplement system and circuit diagrams.

Many engine performance problems can be traced to defects in wiring and connectors. Engine electronic systems operate on low voltage and current. Any excess resistance due to a poor connection or a worn wire can upset system operation. Wiring and connector repair can cure many engine electrical problems and avoid unnecessary replacement of expensive parts.

REVIEW QUESTIONS

1. Mechanic A says that many carmakers' wiring diagrams show wire circuit numbers, connectors, and color codes, as well as schematic operation of circuit devices. Mechanic B says that many carmakers' wiring diagrams show the general arrangement or location of wires and parts on the vehicle. Who is right?

a. A only
b. B only
c. both A and B
d. neither A nor B

2. Mechanic A says that circuit numbers indicate wire size, or gauge. Mechanic B says that wire gauge can be listed in either American Wire Gauge (AWG) or metric specifications. Who is right?

a. A only
b. B only
c. both A and B
d. neither A nor B

3. Mechanic A says that a case ground is a ground connection directly to the engine or chassis. Mechanic B says that many lamps and gauges have remote grounds. Who is right?

a. A only
b. B only
c. both A and B
d. neither A nor B

4. The symbol shown here represents:

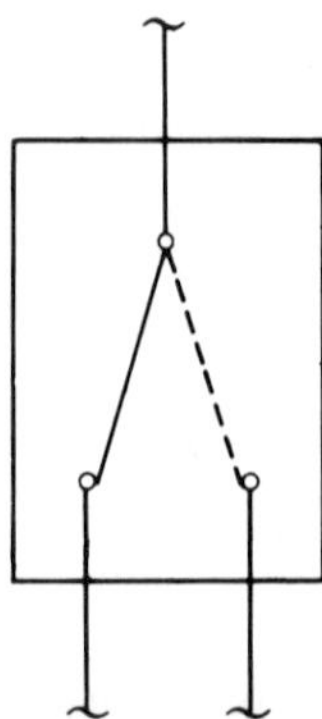

a. an spst switch
b. a relay
c. an spdt switch
d. a solenoid

5. The symbol shown here represents:

a. a ground connection
b. battery voltage supply
c. a connector
d. a wiring splice

6. The symbol shown here represents:

a. a fuse
b. a circuit breaker
c. a bulkhead connector
d. a wiring splice

7. Current flow, or power distribution, diagrams show all of the following EXCEPT:

a. current flow direction
b. voltage supply
c. circuit numbers
d. mechanical installation of circuit parts.

8. Mechanic A says that fuses are rated for different system voltage levels. Mechanic B says that all SFE fuses are the same diameter, but the length varies with the rating. Who is right?

a. A only
b. B only
c. both A and B
d. neither A nor B

9. Mechanic A is installing a 20-gauge fusible link in a 16-gauge circuit wire. Mechanic B is installing a 16-gauge fusible link in a 12-gauge circuit wire. Who is doing the job correctly?

a. A only
b. B only
c. both A and B
d. neither A nor B

10. You can insulate a wiring splice or repair damaged insulation by using:

a. epoxy sealant
b. heat shrink tubing
c. a replacement wiring harness jacket, or cover
d. all of the above.

11. Mechanic A says you can solder electrical wire correctly with acid-core solder. Mechanic B says that you can solder electrical wire correctly with solid-wire solder and acid-paste flux. Who is right?

a. A only
b. B only
c. both A and B
d. neither A nor B

12. Mechanic A says that you can repair a damaged wire in a molded, multiwire connector by cutting out the damaged wire and splicing in a new wire with a single bullet connector. Mechanic B says that you can release individual terminals from many hard-shell plastic connectors with a small screwdriver. Who is right?

a. A only
b. B only
c. both A and B
d. neither A nor B

13. Many late-model GM cars use special connectors, called:

a. molded connectors
b. Weather-Pack connectors
c. AMP connectors
d. inverted-tang connectors.

14. The symbol shown here represents:

a. an ignition coil
b. a capacitor
c. a transformer
d. a battery

15. The symbol shown here represents:

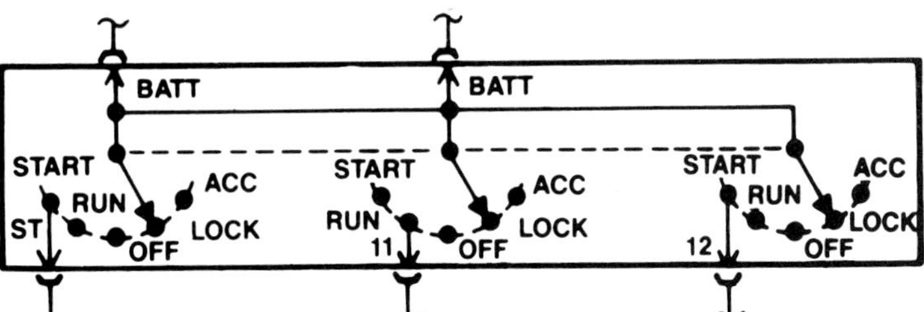

a. a bulkhead connector
b. an ignition switch
c. a fuse block
d. an ignition control module.

9 BATTERY TESTING AND SERVICE

INTRODUCTION

A weak or discharged battery can cause problems in all parts of a car's electrical system. Battery inspection should be the first step of any electrical service job. If battery cable connections are corroded or loose, they should be cleaned and tightened before proceeding with any other electrical tests. In fact, the *first steps* of every procedure that you follow for ignition, starting, charging, lighting, accessory, and other electrical service will be to *check battery voltage and state of charge*.

This chapter presents basic procedures for:

- Battery safety
- Battery inspection and cleaning
- Battery testing
- Battery charging
- Battery replacement

GOALS

After studying this chapter, you should be able to do the following jobs in twice the flat-rate labor time or less:

1. Explain and practice the necessary safety precautions for battery service.

2. Inspect batteries, cables, and mounting hardware for damage and deterioration; clean or replace as required.

3. Perform battery specific gravity, load (capacity), and 3-minute charge tests and evaluate the results.

4. Charge a battery in or out of the vehicle by slow and fast methods.

5. Select and install an appropriate replacement battery for a specific vehicle. Activate it if it is dry charged.

6. Using a booster battery and jumper cables, properly jump start a vehicle with a discharged battery.

BATTERY SAFETY

Review the electrical safety precautions in chapter 1 of this volume and observe the following specific precautions for battery service:

1. A battery gives off explosive hydrogen (H_2), particularly during charging. Keep cigarettes, sparks, and open flame away from a battery at all times. Operate charging equipment only in well-ventilated areas.

2. To avoid sparks and possible explosion, be sure a battery charger is *off* before connecting or disconnecting charger cables at a battery.

3. Never short circuit the two terminals of a battery and always keep metal tools off the top of a battery to avoid sparks and possible explosion. Remove rings, wristwatches, and other metal jewelry before servicing a battery.

4. Battery electrolyte contains sulfuric acid (H_2SO_4), which is corrosive. If electrolyte spills on a vehicle, wash it off with a solution of baking soda and water. Wash your hands immediately if electrolyte gets on them.

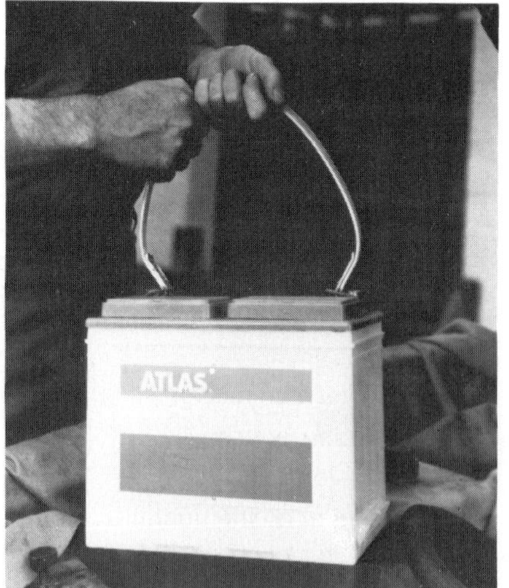

Figure 9-1. Carry or lift a battery with a carrier or lifting strap. (Atlas)

5. Wear safety glasses or goggles when servicing a battery. Keep electrolyte away from your face and eyes. If electrolyte contacts your eyes, flush them with water for 5 minutes, then with a mild solution of baking soda and water. See a doctor immediately.

6. Batteries are heavy and awkward, Always use a battery carrier or lifting strap, figure 9-1, to lift and carry a battery.

7. When disconnecting battery cables, always *disconnect the ground (usually −) cable first. Connect the ground cable last*. Always be sure that battery polarity is correct when connecting cables.

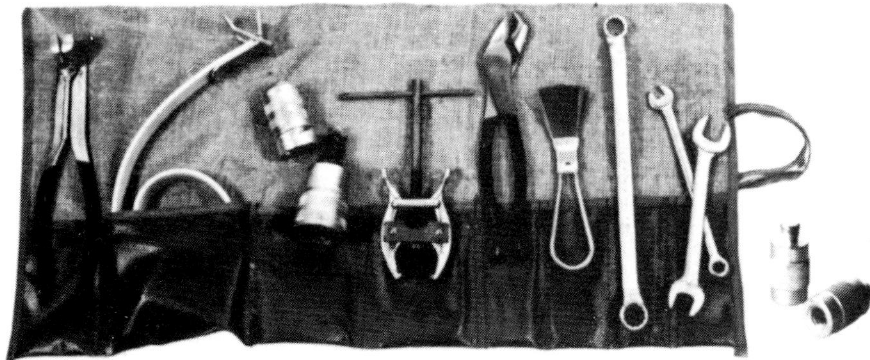

Figure 9-2. These are the basic tools and supplies for battery service. (Atlas)

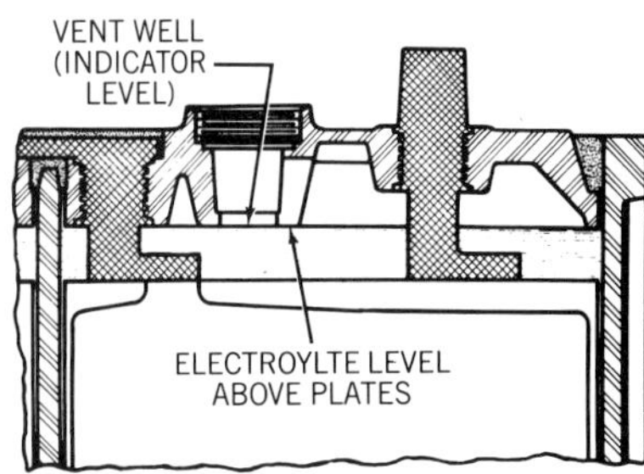

Figure 9-3. Electrolyte level must be above the plates in each cell, but not above the indicator level. (Delco-Remy)

Figure 9-4. Remove corrosion from battery terminals and holddown hardware. (Atlas)

Figure 9-5. Inspect battery cables for worn insulation and defective connectors. (Atlas)

8. Disconnect both battery cables before charging a battery in a vehicle. Charging current through the electrical system may damage the alternator.

9. Vehicles with engine control computers and other electronic systems with random access memory (RAM) need continuous battery voltage to retain electronic memory. Disconnecting the battery will erase data stored in RAM. You must reset electronic radios, clocks, door locks, and other devices after reconnecting the battery.

10. Do not use more than 16 volts to jump start a vehicle with electronic fuel injection or electronic engine controls. High voltage may damage electronic parts.

11. Battery electrolyte in a fully charged battery will not freeze until temperatures drop below −60° F (−51° C). Electrolyte in a discharged or partly charged battery, however, can freeze at temperatures of 18° F (−7.7° C). Cold winter weather can freeze a weak battery. Before jump starting or charging a battery in cold weather, check the electrolyte for signs of ice crystals or slush. If you see these signs of freezing, or if electrolyte level is below the tops of the plates, let the battery warm at room temperature for a few hours before charging or jump starting. Forcing current through frozen electrolyte can rupture or explode the battery.

BATTERY INSPECTION AND CLEANING

Even maintenance-free batteries need periodic inspection and cleaning to ensure that they are in good working order. If a vehicle charging system is working properly and electrical loads are not excessive, inspection and cleaning may be the only services any battery needs. To do these jobs, you will need the following equipment and tools, figure 9-2.

- A cleaning solution of baking soda and water, or ammonia
- Stiff-bristled cleaning brushes
- Terminal pliers and wrenches and perhaps a terminal spreader and puller
- Terminal and connector scraping and cleaning tools
- A battery carrier or lifting strap
- Protective coating for the battery terminals (jelly or spray)

Battery Inspection

Complete battery inspection consists of the following eight steps but takes only a couple of minutes to do.

1. If the battery has removable cell caps, check the electrolyte level. It should be above the tops of the plates or at the split-ring indicator level in each cell, figure 9-3. Add water to raise the electrolyte level, if necessary. Use mineral-free drinking water or distilled water. Do not overfill the battery.

2. Check for missing or damaged cell caps; replace as required.

3. Check battery terminals, cable connectors, and metal holddown parts for acid corrosion, figure 9-4. Clean as required.

4. Check the cables for broken or corroded wire strands, worn insulation, and defective connectors, figure 9-5. Replace defective parts.

5. Check battery case and cover for dirt, grease, or electrolyte condensation that could cause voltage to leak to ground. Clean battery as necessary.

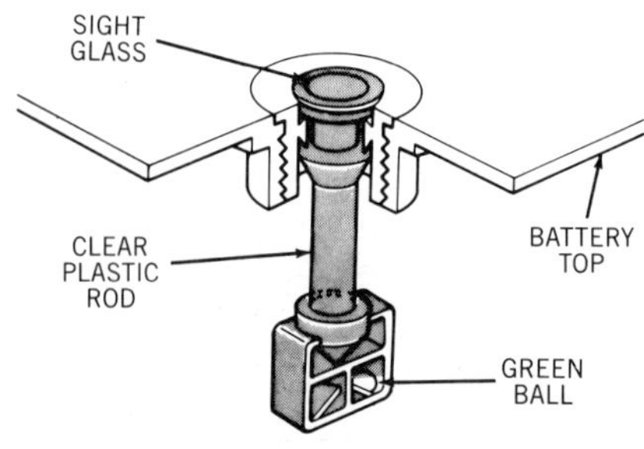

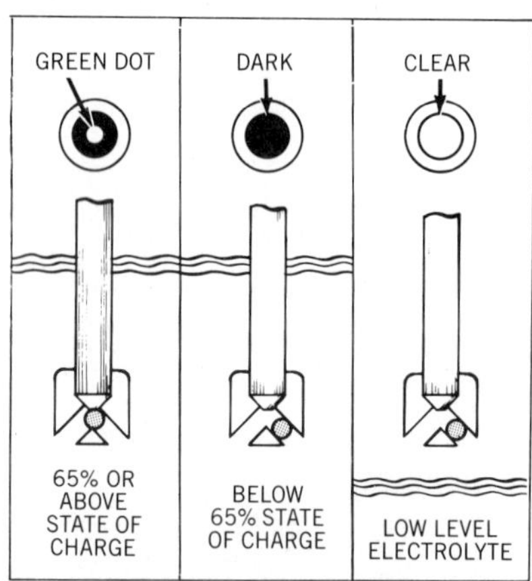

Figure 9-6. The color of a built-in hydrometer indicates the general state of charge. (Chrysler) (Delco-Remy)

6. Inspect the battery for cracks, loose terminals, and other damage. Replace a damaged battery.

7. Check the battery carrier (tray), holddown parts, and heat shields for looseness or improper installation. Tighten or replace loose or damaged parts.

8. If the battery has a built-in hydrometer (state-of-charge indicator), check its color indication, figure 9-6, for general battery condition.

Battery Cleaning

Some dirt and corrosion naturally collect on a battery from two general sources:

1. High temperature and air movement under the hood cause dirt and grease to collect on any flat surface.

2. Normal battery gassing (hydrogen release) and water evaporation carry electrolyte vapors out of the battery. The vapors condense on the battery top and contain a small amount of sulfuric acid. As acid vapors condense over a period of time, they corrode metal parts.

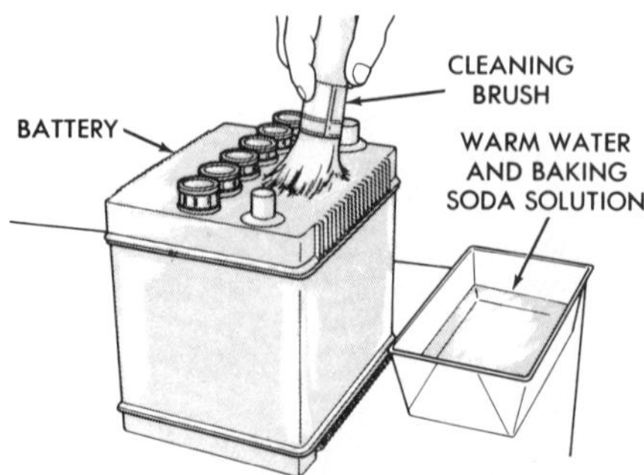

Figure 9-7. Baking soda and water, or ammonia, will neutralize and remove corrosion. (Chrysler)

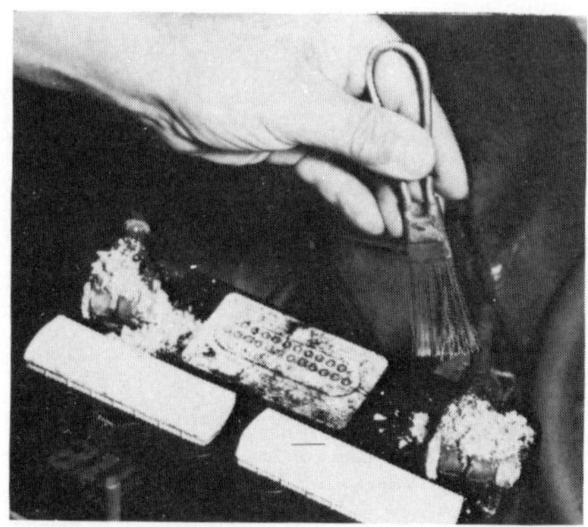

Figure 9-8. Use a stiff-bristled brush to remove corrosion from the outside of a connector or holddown hardware.

Dirt and corrosion cause two general problems:

1. Dirt and grease form a conductive film that causes voltage to discharge slowly to ground or between the + and − battery terminals. Electrolyte condensation adds to this discharge action because it, too, is conductive. The same electrochemical self-discharge that occurs inside a battery will occur outside when electrolyte condenses on a battery top.

2. Electrolyte condensation contains corrosive H_2SO_4, which eats away the metal of battery terminals, cable connectors, and holddown parts. Corrosion on battery terminals and cables adds resistance to the entire electrical system. In extreme cases, corrosion between battery terminals and cables can add enough circuit resistance to drop 12 volts across the cable connection and leave *no* voltage for the electrical system.

Periodic battery cleaning eliminates these two problems of voltage leakage (discharge) and circuit resistance. Thorough battery cleaning consists of the following steps:

> **Caution.** Do not let baking soda or ammonia enter the battery cells. These solutions will neutralize the acid in electrolyte and destroy a battery.
>
> Keep dissolved corrosion and cleaning solutions off painted surfaces and rubber parts. Dissolved acid will harm paint and rubber.

1. Wash the battery top, case, and holddown parts with a mixture of baking soda and water or with household ammonia, figure 9-7. These solutions neutralize acid and dissolve corrosion.

2. Remove heavy corrosion with a stiff-bristled brush, figure 9-8. Do not splash corrosion or the cleaning solution onto painted surfaces.

3. After neutralizing acid and corrosion with baking soda or ammonia, wash the battery with detergent and water to remove dirt. Rinse with clear water from a hose or bucket.

4. Dry the battery, the cables, the holddown parts, and adjoining vehicle parts with a clean cloth or low-pressure compressed air.

5. Cleaning the outside of battery terminals and cables often does not remove corrosion that forms between the cable connectors and terminals. *Starting with the ground cable*, remove the cable connectors from the terminals as follows, figure 9-9:

- **a.** On a side-terminal battery, use a wrench to remove the capscrews that secure the cables to the terminals.
- **b.** On a top-terminal battery, use a wrench or battery pliers to loosen the nut on the cable connector bolt, or use pliers to release spring-type connectors. Use a puller to remove a cable that is stuck to a post, figure 9-10. Do not pry or hammer on a stuck cable connector.

6. Wash the battery terminals and cable connectors with a baking soda solution or ammonia to remove all corrosion. Use a spreading tool to open the connector for a top-terminal battery, figure 9-11.

7. Scrape battery posts and the insides of cable connectors with wire brushes that have internal and external bristles, figure 9-12. Remove corrosion from side-terminal connectors with a stiff-bristled brush.

8. After cleaning cable connectors and battery terminals, dry them with a clean cloth or low-pressure compressed air.

9. Remove and clean corroded hold-down parts with the same methods used for battery cables and terminals.

10. Starting with insulated (positive, or "hot") cable, reconnect the battery cables and reinstall holddown parts securely.

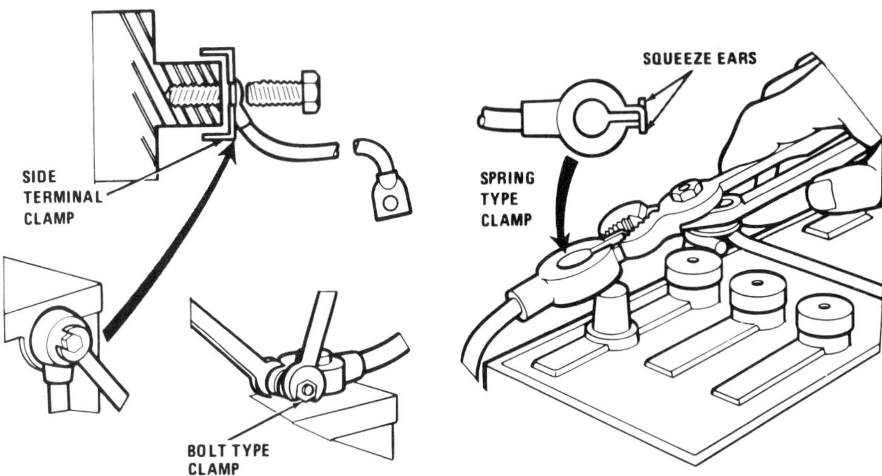

Figure 9-9. Use a wrench of the correct size to loosen battery cable connectors. Use pliers to release spring-type cable clamps. (Chrysler)

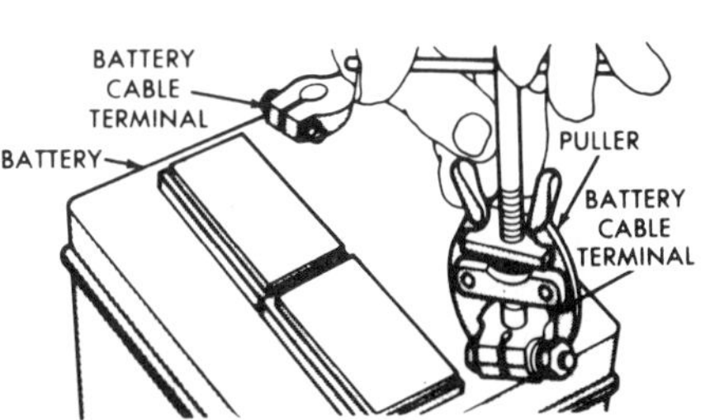

Figure 9-10. Use a small puller to remove a cable. Do not hammer or pry on the connector. (Chrysler)

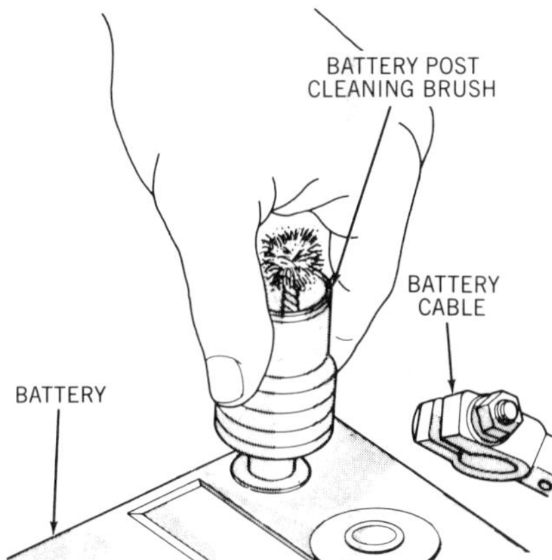

Figure 9-11. Use battery pliers or a spreading tool to open the connector for a top-terminal battery. (Atlas)

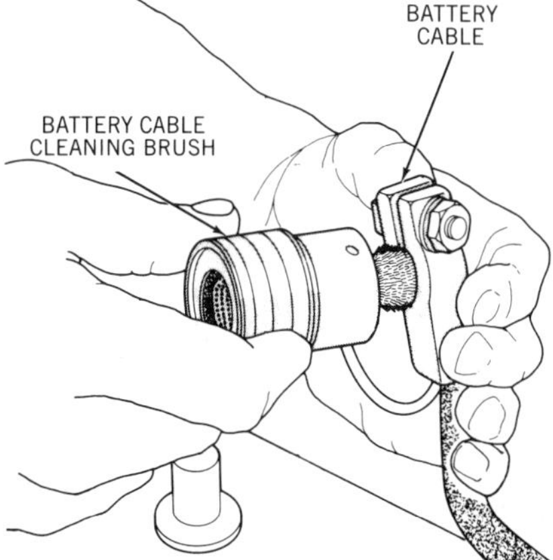

Figure 9-12. Use a special brush with internal and external bristles to clean battery terminals and connectors. (Chrysler)

BATTERY TESTING

For a battery to operate properly, it must be fully charged and have the capacity to deliver current needed by the starter motor and other electrical system parts. You can use a load (capacity) test to measure the current-delivering ability of any battery, maintenance free or vent-cap type. If a battery has removable cell caps, you can test its state of charge by measuring electrolyte specific gravity with a hydrometer. You can use a 3-minute charge test to determine if a battery can be recharged or if it is too badly sulfated to accept a charge. Do not use a 3-minute charge test on a maintenance-free battery because the test results will not be reliable.

Battery Load (Capacity) Test

A load (capacity) test indicates how well a battery will deliver current while maintaining enough voltage to operate the ignition system. A load test is the only way to test a maintenance-free battery and the preferred way to test the operating ability of any battery in a late-model vehicle. A battery must be at least one-half to three-quarters charged for an accurate load test.

Chapter 5 explains load test pro-

Figure 9-13. Remove the coil high-voltage cable from the distributor cap and ground it with a jumper wire to disable the ignition.

Figure 9-14. Disconnect the ignition voltage feed connector from the distributor to disable a Delco HEI system.

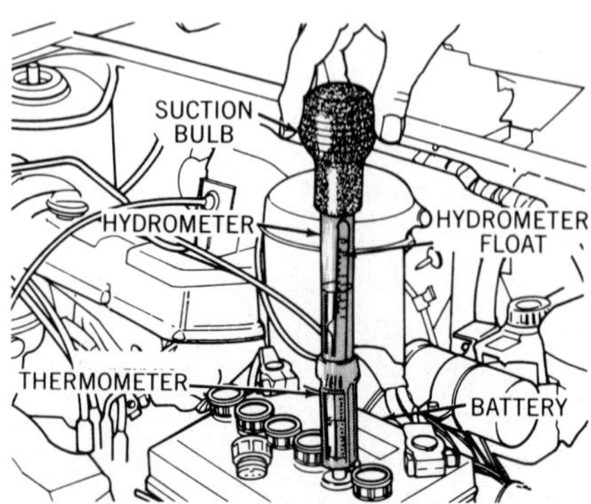

Figure 9-15. Do not let the indicator (float) touch the top or sides of the hydrometer tube when measuring specific gravity. (Chrysler)

cedures, in detail, as part of comprehensive engine testing. Refer to those instructions for basic load testing. If you do not have the necessary test equipment, you can use the following alternative procedure to determine general battery capacity.

Alternative Load Test. If you do not have an ammeter or a battery-starter tester, you can make a general test of battery capacity by using the starter motor to load the battery, as follows:

1. Connect a voltmeter to the battery + and − terminals as explained previously.

2. Disable the ignition system to keep the engine from starting:

a. Remove the coil high-voltage cable from the distributor cap and ground it with a jumper wire, figure 9-13.

b. With a Delco HEI system or similar electronic ignition, disconnect the ignition primary (battery voltage) connector from the distributor, figure 9-14.

3. Using the ignition switch, crank the engine for 15 seconds while watching the voltmeter.

Voltage should remain above 10 volts (9.6 volts, minimum) for a 12-volt battery during cranking. If voltage was below 9.6 volts, the battery or the starting system may be faulty. You must test the battery and the starting system individually to find the problem. If voltage was 10.0 volts or more with the cranking load for 15 seconds, the battery is good. If voltage was between 9.6 and 10.0 volts, recharge and retest the battery before more electrical or engine testing.

Specific Gravity Test

A hydrometer is an instrument that measures the specific gravity of a liquid. Battery hydrometers are calibrated to measure the specific gravity of electrolyte. Because electrolyte specific gravity indicates a battery's state of charge, you can use this test for a vent-cap battery. Obviously, you can't use it on a maintenance-free battery because you can't get to the electrolyte.

Before testing specific gravity, remove the cell caps and check electrolyte level. If electrolyte level is low, add water to the battery and charge it at about 5 amperes for 5 to 15 minutes to mix the water with the electrolyte before testing.

Test specific gravity as follows:

1. Remove the cell caps and lay them carefully on the battery top.

2. Insert the hydrometer tube into either end cell and draw in enough electrolyte to float the indicator without letting it touch the top or sides of the tube, figure 9-15. The indicator must float freely for an accurate reading.

3. If the hydrometer has a built-in thermometer, fill the hydrometer with electrolyte and empty it several times to stabilize the temperature before reading the specific gravity.

4. With the hydrometer at eye level, read the specific gravity on the float, figure 9-16. Read electrolyte temperature on the hydrometer thermometer or a separate thermometer.

5. Return the electrolyte to the cell from which it was taken and repeat these steps with the other battery cells, in order.

The electrolyte in all cells of a fully charged battery will have specific gravity of 1.260 to 1.280 at 80° F (26.7° C). If electrolyte temperature is higher or lower, correct the specific gravity reading as follows:

- Every 10° above 80° F—*add* 0.004 to the specific gravity measurement.
- Every 10° below 80° F—*subtract* 0.004 from the specific gravity measurement.

Many hydrometers have temperature correction tables printed on their sides. Here are two examples of how to correct a specific gravity reading for high or low temperature.

1. *High temperature*

Indicator reading = 1.250
Electrolyte temperature = 120° F
Temperature correction = 40° F
4 × 0.004 = 0.016
1.250 + 0.016 = 1.266

2. *Low temperature*

Indicator reading = 1.250
Electrolyte temperature = 20° F
Temperature correction = 60° F
$6 \times 0.004 = 0.024$
$1.250 - 0.024 = 1.226$

Recharge any battery with a corrected specific gravity below 1.230. If specific gravity varies more than 50 points (0.050) between any two cells, the battery is defective and must be replaced.

Uniformly low specific gravity readings for all cells can indicate two things:

1. The battery has reached the end of its useful life and needs to be replaced.

2. The battery is discharged because of a problem elsewhere in the electrical system.

A discharged battery can be due to something as simple as a loose drive belt for the alternator. It also can be due to a problem in the charging system, a defective starting system that causes high current draw, or a continuous current drain caused by some other electrical problem. In any case, a discharged battery is a clue to test the charging, starting, and other electrical systems for problems that might cause battery failure. Recharging or replacing a battery is not a cure for other electrical problems.

Three-Minute Charge Test

A 3-minute charge test indicates if a discharged battery can be recharged or if it is too badly sulfated to accept a charge. This test is not reliable for maintenance-free batteries, however, because of their plate construction.

The 3-minute charge test uses a fast battery charger to pass high current through a battery for 3 minutes. High current will loosen sulfate deposits from the plates. If a battery is too badly sulfated to accept a charge, however, current will not loosen the deposits, and high voltage will be measured across the battery terminals.

Caution. If voltage higher than 16 volts is recorded early in the test, stop the test immediately to avoid battery explosion or damage to the charger.

1. Disconnect both cables from the battery if it is in the vehicle.

2. Connect the battery charger + lead to the battery + terminal and the − lead to the − terminal, figure 9-17.

3. Connect a voltmeter with the + lead on the battery + terminal and the − lead on the − terminal.

4. Turn on the battery charger and adjust it for a high charging rate of no more than 40 amperes. If the charger has a timer, set it for 3 minutes.

5. After 3 minutes, read the voltmeter and turn off the charger.

If voltage is 15.5 volts or less, the battery can be recharged. If voltage is higher than 15.5 volts, the battery must be replaced.

BATTERY CHARGING

An alternator will *maintain* a battery in a fully charged condition, if:

- The vehicle charging system and other electrical system parts are in good condition.
- The battery has the right current capacity for the vehicle.
- The battery has not neared the end of its useful life and lost its ability to retain a charge.

An alternator will not recharge a badly discharged battery because the alternator must supply current to other electrical systems, as well as charging current to the battery.

A battery charging rate is measured in amperes of current for a period of time. For example if you charge a battery with 10 amperes for 4 hours, you apply a 40-ampere-hour charging rate. If an alternator does not deliver the charging rate that a discharged battery needs, the battery will remain less than fully charged. You often must recharge a battery with a battery charger to maintain its proper operation or to test and service other parts of the electrical system.

Generally, you can charge a battery at any current from 3 to 50 amperes for the time necessary to restore it to full charge. Low current requires longer

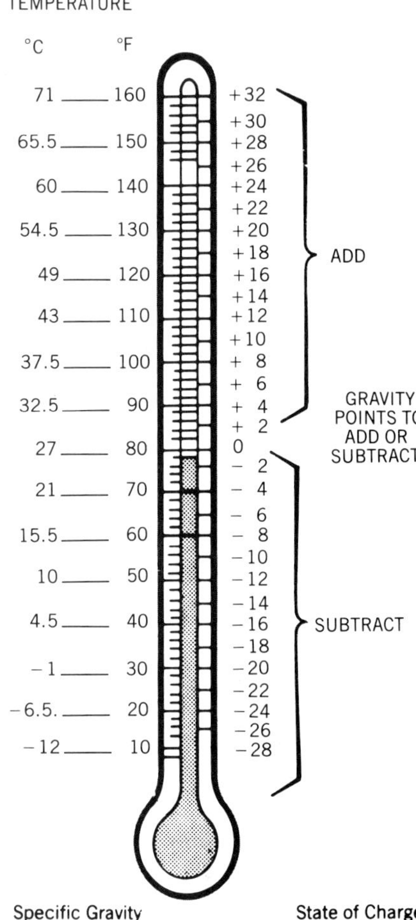

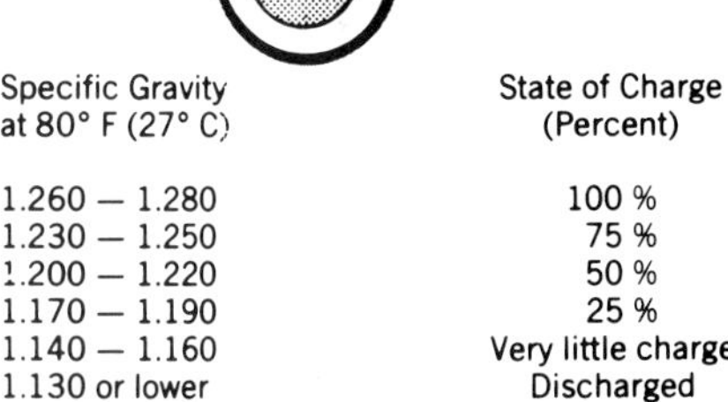

Specific Gravity at 80° F (27° C)	State of Charge (Percent)
1.260 — 1.280	100 %
1.230 — 1.250	75 %
1.200 — 1.220	50 %
1.170 — 1.190	25 %
1.140 — 1.160	Very little charge
1.130 or lower	Discharged

Figure 9-16. Specific gravity readings corrected to 80° F (27° C) indicate the state of charge.

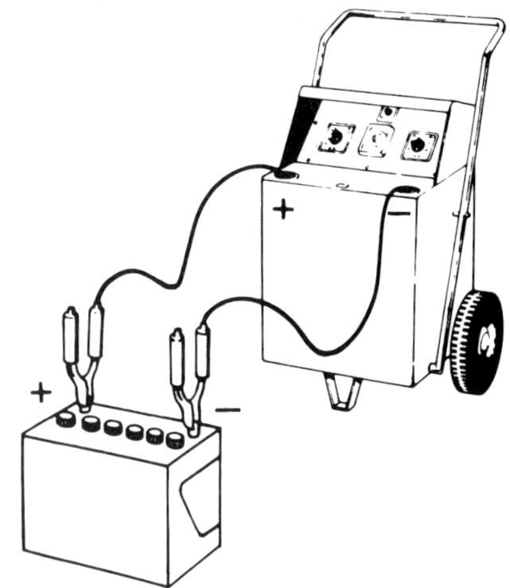

Figure 9-17. Be sure the charger is off. Then connect the + lead to the battery + terminal and the − lead to the − terminal. (Ford)

Battery Charging Guide (6-Volt and 12-Volt Batteries) Recommended Initial Charging Rate and Time for Fully Discharged Condition (Constant-Voltage, Taper-Rate Charger)

Reserve Capacity Rating	*20-Hour Rating*	*5 Amperes*	*10 Amperes*	*20 Amperes*	*30 Amperes*	*40 Amperes*
75 Minutes or less	50 Ampere-Hours or less	10 Hours	5 Hours	2½ Hours	2 Hours	
Above 75 To 115 Minutes	Above 50 To 75 Ampere-Hours	15 Hours	7½ Hours	3¼ Hours	2½ Hours	2 Hours
Above 115 To 160 Minutes	Above 75 To 100 Ampere-Hours	20 Hours	10 Hours	5 Hours	3 Hours	2½ Hours
Above 160 To 245 Minutes	Above 100 To 150 Ampere-Hours	30 Hours	15 Hours	7½ Hours	5 Hours	3½ Hours
Above 245 Minutes	Above 150 Ampere-Hours		20 Hours	10 Hours	6½ Hours	5 Hours

Figure 9-18. Typical charging rates for fully discharged batteries of different current capacities.

charging time, which equals a slow-charging rate. High current requires less charging time, which equals a fast-charging rate. Battery chargers used in service stations and garages can charge at slow and fast rates. Small "trickle chargers" (the kind owned by many motorists) charge only at slow rates.

Fast Charging vs. Slow Charging

When time is available, slow charging at 5 to 15 amperes is better than fast charging. Slow charging minimizes battery heat and plate damage due to high charging current. Slow charging also promotes a more gradual and complete electrochemical recombination of electrolyte and plate materials. But, you can satisfactorily fast charge a battery with high current as long as:

- Electrolyte temperature does not exceed 125° F (52° C).
- Electrolyte does not bubble or spew from the vents.
- Battery gassing is not noticeably excessive.
- Battery terminal voltage does not exceed 15.5 volts.

Additionally, follow these precautions for fast charging:

1. Monitor electrolyte temperature in a vent-cap battery with a thermometer. Feel the case of a maintenance-free battery to estimate the temperature. Stop charging if temperature exceeds 125° F (52° C).
2. Reduce the charging current if battery terminal voltage exceeds 15.5 volts.
3. Do not charge a battery that has failed a 3-minute charge test or is badly sulfated.
4. Do not charge a battery with plate or separator damage or a cracked case.

Most professional battery chargers deliver gradually decreasing charging current as the battery state of charge increases to protect the battery from overcharging. When possible, follow a period of fast charging with a period of slow charging at low current to stabilize the battery and ensure that it is fully charged.

Charging a Battery

Several variable factors affect charging rates such as:

- *Battery capacity* —A high-capacity battery (100 ampere-hours or 500 cold-cranking amperes, for example) requires twice the charging time that a battery with half that capacity needs.
- *Battery state of charge* —A completely discharged battery requires twice the charging rate that a half-charged battery needs
- *Electrolyte temperature* —A battery at 0° F (−18° C) needs two hours more charging time than a battery at 80° F (27° C).
- *Battery age and condition* —A battery that is several years old or that has been used for severe service will need 50 percent more charging rate in ampere-hours than a new battery.

Although we cannot list exact charging specifications, the guidelines in figure 9-18 are typical charging rates for completely discharged batteries of various capacities.

The following procedure includes instructions to check the specific gravity of a vent-cap battery to determine its state of charge. You cannot do this with a maintenance-free battery, but you can estimate the charging rate on the basis of load test specifications:

- If the load test specification is 200 amperes or less, a 50-ampere-hour charging rate should restore a full charge.
- If the load test specification is more than 200 amperes, a 75-ampere-hour charging rate should restore a full charge.

Some calcium-grid maintenance-free batteries may not accept high charg-

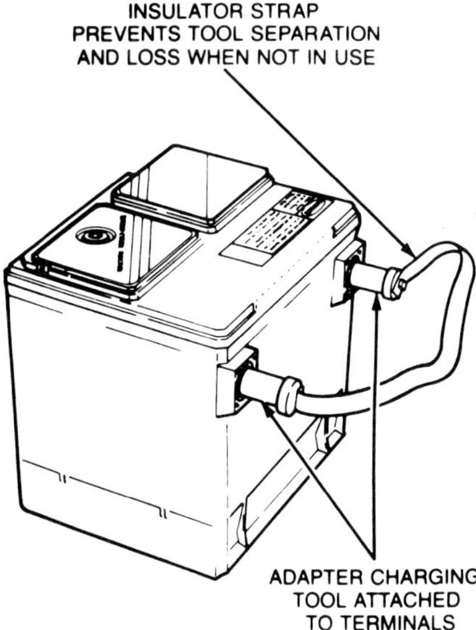

Figure 9-19. Connect adapters to a side-terminal battery for testing and charging. (Chevrolet)

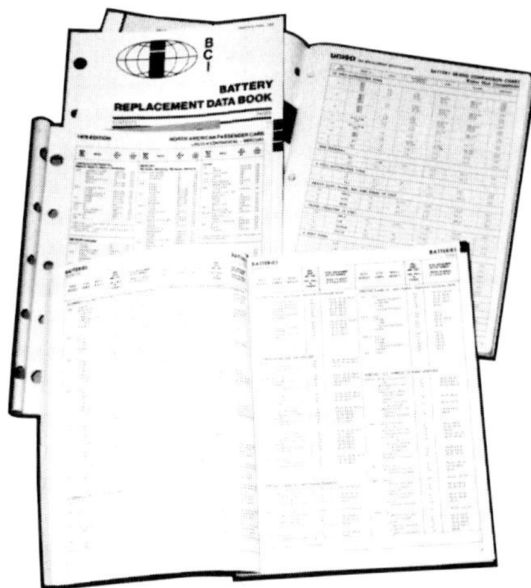

Figure 9-20. Service manuals and parts books contain battery capacity and size specifications for different vehicles.

ing current when completely discharged. If the load test indicates that such a battery is completely discharged, monitor the initial charging rate and terminal voltage closely. If terminal voltage exceeds 15.5 volts with a high charging current, reduce the current and charge the battery for several hours with low current that keeps terminal voltage below 15.5 volts.

After applying these charging rates, repeat the load test. Continue charging for another 50 to 75 ampere-hours if test results are still below specifications.

Before charging a battery, be sure the ac receptacle into which the charger is plugged is delivering full current and voltage. Do not plug the charger into a heavily loaded circuit. A 20-percent ac voltage drop (120 to 96 volts) will reduce the charger output almost 40 percent. If you use an extension cord, be sure it is number 14 or heavier wire and less than 25 feet long to avoid excessive voltage drop. Whether you charge a battery at a fast-charging or a slow-charging rate, follow these general steps:

1. Observe the safety precautions listed at the beginning of this chapter, especially:

- **a.** Wear eye protection and charge the battery in a well-ventilated area. Keep sparks or flame away from the battery.
- **b.** Do not charge a battery that shows any signs of freezing.
- **c.** Be sure the charger is off before connecting or disconnecting its leads at the battery.

2. Check the battery electrolyte level:

- **a.** For a vent-cap battery, remove the cell caps and be sure the electrolyte is above the plates. Add water if necessary.
- **b.** For a maintenance-free battery, check the built-in hydrometer, if so equipped, to be sure electrolyte level is correct. Otherwise, hold a shop light behind the battery and view the electrolyte level through the translucent plastic case.

3. Disconnect the battery cables to avoid damaging the alternator and other electronic parts. Remove the battery from the vehicle if you wish.

4. If necessary, connect charging connector adapters to a side-terminal battery, figure 9-19.

5. Leave the cell caps in place on a vent-cap battery. Be sure all vents are open. Place a damp cloth over the battery top to reduce hydrogen escape.

6. With the charger off, connect the + cable to the battery + terminal and the – cable to the battery – terminal.

7. Adjust the charger for the desired charging current and time and turn it on.

8. On a vent-cap battery, check the electrolyte specific gravity and temperature periodically during charging. Stop charging if the temperature exceeds 125° F (52° C). Estimate the temperature of a maintenance-free battery by touching the case.

9. Use a voltmeter to check voltage across the battery terminals periodically. If voltage exceeds 15.5 volts, reduce the charging current.

10. Stop charging when all cells are gassing freely and specific gravity is 1.260 to 1.280 for three straight hours, corrected for temperature.

11. Repeat the battery load test on a maintenance-free battery and continue charging if results are below specifications.

12. After charging, wash the battery top with baking soda and water to remove electrolyte acid condensation due to gassing. Rinse with fresh water and dry the battery.

13. Reinstall the battery, if removed, and reconnect the battery cables. *Connect the insulated (positive, or "hot") cable first, then the ground cable.*

BATTERY REPLACEMENT

Any replacement battery for a vehicle must have a current capacity equal to, or higher than, the original battery. Installing a low-capacity battery in a car with a heavy electrical load will shorten battery life and overwork the charging system. Additionally, a replacement battery must be the same physical size and group number as the original battery. Battery capacity and group number specifications are listed in vehicle service manuals and parts books, figure 9-20.

The Battery Council International (BCI) replacement data books are the source for all battery ratings published in parts books. Most parts books list cold-cranking, reserve-capacity, and ampere-hour ratings. The cold-cranking rating is the most common specification for late-model automobile batteries. However, the reserve-capacity rating is equally important for vehicles with computer memory systems that place a constant current drain on the battery.

Battery Removal and Installation

Since 1956, all domestic cars and light trucks have used negative-ground electrical systems. The battery − terminal is grounded to the vehicle engine or chassis. Some imported cars as late as 1969, however, had positive-ground electrical systems. Although the − terminals of batteries on late-model cars are the ground terminals, it is good practice to mark both battery cables before disconnecting them to ensure that a new battery is installed with the correct polarity.

Remove and replace a battery as follows:

1. Review and follow the safety precautions at the beginning of this chapter.

2. Mark the battery cables before disconnecting them from the old battery.

3. Using correctly sized wrenches for the cable connectors, *disconnect the ground (−) cable first*. Use a cable puller to remove a connector stuck to a post-type terminal.

4. Then disconnect the insulated, or "hot," (+) cable.

5. Remove the battery holddown hardware and any heat shields that require removal, figure 9-21.

6. Using a battery carrier or lifting strap, remove the battery from the vehicle.

7. Clean and remove corrosion from cables, holddowns, battery trays, and heat shields before installing the new battery.

8. Using a battery carrier or lifting strap, install the battery in the vehicle.

9. Reinstall all holddown hardware and heat shields securely.

10. *Connect the insulated, or "hot," (+) cable first*, then the ground (−) cable. Be sure connectors on post terminals are flush with, or slightly below, the tops of the posts.

11. To reduce corrosion, apply petroleum jelly or anticorrosion paste or spray to the connectors after installation.

BATTERY DIAGNOSIS FOR VEHICLE ELECTRONIC SYSTEMS

With the increasing use of automotive computers and electronic sensors, the battery has become more than just a device that provides power to start the vehicle. Low battery voltage can affect the operation of an electronic control module, causing various driveability problems. Here are some examples of the possible problems resulting from low battery voltage:

- The computer may set false trouble codes.
- If the computer contains a backup mode program, low battery voltage can cause the computer to switch into that program until the ignition is turned off. While the vehicle will continue to run in backup mode, driveability will suffer.
- The computer may attempt to solve the low voltage problem by adjusting fuel injector timing to obtain high idle rpm as a way to increase the charging current from the alternator.

An excessive electrical drain will affect the battery of a computer-equipped vehicle in one of two ways:

1. The battery will be very low or completely dead after sitting overnight.

2. The battery will gradually lose its power over two or three days.

The following procedures are provided by General Motors and Ford Motor Company for their vehicles and should be used in such cases to determine if the battery failure is caused by an excessive electrical drain. Other manufacturers may have their own procedures and specifications to be followed. Check the service literature for the make and model vehicle to be serviced.

GM Vehicles

General Motors recommends a battery current drain test and a test for abnormal ECM power.

Current Drain Test

1. Raise and support the hood, then remove the underhood lamp bulb, if so equipped.

2. Turn the ignition off.

3. Charge the battery until it is fully charged.

4. If the car has an electronic load control device, the pump exhaust valve will cause a 0.5-ampere drain for 4 to 6 minutes after the ignition is turned off. Locate the pump and disconnect the exhaust valve connector, figure 9-22.

5. Disconnect the battery negative cable and connect an ammeter capable of handling at least 2 amperes in series between the cable and battery terminal. (A digital volt-ohmmeter also can be used if set to the DCA scale).

6. Note the amperage draw. With the ammeter on the 2-ampere (2,000 mA) setting, switch the selector switch down the scale one notch at a time until you have an accurate reading of the amperage drain. With the key off, the vehicle doors closed and all accessory systems off, the amperage draw should not exceed 0.1 ampere.

7. If using a digital volt-ohmmeter, disconnect it and reconnect the battery negative cable before opening the doors or operating any accessory system. If it is not disconnected, the fuse in the test instrument will blow.

The current drain from a normal parasitic load on the battery should be between 10 and 30 mA. If a reading above this is obtained, perform an ECM test to determine if the electronic control module is the cause of the excessive current drain. If the ECM test is satisfactory, you should check each of the electrical subsystems and

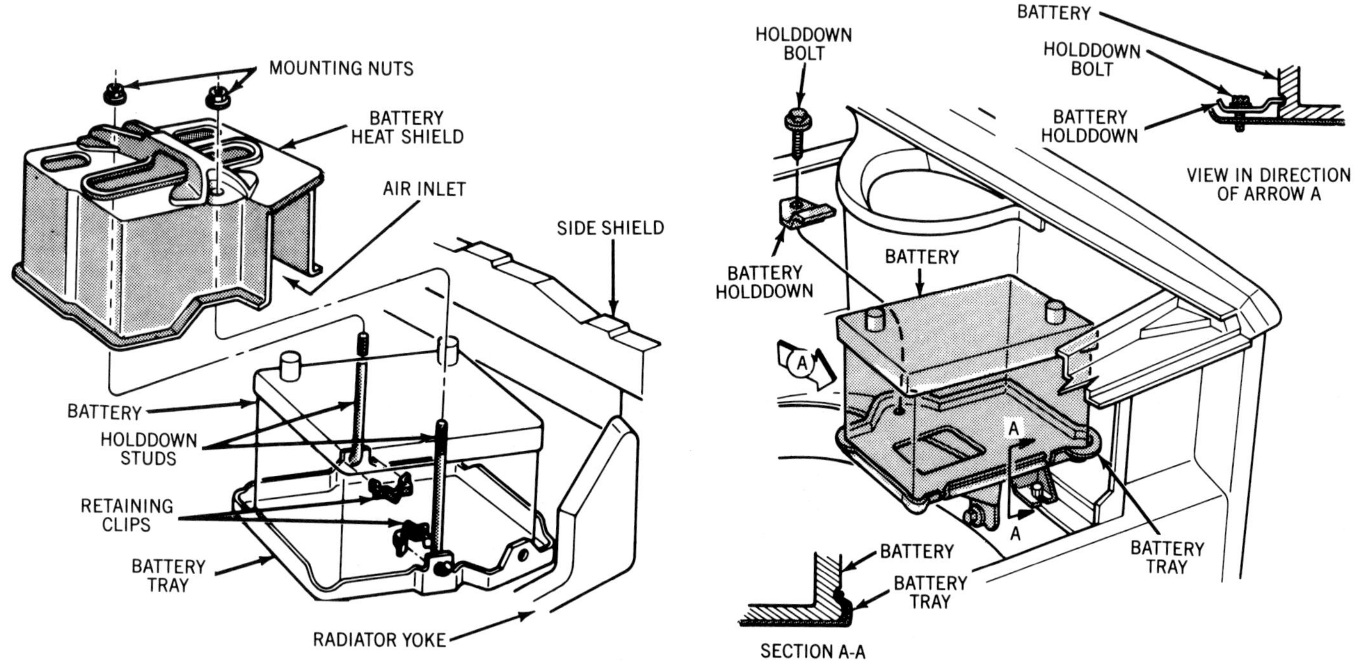

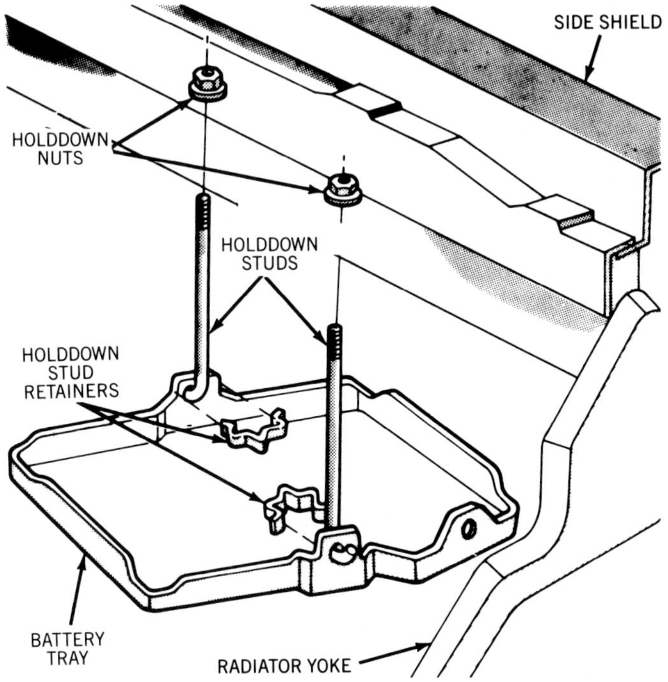

Figure 9-21. Remove holddown hardware and some heat shields to replace a battery. Be sure to reinstall all of these parts securely. (Chrysler)

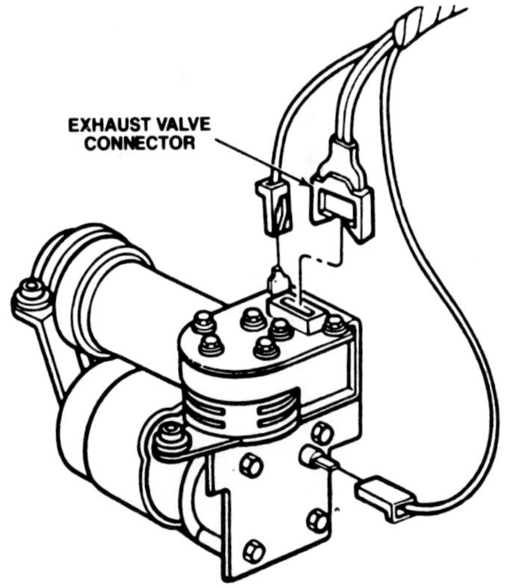

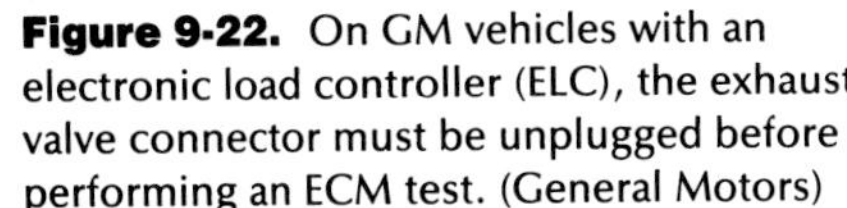

Figure 9-22. On GM vehicles with an electronic load controller (ELC), the exhaust valve connector must be unplugged before performing an ECM test. (General Motors)

components to find the cause of the problem.

ECM Test. When excessive amperage draw is found during a current drain test, the engine computer (the ECM) may be the cause. Under certain conditions, the ECM can remain powered up with the ignition off. If this happens, the battery will go dead quickly. To solve the problem, the defective ECM should be replaced. To determine if the ECM is at fault:

1. Make sure the test equipment has been disconnected and the battery negative cable has been reconnected.

2. Connect a scanner tool to its diagnostic connector on the vehicle.

3. Turn the ignition switch on for 5 seconds, then shut it off. While the switch is on, the scanner will read data. When the switch is turned off, the scanner should stop reading data.

4. If the scanner is still displaying

data 10 seconds after the ignition has been turned off, the ECM is remaining powered up and should be replaced.

5. If the ECM test is satisfactory, the electronic load controller (ELC) may be defective:

a. Disconnect the battery negative cable and connect an ammeter capable of handling at least 2 amperes in series between the cable and the battery terminal.

b. Turn the ignition on. Record the current draw and time, then shut the ignition off.

c. Wait 10 minutes and recheck the ammeter. If the ELC has not turned off after 10 minutes, it is defective.

6. If a new ECM is installed, repeat this procedure to make sure that the problem has been corrected.

Ford Current Drain Test

Ford recommends the use of a test lamp instead of a voltmeter. The voltmeter will react to the normal parasitic loads; a test lamp will generally indicate only current drains large enough to cause a problem. You also can perform this test with an ammeter. If an ammeter is used, the permissible current drain must be less than 0.5 ampere. If the drain exceeds 0.5 ampere, perform step 5a to locate the malfunctioning circuit.

1. Raise and support the hood, then remove the underhood lamp bulb, if so equipped.

2. Charge the battery until it is fully charged.

3. Disconnect the battery negative cable. Connect the test lamp between the disconnected cable and the battery negative terminal. The test lamp should not light.

4. Turn the ignition key on, then off. The test lamp should not light.

5. If the test lamp lights in step 4 on other vehicles with electronic air suspension or load leveling systems, unplug the connector at the system module. Some system modules will remain powered up for as long as 70 to 90 minutes after they have been disconnected. Check the carmaker's specifications and wait the required length of time before continuing the test.

a. If the test lamp remains on after the modules are disconnected, remove the fuses and starter relay leads one at a time until the problem circuit is found.

b. If the test lamp goes out after disconnecting the modules, reconnect all modules. Wait the required length of time as stated above and then repeat the test. Unplug one module at a time until the problem circuit is located.

JUMP STARTING A VEHICLE

Often, you may have to use a booster battery and jumper cables to jump start a vehicle with a weak or discharged battery. Jump starting a vehicle is preferable to push starting for several reasons.

1. Push starting a vehicle can draw gasoline through combustion chambers and flood a catalytic converter. This causes converter overheating and damage.

2. Most vehicles with automatic transmissions cannot be push started because most transmissions have no pump to develop hydraulic pressure and turn the engine.

3. Push starting a diesel engine is dangerous because:

a. The injection system may operate without lubrication from the fuel.

b. Extra fuel in the cylinders may explode and damage the engine when it ignites.

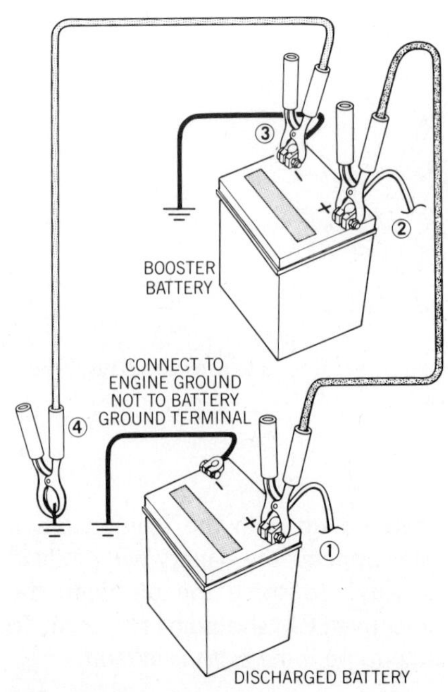

Figure 9-23. Connect the booster battery to the disabled battery like this.

Follow these steps to jump start most vehicles safely. They will avoid damage to the starting system and prevent sparks that might cause a battery explosion.

> **Caution.** Do not use more than 16 volts to jump start a vehicle with electronic fuel injection or electronic engine controls. High voltage may damage electronic parts.
>
> Do not let the bumpers or other parts of the two vehicles touch. Touching will create an electrical connection and possible arcing.

1. Engage the parking brake of the disabled vehicle, put the transmission in neutral or park, and turn off the ignition switch and all accessories.

2. Attach one end of one jumper cable to the insulated, or "hot" (usually +), cable of the discharged battery, figure 9-23.

3. Attach the other end of the same cable to the corresponding terminal (usually +) of the booster battery.

4. Attach one end of the other cable to the ground terminal (usually −) of the booster battery.

5. Attach the other end of the second cable to *an engine ground* on the disabled vehicle as far from the battery as possible. *Do not connect this cable to the battery ground (−) terminal.*

6. Turn on the ignition and try to start the engine of the disabled vehicle. If it does not start immediately and the booster battery is in another vehicle, start and run the engine at fast idle to avoid excessive current drain.

7. After the engine of the disabled vehicle starts, *disconnect the ground jumper cable connection from its engine first.* Then disconnect the ground jumper cable from the booster battery.

8. Disconnect the positive jumper cable from the booster battery; then disconnect its other end from the other battery.

Jump Starting a GM Diesel 2-Battery System

Many GM diesel automobiles have two 12-volt batteries in parallel to provide high cranking current, figure 9-24. The two batteries still provide 12 volts but cranking current over 900 amperes as required by some of the diesel engines. You can jump start a GM diesel with two batteries by using the previous procedure with some changes.

First, be sure the booster battery has the cranking current capacity needed by the diesel engine. A small-capacity battery may not provide the amperage to crank a diesel, particularly in cold weather.

In the 2-battery installation, the positive (+) terminal of one battery is

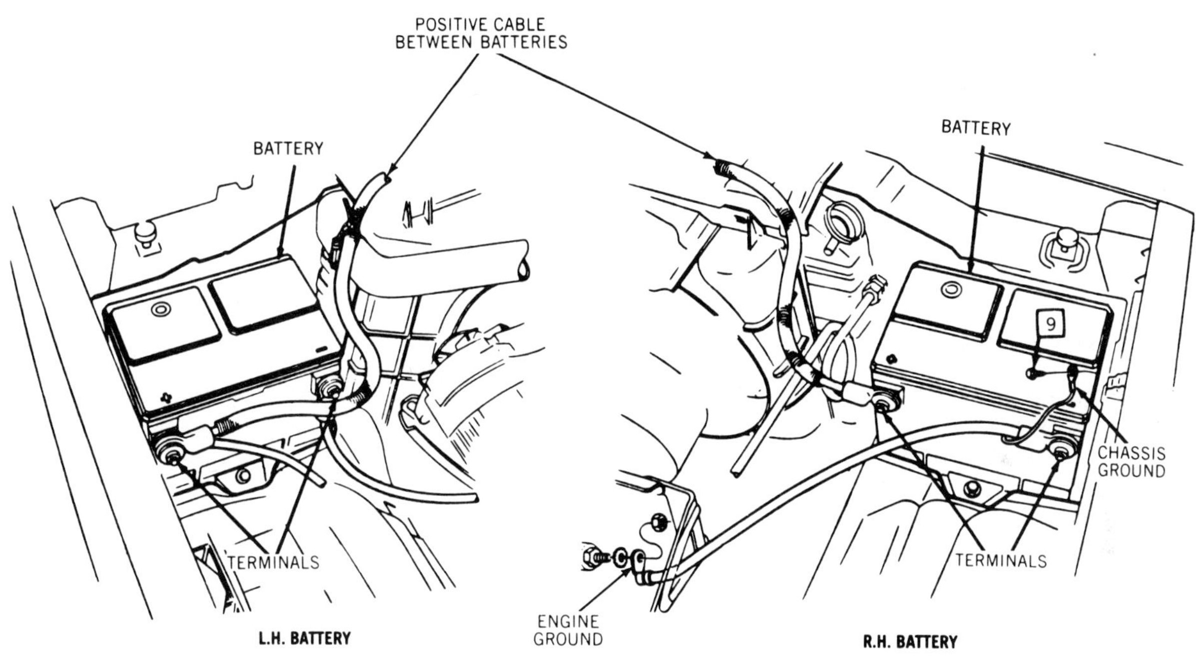

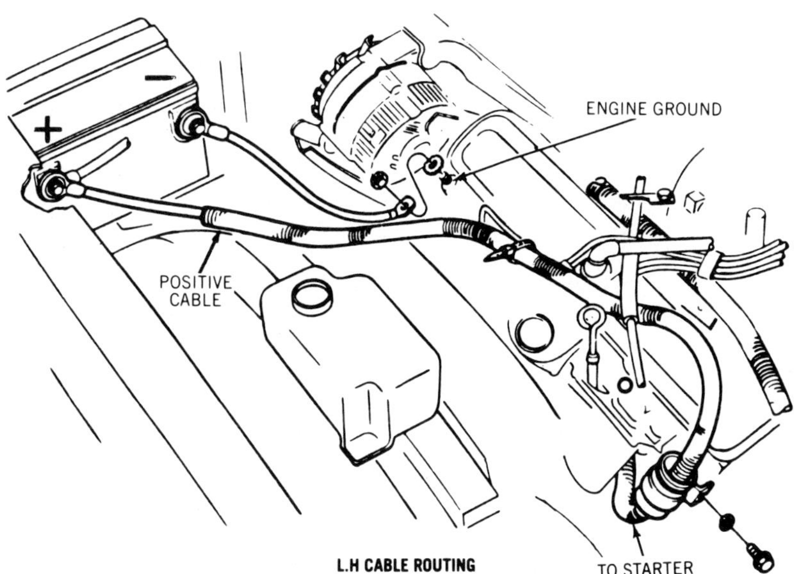

Figure 9-24. Many GM diesel cars have two 12-volt batteries in parallel to provide high cranking current.

connected to the starter motor solenoid, figure 9-24. In step 2 of the jump-starting procedure, connect the cable from the booster battery + terminal to the diesel battery + terminal that is connected to the starter. This minimizes resistance in the starting circuit. In step 5 of the jump-starting procedure, connect ground jumper cable to a ground point on the diesel engine that is at least 18 inches from either battery.

Before trying to start the disabled diesel engine, turn the ignition switch on for about 30 seconds to let the glow plugs warm the combustion chambers. Start the engine in the booster vehicle and run it at fast idle before trying to start the diesel engine. Follow all other steps in the previous jump-starting procedure as you would for any other vehicle.

SUMMARY

Battery testing is the first step of every electrical service procedure. If a battery is undercharged, it must be charged before testing or servicing any other electrical system. Similarly, loose or corroded cables and connections must be cleaned and tightened. Complete battery service consists of:

1. Inspecting the battery for damage, low electrolyte level, dirt, and corrosion.

2. Removing corrosion from the battery with ammonia or baking soda and water and cleaning off dirt with detergent.

3. Testing a battery's state of charge and ability to deliver current by one of the following methods:

- **a.** A load test on any battery
- **b.** Electrolyte specific gravity on vent-cap batteries
- **c.** A 3-minute charge test on vent-cap batteries.

4. Charging a battery at either a slow- or a fast-charging rate.

5. Selecting and installing a replacement battery.

6. Jump starting a vehicle with a dead battery, when necessary.

REVIEW QUESTIONS

Multiple Choice

1. Mechanic A says that you should always disconnect the battery ground cable first when removing battery cables. Mechanic B says that you should always reconnect the ground cable last. Who is right?

- **a.** A only
- **b.** B only
- **c.** both A and B
- **d.** neither A nor B

2. To neutralize spilled battery electrolyte or remove battery corrosion, you can use:

- **a.** commercial cleaning solvent
- **b.** trichloroethylene
- **c.** detergent and water
- **d.** baking soda and water.

3. Electrolyte in a discharged battery can freeze at temperatures near:

- **a.** 18° F
- **b.** 32° F
- **c.** −32° F
- **d.** −60° F.

4. The specific gravity of the electrolyte in a vent-cap battery is 1.260 at 50° F during testing. The true (corrected) specific gravity is:

- **a.** 1.272
- **b.** 1.275
- **c.** 1.248
- **d.** 1.245.

5. The specific gravity of the electrolyte in a vent-cap battery is 1.250 at 120° F during testing. The true (corrected) specific gravity is:

- **a.** 1.266
- **b.** 1.234
- **c.** 1.275
- **d.** 1.280.

6. Mechanic A says that you can use a 3-minute charge test only on a maintenance-free battery. Mechanic B says that a capacity (load) test is the best indication of the cranking ability for any battery. Who is right?

- **a.** A only
- **b.** B only
- **c.** both A and B
- **d.** neither A nor B

7. Before charging a battery in a vehicle, Mechanic A disconnects both battery cables. Mechanic B connects the charger to line voltage (AC power receptacle) before connecting it to the battery. Who is doing the job correctly?

- **a.** A only
- **b.** B only
- **c.** both A and B
- **d.** neither A nor B

8. Generally, you can fast charge a battery as long as all of the following conditions exist EXCEPT:

- **a.** battery terminal voltage remains above 16 volts.
- **b.** the battery does not gas excessively
- **c.** battery temperature stays below 125°F
- **d.** electrolyte does not spew from the vents.

9. Mechanic A says that if a battery load-test specification is more than 200 amperes, a 75-ampere-hour charging rate should restore a full charge. Mechanic B says that a battery at low temperature needs more charging time than a battery at high temperature. Who is right?

- **a.** A only
- **b.** B only
- **c.** both A and B
- **d.** neither A nor B

10. When selecting a replacement battery, you should check any, or all, of the following specifications EXCEPT the:

- **a.** ampere-hour rating
- **b.** reserve capacity
- **c.** load-test current
- **d.** cold-cranking rating.

11. When jump starting a car with a dead battery, Mechanic A makes his first cable connection to the "hot" (+) cable terminal of the discharged battery. Mechanic B makes his last connection to the ground cable terminal of the discharged battery. Who is doing the job correctly?

a. A only
b. B only
c. both A and B
d. neither A nor B

12. A hydrometer test shows that the specific gravity of a vent-cap battery at 80° F is 1.220. This indicates that a battery is:

a. 100% charged
b. 5% charged
c. 50% charged
d. discharged.

13. If the reserve-capacity rating of a battery is 160 to 245 minutes, the recommended charging time at 10 amperes is:

a. 10 hours
b. 15 hours
c. 20 hours
d. 7-1/2 hours.

14. If the ampere-hour rating of a battery is 85 ampere hours, the recommended charging time at 20 amperes is:

a. 5 hours
b. 7-1/2 hours
c. 10 hours
d. 2-1/2 hours.

15. Mechanic A says that some diesel automobiles have two 12-volt batteries connected in parallel to provide high cranking current. Mechanic B says that the two 12-volt batteries in parallel also provide 24 volts of cranking voltage. Who is right?

a. A only
b. B only
c. both A and B
d. neither A nor B

10 STARTING SYSTEM TESTING AND SERVICE

INTRODUCTION

Hard-starting complaints are some of the most common engine performance problems. The cause of an engine starting problem may be in the parts of the starting system; it may be a high-resistance fault in the circuit; or it may be a discharged battery. Starting problems also may be caused by starting circuit effects on other vehicle systems. For example, starting system problems can reduce ignition voltage so that the engine may not fire and run even if it cranks. The starting system also can affect the battery state of charge and battery life.

Chapter 5 explained the general performance tests for battery load (capacity) and for cranking speed, current draw, and voltage drop. If a battery and starting system pass these tests, they are in good working order. But, if these area tests indicate starting system problems, you can locate the cause with the specific tests in this chapter. Problems may exist in the electrical connections and the switches, relays, and solenoids of the system. Problems also may exist in the motor, itself. This chapter also outlines starter motor testing and repair.

GOALS

After studying this chapter, you should be able to do the following jobs in twice the flat-rate labor time or less:

1. Analyze starting system problems for probable causes. Inspect system parts and repair or replace as required.

2. Perform basic voltage-drop, resistance, or current tests on starting circuits using standard test equipment. Interpret the results.

3. Test and adjust starting safety switches on manual and automatic transmissions.

4. Remove a starter motor from vehicle; disassemble, test, repair, reassemble, and install it. Verify proper operation after installation.

STARTING SYSTEM SAFETY

When you service the starting system, you will be working with batteries, electrical systems, and running engines. Follow the basic safety precautions in chapter 1, as well as specific precautions for engine, battery, and test equipment safety. Remember particularly that a battery can explode if exposed to sparks or open flame.

The starter motor draws several hundred amperes of current. A short in the motor circuit or at the battery can cause a large spark. A high-current short circuit can damage parts of the starting circuits, as well as other electrical systems.

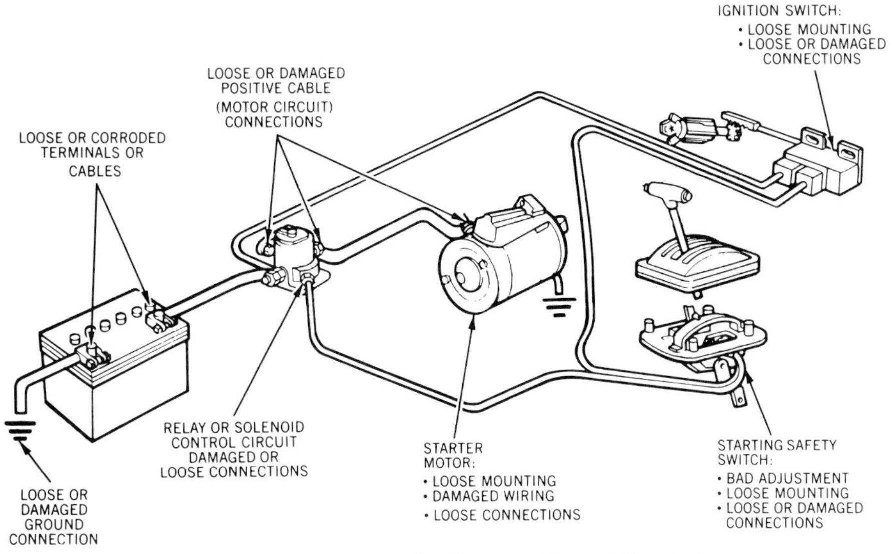

Figure 10-1. Inspect the starting system for these problems. (Chrysler)

STARTING SYSTEM INSPECTION

The general engine inspection in chapter 5 will reveal many obvious starting system problems. Also, when checking a starting system, inspect the following key points, figure 10-1:

1. Check the battery for loose or corroded terminals and damaged cables.

2. Check battery ground cable con-

Symptom	Possible Cause	Correction
Starter spins but does not crank the engine	1. Broken starter drive 2. Worn flywheel ring gear	1. Replace the starter drive. 2. Replace ring gear.
Starter cranks the engine slowly	1. Discharged battery 2. High resistance in the starting system 3. Defective starter motor 4. Engine mechanical problem	1. Recharge or replace the battery. 2. Do voltage-drop tests for high resistance. 3. Repair or replace the starter. 4. Test engine mechanical condition.
Starter does not crank the engine	1. Discharged battery 2. Open control circuit 3. Open motor circuit 4. Defective relay or solenoid	1. Recharge or replace the battery. 2. Do voltage-drop tests to locate the open circuit. 3. Do voltage-drop tests to locate the open circuit. 4. Repair or replace the relay or solenoid.

Figure 10-2. This table lists general starting system problems, their possible causes, and cures.

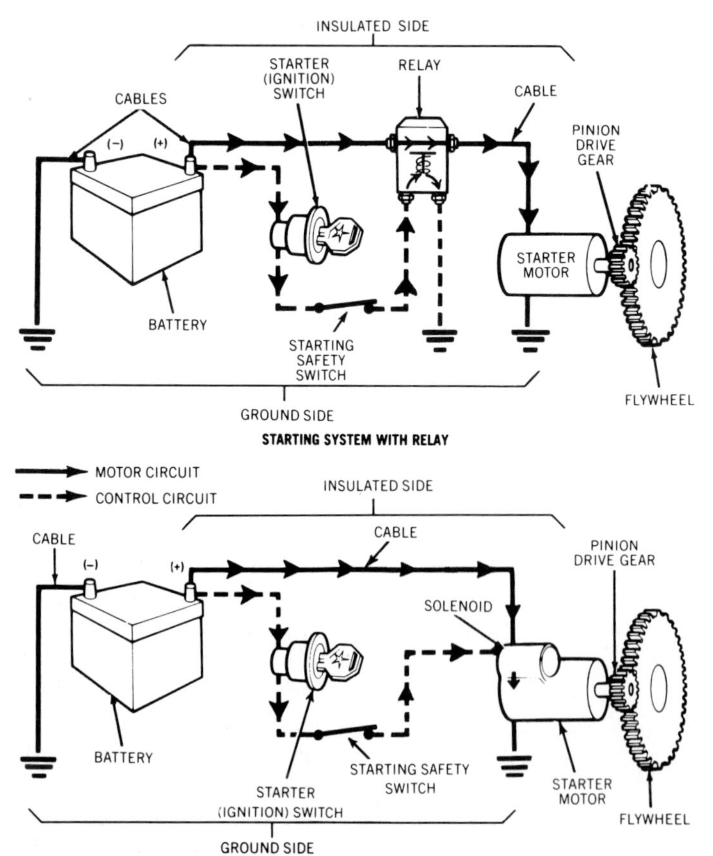

Figure 10-3. For testing, we divide the motor circuit into the insulated side and the ground side. (Chrysler)

nections to the engine and body for looseness.

3. Check battery positive (+) cable connections to the relay or solenoid and the motor for looseness and damage.

4. Inspect connections at the ignition switch, the starting safety switch, and the relay or solenoid for damaged terminals and wiring.

5. Check the switches, themselves, for loose mounting and damaged contacts.

6. Check starting safety switch adjustment.

7. Inspect the starter motor for loose mounting bolts, damaged wiring, and loose connections.

Listening to the starter motor operation can direct you to specific tests and repairs. Figure 10-2 lists general starter motor symptoms, possible causes, and repairs.

STARTING CIRCUIT TESTING

The starting system consists of the control circuit and the motor circuit, figure 10-3. The control circuit contains the ignition switch, the starting safety switch, and the control side (coil) of the starter relay or solenoid. The motor circuit conducts several hundred amperes of current to the starter motor for a few seconds when the control circuit energizes the relay or the solenoid.

For pinpoint testing, we divide the motor circuit into two parts, figure 10-3:

1. The insulated side of the circuit starts at the battery + terminal and contains the cable, or cables, between the battery and the relay or solenoid. It also includes the relay or solenoid high-current contacts and the motor armature and field windings.

2. The ground side of the circuit starts at the motor ground connection to the engine. It includes the low-voltage ground path through the frame or body and the battery ground cable and connections. The ground side ends at the battery − terminal.

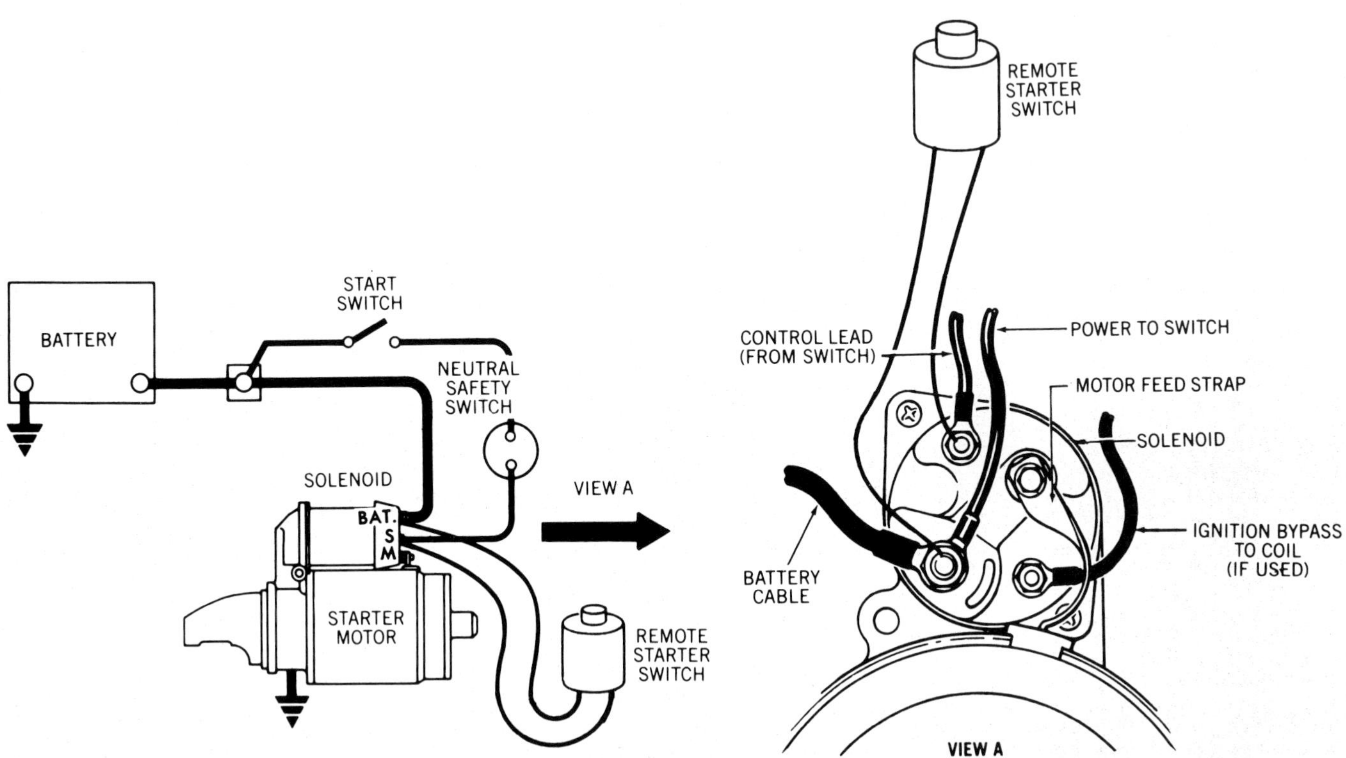

Figure 10-4. Remote starter switch connections for a solenoid-operated starter without a relay. (Delco-Remy) (Ford)

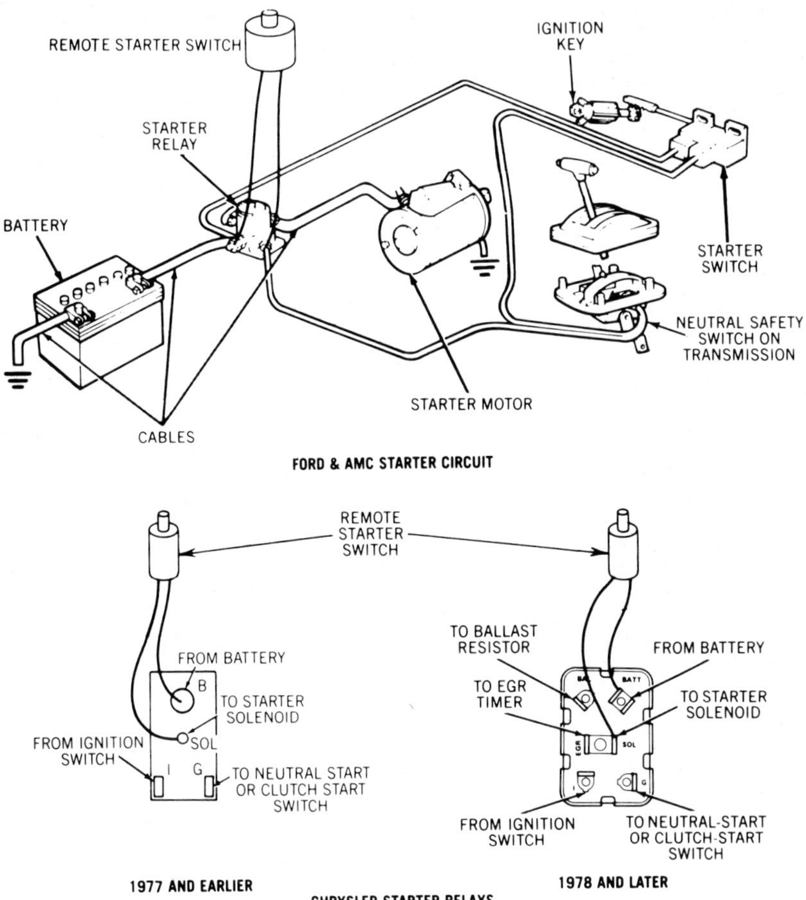

Figure 10-5. Remote starter switch connections for starting systems with relays. (Chrysler)

High resistance in the control circuit or either side of the motor circuit can cause starting system problems. The cranking current draw, voltage drop, and speed tests in chapter 5 will lead you to the circuit resistance tests explained below. If cranking voltage is below specifications (usually 9.6 volts, minimum) during the area tests, check the individual circuits for high resistance. If cranking voltage is within specifications, but the motor cranks slowly, you should remove and test the motor as explained later in this chapter.

You will use a remote starter switch and a voltmeter or volt-ampere or battery-starter tester for these circuit tests. You also may need a jumper wire for some systems. Disable the ignition system as explained in chapter 5. Connect the remote starter switch as shown in figures 10-4 and 10-5. To test the starting system of a diesel engine, disconnect the injection pump fuel shutoff solenoid, figure 10-6.

The following procedures refer to test diagrams for typical GM, Ford, and Chrysler starting systems. These diagrams also represent common starting systems for imported vehicles. Using

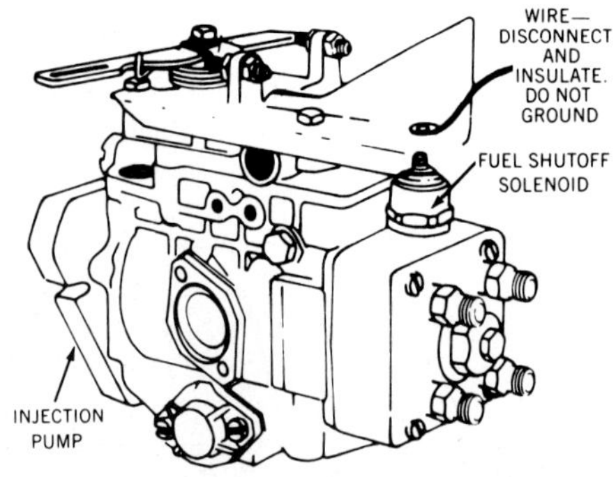

Figure 10-6. Disable a diesel injection pump by disconnecting the fuel shutoff solenoid. (Volkswagen)

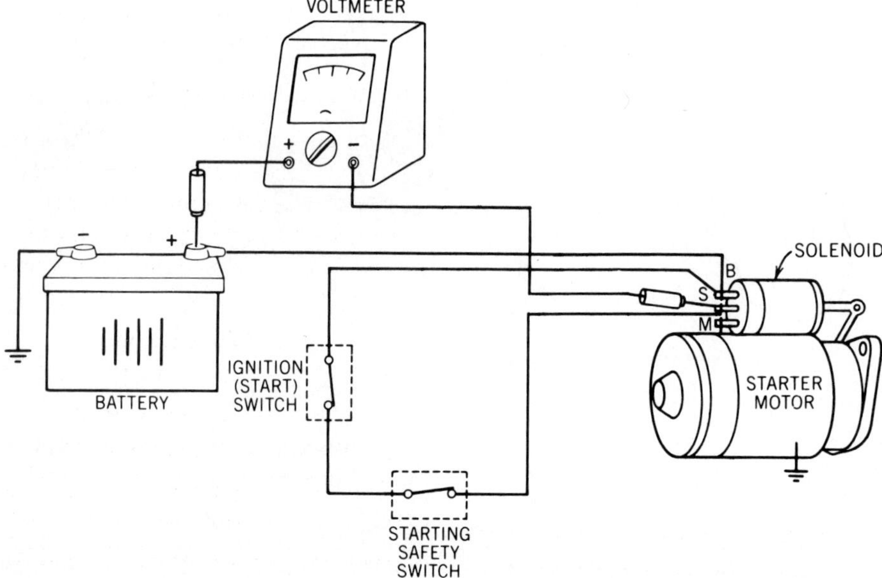

Figure 10-7. Basic GM control circuit test.

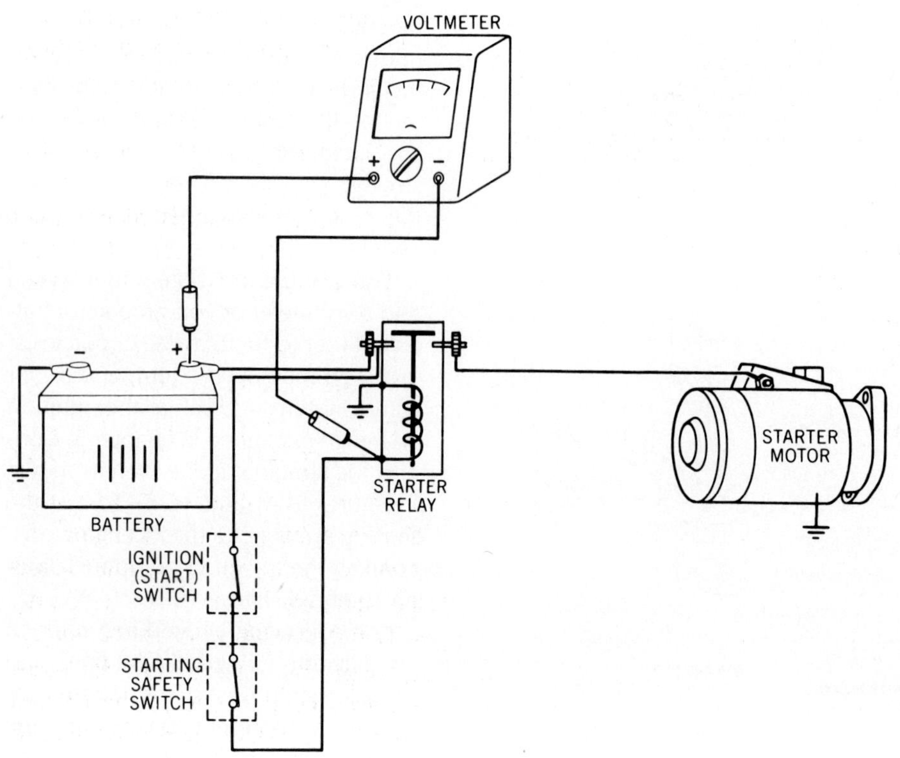

Figure 10-8. Basic Ford control circuit test.

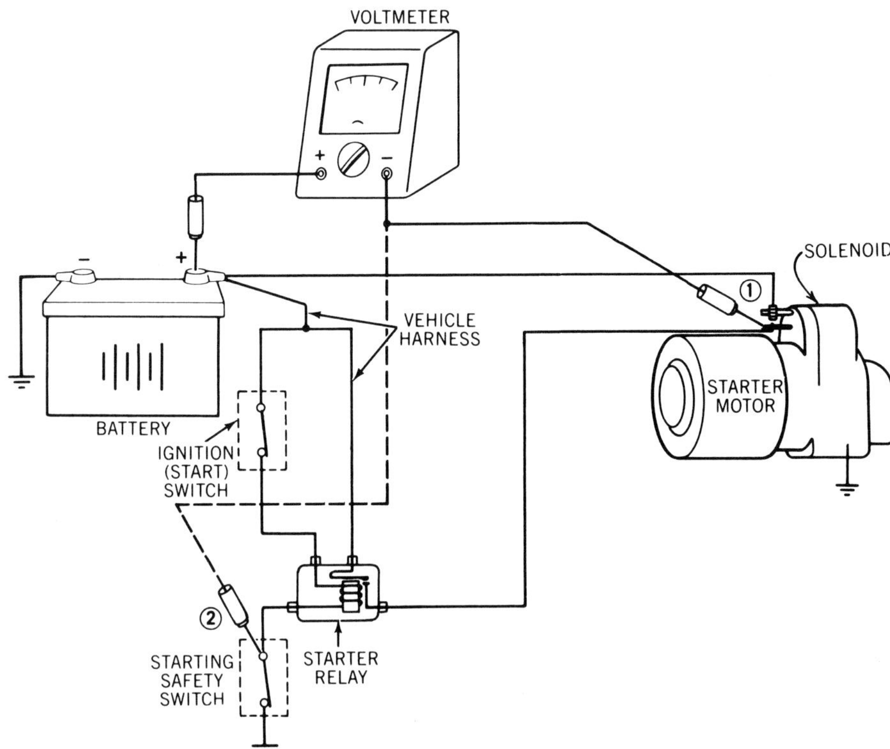

Figure 10-9. Basic Chrysler control circuit test.

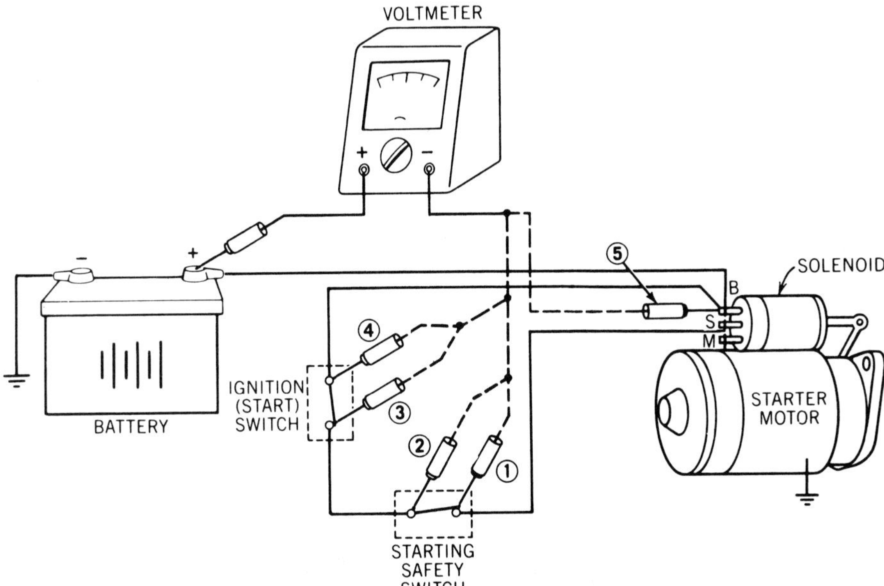

Figure 10-10. GM control circuit pinpoint tests.

a carmaker's electrical diagrams and specifications, as well as these illustrations, you can apply these tests to any vehicle.

Caution. Do not crank the engine of a vehicle with a catalytic converter for more than 15 seconds at a time to avoid drawing fuel into the catalyst. If the converter-equipped vehicle has air injection, disconnect the air hose from the pump or the air control valve to avoid adding air to unburned fuel in the exhaust system.

Control Circuit Test

Disable the ignition system or the diesel injection pump and use the vehicle ignition switch to crank the engine. Do not use a remote starter switch when testing voltage drop in the starter control circuit.

1. Refer to the following figures:
   a. GM systems—figure 10-7
   b. Ford systems—figure 10-8
   c. Chrysler systems—figure 10-9

2. Connect the voltmeter + lead to the battery + terminal.

3. Connect the voltmeter – lead to the switch terminal at the relay or the solenoid.

4. Crank the engine with the ignition switch and note the voltmeter reading.
   a. If voltage drop is within the carmaker's specifications (usually 2.5 volts or less), the control circuit is okay.
   b. If voltage drop is higher than specifications, proceed with steps 5 through 8 to locate the high resistance.

5. Refer to the following figures:
   a. GM systems—figure 10-10
   b. Ford systems—figure 10-11
   c. Chrysler systems—figure 10-12

6. Leave the voltmeter + lead connected to the battery + terminal.

7. While cranking the engine, touch the voltmeter – lead to the points shown in sequence in the diagrams.

8. When the voltage drop decreases to within specifications, you have found the high resistance. Repair or replace wiring or parts as needed.

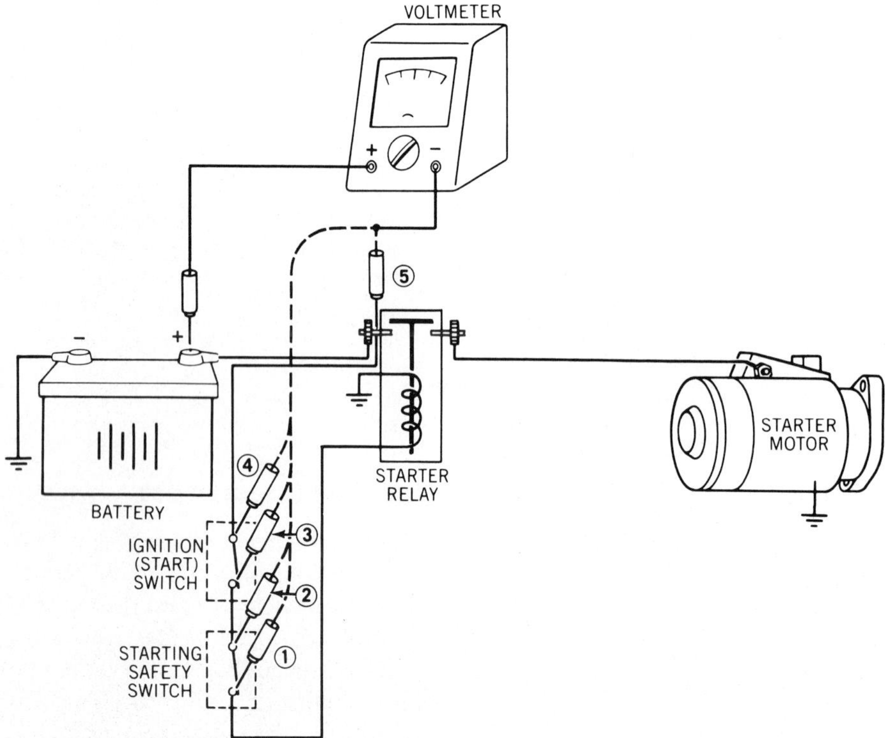

Figure 10-11. Ford control circuit pinpoint tests.

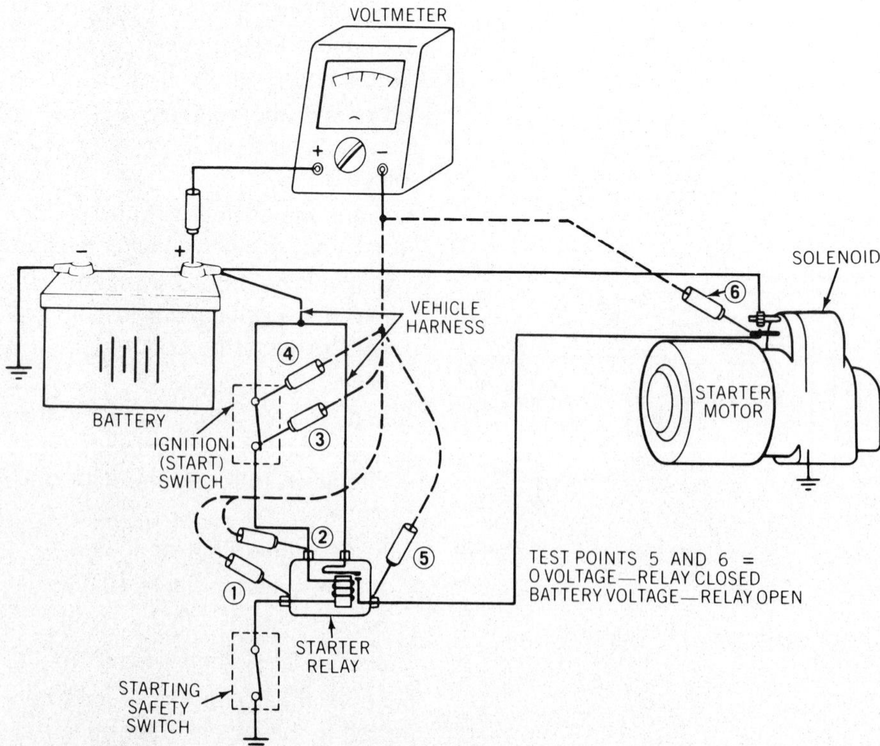

Figure 10-12. Chrysler control circuit pinpoint tests.

Motor Insulated Circuit Test

Disable the ignition system or the diesel injection pump and use a remote starter switch to crank the engine.

1. Connect the voltmeter + lead to the battery + post or terminal nut, figure 10-13. *Do not connect it to the + cable.* If you connect the meter to the cable, you will miss a possible high-resistance connection between the cable and the battery terminal.

2. Connect the voltmeter − lead to the motor terminal of the starter. If the motor terminal is not exposed, connect to the motor battery terminal, figure 10-13.

3. Crank the engine with the remote starter switch and note the voltmeter reading.

a. If voltage drop is within the carmaker's specifications (usually 0.2 to 0.6 volt), the insulated side of the motor circuit is okay.

b. If voltage drop is higher than specifications, proceed with steps 4 through 7 to locate the high resistance.

4. Refer to the following figures:

a. GM systems—figure 10-14

b. Ford systems—figure 10-15

c. Chrysler systems—figure 10-16

5. Leave the voltmeter + lead connected to the battery + terminal.

6. While cranking the engine, touch the voltmeter − lead to the points shown in sequence in the diagrams.

7. When the voltage drop decreases to within specifications, you have found the high resistance. Repair or replace wiring or parts as needed.

The last test point for each system measures voltage drop between the battery + post and the + cable connector. With the voltmeter + lead on the battery post or terminal, touch the − lead to the cable connector. Corrosion between the battery terminal and the cable connector often causes very high resistance in the motor circuit.

Motor Ground Circuit Test

Disable the ignition system or the diesel injection pump and use a remote starter switch to crank the engine.

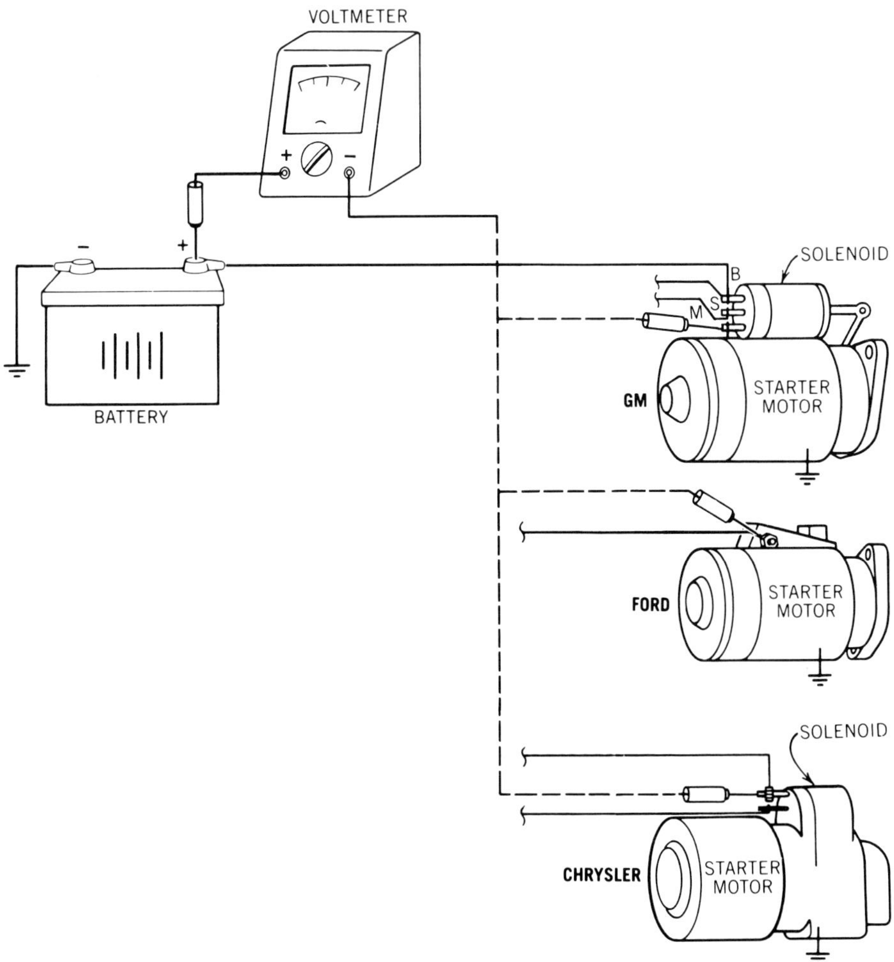

Figure 10-13. Basic motor insulated circuit test.

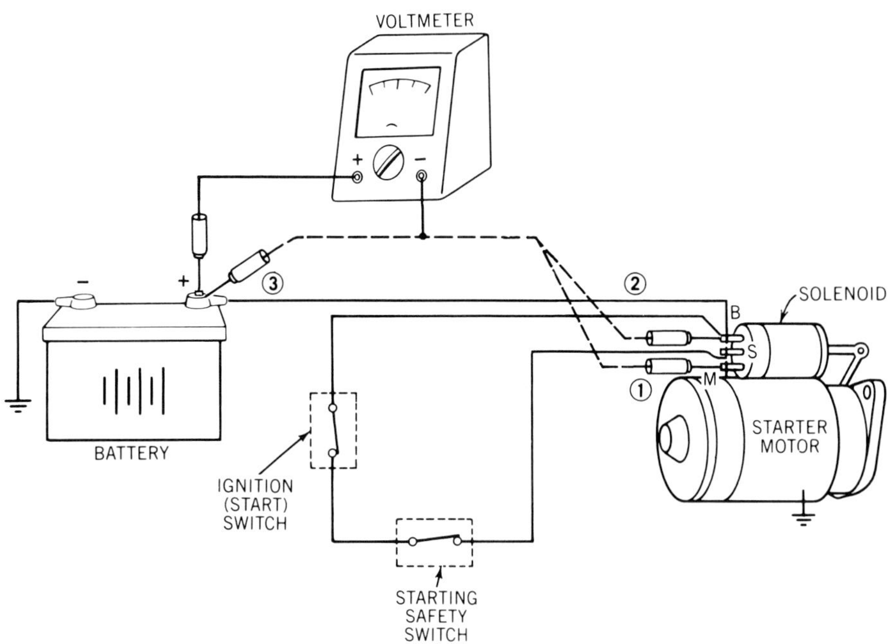

Figure 10-14. Pinpoint tests for the GM motor insulated circuit.

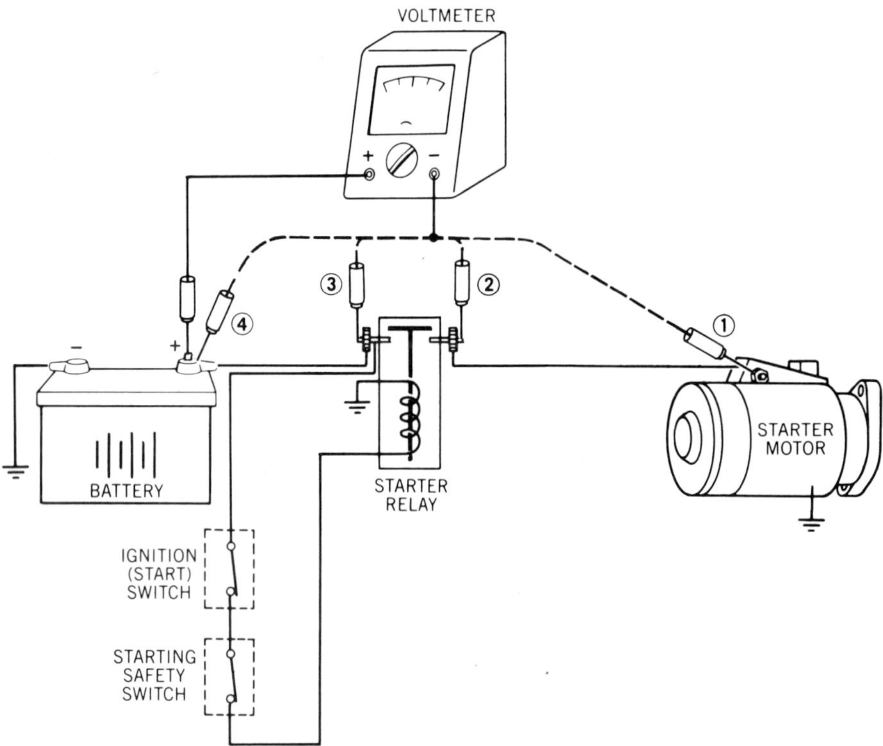

Figure 10-15. Pinpoint tests for the Ford motor insulated circuit.

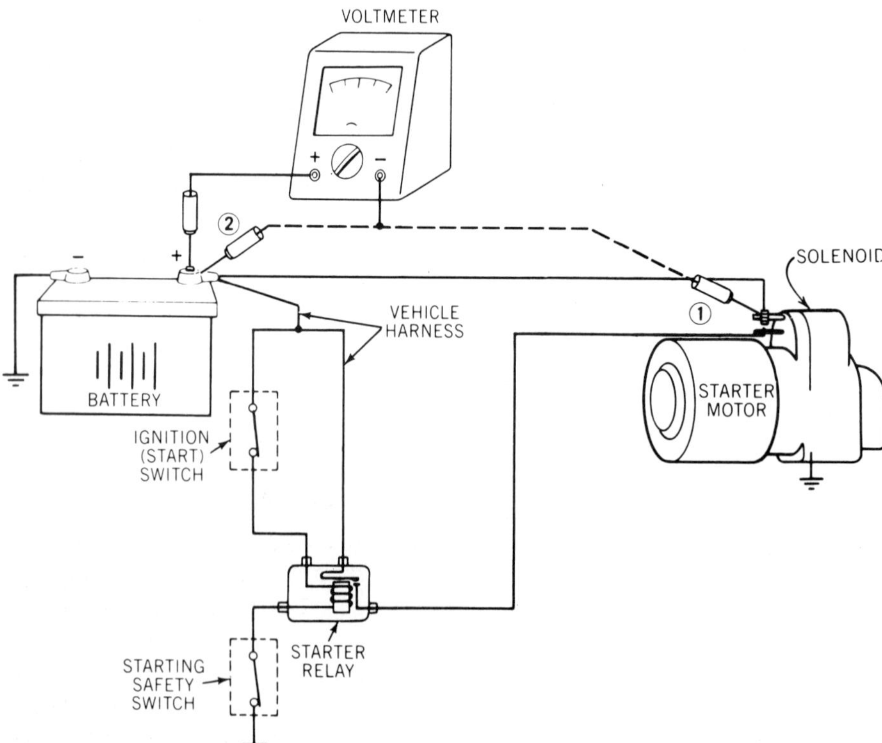

Figure 10-16. Pinpoint tests for the Chrysler motor insulated circuit.

1. Connect the voltmeter + lead to the starter motor housing (the field frame or either end frame), figure 10-17.

2. Connect the voltmeter − lead to the battery − post or terminal nut. *Do not connect it to the − cable.* If you connect the meter to the cable, you will miss a possible high-resistance connection between the cable and the battery terminal.

3. Crank the engine with the remote starter switch and note the voltmeter reading.

a. If voltage drop is within specifications (usually 0.1 to 0.3 volt), the ground side of the motor circuit is okay.

b. If voltage drop is higher than specifications, proceed with steps 4 through 6 to locate the high resistance.

4. Leave the voltmeter − lead connected to the battery − terminal.

5. While cranking the engine, touch the voltmeter + lead to the points shown in sequence in figure 10-17.

6. When the voltage drop decreases to within specifications, you have found the high resistance. Clean and tighten the ground connections or replace cables as required.

The last test point in figure 10-17 measures voltage drop between the battery − post and the − cable connector. With the voltmeter − lead on the battery post or terminal, touch the + lead to the cable connector. Corrosion between the battery terminal and the cable connector often causes very high resistance in the ground side of the motor circuit.

Starter Current Draw Test

The general tests in chapter 5 include a measurement of starter current while cranking the engine. This test is usually enough to detect current draw problems in the starting system. However, you can do an alternative current draw test to verify the results of the general test or to check current draw more precisely. You will need a voltmeter, an ammeter, and a variable resistance carbon pile.

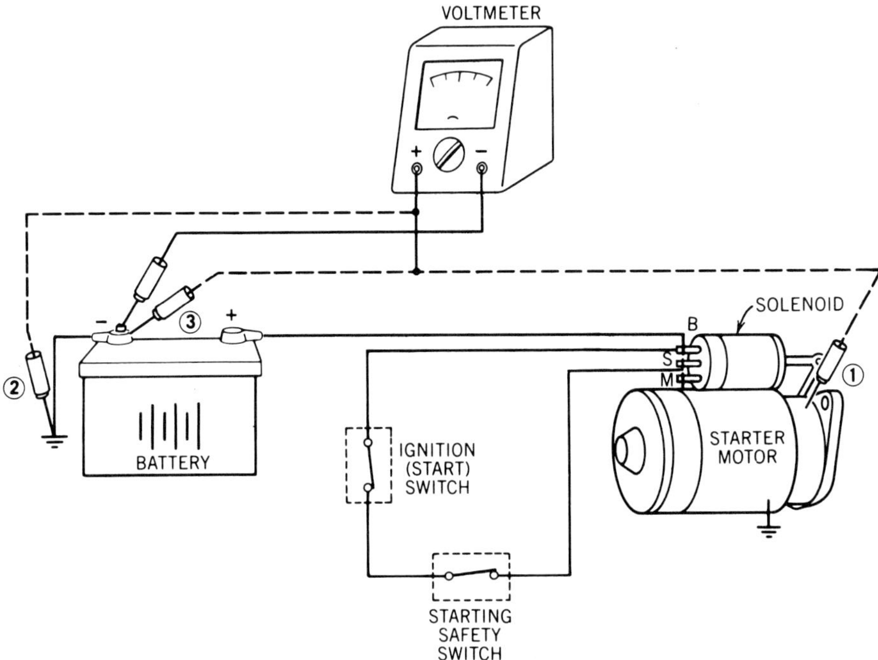

Figure 10-17. Motor ground circuit test.

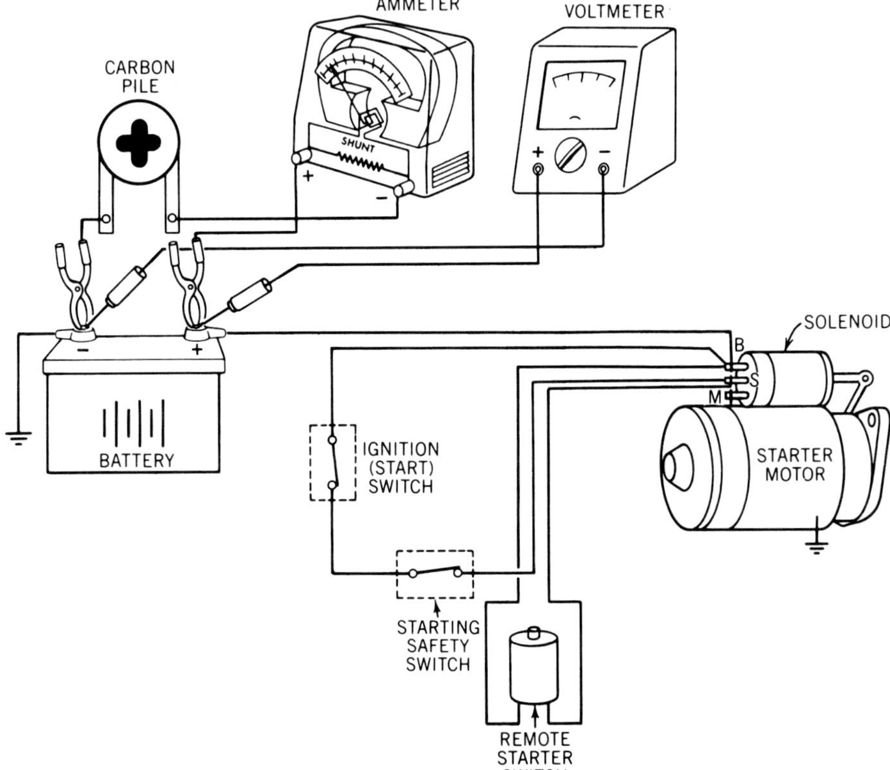

Figure 10-18. Starter current draw test, using a carbon pile.

1. Connect the test equipment to the vehicle according to the manufacturer's directions or as follows, figure 10-18:

a. Connect the voltmeter + lead to the battery + terminal and the − lead to the battery − terminal.

b. Connect the ammeter + lead to the battery + terminal and the − lead to one lead of the carbon pile.

c. Set the carbon pile to maximum resistance (open) and connect the other carbon pile lead to the battery − terminal.

2. Disable the ignition system or the diesel injection pump and use a remote starter switch to crank the engine.

3. Crank the engine and note the voltmeter reading.

4. With the engine stopped, adjust the carbon pile until the voltmeter reading is the same as the reading in step 3.

5. Note the exact ammeter reading and return the carbon pile to its open position.

6. Compare the ammeter reading to the carmaker's current draw specifications.

a. *High current draw* indicates a short in the starter motor or a binding engine or starter motor.

b. *Low current draw* indicates high resistance in the starting circuits, a discharged battery, or a worn engine (if cranking speed and voltage are normal).

STARTING SAFETY SWITCH ADJUSTMENT

A no-start problem can be caused by a starting safety switch that is not properly adjusted. If the switch does not close the control circuit when the transmission is in park or neutral or the clutch is disengaged, the starter cannot operate. Just as importantly, if the safety switch closes when the transmission is engaged, the engine may start and move the vehicle when it is not controlled properly.

Mechanical blocking devices as used on many GM and Ford cars are not electrical switches and need no adjustment. Transmission-mounted safety switches on most Chrysler and AMC cars are not adjustable. If the switch does not work, install a known-good switch and retest the circuit. If the good switch does not work right, the internal transmission linkage must be adjusted or repaired. Ford transmission-mounted switches are adjustable.

Use a voltmeter, an ohmmeter, or a test lamp and the basic electrical troubleshooting methods in chapter 4 to test switch operation. The following procedures explain typical adjustments for floor-mounted and column-mounted switch adjustments. Use these guidelines and carmakers' specific instructions to adjust starting safety switches.

CAUTION. Disconnect the battery ground cable before adjusting a switch to avoid accidental switch operation.

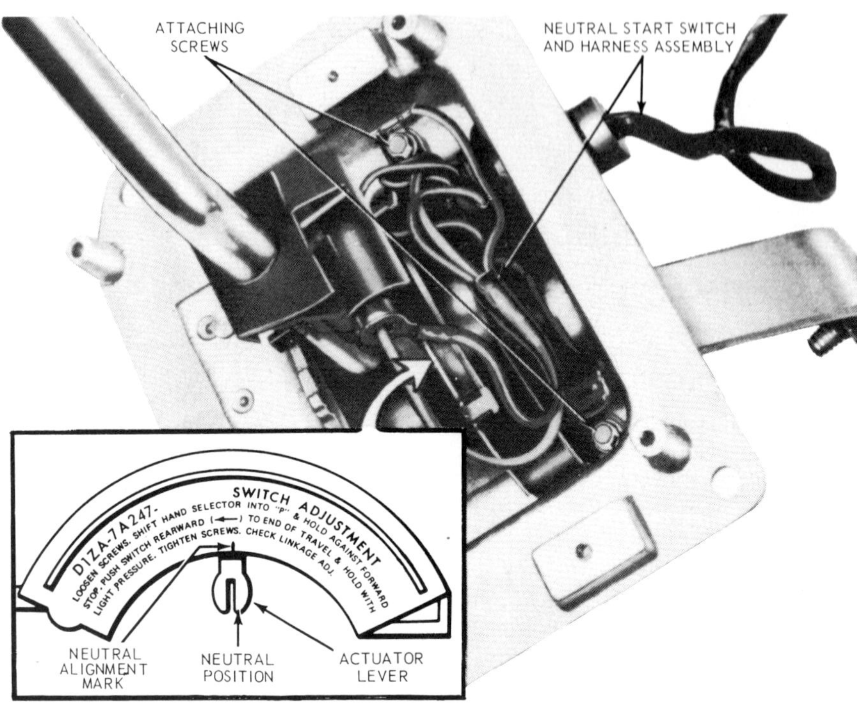

Figure 10-19. Typical Ford floor-mounted transmission switch adjustment. (Ford)

Floor-Mounted Switch Adjustment

Many Ford products and some other cars have safety switches on the floor-mounted automatic transmission selector, or console. Adjust these Ford switches as follows. The methods are typical for other cars, but check the manufacturer's directions.

1. Remove the shift lever handle, the console housing, and the pointer backup shield for access to the switch.

2. Loosen the two screws holding the switch to the housing, figure 10-19.

3. Move the shift lever to the forward stop of the park position and hold it there.

4. Move the switch rearward as far as it will go.

5. Tighten the two screws loosened in step 2.

6. Operate the starter to verify that the engine starts only when the transmission is in park or neutral. Repeat the adjustment if necessary.

7. Reinstall parts removed in step 1.

8. If the switch is mounted on the transmission, loosen the two attaching bolts, figure 10-20. With the shift lever in neutral, insert a gauge pin through the three holes of the switch as shown. Tighten the switch bolts and remove the pin.

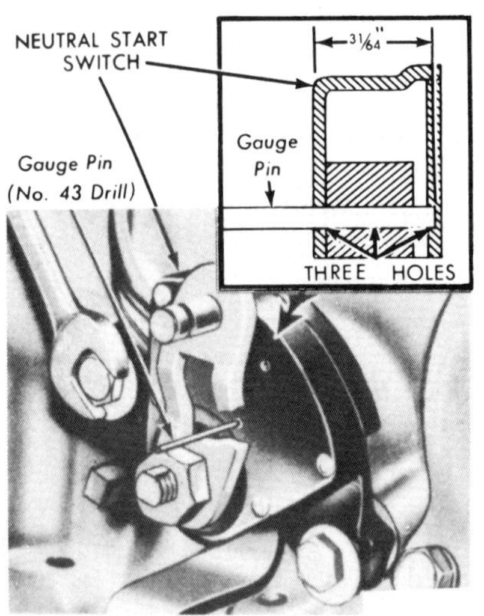

Figure 10-20. Typical Ford transmission-mounted switch adjustment. (Ford)

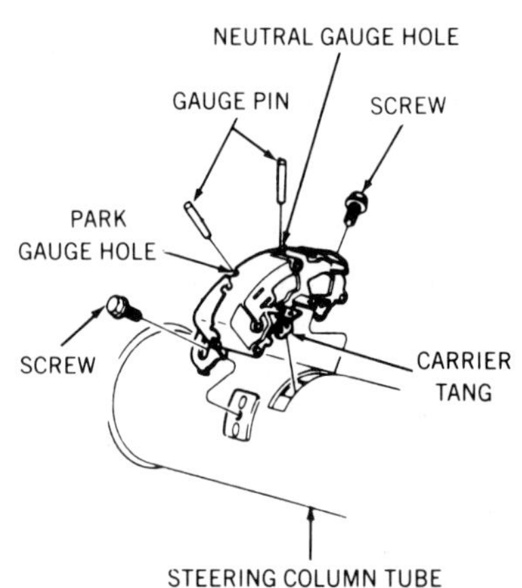

Figure 10-21. Adjust some column-mounted switches by using a gauge pin to align two holes. (Buick)

Column-Mounted Switch Adjustment

Many cars have starting safety switches mounted on the steering column at the transmission gear selector. GM and Ford switches are typical of this kind. They require a gauge pin to align two holes for adjustment, as follows:

1. Loosen the switch mounting screws, figure 10-21.

2. Move the switch until the two alignment holes line up.

3. Insert a gauge pin of the specified size into the holes.

4. Tighten the switch mounting screws and remove the gauge pin.

5. Operate the starter to verify that the engine starts only when the transmission is in park or neutral. Repeat the adjustment if necessary.

STARTER MOTOR SERVICE

The preceding starting system tests may isolate a problem in the system wiring, switches, relays, or solenoids. They also may identify a problem in the motor. When you isolate a problem to the motor, you may simply replace it. Equally or more often, you may remove the motor for further testing and repair.

Starter motor service can be an important part of overall engine performance service. GM and many other carmakers cover starter motor repair under new-vehicle warranty. This is often the job of an engine performance technician.

Starter motor problems fall into three general categories:

1. Mechanical problems in the starter drive
2. Electrical problems in the motor fields or armature and their connections
3. Electrical or mechanical problems in a motor solenoid.

The following sections outline starter drive tests, motor removal and replacement, bench testing, and motor overhaul. The chapter ends with an illustrated procedure for repairing a typical Delco-Remy solenoid-operated starter. You can use this procedure as a guide to service a Delco-Remy or similar starter motor. It also indicates the general sequence for repairing any starter motor.

Starter Drive Test

If starter motor spins but does not engage and crank the engine, the problem almost certainly is in the starter drive. However, a drive may only slip occasionally or fail under certain conditions, such as very cold weather when cranking loads are greatest.

There are three common kinds of starter drives:

1. Solenoid-operated, direct drive
2. Solenoid-operated, reduction drive
3. Movable pole shoe.

All of these drives use a 1-way (overrunning) clutch to release the starter pinion from the flywheel when the engine starts. The pinions and 1-way clutches are similar on all of these drives, figure 10-22. You can use the following basic test to check all of them. Observe the earlier precaution for cranking converter-equipped vehicles. Proceed as follows:

1. Connect a remote starter switch to the relay or solenoid. Use a switch with leads long enough to reach to the driver's seat.
2. If the relay or the solenoid has an ignition bypass terminal, disconnect the bypass wire to keep the engine from starting.
3. Crank the engine with the remote switch and listen to the starter.
4. If the starter engages properly, reconnect the ignition bypass wire (if disconnected in step 2) and turn the ignition switch to the run position.
5. Crank the engine again with the remote switch. As soon as the engine starts, turn the ignition switch off and continue cranking the engine with the remote switch. Repeat this step two or three times.

Rapid cycling of the starter motor usually will uncover an intermittent problem in the starter drive. If the starter repeatedly cranks the dead engine, the drive clutch is okay. If the motor spins without cranking the engine, the drive clutch is bad.

Most starter drive failures are due to a slipping 1-way clutch or improper engagement of the pinion with the flywheel ring gear. Occasionally, the spring in the solenoid or the drive may break. This can prevent full disengagement of the pinion when the engine starts. The flywheel will usually throw the pinion out of the ring gear but damage the drive in the process.

To check for damaged pinion or flywheel ring gear teeth, you usually must remove the motor from the engine. If the engine has an inspection plate on the bellhousing or a removable flywheel cover, you may be able to inspect the pinion and ring gear without removing the motor. Figure 10-23 shows normal gear wear patterns and damaged gears caused by incorrect pinion engagement. Loose starter mounting bolts also can lead to damaged pinion and ring gear teeth.

You can replace the starter drive by removing the motor drive end frame and separating the drive from the armature or the drive gear shaft. You do not need to disassemble the entire mo-

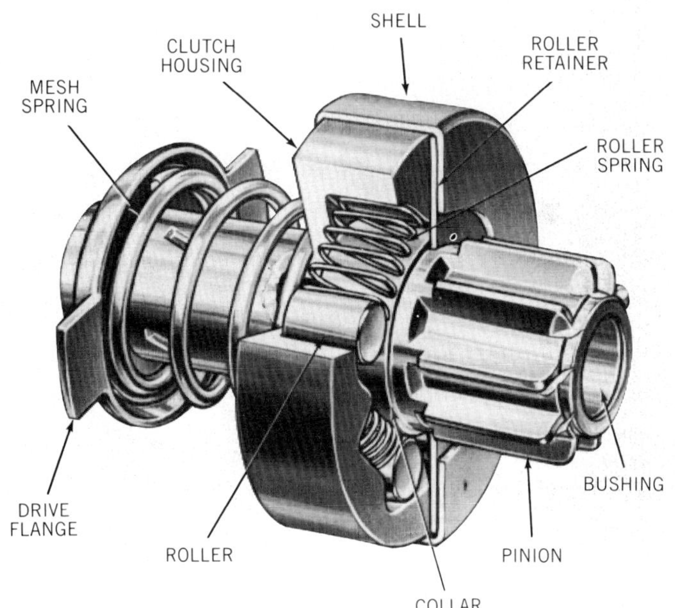

Figure 10-22. Most starter drives have similar 1-way roller clutches. (Ford)

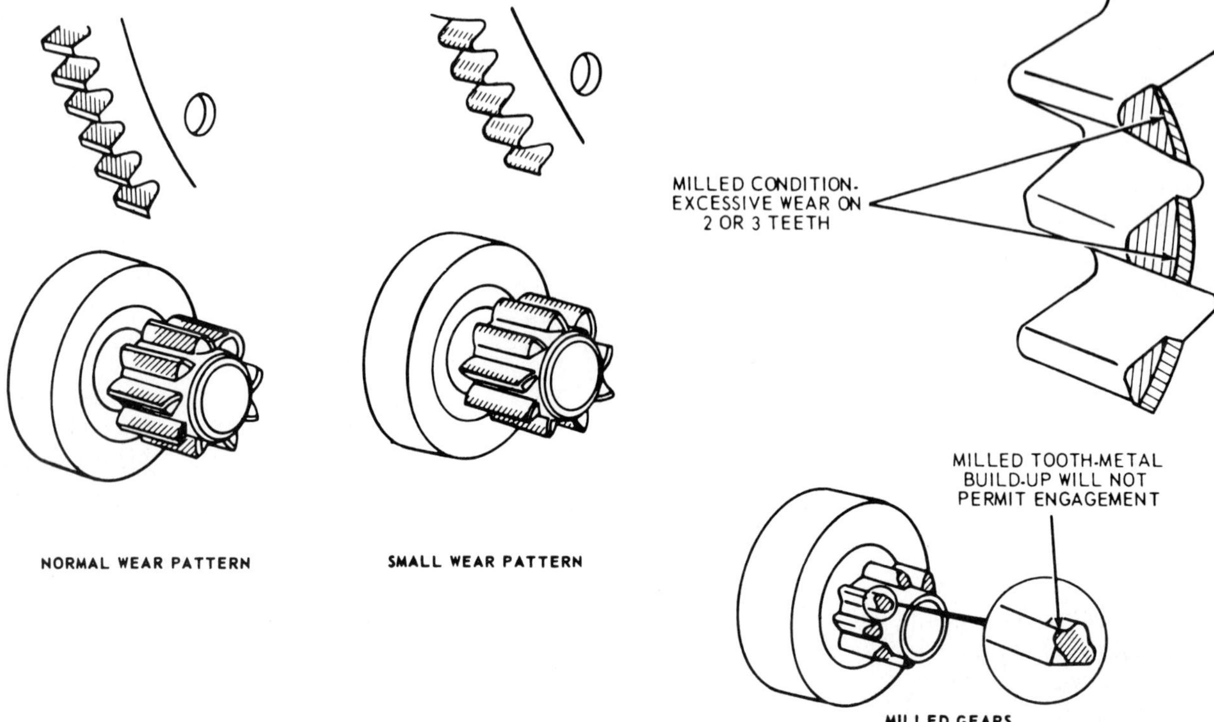

Figure 10-23. Starter pinion and flywheel ring gear wear patterns. (Ford)

tor. The illustrated procedure at the end of this chapter shows typical starter drive replacement.

Starter Motor Removal

Starter removal is not difficult, but getting access to the motor often is. We can't give you detailed instructions to remove every starter on every vehicle. The following general steps, however, will help you use the carmaker's procedures to do the job efficiently.

In most cases, you must raise the vehicle on a hoist and remove the motor from underneath. You may have to turn the steering to one side or the other for clearance. Remove the starter as follows:

1. Disconnect the battery ground cable.

2. Raise the vehicle on a hoist or with safety stands to a convenient height.

3. Inspect the starter installation to determine what vehicle parts must be removed or loosened for access, figures 10-24 and 10-25. These may include:

a. Starter support brackets and heat shields

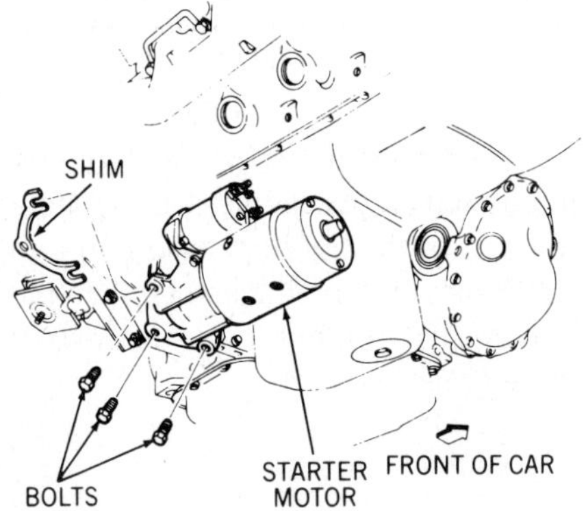

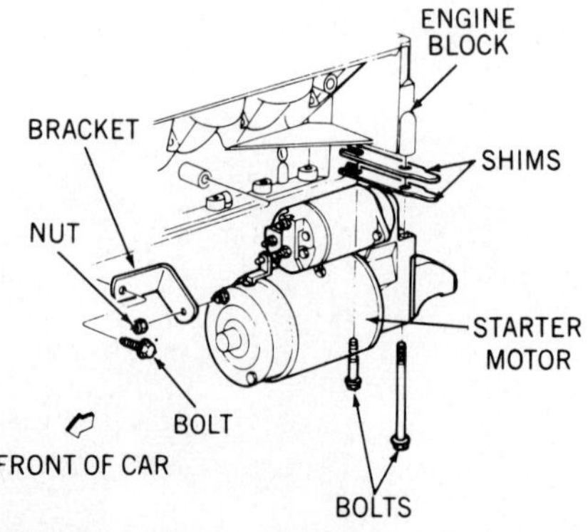

Figure 10-24. Typical starter motor installations with shims. (Cadillac)

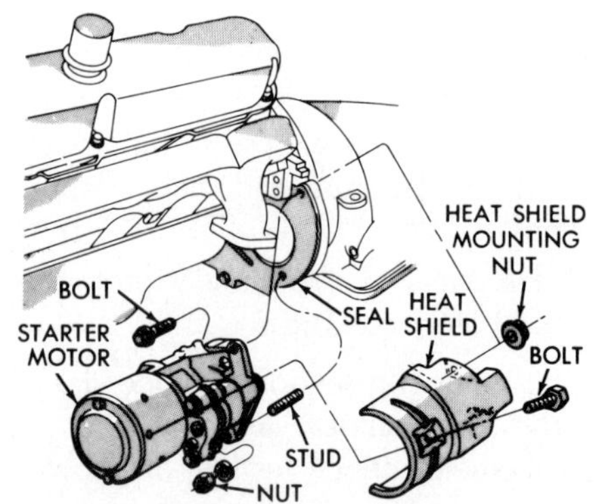

Figure 10-25. Typical starter motor installation with a heat shield. (Chrysler)

b. Exhaust pipes
c. The steering idler arm
d. Steering linkdge
e. Transmission oil cooler lines
f. Air injection tubes to the catalytic converter

4. Loosen or remove parts as needed for access.
5. Disconnect all wires from the motor or the solenoid. Tag the wires for correct reconnection.
6. Remove any support brackets and heat shields.
7. Remove the starter mounting bolts from the engine and remove the starter.
8. If the starter is installed with shims between the drive end frame and the engine, count the shims and note their thickness. Save them for reinstallation. Many starters need shims to establish proper pinion and flywheel engagement.

Starter Motor Tests

Complete starter motor repair includes a no-load operating test of the motor and electrical tests for open and short circuits. Tests using a self-powered test lamp or an ohmmeter are based on the troubleshooting principles you learned in chapter 4. The illustrated procedure at the end of the chapter includes some of the tests. The following sections summarize all of these tests for any motor.

No-Load Test. The no-load test is done with the motor clamped in a large vise or mounted on a test stand. The test does just what the name indicates. It operates the starter with no cranking load. The motor must meet the manufacturer's specifications for speed and current draw. If a motor does not meet no-load specifications, it can't crank an engine properly. The no-load test will help you identify open or shorted windings, worn brushes, a bent or worn shaft, and worn bushings.

You will need a 12-volt battery, a switch, a voltmeter, an ammeter, and a mechanical rpm indicator for this test. These are included in most starter test stands. A carbon pile is optional if you want to regulate battery voltage exactly. Connect the equipment to a Ford movable pole shoe starter as shown in figure 10-26. Connect the equipment to a solenoid-operated starter as shown in figure 10-27. Connect the voltmeter to the solenoid battery terminal if the motor terminal is not exposed, as on some Chrysler motors.

Close the switch to operate the motor. Note the current draw and the motor speed and compare them to specifications. Figure 10-28 lists common no-load motor problems and their possible causes.

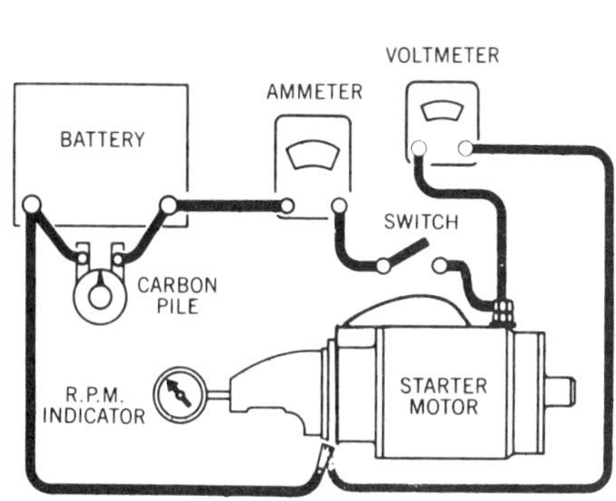

Figure 10-26. Circuit connections for a no-load test of a movable pole shoe starter. (Delco-Remy)

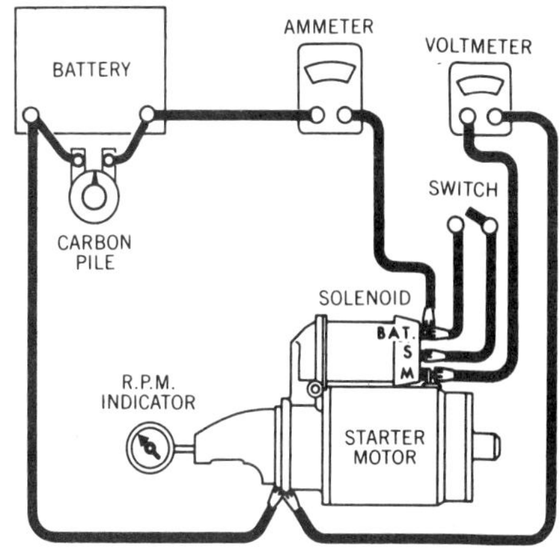

Figure 10-27. Circuit connections for a no-load test of a solenoid-operated starter. (Delco-Remy)

Test Result	*Possible Cause*
Low speed, high current draw	1. Binding or worn motor bushings 2. Bent armature shaft 3. Loose pole shoes dragging on the armature
Low speed, low current draw	1. High resistance in the motor 2. Open or grounded field winding 3. Worn brushes, weak or broken brush springs 4. Shorted or grounded armature 5. Worn commutator
High speed, high current draw	1. Shorted field windings
Motor does not operate but draws high current	1. Insulated circuit connections are grounded inside the motor 2. Bushings are completely frozen
Motor does not operate but draws low current	1. Open field or armature circuit 2. Loose or broken brushes 3. Worn commutator

Figure 10-28. Starter no-load troubleshooting table.

Armature Growler Test. A growler is an ac electromagnet used to test any kind of motor armature for shorted windings. Alternating current at 60 H reverses the magnetic field of t growler 120 times per second. W you place an armature on the grr

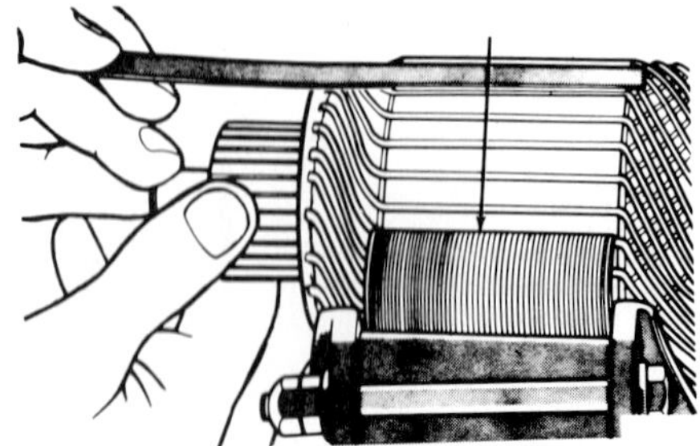
Figure 10-29. Use a growler and a hacksaw blade to test an armature for short circuits. (Chrysler)

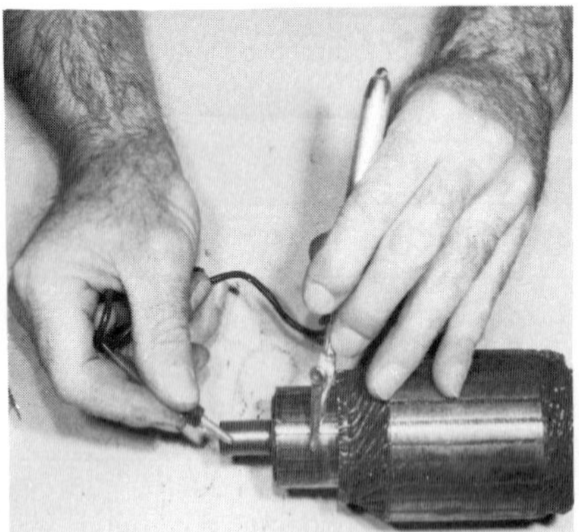
Figure 10-30. Testing for a grounded armature with a test lamp.

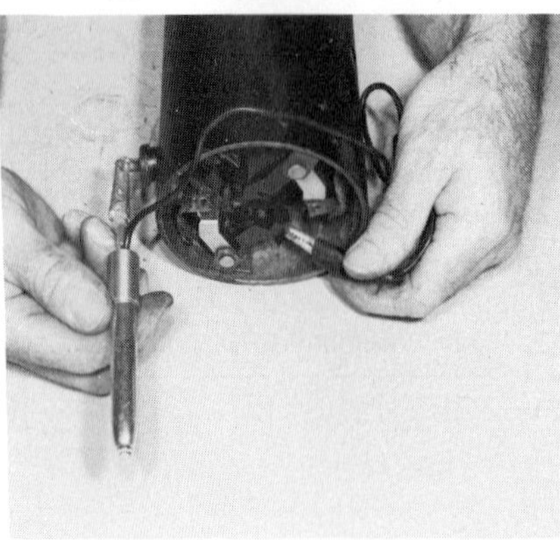
Figure 10-31. Testing for open field windings with a test lamp.

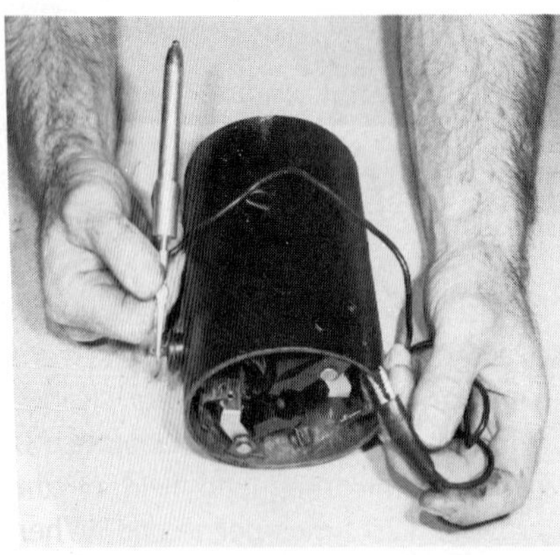
Figure 10-32. Testing for grounded field windings with a test lamp.

the alternating magnetic field causes a growling noise.

To test an armature for short circuits between the windings, place it on the growler and hold a hacksaw blade lengthwise across its surface, figure 10-29. Slowly rotate the armature. If the hacksaw blade jumps or vibrates over any section of the armature core, the windings are shorted at that point.

Armature Ground Test. If any armature windings are grounded to the shaft, the motor will not develop full power. A grounded winding reduces current and the magnetic field of the armature. Windings are connected to the commutator, which is insulated from the shaft. Any grounded winding will establish continuity between the commutator and the shaft.

Place one lead of an ohmmeter or a self-powered test lamp on the armature shaft. Touch the other lead to *each* commutator segment, figure 10-30. If the lamp lights or the ohmmeter shows continuity, the armature is grounded.

Field Coil Tests. Grounded or open windings in the field coils also will reduce motor cranking power. To check for open field windings, place one lead of an ohmmeter or a self-powered test lamp on the insulated motor brush or the motor connector, figure 10-31, and the other lead on the field coil ground connection. The lamp should light or the meter should show continuity. If not, the field coil windings are open. Four-pole motors have two insulated and two ground brushes. Repeat the test for the second set of brushes.

To check for grounded windings, disconnect the field ground leads. On many motors, you will have to disconnect soldered connections. Connect the ohmmeter or test lamp between the motor connector and the field frame, figure 10-32. The lamp should not light, or the meter should show an open circuit. If not, the field windings are grounded to the frame.

Motor Brush Replacement. Brush replacement is a common part of motor repair. Replace any brush that is worn to one-half its original length (about 3/8 inch worn length) or that is cracked or chipped. As a brush wears, brush spring tension decreases. A worn brush can bounce or arc on the commutator, which reduces armature current.

Match the contour of the brush face to the commutator when you install new brushes. If brushes are reversed, the reduced contact area will reduce armature current and cause arcing.

Delco-Remy Pinion Clearance Check. Pinion travel, or the clearance between the pinion and its retainer on the shaft, is important for proper drive engagement. Delco-Remy provides the following procedure to check this clearance on its solenoid-operated motors. Other manufacturers may provide similar instructions.

The starter must be off the engine to check pinion clearance. Clearance is not adjustable. If it is out of limits, the shift fork, the pivot pin, or the drive must be replaced.

1. Disconnect the motor connector from the solenoid. Insulate the connector.

2. Connect a battery from the solenoid switch terminal to the motor frame, figure 10-33.

3. Touch a jumper wire from the solenoid motor terminal to the motor frame. This energizes the solenoid to engage the drive. The drive will remain engaged until the battery is disconnected, but the motor will not rotate.

4. Push the pinion toward the commutator to remove endplay.

5. Use feeler gauges to measure the clearance between the pinion and its retainer, figure 10-34. Clearance should be 0.010 to 0.140 inch.

6. Disconnect the battery.

Starter Motor Installation

Starter installation involves the same steps as removal. You must reinstall any vehicle parts removed for starter access. Connect the battery cable and all other wiring to the motor *before* reconnecting the battery ground ca-

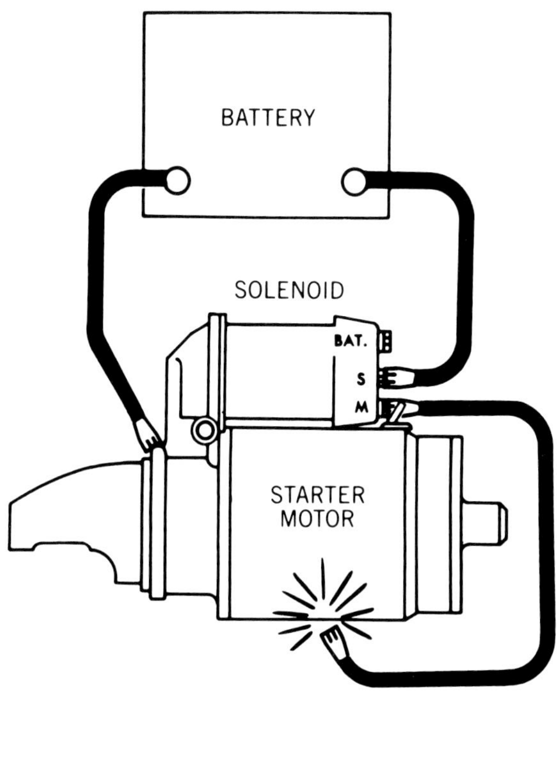

Figure 10-33. Circuit connections for checking Delco-Remy pinion clearance. (Delco-Remy)

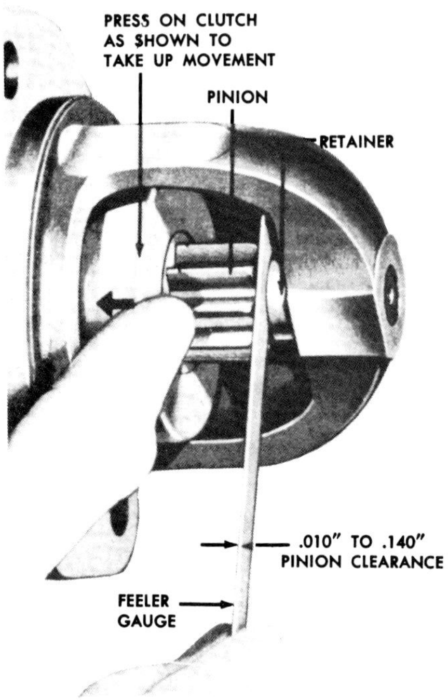

Figure 10-34. Measure pinion clearance with a feeler gauge. (Delco-Remy)

ble. Connect the ground cable last to prevent an accidental short circuit during motor installation.

There are two key points to be aware of during starter motor installation:

1. Many motors have drive end frames of different lengths for different engine and transmission combinations. When installing a new or rebuilt motor, be sure the drive end frame length, or "snout length," matches the original motor. Sometimes, the differences in length are very slight. A motor may attach to an engine perfectly, but if the snout length is wrong, the pinion will not engage the flywheel correctly.

2. Many motors require shims between the motor mounting flange and the engine for correct pinion and flywheel clearance, figure 10-24. If you removed shims when you removed the motor, reinstall the same number and thickness of shims in the original location when you install the motor. You may have to add or subtract shims to adjust pinion engagement. Motor mounting positions vary, but the following general rules will help you with shim adjustment:

a. If the pinion chatters or does not disengage from the flywheel, add shims to the lower or the inner motor mounting position.

b. If the pinion does not engage the flywheel completely, add shims to the upper or the outer motor mounting position.

STARTER MOTOR OVERHAUL

The following illustrated procedures show the basic steps to disassemble, repair, and reassemble these common starter motors:

1. Delco-Remy solenoid-operated starter

2. Chrysler permanent-magnet, gear-reduction (PMGR) starter

3. Delco-Remy (PMGR) starter

Each manufacturer builds these starter motors in various sizes for different engines, but basic designs and overhaul sequences are similar. These procedures will help you service these particular starters, and they illustrate the basic repair steps that are typical for other starters of similar designs.

DELCO STARTER OVERHAUL

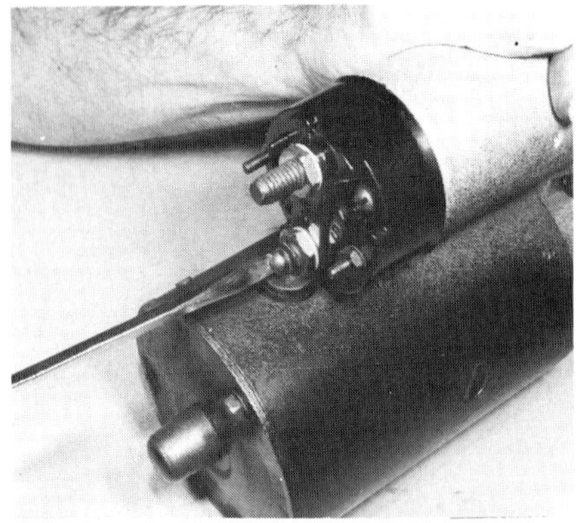

1. Remove the screw that connects the starter motor field coil connector to the solenoid motor (M) terminal. If necessary, you can replace just the solenoid without complete disassembly.

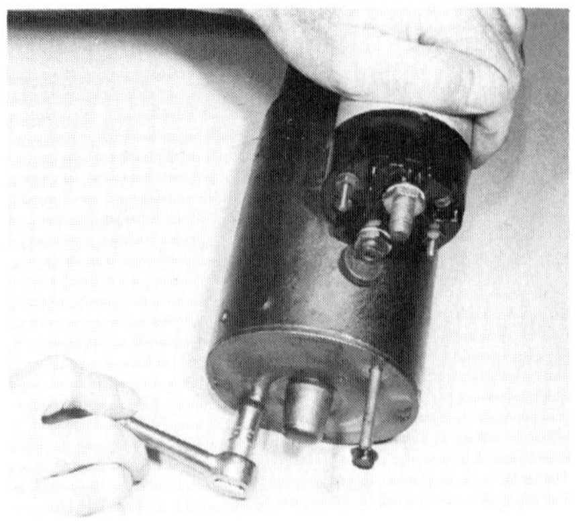

2. Remove the two long through bolts from the commutator end frame.

3. Remove the end frame and inspect the bushing (A) for wear or damage. Note the fiber or leather packing washer (B) on the end of the shaft.

4. If the end frame bushing is damaged, remove it with a special puller. Drive a new bushing in place with a wooden block and a hammer. Lubricate according to the manufacturer's instructions.

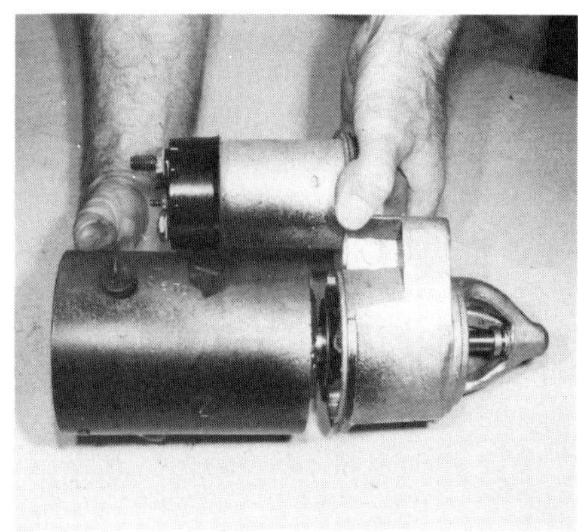

5. Turn the field frame slightly clockwise (viewed from the commutator end) and slide it off the armature shaft.

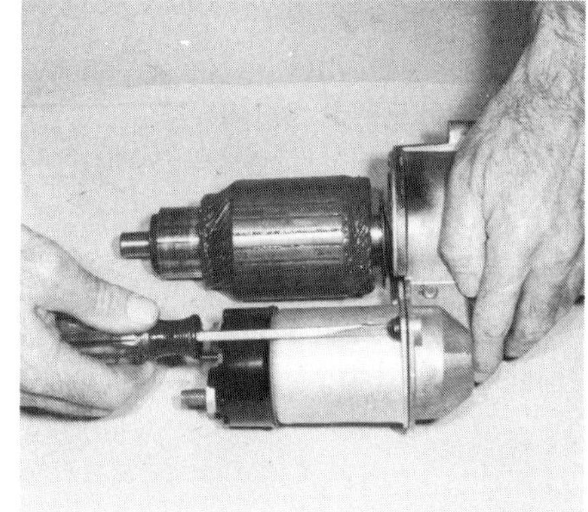

6. Loosen the two screws that hold the solenoid body to the drive end frame. Then hold the solenoid with one hand to retain the plunger spring as you remove the screws.

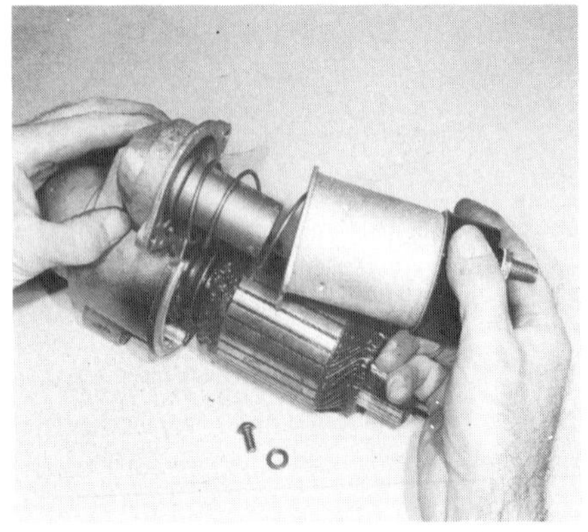

7. Remove the solenoid carefully to keep the plunger spring from flying out or popping the soleniod across the workbench.

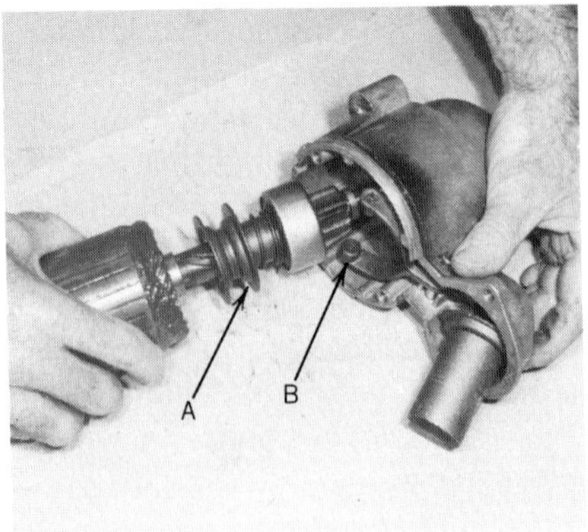

8. Tilt the fork and the armature shaft to disengage the drive collar (A) from the fork (B) and remove the armature shaft. On a 10-MT or larger starter, remove the center bearing support from the drive end frame first.

9. If the fork pivot pin is held by a snapring, use snapring pliers to remove it. If the pivot pin is pressed in, remove it with a drift punch and hammer.

DELCO STARTER OVERHAUL (*Continued*)

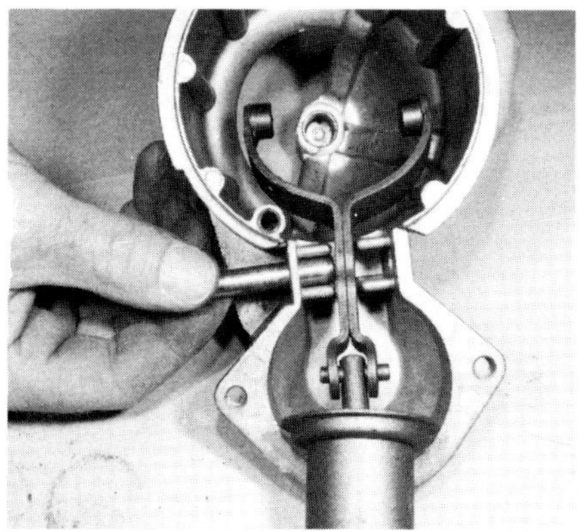

10. Remove the pivot pin to remove the shift fork and solenoid plunger.

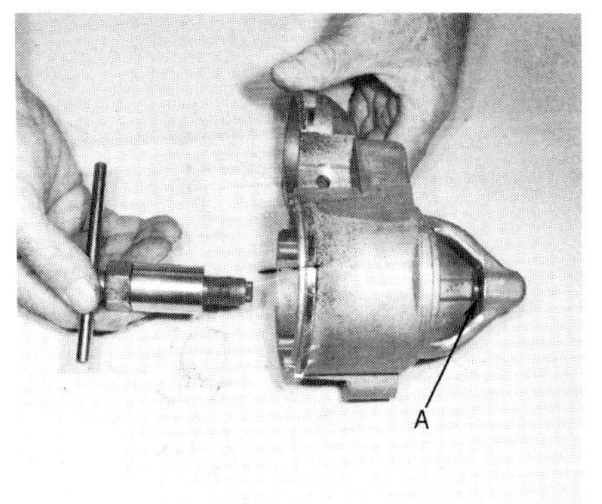

11. Inspect the drive end bushing (A) for wear or damage. If defective, remove it with the same puller you used for the commutator end frame bushing.

12. To remove the starter drive, use a deep socket and a soft mallet to tap the pinion stop collar toward the drive pinion gear.

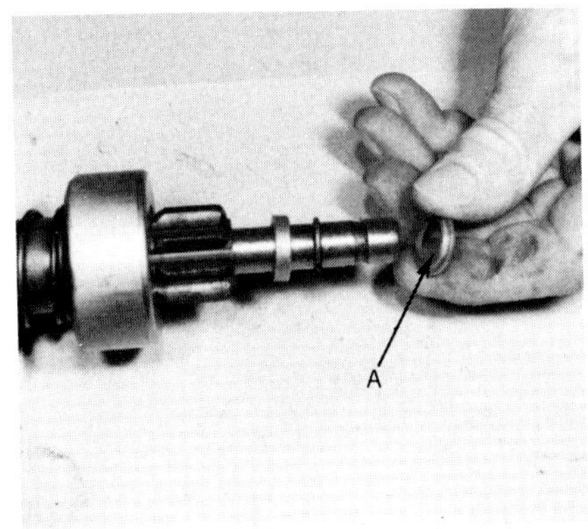

13. Note the direction in which the thrust collar (A) fits on the shaft and then remove it.

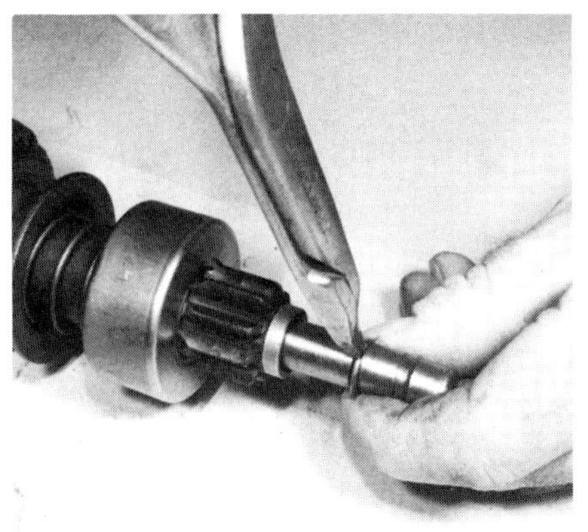

14. Use snapring pliers or two small screwdrivers to remove the retaining ring from the shaft groove. Then slide it off the shaft.

15. Now slide the entire drive assembly and the pinion stop collar off the armature shaft.

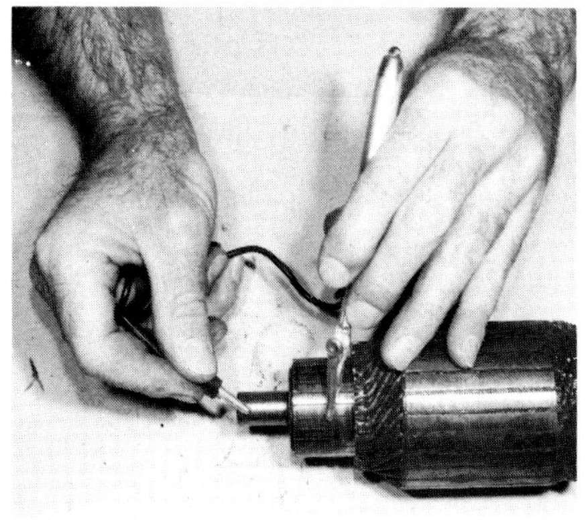

16. Use an ohmmeter or a self-powered test lamp with the leads placed as shown to check for a grounded armature. If the lamp lights or the meter shows low resistance, the armature is grounded to the shaft.

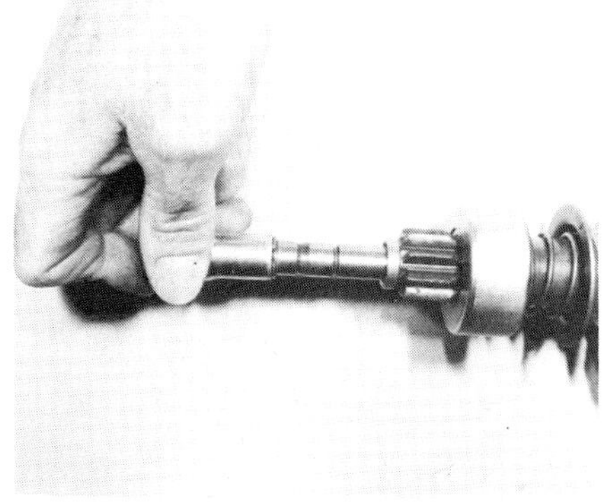

17. Install a new drive assembly and pinion stop collar on the shaft. Use a deep socket and mallet to tap the new retaining ring into the shaft groove.

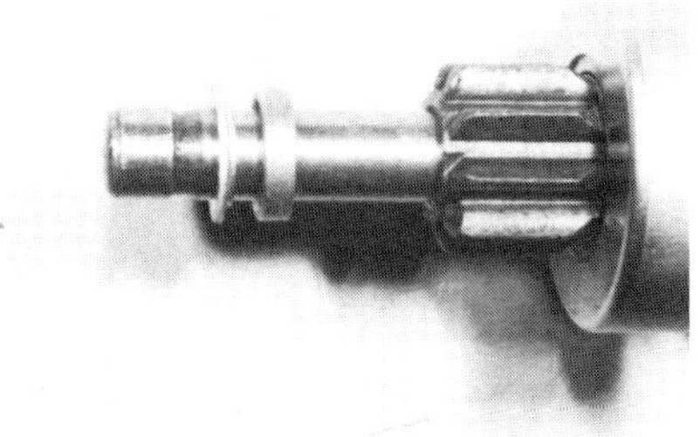

18. Put the thrust collar on the shaft with the flange toward the end. Then use two pair of pliers to squeeze the stop collar and thrust collar together over the retaining ring.

DELCO STARTER OVERHAUL (*Continued*)

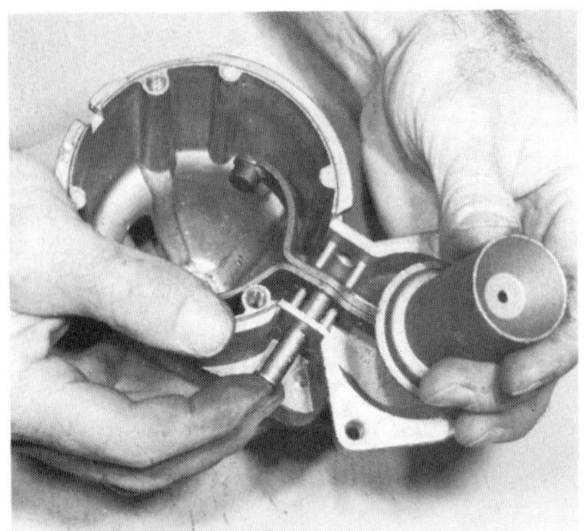

19. Reinstall the solenoid plunger and shift fork assembly. Secure the fork pivot pin with the snapring or drive it in place with a drift punch and hammer.

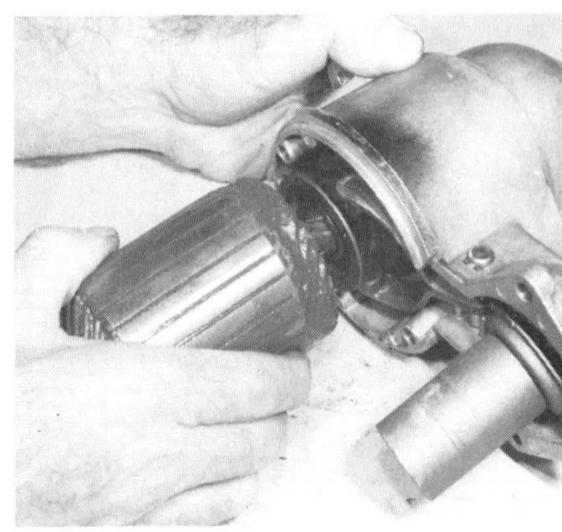

20. Engage the starter drive collar with the fork and install the armature shaft. Set the drive end and shaft assembly aside while you service the solenoid and field frame.

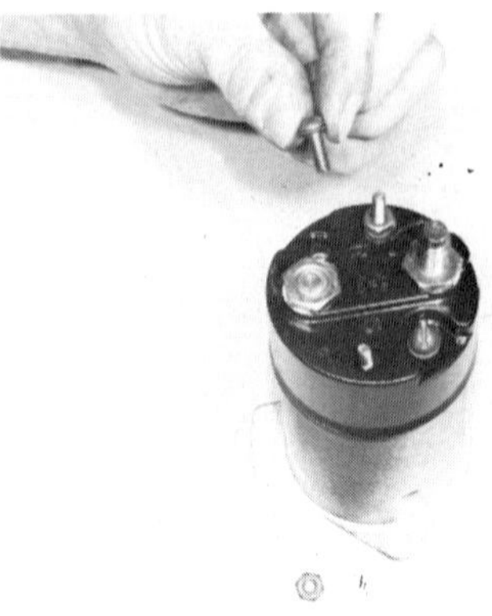

21. You may replace the solenoid as a complete unit or disassemble it to replace the contact ring and other parts. Remove the end cap screws and terminal nuts.

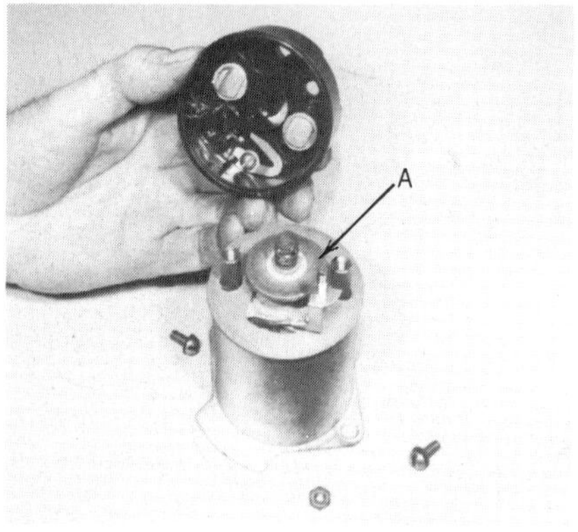

22. Remove the end cap to expose the contact ring, or washer (A), and the terminal connections.

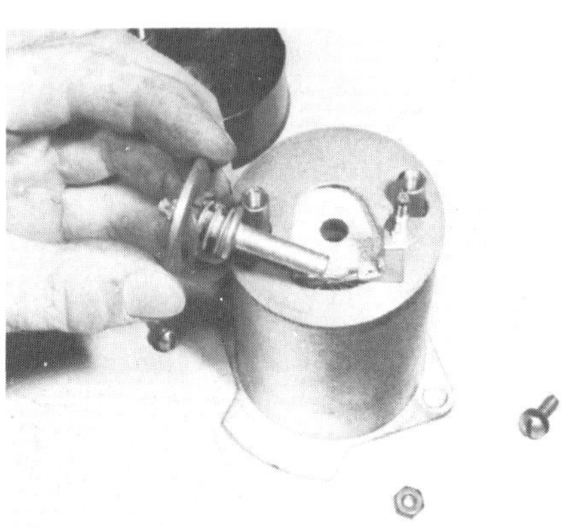

23. Remove and replace the contact ring and shaft on this soleniod as a unit. On some solenoids, you can remove the ring from the shaft and turn it over for reuse.

24. Put the solenoid spring around the plunger in the drive end frame and install the solenoid. Hold the solenoid in place with one hand and install the two screws.

25. Tighten the two solenoid screws alternately and equally. Move the armature shaft by hand to check for a binding solenoid or fork.

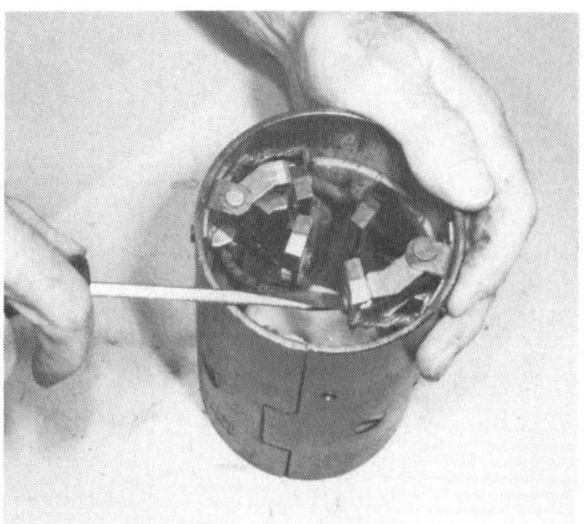

26. If brushes are cracked or worn to one-half original length or less, replace them. Remove the screws from the brush holders and disconnect the leads from ground or field connectors.

27. Do not remove the brush holder pivot pin, or the brush holder pieces will fall out like this.

DELCO STARTER OVERHAUL (*Continued*)

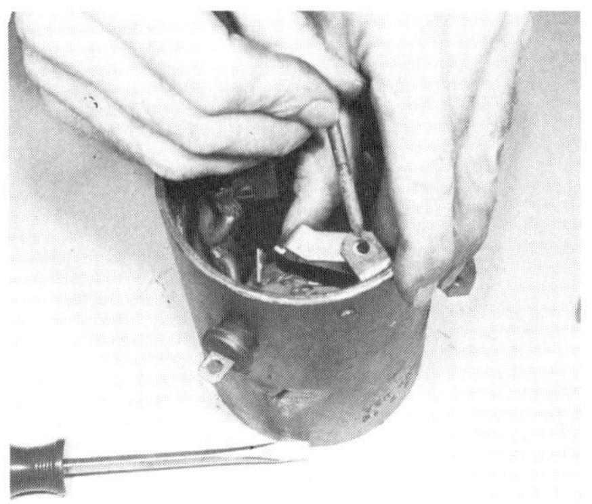

28. If you must replace the brush holders, do one at a time. Use the other holders as a guide to help you install the pivot pins and other pieces correctly.

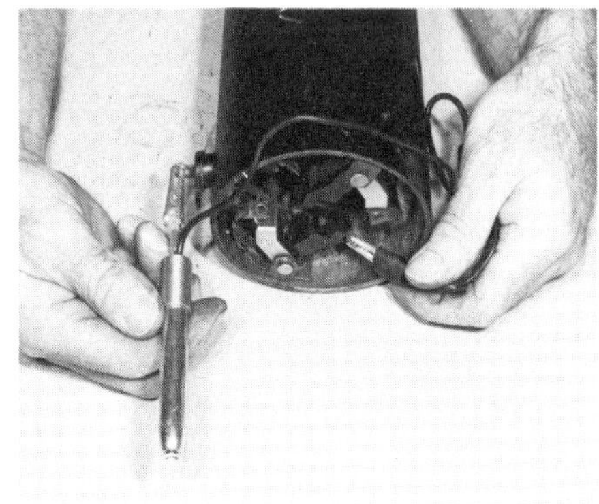

29. Test for field coil open circuits by connecting one lead of a self-powered test lamp to the motor solenoid connector and the other lead to each insulated brush terminal. The lamp should light. If not, the coils are open.

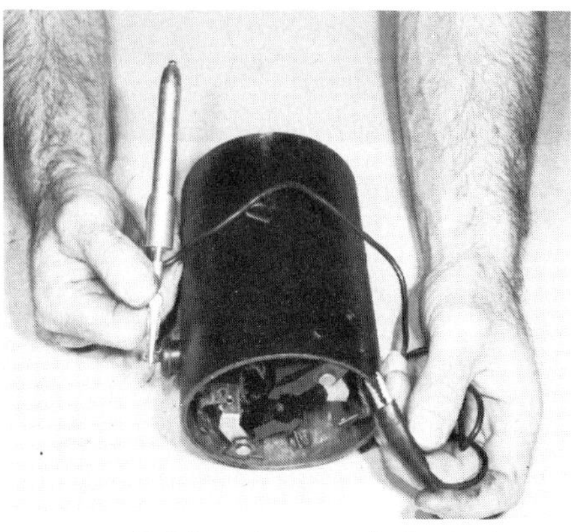

30. Test field coil grounded windings by disconnecting the field ground leads. Then connect one self-powered lamp lead to the frame and the other to the motor connector. The lamp should not light. If it does, the coils are grounded.

31. Carefully fit the field frame over the armature shaft. Hold the brushes away from the commutator and slowly slide the commutator into the brushes. Do not nick the brushes or bend the holders.

32. Three of these four brushes are installed wrong. The contour does not match the commutator. Unscrew and reverse the brushes or fit new brushes with the proper contour.

33. For final assembly, align the dowel pin (A) in the field frame with the hole in the drive end housing and slide the two sections together.

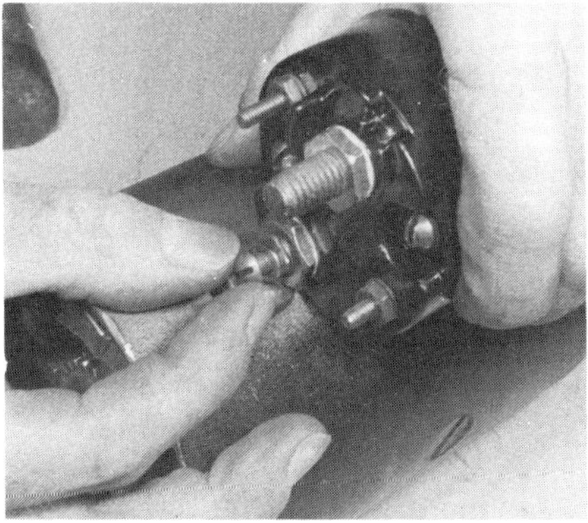

34. Install the screw to connect the motor field coil connector to the solenoid motor (M) terminal.

35. Install the packing washer on the commutator end of the shaft and install the commutator end frame.

36. Carefully slide the through bolts through the motor and thread them into the drive end frame. Be careful not to crossthread them. Tighten the bolts with a wrench and the overhaul is done.

CHRYSLER-BOSCH PMGR STARTER MOTOR

1. Unscrew the brush terminal nut (arrows). Disconnect the terminal and the solenoid stud. Then remove the washer.

2. Hold the solenoid and remove the mounting screws (arrows). Spring pressure will force it from the drive frame housing.

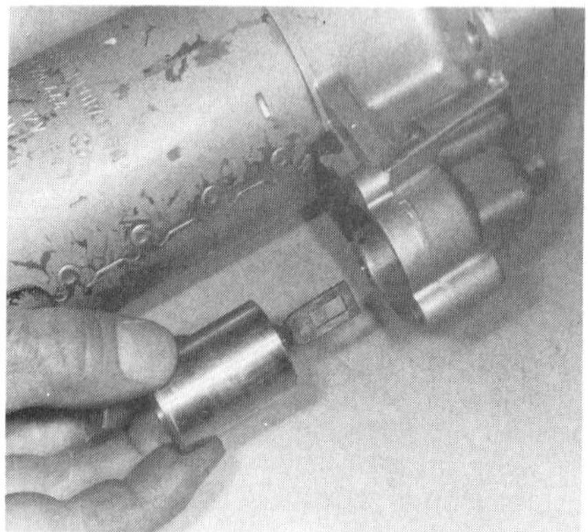

3. Disconnect the solenoid plunger from the shift fork and remove the plunger from the starter housing.

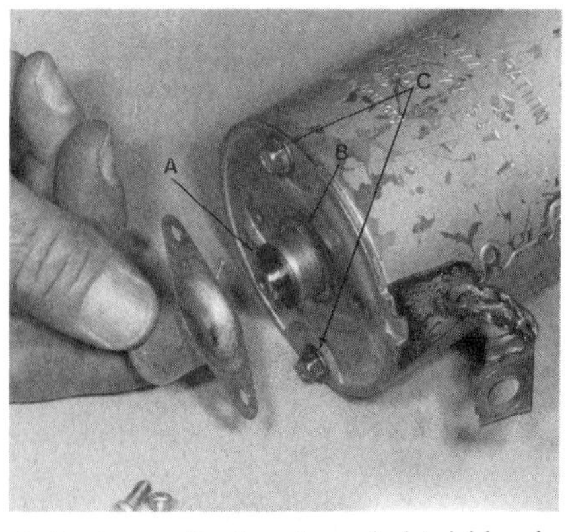

4. Remove the brush end shield bushing cap. Remove the armature shaft C-washer (A) and flatwasher (B). Then remove the through bolts (C).

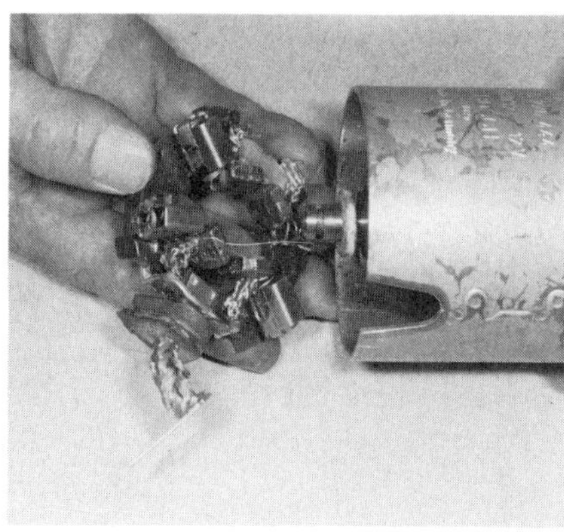

5. Remove the end cap. Then pry the brush terminal insulator from the starter frame. Remove the brush plate assembly.

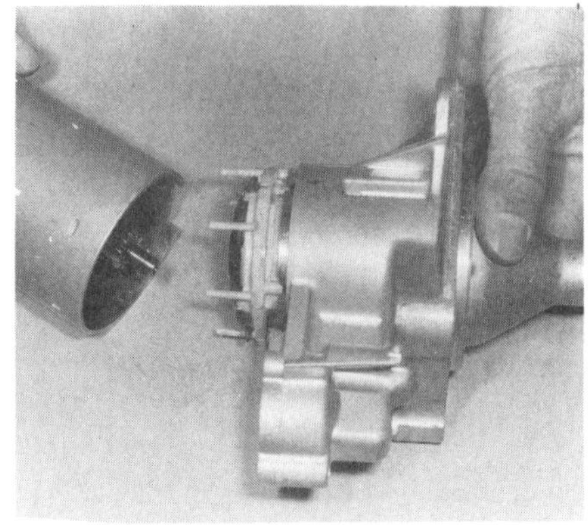

6. Magnetic force makes it difficult to separate the field frame from the drive end frame. The armature will stay in the field frame.

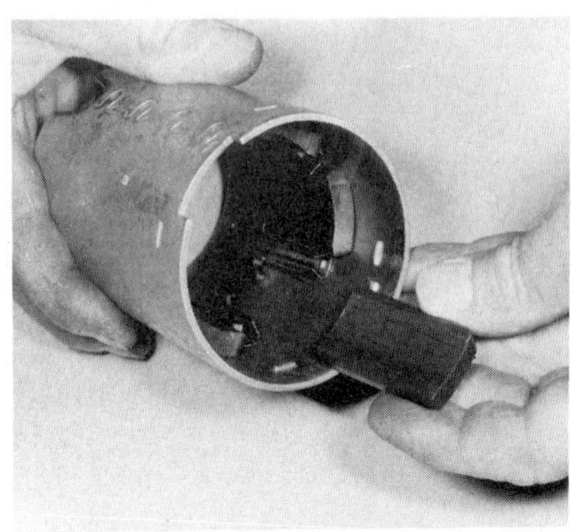

7. Permanent magnets in this starter motor are replaceable. Remove them by sliding the magnets from their retaining clips. Do not drop them.

8. Remove the rubber seal (arrow) from the drive end frame and slide the drive gear train assembly out of the end frame.

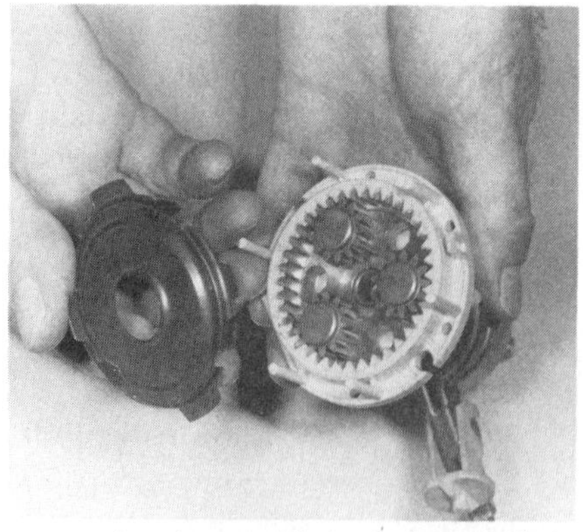

9. Unclip the metal dust plate. When removed, it exposes the planetary gear assembly, which can be serviced.

CHRYSLER-BOSCH PMGR STARTER MOTOR (*Continued*)

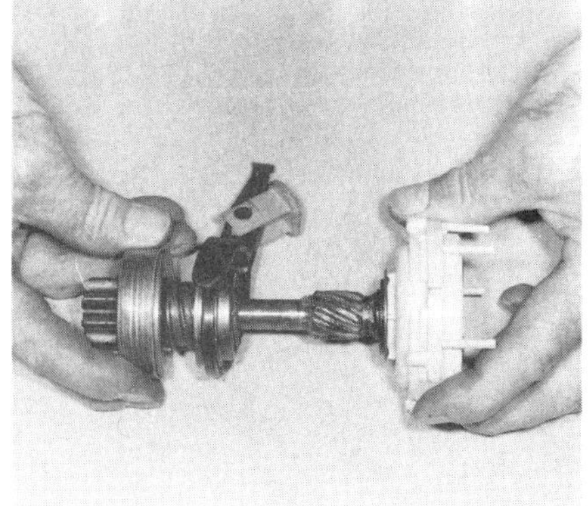

10. Press the stop collar from the snapring with a suitable socket. After removing the snapring and collar, slide the clutch assembly off the output shaft.

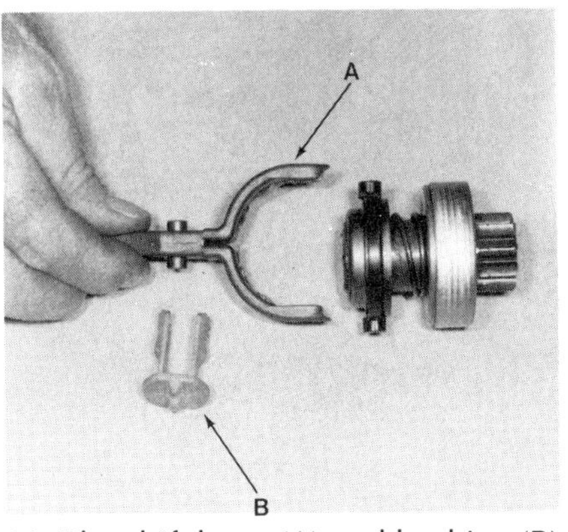

11. The shift lever (A) and bushing (B) are plastic. Remove them from the clutch assembly only if replacement is necessary.

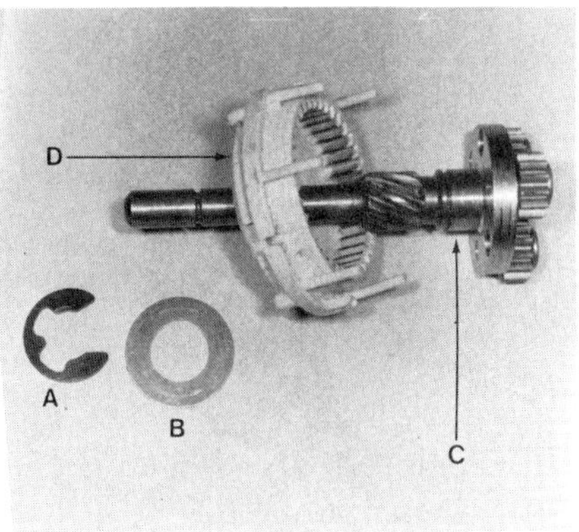

12. Remove the C-clip (A) and washer (B) from the groove (C). Then remove the output shaft and planetary gears from the annulus (ring) gear (D).

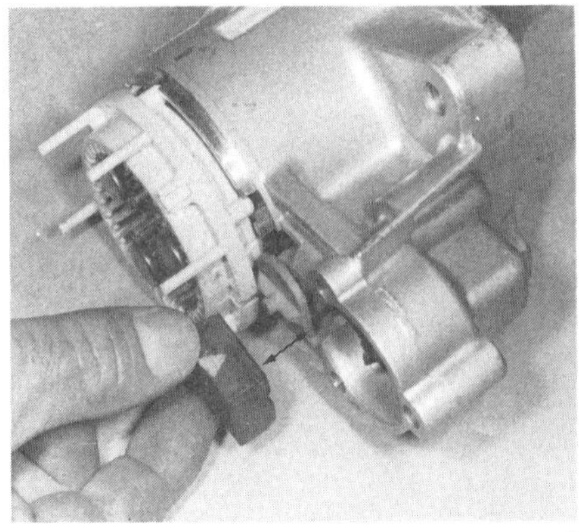

13. Reinstall the drive gear train and shift lever assembly in the drive end frame. Press rubber seal firmly into place (arrow).

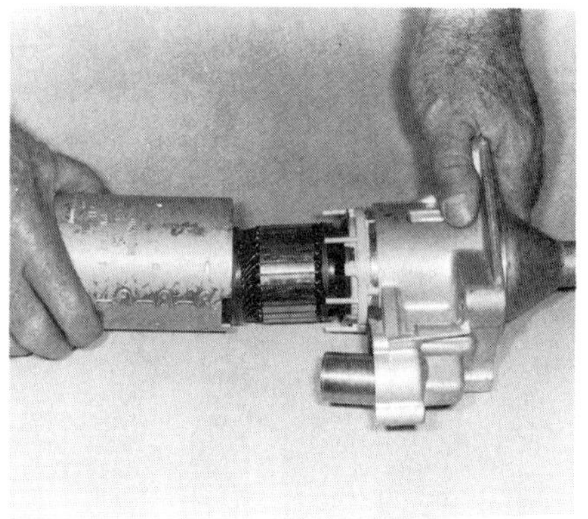

14. Engage the armature in the drive train. Hold the field frame firmly when reinstalling or magnetic attraction will snap it out of your grasp.

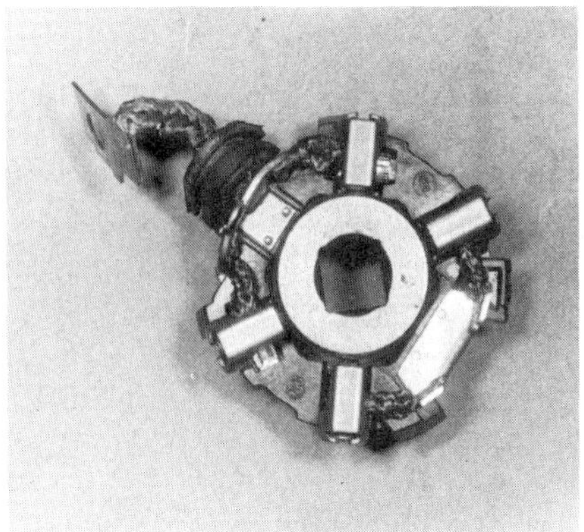

15. Insert a suitable size socket as shown to hold the brushes in place for reinstallation over the armature shaft.

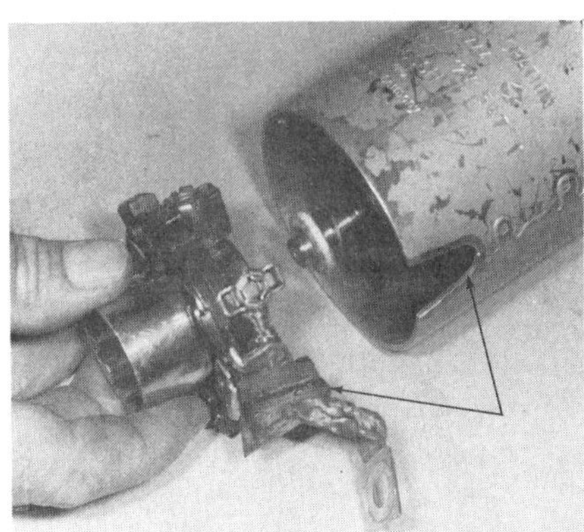

16. Grasp the brush assembly by the socket and fit it in place. Then remove the socket. Fit the brush terminal insulator in place securely (arrows).

17. Install the end shield and tighten the through bolts securely. Install the flatwasher, C-clip, and bushing cap.

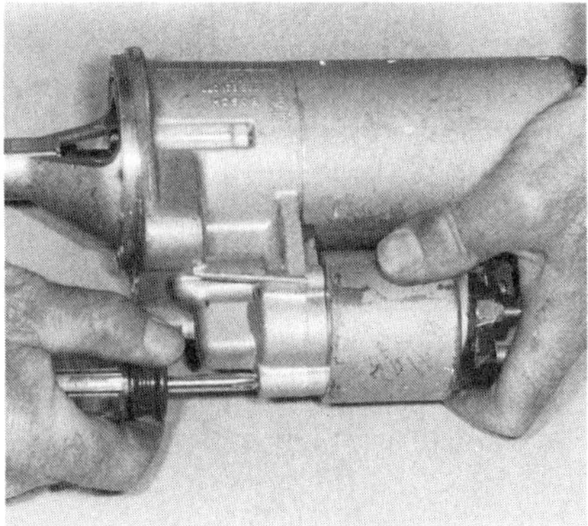

18. Connect the solenoid to the shift fork. Then fit the housing to the drive end frame and hold in place to overcome spring tension while tightening the screws.

DELCO-REMY PMGR STARTER MOTOR

1. Remove the nut holding the brush terminal to the solenoid stud. Disconnect the terminal from the stud (arrows).

2. Remove the two through bolts holding the end cap to the field frame. If cap does not come off easily, tap on its ears with a soft-faced hammer.

3. Remove the end cap. Note that there are six locating dowels in the cap that must align and engage the brush holder during reassembly.

4. Disengage the brush terminal insulator from the field frame; then slide the brush holder from the armature shaft and remove it.

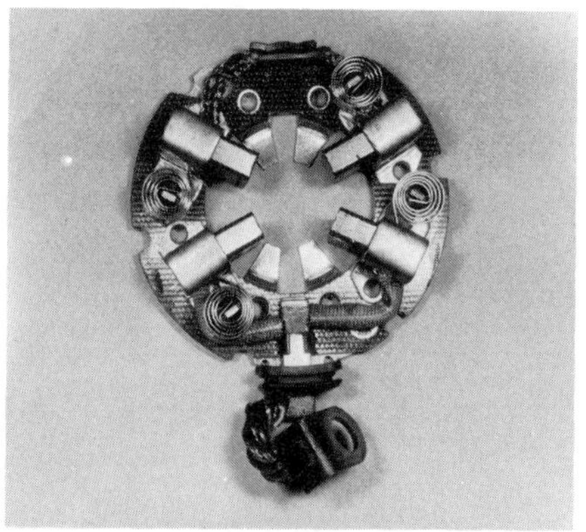

5. The brushes appear to be removable but they are not. If one or more brushes require replacement, the entire assembly must be replaced.

6. Pull the field frame from the drive end housing. This unit does not have the intense magnetic attraction of the Chrysler-Bosch PMGR starter.

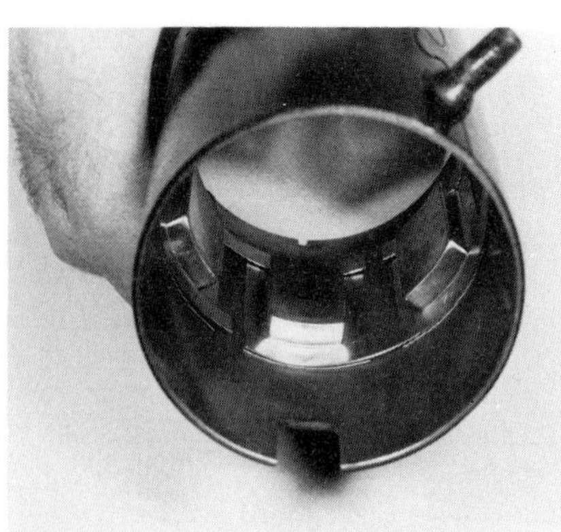

7. The permanent magnets in this starter frame are not serviceable. If defective, the entire assembly must be replaced.

8. Remove the armature from the drive end frame. Further disassembly is not possible because of the solenoid drive housing and geartrain design.

9. To reinstall the brush holder, pry back each spring and slide the brush past it. When the holder is reinstalled, push each brush in position and its spring will snap in place.

REVIEW QUESTIONS

1. The insulated side of a starter circuit:
- **a.** starts at the battery + terminal
- **b.** contains the motor armature and field windings
- **c.** contains the cables and the relay or solenoid high-current contacts
- **d.** all of the above

2. The ground side of a starter circuit:
- **a.** ends at the battery + terminal
- **b.** starts at the battery connection
- **c.** includes the low-voltage ground path through the frame or body
- **d.** includes the field windings

3. Mechanic A says that when you perform starting system tests on a converter-equipped car, you should keep engine cranking time to 15 seconds or less. Mechanic B says that you should disconnect the air injection hose from the pump or the check valve. Who is right?
- **a.** A only
- **b.** B only
- **c.** both A and B
- **d.** neither A nor B

4. When doing a motor insulated circuit test, you can test voltage drop between a battery post and a cable connector by:
- **a.** placing the voltmeter − lead on the battery post and the + lead on the cable connector
- **b.** placing the voltmeter + lead on the battery post and the − lead on the cable connector
- **c.** placing the voltmeter + lead on the cable connector and the − lead on ground
- **d.** placing the voltmeter − lead on the battery post and the + lead on ground

5. When doing a starter current draw test, you need:
- **a.** a voltmeter and an ammeter
- **b.** a variable resistance carbon pile
- **c.** both a and b
- **d.** a voltmeter and an ohmmeter

6. When doing a starter current draw test, high current draw usually indicates:
- **a.** a short in the starter motor or a binding engine or starter motor
- **b.** a discharged battery
- **c.** high resistance in the starting circuits
- **d.** a battery cable corrosion

7. When doing a starter current draw test, low current draw usually indicates:
- **a.** low resistance in the starting circuits
- **b.** a short in the starter motor
- **c.** a discharged battery or high resistance in the starting circuits
- **d.** a binding starter motor

8. Chrysler and AMC transmission-mounted starting safety switches:
- **a.** are not adjustable
- **b.** can be checked with a carbon pile
- **c.** can be checked with an ammeter
- **d.** both b and c

9. If a starter motor spins but does not engage and actually crank the engine, the problem is most likely the:
- **a.** solenoid
- **b.** ring gear
- **c.** field windings
- **d.** starter drive

10. The first step in removing a starter from an automobile is:
- **a.** turning the steering wheel to get adequate clearance
- **b.** removing the starter mounting bolts
- **c.** removing the battery ground cable
- **d.** disconnecting the transmission oil cooler lines

11. Shims may be used between the starter drive end frame and the engine mounting surface to:
- **a.** provide a strong mounting surface
- **b.** establish the proper starter pinion gear and flywheel engagement
- **c.** keep the end frame flanges from breaking
- **d.** keep dust and dirt out of the starter

12. A starter no-load test:
- **a.** is performed with the starter in the car and the transmission in neutral
- **b.** can be used to identify starter drive problems
- **c.** is used to measure starter performance outside the normal operating conditions
- **d.** can be used to identify a discharged battery

13. An armature growler is used to test an armature for shorted windings. This is done by:
- **a.** placing the armature in the growler and holding a hacksaw blade lengthwise across its surface
- **b.** placing the armature in the growler and holding a hacksaw blade on each of the armature ends
- **c.** placing the armature perpendicular to the growler
- **d.** placing the armature in the growler and holding a hacksaw blade only on the commutator end

14. A test for a grounded armature is:
- **a.** placing a hacksaw blade across the armature surface
- **b.** placing a test lamp lead on the armature shaft and placing the other lead on each of the commutator segments
- **c.** placing an ohmmeter lead on the armature shaft and placing the other lead on the field windings
- **d.** both b and c

15. A field coil ground lead has been disconnected from the starter frame. One lead of the test lamp has been connected to the motor connector and the other lead to the field coil frame. The lamp lights. The problem is:
- **a.** the armature is grounded
- **b.** the commutator is grounded
- **c.** the field windings are shorted
- **d.** the field windings are grounded to the frame

16. Starter motor brushes should be replaced if they are worn to:
- **a.** 80 percent of their original length
- **b.** one-half of their original length
- **c.** three-fourths of their original length
- **d.** 60 percent of their original length

17. New brushes have been incorrectly installed in a starter—the contour of the brushes does not match the contour of the commutator. The result will be:
- **a.** brush bouncing and arcing
- **b.** reduced current to the armature
- **c.** a shortened armature
- **d.** both a and b

18. Which of the following is not a common kind of starter drive?

a. Solenoid-operated, direct drive

b. Solenoid-operated, reduction drive

c. Renix drive

d. Movable pole shoe

19. Most starter drive problems are due to:

a. a slipping 1-way clutch

b. a broken spring

c. a broken ring gear

d. a binding shaft

20. A starter will only crank an engine slowly. Which of the following is *not* a probable cause?

a. a discharged battery

b. high resistance in the starting system

c. a broken starter drive

d. an engine mechanical problem

11 CHARGING SYSTEM TESTING AND SERVICE

INTRODUCTION

A charging system problem can disable an entire vehicle. The battery is the power source for the complete electrical system, but the charging system must maintain the proper battery state of charge. The charging system also must deliver additional current for heavy electrical loads, and it must maintain a regulated system voltage. Proper current supply and regulated voltage are both essential for correct engine performance.

Low charging current or low voltage, or both, will result in an undercharged battery. Low current and low voltage cause faulty operation of lighting and accessory systems. Low voltage can cause electronic systems to malfunction. Engine computers require a steady voltage within specific limits for proper operation.

High charging voltage can cause early failure of lamp bulbs. High voltage also causes higher than normal current. This can overheat and overcharge a battery. High voltage and the resultant high current can overheat and damage an ignition coil and burn the contacts of a breaker-point ignition. Most importantly in late-model cars, high voltage and current can damage electronic parts. High voltage can break down diodes, transistors, and other devices. A charging system malfunction can destroy an expensive computer.

Chapter 5 explained basic tests for battery load (capacity) and for charging system output. If a battery and charging system pass these general tests, they probably are in good working order. If voltage and charging current are out of limits, the problem may be in the battery or the charging system.

GOALS

After studying this chapter, you should be able to do the following jobs in twice the flat-rate labor time or less:

1. Analyze abnormal charging system symptoms for possible causes. Inspect the system for damage and defects; repair or replace as required.

2. Perform basic resistance, output, voltage, field current, and regulator tests using voltmeters, ammeters, ohmmeters, multimeter equipment, or an oscilloscope.

3. Perform specific charging system tests according to carmakers' procedures and evaluate the results.

4. Remove an alternator from a vehicle; disassemble, test, repair, reassemble, and reinstall it. Verify proper operation.

CHARGING SYSTEM SAFETY

When you service the charging system, you will be working with batteries, electrical systems, and running engines. Follow the basic safety precautions in chapter 1, as well as specific precautions for engine, battery, and electrical safety. Remember that a battery can explode if exposed to sparks or flame.

Observe these particular points:

1. Keep the ignition switch off except when instructed to turn it on during specific procedures.

2. The alternator output (BAT) terminal has battery voltage present at all times, even with the engine off. Never ground the BAT terminal, intentionally or accidentally.

3. Disconnect the battery ground cable before disconnecting any wires from the alternator.

4. Be sure all alternator and regulator terminals are connected correctly.

5. Be sure the battery and the alternator are connected with the same electrical polarity. Late-model vehicles have negative ground systems. Never reverse battery cable connections. Reversed polarity will destroy alternator diodes.

6. Never run an alternator with unregulated voltage or without an external load connected to the output circuit.

Symptom	*Possible Cause*	*Correction*
Ammeter or voltmeter needle flutters or warning lamp flickers	1. Loose wiring connections 2. Intermittent high resistance in field circuit (for example, worn brushes) 3. Defective regulator	1. Inspect and repair wiring. 2. Test field circuit; repair alternator. 3. Test and repair or replace regulator.
Ammeter shows discharge, voltmeter shows low voltage, indicator lamp stays lit, or battery is continuously discharged	1. Loose or worn drive belt 2. Corroded or loose battery cables 3. Defective battery 4. Battery too small (under capacity) 5. Loose or damaged system wiring 6. Alternator output is low 7. Defective regulator	1. Adjust or replace drive belt. 2. Clean, tighten, or replace cables. 3. Test and charge or replace battery. 4. Replace battery with correct model. 5. Repair wiring. 6. Test and repair alternator. 7. Test and repair or replace regulator.
Ammeter shows continuous charge, voltmeter shows high voltage, or battery is overcharged	1. Defect in system wiring 2. Poor ground at regulator 3. Regulator out of adjustment 4. Defective regulator	1. Test and repair wiring. 2. Repair regulator ground connection. 3. Test and adjust regulator. 4. Test and repair or replace regulator.
Indicator lamp is on when engine is off	1. Shorted positive diode in alternator	1. Test and repair alternator.
Squealing or rumbling noise from alternator	1. Loose or worn drive belt 2. Worn alternator bearings 3. Loose or damaged stator 4. Loose or damaged pulley	1. Adjust or replace drive belt. 2. Repair alternator. 3. Repair alternator. 4. Repair alternator.
Whining noise from alternator	1. Shorted diode	1. Test and repair alternator.

Figure 11-1. Charging system troubleshooting table.

CHARGING SYSTEM INSPECTION

The area tests in chapter 5 will lead you to the battery tests in chapter 9 and the charging system tests in this chapter. The instrument panel meter or lamp and other vehicle symptoms also may lead you to test and service the charging system. Before inspecting or testing the charging system, check the battery as explained in chapter 9. If the battery is worn out, charging system voltage and current may be out of normal limits with no fault in the alternator or the regulator.

Figure 11-1 lists common symptoms of charging problems, along with possible causes and cures. Begin charging system service by checking the symptoms in this table and the following inspection points:

1. Inspect the alternator drive belt for wear, damage, and looseness. More battery and charging system problems are due to defective drive belts than any other cause.

2. Check the alternator and regulator for loose mounting bolts. Tighten or replace as required.

3. Inspect all wiring and connections at the alternator, the regulator, and vehicle harnesses for damage, corrosion, and looseness. Repair as required.

Charging System Protection

The wiring for almost every charging system includes one or more fusible links (fuse links). High current or voltage surges due to charging system problems can damage electrical and electronic parts. Electrical problems elsewhere in the vehicle can overload and damage charging system parts. Fusible links protect the entire system from current overloads.

Carmakers use fusible links to protect charging systems because they can be installed in the wiring close to the voltage and current source: the alternator, figure 11-2. When you inspect the charging system, always check the fusible links. Blistered or burned insulation indicates a blown link. Replace a damaged fusible link as explained in chapter 8.

CHARGING SYSTEM TESTING

The charging system has two circuits: the output circuit and the field circuit. Testing the charging system is similar to testing the starting system and other electrical circuits. You will check for high resistance (excessive voltage drop) in the insulated (hot) and the ground sides of both circuits. On most

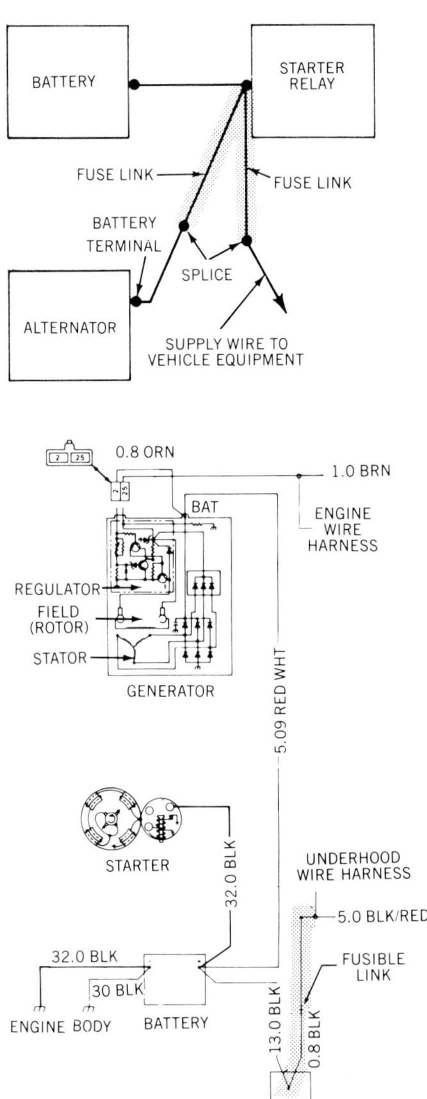

Figure 11-2. These wiring diagrams show typical fusible link installations for charging systems. (Ford) (Cadillac)

systems, you also will check some, or all, of the following:

- Current output
- Maximum regulated voltage
- Field current draw

Carmakers' test procedures vary because of alternator and regulator construction, but all are based on common principles. To test any charging system, follow the carmaker's instructions for the particular vehicle year and model. You also will need the manufacturer's specifications for a particular system. Remember that one basic model of alternator can be built with different current ratings. A small car with few accessories may use a 35- or 40-ampere unit. A large car with many accessories may use a 60- or 80-ampere alternator of the same model. Field current and regulated voltage specifications also can vary from model to model, even though they are all within a general range. Alternator and engine combinations can affect test speeds because of different ratios between crankshaft pulleys and alternator pulleys. Also, an alternator with a delta stator may require some tests at slightly higher speed than one with a Y-type stator.

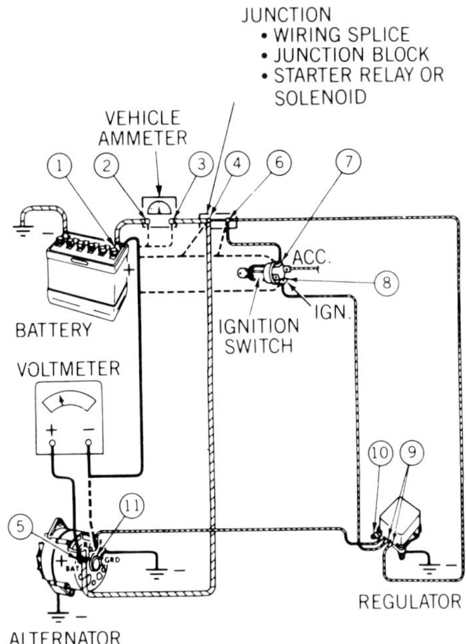

Figure 11-3. Measure voltage drop across the complete circuit to determine resistance. Numbers indicate test points to isolate high resistance. (Delco-Remy)

The following sections explain the principles of basic charging system tests. These are followed by examples of test instructions for several late-model systems with electronic voltage regulators. These procedures will help you use and understand carmakers' instructions quickly and accurately. They also are guidelines for general testing when manufacturer's procedures are unavailable.

Unless otherwise specified, test the charging system with the engine at normal operating temperature. This ensures that the alternator, the regulator, and the battery are at the right temperatures for specified test values.

Test Equipment Use

You can test a charging system with a separate voltmeter, an ammeter, a carbon pile, and other equipment. Most often, these will be combined in multimeter equipment such as a volt-amp tester, a battery-starter tester, or an oscilloscope console. The following sections describe general meter connections to help you understand test principles. Always follow the equipment maker's instructions for specific connections.

Output Circuit Resistance Tests

High resistance in the output circuit can reduce charging voltage and current output. The resistance may be on the insulated (hot) side or the ground side of the circuit. As with most voltage-drop tests, start by measuring the drop across one entire side of the circuit.

Test the output voltage drop with the engine running at 1,500 and 2,000 rpm and the alternator delivering a specified current, typically about 20 amperes. You can connect a carbon pile across the battery or use vehicle accessories to load the circuit.

Connect the voltmeter + lead to the output (BAT) terminal of the alternator. Connect the meter − lead to the battery + terminal, figure 11-3. Depending on circuit design, some instructions may specify connecting the meter − lead to a field terminal of the alternator or the regulator also, figure 11-4. Above idle, field current comes from alternator output. Therefore, the output circuit includes connections to the battery and connections to the field.

If voltage drop is within limits, resistance is okay on the hot side of the output circuit. Specifications vary, but typical voltage drop should be less than 1 volt. Typically, it should be 0.7 volt with an instrument panel ammeter; 0.4 volt with an indicator lamp. If voltage drop is high, move the meter − lead through successive test points, toward the BAT terminal, to pinpoint the high resistance, figure 11-3.

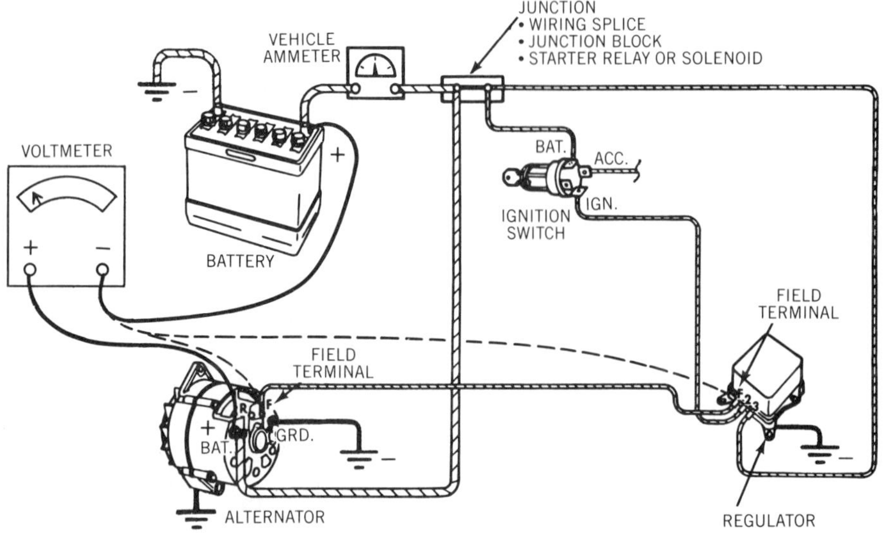

Figure 11-4. The output circuit includes field connections. Check them at the regulator and at the alternator. (Delco-Remy)

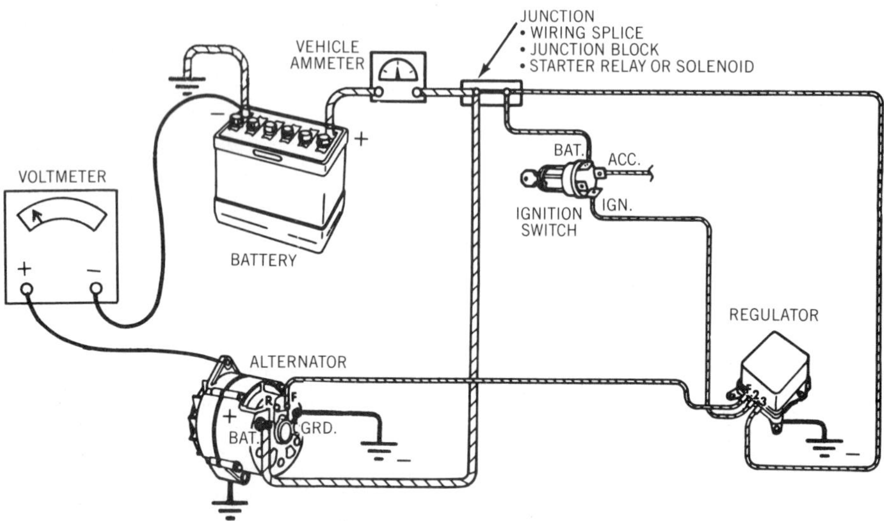

Figure 11-5. Measure voltage drop to determine ground resistance. (Delco-Remy)

Test voltage drop across the ground side of the circuit by connecting the voltmeter + lead to the alternator frame and the − lead to the battery − terminal, figure 11-5. Typical voltage drop should be 0.2 volt or less. If voltage drop is higher than specifications, move the meter + lead through successive ground connections toward the battery to locate the high resistance.

Some carmakers call for a separate voltage-drop test of the field circuit to be made at the regulator. All of these resistance tests are examples of basic voltmeter use that you learned in chapter 4. High resistance in the output circuit often is due to loose or corroded connections and damaged wiring. Fix any high-resistance problems before proceeding with other system tests.

Output Current Tests

You can test current output in one, or both, of two ways. Some carmakers specify one of these methods.

For the first method, connect a carbon pile across the battery to load the alternator output circuit. Connect a voltmeter between the battery + terminal and ground, figure 11-6. Connect an ammeter + lead to the alternator BAT terminal and the − lead to the battery + terminal or to a junction for the alternator output as shown in figure 11-6. If you use an inductive ammeter, follow the equipment maker's directions for connection.

Turn on the ignition and read the rate of discharge on the ammeter. This is field current and ignition current draw. Then, start and run the engine at specified test speed and adjust the carbon pile for a steady 15 volts of system voltage or for the highest possible current. Read the ammeter and add this reading to the previous one. Compare the total current to alternator maximum output specifications. Most manufacturers allow ±10-percent or ±10-ampere tolerance on the rated maximum current.

For the second test method, you must bypass the voltage regulator to apply full current to the alternator field. Some carmakers recommend this method instead of the previous test. If any system fails the first current output test above, you don't know if the cause is in the alternator or the regulator. Bypassing the regulator lets you check unregulated current output and isolate the problem to the alternator or the regulator.

Caution. When applying full field current to the alternator, do not operate the system above the specified test speed. Unregulated field current at high speed will produce high voltage that can damage charging system parts and electronic components.

Use the same test equipment connections as used for the first current output test, figure 11-6. If the regulator is mounted remotely from the alternator, you must bypass it with a jumper wire, figure 11-7. If the regulator is a solid-state unit, mounted on or inside the alternator, manufacturers provide different ways to bypass it. With the regulator bypassed and full current to the field, run the engine at the specified speed and adjust the carbon pile for

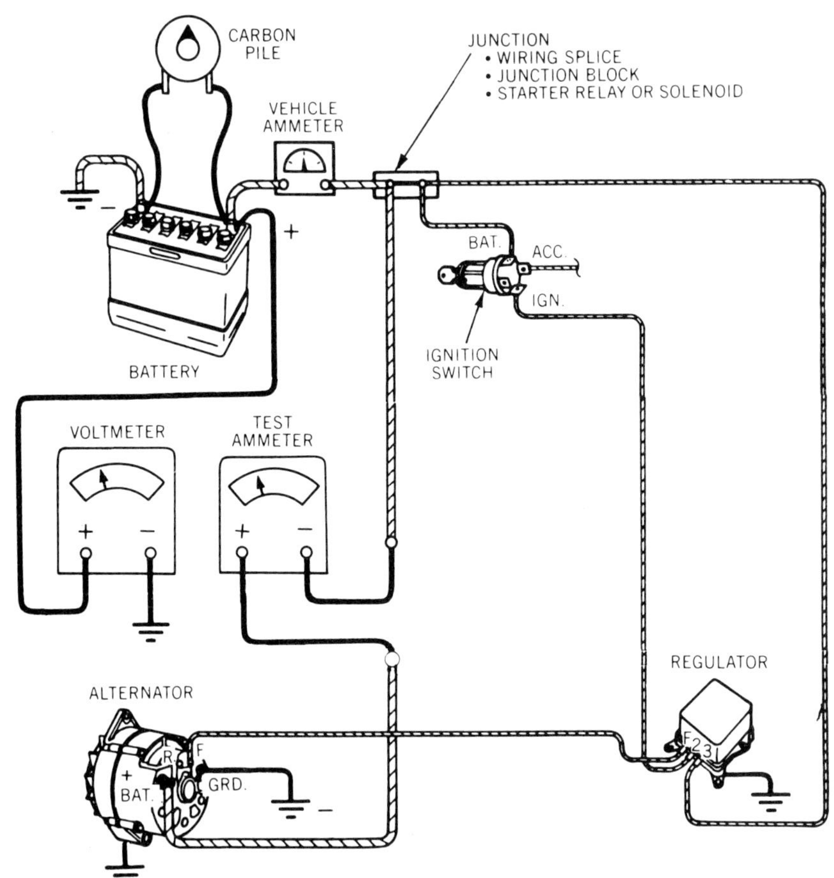

Figure 11-6. Basic output current test connections. (Delco-Remy)

maximum current at a specified voltage (usually about 15 volts). Compare the ammeter reading to specifications. If current is out of limits, the alternator is bad. If current is within limits, the regulator may be bad.

Charging Voltage Tests

Some carmakers recommend a maximum regulated voltage test instead of a current output test. This test avoids possible damage to electronic parts due to high current. Be sure the system has passed the output circuit resistance tests before testing maximum voltage.

Connect a voltmeter between the battery + terminal and ground, figure 11-8. Use a carbon pile across the battery or turn on the headlamps and air conditioner to load the output circuit. Run the engine at the specified speed (usually about 1,500 to 2,000 rpm)

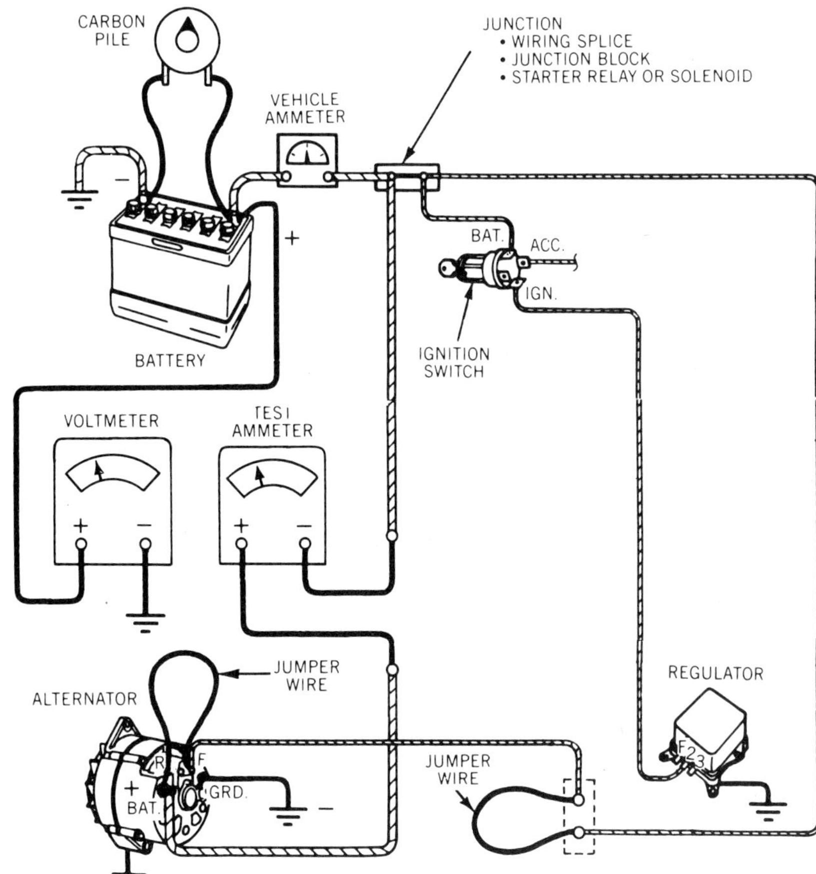

Figure 11-7. Depending on circuit connections, you can bypass the regulator at the regulator connector or at the alternator. (Delco-Remy)

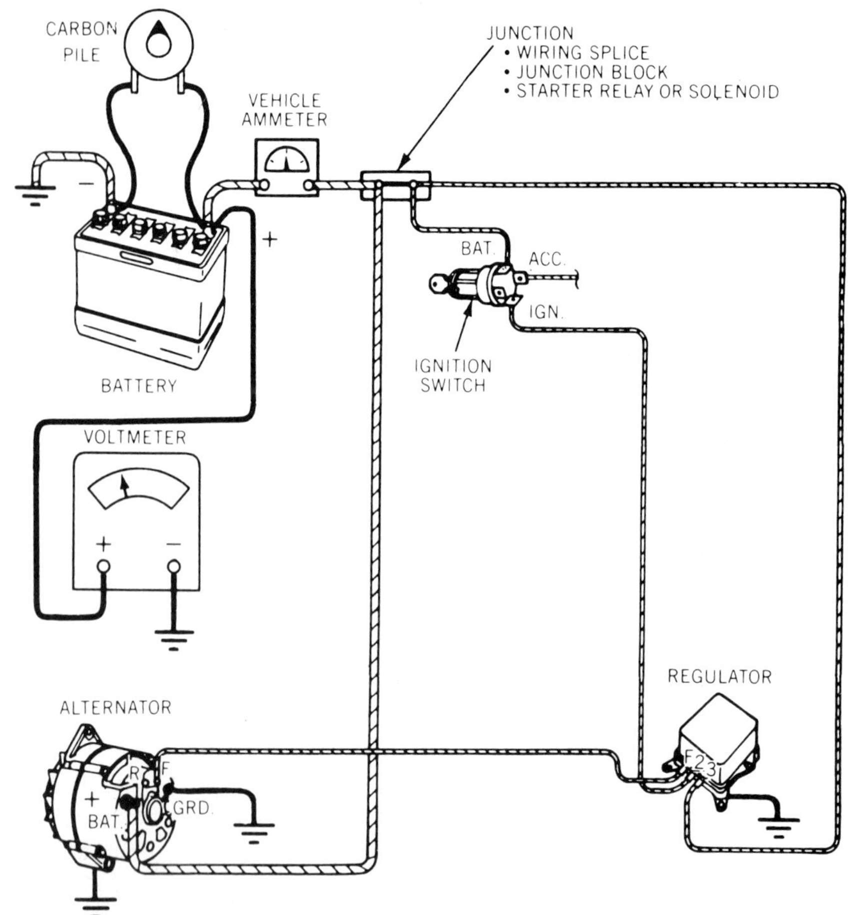

Figure 11-8. Basic charging voltage test connections. (Delco-Remy)

and read the voltmeter. If voltage is below 13 volts, bypass the regulator and repeat the test. Do not exceed the specified test speed with the regulator bypassed. If voltage rises to about 16 volts, the regulator probably is bad. If voltage remains low, the alternator is bad, or there is a high current load or undetected resistance somewhere in the system.

Test the voltage regulator setting by connecting a voltmeter between the alternator BAT terminal and ground or between the battery + terminal and ground. Run the engine at 1,500 to 2,000 rpm with the headlamps on low beam and read the voltmeter. Turn off the headlamps, increase engine speed slightly (about 200 rpm), and read the voltmeter again. With an electronic regulator, charging voltage at both test conditions should be within specifications and approximately the same. With an older double-contact electromechanical regulator, voltage at the first condition should be the setting of the series contacts in the regulator. Voltage at the second condition should be the setting of the upper, grounding contacts.

Field Current Tests

The alternator field coil is a resistor with a fixed resistance value. It should draw a specified amount of current. If current draw is high, there is a short somewhere in the field circuit. If current draw is low, there is high resistance in the circuit.

Test field current draw with the engine off. Specific procedures may tell you to bypass the regulator or the indicator lamp circuit. Some procedures may tell you to connect a jumper wire from the battery to a specific alternator field terminal.

Connect an ammeter in one of two ways, depending on circuit design, figure 11-9:

1. Meter + lead to the battery + terminal and the − lead to the alternator field.
2. Meter + lead to the alternator field wire and the − lead to the field terminal. In most cases, this method works only for systems with an instrument panel ammeter.

Connect an inductive ammeter according to manufacturers' instructions, usually around either battery cable.

Turn on the ignition and read field current draw on the ammeter. If field current is out of limits, test the alternator or the regulator. If field current is within limits, but you suspect a problem, turn the alternator pulley by hand while reading the ammeter. If the reading fluctuates, the brushes and sliprings are not making uniform contact.

Although alternator brushes do not carry high current and suffer the wear of generator or motor brushes, carbon residue can build up on sliprings of high-mileage alternators. Often simply cleaning the sliprings and installing new brushes will restore low or irregular field current to specifications.

Delco-Remy 10-SI System Testing

Delco-Remy alternators with integral electronic regulators are standard equipment on vehicles from GM and several other carmakers. Delco-Remy recommends the following series of tests for model 10-SI alternators. Tests are similar for 15-SI and 27-SI systems.

Circuit Continuity and Voltage Tests

1. Connect your voltmeter − lead to ground, figure 11-10. The alternator frame or mounting bracket is a good ground point.
2. Turn the ignition on; do not start the engine.
3. Touch the voltmeter + lead to the following points, in sequence, figure 11-10.
   a. The alternator BAT terminal

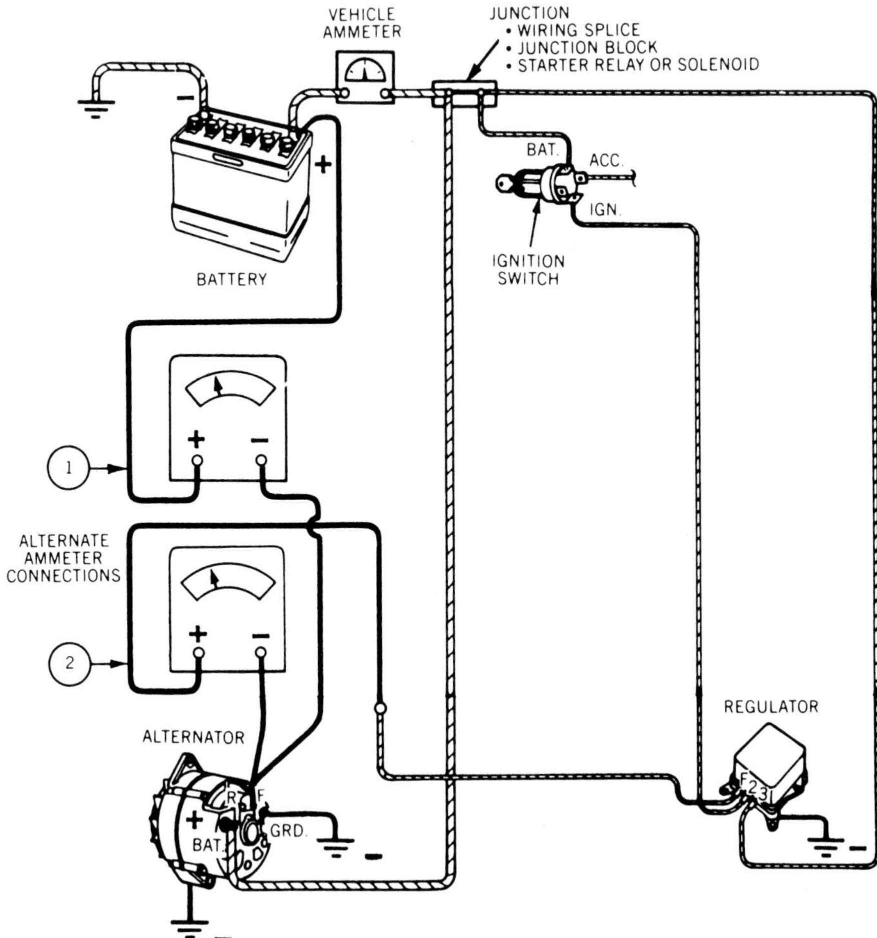

Figure 11-9. Connect the ammeter in one of these two ways to measure field current. (Delco-Remy)

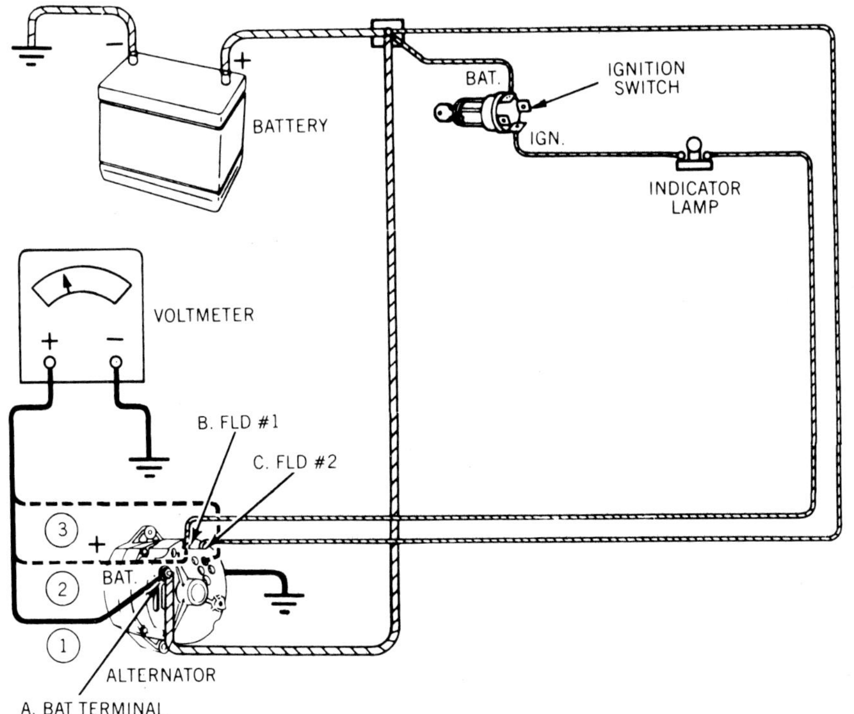

Figure 11-10. Touch the meter + lead to the numbered points in sequence to check circuit voltage and continuity. (Delco-Remy)

b. The alternator field no. 1 terminal

c. The alternator field no. 2 terminal

4. Read the voltmeter at each point. Zero voltage at any test point indicates an open circuit between that point and the battery + terminal. Repair any open circuit before proceeding with other tests.

5. Connect your voltmeter across the battery, figure 11-11, and run the engine at 1,500 to 2,000 rpm. System voltage should be between 12.6 and 15.5 volts. If voltage is out of limits, remove the alternator for further testing.

Current Output Test

1. Disconnect the battery ground cable and the alternator BAT terminal output connection.

2. Connect your ammeter + lead to the alternator BAT terminal and the − lead to the disconnected output wire, figure 11-12.

3. Reconnect the battery ground cable.

4. Connect a carbon pile across the battery and adjust it for maximum current load. Or, turn on the high-beam headlamps, the air conditioner, and other accessories for maximum current load.

5. Run the engine at the specified test speed (usually 1,500 to 2,000 rpm) and adjust the carbon pile for the highest possible ammeter reading.

6. Read the ammeter and proceed as follows:

   a. If the ammeter reading is within 10 amperes of the alternator rating, current output is okay.

   b. If the ammeter reading is within 10 amperes of the rated output, but the indicator lamp stays on, remove the alternator to test the rectifier bridge (diodes) and the field diode trio.

   c. If the current is not within 10 amperes of rated output, proceed with steps 7 and 8.

7. Bypass the regulator to apply full field current to the alternator, as follows:

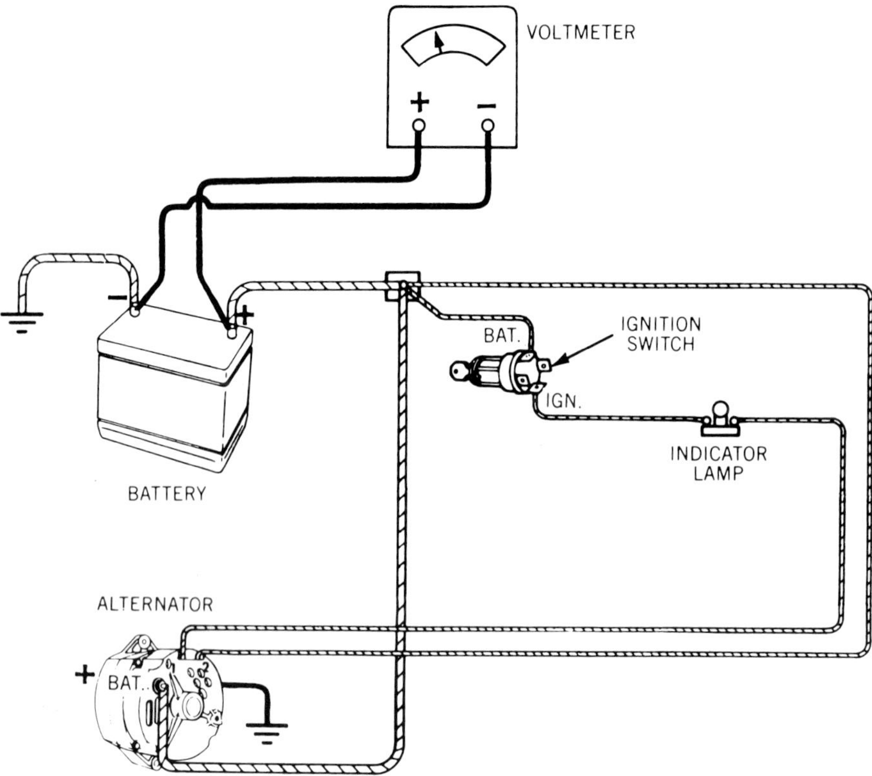

Figure 11-11. Measure regulated charging voltage at the battery. (Delco-Remy)

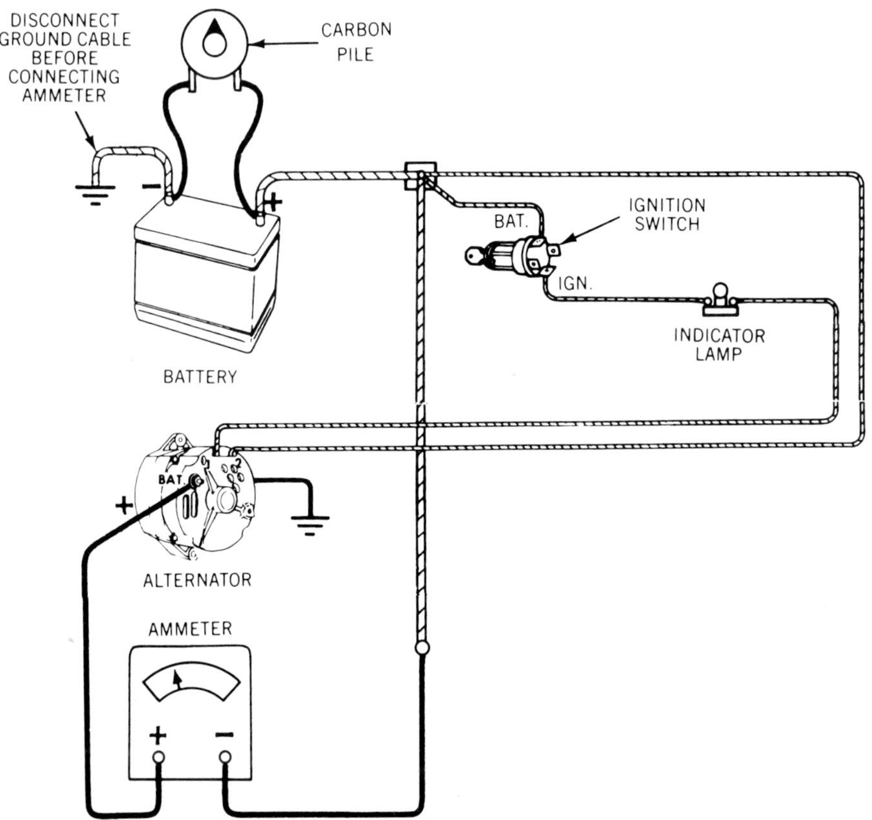

Figure 11-12. Delco-Remy 10-SI current output test. (Delco-Remy)

a. Insert a small screwdriver through the test hole in the rear of the alternator, figure 11-13.

b. Touch the regulator grounding tab and the alternator frame with the blade.

8. Repeat steps 4 and 5 and read the ammeter. If the reading is now within specifications, remove the alternator to replace the regulator. If the reading is still not within specifications, remove the alternator to test and service the stator, the rotor, or the rectifier bridge.

Delco-Remy CS System Testing

Delco-Remy recommends the following general diagnosis procedure for CS-series charging systems to locate the cause of an undercharge or overcharge condition, or an indicator lamp that does not operate properly. On vehicles so equipped, the indicator lamp will come on when the ignition is switched on, and go out once the engine starts.

Diagnosis Procedure. Figures 11-14 and 11-15 show the basic wiring diagrams for the two circuits used.

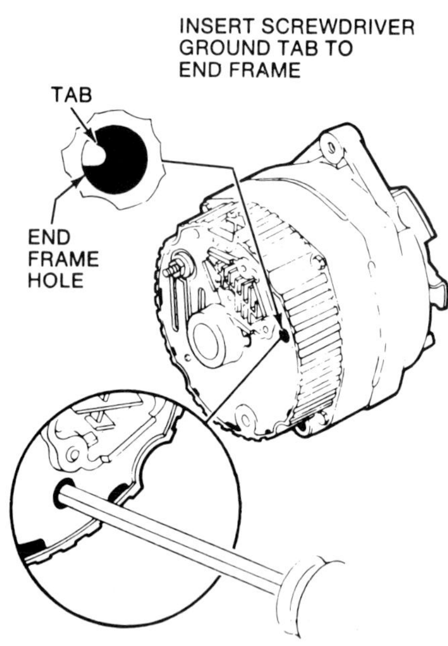

Figure 11-13. Bypass the Delco-Remy integral receptor by grounding it with a screwdriver. (Cadillac)

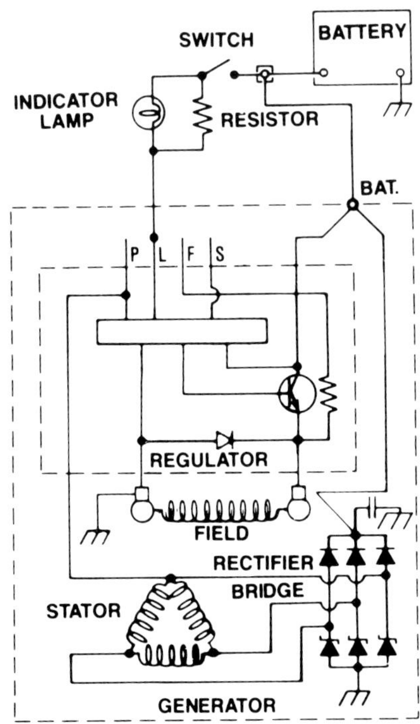

Figure 11-14. The basic CS charging system circuit using the L terminal. (GM)

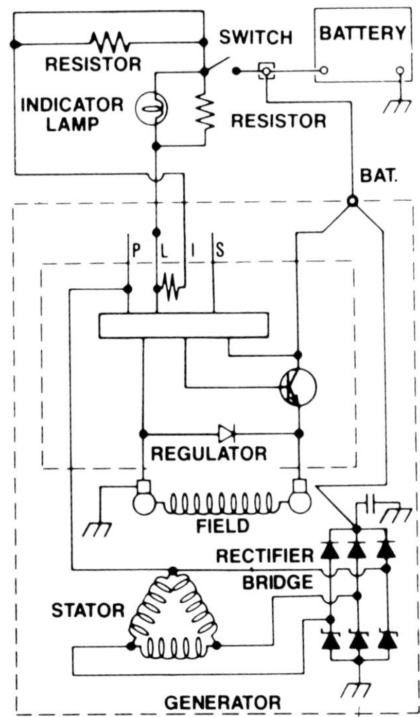

Figure 11-15. The basic CS charging system circuit using the L and I terminals. The I terminal may connect to the switch either through the resistor or directly. (GM)

If the vehicle does not have an indicator lamp, omit steps 4 through 6.

1. Test and charge the battery as required before checking the charging system.

2. Check drive belt tension and correct as required.

3. Visually inspect all circuit wiring for loose connections and other defects. Correct as required.

4. Turn the ignition switch on, but do not start the engine. The indicator lamp should light.

5. If the indicator lamp does not light in step 4, unplug the harness connector at the alternator and ground the L terminal lead in the harness with a 5-ampere fused jumper wire.

- **a.** If the lamp lights, repair or replace the alternator as required.
- **b.** If the lamp does not light, there is an open in the circuit between the grounded L terminal lead in the harness and the ignition switch. Locate and repair as required.

6. Reconnect the alternator harness at the alternator. Start the engine and run at a moderate speed (approximately 2,000 rpm). If the indicator lamp does not go off, shut the engine off. Turn the ignition switch back on and unplug the harness connector at the alternator.

- **a.** If the lamp goes off, repair or replace the alternator as required.
- **b.** If the lamp remains on, check for a grounded L terminal wire in the harness.

7. If the indicator lamp comes on during normal operation, or if the battery is consistently undercharged or overcharged:

- **a.** Unplug the wiring harness at the alternator.
- **b.** Turn the ignition switch on but do not start the engine.
- **c.** Connect a voltmeter between ground and the L terminal in the wiring harness. If the I terminal is used, jumper the voltmeter lead to the I terminal.
- **d.** Note the voltmeter scale. If no reading is shown, check for an open or grounded circuit between the harness terminal and the battery. Correct as required.

8. Reconnect the harness to the alternator. Make sure all electrical accessories are off. Start the engine and run at a moderate speed (approximately 2,000 rpm).

9. Measure battery voltage. If it exceeds 16.0 volts, repair or replace the alternator as required.

10. Connect an ammeter in the alternator output circuit (at the BAT terminal). Connect the voltmeter across the alternator and load the battery with a carbon pile until maximum ampere output is obtained with output voltage at 13.0 volts or greater. If the charging current is not within 15 amperes of the rated output, repair or replace the alternator as required.

CS System Trouble Codes. The electronic control module (ECM) on General Motors vehicles that also have a body computer module (BCM) can set a single trouble code (EO16) whenever system voltage is out of range (too low or too high). The ECM monitors system voltage on its ignition input circuit. On the Cadillac Allanté, the code will set if system voltage drops under 10 volts or exceeds 16 volts for 5 seconds or more with the engine running at 800 rpm or more. When the code is set, all system solenoids are disabled to prevent ECM damage and indicator lamps are illuminated to warn the driver. Other GM vehicles with a BCM respond in the same manner, but the code sets only when an excessive voltage condition exists.

The BCM also can set four trouble codes:

1. Code B409—charging system failure

2. Code B410—charging system problem

3. Code B411—battery voltage too low

4. Code B412—battery voltage too high

A B409 code can be set under a variety of conditions:

• When the voltage regulator sees an open or shorted field circuit, an overvoltage or undervoltage condition, or a broken drive belt, it causes the alternator to pull the I terminal circuit low. As soon as the BCM sees the circuit pulled low for 12 seconds above a specified engine speed (usually 500 or 600 rpm), it sets the code.

• When there is an open or a short to ground in the I terminal circuit, the alternator ceases operation, because the 12-volt signal from the BCM never reaches the I terminal.

• When there is an open or a short to ground in the F terminal circuit, the BCM does not receive its pulse width modulation (PWM) regulation signal from the alternator F terminal. This also can result from an open in the I terminal circuit. Since the alternator is not working, there is no PWM signal to send the BCM on the F terminal circuit.

Although it indicates a general charging system problem, a B410 code is more complicated, because it sets whenever the "alternator malfunction" flag is set in the BCM. The alternator malfunction flag is a condition based on engine speed, battery voltage, and commanded output of the alternator. It works as follows:

• Whenever the alternator is at maximum output, engine speed exceeds a calibrated value and battery voltage is above 16 volts or below 10 volts, a counter in the computer starts testing for a problem.

• Whenever the alternator is under full output but battery voltage is under a specified threshold, the counter also looks for a problem.

• If the counter exceeds a calibrated limit, it sets the alternator malfunction flag.

• Once the malfunction flag has been set, it can be cleared only after the alternator comes back within the specified regulation and battery voltage exceeds a specified limit.

Codes B411 and B412 are similar to the EO16 code set by the ECM in that they are set under the following conditions:

• The ignition is on.

• Engine speed exceeds 800 rpm.

• Reference voltage at the BCM ignition input is below 10.6 volts (code B411) or above 16 volts (code B412).

• All conditions exist continuously for at least 5 seconds.

When all of these conditions are present, the cruise control is disengaged until they come back within acceptable limits. Since the conditions which cause either of these codes to set might also cause false codes to set, these codes should be corrected before attempting to correct any other codes that might be present. Under certain circumstances, correcting a code B411 or B412 may also remove another code that is present. When this happens, the other code is a false one.

Direct testing of the voltage-regulating circuitry is generally not required unless one or more of the above codes are displayed. When the service diagnostic code mode is entered, ECM codes will be displayed first, followed by BCM codes. Stored codes are displayed in numerical order for 2 seconds. If no codes are stored, the display will flash a NO ECM CODES or NO BCM CODES message.

Entering the service diagnostic mode is essentially the same on vehicles currently using the system, but the display units differ:

• *Buick*—Codes are displayed on the graphic control center (GCC). To enter the diagnostic mode, turn the ignition on. Touch the OFF and WARM pads on the climate control page at the same time and hold depressed until the service mode page is brought up on the CRT or a double beep is heard. To exit the diagnostic mode, depress the EXIT pad.

• *Cadillac*—Codes are displayed on the climate control driver information center (CCDIC). To enter the diagnostic mode, turn the ignition on. Touch the OFF and WARM pads on the climate control panel at the same time and hold depressed until the segment check appears on the instrument panel cluster (IPC) and the CCDIC. To exit the diagnostic mode, depress the RESET button on the driver information center.

• *Oldsmobile*—Codes are displayed on the electronic climate control (ECC) panel. To enter the diagnostic mode, turn the ignition on. Touch the OFF and WARM pads on the ECC panel at the same time and hold depressed until the segment check appears on the instrument panel cluster (IPC) and the EEC. To exit the diagnostic mode, depress the BI-LEV button on the ECC.

A trouble code indicates a problem within the system. Individual circuits must be tested to pinpoint the exact cause. Since circuit tests and connector terminals vary according to car model and manufacturer, you will need to consult the proper service publications for detailed test information.

Ford-Motorcraft System Testing

Ford recommends the following voltmeter tests for Motorcraft alternators with either solid-state or electromechanical regulators. Ford does not recommend current output tests for vehicles with electronic ignition or electronic regulators. You can measure charging current with an inductive ammeter, however.

Charging Voltage and Regulator Tests

1. Connect a voltmeter across the battery, figure 11-16.

2. With the ignition off, note the battery voltage reading (approximately 12 to 12.6 volts). If the battery is discharged, charge it with a battery charger before testing alternator charging voltage.

3. Run the engine at 1,500 rpm with no accessory load on the electrical system and read the voltmeter. Voltage should be 1 to 2 volts above battery voltage.

4. Run the engine at 2,000 rpm and turn on the high-beam headlamps, the air conditioner, and other accessories to load the system. Read the voltme-

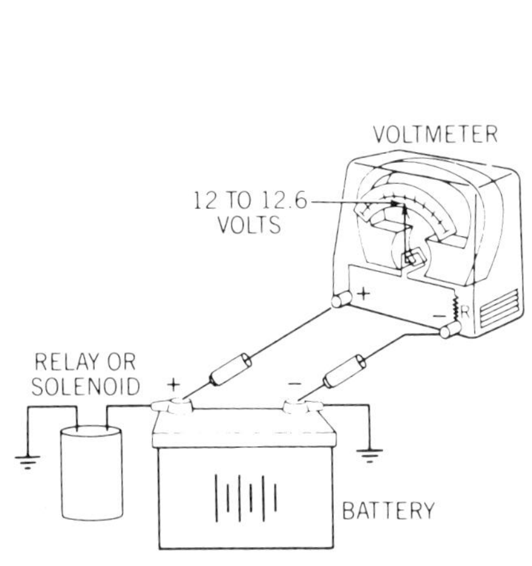

Figure 11-16. Measuring battery voltage.

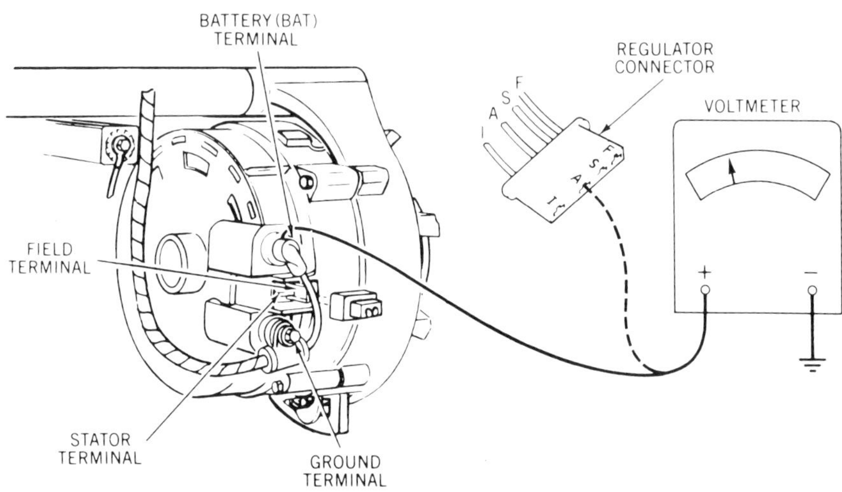

Figure 11-17. Test for battery voltage at the alternator and at the regulator A terminal. (Ford)

ter. Voltage should be about 0.5 volt above open-circuit battery voltage.

5. Compare voltage readings in steps 2, 3, and 4 to the normal values below:

- **a.** Open-circuit battery voltage—12 to 12.6
- **b.** No-load charging voltage—1 to 2 volts above open-circuit battery voltage
- **c.** Loaded charging voltage—0.5 volt above open-circuit battery voltage.

If voltmeter readings are out of limits, proceed with steps 6 through 14.

6. If charging voltage is higher than normal, check for high resistance at all ground connections for the alternator, the regulator, and the battery. Repeat steps 1 through 5.

7. If voltage is still higher than normal, disconnect the regulator connector and repeat steps 1 through 5. If this fixes the problem, replace the regulator.

8. If voltage remains higher than normal with the regulator disconnected, test for a short circuit in the wiring harness between the alternator and the regulator.

9. Test for battery voltage at the alternator BAT terminal and the regulator connector A terminal, figure 11-17. If no voltage is present, the circuit is open.

10. If loaded charging voltage is not at least 0.5 volt above battery open-circuit voltage, disconnect the regulator connector and connect an ohmmeter between the F terminal of the connector and ground to measure field resistance, figure 11-18. Resistance should be 4 to 250 ohms. If it is less than 4 ohms, the field circuit is grounded. If it is more than 250 ohms, the field circuit is open.

11. Disconnect the regulator connector and connect a jumper wire between the connector A and F terminals, figure 11-19. Repeat step 4. If loaded voltage is 0.5 volt or more above battery voltage, the regulator or the wiring harness is bad.

12. If a low-voltage condition still exists, remove the jumper from the regulator connector and leave the connector disconnected.

13. Connect a jumper wire between the alternator BAT and field (FLD) terminals, figure 11-20, to bypass the regulator and apply full field current.

14. Repeat step 4 to measure loaded system voltage. Read the voltmeter and proceed as follows:

- **a.** If loaded voltage is now within limits, repair the wiring between the alternator and the regulator.
- **b.** If voltage is still low, move the voltmeter + lead to the BAT terminal. If you read battery voltage, repair the alternator. If you read no voltage, repair the BAT output wiring.

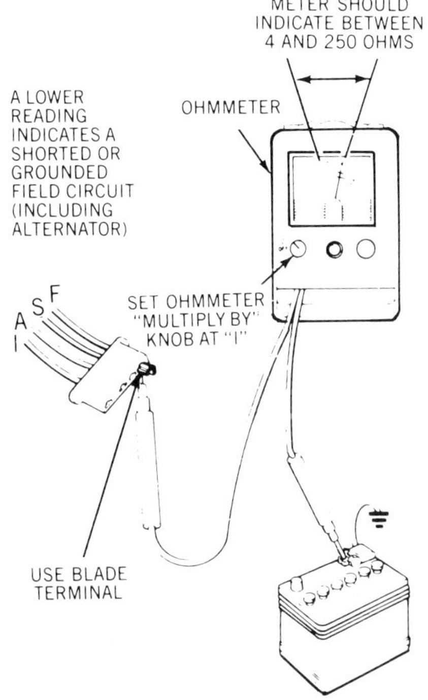

Figure 11-18. Measuring field resistance of a Motorcraft charging system. (Ford)

Regulator Circuit Test. Ford regulator circuit tests vary depending on whether the vehicle has an indicator lamp or an instrument panel ammeter. With the ignition off, disconnect the regulator connector and proceed as follows:

1. For an ammeter circuit, connect your voltmeter + lead to the connector S terminal and the − lead to ground, figure 11-21.

2. Turn the ignition on; *do not start the engine*. Read the voltmeter:

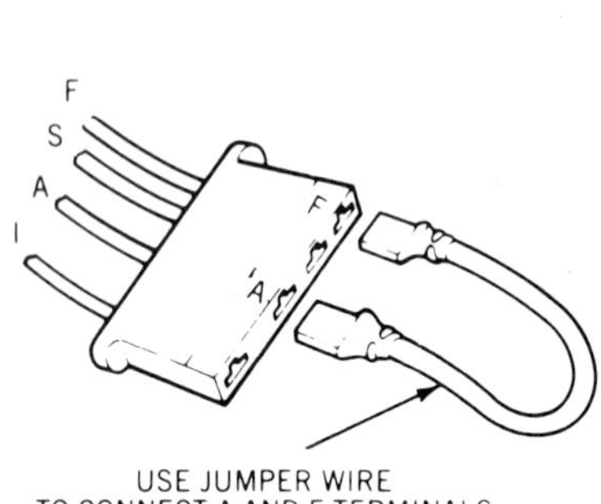

Figure 11-19. Applying full field current to the alternator rotor at the regulator connector. (Ford)

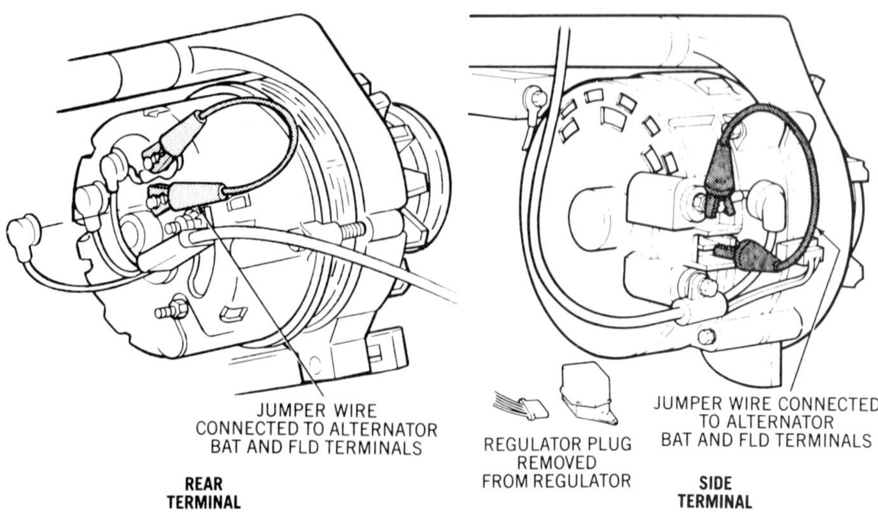

Figure 11-20. Applying full field current to the rotor at the alternator. (Ford)

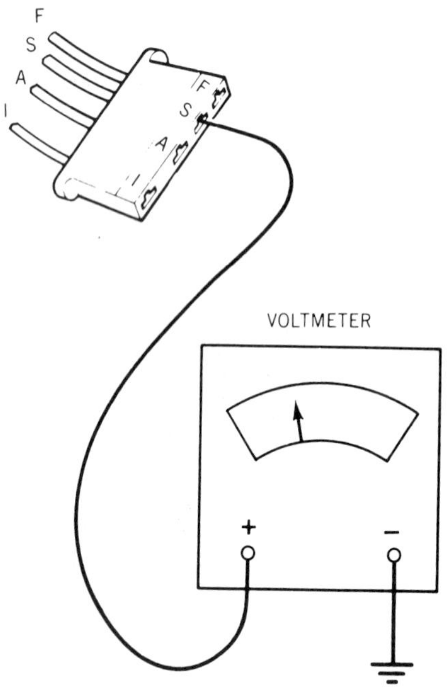

Figure 11-21. Regulator circuit test for a Motorcraft system with an ammeter. (Ford)

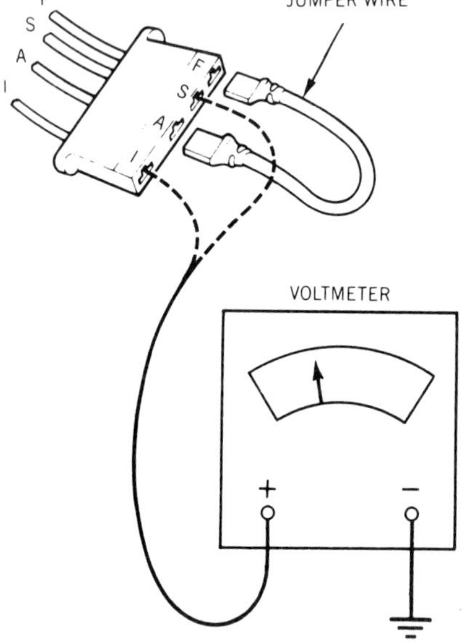

Figure 11-22. Regulator circuit test for a Motorcraft system with an indicator lamp. (Ford)

a. If the meter shows approximately battery voltage, the regulator circuit is okay.

b. If the meter shows no voltage, find and repair the open circuit in the wiring from the ignition switch to the connector S terminal.

3. For an indicator lamp circuit, connect a jumper wire between the connector A and F terminals, figure 11-22.

4. Start and idle the engine.

5. Connect your voltmeter − lead to ground and alternately touch the + lead to the S and I terminals of the connector, figure 11-22. Read the meter:

a. If the I-terminal reading is about 13 volts and the S-terminal reading is about 6 or 7 volts, the regulator circuit is okay.

b. If either reading is zero, repair the wiring to that terminal.

c. If voltage is within limits but charging voltage is not, replace the regulator and repeat the charging voltage tests.

Diode Tests. The voltmeter readings at the S terminal of the regulator connector (step 5 of the previous test) can indicate alternator diode condition, as follows. Disconnect the electric choke before making these tests:

- If the meter shows approximately one-half of battery voltage, the diodes are okay.
- If the meter shows approximately 1.5 volts less than battery voltage, the alternator has a shorted + diode.
- If the meter shows approximately 1.5 volts, the alternator has a shorted − diode or a grounded stator winding.
- If the meter shows approximately 1.0 to 1.5 volts *less than half battery voltage,* the alternator has an open + diode.
- If the meter shows approximately 1.0 to 1.5 volts *more than half battery voltage,* the alternator has an open − diode.

Ford-Motorcraft IAR System Testing

Ford recommends the following circuit test procedure for charging systems with IAR alternators. A field circuit drain check is also recommended to determine the cause of current drain through the alternator field circuit when the ignition is off.

Charging Circuit Test

1. Connect a voltmeter between the battery terminals and take a battery voltage reading.

2. Make sure all accessories are off. Turn the ignition on and start the engine. Run the engine at 1,500 rpm and note the voltmeter reading.

- **a.** If the reading has increased by more than 2 volts over the reading in step 1, go to step 13.
- **b.** If the reading has not increased, go to step 5.
- **c.** If the reading has increased by less than 2 volts, continue with step 3.

3. Turn all electrical accessories on (high position), then increase engine speed to 2,000 rpm and note the voltmeter reading.

- **a.** If the reading does not increase by 0.5 volt, go to step 5.
- **b.** If the reading increases by 0.5 volt or more, proceed with step 4.

4. Shut the engine off. Connect a 12-volt test lamp in series with the battery positive cable to check for a current drain. Locate drains and correct as required.

5. With the ignition switch off, disconnect the harness connector at the regulator on the back of the alternator.

6. Connect an ohmmeter between the F terminal and the A terminal screws on the regulator, figure 11-23.

- **a.** If the ohmmeter reading is 2.4 ohms or less, remove the alternator from the engine and replace the regulator. You should also check the alternator for a shorted field circuit or rotor.
- **b.** If the ohmmeter reading exceeds 2.4 ohms, proceed with the test sequence.

7. Reconnect the harness connector to the regulator. Connect the voltmeter – lead to the rear housing of the alternator. Briefly touch the voltmeter + lead to the A terminal screw on the regulator and note the meter reading.

- **a.** If the voltmeter reading is the same as in step 1, continue testing.
- **b.** If the voltmeter reading differs from that in step 1, repair the A circuit wiring.

8. Make sure the ignition switch is off. Connect the voltmeter between the F terminal on the regulator and ground.

- **a.** If the voltmeter reading is the same as in step 1, continue testing.
- **b.** If there is no voltmeter reading, look for an open or grounded field circuit in the alternator.

9. Leave the voltmeter connected as in step 8. Turn the ignition switch on but do not start the engine.

- **a.** If the regulator F terminal voltage is 1.5 volts or less, repair the wiring between the alternator and starter relay.
- **b.** If the F terminal voltage exceeds 1.5 volts, continue testing.

10. Turn the ignition switch off. Disconnect the harness connector from the regulator.

11. Install a jumper wire between the harness connector and regulator A terminals. Install another jumper wire between the alternator housing and regulator F terminal, figure 11-24.

12. Turn the ignition on. Start the engine and run at idle. Connect the voltmeter – lead to the battery – terminal. Touch the voltmeter + lead first to the S terminal and then to the I terminal in the harness connector and note each reading.

- **a.** If the reading at the S terminal is about one-half that shown at the I terminal, remove the jumper wires. Shut the engine off, install a new regulator and reconnect the regulator plug.
- **b.** If the voltage at the S terminal is more or less than about one-half

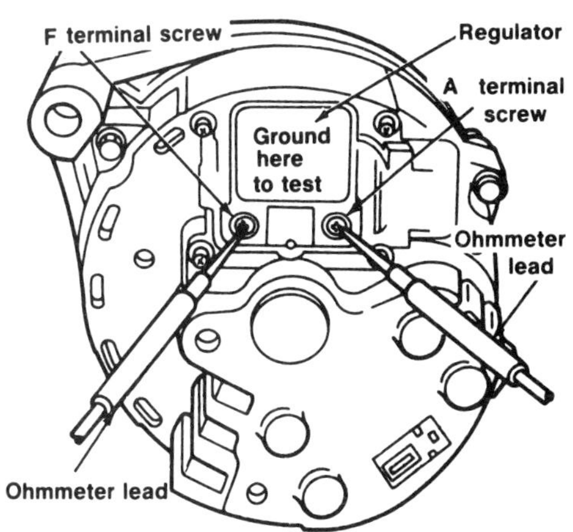

Figure 11-23. Motorcraft IAR under-voltage test points. (Ford)

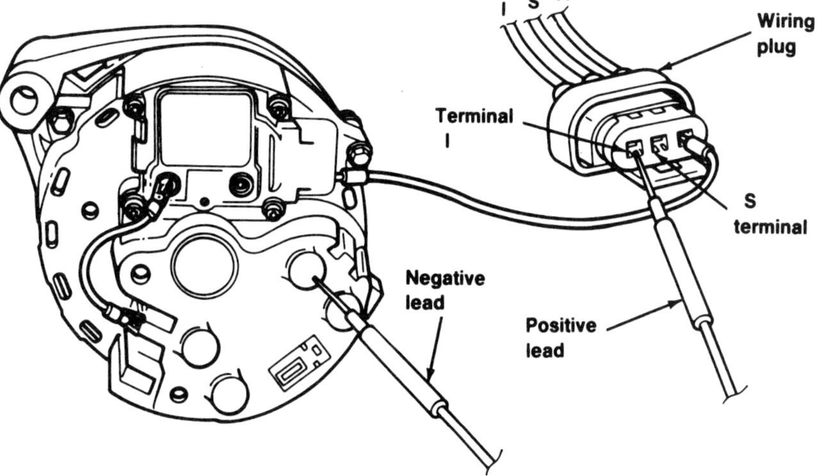

Figure 11-24. Motorcraft IAR regulator S or I circuit test points. (Ford)

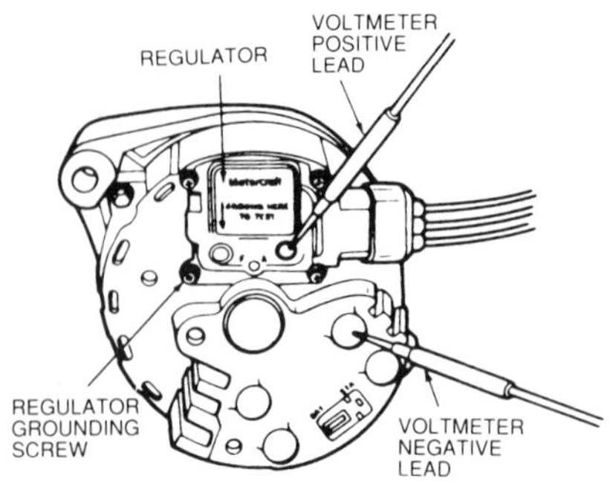

Figure 11-25. Motorcraft IAR over-voltage test points. (Ford)

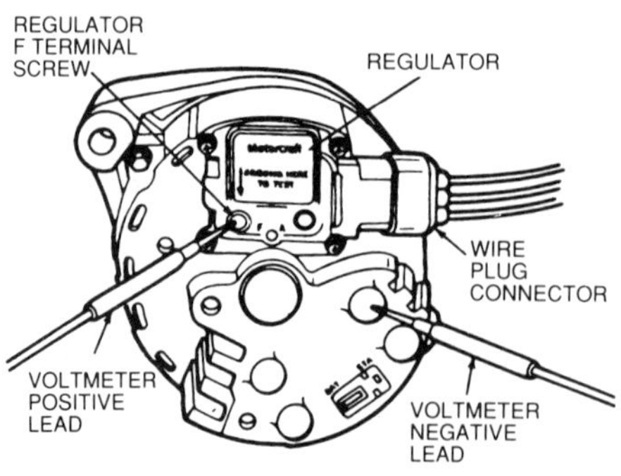

Figure 11-26. Motorcraft IAR I terminal field circuit drain test points. (Ford)

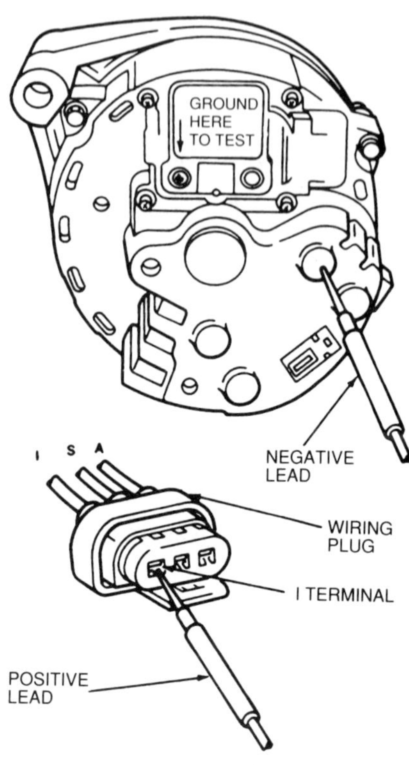

Figure 11-27. Motorcraft IAR I terminal field circuit drain test points. (Ford)

that shown at the I terminal, remove the alternator for further testing.

13. Turn the ignition on but do not start the engine. Connect the voltmeter − lead to the rear housing of the alternator. Connect the voltmeter + lead first to the starter relay output terminal, then to the regulator A terminal, figure 11-25. Record the difference in voltmeter readings.

a. If the readings differ by 0.5 volt or less, continue testing.

b. If the readings differ by more than 0.5 volt, repair the A circuit wiring.

Field Circuit Drain Check. Perform all steps with the voltmeter − lead connected to the rear housing of the alternator.

1. Make sure the ignition is off. Probe the regulator F terminal with the voltmeter + lead, figure 11-26.

a. If battery voltage is shown, the system is functioning properly and no further testing is required.

b. If the reading is less than battery voltage, continue testing.

2. Disconnect the harness connector at the regulator. Probe the connector I terminal with the voltmeter + lead, figure 11-27. If any voltage reading is shown, trace the I circuit lead between the connector and the ignition switch to find and correct the cause of the drain.

3. If no voltage reading is obtained in step 2, probe the connector S terminal with the voltmeter positive lead.

a. If no voltage is obtained, install a new regulator and repeat this step. If there is still no voltage shown, the alternator rectifier assembly is defective.

b. If voltage is now shown, trace the S circuit lead to find and correct the cause of the drain.

Chrysler System Testing with Electronic Regulator

For charging systems with electronic regulators, Chrysler recommends a circuit resistance test, a current output test, a field current draw test, and a regulator test.

Circuit Resistance Test

1. Disconnect the battery ground cable before connecting test equipment.

2. Disconnect the output connector from the alternator BAT terminal.

3. Connect your ammeter + lead to the BAT terminal and the meter − lead to the disconnected wire, figure 11-28.

4. Connect your voltmeter + lead to the output wire and the − lead to the battery + terminal, figure 11-28.

5. Disconnect the green field connector from one alternator field (FLD) terminal and insulate the wire.

6. Connect a jumper wire between the alternator FLD terminal and ground to apply full field current.

7. Connect a carbon pile across the battery and reconnect the battery ground cable.

Caution. To avoid high voltage, reduce engine speed to idle immediately after starting.

8. Start and idle the engine.

9. Adjust the carbon pile to maintain a steady 20 amperes of charging current at the battery.

10. Read the voltmeter to measure voltage drop across the insulated side of the output circuit:

a. If the voltage drop is 0.7 volt or less, circuit resistance is normal.

b. If the voltage drop is higher than 0.7 volt, move the voltmeter + lead through the circuit from the BAT connector toward the battery to locate the high resistance.

Current Output Test. Connect your test equipment as in steps 1 through 7 of the circuit resistance test, but connect the voltmeter + lead to the alternator BAT terminal and the − lead to

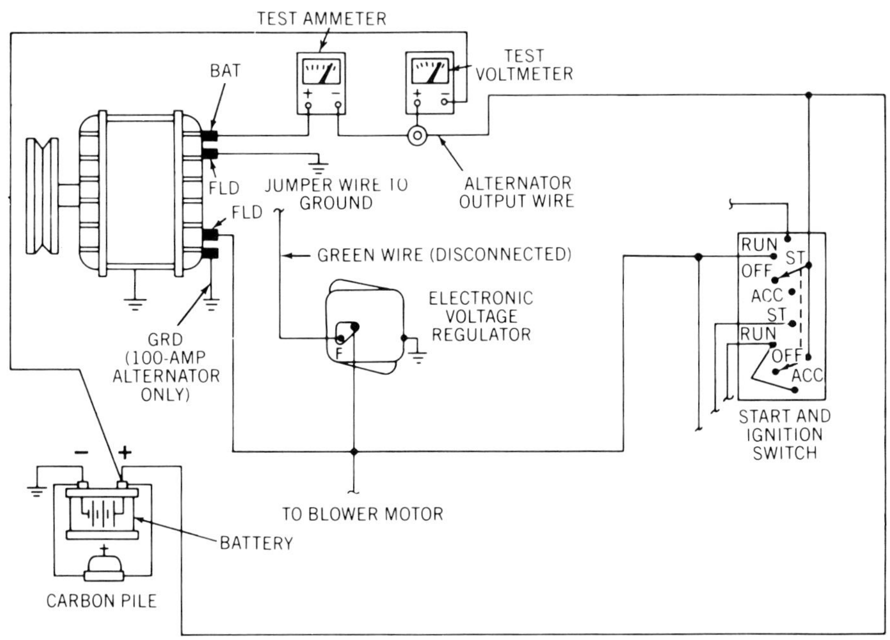

Figure 11-28. Chrysler charging circuit resistance test connections. (Chrysler)

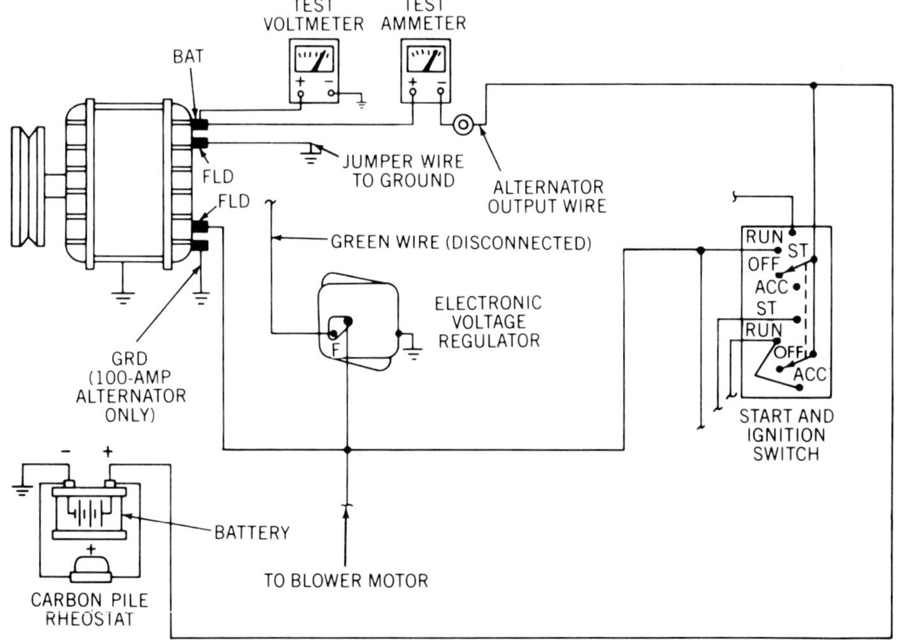

Figure 11-29. Chrysler charging current output test connections. (Chrysler)

ground, figure 11-29. Proceed as follows:

> **Caution.** To avoid high voltage, reduce engine speed to idle immediately after starting.

1. Start and idle the engine.
2. Adjust the carbon pile for maximum current and the engine speed as follows:
   a. 100- or 117-ampere alternators—900 rpm and 13 volts
   b. All other alternators—1,250 rpm and 15 volts

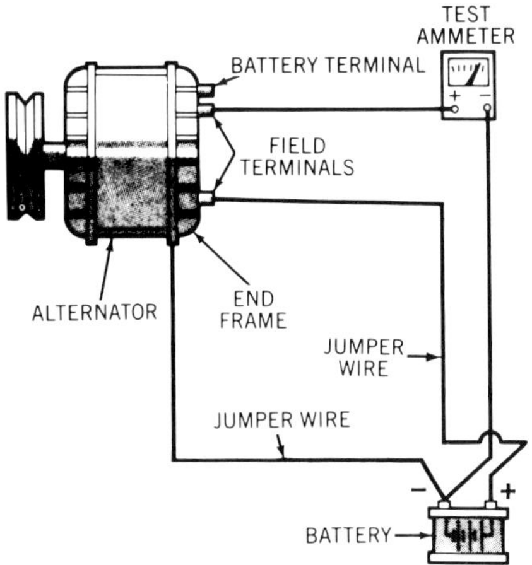

Figure 11-30. Chrysler field current draw test connections. (Chrysler)

3. Read the ammeter and compare output current to specifications. If current is within specifications, the alternator and the regulator are okay. If current is not within specifications, proceed with field current and voltage regulator tests.

Field Current Draw Test

1. Disconnect the battery ground cable.
2. Disconnect both field leads from the alternator, figure 11-30.
3. Connect your ammeter + lead to one FLD terminal on the alternator and the – lead to the battery – terminal.
4. Connect a jumper wire from the other FLD terminal to the battery + terminal.
5. Reconnect the battery ground cable and read the ammeter. You are applying full field current to the rotor.
6. Rotate the alternator pulley by hand and watch for variations in the current reading. Small variations are normal. Large variations indicate problems with the brushes or slip-rings.
7. If current is within specifications, the field is okay. If current is out of limits, remove the alternator for further testing.

Voltage Regulator Test. For a general test of Chrysler's isolated field

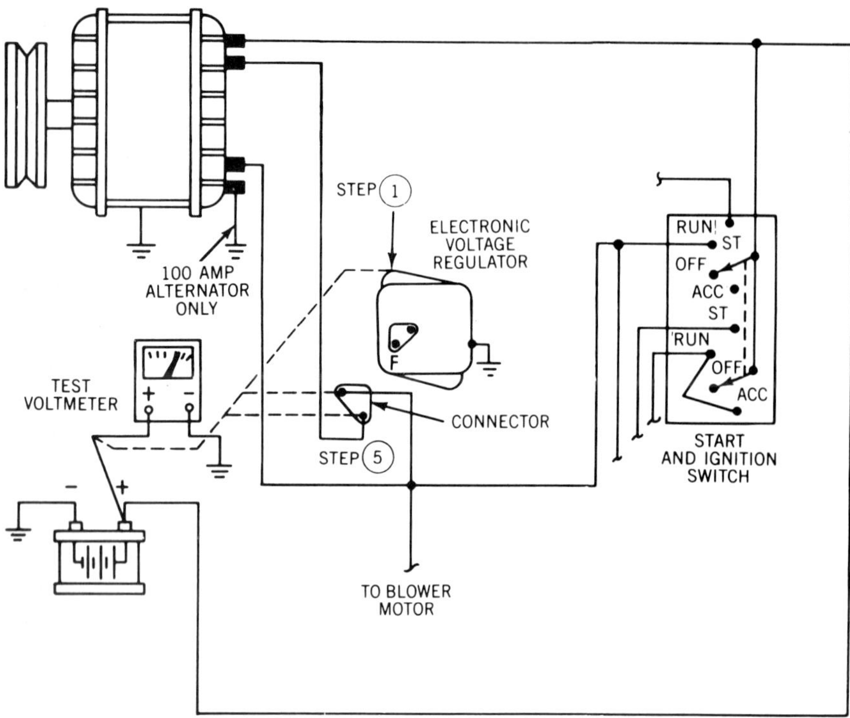

Figure 11-31. Chrysler voltage regulator test connections. (Chrysler)

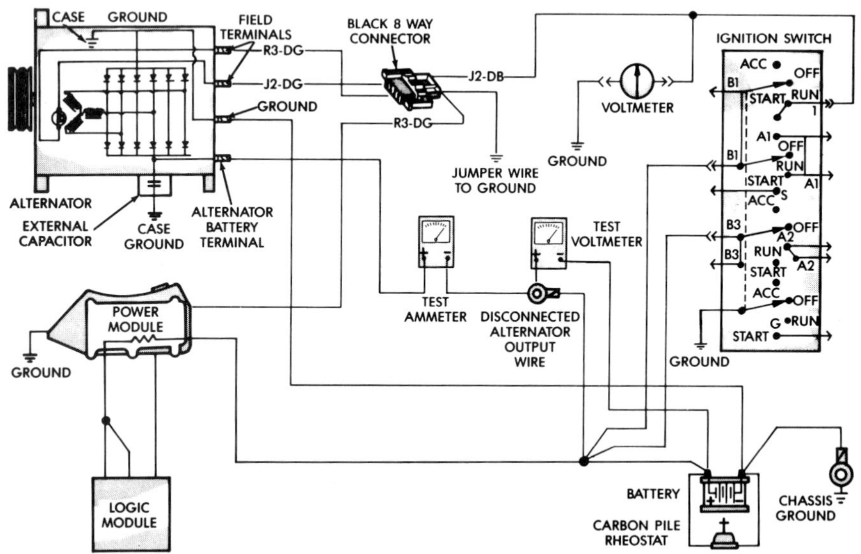

Figure 11-32. When checking the Chrysler computer-controlled charging system, ground terminal R3 at the black connector. (Chrysler)

electronic regulator, measure charging voltage across the battery at 1,250 engine rpm with all accessories off. If voltage is within specifications, the regulator is okay. If voltage is out of limits, proceed as follows:

1. Connect the voltmeter + lead to the regulator housing and the − lead to ground, figure 11-31.

2. With the engine running, read the voltage drop (usually 0.1 to 0.2 volt) to measure ground resistance.

3. If voltage drop is high, clean and tighten the ground connection. If regulated charging voltage is then above specifications, replace the regulator.

4. Turn the ignition off and disconnect the connector from the regulator.

5. Turn the ignition on and measure voltage between the red and the green wire terminals of the connector and ground, figure 11-31. Interpret voltmeter readings as follows:

- **a.** No voltage at either terminal indicates an open circuit in the wiring.
- **b.** If battery voltage is present at both terminals but charging system voltage remains out of limits, replace the regulator.

Chrysler Computer-Controlled Charging System Testing

The circuit resistance and current output tests used with this computer-regulated system are similar to those tests on systems using external transistorized regulators. There are, however, two differences that should be noted:

1. The field circuit should be grounded at terminal R3 on the dash side of the black 8-way connector, figure 11-32. This replaces step 5 of the resistance and current tests for those Chrysler systems using a separate transistorized regulator. Be sure not to ground the blue wire at terminal J2.

2. The maximum circuit drop allowed across the output circuit is 0.05 volt instead of the 0.7 volt for a circuit using a separate regulator.

Direct testing of the voltage regulating circuitry is generally not required unless the computer detects a problem in the charging system and records fault codes in the system memory. Some of the codes involved will turn on the POWER LIMITED, POWER LOSS, or CHECK ENGINE lamp in the instrument cluster; others will not. Charging system fault codes can be checked with a special tester (described in chapter 4) that plugs into the engine compartment diagnostic connector. If the special tester is not available, fault codes can be checked as follows:

1. Make sure the battery has a full charge before attempting to test the charging system. If it does not, charge as required.

2. Turn the ignition switch on-off-on-off-on within a 5-second period to enter the diagnostic mode.

3. Watch the POWER LIMITED, POWER LOSS, or CHECK ENGINE lamp in the instrument cluster. It should come on for two seconds as a bulb and system check.

4. If fault codes are recorded in the system memory, the lamp will display the 2-digit numbers as a series of flashes, with longer pauses between the digits of each code.

The following table lists the charging system fault codes. If more than one code has been recorded, the lamp will remain off for four seconds between the codes. Any codes obtained other than those listed in the table deal with problems in other parts of the engine control system.

Code	*Lamp On*	*Fault*
16	Yes	Battery sensing voltage is below 4 or between 7.5 and 8.5 volts for more than 20 seconds
41	No	There is a problem in the field circuit or improper field control
44	No	The battery temperature sensing signal is less than 0.04 volt or greater than 4.9 volts
46	Yes	Charging system voltage exceeds the desired regulating voltage by 1 volt for more than 20 seconds
47	No	Charging system voltage is less than the desired regulating voltage by 1 volt for more than 20 seconds

A fault code indicates a problem within the system. Individual circuits must be tested to pinpoint the exact cause. Since circuit tests and connector terminals vary somewhat between car models according to the engine control system functions, you will need to consult the proper service publications for detailed test information.

Charging System Tests with an Oscilloscope

The same oscilloscope that you use for ignition testing can be used to analyze alternator voltage. Alternator scope patterns will help you detect a bad diode or rectifier bridge and problems in the stator or rotor without removing the alternator.

1. Connect the scope + voltage lead to the battery + terminal or to the alternator BAT terminal. Use a jumper if necessary to connect the scope without disconnecting the output terminal.

2. Connect the scope – voltage lead to ground. If the scope includes an ammeter, you can connect it to measure charging current while analyzing the scope patterns.

3. Adjust the scope to the low-voltage scale (usually 30 or 50 volts). Start and run the engine at a steady 1,500 to 2,000 rpm.

4. Compare the scope pattern to figures 11-33 through 11-41.

Figure 11-33 shows a normal alternator pattern, which is a steady rectified voltage trace with a slight ac ripple. Figures 11-34 through 11-41 show various abnormal patterns caused by faulty diodes or stator or rotor problems. Actual patterns may vary slightly from one scope to another and from car to car. These illustrations and your experience will help you recognize specific defects shown on your scope.

Generally, a shorted diode affects the pattern more than an open diode does. An open diode affects only one output phase. A shorted diode affects two phases by letting current flow back through the stator to oppose the next phase output.

ALTERNATOR SERVICE

The preceding charging system tests may isolate a problem in the system wiring and switches, a relay, the regulator, or the alternator. When you isolate a problem to the alternator, you may replace it, or you may remove it for further testing and repair. Many alternator repairs such as replacing a rectifier bridge or brushes and cleaning sliprings are simple and can be done quickly and economically. Additionally, alternator repair is covered under new-vehicle warranty by some carmakers.

Alternator service includes some or all of the following:

- Alternator removal
- Alternator overhaul or repair
- Alternator bench tests as part of repair
- Alternator installation

Alternator Removal

As with a starter motor, alternator removal is not difficult, but getting access to the alternator can be. You may have to remove or relocate other accessories, such as:

- The air pump
- The air conditioning compressor and hoses
- The power steering pump
- The air cleaner and air intake ducts
- A vacuum pump on a diesel engine

Tag all wires and drive belts that you remove so that you will reinstall them correctly. Observe all engine and electrical safety precautions.

1. Disconnect battery ground cable.

2. Remove or loosen and relocate engine accessories as required for alternator access.

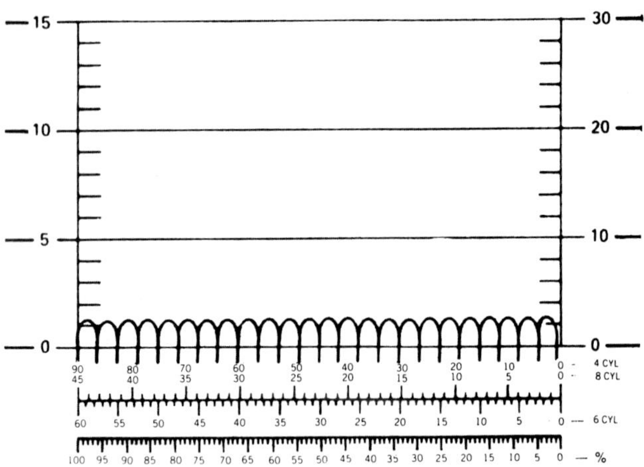

Figure 11-33. Typical normal alternator oscilloscope pattern.

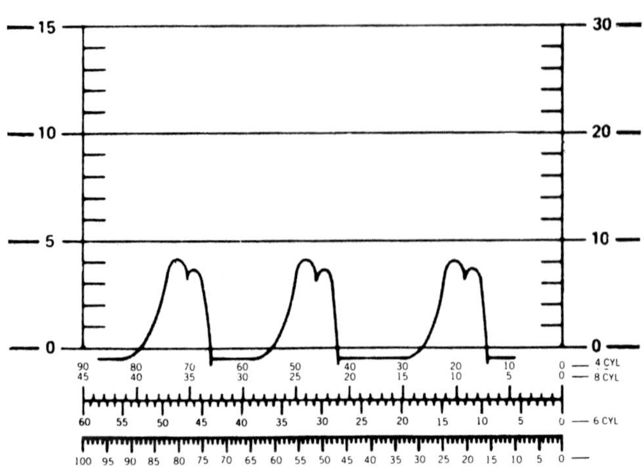

Figure 11-36. Two open diodes of the same polarity.

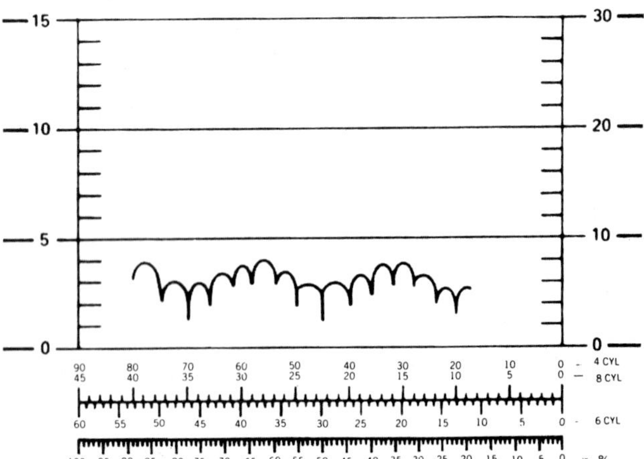

Figure 11-34. Weak diode creating high resistance.

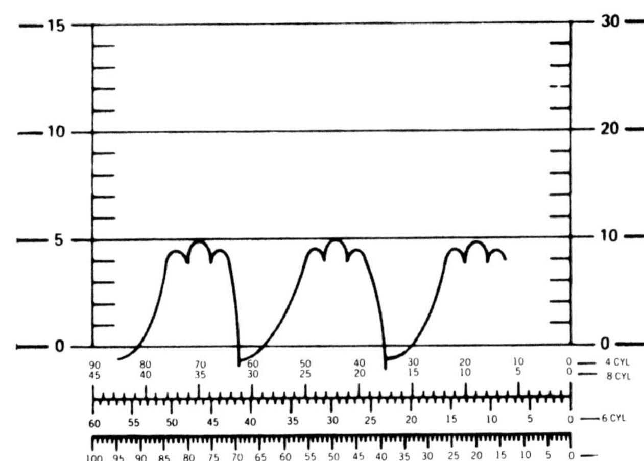

Figure 11-37. Two open diodes connected to different windings.

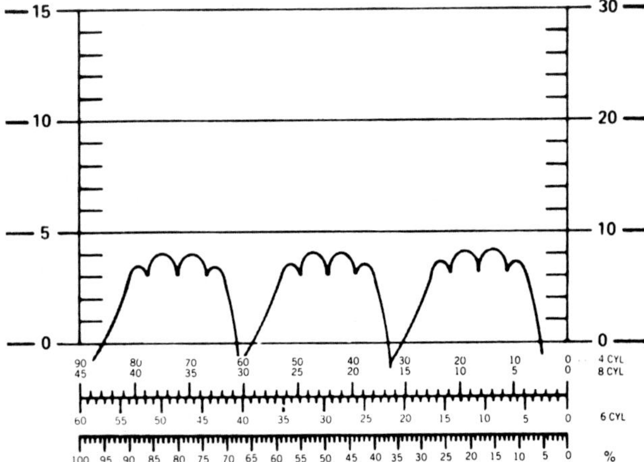

Figure 11-35. One open diode.

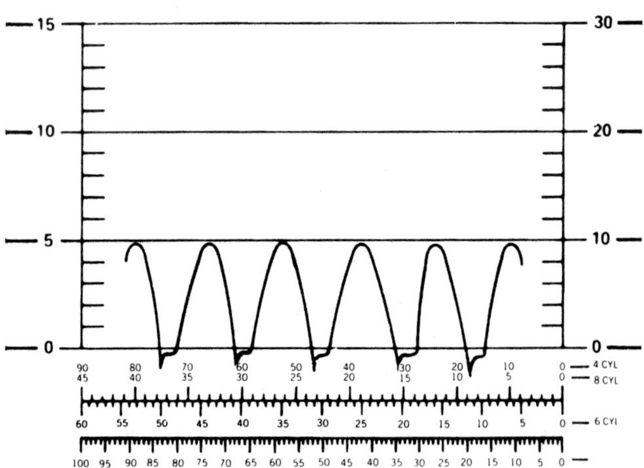

Figure 11-38. Two open diodes, one positive and one negative.

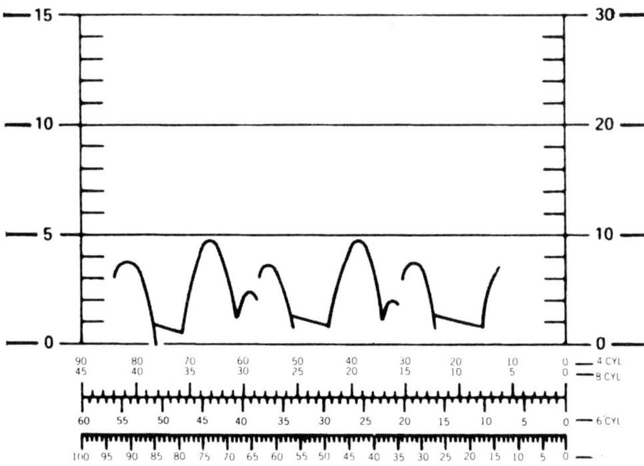

Figure 11-39. One shorted diode.

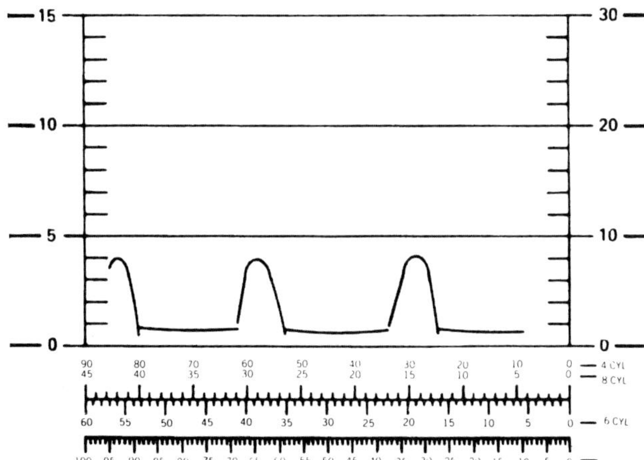

Figure 11-40. Two shorted diodes of the same polarity.

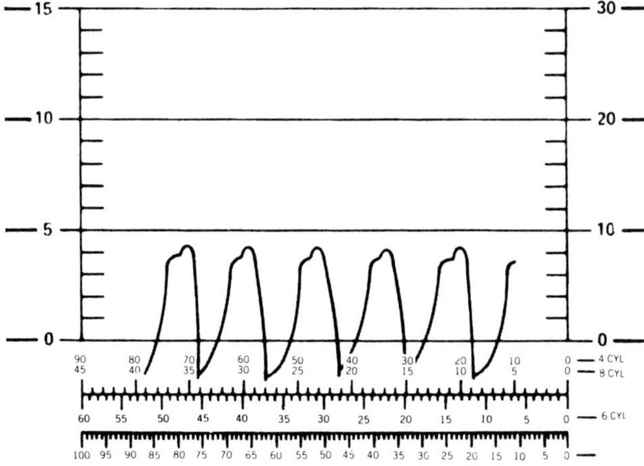

Figure 11-41. Shorted stator windings.

3. Disconnect all wires from the alternator. Tag the connectors for reinstallation.

4. Loosen the adjusting bolt that holds the alternator to its adjusting bracket, figure 11-42.

5. Loosen the lower mounting bolt on the alternator.

6. Carefully push the alternator toward the engine until you can remove the drive belt from the pulleys.

7. Remove the alternator adjusting and mounting bolts. Note the location of washers and spacers, figure 11-43. Remove the alternator.

Alternator Bench Tests

You can use a group of standard tests on all alternators to pinpoint problems and to determine if an alternator can be repaired. You will use an ohmmeter or a self-powered test lamp and make these tests during alternator disassembly. These tests consist of:

- Stator continuity and stator ground
- Rotor continuity and rotor ground
- Test for a shorted capacitor
- Tests for shorted or open diodes or rectifier

An ohmmeter that reads low or zero resistance or a test lamp that lights indicates continuity. An ohmmeter that reads high or infinite resistance or a test lamp that does not light indicates an open circuit or high resistance. A good diode will show continuity with an ohmmeter or test lamp connected in one direction and an open circuit with the meter or lamp connected in the other direction.

Although a test lamp will work to test for short circuits in the rotor or stator, an ohmmeter is more reliable. A high-resistance short may not cause the lamp to light but will still cause alternator problems. Set your ohmmeter on the X10,000 scale to test for high-resistance shorts. The illustrations that follow show both ohmmeters and test lamps. Test principles are the same with either device.

You can make the tests at several convenient points during alternator

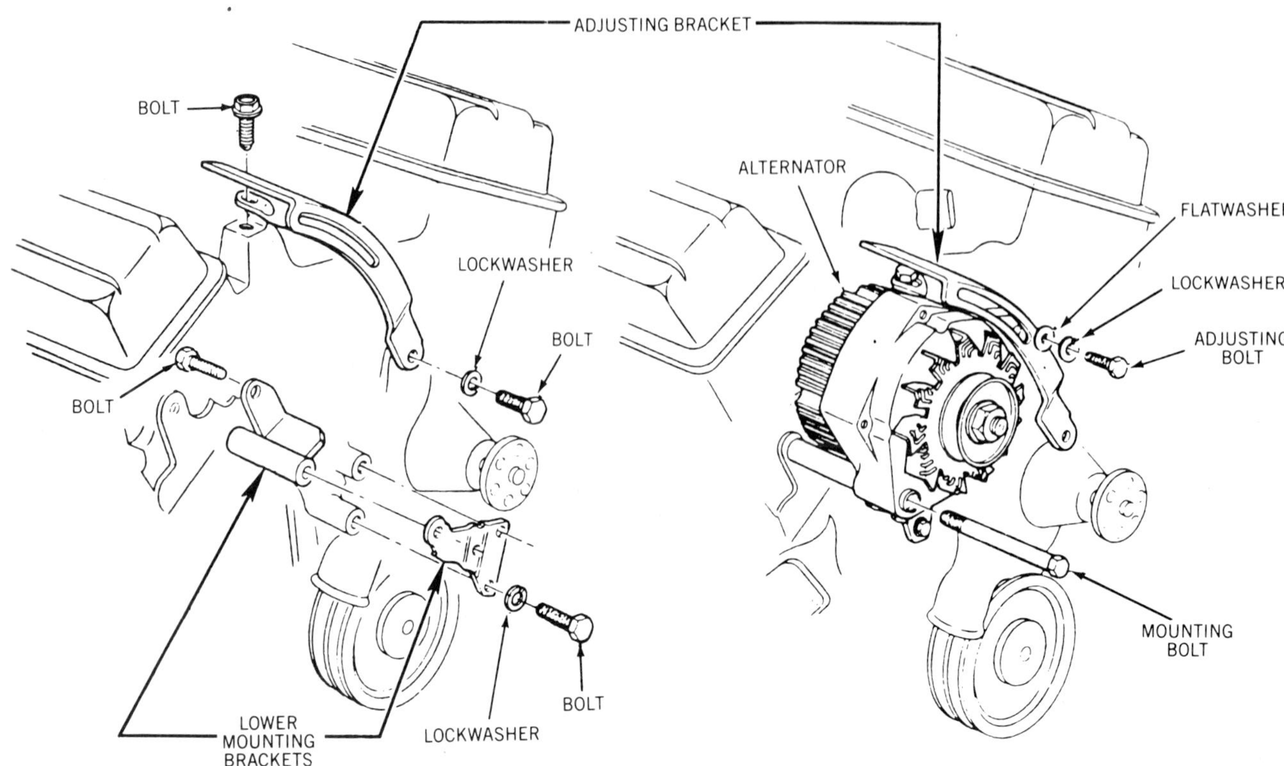

Figure 11-42. Typical alternator installation on a V-type engine. (Chevrolet)

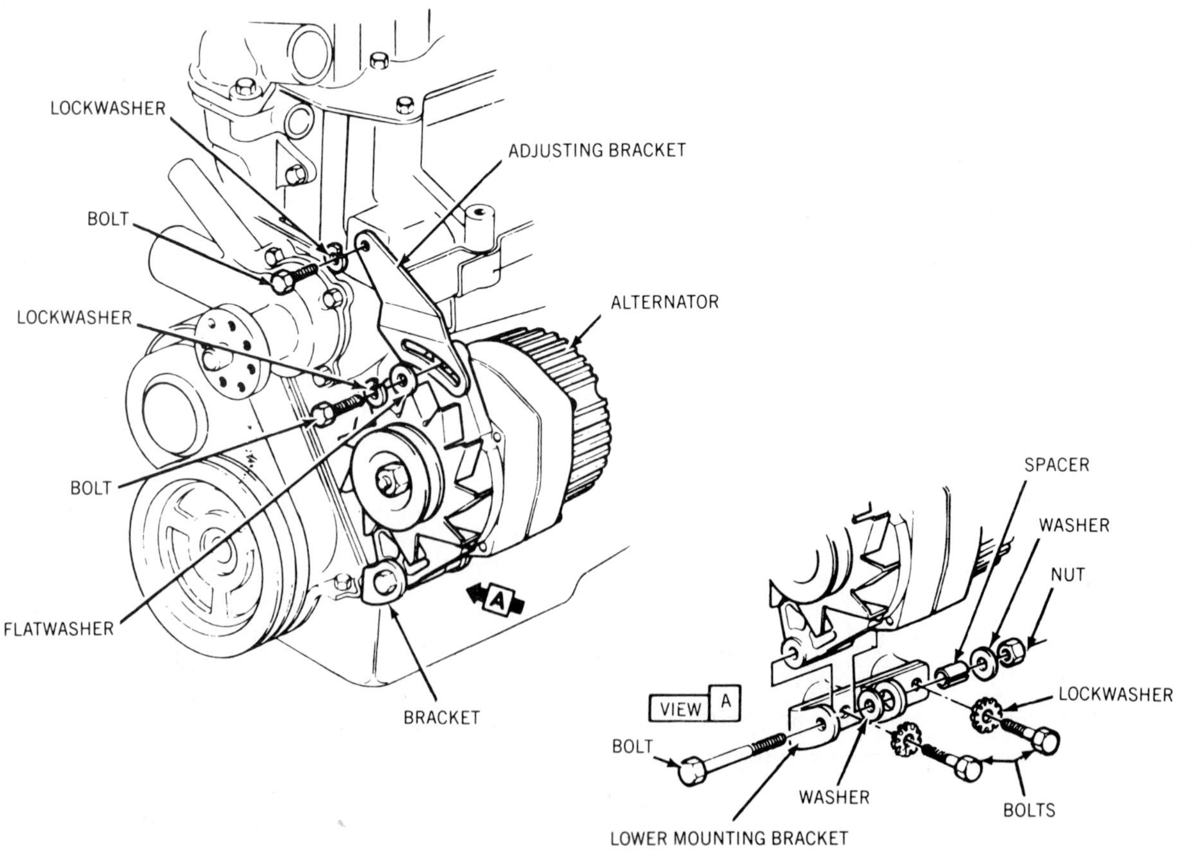

Figure 11-43. Typical alternator installation on an inline engine. (Chevrolet)

repair. The overhaul sequence at the end of this chapter contains examples of all of them. Use the following procedures to test the stator, the rotor, and the rectifiers of any alternator. These procedures also include instructions for diode replacement.

Stator Continuity and Ground Tests. Continuity must exist through all windings of the stator. The windings must not be grounded to their frame.

Test continuity by connecting one ohmmeter or test lamp lead to one stator winding or to the neutral junction. Touch the other lead to each remaining stator lead, figure 11-44. The lamp should light or the meter should show continuity.

Test for a grounded stator by connecting one ohmmeter or test lamp lead to the stator frame. Touch the other lead to each stator connection, figure 11-44. The lamp should not light; the meter should show infinite resistance at each point.

Rotor Continuity and Ground Tests. Continuity must exist through the rotor winding and the sliprings. The winding and sliprings must not be grounded to the shaft.

Test continuity by touching one ohmmeter or test lamp lead to one slipring. Touch the other lead to the other slipring, figure 11-45. The lamp should light or the meter should show continuity.

Test for a grounded rotor by touching one ohmmeter or test lamp lead to one slipring. Touch the other lead to the shaft, figure 11-45. The lamp should not light; the meter should show infinite resistance.

Capacitor Test. If the alternator includes a capacitor for radiofrequency interference (RFI) suppression or to absorb voltage surges, test it for a short circuit. Connect one ohmmeter or test lamp lead to the mounting clip. Connect the other lead to the pigtail connector, figure 11-46. The lamp should not light; the meter should show infinite resistance.

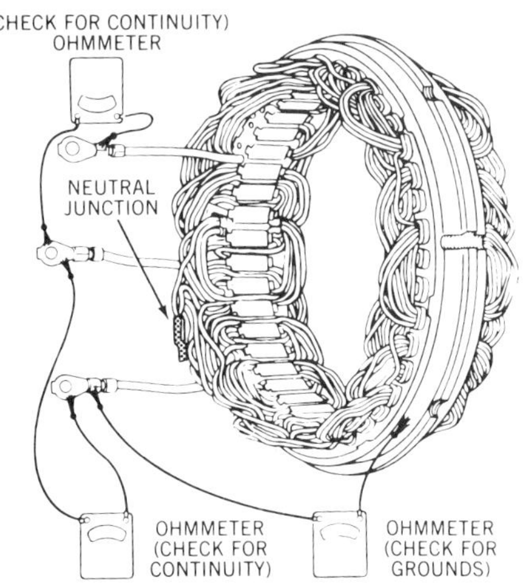

Figure 11-44. Stator continuity and ground test points. (Cadillac)

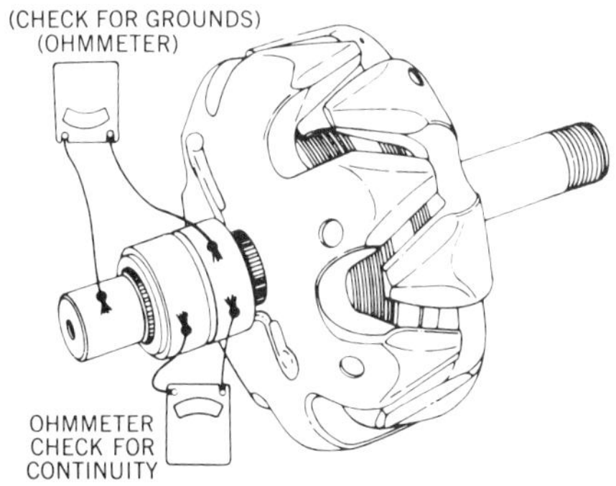

Figure 11-45. Rotor continuity and ground test points. (Cadillac)

Diode Test and Replacement. A diode must pass current in one direction and block it in the other. If a diode is shorted, it will pass current in both directions. If it is open, it will block current in both directions. To test a diode for an open or a short circuit, touch one ohmmeter or test lamp lead to the diode lead and the other meter or lamp lead to the heat sink or alternator frame, figure 11-47. Then reverse the leads. The lamp should light or the meter should show continuity in one direction but not in the other.

If continuity is indicated in both directions, the diode is shorted. If continuity is not indicated in either direction, the diode is open.

You can replace individual diodes

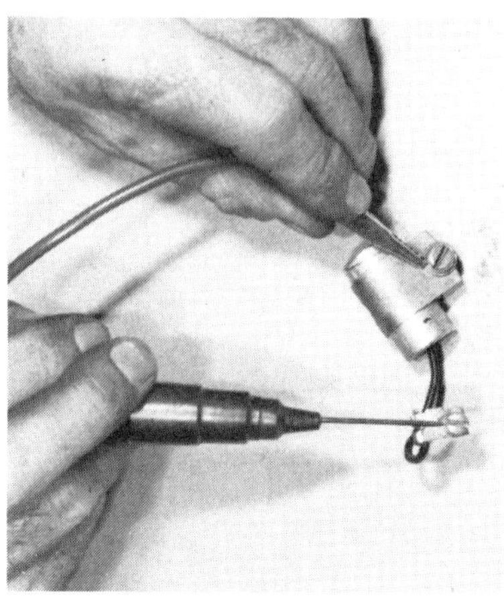

Figure 11-46. Test an alternator capacitor for a short circuit.

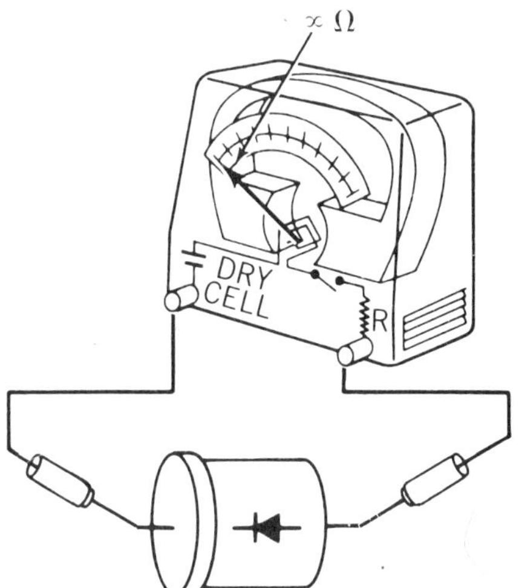

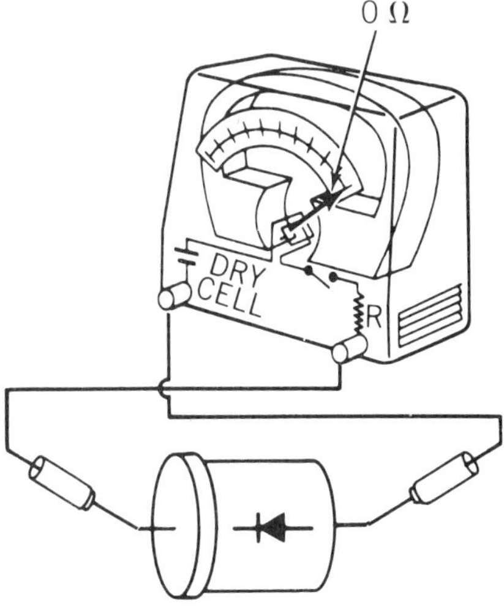

Figure 11-47. Diode tests with an ohmmeter.

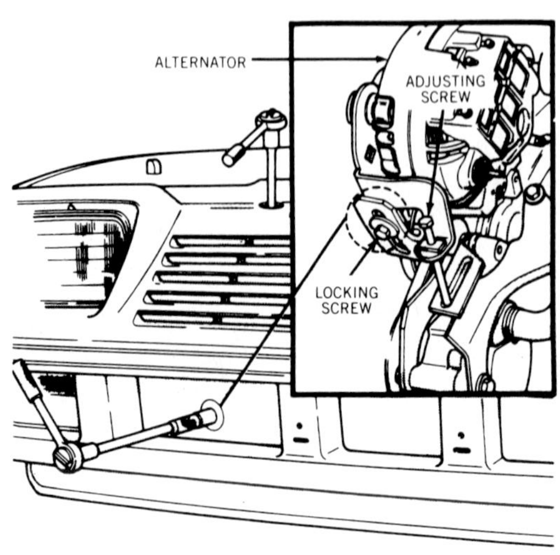

Figure 11-49. Turn the adjusting screw to set alternator belt tension. (Chrysler)

Figure 11-48. Test belt tension with a strand tension gauge.

on older alternators by pressing them out of the end frame or heat sink with special tools. Some diodes are screwed into the end frame or heat sink. Remove these with a wrench.

Most late-model alternators have diodes built into rectifier bridges. Test points vary depending on bridge construction. Follow the carmaker's instruction to test a rectifier bridge, but use the same testing principles outlined above. The overhaul procedure at the end of the chapter includes examples of testing a Delco-Remy rectifier bridge and a field diode trio.

Alternator Installation

Before installing an alternator, spin the shaft by hand to be sure there is no interference between the stator and the rotor or between the fan and the end frame. Inspect the drive belt, or belts. If any belt is worn or damaged, install a new one. The water pump belt on most engines also drives the alternator. Use a strand tension gauge to adjust tension accurately, figure 11-48. Be sure the battery ground cable is disconnected; install the alternator as follows:

1. Put the alternator on the mounting bracket and adjusting bracket. Loosely install the mounting and adjusting bolts and any required spacers and washers, figures 11-42 and 11-43.

2. Install the alternator belt, or belts, and pull the alternator away from the engine by hand to keep the belts in place.

3. Tighten the mounting and adjustment bolts just enough to maintain belt tension.

4. Carefully pry the alternator away from the engine with a wooden hammer handle or pry bar. Pry against thick parts of the drive end frame.

5. Tighten the lower mounting bolt, then the upper adjusting bolt to keep from distorting the alternator housing and binding the bearings.

6. Torque the bolts to specifications and check belt tension with a gauge.

7. Reinstall other engine accessories removed for alternator access. Check tension of all belts with a tension gauge.

8. Be sure that battery polarity matches alternator polarity (negative ground on late-model cars and light trucks).

9. Connect all wires to the alternator.

10. Connect the battery ground cable.

Many late-model cars have adjusting bolts or wrench attachment points to position the alternator and adjust belt tension, figure 11-49. Follow the carmaker's instructions for alternator adjustment, belt tension, and bolt torque. After installing the alternator, verify correct operation with regulated voltage and current output tests as specified by the carmaker.

ALTERNATOR OVERHAUL

The following illustrated procedures show the basic steps to disassemble, test, repair, and reassemble four popular types of alternators:

- GM Delco-Remy 10-SI alternator
- Ford Motorcraft IAR alternator
- Chrysler's new corporate alternator
- GM Delco-Remy CS-144 alternator

These alternators are the most common original-equipment units on late-model GM, Ford, and Chrysler vehicles. These overhaul sequences are typical of the steps you will use to service the different variations of these basic alternators. These procedures also are typical of the overhaul sequences you can follow for imported-car alternators and older domestic units.

DELCO 10-SI ALTERNATOR OVERHAUL

1. Draw a chalk mark across the end frames for alignment during reassembly. Then remove the four through bolts that hold the end frames together.

2. Separate the drive end frame and rotor from the slipring end frame and stator. If they don't come apart easily, pry carefully with a screwdriver at the stator slots (A).

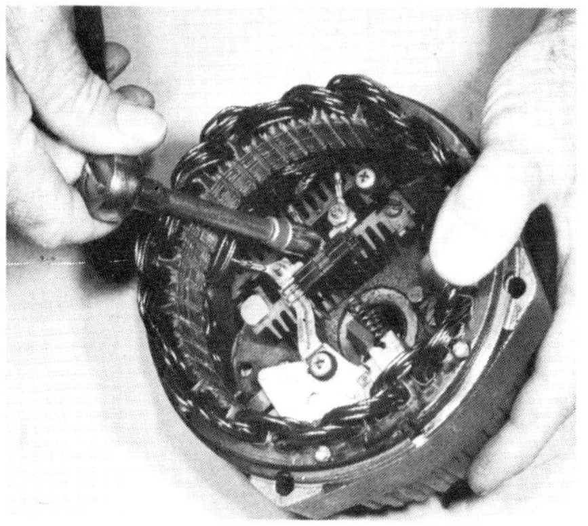

3. Remove the three nuts that hold the stator leads and the diode trio to the rectifier. Brushes and springs will pop out of the brush holder when you remove the stator.

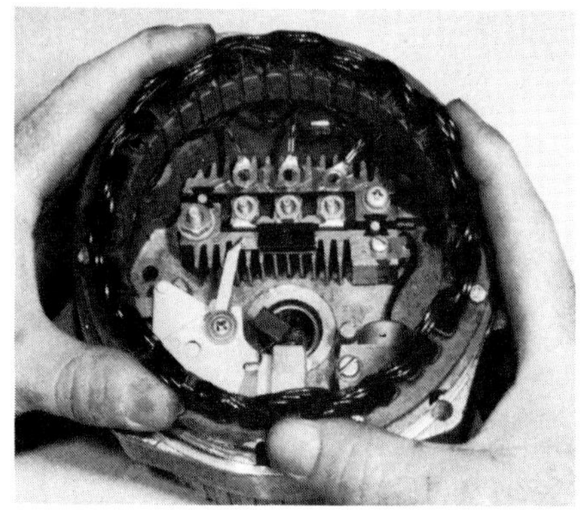

4. Lift the three stator leads off the rectifier terminals and remove the stator from the slipring end frame.

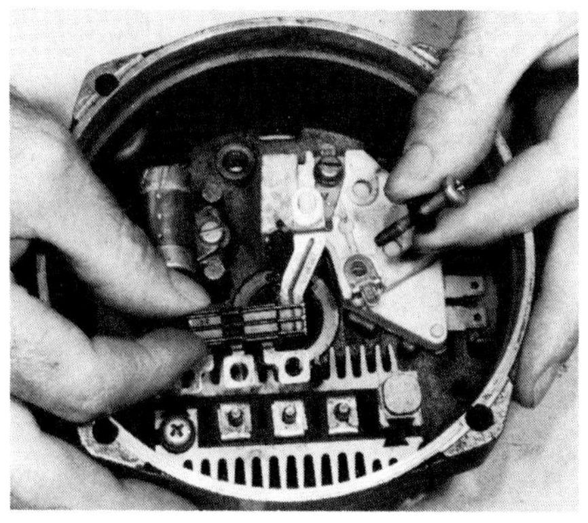

5. Remove the screw that holds the diode trio to the brush holder and regulator assembly. Note that the insulated washer goes between the screw head and the diode trio lead.

6. Disconnect the capacitor lead from the rectifier insulated heat sink (A) and remove the capacitor mounting screw and the capacitor.

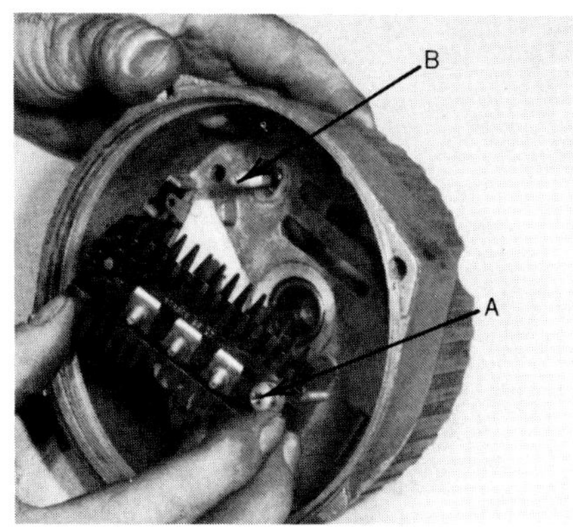

7. Remove the rectifier ground screw (A) and the BAT terminal nut and washer (B). Remove the rectifier. Keep track of all screws, nuts, and washers and their locations.

8. Remove the two mounting screws from the brush holder and regulator. Note the positions and locations of insulating sleeves and washers for correct reassembly.

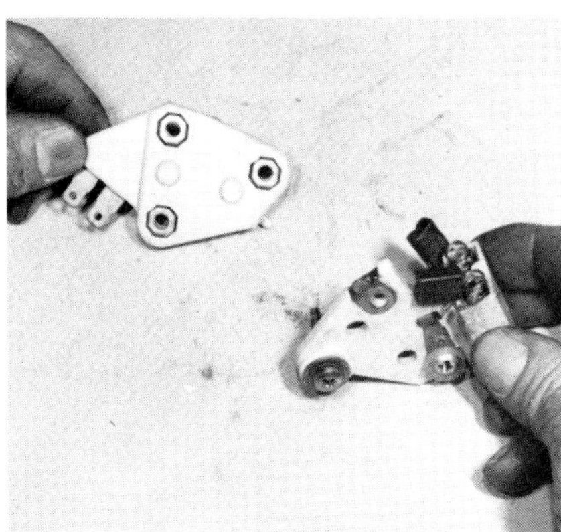

9. Separate the regulator from the brush holder. If the brushes are chipped or worn to half their original length or less, replace them.

DELCO 10-SI ALTERNATOR OVERHAUL (*Continued*)

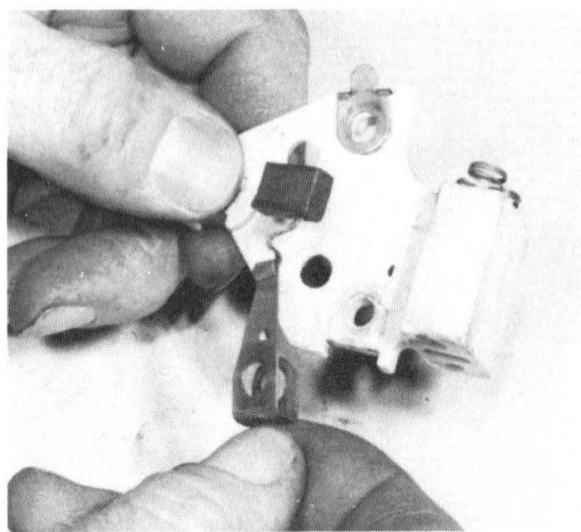

10. Slip the brush clip off the brush holder and install a new clip-and-brush assembly. Both assemblies are the same.

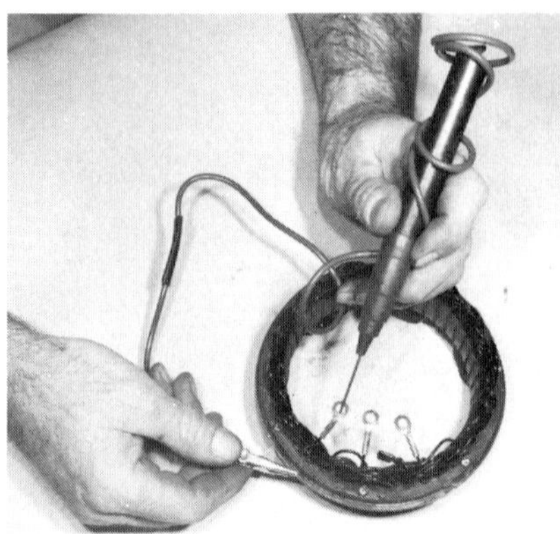

11. Now is a good time for ohmmeter or test lamp checks. Connect one test lead to the stator frame and the other to *each* stator lead. If the lamp lights or the meter shows low resistance at any point, the stator is grounded.

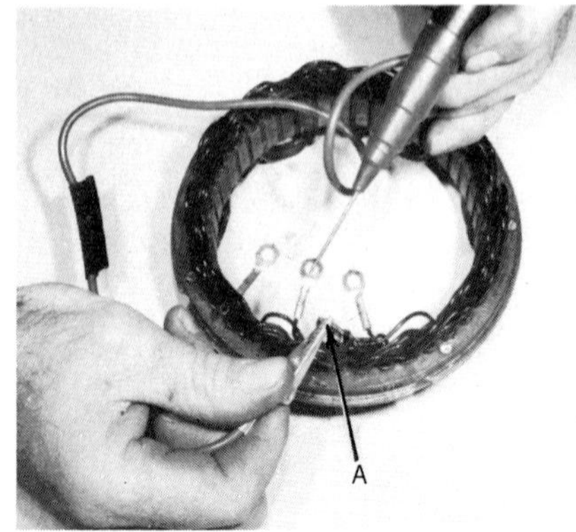

12. Connect one test lead to the stator neutral junction (A). Touch the other lead to each of the three leads to check stator continuity. If the lamp *doesn't* light or the meter *doesn't* show low resistance at all points, the stator is open.

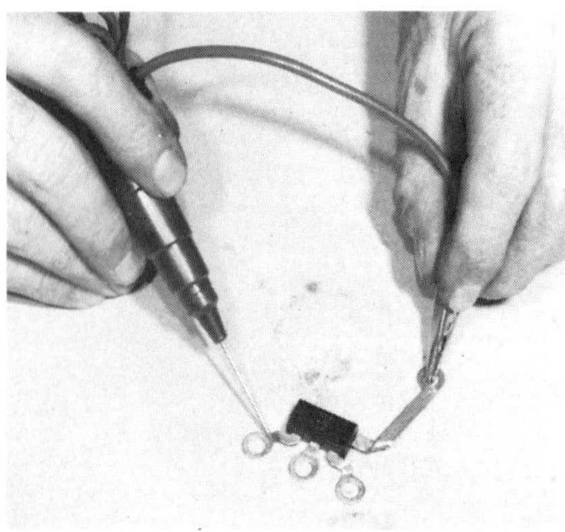

13. Touch one test lead to the diode trio brush holder terminal and the other lead to each of the other three terminals. Note whether or not the lamp lights or whether the meter reads high or low.

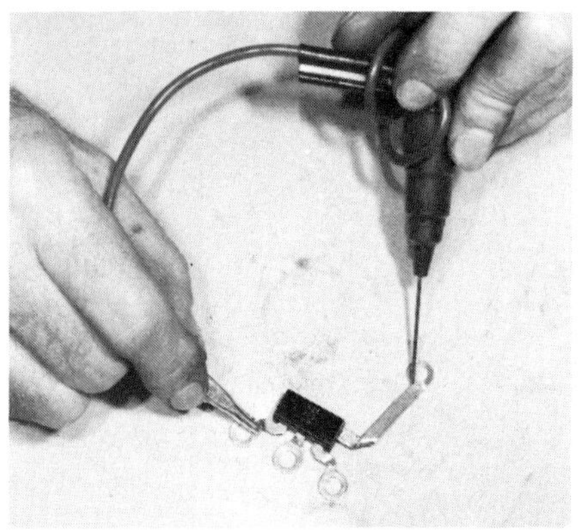

14. Reverse the test leads and repeat step 13. If the lamp lights in both directions or fails to light in one direction at any point, the trio is bad. Touch the test leads to each pair of the three terminals. If the lamp doesn't light or the meter reads high, the trio is bad.

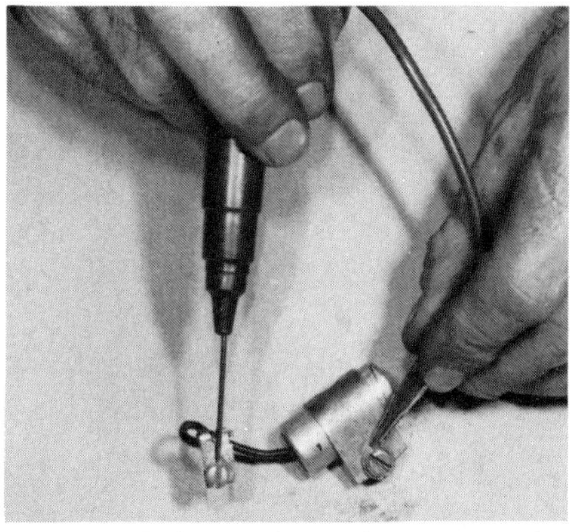

15. Check for a shorted capacitor by touching one test lamp lead to the pigtail terminal and the other to the case or mounting bracket. The lamp should not light. If it does, the capacitor is shorted.

16. Check for a grounded rotor by touching one test lead to the shaft and the other to either slipring. If the lamp lights or the meter reads low resistance, the rotor winding is grounded.

17. Check for rotor continuity by touching the test leads to both sliprings. If the lamp does not light or the meter reads high resistance, the rotor has an open circuit.

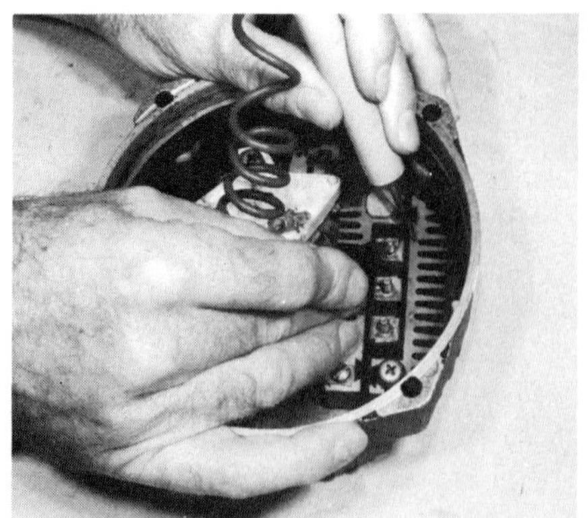

18. Test the rectifier in or out of the alternator. Touch one test lead to the ground heat sink and the other to the base of each diode. Do *not* touch the threaded stud. Note whether or not the lamp lights or whether the meter reads high or low.

DELCO 10-SI ALTERNATOR OVERHAUL *(Continued)*

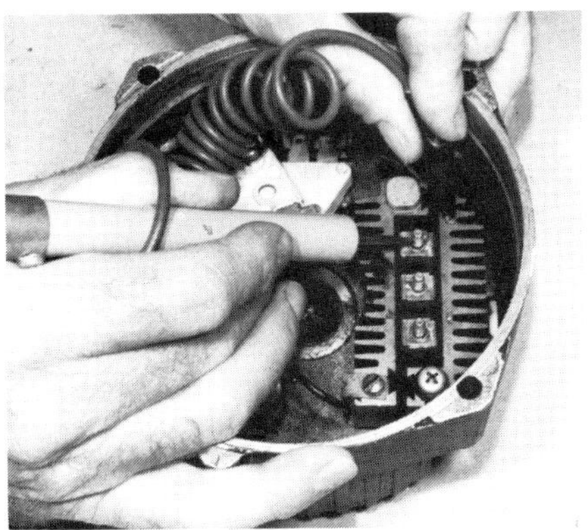

19. Reverse the test leads and repeat step 18 at each diode. If the lamp lights in both directions or fails to light in either direction at any point, or if the meter reads the same in both directions, the rectifier is bad.

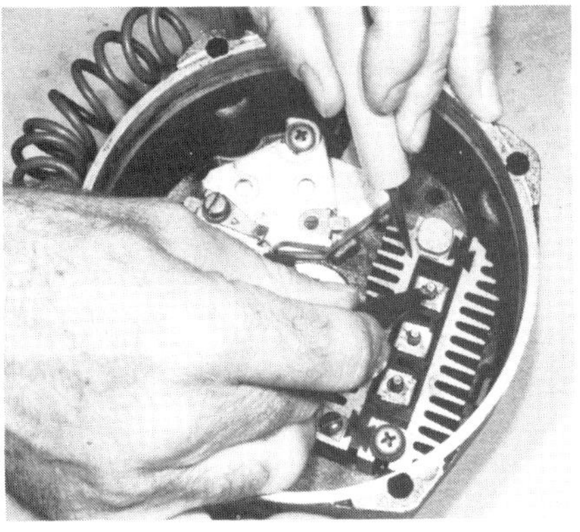

20. Repeat steps 18 and 19 between each diode and the insulated heat sink for a total of 12 separate test lead connections. Replace the rectifier if it is out of limits at any point.

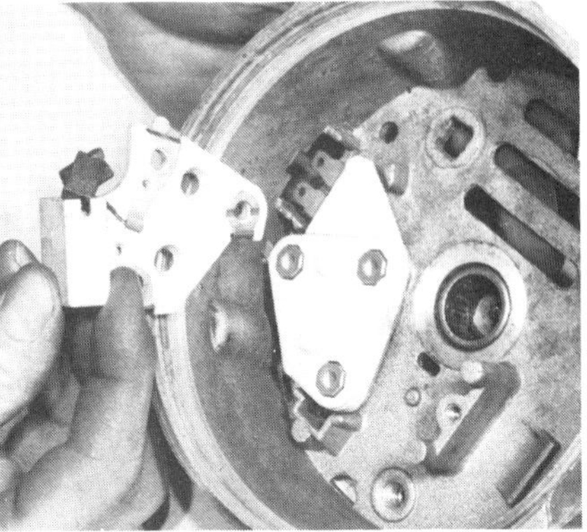

21. Start reassembling the slipring end frame by placing the regulator in position. Install the brush holder on the regulator with screws and insulators in the original locations.

22. Tighten the screws carefully to avoid insulator damage. This screw may or may not have an insulating washer. The brush holder screw may or may not have an insulator sleeve. Insulator applications vary among models.

23. Install the brush springs and brushes in the holder and carefully press them into their slots. Insert a toothpick or drill bit (A) through the end frame and brush holder to hold the brushes for rotor installation.

24. Install the BAT terminal from outside the end frame. Be sure the square shoulder of the insulator is properly seated in the hole.

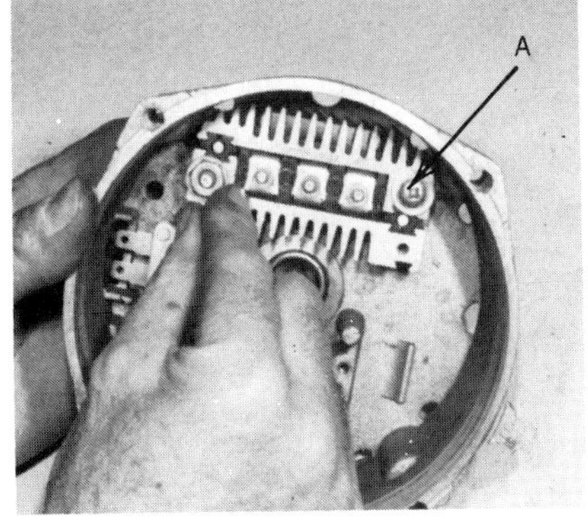

25. Install the rectifier over the BAT terminal. Install the ground screw and insulator (A). Then install the nut on the BAT terminal. Tighten the BAT terminal nuts carefully.

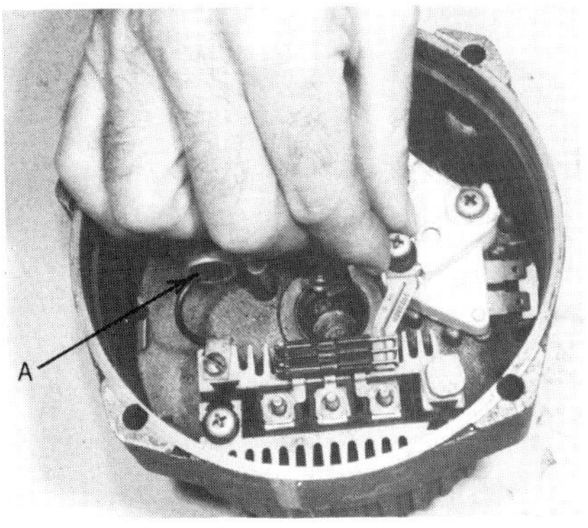

26. Install the capacitor (A). Then install the diode trio with three leads over the rectifier terminals. Install the brush holder and regulator terminal with the required insulator.

27. Align the notches in the stator frame with the bolt holes in the end frame and install the stator. Place the three stator leads on the rectifier terminals and install the terminal nuts.

DELCO 10-SI ALTERNATOR OVERHAUL *(Continued)*

28. To service the drive end frame, carefully clamp the rotor in a vise and remove the nut, washer, pulley, and fan from the shaft. You will need an impact wrench to do this on some alternators.

29. Remove the collar from the rotor shaft and then withdraw the rotor from the end frame.

30. To replace the drive end frame bearing, remove the three screws from the bearing retainer. Then remove the retainer.

31. Support the end frame hub on a suitable fixture and drive or press the bearing out of the hub.

32. Install the bearing by tapping with a wooden block and a hammer or by pressing with a collar over the outer race. Don't apply pressure to the inner race. Pack the bearing one-quarter full of the specified lubricant.

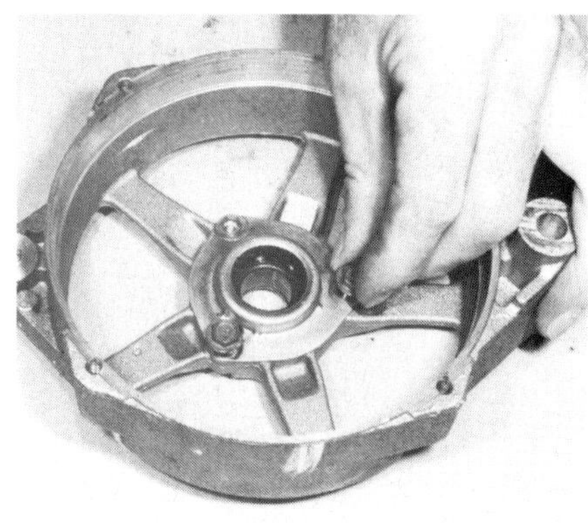

33. If the felt seal is hardened or worn, replace the retainer. Install the retainer and tighten the three screws.

34. Install the collar, the fan, the pulley, the washer, and the nut in that order. Tighten the nut to the specified torque value.

35. Align the chalk marks made at disassembly and assemble the end frames to each other.

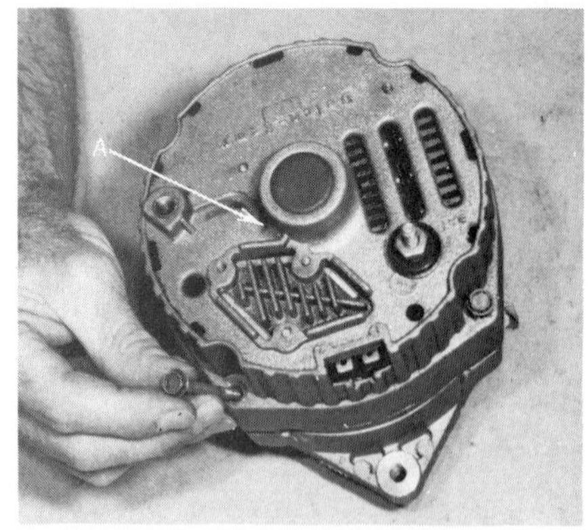

36. Remove the drill bit or wire (A) that holds the brushes away from the sliprings. Install and tighten the four through bolts, and the overhaul is done.

MOTORCRAFT IAR ALTERNATOR

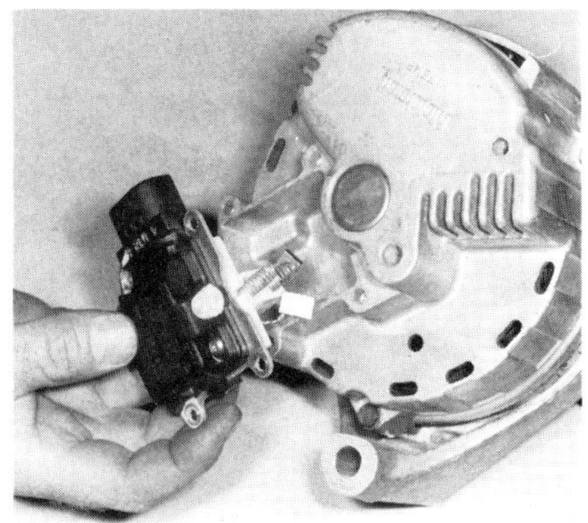

1. Remove the screws holding the brush and regulator assembly to the rear end frame. Remove the brush and regulator assembly.

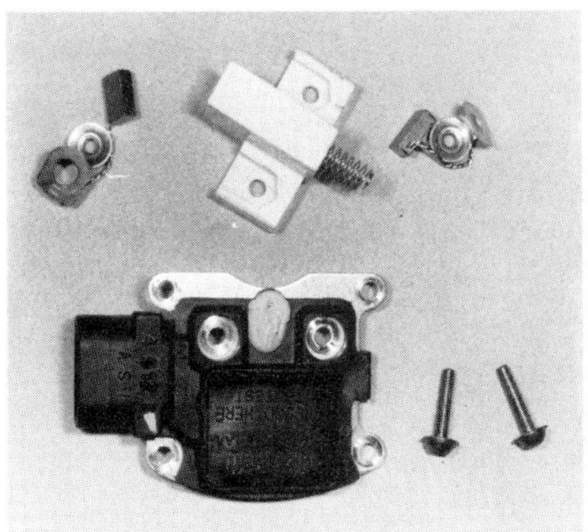

2. Two screws hold the brush holder and regulator together. Each brush lead has an integral retaining nut for its screw. Separate the two units.

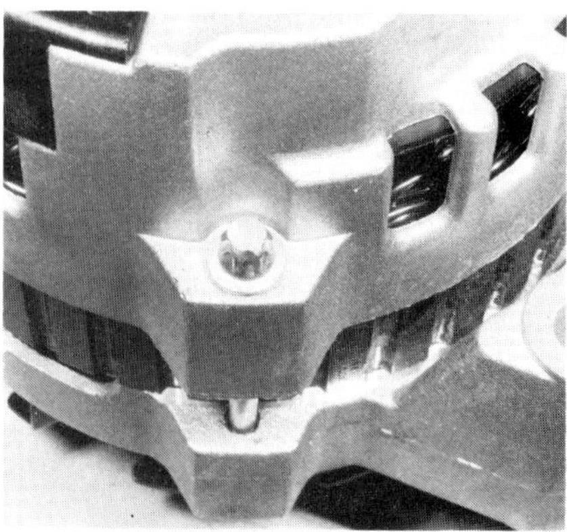

3. The through bolts have external Torx heads. If you do not have the right Torx socket, you can remove them with a 6-point socket.

4. Draw an alignment mark (arrows) across both end frames and the stator for reassembly reference. Then remove the through bolts.

5. Pull the end frames apart with a rotating motion. The drive end frame and rotor will separate from the rear end frame and stator.

6. To service the rotor, remove the pulley nut, washer, pulley, and fan. Remove spacer sleeve (arrow) from rotor shaft.

7. Use a press (do not hammer) to remove the rotor from the drive end frame. Refer to chapter text for testing.

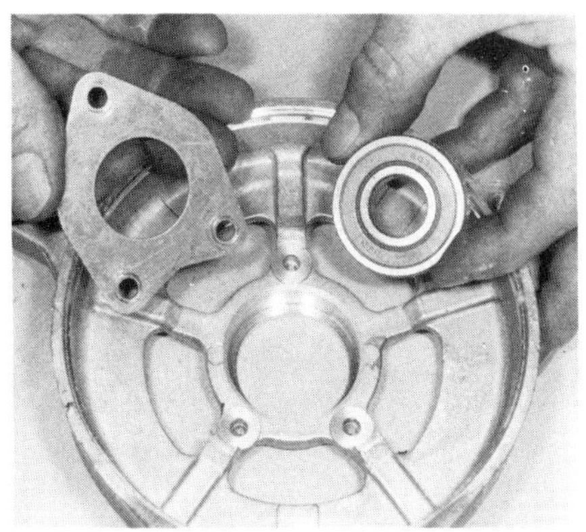

8. Three screws and a retainer hold the drive end bearing in place. Push the slip-fit bearing from the end frame by hand.

9. The stator leads (arrows) are soldered to the rectifier assembly. Unsolder the connections to remove the stator.

MOTORCRAFT IAR ALTERNATOR (*Continued*)

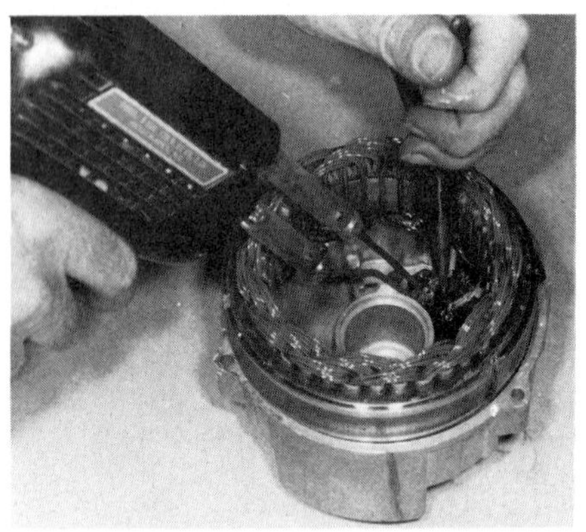

10. To unsolder stator leads, hold the lead with needlenose pliers, heat with a soldering iron, and pull the lead off the rectifier terminal. Repeat to unsolder other leads.

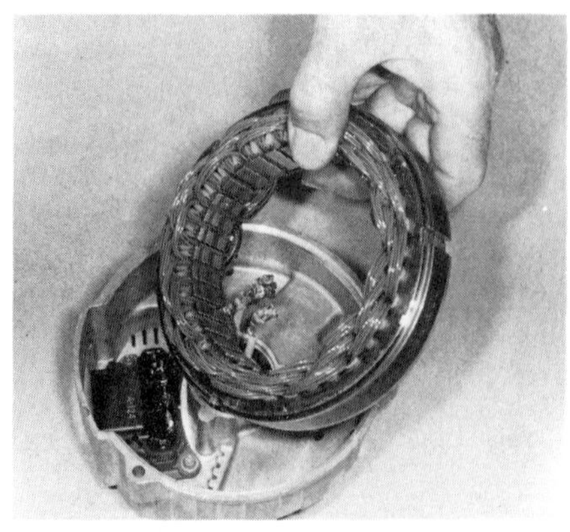

11. After all leads are unsoldered, carefully remove the stator from the end frame. Refer to chapter text for testing.

12. Torx head screws hold the rectifier assembly to the end frame. Use a suitable Torx driver for screw removal to prevent screw head damage.

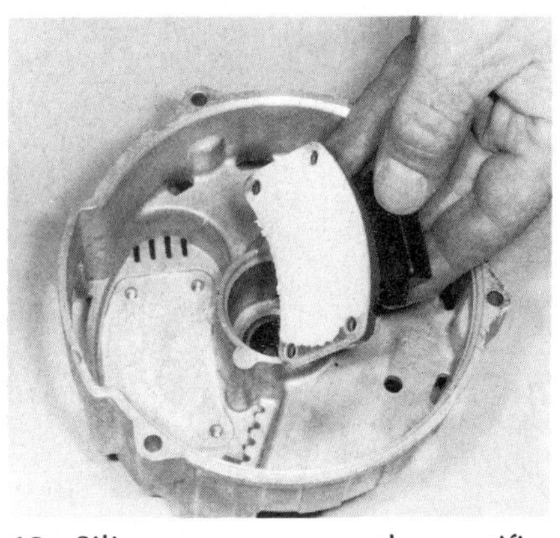

13. Silicone grease on the rectifier base dissipates heat. If wiped from end frame, recoat during reassembly.

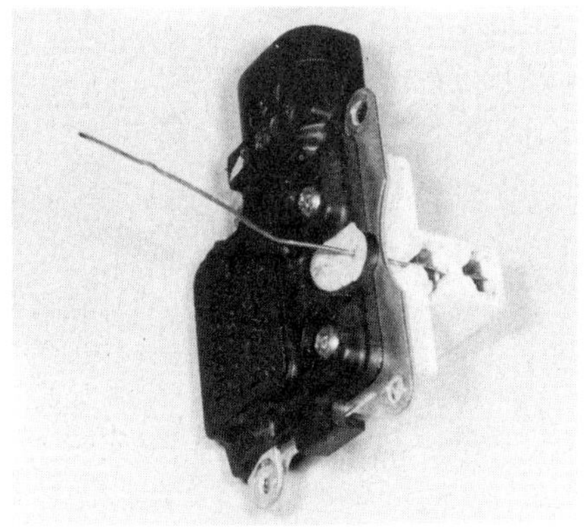

14. Reattach the brush holder to the regulator and insert a bent paper clip as shown (or a plasic toothpick) to retain the brushes for reinstallation.

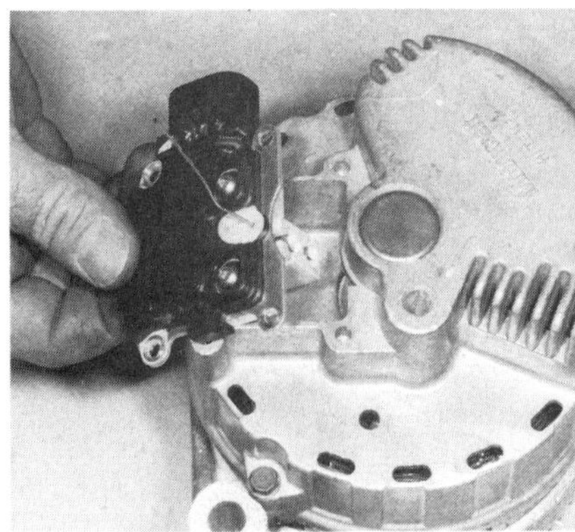

15. After reassembling the alternator, install the brush holder and regulator assembly to the end frame. Then remove clip so brushes will contact rotor sliprings.

CHRYSLER 40/90 ALTERNATOR

1. Remove the nut holding the dust cover to the brush holder assembly stud. Remove the cover.

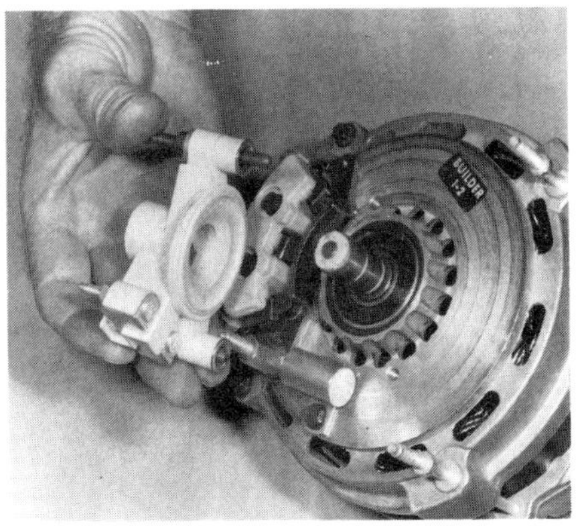

2. Use a deep socket to remove the two brush holder retaining screws. Carefully lift the brush holder from the rotor shaft.

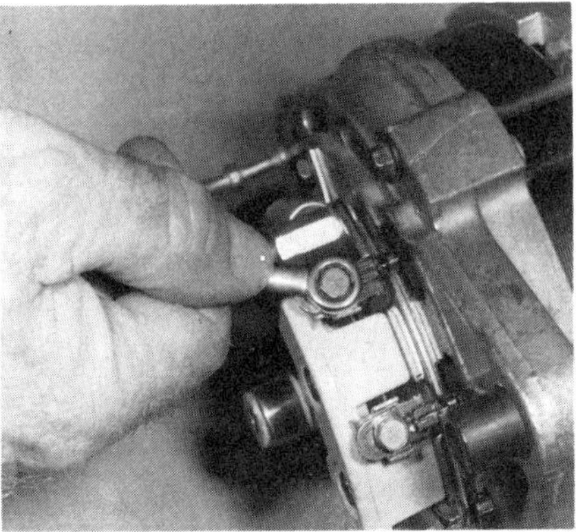

3. Three screws hold the stator leads to the rectifier assembly posts. Remove the screws and separate the leads from the posts.

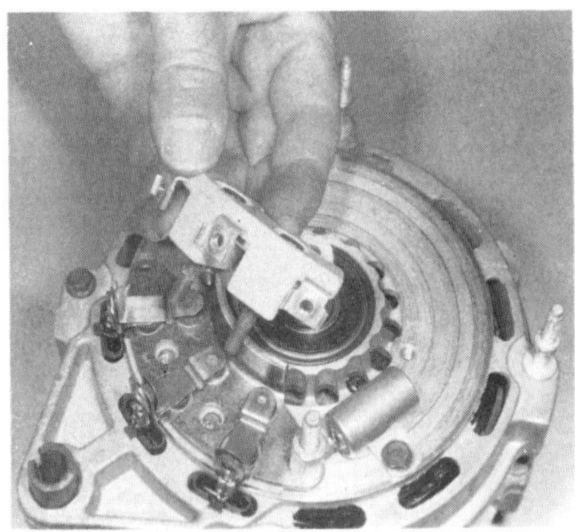

4. Remove the two screws holding the rectifier assembly to the alternator end frame. Remove the rectifier assembly.

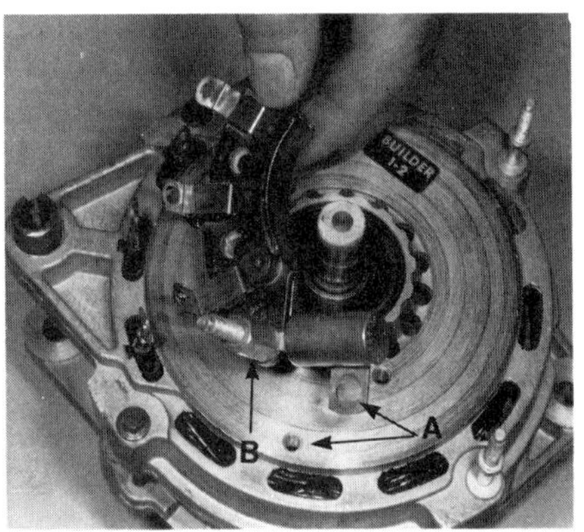

5. Remove the capacitor mounting screw (A) to lift off the capacitor and rectifier assembly. To separate the two, remove the large nut (B).

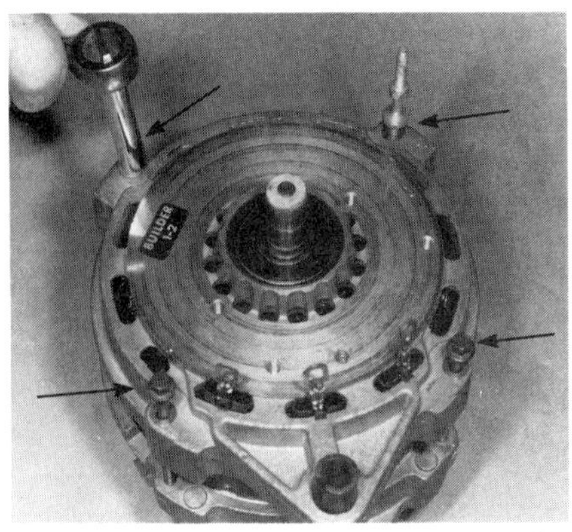

6. Remove the four through bolts holding the alternator together (arrows). The two bolts with stud ends require a deep socket.

7. Pry points for separating the stator and drive end frames are provided on each side of the alternator.

8. Separate the end frame and stator by guiding the stator leads through the end frame insulator holes (arrows).

9. Check the stator lead insulator condition. If damaged, unsnap the insulation and remove it from the end frame.

CHRYSLER 40/90 ALTERNATOR (*Continued*)

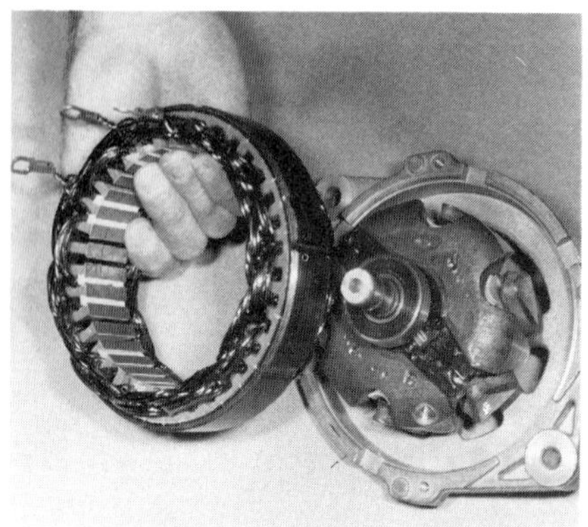

10. Separate the stator from the end frame. Refer to chapter text for testing procedures, if necessary.

11. Remove the nut and washer holding the drive pulley and fan; then remove the pulley and fan. The front bearing chamber fits against the rotor during reassembly.

12. Press the rotor out of the drive end frame. Remove the inner bearing spacer (arrow) from the rotor shaft. See chapter text for testing.

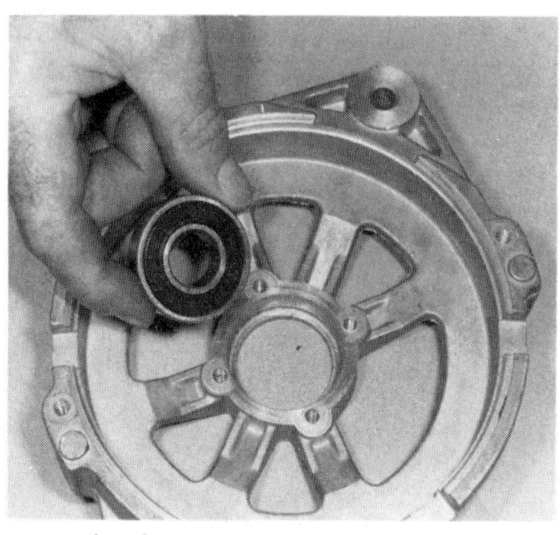

13. The front bearing can be replaced by removing the retaining capscrews and pressing (do not hammer) the bearing from the drive end frame. Press a new bearing in place and install the capscrews.

14. Handle the shaft assembly carefully. Ceramic slipring end (A) can be damaged easily, as can plastic termination plate (B).

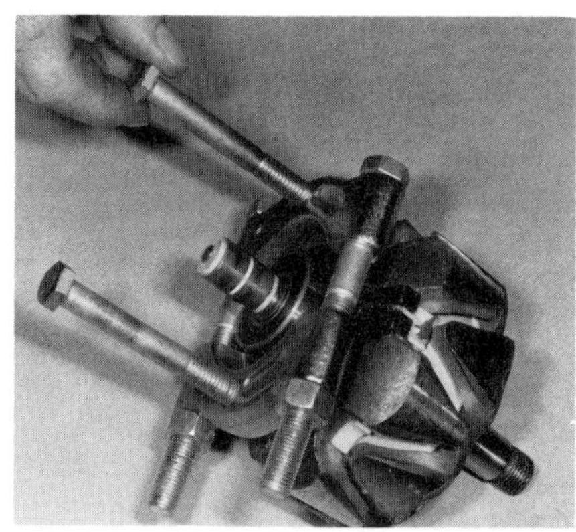

15. If bearing removal is necessary, use a universal bearing puller as shown. A special jig is required to install the new bearing.

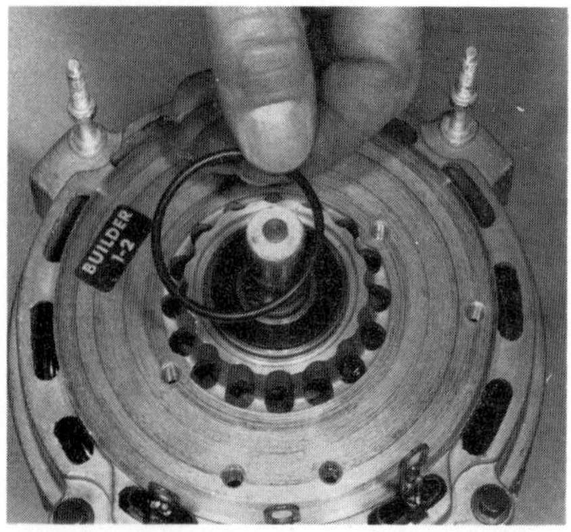

16. Reassemble the end frames and stator. Install and tighten the through bolts. Install a new O-ring as shown on the rectifier end frame.

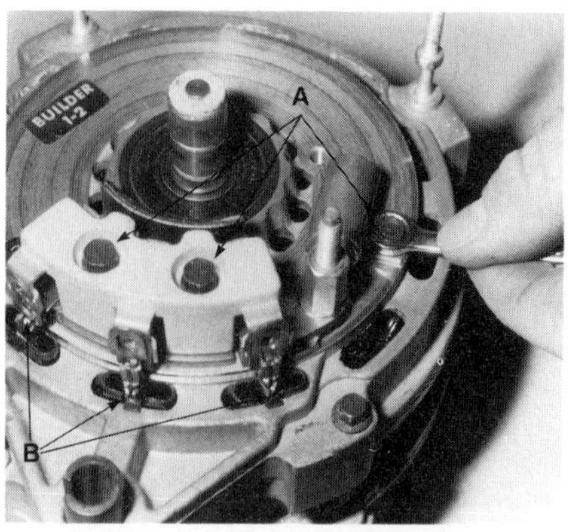

17. Reinstall the rectifier-capacitor assembly with screws (A). Reinstall the stator-to-rectifier screws (B).

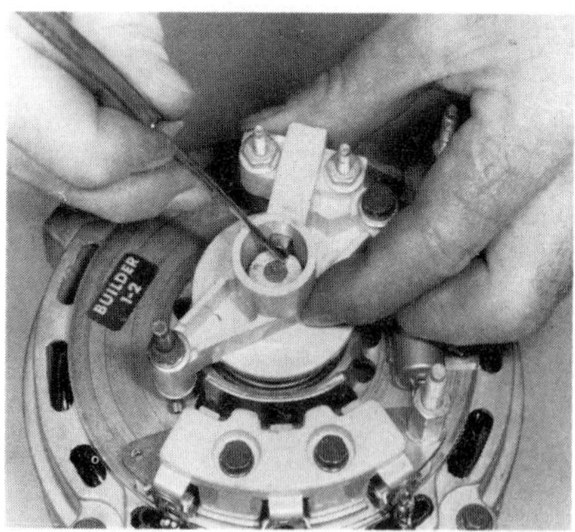

18. Depress brushes with a pointed tool and fit the holder over the rotor sliprings. Tighten brush holder screws. Then reinstall the dust cover with its retaining nut.

DELCO-REMY CS-144 ALTERNATOR

1. Draw chalk marks across both end frames and the stator to serve as reassembly reference points.

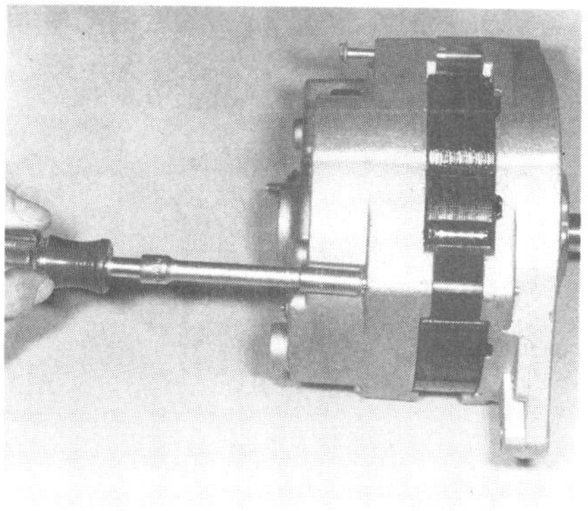

2. Loosen and remove each of the through bolts holding the alternator end frames together.

3. Separate the front end frame from the stator assembly by gently tapping on the alternator ears with a soft-faced hammer.

4. Remove the front end frame and spacer (A). Bearing (B) can be removed, but if worn or defective, replace the end frame and bearing as a unit.

5. Carefully separate the rotor from the stator and rear end frame assembly with a slow, rotating motion.

6. If the rotor bearing (A) requires replacement, press the old bearing off the shaft; then press on a new one. Do not pry or hammer bearing.

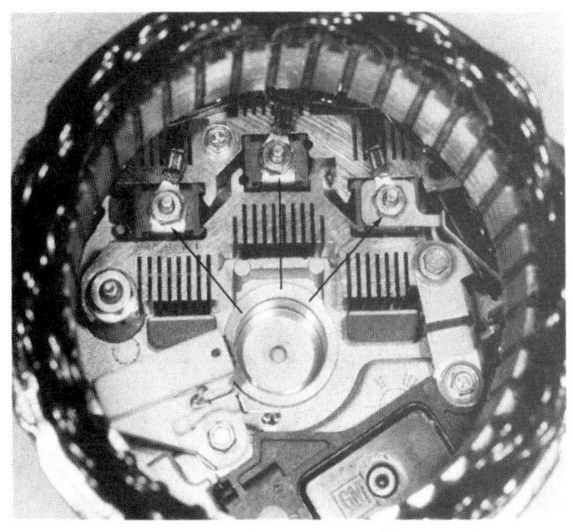

7. To separate the stator from the rear end frame, remove the three nuts (arrows) and disengage the stator leads from the studs.

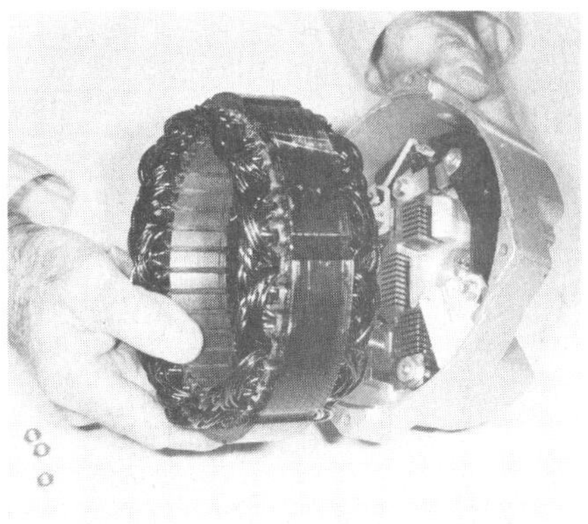

8. Carefully work the stator out of the rear end frame. Do not damage the stator leads in this step. Refer to chapter text for testing.

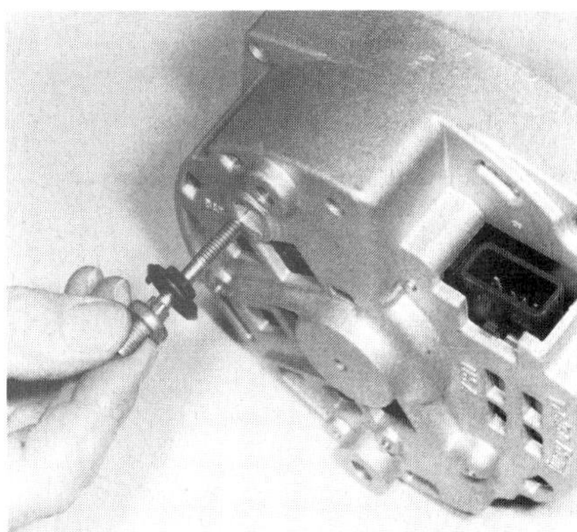

9. Unscrew and remove the BAT terminal with its insulator. Check the condition of the insulator and replace as required.

DELCO-REMY CS-144 ALTERNATOR (*Continued*)

10. Remove the screw holding the brush holder and one end of the regulator to the end frame.

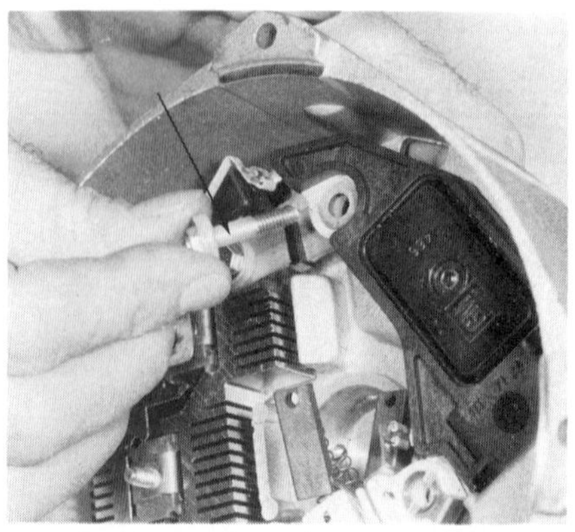

11. Remove the other regulator retaining screw. Note insulator on screw (arrow). If the insulator is damaged, replace it during reassembly.

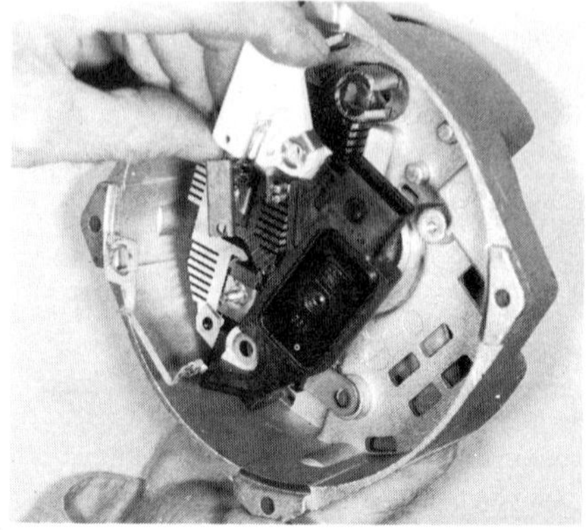

12. Carefully lift the brush holder and regulator assembly from the rear end frame.

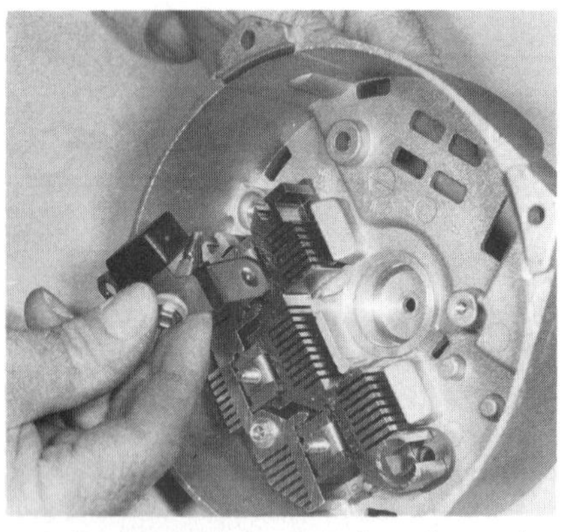

13. Remove the capacitor from the rear end frame. The retaining screw also has an insulator. Inspect it and replace if necessary.

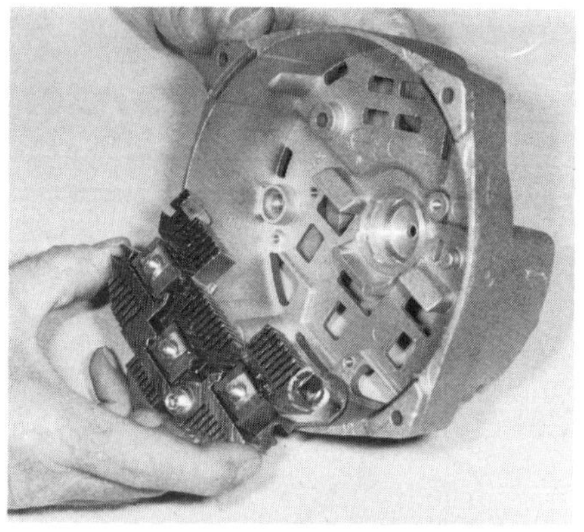

14. Remove the rectifier assembly from the rear end frame. The rectifier assembly looks as if it comes apart, but do not try to separate it. The unit is serviced as an assembly.

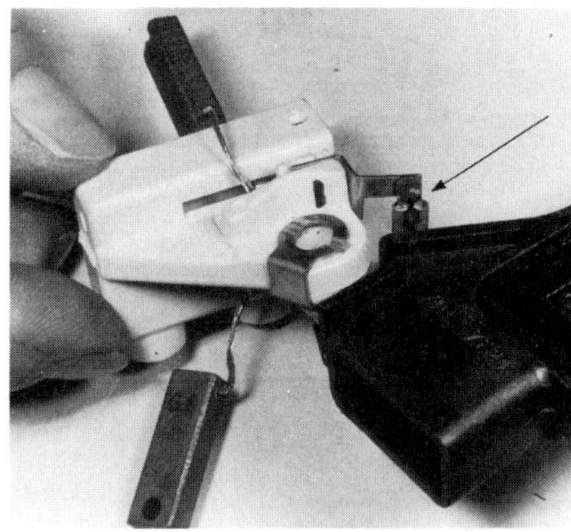

15. The brush holder and regulator assembly can be separated by uncrimping and unsoldering the connector (arrow).

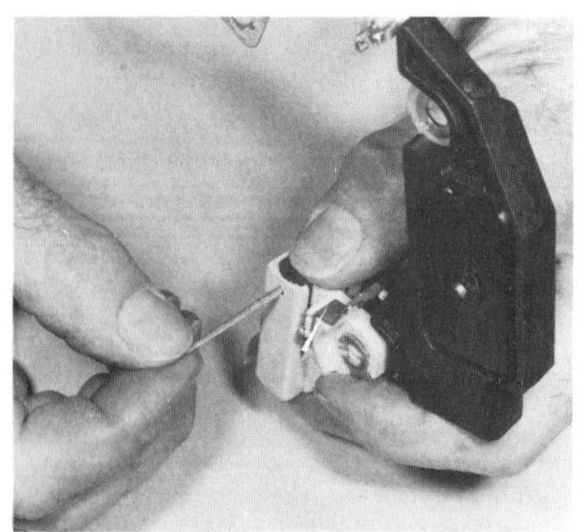

16. Before reassembling the alternator, press the brushes into the holder and slide a toothpick in the housing as shown to hold the brushes in place. Do not forget to remove it once the stator is installed.

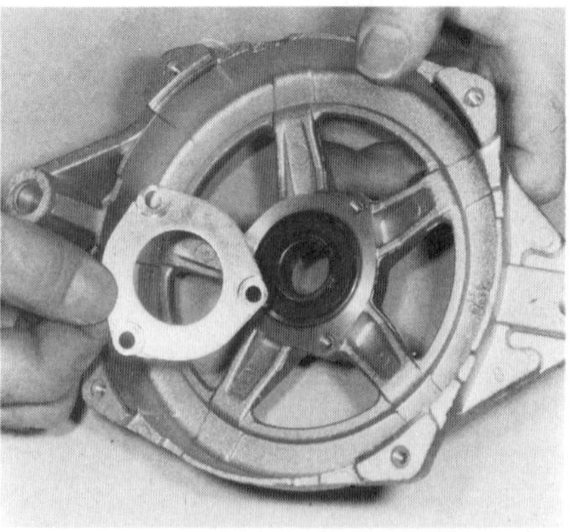

17. You can service the front end bearing by removing the bearing cover and pressing the bearing from the end frame.

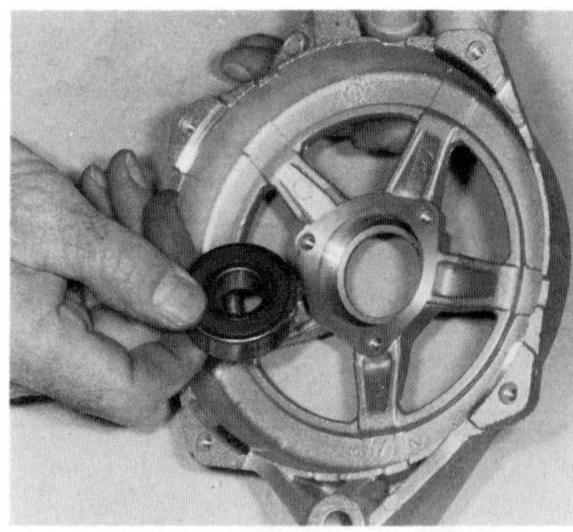

18. Clean bearing bore and press a new bearing in place. Bearing must be installed facing as shown. The alternator is now ready for reassembly.

REVIEW QUESTIONS

1. The alternator output (BAT) terminal has battery voltage present:

a. at all times
b. when the igntion key is in the On position
c. when the regulator allows it
d. only when the battery is discharged

2. A vehicle seems to have a charging system problem. The first test you will make is:

a. wiring and connection tests
b. fusible link inspection
c. a test of the battery
d. a test of the output circuit

3. A vehicle's charging system indicator lamp is on when the engine is off. The probable cause is:

a. a defective voltage regulator
b. a shorted positive diode in the alternator
c. worn brushes
d. a bad ground

4. High resistance in an alternator output circuit is often due to:

a. a discharged battery
b. a shorted diode
c. loose or corroded conections or bad wiring
d. a bad regulator

5. You are making general alternator charging system tests. The manufacturer's instructions say to put a load on the alternator output circuit. You will use:

a. a voltmeter
b. an ammeter
c. a volt-amp tester
d. a carbon pile

6. Some carmakers recommend a charging voltage test rather than an output current test because:

a. the test pinpoints a bad regulator more quickly
b. the test avoids possible high current damage to electronic components
c. the test avoids the use of a carbon pile
d. you don't have to do an output circuit resistance test first

7. An alternator field current test is used to determine:

a. if current draw is high
b. if current draw is low
c. if there is a short in the field current
d. all of the above

8. An oscilloscope is most useful for testing for:

a. a bad regulator
b. bad alternator diodes
c. worn brushes
d. defective sliprings

9. A stator continuity test:

a. identifies a grounded wire
b. can be made with a test lamp or ohmmeter
c. assures that all stator windings have continuity
d. both b and c

10. A test lamp lead is touching an alternator rotor slipring. The other lead is touching the rotor shaft. The lamp lights. What does the test show?

a. The rotor is grounded.
b. The sliprings are worn.
c. The stator is grounded.
d. The rotor has continuity.

11. An open diode will:

a. block current in one direction
b. allow current to pass in both directions
c. block current in both directions
d. allow current to pass in one direction

12. A shorted diode will:

a. allow current to pass in both directions
b. block current in one direction
c. allow current to pass in one direction
d. block current in both directions

13. A vehicle is equipped with a voltmeter. The voltmeter needle flutters when the engine is running. A possible cause is:

a. a bad diode
b. low alternator output
c. loose wiring connections
d. battery that is too small

14. The most common cause of charging system problems is:

a. loose or bad alternator drive belts
b. a grounded stator
c. worn brushes
d. a bad voltage regulator

15. Fusible links are used to:

a. provide a path for alternator output
b. protect the charging system from low current situations
c. protect the charging system from voltage surges
d. both b and c

16. When making an alternator output voltage drop test the typical current is approximately:

a. 30 amperes
b. 20 amperes
c. 10 amperes
d. 40 amperes

17. You have completed a test lamp test of an alternator capacitor. The lamp did not light. The capacitor is:

a. bad
b. shorted
c. open
d. good

18. You are testing a Delco-Remy 10-SI alternator. You have connected a voltmeter across the battery and the engine is running at 1,500 rpm. The system voltage should be:

a. between 12.6 and 15.5 volts
b. about 16 volts
c. less than 12.6 volts
d. exactly 13.2 volts

19. You discover that someone has reversed the connections of a vehicle's battery cables. Although there are no visible problems you know that this situation can:

a. cause unregulated voltage
b. destroy alternator diodes
c. produce CEMF
d. affect brush life

PART FOUR
IGNITION SYSTEM TESTING AND SERVICE

INTRODUCTION

Ignition system service has always been a primary part of a traditional engine tuneup. Changing the plugs, points, and condenser and setting the timing could often restore new-car performance to a high-mileage engine. Ignition service is not quite that simple anymore.

Although breaker-point ignitions have not been used in many new cars for over a decade, a large number of vehicles still on the road have these systems and require regular service. Moreover, if you understand the basic service procedures for an electromechanical breaker-point ignition, you can apply comparable skills to electronic systems.

Chapters 12 and 13 cover ignition service in terms of the low-voltage primary and the high-voltage secondary circuits. The modern tuneup technician is a skilled electronic troubleshooter, and these two chapters concentrate on the testing and diagnostic principles of ignition service. By emphasizing the fundamental similarities of ignition test procedures, these chapters will help you apply basic service principles to all systems, breaker-point and electronic alike. Chapter 13 also contains inspection, repair, and adjustment instructions for secondary-circuit components common to all ignitions.

Distributor service has always been a critical part of overall ignition service. Chapter 14 contains instructions for distributor service in the engine, as well as for distributor removal, cleaning, and installation. You can do many test and service operations on a distributor using a distributor test machine, and this chapter outlines these procedures. Chapter 14 also has several photo sequences as examples of step-by-step service procedures for different distributors.

Along with proper air-fuel ratio control, ignition timing is the most important engine-operating factor for emission control, fuel economy, and top performance. Chapter 15 covers the basic procedures for timing adjustment and for spark advance testing.

12 IGNITION PRIMARY CIRCUIT TESTING AND SERVICE

INTRODUCTION

Because the ignition system is a major combustion control system, you must know how ignition operation and adjustment affect other systems and overall engine operation. There is an old saying that 80 percent of all carburetor problems are in the distributor. Whether this extreme statement is true or not, it emphasizes the importance of proper ignition service. Some engineers and mechanics consider the ignition system to be the most important emission control system on an engine. The number of ignition modifications that engineers have made to control emissions, and the effects of ignition misadjustment or misfire on emissions support this view. If an engine is to run properly, the ignition must work correctly from the battery supply voltage at the ignition switch to the spark at the plugs. One fault in the primary or the secondary circuit will seriously affect performance, emissions, economy, and driveability.

To describe ignition operation, we divide the system into the low-voltage primary circuit and the high-voltage secondary circuit, figure 12-1. This chapter and the next use this same division to explain service procedures. This is an effective approach to ignition service because it emphasizes that voltage tests of a primary circuit will detect the same kinds of problems in breaker-point and electronic systems. It also points out that secondary circuit services are basically the same for breaker-point and electronic systems.

The distinction between primary and secondary circuit services is not absolute, however. You must remember your total-system approach to service. For example, the coil is part of both circuits, but you will learn to test it as part of primary circuit service. The distributor contains both primary and secondary circuit parts. Timing adjustment sets primary circuit switching, which controls secondary ignition. Because of their importance to performance and emission control, distributor service and ignition timing are covered in separate chapters.

This chapter and the next three complement each other. In this chapter, you will learn to test the primary circuits of breaker-point and electronic ignitions and to replace key parts.

GOALS

After studying this chapter, you should be able to do the following jobs in twice the flat-rate labor time, or less, using carmakers' instructions when necessary:

1. Make voltage-drop, available-voltage, and resistance tests on breaker-point and electronic ignition primary circuits. Interpret results and make necessary repairs.

2. Test coil current draw, resistance, and polarity. Interpret the results.

3. Measure and adjust ignition dwell, using a tach-dwellmeter.

4. Using an engine analyzer oscilloscope, test breaker-point and electronic primary circuits, analyze waveforms, and repair as required.

5. Remove, replace, and adjust primary circuit components of electronic ignitions.

IGNITION SERVICE SAFETY

Review the precautions in chapter 1 for ignition and electrical system safety and pay particular attention to the following points.

1. Do not disconnect battery terminals with the ignition switch on. High-voltage surges can damage electronic parts.

2. Do not disconnect or connect electrical connectors with the ignition on unless *specifically* directed to do so. Breaking or making a connection can cause a high-voltage surge that may damage electronic parts.

3. Do not ground the TACH terminal on Delco-Remy high-energy ignition (HEI) systems or you will destroy the module.

4. Do not short circuit the ignition primary circuit to ground without resistance. High current will burn wiring and connectors and destroy electronic parts.

5. Do not allow a secondary voltage arc near fuel system components. Explosion may result.

6. An early step of every ignition test procedure is to test for secondary voltage, delivered by the coil. Carmakers' instructions tell you to do this in one of two ways.

a. Remove a secondary cable from a spark plug or remove the coil cable from the distributor cap. Hold the disconnected end about 1/4 inch from an engine ground while cranking the engine and watch for a spark, figure 12-2.

b. Disconnect a cable from a spark plug and connect it to a spark plug simulator, or test plug, figure 12-3. Clamp the test plug to an engine ground, crank the engine, and watch for a spark between the center electrode and the shell.

The procedures in this chapter contain the individual manufacturers' recommendations. However, you can use a test plug reliably and safely to check the secondary voltage of any ignition. Many high-voltage electronic ignitions require the use of a test plug to avoid high open-circuit voltage that could damage the coil.

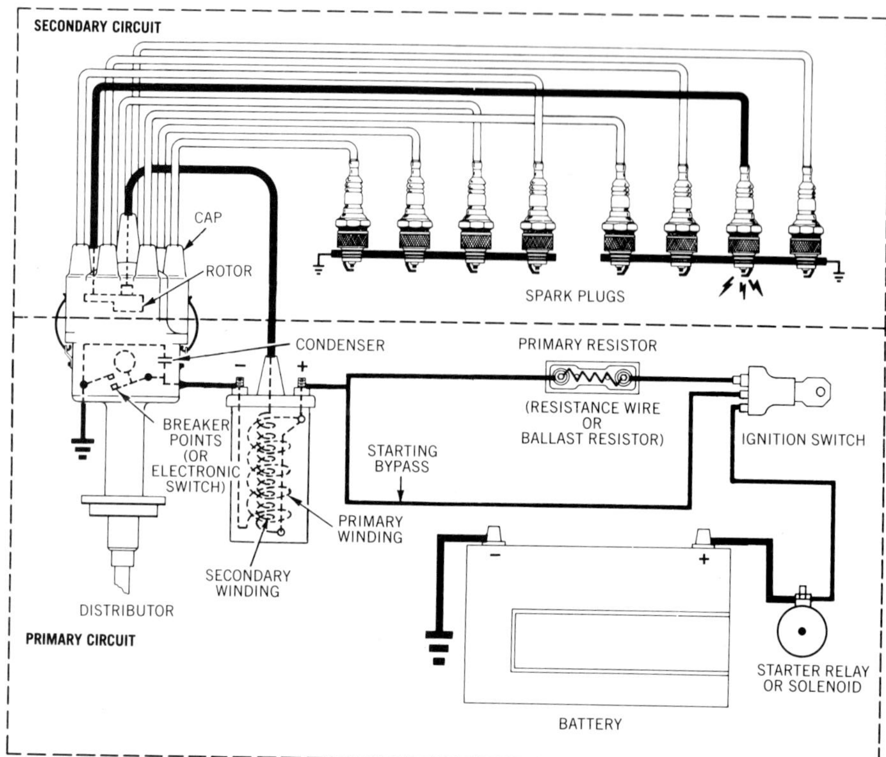

Figure 12-1. Each part of the primary and secondary circuits must work right for the complete system to operate properly. (Prestolite)

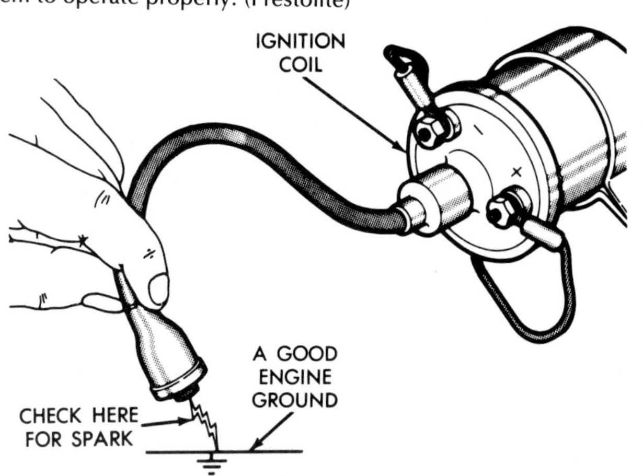

Figure 12-2. Hold the coil wire or a spark plug cable about 1/4 inch from ground while cranking the engine to check for secondary voltage. (Chrysler)

THE IGNITION SYSTEM AND OVERALL ENGINE OPERATION

Figure 12-4 is a general table that lists engine performance complaints, possible ignition problems that could cause the symptoms, and general cures. If the performance complaints look familiar, they should. Many are the same

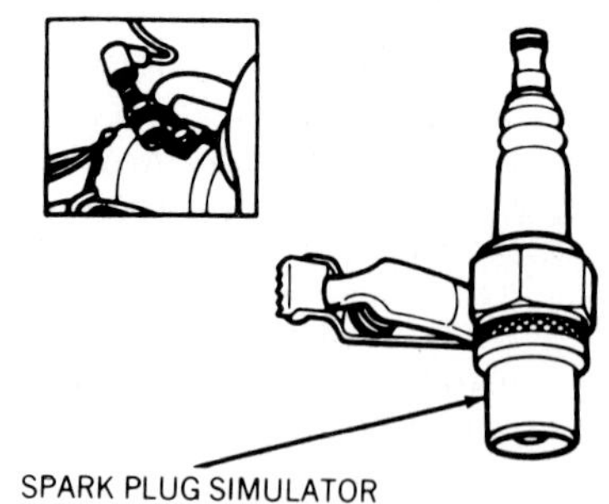

Figure 12-3. Clamp the test plug to ground, connect a spark plug cable, and crank the engine to check for secondary voltage. (Ford)

SYMPTOM—Engine cranks normally but will not start and run

Possible Cause	*Correction*
1. Open or grounded primary circuit	1. Check all primary wiring and connections. Inspect and test the coil, condenser, breaker points, distributor pickup coil or switch, and ignition switch.
2. Coil shorted or grounded	2. Test and replace the coil.
3. Points burned or not opening	3. Adjust or replace the points. Check voltage and current.
4. Wrong basic ignition timing	4. Adjust the timing.
5. Fouled spark plugs	5. Replace the spark plugs.
6. Secondary voltage leak	6. Inspect the coil, distributor cap and rotor, and spark plug cables. Replace as necessary.
7. Missing signal from the distributor pickup or engine crankshaft sensor	7. Troubleshoot the distributor pickup or crankshaft sensor circuit, following the carmaker's procedures.

SYMPTOM—Engine runs, but one cylinder misfires

Possible Cause	*Correction*
1. Bad spark plug	1. Locate and replace the bad plug.
2. Bad distributor cap terminal	2. Replace the distributor cap.
3. Loose or bad spark plug cable	3. Tighten or replace the cable.

SYMPTOM—Engine runs, but various cylinders misfire

Possible Cause	*Correction*
1. Breaker points are dirty, worn, or out of adjustment	1. Clean and adjust or replace the points.
2. Bad condenser	2. Test and replace the condenser.
3. Intermittent open circuit in distributor pickup coil or switch	3. Test and repair or replace the pickup coil or switch.
4. Intermittent open circuit in the ignition module	4. Test and replace the module as necessary.
5. Intermittent open or short circuit in ignition coil	5. Test and replace the coil if necessary.
6. Spark advance mechanisms not working properly	6. Test and repair the spark advance devices.
7. Leaking secondary wiring or distributor cap	7. Inspect, test, and replace spark plug cables and cap as necessary.
8. Intermittent open circuit or high resistance in primary wiring or connections	8. Inspect, test, and repair primary wiring and connections.
9. Fouled spark plugs	9. Replace spark plugs and correct the fouling problem.

SYMPTOM—Engine backfires

Possible Cause	*Correction*
1. Incorrect timing or spark advance	1. Test and adjust timing and advance as necessary.
2. Wrong spark plug heat range	2. Install correct plugs.
3. Ignition crossfire	3. Check the routing of spark plug cables. Inspect the cables for leakage.

SYMPTOM—Engine knocks or pings

Possible Cause	*Correction*
1. Incorrect timing or spark advance	1. Test and adjust timing and advance as necessary.
2. Wrong spark plug heat range	2. Install correct plugs.
3. Breaker points out of adjustment	3. Adjust or replace the points.
4. Low-octane gasoline	4. Use higher octane gasoline.
5. Detonation sensor not working	5. Test and replace the sensor.

SYMPTOM—Engine has low power and poor performance

Possible Cause	*Correction*
1. Incorrect timing or spark advance	1. Test and adjust timing and advance as necessary.
2. Breaker points out of adjustment	2. Adjust or replace the points.
3. Engine with electronic spark timing is running on base timing from the ignition module, without computer control of spark advance	3. Troubleshoot the electronic spark timing and engine computer, following the carmaker's procedures.

SYMPTOM—Breaker points burn or oxidize

Possible Cause	*Correction*
1. High charging system voltage	1. Test and repair the charging system.
2. Long dwell angle	2. Adjust the dwell.
3. High resistance in the condenser	3. Test and replace the condenser.
4. Weak point spring tension	4. Replace points and check spring tension.
5. Oil vapor collecting on the points due to overlubricated distributor or high crankcase pressure	5. Clean the distributor and service the PCV system.

Figure 12-4. This table lists possible ignition causes of common performance complaints.

symptoms with which you start troubleshooting fuel and emission control systems. While this table lists only the possible ignition causes for each complaint, it shows that several faults, or faults in several systems, can cause the same general problem.

PRIMARY CIRCUIT VOLTAGE AND RESISTANCE TESTS

Primary circuit voltage directly affects secondary ignition voltage. If you lose 1 volt in the primary circuit, you can lose 1 kilovolt (1,000 volts) in the secondary. Low primary voltage can be caused by low source voltage from the battery or charging system. More often, however, it is caused by high resistance in the primary circuit. Low source voltage can be due to:

- A discharged battery
- Low charging voltage
- High current draw at the starter motor.

High primary resistance can be due to:

- A bad coil, points, condenser, or ignition module
- A bad ballast resistor
- A bad ground at the ignition module
- Loose, damaged, or corroded primary wiring and connectors.

The following procedures allow you to test available voltage, voltage drop, and resistance at key points in the primary circuit. These procedures are for breaker-point systems and the most common electronic ignitions from Chrysler, GM, and Ford. Test principles are similar for other systems, but refer to the carmaker's specific procedures for accurate troubleshooting.

Remember, when you use a voltmeter to measure available voltage or voltage drop, you are also checking resistance. High resistance will lower the voltage at a point in a circuit. High voltage drop across a component or connection indicates high resistance.

Before testing the ignition, be sure the battery is fully charged. Disable the secondary circuit, if required, as explained in each procedure. Be sure that the transmission is in neutral or park and that the parking brake is set.

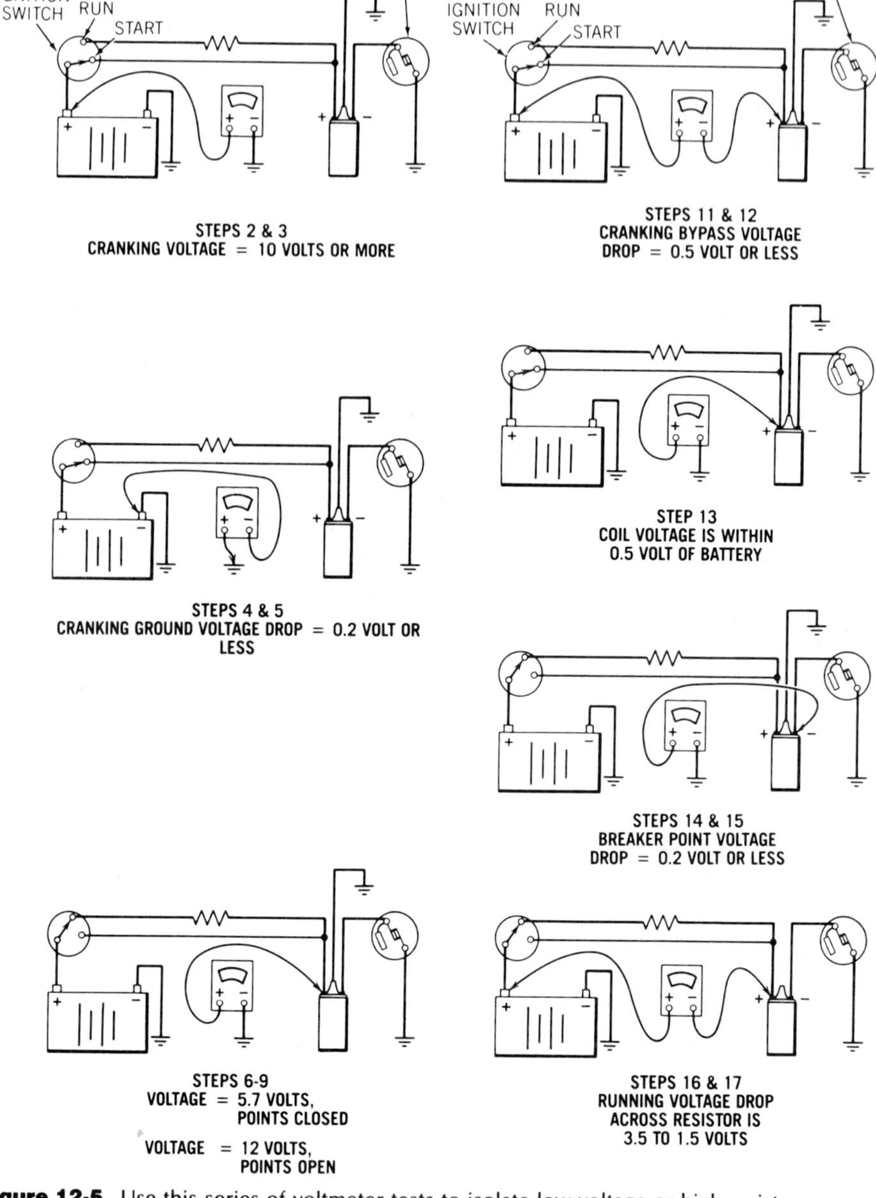

Figure 12-5. Use this series of voltmeter tests to isolate low-voltage or high-resistance problems in a breaker-point primary circuit.

Breaker-Point Primary Circuit Voltmeter Tests

Refer to figure 12-5, which shows the six test points for the following procedure. All instructions are for a negative-ground electrical system. Observe the correct polarity as shown in the drawings when you connect your voltmeter for each test point.

1. Remove the secondary cable from the center tower of the distributor cap and connect it to ground with a jumper wire, figure 12-6.

2. Connect the voltmeter + lead to the battery + terminal or post, *not* to the cable. Connect the − lead to ground.

3. Crank the engine and watch the voltmeter. If voltage is 10 volts or more, the battery is supplying adequate voltage. If voltage is below approximately 9.5 volts, test the starter motor current draw and do a battery load test to check

Figure 12-6. Ground the coil secondary lead during primary circuit testing.

battery capacity. Also check for high resistance in the battery ground connection as directed in the next step.

4. Connect the voltmeter + lead to ground. Connect the − lead to the battery − terminal or post, *not* to the cable.

5. Crank the engine and watch the voltmeter. The voltage drop should be 0.2 volt or less. If voltage is above 0.2 volt, look for high resistance in the ground connection. Clean and tighten the battery ground cable connections or replace the cable.

6. "Bump" the engine with the starter motor or remote starter switch until the points are closed. Turn the ignition on.

7. Connect the voltmeter + lead to the coil + terminal and the − lead to ground.

8. Note the meter reading. It should be 5 to 7 volts (about one-half of battery voltage). A lower reading indicates high resistance in the primary circuit.

9. Bump the engine with the starter to open the points or open them by hand. Leave the meter connected as in step 8 and note the reading. It should equal battery voltage (about 12 volts). Lower voltage indicates a high-resistance short to ground between the battery and the coil.

10. If voltage in steps 8 and 9 is not within limits, test primary (ballast) circuit connections, resistor, and coil.

11. To test the primary resistor bypass circuit, connect the voltmeter + lead to the battery + terminal (not the cable clamp) and the − lead to the coil + terminal.

12. Crank the engine and watch the voltmeter. Voltage drop should be 0.5 volt or less. If it is higher, there is high resistance in the bypass circuit. Test the circuit wiring and connections from the battery through the ignition switch or relay to the coil.

13. You also can check the primary resistor bypass circuit by connecting the voltmeter + lead to the coil + terminal and the − lead to ground while cranking the engine. Voltage should be within 0.5 volt of battery voltage measured in step 3.

14. To test voltage drop (resistance) across the breaker points, connect the voltmeter + lead to the coil − terminal or the distributor primary terminal. Connect the meter − lead to ground.

15. Bump the engine with the starter until the points are closed and read the meter. Voltage drop should be 0.2 volt or less (ideally 0.1 volt). If it is higher, look for burned or pitted points, a poor ground connection, and loose primary wiring or connectors.

16. To test voltage drop across the primary resistor with the engine running, reconnect the coil secondary cable to the distributor. Connect the voltmeter + lead to the battery + terminal and the − lead to the coil + terminal.

17. Start and run the engine at 1,500 rpm and read the meter. Voltage drop should be 1.5 to 3.5 volts depending on the vehicle. If the reading is higher or lower than this range, test the resistor and its circuit for short circuits or high resistance.

Chrysler Electronic Ignition Primary Voltage and Resistance Tests—Magnetic-Pulse Distributors

Use the following basic procedure to test Chrysler magnetic-pulse electronic ignitions from 1972 to the late 1980's. If the ignition is part of an electronic lean-burn (ELB) or electronic spark control (ESC) system, you must use Chrysler test procedures for the specific year and model. Early Chrysler systems have a dual ballast resistor. Many later systems have only a single ballast resistor. The accompanying illustrations show a dual resistor. For a single-resistor system, disregard the 5.0-ohm auxiliary resistor circuit.

Most late-model Chrysler engines use ignitions with Hall-effect distributor switches or direct ignitions without distributors. Mitsubishi engines in Chrysler vehicles use ignitions with magnetic-pulse distributors or optical

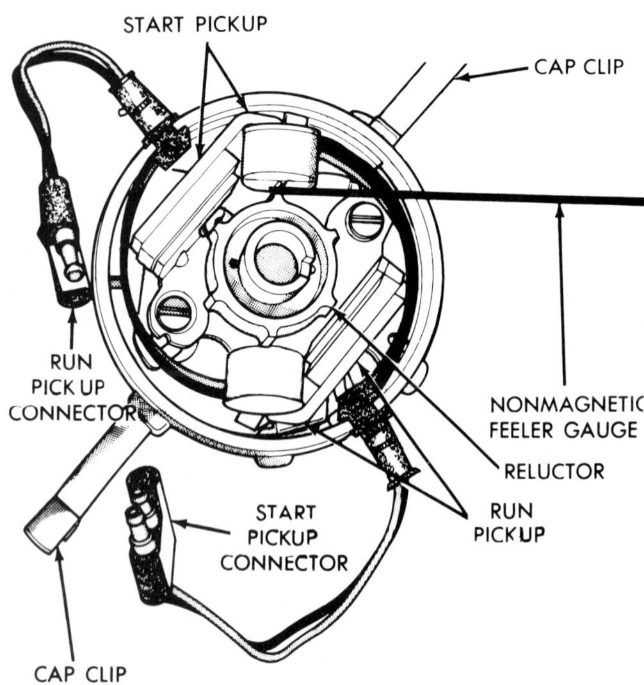

Figure 12-7. Some Chrysler distributors have dual pickups with different air gaps. Adjust each separately. (Chrysler)

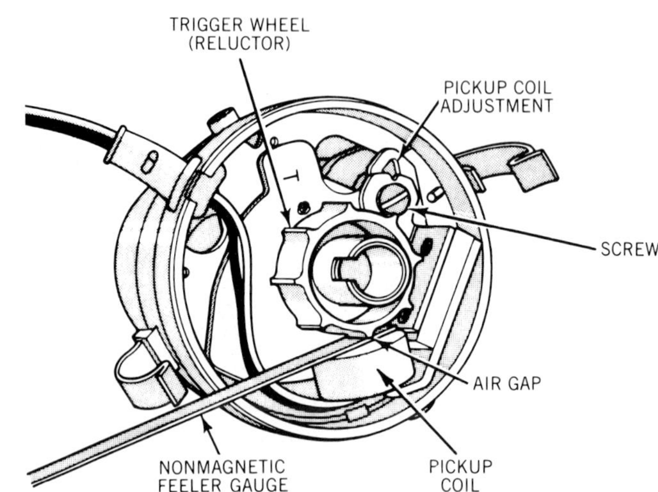

Figure 12-8. Adjust the air gap by moving the pickup coil on the breaker plate. Use nonmagnetic (brass) feeler gauges. (Chrysler)

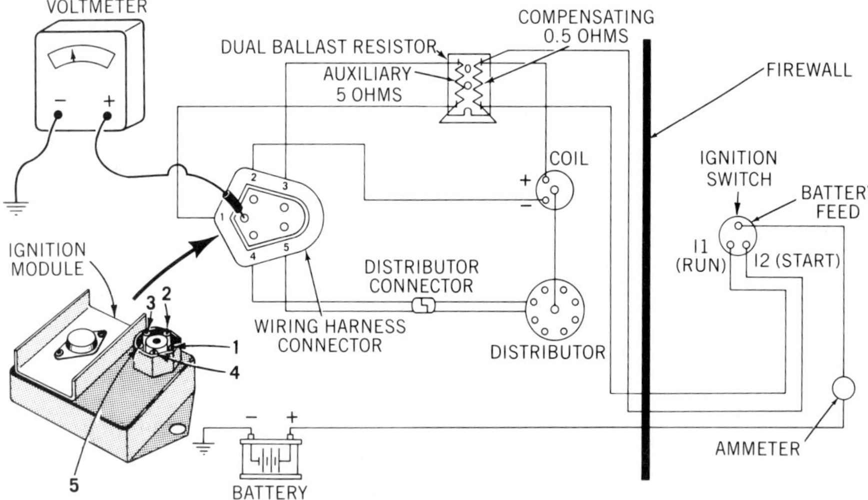

Figure 12-9. Make your voltmeter and ohmmeter tests at the ignition module harness connector. (Chrysler)

distributors. Test procedures vary slightly for different models and years. One example of testing a Hall-effect distributor system follows these pages on testing the basic magnetic-pulse system.

Chrysler Air Gap Check and Adjustment. Although an electronic distributor has no breaker-point rubbing block and cam to wear, the distributor bushings can wear from simple rotation on a high-mileage distributor. Also, the breaker plate and vacuum advance mechanism may wear and become loose. All of these things can change the air gap between the trigger wheel (reluctor) and the pickup coil.

Use nonmagnetic (brass) feeler gauges to check and adjust the air gap. The pickup coil magnet will attract a steel gauge, which makes accurate adjustment impossible. Chrysler specifies different air gaps for different years and models. Check the specifications.

If the distributor has two pickup coils, figure 12-7, check and adjust the air gap for each.

1. Check the air gap between the trigger wheel (reluctor) and the pickup coil by bumping the engine with the starter to align one wheel tooth with the pickup coil.

2. Place a nonmagnetic feeler gauge of the specified thickness between the reluctor tooth and the pickup coil pole piece, figure 12-8.

3. If clearance is too loose or too tight, loosen the pickup coil lockscrew and use a screwdriver blade in the adjustment slot, figure 12-8, to shift the breaker plate so the feeler gauge *just contacts* the wheel tooth and the pickup coil. Tighten the lockscrew.

4. Check the adjustment with the specified "no-go" feeler gauge. It should not fit into the air gap.

5. Apply vacuum to the advance diaphragm to move the breaker plate through its full travel. Be sure the pickup coil does not hit the trigger wheel (reluctor) tooth.

Chrysler Electronic Ignition Circuit Tests

Caution. Be sure the ignition switch is off whenever you disconnect or connect the ignition module connector. Do not touch the switching transistor on the ignition module when the ignition is on. High voltage is present.

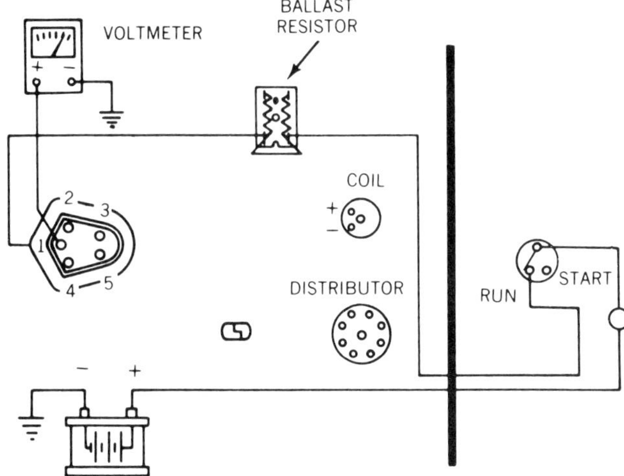

Figure 12-10. Chrysler ignition circuit for connector terminal 1. (Chrysler)

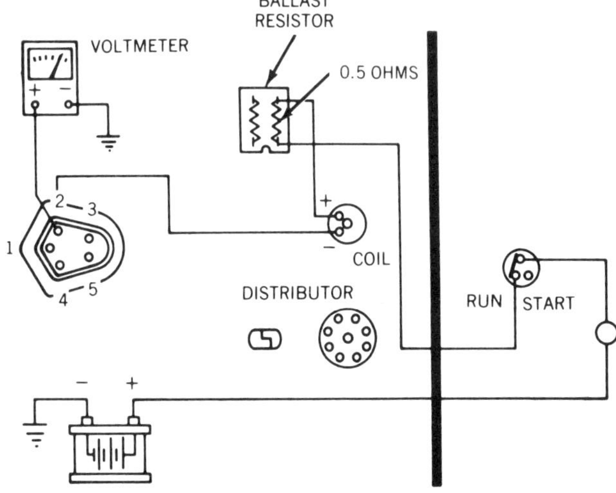

Figure 12-11. Chrysler ignition circuit for connector terminal 2. (Chrysler)

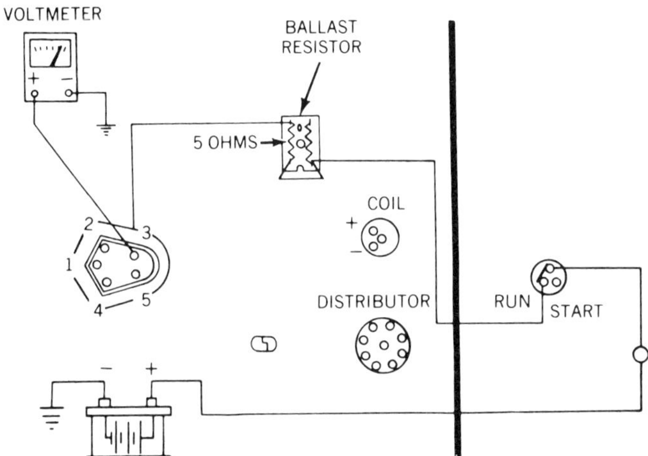

Figure 12-12. Chrysler ignition circuit for connector terminal 3. (Chrysler)

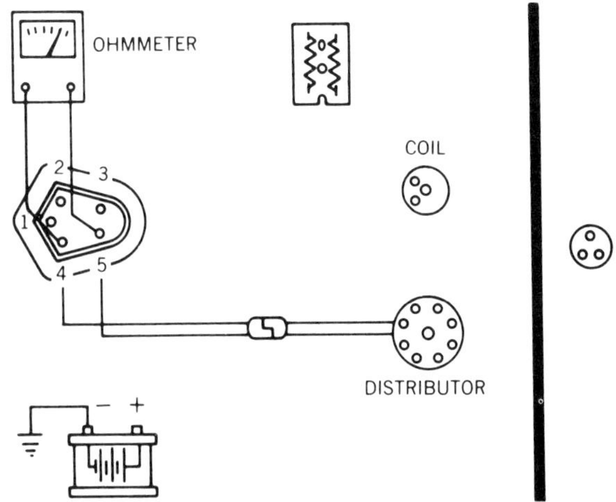

Figure 12-13. Chrysler ignition circuit for connector terminals 4 and 5 (pickup coil). (Chrysler)

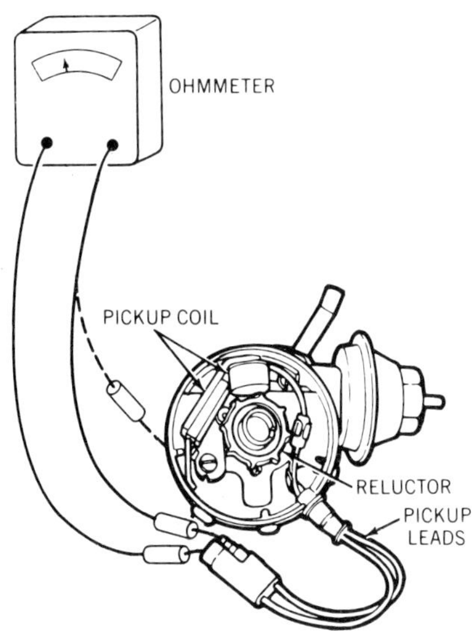

Figure 12-14. Test pickup coil resistance and for a grounded pickup coil at the distributor. (Chrysler)

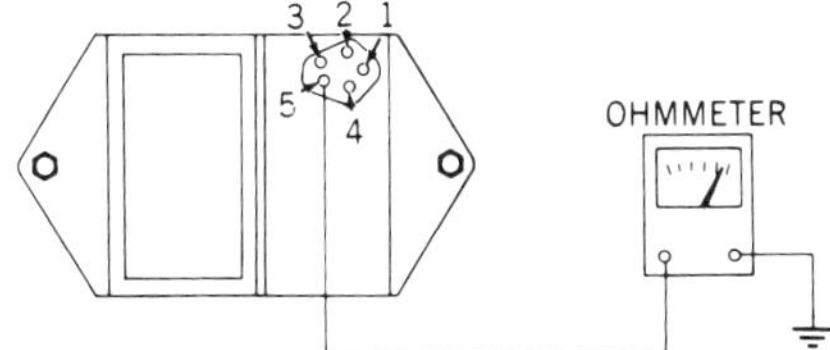

Figure 12-15. Check ignition ground continuity at module pin 5. (Chrysler)

1. With the ignition off, disconnect the vehicle harness connector from the ignition module.

2. Turn the ignition on and connect your voltmeter – lead to ground and the + lead to cavity 1 of the harness connector, figure 12-9, not to the module.

3. Read the voltmeter. Voltage should be within 1 volt of battery voltage. If it is not, test and repair the circuit shown in figure 12-10.

4. Move the voltmeter + lead to harness connector cavity 2 and read the voltmeter. If voltage is not within 1 volt of battery voltage with the ignition on, test and repair the circuit shown in figure 12-11.

5. Move the voltmeter + lead to harness connector cavity 3 and read the voltmeter. If voltage is not within 1 volt of battery voltage with the ignition on, test and repair the circuit shown in figure 12-12.

6. Turn the ignition switch off.

7. Connect an ohmmeter between cavities 4 and 5 of the harness connector, figure 12-13, to test pickup coil resistance. It should be 150 to 900

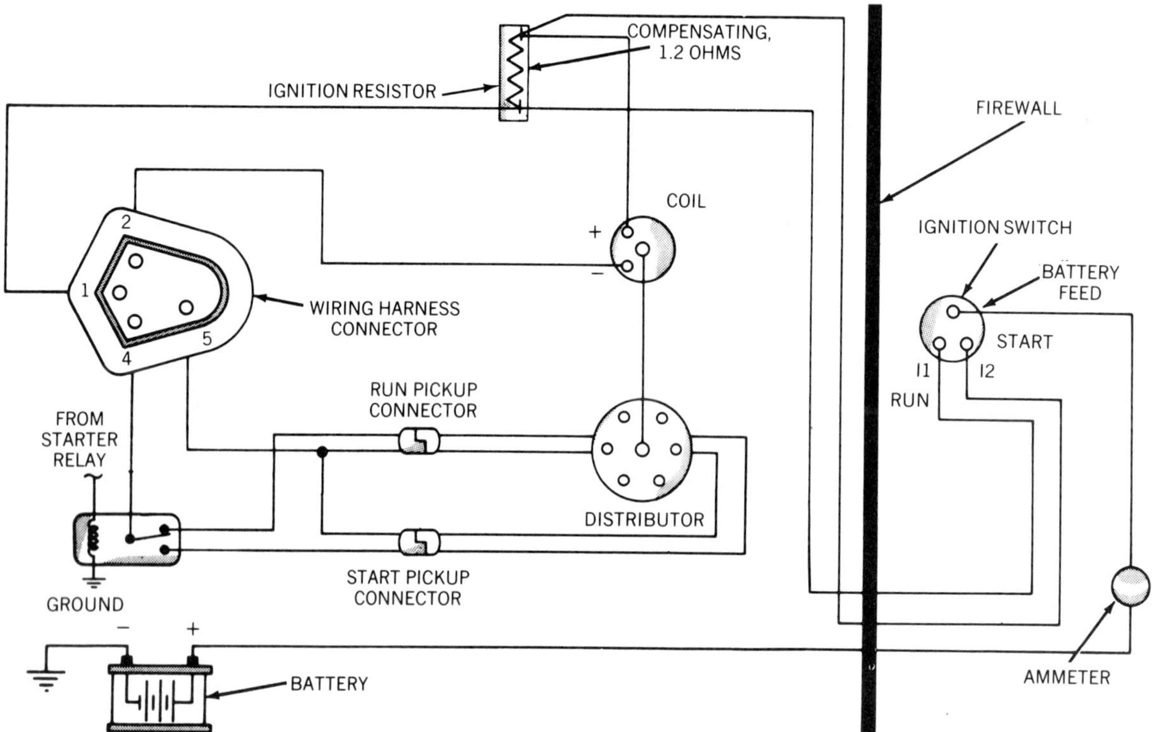

Figure 12-16. Late-model Chrysler magnetic pulse ignitions have a single ballast resistor and dual pickup coils. Test principles are similar to those for earlier systems, however. (Chrysler)

ohms. If it is, go to step 9; if not, continue with step 8.

8. Disconnect the connector at the distributor and measure resistance through the pickup coil at the distributor, figure 12-14. If it is 150 to 900 ohms, repair the wiring from the harness to the distributor. If resistance is out of limits, replace the pickup coil.

9. Connect one ohmmeter lead to ground and the other alternately to both sides of the pickup coil lead at the distributor. The ohmmeter should show infinite resistance at both points to indicate that the pickup coil is not grounded. If not, replace the pickup coil.

10. With one ohmmeter lead connected to ground, connect the other to ignition module pin 5, figure 12-15. The meter should show continuity (0 to 1 ohm). If not, clean the module mounting points and tighten the mounting screws and repeat the test. If resistance is still high, replace the module.

11. If the system still does not operate right, test the ballast resistor and the coil as explained later in this chapter. Many mid-1980's Chrysler 6- and 8-cylinder ignitions have distributors with two pickup coils and only a single ballast resistor. Although specific test procedures vary slightly, they are based on the principles of the preceding steps. Figure 12-16 is a diagram of a typical late-model Chrysler ignition. Notice that the system has a pickup coil relay and that terminal 3 has been removed from the harness connector. The ignition module of most current Chrysler ignitions is inside the spark control computer housing. Tests are made at the 10-wire and 12-wire connectors to the computer.

Chrysler Hall-Effect Switch Tests

Chrysler engines with fully integrated engine control systems have the ignition module built into the power section of the engine computer. This is in the power module on most systems before 1988 and in the single-module engine controller (SMEC) or single-board engine controller (SBEC) on most 1988 and later vehicles. Primary circuit switching is similar to earlier Chrysler electronic ignitions, but most late-model Chrysler ignitions are variable-dwell systems. They have no ballast resistor.

Chrysler driveability test procedures give instructions for voltmeter and ohmmeter tests of the primary circuit. These tests are made at the terminals of the computer harness connectors, not at a single ignition module connector. Test points vary slightly at connector terminals for different years and models. Therefore, we will not try to give you a single test procedure here. You should refer to Chrysler wiring diagrams and instructions for the specific vehicle you are testing.

You can, however, make some quick tests on the distributor Hall-effect switch, or switches, to see if it can send a timing signal to the computer. Some distributors have a single Hall-effect switch; others have two. If the distributor has two, check Chrysler procedures to determine which switch to test for specific driveability problems or fault codes.

This Hall-effect switch is particularly useful for troubleshooting no-start problems. Perform this test as follows. Be sure the computer harness connectors are connected.

1. Disconnect the vehicle harness connector from the distributor connector and put a jumper wire between cavities 2 and 3 of the vehicle harness connector, figure 12-17A.

2. Disconnect the coil cable from the distributor cap, turn the ignition on, and hold the end of the coil near a good ground.

3. Make and break the jumper wire connection several times at either cavity 2 or 3 and watch for a spark from the coil cable.

- **a.** If spark is present, the computer is capable of receiving a Hall-effect signal and firing the ignition coil.
- **b.** If spark is not present, an open circuit may exist in the wiring from cavity 2 or 3 to the computer, or the computer may be at fault. Use an ohmmeter and a wiring diagram to check wire continuity.

4. If spark is present in step 3, connect a voltmeter between cavity 1 of the harness connector and ground, figure 12-17B. Voltage should be 7 volts or more.

- **a.** If voltage is not present, trace and repair the wiring. If wiring is okay, the computer may be at fault.
- **b.** If the required voltage is present, the Hall-effect switch is probably bad.

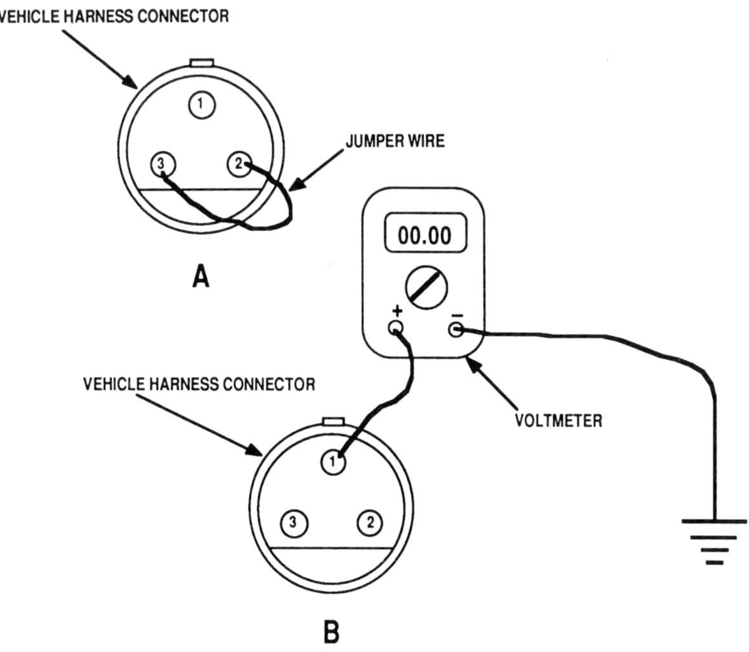

Figure 12-17. Chrysler distributor Hall-effect test points.

GM-Delco High-Energy Ignition (HEI) Primary Voltage and Resistance Tests

The following tests are for basic HEI systems with magnetic-pulse signal generators. The illustrations show test points for systems with the coil in the distributor cap and with the coil mounted separately. Before testing the ignition primary circuit, check the fuel delivery and be sure the battery cranking voltage is 9.5 volts or more. Inspect the primary wiring and connectors for damage and looseness.

> **Caution.** Do not operate the HEI system with an open secondary circuit. High secondary voltage may damage the coil or cause a short circuit in the distributor.
>
> Do not ground the TACH terminal during testing or the module will be damaged.

Delco recommends that you test for secondary voltage, or spark at the plugs, with a spark tester, or spark plug simulator, figure 12-3. Connect the grounding clip to an engine ground and connect one spark plug cable to the tester. Crank the engine and watch for a spark between the tester center electrode and shell. This indicates adequate secondary voltage capability. After the preliminary tests, check the HEI system as follows:

1. Be sure that the four distributor cap latches are secure and that the coil cover screws are tight on systems with the coil in the distributor.

2. Connect the voltmeter + lead to the BAT terminal at the distributor and the − lead to ground, figure 12-18.

3. Turn the ignition to start (crank the engine) and read the voltmeter. It should indicate 7 volts or more. If cranking voltage is below 7 volts, battery voltage is low or there is high resistance in the primary circuit. If cranking voltage is zero, the primary circuit is open.

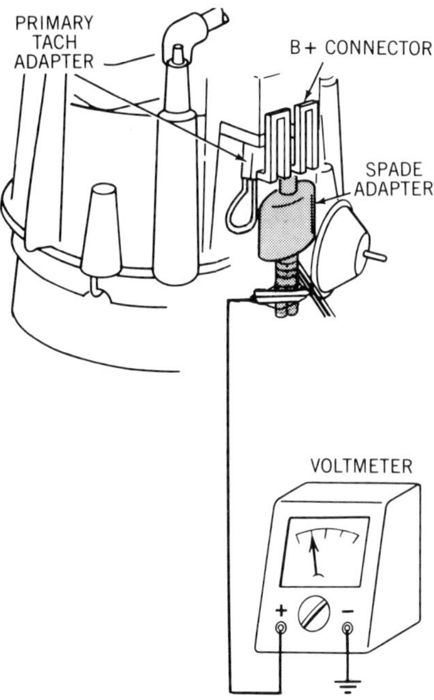

Figure 12-18. Check HEI primary cranking voltage at the BAT terminal.

4. Move the voltmeter + lead to the TACH terminal, figure 12-19, turn the ignition on, do not crank the engine. Voltage should be at least 10 volts (ideally within 0.5 volt of battery voltage). If voltage is 1 volt or less, the

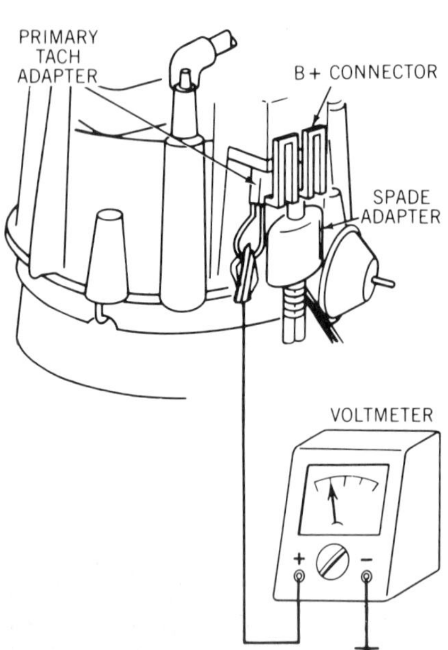

Figure 12-19. Check HEI primary run voltage at the TACH terminal.

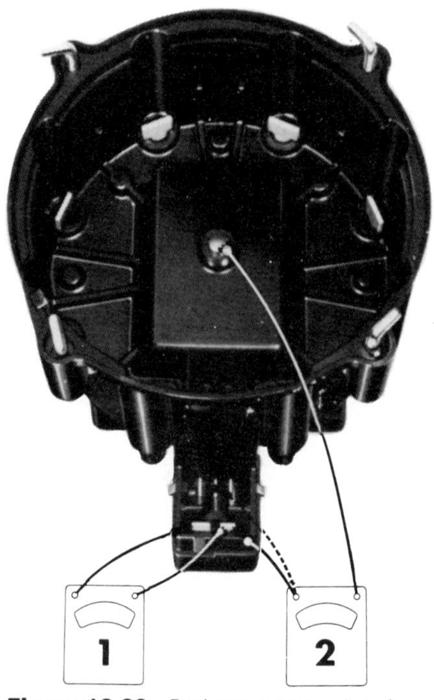

Figure 12-20. Resistance test points for an HEI built-in coil. (Delco-Remy)

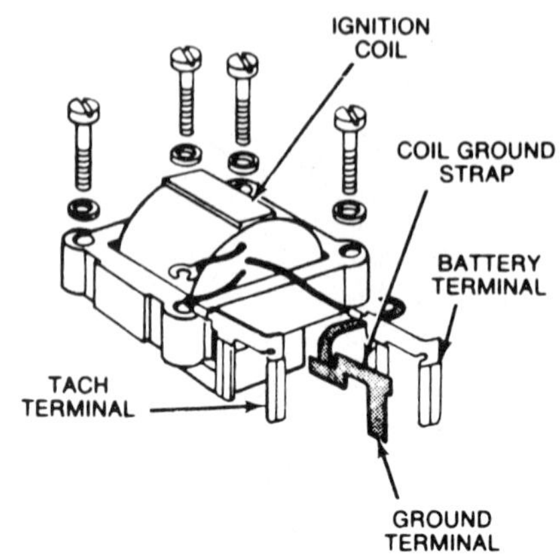

Figure 12-21. Integral HEI coils have a ground terminal and strap that must be installed correctly.

coil is probably bad. If voltage is between 1 and 10 volts, test or replace the ignition module according to the carmaker's procedures.

5. If voltage was out of limits in step 3 or 4, use the carmaker's wiring diagrams to trace and repair the primary circuit. Then proceed to test the coil.

6. Remove the cap and coil from a distributor with a built-in coil by turning the four cap latches.

7. Test the resistance of built-in coils, figure 12-20, as follows:

a. Connect an ohmmeter as shown for test point 1 to measure primary resistance. It should be 0 to 1 ohm on 1980 and earlier models or 0 to 2 ohms on 1981 and later models.

b. Connect on ohmmeter as shown for test point 2 to measure secondary resistance. Measure between the cap button and the TACH terminal, and then between the cap button and the GROUND terminal. Replace the coil if *both* readings are infinite or do not match secondary resistance specifications, which vary for different years and models. All integral HEI coils builkt since 1978 have a coil ground strap with a terminal that fits into the distributor connector, figure 12-21. The other end of the ground strap is fastened to one of the mounting screws. Be sure the ground strap and terminal are installed correctly when replacing a coil.

8. Test the resistance of separate HEI coils, figure 12-22, as follows:

a. Connect an ohmmeter as shown for test point 1 to measure primary resistance. It should be 0 to 1 ohm.

b. Connect an ohmmeter as shown for test point 2 to measure secondary resistance. Resistance specifications vary for different years and models.

c. Connect an ohmmeter as shown for test point 3 to check for a grounded coil. The meter should show infinite resistance. Note the terminal shown on the late-model coil in figure 12-22, for which internal connection problems have existed. Wiggle the terminal while testing if you suspect an intermittent secondary problem.

9. Test the HEI pickup coil (pole piece), figure 12-23, as follows:

a. Disconnect the pickup coil leads from the ignition module in the distributor. On 5- and 7-pin modules, be sure to disconnect only the pickup coil.

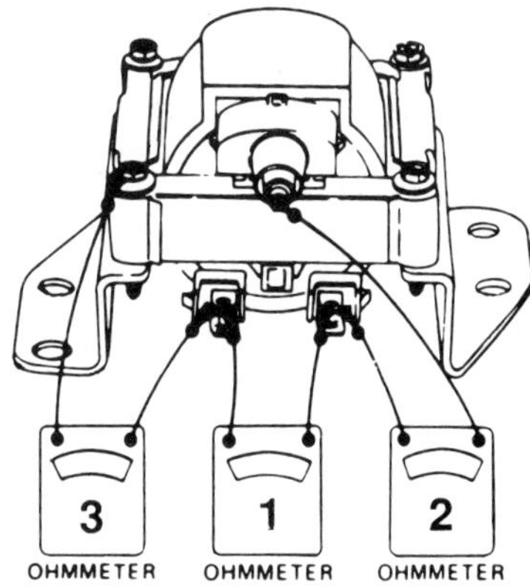

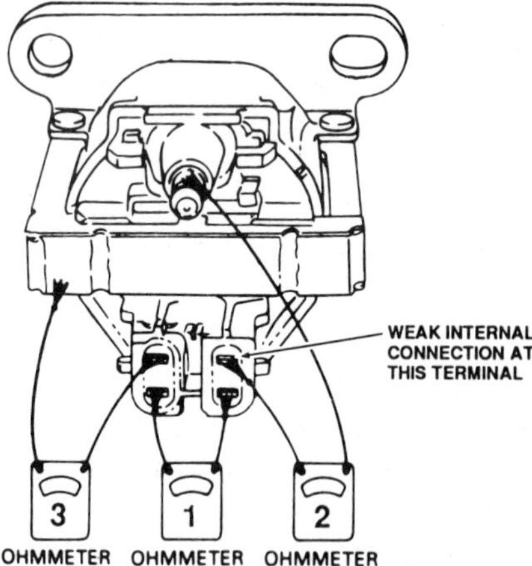

Figure 12-22. Check primary and secondary windings of remote HEI coils with an ohmmeter. Note the problem area in late-model coils as shown above.

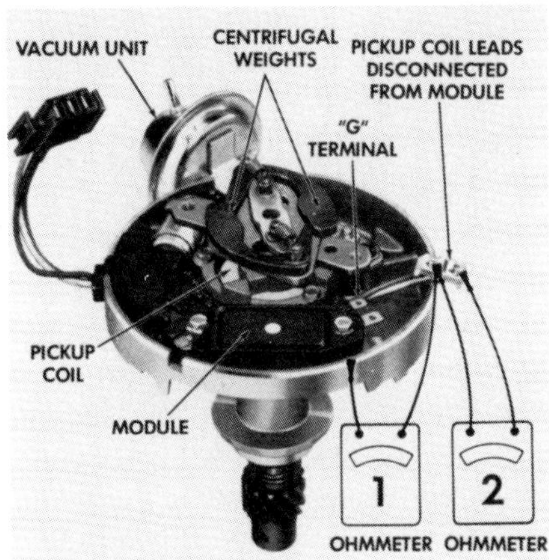

Figure 12-23. Test pickup coil resistance and for a grounded pickup coil at the distributor. (Delco-Remy)

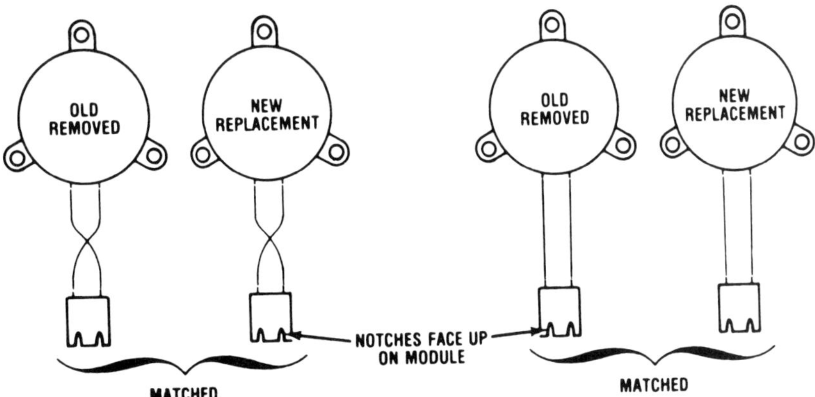

Figure 12-24. Check HEI pickup coils to be sure you install the correct replacement.

b. Connect an ohmmeter as shown for test point 1 and a hand-operated vacuum pump to the vacuum advance diaphragm.

c. Operate the vacuum advance unit through its full range and read the ohmmeter. Resistance between either pickup coil lead and the distributor (ground) must be infinite.

d. Connect the ohmmeter for test point 2 (across the pickup coil leads) and again operate the vacuum advance. Resistance must be within the carmaker's specifications.

e. Replace the pickup coil if resistance is out of limits for either test pint 1 or 2. Replace the vacuum advance unit if it does not operate throughout its full range without leaking. You can check the pickup coil signal voltage by connecting a digital voltmeter across the pickup coil leads. Set the emter on the low ac scale and crank the engine. The meter should read more than 300 millivolts (0.3 volt).

10. General Motors has used more than 21 different pole pieces for pickup coils in HEI distributors since 1975. When replacing a pole piece, you must be sure to install the correct one. If a pole piece with reversed magnets or reversed leads is installed, the signal to the ignition module will be out of phase with engine tdc. The engine may not start, or it may be impossible to adjust timing correctly. Make these two checks for a replacement pole piece:

a. Place the top of the old pole piece against the top of the replacement. The magnets should repel each other. They definitely should not attract.

b. Lay the pickup connector leads out with the notches in the connector up, figure 12-24. If the leads of the old pole piece are crossed, the leads of the new one should be crossed. If the leads of the old pole piece are parallel, the leads of the new one should be parallel.

11. Test the HEI module current draw with a low-scale, series-connected ammeter as follows:

a. Disconnect the battery + lead (large pink wire) from the distributor, figure 12-25.

b. Connect the ammeter + lead to the pink wire connector and the – lead to the distributor terminal.

c. Operate the ignition and engine in the following conditions and read the ammeter.

- Key on, engine off—0.1 to 0.2 amp
- Engine cranking—0.5 to 1.5 amps
- Engine idling—0.5 to 1.5 amps
- Engine at 2,000 to 2,500 rpm—1.0 to 2.8 amps

Testing HEI with Electronic Spark Timing

Figure 12-25. To measure HEI coil current draw, connect an ammeter to the large pink wire at the BAT connector terminal.

Most GM engines since 1981 with fully integrated electronic control systems use HEI systems with electronic spark timing (EST) control by the engine computer. EST problems can cause several symptoms, such as:

- The engine may not start.
- The engine may die but then restart.
- The engine may start and run on base timing with little or no spark advance.

Because of the variety of ignitions used by GM, both distributor-type and DIS, you should check wiring diagrams and test procedures for a specific car or truck when troubleshooting an EST problem. The basic tests in the following sections, however, will help you diagnose HEI-EST problems.

Code 42 and EST Problems. Trouble code 42 is the most common GM code for EST problems. It should appear for any of the following conditions:

- An open or grounded EST line (A grounded EST line should cause the engine to die and not start.)
- An open or grounded bypass line
- An open or grounded "set-timing" connector in the bypass line
- A PROM not fully seated in the computer
- A 1981 ignition module installed in a later system

Figure 12-26 is a basic HEI-EST diagram that shows the EST connections at the HEI module 4-wire connector.

- The HEI reference circuit (430) is a 5-volt square wave signal that goes from pin R of the HEI module to the computer. This is the basic pickup coil timing signal that indicates crankshaft position.
- The bypass circuit (424) is a 5-volt signal that goes from the computer to the HEI module pin B. This signal changes the control of the ignition primary circuit from the HEI module to the computer for electronic spark advance.
- The EST signal (423) is a 5-volt square wave that goes from the computer to pin E of the HEI module. This circuit triggers the module when the computer is controlling timing. The computer does not know what the actual timing is, but it knows when it gets a reference signal from the HEI module. The reference signal is the base point from which the computer advances or retards timing. If base timing is incorrect, the entire spark curve will shift by the amount that base timing is off.

The table in figure 12-27 summarizes some of the problems that can be caused by a grounded or open wire in the EST, the HEI reference, or the 5-volt bypass circuit from the ignition module to the computer. Figure 12-27 applies to *carbureted engines only*. Similar problems may or may not occur on fuel-injected engines. Refer to GM test procedures for specific late-model, fuel-injected engines.

EST Quick Test. You can make a quick test to see if the engine computer is providing EST control by running the engine at about 2,000 rpm and then interrupting EST operation. If engine speed then drops, it means that the computer was advancing timing and that spark advance has now returned to base timing. Make this test in either of two ways:

1. With engine speed at about 2,000 rpm, install a jumper between pins A and B in the ALDL test connector (the top two terminals on the left of the 12-pin connector, figure 12-28).

2. Check the vehicle emission control information (VECI) decal for the location of the "set-timing" connector. Then open the connector with the engine running.

If rpm does not change with either of these two methods, the computer EST function was not working. Follow GM diagnostic procedures to troubleshoot the EST function. The second method outlined above will cause code 42 to appear in computer memory, which must be cleared after service.

HEI Module Test. You can test the HEI module to see if it can respond to a pickup coil signal and switch the primary circuit, as follows:

1. Connect an HEI spark tester to the ignition coil in one of two ways:

a. For a remote coil, connect a spark plug cable from the coil tower to the tester. Then clamp the tester to ground.

b. For an integral coil, cut a spark plug boot as shown in figure 12-29. Install the boot on the tester and then connect the tester to the coil center terminal and ground as shown.

2. Connect a test lamp to battery positive (+) or to a voltage source of 1.5 to 8.0 volts.

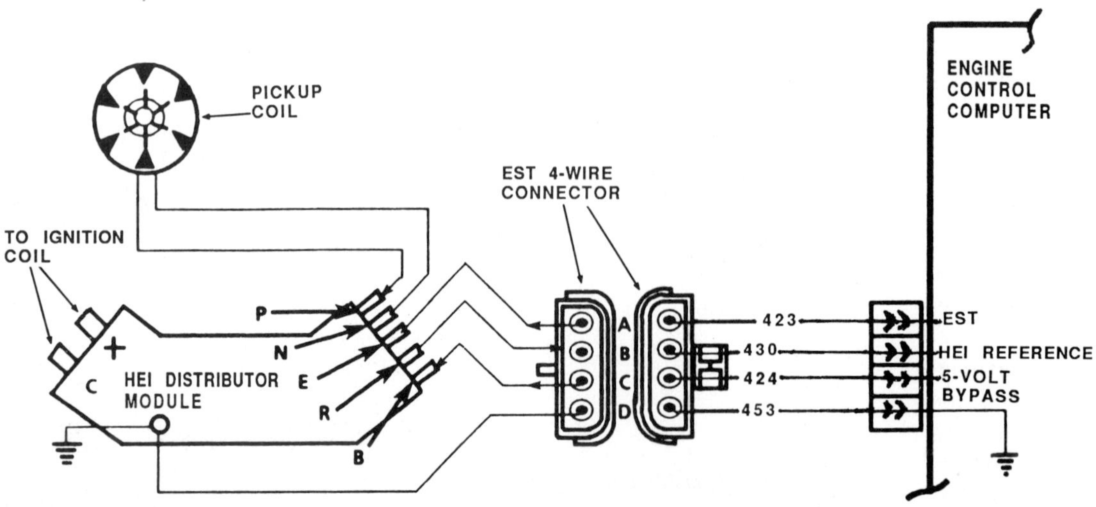

Figure 12-26. This diagram is typical of all GM HEI systems with electronic spark timing (EST).

GM HEI EST Circuits (Carbureted Engines—Most Systems)

	GROUNDED Engine Off or Cranking	GROUNDED Engine Running	OPEN Engine Off or Cranking	OPEN Engine Running
EST MODE (HEI Pin E, ECM Pin 12) *4-wire Pin A*	No fault when cranking. At start, voltage on EST line does not operate transistor. Engine stalls, will not run. Then sets Code 42.	Engine stalls. Will not restart. Sets Code 42.	Engine starts and runs on pickup coil (HEI mode) and base timing. Sets Code 42.	Engine stalls; then restarts. Runs on pickup coil (HEI mode) and base timing. Sets Code 42. (NOTE: Some 1982 and earlier systems may not stall; may not set codes.)
HEI REF Signal (HEI Pin R, ECM Pin 10) *4-wire Pin B*	Sets Code 12. Engine does not start.	Sets Code 12 or 41. Engine continues to run.	Engine starts, stays in HEI (module) mode, removes 5-volt bypass, dumps AIR to atmosphere. Sets Code 41.	Engine continues running, stays in HEI (module) mode, removes 5-volt bypass, dumps AIR to atmosphere. Sets Code 41.
5-VOLT BYPASS (HEI Pin B, ECM Pin 11) *4-wire Pin C*	Engine starts and stays in HEI (module) mode and continues to run on base timing. Sets Code 42.	Shifts to HEI (module) mode and engine continues to run on base timing. Sets Code 42.	Engine starts and stays in HEI (module) mode and continues to run on base timing. Sets Code 42.	Shifts to HEI (module) mode and engine continues to run on base timing. Sets Code 42.

NOTE: HEI ground is module Pin D, 4-wire Pin D, and ECM Pin 13.

Figure 12-27. Troubleshooting table for HEI-EST circuits on most GM carbureted engines. (Layne © 1991)

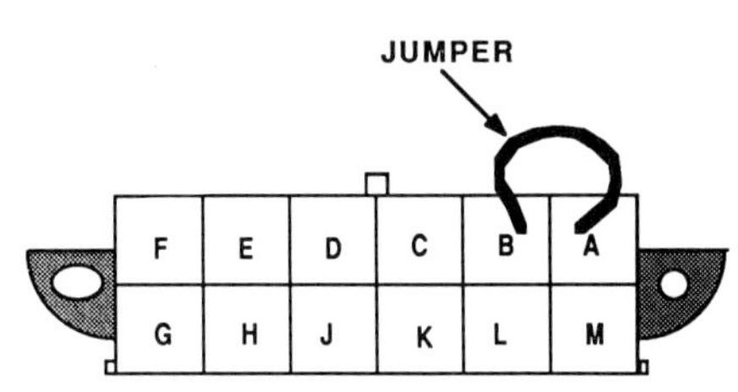

Figure 12-28. Use a jumper to connect terminals A and B in the ALDL connector for the "field service" test mode.

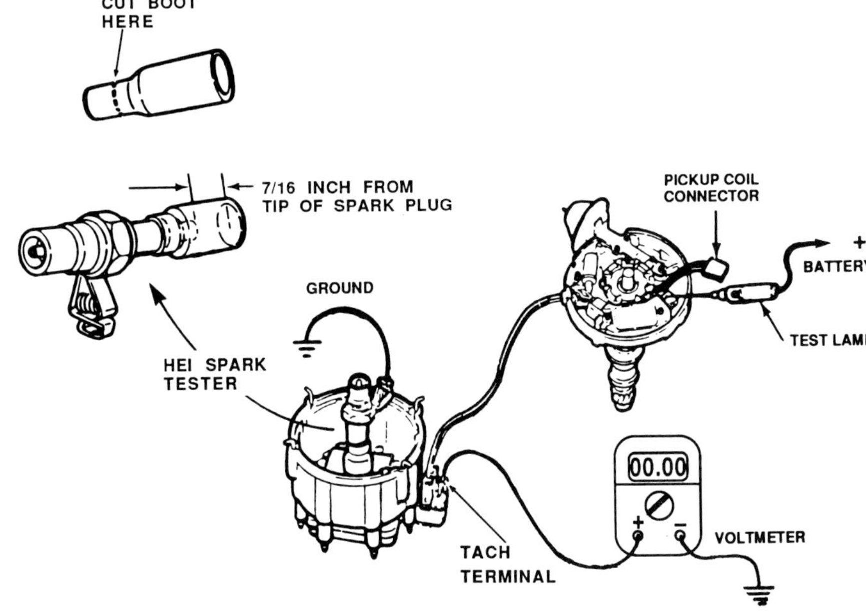

Figure 12-29. Test the ignition module's ability to respond to a pickup coil signal and fire the coil with a setup like this.

3. Disconnect the pickup coil leads (usually marked P and N) from the module. Leave all other module leads connected.

4. Set a voltmeter on the 20- or 40-vdc scale and connect it to the distributor cap TACH terminal.

5. Turn the ignition on and tap the test lamp against the module P terminal for 3 seconds or less, figure 12-29. If the module terminal is not marked P, tap the L or H terminal.

6. Watch the voltmeter. The reading should drop from battery voltage to about 7 to 9 volts each time you tap the P terminal. If it does not, the module may be bad, or the coil may be open.

7. If the voltmeter responds as specified in step 6, watch the spark tester for spark as you repeatedly tap the P terminal. If the tester does not fire, the coil probably is bad. If the tester fires regularly, the pickup coil may have an intermittent fault.

General Motors DIS Testing

GM's first distributorless, or direct, ignition system (DIS) was used on some fuel-injected Buick-built V-6 engines in 1984. Since then GM DIS installations have expanded to most of the carmaker's engine families. The following sections outline some basic tests for the following DIS variations:

- Computer-controlled coil ignition (C31) on Buick-built V-6 engines.
- DIS on Chevrolet-built 4-cylinder and V-6 engines, as well as on Pontiac-built 4-cylinder engines.
- Integrated direct ignition (IDI) on Oldsmobile-built 4-cylinder engines.

Magnetic Pickup (Reluctance) Crankshaft Sensor Testing. You can test the crankshaft sensors of DIS and IDI systems by disconnecting the 2-wire connector at the ignition module and measuring the resistance of the ac voltage signal with a digital volt-ohmmeter (DVOM). Specifications are as follows:

- DIS sensor—900 ohms to 1200 ohms resistance or 200 mvac at cranking rpm
- IDI sensor—600 ohms to 900 ohms resistance or 250 mvac at cranking rpm

Hall-Effect Crankshaft and Camshaft Sensor Testing. Buick-built V-6 engines have Hall-effect crankshaft and camshaft sensors that send a digital (high-low) square wave signal to the C3I ignition module. The crankshaft sensor is on the front of the engine, next to the harmonic balancer, figure 12-30, to provide cylinder tdc timing pulses. These are equivalent to the ignition tachometer signal. The 3.8-liter engines with sequential fuel injection also have camshaft sensors to provide a number 1 cylinder signal for ignition and fuel injection synchronization. The 3.0- and 3.3-liter engines do not have camshaft sensors. They use a combination sensor on the front cover, figure 12-31, that sends a crankshaft tdc signal and a synchronizing signal. Because these engines do not have sequential injection, they do not need a camshaft signal. The synchronizing signal indicates cylinder position and takes the place of a camshaft signal.

On late-model 3.8-liter (3800) V-6 engines, the "fast-start" C3I system uses a dual crankshaft sensor and a separate camshaft sensor. The outer crankshaft sensor works with an 18-blade trigger ring and sends an 18-pulse-per-revolution (18X) signal to the C3I module. The inner crankshaft sensor works with an unevenly spaced trigger ring and sends a 3-pulse-per-revolution (3X) signal to the C3I module. Each pulse is a different time length, figure 12-32. The C3I module

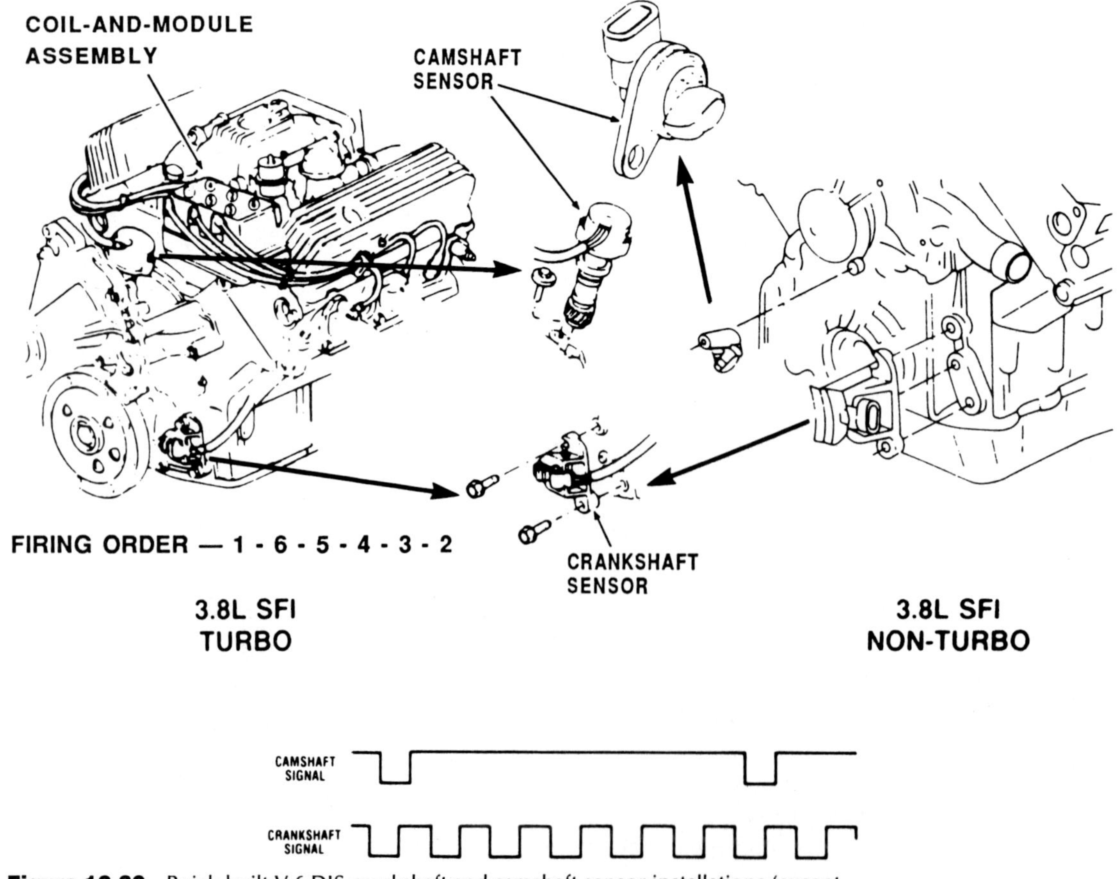

Figure 12-30. Buick-built V-6 DIS crankshaft and camshaft sensor installations (except 3800 "fast-start" system).

compares the uneven 3X signals to the 18X clock signals to determine which coil to fire. The module's circuitry also divides the 18X signal by 6 and sends a fuel control signal to the engine computer. Additionally, the module processes the camshaft sensor signal and sends it to the computer.

All C3I systems need a crankshaft sensor signal to start the engine. If a 3.8-liter C3I system loses the camshaft signal while running, it will continue to run. When it is stopped, however, it will not restart unless it is a "fast-start" system. These systems will start if the camshaft signal is open; if it is shorted or grounded, however, they will not restart. To identify a camshaft sensor problem on a "fast-start" 3800 engine, disconnect the camshaft sensor if the engine does not start. If it then starts, the sensor is probably grounded or shorted.

Referring to vehicle wiring diagrams, use these basic tests to check crankshaft and camshaft Hall-effect sensors on GM systems:

1. Check system supply (battery) voltage to the C3I module and to the sensor.

2. Verify that the module and all sensors have good ground connections.

3. If the engine does not start, tap the crankshaft sensor with a screwdriver handle. If the engine then starts, the sensor may have an intermittent open or short circuit.

Crankshaft sensors are not adjustable for ignition timing, but they must be aligned correctly for proper operation. If a crankshaft sensor is damaged due to misalignment or a bent trigger ring, it can cause hesitation, sag, stumbling, or dieseling.

If a C3I V-6 does not start or dies immediately after starting, it may be due to intermittent or permanent failure of the crankshaft sensor. The crankshaft sensor may fail because of engine vibration or damage from road debris, such as rocks. Early symptoms may appear as a misfire, engine stumble, or an intermittent no-start condition.

Inspect the sensor for damage from both above and below the engine. Inspect the trigger ring vanes on the harmonic balancer for shiny spots, which may indicate that the vanes are hitting the sensor. Check the magnet and the sensor body for cracks or broken sections. Replace the sensor if you find any of this damage. Inspect the sensor connector for damage, corrosion, and poor electrical connections.

If you have to replace a crankshaft sensor, you must set the gap between the interrupter vanes and the sensor precisely. Use Kent-Moore tool J-37089, or its equivalent, figure 12-33, for all dual-slot sensors. Earlier single-slot sensors have been superseded by dual-slot sensors as service replacements. Figure 12-33 shows adjustment of the dual-slot sensor using tool J-37089. The harmonic balancer must be removed from the engine for sensor adjustment.

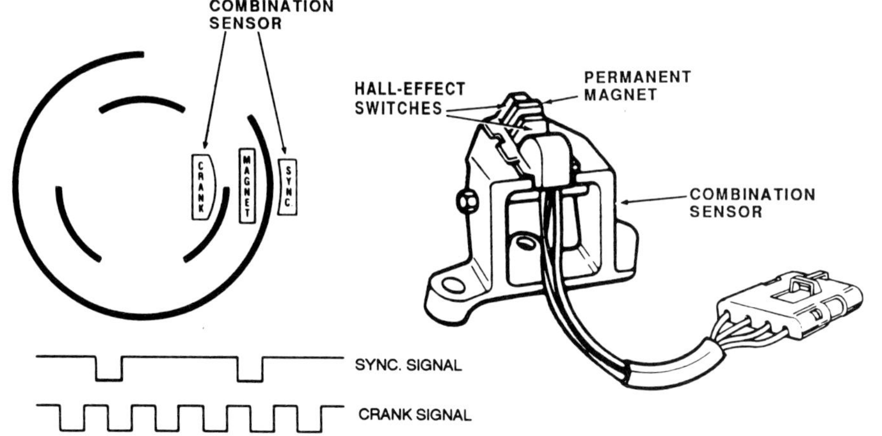

Figure 12-31. Combination sensor used on 3.0- and 3.3-liter Buick V-6 engines.

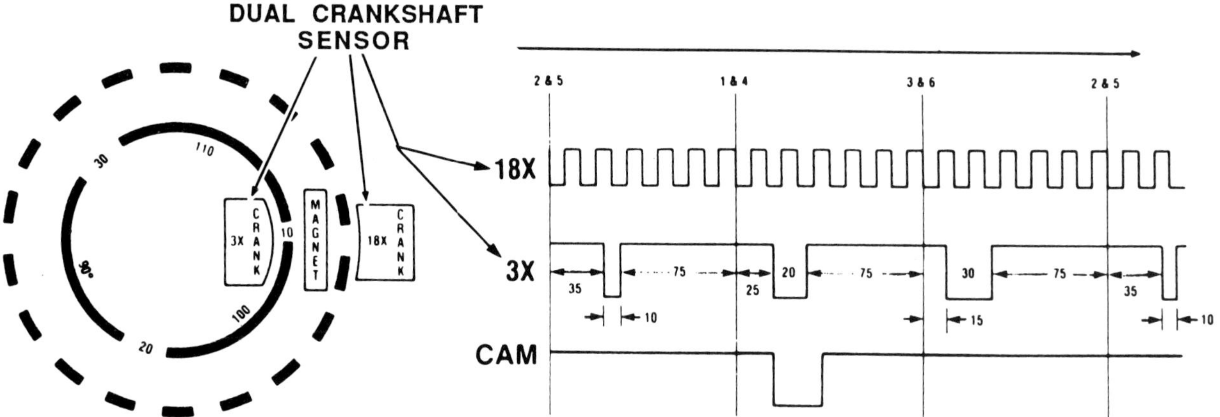

Figure 12-32. Crankshaft sensor signals for the Buick 3800 "fast-start" ignition. The camshaft signal, used for fuel injection synchronization, is shown for reference.

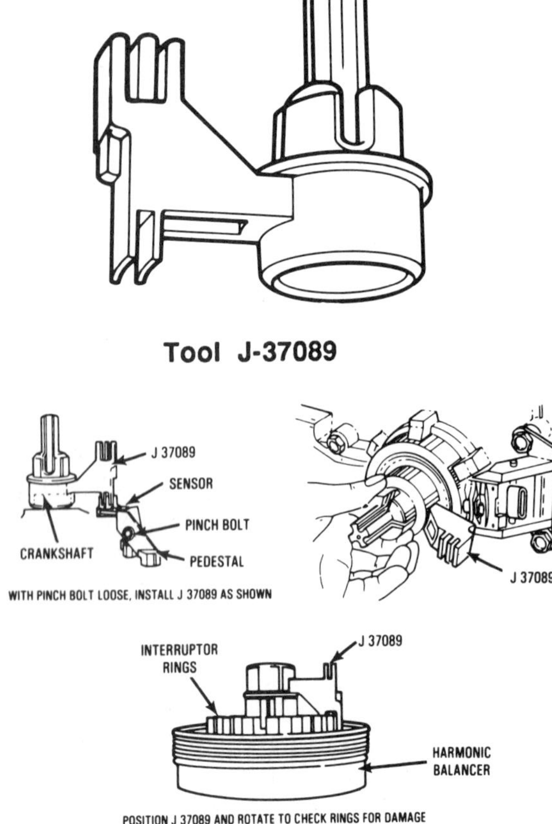

Figure 12-33. Crankshaft sensor alignment tool J-37089 required for Buick V-6 DIS.

Coil Testing. You can test the coils or coil packs on GM direct ignitions by removing them from the engine and checking the resistance of the primary and secondary windings with an ohmmeter.

C3I coil packs for Buick V-6 engines are one of three different types:

- *Type 1* is used on 1984 and later turbocharged 3.8-liter V-8 engines, on 1985 and later 3.0-liter V-6 engines, and on some 1986 and later 3.8-liter V-6 engines with sequential fuel injection (SFI). This type has all three coils molded into a single pack, figure 12-34. If one coil malfunctions, the entire coil pack must be replaced.
- *Type 2* is used on some 1986 and later 3.8-liter SFI V-6 engines. This type is similar to type 1, but each coil is separate and can be replaced individually.

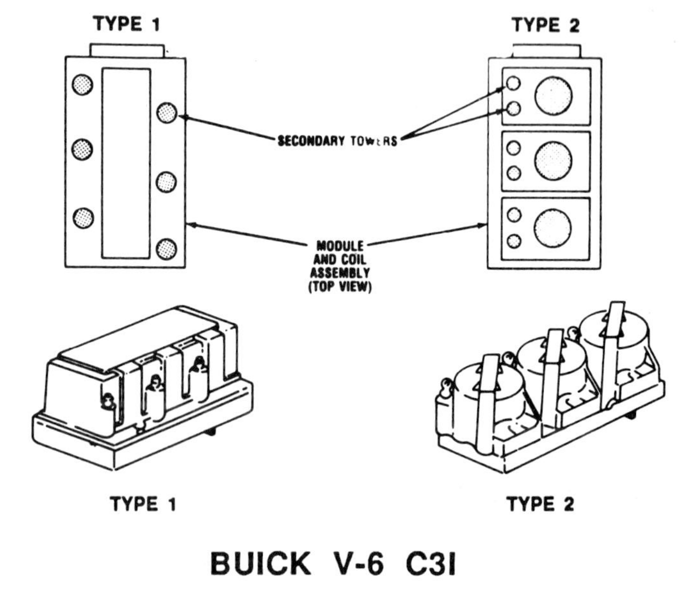

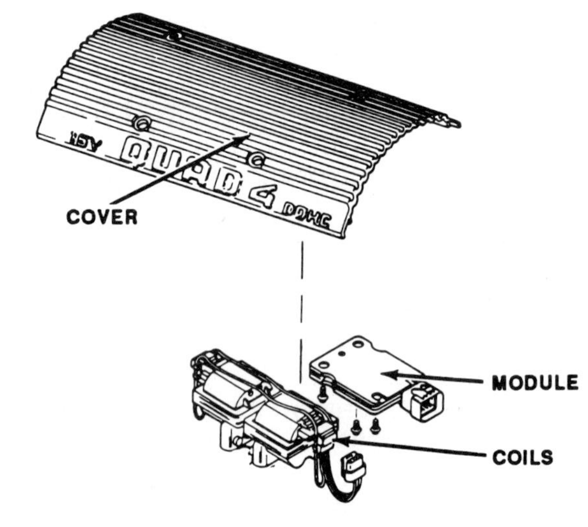

Figure 12-34. Coil packs used on General Motors various DIS installations.

• *Type 3* is used on 1988 and later 3800 V-6 engines. The type 3 coil pack is similar to, and can be interchanged with, a type 1 coil pack. The module differs electronically, however, and the connector plugs are not compatible with other systems.

Resistance specifications for all GM DIS coils, figure 12-34, are as follows:

Primary Winding	*Secondary Winding*
C3I type 1	
0.35 Ω to 1.5 Ω	10k Ω to 14k Ω
C3I type 2	
0.35 Ω to 1.5 Ω	5k Ω to 7k Ω
C3I type 3	
0.35 Ω to 1.5 Ω	10k Ω to 14k Ω
DIS and IDI	
0.35 Ω to 1.5 Ω	5k Ω to 7k Ω

Spark Testing. After testing coil resistance, test the secondary firing capability of each coil. Connect the coil end of one spark plug cable to an HEI spark tester and then clamp the tester to the coil tower to put the tester in series with both plugs for that coil, figure 12-35. Start the engine and watch the tester. A strong spark indicates proper coil firing. A weak or intermittent spark indicates a bad coil. Let the engine idle for 10 minutes. If the spark remains strong, the coil is okay.

Power Balance Testing. You can do a power balance test on a GM direct ignition system by using short lengths of vacuum hose and a 12-volt test lamp. The vacuum hose must have a high carbon content so that it will conduct secondary voltage.

Install a short piece of vacuum hose between each coil terminal and its spark plug cable or between each spark plug terminal and its cable, figure 12-36. The vacuum hose should be long enough to be exposed when the plug cable is connected, but short enough so that it does not touch any metal object. Access to the coil towers on 4-cylinder engines may be difficult, so install the vacuum hoses at the spark plugs if this is the case.

Using a 12-volt test lamp connected to ground, touch each vacuum hose to short the cylinder. Note the engine speed as you short each cylinder. Then slowly move the test lamp probe away from the hose and note the strength of the arc between the vacuum hose and the test lamp. As with any power balance test, engine speed will decrease the least on a cylinder with a problem. Interpret the strength of the arc between the hose and the lamp probe as explained in the next paragraphs.

• For vacuum hoses placed at coil towers, touch the hoses with the lamp until you find little or no rpm drop. Then move the probe away from the hose and note the arc. If the arc jumps a long distance to the grounded test lamp, suspect an open circuit between the coil and the plug. Look for an open plug cable or plug or a very wide plug gap. If the arc jumps only a short distance to the grounded test lamp, suspect a short circuit between the coil and the plug. Look for a shorted cable or plug, or a fouled or cracked spark plug.

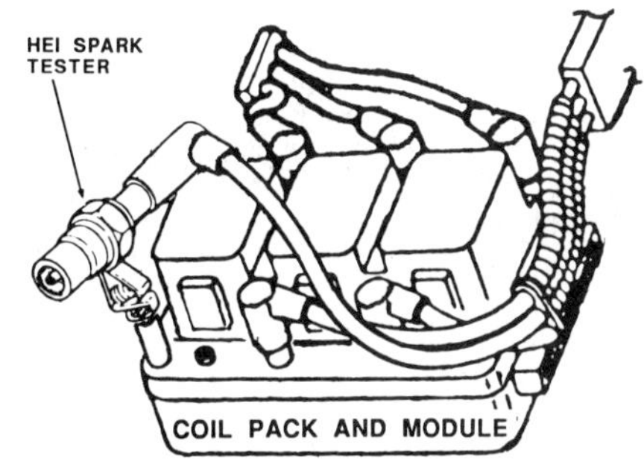

Figure 12-35. Connect an HEI spark tester to one coil circuit like this to test coil firing voltage.

• For vacuum hoses placed at spark plugs, touch the vacuum hoses with the test lamp until you find little or no rpm drop. Then move the probe away from the hose and note the arc. If the arc jumps a long way to the test lamp, suspect an open plug or a very wide plug gap. If the arc jumps only a short distance to the test lamp, suspect a shorted or fouled plug, or a nearly open spark plug cable. if there is no arc, suspect an open circuit in the plug cable or the coil tower. If the strength of the arc on a cylinder with little rpm drop seems to be the same as the arc on cylinders with normal rpm drop, suspect a fuel system or compression problem.

Motorcraft (Ford) Breakerless and Dura-Spark Ignition Primary Voltage and Resistance Tests

The following tests are for basic Motorcraft breakerless and Dura-Spark

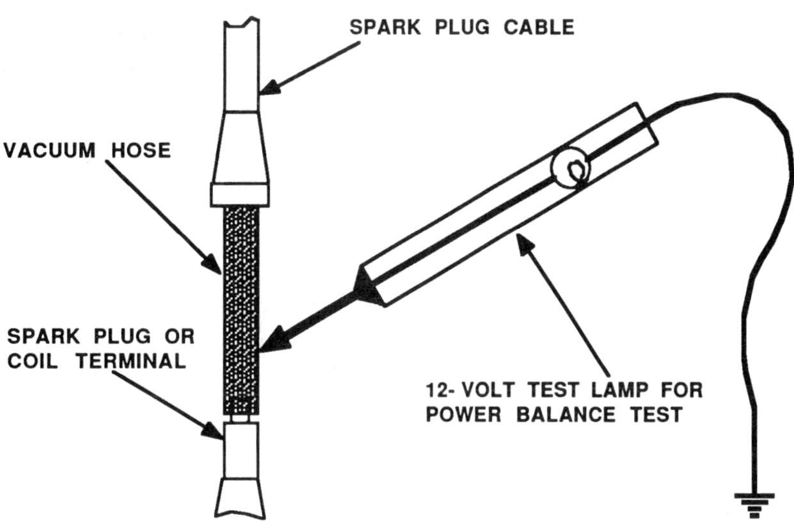

Figure 12-36. Connect conductive vacuum hose to the plug cables and use a probe light to short each cylinder for a DIS power balance test.

systems with magnetic-pulse signal generators. Before testing the ignition primary circuit, check the fuel delivery and be sure the battery cranking voltage is 9.5 volts or more. Inspect the primary wiring and connectors for damage and looseness.

Caution. Do not operate the Dura-Spark I system with an open secondary circuit. High secondary voltage may damage the coil or cause a short circuit in the distributor.

When testing for spark at a plug, do not disconnect cylinders 1 or 8 on a V-8 engine, 3 or 5 on an inline six, 1 or 4 on a V-6, and 1 or 3 on a 4-cylinder engine. The pickup coil is directly below these cable terminals in the distributor and high voltage may arc to the pickup coil.

You can test for secondary voltage, or spark at the plugs, by disconnecting a cable and holding it 1/4 inch from ground while cranking the engine. It is better, however, to use a spark tester, or spark plug simulator, figure 12-3. Connect the grounding clip to an engine ground and connect one spark plug cable to the tester. Crank the engine and watch for a spark between the tester center electrode and shell. This indicates adequate secondary voltage capability.

Since 1974, Ford has steadily developed and modified its basic electronic ignition. You can test a specific year and model most efficiently if you have Ford's test procedures for that vehicle. But, if you know the basic Ford circuit arrangement, color coding, and module connector pin numbers you can make basic system tests and follow the specific instructions more easily.

Ford Ignition Wire Color Codes. Ford's basic ignition wire color codes for breakerless and Dura-Spark systems are as follows:

- Red = Voltage supply, running
- White = Voltage supply cranking
- Green = Primary current from the coil – terminal to the module (ground side of coil)
- Orange & Purple = Distributor pickup coil (stator) signals to the module

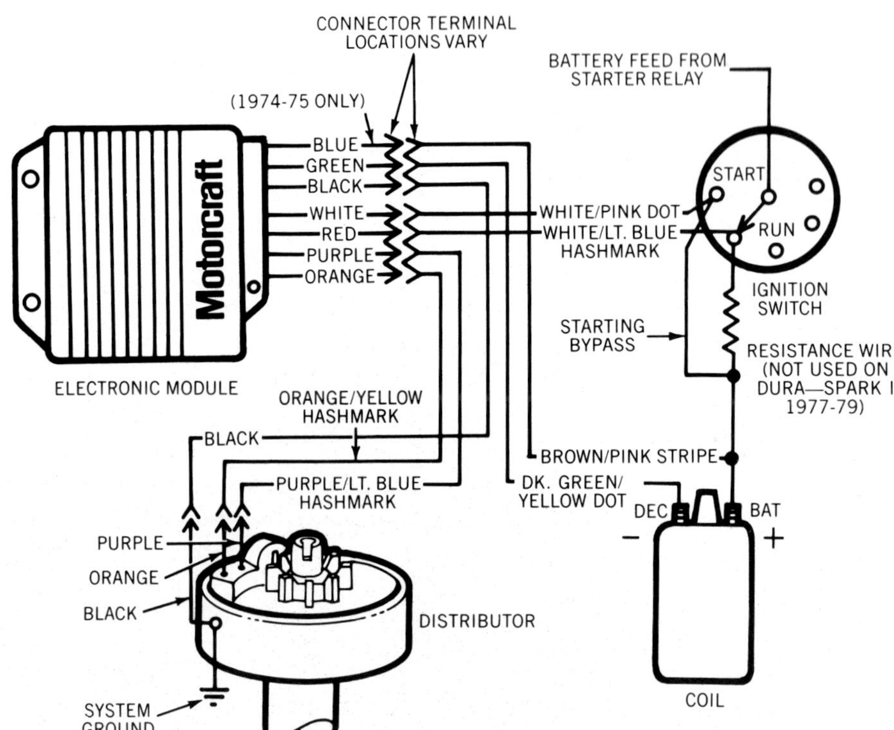

Figure 12-37. Ford solid-state and Dura-Spark ignitions use this basic circuit. Wire arrangements in connectors vary from year to year. (Ford)

- Black = System ground (module to distributor body)
- Blue = System protection (1973-75 systems only)

These color codes refer to the wiring on the *vehicle harness side* of the ignition module connectors. Some wire colors change at the module connectors. For example, the red harness wire connects to the white module wire, and the white harness wire connects to the red module wire. Figure 12-37 is a simplified diagram of Ford Solid-State and Dura-Spark electronic ignitions from 1973 to the present. You will make voltage and resistance tests on the *harness side* of the connectors, unless directed otherwise. Many Ford test procedures recommend that you make voltage tests with the igniton module connectors connected to the harness. Connect your voltmeter to the wiring by carefully inserting a small straight pin through the wire insulation, figure 12-38.

Caution. Do not let the pin touch ground during testing. A short circuit may damage wiring or electronic components.

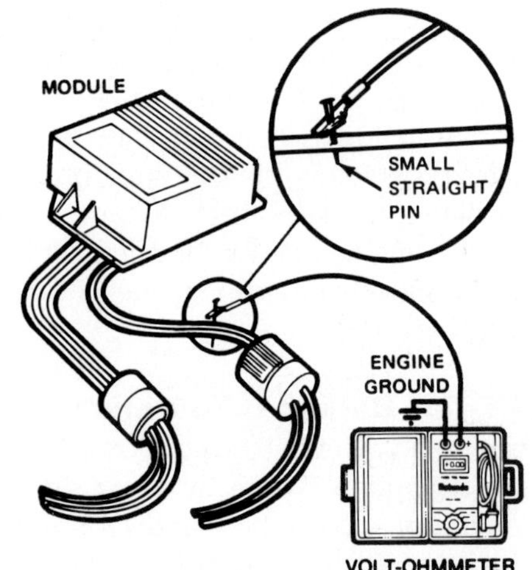

Figure 12-38. Use a straight pin to pierce the ignition wiring and connect your voltmeter. (Ford)

Ford Ignition Module Connectors and Module Color Codes. Ford has relocated the blades (pins) and sockets in ignition module connectors from year to year to prevent installation of the wrong module in a particular ignition system. Figure 12-39 shows the blade and socket numbers and the wire colors for Ford ignition systems of different years. These are the two connectors for the basic ignition circuit,

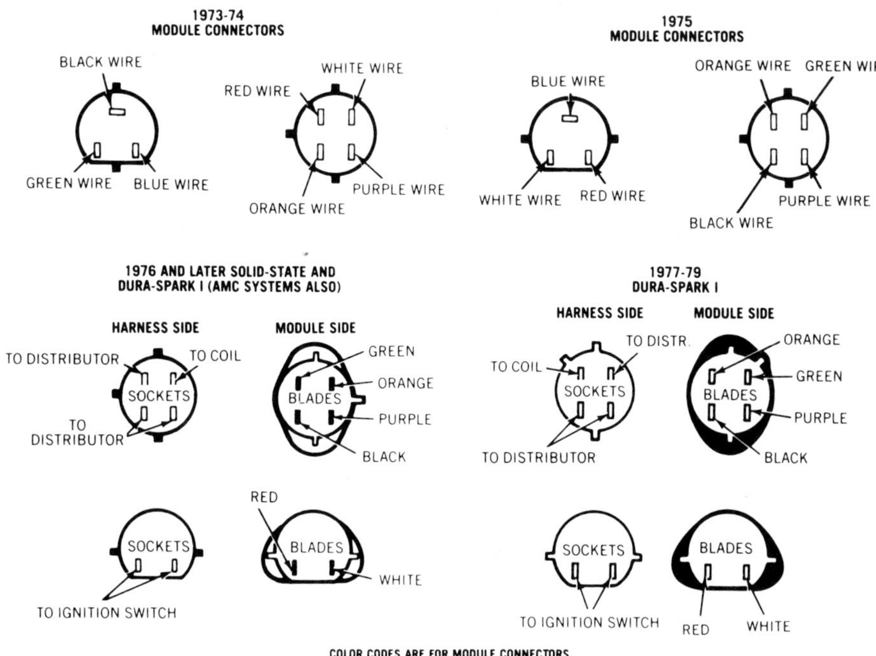

Figure 12-39. Ford Solid-State and Dura-Spark ignition module connectors. Third connectors for dual-mode systems are not shown. (Ford)

Ignition Type	*Grommet Color*
1973–74 Solid-State	Black (7 wires)
1975 Solid-State	Green (7 wires)
1976 Solid-State	Green (6 wires)
1976 & later Dura-Spark II	Blue
1979 & later Dura-Spark II with "start retard"	White
1978 & later Dura-Spark II with "dual mode"	Yellow
1977–79 Dura-Spark I	Red
1979 & later Dura-Spark III with EEC-III systems	Brown
1981 & later Dura-Spark "universal ignition module" with MCU-D systems	Yellow

Figure 12-40. Ford solid-state and Dura-Spark ignition module color codes.

not for the third connectors of dual-mode, or cranking-retard subsystems.

You can use the color of the wire grommet of the ignition module for the main connectors to help identify the type of ignition system on various Ford vehicles, figure 12-40.

Ford Primary Circuit Voltage and Resistance Tests. By following the wire color codes, you can check voltage and resistance at the harness connectors under the following conditions:

1. Key on, engine off, *voltage* should be:

   a. Coil BAT terminal to ground (module connected, coil DEC terminal grounded) = 5 to 8 volts. This test does not apply to Dura-Spark I because it has no ballast resistor.

   b. Harness red wire to ground = within 1 volt of battery voltage.

   c. Harness green wire to ground = within 1 volt of battery voltage.

2. Key on, engine cranking, *voltage* should be:

   a. Harness white wire to ground = within 1 volt of battery voltage.

   b. Distributor harness orange wire to purple wire (pickup coil signal) = 0.5 volt ac or any dc voltage pulse.

   c. Dura-Spark I coil BAT terminal to ground with coil DEC terminal grounded = 6 volts or more.

3. Key off, *resistance* should be:

   a. Between orange and purple distributor harness wires, figure 12-41, (pickup coil resistance) = 400 to 800 ohms

   b. Between *both* the orange wire and ground and the purple wire and ground = 70 kilohms or more (infinity).

   c. Between black distributor harness and ground (ground continuity) = 1 ohm or less.

   d. Between the red harness wire and the coil center tower (primary continuity through the coil) = specifications vary; check for each model (generally 7,000 to 10,000 ohms).

   e. Between the green and red harness wires (green and ground on Dura-Spark II) = approximately 4 ohms.

   f. Between the red harness wire and the coil + (BAT) terminal (except Dura-Spark I) = ballast resistor resistance; specifications vary from 0.6 to 2.0 ohms for different models. High resistance equals an open resistance wire.

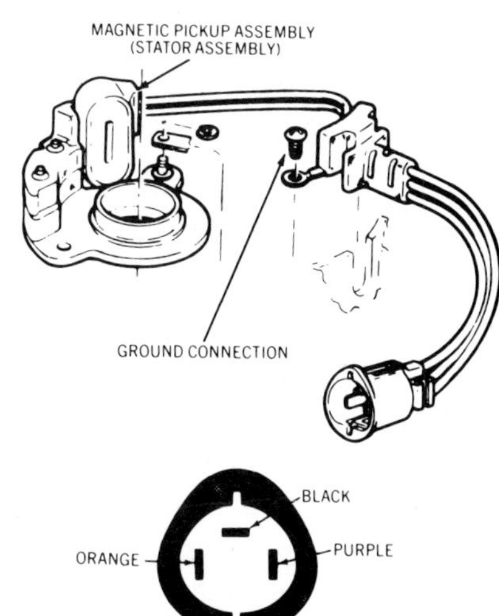

Figure 12-41. Test pickup coil resistance and for a grounded pickup coil at the distributor. Also check system ground continuity at the black wire. (Ford)

g. Across the coil primary terminals (BAT and DEC) = coil primary winding resistance.

Testing Ford Dura-Spark, TFI, and DIS Ignitions Used with Electronic Engine Controls

Figures 12-42 through 12-46 are wiring diagrams for the following Ford ignitions used with electronic engine controls:

- Dura-Spark II and Dura-Spark III, the latter used with EEC-III systems
- Thick-film integrated (TFI), TFI with computer-controlled dwell (CCD), and TFI with closed-bowl distributor used with EEC-IV systems
- The distributorless ignition system (DIS) used on dual-plug 2.3-liter engines
- The distributorless ignition system (DIS) used on 3.8-liter supercharged V-6 engines
- Electronic distributorless ignition system (EDIS) used on other V-6 engines

You can use these diagrams and Ford test procedures to help troubleshoot various ignition problems. We can't provide all of Ford's detailed pinpoint tests for ignition systems in this volume. Most, however, are based on the principles of testing Ford ignitions, outlined in the previous section.

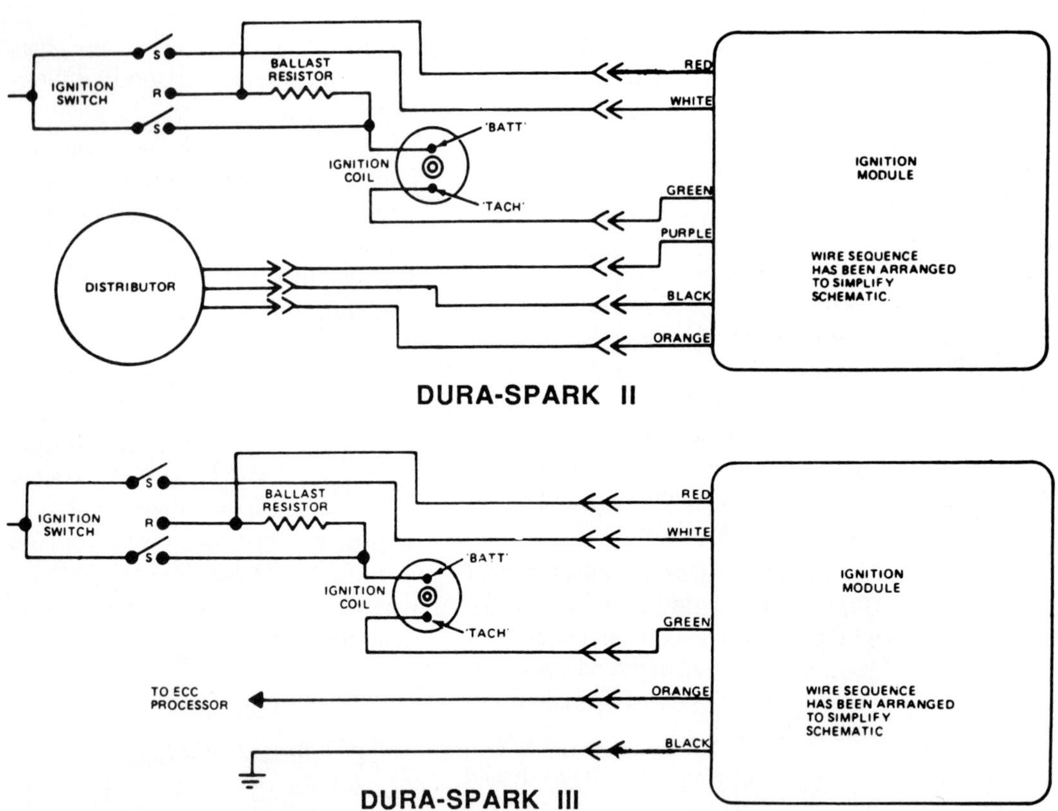

Figure 12-42. These diagrams are typical of Ford's Dura-Spark II and Dura-Spark III systems, used with early electronic engine control systems.

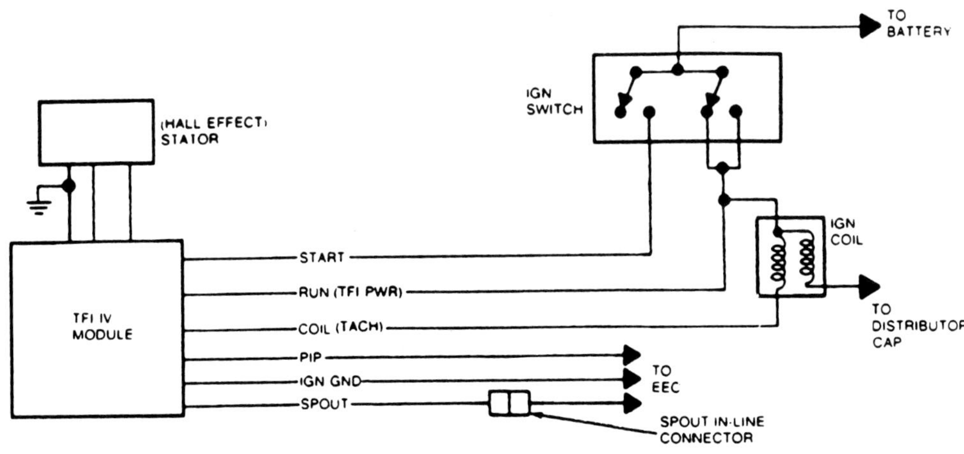

BASIC TFI IGNITION

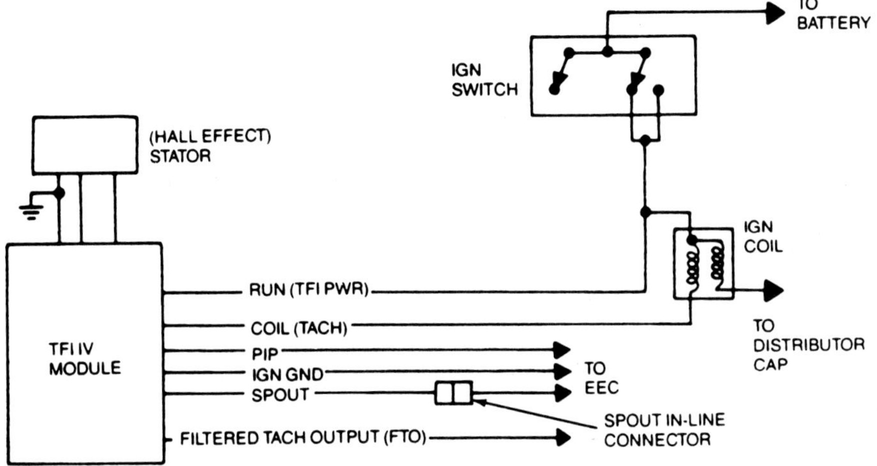

TFI IGNITION WITH COMPUTER-CONTROLLED DWELL (CCD)

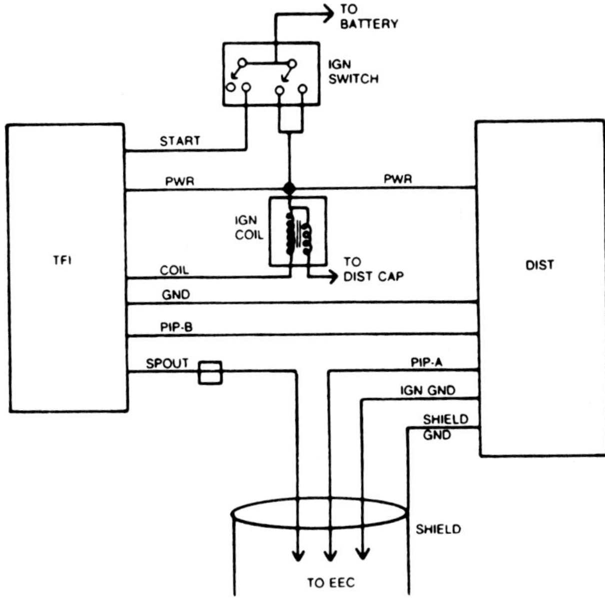

TFI IGNITION WITH CLOSED-BOWL DISTRIBUTOR (REMOTE MODULE)

Figure 12-43. These diagrams show three versions of Ford's thick-film integrated (TFI) ignition systems.

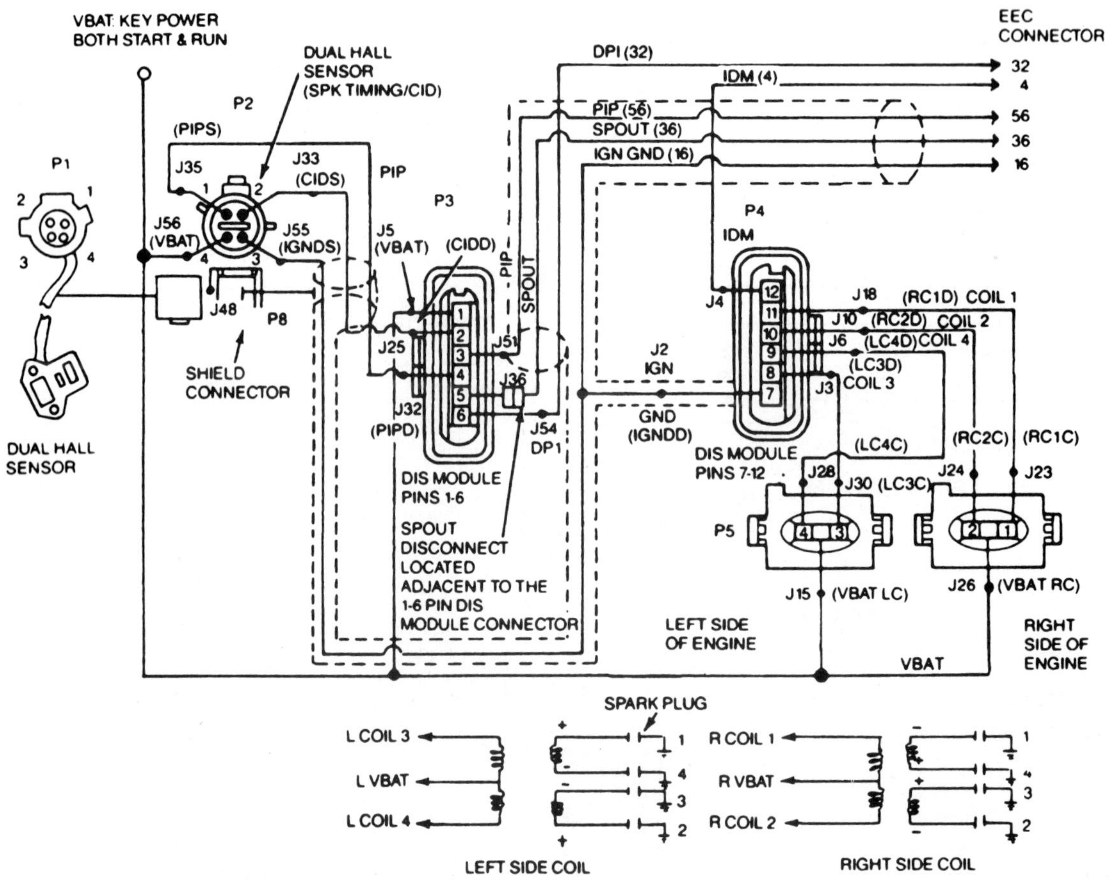

CONNECTORS ARE SHOWN LOOKING INTO THE WIRING HARNESS.
IGN GND IS A LOW CURRENT REFERENCE FOR THE EEC IV AND DIS MODULE.
IT IS CONNECTED TO BATTERY NEGATIVE (GROUND) VIA THE DIS MODULE METAL BASE PLATE. PROPER SYSTEM OPERATION DEPENDS ON A LOW RESISTANCE PATH.

FORD DIS INSTALLATION FOR 2.3-LITER ENGINE WITH DUAL PLUGS

Figure 12-44. Ford's distributorless ignition system (DIS) used on dual-plug 2.3-liter engines in Ranger trucks. The DIS installation on 1991 and later 4-cylinder Mustangs is similar.

DISTRIBUTOR HALL-EFFECT SWITCH TEST

Several late-model distributors from GM, Ford, Chrysler, and other carmakers have Hall-effect switches. These switches may be used in place of magnetic-pulse generators for primary trigger signals. Or, as in many HEI distributors, the Hall-effect switch may be used along with the magnetic pickup coil.

You can test most Hall-effect switches by connecting a 12-volt battery across the + and − source voltage terminals of the switch and a voltmeter across the signal-voltage and − terminals, figure 12-47. Then insert a knife blade or steel feeler gauge between the magnet and the Hall switch. Voltmeter readings for a good Hall-effect switch should be as follows:

- Without knife blade or feeler gauge, less than 0.5 volt.
- With knife blade or feeler gauge inserted and touching the magnet, within 0.5 volt of battery voltage.

PRIMARY (BALLAST) RESISTOR CIRCUIT TESTS

You have already learned some resistance and voltage drop tests for ballast resistor circuits in the previous procedures. When you check this circuit, you must check both the starting bypass branch and the resistor itself. The most common problems in either branch are high resistance or an open circuit. Two common symptoms can lead you to check for open circuits:

1. The engine cranks normally, almost starts, and then dies. This often means an open resistor. The ignition works on the bypass circuit and almost fires, but when the key switches to the run position, the circuit is open.

2. The engine cranks but does not fire until you release the key and it switches to the run position. This usually means that the bypass circuit is open. While

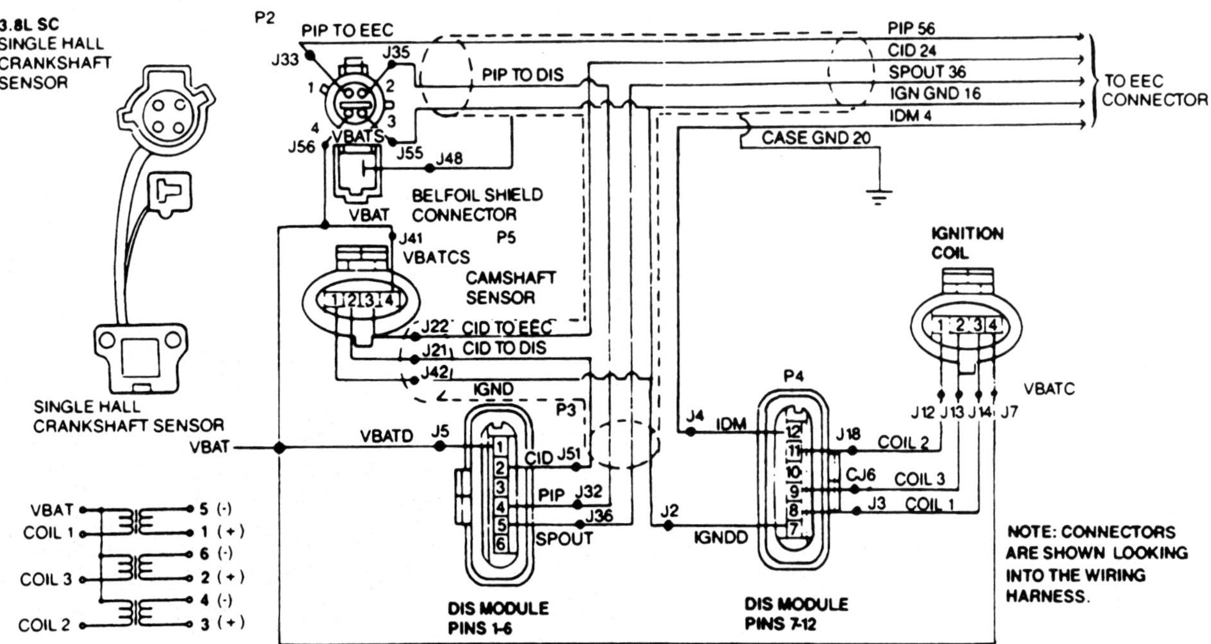

FORD DIS INSTALLATION FOR 3.8-LITER SUPERCHARGED V-6 ENGINE

Figure 12-45. Ford's distributorless ignition system (DIS) used on 3.8-liter supercharged V-6 engines.

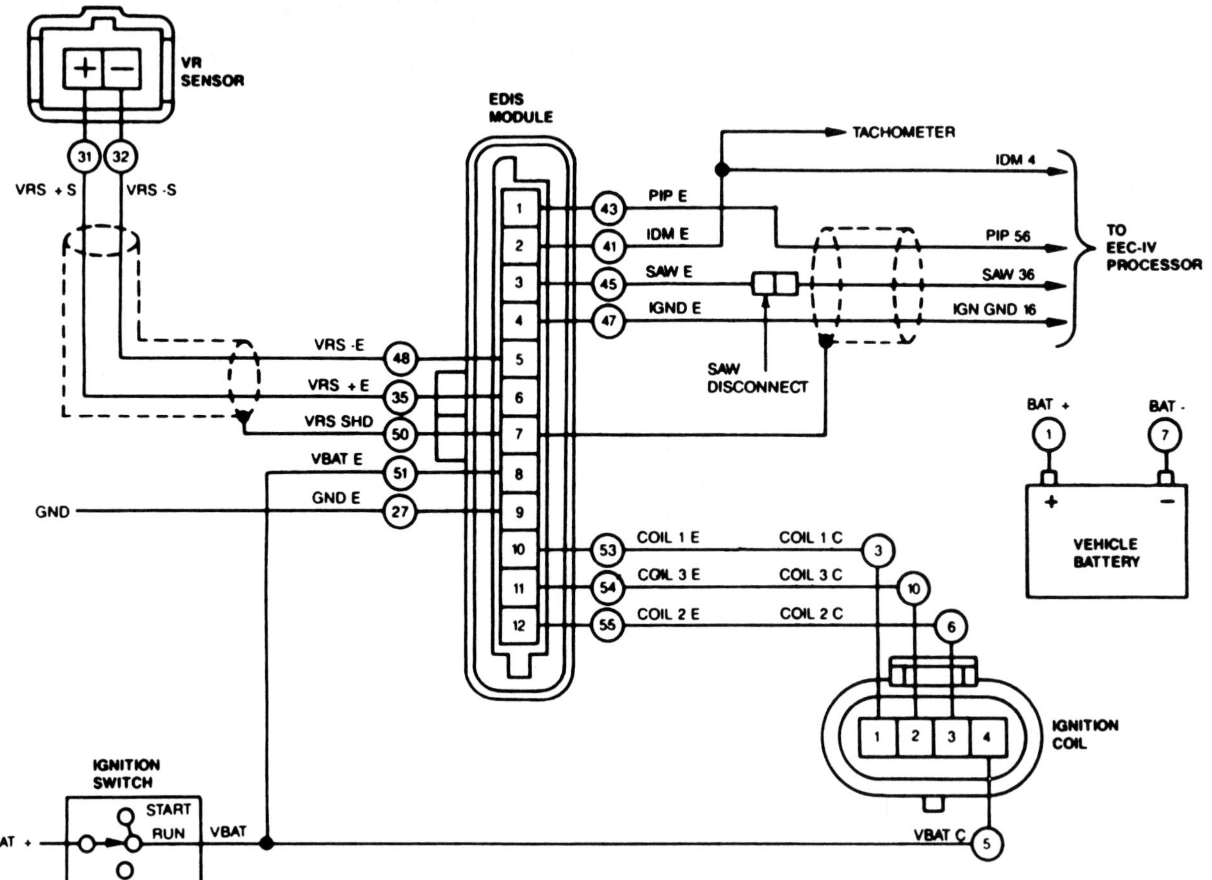

FORD EDIS INSTALLATION FOR V-6 ENGINES

Figure 12-46. Ford's distributorless ignition system (DIS) used on some 1991 and later V-6 engines. The 4.1-liter truck engine is shown.

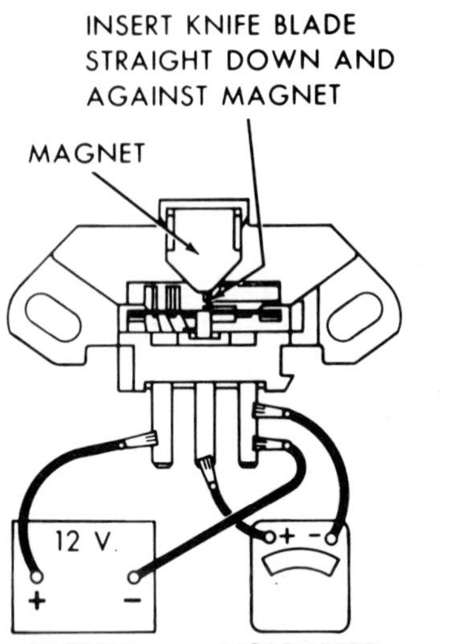

INSERT KNIFE BLADE STRAIGHT DOWN AND AGAINST MAGNET

MAGNET

12 V.

BATTERY

VOLTMETER

Figure 12-47. Apply input voltage to a Hall-effect switch and test its output by inserting a knife or feeler gauge against the magnet. (Delco-Remy)

the engine is still turning as you release the key to the run position, the circuit closes through the resistor, and the engine starts.

You can measure voltage drop across the resistor and the bypass circuit as you have learned in the previous procedures. Measuring voltage drop lets you check for open circuits or high resistance. You also can test the resistor with an ohmmeter. Remember to make any ohmmeter tests with the ignition off or the resistor disconnected from the circuit.

A ballast resistor, such as the traditional Chrysler type, is easy to test with an ohmmeter because both terminals are close together on the ceramic block, figure 12-48. Many cars from GM, Ford, and other makers have ballast resistors that are a length of resistance wire in the vehicle wiring harness, figure 12-49. These often are hard to check with an ohmmeter because one end is connected to the ignition switch and the other to the coil in the engine compartment. In this case, use the voltage-drop tests you learned previously. Check the carmaker's wiring diagrams. If the resistance wire connectors are accessible in the engine compartment, use your ohmmeter to test the resistor.

To replace a unit-type (Chrysler-type) resistor, disconnect the wires from either end and remove the resistor. To replace a length of resistance wire, refer to the carmaker's wiring diagrams and procedures to disconnect the defective wire from the circuit and install a new one in its place, figure 12-49. Replacement resistor wires must be the exact length and exact resistance specified by the carmaker.

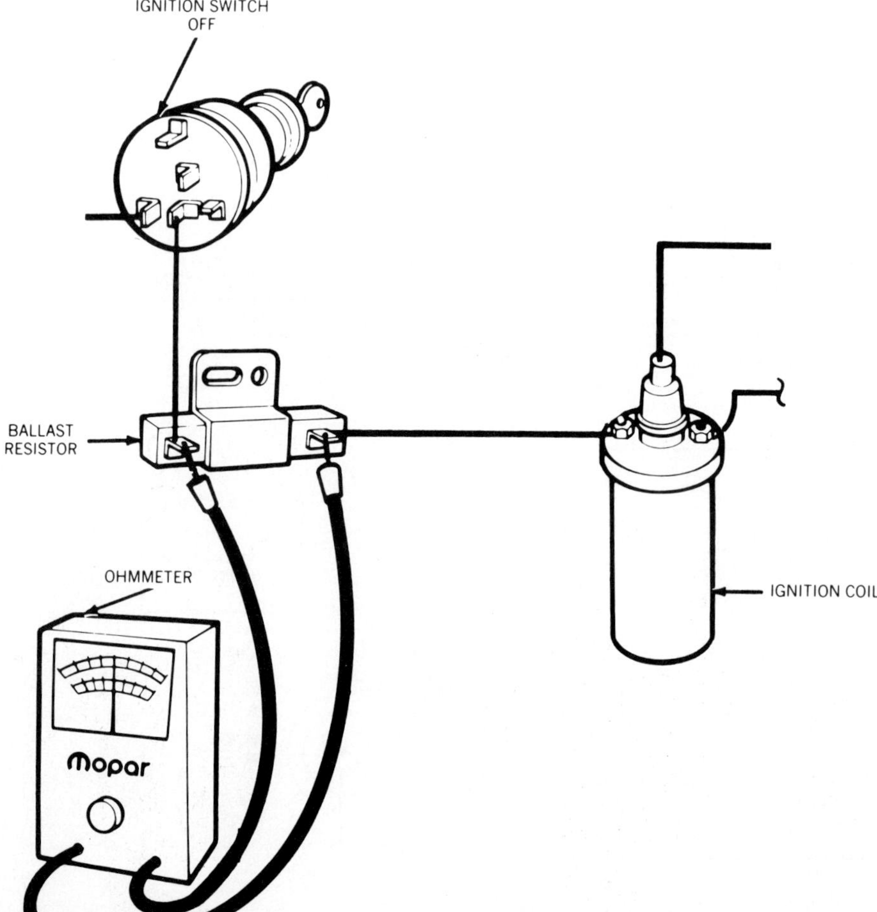

Figure 12-48. Measure resistance across the terminals of a unit ballast resistor like this Chrysler resistor. (Chrysler)

IGNITION COIL TESTS

Several of the ignition circuit tests that you learned earlier included coil winding resistance tests. You can check primary and secondary winding resistance of any coil with an ohmmeter. Other standard coil tests include current draw, polarity, and output voltage. You can test polarity and output voltage most easily with an oscilloscope, as explained later in this chapter. The following sections outline basic resistance, current draw, and polarity tests.

When you do any coil test, be sure that the coil is at normal operating temperature before testing. Temperature has a direct effect on resistance and performance. Also before testing, inspect the coil for loose or corroded connections, a dented or cracked housing, a cracked or burned center tower, and oil leakage. Replace any damaged coil.

Coil Resistance Tests

Test the primary winding resistance by disconnecting the coil from the igni-

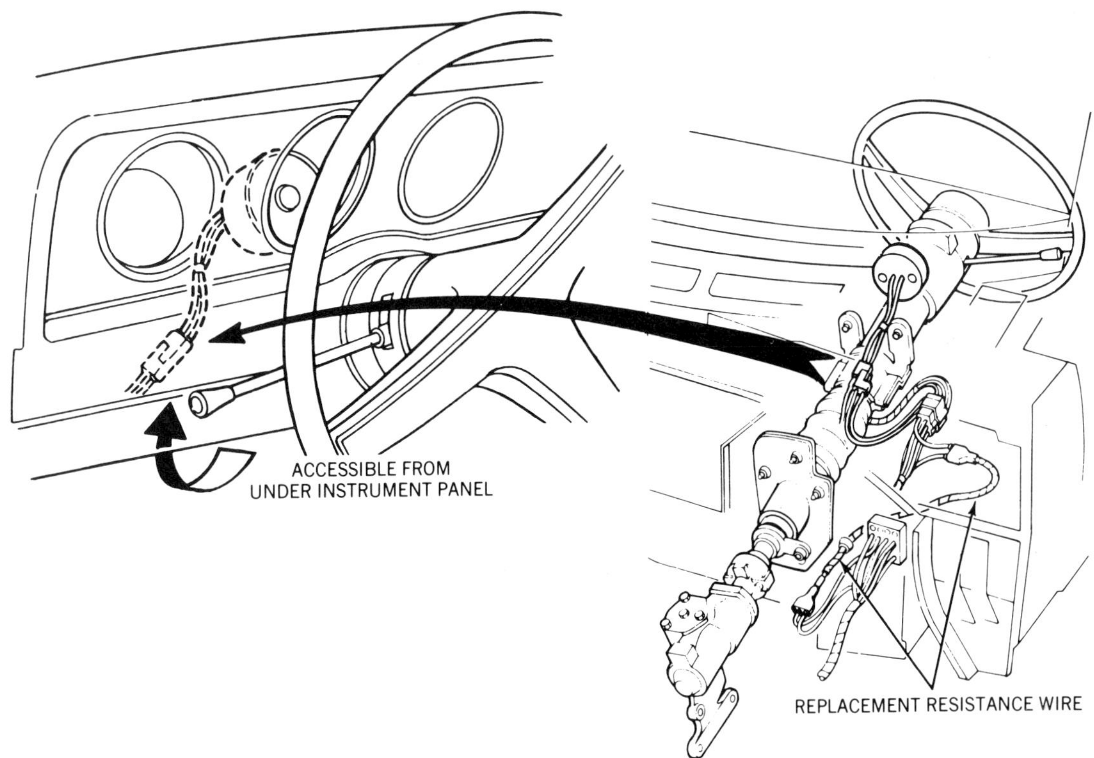

Figure 12-49. The ballast resistor of many cars is a length of resistance wire, which is hard to check with an ohmmeter. (Ford)

tion circuit and connecting your ohmmeter across the primary terminals, figure 12-50. Set the ohmmeter on the lowest scale and compare the reading to manufacturer's specifications. If the reading is above or below specifications, replace the coil.

To test the secondary winding, set the ohmmeter on the highest scale. Connect one meter lead to the center tower (secondary terminal) and touch the other lead alternately to both primary terminals, figure 12-51. Compare the lower reading to the maker's specifications. If the reading is above or below specifications, replace the coil.

To test the coil for shorted windings, set the meter on the lowest scale. Touch one lead to the coil case and the other to either primary terminal, figure 12-52. The meter should indicate infinite resistance. A reading lower than infinity means that the windings are shorted to the coil case and the coil must be replaced.

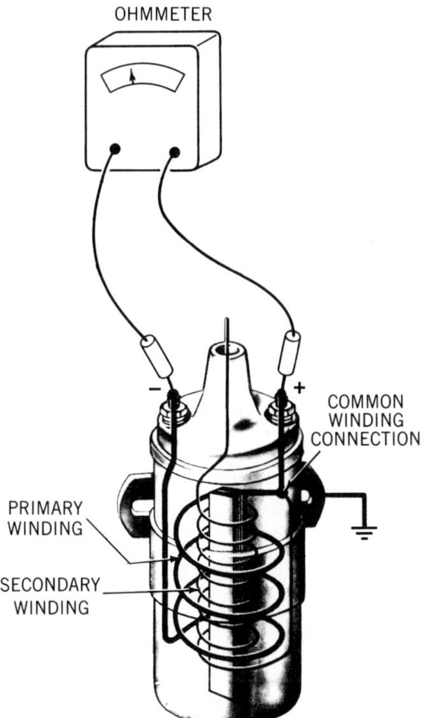

Figure 12-50. Measure coil primary resistance across the primary terminals. (Bosch)

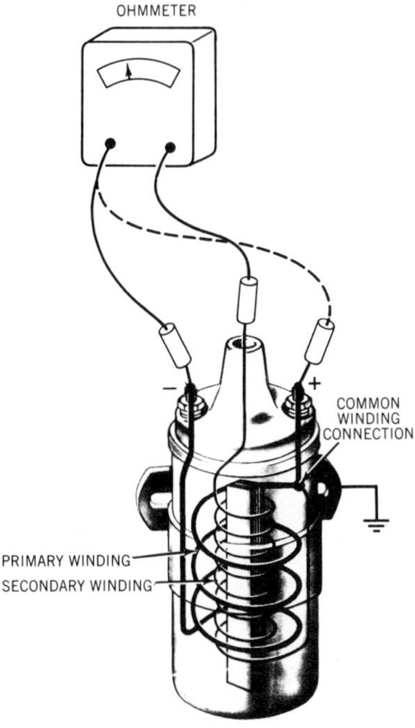

Figure 12-51. Measure coil secondary resistance across the center tower and both primary terminals. (Bosch)

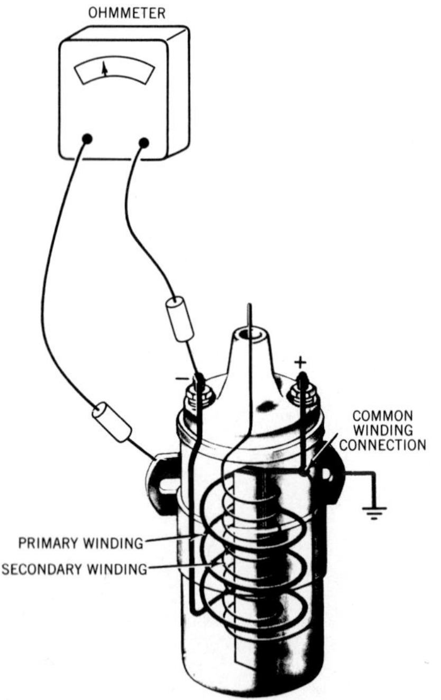

Figure 12-52. Test a coil for a short circuit to ground between either primary terminal and the case. (Bosch)

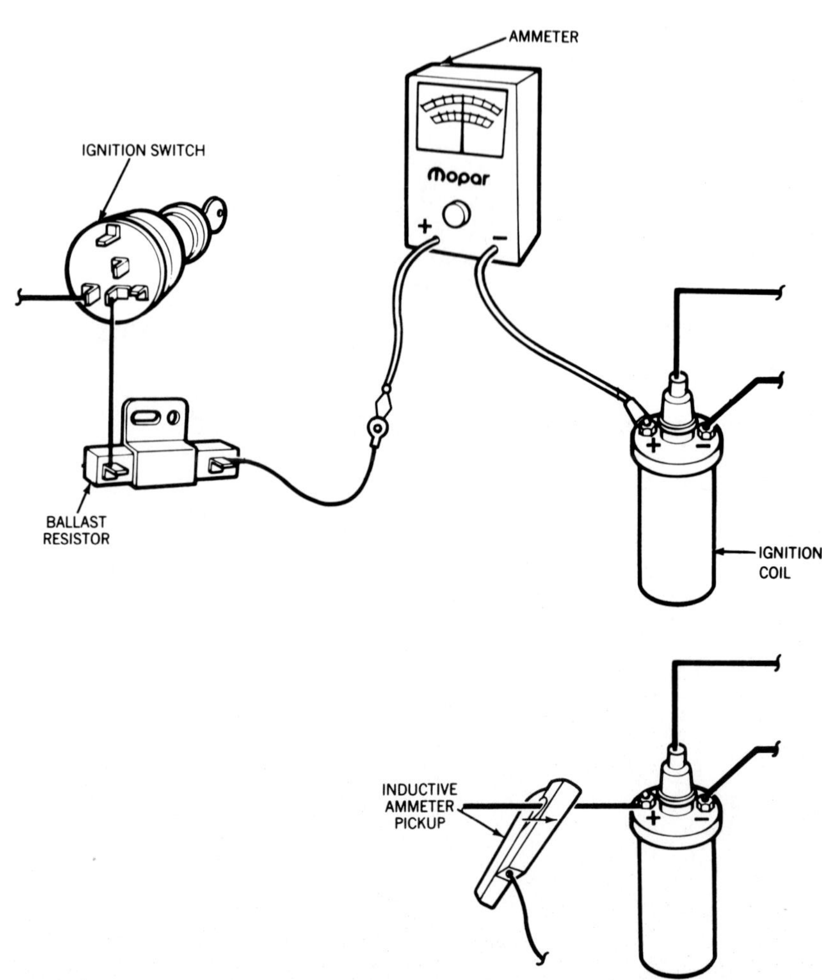

Figure 12-53. Connect an ammeter in either of these ways to test coil current draw. (Chrysler)

Coil Current Draw Test

Several carmakers publish current draw specifications for the coil. The Delco HEI test procedures in the previous section included current draw tests. Current specifications may be with the engine running, cranking, or with the ignition on and breaker points closed. To test current draw, connect an ammeter − lead to the coil BAT, or +, terminal. Connect the ammeter + lead to the + primary wire, disconnected from the coil + terminal, figure 12-53. If you have an inductive ammeter, place the inductive pickup around the primary wire to the coil + terminal. Leave the wire connected to the coil. Point the arrow on the meter pickup in the direction specified in the ammeter instructions (usually toward the coil).

Start or crank the engine, or turn the ignition on and close the points as directed by the manufacturer's procedures. Read the ammeter. No current reading means the primary circuit is open. The engine will not run in this case.

Current draw higher than specifications can be caused by:

- A short circuit in the coil or ballast resistor
- An incorrect coil or ballast resistor installed on the car.

Current draw lower than specifications can be caused by:

- A discharged battery
- High resistance in the coil primary winding
- Loose or corroded primary connections or high resistance elsewhere in the primary circuit.

Coil Polarity Test

It doesn't happen too often, but a coil can be connected backwards in an ignition system. In a negative-ground electrical system, if the coil negative terminal is connected to the battery voltage wire and the positive terminal is connected to the distributor or the ignition module, coil polarity is reversed, figure 12-54. Reversed polarity also can be caused by connecting a battery backwards, which also will damage the alternator, or by a manufacturing defect in the coil. If coil polarity is reversed, the engine will start and run, but probably will misfire under load or at high speed. This happens because maximum secondary voltage is reduced. If required voltage (high speed or heavy load) is higher than coil available voltage, the spark plugs will not fire.

You can check coil polarity easily with an oscilloscope, which you will

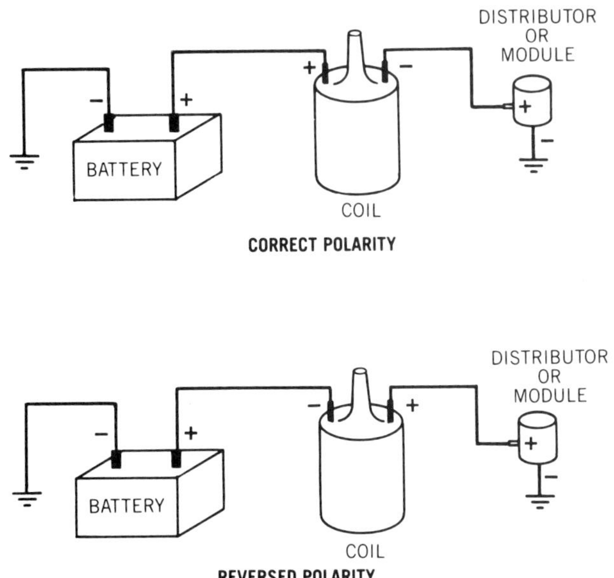

Figure 12-54. Correct and reversed coil polarity.

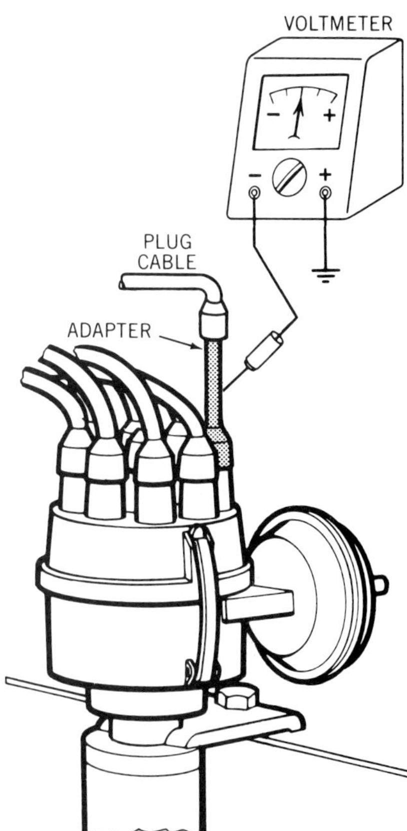

Figure 12-55. If the voltmeter reads upscale (toward + voltage), coil polarity is correct.

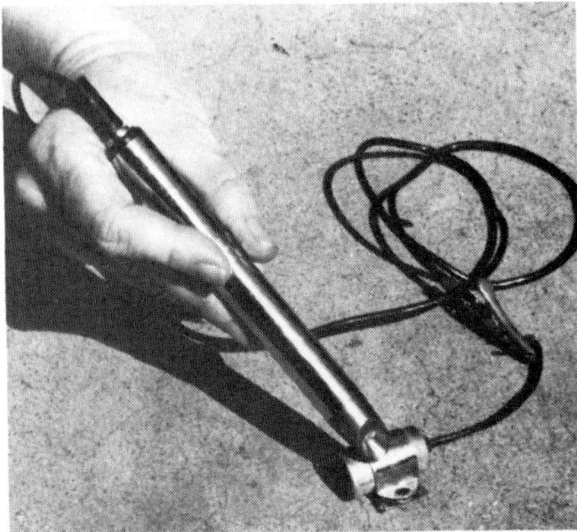

Figure 12-56. If the self-powered lamp lights between the condenser lead and the case, the condenser is shorted.

learn to do later. You also can check it with a voltmeter.

Voltmeter Polarity Test. For this test, you need a negative-reading voltmeter. That is, the meter dial must be graduated on the negative side of zero, as well as on the positive side. If the voltmeter is not negative reading, it can be damaged during the test.

1. Install an adapter with an exposed terminal into any spark plug cable, either at a spark plug or at the distributor, figure 12-55.

2. Connect the voltmeter + lead to an engine ground. Set the meter on its highest voltage scale. Start and idle the engine.

3. Briefly touch the voltmeter − lead to the adapter terminal in the plug cable and read the meter. If the needle moves upscale (toward +), coil polarity is correct. If the needle moves downscale (toward −), coil polarity is reversed.

CONDENSER TESTS

You can test the condenser of a breaker-point ignition system with a special condenser tester that uses ac and dc voltage to test condenser resistance and insulation quality and that measures actual capacity in microfarads. If you have such a tester, follow the manufacturer's directions for use. Later in this chapter you will learn to analyze condenser operation from oscilloscope patterns.

You can make a quick test of a condenser to check for an internal short circuit (voltage leakage through the insulation) with a self-powered test lamp. Connect one lead of the test lamp to the condenser lead, or pigtail, and the other lead to the condenser case, figure 12-56. If the lamp lights, the condenser insulation has broken down and the condenser is shorted.

Breaker-point wear can be a clue to condenser condition. Badly pitted or burned points can indicate a shorted or leaking condenser. Heavy pitting on the ground point can indicate an under-capacity condenser. Heavy pitting on the insulated point can indicate an over-capacity condenser. Remember, though, that point wear can be caused by other problems, as well. These include high system voltage, misalignment, and dirt or grease on the contact surfaces. Nevertheless, burned or pitted points should lead you to check the condenser, as well as other parts of the primary circuit.

BREAKER-POINT DWELL TEST AND ADJUSTMENT

Measuring and adjusting the dwell angle on a breaker-point ignition is an important and basic service because it affects coil saturation and secondary voltage, as well as ignition timing. Later in this chapter, you will learn to measure and analyze the dwell angle with an oscilloscope. In a later chapter, you will learn to adjust timing. You must always adjust breaker-point dwell before setting the timing because as you increase or decrease the dwell angle, you change the time at which the points open in relation to crankshaft angle. The point-opening time *is* ignition timing.

A tach-dwellmeter, figure 12-57, is a simple piece of test equipment that measures both engine speed and dwell angle. It is basically a voltmeter with internal circuits that translate the interruptions of primary voltage into rpm and dwell angle measurements. A tach-dwellmeter measures rpm as well as dwell angle because you often must check dwell variations at different engine speeds. You also can use the meter as a tachometer for basic idle speed adjustments.

Most tach-dwellmeters show an average dwell reading for all cylinders. They do not show individual cylinder dwell angles. An oscilloscope or a distributor tester shows individual cylinder dwell, but a tach-dwellmeter is fine for on-car dwell measurement and adjustment.

A tach-dwellmeter has negative and positive test leads, as a voltmeter does. Connect the negative lead to an engine ground and connect the positive lead to the coil + or − terminal, figure 12-57, depending on the instructions for the meter. Some coils need adapters for meter connections, figure 12-58. Set the cylinder selector knob on the meter for the number of cylinders in the engine you are testing, usually 4, 6, or 8. Set the function selector knob for tachometer or dwell measurement. The meter has several scales for high and low rpm and for dwell angle for 4-, 6-, and 8-cylinder engines.

For a basic dwell measurement, start and idle the engine and read both rpm and dwell measurements. Most carmakers allow a slight amount of dwell variation as engine speed changes. In fact, dwell will decrease as engine speed increases for a Ford, Chrysler, or other distributor with a side-pivot breaker plate as the vacuum advance operates. Allowable dwell variation is less for Delco distributors and others with center-pivot breaker plates.

Always check the manufacturer's specifications for dwell angle and allowable variation. For example, an 8-cylinder engine might have a dwell specification of 30° ± 2° or 28° to 32° at a given speed. The speed may be given in engine rpm or distributor rpm. If it is in distributor rpm, you must double it for test and adjustment in the engine. (For example, 2,000 rpm distributor speed = 4,000 rpm engine speed.)

Set the dwell for a new set of points to the lower end of the specified range. If the specification is 28° to 32°, adjust new points to 28° or 29°. This allows for rubbing block wear, which will *increase* the dwell angle during several thousand miles of operation.

Measure the dwell angle at the speed specified by the carmaker. Normally, you will adjust the dwell at cranking or idle speed. Delco V-6 and V-8 breaker-point distributors have windows in the cap that let you adjust

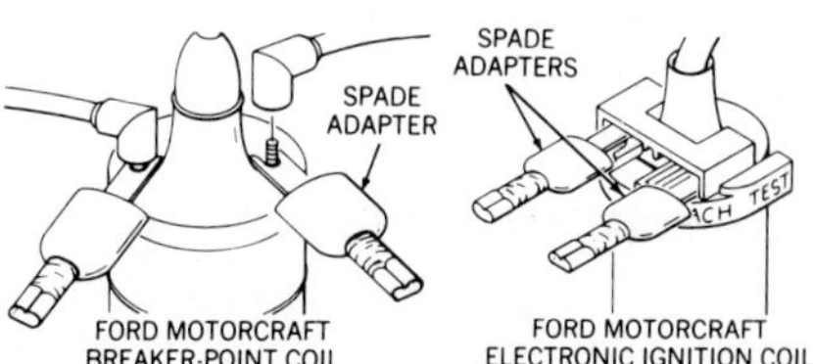

Figure 12-58. Use adapters to connect the tach-dwellmeter to Ford and similar coils.

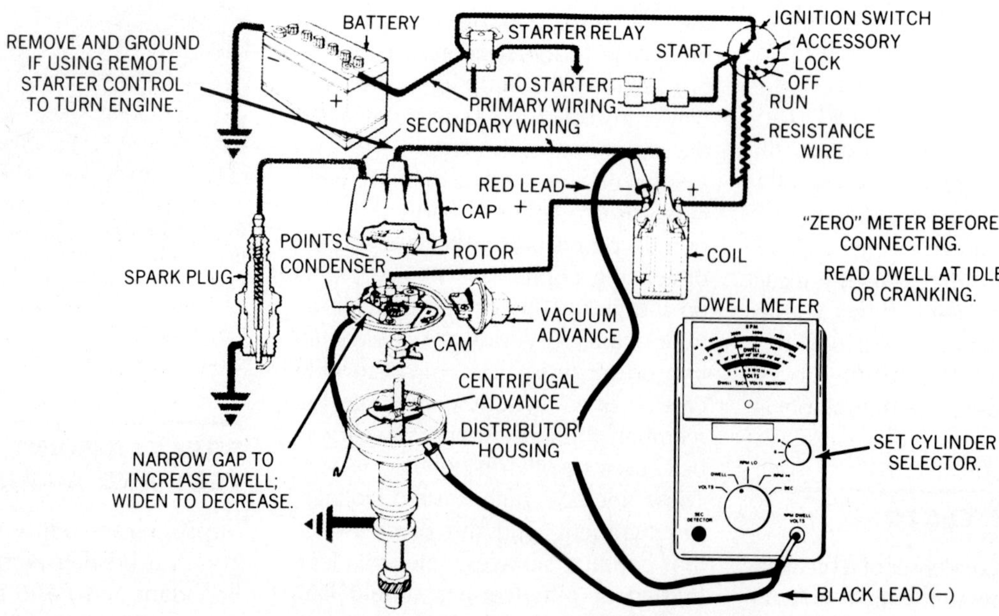

Figure 12-57. Connect the dwellmeter according to the manufacturer's instructions to measure dwell. (Ford)

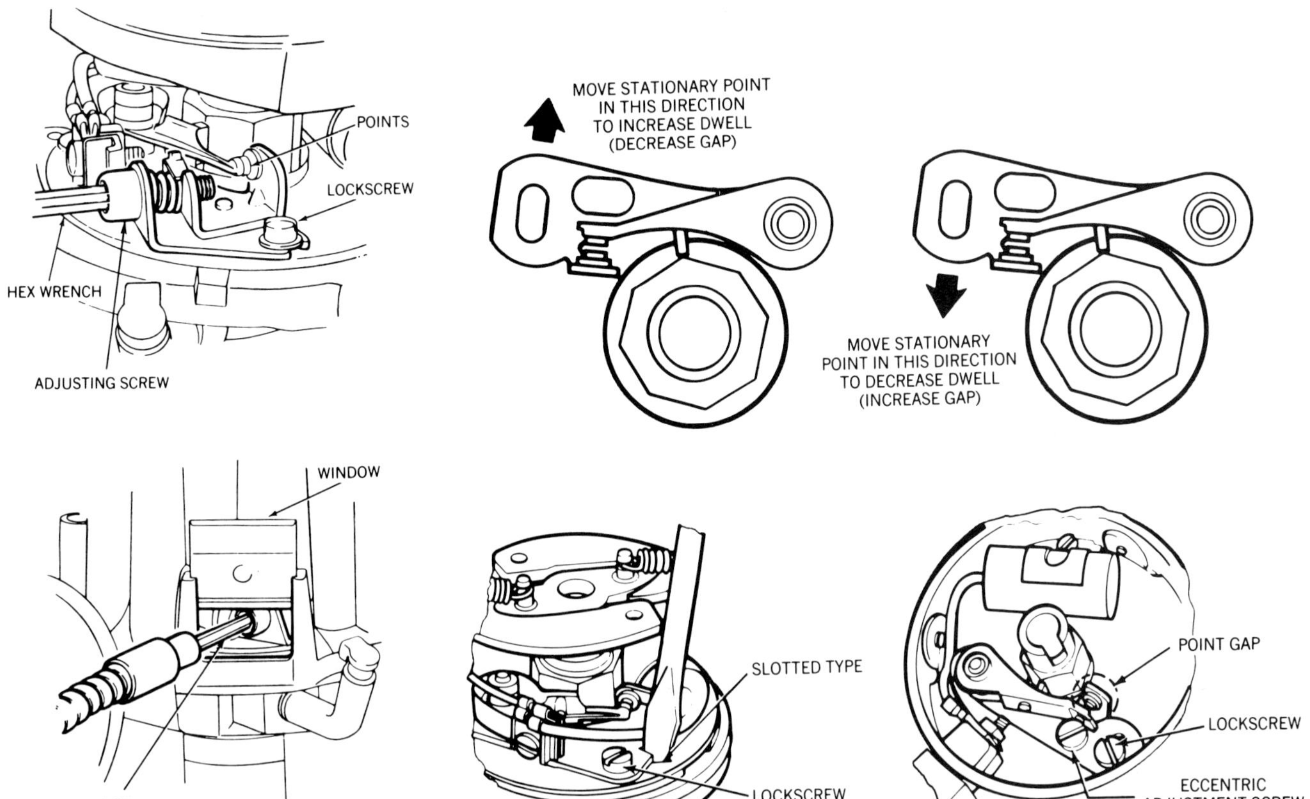

Figure 12-59. Use a hexhead (Allen) wrench through the cap opening to set dwell on a Delco external-adjustment distributor. (Delco-Remy)

Figure 12-60. For an internal-adjustment distributor, move a screw or a slot in the point base to adjust dwell. (Chrysler)

dwell with the engine running, figure 12-59. Raise the metal tab in the side of the cap and insert a 1/8-inch hexhead (Allen) wrench into the adjusting screw on the ground breaker point. With the engine running at the specified speed, turn the screw clockwise to increase dwell or counterclockwise to decrease dwell.

You can measure dwell on other distributors with the engine running, but you must remove the cap to adjust it. Loosen the lockscrew or holddown screw on the point assembly and move the ground point by turning an adjustment screw, figure 12-60, or by moving an adjustment slot with a screwdriver. A good tach-dwellmeter will give an accurate dwell measurement at engine cranking speed. Crank the engine with the ignition switch or a remote starter switch and read the meter. Adjust the point assembly until dwell is within specifications and tighten the lockscrew. Then reinstall the distributor cap and check the dwell adjustment with the engine running at the specified speed.

If dwell is not within specifications at a given engine speed, repeat the adjustment. If dwell varies more than allowed by the manufacturer at a given speed or at different speeds, remove the distributor for test and overhaul in a distributor tester.

Chapter 14 covers breaker-point installation and adjustment in more detail as part of complete distributor service.

PRIMARY CIRCUIT OSCILLOSCOPE TESTS

Chapter 4 introduced you to the engine analyzer oscilloscope. Now you will learn to use it to test the ignition primary circuit. In the next chapter, you will learn to use it to test the secondary circuit.

An oscilloscope, or scope, is really a voltmeter that displays a changing voltage picture (a pattern or waveform) over a period of time. It is the best device for analyzing dynamic voltage changes while a circuit is operating. Most scopes are mounted in complete engine analyzers that also contain voltmeters, ohmmeters, ammeters, tach-dwellmeters, vacuum and pressure gauges, timing lights, and exhaust analyzers. The analyzer has multiple test leads that you connect to various points on an engine for different tests, figure 12-61. Often, you will have all leads connected at once for comprehensive testing. To check only the ignition system, you must have the following minimum oscilloscope connections:

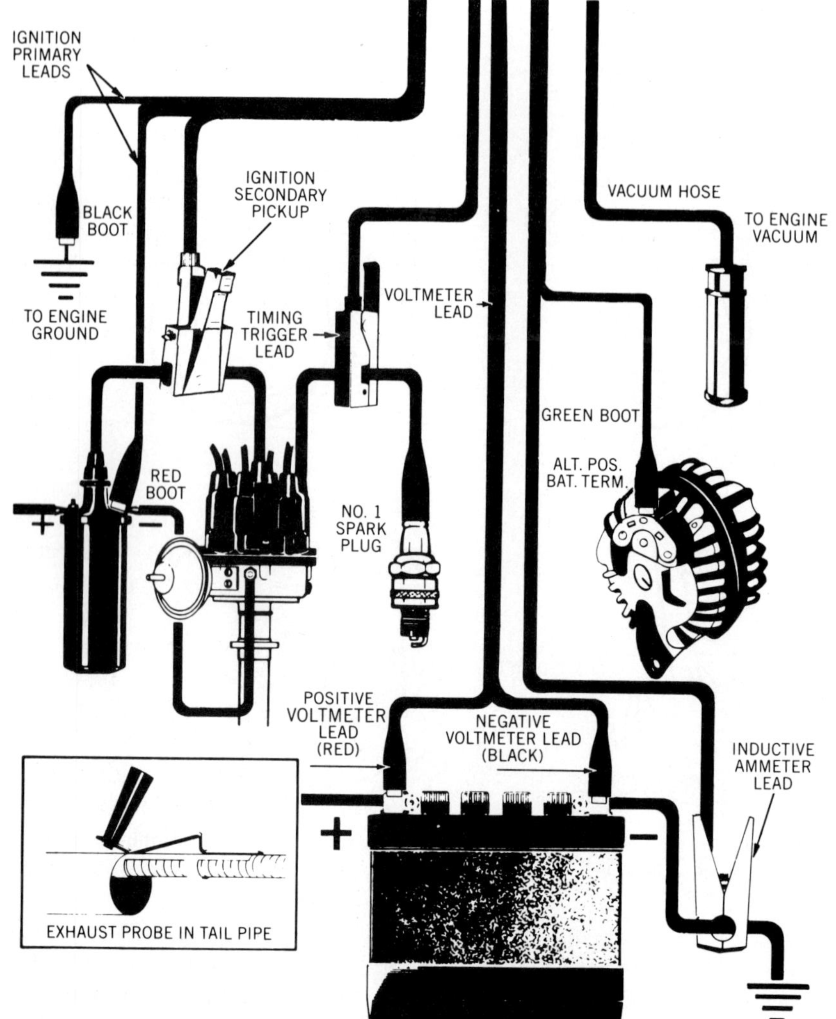

Figure 12-61. These are typical test lead connections for an oscilloscope engine analyzer. (Sun)

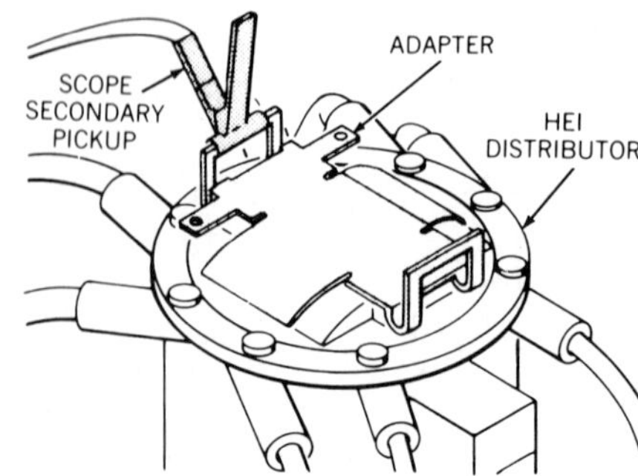

Figure 12-62. Delco HEI coils require special secondary pickups.

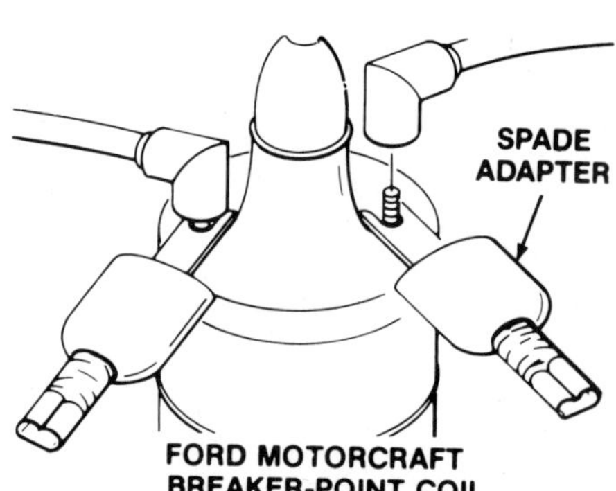

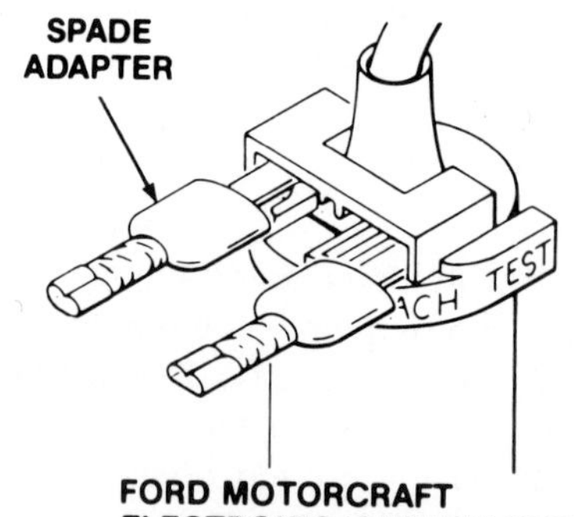

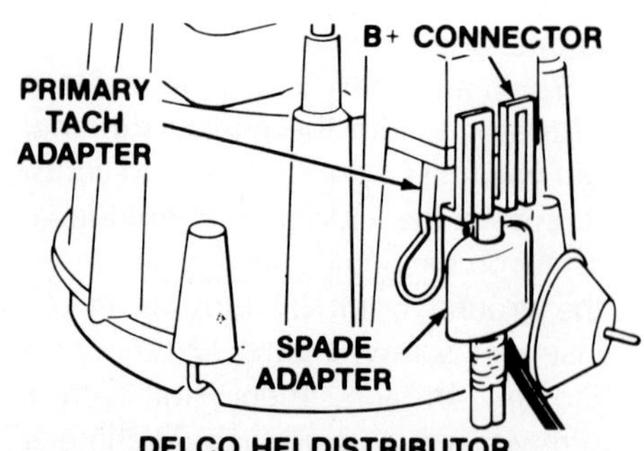

Figure 12-63. Some primary circuits also require test lead adapters.

1. Analyzer ground connection to engine ground to complete all circuits

2. Scope primary connection to the ignition primary circuit at the coil + or − terminal, depending on equipment design

3. Scope high-voltage connection to the coil secondary terminal or cable

4. Scope high-voltage connection to the number 1 spark plug for timing signals and proper firing order waveform sequence.

Most late-model scopes have inductive pickups for the high-voltage connections, figure 12-61. Older models have series connectors, and these may not work on many electronic ignitions. Some ignitions require adapters for the test lead connections. Delco HEI systems with coils in the distributors require an inductive adapter for the coil secondary test lead, figure 12-62. Ford and Delco HEI systems require adapters for the primary connections to the coil, figure 12-63.

The oscilloscope cathode ray tube (CRT) has vertical voltage scales on each side, usually graduated from 0 to 20 or 25 and from 0 to 40 or 50, figure 12-64. A range selector switch lets you choose low- and high-voltage scales. For primary circuit displays, you will use the low-kilovolt scale or a 0- to 500-volt scale, according to analyzer directions. The horizontal scales at the bottom of the CRT show ignition dwell in degrees, and often time in milliseconds and percentage to measure duty cycles.

Primary and Secondary Oscilloscope Patterns

Primary and secondary patterns, or waveforms, both show ignition voltage in three basic sections: firing, intermediate, and dwell, figure 12-65. Because of the different voltage ranges of the scope, primary and secondary patterns appear about the same size even though primary voltage ranges from 0 to 300 or 400 volts and secondary voltage ranges from 0 to 20,000 volts (20 kV) and higher.

The firing section shows what happens when primary current to the coil stops. The rapid oscillations in the primary pattern are due to the rapid charging and discharging cycle between the coil and the condenser or the ignition module. Voltage does not stop abruptly when the primary circuit opens, but oscillates between inductive and capacitive components. Some electronic ignitions, such as Delco HEI systems, do not display these primary oscillations, however.

In the firing section of a secondary pattern, voltage rises to a peak as the plug gap ionizes and the plug fires. It then immediately drops to a lower "burn voltage" as the plug continues to fire. This is usually called the "spark line." It also represents spark plug "burn time."

At point B in figure 12-65, coil voltage is no longer high enough to maintain a spark at the plug, and the firing section ends. The intermediate section extends from B to C and is the time between the end of the spark and the start of the dwell period when the primary circuit closes. During the intermediate period, remaining energy in the coil dissipates as gradually decreasing voltage oscillations. The shape of the oscillations indicates the condition of the coil (and the condenser in a breaker-point system).

Chrysler electronic ignitions do not have a distinct intermediate section. Coil oscillations start at the end of the firing section (end of sparkline) and gradually diminish to a straight voltage line that ends at the next vertical firing line.

The dwell section of the waveform begins at point C in figure 12-65 and extends to point D at the right. At that point, the firing section starts again (point A). During the dwell section, the primary circuit is closed, and current flows to the coil. The dwell section begins with a short downward line. In a breaker-point system, this is the "points-close" signal. In an electronic system, the downward line is the point where the module power transistors turn on primary current. The rest of the dwell section is a flat, almost horizontal line that indicates primary current to the coil until the primary circuit opens again

The dwell sections vary for variable-dwell electronic ignitions, such as Delco HEI, Prestolite, and some Ford systems. The hump, or ripple, toward the end of the dwell section indicates that current-limiting circuits in the module have reduced primary current.

Control keys on the oscilloscope console allow you to select one of three basic patterns for either the primary or the secondary circuit. These are:

1. The *superimposed pattern* superimposes all cylinder waveforms upon each other so that there appears to be only one pattern, figure 12-66. This gives you a quick way to check the uniformity of all the waveforms.

2. The *parade pattern* displays the individual cylinder waveforms side by side, from left to right in firing-order sequence, figure 12-67. This compresses each individual pattern horizontally, but allows you to compare voltage levels.

3. The *raster pattern* stacks the individual cylinder patterns one above the other from bottom to top in firing-or-

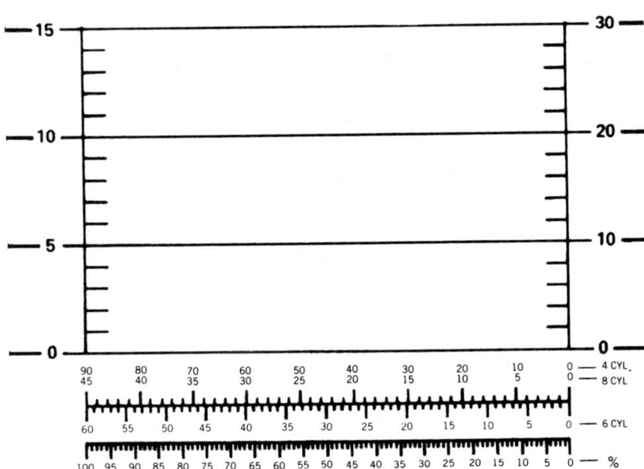

Figure 12-64. An oscilloscope screen, or CRT, has vertical voltage scales and horizontal dwell scales. It also may have time and percentage scales.

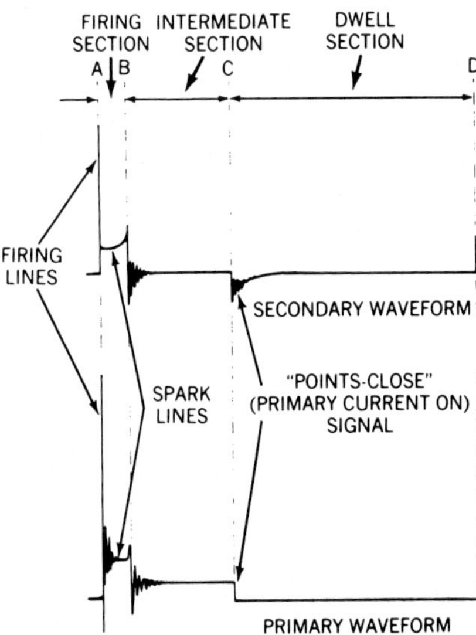

Figure 12-65. Primary and secondary patterns (waveforms) both have three corresponding sections: firing, intermediate, and dwell.

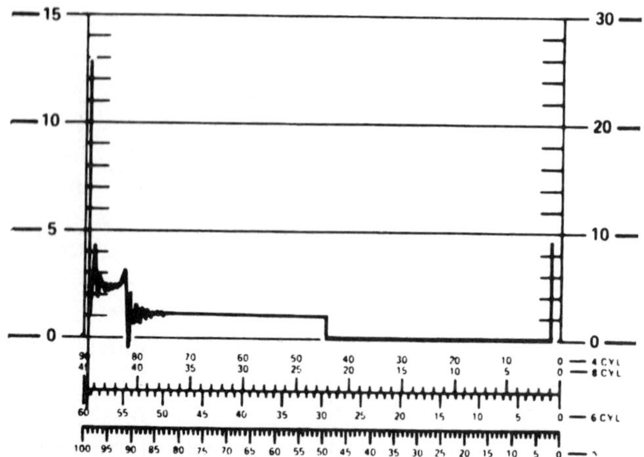

Figure 12-66. Typical primary circuit superimposed pattern for breaker-point and many electronic ignitions.

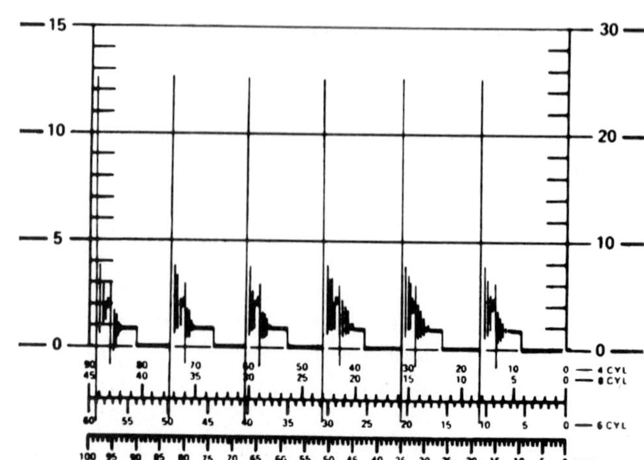

Figure 12-67. Typical primary circuit parade pattern.

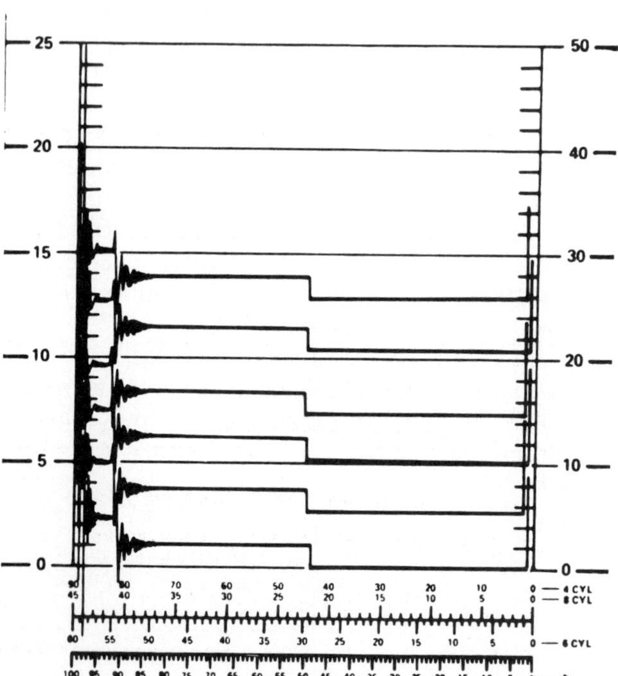

Figure 12-68. Typical primary circuit raster pattern for a breaker-point and many electronic ignitions.

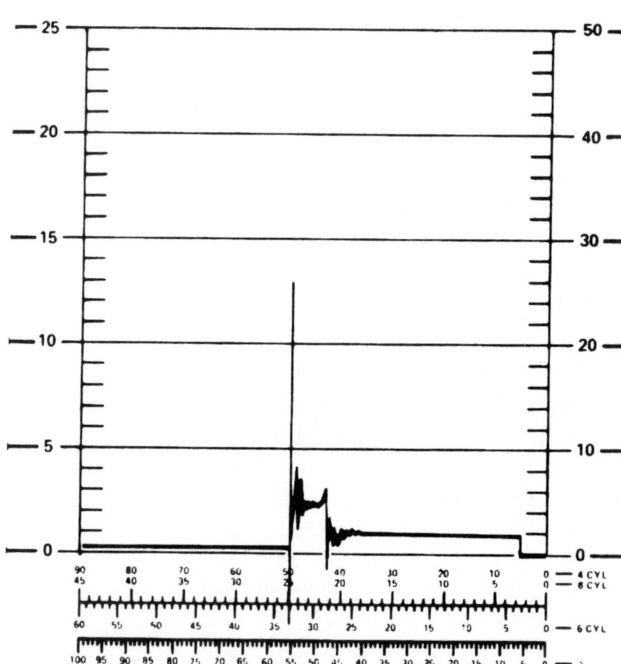

Figure 12-69. This primary circuit superimposed pattern is shifted to the right.

der sequence, figure 12-68. This shows the full horizontal pattern for each cylinder so that you can compare the spark line (burn time), dwell, and other time-related events.

These three displays for either the primary or the secondary circuit give you six basic patterns to use for diagnosis and service. Most modern scopes also allow you to shift the display about halfway across the screen, figure 12-69, for closer examination of the firing and intermediate sections. Some scopes also allow you to select the waveform for one cylinder and separate it from a superimposed or parade pattern for close analysis, figure 12-70. You also may be able to expand part of a pattern and break it into millisecond increments to analyze functions such as dwell or burn time, figure 12-71.

Primary Pattern Analysis

Each section of any pattern is created by a different part of the circuit. If a part is defective or misadjusted, the pattern will be abnormal. You can compare an abnormal pattern to a normal pattern to identify and repair a fault. Primary patterns indicate the condition of the entire primary circuit, including the ignition switch, the pri-

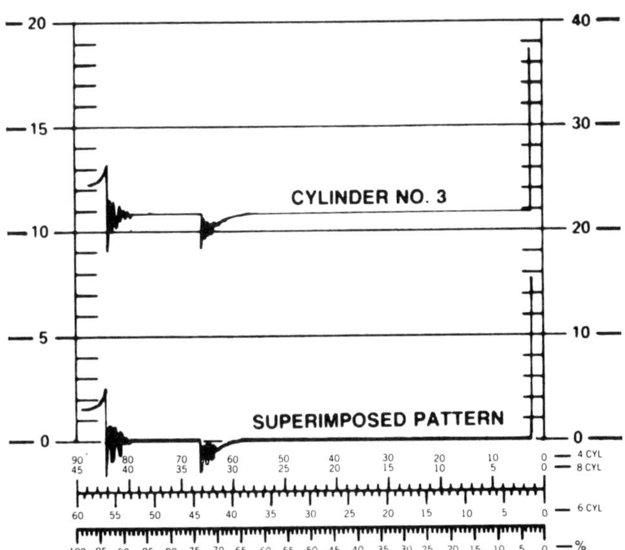

Figure 12-70. Cylinder no. 3 is selected from this superimposed pattern.

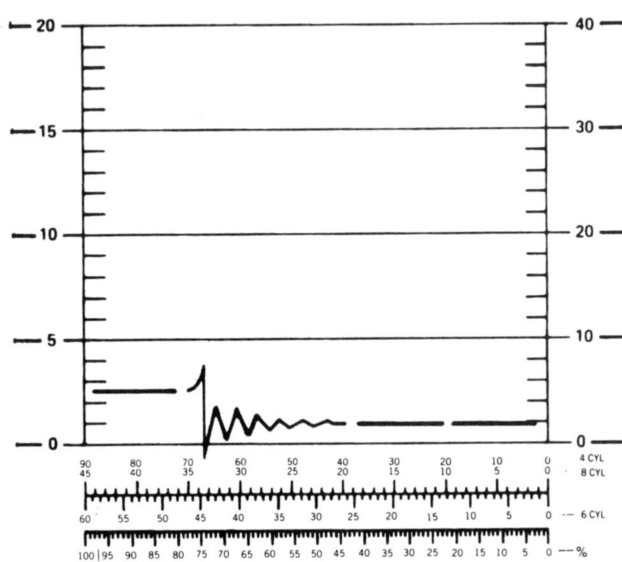

Figure 12-71. This pattern is expanded and broken into 1-millisecond (1-ms) divisions for close analysis.

mary (ballast) resistor, the coil primary winding, the points and condenser, or the ignition module and pickup coil. Make the primary circuit tests at idle with a fully warm engine.

Superimposed Patterns. Figure 12-66 shows a normal primary superimposed pattern for breaker-point ignition. Motorcraft electronic ignitions display similar patterns, except that the dwell section for Dura-Spark I will be shorter and will vary with engine speed. Figure 12-72 shows normal primary superimposed patterns for Chrysler, Delco HEI, and Prestolite electronic ignitions. In a breaker-point ignition, the condenser oscillations should start high and taper off through the firing section. There should be at least five distinct coil oscillations in the intermediate section, with the first being largest.

Figures 12-73 through 12-77 are abnormal primary superimposed patterns that show the following problems and summarize the most likely causes:

- Reversed coil polarity (or reversed primary scope leads), figure 12-73
- Breaker-point primary circuit high resistance, figure 12-74

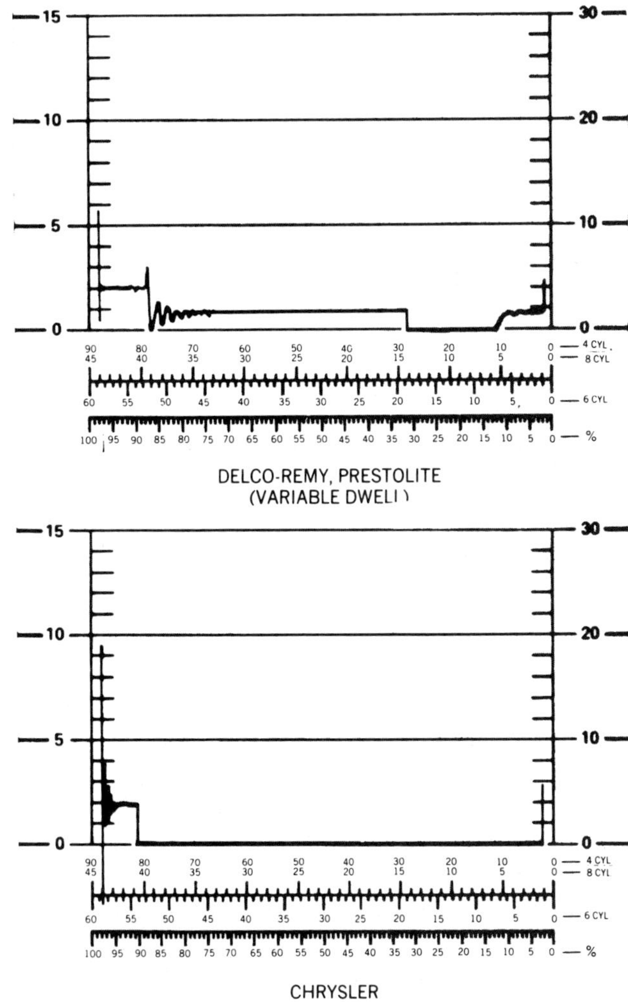

Figure 12-72. Normal primary superimposed patterns for electronic ignitions: variable dwell, top, fixed dwell, bottom.

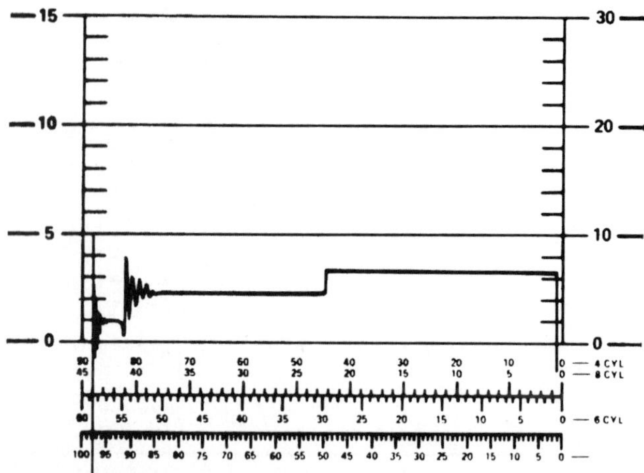

Figure 12-73. This upside-down primary pattern indicates reversed coil polarity. *Causes:* Battery cables or coil primary connections reversed. Scope primary leads reversed.

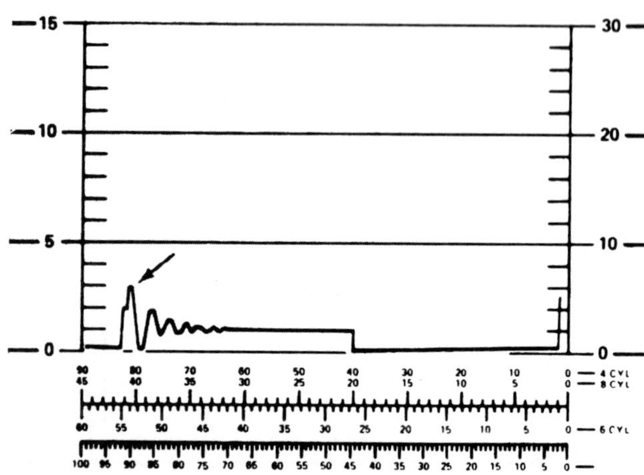

Figure 12-74. Breaker-point primary circuit high resistance. Firing oscillations are reduced or missing. *Causes:* High resistance in the primary circuit, such as the ignition switch, the ballast resistor, or any wiring connections. Low battery or charging voltage also can cause this.

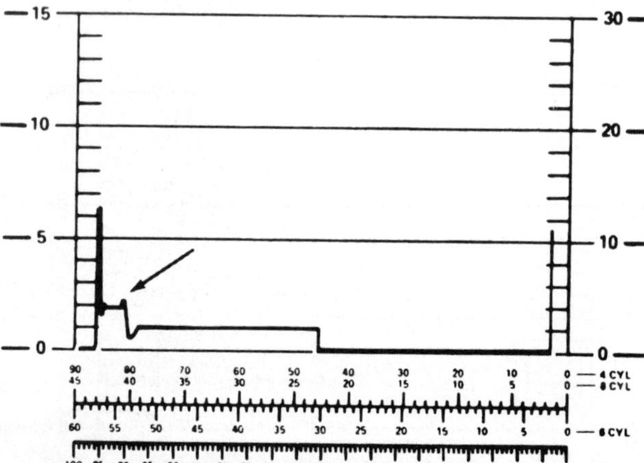

Figure 12-75. Reduced coil and condenser oscillations in the firing or intermediate section of a breaker-point ignition. *Causes:* Short circuit in the coil primary winding or in the condenser (condenser leakage).

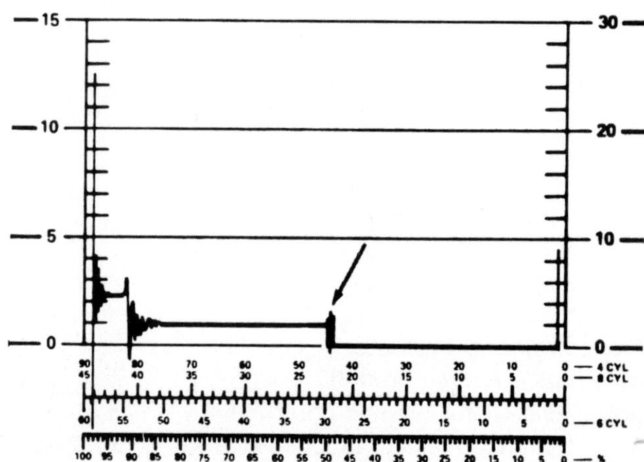

Figure 12-76. Broken lines or "hash" marks at the points-close signal indicate breaker-point bounce. This is most noticeable at medium and high speed.
Causes: Weak point spring. Dry or dirty distributor cam lobes. Pitted, misaligned, or loose points.

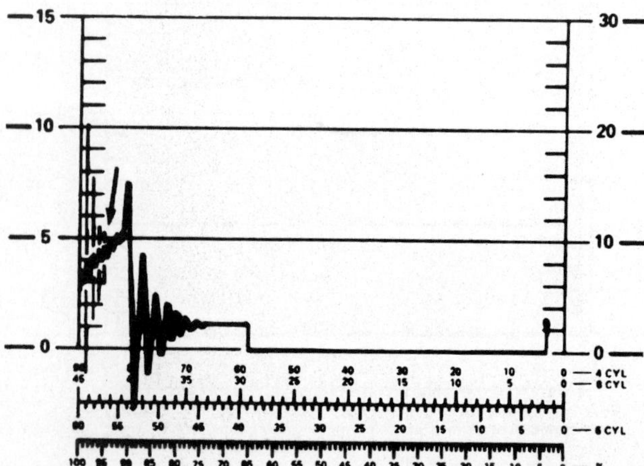

Figure 12-77. This pattern shows a bright sloping line or "blip" of light at the points-open signal that indicates breaker-point arcing.
Causes: Burned, misaligned, or pitted points. Primary voltage is too high. Bad condenser or ballast resistor.

- Reduced coil-condenser oscillations in a breaker-point circuit, figure 12-75
- Breaker-point bounce, figure 12-76
- Breaker-point arcing, figure 12-77.

Raster Patterns. Figure 12-68 shows a normal primary raster pattern for a breaker-point ignition. Figures 12-78 through 12-80 show normal primary circuit raster patterns for electronic ignitions. The raster pattern is useful for comparing time-related events among cylinders. Figure 12–81 shows a dwell variation at the points-

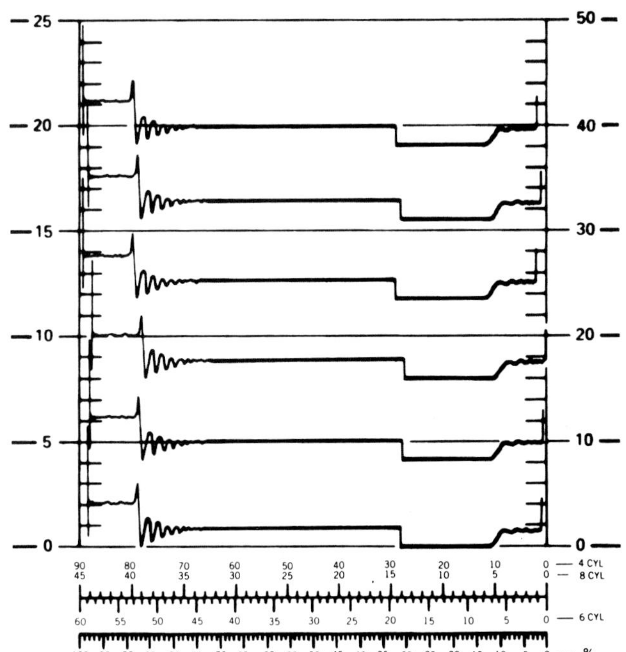

Figure 12-78. Normal primary raster pattern for Delco, Prestolite, and other variable-dwell electronic ignitions.

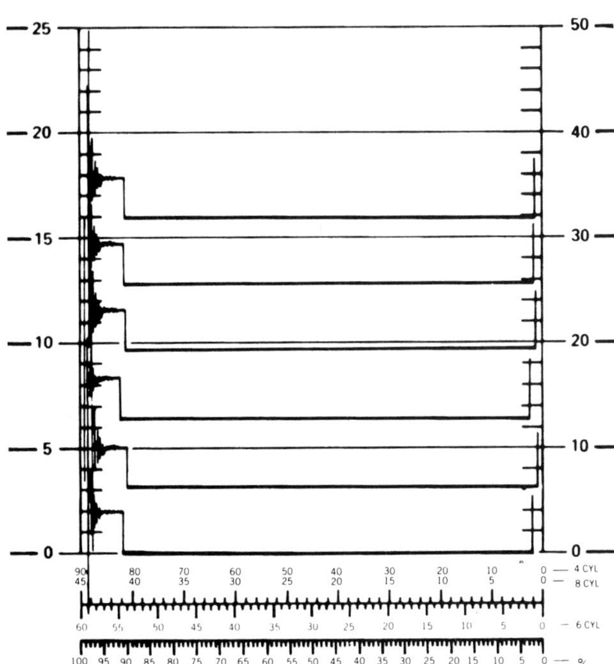

Figure 12-79. Normal primary raster pattern for Chrysler electronic ignitions.

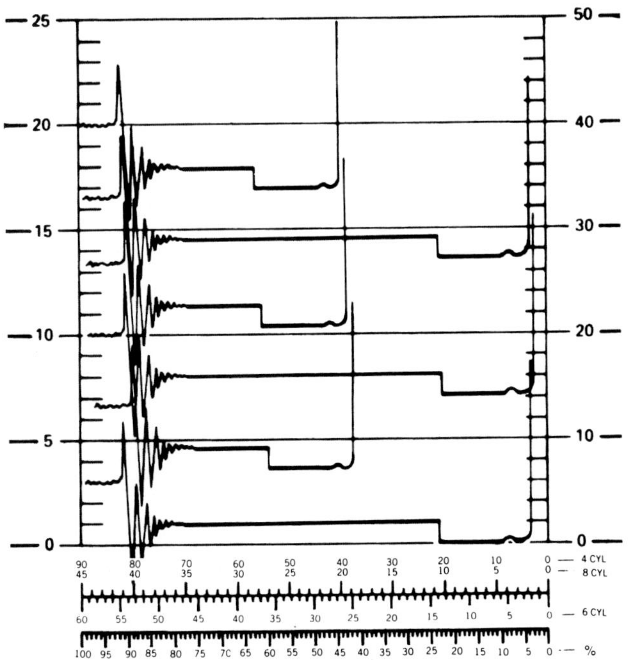

Figure 12-80. Normal primary raster pattern for uneven-firing V-6 electronic ignitions.

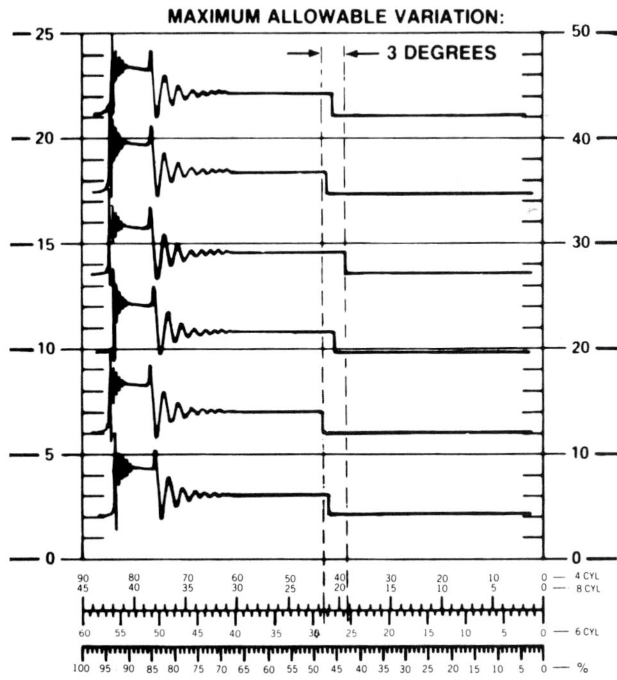

Figure 12-81. If the points-close signals are out of line by more than 3 degrees, there is too much dwell variation. There are two common symptoms and their related causes:
Symptom 1: Points-close signals are uneven but steady.
Cause 1: Worn distributor cam lobes.
Symptom 2: Points-close signals are uneven and fluctuating.
Cause 2: Worn distributor bushings, loose breaker plate, bent distributor shaft, loose timing chain.

close signal for a breaker-point ignition. If the points-close signal varies by more than 3 degrees among cylinders but is steady at a constant speed, the distributor cam lobes probably are worn. If the signal varies and is unsteady, the engine may have a loose timing chain; or the distributor may have a bent shaft, worn bushings, or a loose breaker plate.

Parade Patterns. Figure 12-82 shows an abnormal primary parade pattern for a breaker-point ignition. Compare this to the normal pattern in figure 12–67. The parade display is more useful for checking the secondary than the primary circuit, but you can use a primary parade pattern to check firing voltages among cylinders. If the first oscillation in the firing section is higher or lower for some cylinders, it can indicate the following problems:

- At idle—pitted or misaligned breaker points
- At medium or high speed—point bounce
- At all speeds—worn distributor cam lobes.

Primary Circuit Dwell Measurement

Primary superimposed and raster patterns allow close analysis of the basic dwell angle and dwell variation among cylinders. To display the basic dwell angle, select the primary superimposed pattern and adjust the pattern width to fill the scope CRT exactly from left to right. Read the dwell angle on the horizontal dwell scale for a 4-, 6-, or 8-cylinder engine, figure 12-83, and compare it to specifications at a given rpm.

Dwell should be constant on breaker-point and fixed-dwell electronic ignitions, but manufacturers allow some variation on breaker-point systems. On a superimposed pattern, you can see dwell variation as a slight offset in the display, figure 12-84. If the variation exceeds 3 degrees, use the raster pattern to look at the dwell variation for individual cylinders, figure 12-81. Accelerate the engine and watch the dwell section on either the superimposed or the raster pattern. Remember that dwell will decrease 3 to 6 degrees on a distributor with a side-pivot breaker plate as vacuum advance is applied. Dwell variation among cylinders should remain within 3 degrees at any engine speed.

At this point, we must clarify two terms: "dwell variation" and "cam lobe variation." If dwell variation exists *among the cylinders*, it probably is due to cam lobe wear or shaft and bushing wear. If dwell varies uniformly for all cylinders as speed changes, it is normal for a side-pivot breaker plate. However, erratic dwell variation or

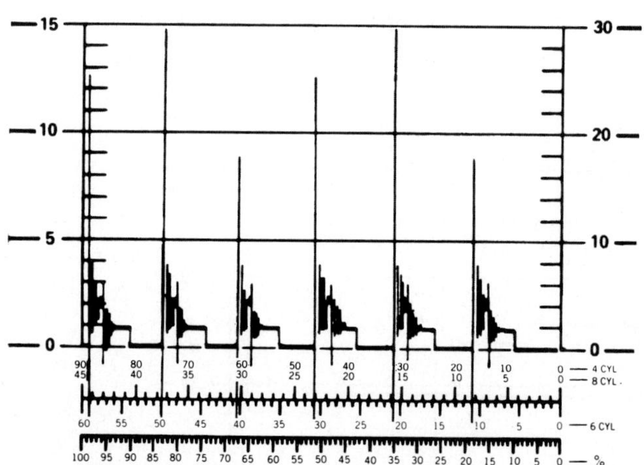

Figure 12-82. These uneven firing voltages can be caused by:
At Idle: Pitted or misaligned breaker points.
At Medium to High Speed: Breaker-point bounce or float.
At All Speeds: Worn distributor cam lobes or other timing variation.

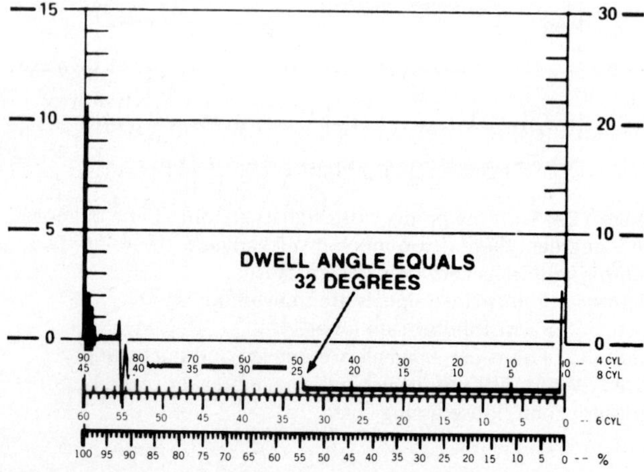

Figure 12-83. Use the primary superimposed pattern to measure dwell accurately.

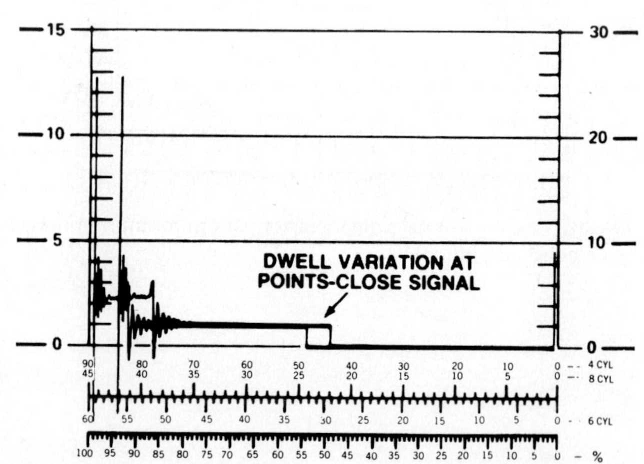

Figure 12-84. Dwell variation at the points-close signal of a superimposed pattern. Compare this display to figure 12-81.

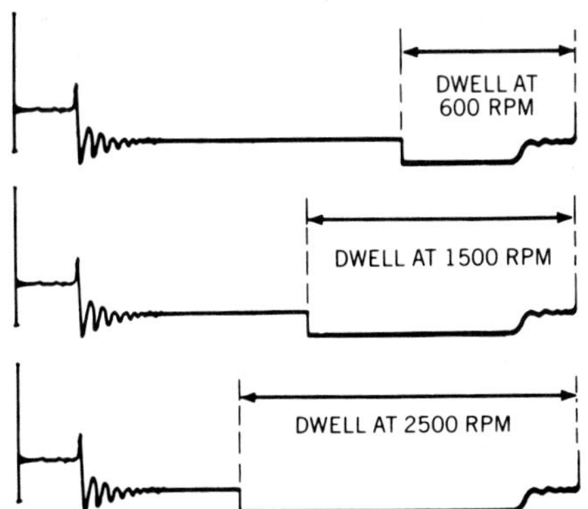

Figure 12-85. Normal dwell variation for a variable-dwell electronic ignition.

dwell variation with changing speed for a center-pivot breaker plate can indicate a worn distributor shaft, breaker plate, or vacuum advance mechanism.

You can use the primary superimposed pattern as a dwellmeter to adjust breaker-point dwell as explained earlier. For a Delco external-adjustment distributor, run the engine at idle and watch the pattern dwell section as you adjust the points. Many scopes will display a primary pattern at cranking speed that you can use to adjust an internal-adjustment distributor.

The dwell period for an electronic ignition is controlled by the ignition module, and you cannot adjust it. Dwell variation on a fixed-dwell system, however, can be used to identify a bent distributor shaft, worn distributor bushings, or a worn timing chain. These problems will cause the same kind of dwell variation for breaker-point and electronic systems.

The dwell measurement for a variable-dwell electronic ignition will increase with engine speed and may fluctuate at a constant speed because dwell is controlled electronically by the module, not by the distributor. Figure 12-85 shows the changing dwell period for a Delco HEI system. If dwell does not increase as speed increases, the ignition module may be faulty.

ELECTRONIC PRIMARY CIRCUIT SERVICE

The primary circuits of electronic ignitions do not need periodic service as breaker-point primary circuits do. Electronic primary circuit service consists of identifying and fixing problems in the following three general areas, if they occur:

1. *High resistance or open or grounded circuits in primary wiring and connectors.* You can locate such problems with common voltmeter and ohmmeter electrical tests. You can fix many primary circuit problems simply by cleaning and repairing wiring or connectors.

2. *Distributor signal generator (pickup coil) problems.* The voltmeter and ohmmeter tests and oscilloscope tests that you learned earlier will help you identify an open or grounded circuit in an electronic distributor pulse generator, or pickup coil. The Chrysler ignition test procedures outlined air gap adjustment between the trigger wheel (reluctor) and pickup coil. Other than this adjustment on some Chrysler distributors, electronic distributor parts are not adjustable.

3. *Ignition module replacement.* Your voltmeter, ohmmeter, and oscilloscope tests will help you identify a bad ignition module. Modules cannot be adjusted; they can only be replaced. Replacing a module mounted separately from the distributor requires simply disconnecting the harness connectors, removing the mounting screws, and installing and connecting the new module. Overhaul procedures in a later chapter explain replacement of Delco HEI modules. All modules provide the ground connection for the primary circuit. Many intermittent faults and hard-starting problems are caused by high-resistance or open ground connections at the module. Whenever you replace an ignition module, follow the manufacturer's instructions to ensure a good ground connection.

SUMMARY

Voltage in the primary circuit and primary circuit operation directly affect secondary ignition operation. Low primary voltage can be caused by low voltage from the battery or charging system, but more often it is caused by high resistance in the primary circuit. Voltmeter, ohmmeter, and oscilloscope tests will help you find causes of low primary voltage such as a discharged battery or low charging voltage or a bad coil, points, condenser, ballast resistor, ignition module, or a poor ignition system ground.

Oscilloscope testing is an important method of ignition diagnosis. The scope paints a moving picture of ignition voltage on a cathode ray tube (CRT). By using the basic superimposed, raster, and parade patterns, you can analyze voltage levels and study changing voltage over a period of time. The scope also is an accurate dwellmeter for adjusting breaker-point dwell and diagnosing distributor condition.

Breaker-point primary circuits require periodic adjustment and replacement of points and condenser. Electronic primary circuits require troubleshooting when problems occur. Most electronic primary circuit service consists of identifying and fixing problems of high resistance or poor ground connections, defective signal generators (pickup coils), and bad ignition modules.

REVIEW QUESTIONS

Multiple Choice

1. Before servicing an electronic ignition system, mechanic A grounds the TACH terminal. Mechanic B disconnects and grounds the battery voltage, or BAT, connector. Who is doing to the job correctly?

a. A only
b. B only
c. both A and B
d. neither A nor B

2. An engine runs, but one cylinder misfires continually. All of the following can cause this problem EXCEPT:

a. a fouled spark plug
b. a bad distributor cap terminal
c. an open circuit in the distributor pickup coil
d. an open circuit in the spark plug cable

3. High resistance in the ignition primary circuit can be caused by any of the following EXCEPT:

a. a bad coil
b. high charging voltage
c. a defective ballast resistor
d. a poor ground at the ignition module

4. When testing a breaker-point primary circuit during cranking (START), voltage drop across the starting bypass circuit should be no more than:

a. 0.2 volt
b. 2.0 volts
c. 0.5 volt
d. 1.5 to 3.5 volts

5. Mechanic A says that with the ignition on, the breaker points closed, and the engine not running, voltage drop across the ballast resistor should be 5 to 7 volts. Mechanic B says that with the ignition on, the breaker points open, and the engine not running, voltage at the coil + terminal should equal battery voltage. Who is right?

a. A only
b. B only
c. both A and B
d. neither A nor B

6. Mechanic A says that you must use a steel feeler gauge to check the air gap between the trigger wheel and the pickup coil and magnet in some distributors. Mechanic B says that you can check pickup coil resistance in a magnetic-pulse distributor with a voltmeter across the pickup coil connector terminals. Who is right?

a. A only
b. B only
c. both A and B
d. neither A nor B

7. Chrysler's original electronic ignition system has:

a. no ballast resistor
b. a single 5.0-ohm ballast resistor
c. a single 0.5-ohm ballast resistor
d. a dual ballast resistor

8. The Delco-Remy high energy ignition (HEI) system has:

a. no ballast resistor
b. a single 5.0-ohm ballast resistor
c. a single 0.5-ohm ballast resistor
d. a dual ballast resistor

9. Mechanic A says that the primary circuits of all electronic ignitions are grounded through the ignition module. Mechanic B says that Ford Dura-Spark ignitions are grounded at the distributor. Who is right?

a. A only
b. B only
c. both A and B
d. neither A nor B

10. A primary (ballast) resistor is NOT used in Ford's:

a. Dura-Spark II ignition
b. Dura-Spark III ignition
c. thick-film integrated (TFI) ignition
d. all of the above

11. An engine cranks normally, almost starts, and then dies as you release the ignition key to the run position. This often indicates:

a. an open ballast bypass circuit
b. an open ballast resistor
c. low battery voltage
d. any of the above

12. All of the following can cause current draw at the ignition coil that is *lower* than specifications EXCEPT:

a. a short circuit in the ballast resistor
b. a discharged battery
c. high resistance in the coil primary winding
d. loose connections in the primary circuit

13. Mechanic A says that breaker-point dwell will decrease as the vacuum advance moves a side-pivot breaker plate. Mechanic B says that you should double the distributor rpm specifications to test dwell at a specific engine speed. Who is right?

a. A only
b. B only
c. both A and B
d. neither A nor B

14. Mechanic A adjusts a new set of breaker points to a narrow gap to allow the proper dwell as the rubbing block wears. Mechanic B adjusts the dwell angle for a new set of breaker points to the lower end of the specified dwell range. Who is doing the job correctly?

a. A only
b. B only
c. both A and B
d. neither A nor B

15. In an oscilloscope pattern, the peak voltage at the time the spark plug ionizes and first fires is usually called the:

a. spark line, or period
b. dwell period
c. firing line, or point
d. burn time, or point

16. When individual cylinder patterns are stacked one above the other in an oscilloscope display, it is called a:

a. superimposed pattern
b. parade pattern
c. shifted pattern
d. raster pattern

17. If an oscilloscope pattern is upside down, ignition coil polarity may be reversed, or:

a. distributor pickup coil polarity is reversed
b. the ignition module has a high-resistance ground connection
c. the oscilloscope primary leads are connected backwards
d. the ignition coil is shorted

18. Mechanic A says that dwell variation that occurs when a vacuum advance unit operates is the same as cam lobe variation seen in an oscilloscope pattern. Mechanic B says that erratic dwell variation can indicate a loose breaker plate. Who is right?

- **a.** A only
- **b.** B only
- **c.** both A and B
- **d.** neither A nor B

19. The primary ignition pattern shown at the right usually indicates:

- **a.** improper coil-condenser oscillations
- **b.** reversed coil polarity
- **c.** high primary circuit resistance
- **d.** breaker-point arcing

20. The primary ignition pattern shown at the right indicates:

- **a.** excessive cam lobe wear
- **b.** normal dwell variation for an electronic ignition
- **c.** an uneven-firing V-6 engine
- **d.** uneven spark plug firing voltage

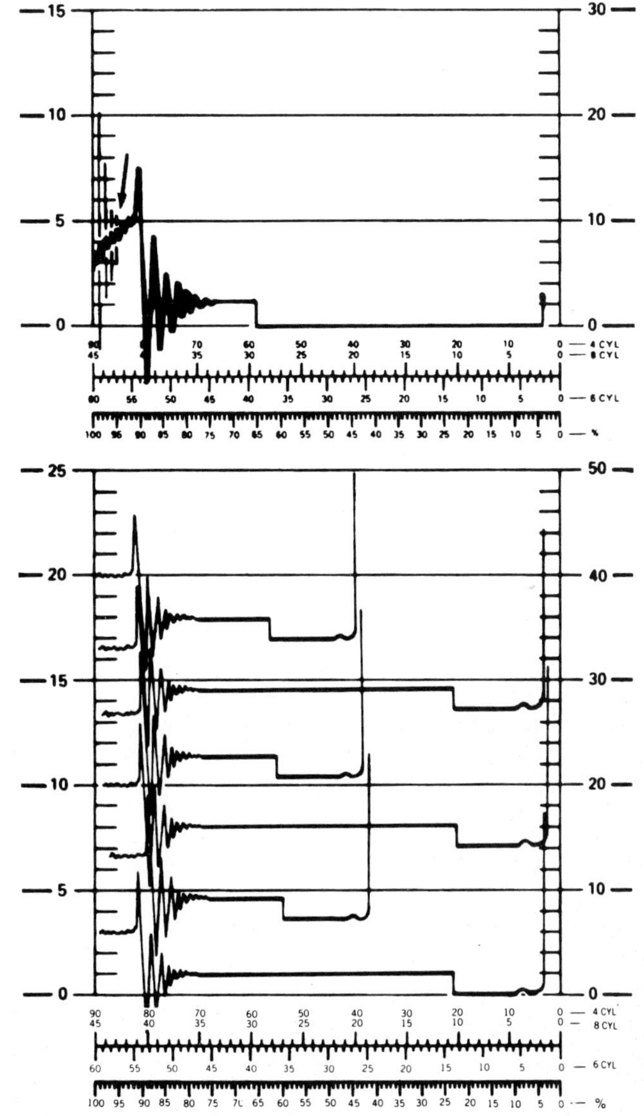

21. An engine cranks normally but will not start. All of the following ignition problems may cause this EXCEPT:

- **a.** an open circuit in a spark plug cable
- **b.** a short circuit in the distributor pickup coil
- **c.** a short circuit in the ignition coil
- **d.** a secondary voltage leak

22. An engine runs but backfires repeatedly. All of the following are common causes of this EXCEPT:

- **a.** incorrect timing or spark advance
- **b.** wrong spark plug heat range
- **c.** ignition crossfire
- **d.** low-octane gasoline

23. With the engine running, voltage drop across a ballast resistor is usually:

- **a.** 0.2 to 0.5 volt
- **b.** 1.5 to 3.5 volts
- **c.** 6.0 to 7.5 volts
- **d.** 7.5 to 9.6 volts

24. Mechanic A says that the air gap between the pickup coil and the trigger wheel should be checked through the full range of vacuum advance on a distributor that requires this adjustment. Mechanic B says that air gap always affects ignition dwell time. Who is right?

- **a.** A only
- **b.** B only
- **c.** both A and B
- **d.** neither A nor B

25. Mechanic A says that a Delco-Remy HEI system should have full battery or charging voltage at the ignition coil. Mechanic B says that HEI system dwell should vary with engine speed. Who is right?

- **a.** A only
- **b.** B only
- **c.** both A and B
- **d.** neither A nor B

13 IGNITION SECONDARY CIRCUIT SERVICE

INTRODUCTION

The primary circuit of any ignition system must be in good operating condition for the secondary circuit to deliver required voltage. In the last chapter, you learned that a loss of one volt in the primary circuit can reduce secondary voltage by several thousand volts. You also learned to test, adjust, and repair the primary circuit for proper operation.

This chapter continues with ignition services for secondary circuit components, including:

- Cap, rotor, and cable service
- Secondary circuit oscilloscope tests
- Spark plug service.

GOALS

After studying this chapter, you should be able to do the following jobs in twice the flat-rate labor time or less, using the carmaker's procedures:

1. Inspect and test distributor caps, rotors, and spark plug cables and make necessary replacements.

2. Use an oscilloscope engine analyzer to test secondary circuit operation. Interpret scope patterns and make necessary adjustments or repairs.

3. Remove spark plugs, analyze plug condition, and install correct replacements.

IGNITION SECONDARY CIRCUIT INSPECTION AND OHMMETER TESTING

The secondary circuits of breaker-point and electronic ignitions are basically identical, but some parts look different on some electronic systems. The secondary components, figure 13-1, are the:

- Coil secondary winding and high-voltage terminal (coil tower)
- The high-voltage spark plug cables and coil-to-distributor cable
- The distributor cap and rotor
- The spark plugs.

You can find many secondary problems with a simple visual inspection and some ohmmeter tests. Later in the chapter, you will learn to test the secondary circuit with an oscilloscope.

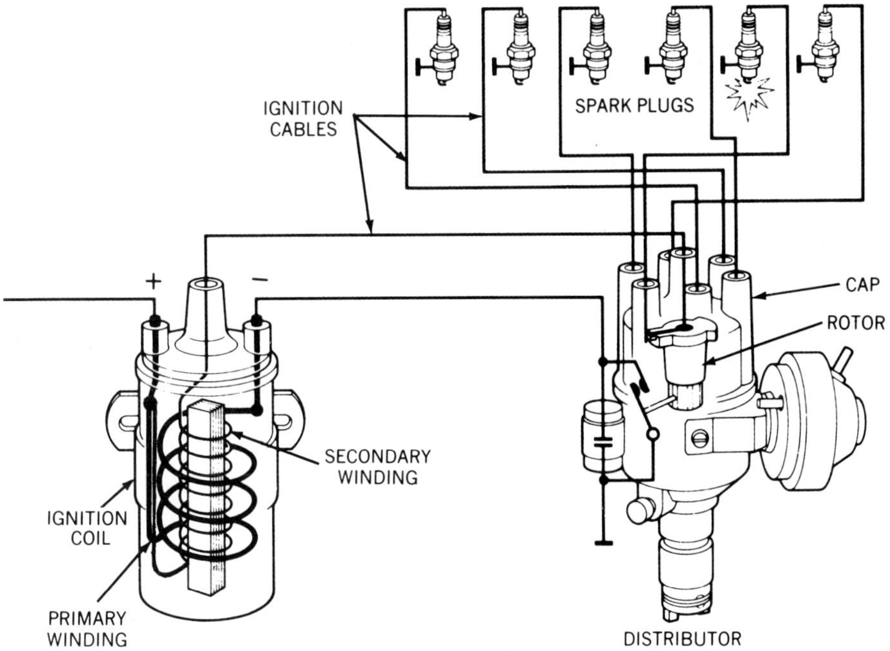

Figure 13-1. Proper secondary circuit operation depends on primary circuit performance. (Bosch)

Caution. Handle spark plug cables carefully. Twisting or pulling a cable can damage it internally and cause high resistance or high-voltage leakage to ground. High secondary voltage can arc across a wide air gap or leak through damaged insulation. Do not pull on cables to disconnect them. Grasp the cable boot and twist gently as you remove the cable. Do not hold an exposed cable end near your body and be careful handling cables when the engine is cranking or running.

Secondary Circuit Inspection

Inspect the secondary circuit components as follows:

1. Inspect all ignition (spark plug) cables and boots for looseness and for cracked, brittle, or damaged insulation, figure 13-2.

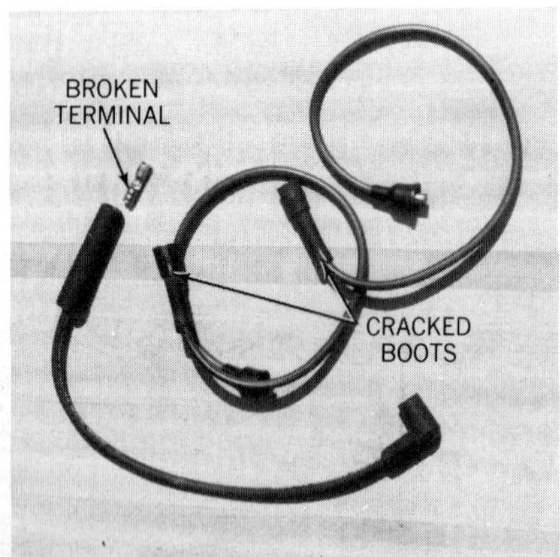

Figure 13-2. Inspect spark plug cables for damage like this. Handle cables carefully; do not jerk them.

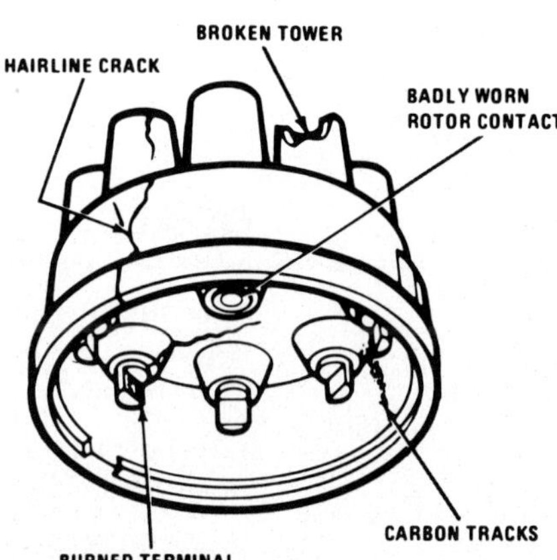

Figure 13-3. Replace the distributor cap if it has any of these defects. (Chrysler)

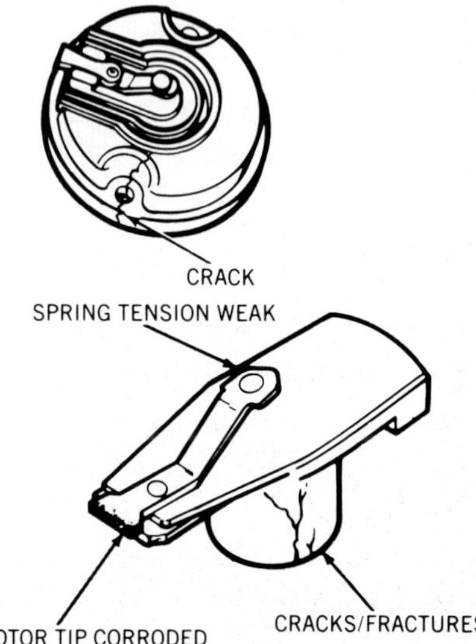

Figure 13-4. Inspect rotors for carbon tracks, cracks, and corrosion like this. (Chrysler)

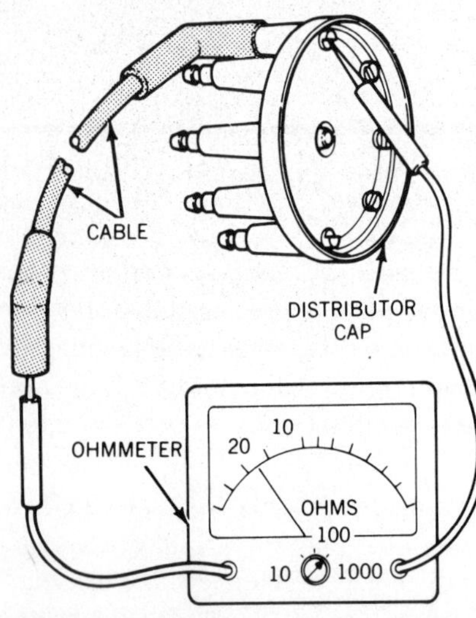

Figure 13-5. Measure spark plug cable resistance *through the cap terminal.* (Ford)

2. Inspect cable connections at spark plugs and at the distributor cap for signs of high-voltage arcing. This usually appears as a carbon track on the cap terminal or the plug insulator. An open wire or very high resistance also can cause green-colored corrosion at the cap terminal.

3. Inspect the distributor cap and replace it if you find any of the defects shown in figure 13-3:

a. Carbon tracks from high-voltage arcing or cracks of any kind

b. Burned, corroded, or broken terminals inside the cap

c. A loose or broken carbon button inside the coil tower

d. A damaged coil connection or signs of high-voltage arcing between the coil and the cap of a Delco HEI system.

4. Inspect the distributor rotor and replace it if you find any of the defects shown in figure 13-4:

a. Carbon tracks or cracks of any kind on the body

b. Bent, broken, or burned electrodes, electrode tips, or coil tower contact strip

c. A broken or worn locating lug or mounting screw holes.

Distributor Cap and Ignition Cable Ohmmeter Testing

In the preceding chapter, you learned to use an ohmmeter to test the ignition coil. You also can use an ohmmeter to check the resistance of the ignition (spark plug) cables and their connections to the distributor cap. Suppression-type, or TVRS, ignition cables are made with a carbon conductor that has high resistance of several thousand ohms per foot. Nevertheless, resistance higher than specifications can cause performance problems such as misfire, higher burn voltage, and reduced burn time. Resistance can increase because of:

- A loose or corroded connection at a plug of distributor cap terminal
- A broken terminal on either end of the cable
- A broken conductor inside the cable due to pulling, kinking, or rough handling
- A breakdown of the conductor or an increase in resistance due to heat or vibration.

Because cable resistance is normally high, you must use one of the high scales on your ohmmeter for accurate measurement. Proceed as follows:

1. Remove the cap from the distributor with the cables attached.

2. Disconnect one cable from its spark plug.

3. Connect one ohmmeter lead to the plug terminal of the cable and the other lead to the corresponding terminal *inside* the distributor cap, figure 13-5.

4. Read the ohmmeter. If resistance is within specifications, the cable and its connection to the cap terminal are good.

5. If resistance is out of limits, disconnect the cable from the cap and read the resistance of just the cable, figure 13-6. If resistance is still out of limits,

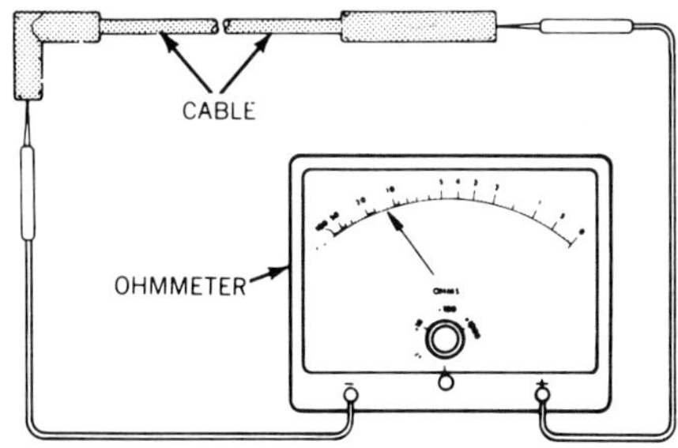

Figure 13-6. If resistance is high when measured through the cap, test just the cable to isolate the problem. (Ford)

Carmaker (1975 & later, unless otherwise listed)	*Maximum Resistance in Ohms*
General Motors	20,000
Ford	5,000 per inch
Chrysler	
1975–82 and 1983 & later 6-, 8-cyl	50,000
1983 & later 4-cyl	3,000 to 7,200 per foot
AMC-Renault	
Electrofill cables	200 per foot (600 per meter)
15″ (380-mm) cables	1,000 to 10,000
15–25″ (380–630 mm)	4,000 to 15,000
25–35″ (630–900 mm)	6,000 to 20,000
More than 35″ (900 mm)	8,000 to 25,000
Mitsubishi-Chrysler imports	22,000
Toyota and Honda	25,000
Datsun-Nissan	30,000
Mazda	16,000

Figure 13-7. These are maximum resistance specifications for some manufacturers' spark plug cables, measured through the cap.

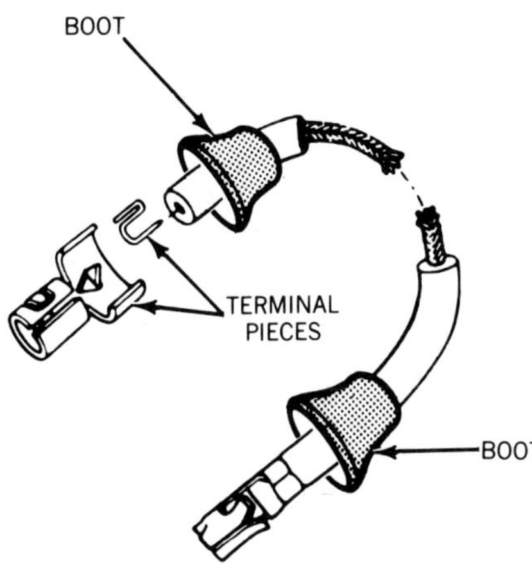

Figure 13-8. Typical installation methods for spark plug cable terminals. (Delco-Remy)

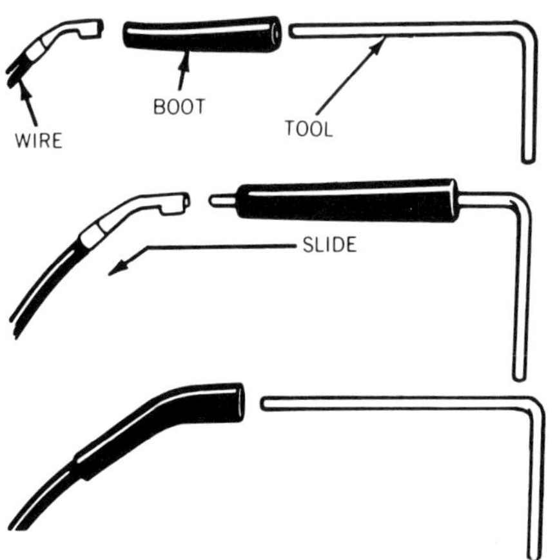

Figure 13-9. Use a smooth metal or plastic rod to slide a boot over a cable terminal, if necessary. (Ford)

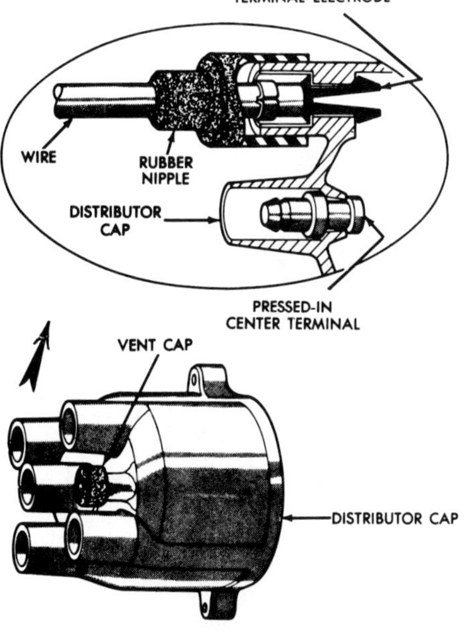

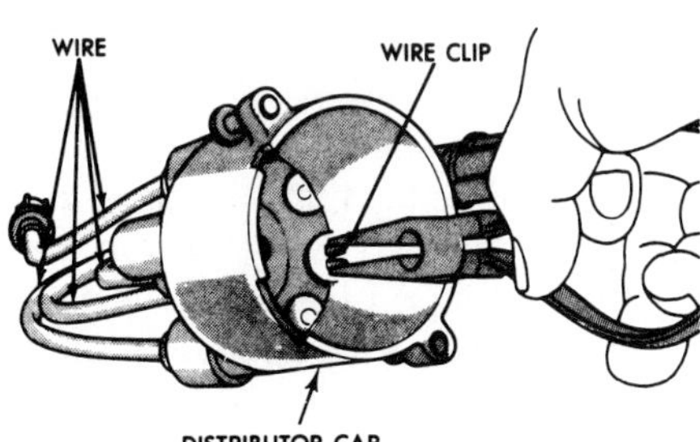

Figure 13-10. Chrysler was the first domestic carmaker to use positive-locking spark plug cables. (Chrysler)

replace the cable. If cable resistance is okay, clean or tighten the connection to the cap or replace the cap.

6. Repeat steps 2 through 5 for each plug cable and for the coil cable.

Figure 13-7 lists spark plug cable resistance specifications for late-model vehicles from several carmakers.

DISTRIBUTOR CAP, ROTOR, AND CABLE REPLACEMENT

Replacing distributor caps and rotors and ignition cables are common service jobs. The following procedures will help you do these jobs quickly and accurately.

Ignition Cable Replacement

All breaker-point ignitions and most electronic systems use ignition cables with a 7-mm diameter. Some Delco HEI, Ford (Motorcraft) Dura-Spark, and a few other electronic systems use 8-mm cables. The 8-mm insulation provides extra dielectric resistance for high-voltage systems where secondary voltage can be as high as 50 kV. Most 7-mm and 8-mm cables are not interchangeable. Do not try to install 7-mm cables on systems that require 8-mm cables.

Many ignition cables are made in predetermined lengths with molded boots and terminals at each end. Other cables are sold in bulk rolls or in average lengths without terminals or boots installed. When you install preassembled cables, you simply remove the old one and install the new one on the distributor cap and spark plug. If the cable is not preassembled, you must cut it to length, slide the boots on each end, and install the terminals with an appropriate crimping tool, figure 13-8.

Normally, put the boots on the cable before the terminals. Do not try to pull them over the terminals, or you may tear them. If you must replace a damaged boot or install a boot over a cable terminal, use a smooth plastic or metal rod as shown in figure 13-9 to avoid damaging the boot.

Many late-model Chrysler 4-cylinder engines have spark plug cables with locking terminals, figure 13-10. You must remove the distributor cap to disconnect these cables from the cap. Do not try to pull on the cables, or you will damage the terminals and the cap.

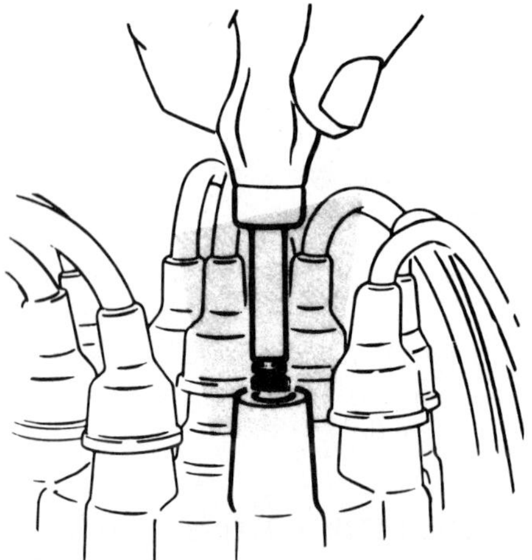

Figure 13-11. Use a small round brush to remove light corrosion inside a distributor cap tower. (Delco-Remy)

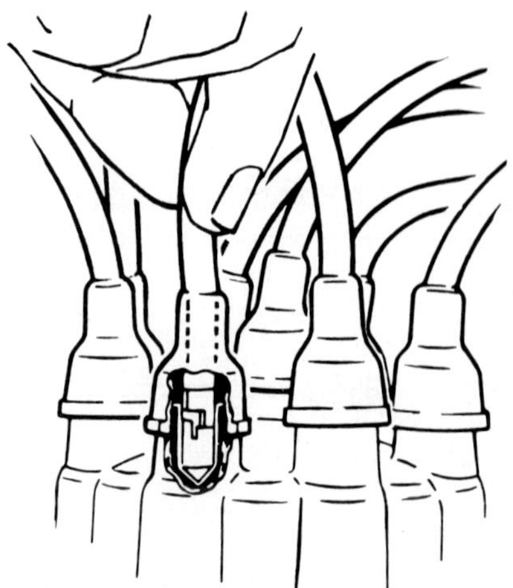

Figure 13-12. Press firmly to be sure the cable terminal seats in the tower. Squeeze the boot to release trapped air. (Delco-Remy)

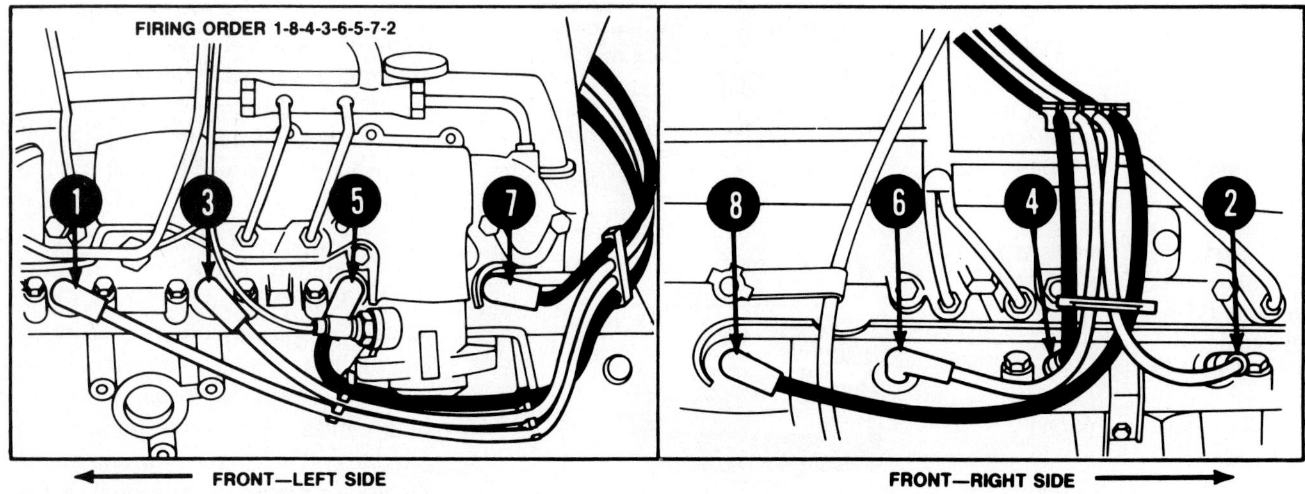

Figure 13-13. Typical spark plug cable routing to avoid crossfiring. (Champion)

When you install any ignition cables, follow this sequence to be sure you install them in the right order:

1. Remove the cap from the distributor if it has locking terminals. Otherwise, leave the distributor cap in place and start at the front of the engine in firing-order or cylinder-number sequence.

2. Remove one cable from the cap and the spark plug.

3. Select the correct length replacement cable or cut one to the length of the old cable and install the boots and terminals.

4. Inspect the cap tower for dirt, corrosion, or cracks. Clean it, figure 13-11, or replace it if necessary.

5. Install the new cable in the cap with the terminal seated firmly on or inside the tower, figure 13-12. Be sure the boot fits firmly on the tower. If necessary, squeeze it to remove air and let it fit securely.

6. Install the other end of the cable on the spark plug. Be sure that the terminal and the boot fit securely.

7. Repeat steps 2 through 6 for the other cables, including the coil cable. Route the cables in their original locations and secure them in their holders. To avoid crossfiring, do not route cables in firing-order sequence next to each other, figure 13-13. Keep the cables away from other electrical wiring wherever possible and do not let them touch exhaust manifolds.

Distributor Cap and Rotor Replacement

Most distributor caps are held to the distributor by two spring clips. The cap has a locating lug to align it correctly, figure 13-14. Most rotors simply slip onto the top of the shaft and also have lugs to align them. Some systems, particularly Ford Dura-Spark and Delco ignitions, have parts that are unique. These systems have large distributor caps and special rotors for wide spacing between secondary terminals. This

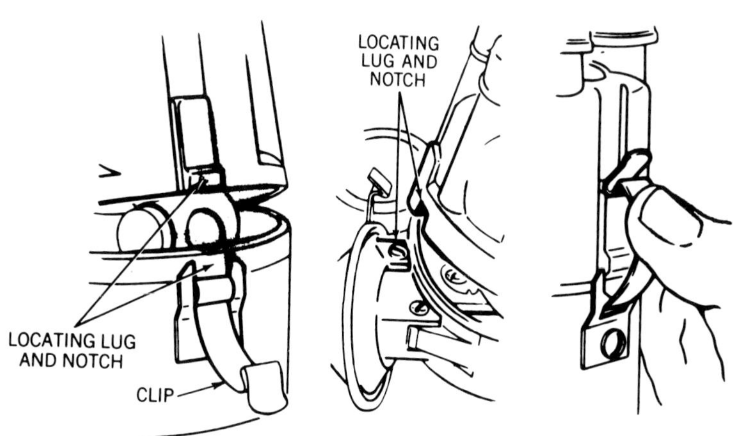

Figure 13-14. Typical distributor cap latches and locating lugs. (Delco-Remy)

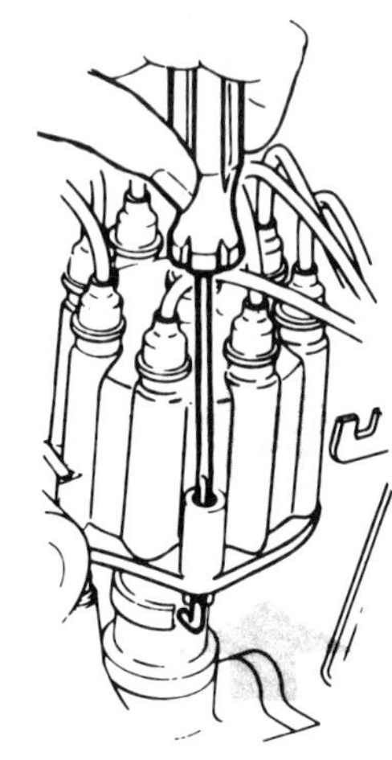

Figure 13-15. Release the spring-loaded latches on Delco-Remy distributor caps with a small screwdriver. (Delco-Remy)

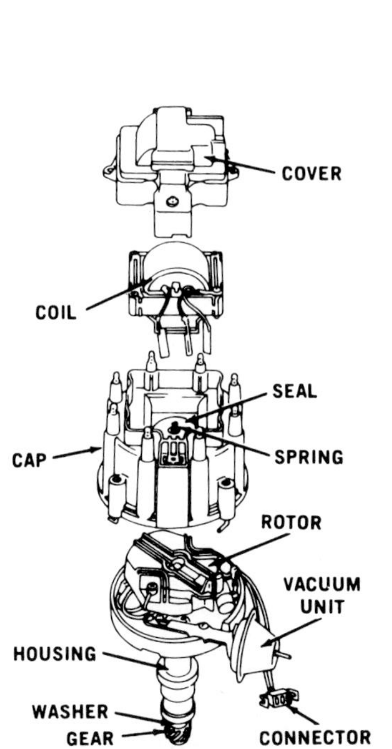

Figure 13-16. The ignition coil is in the cap of most Delco HEI distributors. (Delco-Remy)

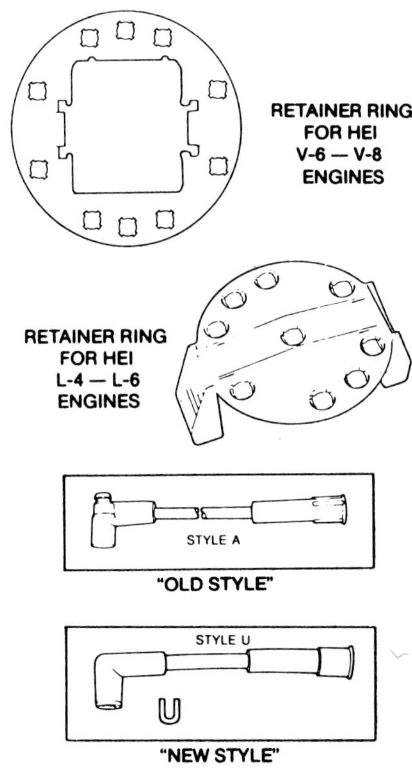

Figure 13-17. Typical Delco-Remy HEI cable retainers and cables. (Champion)

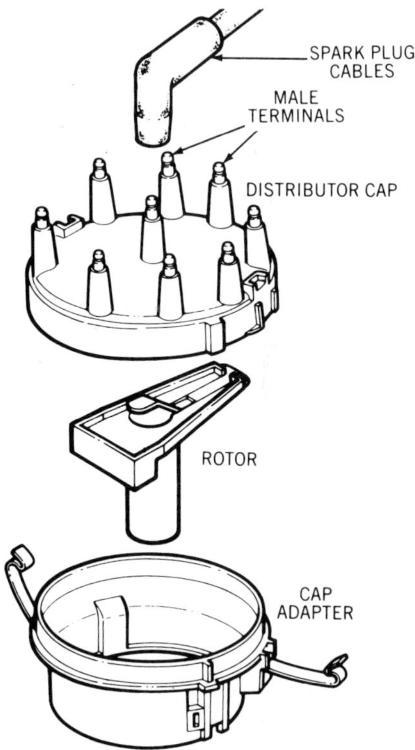

Figure 13-18. Many Ford Dura-Spark distributors have cap adapters like this. (Ford)

helps to prevent crossfiring with secondary voltage as high as 50 kV. These parts are no problem, however, if you recognize their features.

Cap Replacement. You replace most caps simply by unclipping the old one and installing the new one. Delco caps are secured with spring-loaded latches that you press and turn with a screwdriver, figure 13-15. V-8 and V-6 breaker-point caps have two; HEI caps have four, figure 13-16. You must disconnect the ignition feed wire and module connector from a Delco HEI cap and remove the coil to replace the cap. Caps for Delco HEI and most Ford Dura-Spark systems have male cable terminals. Many Delco caps also have cable retainers to help hold the cables in place, figure 13-17.

Many Ford Dura-Spark distributors have an adapter ring between the cap and the distributor body, figure 13-18. Do not try to remove the cap and adapter as a unit because the adapter will jam on the rotor. Remove the upper distributor cap first; then remove the rotor before removing the adapter. Some adapters are held to the distributor body by spring clips. Others are held by screws inside the ring to en-

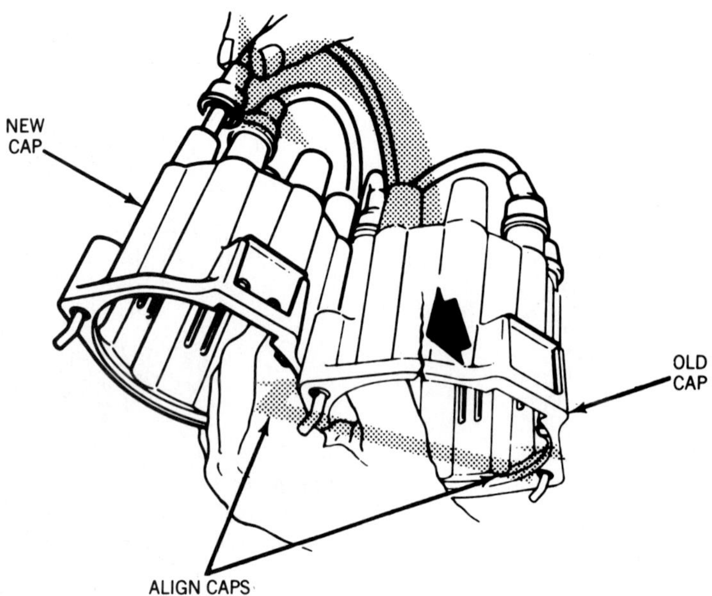

Figure 13-19. Hold the old and the new cap side by side and change cables one at a time. (Delco-Remy)

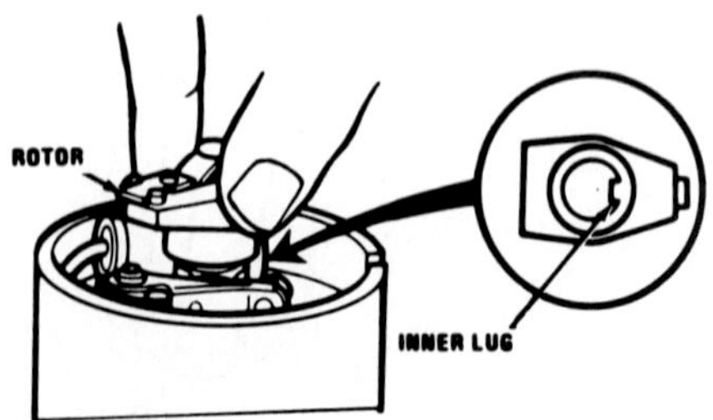

Figure 13-21. Align the inner lug with the notch or flat spot in the shaft and press the rotor into place. (Chrysler)

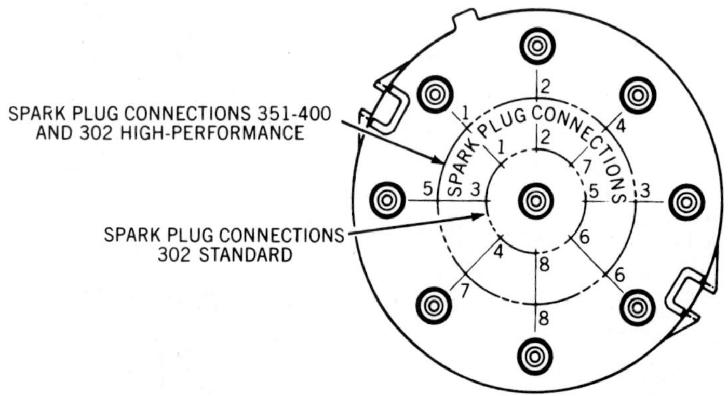

Figure 13-20. The numbers on this 2-level Dura-Spark cap are engine cylinder numbers, not firing order. (Ford)

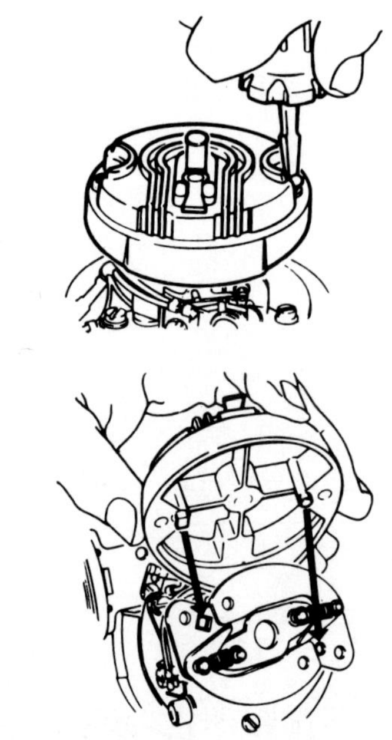

Figure 13-22. Delco V-8 breaker-point rotors and HEI rotors have locating lugs for alignment and are held in place by two screws. (Delco-Remy)

sure that you remove the rotor before the adapter.

When you replace any distributor cap, you must be sure to install the spark plug cables in firing order. A simple way to do this is to hold the new cap next to the old one with the locating lugs or clips aligned in parallel positions. Then switch the cables from the old to the new cap, one at a time in firing order, figure 13-19.

Some Ford 8-cylinder Dura-Spark distributors have caps and rotors with terminals on two levels. (Ford calls them "bilevel" caps and rotors.) The cap terminals are *not* in firing-order sequence. The 2-level cap and rotor place terminals in firing order 135 degrees away from each other in the cap. This also helps to prevent high-voltage crossfiring within the distributor. These caps have cylinder numbers molded into their tops, figure 13-20. There are two rings of numbers, which indicate engine cylinder numbers, *not* firing order. The outer ring is for 351- and 400-cid (5.7- and 6.5-liter) V-8's. The inner ring is for most 302-cid (5.0-liter) V-8's, except high-performance models. High-performance 302 engines have 351 camshafts and use the 351 firing order. All of this means that you must be very careful when replacing the cap or ignition cables on a Ford V-8 Dura-Spark distributor. Trace each cable from the cap to its cylinder to be sure it is installed in the right order.

Rotor Replacement. Replace most distributor rotors by lifting the old one off the shaft, aligning the lug in the new one with the shaft slot, and pressing it into place, figure 13-21. Be sure it is seated securely on the shaft or it may hit the cap when the engine is running.

Occasionally, you may find a rotor that is stuck to the distributor shaft. You may be able to remove it by prying carefully under opposite sides with two screwdrivers. If this doesn't work immediately, do not force it, or you will

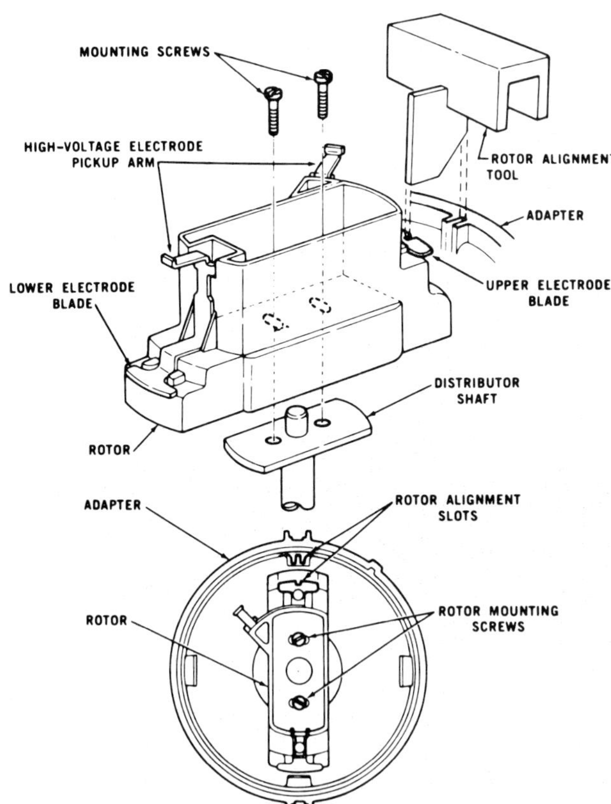

Figure 13-23. Ford EEC-I Dura-Spark rotor alignment. (Ford)

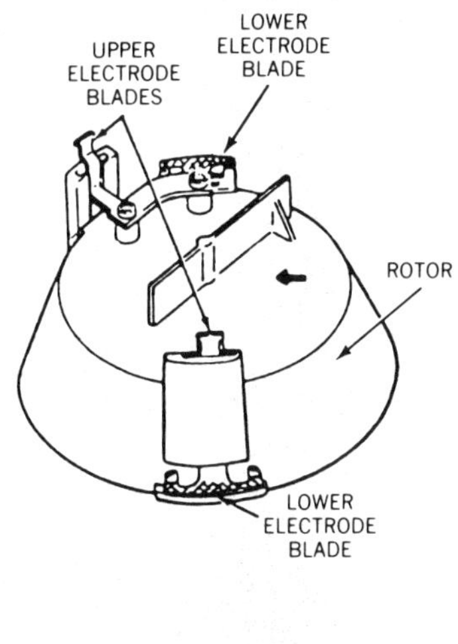

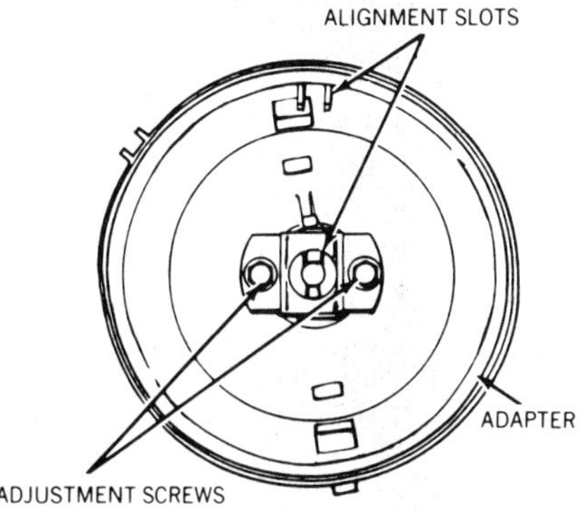

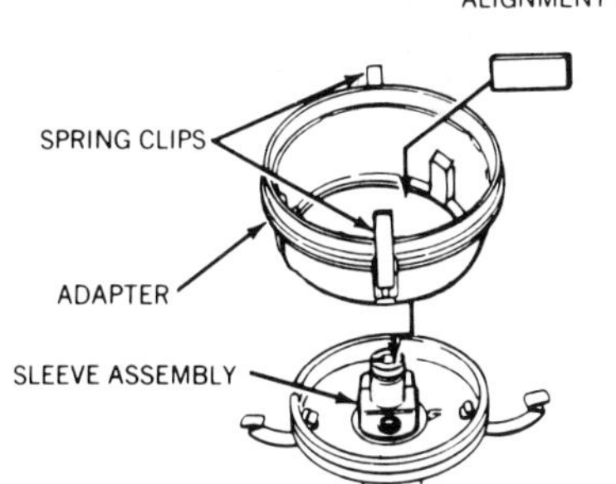

Figure 13-24. Ford EEC-II & III Dura-Spark rotor replacement and sleeve alignment. (Ford)

break the aluminum distributor body. If the rotor is stuck tightly to the shaft, you can crack it to remove it by tapping gently with a hammer and small punch or chisel. Be sure to remove all broken rotor pieces from the distributor and inspect the shaft for damage before installing a new rotor.

All Delco rotors (except 4- and 6-cylinder breaker-point types) and some Ford Dura-Spark rotors are held to the distributor shaft by two screws, figure 13-22. Delco rotors have two lugs to align them with the shaft and cap.

Ford has two different distributor designs for Dura-Spark caps and rotors used with early EEC systems. Complete distributors may be interchangeable from one model year to another, but individual parts are not. Each kind of Ford rotor or adapter requires special alignment procedures.

Early Dura-Spark I rotors are held to the distributor shaft by two screws in slotted holes, figure 13-23. Replace and align one of these rotors as follows:

1. Remove the distributor cap and crank the engine to align the slot in the rotor upper coil electrode with the slot in the cap adapter.

2. Remove the rotor screws and the rotor. Do not turn the crankshaft after removing the rotor or rotor alignment and retiming will be difficult.

3. Install the new rotor and the screws but do not tighten the screws.

4. Slide the rotor alignment tool into the slots on the distributor adapter and the rotor electrode to align them.

5. Tighten the rotor screws and remove the alignment tool. Reinstall the cap.

Later 2-level Dura-Spark rotors used with EEC III systems have spring clips to hold them to the shaft and do not usually need special alignment. The rotor sleeve on the shaft, however, is adjustable if necessary, figure 13-24. Replace these rotors as follows:

1. Rotate the engine to align the specified timing marks. Be sure number 1 piston is on the compression stroke or the rotor can be misaligned 180 degrees if you adjust the shaft sleeve.

2. Remove the distributor cap. Hold the rotor by the lifting tab on the top and remove the rotor.

3. Verify that the slots on the shaft sleeve are in line with the alignment slot in the distributor adapter. Align the arrow on the new rotor with the large slot in the sleeve and press the rotor into place.

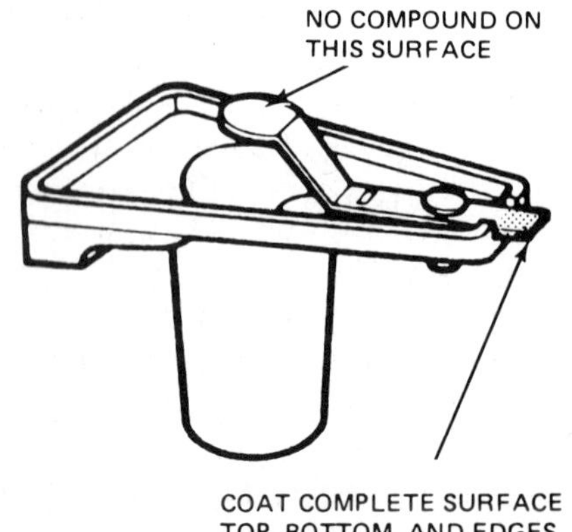

Figure 13-25. These are typical locations for silicone grease on some Ford rotors. (Ford)

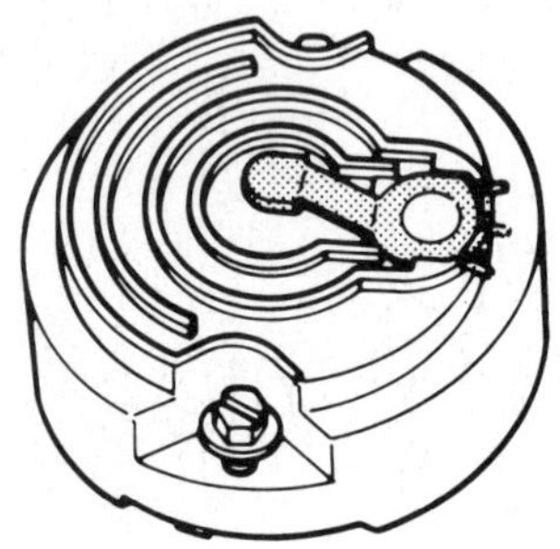

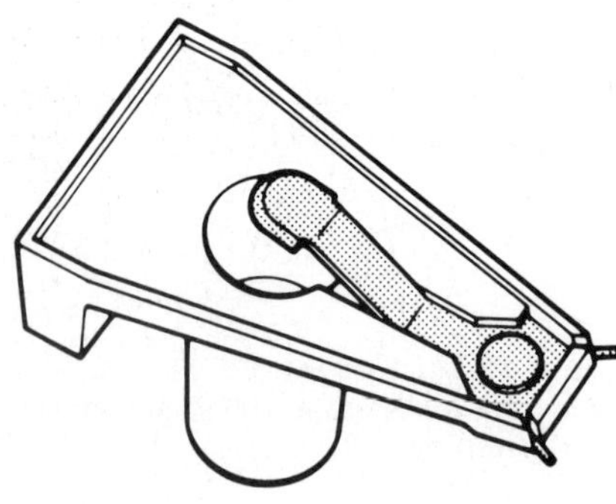

Figure 13-26. These multiple-point rotors suppress RFI without silicone grease. (Ford)

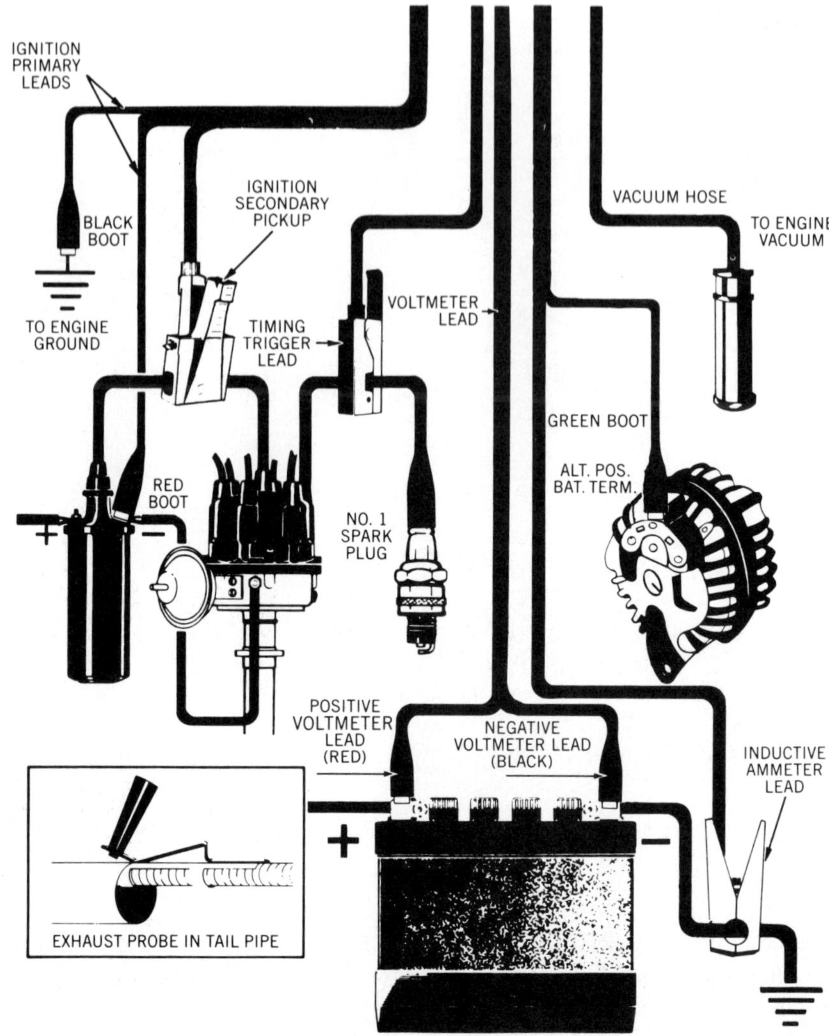

Figure 13-27. Typical test lead connections for an oscilloscope engine analyzer. (Sun)

This normally aligns the rotor satisfactorily. If you have removed the distributor, the cap adapter, or the sleeve, loosen the two screws on the sleeve and insert the alignment tool into the slots on the sleeve and the adapter. Then tighten the sleeve screws, remove the alignment tool, and install the rotor.

Since the late 1970's Ford and some other carmakers have used silicone grease on cap and rotor terminals to suppress radiofrequency interference. After a few hundred miles of operation, the silicone grease forms white crystals that look like corrosion on the rotor and cap terminals. This is normal and does not harm ignition operation. Do not remove the grease.

When you replace one of these caps or rotors, apply a thin coat of silicone grease to the specified areas of the rotor electrode, figure 13-25, or the cap terminals. Use Ford D7AZ-19A331-A, Dow Corning 111, or General Electric G-627 grease or equivalent. Because Ford cap and rotor designs vary, the locations for the grease are different for various parts.

In 1981 Ford began using some rotors with multiple-point electrodes, figure 13-26. These reduce ignition radiofrequency interference without the use of silicone grease.

SECONDARY CIRCUIT OSCILLOSCOPE TESTS

In chapter 12, you learned to use the oscilloscope to test and service the primary circuit. Now you will learn to use it to test and service the secondary circuit.

Follow the equipment maker's directions to connect the scope, or analyzer, leads to the engine, figure 13-27. Although you may have all of the analyzer leads connected, you only need the same minimum connections to test the secondary circuit that you used for the primary circuit:

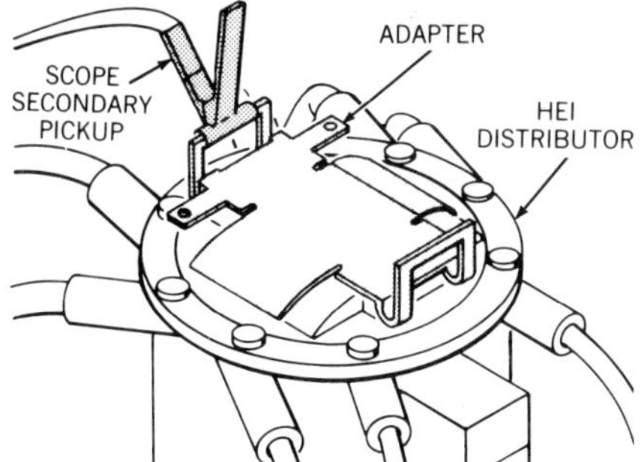

Figure 13-28. Delco HEI coils require special secondary pickups.

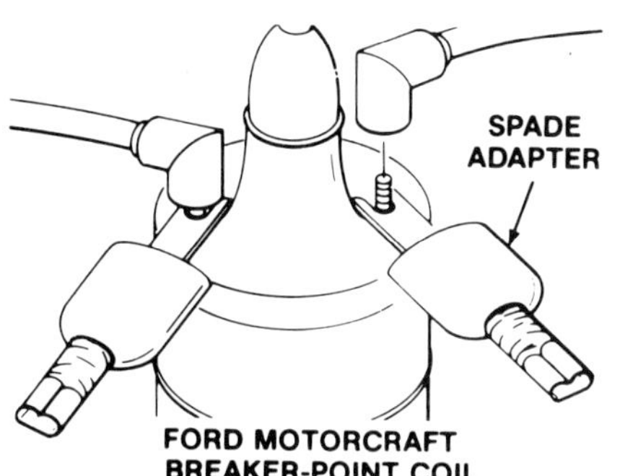

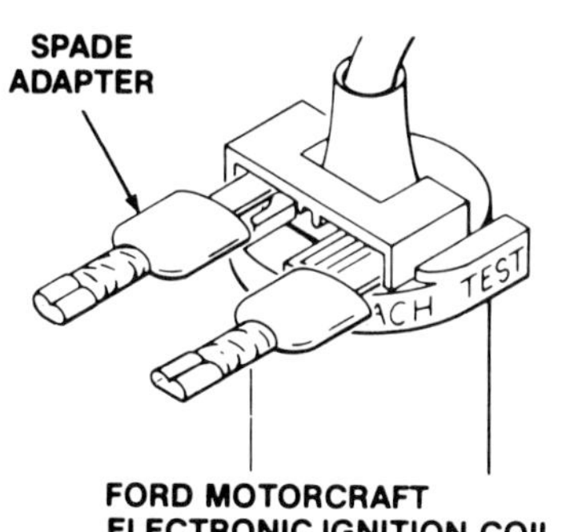

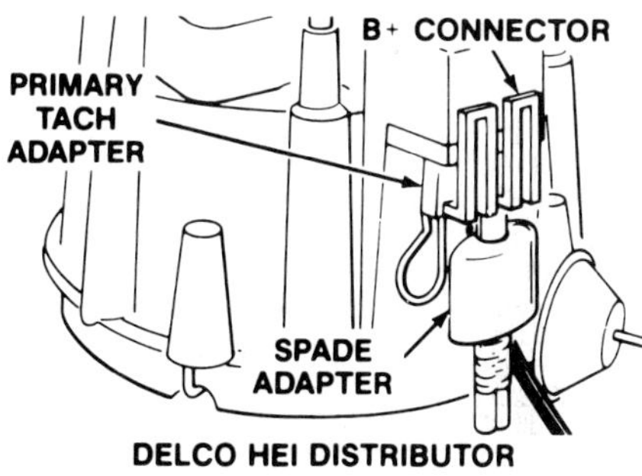

Figure 13-29. Some primary circuits also require test lead adapters.

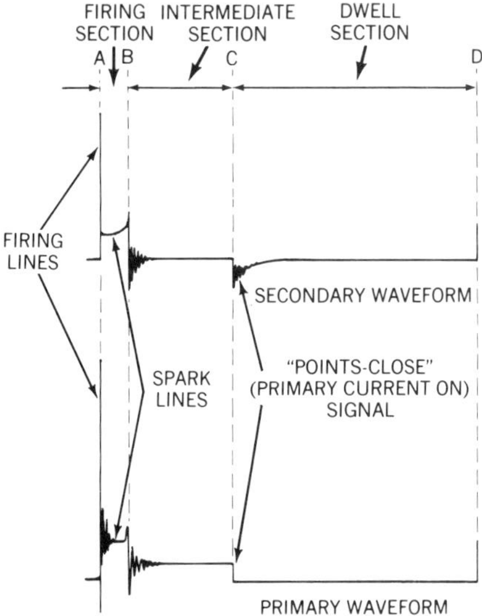

Figure 13-30. Primary and secondary patterns have corresponding sections.

1. Analyzer ground connection to engine ground to complete all circuits
2. Scope primary connection to the ignition primary circuit at the coil + or − terminal, depending on equipment design
3. Scope high-voltage connection to the coil secondary terminal or cable
4. Scope high-voltage connection to the number 1 spark plug for timing signals and proper firing-order waveform sequence.

Use adapters as required for coil primary and secondary connections, figures 13-28 and 13-29. To view the secondary circuit patterns, you will use the kilovolt (kV) scales on either side of the scope CRT, or screen.

Secondary Oscilloscope Patterns

Like the primary pattern, the secondary pattern is divided into three basic sections: firing, intermediate, and dwell, figure 13-30. The secondary pattern displays the high voltages in the circuits from the coil, through the distributor, to each spark plug. The time events are the same in both the primary and the secondary circuits, however.

The firing section of a secondary pattern begins as voltage rises to a peak at the plug gap and the plug fires. It then drops to a lower burn voltage as the plug continues to fire. The intermediate section begins as the spark stops and extends until the dwell section begins. The secondary intermediate section shows voltage oscillations as remaining coil energy dissipates in the circuit. The dwell section begins as the points close or as the ignition module turns on to close the primary circuit and recharge the coil.

As with primary circuit patterns, you have three basic scope displays for secondary circuit patterns:

1. The *superimposed pattern* places all cylinder waveforms upon each other so that there appears to be only one pattern.
2. The *parade pattern* displays the individual cylinder waveforms side by side, from left to right in firing-order sequence.
3. The *raster pattern* stacks the individual cylinder patterns one above the other from bottom to top in firing-order sequence.

Also, as with primary patterns, most modern scopes allow you to shift the secondary pattern on the screen, expand the pattern, or break it into millisecond increments. You also can select the waveform for one cylinder and separate it from a superimposed or parade pattern. This is particularly useful to analyze the plugs, cables, and internal condition of individual cylinders.

Breaker-Point and Electronic Pattern Differences

Because the secondary circuits of breaker-point and electronic ignitions are fundamentally identical, secondary scope patterns for all systems are similar. Figures 13-31 through 13-35 show typical secondary superimposed patterns for breaker-point and several electronic ignitions. Chrysler systems do not have a distinct intermediate section. Coil oscillations start at the end of the firing section (end of sparkline) and gradually diminish to a straight voltage line that ends at the next vertical firing line.

In a breaker-point system, the os-

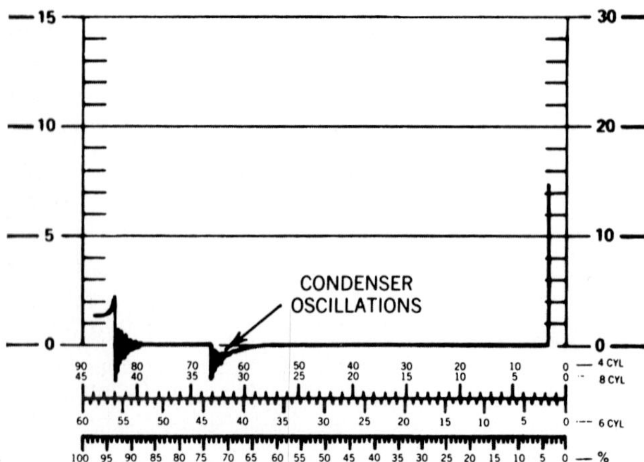

Figure 13-31. Normal breaker-point secondary superimposed pattern. Most electronic ignitions are similar.

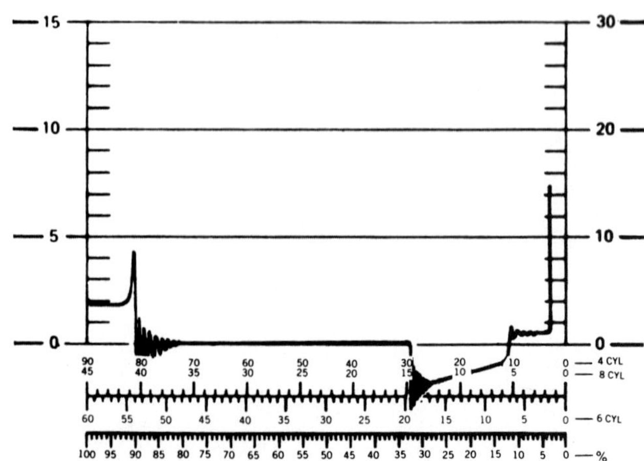

Figure 13-32. Normal Delco HEI secondary superimposed pattern. Other variable-dwell electronic ignitions are similar.

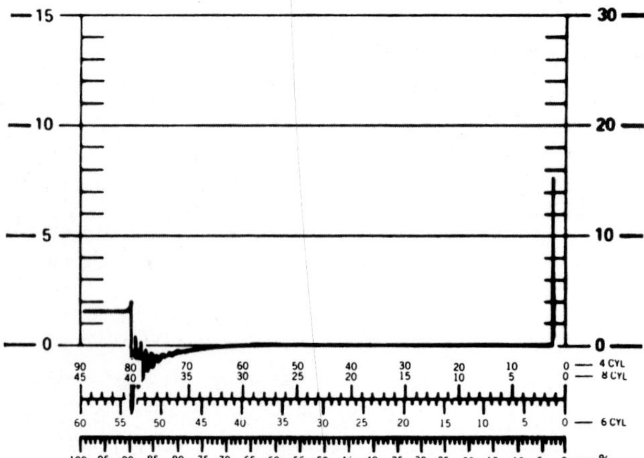

Figure 13-33. Normal Chrysler electronic secondary superimposed pattern. Note the absence of a distinct "dwell" section.

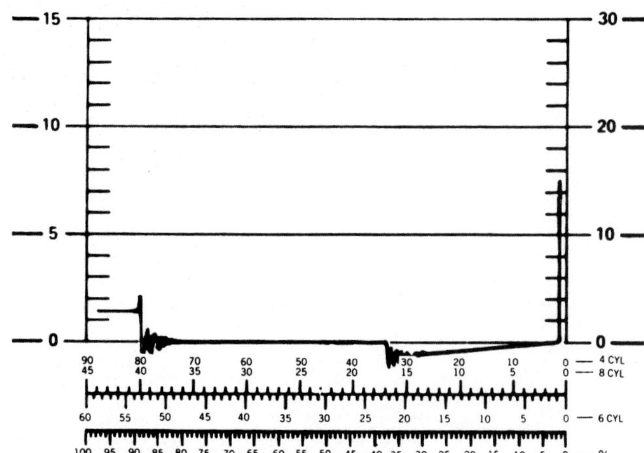

Figure 13-34. Normal Ford Dura-Spark I and TFI secondary superimposed pattern.

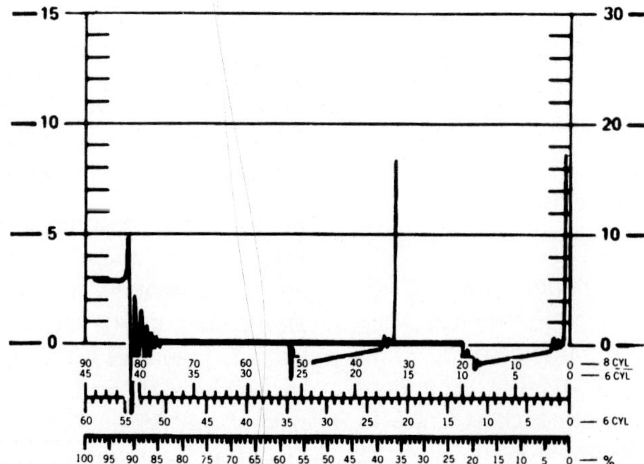

Figure 13-35. Normal secondary superimposed pattern for an uneven-firing V-6.

cillations at the points-close signal as the dwell starts are caused by the condenser, figure 13-31. They also reflect breaker-point action and are useful for testing point and condenser condition. The similar oscillations in some electronic systems are caused by circuits in the control module.

The dwell sections will vary for variable-dwell electronic ignitions, such as Delco HEI, Prestolite, and some Ford Motorcraft systems, figure 13-32. The hump, or ripple, toward the end of the dwell section for these systems indicates that current-limiting circuits in the module have reduced primary current.

The secondary patterns for uneven-firing V-6, figure 13-35, engines will have varying horizontal spacing, as the primary patterns do.

Secondary Pattern Analysis

Ignition secondary patterns show the condition of the coil secondary winding, the coil high-voltage cable or connection to the distributor, the distributor cap and rotor, the spark plug cables, and the spark plugs. Also, because mechanical conditions inside the cylinders affect secondary circuit resistance (the voltage needed to fire the plugs and maintain a spark), secondary patterns are useful for analyzing such things as:

- Air-fuel ratios
- Stuck or burned valves
- Cylinder compression.

If any pattern is upside down, it indicates reversed coil polarity, reversed battery connections, or reversed scope primary leads as an upside down primary pattern does.

Superimposed Pattern Analysis. Figure 13-36 shows a secondary superimposed pattern that is jumping up and down on the screen. This pattern shows a problem that is shared by all cylinders. In this case, the jumping pattern is probably caused by a loose connection or corrosion in the coil high-voltage cable at the coil or at the distributor cap. It also might be caused by an intermittent open circuit in the coil secondary winding or cable.

The secondary superimposed pattern also is a starting point to isolate problems in just one, or a few, cylinders. If all cylinders are operating normally, the superimposed patterns should appear as shown in figures 13-31 through 13-35. The individual traces for all cylinders will fit together to form one uniform pattern.

If one or more cylinders are not operating properly, however, part of the superimposed pattern will be out of register. Such abnormalities are most noticeable in the firing section and the intermediate section. Figure 13-37 shows a superimposed pattern in which one cylinder has an abnormally long sparkline and short coil oscillations. Figure 13-38 shows a pattern in which one cylinder has a high firing line, no sparkline, and large coil oscillations. You can now use a raster pattern or a parade pattern to isolate the cylinder, or cylinders, in which the problems exist. We will look at firing line and sparkline analysis in more detail when we review raster and parade patterns.

The points-close signal in a breaker-point secondary superimposed pattern can show you problems in the points or the primary wiring. A normal points-close signal should have at least three

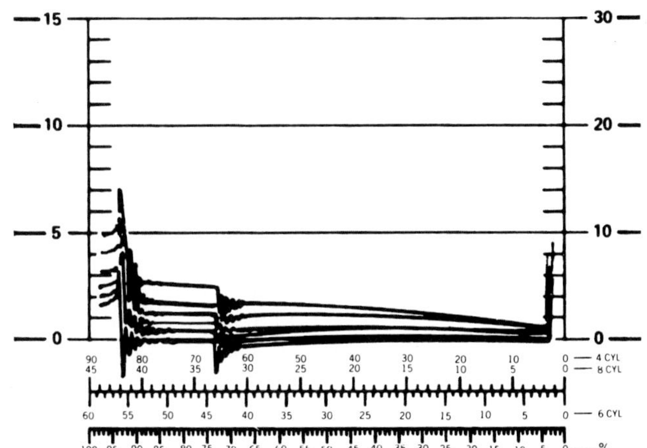

Figure 13-36. This secondary superimposed pattern is jumping up and down on the screen.
Causes: Loose connection or corrosion in the coil high-voltage cable at the coil or the distributor cap. An intermittent open in the coil secondary winding.

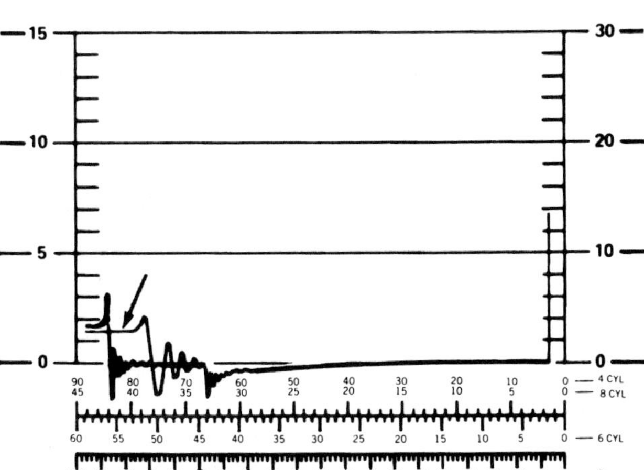

Figure 13-37. This secondary superimposed pattern has one long, low sparkline, showing long burn time and short coil oscillations.
Causes: Fouled spark plug. Shorted or low-resistance plug cable. Crossfiring inside distributor.

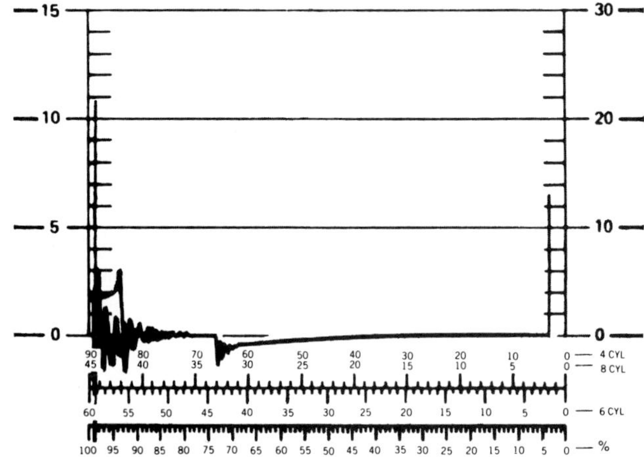

Figure 13-38. This secondary superimposed pattern has one cylinder with no sparkline and large coil oscillations.
Causes: Disconnected or open spark plug cable.

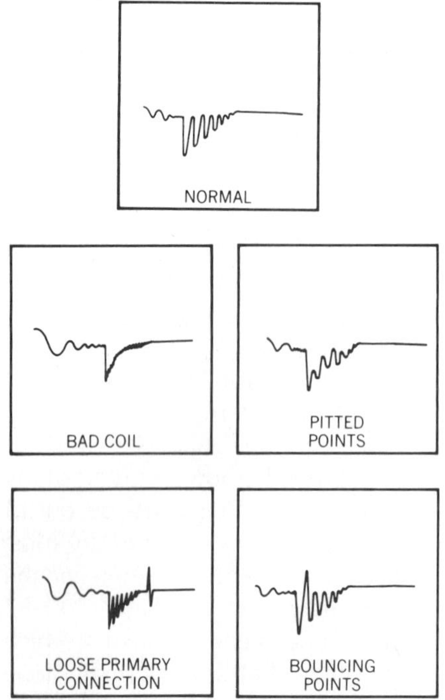

Figure 13-39. Here are four abnormal points-close signals that indicate various defects. A normal signal is shown for comparison.

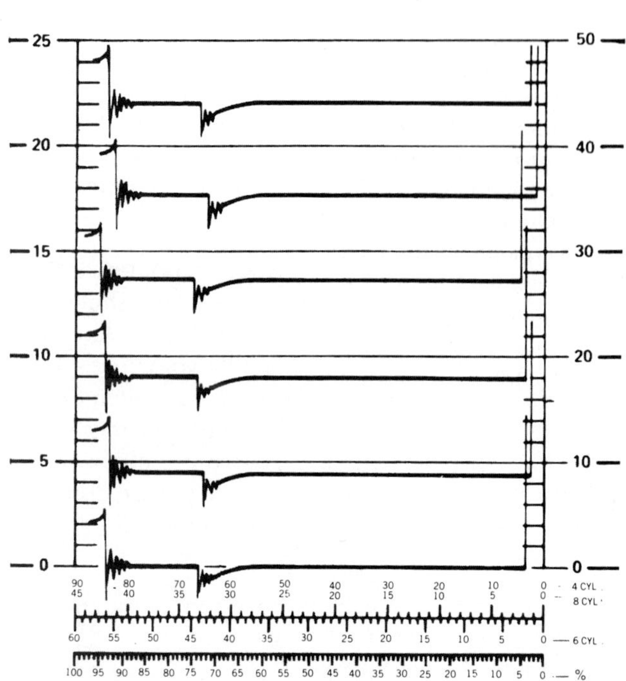

Figure 13-40. Typical normal secondary raster pattern for a breaker-point ignition. Most electronic systems are similar.

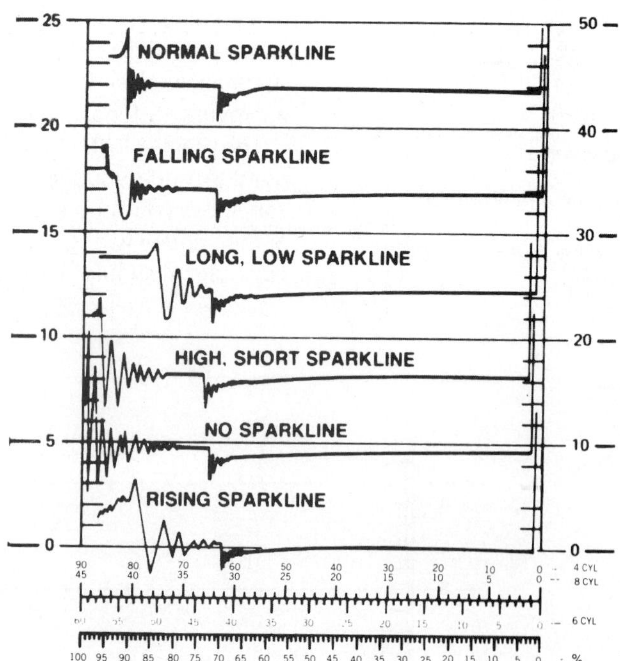

Figure 13-41. This secondary raster pattern highlights the defects shown in abnormal superimposed patterns.

downward oscillations, all below the dwell line, and the first should be longest. Figure 13-39 shows a normal points-close signal and four abnormal signals that indicate different problems.

Raster Pattern Analysis. Figure 13-40 shows a secondary raster pattern with all cylinders operating normally. This is for a breaker-point system, but most electronic ignitions display similar patterns. The raster pattern is particularly useful for checking the sparklines in the firing section of each cylinder. Figure 13-41 is a raster display that shows one pattern with a normal sparkline and five patterns with abnormal sparklines. Figure 13-42 lists the possible causes of these and other abnormal sparklines.

Parade Pattern Analysis. Figures 13-43 through 13-46 show normal secondary parade patterns for breaker-point and electronic ignitions. Parade patterns are most useful for comparing spark plug firing voltages for individual cylinders and for checking coil output voltage. Figures 13-47 through 13-49 are parade patterns that show high, low, and uneven firing voltages and list the possible causes.

SPARKLINE ANALYSIS			
Pattern Defect	**Possible Causes**	**Pattern Defect**	**Possible Causes**
Falling sparkline	High resistance between the distributor cap and the plug due to corrosion in the cap or on the cable terminals.	No sparkline	Disconnected or open spark plug cable.
Long, low sparkline	1. Fouled spark plug. 2. Shorted or low-resistance plug cable. 3. Crossfiring inside distributor.	Rising sparkline	1. At low speed— sticking valve. 2. With increasing speed—floating valve. 3. At all speeds—burned valve, wide plug gap, or intake manifold leak.
High, short sparkline	1. Wide spark plug gap. 2. High resistance in a spark plug cable or cable connections.	Jumping sparkline	Sticking or worn valve that causes an uneven air-fuel mixture in the cylinder.

Figure 13-42. These are typical causes of abnormal sparklines seen in secondary ignition patterns.

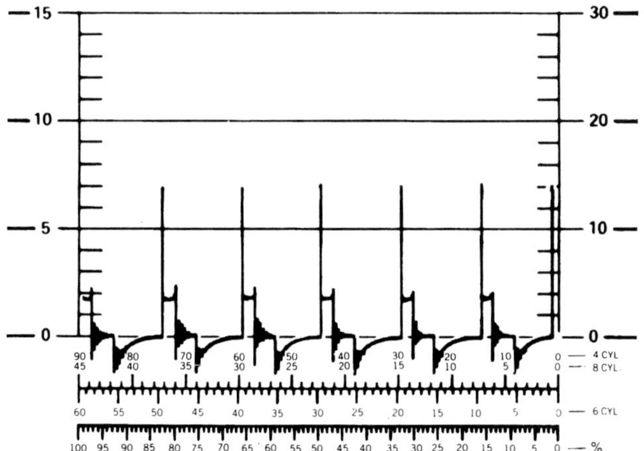

Figure 13-43. Normal secondary parade pattern for breaker-point and many electronic ignitions.

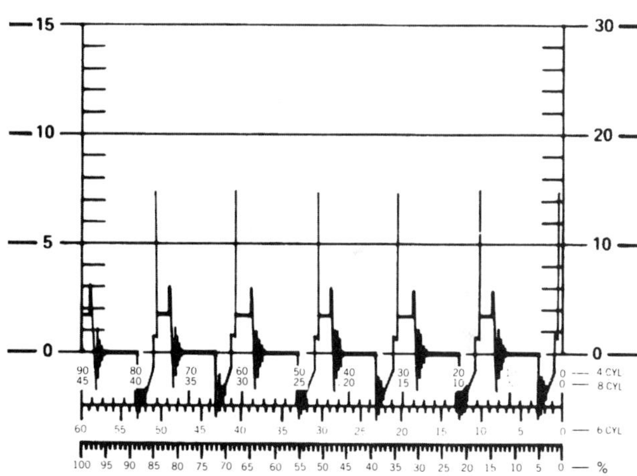

Figure 13-44. Normal secondary parade pattern for Delco HEI and other high-voltage, variable-dwell electronic ignitions.

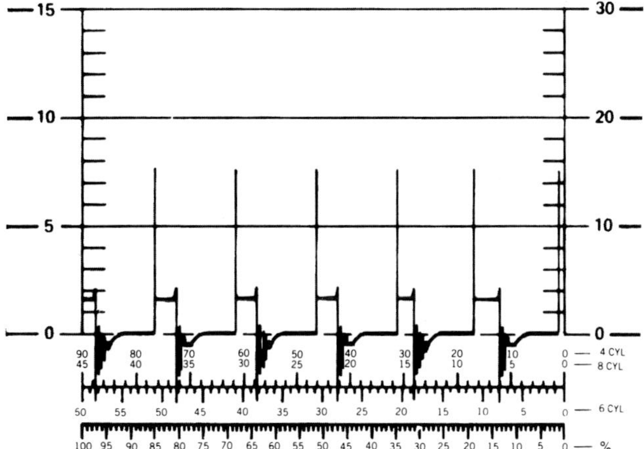

Figure 13-45. Normal secondary parade pattern for Chrysler electronic ignitions.

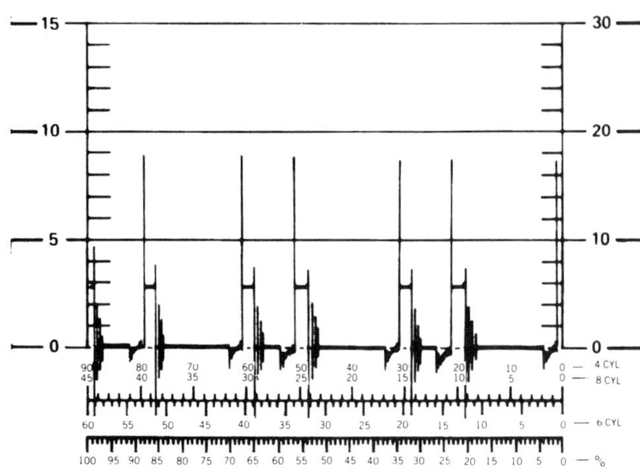

Figure 13-46. Normal secondary parade pattern for an uneven-firing V-6.

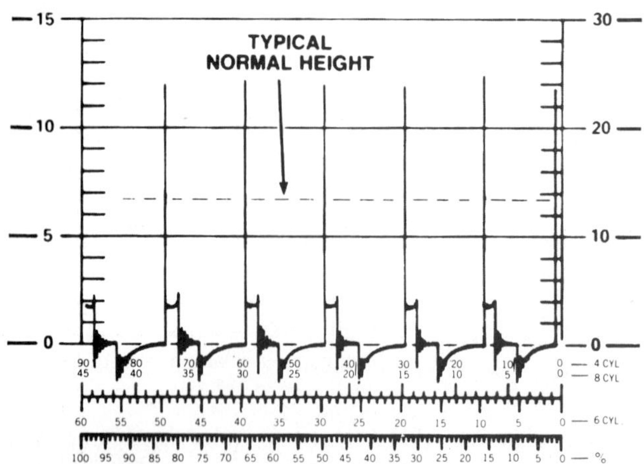

Figure 13-47. This secondary parade pattern has high firing lines for all cylinders.
Causes: Worn spark plugs. Wide plug gaps. Wide rotor air gap. A break in the coil high-voltage cable. Late ignition timing. A lean air-fuel ratio.

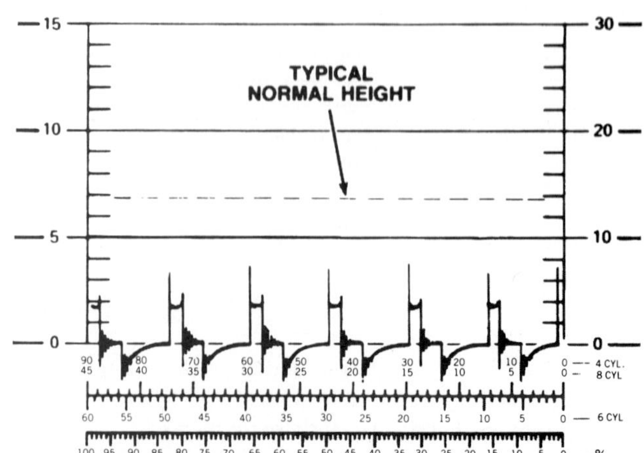

Figure 13-48. This secondary parade pattern has low firing lines for all cylinders.
Causes: Fouled spark plugs. Narrow plug gaps. Low compression. Late ignition timing. A rich air-fuel ratio.

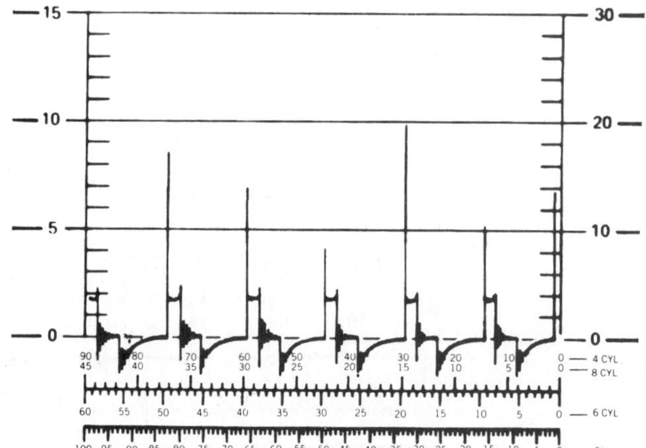

Figure 13-49. This secondary parade pattern has uneven firing lines.
Causes: Any of the causes listed for figures 13-47 and 13-48 can exist in an individual cylinder and cause an abnormally high or low firing voltage. Also, a bent distributor shaft or worn bushings can cause uneven firing voltages.

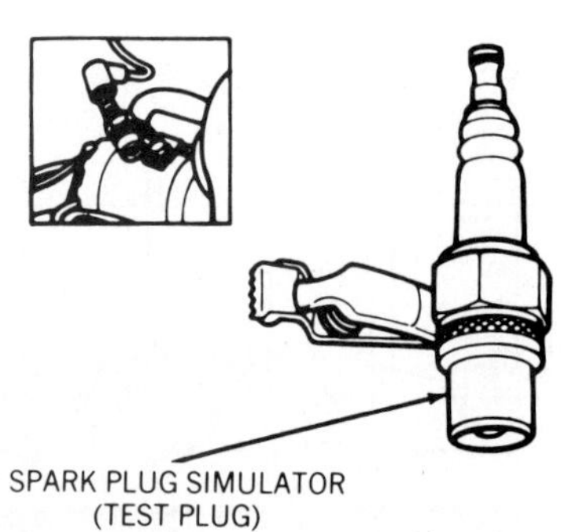

Figure 13-50. Use this spark tester to check secondary voltage capability of a high-voltage electronic ignition system. (Ford)

Special Secondary Tests

You can use secondary scope patterns for special tests of ignition and engine operation as explained below.

Coil Output and Secondary Leakage Tests. For a breaker-point ignition, you can test coil output (maximum available voltage) by disconnecting a cable from one spark plug with the engine at idle and observing open-circuit voltage on a secondary parade pattern. However, you must use a spark tester, or spark plug simulator, figure 13-50, to test high-voltage electronic ignitions, such as Delco HEI systems. The spark tester is a modified spark plug with the ground electrode removed and a clamp attached for grounding to the engine. You can use a spark tester reliably with *all* ignitions, and it is recommended as standard practice.

Clamp the spark tester to an engine ground and connect a spark plug cable to it. Don't use the number 1 cable to which the scope trigger lead is attached. Select the secondary parade pattern on the high-kV scale. Figure 13-51 shows this pattern with open-circuit voltage displayed on the number 6 cylinder. Minimum coil output voltage for most ignitions should be 20 to 25 kV, or slightly higher for some electronic systems. Lower voltage indicates:

- A bad coil
- A leaking condenser
- A bad ballast resistor
- Insufficient dwell
- Other high primary resistance
- Secondary circuit leakage.

More importantly, compare the maximum available voltage to the plug-firing voltage for the other cylinders. Firing voltage is *required* voltage. Maximum coil output is *available*

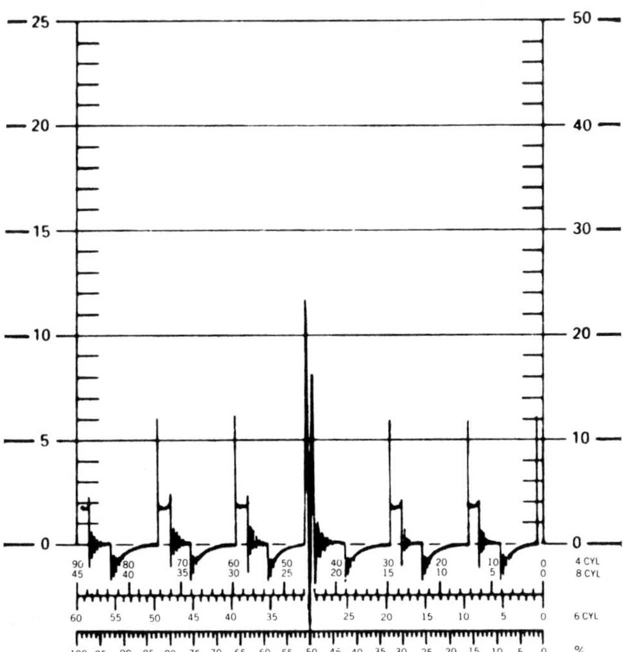

Figure 13-51. This secondary parade pattern shows maximum coil output (open-circuit voltage) for one cylinder.

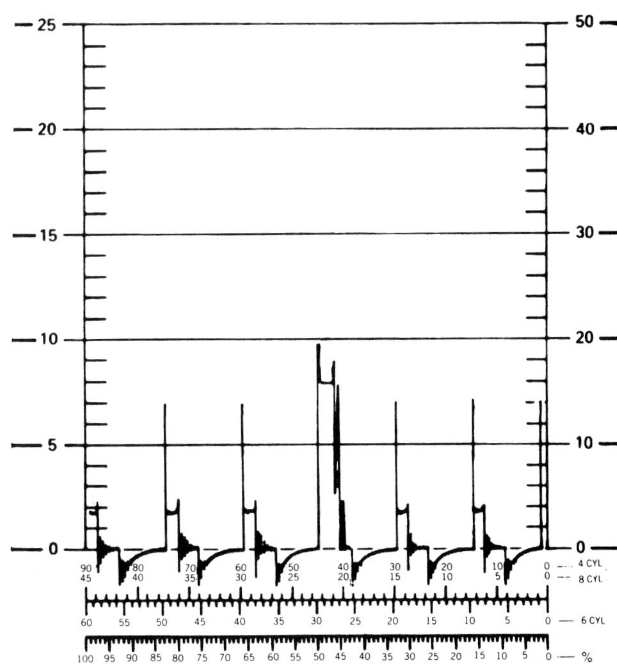

Figure 13-52. Any trace of a sparkline in the available-voltage display indicates secondary voltage leakage, usually inside the distributor.

voltage. The difference between the two is *reserve* voltage. If available voltage falls below required voltage under any condition, a misfire occurs. A well-tuned ignition system on a sound engine should have a *60-percent voltage reserve* under most conditions. For example, if the plug-firing required voltage is 8 kV, maximum available voltage from the coil should be about 24 kV.

You also can check coil available voltage during cranking by connecting the coil high-voltage cable to the spark tester and looking at a secondary parade or superimposed pattern while cranking the engine. The available voltage should be about the same during cranking as it is with the engine running. If it is not, check for the problems listed above, as well as a weak or defective battery and high resistance in the starting bypass circuit.

Notice that the negative-voltage oscillations for open-circuit voltage in figure 13-51 extend about half the distance below the 0-voltage line as the positive oscillations extend upward. That is, if positive voltage is 20 kV, negative voltage should be about 10 kV. If the negative oscillations are shorter, the secondary circuit has poor insulation or voltage leakage to ground at some point. If the condition is common to all cylinders, the cause probably is:

- Poor insulation at the coil tower
- A bad coil high-voltage cable
- A bad distributor cap or rotor.

If the problem occurs with one or a few, but not all, cylinders, the cause is usually voltage leakage at the distributor cap or spark plug cables. These problems are most often caused by a carbon track or dirt that forms a high-resistance short to ground.

If the coil open-circuit voltage spike shows any trace of a sparkline and few oscillations, figure 13-52, there is crossfiring or voltage leakage inside the distributor cap.

Rotor Air Gap Test. The gap between the rotor tip and the plug terminals in the distributor cap creates resistance in the secondary circuit. Engineers calculate this resistance as a design factor in the ignition system. If it is higher than intended, it can raise required voltage and reduce voltage reserve. You can't measure the resistance directly, but you can measure the voltage required to bridge the air gap.

Disconnect one cable from a spark plug and ground it with a jumper wire. Then select the secondary parade pattern on the low-kV scale. The height of the firing line for the grounded cylinder shows the voltage needed to jump the rotor air gap, figure 13-53. It may be as low as 3 kV for most breaker-point ignitions or as high as 6 to 8 kV for high-energy electronic systems. Experience will teach you the normal air gap voltage for different systems. If it is excessively high compared to plug-firing voltages, replace the cap and rotor.

You can use rotor air gap voltage measurements for other distributor tests also. Attach a hand-operated vacuum pump to the vacuum advance diaphragm and operate it through its full range as you watch the air gap voltage. With full vacuum advance, air gap voltage should be no more than twice what it is with no advance. If it is higher, look for a bad rotor or cap, a bent or worn distributor shaft or breaker plate, or an incorrect dwell setting on a breaker-point system.

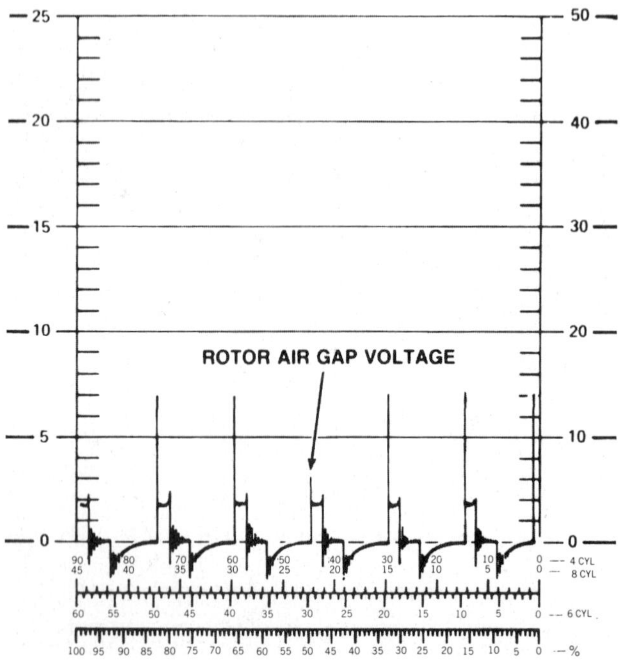

Figure 13-53. The short firing line (spike) is rotor air gap voltage.

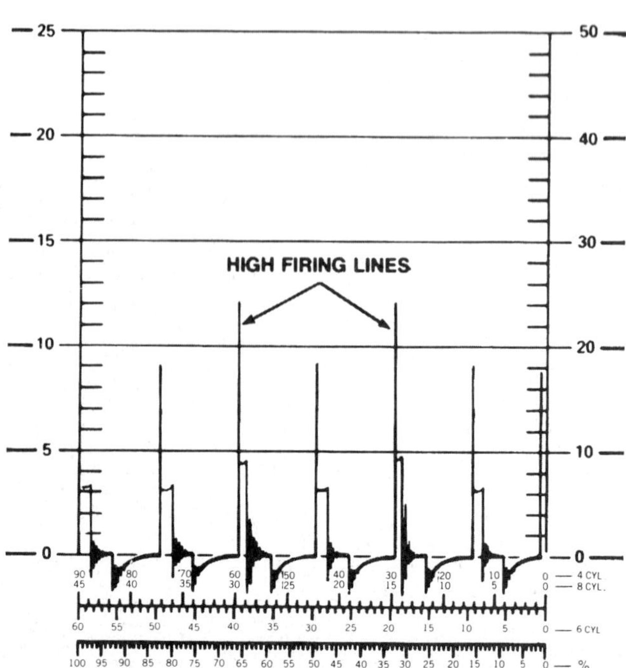

Figure 13-54. Snap-acceleration test with abnormally high firing voltages for two cylinders.

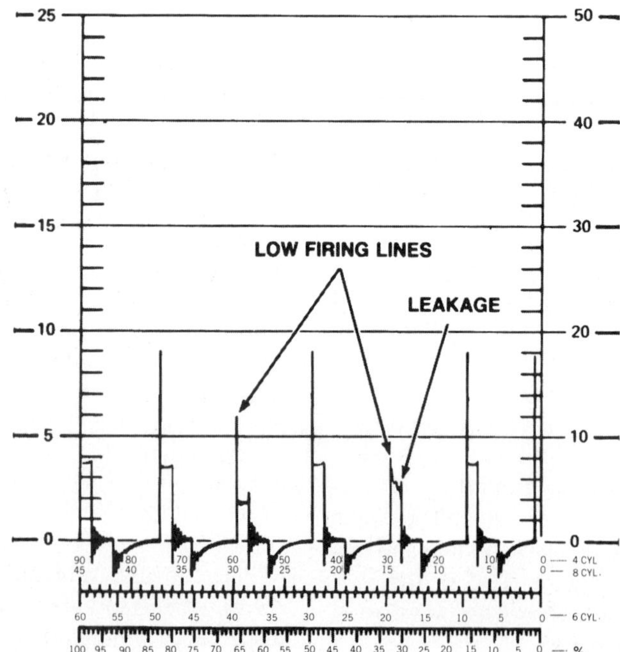

Figure 13-55. Snap-acceleration test with abnormally low firing voltages for two cylinders.

Check rotor air gap for two cylinders 180 degrees from each other in the distributor cap. If air gap voltages differ by more than 2 or 3 kV, the distributor bushings may be worn or the shaft may be bent or worn.

Snap Acceleration Test. With this test, you can locate worn spark plugs by placing a brief load on the engine. Select the secondary parade pattern on the low-kV scale. Momentarily snap the accelerator open and release it immediately. Observe the height of the firing lines at the instant that you snap the accelerator open. They should jump abruptly and then return to the previous level at idle. High firing lines for one or more cylinders, figure 13-54, indicate:

- Worn spark plugs or wide plug gaps
- Broken or high-resistance plug cables
- Defective resistor-type spark plugs
- A worn or cocked distributor cap or a loose distributor shaft
- Lean air-fuel ratio.

Low firing lines or little increase in firing voltage, figure 13-55, indicates:

- Fouled plugs or narrow plug gaps
- Secondåry voltage leakage to ground
- Rich air-fuel ratio.

SPARK PLUG SERVICE

The spark plugs are the final parts in all ignition secondary circuits. If the plugs are not in good condition and do not fire properly, the rest of the circuit cannot do its job. Carmakers recommend plug service as often as every 5,000 miles (8,000 km) or as seldom as every 50,000 miles (80,000 km). Spark plug service intervals depend on:

- Engine design
- Driving habits
- The kind of gasoline used
- The emission devices on the engine.

Plugs in low-compression engines that burn unleaded gasoline generally last longer than plugs in high-compression engines that burn leaded fuel.

Spark plug service should be a straightforward job of removing and inspecting the old plugs and gapping and installing new ones. Accessories and emission devices on many engines, however, make access to the plugs difficult. Some carmakers give you specific instructions for removing accessories or other steps to get clear access to the plugs. Others do not. On many engines, you must loosen and relocate the air conditioning compressor, the power steering pump, the alternator, the air pump, or some other accessory for access to one or more plugs. Whenever you relocate an accessory for plug service, be careful of the wiring and plumbing. Pumps and compressors have heavy hoses that are bulky and that can be torn loose through rough handling.

Some spark plugs may be most accessible from below the engine. You must raise such vehicles on a hoist or with a jack and safety stands and go below the engine to change the plugs.

Tool companies make a variety of special wrench extensions, adapters, and U-joint sockets to make plug replacement easier. In all cases, however, you will need two basic spark plug sockets:

- A 13/16-inch deep socket for 14-mm gasketed and 18-mm tapered-seat plugs
- A 5/8-inch deep socket for 14-mm tapered-seat plugs.

Spark plug sockets are made for either a 3/8- or a 1/2-inch drive ratchet and usually have a rubber insert to hold and cushion the plug insulator. A length of rubber or nylon tubing with an inside diameter that fits over the plug terminal or insulator is another handy tool to hold a plug for removal and installation. Force the tubing over the plug, figure 13-56, and turn it to turn the plug if you can't reach the plug with your fingers.

Figure 13-56. Use a length of nylon or rubber tubing to hold a plug that is hard to reach for installation.

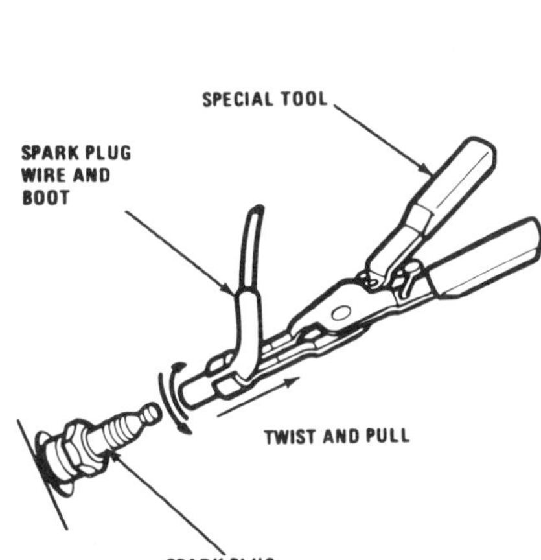

Figure 13-57. Grip the cable with insulated pliers for extra gripping power and to avoid damage. (Chrysler)

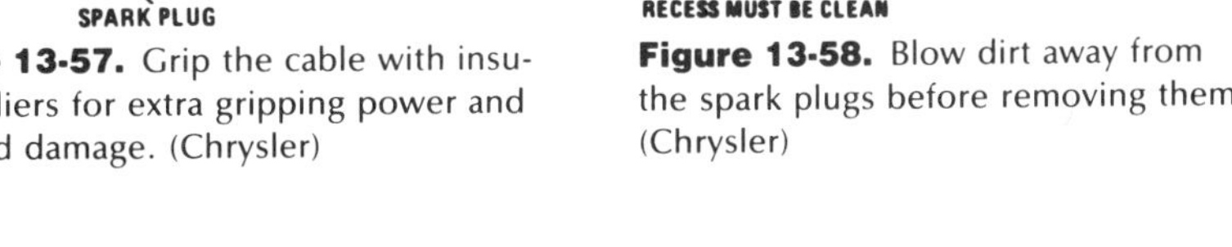

Figure 13-58. Blow dirt away from the spark plugs before removing them. (Chrysler)

Spark Plug Removal

After gaining access to the spark plugs, follow these guidelines to remove them:

1. Disconnect the cables from the plugs by grasping the boots and twisting gently while pulling. Do not jerk on the cables, or you will damage them. Use insulated spark plug cable pliers, figure 13-57, for a better grip.

2. Loosen each plug one or two turns and then blow dirt away from the plugs with compressed air, figure 13-58.

3. Remove the plugs and place them in a tray or holder in cylinder-number sequence, figure 13-59.

4. If the plugs have gaskets, be sure that the old gaskets come out with the plugs. If the engine has spark plug tubes (such as older 6-cylinder Chryslers), gasket-type plugs are installed without gaskets.

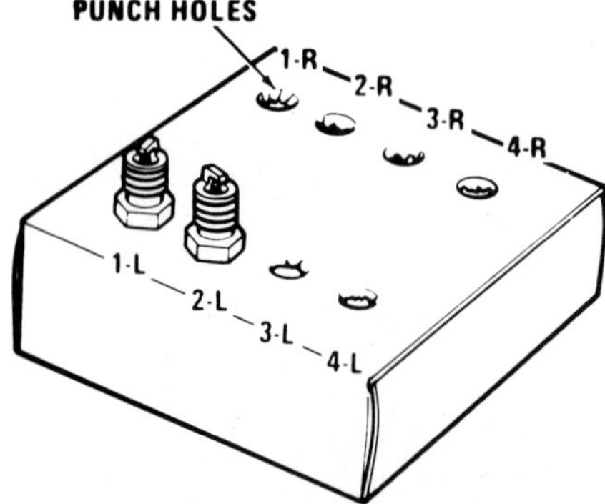

Figure 13-59. Place the used plugs in a numbered holder for analysis. (Chrysler)

Figure 13-60. A normal spark plug in good condition. (Champion)

Figure 13-61 A worn spark plug of the right heat range and normal operating conditions. (Champion)

Figure 13-62. An oil-fouled spark plug. (Champion)

Figure 13-63. A carbon-fouled spark plug. (Champion)

Figure 13-64. An ash-fouled spark plug. (Champion)

Spark Plug Diagnosis

Looking at the firing ends of the plugs can tell you a lot about plug-operating conditions and about general engine condition. The following sections and photos explain the basic points of "reading" spark plugs.

Normal Condition. Figure 13-60 shows a plug with light-brown or gray deposits that accumulate in normal use. There is little electrode wear, and the deposits are not heavy. This plug could be cleaned, regapped, and reinstalled for satisfactory service.

Worn Condition. The plug shown in figure 13-61 has the same light-brown or gray deposits as the plug in figure 13-60. However, the rounded and worn electrodes and insulator indicate that this plug is worn out. This plug can require twice the firing voltage of a plug in good condition and will probably misfire under even a light load. It should be replaced.

Oil-Fouled Condition. Figure 13-62 shows a plug with damp, oily deposits on the electrodes and insulator. This can be caused by oil entering the engine past piston rings or valve guides in a high-mileage engine. It also can happen in a rebuilt engine before the rings seat and establish normal oil control. A defective PCV system also can cause oil to enter the combustion chamber. A cheap cure for oil-fouled plugs is to install plugs that are one or two heat ranges hotter to help burn off the deposits. However, this treats the symptom, not the cause. Locate the cause of oil fouling and recommend repairs to the car owner.

Carbon-Fouled Condition. Figure 13-63 shows a plug with soft, black, sooty deposits. These can accumulate on a plug that is too cold for the engine. Be sure that you install plugs of the correct heat range. A rich air-fuel mixture, a stuck choke, or a clogged air filter also can cause carbon fouling. Cold-engine operation due to stop-and-go driving also causes carbon fouling, as can a defective thermostat, a bad manifold heat control valve, retarded timing, low compression, or bad plug wires or distributor cap that lead to prolonged misfiring.

Ash-Fouled Condition. Figure 13-64 shows a plug with light brown or white deposits that come from some oil or fuel additives. These usually are nonconductive, but large amounts on the plugs and in the combustion chamber can mask the spark and cause misfiring. Clean and regap the plugs or replace them more often to counteract this condition.

Splash-Fouled Condition. Figure 13-65 shows a plug with deposits that have broken loose from the piston or valves and splashed onto the plug. Such deposits may break loose immediately after a tuneup. Clean and regap the plugs before recommending replacement.

Figure 13-65. A splash-fouled spark plug. (Champion)

Figure 13-66. A spark plug with mechanical damage from an object in the combustion chamber. (Champion)

Figure 13-67. A spark plug with gap bridging. (Champion)

Figure 13-68. A spark plug with a glazed insulator. (Champion)

Figure 13-69. An overheated spark plug. (Champion)

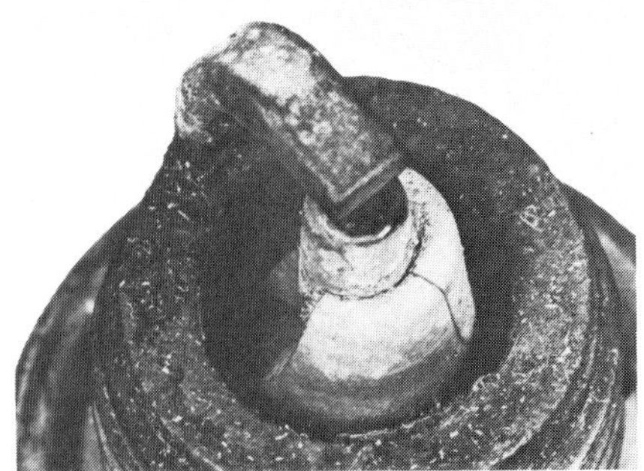

Figure 13-70. A spark plug with detonation damage. (Champion)

Mechanical Damage. Figure 13-66 shows a plug with damage from a foreign object in the combustion chamber. This also can happen if a plug of the wrong reach or the wrong tip length is installed in some engines. Find the cause of the damage before installing new plugs.

Gap Bridging. Figure 13-67 shows a plug with deposits similar to splash fouling, except that these have bridged the gap and short circuited the plug. A secondary scope pattern will show this condition as a low firing line, or no firing line, for this cylinder.

Glazed Insulator. Figure 13-68 shows a plug with yellow or tan deposits caused by hard acceleration and increased plug temperatures. Normal deposits that would usually burn off or flake off now fuse into a hard, conductive coating that causes a misfire. Glazed plugs should be replaced. Plugs that are one heat range colder may prevent future glazing.

Overheated Condition. Figure 13-69 shows a plug with a clean, white insulator tip that indicates overheating. The difference between a glazed plug and an overheated one is a difference of degree. Temperatures for this plug are too high for all operating conditions. If the heat range is hotter than recommended, overheating like this may occur. It also can be caused by overly advanced timing, lean air-fuel ratios, or an overheated engine.

Detonation and Preignition. Figures 13-70 and 13-71 show plugs with damage similar to the mechanical damage in figure 13-66. These conditions were caused by spark knock, or the untimed or uncontrolled explosion of the air-fuel mixture. You can't tell from the plug condition whether the problem is preignition (explosion before the spark) or detonation (uncontrolled explosion after the spark), but this kind of damage should alert you to look for other engine problems before installing the new plugs.

Figure 13-71. A spark plug with preignition damage. (Champion)

Cracked Insulator. Figure 13-72 shows a plug with a hairline crack in the insulator. When this plug was removed from the engine, it appeared carbon fouled, like the plug in figure 13-63. The plug was cleaned to reveal the crack. The carbon fouling was not due to cold operation or a rich air-fuel mixture. It occurred because the crack caused the plug to short circuit and create a misfire. The crack was prob-

Figure 13-72. A cracked insulator will cause a misfire.

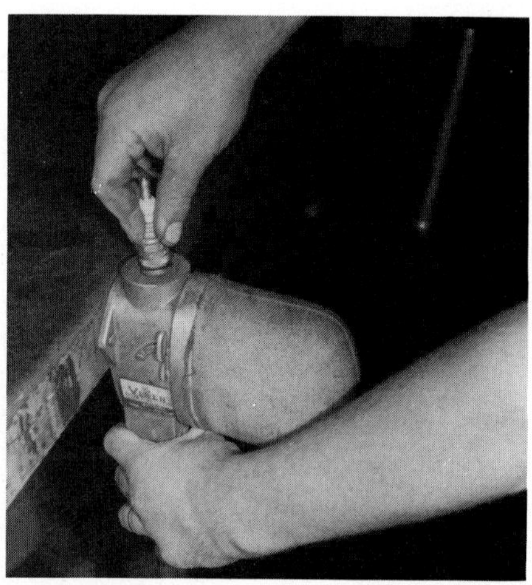

Figure 13-73. Use a spark plug blaster to clean a used plug that is still usable.

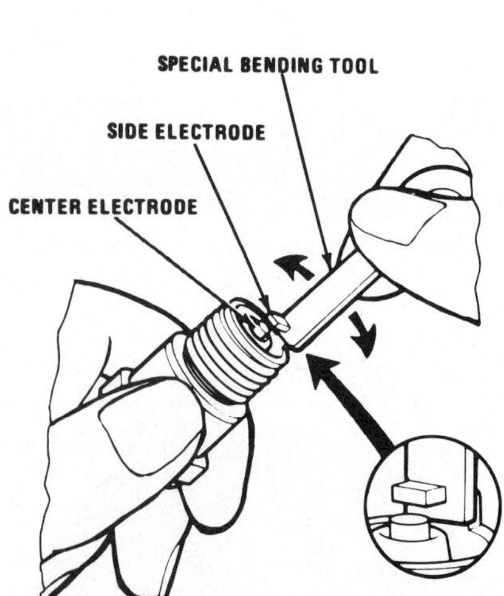

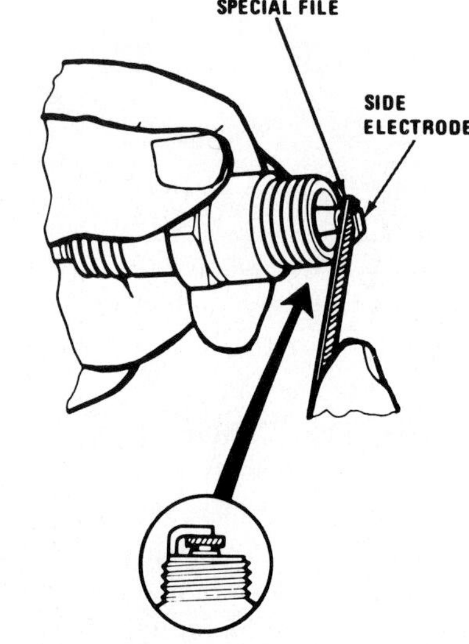

Figure 13-74. File the electrodes lightly before setting the gap on a used plug. (Chrysler)

ably caused by preignition or detonation. This is most likely to occur in an older high-compression engine operating on low-octane, low-lead fuel.

Spark Plug Cleaning, Gap Adjustment, and Installation

Installing new plugs is usually more economical and ensures better performance than cleaning and regapping old ones. However, used plugs that are not badly worn can be cleaned in sandblasting equipment. The gap of either a new or a used plug must be checked and adjusted before installation.

Carmakers and plug manufacturers specify spark plugs by part number for specific engines. The part number indicates the plug diameter, reach, basic gap (wide or narrow), heat range, tip design, and other features such as a resistor or series gap. Always check the part number specifications of the vehicle and plug manufacturer before installing replacement plugs. Do not trust that the last person who installed plugs in a particular engine chose the correct ones.

Some manufacturers list two or three heat ranges for different operating conditions. If a vehicle is used for short trips and stop-and-go driving, a slightly hotter range may work best. If the vehicle is used for extended highway driving, particularly in hot weather, a colder range may be the best choice. In any case, heat ranges should not vary more than one or two numbers from a basic recommendation. Regardless of the heat range, all other plug specifications must match the carmaker's original design specifications.

Spark Plug Cleaning. Although spark plug cleaning is often called "sandblasting," the abrasive material is not sand. The proper material for cleaning plugs is an aluminum oxide abrasive that is similar to the plug insulator material. Ordinary blasting sand may stick to the plug insulator and form a conductive coating when exposed to combustion temperatures. This will cause the plug to misfire.

Clean used spark plugs as follows:

> **Caution.** Do not soak spark plugs in lacquer thinner or solvent. Solvents can penetrate the ceramic insulator and form conductive deposits when the plugs are reinstalled and heat up. Such deposits cause short circuits and misfiring.

1. Wipe dirt, oil, and grease from the plugs with a clean cloth. Blow loose deposits away with compressed air.
2. Place the plug in a spark plug blaster, figure 13-73, and rotate the plug while applying short bursts of abrasive material.
3. After blasting, use compressed air to remove all traces of abrasive from the plug.
4. Clean the threads and gasket or seat area with a soft-bristled wire brush.
5. Carefully file the center electrode to get a flat, shiny surface, figure 13-74.

Setting the Plug Gap. After cleaning a used plug, you must readjust the gap. Even new plugs must be gapped before installation. Tool companies make a variety of plug-gapping tools, figure 13-75, that make gap adjustment easy and ensure an accurate setting. The simplest gapping tool is a round wire feeler gauge and a bending

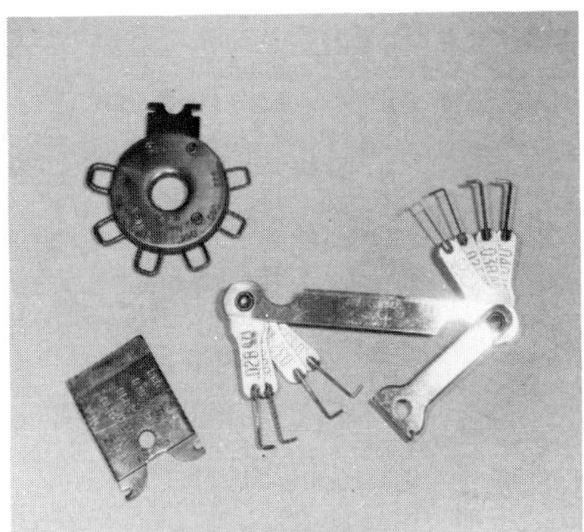

Figure 13-75. Any of these tools will make plug gapping easy and accurate.

tool to bend the ground electrode. A flat feeler gauge may give you an inaccurate measurement on a used plug. Measure the gap with a round feeler gauge, figure 13-76. Bend the side electrode, figure 13-77, to adjust the gap. Do not try to bend or hammer the center electrode to set the gap.

Many electronic ignitions use plugs with gaps that range from 0.045 to 0.080 inch. Manufacturers make plugs with an initial gap within this range for accurate adjustment. Do not try to bend a narrow-gap plug electrode to a wide gap setting. Do not try to close the gap of a wide-gap plug for use in an engine that requires a plug gap of 0.030 or 0.035 inch.

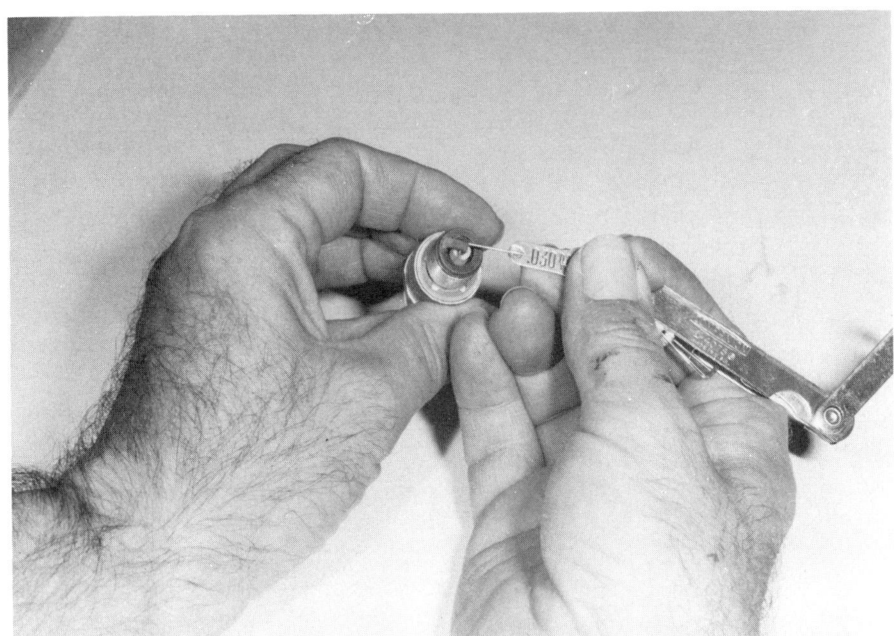

Figure 13-76. Measure the plug gap with a round feeler gauge.

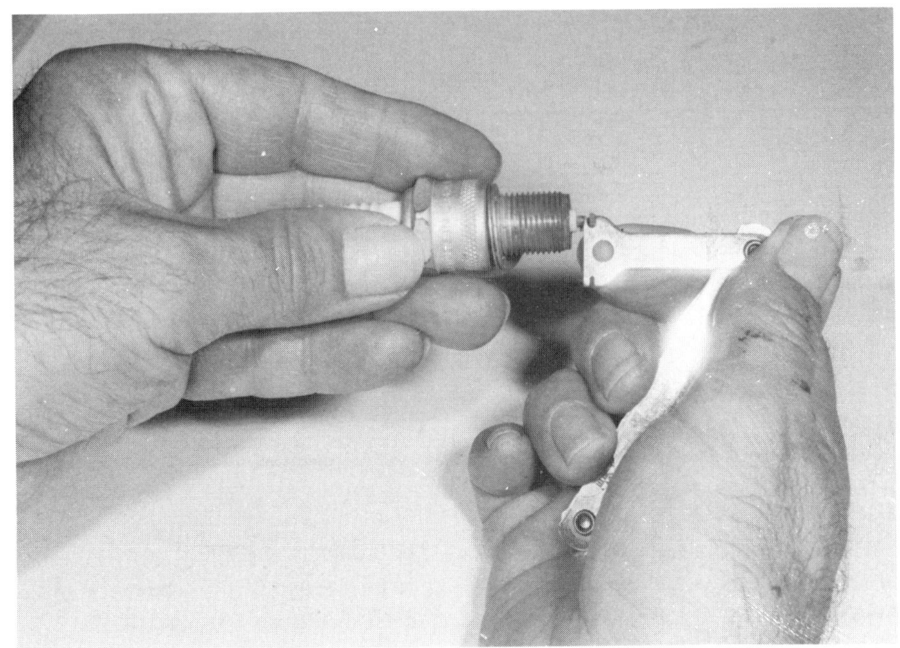

Figure 13-77. Bend the side electrode to set the gap.

Spark Plug Installation. Whether you are installing new or used plugs, follow these basic steps. Some carmakers and plug manufacturers recommend a thread lubricant or antiseize compound for plugs in some engines, particularly those with aluminum heads. Follow these recommendations when installing plugs. Use only specific antiseize compounds recommended by the carmaker. Some may not be compatible with aluminum heads:

1. Remove dirt and grease from the engine plug seats with a clean cloth. Do not let dirt fall into the combustion chambers.

2. If the engine requires spark plugs with gaskets, be sure the gaskets are in good condition and correctly installed on the plugs. Be sure there is only one gasket on each plug.

3. If the engine requires gasket-type plugs in metal cylinders, remove the gaskets before installing the plugs.

4. Be sure that tapered seats of cylinder heads and plugs are clean and free from nicks.

5. Install the plugs into the engine by hand. Use nylon or rubber tubing on the plug terminals or insulators to turn hard-to-reach plugs. Be sure the plugs turn freely in their holes and are not cross threaded. If the plugs do not turn easily, you may have to clean the cylinder head threads with a thread chaser, figure 13-78.

6. If possible, use a torque wrench to tighten the plugs to the values listed in figure 13-79. If the plugs have used gaskets or if you use an antiseize com-

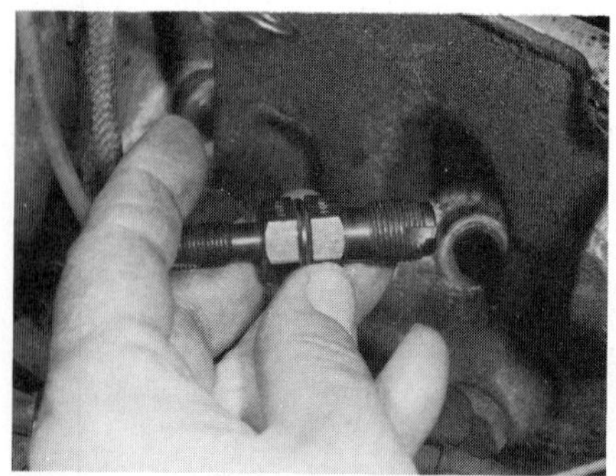

Figure 13-78. Use a thread chaser to clean plug threads in cylinder heads.

PLUG TYPE	CAST-IRON HEAD Foot-Pounds	CAST-IRON HEAD Newton-Meters	ALUMINUM HEAD Foot-Pounds	ALUMINUM HEAD Newton-Meters
14-MM GASKETED	25-30	34-40	15-22	20-30
14-MM TAPERED SEAT	7-15	9-20	7-15	9-20
18-MM TAPERED SEAT	15-20	20-27	15-20	20-27

Figure 13-79. Torque values for spark plug installation.

pound, reduce the torque values slightly.

7. If you can't use a torque wrench, install the plugs finger tight. Then tighten 14-mm gasketed plugs an additional 1/4 turn. Tighten 14-mm and 18-mm tapered-seat plugs only an additional 1/16 turn. Tapered-seat plugs require less torque. Do not overtighten any spark plug, particularly in an aluminum cylinder head.

SUMMARY

Proper service of the secondary ignition circuit is critical for good engine performance. It begins with accurate timing adjustment. Spark-timing control systems must work properly for the best combination of performance, economy, and emission control. You can test and service all spark advance control systems with the guidelines in this chapter and the carmaker's specific procedures.

Secondary circuit oscilloscope tests use the same basic scope displays—superimposed, raster, and parade patterns—that are used for primary circuit troubleshooting. Secondary patterns, however, let you analyze the spark plug circuits for each cylinder. Because engine mechanical conditions affect cylinder resistance and firing conditions, secondary scope tests also let you isolate problems such as incorrect air-fuel ratios, low compression, or bad valves.

Testing and replacing distributor caps, rotors, and spark plug cables involve visual inspection and ohmmeter tests. Most cables, caps, and rotors are easy items to replace, but you must be aware of the unique designs for some electronic ignition parts.

Spark plug condition can tell you a lot about engine and ignition system condition if you learn to "read" plugs accurately. Changing spark plugs on many engines is complicated by crowded engine compartments, but you will use the same principles of plug removal, gapping, and installation on all vehicles.

REVIEW QUESTIONS

Multiple Choice

1. An engine misfires on several cylinders at random. Mechanic A says that a loose, corroded, or otherwise bad coil high-voltage lead can cause this. Mechanic B says that a carbon track on the distributor rotor can cause this problem. Who is right?

- **a.** A only
- **b.** B only
- **c.** both A and B
- **d.** neither A nor B

2. Mechanic A finds a spark plug cable with a loose spark plug terminal. He says this can be caused by improper handling of the plug cables. Mechanic B says that high coil voltage can cause a terminal to separate from the cable. Who is right?

- **a.** A only
- **b.** B only
- **c.** both A and B
- **d.** neither A nor B

3. When checking the resistance in a complete secondary circuit for one cylinder, you should begin by using your ohmmeter to measure resistance between which of the following points:

- **a.** the spark plug ground electrode to the cable terminal inside the distributor cap.
- **b.** the spark plug terminal of a cable to the cable terminal inside the distributor cap.
- **c.** the spark plug terminal of a cable to the terminal at the other end of the cable
- **d.** the spark plug center electrode to the cable terminal inside the distributor cap

4. Several electronic ignition systems use which of the following kinds of spark plug cables:

- **a.** 7-mm, copper core
- **b.** 8-mm, copper core
- **c.** 8-mm, TVRS conductor
- **d.** 8-mm, silicon conductor

5. Mechanic A says that the numbers on Ford bilevel Dura-Spark I distributor caps indicate cylinder numbers. Mechanic B says that firing orders are different for various Ford V-8 engines. Who is right?

- **a.** A only
- **b.** B only

c. both A and B
d. neither A nor B

6. Mechanic A says that silicone grease used on some distributor rotors suppresses radiofrequency intereference (RFI). Mechanic B says that silicone grease prevents high-voltage arcing inside the distributor. Who is right?

a. A only
b. B only
c. both A and B
d. neither A nor B

7. An oscilloscope secondary parade pattern is an easy way to compare:

a. dwell variation
b. voltage reserve and required voltage
c. peak firing voltage among cylinders
d. all of the above

8. Mechanic A says that secondary oscilloscope patterns indicate the resistance of spark plugs, cables, and other secondary circuit parts. Mechanic B says that mechanical conditions inside the cylinder affect secondary resistance. Who is right?

a. A only
b. B only
c. both A and B
d. neither A nor B

9. The secondary pattern shown here probably indicates:

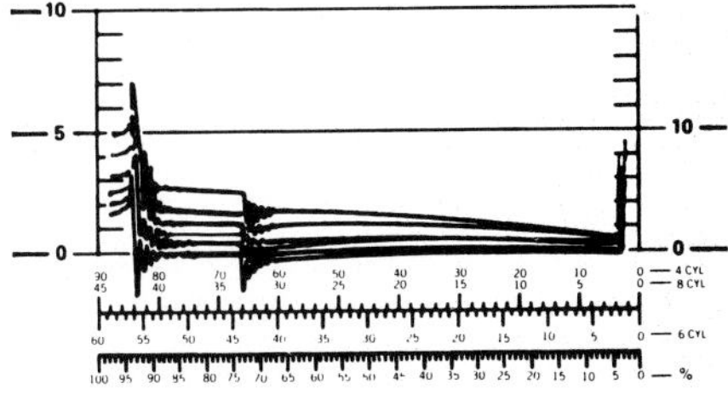

a. fouled spark plugs
b. an open, loose, or corroded plug cable
c. a corroded or loose coil high-voltage cable
d. a lean air-fuel mixture

10. The secondary pattern shown here probably indicates:

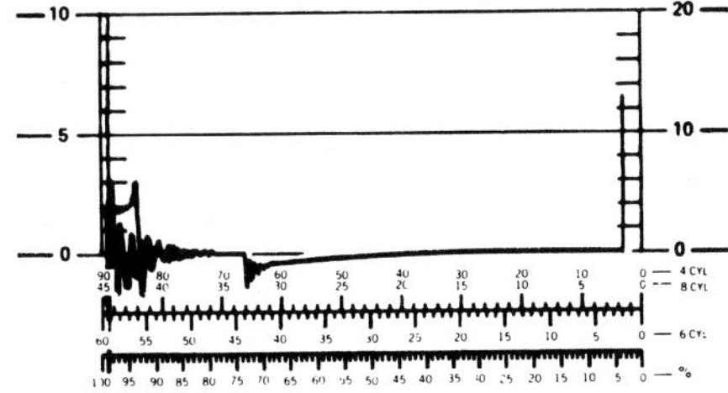

a. a fouled spark plug
b. an open spark plug cable
c. a loose coil connection
d. a rich air-fuel mixture

11. The high firing line (spike) in the secondary parade pattern shown here indicates:

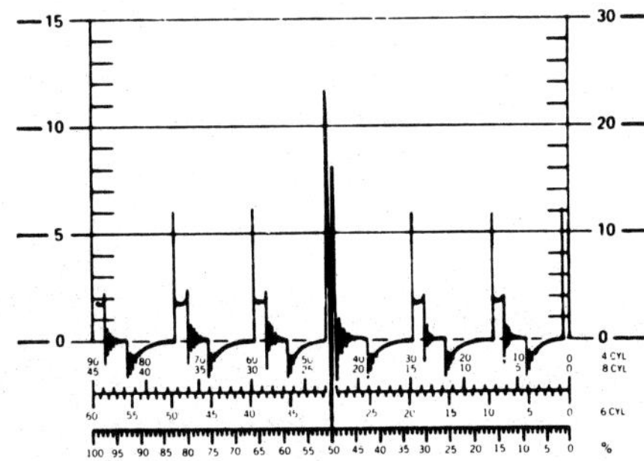

a. a fouled spark plug
b. a loose coil conection
c. coil open-circuit voltage
d. low compression

12. Mechanic A says that you should use a modified spark plug (spark tester, or simulator) to check available voltage on many electronic ignitions. Mechanic B says that you cannot use this kind of tester on a breaker-point ignition. Who is right?

a. A only
b. B only
c. both A and B
d. neither A nor B

13. Mechanic A says that available voltage is the voltage required to fire a spark plug under various conditions. Mechanic B says that the voltage needed to bridge the rotor air gap may be 3 to 8 kV, which is part of required voltage. Who is right?

a. A only
b. B only
c. both A and B
d. neither A nor B

14. A blackened, soot-covered spark plug may be the result of:

a. a heat range that is too hot
b. lead additives in gasoline
c. a rich air-fuel mixture
d. high compression due to combustion chamber deposits

15. If a spark plug heat range is too hot, it may cause:

a. splash fouling
b. oil fouling
c. gap bridging
d. detonation

16. When installing a 14-mm, tapered-seat spark plug, mechanic A torques it to 30 foot-pounds. Mechanic B applies antiseize compound to the threads of a plug used in an aluminum head. Who is doing the job correctly?

a. A only
b. B only
c. both A and B
d. neither A nor B

17. Mechanic A says that you can adjust the gap of any spark plug to 0.050 inch. Mechanic B says that you also can file the center electrode to increase plug gap. Who is right?

a. A only
b. B only
c. both A and B
d. neither A nor B

14 DISTRIBUTOR SERVICE

INTRODUCTION

Distributor service is an essential part of a traditional tuneup for an engine with breaker-point ignition. It is equally important for most electronic ignitions. While you don't have to replace and adjust breaker points in an electronic distributor, you may have to replace a pickup coil or a trigger wheel or adjust centrifugal and vacuum advance mechanisms. You can test spark advance and pickup coil operation accurately with the distributor in the engine. Often, however, you can replace and adjust these parts most easily and accurately with the distributor out of the engine.

This chapter contains basic procedures for condenser and breaker-point replacement and for dwell adjustment with a distributor in or out of an engine. Several photo procedure sequences illustrate the principles of distributor overhaul. This chapter also includes instructions for using a distributor tester to adjust dwell, to test for distributor wear, and to test and adjust advance mechanisms.

GOALS

After studying this chapter, you should be able to do the following jobs in twice the flat-rate labor time or less, using the carmaker's procedures:

1. Install and adjust breaker points and install a condenser with the distributor in the engine.
2. Remove a breaker-point or an electronic distributor from an engine.
3. Overhaul a breaker-point or an electronic distributor. Replace points, condenser, or other defective parts.
4. Using a distributor tester, test and adjust dwell, centrifugal advance, and vacuum advance and test for distributor wear. Replace or adjust parts as required.
5. Install a distributor in an engine.

BREAKER-POINT AND CONDENSER INSTALLATION AND ADJUSTMENT

Breaker-point ignitions require periodic replacement of the contact points. Continual opening and closing of the primary circuit causes wear on the point rubbing block, the point spring, the distributor cam, and the points themselves. All of these actions cause the dwell angle and timing to change during several thousand miles of operation, figure 14-1. Additionally, the current flow stops and starts across the points as they open and close. This always causes some arcing and pitting on the point contact surfaces, which also affects dwell and timing.

Adjusting the dwell angle will compensate for minor wear and pitting of the points, but eventually they will need replacement. When points are burned or pitted, it is difficult to clean and file them to restore the correct gap and dwell. Most points have a hardened tungsten surface. When you file the points, you may remove some of the pits. You also will probably leave some of the file metal on the hard tungsten. The points will then burn and wear out more rapidly. In an emergency, you can file the point surfaces and readjust the gap and dwell, but it is more efficient to replace the point set. It also is common practice to replace the condenser along with the points.

Condensers do not wear mechanically as points do, but heat, moisture, and dirt may damage the condenser connections or cause a condenser to break down. If you can check the capacitance and leakage of a condenser on a condenser tester and it proves to be in good condition, you need not replace it. If you cannot test the condenser, it is usually better to replace it along with the points.

Breaker-point assemblies are mounted to the breaker plate by two screws or by one screw and a locating pin, figure 14-2. Even if the replacement point set is assembled and aligned at the factory, you should check alignment and spring tension. Two-piece point sets always require alignment and spring tension adjustment for accurate installation and long life.

Use a spring scale, hooked on the movable point arm to measure spring tension, figure 14-3. Do not let the scale arm drag on the distributor body, or you will get an inaccurate reading. Adjust spring tension by moving a long notch in the end of the spring back and forth on the spring retainer, figure 14-3, before tightening the screw. If spring tension is too low, the points will bounce at high speed. If tension is too high, it will accelerate wear on the rubbing block and cam.

Use an aligning tool to center the contact surfaces of the fixed and movable breaker points. Bend the stationary contact point, not the movable one, figure 14-4. The point surfaces should contact in their exact centers for long life and minimum pitting. Alignment

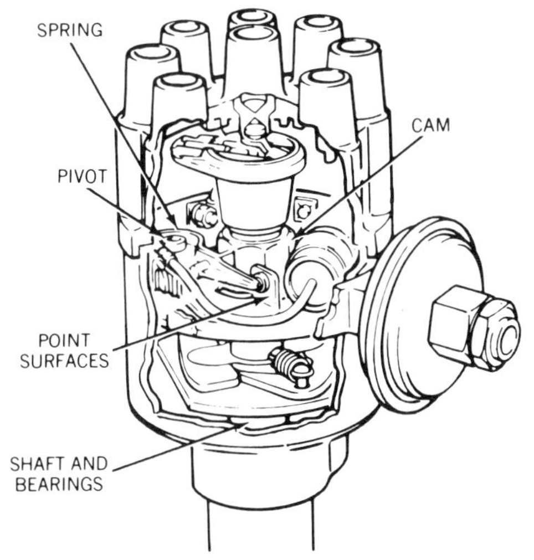

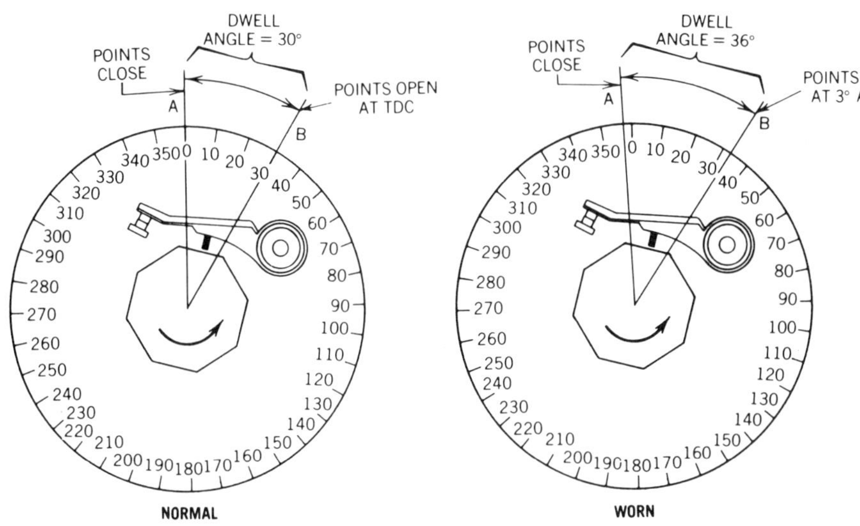

Figure 14-1. Normal wear in a breaker-point distributor causes dwell and timing to change.

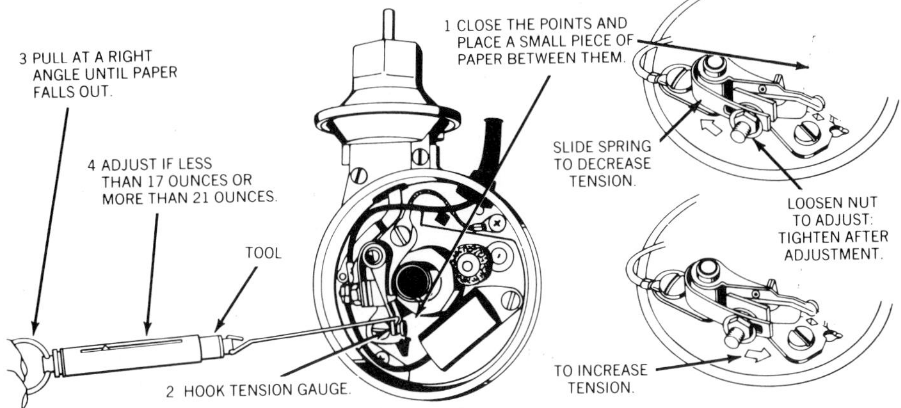

Figure 14-3. Pull gently on the movable point arm with a scale to measure spring tension. (Ford)

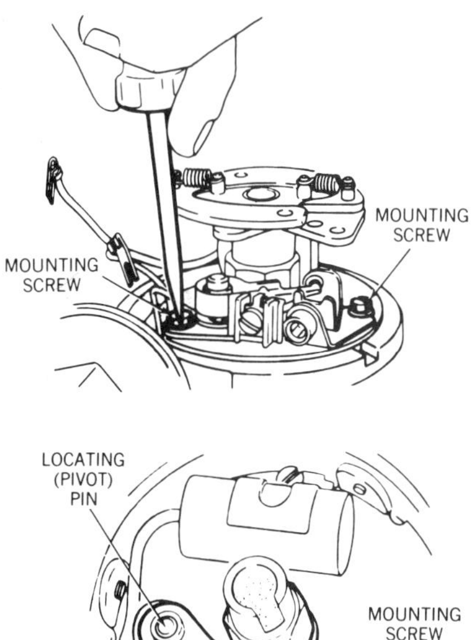

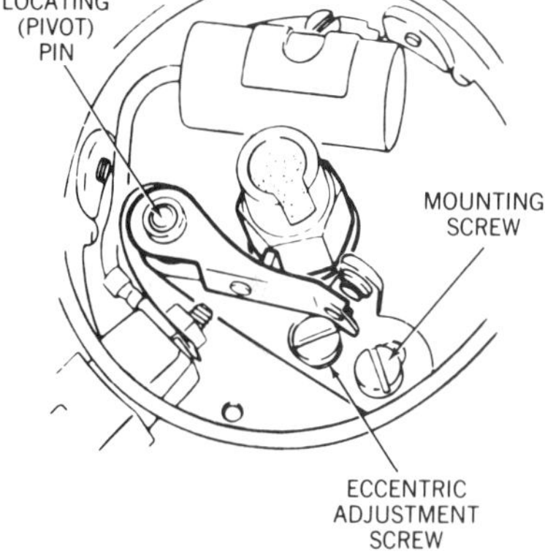

Figure 14-2. Points are mounted to the breaker plate by two screws or by one screw and a locating pin. (Delco-Remy)

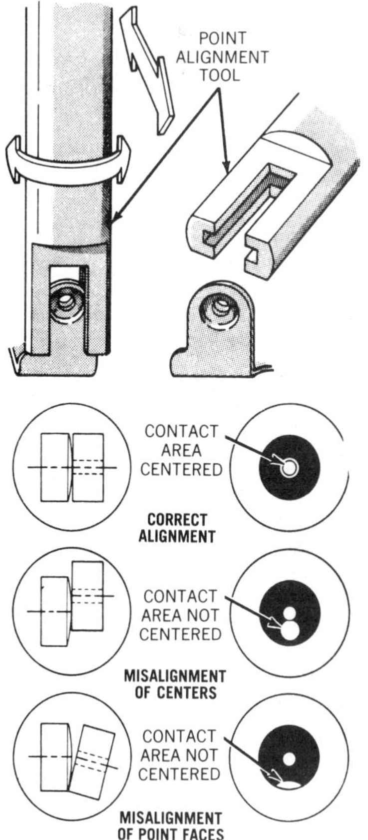

Figure 14-4. Bend the stationary, or ground, point with an alignment tool. (Ford)

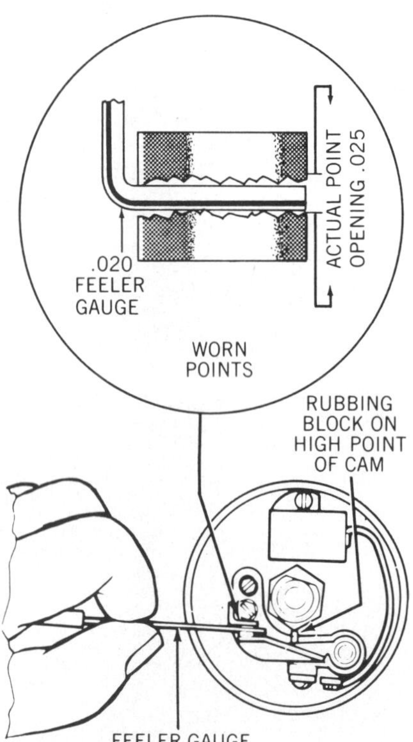

Figure 14-5. With the rubbing block on the high point of a cam lobe, you can adjust new points with a feeler gauge.

PRIMARY LEADS

LOCATING PIN

PIVOT POST

ECCENTRIC SCREW

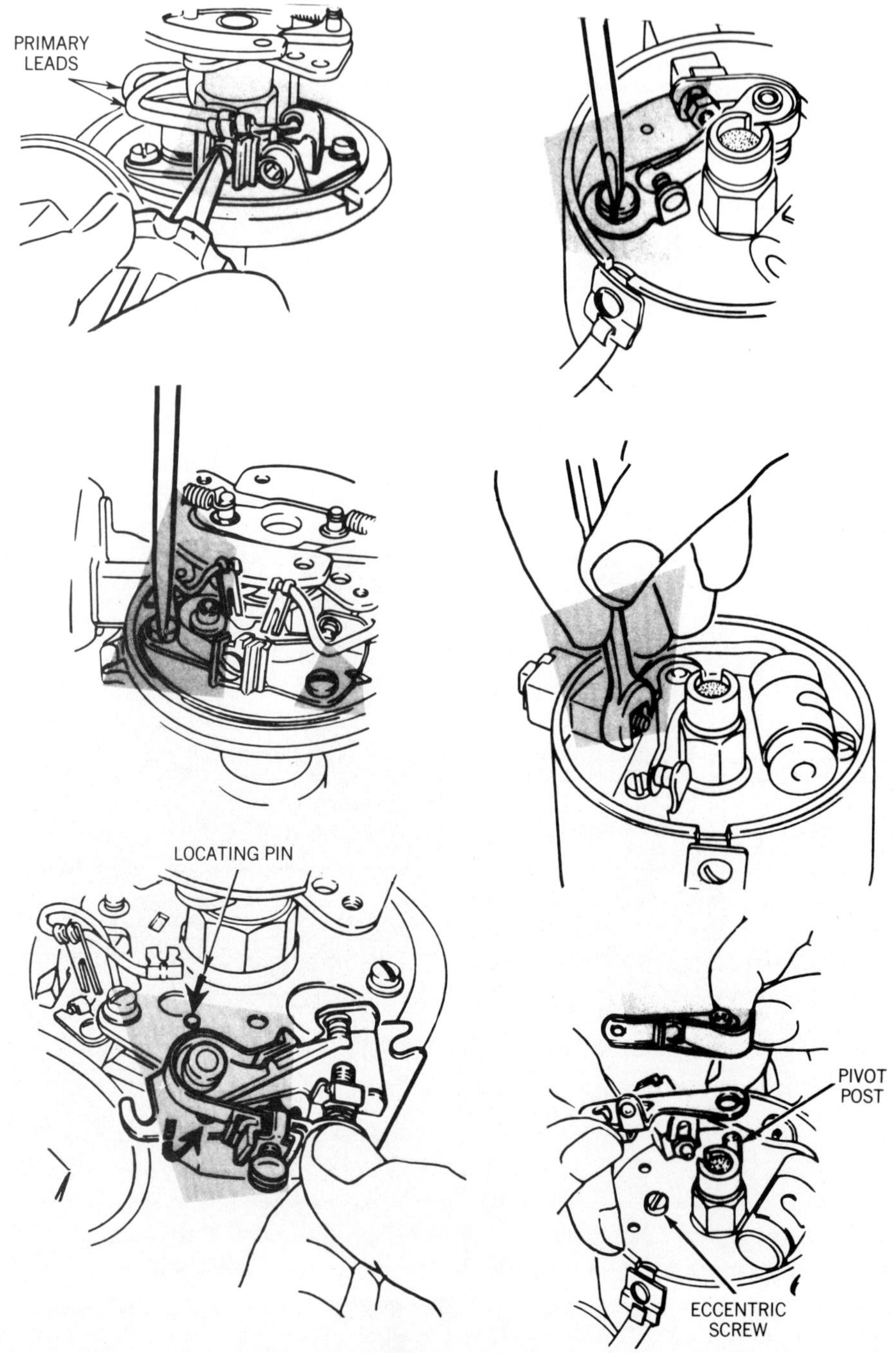

Figure 14-6. Typical breaker-point installations. (Delco-Remy)

also ensures accurate dwell adjustment. You can align a new set of points by operating the distributor in a distributor tester or by cranking the engine with the ignition on and watching the arc between the points. Bend the stationary point until the arc is centered on the contact surfaces for correct alignment.

You can replace breaker points with the distributor in or out of the engine. The photo procedures later in this chapter show both methods for different distributors. It is usually more accurate, and easier in the long run, to remove the distributor for service. This also gives you the opportunity to test distributor operation and spark advance in a distributor tester.

You can adjust new, properly aligned points by measuring the gap with a specified feeler gauge, figure 14-5. Be sure that the feeler gauge is clean and free from oil and grease to avoid leaving dirt on the points. Do not try to adjust used points with a feeler gauge because the uneven surfaces cause inaccurate measurements, figure 14-5. After adjusting the gap with a feeler gauge, always check the dwell with a dwellmeter, an oscilloscope, or a distributor tester to ensure accurate adjustment. Remember that dwell will increase (point gap will decrease) as the point rubbing block wears. To allow for this wear, adjust dwell toward the minimum end of the specification range when you install new points. That is, if the dwell specification is 28° to 32°, set the dwell at 28° or 29° when you install the points to allow for rubbing block wear.

If you adjust the dwell by setting the point gap, adjust the gap toward the *wide end* of the specification range. That is, if the gap specification is 0.019 to 0.021 inch, adjust a new set of points to 0.020 or 0.021. The gap will decrease and the dwell will increase as the rubbing block wears, but dwell will

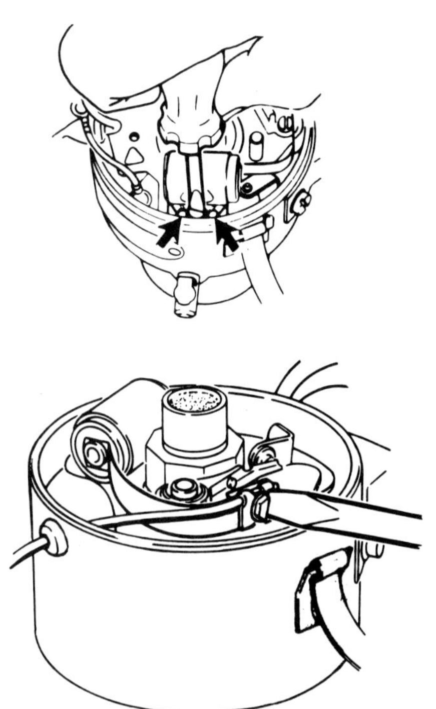

Figure 14-7. Install the condenser and connect the condenser lead as shown here. (Delco-Remy)

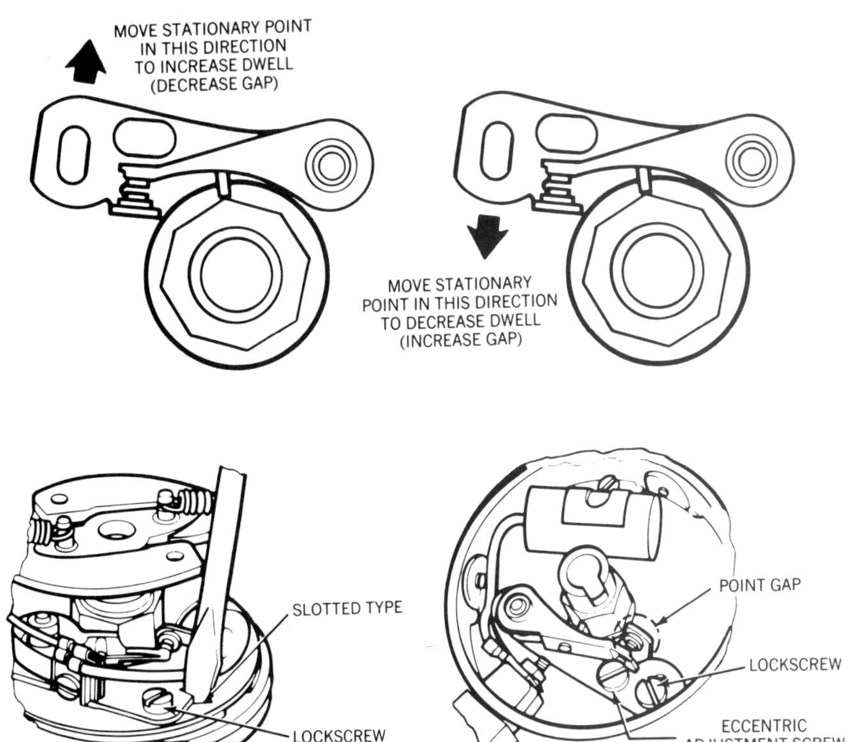

Figure 14-8. Adjust dwell by moving the breaker-point base at a slot or by turning an eccentric adjustment screw. (Chrysler)

stay within the specification range.

Whether you install and adjust points and condenser with the distributor in or out of the engine, the following general procedure outlines the steps you will follow:

1. Remove the distributor cap and rotor.

2. Using a remote starter switch, or with the distributor in a tester or bench vise, rotate the shaft so that the breaker-point rubbing block is *exactly* on the high point of a cam lobe, figure 14-5.

3. Loosen or remove the condenser mounting screws and the point mounting screws. Disconnect the primary lead and the condenser lead from the point insulated terminal and remove the old points and condenser. Figure 14-6 shows typical installations.

4. Install the new point set on the breaker plate. Do not tighten the mounting screws, except on a Delco V-8 distributor.

5. Install the new condenser on the distributor and place the pigtail connector or strap on the insulated terminal at the movable point arm, figure 14-7.

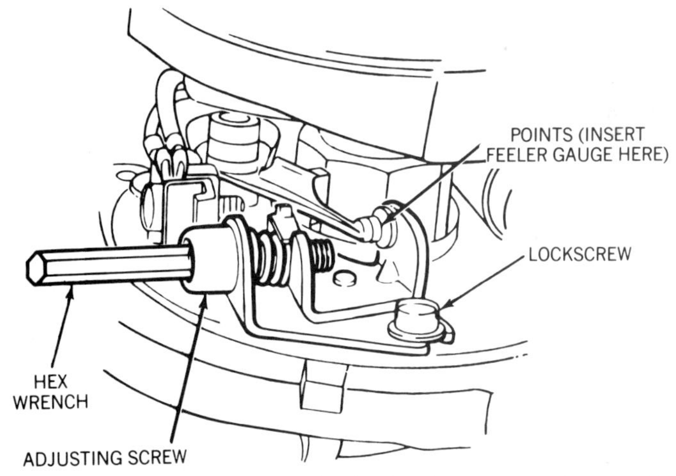

Figure 14-9. Turn the adjusting screw with a hexhead wrench to adjust dwell on a Delco external-adjustment distributor. (Chrysler)

6. Be sure that the condenser lead is not grounded to the distributor body and tighten the condenser mounting screw.

7. Install the primary wire lead on the point insulated terminal. Tighten the terminal screw or nut if required.

8. With the point rubbing block on the high point of a cam lobe, place a feeler gauge of the specified thickness between the points, figure 14-5.

9. Turn the point adjusting screw or use a screwdriver to move the adjusting slot, figure 14-8, while gently sliding the feeler gauge back and forth between the points. Be sure to hold the gauge squarely in the gap. Do not twist it or let it drag on the distributor body. Adjust Delco V-8 points by using a hexhead (Allen) wrench to turn the adjusting screw on the back of the ground point, figure 14-9.

10. When the feeler gauge passes through the gap with a light drag and does not force the points open as it enters the gap, tighten the point mounting lockscrew. Recheck the adjustment after you tighten the screw.

11. Check the point adjustment by measuring actual dwell in a distributor tester or with a dwellmeter or an oscilloscope when the distributor is in the engine. Readjust the points, if necessary, for an accurate dwell adjustment.

BREAKER-POINT SERVICE WITH THE DISTRIBUTOR IN THE ENGINE

The following photo procedure shows typical steps for installing and adjusting breaker points with the distributor in the engine. After dwell adjustment, set the initial timing according to the carmaker's procedures.

Because many imported cars continued to have breaker-point ignitions for one or two years after domestic makers switched to electronic systems, we are using a 4-cylinder distributor from a Honda Civic as our example. It has the features of most basic breaker-point distributors, and you can use this general sequence for similar systems.

HONDA BREAKER-POINT REPLACEMENT

1. This primary connector is a spade terminal on the side of the distributor. Unclip it; then remove the distributor cap with cables attached.

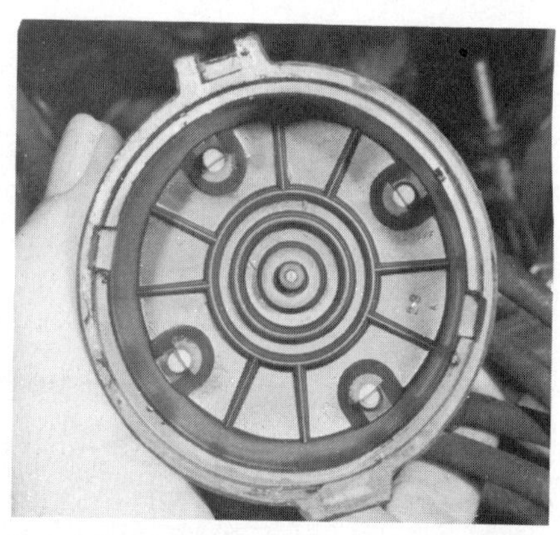

2. Inspect the cap for cracks, carbon tracks, corrosion, and other damage. Wipe it clean with a clean cloth. Use no solvent.

3. Pull the rotor straight up and off the shaft to remove it. Inspect it for the same possible defects as the cap.

4. Loosen the connector screw and slip the primary lead connector off the insulated breaker-point terminal.

5. Loosen the two mounting screws (A) that hold the points to the breaker plate. Remove the old point assembly.

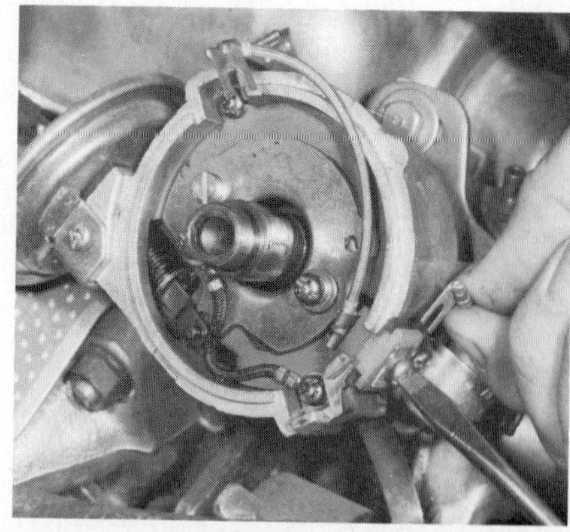

6. Whether the condenser is inside or outside the distributor, disconnect its pigtail lead from the primary terminal. Then remove the mounting screw and the condenser.

HONDA BREAKER-POINT REPLACEMENT (*Continued*)

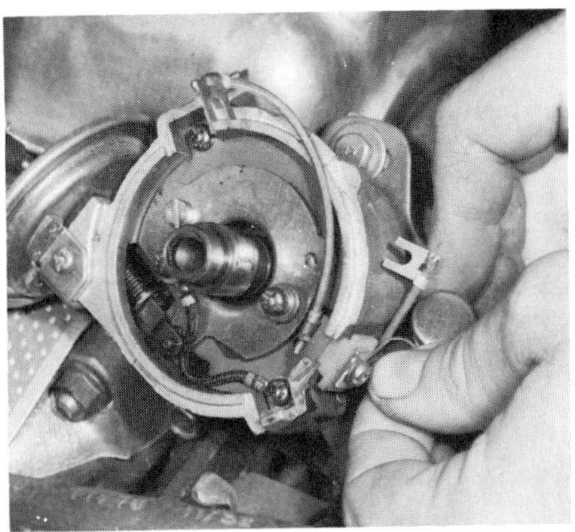

7. Mount the new condenser on (or inside) the distributor and connect its pigtail lead to the primary terminal.

8. Apply a thin film of lubricant to the distributor cam lobes. You also can put a small amount on the point rubbing block.

9. Install the new point assembly and the two mounting screws (A). Be sure the ground wire (B) is in good shape. Don't tighten the screws yet.

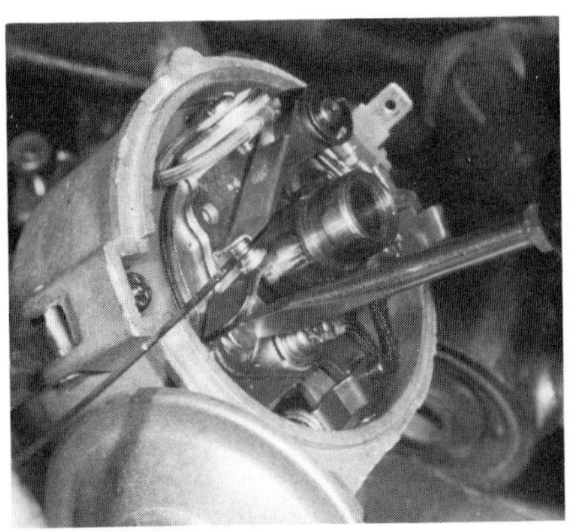

10. Turn the crankshaft *or* the distributor to put the point rubbing block on the high point of a cam lobe. Use a feeler gauge to set the gap by turning an eccentric adjustment screw.

11. Tighten the point mounting screws after adjusting the gap. Then, connect your dwellmeter and verify the dwell adjustment while cranking the engine.

12. After installing the rotor and cap (A), connect your timing light and adjust the initial timing.

DISTRIBUTOR REMOVAL

The details of distributor removal vary from engine to engine, but there are several principles that you can follow to make the job easy and troublefree. As with spark plug service, you may have to loosen and relocate some engine accessories for access to the distributor. On many engines, you will have to remove the air cleaner or air intake ducts. Tag any wires or vacuum lines that you disconnect so that you can reinstall them correctly.

Before removing the distributor, release the clips or screws for the cap and set the cap aside with the spark plug cables attached. For a Delco HEI distributor with the coil in the cap, disconnect the cables from the cap but leave them attached to the cable retainer, figure 14-10. Set the cables and retainer aside and remove the cap and coil with the distributor.

Figure 14-10. Remove the distributor cap or the spark plug cables and retainer before removing the distributor from the engine.

Next, establish reference marks for correct reinstallation, figure 14-11. Mark the rim of the distributor body with a pencil, chalk, or a light scribe line in line with the rotor tip. Make another pair of matching marks in line with each other on the base of the distributor and the engine block. If you can't reach the distributor base to mark it, note the position of the vacuum advance unit or the primary connector in relation to a point on the engine.

Many mechanics use one of the following methods to establish distributor location reference points.

1. Crank the engine to align the timing marks at tdc for the number 1 cylinder. Then mark the distributor body in line with the rotor tip and make reference marks on the distributor base and engine block. This takes a few extra seconds, but it helps you set static timing when you reinstall the distributor.

2. Crank the engine so the distributor rotor points in the same direction on any engine before removing the distributor. That is, crank the engine so the rotor is in line with the crankshaft and pointing to the front of the engine. Or crank the engine to align the rotor perpendicular with the crankshaft. This method establishes a habit that makes distributor alignment instinctive.

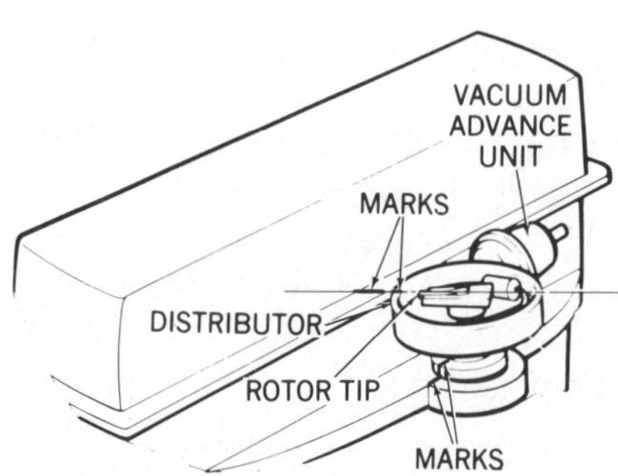

Figure 14-11. Make alignment marks as shown here before removing the distributor. (Chrysler)

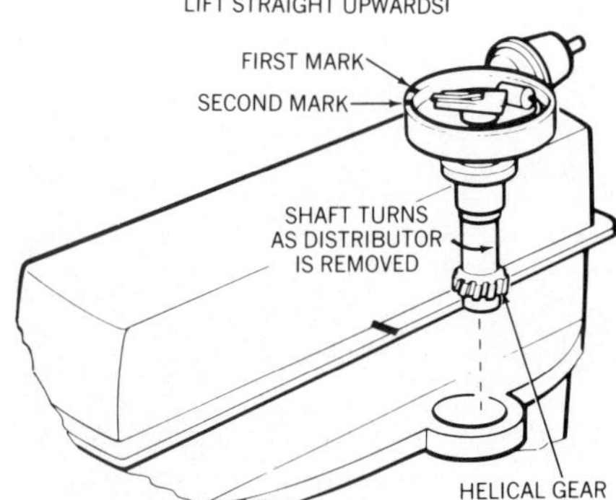

Figure 14-12. Make a second mark on the distributor rim as the rotor turns when you remove the distributor. (Chrysler)

Many distributors are driven by a helical gear, or bevel gear, that mates with a spiral gear on the camshaft. When you remove the distributor, the shaft will turn as the distributor gear slides out of the camshaft gear. The rotor tip will move about 1/4-inch opposite to its normal rotation, measured in relation to the distributor rim. As you remove such a distributor, stop when the gear slides free of the camshaft and make a second mark on the distributor rim, in line with the rotor tip, figure 14-12. Use this second mark to align the rotor when you reinstall the distributor. When the distributor seats in the engine, the rotor tip should turn into alignment with the first mark. These extra steps will save you time and trouble when you reinstall the distributor and adjust the timing.

Most distributors are mounted to the engine with one holddown capscrew and a small clamp. After removing the capscrew and clamp, figure 14-13, loosen the distributor carefully. Do not jerk it from the engine. Carelessness will cause you problems.

Heat, dirt, and varnish deposits may cause the distributor body to stick in the engine. Gently twist and rock the distributor to loosen it. If that does not work, squirt penetrating oil at the distributor base where it mounts on the engine and again twist gently. On many distributors, you can grip the body with an oil filter wrench for extra leverage as you twist. As a final resort, tap the

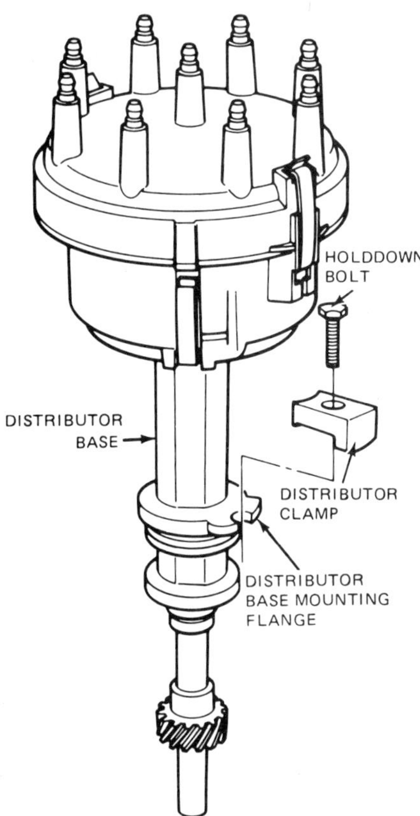

Figure 14-13. Typical distributor holddown bolt and clamp. (Ford)

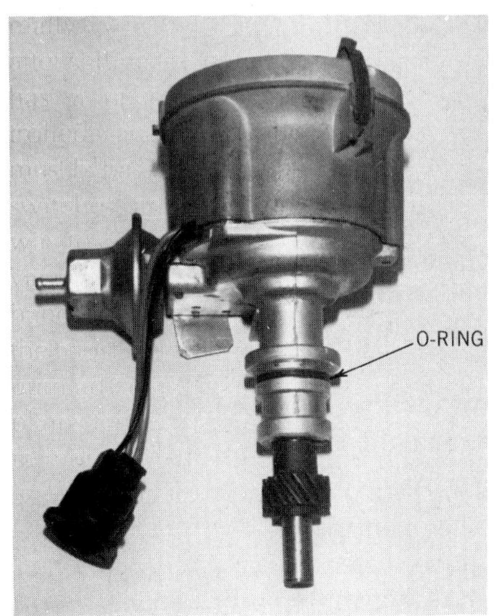

Figure 14-14. Many distributors have O-rings or metal sealing rings. Inspect for damage. Be sure the ring is in good condition for reinstallation.

Figure 14-15. Disconnect the primary lead from the coil or the ignition pickup connector from the distributor.

Figure 14-16. Disconnect the ignition feed wire connector from a Delco HEI distributor like this.

distributor body gently with a small plastic hammer. To remove the distributor from some GM 4-cylinder engines, you must first remove the fuel pump. This is necessary for clearance to remove the distributor.

Some distributors have a metal sealing ring or a rubber O-ring where the base joins the engine. For example, you will find metal rings on Chrysler V-8 distributors and rubber O-rings, figure 14-14, on many Ford engines and several imports. Do not lose these sealing rings when you remove the distributor. If a rubber O-ring is damaged, install a new one. A damaged O-ring will cause an oil leak at the distributor base.

Several engines have an intermediate shaft from the distributor to the oil pump. The shaft should stay in the engine when you remove the distributor, but if it has a lot of sludge buildup, it may pull out with the distributor. Ford small-block V-8's have such a shaft. If the shaft comes out with the distributor, you can reinstall it easily later. If the shaft lifts part way out of the engine and then falls into the engine, you have a problem. It usually falls into the front of the oil pan or the timing cover, which means that you must remove the pan or timing cover to retrieve it.

To avoid such a problem on a Ford V-8 or similar engine, raise the distributor only an inch or two from the engine and look underneath to see if an oil pump shaft is attached. If a shaft is coming out with the distributor, grab it with long-nosed pliers and lift it out of the engine. Do not try to force it back. To reinstall such a shaft, hold it firmly with a gripping tool and insert it into the engine to engage the oil pump or other accessory. Be sure it is seated completely before reinstalling the distributor. If you are unsure about the drive gear arrangement and the use of an intermediate shaft on any distributor, check the manufacturer's installation drawings and remove the distributor cautiously.

Use the following basic steps, along with the preceding precautions, to remove any distributor:

1. Before removing the distributor, align the rotor in the desired position and make your alignment marks, figure 14-11. Then be sure the ignition switch is off. It is good practice to disconnect the battery ground cable to keep the engine from being cranked while the distributor is out.

2. Release the cap clips or screws and remove the cap. Set it aside with cables attached. On Delco HEI systems, disconnect the cables and retainers from the cap so you can remove the cap with the distributor.

3. On a breaker-point distributor, disconnect the primary lead from the distributor or the distributor side of the coil, figure 14-15. It is usually easier to disconnect it from the coil and remove the short lead with the distributor.

4. On an electronic distributor, disconnect the signal generator (pickup coil) connector at the distributor. On a Delco HEI system, disconnect the ignition feed wire from the distributor cap, figure 14-16. Unlatch the connector with a small screwdriver.

Figure 14-17. Remove the holddown bolt and clamp.

Figure 14-18. Lift the distributor carefully up and out of the engine.

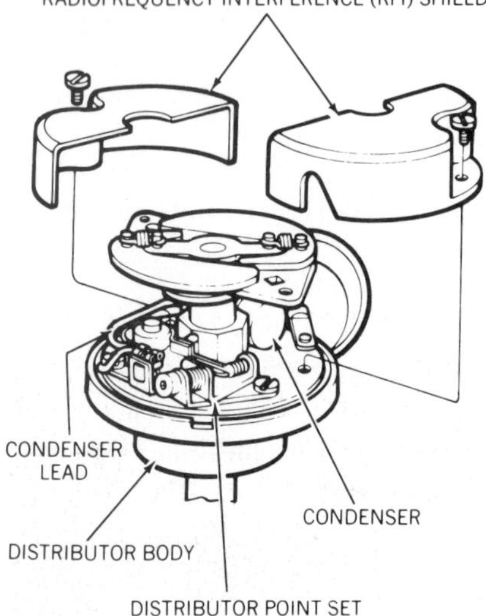

Figure 14-19. Many Delco-Remy breaker-point distributors have RFI shields that you must remove to adjust dwell. (Chrysler)

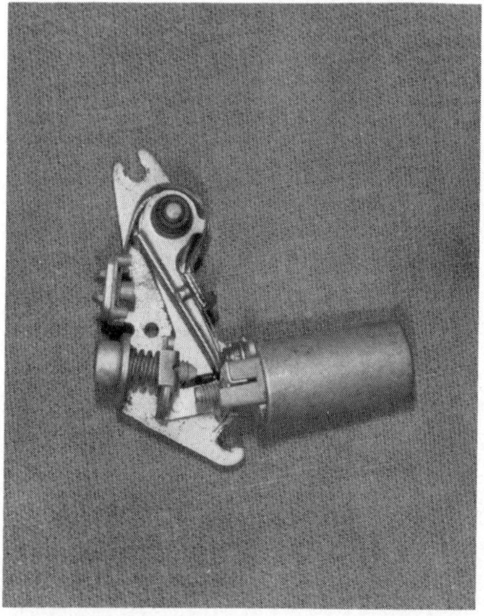

Figure 14-20. You can eliminate an RFI shield by installing a point set with the condenser mounted directly to the points.

5. Disconnect all vacuum lines from the distributor.

6. Use a distributor wrench or a socket and extension with a ratchet to remove the distributor holddown bolt (capscrew) and clamp, figure 14-17.

7. Loosen and remove the distributor, figure 14-18. Do not let dirt fall into the engine.

> **Caution** Do not let solvent enter the vacuum advance unit or contact electronic pickup coils. Keep solvent out of the shaft and bushing areas. Solvent will damage vacuum diaphragms and some electronic parts. Solvent will remove lubricant from permanently lubricated bushings.

8. Clean dirt and grease from the distributor body with solvent and a soft brush. Then install the distributor in a bench vise with padded jaws or in a distributor tester for service.

DISTRIBUTOR OVERHAUL

This section contains photo procedures that show overhaul steps for five typical distributors:

1. A Ford (Motorcraft) 6-cylinder Dura-Spark distributor with centrifugal and vacuum advance. This is typical of many distributors with electronic pickup coils (magnetic pulse generators). This photo sequence includes steps that show a variation in Ford distributor design. These steps also illustrate that while component shapes may vary, service principles remain the same.

2. A Delco 8-cylinder HEI distributor with centrifugal and vacuum advance. Later models used with electronic spark control do not have these mechanical advance devices and have modules with 5- or 7-wire connectors.

3. A Ford (Motorcraft) TFI-IV distributor used with the 2.3-liter HSC 4-cylinder engine. Conventionally mounted on the engine, the distributor is driven by the cam gear and has an integral ignition module, a Hall-effect switch and stator assembly, and a fixed octane adjustment provision.

4. A Delco 4-cylinder HEI-EST distributor containing both a magnetic pulse generator and a Hall-effect switch.

5. A Chrysler optical distributor. Introduced on 1987 3.0-liter V-6 engines, this distributor contains two LED's and photodiodes with a dual-track timing disc. It represents the first OEM use of an optical distributor on production cars.

Construction and adjustment details vary among distributors, but these photo procedures show the general steps common to most. Use these examples, along with the carmaker's specific procedures to service a distributor.

Some of these overhaul procedures can be done with the distributor in the engine. To do them with the distributor removed, mount the distributor in a vise with padded jaws or in a distributor tester. The section immediately after the photo procedures contains instructions for testing and adjusting a distributor in a tester. You can use several of the test procedures to check a distributor before overhaul to identify wear and determine how much work needs to be done. If you have access to a distributor tester, review these instructions before overhauling a distributor.

Many Delco 8-cylinder breaker-point distributors made in the early and mid-1970's had a 2-piece radio-frequency interference (RFI) shield inside the distributor, figure 14-19. The shield prevents RFI signals from the condenser lead from interfering with vehicle radios. You must remove the shield to adjust the dwell and reinstall it after dwell adjustment is complete. You can eliminate the RFI shield by installing a point set with the condenser attached directly to the ground point, figure 14-20.

FORD (MOTORCRAFT) DURA-SPARK DISTRIBUTOR OVERHAUL

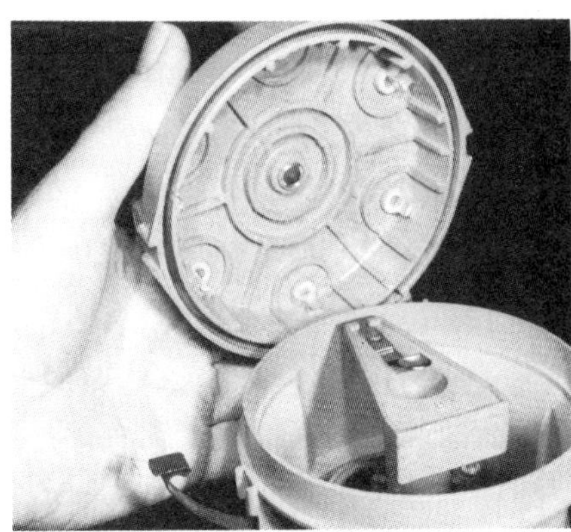

1. Ford has built Dura-Spark distributors in several forms. This is a Dura-Spark II. Start by removing the cap.

2. Remove the rotor next, before trying to remove the adapter ring. Some Dura-Spark I rotors are held by screws.

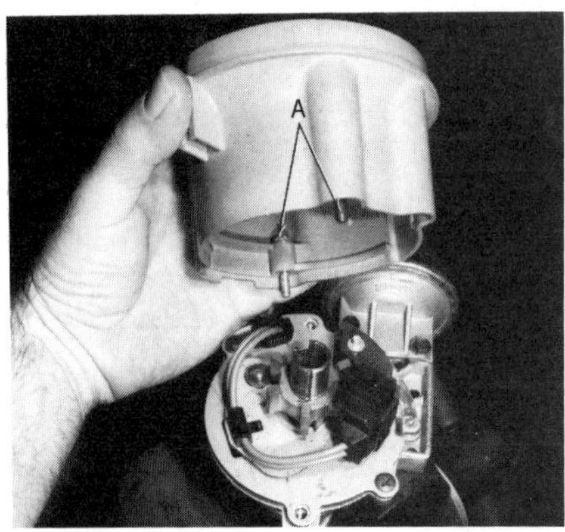

3. Loosen the internal screws (A) to remove the adapter. Leave them in the adapter. Some adapters are held by external clips.

FORD (MOTORCRAFT) DURA-SPARK DISTRIBUTOR OVERHAUL (*Continued*)

4. Use two screwdrivers to carefully pry the trigger wheel (armature) from the sleeve. If it doesn't come off easily, stop.

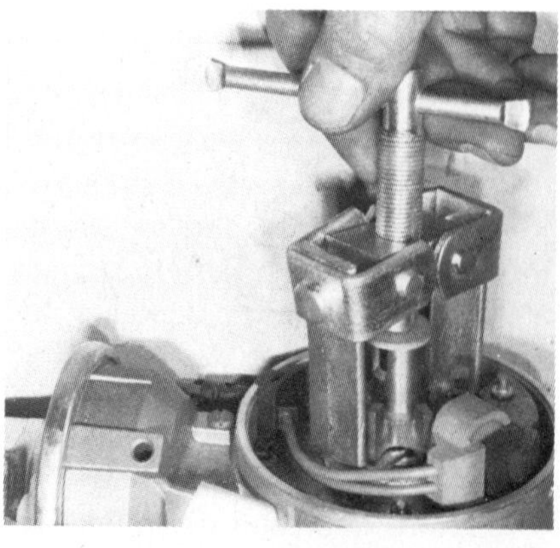

5. To remove a stubborn armature, use a small 2-jaw puller. We changed distributors to show you this step.

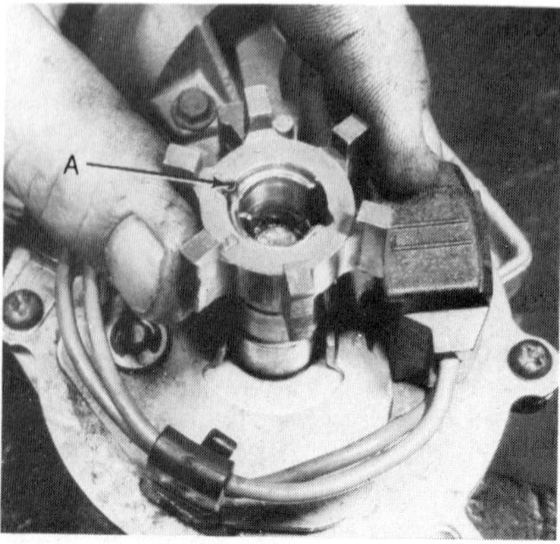

6. As you remove the armature, notice the rollpin (A) that locks it to the sleeve shaft. Note the position and don't lose the pin.

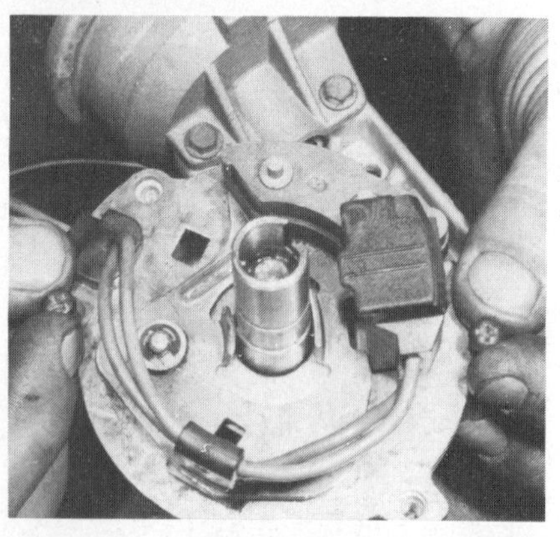

7. Remove the two screws that hold the pickup coil (stator) and its plate to the distributor body. This stator and plate are a single assembly.

8. Disengage the pin on the bottom of the stator from the vacuum advance linkage and remove the stator and plate assembly.

9. Let's look at four photos of an earlier Ford Solid-State distributor (steps 9 to 12) for comparison. Remove this wire retaining ring to remove the stator.

10. Be careful not to lose the spring clip that holds the vacuum advance link to the stator when you remove it.

11. This Ford distributor shows a critical point for many electronic ignitions. The primary lead mounting screw is the system ground for the entire ignition. Be sure it is clean and tight at reassembly.

12. Separate the primary lead grommet from the distributor and remove the stator and the primary lead as an assembly.

FORD (MOTORCRAFT) DURA-SPARK DISTRIBUTOR OVERHAUL (*Continued*)

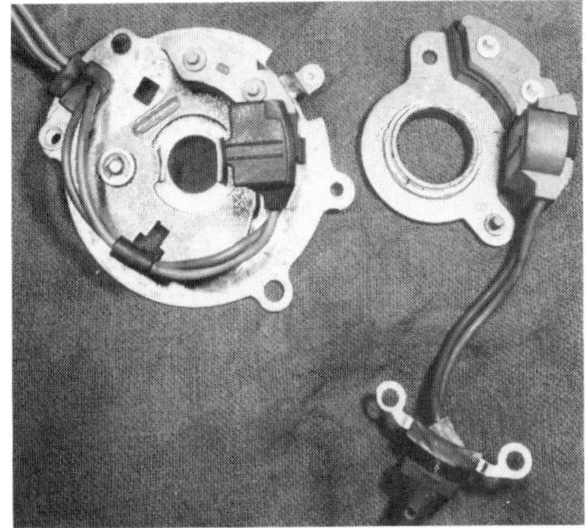

13. Compare the Dura-Spark baseplate and stator on the left with the earlier stator on the right. Functionally, they are identical, but they are built differently.

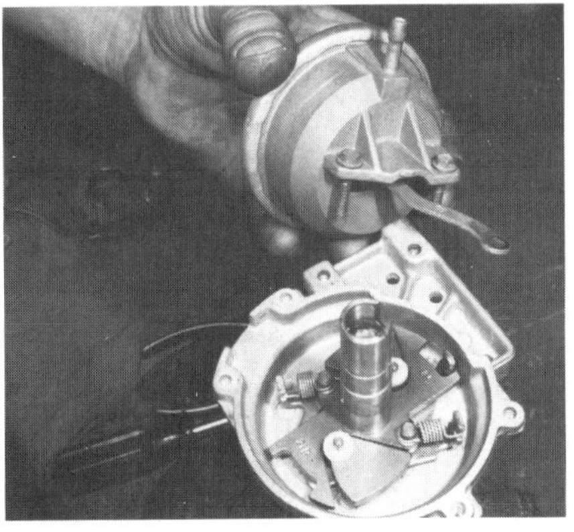

14. After removing the Dura-Spark stator and baseplate, remove the two screws that hold the vacuum advance unit.

15. Mark the advance weights and springs for correct reassembly, then remove them. Notice the nylon bushings on the weight pivot pins.

16. Remove the lubricating wick and retaining snapring from the center of the rotor sleeve.

17. Remove the sleeve and then the weights for cleaning. Don't lose or mix up the nylon bushings on the weights.

18. If you want to remove the shaft, remove this rollpin from the drive gear. Replace the O-ring on the distributor body if it is worn or damaged.

19. The tanged washer may lift out with the sleeve or stay on the weight base. Remove it for cleaning and reinstall it along with the weights.

20. Clean and reinstall the sleeve with light lubricant on the shaft. Be sure the rubber stop (at the screwdriver) is in place, or you will get too much centrifugal advance.

21. Reinstall the retaining snapring and the lubricating wick in the sleeve. Put two or three drops of oil on the wick.

FORD (MOTORCRAFT) DURA-SPARK DISTRIBUTOR OVERHAUL (*Continued*)

22. Lightly lubricate the pivot pins and spring ends and reinstall the springs in the correct locations on the weight base.

23. Connect the vacuum link to the stator pin and place the vacuum unit and the stator and plate assembly on the distributor. The grommet tab (A) is the system ground.

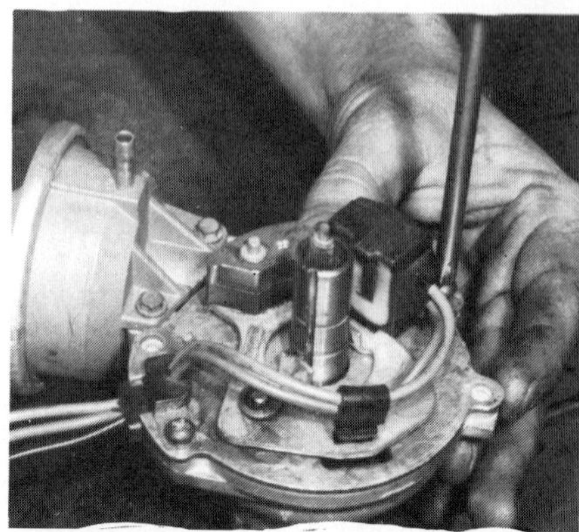

24. Install two hexhead screws to secure the vacuum unit and the two screws that hold the stator and plate.

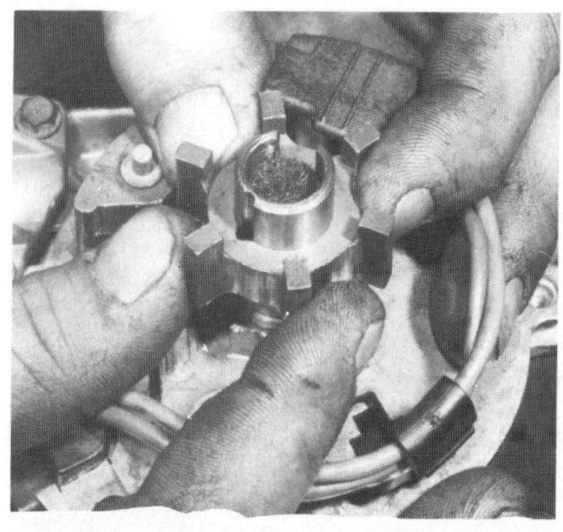

25. Place the armature on the shaft with the rollpin slots aligned. Be sure the armature is seated on the base of the sleeve.

26. Use long-nosed pliers to put the rollpin in its slot. Then gently tap it into place with a hammer and small punch.

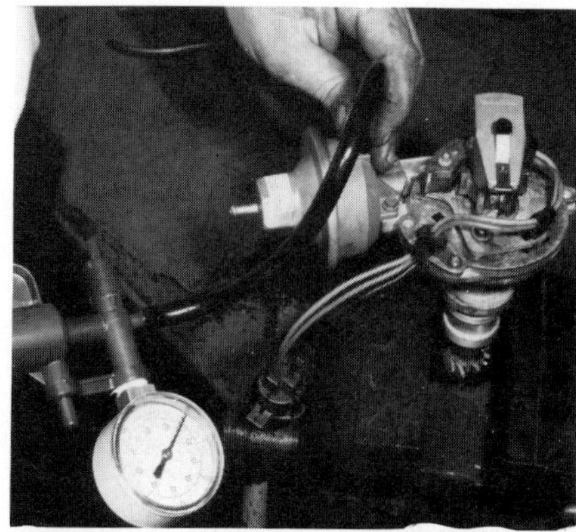

27. Install the rotor and use your hand-operated vacuum pump to check the advance and the retard vacuum diaphragms.

DELCO 8-CYLINDER HEI DISTRIBUTOR OVERHAUL

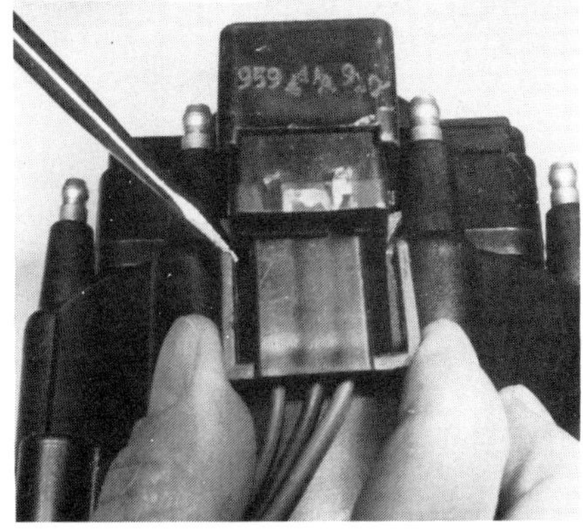

1. Begin the HEI overhaul by disconnecting the module connector from the cap terminal. Use a screwdriver to carefully release the connector tabs.

2. Release the four spring latches and remove the cap. Inspect the cap thoroughly for signs of crossfiring. HEI units with separate coils are similar inside the distributor.

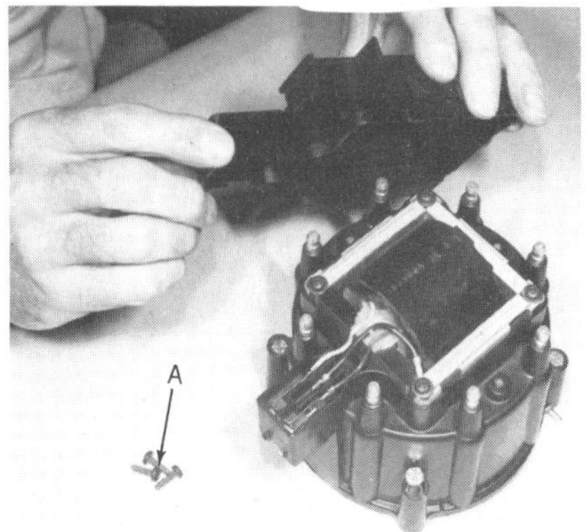

3. Remove the three screws (A) that hold the coil cover, then remove the cover.

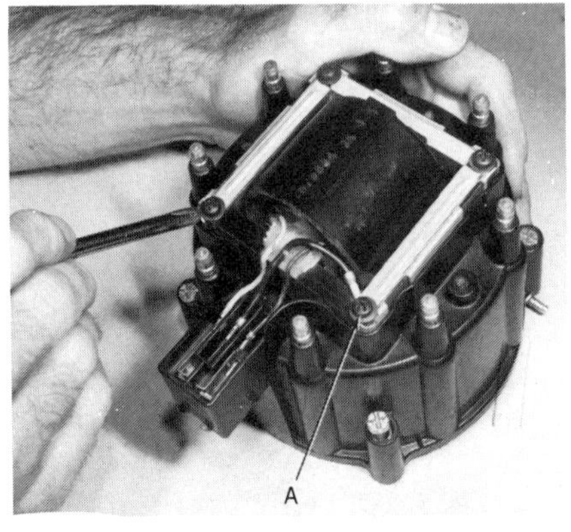

4. Remove the four screws that secure the coil. Note the ground lead at screw (A). The coil also is grounded at the screw near the screwdriver.

5. After removing the coil, take out the rubber seal. Then lift out the carbon button and spring (A).

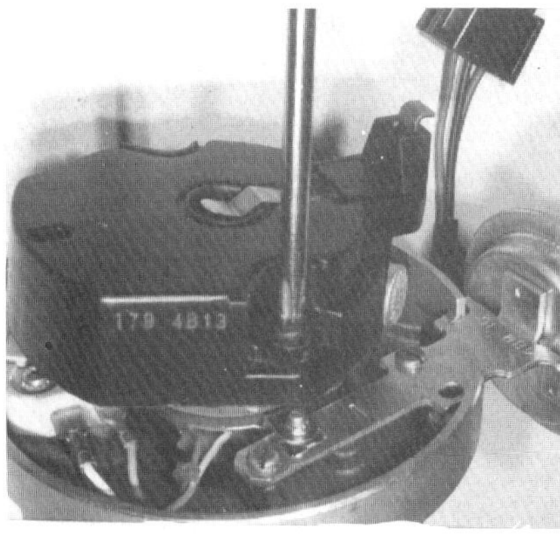

6. Returning to the distributor body, remove the two screws that hold the rotor. Inspect the rotor carefully for signs of crossfiring and "burn through."

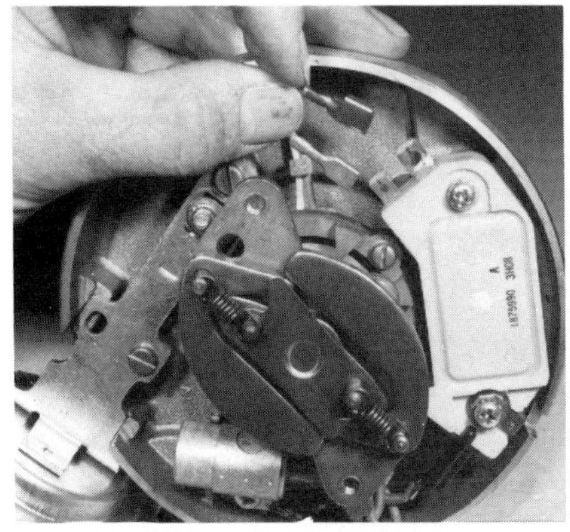

7. Disconnect the pickup coil leads from the ignition module. Use an ohmmeter to check pickup coil resistance across these leads.

8. Disconnect the connector at the other end of the module from the B + (battery voltage) and C (ground) terminals. Slip the wiring grommet out of its slot in the body.

9. Remove the two module screws and the module. Later units with electronic timing have 5 or 7 pins. Apply silicone grease to the module and mounting base at reassembly.

DELCO 8-CYLINDER HEI DISTRIBUTOR OVERHAUL (*Continued*)

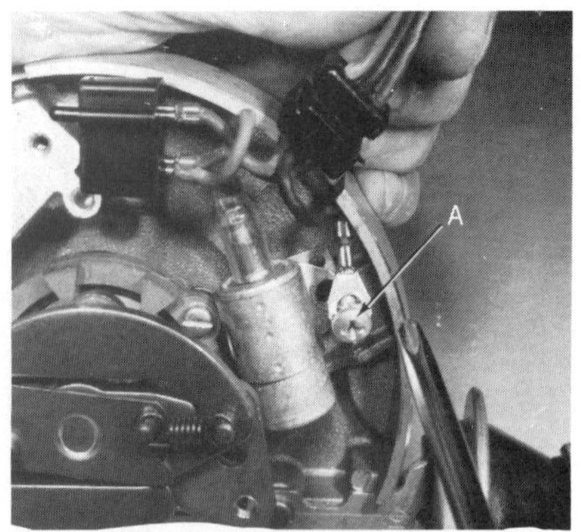

10. This capacitor suppresses radio interference; it is not a primary circuit condenser. Remove the mounting screw (A).

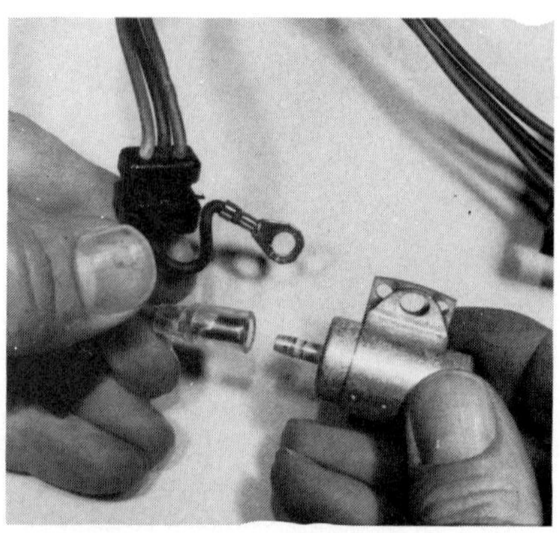

11. Disconnect the capacitor from the primary lead. Test the lead for continuity and inspect it for damage. Test the capacitor for short circuits also.

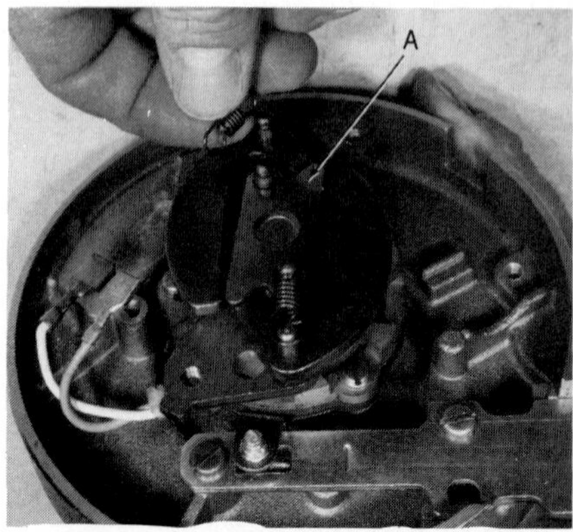

12. Mark the spring and weight locations for correct reassembly. Then remove the springs, the weight retainer (A), and both weights.

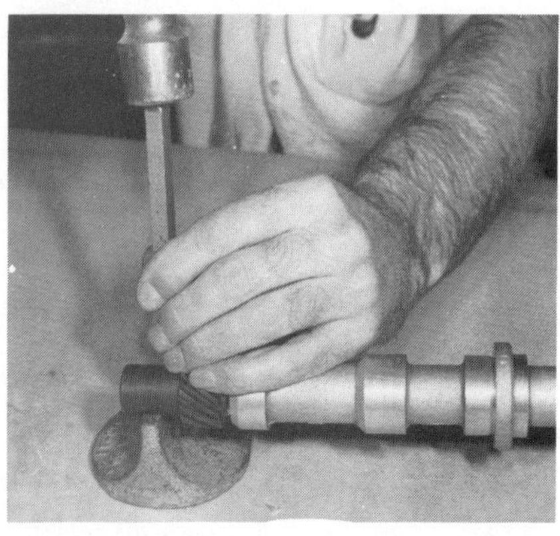

13. To remove the pickup coil, you must remove the shaft. Start by removing the rollpin from the drive gear with a hammer and small punch.

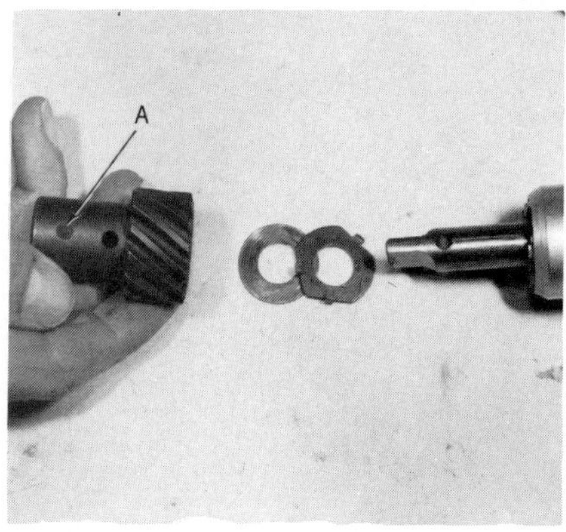

14. Note the positions of the gear, the shim, and the tanged washer as you remove them. The dimple (A) at the end of the gear must align with the rotor tip when you reassemble them.

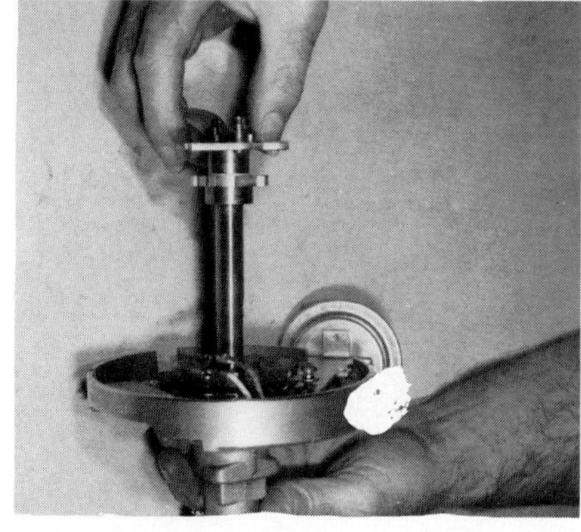

15. Carefully slide the shaft and trigger wheel (timer core) assembly from the distributor. Do not nick the bushings as you withdraw the shaft.

16. Slide the shaft out of the trigger wheel and weight base assembly. Inspect the trigger wheel bushing and upper distributor bushing for wear.

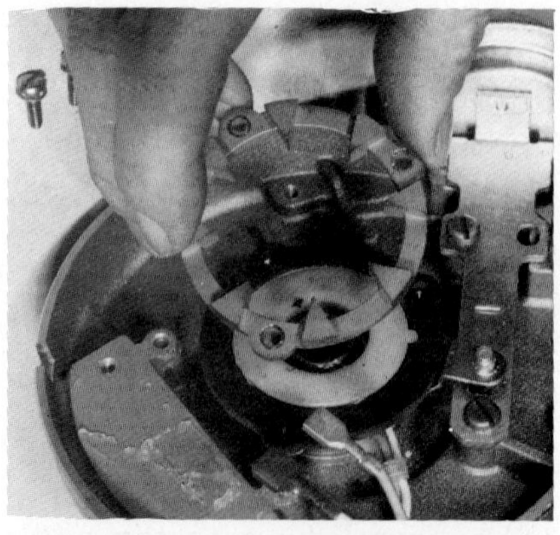

17. Remove three screws and then remove the toothed pole piece and the rubber gasket beneath it.

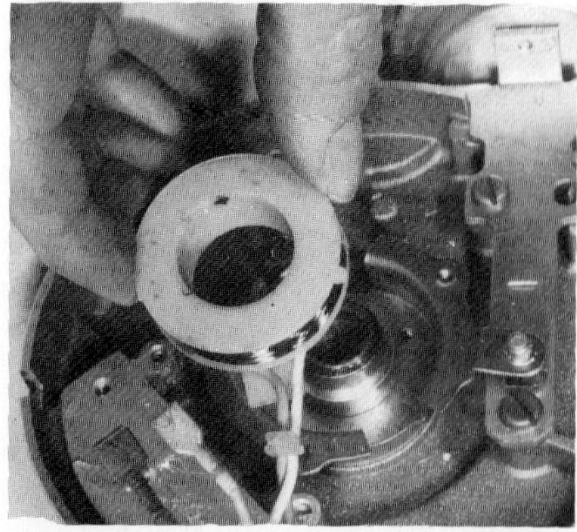

18. Carefully lift the pickup coil out of the retainer. Replace it if resistance is out of limits or if shorted to distributor body.

DELCO 8-CYLINDER HEI DISTRIBUTOR OVERHAUL (*Continued*)

19. This small wave washer holds the pickup coil retainer to the upper bushing. Carefully pry it from its slot.

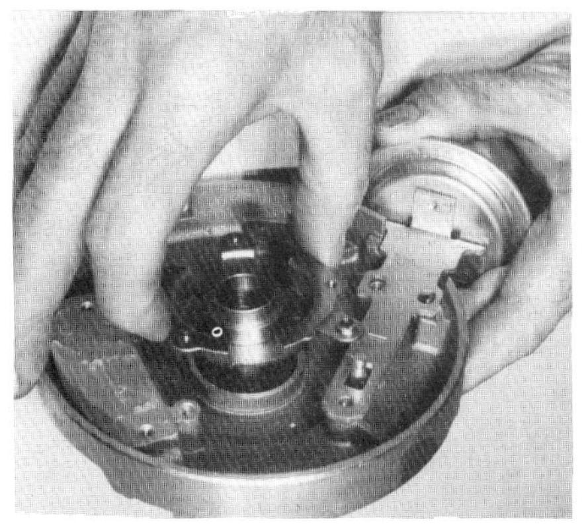

20. Slide the pickup coil retainer off the vacuum link and lift it from the distributor. Remove the vacuum advance unit at this time by removing two screws.

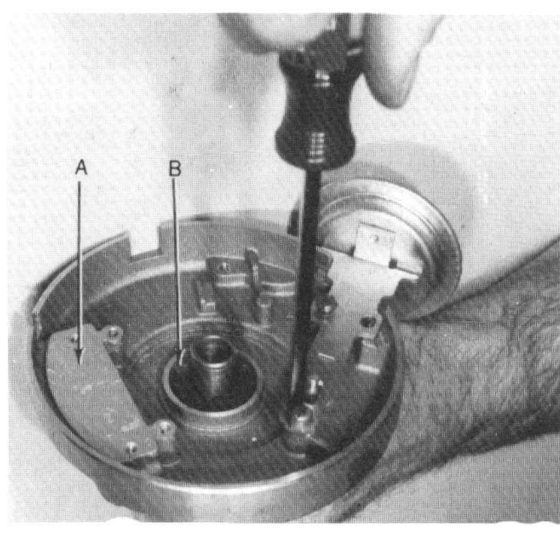

21. Put silicone grease on the module mounting base (A). Lubricate or replace the felt washer (B). Reinstall the vacuum advance unit.

22. Reinstall the pickup retainer and its wave washer. Be sure the vacuum unit works freely. Install the pickup coil and rubber gasket, then the pole piece and three screws.

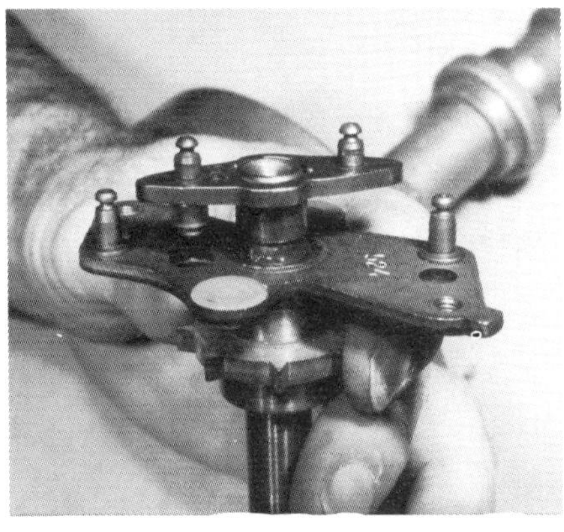

23. Apply a thin film of grease to the upper end of the shaft. Then reassemble it to the trigger wheel and weight base.

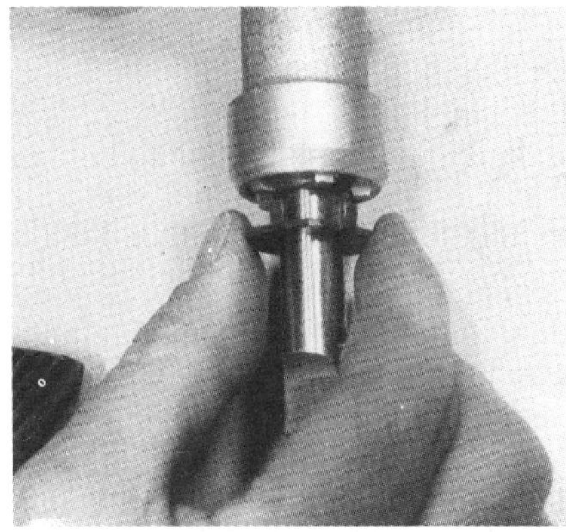

24. Carefully slide the shaft assembly into the housing without hitting the bushings. Put the tanged washer, the shim, and the gear on the shaft.

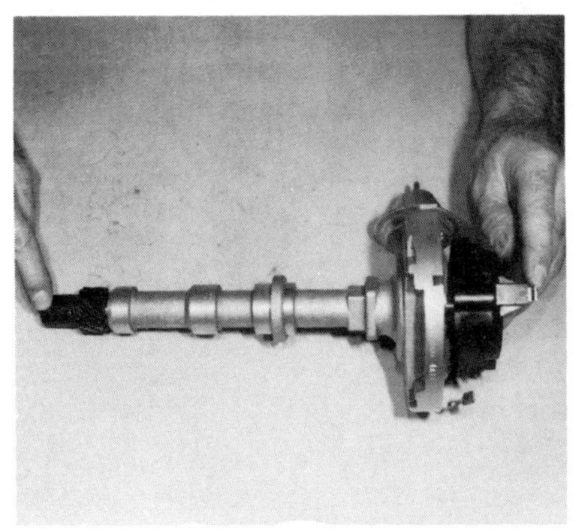

25. Set the rotor in position on the weight base and align its tip with the dimple on the drive gear. Then drive the rollpin through the gear and shaft.

26. Install the weights in the original positions. Then install the weight retainer and assemble the springs in the correct positions. Lightly lubricate the weights.

27. Connect the RFI capacitor to the primary lead. Slide the grommet into the housing slot and install the capacitor screw.

DELCO 8-CYLINDER HEI DISTRIBUTOR OVERHAUL (*Continued*)

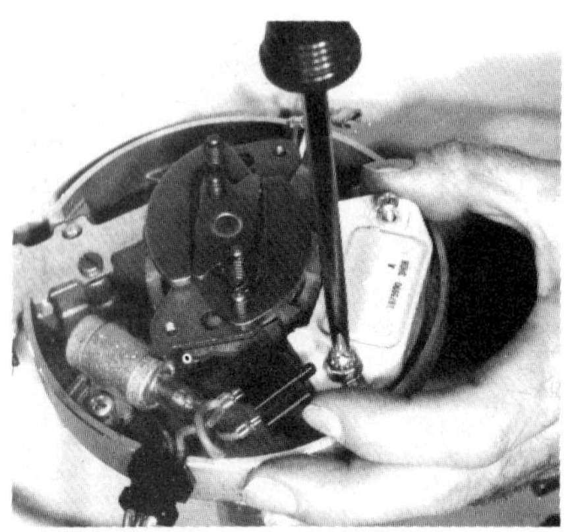

28. Apply silicone grease to the bottom of the module and then install it with two screws. Connect the primary lead to one end and the pickup coil leads to the other.

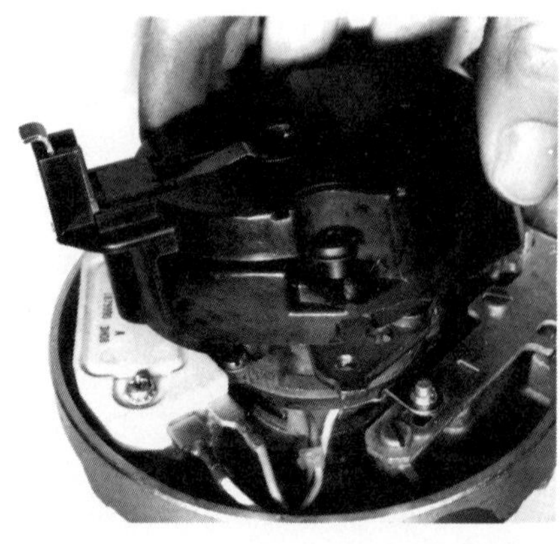

29. Align the locating lugs on the bottom of the rotor with the holes in the weight base. Install the rotor and secure it with two screws.

30. Install the carbon button, spring, and rubber seal in the cap. Install the coil and secure it with four screws. Be sure the ground connection (A) is clean and tight.

MOTORCRAFT TFI DISTRIBUTOR

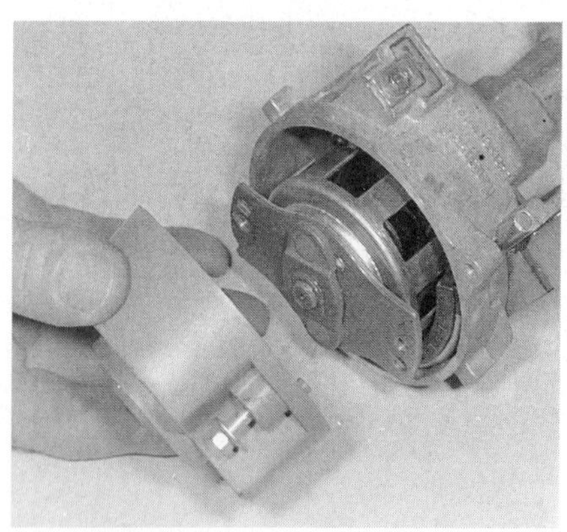

1. The rotor attaches to the distributor shaft with two screws. Square and round locating pegs on the rotor are used to align the rotor. Remove the screws and rotor.

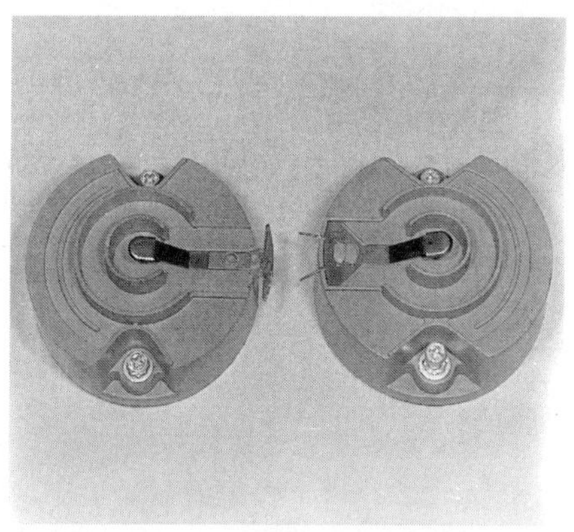

2. Replace wire-terminal (cat whisker) rotors (right) with the blade type (left). Be sure to coat the blade edge with silicone grease.

3. Two screws hold the TFI module to the distributor base. Remove the screws and carefully work the module free to avoid terminal damage.

MOTORCRAFT TFI DISTRIBUTOR (*Continued*)

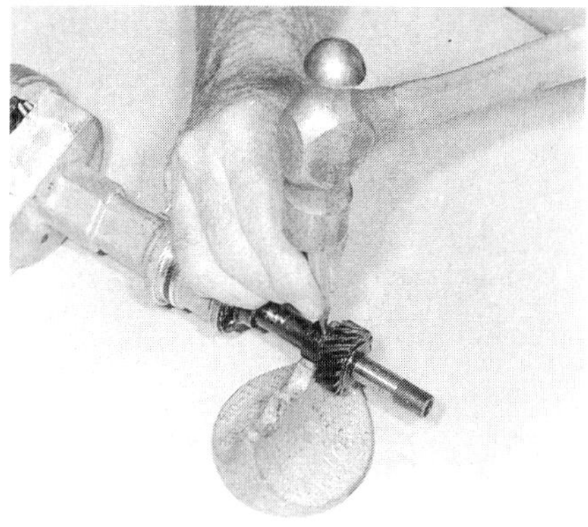

4. Further disassembly requires removal of the drive gear rollpin. Use the proper tools to avoid gear or shaft damage.

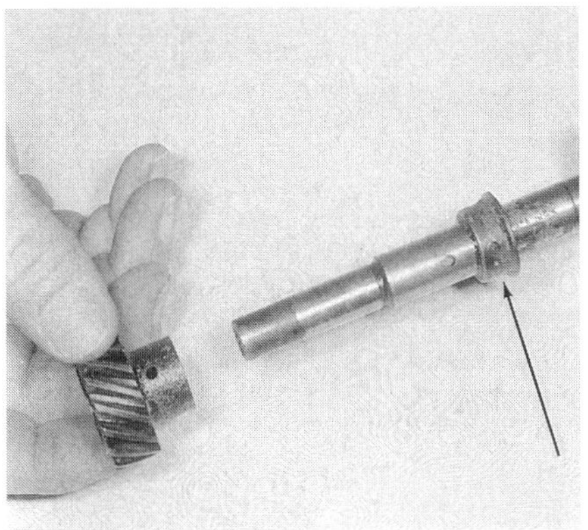

5. When the rollpin is removed, the drive gear will slide off the end of the shaft. Remove the second rollpin (arrow) from the bushing.

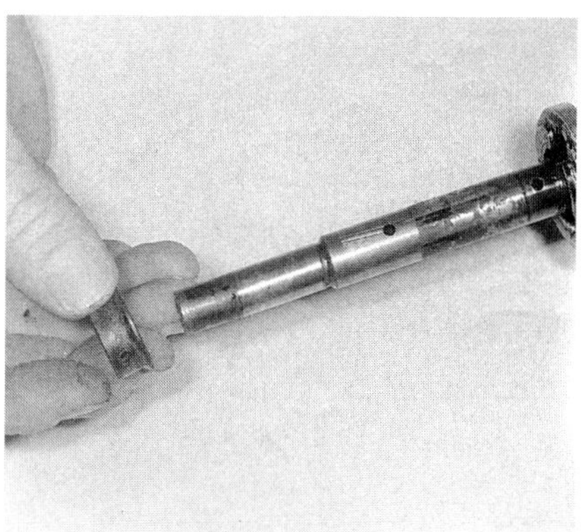

6. The bushing rollpin hole is offset 90 degrees from the drive gear rollpin hole. Slide the bushing off the shaft.

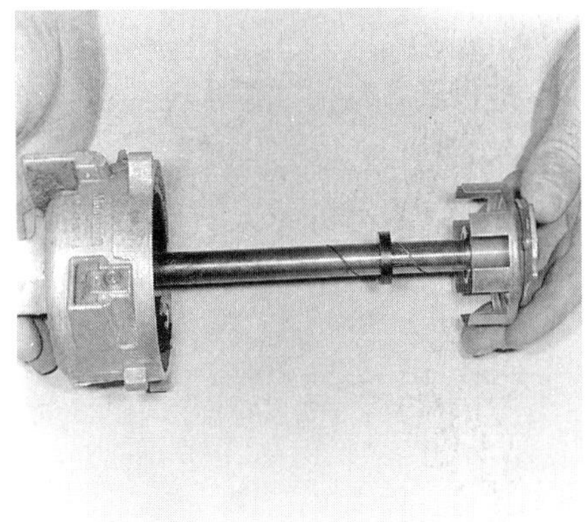

7. Deburr the shaft with crocus cloth or a fine file to prevent damage to the housing bushings when removing the shaft. Then carefully slide the shaft from the housing.

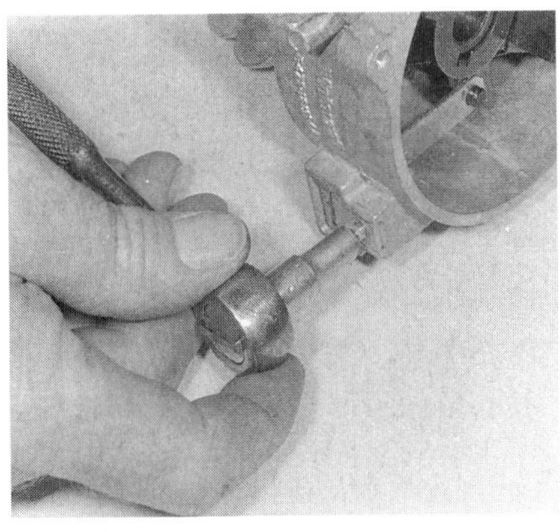

8. Remove the screw holding the octane selector rod in place against the distributor housing.

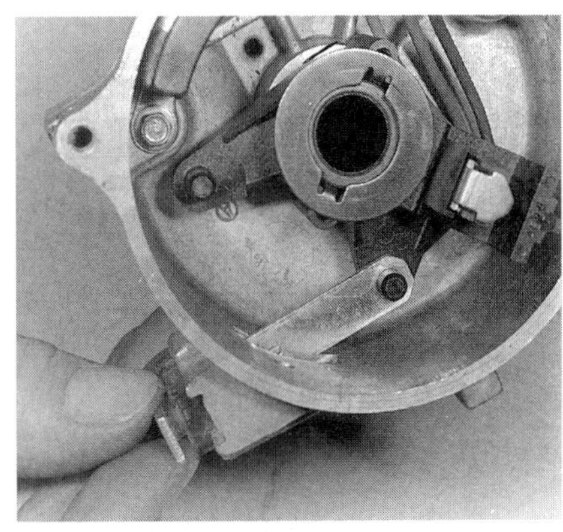

9. Disengage the octane selector rod from the Hall-effect pickup assembly by tilting the rod until it pops off the pickup stud.

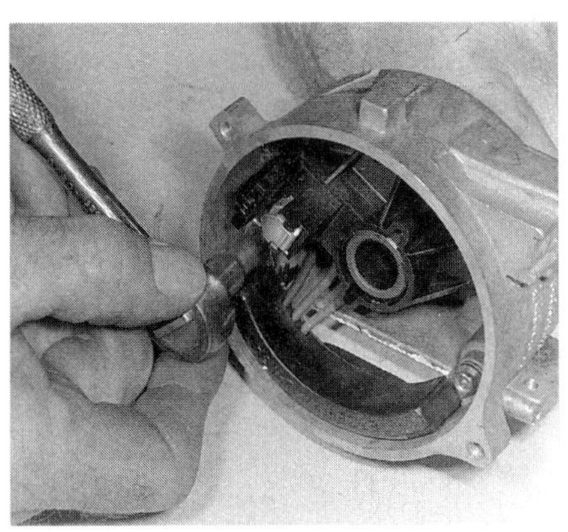

10. Remove the two capscrews holding the Hall-effect pickup assembly to the base of the distributor housing.

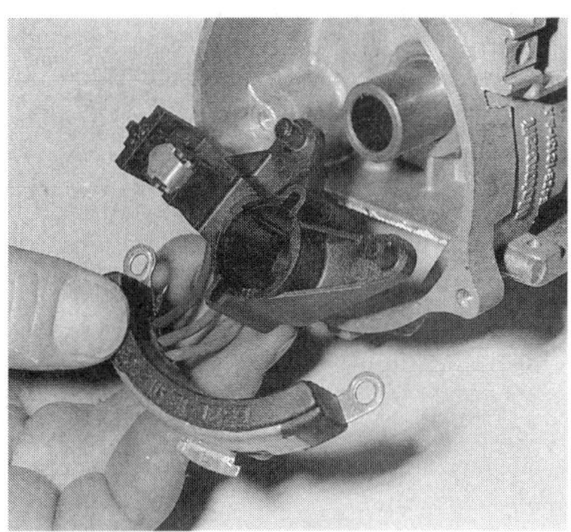

11. Remove the Hall-effect pickup assembly. Do not try to disassemble the pickup. It is serviced as an assembly.

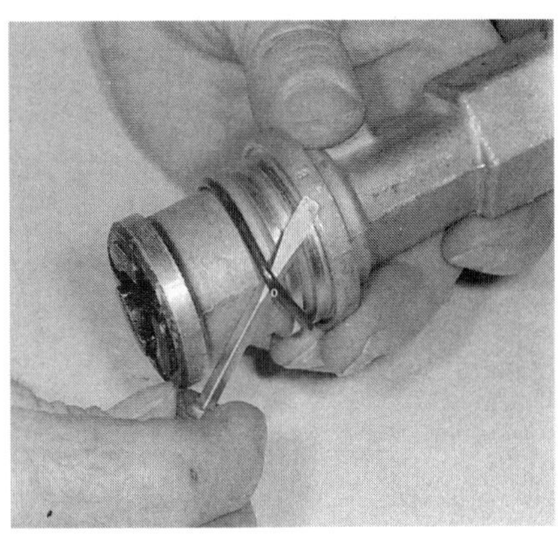

12. Remove the housing O-ring and install a new one. Reassemble the distributor by reversing the disassembly steps.

DELCO-REMY 4-CYLINDER HEI-EST DISTRIBUTOR

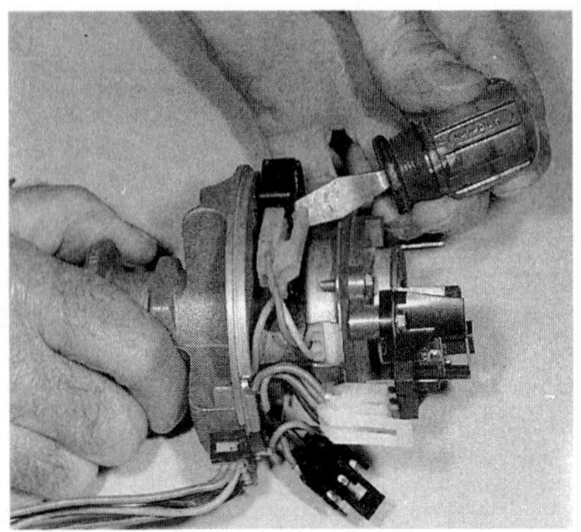

1. With the cap and rotor removed, unclip and disconnect all electrical connections at the module and Hall-effect switch.

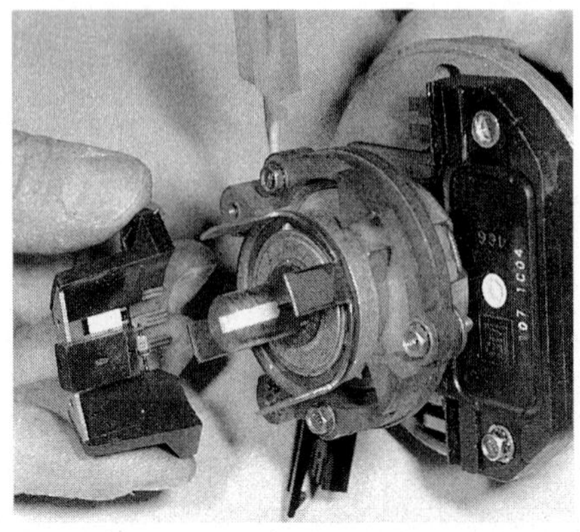

2. Remove the screws holding the Hall-effect switch to the pickup coil assembly. Remove the Hall-effect switch.

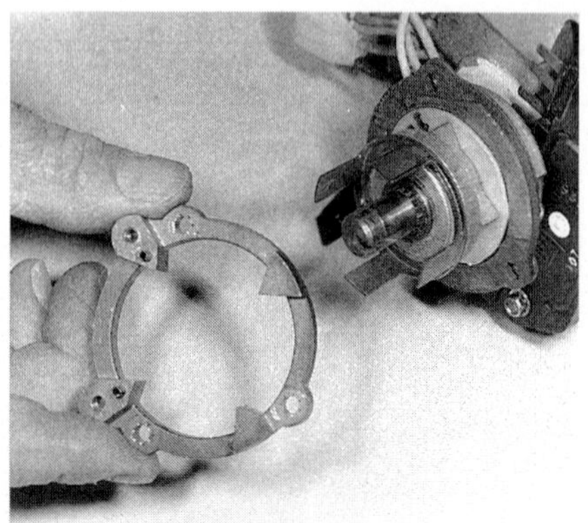

3. Three screws hold the pole piece and magnetic pickup to the distributor base. Remove the pole piece.

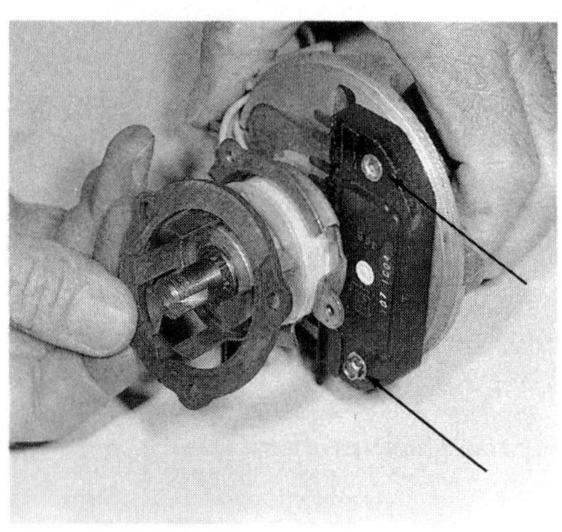

4. Remove the magnet from the distributor base. Remove the two screws (arrows) holding the HEI-EST module to the distributor base.

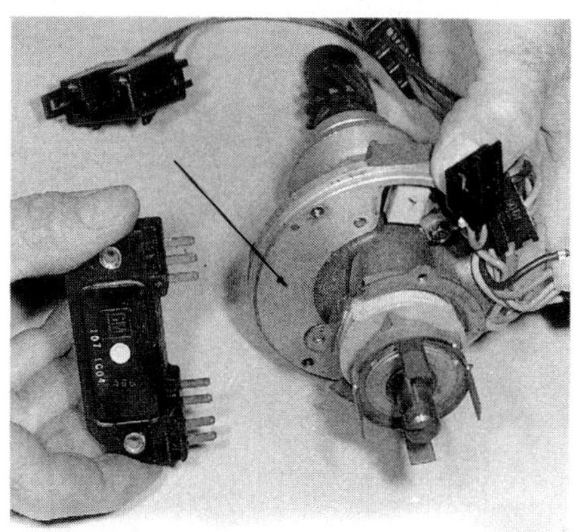

5. Remove the HEI-EST module from the distributor base. If a new module is being installed, clean the silicone grease from the mounting pad on the base (arrow).

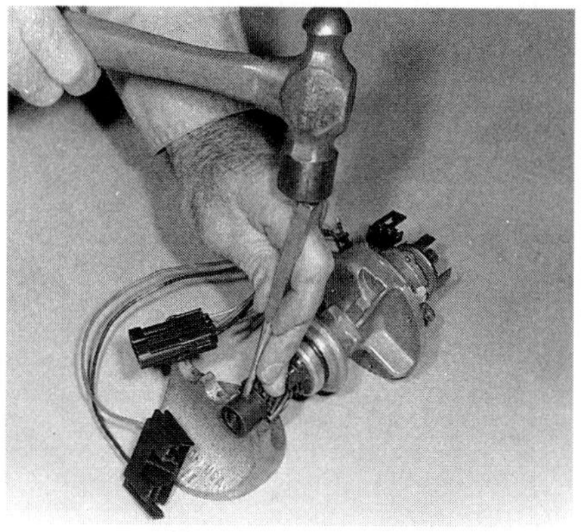

6. Further disassembly requires removal of the distributor shaft drive gear. Use the proper tools to remove the rollpin to prevent shaft or gear damage.

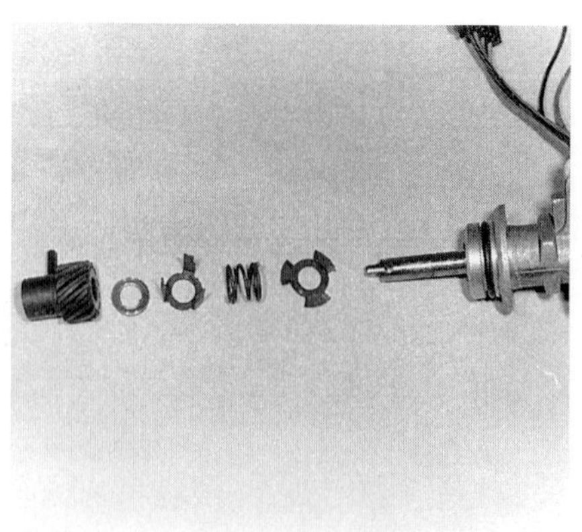

7. With the rollpin removed, slide the drive gear and the components shown from the distributor shaft. Reinstall all in the order shown.

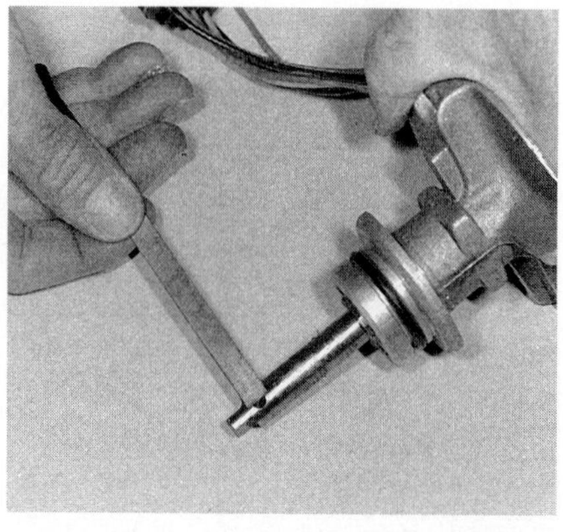

8. Deburr the shaft with crocus cloth or a fine file to prevent damage to the housing bushings when removing the shaft.

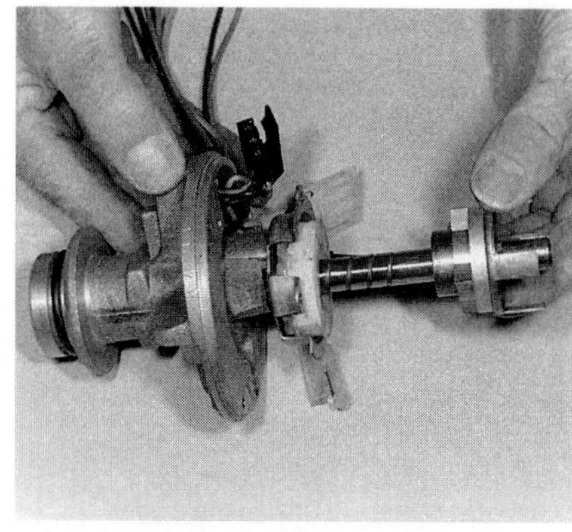

9. Carefully pull the shaft with the Hall-effect shutter and trigger wheel from the distributor housing.

DELCO-REMY 4-CYLINDER HEI-EST DISTRIBUTOR (*Continued*)

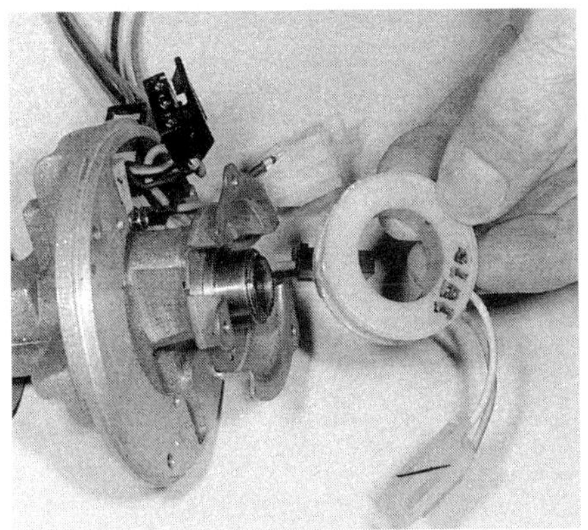

10. Remove the magnetic pulse pickup coil from its cup on the distributor base.

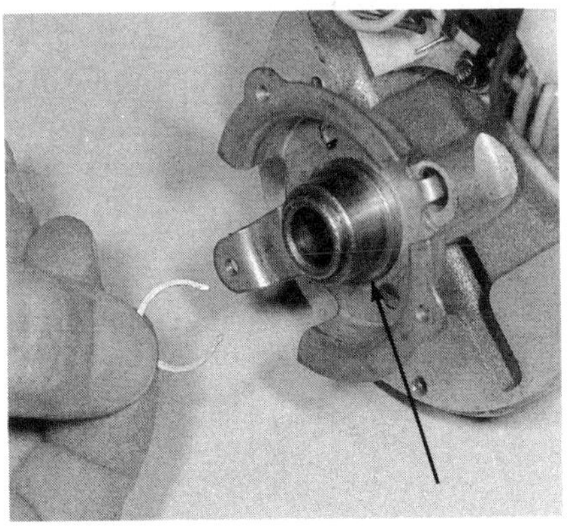

11. To remove the cup from the base, expand and remove this snapring from the cup assembly (arrow).

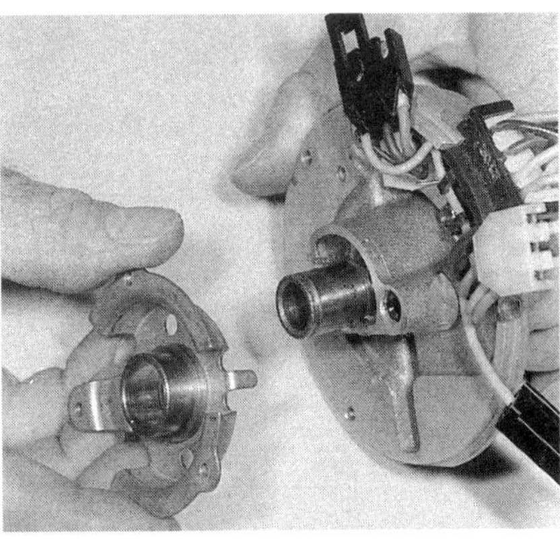

12. Slide the distributor cup off and remove it from the base to expose the lubricant chamber.

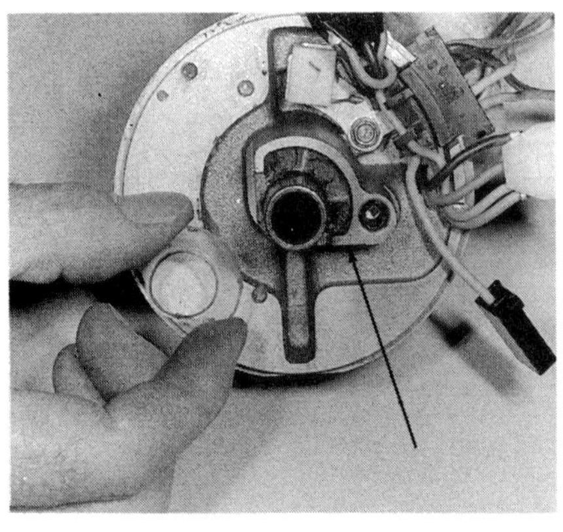

13. Carefully pry this small plastic cover off the lubricant chamber (arrow). Clean out old lubricant, fill the cavity with fresh lubricant, and reinstall the cover.

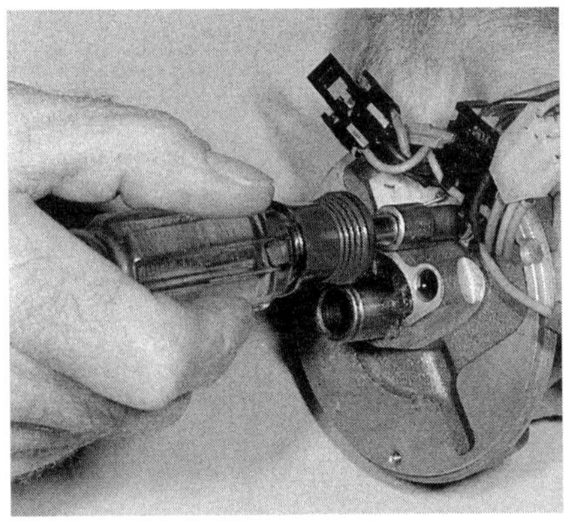

14. If the wiring harness assembly requires removal, remove the capscrew holding it to the distributor base.

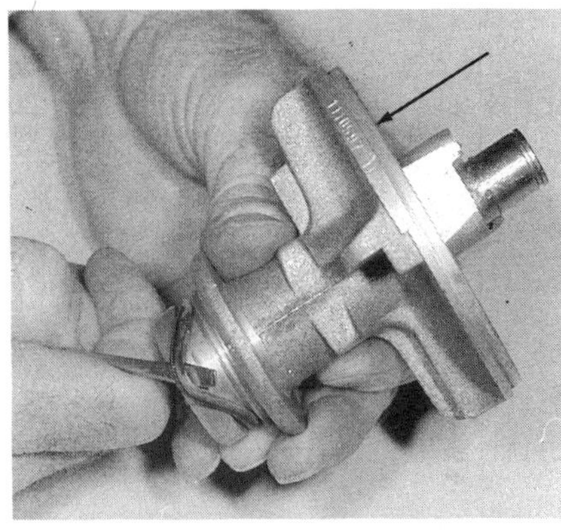

15. Remove the housing O-ring and install a new one. Note the distributor part number stamped on the edge of the housing (arrow).

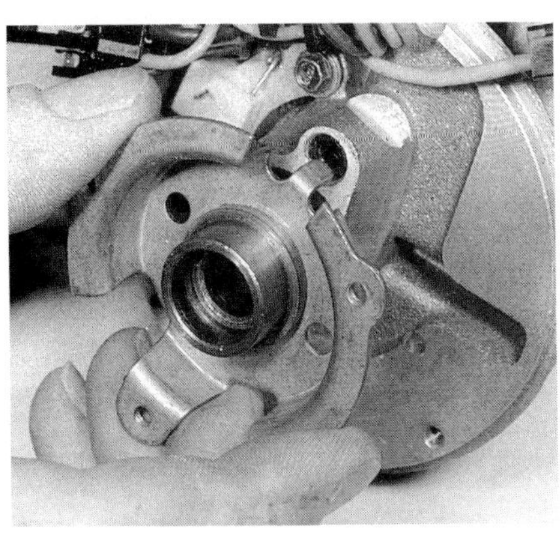

16. When reinstalling the distributor cup to the housing, make sure the tang engages the hole as shown. Then reinstall the snapring.

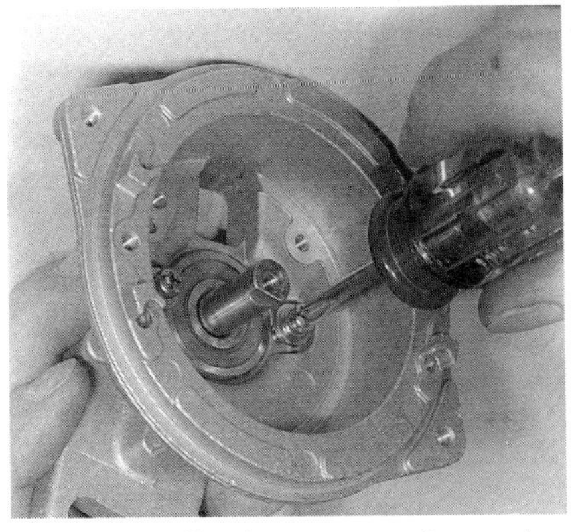

17. Reinstall the magnetic pulse pickup coil. Then carefully slide the distributor shaft through the housing bushings.

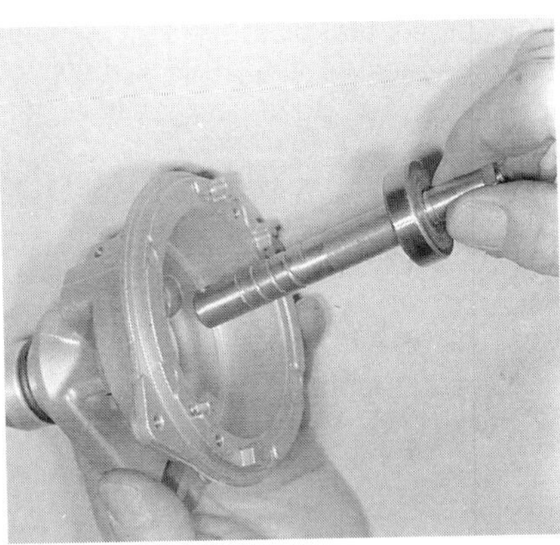

18. Reinstall the distributor shaft components in the order shown in step 7. Then fit the drive gear in place and reinstall the rollpin.

DELCO-REMY 4-CYLINDER HEI-EST DISTRIBUTOR (*Continued*)

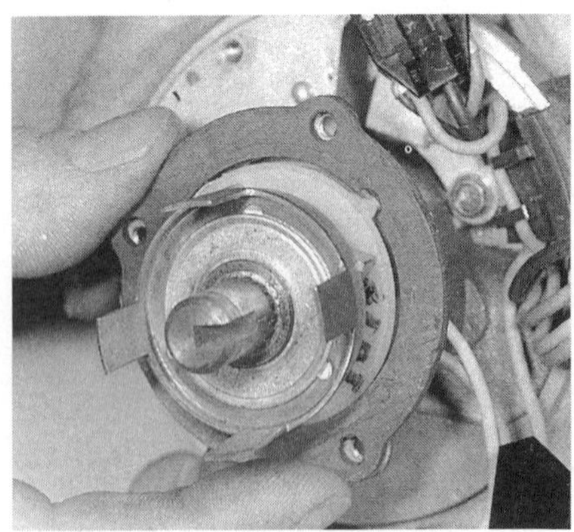

19. Reinstall the magnet and align its mounting holes with those of the distributor cap.

20. Reinstall the pole piece with screws finger-tight. Pole piece must be properly aligned at arrows to produce proper signal and prevent distributor damage. Special alignment tools are available.

21. Wipe the HEI-EST module base and distributor mounting pad with silicone grease before reinstalling the module.

CHRYSLER OPTICAL DISTRIBUTOR

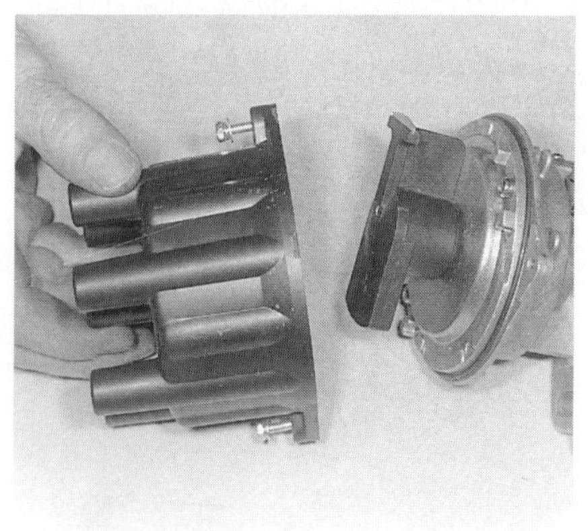

1. Remove the distributor cap. The cap attaches to the base with two screws. Screws do not come out of the cap.

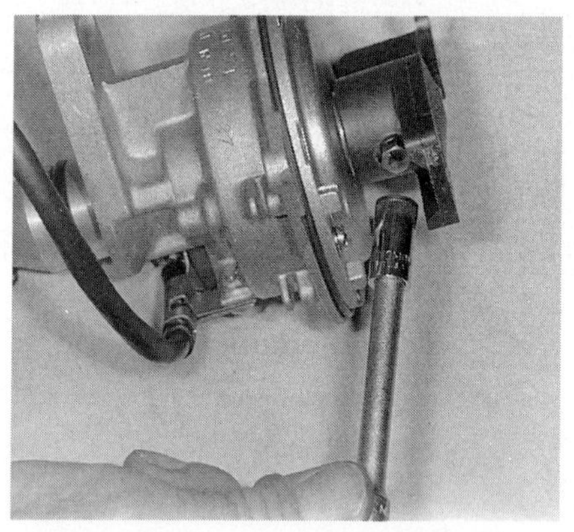

2. Remove the screw holding the rotor to the distributor shaft before trying to remove the rotor.

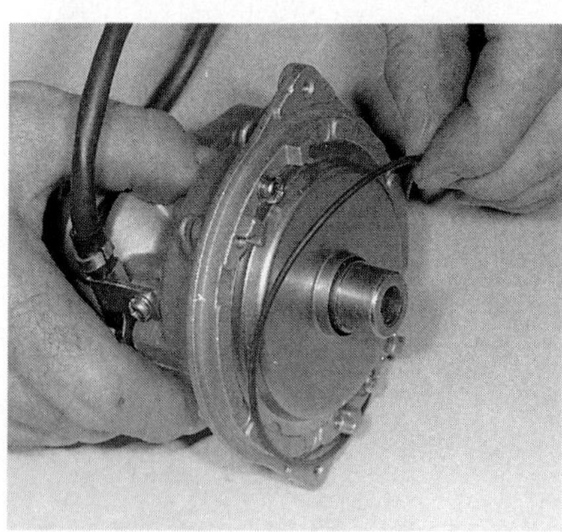

3. A thin O-ring is used as a dust seal between the cap and the housing. Carefully remove the O-ring. It can be reused if not damaged.

CHRYSLER OPTICAL DISTRIBUTOR *(Continued)*

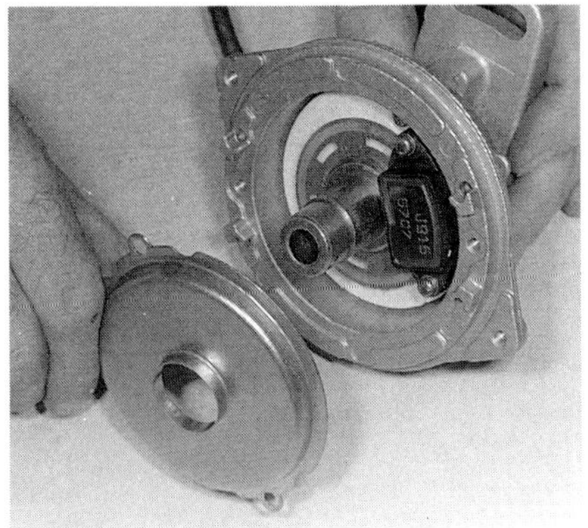

4. Remove the dust cap from the housing. The cap serves two purposes: it keeps out contamination and reduces electromagnetic interference (EMI).

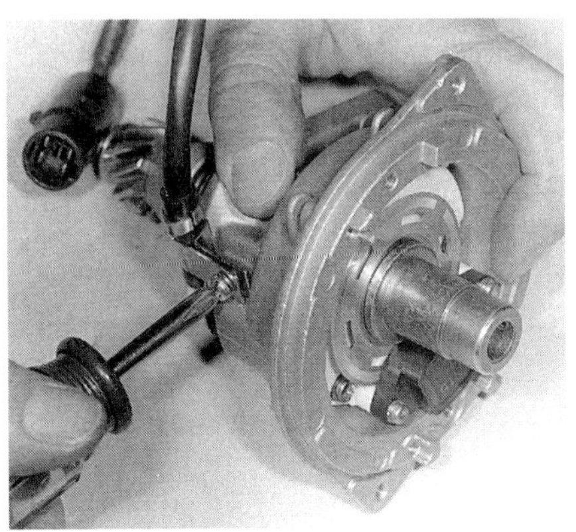

5. Remove the wiring harness screw. Then disconnect the harness from the optical sensing unit.

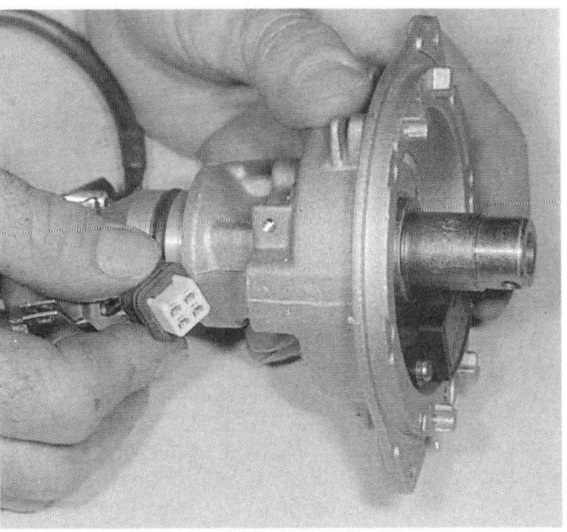

6. Carefully work the harness plug and weatherproof seal from its location in the base of the distributor.

7. Remove the screw in the center of the shaft (A). The optical module screws (B) are sealed with epoxy and should not be removed.

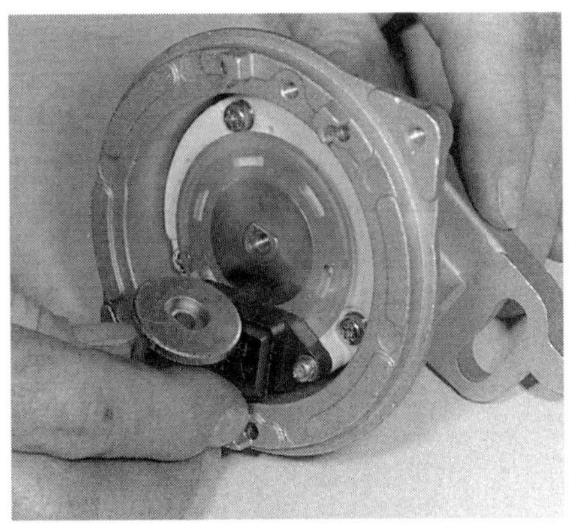

8. After removing the screw, remove the upper bushing to expose the disc and spacer assembly.

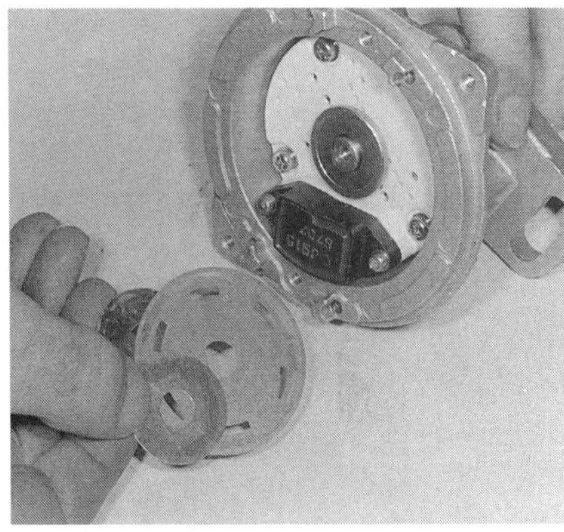

9. Carefully remove the disc and spacer assembly. Note the relationship so you can replace the assembly with the components in the same positions as removed.

10. Remove the lower bushing from the housing. The shaft and bushing both have a flat on one side that must engage for reassembly.

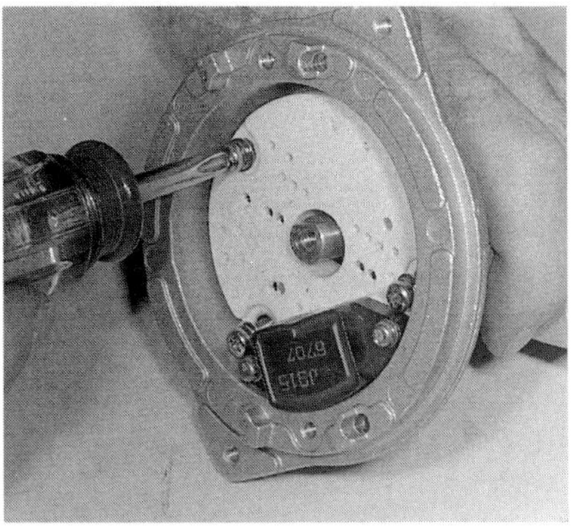

11. Remove the three screws holding the optical module unit to the distributor base. The module unit is serviced as an assembly.

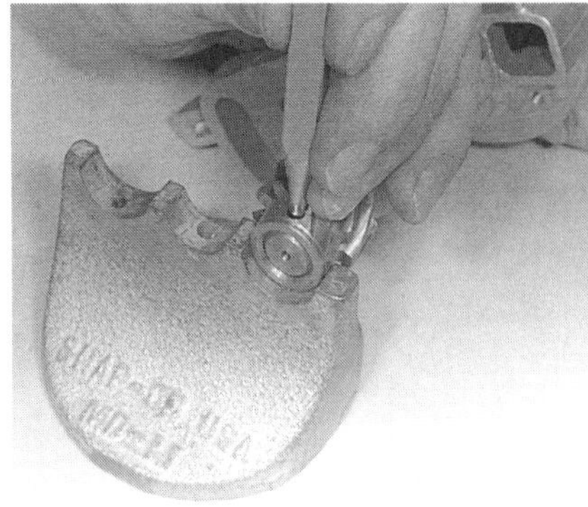

12. To remove the distributor shaft from the housing, use the appropriate tools to remove the drive gear rollpin.

CHRYSLER OPTICAL DISTRIBUTOR *(Continued)*

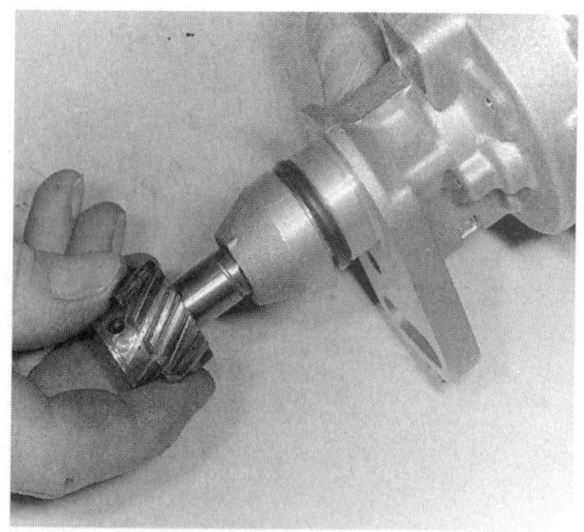

13. Remove the drive gear. It is not necessary to deburr the end of the shaft before removal, because the housing has no bushings.

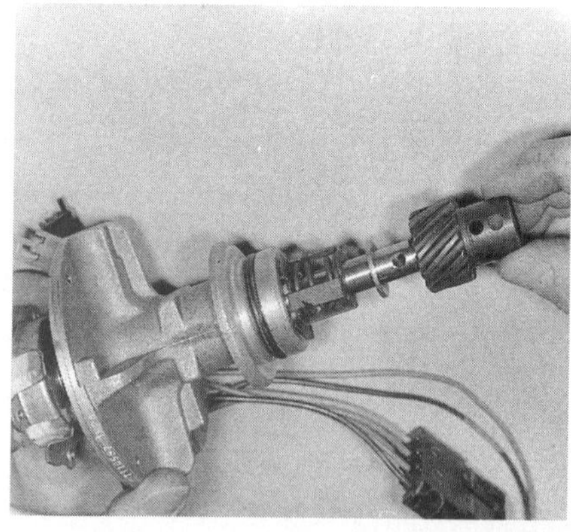

14. Remove the two screws holding the distributor shaft and bearing assembly in the housing.

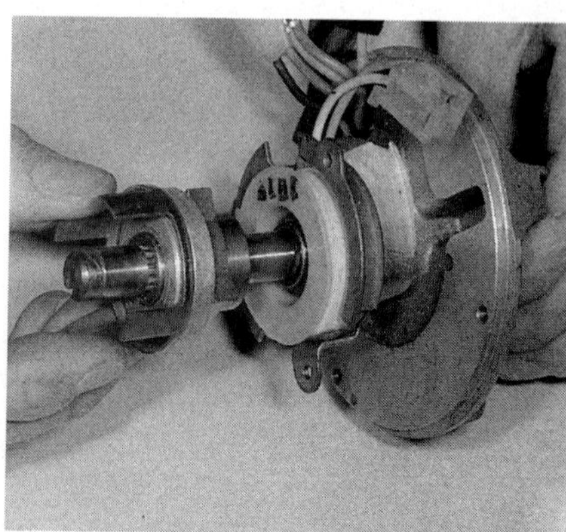

15. Remove the shaft and bearing assembly. Replace the housing O-ring. Reverse disassembly steps to reassemble the distributor.

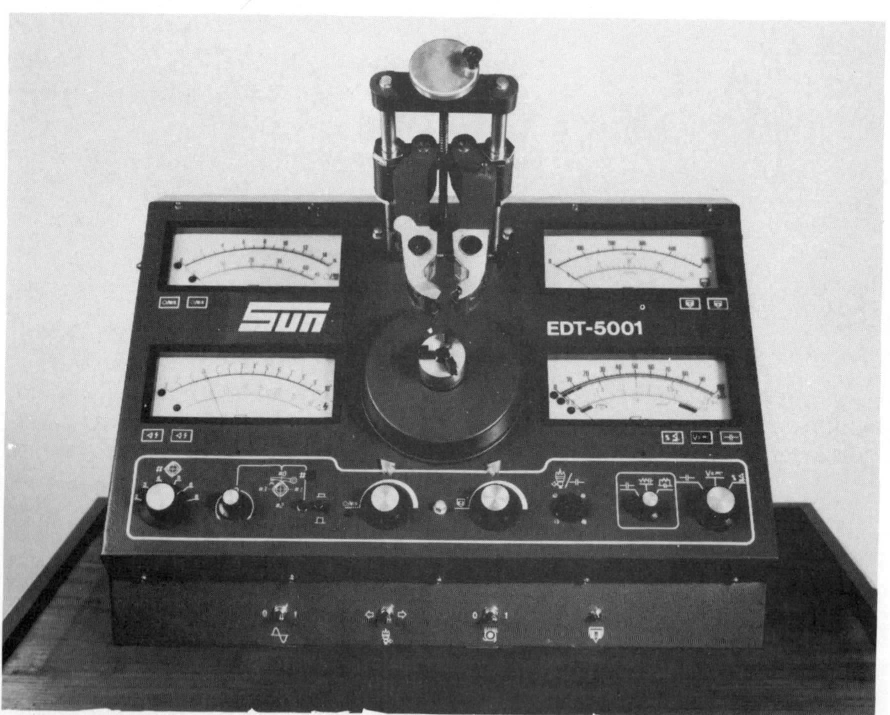

Figure 14-21. This tester allows thorough overhaul and adjustment of any distributor. (Sun)

DISTRIBUTOR SERVICE IN A DISTRIBUTOR TESTER

A distributor tester, figure 14-21, also is called a synchrograph or a synchroscope. Although designed originally to service breaker-point distributors, the tester is a useful instrument to check any distributor for:

- Shaft and bushing wear, and cam wear on a breaker-point distributor
- Centrifugal operation at all points on an advance curve
- Vacuum advance operation at various vacuum levels and speeds
- Dwell angle and dwell variation on a breaker-point distributor
- Breaker-point alignment, bounce, and spring tension

Unfortunately, distributor testers are not as common in shops as they were a decade ago. If you have access to a distributor tester, you can use the following sections as an outline of the tests and adjustments that you can perform by using one of these machines. Refer to the tester operating instructions for detailed procedures.

Tester Controls and Operation

The tester has a reversible, variable-speed electric motor to rotate a distributor clockwise or counterclockwise, as required. You mount the distributor in a fixture, which is adjustable for height, and clamp the shaft or drive gear in the chuck, figure 14-22. The tester contains a dwellmeter, a tachometer, and usually has a vacuum pump and gauge. It also may have an advance meter, an ohmmeter, and a condenser tester or ignition module tester. The tester has a 12-volt power supply or connections for a 12-volt

Figure 14-22. Position the tester clamp to hold the largest machined area on the distributor body. Tighten the chuck.

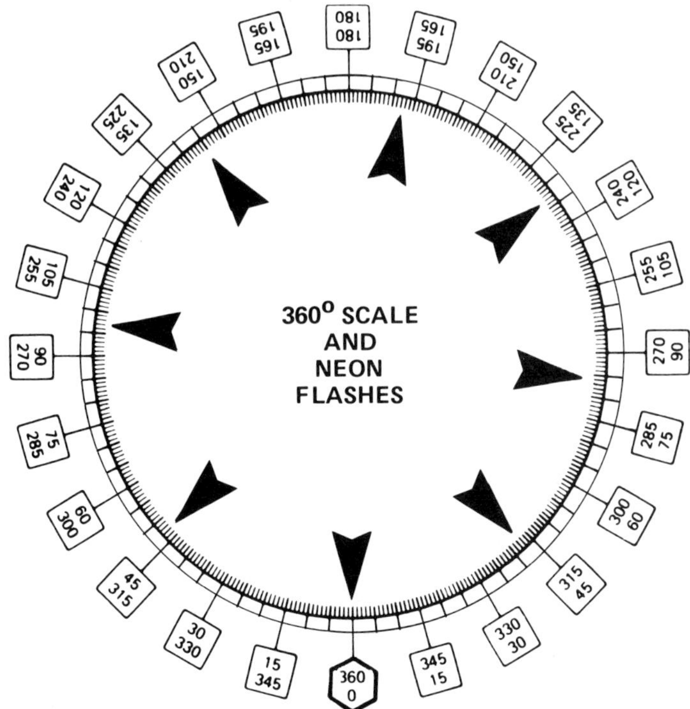

Figure 14-23. Typical distributor tester synchronizing scale. Each neon arrow is a timing signal for one cylinder.

battery to supply primary dc voltage to the distributor.

The unique feature of a distributor tester is the stroboscopic synchronizing scale. This is a ring around the rotating chuck, with each degree marked on a 360-degree scale. Whenever the points open or the pickup coil generates a voltage signal, it triggers an arrow-shaped neon light inside the ring, figure 14-23. A light will flash for each lobe on the distributor cam or each tooth on the trigger wheel during each shaft revolution. The flashes represent ignition for each cylinder in firing-order sequence. At medium and high speeds, the stroboscopic action of the lights makes the arrows appear continuously lit. It is similar to the effect a timing light has on timing marks.

You can rotate the scale ring to align the 0-degree mark with any arrow. For a normal 8-cylinder distributor, the arrows should be equally 45 degrees apart, figure 14-23. They will be 60 degrees apart for an even-firing 6-cylinder engine and 90 degrees apart for a 4-cylinder engine. Spacing varies depending on the ignition intervals of the engine. You will use the arrow positions to check for dwell and timing variation and distributor shaft and cam wear. When you operate the distributor centrifugal and vacuum advance or retard mechanisms, the arrows appear to move around the scale and you can check spark advance to the exact degree.

Distributor Installation in the Tester

You may want to do a complete overhaul, as well as run the distributor in the tester. To remove the distributor shaft, however, you must have the distributor out of the tester. Before mounting the distributor in the tester, clean it with a cloth dampened in solvent. If it is very dirty, you may want to clean it in a solvent tank. If so, remove the vacuum unit and all electronic parts first. Inspect the distributor for the following problems and fix them during service:

- Check the shaft for obvious endplay or sideplay. Rotate the shaft by hand and feel for binding.
- Check the cam lobes for wear and cracks.
- Check the breaker plate or pickup coil plate for looseness or worn bearings.
- Check the primary wiring for worn insulation, broken connectors and other damage or causes of high resistance.

- Check for inadequate lubrication of the shaft, advance weights, and breaker-point cam.
- Look for signs of crossfiring and point pitting that indicate too much lubrication.

Install the distributor in the tester and connect the primary circuit or pickup coil leads according to the tester operating instructions.

Breaker-Point Distributor Service

You can make the following tests and adjustments while servicing a breaker-point distributor in a distributor tester. Usually, it is easiest to do these jobs in this order:

- Install the points and check the alignment of the contact surfaces and the rubbing block on the cam.
- Check and adjust point spring tension.
- Test point resistance and fix high-resistance problems.
- Adjust dwell and check dwell variation.
- Test for point bounce and float.

Check cam, bushing, and shaft wear.

Electronic Distributor Service

Distributor testers were invented for servicing breaker-point distributors, but they are equally useful for testing and adjusting many electronic distributors. Many testers have ohmmeters that you can use to test pickup coil resistance. Often, a pickup will develop an intermittent open or short circuit due to vibration at high speed or movement of the vacuum advance mechanism. You can't check for these intermittent conditions easily with the distributor in the engine, but you can with a distributor tester. Operate the distributor at varying speeds and through the full range of vacuum advance while testing the pickup coil. Also, wiggle the pickup connector wires while checking coil resistance and testing for short circuits.

Some testers also can check ignition module operation similarly to the volt-ohmmeter checks you can make of the vehicle system. These are particularly useful for testing modules built into the distributor, such as Delco HEI and Ford TFI systems.

Obviously, you don't have breaker points to align and adjust or cam lobes to wear in an electronic distributor. Shafts and bushings can wear, however, and affect trigger wheel alignment with the pickup coil. This can cause timing to vary among the cylinders. The preceding tests for shaft and bushing wear on a breaker-point distributor work equally well on an electronic unit.

The following sections contain instructions for testing and adjusting centrifugal and vacuum advance in a distributor tester. Because many electronic distributors have advance mechanisms identical to those in breaker-point distributors, the procedures apply to both kinds of distributors.

Some distributors used with electronic engine control systems simply "distribute" secondary voltage. They have no spark advance mechanisms, and some do not have a signal generator (pickup coil). If the distributor has a signal generator (pickup coil), you can test for shaft and bushing wear. Other tests do not apply.

Testing Spark Advance

You can test the spark advance mechanisms with the distributor in the engine, but it is difficult (often impossible) to adjust them accurately. A distributor tester allows you to test *and adjust* centrifugal and vacuum advance with precision.

Spark advance specifications, or curves, can be given in engine speed and degrees or in distributor speed and degrees. You will need specifications in distributor speed and degrees for use with a tester. If you have them in engine speed and degrees, divide them in half. Vacuum specifications in inches of mercury are the same for use in a tester or on the engine.

Centrifugal specifications begin with a starting speed of zero (or 0° to 1°) advance. Advance should start at this speed, and there must be no advance below the speed. The specifications then list several intermediate speeds with increasing degrees of advance. Maximum advance is usually listed at about 2,000 to 2,200 distributor rpm. There should be no more centrifugal advance above the specified maximum speed.

Vacuum advance specifications are given in various degrees with specific vacuum levels. You check vacuum advance *below* the starting speed for centrifugal advance. This keeps centrifugal advance from being added to vacuum advance and producing false indications. Figure 14-24 is an example of typical centrifugal and vacuum advance specifications for several common distributors.

Centrifugal Advance Test and Adjustment

1. Operate the distributor in the correct direction (clockwise or counterclockwise) at about 200 rpm.

2. Align the zero point on the synchronizing degree ring with one of the flashing arrows, figure 14-25.

3. Slowly accelerate the distributor to the specified starting speed for centrifugal advance.

4. Note the position of the arrow on the degree ring. It should remain at 0 or advance no more than 1 degree.

5. Accelerate the distributor to the speed for the first stage of advance and note the arrow position. All arrows should advance clockwise if the distributor turns counterclockwise, or they should advance counterclockwise if the distributor turns clockwise, figure 14-25.

6. Repeat step 5 for the other points on the advance curve and record the advance reading at each one.

7. After recording maximum advance at the specified maximum speed, increase distributor speed another 200

Distributor Part No.	*Centrifugal Advance in Distributor Degrees at Dist. RPM* Start	Intermediate (1)	Intermediate (2)	Intermediate (3)	Intermediate (4)	Maximum	*Vacuum Advance (Max. Dist. Deg at in. Hg)*
1112163 (Delco)	0-2.5 @ 600	7.0-10 @ 900	8-10 @ 1000	—	—	14-16 @ 2200	10° @ 13-15"
1112958 (Delco)	0-3.0 @ 550	1.5-5.0 @ 600	9-12 @ 1000	—	—	15-18 @ 2300	13° @ 16-18"
D50F-BA (Ford)	0.1.5 @ 500	4-5 @ 750	5-7 @ 1000	6-8 @ 1250	9-10 @ 1500	10-12 @ 2000	6°-8° @ 10"
D70F-AH (Ford)	0-1 @ 500	1-3 @ 750	4-6 @ 1000	5-7 @ 1250	8-9 @ 1500	10-13 @ 2000	8°-9° @ 10"
3666781 (Chrys)	1-3 @ 500	10-12 @ 800	—	11-13 @ 1250	—	14-16 @ 1800	9°-12° @ 15"
3755468 (Chrys)	1-4 @ 650	8-10 @ 800	9-11 @ 1000	12-14 @ 1250	—	14-16 @ 2000	9°-12° @ 15"

Figure 14-24. Typical centrifugal and vacuum advance specifications for use with a distributor tester.

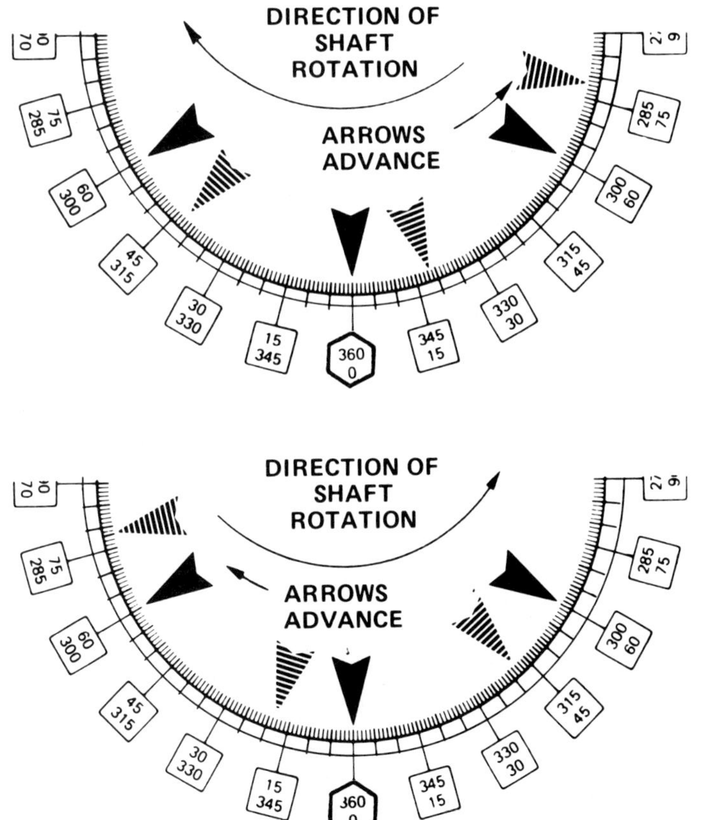

Figure 14-25. Synchronizing arrows advance opposite to distributor rotation.

or 300 rpm. There should be no more advance.

8. Decrease distributor speed in steps to each speed listed in the specifications and again record the advance at each point. It should be the same as you recorded with increasing speed.

Centrifugal advance should be even and constant throughout the distributor speed range at both increasing and decreasing speeds. Here are some symptoms and possible causes of incorrect advance:

- Too much advance, too fast = weak or broken springs
- Too little advance, too slowly = too much spring tension or a sticking or frozen advance weight
- No advance = frozen advance weights
- Erratic advance = sticking advance weights or uneven spring tension
- Too much advance at high speed = a broken stop on the centrifugal mechanism.

Cleaning, lubrication, and adjustment of the springs and weights will cure most problems. If you cannot get the advance within specifications after servicing the distributor, replace the weights and springs or the complete distributor.

You can adjust advance spring tension on many bowl-type distributors by reaching through the breaker plate with a small screwdriver and bending the spring anchors, figure 14-26. Bend them inward to decrease tension and increase advance. Bend them outward to increase tension and decrease advance. On some Chrysler electronic distributors, the springs are anchored on eccentric pins that you must turn with a special tool. Replace the springs to correct advance problems on a Delco distributor (breaker-point or HEI) with the weights on the top of the shaft.

Vacuum Advance Test and Adjustment. Before testing vacuum advance or retard operation, test the vacuum diaphragm for leakage. Apply 15 to 20 in. Hg of vacuum, hold it for 30 to 60 seconds and note the vacuum gauge reading. Decreasing vacuum indicates a leaking diaphragm, which

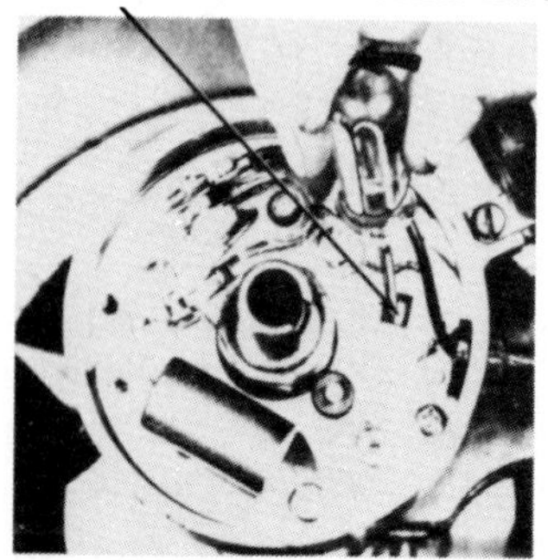

Figure 14-26. Adjust centrifugal advance springs on many distributors by bending the anchors. (Ford)

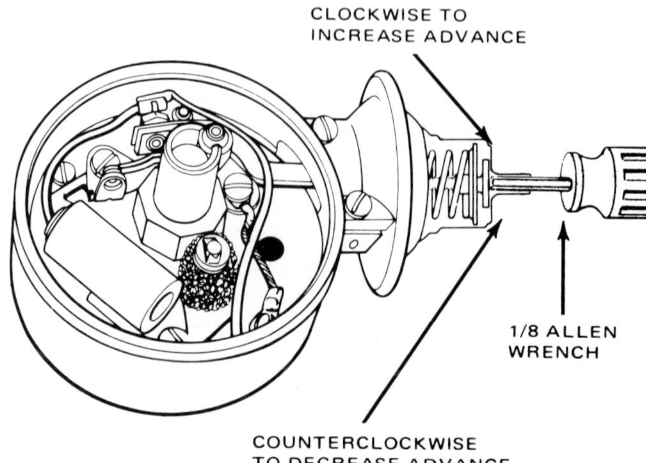

Figure 14-27. Many Ford and other vacuum units can be adjusted with a hexhead (Allen) wrench. (Ford)

must be replaced. Test both diaphragms of a dual-diaphragm distributor.

1. Operate the distributor in the correct direction (clockwise or counterclockwise) at about 200 rpm or a specified speed below the centrifugal advance starting point.

2. Align the zero point on the synchronizing degree ring with one of the flashing arrows, figure 14-26.

3. Connect the tester vacuum line to the distributor and adjust the vacuum pump for 0 in. Hg.

4. Slowly increase vacuum and note the gauge reading when the synchronizing arrows start to advance. Compare it to specifications for the vacuum advance starting point.

5. Increase vacuum to the next specified level and record the advance reading.

6. Repeat step 5 for other points on the vacuum advance curve and record the advance at each one.

7. After recording maximum advance at maximum vacuum, decrease vacuum in steps and again record the advance at each point. It should be the same as you recorded with increasing speed.

8. If the distributor has a vacuum-retard diaphragm, apply the specified amount of vacuum and note the amount of retard shown on the synchronizing scale. If vacuum-retard operation is not correct, replace the vacuum assembly. Retard diaphragms are not adjustable, and you can't get correct vacuum advance with a defective retard diaphragm.

Like centrifugal advance, vacuum advance should be smooth and steady at all points in the specifications. If advance is erratic, you may have a loose or a sticking breaker plate or diaphragm linkage. If advance is steady but out of limits, adjust or replace the diaphragm assembly.

You can adjust some Ford, Chrysler, Prestolite, and Bosch distributors by inserting a 1/8-inch hexhead (Allen) wrench through the vacuum hose nipple and turning an adjustment screw, figure 14-27. Turn the screw clockwise to increase advance, counterclockwise to decrease it.

DISTRIBUTOR LUBRICATION AND CLEANING

Throughout the previous overhaul and adjustment procedures we have referred to cleaning and lubricating parts of the distributor. These are essential steps, but you must do them with care.

In normal operation, dirt, grease, varnish, and rust accumulate inside a distributor. These kinds of contamination can cause crossfiring, incorrect spark advance, shaft and bushing wear, accelerated breaker-point wear, and primary high resistance or short circuits. Lack of lubrication also causes wear on shafts, bushings, cam lobes, points, and advance mechanisms. Here are guidelines for cleaning and lubricating a distributor.

Caution. Do not get solvent inside a vacuum advance unit. It can cause the diaphragm to leak. Do not get solvent on pickup coils or other electronic parts. Solvent residue can cause either high resistance or a short circuit for signal voltage. Remove electronic parts before cleaning a distributor.

Carefully remove grease from a distributor with a little solvent and a small brush. Remove dirt with a clean, lint-free cloth. Wipe off any solvent residue with a dry cloth. Be sure that point contact surfaces and wiring connections are clean and dry.

The most important point about distributor lubrication is *not to overlubricate*. Centrifugal force of the rotating shaft and advance weights can throw grease and oil onto breaker points, electronic connections, and cap and rotor terminals. *After* you install new points and adjust the dwell, put a light film of distributor lubricant on the cam and rubbing block, figure 14-28. Some distributors have wicks to lubricate the cam. If the wick is replaceable, install a new one. If the wick is not replaceable, apply *one drop* of oil. *One small drop* of light oil, such as penetrating oil, on the following locations, figure 14-29, will ensure adequate lubrication:

- The breaker-point pivot pin
- The advance weight pivot pins
- The ends of the advance weight springs
- The breaker plate bushing or pivot point.

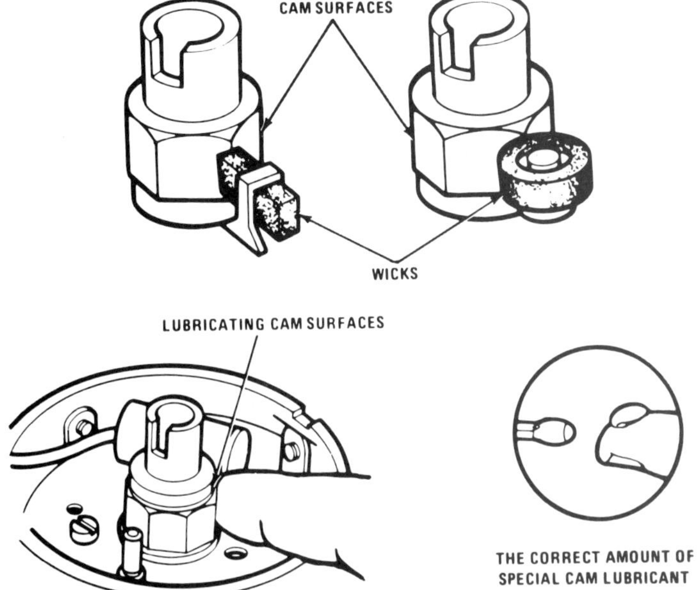

Figure 14-28. Typical distributor cam lubrication methods. (Chrysler)

Many distributors have a felt wick in the top of the rotor sleeve, beneath the rotor. Two or three drops of oil on the wick, figure 14-29, provide lubrication between the sleeve and the distributor shaft.

You may work on an older distributor that has an oil cup or a grease cup for shaft and bushing lubrication, figure 14-30, but most late models do not have these. Nevertheless, when you overhaul a distributor, put a light coating of grease or motor oil on the shaft bushings. Before reinstalling a distributor, apply a few drops of oil to the shaft through an opening or clearance in the drive end.

Many electronic distributors require special lubricants to ensure proper electrical connections or to suppress radiofrequency interference (RFI). Manufacturers require specific silicone lubricants for applications such as the following:

- Many Ford Dura-Spark distributors require silicone grease on the rotor tips and cap terminals for RFI suppression, figure 14-31.
- Delco HEI distributors require silicone grease between the module and its mounting base in the distributor. This grease also aids heat dissipation.
- Ford, AMC, and several other carmakers require that primary connectors at the ignition module and at the

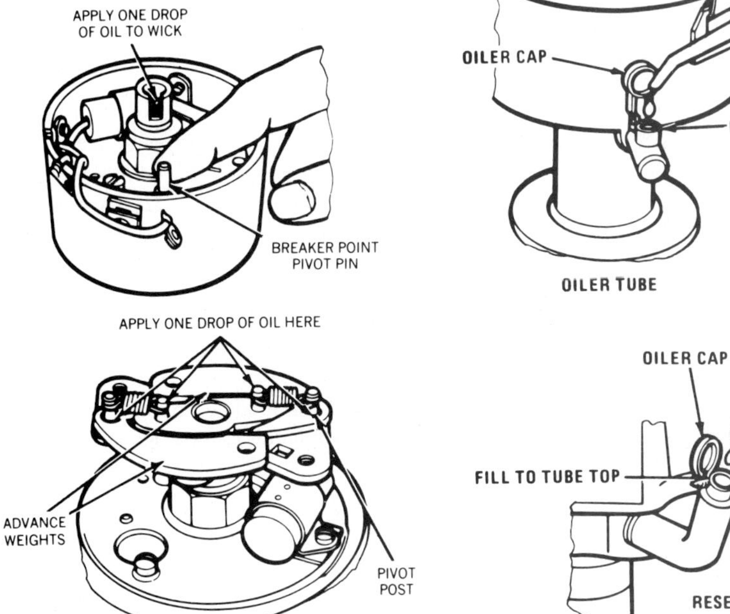

Figure 14-29. Apply a small amount of oil to these points in a distributor. (Chrysler)

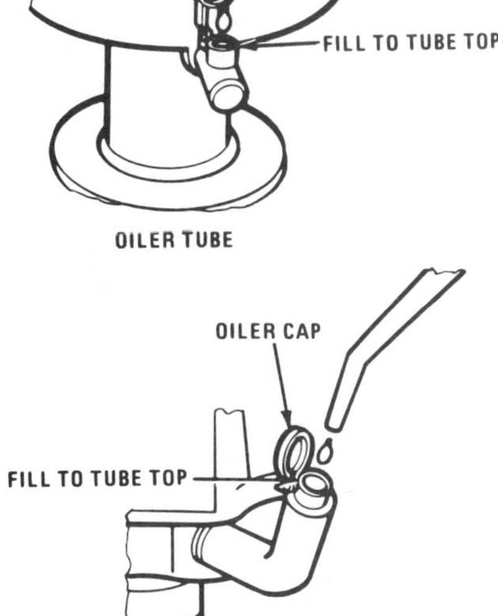

Figure 14-30. Many older distributors have oil cups for lubrication. (Chrysler)

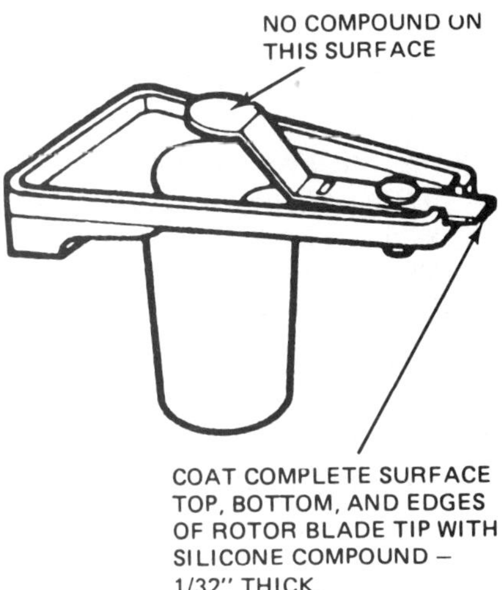

Figure 14-31. These are typical locations for silicone grease on some Ford rotors. (Ford)

distributor be filled with silicone lubricant to keep out moisture and corrosion that can interfere with voltage signals.

DISTRIBUTOR INSTALLATION

After servicing a distributor, reinstall it by following these basic steps. Refer to figures 14-11 through 14-14.

1. Be sure the ignition switch is off or the battery ground cable disconnected.

2. Install the rotor on the shaft and align its tip with your reference mark on the distributor body.

3. Align your other reference mark on the distributor body with the reference mark, or point, on the engine.

4. Be sure that any required gasket or O-ring is in place and carefully insert the distributor into the engine.

5. Engage the distributor drive with the engine camshaft or distributor driveshaft. If the distributor shaft turns the oil pump, be sure it engages with the pump drive.

6. Be sure the distributor is properly seated in the engine and engages with its driveshaft (and the oil pump, if required). Be sure the rotor stays aligned with its reference mark. You may have to:

- **a.** Wiggle the distributor shaft slightly or turn the engine crankshaft by hand or with the starter to engage the distributor.
- **b.** If a helical drive gear on the distributor turns the shaft out of position as you install it, raise the distributor and rotate its shaft backwards 15 to 20 degrees to align the right gear teeth. This is about 1/4-inch of rotor tip movement in relation to the distributor rim.

7. Install the holddown bolt (capscrew) and clamp. Tighten the bolt enough to allow distributor movement with some resistance.

8. Connect the distributor primary lead to the coil or the pickup coil connector to the module harness.

9. Connect the distributor vacuum line, or lines, if so equipped.

10. Install the distributor cap and spark plug cables.

11. Reconnect the battery ground cable if disconnected.

SUMMARY

Distributor service is a critical part of ignition system service. Simple replacement of breaker points and condenser often can be done with the distributor in the engine. Complete distributor service is usually easier and more accurate, however, with the distributor removed.

A distributor tester allows you to replace the points and condenser of a breaker-point distributor and adjust dwell precisely. A tester also allows you to test centrifugal and vacuum advance mechanisms of breaker-point and electronic distributors with equal precision.

Many problems of shaft, bushing, and breaker plate (pickup plate) wear are common to all distibutors and must be corrected for optimum engine performance.

REVIEW QUESTIONS

Multiple Choice

1. Breaker points must be adjusted or replaced at scheduled intervals because:

- **a.** the rubbing block wears and dwell changes
- **b.** the point surfaces wear and can burn
- **c.** the point spring can wear and lose tension over a period of time.
- **d.** all of the above

2. When installing a new set of breaker points, you should make the following checks and adjustments EXCEPT:

- **a.** point spring tension
- **b.** point alignment
- **c.** point bounce
- **d.** point dwell

3. Mechanic A says that high tension on the breaker-point spring will cause point bounce at high speed. Mechanic B says that high spring tension will increase cam and rubbing block wear. Who is right?

- **a.** A only
- **b.** B only
- **c.** both A and B
- **d.** neither A nor B

4. Mechanic A adjusts the gap on a new set of points with a feeler gauge while the rubbing block is on the high point of the distributor cam. Mechanic B adjusts the dwell, using a dwellmeter while cranking the engine. Who is doing the job correctly?

- **a.** A only
- **b.** B only
- **c.** both A and B
- **d.** neither A nor B

5. Mechanic A adjusts dwell on a Delco V-8 breaker-point distributor with a hex-head (Allen) wrench while the engine is idling. Mechanic B checks dwell adjustment before the final ignition timing adjustment. Who is doing the job correctly?

- **a.** A only
- **b.** B only
- **c.** both A and B
- **d.** neither A nor B

6. Before removing a distributor, Mechanic A cranks the engine to align the timing marks at tdc for number 1 cylinder. Mechanic B cranks the engine to align the rotor tip parallel with the engine centerline. Who is doing the job correctly?

- **a.** A only
- **b.** B only
- **c.** both A and B
- **d.** neither A nor B

7. When removing a distributor, an intermediate drive shaft for the oil pump comes out of the engine with the distributor. You should:

a. twist the distributor gently to allow it to drop back into the engine

b. grasp the shaft with pliers or a similar tool, remove it and install a new shaft when you reinstall the distributor

c. grasp the shaft with pliers or a similar tool, remove it, and clean it for reinstallation if it is in good shape

d. carefully reinstall the distributor and do not attempt to remove it

8. You can do all of the following jobs on an electronic distributor, using a distributor tester EXCEPT:

a. check shaft and bushing wear

b. check dwell

c. check and adjust centrifugal advance

d. check and adjust vacuum advance

9. Before removing a distributor from an engine, it is good practice (for safety and other reasons) to:

a. remove an electric fuel pump fuse

b. disconnect the coil battery voltage (BAT) feed wire

c. disconnect the battery ground cable

d. disconnect the ignition module from the engine wiring harness

10. Excessive dwell variation among cylinders can be caused by any of the following EXCEPT:

a. high resistance in the ignition pickup coil

b. a loose breaker plate

c. a worn distributor shaft or bushings

d. worn distributor cam lobes

11. To test distributor spark advance in a distributor tester, Mechanic A says that you need centrifugal advance specifications in distributor degrees at distributor rpm. Mechanic B says that you need vacuum advance specifications in distributor degrees at various vacuum levels, with no regard to distributor speed. Who is right?

a. A only

b. B only

c. both A and B

d. neither A nor B

12. Before testing distributor vacuum advance in a distributor tester, you should check

a. initial timing

b. centrifugal advance

c. vacuum diaphragm leakage

d. all of the above

13. A distributor may require silicone grease at any of the following points EXCEPT:

a. between a distributor-mounted module and its base in the distributor

b. on the rotor tip and cap electrodes (terminals)

c. on primary connectors for the pickup coil

d. on the trigger wheel teeth

14. Mechanic A says that when you reinstall a distributor, you may have to rotate the shaft to allow for the angle of helical teeth on the drive gear. Mechanic B says that you cannot adjust static timing on an electronic distributor. Who is right?

a. A only

b. B only

c. both A and B

d. neither A nor B

15 IGNITION TIMING

INTRODUCTION

In the preceding three chapters, you have learned the principles of testing and servicing the ignition primary and secondary circuits and their parts. You have also learned the importance of distributor service and how to do it accurately. This chapter covers the final steps of complete ignition service: ignition timing and spark advance testing in the engine.

The chapter begins with basic instructions for distributor installation and static timing adjustment so that an engine will start and run. It continues with procedures for checking and adjusting initial timing and testing spark advance control systems.

Timing adjustment is the final step of complete ignition service and *one* of the last steps of comprehensive engine service. You may, however, check timing and spark advance at several stages during engine performance troubleshooting. A timing check may be an early step of an overall engine test to isolate the cause of a driver's complaint. You learned that when you check and adjust breaker-point dwell on the engine, you also must readjust timing. When you studied distributor service, you learned to test and adjust centrifugal and vacuum advance mechanisms in a distributor tester.

Centrifugal and vacuum advance mechanisms may operate perfectly in a distributor tester, but ignition timing can be incorrect if the basic setting is wrong or if vacuum advance control devices on an engine do not work right. You will learn to test and adjust these systems in this chapter.

There are three critical engine-operating factors for best performance, economy, and emission control:

1. Engine mechanical condition
2. Air-fuel ratio
3. Ignition timing.

Engineers may argue about which of the three is the most critical, but from a service standpoint, you can consider them equal. Ignition timing from the initial (basic) setting through the full range of advance for different speeds and loads has a major effect on power, fuel consumption, and exhaust emissions. Timing must be correct at every point in an engine's operating range.

GOALS

After studying this chapter, you should be able to do the following jobs in twice the flat-rate labor time or less, using the carmaker's procedures.

1. Install a breaker-point or an electronic distributor and adjust the static timing so the engine will start and run.
2. Check and adjust basic ignition timing, using a timing light or a magnetic timing indicator.
3. Test centrifugal, vacuum, and electronic spark advance according to manufacturer's specifications with the distributor in the engine.

STATIC TIMING ADJUSTMENT

You can adjust ignition timing statically (with the engine off) or dynamically (with the engine running). You will probably use the static timing method only when you install a distributor for a basic setting so the engine will start. Chapter 14 ended with instructions for distributor installation. The following procedures for static timing continue from that point.

When you reinstall a distributor in an engine, the engine should start and run if it was not cranked while the distributor was out and if you align your reference marks correctly. If the engine was cranked while the distributor was out or if you are installing a distributor in a rebuilt engine, you must set the static timing. The first steps are the same for breaker-point and electronic distributors, but final adjustment steps vary slightly.

Distributor Alignment

The following steps are based on aligning the distributor at top dead center (tdc) for the number 1 cylinder. For routine service, you can install the distributor, using the reference marks you made when you removed it. These may not be at tdc for the number 1 cylinder, but timing should be close enough to let you start and run the engine. If timing is off, or if you install a distributor without reference marks made at removal, use the following procedure to align the distributor at tdc for number 1 cylinder. Then adjust static timing.

1. Rotate the engine to bring the number 1 piston to top dead center on the compression stroke. This is easiest with the spark plugs removed to relieve compression. If the engine is timed to any other cylinder (number 8 or number 4), use that cylinder for static timing.

2. Align the timing marks at tdc or the specified basic timing setting, figure 15-1.

3. Put the rotor on the distributor shaft and align the electrode tip with a reference line on the body that corresponds to the number 1 spark plug terminal.

4. Install the distributor in the engine so that the rotor points to the number 1 cylinder reference line, figure 15-2.

5. Be sure the distributor is correctly installed and seated as explained in chapter 14. Check a repair manual drawing that shows vacuum diaphragm and distributor cap clip positions in relation to the engine. Distributor position is critical on many engines to allow enough movement to adjust timing.

6. Install the distributor holddown bolt (capscrew) and clamp. Do not tighten the bolt.

7. Rotate the distributor body in the direction of rotor rotation so that the breaker points are closed or so that the pickup coil pole piece is aligned with one tooth of the trigger wheel, figure 15-3.

8. Continue with final static timing adjustment in accordance with the following procedures for breaker-point or electronic distributors.

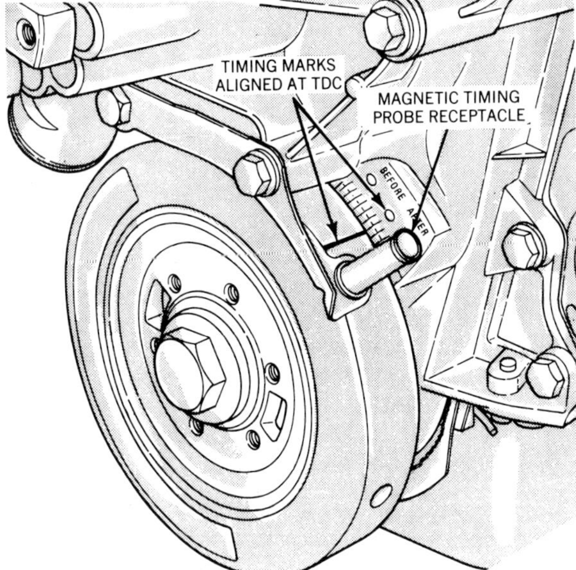

Figure 15-1. Align the engine timing marks to set static timing. (Chrysler)

Breaker-Point Static Timing

1. Connect one lead of a 12-volt test lamp to the distributor terminal on the coil (usually −) or to the distributor primary terminal. Connect the other test lamp lead to ground, figure 15-4.

2. Turn the ignition switch on and watch the test lamp. It should not light if the points are closed.

3. Carefully turn the distributor body opposite to rotor rotation while watch-

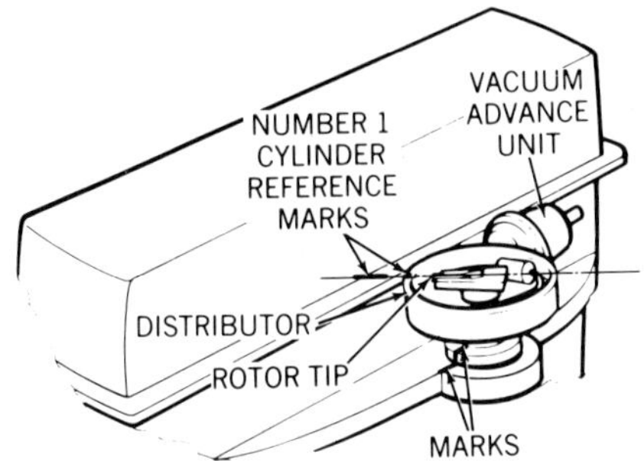

Figure 15-2. Use your alignment marks on the distributor rim and the body for distributor installation and static timing. (Chrysler)

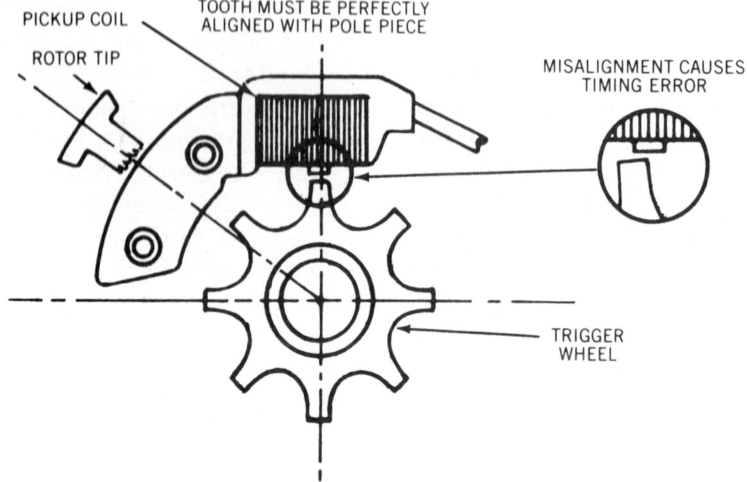

Figure 15-3. Align a trigger wheel tooth with the pickup coil to set static timing for an electronic distributor. (Ford)

ROTATE SLOWLY, OPPOSITE TO ROTOR ROTATION TO SET STATIC TIMING

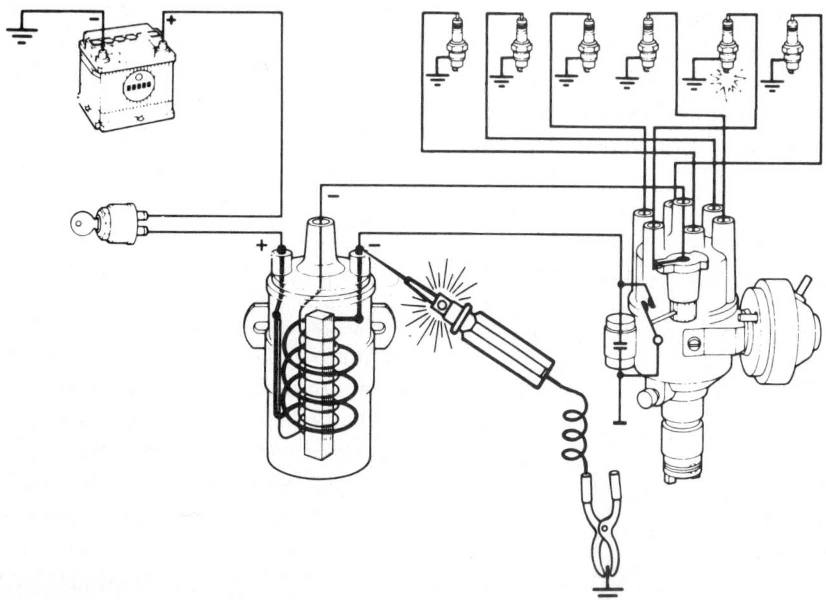

Figure 15-4. Connect a 12-volt test lamp across the distributor to set static timing for a breaker-point ignition. (Bosch)

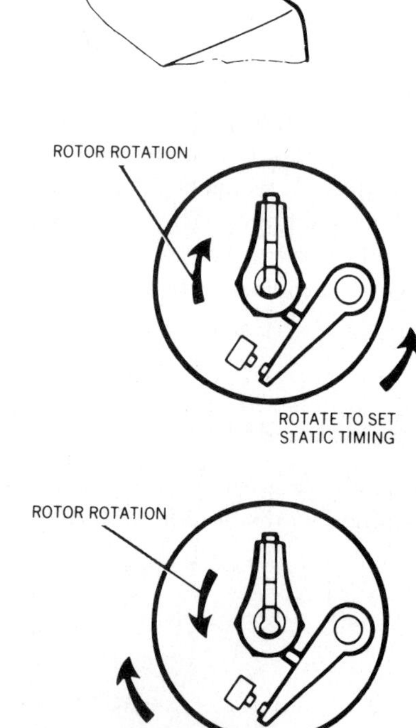

Figure 15-5. Turn the distributor opposite to rotor rotation until the lamp lights. (Chrysler)

ing the lamp, figure 15-5. When the lamp lights, the points have opened. This is your static timing setting.

4. Tighten the holddown bolt and clamp. Timing should be close enough for the engine to start and run.

5. Check and adjust timing at the carmaker's specified speed with a timing light or magnetic timing meter.

If you don't want to use a test lamp for static timing of a breaker-point distributor, turn the ignition on with the points closed. Slowly turn the distributor opposite from rotor rotation until an arc jumps across the point contact surfaces. Tighten the holddown bolt.

Electronic Distributor Static Timing

After aligning the trigger wheel and the pickup coil, tighten the distributor holddown bolt and clamp. Often this will set the timing close enough to start the engine. If not, proceed as follows.

1. Turn the ignition switch off and install the distributor rotor, cap, and plug cables. Be sure all connections are secure.

2. Loosen the distributor holddown bolt and turn the distributor slightly in the direction of rotor rotation.

3. Connect an inductive timing light to the number 1 spark plug cable.

4. Turn the ignition switch on.

5. Carefully turn the distributor body opposite to rotor rotation until the tim-

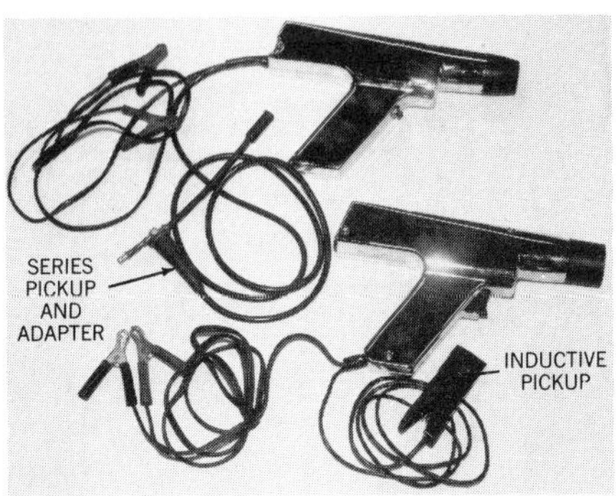

Figure 15-6. Typical dc timing lights; one with a series pickup, the other with an inductive pickup.

Figure 15-7. Use an adapter to connect a timing light with a series pickup. Do not puncture a plug cable.

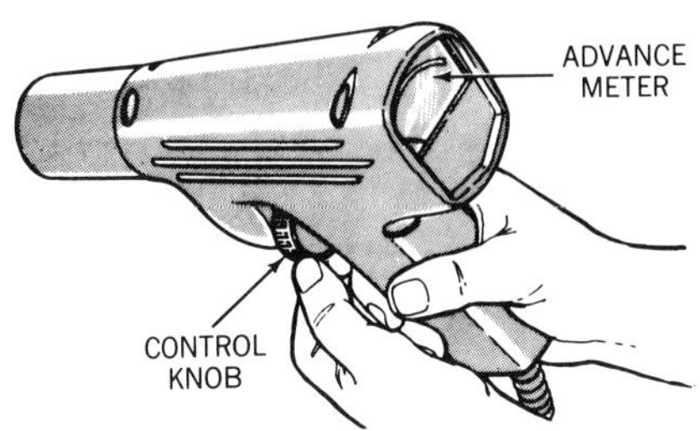

Figure 15-8. An adjustable timing light has a control knob and a timing advance meter. (Chrysler)

ing light flashes. Tighten the hold-down bolt.

6. Turn the ignition off and disconnect the timing light.

7. Check and adjust timing at the carmaker's specified speed with a timing light or magnetic timing meter after you start the engine.

IGNITION TIMING AND SPARK ADVANCE TEST AND ADJUSTMENT

Most engines require dynamic timing test and adjustment with a stroboscopic timing light or a magnetic timing meter. Additionally, most engines require that you set timing in relation to top dead center on the compression stroke for the number 1 cylinder. A few engines, however, require timing adjustment in relation to another cylinder, such as number 8 or number 4. You will set the basic timing on most engines at the normal slow-idle (curb-idle) speed, but a few carmakers specify timing test and adjustment at higher speeds.

Timing Test Equipment

To test and adjust timing and spark advance, you will need some or all of the following equipment:

- A tachometer to measure engine speed
- A timing light (with or without an advance meter)
- A magnetic timing meter
- A hand-operated vacuum pump and gauge to test vacuum advance diaphragms, solenoids, and valves.

You have learned to connect and use a dwellmeter to adjust ignition dwell. Most meters for ignition service are combination dwellmeters and tachometers. They are usually called "tach-dwell" or "dwell-tach" meters. You connect such a meter to the ignition system as explained in chapter 12. You will use the tachometer to check timing and advance at different engine speeds. Chapter 4 on basic test equipment introduced the hand-operated vacuum pump and the timing light. You can use the basic test procedures for vacuum devices outlined in chapter 4 to check vacuum diaphragms, solenoids, and valves.

Timing Lights. The simplest timing lights are powered by dc voltage from the car's battery, figure 15-6. You connect the red lead to the battery positive (+) terminal and the black lead to the negative (−) terminal. Most modern timing lights have inductive pickups that you simply clamp around the spark plug cable. Many electronic ignitions require this kind of light. The ignition may not operate properly with an older light that has a series pickup. If you use a timing light with a series pickup, you must connect the light to an adapter at the distributor cap or the spark plug terminal, figure 15-7.

Caution. Never puncture a spark plug cable or boot to connect the light. This will cause secondary voltage leakage.

Some timing lights are built into an engine analyzer console and do not need power from the car battery. You simply connect the timing pickup lead to the spark plug cable. This is usually the same lead that provides the timing trigger signal for the scope. All timing lights that are part of analyzer consoles, and many portable lights, are adjustable. These are often called "power timing lights." An adjustable light has a timing advance meter, built into the back of the light, figure 15-8, or on the analyzer console.

An adjustable timing light has a control knob that you use to adjust the light so that it flashes before or after the plug fires. When you check spark advance with a nonadjustable light, the light flashes so that the timing marks appear to advance along the degree scale from tdc. At full advance, the pointer may appear to be completely off the scale, figure 15-9, and you can't be sure if advance is 20, 25, 30, or 35 degrees btdc. As you rotate the control knob of an adjustable light, you bring the timing marks back to tdc alignment as the light flashes at high engine speed. The meter on the analyzer console or on the light then shows the exact amount of advance for which the light has compensated, figure 15-10. An adjustable timing light also lets you read the spark advance much more accurately than you can from the engine timing marks.

Magnetic Timing. Most late-model engines have pickup sockets for magnetic timing meters. Carmakers use

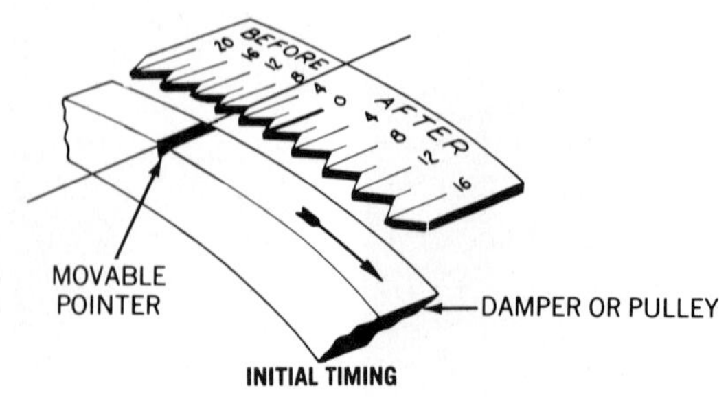

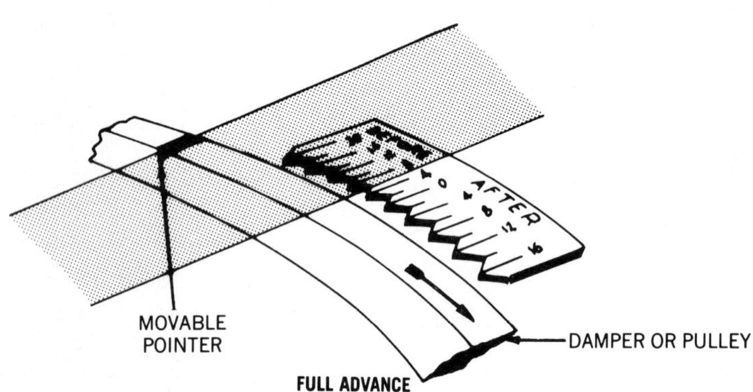

Figure 15-9. You can watch timing advance with a nonadjustable light, but you can't measure the exact number of degrees. (SAE)

Figure 15-10. After you adjust the timing light control knob to tdc or the basic timing setting, the meter displays the exact amount of advance. (Bosch)

these to adjust timing with automatic test equipment at the factory. Magnetic timing provides a more accurate timing signal than timing marks and a stroboscopic light can. The magnetic timing pickup socket may be on a bracket as part of the optical timing marks at the front of the crankshaft, figure 15-1, or it may be in the side of the block, figure 15-11, or in the bellhousing.

A magnetic timing meter works like the magnetic-pulse generator of an electronic ignition. The meter probe is a magnetic pickup coil, similar to the one in a distributor. When you insert the probe in the engine socket, a notch, a tab, or a magnetic particle on the crankshaft damper, on the crankshaft itself, or on the flywheel changes the magnetic reluctance around the pickup coil as it rotates past the probe. Either a notch, a tab, or a magnetic particle works equally well because the pickup simply senses a *change* in reluctance. This signal indicates the crankshaft position.

Magnetic timing adjustment is more accurate than optical adjustment with a timing light and marks because you get the crankshaft position signal directly from the crankshaft instead of from the distributor as the light flashes. Looseness in engine timing chains and distributor drive gears can actually delay ignition timing a few degrees after tdc in relation to crankshaft position. Magnetic timing adjustment also eliminates the optical viewing error that can occur as you look at timing marks from different angles.

The notch or tab on the crankshaft, the damper, or the flywheel of many engines is offset from the tdc position of the number 1 cylinder. The offset can be 10, 15, 32, 67, 90, or some other number of degrees, depending on engine design. Many magnetic timing meters require that you set the off-

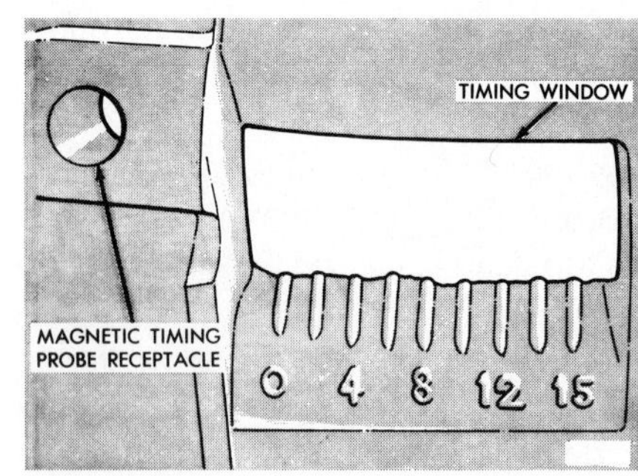

Figure 15-11. Timing meter sockets on some engines are in the block or in the bellhousing. (Chrysler)

set specification on the meter before use. You will find the specifications for different engines in the equipment operating instructions, figure 15-12, or in the carmaker's specifications.

Basic Timing Test and Adjustment

Before you check and adjust initial timing, service the primary circuit and be sure dwell is correct on a breaker-point ignition. If you adjust dwell *after* you set the timing, you will change the timing. Also be sure that the engine is at normal operating temperature and that the idle speed is correct.

As with other engine adjustments, you will need the carmaker's instructions to set the timing on a specific year and model of vehicle. Even though all timing adjustments include the same basic principles, do not guess about a particular vehicle. Start by checking the information on the engine tuneup decal. Then refer to more detailed instructions in factory shop manuals or independent service guides. The following guidelines will help you to use the carmaker's procedures easily and accurately.

- If the distributor has a vacuum advance unit, you can determine the direction of distributor rotation (clockwise or counterclockwise) by looking at the diaphragm housing on the distributor. If you point your right or left hand into the side of the distributor in the direction of the diaphragm linkage, you point in the direction of rotor rotation, figure 15-13. That is, the rotor always turns in the direction opposite from which the vacuum linkage moves the breaker plate.
- If the distributor has centrifugal advance weights, you can turn the rotor freely a few degrees in the rotation direction. The advance weights keep you from turning the rotor backwards. You also can determine the rotation of any distributor by cranking the engine with the distributor cap removed or by checking diagrams in tuneup specifications.
- Most carmakers specify that timing be set at normal slow idle, but Oldsmobile (and a few others) specifies that timing be set at speeds above 1,000 rpm on many engines.
- Most carmakers require that you disconnect and plug the distributor vacuum lines before adjusting timing. Some, however, do not. If the distributor has a dual-diaphragm or vacuum-retard unit, you almost always disconnect and plug the retard vacuum line, figure 15-14. Note the position of any vacuum lines that you disconnect so that you reinstall them correctly.
- If you remove an air cleaner to reach the distributor, plug any vacuum lines that you disconnect from the engine.
- If the distributor has an advance or a retard solenoid, follow directions to energize or disconnect the solenoid for specific tests.
- Before timing an engine, clean the timing marks and highlight them with chalk or paint. Many front-wheel-drive (fwd) cars with transverse engines have the marks on the flywheel, figure 15-15, rather than on the damper. If you can't reach the marks through a maze of belts, hoses, and accessories, clean them with choke-cleaning spray. In some vans, trucks, and motor homes, the timing marks are not where they would be on the same engine in a passenger car. They often are at the bottom of the crankshaft damper or on the flywheel.

Manufacturer	*Magnetic Timing Offset Degrees*
AMC	10
Chrysler (6 & 8 cyl)	10
Ford	
8 cylinder	135
Inline 6 cylinder	135
2300-cc, 4-cylinder	52
1600-cc, 4-cylinder	10
General Motors	9.5

Figure 15-12. This is a sample of the kind of offset specifications you will use with a magnetic timing meter.

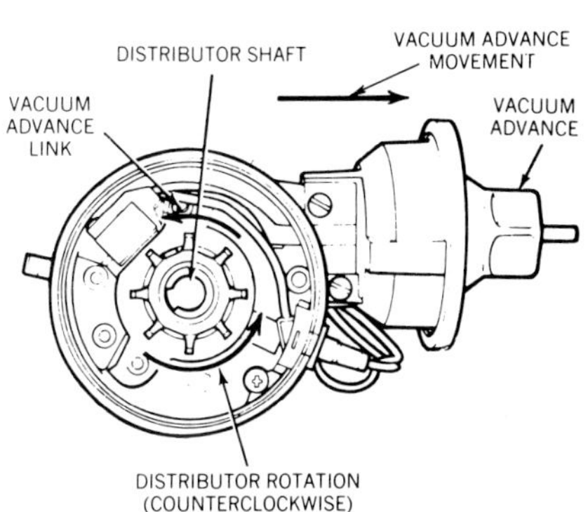

Figure 15-13. The vacuum diaphragm always advances the breaker plate opposite to rotor rotation. (Chrysler)

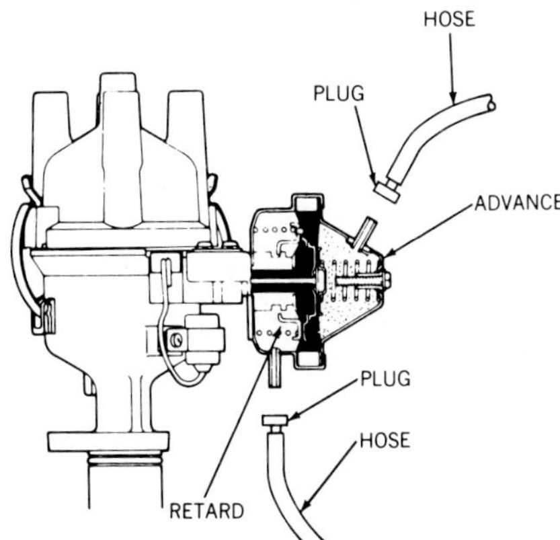

Figure 15-14. Disconnect and plug vacuum hoses according to the carmaker's directions to test and adjust timing, as well as advance and retard. (Chrysler)

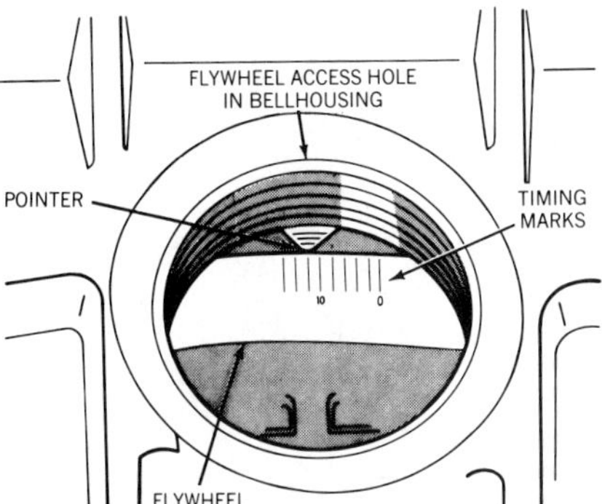

Figure 15-15. Many fwd cars with transverse engines have timing marks on the flywheel. View them through an opening in the bellhousing. (Chrysler)

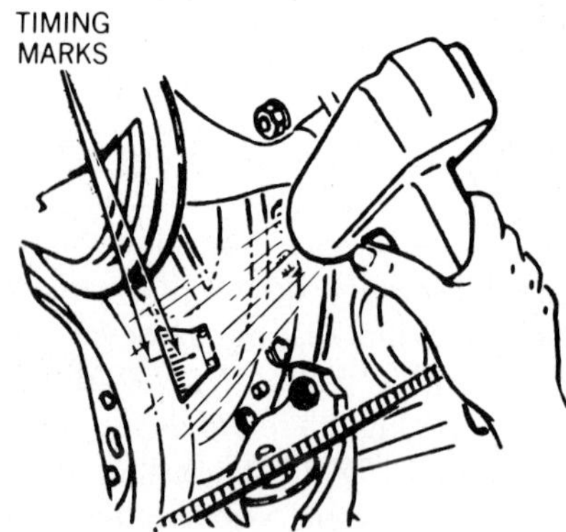

Figure 15-16. Observe the timing mark position as the light flashes. (Delco-Remy)

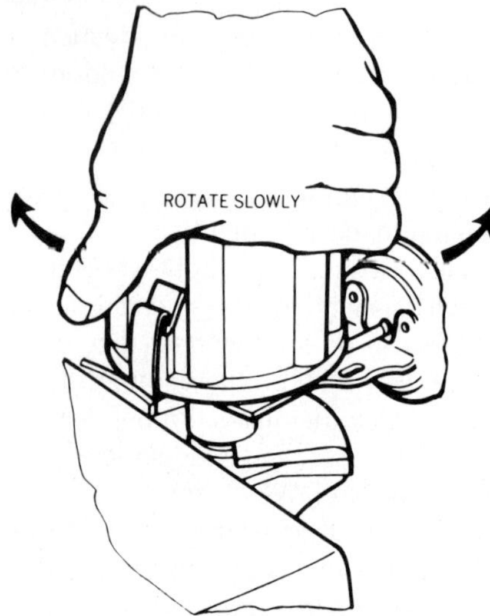

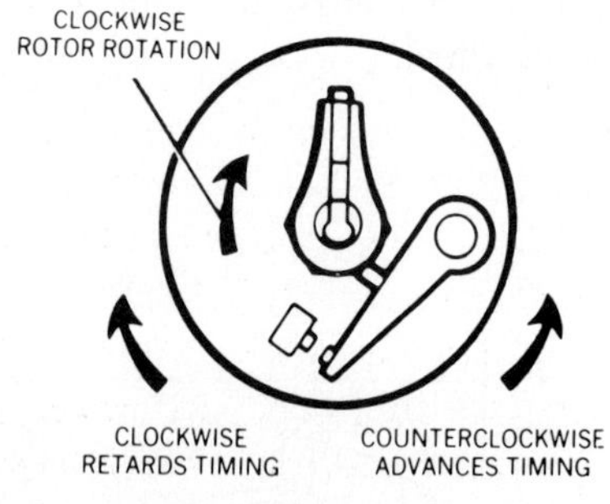

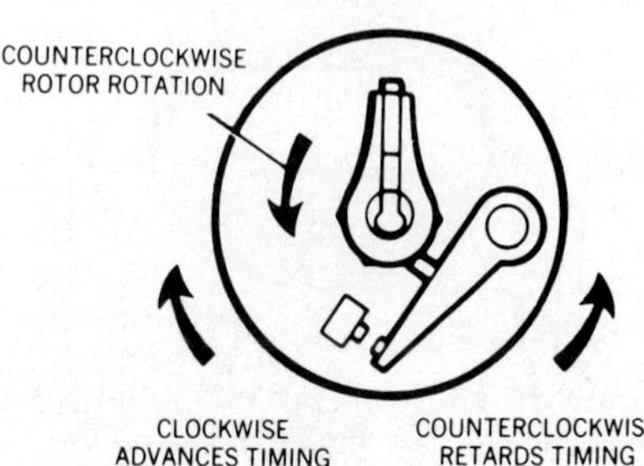

Figure 15-17. These are the directions you turn any distributor to advance or retard timing. (Chrysler)

• If the engine has electronic timing control, you probably will have to disconnect a specified connector at the ignition module or at the distributor to check basic timing. This puts the ignition in a "default" operating mode, where the electronic controls do not work. Some engines also require that you connect a jumper wire between two connector terminals to check timing. On a few engines with electronic timing control, you can neither check nor adjust basic timing. If you remove a distributor from such an engine, you will use a static timing method to reinstall it for a basic timing setting. A later section of this chapter outlines some of the principles used to check timing on engines with electronic controls.

Checking and Adjusting Timing with a Timing Light. Use this basic procedure and the carmaker's instructions to check and adjust basic timing with a timing light.

1. Run the engine to normal operating temperature. Then turn it off and connect a tachometer and timing light to the engine.

2. Set the parking brake, put the transmission in neutral or park, and block the drive wheels. Follow the manufacturer's directions to disconnect vacuum lines and electrical connectors.

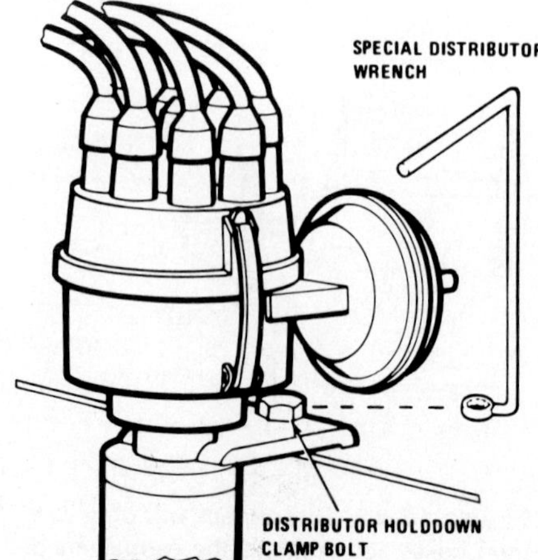

Figure 15-18. Use a special wrench to loosen the distributor holddown bolt. (Chrysler)

Caution. Keep the timing light and tachometer leads away from the rotating fan, accessories, and drive belts.

3. Start and run the engine steadily at the specified speed. If you are checking timing at idle, be sure the idle speed is correct before proceeding.

4. Aim the flashing timing light at the timing marks, figure 15-16, observe their position, and compare it to specifications.

5. If timing is not within specifications, adjust it as follows, figure 15-17:

a. Use a distributor wrench or a U-joint socket and ratchet extension to loosen the distributor holddown bolt, figure 15-18.

b. If timing is retarded, advance it by turning the distributor opposite to rotor rotation.

c. If timing is advanced, retard it by turning the distributor toward rotor rotation.

d. Tighten the holddown bolt and check timing again with the timing light.

6. Reconnect all vacuum hoses and connectors. Check idle speed, and adjust it if necessary, after adjusting timing. If you reset the idle speed, recheck the timing to be sure it is still correct.

A good way to verify the initial timing setting is to reduce engine speed below the normal idle speed and recheck the timing. With the vacuum diaphragm disconnected, timing should not change as you reduce speed. If it retards from your previous setting, the centrifugal advance was operating at the previous engine speed. In this case, you must readjust and recheck timing and idle speed. If timing and centrifugal advance operation are still not within specifications, remove the distributor for service.

Checking and Adjusting Timing with a Magnetic Timing Meter. Checking timing with a magnetic timing meter is actually faster and more accurate than with a timing light. Connect the meter to the engine and install

Distributor Part No.	Centrifugal Advance in Distributor Degrees at Dist. RPM						Vacuum Advance (Max. Dist. Deg at in. Hg)
	Start	Intermediate (1)	Intermediate (2)	Intermediate (3)	Intermediate (4)	Maximum	
1112163 (Delco)	0-2.5 @ 600	7.0-10 @ 900	8-10 @ 1000	—	—	14-16 @ 2200	10° @ 13-15″
1112958 (Delco)	0-3.0 @ 550	1.5-5.0 @ 600	9-12 @ 1000	—	—	15-18 @ 2300	13° @ 16-18″
D50F-BA (Ford)	0-1.5 @ 500	4-5 @ 750	5-7 @ 1000	6-8 @ 1250	9-10 @ 1500	10-12 @ 2000	6°-8° @ 10″
D70F-AH (Ford)	0-1 @ 500	1-3 @ 750	4-6 @ 1000	5-7 @ 1250	8-9 @ 1500	10-13 @ 2000	8°-9° @ 10″
3666781 (Chrys)	1-3 @ 500	10-12 @ 800	—	11-13 @ 1250	—	14-16 @ 1800	9°-12° @ 15″
3755468 (Chrys)	1-4 @ 650	8-10 @ 800	9-11 @ 1000	12-14 @ 1250	—	14-16 @ 2000	9°-12° @ 15″

Figure 15-19. Typical centrifugal and vacuum advance specifications in distributor degrees and distributor rpm.

Distributor Part No.	Centrifugal Advance in Crankshaft Degrees at Engine RPM					
	Start	Intermediate (1)	Intermediate (2)	Intermediate (3)	Intermediate (4)	Maximum
1112163 (Delco)	0-5 @ 1200	14-20 @ 1800	16-20 @ 2000	—	—	28-32 @ 4400
1112958 (Delco)	0-6 @ 1100	3-10 @ 1200	18-24 @ 2000	—	—	30-36 @ 4600
D50F-BA (Ford)	0-3 @ 1000	8-10 @ 1500	10-14 @ 2000	12-16 @ 2500	18-20 @ 3000	20-24 @ 4000
D70F-AH (Ford)	0-2 @ 1000	2-6 @ 1500	8-12 @ 2000	10-14 @ 2500	16-18 @ 3000	20-26 @ 4000
3666781 (Chrys)	2-6 @ 1000	20-24 @ 1600	—	22-26 @ 2500	—	28-32 @ 3600
3755468 (Chrys)	2-8 @ 1300	16-20 @ 1600	18-22 @ 2000	24-28 @ 2500	—	28-32 @ 4000

Figure 15-20. Centrifugal advance specifications in crankshaft (engine) degrees and engine speed.

the probe in the engine timing socket according to the equipment instructions. Adjust the crankshaft offset on the meter, if required.

> **Caution.** Keep the meter probe and leads away from the rotating fan, accessories, and drive belts.

Start and run the engine as you would to check timing with a light, but read the timing from the meter dial. Now you have both hands free to adjust the distributor. You don't have to hold a timing light to check the adjustment.

Spark Timing Advance Tests

Most engines with electronic timing controls have distributors with no centrifugal and vacuum advance mechanisms. There is no distributor spark advance to test. Some manufacturers, however, do have special procedures to check the timing control function of the system computer. These usually are spot checks of advance at a certain engine speed under specified load conditions. The test may include disconnecting and reconnecting a connector and observing a change in timing. You also may have to install a jumper wire between two connector terminals or ground a connector to check electronic advance. To check an electronic timing system, you will need the carmaker's detailed instructions for the specific vehicle. Remember not to disconnect any connector with the ignition on unless *specifically* directed to do so.

Testing and adjusting distributor spark advance on an engine with centrifugal and vacuum advance mechanisms is essential for proper engine service. Because the centrifugal and vacuum mechanisms are the same on breaker-point and electronic distributors that have these controls, test and adjustment procedures are identical.

Spark Advance Specifications. The first things you need for checking and adjusting spark advance are the specifications. Manufacturers publish these in several ways:

- Centrifugal advance in distributor degrees at distributor rpm, figure 15-19
- Centrifugal advance in engine crankshaft degrees at engine rpm, figure 15-20
- Vacuum advance in distributor *or* engine degrees at specified levels of vacuum

Engine	Trans	Year	Distributor Part No.	Advance Degrees@ 2500 rpm	
				Total	Centrifugal
200-cid	AT	1980–81	D9BE-DA	24.3–33.5	6.8–11.0
156-cid	MT	1981–82	4243694	28.2–36.2	10.2–14.2
173-cid	AT	1982	1103485	34.9–42.9	12.9–16.9

Figure 15-21. Spark advance check specifications in engine speed and degrees. Add the basic timing setting to use these specs.

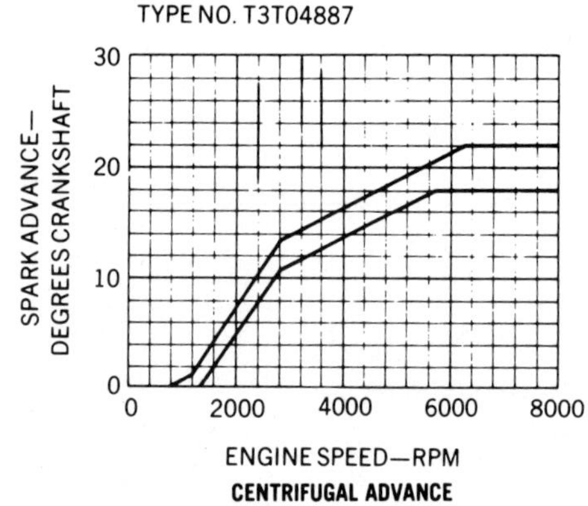

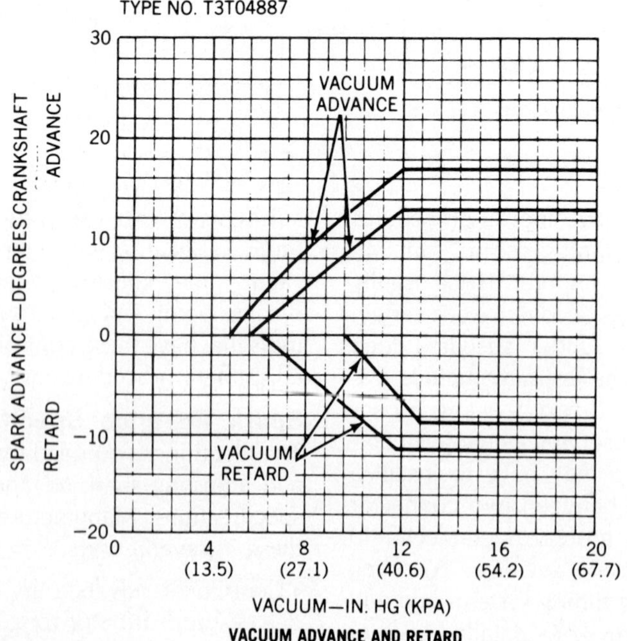

Figure 15-22. These are the centrifugal advance and the vacuum advance and retard curves for one distributor. (Chrysler)

Figure 15-23. Spark advance specifications are listed by distributor part number or model number. You will find the number stamped on the body or on a metal tag. (Delco-Remy)

• Total advance and centrifugal advance at a specified engine cruising speed such as 2,500 rpm, figure 15-21.

Usually, you will find the specifications in table form, as in the examples above. Sometimes, you will find them plotted as a curve on a graph, figure 15-22, which is why they are often called, "spark advance curves."

Carmakers use several different distributors with different centrifugal and vacuum advance curves in the same basic engine for a single model year. Therefore, advance specifications are listed by distributor part number. You will find the part number stamped on the distributor body or on a metal tag attached to the distributor, figure 15-23.

Specifications in distributor degrees and speed are primarily for use with a distributor tester. If you use these to test a distributor in an engine, you must double them because the distributor turns at one-half engine speed. Specifications for total and centrifugal advance at one given engine speed allow you to quickly check overall advance operation. They do not allow you to check the advance at several speeds and with different loads. Subtract the centrifugal advance specification from the total advance to determine vacuum advance.

Whenever you test spark advance with the distributor in the engine, you must add or subtract the basic (initial) timing from the distributor specifications. For example, if distributor centrifugal advance is 25 *crankshaft* degrees btdc at 2,500 engine rpm and initial timing is 8 degrees btdc, add the two numbers to determine spark advance for the engine:

25° distributor advance
+ 8° btdc initial timing =
33° engine spark advance at 2,500 rpm.

If the same distributor is installed in an engine with initial timing 4° atdc (retarded), subtract the initial timing from the distributor advance specification to determine spark advance for the engine:

25° distributor advance
− 4° atdc initial timing =
21° engine spark advance at 2,500 rpm.

Testing Spark Advance with an Adjustable Timing Light. Use the following basic steps and the carmaker's procedures to check centrifugal and vacuum advance on the engine with an adjustable timing light. Normally, you will do this immediately after setting the initial timing. If the car has a speed- or transmission-controlled vacuum advance system, you must bypass the vacuum control valves, switches, and solenoids. The simplest way to do this is with a hand-operated vacuum pump as explained below. A hand-operated vacuum pump allows the most accurate testing of any vacuum advance unit. A later section explains testing vacuum control systems in more detail.

Before or during the spark advance tests, you should check the vacuum advance diaphragm to be sure it is not leaking. Connect a vacuum pump to the distributor diaphragm and apply 15 to 20 in. Hg of vacuum. Close the pump valve and watch the gauge for about 10 seconds, figure 15-24. The vacuum should hold steadily. If vacuum drops, the diaphragm is leaking and you must replace it before you can test vacuum advance accurately. Test both the advance and the retard sides of a dual-diaphragm distributor.

1. Connect your tachometer and timing light to the engine and follow the first three steps of the timing adjustment procedure.

2. Accelerate the engine to the specified speed for checking *total* spark advance.

3. Aim the timing light at the timing marks and adjust the control knob to visually align the marks at the initial timing setting or at tdc.

4. Read the total advance shown on the timing light meter and compare it to specificatons. If you aligned the marks at the initial timing setting, add or subtract the setting from the meter reading to determine distributor advance.

5. To test centrifugal advance, proceed as follows:

a. Disconnect and plug the distributor vacuum advance hose, or hoses.

b. Run the engine steadily at the specified test speed, or speeds, and repeat steps 3 and 4 to check centrifugal advance at each point.

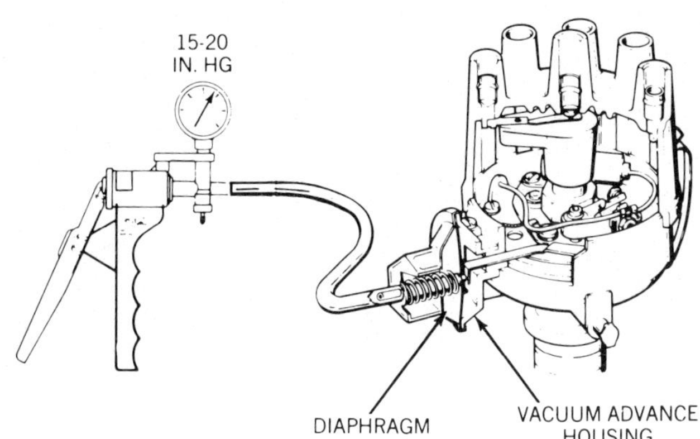

Figure 15-24. Test the vacuum advance diaphragm for leakage before checking spark advance. (Ford)

6. At this point, you can subtract the centrifugal advance readings in step 5 from the total advance readings in step 4 to get a general indication of vacuum advance. Test the vacuum advance, itself, as follows:

a. Connect a hand-operated vacuum pump with a gauge to the distributor vacuum advance unit.

b. Run the engine at idle so that the centrifugal advance mechanism does not operate.

c. Apply the specified amounts of vacuum (in. Hg) to the distributor, align the timing marks with the timing light knob, and read the advance on the meter. If you align the marks at the initial timing setting, add or subtract that number of degrees from the advance reading. If you align the timing marks at tdc, you do not have to add or subtract.

7. Some carmakers specify testing vacuum advance at various engine speeds so that the engine will not stall or run roughly when vacuum advance is applied. To do this:

a. Measure and record the centrifugal advance at the specified speeds as in step 5, above.

b. Use your vacuum pump to apply the specified amount of vacuum at each test speed. Read the total spark advance on the meter.

c. Subtract the centrifugal advance (a) from the total advance (b) to determine exact vacuum advance at the given speeds and amounts of vacuum.

8. If the distributor has a vacuum retard diaphragm (dual-diaphragm unit), run the engine at idle and connect a manifold vacuum line to the retard nipple of the diaphragm.

9. Use the timing light control knob to align the marks at the basic setting or at tdc and read the amount of retard on the timing meter.

Testing Spark Advance Without an Adjustable Timing Light. If you do not have an adjustable timing light, or if you just want a quick check of spark advance operation, do the following:

1. Be sure the engine is at normal temperature.

2. Connect your timing light to the engine. (A tachometer is optional but recommended.)

3. Disconnect and plug the distributor vacuum lines.

4. Run the engine at idle and observe the timing marks with the timing light.

5. Accelerate the engine to approximately 2,500 rpm, or a part-throttle cruising speed, and watch the timing marks as you do so. The timing indication should advance smoothly to indicate that the centrifugal advance is working right.

6. Hold the engine at the steady speed from step 5 and connect the vacuum advance line or a manifold vacuum line to the distributor advance diaphragm. The timing indication should advance more and engine speed should increase slightly to indicate that the vacuum advance is working right.

7. If the distributor has a vacuum retard diaphragm, return the engine to idle speed and connect the retard vacuum line while watching the timing marks. Timing should retard a few degrees and engine speed should drop if the vacuum retard diaphragm is working.

After finishing all vacuum advance and retard tests, be sure the vacuum lines are properly connected to the distributor.

TIMING TEST AND ADJUSTMENT WITH ELECTRONIC ENGINE CONTROLS

With the trend toward direct ignition system (DIS) installation on late-model engines, timing check and adjustment will seldom be required by the 21st century. Nevertheless, millions of cars and trucks with distributors and adjustable base timing will remain on the roads for the next decade.

To test timing on an engine with electronic timing control, start by checking the instructions on the engine decal. For more detailed instructions, check a reference manual. You must identify the year, model, and engine calibration exactly because test procedures vary from year to year and model to model.

The basic rule that applies to all electronic engine control systems is that you must "turn off" the computer to set base timing. That is, you must put the system into open-loop operation or take the computer out of the control loop for ignition timing. The following sections outline the timing test and adjustment principles used by several carmakers. This information will help you to use and understand specific procedures more accurately.

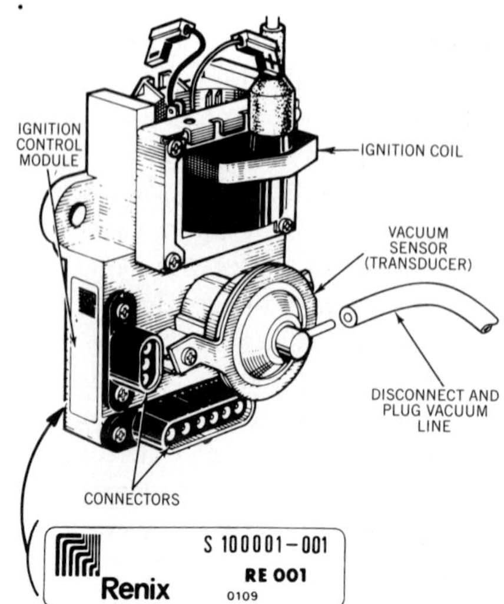

Figure 15-25. Disconnect the vacuum line from the transducer to check timing on AMC-Renault vehicles.

American Motors Electronic Timing (Before 1987)

Timing on AMC engines with Ford Dura-Spark or Delco HEI systems is checked with procedures similar to those used for Ford and GM vehicles. To check timing on AMC-Renault 4-cylinder engines, disconnect the vacuum line from the transducer on the ignition control module, figure 15-25. Then check the basic timing at normal slow (curb) idle. Timing is not adjustable. If it is not within specifications, follow the manufacturer's instructions to locate the defective component.

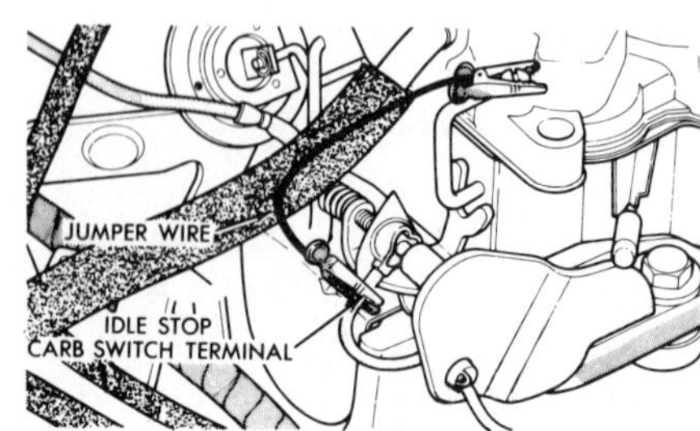

Figure 15-26. Ground the carburetor switch to check timing on most Chrysler products with electronic timing control. (Chrysler)

Chrysler Electronic Timing

When Chrysler introduced its first electronic timing system in 1976, it adopted the common timing test procedure of grounding the carburetor idle-stop switch, figure 15-26. With the switch grounded and the engine running at warmed-up slow (curb) idle, timing should be within 2 degrees of the specification on the engine decal. If timing is out of limits, loosen the distributor holddown bolt and adjust timing as you would for a distributor without electronic timing control.

This method remained the most common way to adjust timing on carbureted Chrysler engines through the mid-1980's. For fuel-injected Chrysler engines, however, the most common way to adjust timing is to disconnect the coolant temperature sensor with the engine warmed up and running. Check and adjust ignition timing on fuel-injected Chrysler vehicles as follows:

1. Set the parking brake and put the transmission in park or neutral.

2. Connect a timing light to the number 1 cylinder or connect a magnetic timing meter. If you use a magnetic timing meter, select −10° offset.

3. Be sure all lights and accessories are off.

4. Start and run the engine until it reaches normal operating temperature. Let engine speed stabilize at idle speed. Run the engine for two minutes after a hot restart to stabilize the temperature.

5. Disconnect the coolant temperature sensor (CTS). The POWER LOSS, POWER LIMITED, or CHECK ENGINE lamp must light.

6. Check the timing and adjust as necessary.

7. Reconnect the CTS and stop the engine.

8. Clear codes from the engine computer caused by disconnecting the CTS.

On 1983–88 engines, you can disconnect and then reconnect the CTS to take the computer out of the timing control loop. If you reconnect the CTS on 1989 and later engines, the computer will go back into closed-loop timing control. Therefore, it is best practice to leave the CTS disconnected during timing adjustment on any Chrysler fuel-injected engine.

For 2.2-liter carbureted Chrysler engines (except in minivans), follow the previous instructions to set timing, but momentarily open the throttle above 1,100 rpm and then release it. Be sure that the carburetor linkage does not bind. Connect a jumper wire between the carburetor idle switch and ground. Then disconnect and plug the vacuum hose at the computer. Check and adjust timing as necessary.

For carbureted minivans, disconnect the 6-way connector at the carburetor. Remove the purple wire from the connector and reconnect the connector. Then check the timing and adjust as necessary.

Ford Electronic Timing

Since the late 1970's, Ford has specified a variety of procedures to test timing on engines with electronic controls. On some engines, timing is adjustable; on others, it is not. Ford's electronic engine conrol (EEC) I and II systems require a special tester to check electronic timing completely. You can test basic timing by removing the PROM from the system computer and then checking the timing marks with your light. You also remove the PROM to check basic timing on EEC III. You can check spark advance by following Ford's procedures to apply vacuum to the manifold pressure sensor, which puts the system into a self-test mode. As part of the self-test, the system checks advance at about 2,000 rpm with no load. You can use your timing light to check the advance at the same time.

Ford's EEC-IV and MCU systems have a diagnostic connector in the engine compartment. In a later chapter, you will learn to use this connector to test for system problems. EEC-IV and MCU systems also have a "computed timing" program as part of the self-test routines built into the vehicle computer. To use this program to check timing, connect an analog voltmeter – lead to pin 4 of the connector. This is the self-test output (STO) terminal, from which the computer pulses service codes. Connect the meter + lead to the battery + terminal. Start the engine and run it at 2,000 rpm for two minutes to warm it up. Then turn it off.

For an EEC-IV system, connect a jumper wire between pin 2 (ground) in the test connector and the separate self-test input (STI) pigtail connector, figure 15-27 (left), which acts as the trigger to start the vehicle self-test. For an MCU system, connect the jumper between pins 2 and 5 in the test connector, figure 15-27 (right). MCU systems do not have a separate STI pigtail connector.

Restart the engine and let it go through the engine-running self-test. The voltmeter will pulse service codes at the end of the test, which takes about one minute. If any code other than 10 or 11 is present, troubleshoot and correct it before proceeding with timing adjustment.

After the last service code is pulsed on the voltmeter, engine speed will accelerate to about 1,500 to 2,000 rpm. At this point, you should read the spark advance on the timing marks. Timing should be 20 degrees advanced from base timing. Therefore, subtract 20 degrees from your reading to determine base timing. Because most Ford engines with electronic

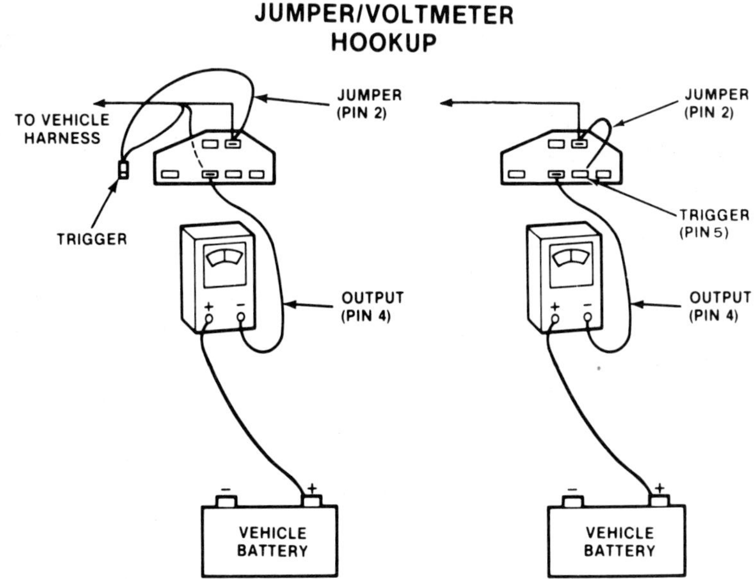

Figure 15-27. Connect a voltmeter and jumper wire in one of these ways (as specified) to the self-test connector on Ford's EEC IV systems. (Ford)

controls are timed at 10° btdc, the computed timing test usually displays 30 degrees of advance if base timing is set correctly.

To check and adjust base timing on an EEC-IV system with Ford's thick-film integrated (TFI) ignition, disconnect the inline connector in the yellow wire with the light green dots. This connector is in the ignition primary harness near the TFI module on the distributor, figure 15-28. It opens the "spark out," or SPOUT, circuit between the ignition module and the system computer. This locks the ignition into base timing with no electronic advance control. You can now use a timing light or magnetic timing meter to check the timing on the marks at the specified engine speed.

To check basic timing on Ford's older dual-mode ignition systems, you must disconnect the third connector at the ignition module, figure 15-29, which connects the module to a vacuum or pressure switch. The switch signals the module to retard timing at idle and some other conditions. Some instructions may specify a jumper across the black and yellow terminals; others do not. After disconnecting the connector, check timing with your timing light. You must follow specific Ford procedures exactly to avoid damaging the system or setting the timing incorrectly.

General Motors Electronic Timing

Like other carmakers, GM has several different procedures to test timing on electronically controlled engines. On most carbureted engines with electronic controls (C-4 and C-3 systems), disconnect the 4-wire ignition primary connector at the distributor to check base timing, figure 15-30. This isolates the ignition module from the system computer. The module then controls timing in relation to engine speed only, and you can check timing at the specified rpm.

You can't disconnect the 4-wire connector to test timing on fuel-injected GM cars because the engine won't run. The computer needs a distributor tach signal to time the injectors. There are three general ways to test timing on fuel-injected GM cars:

1. Put a jumper wire across terminals A and B in the ALDL diagnostic connector.

2. Open a special "set-timing" connector in the distributor wiring harness.

3. Ground a test connector in the engine compartment.

Again, specific procedures vary from year to year and model to model. Check the carmaker's instructions carefully. Don't assume, for example, that jumpering terminals A and B in the ALDL connector will put any GM engine into the base timing setting. On many vehicles, jumpering these two terminals will put the engine into a fixed timing condition that is *not* base timing.

Cadillacs with onboard diagnostics let you check timing by entering test codes into the automatic climate control system. The engine computer then shows the timing on the dashboard display panel.

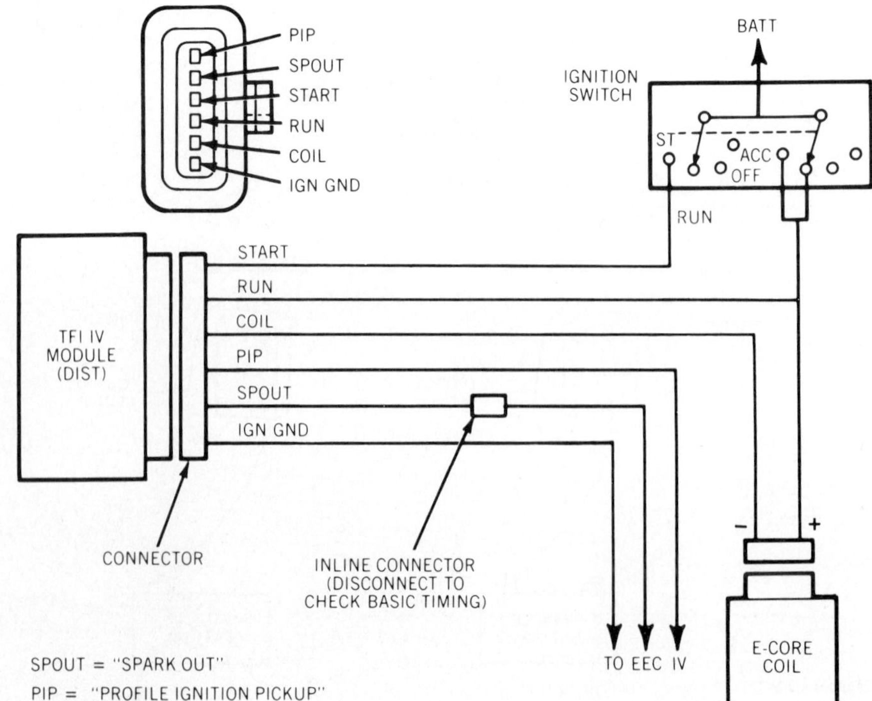

Figure 15-28. Open the inline connector to check basic timing on Ford's EEC IV systems with TFI ignition. (Ford)

VACUUM ADVANCE CONTROL SYSTEM TESTS

The basic tests outlined earlier in this chapter allow you to check the centrifugal and vacuum advance mechanisms of the distributor. They do not test the vacuum advance control devices used as part of emission systems. These systems were used on most automobiles of the 1970's, before electronic timing control became common. You will still find them on several late-model vehicles.

Carmakers developed and modified these systems from year to year and model to model. Therefore, you must have the specifications, diagrams, description, and procedures for a specific vehicle. Volume One of this text outlines the principles and common parts of most vacuum advance control systems. Most have one or more of the following parts:

- A spark delay valve
- A thermostatic vacuum valve
- A vacuum control solenoid

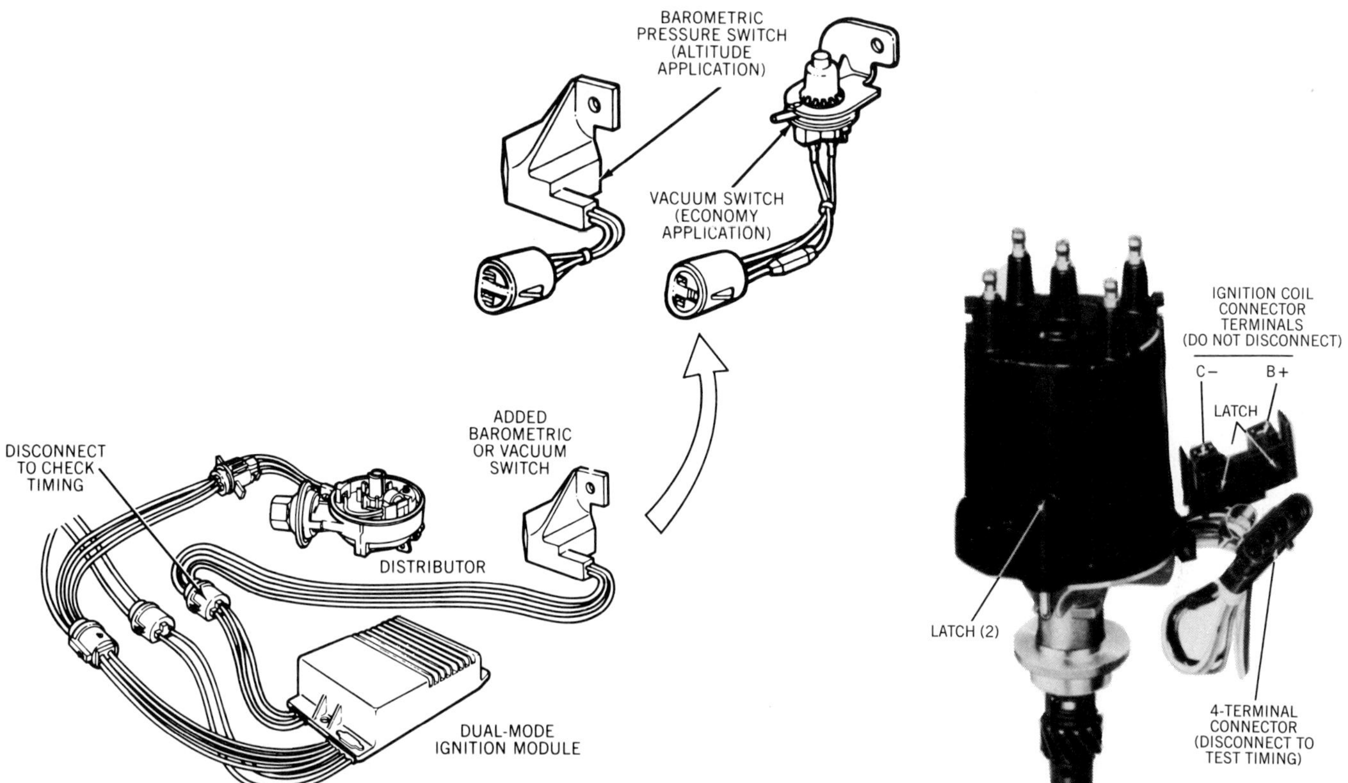

Figure 15-29. Disconnect the barometric or vacuum switch from the ignition module and jumper the module connector to check basic timing on a dual-mode ignition. (Ford)

Figure 15-30. Disconnect the 4-terminal connector at the HEI distributor to set timing on early GM C-4 and C-3 systems. (Delco-Remy)

- A transmission-controlled or a speed-controlled switch to operate a solenoid
- A time-delay, latching, or reversing relay.

Systems with spark-delay valves often have no solenoids or switches and are usually the simplest, figure 15-31. Some, however, have bypass circuits that route vacuum around the delay valve during cold-engine operation, figure 15-32. The bypass may be controlled by a thermostatic switch and solenoid or by a thermostatic vacuum valve.

Speed- and transmission-controlled systems can be plumbed and wired in many ways, but all work on the principle of denying vacuum advance at low speed or in low gears, and applying vacuum advance in high gear or at cruising speed.

You can check any system operation by testing for vacuum at the distributor. It should be delayed or absent under some specific conditions and present during others. You will need a timing light, a tachometer, and a vacuum gauge for most system tests.

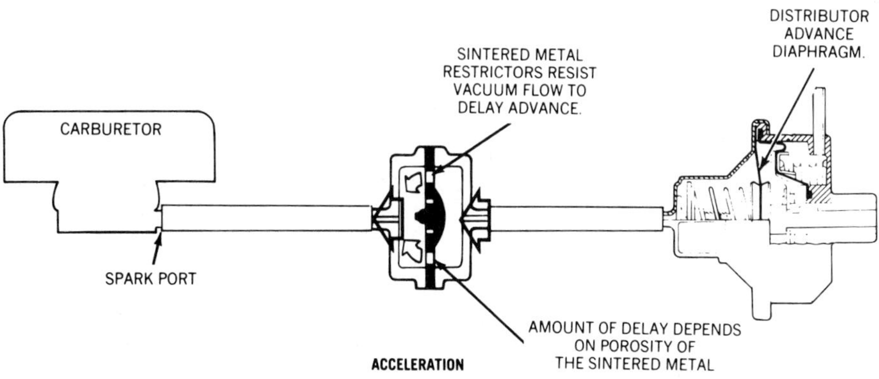

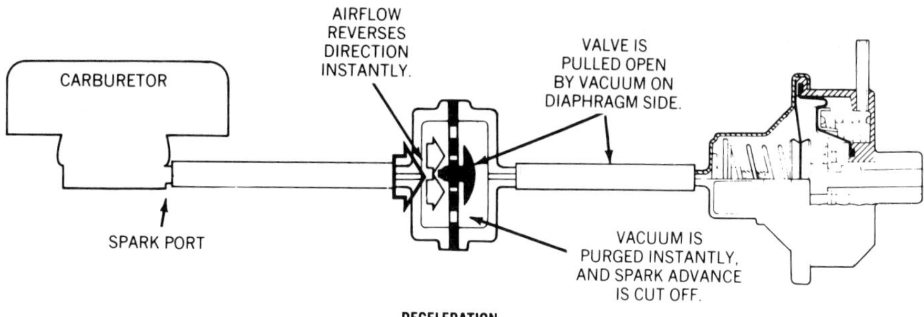

Figure 15-31. Typical spark-delay valve operation. (Ford)

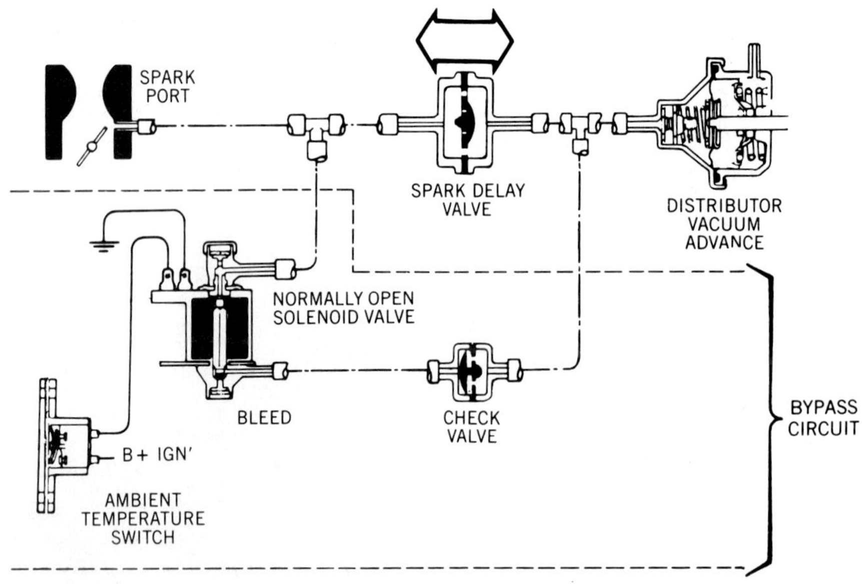

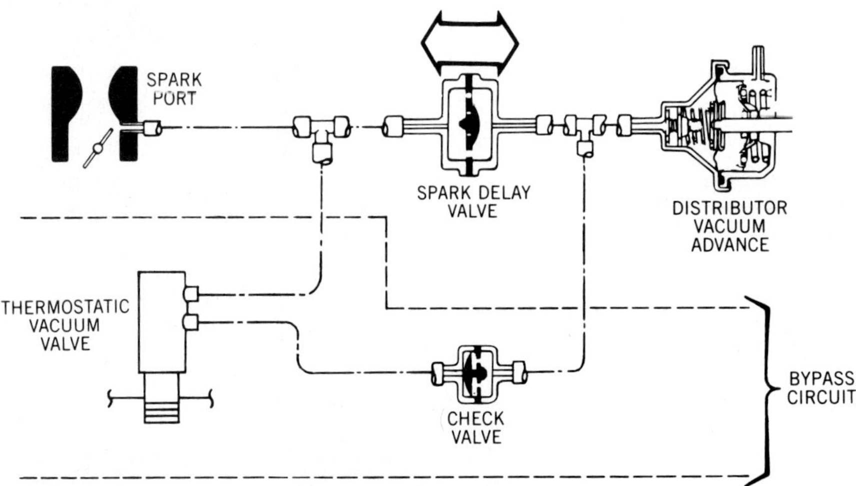

Figure 15-32. A vacuum-delay valve system can have either a thermostatic valve or a solenoid-controlled bypass. The vacuum bypass can occur at high or low temperature, depending on system requirements. (Ford)

Speed- or Transmission-Controlled Vacuum Advance System Test

Use these guidelines and the carmaker's procedures to test a speed- or transmission-controlled vacuum advance system.

> **Warning.** Test a front-wheel-drive vehicle by checking individual system parts or during a road test. Do not operate a front-wheel-drive vehicle on a hoist with the transmission engaged and the wheels turning while working around the engine compartment. Serious injury can result.

For a speed-controlled system or an automatic transmission-controlled system on a rear-wheel-drive car, test the vehicle on the road or raised on a hoist to get distributor vacuum at cruising speed or in high gear. Automatic transmission switches are operated by transmission fluid pressure, which is available in high gear or at wide-open throttle kickdown.

For a manual transmission-controlled system, you also can test the vehicle on the road or on a hoist. On some cars, you can start the engine and shift the transmission through all gears with the clutch disengaged to check for distributor vacuum. These transmission switches are operated by the shift linkage.

If an automatic transmission-controlled system allows vacuum advance in reverse gear, check for distributor vacuum by setting the parking brake, blocking the drive wheels, and shifting the transmission to reverse with the engine at fast idle.

To check for distributor vacuum, use a T-fitting to connect a vacuum gauge into the vacuum line at the distributor or at the distributor port of a vacuum solenoid or valve, figure 15-33. If you are going to road test the car, use a length of hose long enough to route the vacuum gauge into the driver's area. Operate the vehicle in one of the ways outlined above and read the gauge. You should see no vacuum advance at low speed or in low gear. Vacuum advance should occur at cruising speed or in high gear.

To check the actual advance when vacuum is applied, you may need a second person to help you. Connect your timing light and watch the timing marks and the vacuum gauge as the vehicle is operated on a hoist or shifted through the gears. Timing should advance and engine speed should increase as a vacuum reading appears on the gauge.

Vacuum Advance System Component Tests

If the system does not operate correctly for the general tests, you must check the individual parts. The first thing to check is the distributor vacuum diaphragm for leakage, as explained in the timing test procedures. Chapter 4 of this volume has basic test information for vacuum devices. You can use a 12-volt test lamp, a voltmeter, and an ohmmeter to test switches, relays, and solenoids for correct supply voltage and open or shorted circuits. As with any electrical system, you can often fix a problem simply by tightening or cleaning connectors or repairing circuit wiring. The following sections contain more guidelines for component testing.

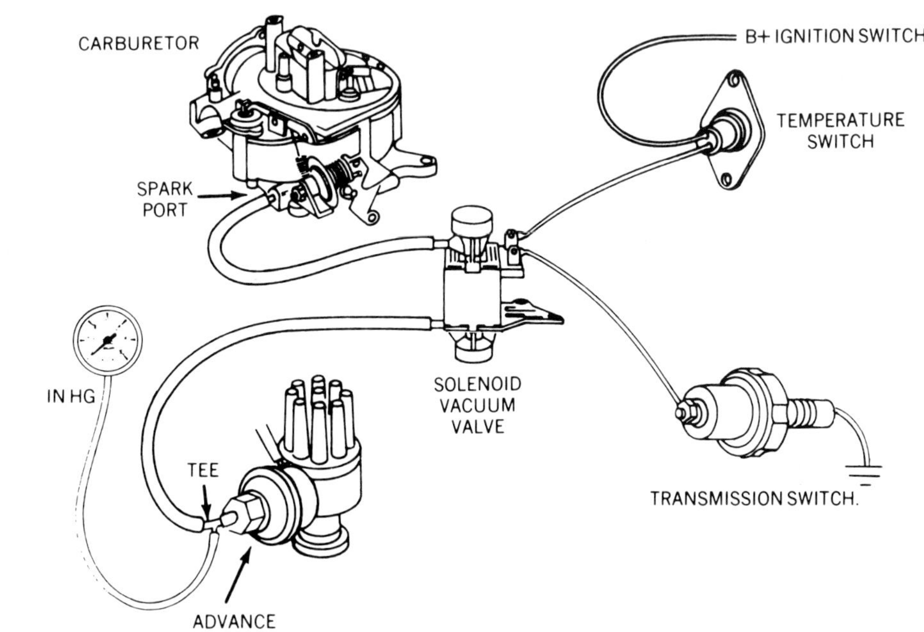

Figure 15-33. Install a vacuum gauge at the distributor or the solenoid outlet to check actual vacuum at the distributor. (Ford)

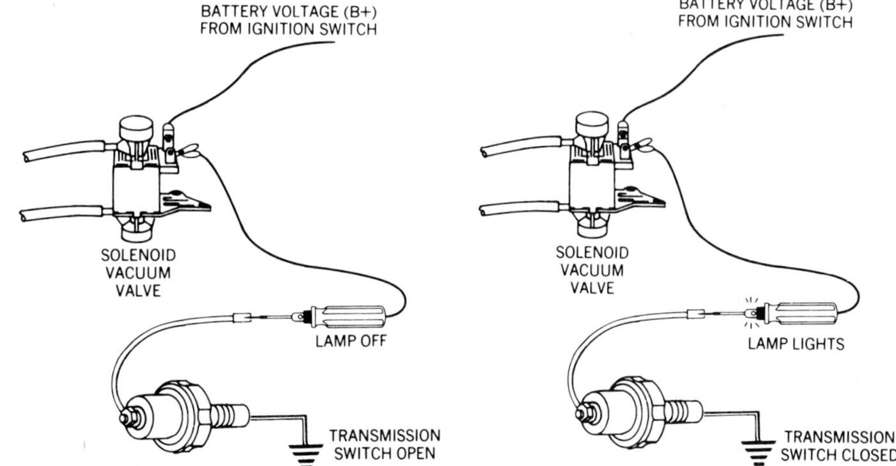

Figure 15-34. After determining if a switch is normally open or closed, connect a test lamp to check switch operation. (Ford)

Transmission Switch and Speed Sensor Tests. All transmission switches are grounding switches, which complete a solenoid circuit to ground. Whether controlled by shift linkage or transmission fluid pressure, most transmission switches work in one of the following ways:

- Normally open in low and intermediate gears to deenergize a solenoid
- Normally closed in low and intermediate gears to energize a solenoid.

After determining if a switch is normally open or normally closed, connect a 12-volt test lamp or a voltmeter in series between the switch connector and its feed wire. You can do this at the switch or at the solenoid, figure 15-34. Turn the ignition on and operate the transmission through all gears. The lamp will light or the voltmeter will show a reading as the switch closes. The lamp will go out or the voltmeter will show no voltage when the switch opens. If you test the vehicle on the road or operating on a hoist, use jumper wires long enough to locate the meter or test lamp in the driver's area.

Use your voltmeter or test lamp to check for applied voltage at the switch wire connector and at the solenoid, figure 15-35.

You can test a speed control switch on a GM system as you would test a transmission switch. Many other speed control switches are sensors, which are magnetic-pulse generators. They are usually mounted in the speedometer cable, figure 15-36. You can test system operation with a road test. You also can use an ohmmeter to check the sensor coil resistance and to check for a short between the sensor coil and its case, figure 15-37.

Vacuum Solenoid Tests. To be sure of its correct operation, you should test a solenoid electrically and for

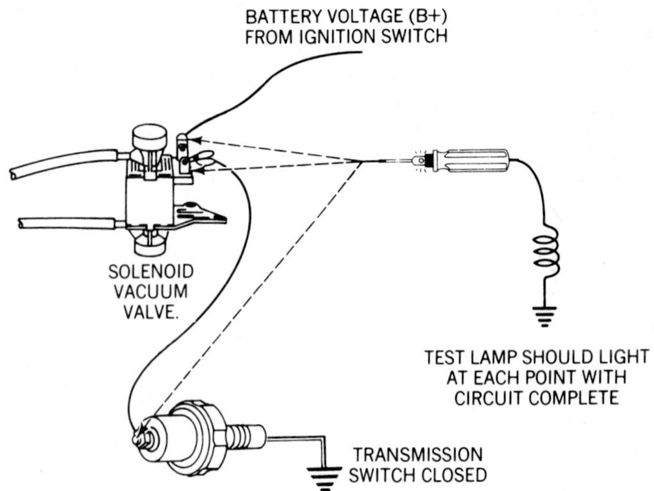

Figure 15-35. With the circuit closed, use a test lamp or voltmeter to check for applied voltage at a switch or a solenoid. (Ford)

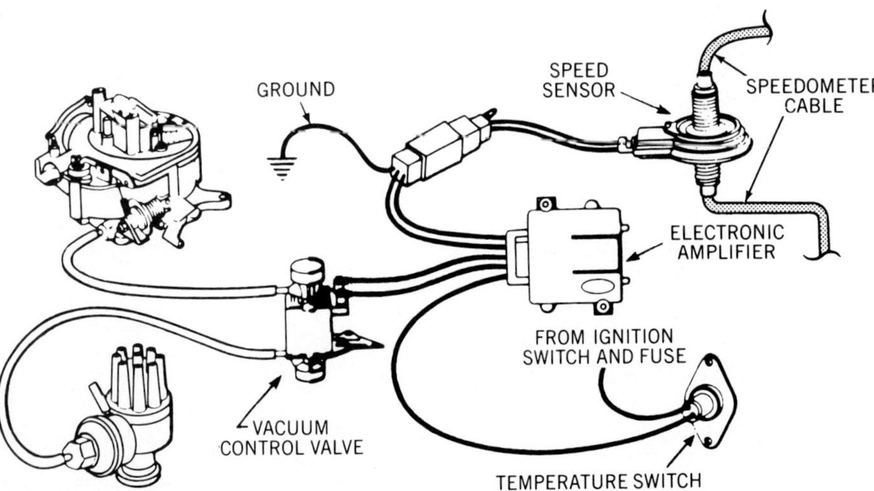

Figure 15-36. This speed sensor is a magnetic pulse generator, mounted in the speedometer cable.

proper vacuum control. Test a solenoid electrically, figure 15-38, as follows:

1. Disconnect the 2-wire harness connector from the solenoid.

2. Connect one solenoid terminal to ground with a jumper wire.

3. Connect another jumper wire to the other solenoid terminal and touch the other end of this jumper to the battery + terminal or another source of battery voltage.

4. Listen for a sharp click as you apply voltage to the solenoid and another click as you remove it. Do this several times to be sure the solenoid operates quickly each time you apply and remove voltage.

5. If you have specifications for the solenoid coil resistance, connect an ohmmeter across both solenoid terminals. If resistance is above specifications, the coil is open or has high internal resistance. If resistance is below specifications, the coil is shorted.

6. Set the ohmmeter on the high scale and connect one lead to the solenoid body or bracket. Connect the other lead alternately to each solenoid terminal. The ohmmeter should show infinite resistance between the body and each terminal. Any continuity indicates a grounded winding.

To test vacuum switching action of a solenoid, you must know if it blocks or passes vacuum when energized or deenergized. Figure 15-39 shows the common designs of 2-port and 3-port vacuum solenoids. To test a 2-port solenoid, disconnect the electrical connector and the vacuum hoses. Apply vacuum with a hand-operated pump or blow air through the distributor port. Either the vent port or the carburetor port should be open as shown in figure 15-39 when the solenoid is deenergized. Then apply battery voltage to energize the solenoid and again apply vacuum or blow air through the distributor port. The opposite port (vent or carburetor) should be open, figure 15-39, when the solenoid is energized.

Energize or deenergize the solenoid to block the inlet (carburetor) port. Use a vacuum pump to apply 15 to 20 in. Hg of vacuum to the inlet port and close the pump valve. Vacuum should remain steady with the inlet port blocked. If it drops, the solenoid has an internal leak.

Three-port solenoids switch vacuum from two different sources (usually between ported and manifold vacuum) to one outlet. Connect a vacuum gauge to the outlet port and alternately energize and deenergize the solenoid as you apply vacuum to each inlet port, figure 15-39. Check for proper vacuum at the outlet and for leakage as you would a 2-port solenoid.

Thermostatic Vacuum Valve Tests.

Thermostatic vacuum valves are known as PVS, TVS, CTO, TIC, and TVV devices. They come in different shapes and sizes, but all are operated by air or coolant temperature. To test one, you must know if it opens or closes a vacuum passage at high or low temperature. Use a vacuum pump and a gauge to test the valve at both high and low temperature. Figure 15-40 shows two examples of testing common thermostatic vacuum valves. Some valves have as many as four or six ports. Test an air temperature valve by heating it with a heat gun. Test a coolant-operated valve with the engine cold and with the engine at normal operating temperature. Do not let the engine overheat, which could harm the catalytic converter. You also can test a coolant-operated valve by removing it from the vehicle and testing it in hot and cold water.

Relay Tests. Use a voltmeter or 12-volt test lamp to check voltage at relay terminals as shown in figure 15-41. Use an ohmmeter to test the relay coil resistance or to check for a short between the relay coil and body, figure 15-41.

Some spark control systems have time-delay relays that do not close for about 20 seconds after voltage is applied through the switch. This allows a vacuum solenoid to apply vacuum spark advance during cold-engine warmup. Test a time-delay relay by connecting a 12-volt test lamp between the relay armature terminal and ground. Apply voltage to the relay switch circuit (usually by grounding the relay coil terminal as a transmission switch would) and count the time until the lamp lights. If it lights immediately or does not light at all, the relay is bad.

A reversing relay closes a solenoid circuit when its switch (control) circuit is open, and vice versa. That is, it reverses the switch operation. Test a reversing relay by again connecting a 12-volt lamp between the armature terminal and ground. Apply voltage to the relay battery terminal with the control circuit (the coil or switch terminal) open. The lamp should light. Close the control circuit, and the lamp should go out.

Vacuum-Delay Valve Tests. Chapter 4 summarizes the tests for vacuum-delay valves, and you also will learn to test them as choke-delay valves during carburetor service. Basically, connect a vacuum pump to the valve inlet and a second gauge to the valve outlet, figure 15-42. Apply the amount of vacuum specified by the manufacturer and measure the time it takes for the vacuum to equalize on both sides of

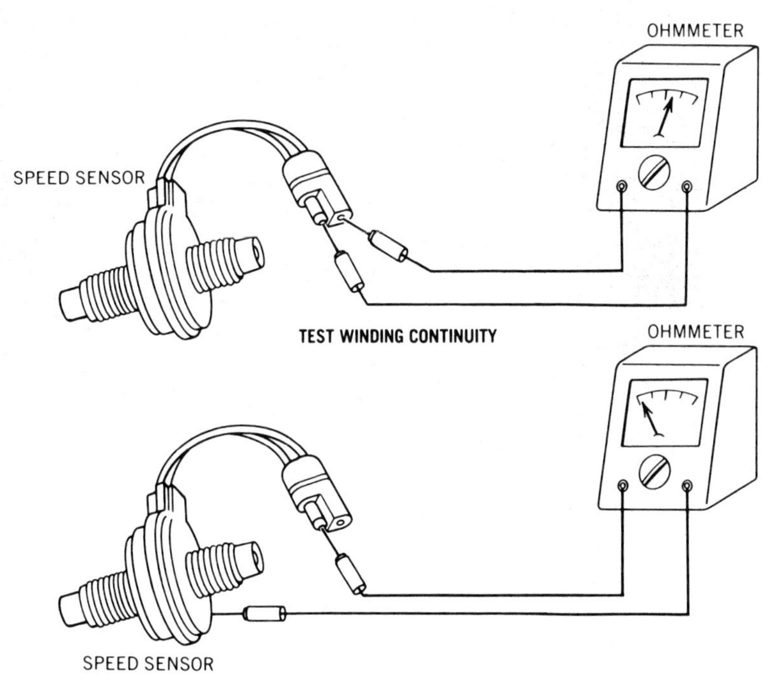

Figure 15-37. Use an ohmmeter to check a speed sensor as you would any coil for resistance (continuity) and for a short circuit to the case.

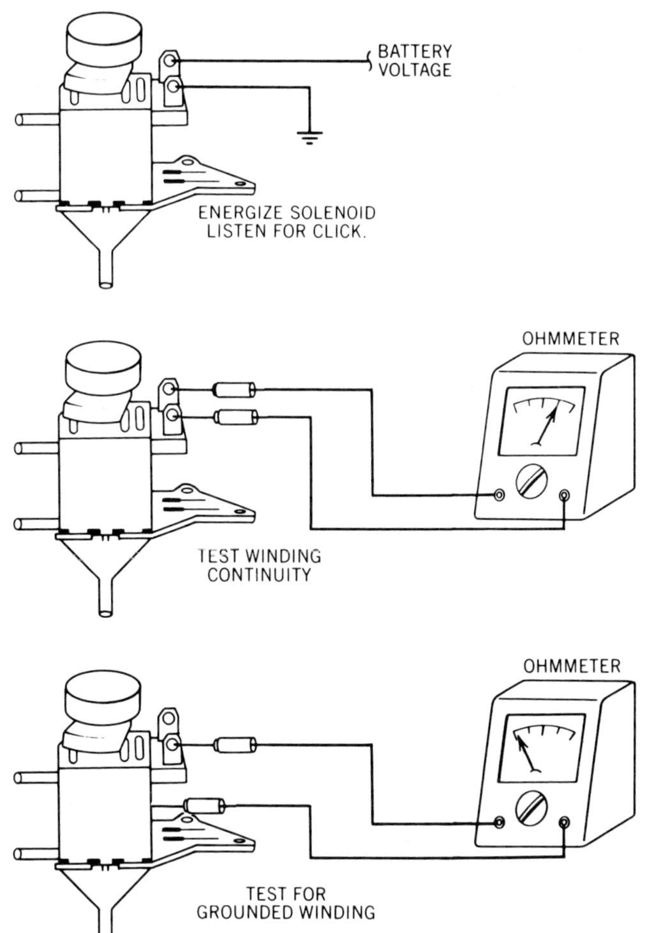

Figure 15-38. Apply battery voltage to test solenoid operation and use an ohmmeter to check the solenoid coil.

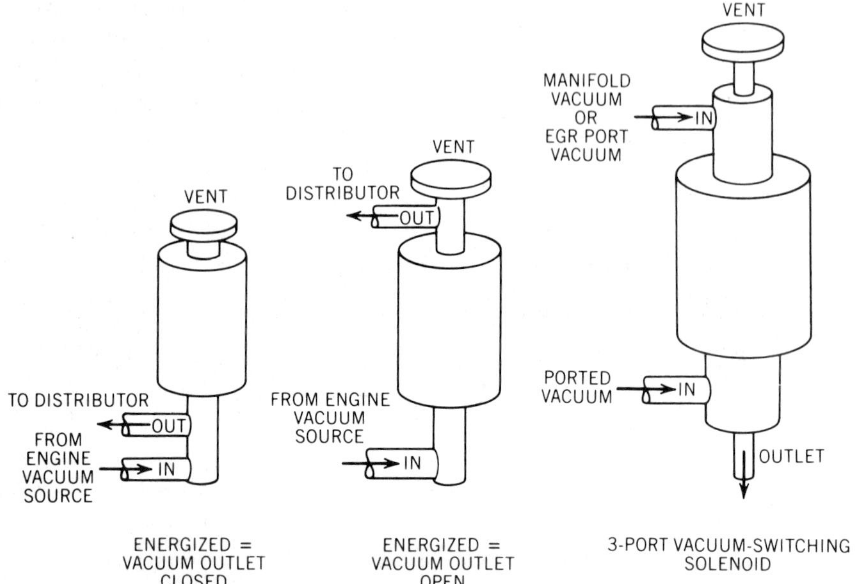

Figure 15-39. Typical 2-port and 3-port solenoids. A 2-port solenoid can be energized or deenergized to block vacuum, depending on design.

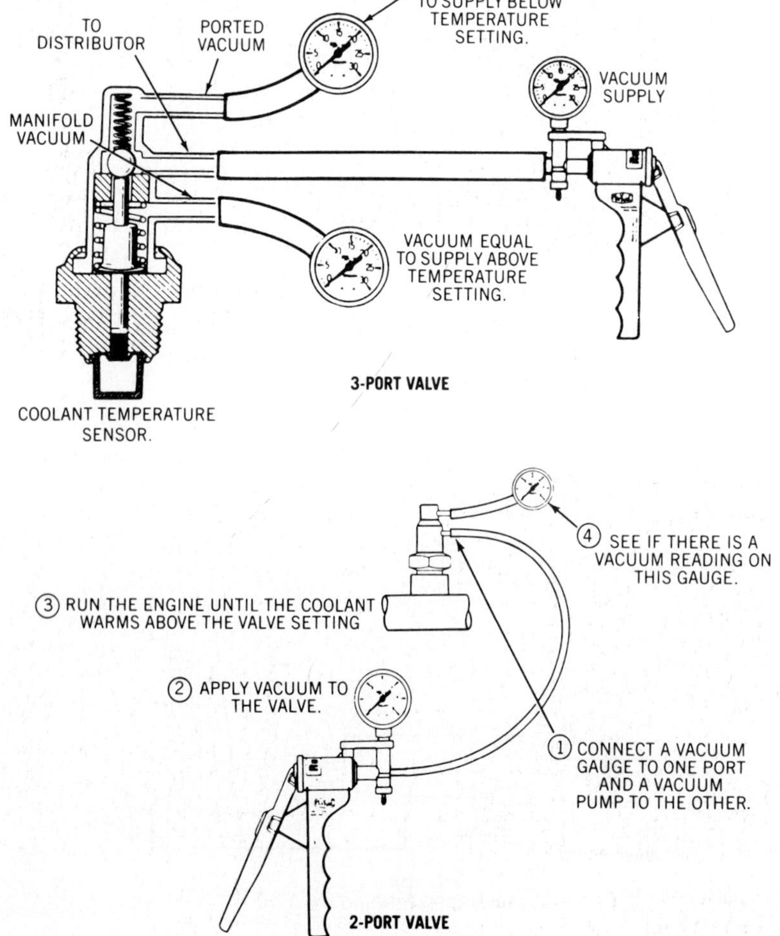

Figure 15-40. Test the operation of a thermostatic vacuum valve at high and low temperature. (Ford)

the valve. Ford gives time specifications for the outlet vacuum to reach 6 or 8 in. Hg with 10 in. Hg applied to the inlet. If the time delay is not within specifications, replace the valve. Ford and some other carmakers recommend that the valves on some engines be replaced at scheduled intervals to avoid problems.

Vacuum-delay valves on some GM and Chrysler engines have a cold-temperature override feature. When valve temperature is below 50° to 60° F (10° to 16° C), a thermostatic disc in the valve opens to allow immediate vacuum advance.

When you replace a vacuum-delay valve, be sure you install it in the right direction. Some are color coded; others have the ports labeled to indicate which goes toward the carburetor and which goes toward the distributor.

SUMMARY

The three critical engine-operating factors are: engine mechanical condition, air-fuel ratio, and ignition timing. Timing adjustment is the final step of complete ignition service and *one* of the last steps of comprehensive engine service. You may, however, check timing and spark advance at several stages during engine performance troubleshooting.

The simplest form of ignition timing is static timing, which you do when you install a distributor in an engine. Dynamic timing requires that you check and adjust the initial timing, using a timing light or magnetic timing meter and a tachometer, at the specified engine speed.

Centrifugal and vacuum advance mechanisms may operate perfectly in a distributor tester, but ignition timing can be incorrect if the basic setting is wrong or if vacuum advance control devices on an engine do not work properly. You can test centrifugal and vacuum advance with the distributor in the engine, using an adjustable timing light and a tachometer.

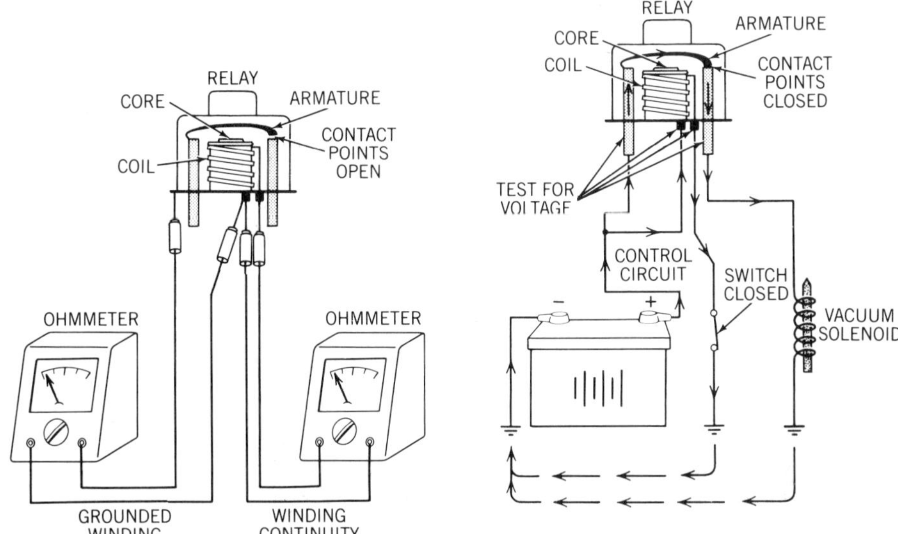

Figure 15-41. Test a relay similarly to a solenoid. Check for voltage at the battery terminal and use an ohmmeter to test the winding.

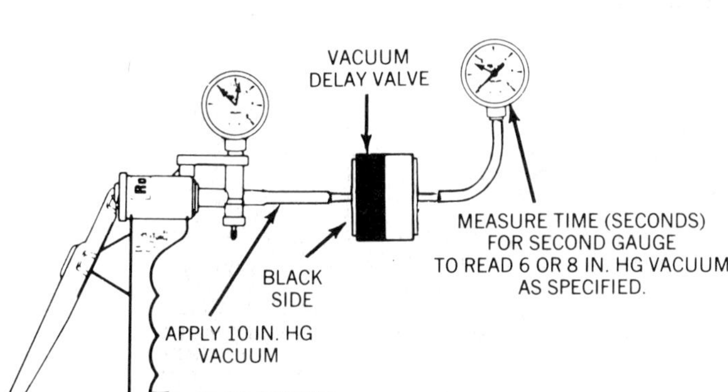

Figure 15-42. Use a vacuum pump and a second gauge to test a vacuum-delay valve. (Ford)

Vacuum advance controls are major emission control systems used on cars of the 1970's to reduce NO_x and HC emissions. While the electrical and vacuum connections vary among systems, they all work on the principle of delaying or denying vacuum advance at low speed or in low gears, and applying vacuum advance in high gear or at cruising speed. You can test the general system operation and the individual parts by using the guidelines in this chapter.

REVIEW QUESTIONS

Multiple Choice

1. Mechanic A says that if you install a distributor correctly and the engine was not cranked with the distributor out, the engine should start and run well enough to adjust timing. Mechanic B says that you can adjust static timing by aligning the timing marks at tdc on the compression stroke for number 1 cylinder. Who is right?

a. A only
b. B only
c. both A and B
d. neither A nor B

2. Mechanic A says that you can set static timing for a breaker-point distributor by connecting a 12-volt lamp across the coil distributor terminal and ground and then turning the distributor until the lamp lights. Mechanic B says you can set static timing by turning the distributor and watching for an arc across the points. Who is right?

a. A only
b. B only
c. both A and B
d. neither A nor B

3. Mechanic A says that you can set static timing for an electronic distributor by connecting an inductive timing light to the number 1 plug cable and then turning the distributor until the light flashes. Mechanic B says you can set static timing by turning the distributor and watching for an arc across the trigger wheel and pickup coil. Who is right?

a. A only
b. B only
c. both A and B
d. neither A nor B

4. A special pickup socket near the timing mark pointer indicates that the engine has provisions for:

a. stroboscopic timing
b. timing advance and retard tests
c. magnetic timing
d. inductive lamp timing

5. Mechanic A says that you must connect an inductive timing light to an adapter at the plug cable. Mechanic B says that you can easily connect a series timing light to a TVRS cable by inserting a small pin through the cable at any convenient point. Who is right?

- **a.** A only
- **b.** B only
- **c.** both A and B
- **d.** neither A nor B

6. A magnetic timing meter works similarly to:

- **a.** a Hall-effect switch
- **b.** a distributor pulse generator
- **c.** an optical pickup
- **d.** a tach-dwellmeter

7. Before checking timing on a specific engine, many magnetic timing meters require you to adjust the:

- **a.** centrifugal advance
- **b.** timing parallax
- **c.** timing offset
- **d.** vacuum retard

8. Mechanic A says that you should adjust dwell on a breaker-point distributor *after* you adjust timing. Mechanic B says that many carmakers specify that you disconnect and plug distributor vacuum lines before adjusting timing. Who is right?

- **a.** A only
- **b.** B only
- **c.** both A and B
- **d.** neither A nor B

9. Before adjusting timing, some carmakers may specify that you do any of the following EXCEPT:

- **a.** adjust the idle speed and mixture
- **b.** disconnect and plug distributor vacuum lines
- **c.** apply manifold vacuum to the EGR valve
- **d.** turn off the air conditioning

10. Mechanic A says that you turn a distributor *opposite* to rotor rotation to advance timing. Mechanic B says that you turn a distributor *toward (in line with)* the vacuum advance unit to advance timing. Who is right?

- **a.** A only
- **b.** B only
- **c.** both A and B
- **d.** neither A nor B

11. With an engine running at 2,500 rpm (or other speed specified for testing total advance) and with no load, the total spark advance is below specifications. You disconnect the vacuum advance line and find very little decrease in advance. This probably indicates that:

- **a.** initial timing is overly advanced
- **b.** centrifugal advance is not working
- **c.** a throttle position transducer is not working correctly
- **d.** vacuum advance is incorrect

12. You test a distributor vacuum advance unit with a hand-operated vacuum pump and find advance within specifications. You also find correct vacuum at the carburetor spark port when you test it with a gauge. But, distributor vacuum advance is not within specifications with the engine running at the specified speed. This can indicate:

- **a.** a broken vacuum line
- **b.** a defective vacuum solenoid
- **c.** a defective transmission control switch
- **d.** any of the above

13. To check and adjust initial (basic) timing on an engine with electronic spark advance control, you may have to:

- **a.** disconnect a vacuum transducer or sensor
- **b.** ground a carburetor idle switch or sensor
- **c.** disconnect a distributor connector
- **d.** any of the above

14. Most transmission-controlled vacuum advance systems have:

- **a.** a normally open transmission switch and a normally deenergized solenoid
- **b.** a normally closed transmission switch and a normally energized solenoid
- **c.** a normally open transmission switch and a normally energized solenoid
- **d.** a normally closed transmission switch and a normally deenergized solenoid

15. Mechanic A says that if an automatic-transmission-controlled vacuum advance system allows advance in reverse, you can check advance with the transmission in reverse, the parking brake set, the wheels blocked, and the engine idling. Mechanic B says that the easiest way to check a transmission-controlled spark system on a front-wheel-drive car is to run the engine with the transmission in high gear while the car is raised on a hoist. Who is right?

- **a.** A only
- **b.** B only
- **c.** both A and B
- **d.** neither A nor B

PART FIVE
FUEL AND EMISSION SYSTEM TESTING AND SERVICE

INTRODUCTION

Complete fuel system service includes more than carburetor or fuel-injection adjustment and repair. Engine fuel metering and air-fuel ratios cannot be correct if problems exist in fuel tanks, pumps, and filters. Proper air induction and air filtration are equally important for good engine performance. Moreover, fuel evaporative emission controls must work properly for efficient engine operation and emission control. Chapter 16 outlines the principles that all carmakers use in their service procedures for these fuel delivery and air induction systems.

Carburetor service has been a central part of engine tuneup since the earliest days of the automobile. Even in the last decade of the 20th century when carburetors are "obsolete," there are millions of carbureted vehicles on the roads that require regular service. Chapter 17 summarizes the basic testing, adjustment, and repair methods that you will use for complete carburetor service. This chapter concentrates on the principles that you will find in carmakers' specific instructions and will help you follow specific procedures more accurately. Chapter 17 ends with a photo sequence that illustrates a typical carburetor overhaul.

The popularity of diesel automobiles rises and falls with the price and availability of crude oil. Even though automotive diesel engine service is not the topical subject that it was in the early 1980's, a lot of automotive diesel engines are regularly installed in light- and medium-duty trucks. The skilled engine performance technician, taking the long view of the trade, will realize that the ability to service diesel engines can pay dividends throughout a career. Chapter 18 covers the basics of testing and adjusting diesel fuel injection systems.

Emission controls have been integral parts of engine combustion control systems for about 30 years. Chapter 19 traces the development of testing and service procedures for these systems. The information in this chapter will help you to develop the basic skills of emission control service and to recognize the similarities among carmakers' specific instructions.

Turbochargers became commonplace on passenger car engines in the early 1980's. Recent years also have seen a rebirth of mechanical superchargers to boost performance of production automobiles. Chapter 20 summarizes the service requirements of these devices as part of the total fuel system.

16 FUEL SYSTEM INSPECTION, TESTING, AND SERVICE

INTRODUCTION

When many people think of fuel system service, they think of carburetor or fuel injection adjustments and repairs. All other parts of the system must be in good condition, however, for the carburetor or injection system to meter fuel correctly. Some parts, such as tanks and lines, require repair only if they are contaminated or damaged. Fuel and air filters and evaporative emission control (EEC) systems, on the other hand, require periodic service. EEC systems and fuel pumps often require troubleshooting and repair to correct fuel system problems. This chapter outlines the principles of inspection, testing, and service for major fuel system components other than the carburetor or injection system.

GOALS

After studying this chapter, you should be able to do the following jobs in twice the flat-rate labor time or less:

1. Inspect the fuel system and all components for damage, deterioration, and leakage. Replace defective parts as required.

2. Test fuel pump pressure and volume. Test operation of electric fuel pumps. Replace defective pumps.

3. Remove and replace various fuel filters. Service diesel fuel filters.

4. Inspect and replace air filters.

5. Test the operation of thermostatically controlled air cleaners. Repair or replace defective parts.

6. Inspect and test the operation of evaporative emission control systems. Repair or replace defective parts. Replace canister air filters.

7. Inspect, test, and repair manifold heat control valves.

FUEL SYSTEM SAFETY

Before servicing any part of a fuel system, review the shop and vehicle safety information in chapter 1. Particularly, observe these precautions:

1. Do not smoke when working on or near any part of a fuel system. Keep sparks, flame, and hot metal away from fuel system parts.

Figure 16-1. Cap or plug disconnected fuel lines to prevent spillage.

2. Work on fuel systems in a well-ventilated area. Clean up any spilled fluids promptly.

3. Cap or plug all fuel lines when disconnected to prevent fuel leakage and to keep dirt out of the system, figure 16-1.

4. Keep the air cleaner installed whenever possible to avoid a backfire or the possibility of dropping something into the engine.

5. Do not use gasoline as a cleaning solvent. Avoid prolonged skin contact with fuels and solvents.

6. Disconnect the battery ground cable before removing and replacing fuel system components on vehicles without electronic engine controls. If it is impractical to disconnect the battery, be sure the ignition is off and use extreme caution when servicing the fuel system.

7. Diesel fuel injection systems retain high pressure between the injection pump and the injectors when the engine is off. Most gasoline injection systems also hold pressure between the fuel pump and the injector body or fuel rail when the engine is off. Follow the carmaker's specific procedures for relieving fuel pressure before loosening any fuel lines or replacing any parts of these systems.

FUEL SYSTEM INSPECTION

When you inspect any part of a fuel system, you will be looking for four basic things: leakage, damage (road damage and rust), looseness, and deteriorated rubber hoses. You will seldom do a complete fuel system inspection from bumper to bumper as a specific service job. However, whenever you are working under the hood or under the car (when it is raised on

Figure 16-2. Replace fuel hoses that are aged, loose, leaking, or otherwise damaged.

Figure 16-3. Tighten carburetor mounting nuts evenly and securely.

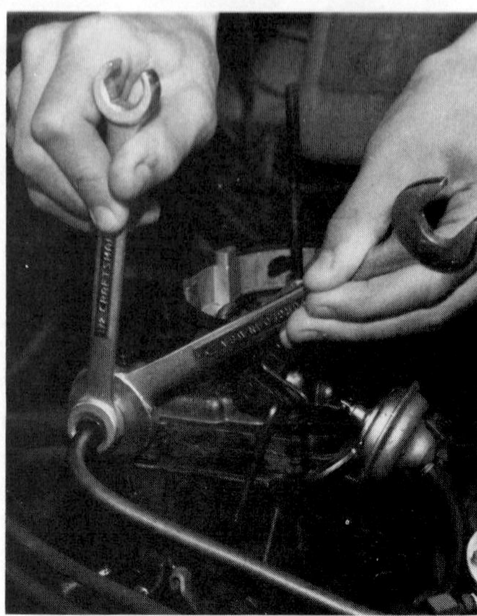

Figure 16-4. If you find leakage at the carburetor inlet, tighten the fittings.

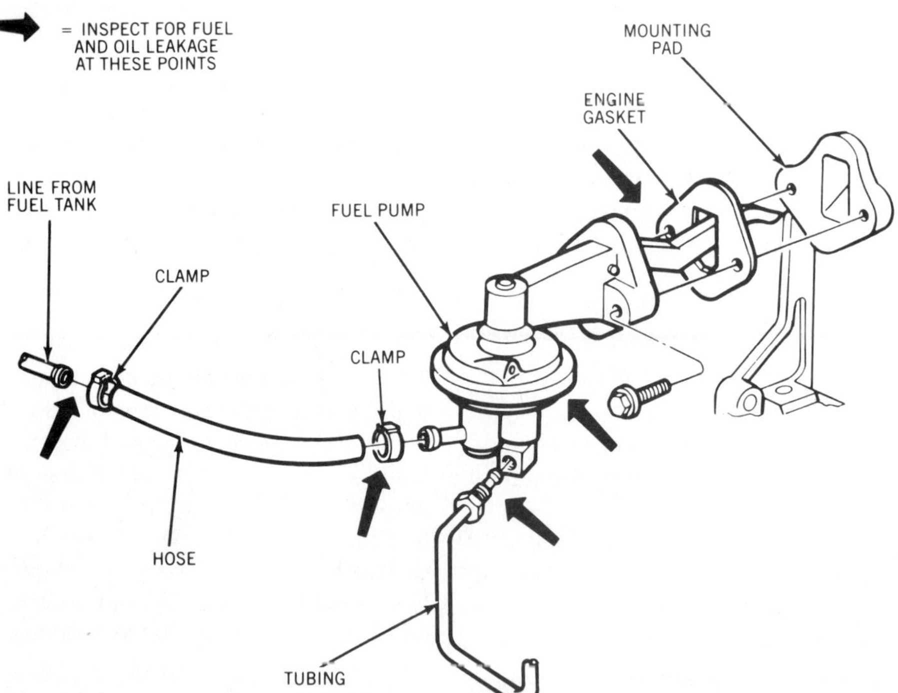

Figure 16-5. Inspect for fuel and oil leakage at a mechanical fuel pump. (Chrysler)

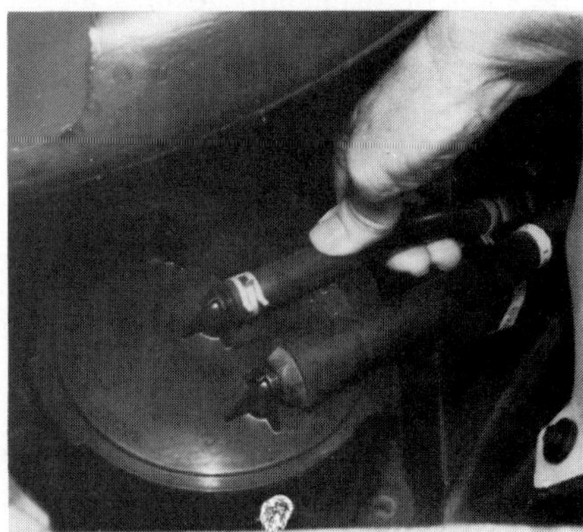

Figure 16-6. Inspect the EEC system hoses for looseness, damage, and incorrect connections.

a hoist), check the general points outlined in the following procedures.

Underhood Fuel System Inspection

1. Check the fuel line fittings at the pump, the carburetor, or the injection system for looseness and leakage.

2. Inspect rubber hoses for leakage, cracks, brittleness, and loose clamps, figure 16-2.

3. Check carburetor or throttle body mounting nuts for looseness, figure 16-3. These are common locations for vacuum leaks.

4. Check the carburetor inlet fitting and gasket areas for leakage. Tighten the fitting and carburetor body screws as necessary, figure 16-4.

5. Inspect all vacuum lines at the carburetor, the manifold, all vacuum switches and valves, and all vacuum-operated devices for leaks, brittleness, oil damage, and proper connections. Use a vacuum diagram, if necessary, to check hose connections.

6. Inspect a mechanical fuel pump for oil leakage at the mounting flange, figure 16-5. Inspect the gasket area for a break or tighten the bolts as necessary.

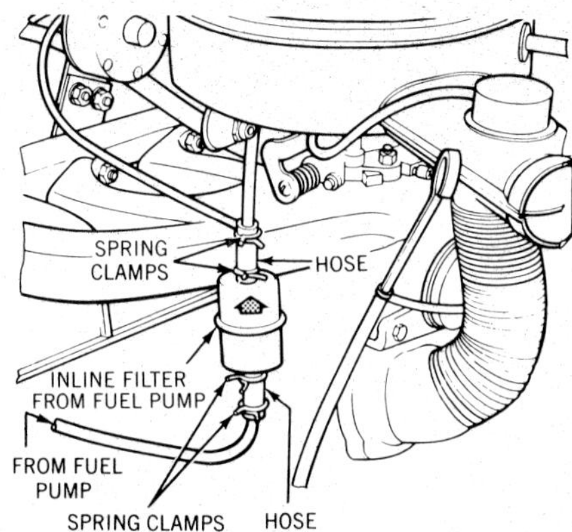

Figure 16-7. Be sure the flow arrow on an inline fuel filter points toward the carburetor or injection system. (Chrysler)

7. Check the breather hole in a mechanical fuel pump body for signs of fuel or oil.

a. If you find fuel, the diaphragm is probably broken, and fuel is passing through it. Check the engine oil for fuel dilution.

b. If you find oil at the breather hole, the diaphragm seal is probably leaking.

8. Check the EEC system vent and purge lines at the vapor canister and the engine for looseness, oil damage, and leakage, figure 16-6.

9. Check vacuum-operated purge valves according to the carmaker's procedures.

10. Check inline fuel filters to be sure they are installed in the right direction, figure 16-7, and that hose connections are tight.

Undervehicle Fuel System Inspection

1. Before raising the vehicle, inspect the fuel filler and cap, figure 16-8, for:

a. The correct cap type, vented or nonvented, according to fuel system design

b. A loose-fitting or damaged cap or leakage around the filler

c. Damage to the gas pump nozzle restrictor on vehicles that require unleaded gasoline.

2. Inspect the fuel tank for loose mountings, dents, cracks, and leakage, figure 16-8.

3. On an EEC-equipped vehicle, inspect the tank vent system for loose connections, kinked lines, and blocked vents, figure 16-9.

4. Inspect the fuel supply and return line connections at the tank for leakage, figures 16-8 and 16-9.

5. Inspect fuel supply and return lines along the length of the chassis for rust, road damage, leakage, and restrictions, figure 16-9.

6. Inspect chassis-mounted electric pumps for loose mountings, loose electrical connectors, and leakage, figure 16-10.

7. If necessary, remove a drain plug or disconnect the line from the pump

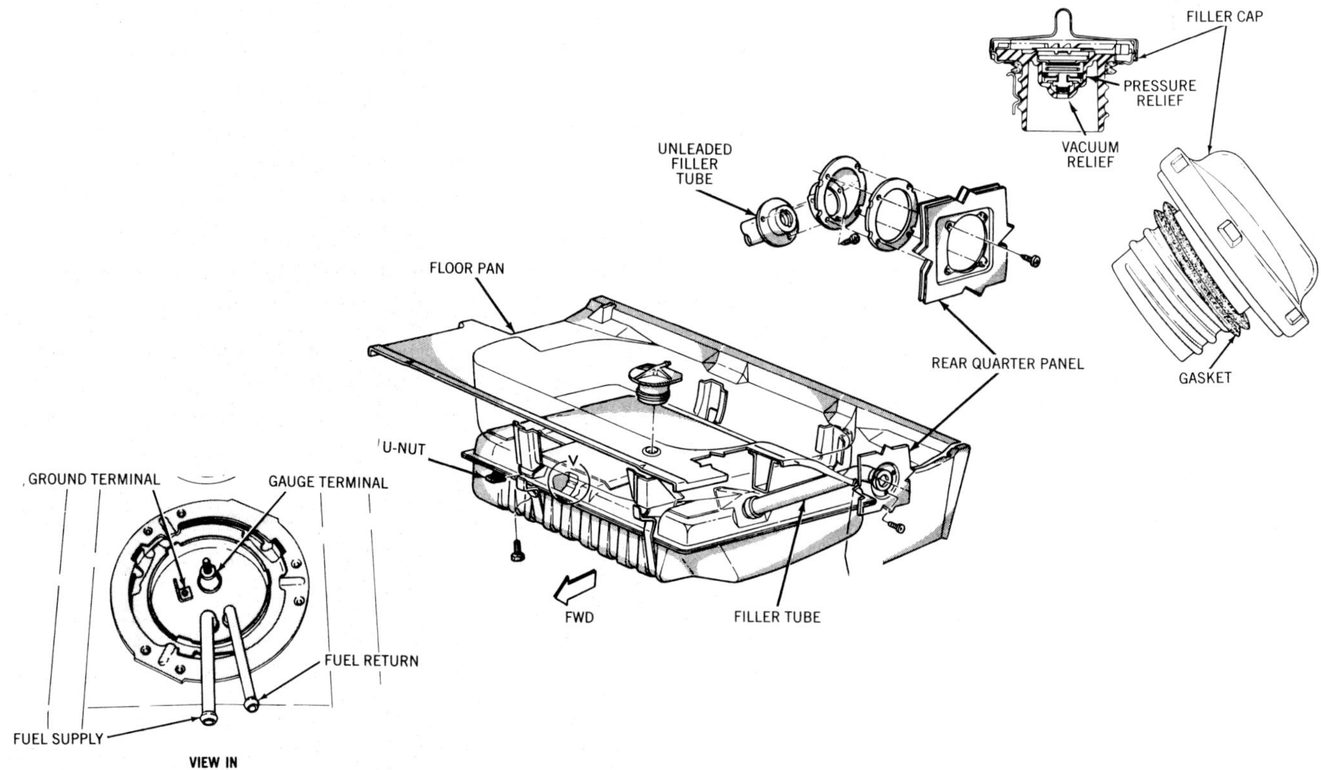

Figure 16-8. Inspect the fuel tank cap for damaged vacuum and pressure valves or gasket. Be sure the fuel filler restrictor is in place on cars requiring unleaded gasoline. (Chrysler)

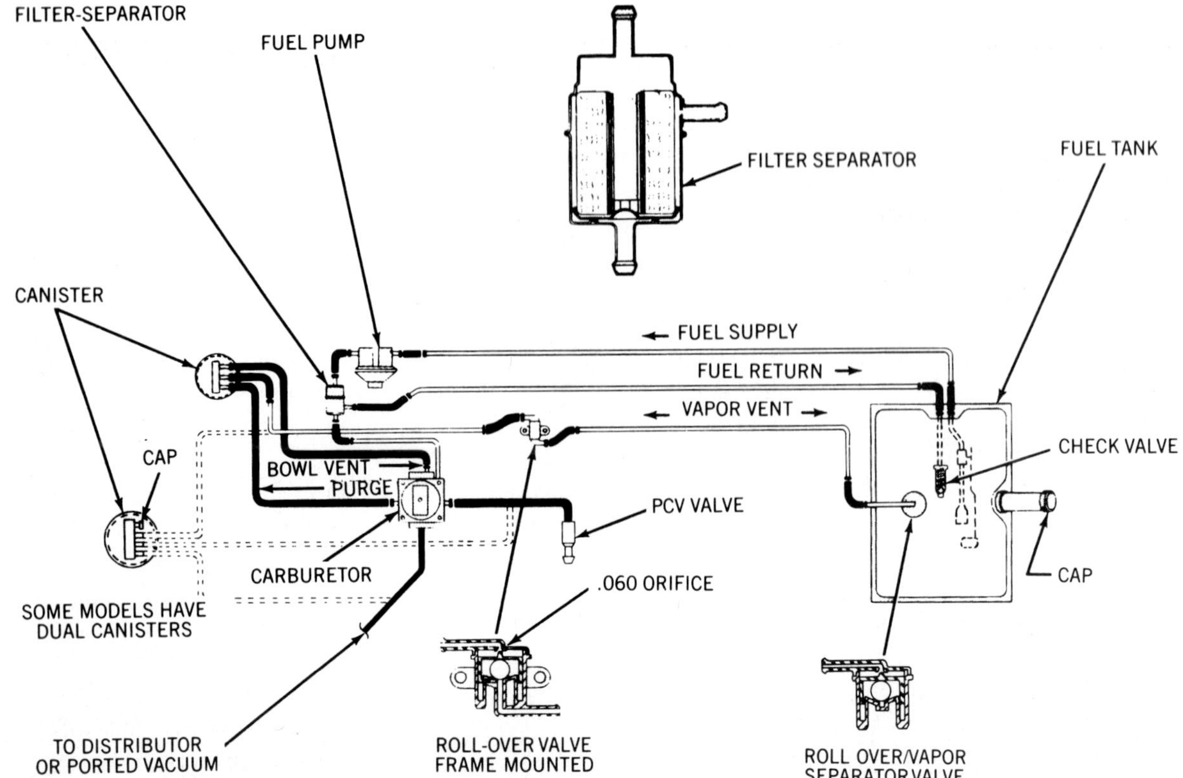

Figure 16-9. This fuel system installation shows typical inspection points for most vehicles. (Chrysler)

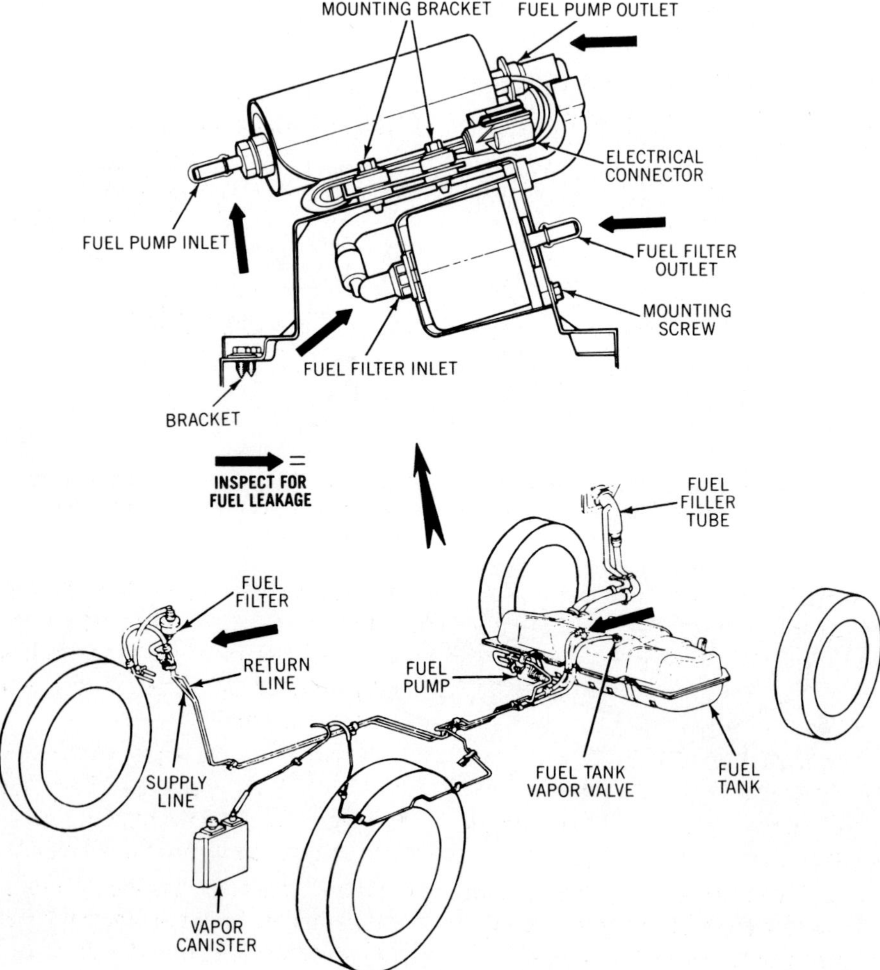

Figure 16-10. Inspect chassis-mounted electric fuel pumps for leakage, loose mountings, and damaged connectors. (Ford)

and drain some fuel into a container to inspect for dirt, sludge, or water. If the fuel is contaminated, drain the tank and remove it for cleaning or replacement.

FUEL TANK AND LINE REPLACEMENT

Fuel Tank Removal and Replacement

Fuel tank removal may not seem at first like engine service, but it can be. With in-tank electric pumps common on many late-model cars, it is often necessary to remove a tank to cure a fuel pump problem. Also, dirt, water, and other contamination in the fuel tank can cause filtration and delivery problems. Such problems can be particularly critical for diesel and gasoline injection systems. Often the only way to cure chronic fuel contamination is to remove and clean the tank.

Fuel tank installations vary from car to car, but the methods for removing and replacing a tank share the common points outlined below. If you can obtain the carmaker's drawing of the tank installation, study it before you start the job to determine the locations

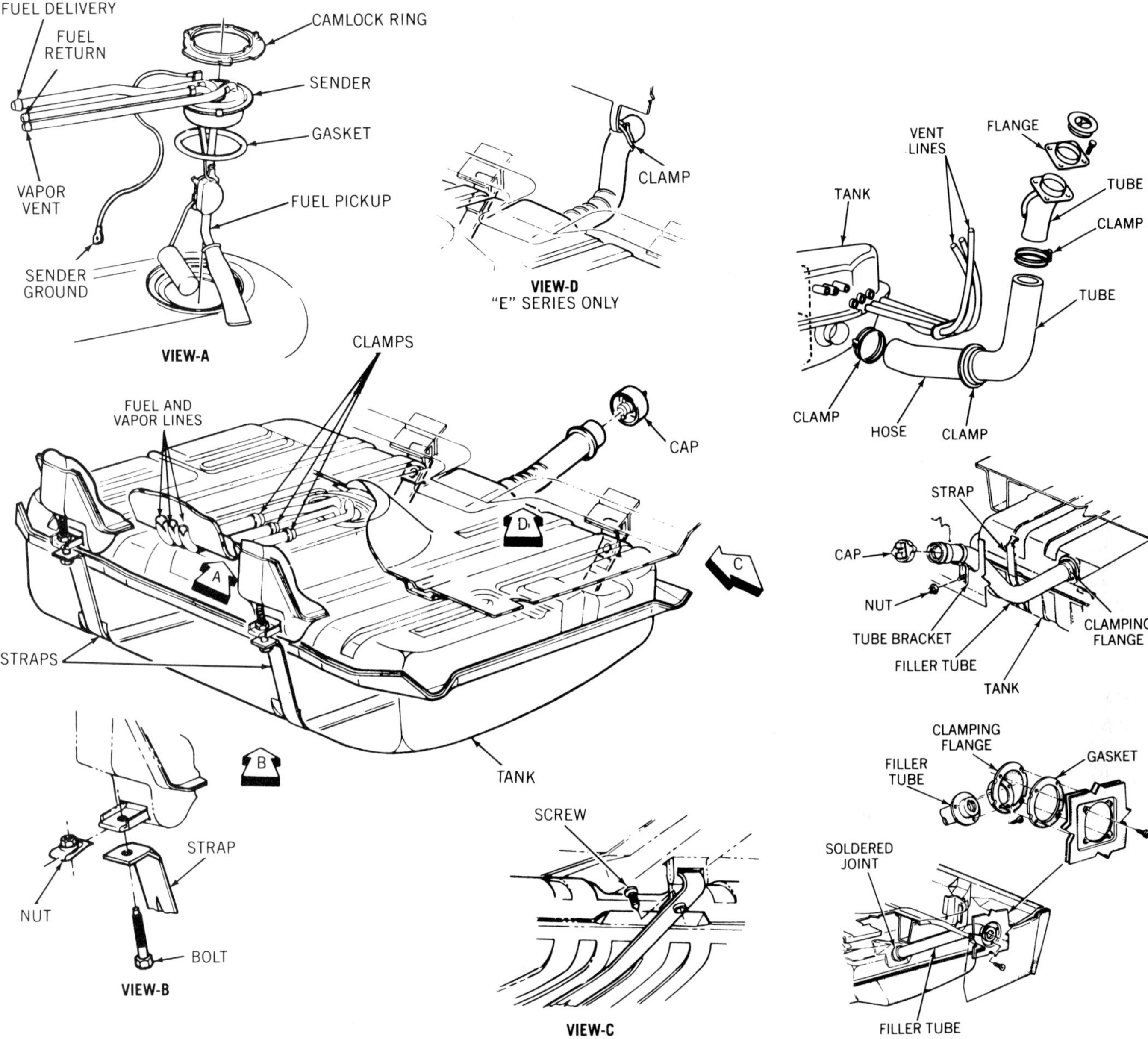

Figure 16-11. To remove a fuel tank, you will have to disconnect electrical connectors, steel fittings, and rubber hoses. (Buick)

Figure 16-12. These are common ways that a filler tube is attached to a tank or car body.

of lines, vent hoses, electrical connectors, and mounting straps.

Warning. Explosive vapors are present in an empty fuel tank. Because of the air and fuel vapor mixture, an empty tank can be more explosive than a full one. Observe all fuel and electrical safety precautions to keep heat, sparks, and flame away from a fuel tank.

1. Remove the filler cap to equalize air pressure in the tank and disconnect the battery ground cable. Depressurize a fuel injection system according to the carmaker's procedures.

2. Most tanks are under the rear of the vehicle. Following the carmaker's lifting instructions, raise the vehicle on a hoist or with a jack and stands to a convenient working height.

3. Drain remaining fuel in the tank into a metal or plastic container by:

a. Removing a drain plug from the tank.

b. Disconnecting the fuel line at the inlet of the fuel pump and using a hand-operated siphon pump to siphon fuel from the tank.

4. If electrical connectors for the fuel gauge sending unit and an in-tank electric pump are accessible, disconnect them, figures 16-10 and 16-11. You may have to wait until the tank is partly lowered from the car to reach some connectors.

5. Disconnect fuel supply and return lines and EEC vapor hoses from the tank, figure 16-11. You may have to remove a vapor separator from the tank.

6. Separate the tank from the filler tube or separate the filler tube from the car body as follows, figure 16-12:

a. If the filler tube is connected to the tank with a hose and clamps, disconnect it at this point.

b. If the tube is soldered to the tank and fastened to the body by a clamping flange, remove the flange screws.

c. If the tube is fastened to the tank by a clamping flange, disconnect it at this point.

Some filler tubes do not have to be removed from the car once they are separated from the tank. This is a good point to check the carmaker's installation drawing.

7. Many station wagon tanks are located in rear fender wells. Remove the wheel and tire for access to such a tank.

8. Remove any shields, shock absorbers, or other chassis parts that may block tank removal.

9. Place a jack or suitable support under the tank to hold its weight as you remove the mounting straps.

10. Loosen the mounting straps and lower the tank far enough to remove any other lines or wires.

11. After removing all connections and clearing obstructions, remove the mounting straps and lower the tank to the ground.

After removing the tank, you can clean it, repair it, or replace it as required. Remove the fuel gauge sending unit, any electric pumps or water separators, and other reusable parts from the old tank. You can clean some tanks by steam cleaning or with special cleaning solutions.

Warning. Do not try to repair a tank by welding. Fuel vapors trapped in the tank may explode.

To install a new or a cleaned tank, first install all reusable parts removed from the original tank. Use new gaskets and install new rubber hoses where required. Use a jack to raise the tank into position under the car and reinstall the mounting straps, filler tube, lines, and other parts.

Fuel Line Replacement

The rigid steel lines and flexible rubber hoses used in fuel systems may need replacement because of damage or deterioration. Rubber hoses can age and become brittle over a length of time. Steel tubing may suffer road damage or corrode from road salt.

Steel fuel and vapor lines are fastened to the chassis with screws and clamps, figures 16-9 and 16-10. Most rigid lines have flexible hoses at the ends that connect to tanks and pumps to absorb vehicle vibration. Many rigid lines can be replaced as preformed assemblies. Others must be cut and formed. Hoses with permanent fittings are usually replaced as assemblies. Hoses secured by clamps are cut to length from bulk material.

A damaged length of steel tubing longer than 12 inches (305 mm) can be cut out of a line and replaced with a comparable length of tubing, spliced into the line with hoses and clamps. Tube ends must be flared or beaded for proper hose and clamp retention, figure 16-13. A damaged length of tubing shorter than 12 inches can be replaced by a length of hose, secured with two clamps. Cut any replacement hose long enough to ensure proper clamping beyond the flared or beaded ends of steel tubing.

Most fuel line fittings are the 45-degree flare type, but some are the compression type. Refer to the instructions in chapter 3 for the selection of fittings and their installation. Always use double-wrapped, brazed steel tubing for fuel lines. *Do not substitute copper tubing for fuel system steel tubing.* Also, always install hoses marked for fuel system use. Vacuum and water hoses, and some plastic tubing, will deteriorate in contact with gasoline or diesel fuel.

Late-model Ford products and some other vehicles have nylon fuel tubing and push-connect fittings to join nylon and steel tubing, figure 16-14. You can disconnect these fittings by removing the retaining clips and sliding the fittings off the lines. When servicing these parts, use only nylon tubing and fittings approved by the manufacturer, figure 16-15. Do not substitute rubber hose and clamps.

FUEL PUMP TESTING

Incorrect fuel pump pressure and volume can cause fuel-metering problems at the carburetor or injection system. If gasoline supply to a carburetor is low, the fuel level in the carburetor

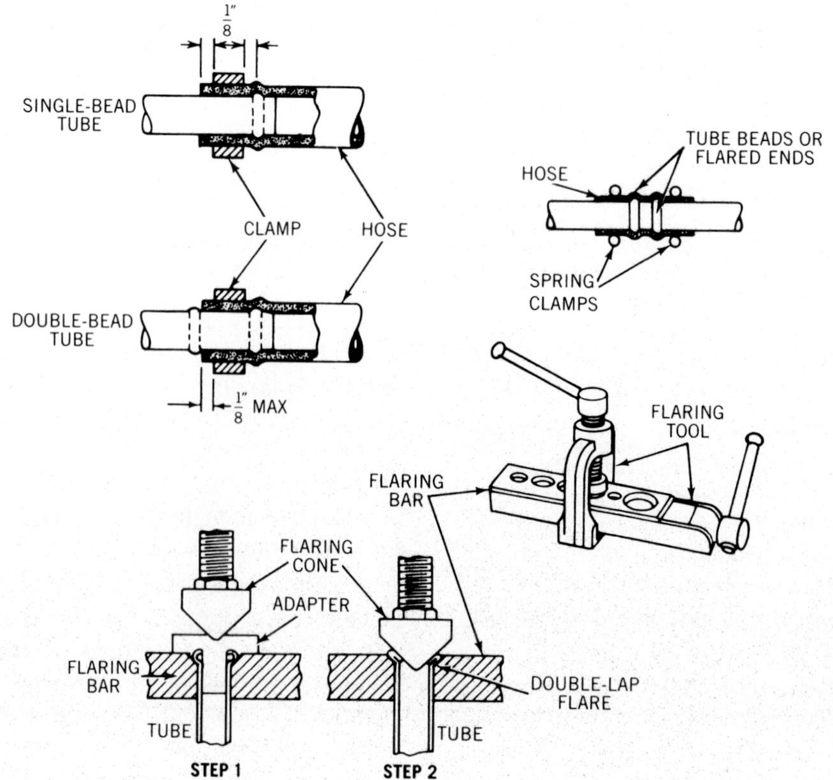

Figure 16-13. You can repair many damaged fuel lines with hoses and clamps as shown here.

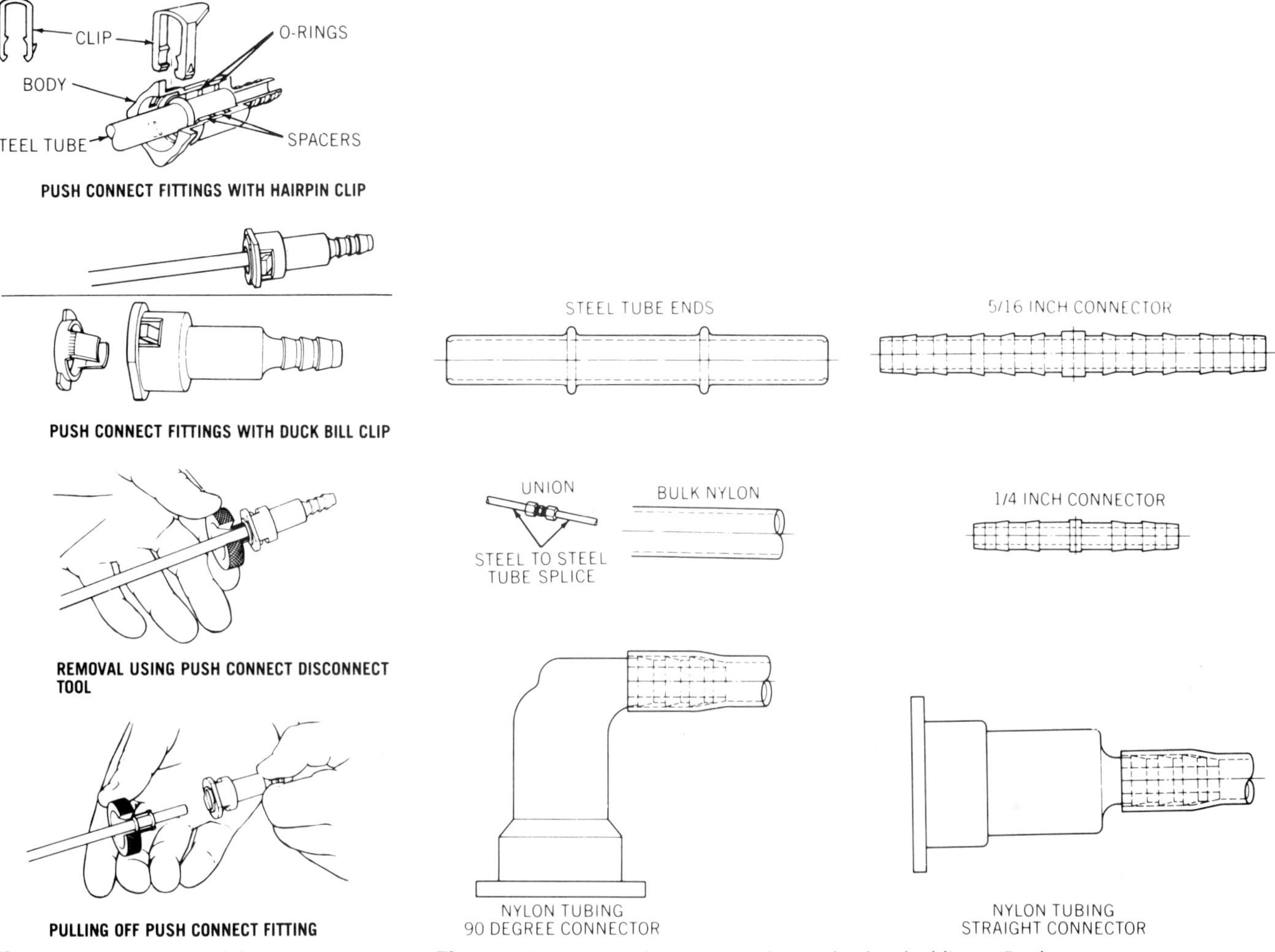

Figure 16-14. Late-model Ford products use these push-connect fuel line fittings. (Ford)

Figure 16-15. Typical connectors for Ford nylon fuel lines. (Ford)

bowl will be low, and the engine can run lean and misfire. If fuel pressure and volume to a carburetor are too high, the float needle can be forced off its seat and the fuel level in the bowl will be too high. The engine will then run rich.

Gasoline and diesel fuel injection systems are not as susceptible to high-pressure or high-volume problems as carburetors are because injection systems always operate with excess fuel and have relief valves, pressure regulators, and return lines to bypass excess fuel back to the tank. Low pressure or low volume, however, can starve an injection system just as it can a carburetor.

All carmakers publish pressure specifications for fuel pumps, and many also supply volume specifications. Pressure specifications are usually in pounds per square inch (psi) at idle or cranking speed. Volume specifications are usually for 1 pint of fuel in a specific number of seconds at idle or at cranking speed. Some volume specifications require only "good volume at cranking speed." This generally means 1 pint in 30 seconds or less at 200 to 400 rpm.

The following pressure and volume test procedures are general methods for checking mechanical and electric pumps on gasoline engines and separate lift or transfer pumps on diesel engines. Late-model diesel and gasoline injection systems may require tests that differ slightly from these.

Depressurize a fuel injection system before opening a line to test a fuel pump.

Fuel Pump Pressure Test

To test the fuel pump on most carbureted engines, you can use the same vacuum-pressure gauge you use to test engine vacuum, which indicates pressures up to about 10 psi. For some fuel-injected engines, you will need a gauge that measures pressures of 30 to 50 psi. For accurate readings, hold the gauge at the height of the carburetor or injector inlet. Use a short connecting hose or length of tubing of 6 inches or less. If the pump has a vapor return line, clamp it closed during testing. Proceed as follows:

1. Remove the air cleaner or air inlet duct for access to the fuel line at the carburetor or injector inlet.

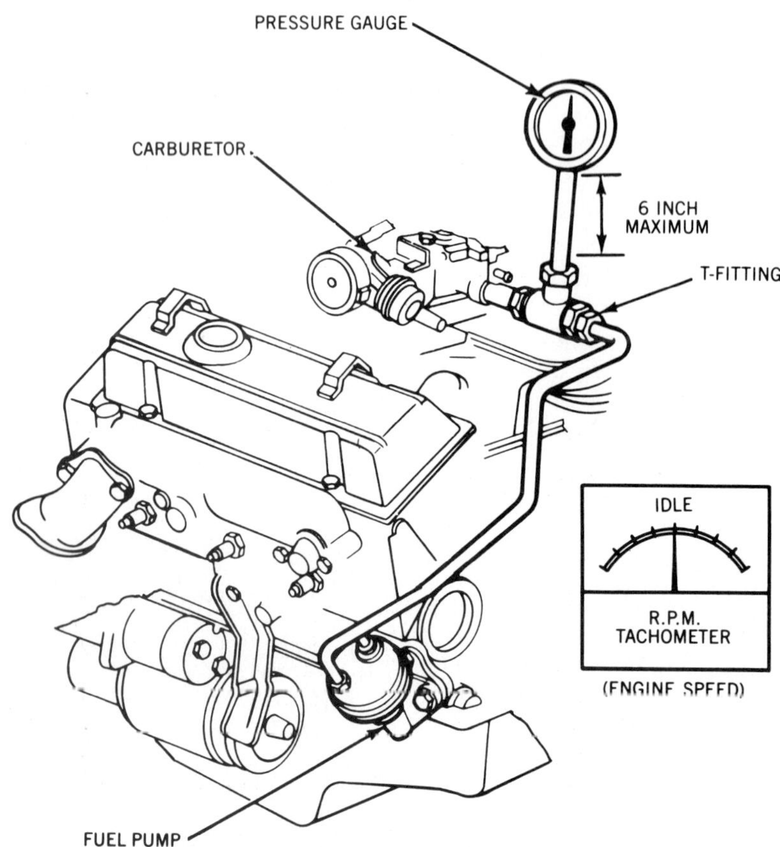

Figure 16-16. For accurate fuel pump pressure testing, install the gauge with a T-fitting at the carburetor inlet. (Chrysler)

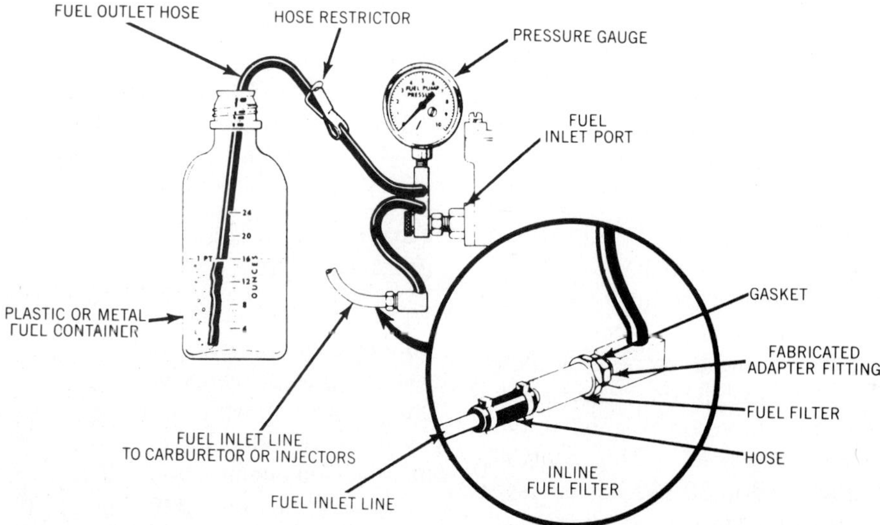

Figure 16-17. You can use this test setup to check pump pressure and volume. (Ford)

2. Disconnect the fuel line at the carburetor or injector inlet and install a tee as shown in figure 16-16.

a. If pressure specifications are below 10 psi, connect the gauge to the tee with a length of fuel hose.

b. If pressure specifications are above 10 psi, connect the gauge with fittings and high-pressure hose or tubing to prevent leakage.

3. Test pump pressure in one of the following three ways, depending on the manufacturer's specifications:

a. If pressure is measured at cranking speed, disable the ignition and crank the engine until the gauge reading stabilizes. Do not crank the engine for more than 30 seconds to avoid overheating the starter motor.

b. If pressure is measured at idle, start the engine and run it at idle.

c. If electric pump pressure is measured with the engine off, turn the ignition to RUN and let the pump operate through its starting cycle, or connect the pump directly to a 12-volt battery for 10 seconds.

4. Compare the gauge reading to specifications. Continue watching the gauge for about 2 minutes. If the pump valves are sealing correctly, pressure should remain stable for several minutes.

5. Disconnect the gauge, remove the tee, reconnect the fuel line, and reinstall the air cleaner.

Fuel Pump Volume Test

For this test, you will need a graduated metal or plastic (not glass) container and the test equipment shown in figure 16-17. You also can use this tester to check pump pressure on many engines. To test volume, proceed as follows:

1. Remove the air cleaner or air inlet duct and disconnect the fuel line at the carburetor or injector inlet.

2. Install the test gauge, hose, and fuel container, figure 16-17. If the pump has a vapor return line, clamp it closed.

3. Open the shutoff clamp on the hose to the fuel sample container.

4. Crank or run the engine for the time specified by the manufacturer. Measure the fuel in the container and compare it to specifications. If air bubbles appear in the sample during testing, the fuel line between the tank and the pump probably has an air leak.

Low pressure and volume are often caused by restricted filters, lines, or the fuel pickup in the tank. Always check the lines and filters for restrictions before replacing a pump to cure a low-pressure problem. In a mechanical pump, a leaking diaphragm, a leaking inlet check valve, a worn

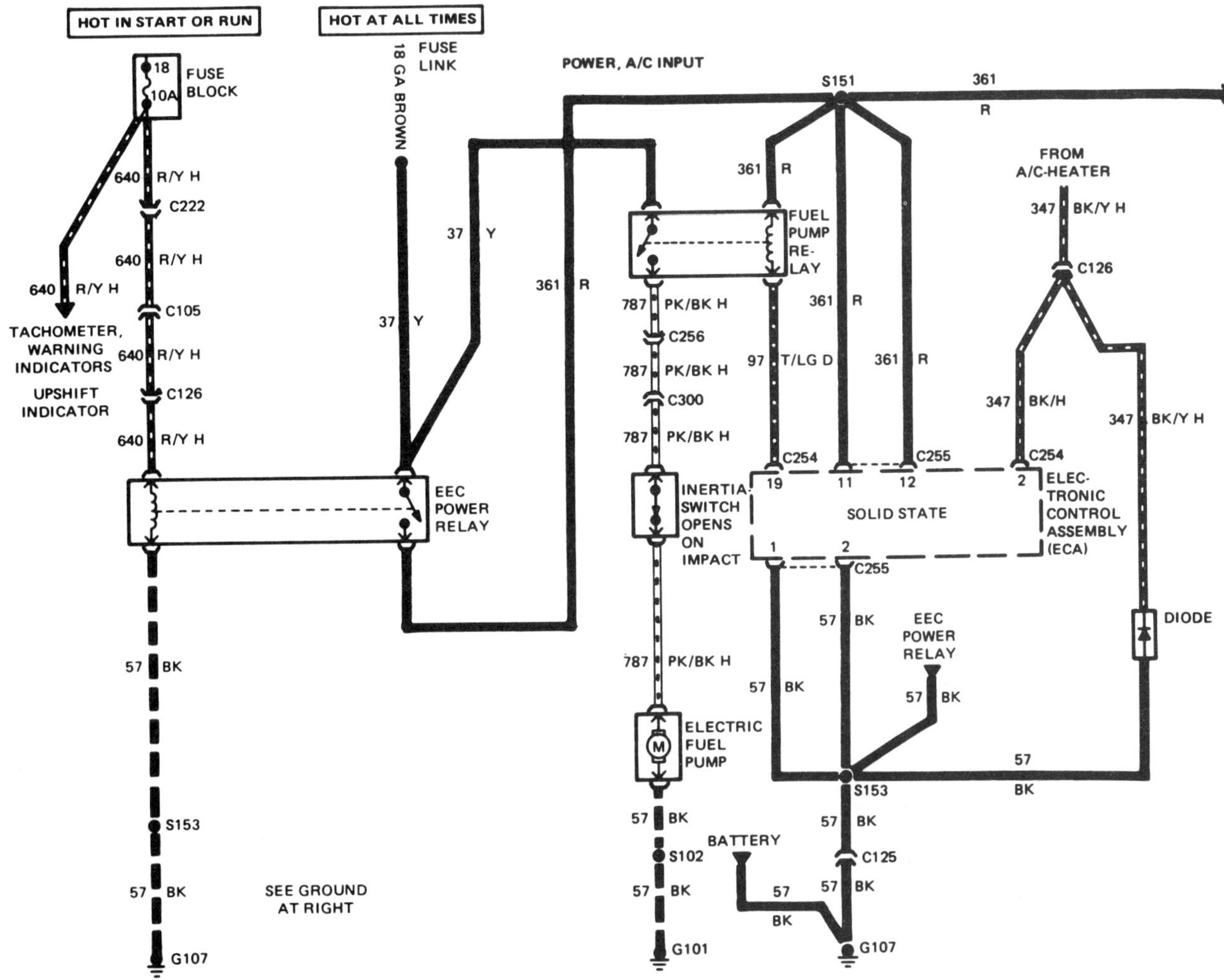

Figure 16-18. To test electric fuel pump operation, refer to a diagram like this one. (Ford)

pushrod or pump linkage, or a worn camshaft drive eccentric also can cause low pressure. High pressure is usually caused by a defective pump check valve or diaphragm spring.

Electric Fuel Pump Testing

To test an electric fuel pump installation accurately, you will need the carmaker's procedures and wiring diagram from a service manual, figure 16-18, to supplement the preceding basic pressure and volume tests. Many carmakers have specific test instructions, particularly to bypass oil pressure switches or relays in the fuel pump circuit.

Besides testing for pressure and volume, you may have to check an electric pump for system voltage, ground connections, and all other electrical connections and switches. Also, circuit connections and color coding can vary between different years and models. These are additional reasons to refer to the carmaker's procedures and diagrams. Don't forget, however, that an inoperable electric fuel pump can be caused by nothing more than a blown fuse. Check this first before doing lengthy electrical tests.

FUEL PUMP REPLACEMENT

A mechanical fuel pump cannot be repaired; it can only be replaced if defective. Repair kits are available for some electric pumps, but most are replaced if defective.

Mechanical Fuel Pump Replacement

Most mechanical fuel pumps are mounted low on the engine where they are driven by the camshaft or an accessory drive shaft. Some are on the side of the timing cover. In most cases, access to the pump is limited by manifolds and engine accessories. Use the following guidelines and the carmaker's instructions to remove and replace a pump:

1. Disconnect the battery ground cable.

2. Remove any belt-driven accessories as needed for access to the pump. You can reach some low-mounted pumps more easily from under the car.

3. Disconnect the pump inlet and outlet lines and plug or cap them, figure 16-19. If the pump has a vapor return line, disconnect and cap it also.

4. Loosen, but do not remove, the pump mounting bolts.

5. Crank the engine until the pump rocker arm or pushrod is on the low point of the camshaft eccentric to release tension from the pump rocker arm or linkage.

6. Remove the pump from the engine, figure 16-20.

7. If the pump is driven by a pushrod, as on Chevrolet small-block engines and some Ford 4-cylinder engines, remove the pushrod and check it for wear and length, figure 16-21. Use a straightedge to check its straightness.

8. Remove old gasket material and dirt from the mounting surface on the engine.

9. Apply gasket cement to both sides of a new pump gasket and place it on the pump flange.

10. Install the pump on the engine in one of the following ways:

a. If the pump rocker arm is driven directly by the camshaft, insert the arm through the mounting hole and be sure it bears correctly on the shaft. Install the mounting bolts.

b. If the pump has a pushrod between the rocker arm and the shaft, put a wad of heavy grease on the shaft end of the pushrod and insert it in the engine, against the eccentric, figures 16-20 and 16-21. The grease will hold it while you install the pump. Place the pump arm against the pushrod and install the bolts.

11. Tighten the mounting bolts alternately and evenly to the specified torque.

12. Reconnect the fuel lines and vapor return line to the pump. Use new clamps on rubber hoses and be careful not to crossthread fittings. If the outlet fitting is hard to install, loosen the line at the carburetor or injector inlet to provide some movement in the line. Tighten all clamps and fittings securely.

13. Reinstall any accessories removed for pump access and reconnect the battery ground cable. Start the engine and check for leaks.

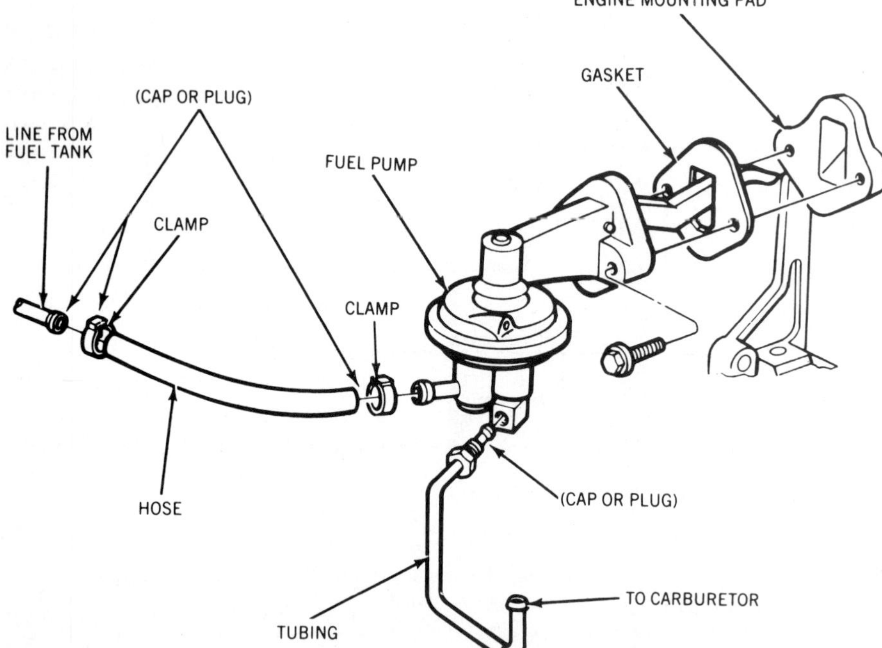

Figure 16-19. Disconnect the lines at the pump and cap the fittings. (Chrysler)

Electric Fuel Pump Replacement

Refer to the carmaker's instructions before servicing an electric fuel pump because there are more variations in electric pump installations than in mechanical pump installations. However, removing and replacing most electric pumps will include one or more of the following tasks:

1. Depressurize a fuel injection system according to the manufacturer's instructions before opening any fuel lines.

2. If necessary, raise the vehicle on a hoist for access to a chassis-mounted or tank-mounted pump. You can re-

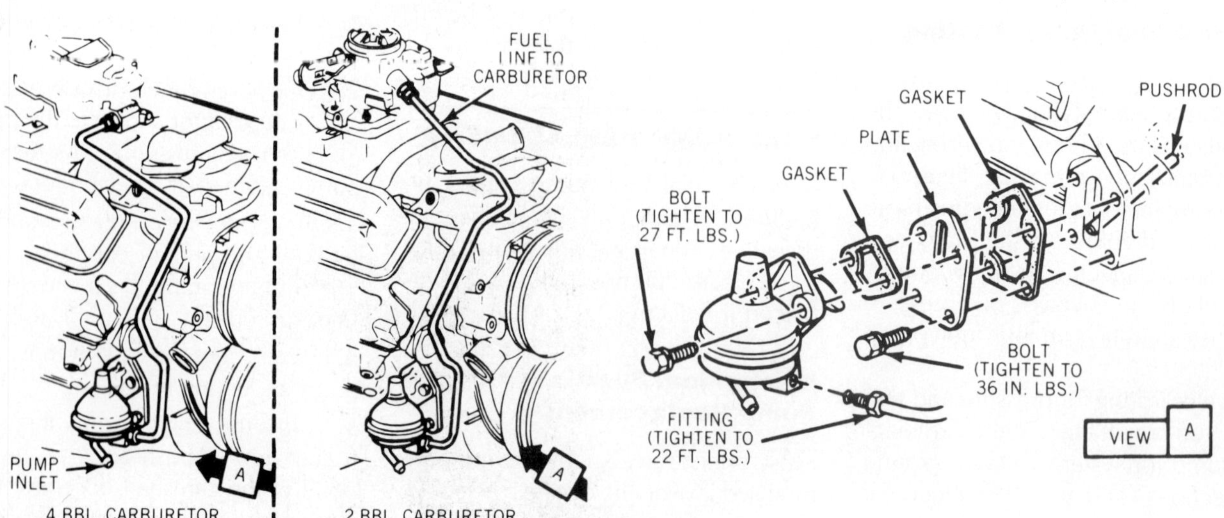

Figure 16-20. Remove the pump from the engine, along with the pushrod if one is used. (Chevrolet)

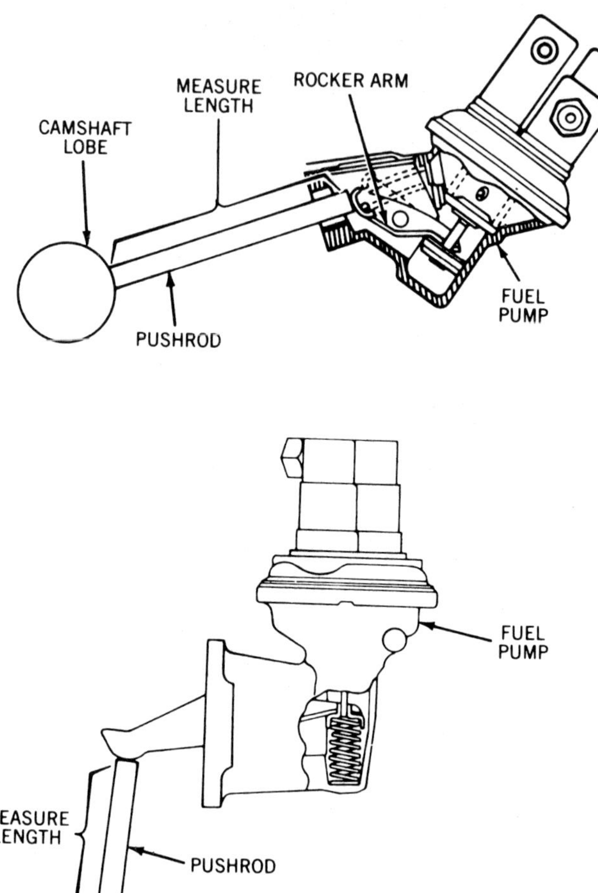

Figure 16-21. Measure the pushrod length and compare it to the manufacturer's specifications. Replace it if it is worn or bent. (Ford)

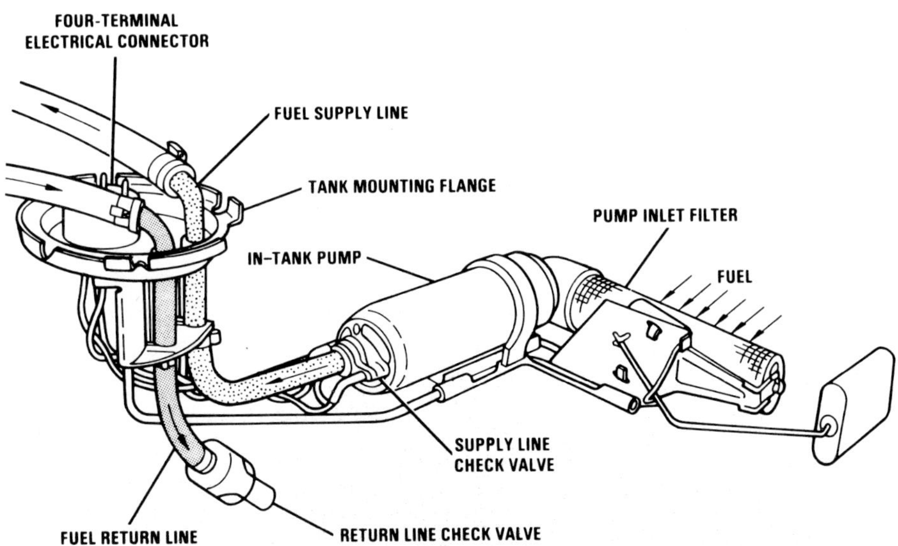

Figure 16-22. You must drop the fuel tank for access to many in-tank pumps. (Chrysler)

move some tank-mounted pumps through an access hatch in the car trunk and a cover on the tank.

3. If necessary, remove the fuel tank for access to an in-tank pump, figure 16-22.

4. Refer to a wiring diagram and test all electrical circuits for continuity and switch or relay operation.

5. Use new gaskets and O-rings where required when installing a pump. It also is a good idea to replace filters when installing a new pump.

FUEL FILTER SERVICE

Gasoline and diesel fuel filters should be replaced at the manufacturer's recommended intervals or whenever system contamination is diagnosed. Diesel fuel conditioners and water separators also may require draining and cleaning.

Gasoline Filter Replacement

Most gasoline filters are the inline type or the carburetor inlet type. Some engines, however, have fuel pump outlet filters or canister-type filters. Depressurize a gasoline injection system before changing the filter.

Inline Filter Replacement. Locate the filter in the fuel line and remove the retaining clamps and hoses. Remove the old filter and install the replacement with the flow arrow pointing toward the carburetor or injection system, figure 16-23. Replace the hoses and clamps if damaged or worn. If the filter has a vapor bypass line, connect it to the new filter. Do not press too hard on a plastic or thin metal filter case. If you distort the filter, it may leak or restrict fuel flow.

Some cars with electric pumps have large-capacity inline filters under the chassis near the tank, figure 16-24. Raise the car on a hoist for access to these filters.

Carburetor Inlet Filter Replacement. GM Rochester carburetors have filters inside the fuel inlet fitting, figure 16-25. Many Ford Motorcraft car-

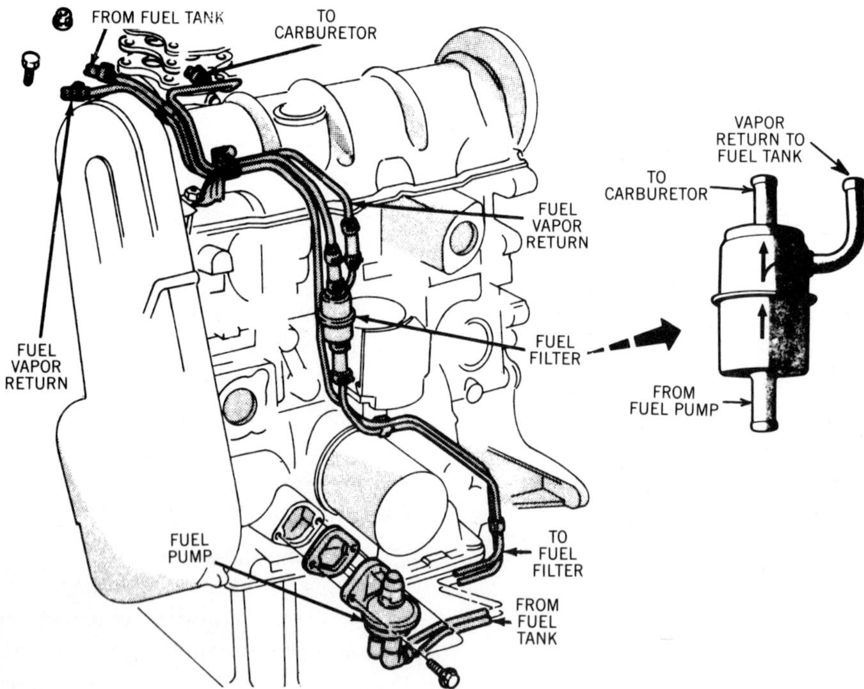

Figure 16-23. Locate the inline fuel filter and install the replacement with the arrow toward the carburetor or injection system. (Chrysler)

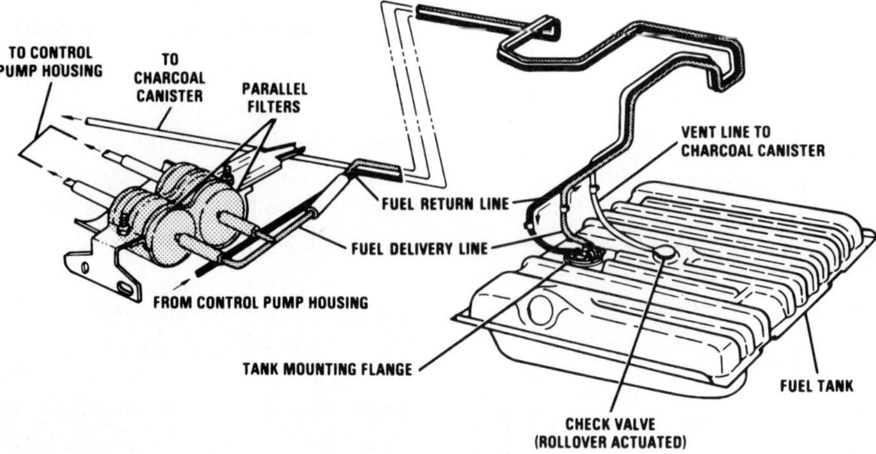

Figure 16-24. Many gasoline injection systems have large inline filters under the vehicle. (Chrysler)

buretors have filters that screw into the carburetor inlet and attach to the fuel line with a hose and clamps, figure 16-26, or with an inverted flare fitting.

To replace a Rochester inlet filter, disconnect the fuel line at the carburetor inlet and remove the inlet fitting. Remove the gaskets, filter, and spring. Install the spring, new filter, and gaskets as shown in figure 16-25. Be sure the filter is in the right direction. If installed backwards, it will restrict fuel flow. If the vehicle requires a filter with a rollover check valve, be sure you install this type. After replacing the filter, install the fuel inlet fittings carefully to avoid crossthreading.

To replace a Motorcraft or other external inlet filter, disconnect the fuel line and unscrew the filter from the carburetor inlet. Install the new filter and reconnect the fuel line. Be careful to avoid crossthreading the fittings.

Fuel Pump Outlet Filter Replacement. Older Chrysler 6-cylinder engines, figure 16-27, and some Cadillacs have filters in the fuel pump outlets. Replace these filters by unscrewing the fitting cap, removing the old filter, and installing a new one.

Canister Filter Replacement. Some engines have large-capacity canister-type fuel filters that look like replaceable oil filters, figure 16-28. Fuel-injected Cadillacs have these filters mounted on the chassis near the tank or on the engine. Some Ford engines have canister filters on the fuel pumps. The location may vary from car to car, but you can replace a canister filter simply by unscrewing the old unit and installing the new one.

Diesel Fuel Filter and Fuel Conditioner Service

Diesel fuel filters are larger capacity than gasoline filters, and many contain water separators. Follow the carmaker's maintenance schedule for filter service. Drain water from the system at recommended intervals or whenever a water-in-fuel lamp indicates the presence of water.

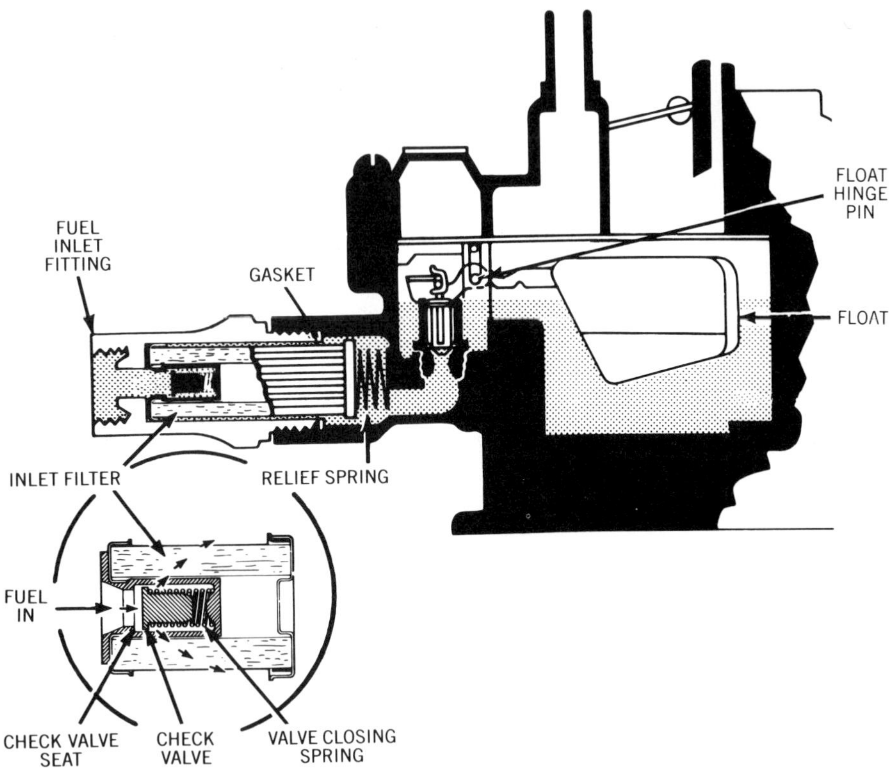

Figure 16-25. Typical Rochester carburetor inlet filter. (Chevrolet)

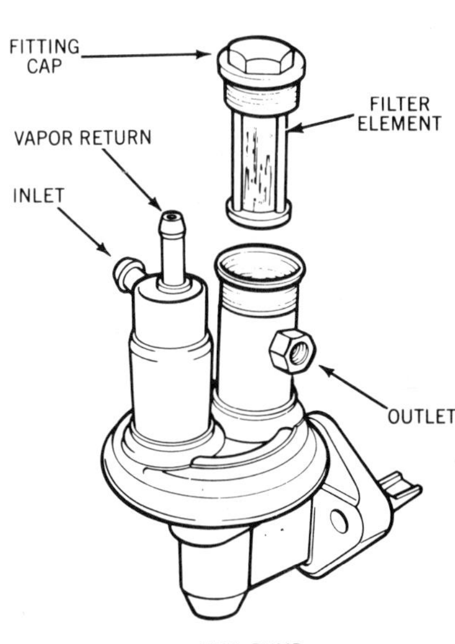

Figure 16-27. Replace a pump outlet filter by unscrewing the fitting cap and removing the filter. (Chrysler)

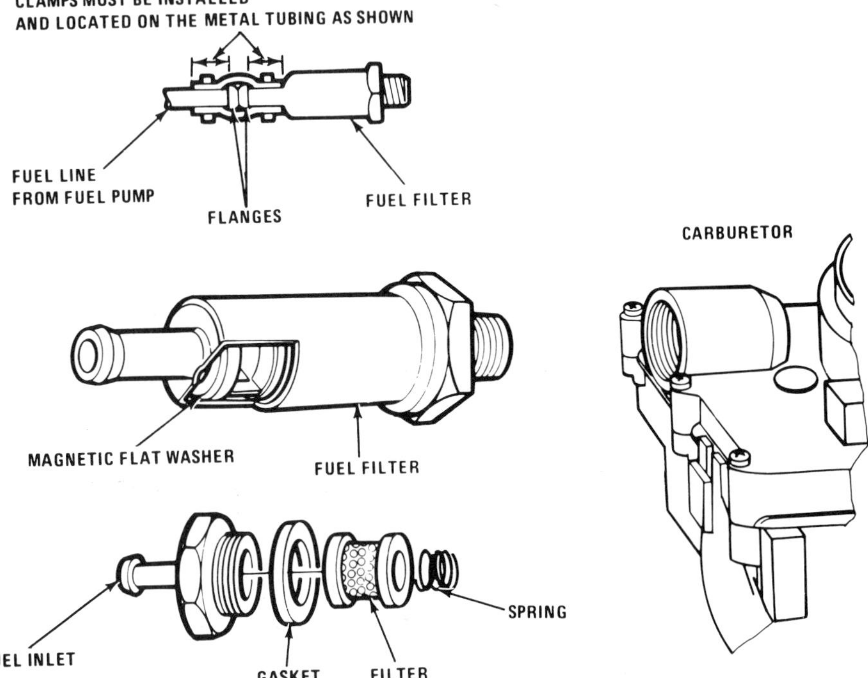

Figure 16-26. Typical Motorcraft carburetor inlet filter. (Chrysler)

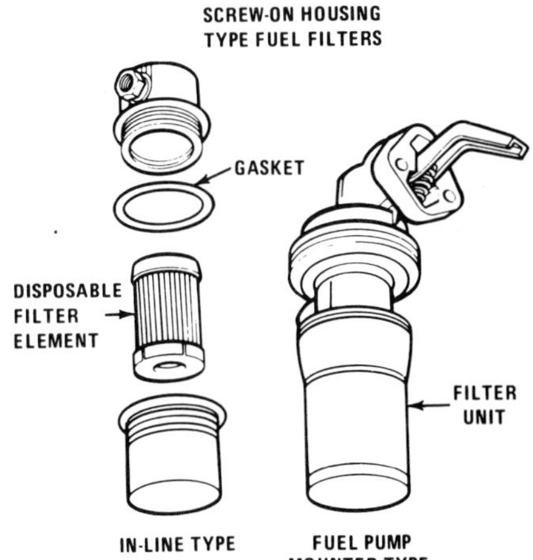

Figure 16-28. Canister fuel filters screw into place like replaceable oil filters. (Chrysler)

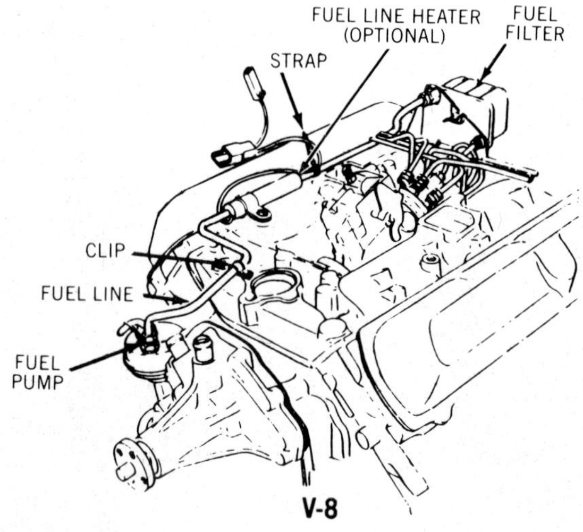

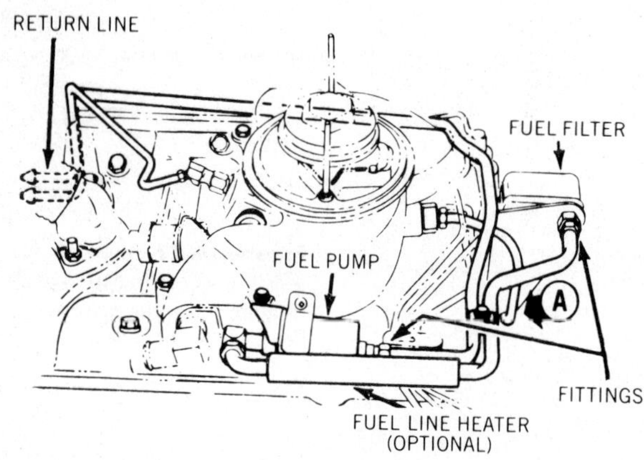

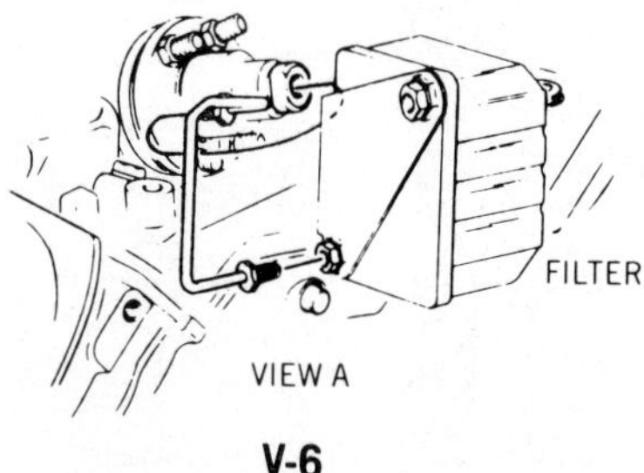

Figure 16-29. GM V-6 and V-8 light-duty diesels have fuel filters like this. (Chevrolet)

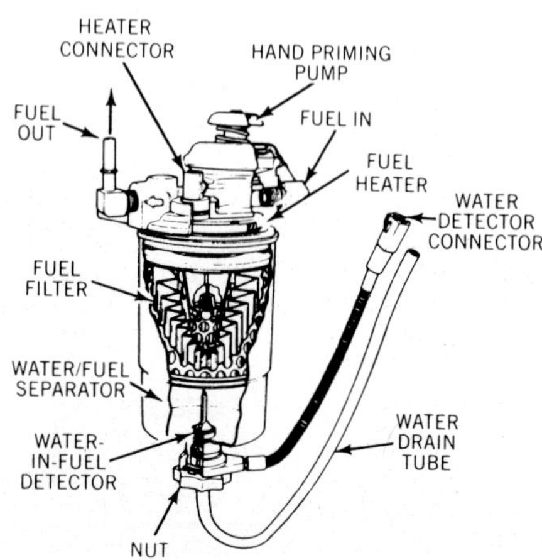

Figure 16-30. Many Ford and Japanese diesel passenger cars have fuel conditioners like this. (Ford)

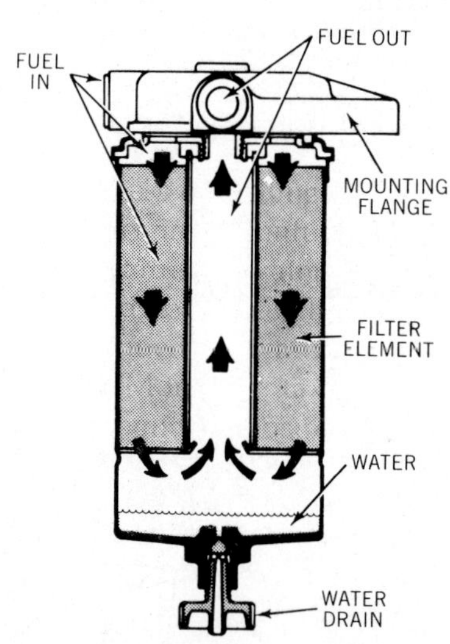

Figure 16-31. Bosch diesel fuel systems have large-capacity fuel filters like this. (Volkswagen)

GM light-duty diesels have a filter mounted on a bracket at the rear of the engine, figure 16-29. Disconnect the fuel lines from the filter with flare-nut wrenches. Remove the old filter from the bracket. Prime the new filter by filling it with diesel fuel and install it on the bracket. Reconnect the fuel lines. If the water-in-fuel lamp on a GM car indicates water in the tank, siphon the tank through the fuel return line at the fuel transfer pump.

Ford diesel cars have a fuel conditioner that contains a filter, a water separator, a heater, and a water indicator, figure 16-30. To drain water from the conditioner, place a container under the water drain hose on the car and open the drain valve on the bottom of the housing 2 1/2 to 3 turns. Operate the priming pump on the conditioner until all water is pumped from the filter and clear fuel flows from the hose. Then close the drain valve.

The Ford diesel fuel filter screws onto the fuel conditioner body like a replaceable oil filter. To change the fuel filter, disconnect the module connector from the water sensor pigtail lead and unscrew the filter with a filter wrench. Remove the water drain valve and sensor from the old filter and install them in a new one. Screw the filter onto the housing hand tight and reconnect the water sensor connector. Prime the system with the pump on the fuel conditioner until fuel flows from the outlet fitting.

Bosch diesel systems have filters with replaceable cartridges. Remove the filter housing from the filter body, remove the old cartridge, and install a new one. Some Bosch filters also have drains on the housings that you can open to remove water from the sys-

Figure 16-32. Replace the air filter if it is dirty or contaminated by PCV blowby.

Figure 16-33. Use a shop light to check for breaks or obstructions in an air filter.

tem, figure 16-31. A Bosch filter also may have a priming pump to remove air from the system. To prime such a system, loosen the vent screw on the filter housing and operate the hand pump until fuel flows from the vent. To prime a Bosch system without a hand pump, loosen the fittings at two injection nozzles and crank the engine with the glow plugs disabled until fuel emerges from the loosened fittings.

AIR FILTER SERVICE

Air filter service is one of the easiest engine maintenance tasks. Filters should be changed at the carmaker's recommended intervals or sooner if dirty.

To replace a filter, remove the air cleaner cover or open the air filter housing. Remove the filter and inspect it for dirt or oil contamination from PCV system blowby, figure 16-32. Several testers are available to measure the air pressure drop across the filter, but it is usually cheaper to replace the filter than to take the time to test it. You can check a filter quickly by holding a shop light inside or behind the filter and looking through it, figure 16-33. If you cannot see the glow of the lamp, the filter is probably restricted, even if it looks clean on the outside. Also use the light to check for holes or cracks. If you find any, replace the filter.

When you replace a filter, install it in the housing, observing any markings for filter direction. Do not overtighten the wingnut holding the cover of a carburetor-mounted air cleaner. Overtightening can distort the carburetor airhorn and cause the choke to stick.

THERMOSTATICALLY CONTROLLED AIR CLEANER SERVICE

Air cleaner service consists of:

- Checking all vacuum hoses and air ducts to be sure they are connected securely and not leaking
- Testing vacuum motors, sensors, or thermostatic bulbs
- Replacing defective parts.

Air Cleaner Inspection, Removal, and Installation

You will have to remove the air cleaner and some of the air ducts to service the carburetor or injection system. Remove, inspect, and install the air cleaner as follows:

1. Check the air ducts and connections for looseness and breaks.

2. When disconnecting PCV and vacuum lines from air cleaners, carefully note their positions for reinstallation. Tag vacuum lines with pieces of tape, if desired, to aid reconnection. Replace worn or broken hoses.

3. Check the gasket at the carburetor airhorn or throttle body. Be sure it fits squarely and is not distorted or broken. A bad gasket at this point can allow unfiltered air into the engine and cause the airhorn to warp when the air cleaner is installed.

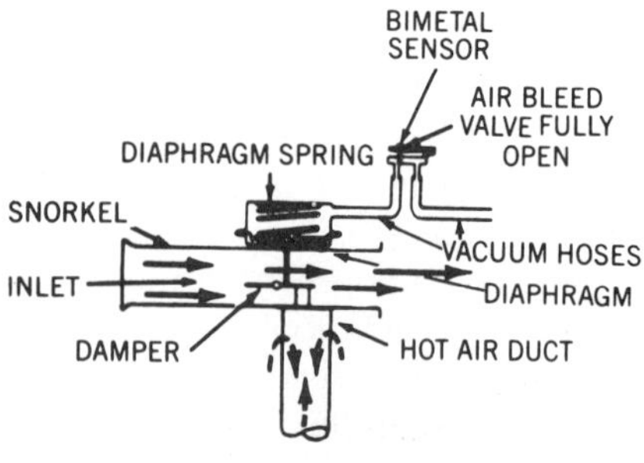

Figure 16-34. With the engine off, a vacuum-operated air cleaner should be in the cold-air position. (GM)

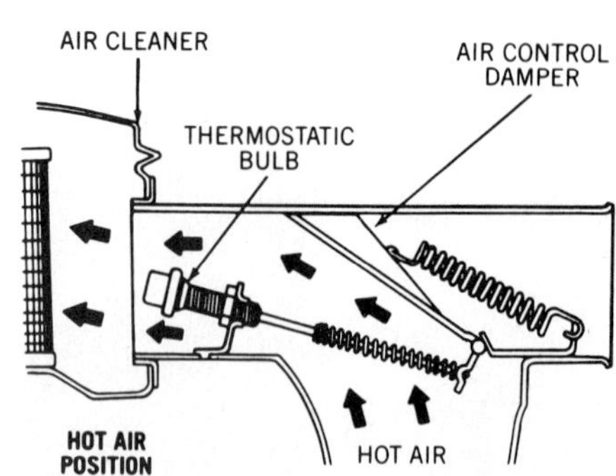

Figure 16-35. With the engine off or cold, a thermostatic bulb air cleaner should be in the hot-air position. (Ford)

Figure 16-36. Check vacuum motor operation with a hand-operated vacuum pump.

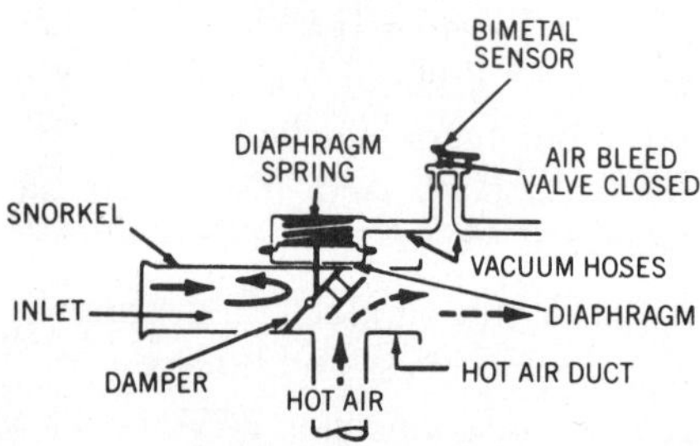

Figure 16-37. With the engine running and cold, vacuum should pull the damper to the hot-air position. (GM)

4. Inspect the air control damper in the air cleaner:

a. On a vacuum-operated air cleaner, the spring in the vacuum motor should hold the damper in the cold-air position when the engine is off, figure 16-34.

b. On a thermostatic bulb air cleaner, the bulb should hold the damper in the hot-air position when the engine is off, figure 16-35.

The linkage should not be loose or rattle, but should move freely when you press on the damper.

Air Cleaner Vacuum Motor Test

1. With the engine off, check all vacuum hoses for damage and incorrect connections.

2. Verify that the damper is in the cold-air position, figure 16-34. It should cover the hot-air inlet and open the cold-air duct, or snorkel.

3. Disconnect the hose from the vacuum motor on the cold-air duct, or snorkel.

4. Attach a hand-operated vacuum pump to the vacuum motor and apply 9 or 10 inches of vacuum, figure 16-36.

5. Check the damper to see if it has moved to the hot-air position, figure 16-37. If it has not, the linkage or the vacuum motor is broken.

6. If the damper moves to the hot-air position, keep vacuum applied to the motor for about 30 seconds and watch the pump gauge. Vacuum should remain steady. If vacuum drops and the damper opens, the vacuum motor diaphragm is leaking.

Vacuum Motor Replacement

Some manufacturers do not supply replacement vacuum motors, and you must replace the whole air cleaner housing if the motor is bad. On many air cleaners, however, you can replace a vacuum motor as follows:

1. Disconnect all air cleaner ducts and hoses and remove the air cleaner from the engine.

2. Disconnect the hose at the vacuum motor that runs to the temperature sensor.

3. Using a 1/16-inch drill, drill out the spot welds or lock tabs that hold the motor to the air cleaner, figure 16-38. On some older Chrysler air cleaners, reach inside the snorkel and bend the locking tab down. You do not have to drill the vacuum motor mounting to replace it.

4. Raise the motor from the housing and tilt it for access to the damper linkage. Unhook the linkage from the damper and remove the motor.

5. Connect the damper linkage to a new vacuum motor and install it as follows:

a. Insert the locking tabs into the slots in the snorkel, align the motor, and install a pop rivet or bend the tabs into place.

b. Drill 7/64-inch holes in the tabs of the motor, place the motor in position, and install it with short

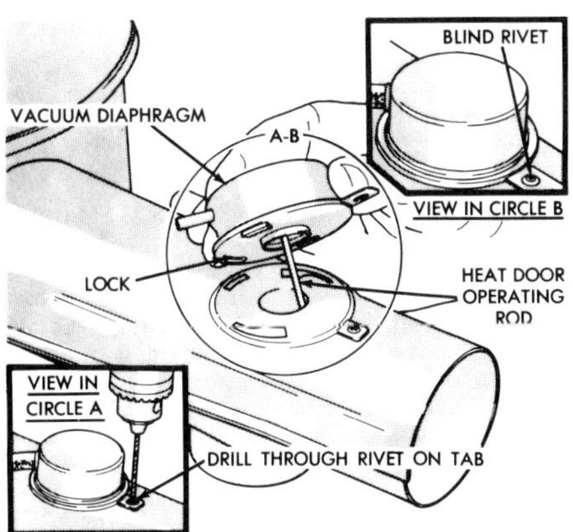

Figure 16-38. Drill out the mounting tabs to remove most vacuum motors. (Chrysler)

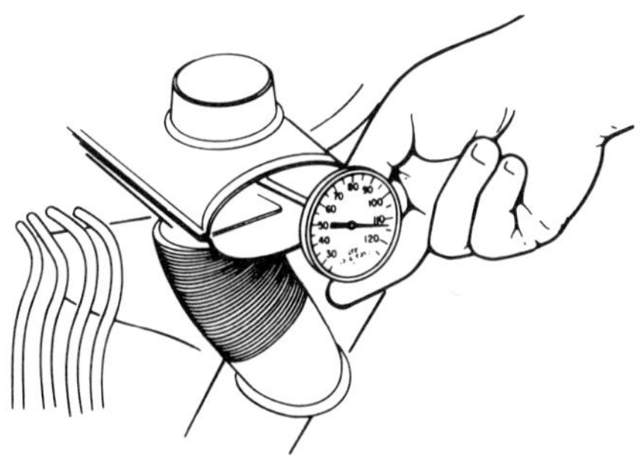

Figure 16-39. Use a thermometer to check intake air temperature at the air cleaner inlet and near the sensor.

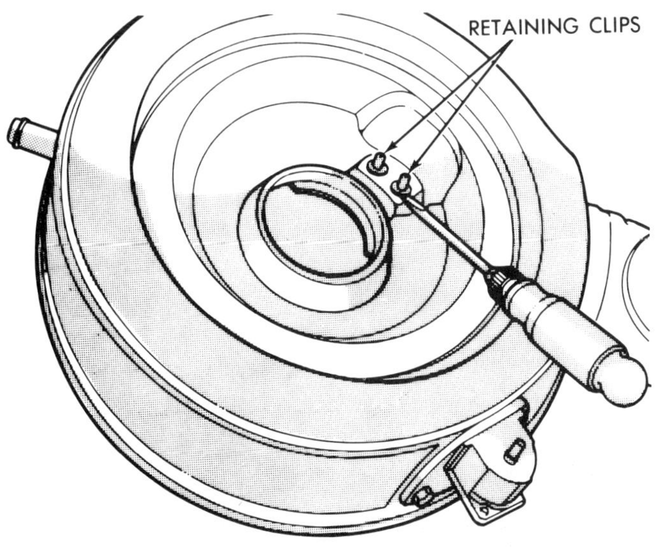

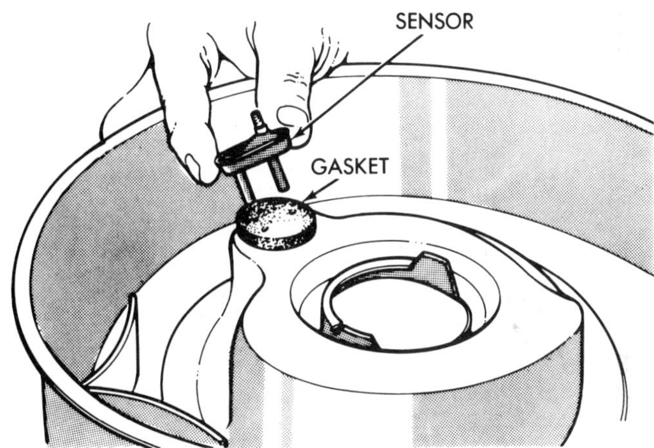

Figure 16-40. Pry out the retaining clips to remove the temperature sensor. (Chrysler)

1/8-inch sheet metal screws. Be sure the screws do not interfere with damper operation.

6. Reinstall the air cleaner and check vacuum motor operation. Connect all hoses and air ducts.

Air Temperature Sensor Test

The bimetal air temperature sensor is in the air cleaner housing. It opens a bleed in the line to the vacuum motor to regulate the air damper position. Temperature specifications for the sensors vary from one carmaker to another, so check the specifications before testing. As a general rule, most sensors are closed at air temperatures below 70° F (21° C), start to open at 85° to 100° F (29° to 38° C), and are fully open at 110° to 120° F (43° to 49° C). Test the sensor as follows:

1. Remove the air cleaner cover and tape a thermometer next to the sensor in the housing, figure 16-39.
2. Disconnect the cold-air duct to the snorkel. Verify that the damper is in the cold-air position. You may want to use a mirror to look inside the duct. Replace the air cleaner cover and start the engine.
3. Verify that the damper moves to the hot-air position as vacuum is applied to the motor when the engine starts.
4. Continue watching the damper as the engine warms up. When it starts to move toward the cold-air position, remove the air cleaner cover and check the thermometer reading.
5. Reinstall the cover and continue watching the damper. When it reaches the full cold-air position, remove the cover again and check the thermometer.

The temperatures when the damper starts to open and when it is fully open should be within specifications. If either temperature is higher than specifications, replace the sensor. If the damper does not open, the problem could be in the sensor or the vacuum motor. Test the vacuum motor before replacing the sensor.

Some air cleaners have a vacuum delay valve in the line before or after the sensor. This valve traps vacuum in the motor and holds the damper in the hot-air position during sudden acceleration with a cold engine. Test the vacuum delay valve as explained in chapter 15.

Air Temperature Sensor Replacement

Replace an air cleaner air temperature sensor as follows:

1. Disconnect all hoses and ducts from the air cleaner and remove it from the engine.
2. Disconnect the vacuum lines from the sensor.
3. Pry out and remove the spring clip, or clips, figure 16-40, that hold the

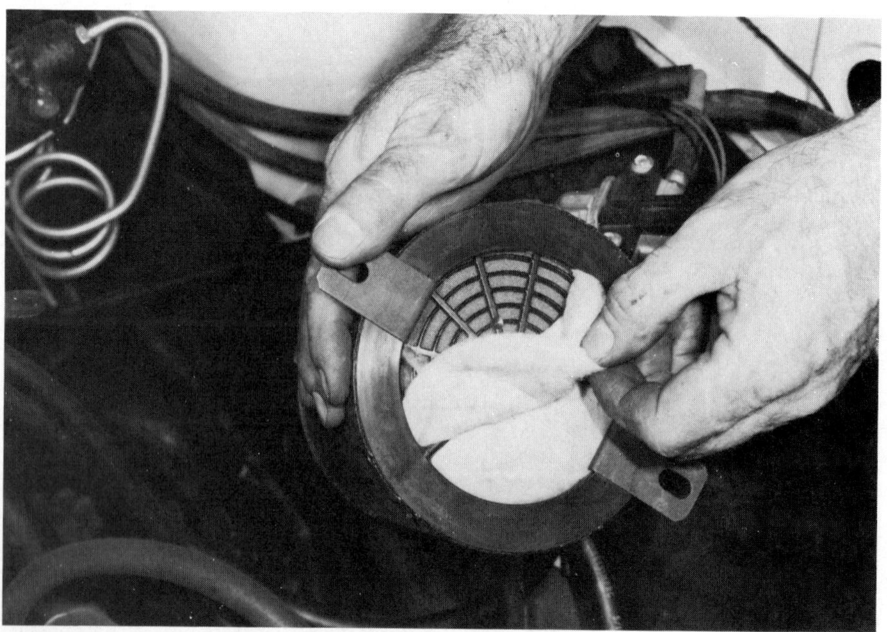

Figure 16-41. Replace EEC canister filters at the carmaker's scheduled intervals.

MANIFOLD VACUUM
DIAPHRAGM ASSEMBLY
SPRING
CANISTER
CARBURETOR BOWL
VALVE SEAT
PURGE VACUUM
GM TYPE 1
CONTROL VACUUM
SPRING
MANIFOLD VACUUM FROM PCV
DIAPHRAGM ASSEMBLY
CARBURETOR BOWL
DIAPHRAGM ASSEMBLY
SPRING
CANISTER
VALVE SEAT
GM TYPE 2

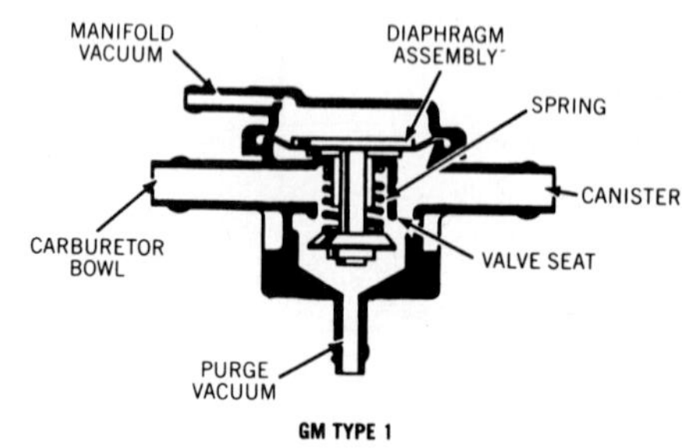

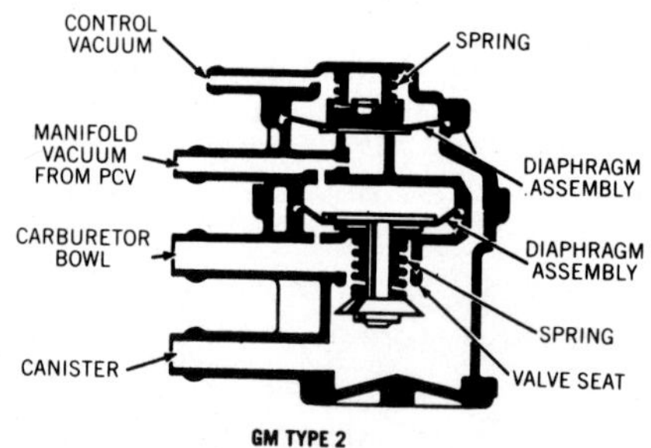

Figure 16-42. Purge valve designs vary. Follow the manufacturer's procedures to test operation. (GM)

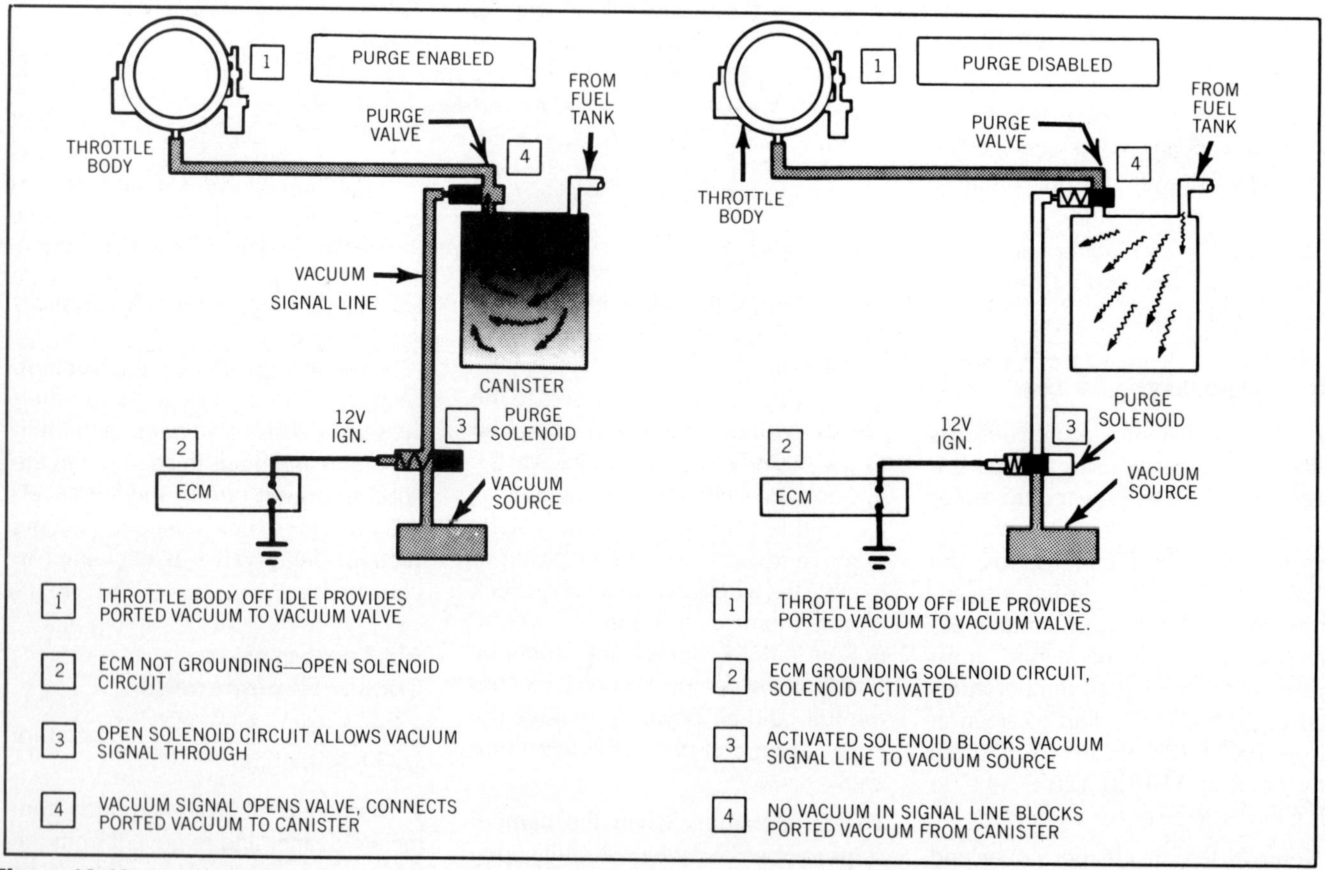

1
PURGE ENABLED
FROM FUEL TANK
PURGE VALVE
4
THROTTLE BODY
VACUUM SIGNAL LINE
CANISTER
12V IGN.
3
PURGE SOLENOID
2
ECM
VACUUM SOURCE
1 THROTTLE BODY OFF IDLE PROVIDES PORTED VACUUM TO VACUUM VALVE
2 ECM NOT GROUNDING—OPEN SOLENOID CIRCUIT
3 OPEN SOLENOID CIRCUIT ALLOWS VACUUM SIGNAL THROUGH
4 VACUUM SIGNAL OPENS VALVE, CONNECTS PORTED VACUUM TO CANISTER
1
PURGE DISABLED
FROM FUEL TANK
PURGE VALVE
4
THROTTLE BODY
12V IGN.
3
PURGE SOLENOID
2
ECM
VACUUM SOURCE
1 THROTTLE BODY OFF IDLE PROVIDES PORTED VACUUM TO VACUUM VALVE.
2 ECM GROUNDING SOLENOID CIRCUIT, SOLENOID ACTIVATED
3 ACTIVATED SOLENOID BLOCKS VACUUM SIGNAL LINE TO VACUUM SOURCE
4 NO VACUUM IN SIGNAL LINE BLOCKS PORTED VACUUM FROM CANISTER

Figure 16-43. Test electronic engine control system purge solenoids according to the carmaker's instructions. (Cadillac)

sensor in the housing. Remove the sensor.

4. Install a new sensor and gasket and insert the spring clips.

5. Reconnect all vacuum hoses, reinstall the air cleaner, and connect all ducts.

Thermostatic Bulb Air Cleaner Test

You can test the thermostatic bulb of older Ford and AMC air cleaners by removing the snorkel from the air cleaner and putting the entire assembly in a pan of hot tap water at 100° F (38° C). Be sure the bulb is completely immersed in water. The damper should be in the hot-air position. Raise the water to 130° to 150° F (54° to 66° C) and allow it to stabilize for 5 minutes. The damper should move to the cold-air position. If damper operation is outside of these temperature limits, replace the entire snorkel, damper, and bulb assembly.

EVAPORATIVE EMISSION CONTROL (EEC) SYSTEM SERVICE

Many EEC vapor storage canisters have replaceable fresh-air filters, which are small foam discs in the bottom of the canister. Carmakers call for replacement of these filters at specified intervals. Replace the filter more often than recommended if the car is driven in dusty conditions.

To replace many canister filters, you can simply reach under the bottom of the canister, release a spring clip that holds the filter, and remove the filter, figure 16-41. On some vehicles, access to the canister is limited by its shape and location. In these cases, disconnect all hoses from the canister and loosen or remove the clamps or brackets that hold the canister to the vehicle. Remove the canister and remove the old filter. Install a new filter and reinstall the canister.

The general fuel system inspection procedure at the beginning of this chapter includes checking the EEC system hoses for damage and incorrect connections. Replace defective hoses and be sure all connections are secure. If you find liquid fuel in the canister, the liquid-vapor separator in the system may not be working. Inspect and test the separator according to the manufacturer's instructions.

The routing of the vapor vent hose from the carburetor to the canister is critical. The hose must be routed in a continuous downward position. Any bends and rises will allow liquid fuel to collect in the hose. This will cause improper canister purging and can create a fire hazard.

Most EEC systems have vacuum-operated purge valves that use manifold or ported vacuum to open the vapor purge line from the canister to the engine, figure 16-42. The purge valve also may close the vapor vent line from the carburetor to the canister when the engine is running. Many electronic engine control systems have a solenoid that controls vacuum to the purge valve to turn off the canister purge under some conditions, figure 16-43.

Incorrect canister purging, whether vacuum or electronically controlled, can upset the air-fuel ratio, increase exhaust emissions, and cause driveability and fuel economy complaints. When troubleshooting any of these problems, test the EEC system purge valve operation according to the carmaker's procedures.

MANIFOLD HEAT CONTROL VALVE SERVICE

Gasoline engines use the following kinds of manifold heat control valves or other methods to preheat the air-fuel mixture of a cold engine:

- Thermostatic manifold heat control valve
- Vacuum-operated manifold heat control valve
- Hot coolant mixture preheating
- Electric grid heater mixture preheating.

If a heat control valve sticks open or heat is otherwise not applied to the mixture of a cold engine, the engine will run poorly when cold and fuel consumption and emissions will increase. If a heat control valve sticks closed or heat is applied to the mixture of a warm engine, acceleration and power will suffer.

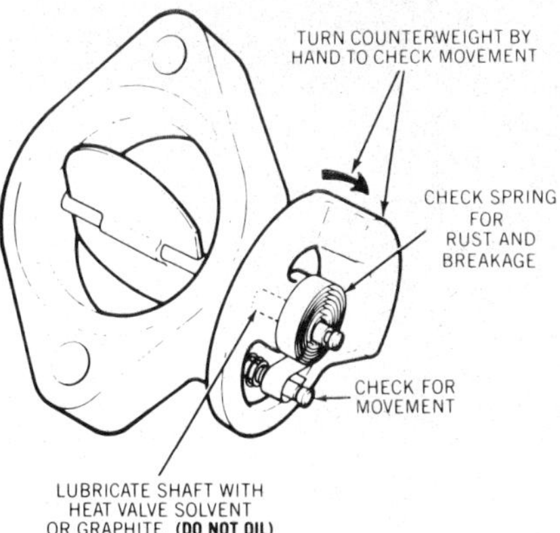

Figure 16-44. Apply heat valve solvent to the shaft and tap the counterweight to free a stuck heat control valve. (Ford)

Thermostatic Heat Control Valve Service

With the engine cold, move the valve counterweight with your hand to verify that the valve moves freely. Inspect the thermostatic spring for rust and damage. If the valve is stuck or moves roughly, apply manifold heat valve solvent to the ends of the shaft, figure 16-44, and tap the counterweight lightly with a small hammer. Do not apply motor oil to the valve because exhaust heat will form carbon deposits from the oil and cause the valve to stick more.

If you cannot free the valve, remove the valve body or the manifold from the engine and try to free it on your workbench, figure 16-45. If you still cannot free the valve, or if it is badly damaged, replace the valve body assembly or the manifold.

Vacuum-Operated Heat Control Valve Service

A vacuum-operated heat control valve directs exhaust to the intake manifold heat passage just as a thermostatic spring-operated valve does. However, the vacuum-operated valve does it more precisely. When the engine is

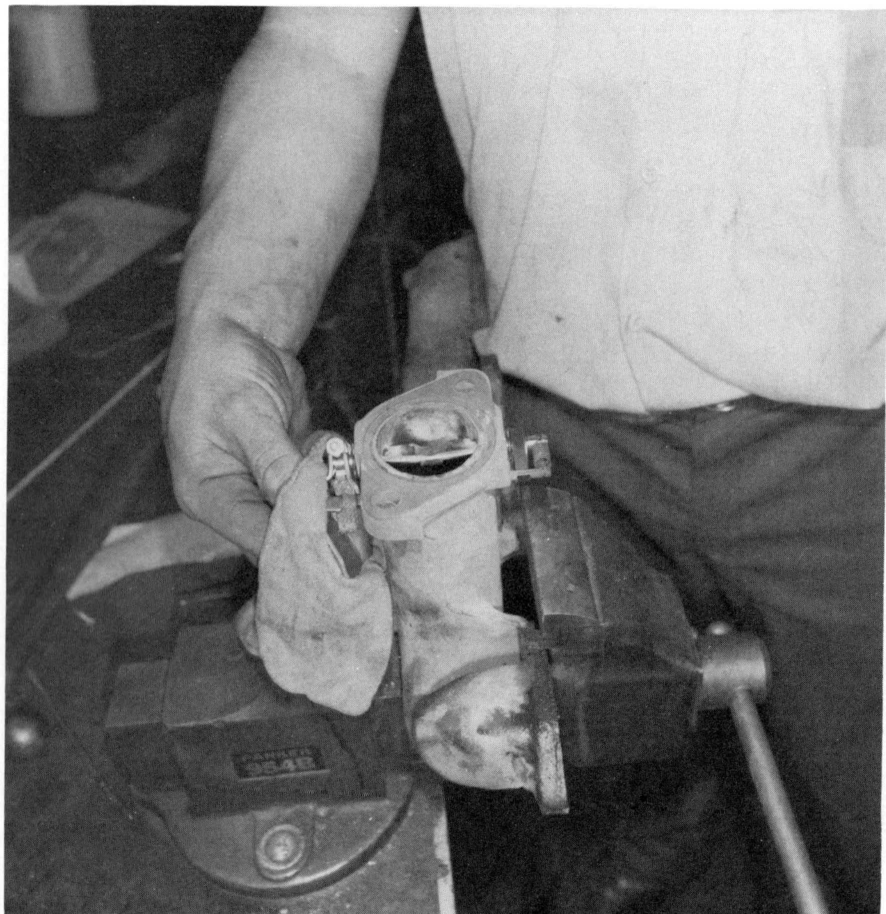

Figure 16-45. If you can't free the heat valve on the car, you will have to remove the manifold.

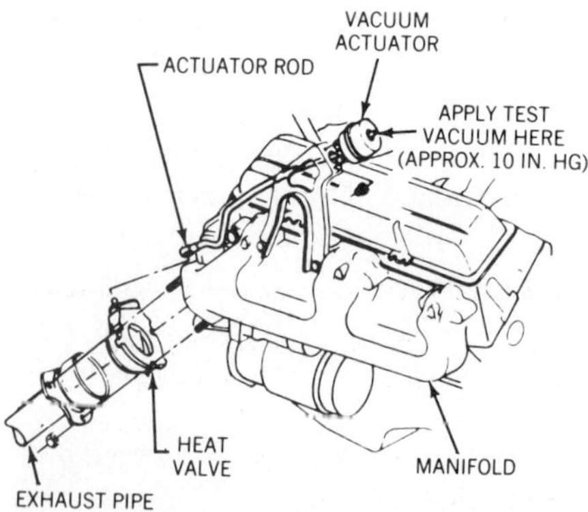

Figure 16-46. Use a vacuum pump to test a heat valve vacuum actuator. (GM)

cold, manifold vacuum passes through a thermostatic vacuum valve to the vacuum actuator for the heat valve. Vacuum closes the heat valve to route exhaust heat to the intake manifold. Inspect and test the system as follows:

1. Check all vacuum hoses for leaks and loose connections. Check the valve linkage for damage.

2. Connect a hand-operated vacuum pump to the valve actuator and apply about 10 inches of vacuum, figure 16-46. The actuator should close the heat valve and the valve and linkage should move freely.

3. Leave vacuum applied to the actuator for one minute and watch the gauge and the heat valve. The heat valve should stay closed with no loss of vacuum.

4. If the valve binds or does not move during the test, service as you would a thermostatic spring-operated valve. If you cannot free it, replace it.

5. If the actuator does not hold vacuum, replace it.

6. Test the switching action of the thermostatic vacuum valve as explained in chapter 4 for basic vacuum valve tests. On a cold engine, the valve must be open. On a warm engine, it must be closed.

7. If the engine has a solenoid vacuum valve instead of a thermostatic vacuum valve, test it according to the carmaker's procedures. Most such solenoids are normally deenergized to open the vacuum line. When the engine warms, the computer energizes the solenoid to block vacuum.

Hot Coolant Mixture Preheating Service

On an engine that uses engine coolant to preheat the air-fuel mixture, check the coolant hoses for leaks or restrictions. With the engine cold, start it and feel the coolant lines at the intake manifold. They should heat quickly. If they do not, or if the mixture overheats on a warm engine, check the cooling system thermostat and other control valves in the system for proper operation.

Electric Grid Heater Service

The electric grid heaters used by GM, Ford, and some other manufacturers since about 1981 are installed between the carburetor or the injection throttle body and the manifold, figure 16-47. The engine computer or a timer controls current to the heater. To test an electric grid heater, you will need the carmaker's specifications and wiring diagram for the system, figure 16-48. The circuit to the heater should be closed when the engine is cold and open when it is warm. Some heaters have a ground connection at the manifold or throttle body base. Check this connection for looseness and broken wiring. With the engine off, open the throttle valve and use a flashlight to inspect the grid heater for damage.

SUMMARY

Inspecting for fuel leakage and damaged or worn lines and hoses is an important part of service for all fuel system components. Some components, such as tanks and lines, do not need service unless they are damaged. Other components, such as air and fuel filters and EEC canister filters are replaced at specified intervals.

Correct fuel pump pressure and vol-

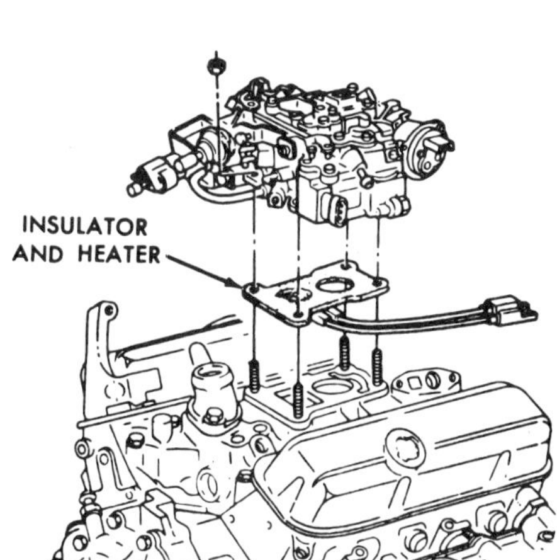

Figure 16-47. Typical electric grid heater installation. (Chevrolet)

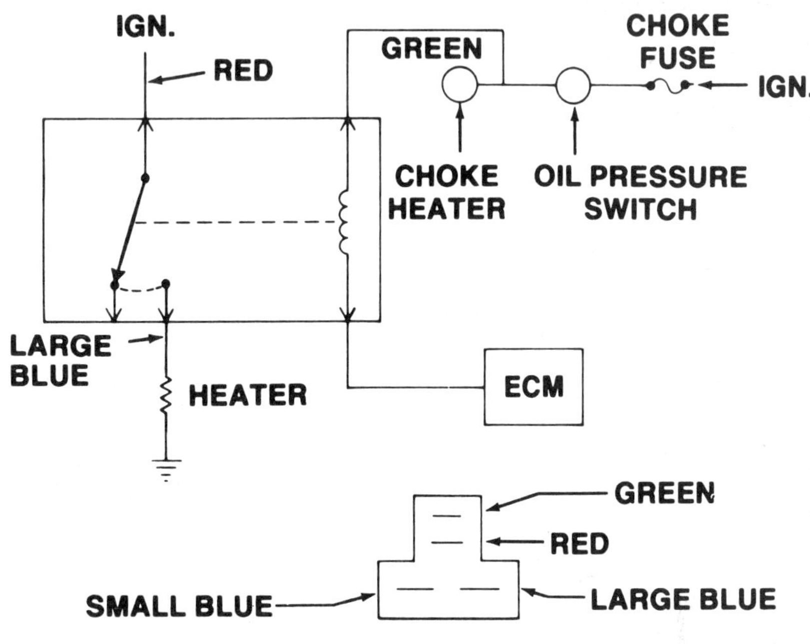

Figure 16-48. Electrical circuits for grid heaters vary from one manufacturer to another. (GM)

ume are important to correct fuel metering and overall system operation. You can test fuel pump pressure with a gauge and delivery volume with a graduated container.

Mixture temperature is important for correct combustion, particularly on late-model, electronically controlled engines. Thermostatically controlled air cleaners regulate intake air temperature. Manifold heat control valves or other preheating devices preheat the air-fuel mixture in the manifold of a cold engine. If either the thermostatic air cleaner or the manifold heat control system fails, economy, emission control, and driveability will suffer.

You can test the operation of air cleaner vacuum motors and sensors with a hand-operated vacuum pump, a gauge, and a thermometer. You can service a manifold heat control valve by cleaning with suitable solvent. You can test vacuum-operated valves with a vacuum pump and electric grid heaters with simple test equipment and the manufacturer's procedures.

Defective fuel pumps, air cleaners, heat control valves, and other parts should be replaced. Few can be repaired.

REVIEW QUESTIONS

Multiple Choice

1. There is evidence that fuel has been leaking from a fuel pump breather hole. Mechanic A says that the diaphragm is probably broken. Mechanic B says that the pump spring may be broken. Who is right?

a. Mechanic A only
b. Mechanic B only
c. Both A and B
d. Neither A nor B

2. Mechanic A says that a partially full fuel tank can be more dangerous than a completely full one. Mechanic B says that a full fuel tank is more dangerous. Who is right?

a. Mechanic A only
b. Mechanic B only
c. Both A and B
d. Neither A nor B

3. A 24-inch section of fuel line must be replaced. Mechanic A says that it will be OK to use a 26-inch section of flexible gas hose and some clamps to make the repair. Mechanic B says that it is better to use a length of copper tubing, two pieces of flexible hose (one for each end), and some clamps to make the repair. Who is right?

a. Mechanic A only
b. Mechanic B only
c. Both A and B
d. Neither A nor B

4. A fuel pump is causing the fuel supply to a carburetor to be too low. Mechanic A says that this will cause the engine to run lean. Mechanic B says that this type of problem can affect a fuel-injected engine in the same way. Who is right?

a. Mechanic A only
b. Mechanic B only
c. Both A and B
d. Neither A nor B

5. To take an accurate fuel pump pressure reading, the pressure-test gauge should be:

a. below the carburetor inlet
b. above the carburetor inlet
c. at the same height as the carburetor inlet
d. connected to the fuel pump outlet

6. Air bubbles appear in the fuel during a fuel pump volume test. Mechanic A says that the pressure is too high, causing the fuel to foam. Mechanic B says that there is probably a leak in the fuel line. Who is right?

- **a.** Mechanic A only
- **b.** Mechanic B only
- **c.** Both A and B
- **d.** Neither A nor B

7. High pressure from a fuel pump is usually caused by:

- **a.** a worn pushrod
- **b.** a defective fuel pump check valve
- **c.** a leaking diaphragm
- **d.** a worn eccentric

8. An electric fuel pump is suspected of being bad. Mechanic A says that the first test to perform is a fuel pump pressure test. Mechanic B says that the electrical circuit should be checked first. Who is right?

- **a.** Mechanic A only
- **b.** Mechanic B only
- **c.** Both A and B
- **d.** Neither A nor B

9. A mechanical fuel pump is being removed. The mounting bolts have been loosened. Mechanic A says that the next step is to remove the bolts completely. Mechanic B says that the pressure on the pump rocker arm should be relieved by cranking the engine before anything else is done. Who is right?

- **a.** Mechanic A only
- **b.** Mechanic B only
- **c.** Both A and B
- **d.** Neither A nor B

10. In a vacuum-operated air cleaner the spring in the vacuum motor should:

- **a.** hold the damper in the cold-air position when the engine is off
- **b.** hold the damper in the cold-air position when the engine is on
- **c.** hold the damper open only at idle
- **d.** hold the damper closed only at idle

11. A vacuum test has been made on an air cleaner vacuum motor. The vacuum reading dropped slowly during a 30-second period. Mechanic A says that this is normal. Mechanic B says that the vacuum motor should be repaired or replaced. Who is right?

- **a.** Mechanic A only
- **b.** Mechanic B only
- **c.** Both A and B
- **d.** Neither A nor B

12. A vapor vent hose from the carburetor to a vapor storage canister must be:

- **a.** routed as close to the intake manifold as possible
- **b.** routed over the heat-control valve to vaporize liquid fuel
- **c.** routed so that it matches the curves in the fuel supply line
- **d.** routed in a continuous, unbent, downward position

Fill in the Blank

13. Most EEC systems use manifold or __________ vacuum to open the vapor purge line from the canister to the engine.

14. A vacuum-operated heat control valve directs exhaust to the __________.

15. The electrical circuit to an electrical grid heater should be __________ when the engine is cold and __________ when it is warm.

16. Diesel fuel injection systems retain high pressure between the __________ and the __________ when the engine is off.

17. When you remove a pressurized fuel filler cap, the air pressure in the tank is __________ to the outside air pressure.

18. Most fuel line fittings are the __________ type.

19. Hoses made for handling vacuum and water will __________ if they are used as fuel hoses.

20. Diesel fuel injection systems aren't as likely to have high-pressure or high-volume fuel problems because injection systems have __________ or __________ to bypass excess fuel to the tank.

21. An air-temperature sensor in an air cleaner opens a bleed in the line to the __________ to regulate the air damper position.

22. Incorrect vapor canister purging can upset the air-fuel ratio and increase __________.

23. On some engines, hot coolant is used to __________ the air-fuel mixture.

17 CARBURETOR SERVICE

INTRODUCTION

Carburetor service can be an automotive specialty by itself and is the subject of several specialized courses. For any service, you will need the carmaker's procedures for carburetor testing, adjustment, and overhaul for a specific year and model of vehicle. You can find these in a factory shop manual or an independent repair manual. Basic instructions for carburetor adjustment are on the tuneup and emission control decal in the engine compartment. Carburetor repair kits also contain detailed instructions for overhaul and adjustment.

This chapter outlines the basic jobs of carburetor service to help you understand and use the manufacturer's procedures. We can divide carburetor service into three general areas:

1. Carburetor test and adjustment on the engine

2. Carburetor removal and replacement

3. Carburetor overhaul.

This chapter follows this sequence to summarize carburetor service.

GOALS

After studying this chapter, you should be able to use manufacturers' procedures to do the following jobs in twice the flat-rate labor time or less:

1. Test and adjust choke operation and choke linkage.

2. Test electric heaters and other choke assist devices; repair or replace as required.

3. Adjust idle speed and mixture by various methods, including propane injection and by using an infrared analyzer.

4. Test general carburetor operation with an infrared analyzer.

5. Adjust carburetor fast idle and deceleration throttle controls.

6. Remove and replace a carburetor, start the engine, and set initial adjustments.

7. Overhaul various carburetors; repair, replace, and adjust parts as required.

THE CARBURETOR AND THE TOTAL COMBUSTION CONTROL SYSTEM

Late-model carburetors have many assist devices that aid driveability and emission control. You must think of the carburetor, its assist devices, and other fuel system parts as a *complete* system. Each part must work properly, or other parts can't do their jobs. Additionally, the carburetor and the rest of the fuel system are part of the complete combustion control system. Carburetor service will not improve engine performance if ignition and emission control devices are not working properly. Before servicing a carburetor to cure an engine problem, check and service the following systems and components:

1. Spark plugs and ignition timing

2. Ignition dwell and advance

3. Engine power balance and compression

4. Fuel filters and fuel pump pressure and volume

5. Air filters and PCV system operation

6. Mechanical valve lifter adjustment.

A WORD ABOUT LATE-MODEL CARBURETORS

Carburetor design has evolved steadily for almost 100 years. At first glance, a modern carburetor hardly resembles a comparable unit of even 40 years ago, yet all carburetors have the basic systems that you learned about in Volume One of this text. Since the late 1970's and early 1980's, manufacturers have changed several carburetor features that affect service. The system functions are the same as they have always been, but the service methods are different from those of the past.

These changes occurred for two reasons:

1. The Federal Environmental Protection Agency (EPA) required that carmakers remove carburetor adjustment points that would allow car owners to "tamper" with adjustments that would upset emission control.

2. The development of electronic engine controls allowed more precise carburetor regulation than traditional mechanical adjustments allowed and made such adjustments unnecessary.

The "tamperproof" and electronic features of late-model carburetors are principally in the following areas:

• Idle mixture screws are adjusted at the factory and sealed, figure 17-1. Routine adjustment as part of engine tuneup is not required. Special procedures are needed to remove the screws and adjust them during overhaul, if necessary.

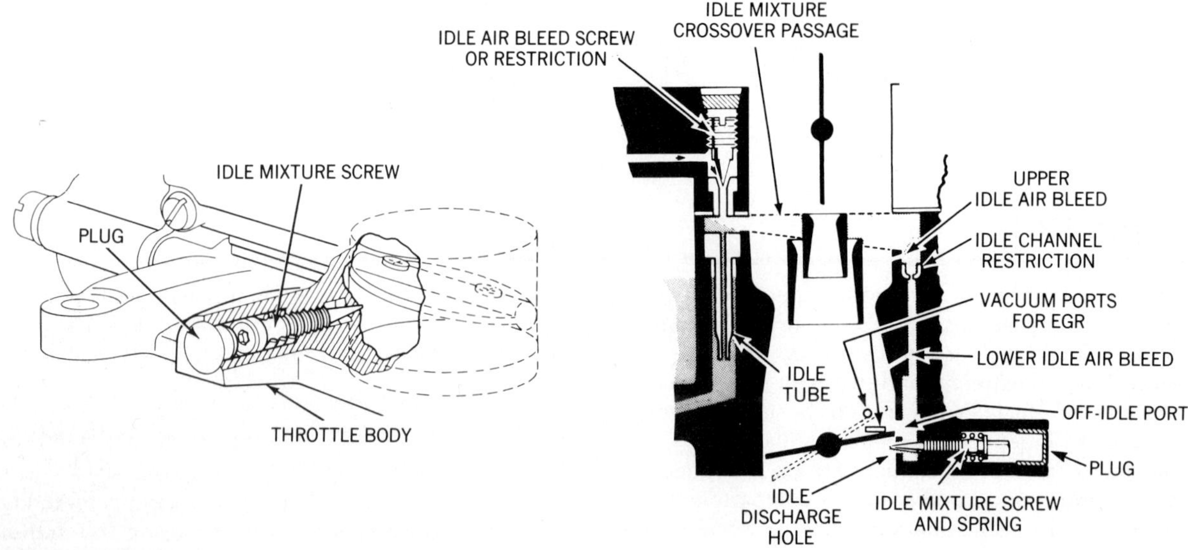

Figure 17-1. Mixture screws on late-model carburetors are adjusted at the factory and sealed. (Ford) (Rochester)

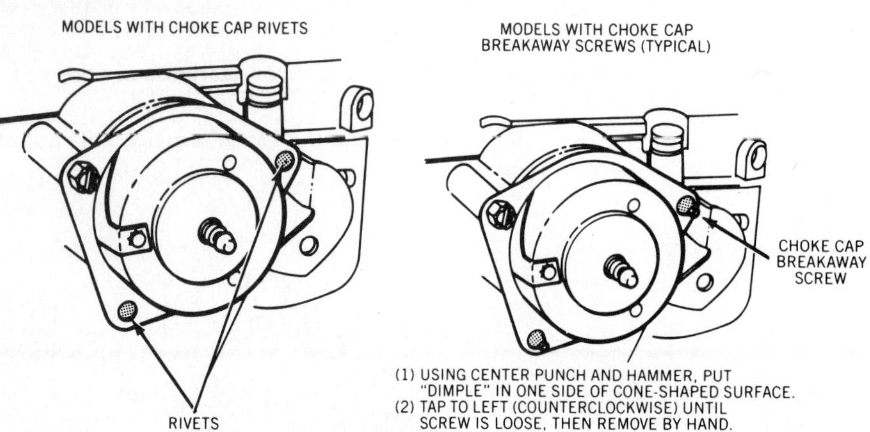

Figure 17-2. Late-model choke covers also are adjusted and sealed at the factory. (Ford)

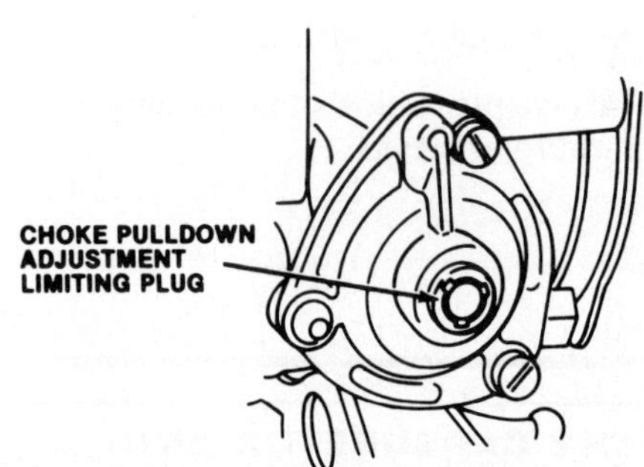

Figure 17-3. Vacuum break pulloff or pulldown adjustment screws are sealed on many late-model carburetors. (Ford)

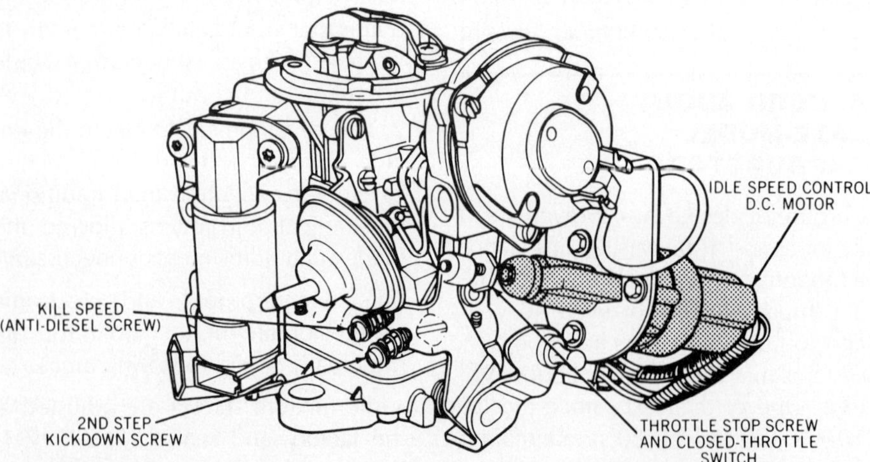

Figure 17-4. This idle speed motor is controlled by the engine computer. Idle speed is not adjustable in the traditional way. (Ford)

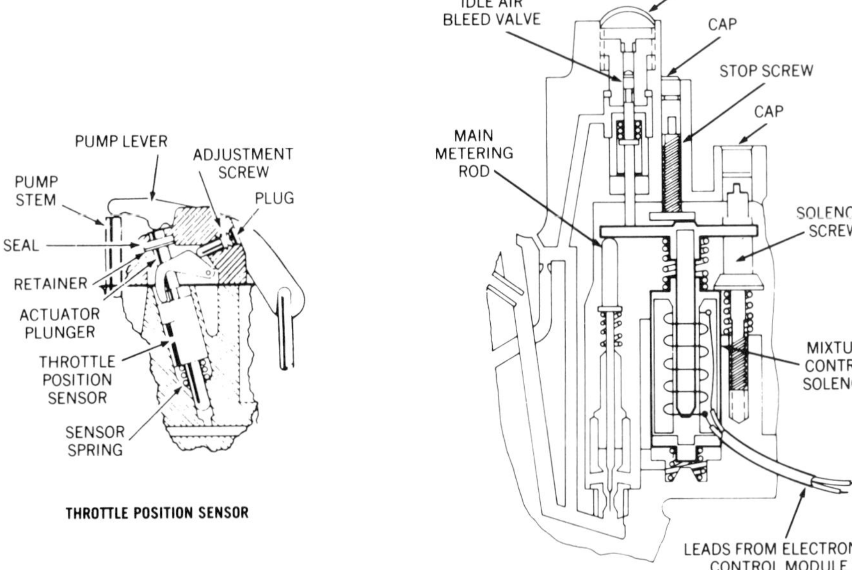

Figure 17-5. The adjustments for this mixture control solenoid and throttle position sensor did not exist on carburetors of a decade ago. (GM)

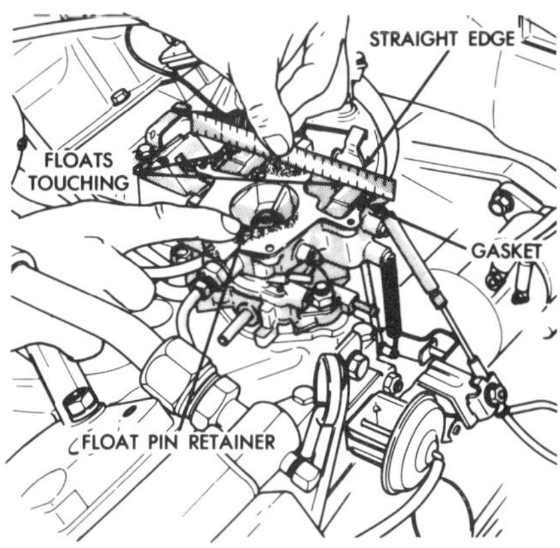

Figure 17-6. On some carburetors, you can remove the airhorn to adjust float level on the car. (Chrysler)

- Choke thermostatic spring housings also are adjusted and sealed at the factory, figure 17-2. These, too, require special procedures if adjustment is needed during overhaul.
- Vacuum breaks (choke pulloff or pulldown devices) on some carburetors have plugs over adjustment screws, figure 17-3. Special procedures, again, are required if adjustment is necessary.
- Idle speed adjustment screws have been replaced by electronic idle speed control motors, figure 17-4. Idle speed is not adjustable, but is controlled by the engine computer.
- Mixture control solenoids and stepper motors control fuel metering in feedback carburetors, and throttle position sensors send signals to engine computers. These have added new adjustment and repair points, figure 17-5, that did not exist a decade ago.

This chapter summarizes carburetor service points for most vehicles built in the last 20 years. Refer to the manufacturer's procedures for specific carburetors, particularly for late-model units on which some traditional choke, idle, and other adjustments are no longer part of routine service.

CARBURETOR TEST AND ADJUSTMENT ON THE ENGINE

When you service a carburetor on an engine, you usually will follow this sequence:

- Choke test and adjustment
- Idle speed adjustment
- Idle mixture adjustment
- Fast-idle adjustment
- Assist device test and adjustment

Float adjustment is not usually part of routine carburetor service, but if float adjustment is not correct, it may be impossible to get other adjustments correct. Float adjustment is part of carburetor overhaul and is summarized later in this chapter. In some cases, however, you may have to remove or partly disassemble a carburetor to check the float settings to get correct idle adjustments or cure main metering problems.

Warning. Never adjust the float level with the engine running. Fire may result.

Some floats can be adjusted with the carburetor on the engine by removing the airhorn, figure 17-6. A few carburetors (particularly some Holley models) have a sight gauge plug in the fuel bowl to check float level, figure 17-7. These carburetors usually have external float adjustment screws. The float level on late-model Rochester Varajet and Quadrajet carburetors can be checked by inserting a gauge through an opening in the top of the bowl, figure 17-8.

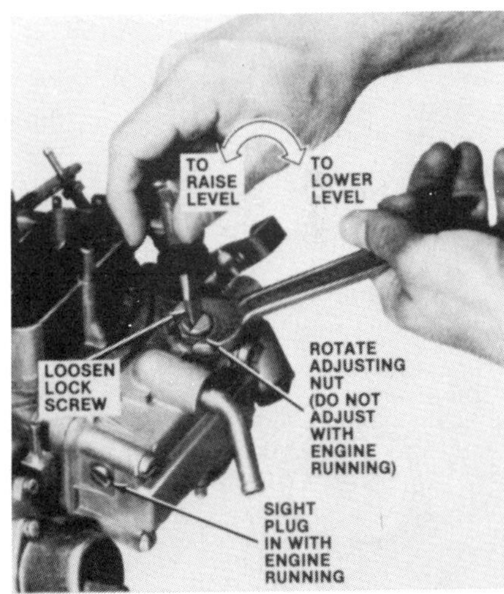

Figure 17-7. Many Holley carburetors have a fuel level sight gauge, or plug, and an external float adjustment screw. (Ford)

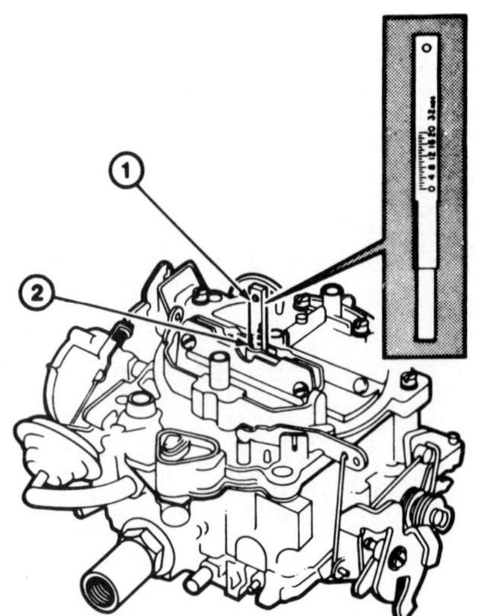

Figure 17-8. You can check float level on many late-model Rochester carburetors by inserting a gauge through a vent opening. (Rochester)

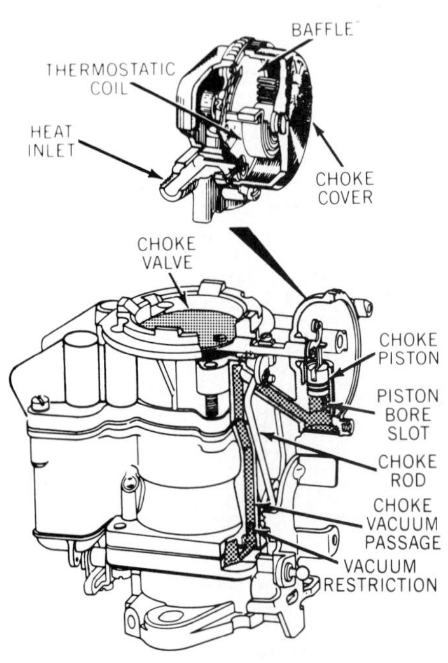

Figure 17-9. An integral choke cover (cap) has an index mark and arrows to indicate rich and lean adjustments. (AMC)

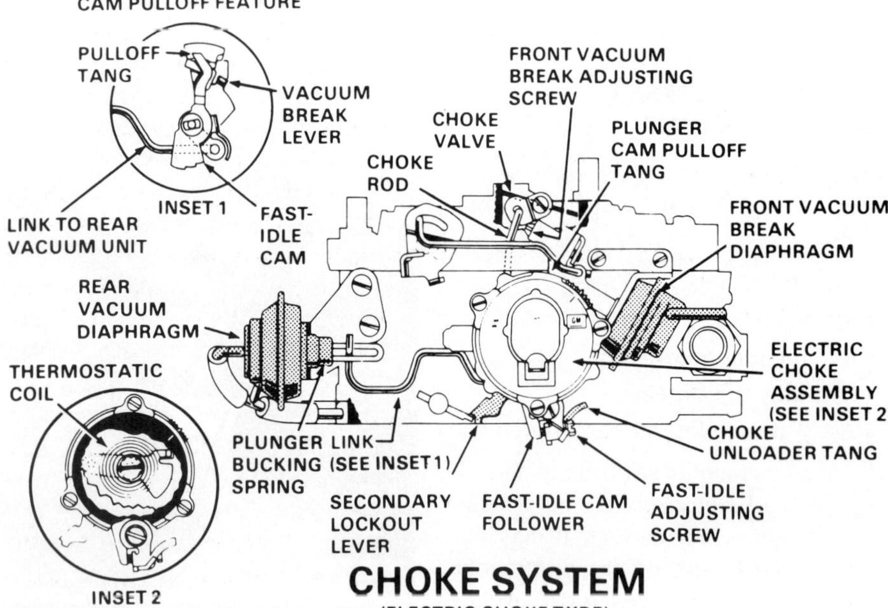

Figure 17-10. Many integral chokes have separate vacuum break diaphragms.

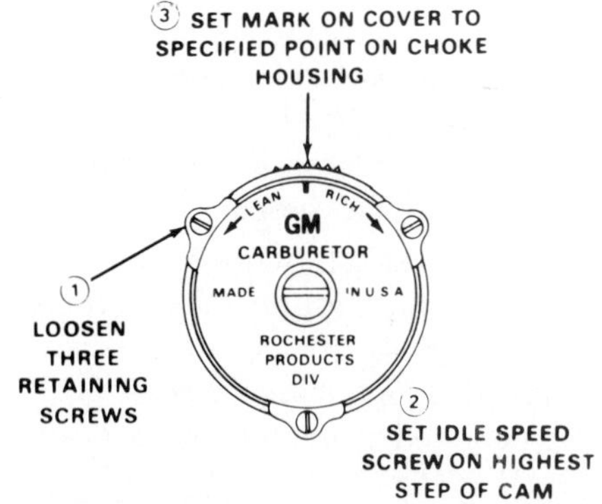

Figure 17-11. Typical adjustment markings for an adjustable choke cap. (Chevrolet)

Automatic Choke Test and Adjustment

Choke service consists of testing and adjusting:

- The closing action of the choke
- The choke unloader (some models)
- The choke vacuum break (pulldown or vacuum kick)
- Electric heaters
- Various choke assist devices.

Choke Closing Adjustment (Integral Choke). The housing for the choke thermostatic spring is attached to the carburetor, figure 17-9. Many contain a vacuum piston for the choke vacuum break opening. Others have a separate vacuum break diaphragm, figure 17-10. Most late-model chokes also have an electric heating element to warm and open the choke quickly to reduce CO emissions.

Most nonsealed choke covers, or caps, have index marks and arrows that show which way to turn the cap for a richer or leaner choke setting, figure 17-11. Some caps with electric heaters do not have the rich and lean arrows. The choke housing body has several notches. The center notch is the "index" mark. If choke adjustment specifications call for an "index" setting, align the cap index mark with the index notch on the housing. If the choke specification calls for a 1R, 2R, or 3R setting, align the cap index mark with the first, second, or third notch in the RICH direction. Similarly, if the specification is 1L, 2L, or 3L, align the cap

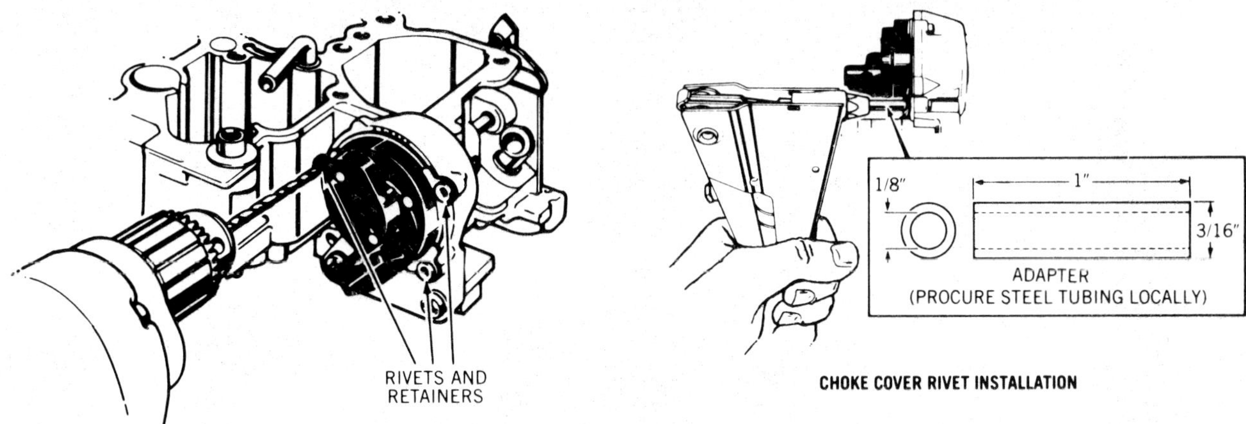

Figure 17-12. Remove rivets on a sealed choke cap with a drill. Install new rivets after adjustment. (Chevrolet)

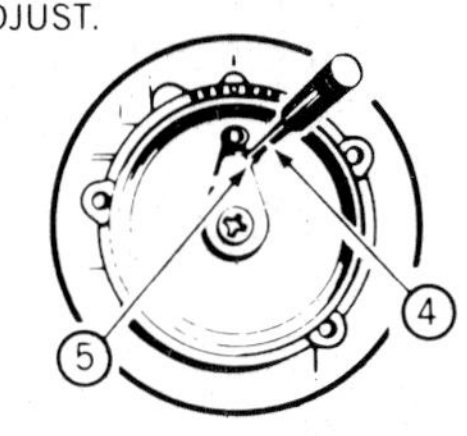

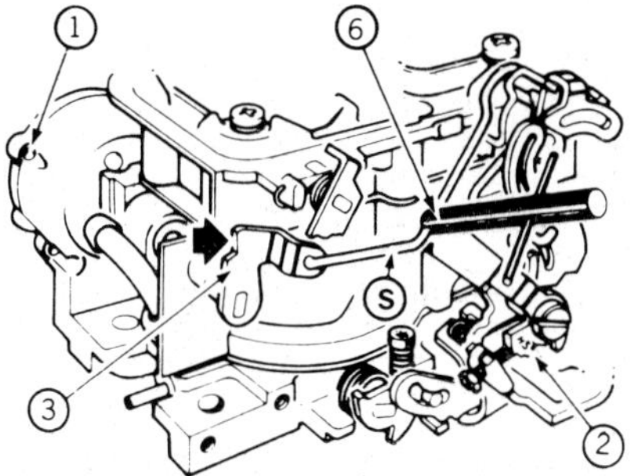

Figure 17-13. Some late-model chokes are adjusted by positioning the choke lever and bending a link. (Rochester)

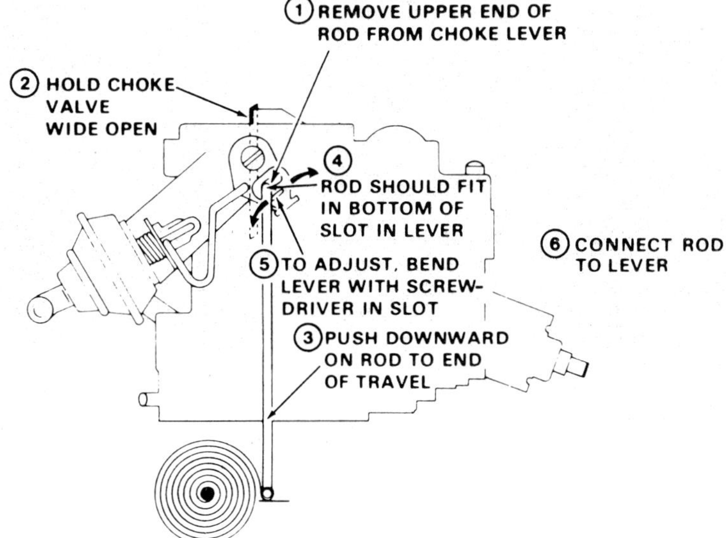

Figure 17-14. Typical adjustment points for a remote, well-type choke. (Rochester)

mark with the appropriate notch in the LEAN direction.

Be sure that the thermostatic spring is cold when making the closing adjustment. Before adjusting the choke thermostatic spring, move the choke plate and linkage by hand to be sure it does not bind or stick. Use an approved choke cleaning spray to remove gum deposits that interfere with choke movement. To adjust the closing tension of an adjustable integral choke, loosen the screws that hold the cover and disconnect the hot-air tube if it is attached to the cover. Then, turn the cover to align its index mark with the specified notch. Finally, tighten the screws and reconnect the hot-air tube.

Late-model nonadjustable integral chokes have rivets or breakaway screws rather than removable screws to secure the choke cover, figure 17-2. You can't adjust the spring tension on these chokes as part of regular maintenance. If choke operation or carburetor overhaul requires a choke adjustment, drill out the rivets (usually with a 1/8-inch drill), figure 17-12, or break out the breakaway screws. On some chokes, you then turn the cap for the correct setting. On Rochester chokes, you put the spring and choke valve in their correct closed positions and bend a link to adjust the choke, figure 17-13. Install new rivets to secure the cap.

Choke Closing Adjustment (Remote Choke). On a remote, or well-type, choke, a rod connects the thermostatic spring to the choke plate on the carburetor, figure 17-14. The spring

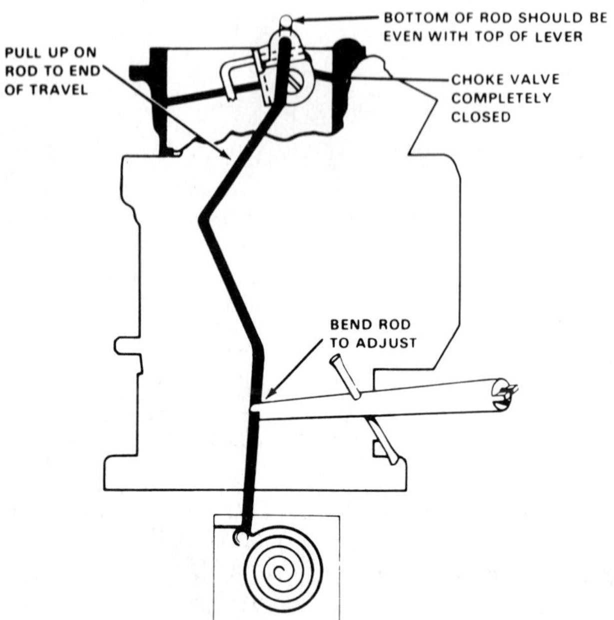

Figure 17-15. Adjust remote choke opening by bending the linkage at specified positions. (Chevrolet)

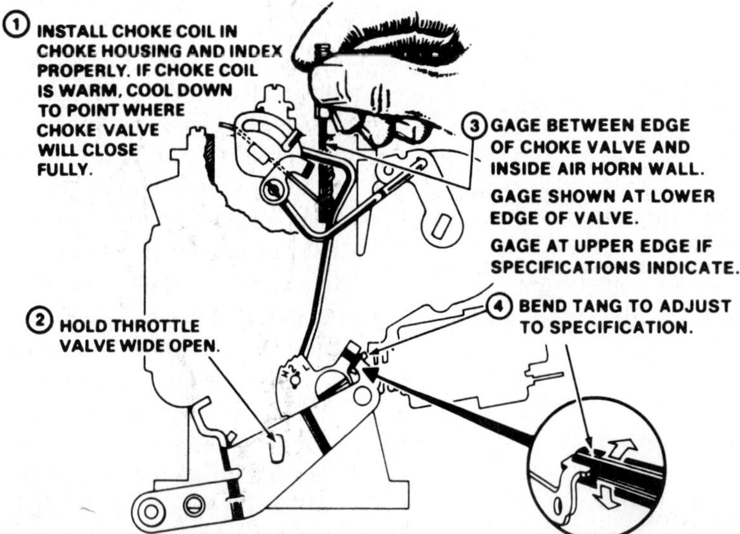

Figure 17-16. Use a drill bit of specified diameter to check choke unloader opening. (Rochester)

may have index marks, similar to those on an integral choke, and an adjustment nut. These are for factory calibration, however, usually not for service adjustment.

Be sure that the thermostatic spring is cold when making the closing adjustment. To adjust a remote choke, disconnect the choke rod from the choke plate linkage and hold the choke plate closed. Firmly but gently, move the rod completely up or down as directed. Note the position of the disconnected rod end in relation to the hole or slot in the plate linkage. Adjustment specifications usually call for the rod end to be one diameter or one-half diameter above or below the linkage hole, or a sliding fit in the linkage slot. Follow the carmaker's instructions to bend the choke rod or the linkage tang to adjust the choke, figure 17-15. Use a bending tool to bend the rod or linkage only where specified.

Choke Unloader Adjustment. If a cold engine floods, opening the throttle fully lets the choke unloader force the choke plate open slightly. This allows more air into the engine to clear the flooding. Many choke unloaders can be adjusted on the engine. Some require that the carburetor be removed. Manufacturers' instructions for adjustment usually include the following steps.

Open the throttle fully to allow the choke to close and the fast-idle linkage to engage. Then close the throttle and be sure the choke stays fully closed. Reopen the throttle and block or hold it fully open so that the unloader opens the choke slightly. On most carburetors, you next insert a drill bit or gauge rod of the specified diameter between the high or low edge of the choke plate, figure 17-16. The drill or gauge should slide through the opening with a slight drag. If it does not, bend the unloader linkage at the specified point to adjust the plate angle. On late-model Rochester carburetors, use a choke angle gauge with a magnetic base that holds it to the choke plate. Figure 17-17 shows a typical unloader adjustment for a late-model Rochester carburetor.

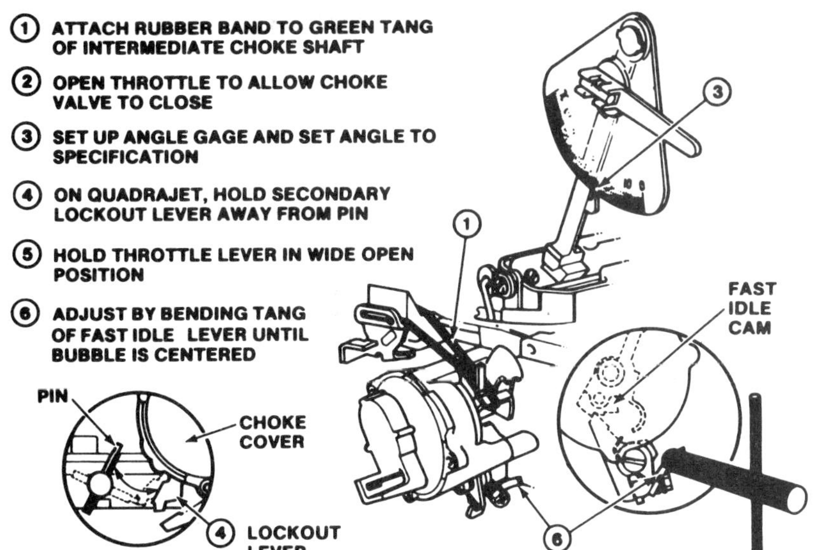

Figure 17-17. Adjust choke unloader opening on late-model Rochester carburetors with an angle gauge. (Rochester)

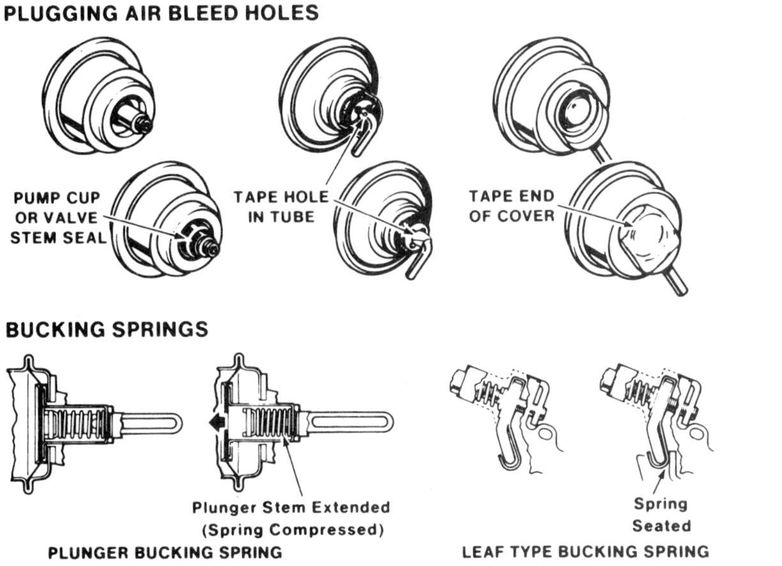

PLUGGING AIR BLEED HOLES

PUMP CUP OR VALVE STEM SEAL

TAPE HOLE IN TUBE

TAPE END OF COVER

BUCKING SPRINGS

Plunger Stem Extended (Spring Compressed)

PLUNGER BUCKING SPRING

Spring Seated

LEAF TYPE BUCKING SPRING

Figure 17-18. Some vacuum breaks have calibrated air bleeds that must be covered to test diaphragm operation. (Rochester)

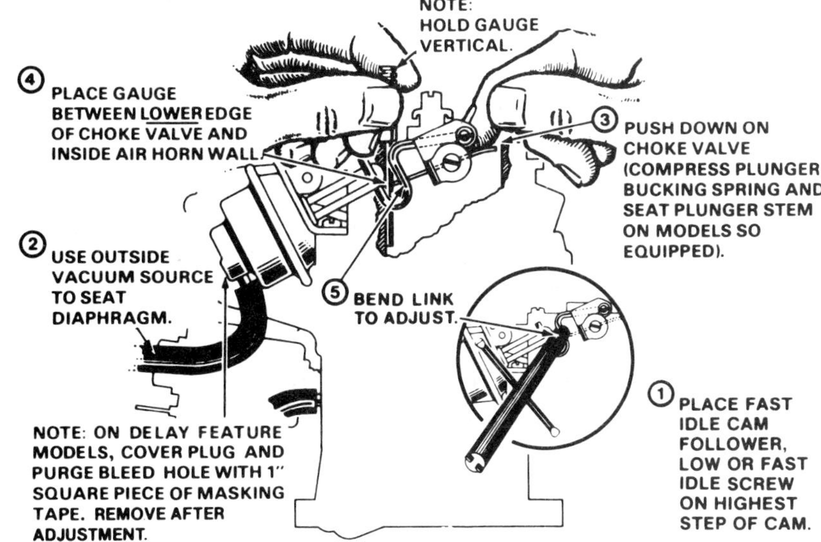

Figure 17-19. Use a drill bit to check vacuum break opening as you did for choke unloader adjustment. (Rochester)

Choke Vacuum Break Adjustment. The choke vacuum break (also called the pulloff, pulldown, or vacuum kick) opens the choke a specified amount after the engine starts. It is either a piston, figure 17-9, or a diaphragm, figure 17-10, operated by manifold vacuum. A vacuum diaphragm can be damaged by heat, oil, or age. If a diaphragm leaks, the vacuum break cannot open the choke properly.

Before trying to adjust a diaphragm vacuum break, disconnect the rod from the diaphragm and apply vacuum (15 to 18 in. Hg) to the diaphragm from the engine or from a hand-operated pump. The diaphragm should pull back to the limit of its travel. If it does not, replace it before adjusting the linkage. Many vacuum breaks have air bleeds to modulate diaphragm action. You can't test the diaphragm condition accurately without first plugging the air bleed, figure 17-18.

As with other adjustments, vacuum break adjustment varies among carburetors but includes these points.

Open the throttle fully to allow the choke to close. Release the throttle and apply vacuum to the diaphragm. (Block the air bleed if necessary.) Measure the choke plate position with a specified drill bit or gauge rod, figure 17-19, or a choke angle gauge, figure 17-20. Adjust the vacuum break opening by bending the linkage or turning an adjusting screw in the diaphragm housing.

A piston-type vacuum break can be damaged by varnish or carbon deposits. You will have to remove the choke cap to inspect and adjust a piston choke. If the piston does not move freely in its bore or is dirty, remove the choke housing (on most such carburetors, the entire airhorn) for cleaning or replacement.

To adjust a piston choke, insert a wire gauge of the specified size between the piston slot and the right-hand slot in the housing, figure 17-21. Turn the piston lever until the gauge is snug in the slot and press lightly on the choke plate to hold it in position. Measure choke plate clearance in the airhorn with the specified drill bit. If clearance is not correct, bend the pis-

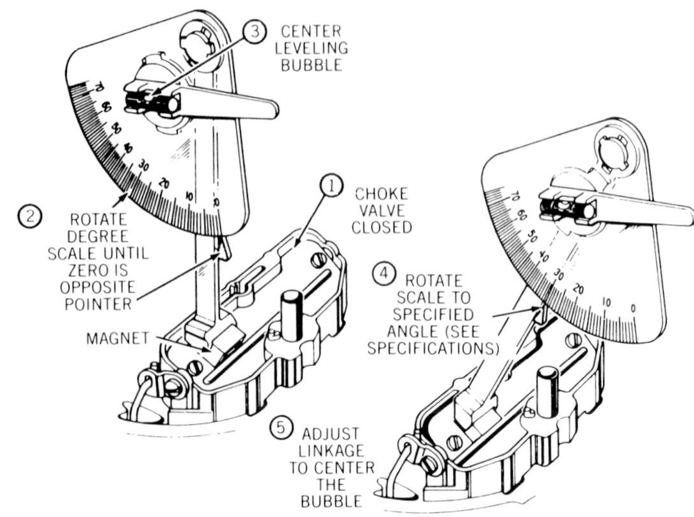

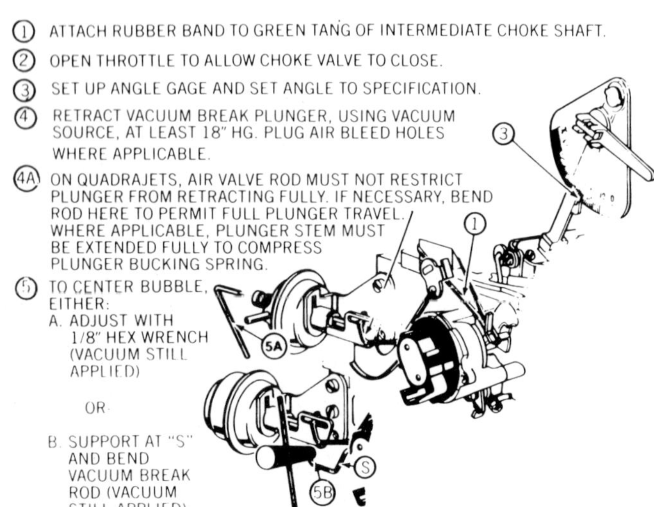

Figure 17-20. Use an angle gauge to set the vacuum break opening on late-model Rochester carburetors. (Rochester)

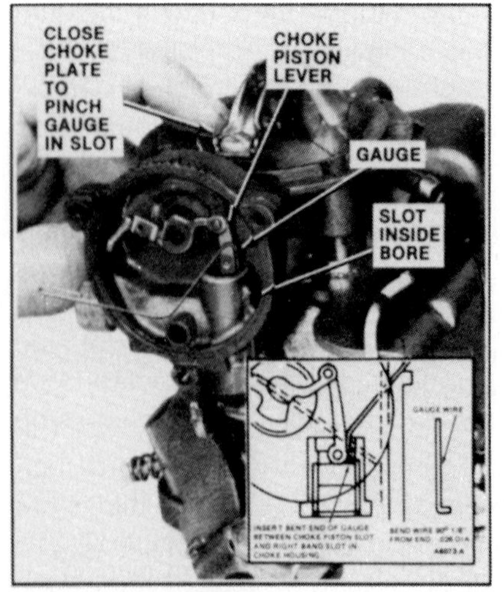

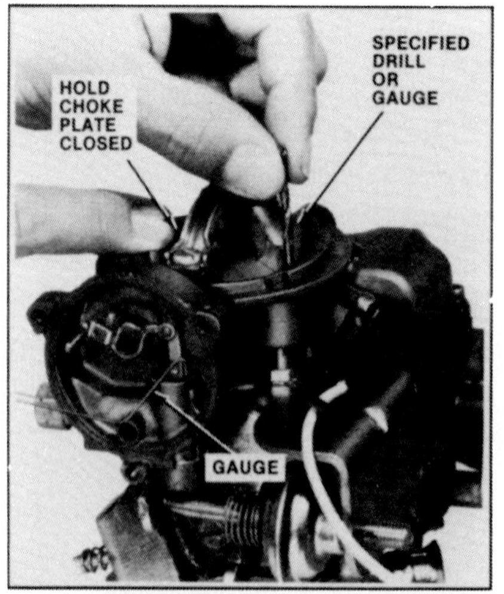

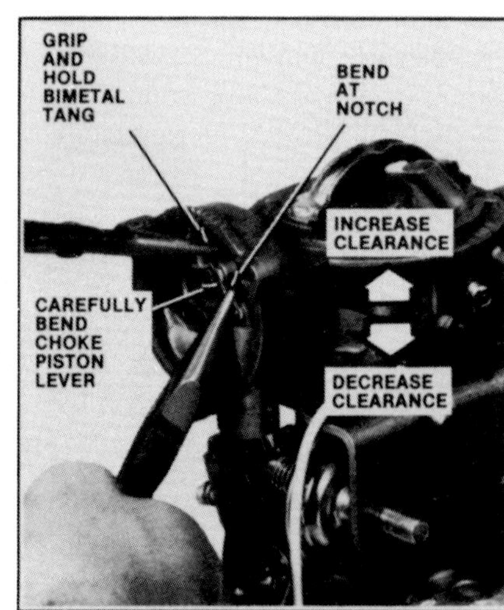

Figure 17-21. Use a wire gauge to measure vacuum break opening on a piston choke. Bend linkage only where specified. (Ford)

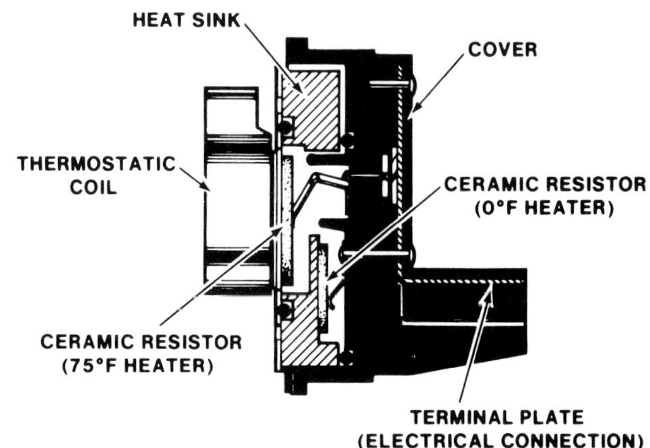

Figure 17-22. Many late-model chokes have 2-stage heaters. (Rochester)

OHMMETER

① TEST CONTINUITY BETWEEN CHOKE TERMINAL AND GROUND STRAP. SHOULD MEET CARMAKER'S SPECIFICATIONS WHEN COLD.

② TEST CONTINUITY BETWEEN GROUND STRAP AND CARBURETOR GROUND. SHOULD BE ZERO OHMS.

12-VOLT TEST LAMP

③ LAMP CONNECTED BETWEEN CHOKE TERMINAL AND CHOKE LEAD CONNECTOR SHOULD GLOW WHEN ENGINE IS RUNNING OR CHOKE SWITCH IS CLOSED.

④ LAMP CONNECTED BETWEEN CHOKE LEAD CONNECTOR AND GROUND SHOULD GLOW WHEN ENGINE IS RUNNING OR WHEN CHOKE SWITCH IS CLOSED.

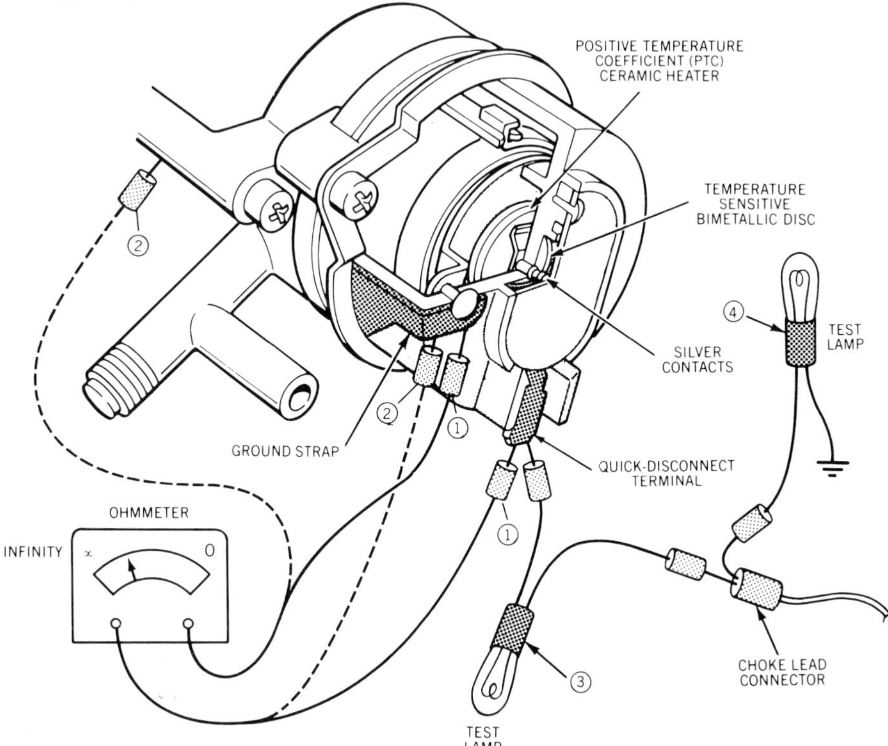

Figure 17-23. These tests for a Ford choke heater are typical for most electric chokes. (Ford)

ton lever (not the piston link) to adjust the opening. Reinstall and adjust the choke cap.

Choke Heater Service. Many cars built since the mid-1970's have electric choke heaters. Some are assist devices that work with hot air to heat the choke spring. Other chokes are "all electric," which means that the heater is the only source of choke heat. Some heaters work at a single temperature; others are 2-stage devices that deliver different amounts of heat, depending on air temperature, figure 17-22. Some heaters operate continuously; others turn on and off at specific temperatures.

You can test the operation of most choke heaters in two basic ways:

1. Use an ohmmeter to check continuity of heating elements and connections with the circuit open.

2. Use a 12-volt test lamp to check circuit operation with the circuit closed.

We can't give you all the test procedures for the variations in choke heaters, but figure 17-23 shows the principles of most choke heater tests. Choke heaters are not adjustable. Replace a defective unit.

Choke Assist Device Testing. Carmakers use a variety of devices to modulate choke operation for the best tradeoff between driveability and emission control. Some principal assist devices are summarized in Volume One. Most of these are vacuum devices that delay choke opening in very cold temperatures or that increase choke opening after a few seconds. None of the preceding choke adjustments is complete without checking assist device operation. Some are adjustable; others must be replaced if defective.

Delay valves on some chokes slow the vacuum break opening for a few seconds. Test a delay valve built into a vacuum diaphragm by applying vacuum to the diaphragm and noting how many seconds are required for the plunger to retract completely. Test a separate delay valve by applying vacuum to the inlet with a second gauge

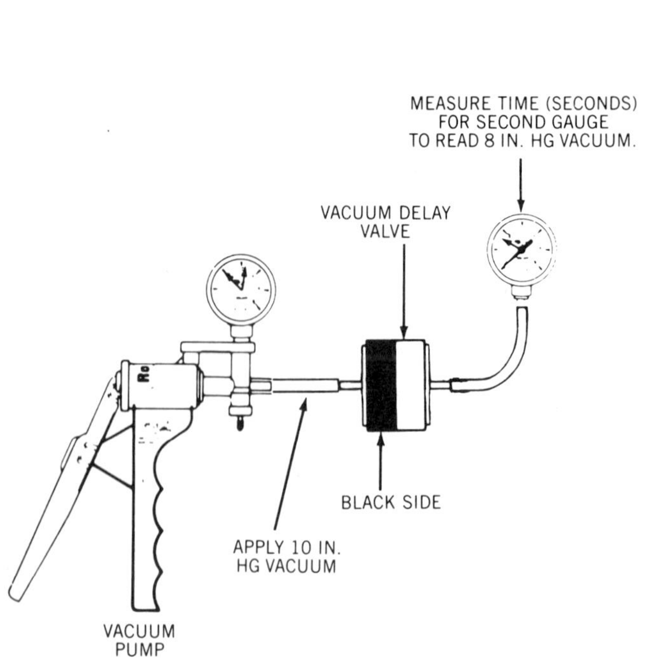

Figure 17-24. Test a vacuum delay valve by checking the time required for both gauges to equalize. (Ford)

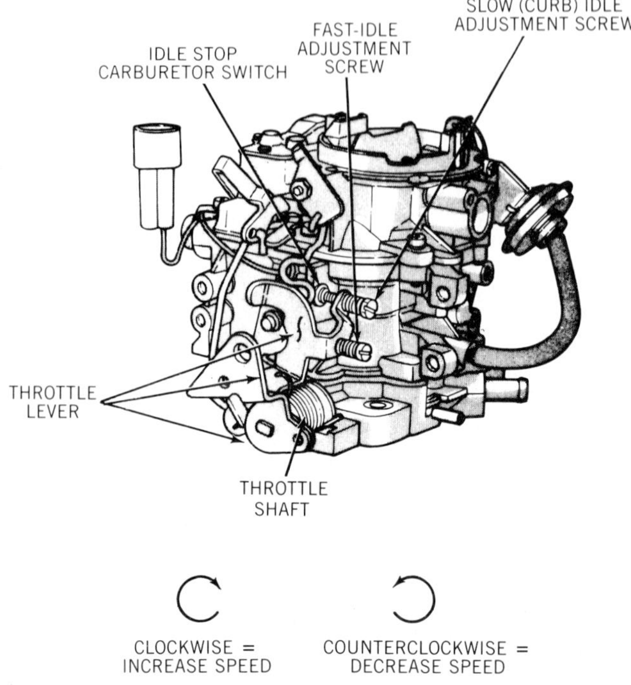

Figure 17-25. Turn the idle speed screw clockwise to increase speed, counterclockwise to decrease it. (Chrysler)

on the outlet, figure 17-24. Note the time needed for the gauge readings to equalize and compare to specifications.

Many carburetors have 2-stage vacuum breaks or two separate vacuum break units, figure 17-10. Each vacuum break must be adjusted separately, according to the carmaker's instructions. Refer to the carmaker's procedures for other tests and adjustments of choke assist devices.

Idle Speed and Mixture Adjustment

Idle speed and mixture adjustments vary considerably for different cars of different years, particularly those with electronic engine controls. The principles explained below will help you use the manufacturer's procedures more easily. Remember that on many late-model engines, the mixture screws are sealed. Adjustment requires carburetor disassembly and is not part of routine service. Also, some carburetors have electronic idle speed control, and traditional speed adjustment is not possible.

Generally, you will adjust idle speed first, then the mixture. After adjusting the mixture, check and readjust the speed. Then, recheck the mixture adjustment. Before any adjustments, do the following:

1. Set the parking brake and block the drive wheels.

2. Run the engine to normal temperature. Be sure the choke is fully open and the throttle linkage is off the fast-idle cam.

3. Connect a tachometer to the engine.

4. Follow the carmaker's instructions to disconnect and plug the EEC canister vapor purge hose to the engine.

5. Follow the carmaker's directions on whether or not to disconnect and plug other vacuum hoses. A few carmakers specify removing the PCV valve and the EEC purge hose from the engine and leaving them open to the atmosphere. You also may have to disconnect and plug an EGR vacuum line.

6. If the engine has a throttle position sensor or an idle switch, you may have to disconnect it or insulate it from the throttle screw for some idle adjustments. Check the engine decal for instructions.

7. If the carburetor has a hot-idle compensator valve, hold or block it closed.

8. If the car has a vacuum-release parking brake, disconnect the hose and plug it.

9. Follow the carmaker's directions regarding air cleaner position (on or off), transmission position (drive or neutral), air conditioning and headlamps (on or off), and other specific engine conditions.

10. Check the engine decal for idle speed specifications and basic instructions.

11. If you are using an infrared exhaust analyzer, place the probe in the tailpipe.

Basic Idle Speed Adjustment. You will still find some engines on which you set the idle speed simply by turning an idle speed screw on the throttle linkage, figure 17-25. Turn the screw clockwise to increase speed, counterclockwise to decrease speed.

Adjust the idle speed on a carburetor with an idle air bypass adjustment by turning the large adjusting screw. This screw opens and closes the bypass passage to regulate idle airflow, figure 17-26. Turn it *clockwise to decrease speed, counterclockwise to increase speed.*

Figure 17-26. Turn an idle air bypass screw counterclockwise to increase speed, clockwise to decrease it.

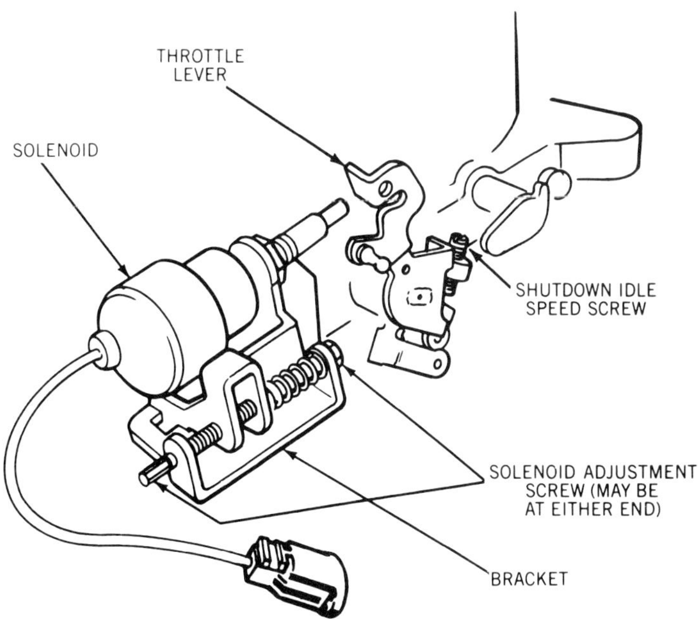

Figure 17-27. Use the bracket adjusting screw to move this throttle solenoid. (Ford)

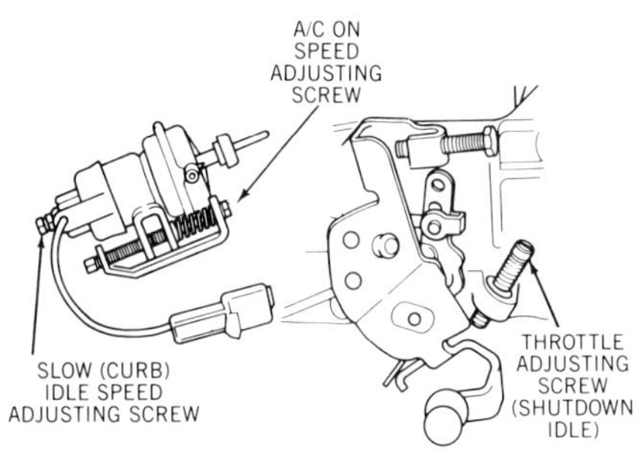

Figure 17-28. This combination solenoid and vacuum diaphragm has two adjustments for idle speed, plus a separate shutdown idle screw. (Ford)

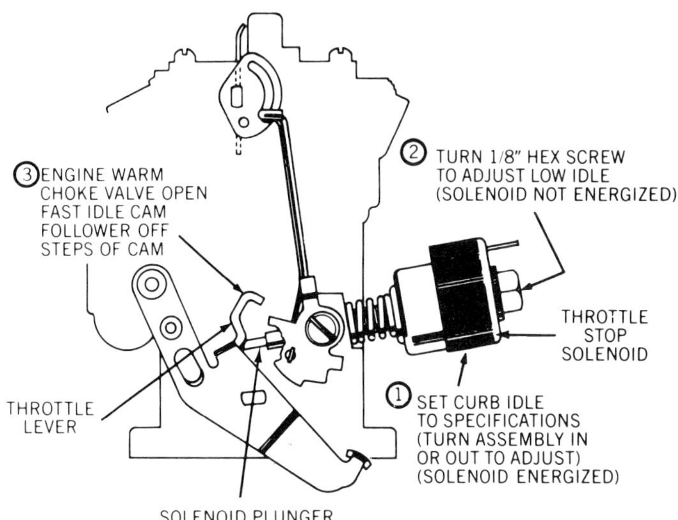

Figure 17-29. Move this solenoid in its bracket to adjust curb idle. Turn the adjusting screw to set shutdown idle. (Chevrolet)

Throttle Stop Solenoid Adjustment. Most late-model cars have throttle stop solenoids, or throttle position solenoids, to set the normal slow-idle speed *and* a lower shutdown idle speed. Slow idle, or **curb idle** is the normal idle speed of a fully warm engine. Late-model engines run at high temperatures and idle speeds close to 1,000 rpm. Also, they often idle with retarded timing. All these factors can cause afterrun, or **dieseling**. The throttle stays open far enough and the engine stays hot enough to draw in air and fuel and ignite the mixture after the ignition is turned off. A throttle stop solenoid prevents this.

When the ignition is on, the solenoid is energized, and the plunger extends to contact the throttle linkage. This holds the throttle in the slow-idle position. When the ignition is turned off, the solenoid is deenergized. The plunger retracts and lets the throttle close farther for a shutdown idle.

There are several kinds of throttle solenoids, each with different adjustment procedures, but we can group them into five basic types:

1. Solenoids adjusted by turning the plunger in or out

2. Throttle linkage with an idle speed screw that is turned in or out to contact the solenoid plunger

3. Solenoids on movable brackets with bracket adjustment screws, figure 17-27

4. Solenoids with an adjusting screw or nut at the rear of the solenoid, figure 17-28

5. Solenoids that are rotated in a bracket after loosening a locknut, figure 17-29.

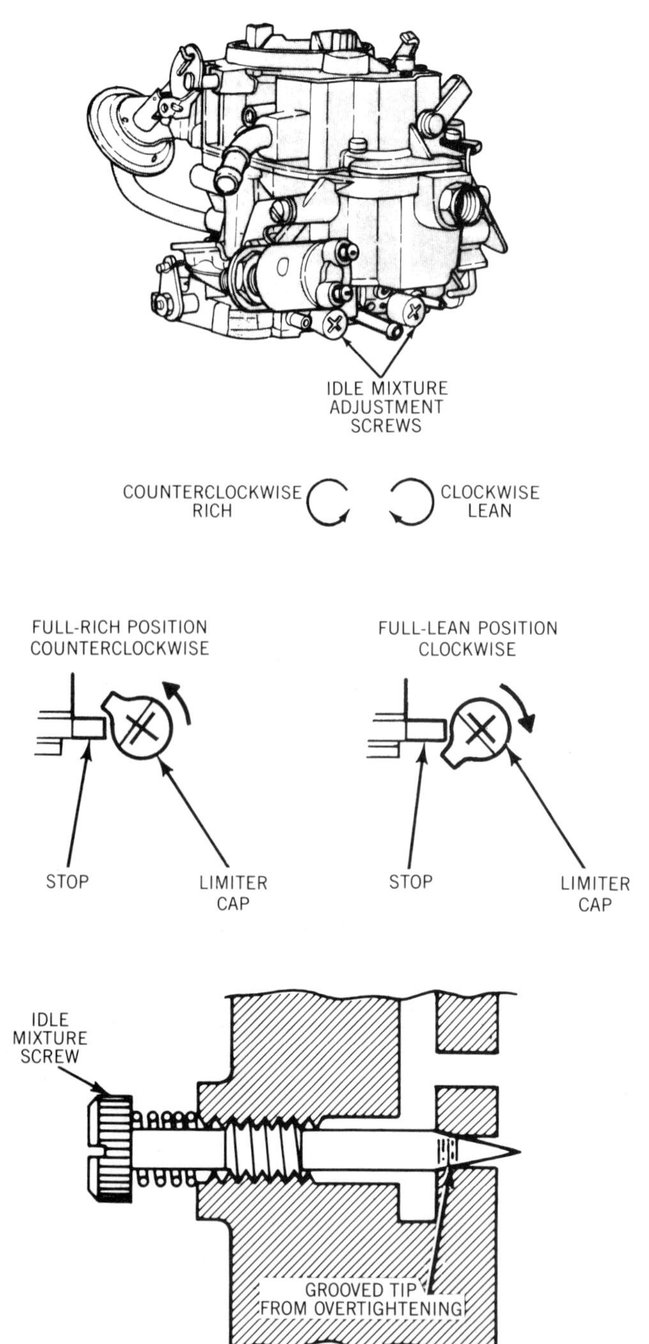

Figure 17-30. Turn mixture screws clockwise for a lean mixture, counterclockwise for a rich mixture. Do not overtighten. (Chrysler)

Figure 17-31. Use this tool to remove limiter caps. (Ford)

Before adjusting idle speed with a solenoid, you may have to check the solenoid operation. Do this by disconnecting the solenoid lead and holding or blocking the throttle partly open with the engine off. Apply battery voltage to the solenoid lead with a jumper wire and verify that the solenoid extends. Remove voltage and verify that the solenoid retracts. If the solenoid does not work right, replace it.

When you adjust idle speed with a solenoid, you will follow the carmaker's procedures and these principles:

1. With the engine running at idle, be sure the solenoid is energized with its plunger extended.

2. Adjust the solenoid plunger, the idle speed screw, the solenoid body, or the bracket screw to get the specified slow-idle speed.

3. Disconnect the solenoid lead to de-energize it and retract the plunger.

4. Adjust the specified screw to get the shutdown-idle speed.

5. Reconnect the solenoid and re-check the slow-idle speed.

These guidelines summarize basic idle speed adjustment with a throttle stop solenoid. Since the late 1970's, however, manufacturers have used more sophisticated throttle positioning devices to control engine speed during:

- Normal slow idle
- Deceleration
- Air conditioning operation
- Combinations of all of these conditions.

Late-model throttle positioners are often combinations of solenoids, vacuum diaphragms, and dashpots. We will examine their adjustment later in this chapter.

Idle Mixture Adjustment. All idle mixture screws are adjusted in the same basic ways, whether they are unrestricted, have limiter caps, or are sealed. To adjust an unrestricted mixture screw, turn it clockwise (inward) to lightly seat it, figure 17-30. Then, turn it counterclockwise (outward) a specified number of turns or until you

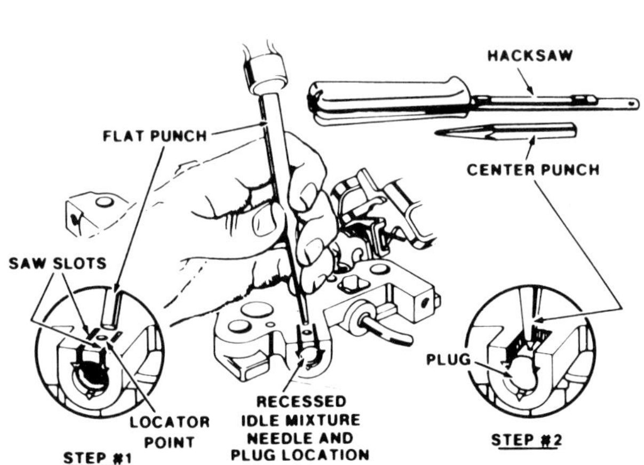

Figure 17-32. Remove some mixture screw plugs by drilling and punching them out from the carburetor base. (Chevrolet)

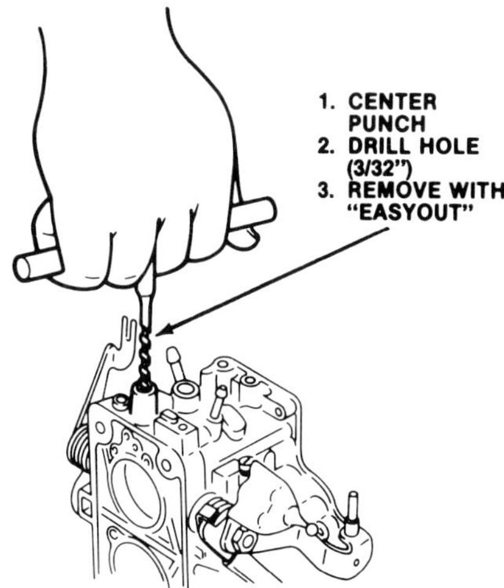

Figure 17-33. Remove some mixture screw plugs with a screw extractor or small slide hammer. (Ford)

reach a specified engine speed. Turning the screw clockwise closes the idle passage and creates a lean mixture. Turning it counterclockwise opens the passage and creates a rich mixture. If you have to remove a cap or plug from a sealed mixture screw for a major adjustment, you will still follow these principles.

The following guidelines apply to all idle mixture adjustments:

- When you seat a mixture screw, do it carefully so that the tapered end just touches the seat. Overtightening will groove the needle and damage the seat, figure 17-30.
- Adjust dual mixture needles of 2- and 4-barrel carburetors equally and alternately in small increments to balance the idle mixture.
- Limiter caps restrict adjustments to one turn or less, figure 17-30. If you can't adjust the mixture with the caps in place, the carburetor may need an overhaul, or other engine problems may exist. If specified by the carmaker and allowed by local regulations, you may remove limiter caps for adjustment when using an infrared analyzer. Use a special tool to remove limiter caps, figure 17-31, and install new caps after adjustment to prevent tampering.

Idle mixture instructions on an engine decal or in a reference manual usually call for one of the three basic adjustment methods below. The lean-best-idle and one-quarter-turn-rich methods are most often used on older engines. Adjust idle speed to the correct rpm setting and then proceed as follows:

1. *Lean-Drop Method*—Adjust the mixture screws equally to get the highest speed and smoothest idle. Readjust the idle speed to specifications. Then continue turning the mixture screws clockwise (leaner) until speed decreases by a specified amount.
2. *Lean-Best-Idle Method*—Adjust the mixture screws equally to get the smoothest idle. Then turn the screws clockwise (leaner) until idle speed decreases. Finally, turn the screws counterclockwise (richer) equally just enough to regain the lost speed.
3. *One-Quarter-Turn-Rich Method*—Adjust the screws clockwise (leaner) past the smoothest idle until idle speed decreases slightly. Then turn the screws counterclockwise (richer) one-quarter turn.

Mixture Adjustment with Sealed (Restricted) Mixture Screws. Idle mixture adjustments are not part of routine service on late-model engines. U.S. Federal regulations require that the screws be sealed or restricted to prevent improper adjustment. You may, however, remove the plugs or caps to adjust idle mixture under two general conditions:

1. To cure a driveability problem that can't be corrected by any other adjustments
2. To readjust idle mixture after carburetor overhaul.

To prevent tampering, most late-model carburetors are built so that you must remove the carburetor from the engine to remove the mixture screw plugs or caps. Tamperproof devices vary slightly, but the following paragraphs explain three common designs and their service methods.

1. Disassemble the throttle body from the rest of the carburetor or turn the carburetor upside down and support the throttle body on a stand. Drill a hole at a 45-degree angle through the base of the throttle body, toward the plug. Use a small punch to knock out the plug. On some carburetors, you must cut two small slits alongside the screw and break away a small area in the carburetor base, figure 17-32.
2. Drill a small hole through the end of the mixture screw plug and use a screw extractor or small slide hammer puller to remove the plug, figure 17-33.
3. On some Motorcraft carburetors, use snapring pliers to remove a 2-piece restrictor cap, figure 17-34. Remove external restrictor caps on other car-

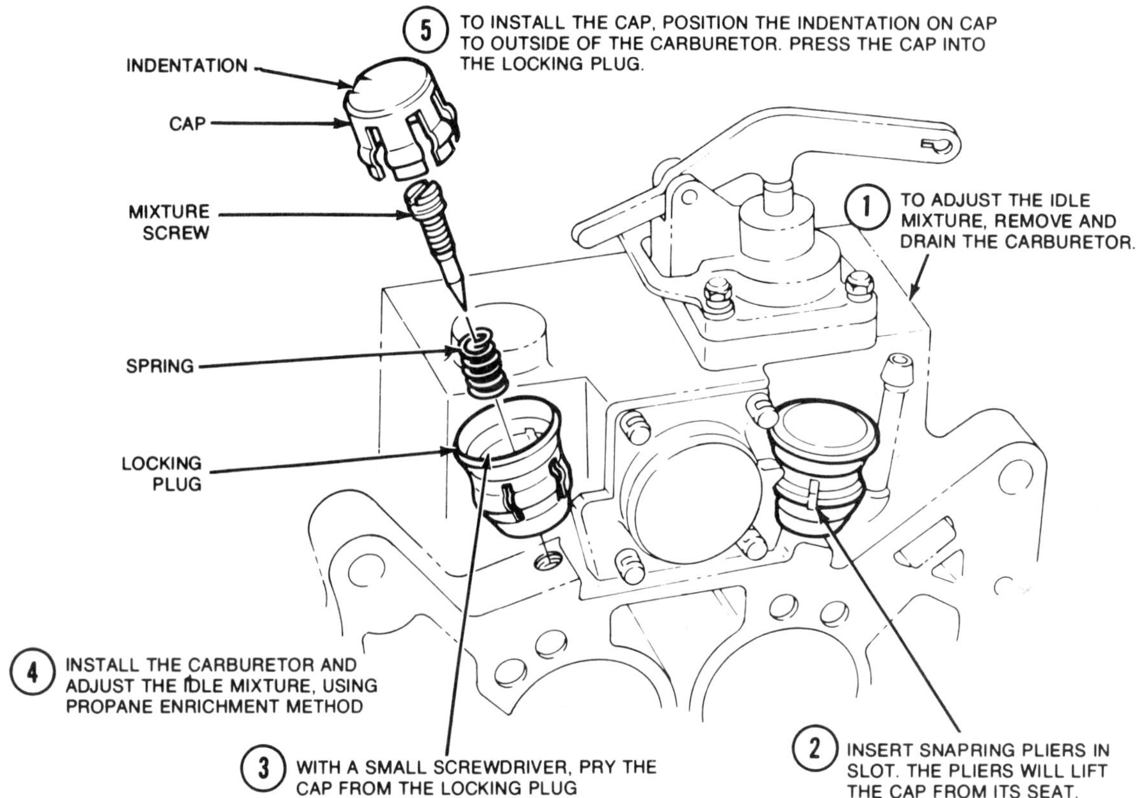

Figure 17-34. Remove these mixture screw covers with snapring pliers. (Ford)

buretors by cutting them off with a hacksaw.

After removing the plug or cap, reinstall the carburetor on the engine and adjust the idle mixture using an infrared analyzer or the propane enrichment method. Most carmakers specify propane enrichment. Following adjustment, install a new plug or cap on the mixture screw according to the manufacturer's instructions.

Mixture Adjustment with an Infrared Analyzer. An infrared exhaust analyzer is not only a valuable diagnostic tool, it is excellent for precise carburetor adjustments. Set up the analyzer as instructed in chapter 5 and proceed as follows:

1. With the engine off, turn the mixture screws to the fully lean position, figure 17-30. If necessary, remove the limiter caps and lightly seat the screws.

2. Turn the screws counterclockwise equally for a rich mixture.

3. Start the engine and adjust idle speed to the specified rpm.

4. Turn the mixture screws clockwise (leaner) in equal 1/16-turn increments to get the smoothest idle with the lowest HC reading. Allow 5 to 10 seconds for HC and CO readings to stabilize after each adjustment.

5. Readjust idle speed as necessary to maintain specified rpm.

6. Gradually adjust the mixture for the lowest possible CO reading without increasing HC. If HC increases, the engine is misfiring.

7. If you cannot get acceptably low HC and CO readings, try these basic corrections.

a. Install a new air filter and readjust for the lowest possible CO reading.

b. Check and adjust ignition timing.

c. Check the PCV system for excessive crankcase vapors to the intake manifold. Service the PCV system as necessary.

d. Change the engine oil to reduce crankcase vapors through the PCV system caused by fuel dilution of the oil.

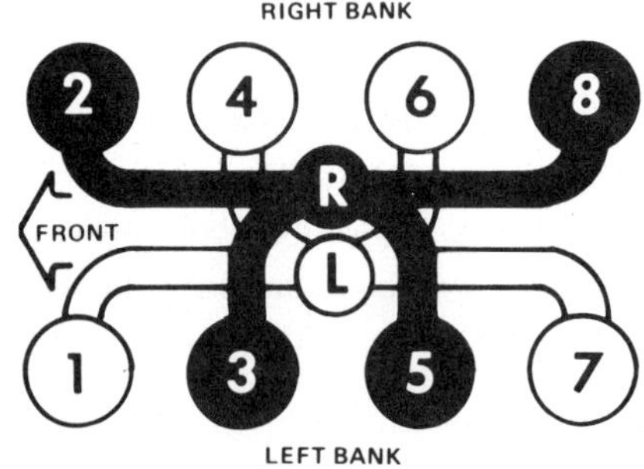

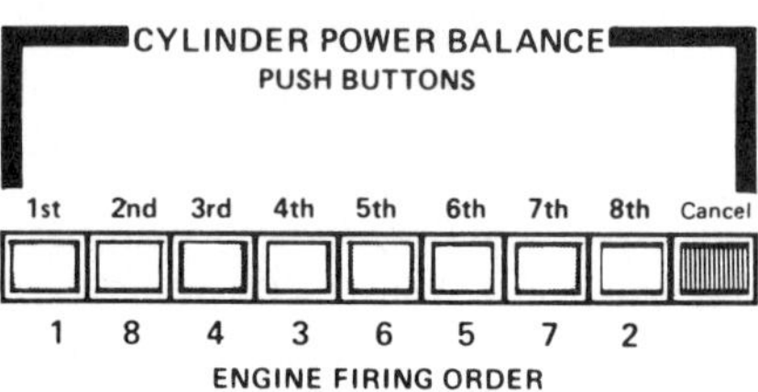

Figure 17-35. Alternately press the odd- and even-numbered buttons to kill cylinders fed from one-half of the carburetor. (Sun)

8. After adjustment, install new limiter caps, if necessary, with the tangs against their fully rich stops, figure 17-30.

If you are using a 4-gas analyzer on a converter-equipped car, CO_2 should be above 10 percent; O_2 should be less than 2 percent. If O_2 is greater, the mixture is lean. If CO is more than 0.5 percent and higher than O_2, the mixture is rich.

Carburetor Balance Adjustment. You can use a variation of the engine analyzer power balance test to be'sure the cylinders of a V-type engine with a 2- or 4-barrel carburetor are getting an equal idle mixture from each side of the carburetor. This allows you to check and adjust carburetor balance and to detect possible problems in individual cylinders.

Caution. When using this procedure on a converter-equipped car, do not kill the cylinders for more than 20 seconds at a time. Allow the engine to run for 30 seconds between adjustments to clear unburned fuel and prevent converter overheating.

Follow the analyzer instructions and these basic steps:

1. Identify each cylinder and match the firing order to the power balance control buttons on the analyzer, figure 17-35.

2. Run the engine at idle and kill the cylinders fed from one side of the carburetor (the odd or even set of pushbuttons in figure 17-35). Note the engine speed.

3. Repeat step 2 for the cylinders fed by the other side of the carburetor and again note the engine speed.

4. Adjust the mixture screws so that the readings in steps 2 and 3 are as close as possible to each other. If engine speeds for each set of cylinders are within 10 rpm, the carburetor is well balanced.

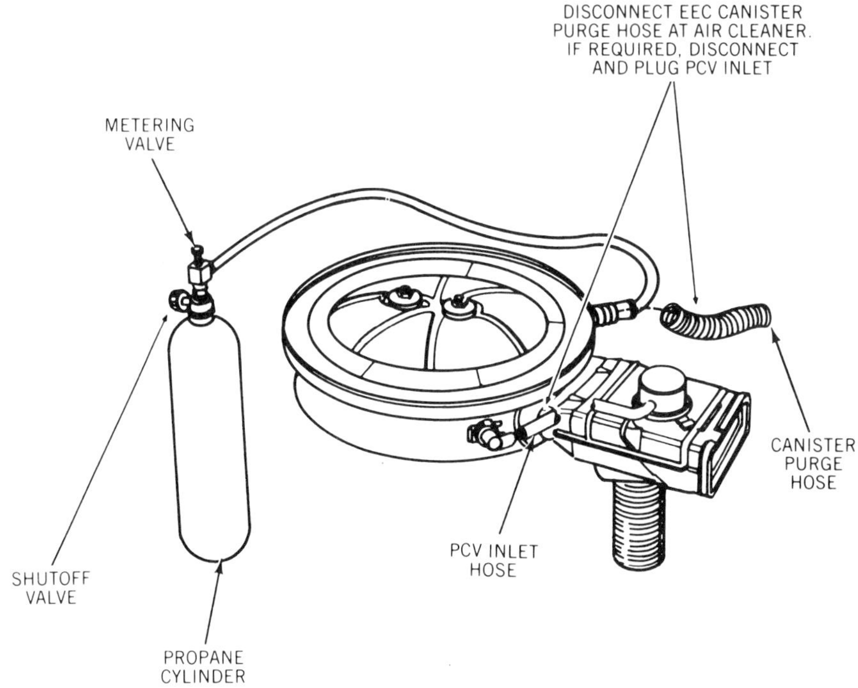

Figure 17-36. Hold the bottle vertical and inject propane at the specified location. (Ford)

Carburetor Adjustment with Propane Injection. Propane injection is a carburetor adjustment method specified by many carmakers from about 1975 until sealed mixture screws were introduced. It also can be used to test mixture adjustment on earlier carburetors and on those with sealed mixture screws. Carmakers' procedures for propane injection differ in their details, but all operate on the principles outlined here.

Catalytic converters on post-1975 cars have reduced exhaust CO to almost immeasurable amounts. Some infrared analyzers may not give precise idle CO readings that you can use for idle adjustment. Propane injection adds a controlled amount of another fuel to the air-fuel mixture and causes idle speed to increase. There is a 3-way relationship of idle speed increase, air-fuel ratio, and the amount of propane injected. The propane amount stays constant during testing. Engineers can calculate the speed increase that a given amount of propane will provide with a specific air-fuel ratio and intake volume. This determines rpm increase specifications. The relationships between speed increase and air-fuel ratio are as follows:

- If speed increase is *less* than specified during testing, the air-fuel ratio is too *rich*. The propane becomes a smaller percentage of total fuel volume and causes a smaller speed increase.
- If speed increase is *more* than specified during testing, the air-fuel ratio is too *lean*. The propane becomes a larger percentage of total fuel volume and causes a greater speed increase.
- Propane injection requires a commercial propane bottle, a shutoff valve, a metering valve, and a length of hose. Many tool companies sell propane injection kits. Carmakers specify propane injection through one of the following locations:
- An EEC hose connection on the air cleaner, figure 17-36
- A PCV hose connection if it does not have a filter restriction
- Through the air cleaner snorkel
- Through a manifold vacuum port at the carburetor base.

Keep the propane bottle vertical, with the valve at the top, for uniform propane flow during testing. The following points will help you use the manufacturer's procedures for propane testing and adjustment.

1. Be sure ignition timing is properly adjusted before testing and adjusting the idle mixture.

2. Connect a tachometer to the engine and run the engine at fast idle to normal operating temperature. Use a dial-type (analog), not digital, tachometer to view the speed changes most accurately.

3. Complete all other preliminary steps for idle adjustment. Doublecheck the

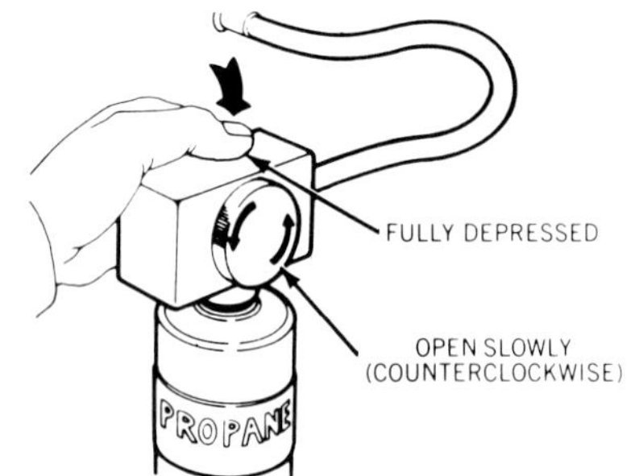

Figure 17-37. This propane injection tool has a combination metering and shutoff valve. Press the plunger to inject gas. (Ford)

carmaker's directions, particularly for transmission gear selector position.

4. Remove the PCV valve from the engine and let it draw fresh air during test and adjustment. This avoids enriching the air-fuel mixture with excessive crankcase blowby.

5. Follow the carmaker's directions to disconnect or reroute:

a. Air injection lines and vacuum hoses

b. EEC purge and vent hoses

c. EGR vacuum lines

d. Air cleaner vacuum hoses

e. Engine vacuum or manifold pressure sensors

6. If the engine has an exhaust gas oxygen (EGO) sensor, disconnect it. Follow directions to ground the engine wiring harness connector (not the sensor) or otherwise put the engine in open-loop operation.

7. With the propane bottle connected as directed, open the shutoff valve. Then gradually open and adjust the metering valve until you get the maximum rpm increase, figure 17-37. Opening the valve more will cause speed to drop as the mixture becomes overly rich.

8. Record the rpm gain and compare it to specifications.

9. If the rpm gain is out of limits, adjust either the mixture screws or the speed screw according to the carmaker's directions to get the specified speed increase.

10. Turn off the propane and check the normal curb idle speed. Follow directions to adjust either the mixture screws or the speed screw.

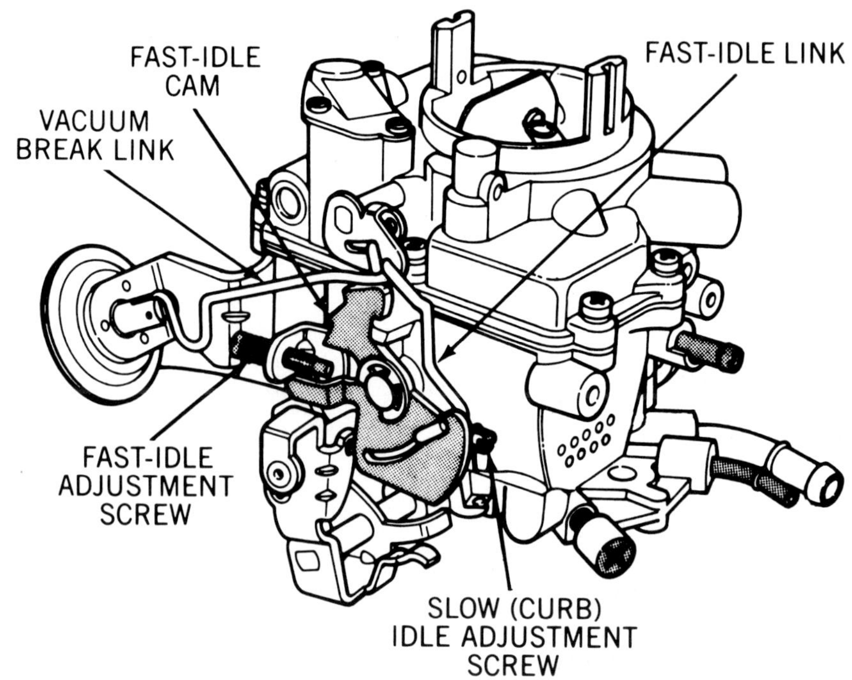

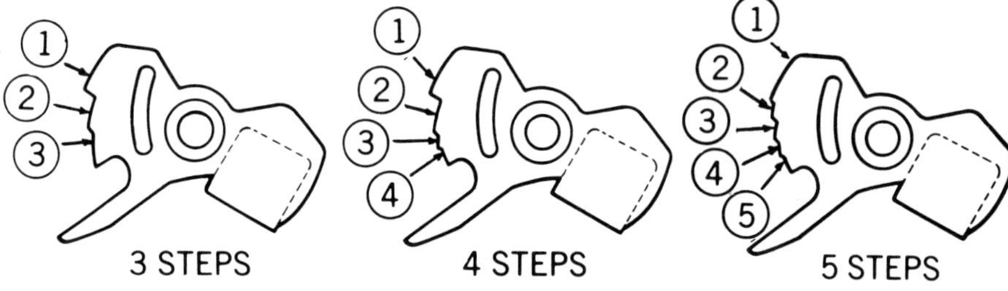

Figure 17-38. Adjust fast idle with the idle speed screw or a separate fast-idle screw on the specified step of the cam. (Chrysler) (Ford)

11. Turn on the propane and adjust the metering valve for the highest speed increase to recheck the enriched rpm.

12. Repeat steps 7 through 11 until the speed increase is within limits. Check and readjust idle speed after each mixture adjustment. Install new mixture limiter caps, if required, after final adjustment.

You can use propane injection to test the idle mixture of older cars and those with sealed mixture screws, *if the car does not have electronic engine controls.* An exhaust gas oxygen (EGO) sensor and engine computer will react to propane enrichment and lean out the idle mixture to cancel any rpm gain. In addition to the instructions above, some carmakers have special test procedures for propane injection with the control system in an open-loop operating mode.

Fast-Idle Speed Adjustment

When the choke closes, its linkage moves a fast-idle cam to contact the idle speed screw or a separate fast-idle screw and open the throttle slightly for fast idle during warmup. The fast-idle cam has several steps, and the specified step must contact the idle speed screw or the fast-idle screw when the choke closes, figure 17-38.

Adjust the slow-idle speed and mixture and the choke before adjusting fast-idle speed. If the carburetor has a separate fast-idle screw, place it on

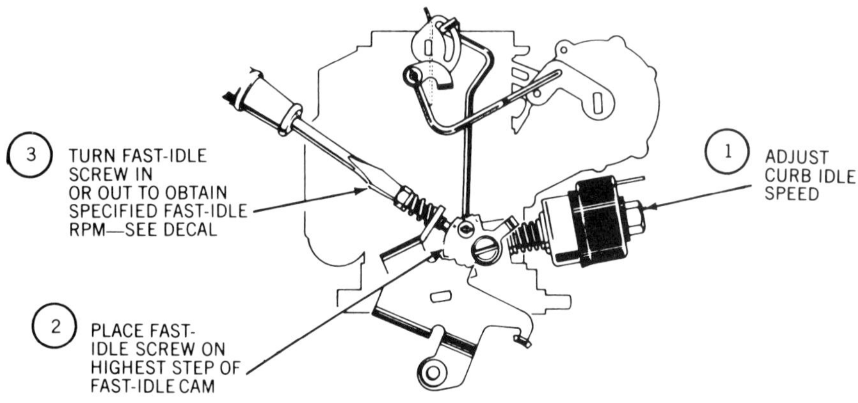

Figure 17-39. Basic fast-idle screw adjustment. (Rochester)

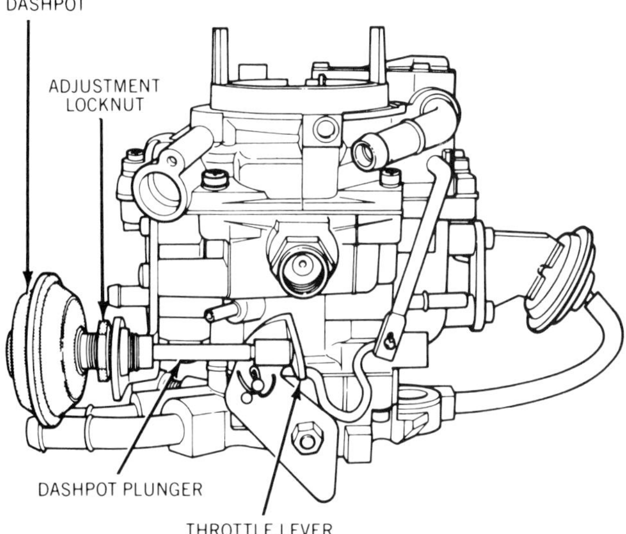

Figure 17-40. Basic dashpot adjustment. (Chrysler)

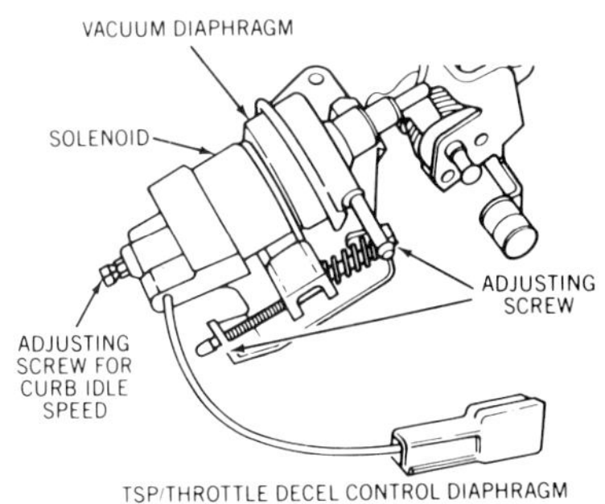

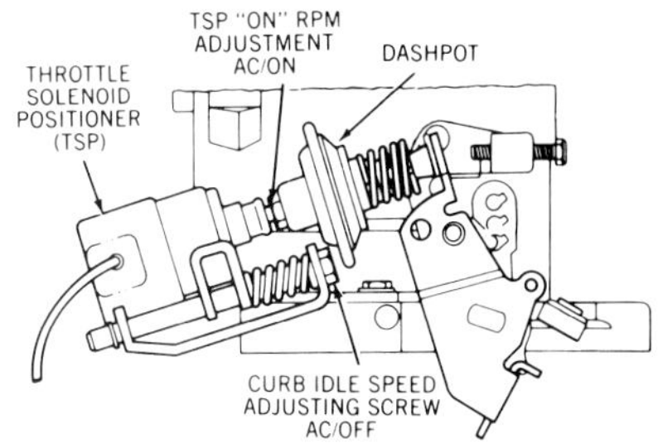

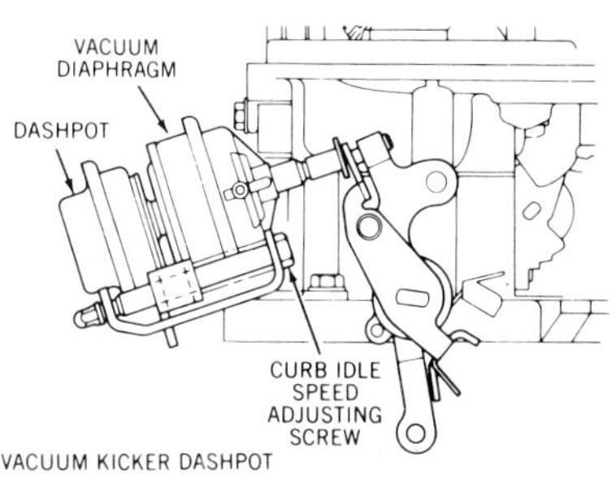

Figure 17-41. Adjustment procedures vary depending on the kind of throttle positioner used. (Ford)

the specified step of the cam and turn it in or out to get the required fast-idle speed, figure 17-39. If the carburetor does not have a fast-idle screw, place the slow-idle screw or the fast-idle cam follower (tang) on the specified step of the cam and check the fast-idle speed. If the choke is adjusted right, its linkage should move the cam so that the correct step contacts the idle speed screw when the choke closes. If it does not, bend the linkage rod so the correct cam step contacts the screw.

Several other adjustments are possible for choke and fast-idle linkage. It is important to understand that choke closing adjustment, vacuum break (pulloff) adjustment, and fast-idle linkage adjustments are all related. Fast-idle speed adjustment is a common part of on-car carburetor service. The other adjustments are usually done as part of major carburetor service. But, you may have to check and adjust all parts of the carburetor linkage to cure a driveability problem. The number and kinds of adjustments vary from one carburetor to another. For complete adjustment, you must follow the carmaker's procedures and adjustment sequence exactly.

Throttle Positioner and Deceleration Control Adjustment

Late-model carburetors use a variety of throttle positioner devices to delay throttle closing on deceleration or to open the throttle slightly to maintain required idle speed. Many carburetors have combination devices that are both a solenoid and a dashpot or a vacuum-operated positioner. Basic throttle stop solenoid adjustment was explained earlier in this chapter.

Deceleration throttle positioners are either solenoids or vacuum diaphragms. You can use basic electrical and vacuum test methods to check them. Deceleration positioners and combination devices require special adjustment procedures.

Adjust a simple dashpot by turning or sliding the dashpot in its mounting bracket after loosening the locknut, figure 17-40. Follow the manufacturer's instructions for extending or retracting the plunger and positioning the throttle lever.

Some carburetors have a combination throttle solenoid and dashpot, a vacuum-operated throttle positioner

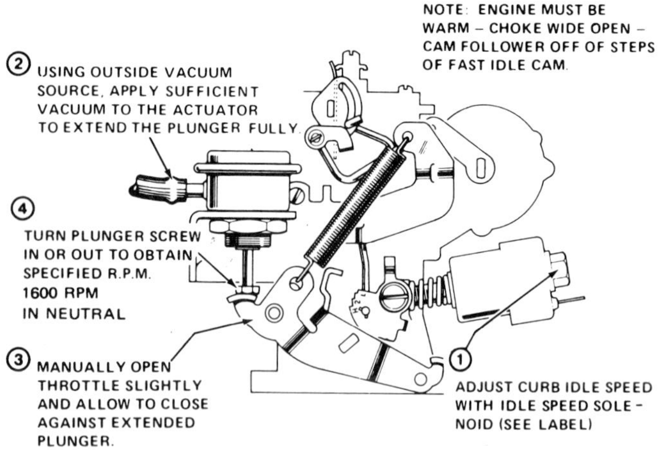

Figure 17-42. Typical adjustment for a vacuum-operated deceleration positioner. (Rochester)

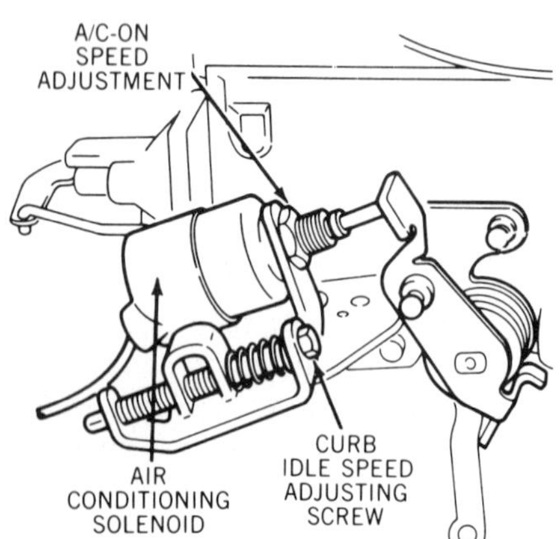

Figure 17-43. This solenoid and bracket have separate adjustments for curb idle and air conditioning throttle position. (Ford)

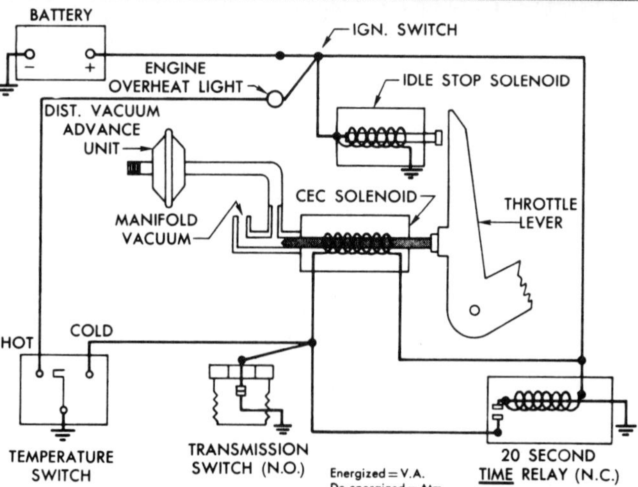

Figure 17-44. The CEC valve used by GM in the early 1970's was a combination throttle deceleration positioner and vacuum advance solenoid. (Chevrolet)

plus dashpot, or a vacuum-operated deceleration positioner. Figure 17-41 shows some of these variations. Figure 17-42 shows typical adjustments for a throttle stop solenoid and a separate vacuum-operated throttle positioner.

Air conditioning throttle solenoids are also common on late-model engines. Some include separate adjustment points for curb idle and for A/C idle speed, figure 17-43.

Adjustment may be needed for any of these devices for normal curb idle, or if the driver complains about sustained high-idle speeds after deceleration. Refer to the manufacturer's procedures to identify, test, and adjust these devices.

Some older GM cars had combi-

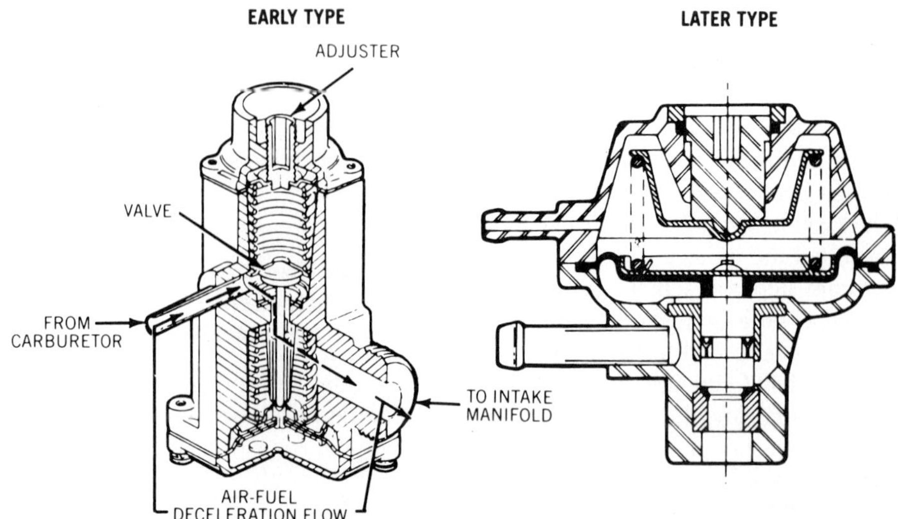

Figure 17-45. Ford's decel valve of the 1970's increased the deceleration air-fuel mixture. (Ford)

Figure 17-46. Air-fuel mixture timing is adjustable on many decel valves.

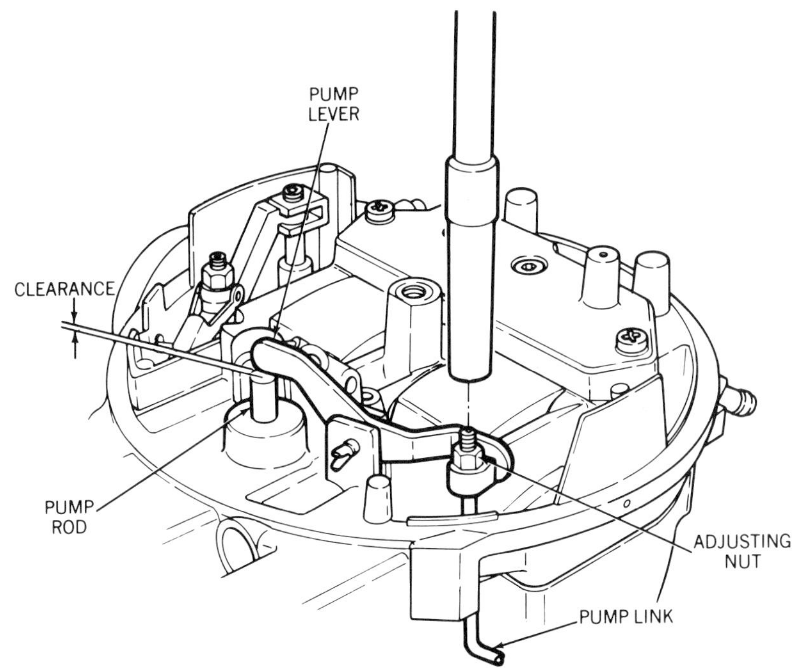

Figure 17-47. Accelerator pump adjustment on the 2700 VV and 7200 VV carburetors. (Ford)

nation emission control (CEC) valves that were both deceleration throttle positioners and vacuum spark advance control valves, figure 17-44.

Deceleration Air-Fuel Valve. Some engines have vacuum-operated valves that pass an extra air-fuel charge to the intake manifold for a few seconds during deceleration. These "decel" valves were used by several carmakers but were most common on Ford 4- and 6-cylinder engines in the 1970's, figure 17-45. You can adjust the spring tension on most of these valves to control the timing for the deceleration air-fuel charge, figure 17-46. Adjustment requires a special tool or a modified screwdriver. Turn the screw clockwise to decrease valve-opening time, counterclockwise to increase it.

Variable-Venturi Carburetor Adjustments

The adjustments you have studied up to this point are typical for most fixed-venturi carburetors. Ford's 2700 VV and 7200 VV carburetors also require several on-car checks and adjustments as part of regular service. Some are similar to those for other carburetors; others are different.

Slow (curb) idle is adjusted as on any other carburetor. The exact methods vary with the kind of throttle positioner used on a specific installation. Because most fuel for the idle mixture comes from the main jets, controlled by the venturi metering rods, idle mixture adjustments are not required. Early models had an idle trim system with adjustable metering screws. These, however, were eliminated in 1978.

Ford's variable-venturi carburetors have cup-type accelerator pumps. Whenever you adjust curb idle speed, you must check and adjust pump lever clearance, figure 17-47.

Because the venturi valves create a variable opening at the top of the barrels, which is closed during cranking, these carburetors do not have conventional choke valves. They do, however, have a choke cap with a thermostatic spring and an electric heater. The choke cap controls the po-

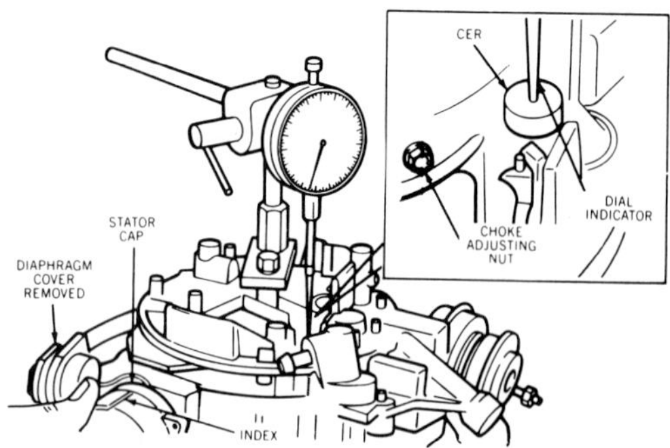

Figure 17-48. Typical CER and CVR adjustment on Ford's variable-venturi carburetors. (Ford)

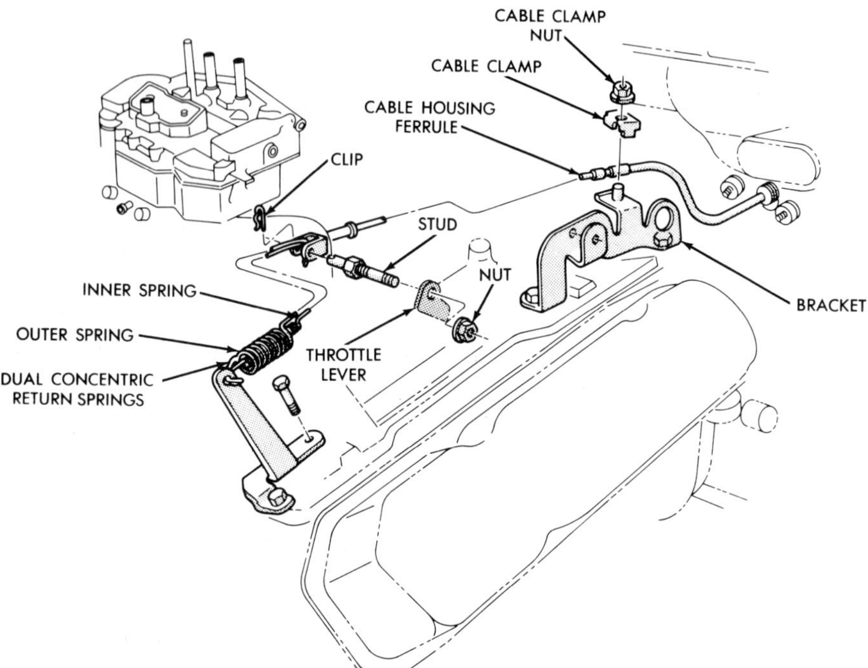

Figure 17-49. For easy reinstallation, tag the throttle linkage parts when you remove them. (Chrysler)

sitions of the control vacuum regulator (CVR) rod and the cold enrichment rod (CER). The CVR controls venturi vacuum, and the CER adds extra fuel to enrich the starting mixture. These devices affect fuel metering throughout the carburetor's operating range. "Choke" adjustment on a variable-venturi carburetor requires a dial indicator to measure CVR rod and CER travel, figure 17-48. Several precise adjustments are required, and adjustment points are sealed with epoxy to prevent tampering on late-model units.

The 2700 VV and 7200 VV carburetors also have fast-idle cams and screws, as well as choke vacuum break diaphragms. The fast-idle linkage controls warmup throttle position. The vacuum diaphragm acts on the fast-idle linkage and the CVR and CER linkage.

As with any other late-model carburetor, the 2700 VV and 7200 VV models have gone through many design variations since their introduction in 1977. You must follow Ford's procedures for specific years and applications for accurate service.

CARBURETOR REMOVAL AND REPLACEMENT

You must remove the carburetor from the manifold to overhaul it or to replace it. Carburetor installations vary from car to car and engine to engine, but you can use the following guidelines to help with these jobs. If the carburetor is mounted on a turbocharger inlet, removal and replacement is basically the same as if it were mounted on the manifold.

Carburetor Removal

1. Disconnect all vacuum, EEC, and PCV hoses from the air cleaner and the carburetor throttle body. If the hoses are not color coded and you do not have a vacuum diagram, tag the hoses and their connections for reconnection.

2. Remove the air cleaner or air intake duct from the carburetor.

3. Disconnect all electrical connectors from choke heaters and carburetor sensors, solenoids, or motors.

4. Disconnect the following devices as required: remote choke linkage, choke hot-air tube, and choke coolant lines.

5. Disconnect the throttle linkage, figure 17-49, and transmission throttle valve linkage, if required.

6. Disconnect the fuel line from the carburetor and plug or cap it.

7. Remove the carburetor mounting nuts or capscrews.

8. Remove the carburetor. If it is stuck to the manifold, tap the throttle body base lightly with a small plastic or rubber hammer to loosen it. Do not pry with a screwdriver.

9. Place a clean cloth in the manifold opening to keep dirt and small parts from falling into the engine.

10. Remove the old mounting gasket and scrape the manifold base and the carburetor base to remove gasket cement and old gasket fragments.

Carburetor Installation

1. Remove the cloth from the manifold and place a new gasket on the manifold base.

2. If the carburetor has an electric grid heater or an EGR valve spacer beneath it, be sure these parts are correctly installed before installing the carburetor.

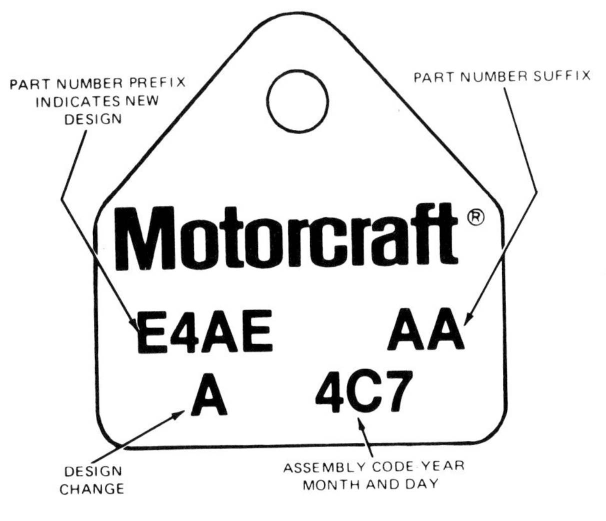

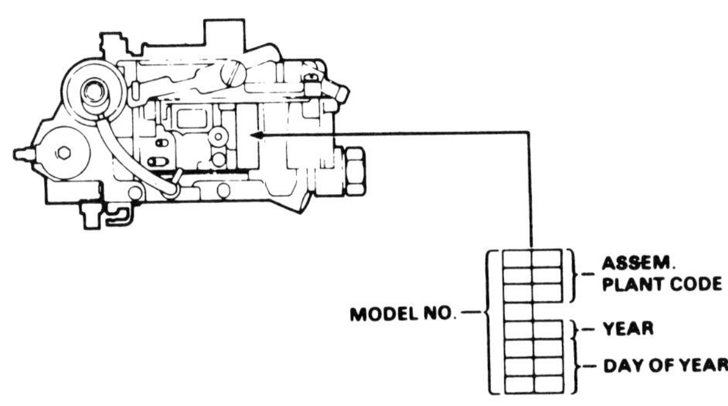

CARBURETOR IDENTIFICATION

VARAJET II MODELS 2SE-E2SE (TYPICAL)

Figure 17-50. Model and part numbers are stamped on the carburetor or on a metal tag. (Ford) (Rochester)

3. Place the carburetor on the manifold and install the nuts or capscrews.

4. Connect the fuel line. Install a new fuel filter if required.

5. Tighten the fasteners alternately and evenly to the specified torque. Do not overtighten.

6. Connect the throttle linkage, figure 17-49.

7. Connect the choke linkage and a hot-air tube or coolant line, if required. Connect the choke heater wiring.

8. Connect all other electrical connectors and all vacuum hoses.

9. Install the air cleaner or air intake duct.

10. Start the engine, check carburetor operation, and make all necessary adjustments.

GENERAL CARBURETOR OVERHAUL

The carburetor adjustments outlined previously are the ones you will use for most routine service, and they will cure most performance problems. If the carburetor is leaking, dirty, or has internal problems, you will have to overhaul it.

Carburetor Identification

The first step of carburetor overhaul is accurate identification. Remember that dozens of part number variations exist for each basic carburetor model. Detail specifications and adjustments will vary, as will specific overhaul steps. You can identify a carburetor accurately by the year and model of vehicle on which it is installed and by the model and part numbers stamped on the carburetor body or on a metal tag attached to a screw, figure 17-50. You will use model and part numbers to get the right overhaul instructions, specifications, and repair kit.

Carburetor Repair Kits

Besides instructions and drawings of the carburetor, a repair kit usually contains:

1. Gaskets

2. An inlet needle and seat

3. An accelerator pump

4. A power valve diaphragm (if used)

5. Accelerator pump check balls

6. Miscellaneous small clips.

A repair kit does not contain jets, metering rods, linkage parts, vacuum diaphragms, and screws. Be careful when removing these parts so that they are not lost or damaged. Many repair kits have gaskets for several variations of a carburetor model. They usually differ in vacuum passage openings. Do not throw the old gaskets away immediately. Save them to match with those in the kit to select the correct replacements.

Carburetor Disassembly

You do not need to separate every screw, nut, and linkage part of a carburetor to clean and overhaul it effectively. For example, you do not have to remove throttles, choke plates, and fast-idle linkage. If you leave some linkage parts assembled, carburetor reassembly and adjustment will be easier.

Note the positions of all linkage parts before disassembly. Follow disassembly instructions carefully, particularly for some late-model carburetors on which certain subassemblies cannot be separated. For example, screws for throttle valves are staked in place on many carburetors, and the valves and shafts cannot be removed. Be sure to remove all vacuum and pump diaphragms, rubber or plastic parts, and electronic parts (mixture control solenoids, stepper motors, and idle speed motors) before putting the carburetor in cleaning solvent. Carburetor cleaners will destroy these parts.

Remove jets carefully with a jet removal tool. Avoid using a screwdriver on carburetor jets. The slightest burr or scratch on a jet can upset fuel metering. Jets for many carburetors are not available as replacement parts, and you must replace the complete carburetor if you damage a jet. Keep primary and secondary jets separate during disassembly, and keep metering rods matched to their respective jets.

Use a removal tool to remove limiter caps from idle mixture screws. If mixture screw removal is necessary from a carburetor with sealed screws, follow the manufacturer's directions, figures 17-32 through 17-34.

Before removing any idle mixture screw, turn it clockwise and count the number of turns needed to lightly seat it. When you reassemble the carburetor, install the mixture screw to this same position.

Be careful not to lose accelerator pump check balls. Note their positions when you remove them.

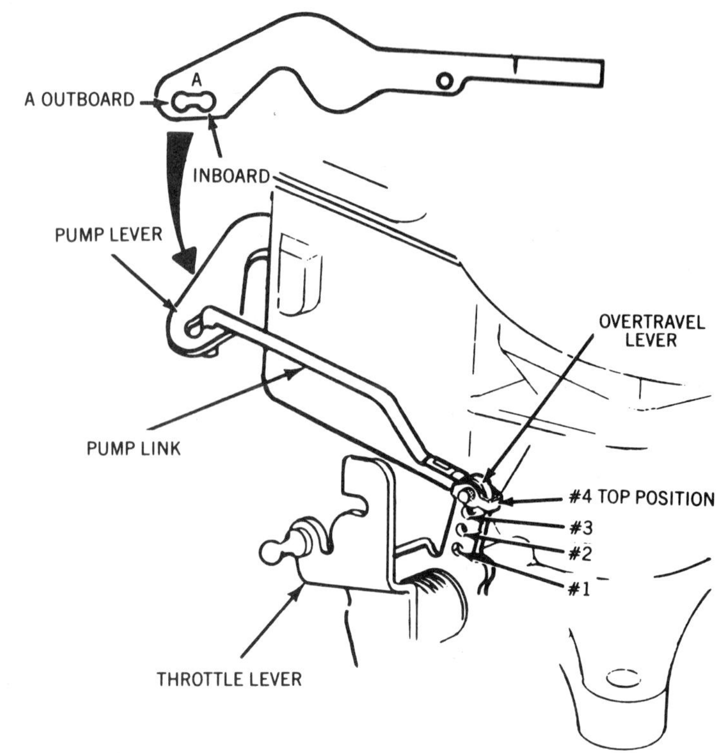

Figure 17-51. Accelerator pump linkage has several adjustment locations. (Ford)

Carburetor Cleaning

Warning. Wear rubber gloves when working with carburetor cleaner. Do not allow cleaner to contact your skin. Carburetor cleaners are toxic and may cause illness.

Put the carburetor parts in a basket or hang them from wires in the cleaner tank. Do not put rubber, plastic, or electronic parts into the cleaner. Some carburetor cleaners act very quickly. Do not leave parts in the cleaner longer than necessary. Caustic cleaners can remove protective coatings from aluminum and zinc parts and make them porous.

After soaking the parts in carburetor cleaner, rinse them thoroughly in running water or common cleaning solvent. Use compressed air to dry the parts and blow out passages. Do not use wires, drills, or other hard objects to clean jets, passages, or other openings. Nicks and scratches can damage them. Do not use sandpaper, a wire brush, or steel wool on carburetor parts. Abrasives will scratch the parts and remove protective coatings.

Carburetor Assembly and Linkage Adjustment

Assemble and adjust the carburetor according to the drawings and instructions in the repair kit or a carburetor overhaul manual. Be careful to install primary and secondary jets in their correct positions. Use a jet tool and do not overtighten. Keep metering rods matched with their jets. When you install accelerator pump check balls, the large one *usually* is for the pump inlet. Install accelerator pump and choke linkages in their original positions or in specified holes or slots. Accelerator pump linkage usually has several positions for the pump link to control pump stroke and fuel delivery, figure 17-51.

Overhaul manuals and repair kit instructions may include a dozen or more specific illustrations and instructions for various adjustments, such as:

- Choke opening and closing
- Vacuum break operation
- Choke unloader
- Accelerator pump stroke
- Secondary throttles and air valves
- Secondary lockout linkage.

Adjustment methods and specifications vary for different carburetor designs and engine applications.

Install mixture screws by turning them in until lightly seated, then backing them out the number of turns that you counted at disassembly. Make the final mixture adjustments when the carburetor is installed. Then replace limiter caps or sealing plugs, if required.

Some carburetor makers specify that certain screws must be tightened in specific sequences to specific torque values. Side-mounted fuel bowls on some Holley carburetors, for example, must be torqued precisely with a torque-limiting screwdriver or nutdriver. Airhorn installation is a critical point on most carburetors. If airhorn screws are tightened unevenly or overtightened, choke valve, metering rod, and accelerator pump linkage may bind. The carburetor may leak, and screw holes may be stripped if

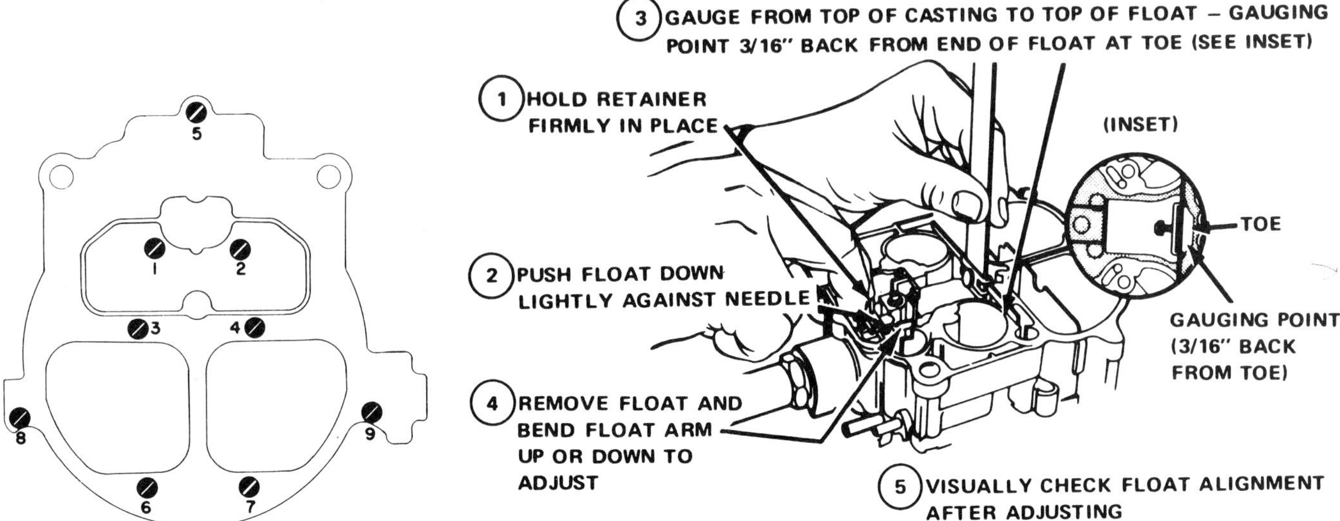

Figure 17-52. This airhorn tightening sequence for a Quadrajet is typical of drawings you will find for many carburetors. (Chevrolet)

Figure 17-53. Typical adjustment methods for a float mounted in the fuel bowl. (Rochester)

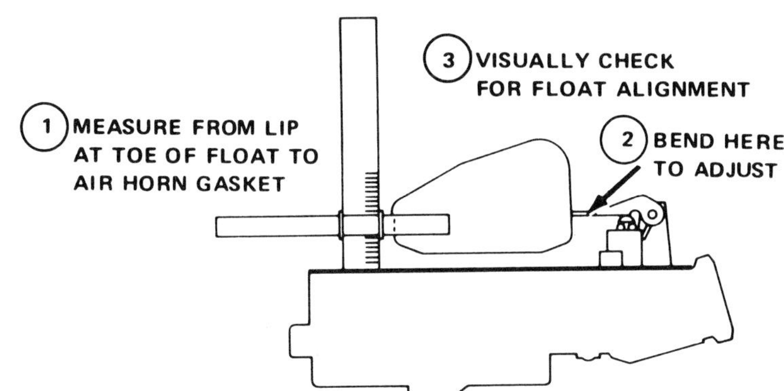

Figure 17-54. Typical adjustment methods for airhorn-mounted floats. (Rochester)

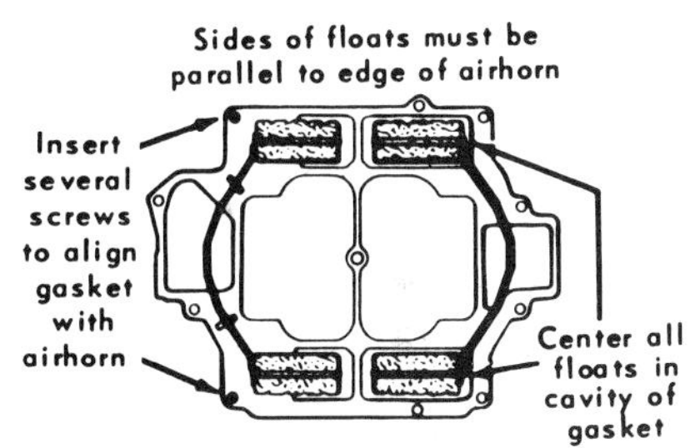

Figure 17-55. The float pontoon must be parallel with the fuel bowl. (Rochester)

the airhorn is not installed properly. Most repair instructions show a specified airhorn tightening sequence, figure 17-52. Some also include screw torque values.

Float Adjustment

Float settings are critical adjustments on all carburetors. Four basic float adjustments are described below. Every carburetor requires float level adjustment. Some carburetors require one or all of the others. Follow the manufacturer's instructions for measuring at certain points and for bending the float arm or tang at specific locations. Instructions also will specify whether float measurements and adjustments are made with or without the gasket in place. When you are adjusting float position to tolerances of + 1/32 inch (0.8 mm), the gasket thickness makes a critical difference.

1. *Float Level*—Float level is the basic float adjustment. Figure 17-53 shows a typical adjustment for a float hung in the fuel bowl. Figure 17-54 shows a typical adjustment for a float hung from the airhorn. Use the float level gauge supplied with the kit or a precision T-scale to measure float position exactly. If the float has two pontoons, adjust both carefully.

2. *Float Alignment*—The float must be parallel with the edges of the fuel bowl or they may rub on the bowl and cause the inlet needle to stick. Some floats have an alignment adjustment, figure 17-55.

3. *Float Drop*—If the float hangs from the airhorn, it may require float drop adjustment. This is separate from float level adjustment and is made by bending a tab, or tang, on the back of the float arm, figure 17-56.

4. *Float Toe*—Float toe is a third adjustment needed for some floats hung from the airhorn, figure 17-57. After adjusting float toe, recheck float level and drop to be sure they are correct.

These final precautions will help you adjust floats carefully and accurately:

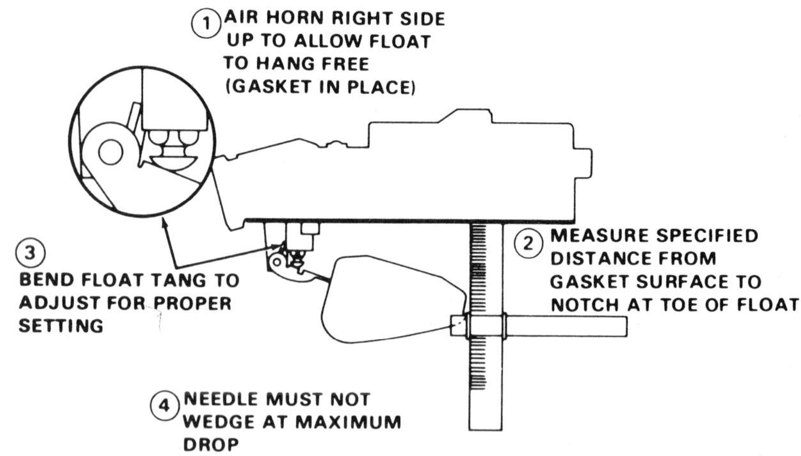

Figure 17-56. Typical float drop adjustment. (Rochester)

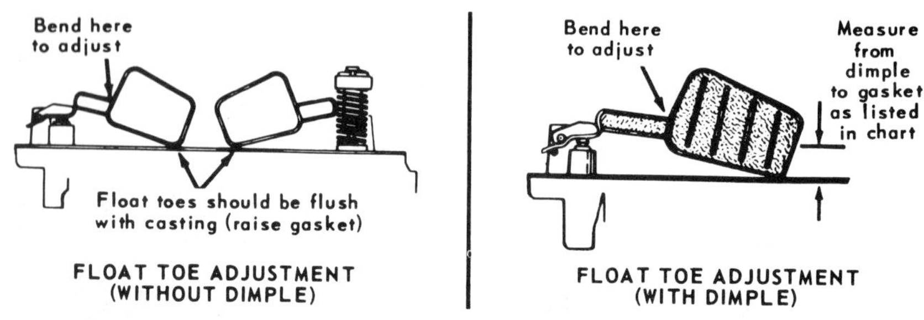

Figure 17-57. Typical float toe adjustment. (Rochester)

1. Before installing a float, check it for buoyancy and leaks according to the manufacturer's instructions. This is very important for plastic floats. Some manufacturers recommend comparing a used float with a new float on a small balance scale to check float weight.

2. Lift the float tang away from the needle when bending an arm or tang. This prevents a false measurement and avoids damage to the needle and seat.

3. Keep the float tang parallel or perpendicular with the needle, as required. Side pressure can cause the needle to stick.

4. For a 2-pontoon float, adjust each separately and equally. Keep them parallel.

5. Don't scratch a float tang during adjustment or the needle may stick.

Typical Carburetor Overhaul

The following photo procedure illustrates the overhaul of a Rochester 4M Quadrajet carburetor. Although exact overhaul steps will vary from one carburetor to another, this sequence shows the typical steps of this job. It also illustrates the general sequence for any carburetor overhaul.

Rochester Quadrajets have been common on GM cars for over 20 years, but many variations exist among their part numbers. Late models have mixture control solenoids, electric chokes, and other differences not shown in this sequence. Nevertheless, this procedure will help you follow the manufacturer's instructions for most carburetor overhauls. Read through this procedure, or any repair instructions, thoroughly before starting the job to help you understand key points and adjustments.

QUADRAJET OVERHAUL

1. Remove the screw securing the upper choke lever to the choke shaft. Then twist the lever to remove it from the rod. Remove the rod later.

2. Remove the screw from the secondary metering rod hanger and lift the rods carefully out of the airhorn. Don't separate the rods from the hanger.

3. Remove the front and rear vacuum breaks and other devices, such as mixture solenoids, throttle solenoids, dashpots, and linkage.

QUADRAJET OVERHAUL (*Continued*)

4. Remove 9 screws that hold the airhorn to the fuel bowl (main body). Don't overlook 2 in counterbores next to the venturi.

5. Remove the bowl vent valve cover, gasket, and spring (arrow) from the airhorn.

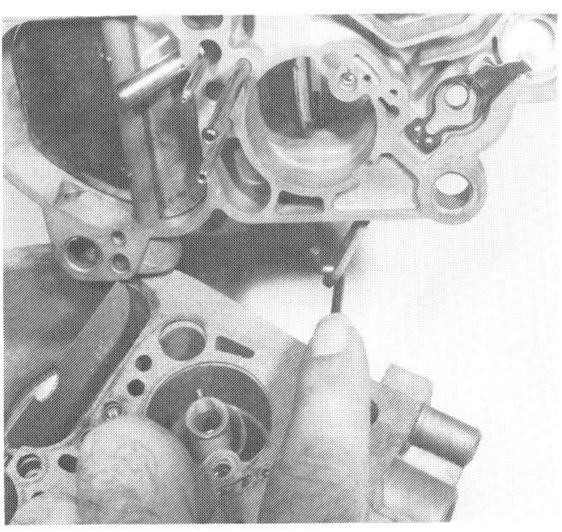

6. Raise the airhorn straight up from the bowl (main body) until the accelerator and secondary bleed tubes clear the body. Then turn the airhorn to disengage the pump rod.

7. Raise the gasket carefully at the corner to remove the accelerator pump plunger. No further airhorn disassembly is needed.

8. Before removing the gasket, disengage the auxiliary (front) power piston from the metering rod. Hold the piston down to remove the hanger.

9. Now, turn the hanger to touch the main (rear) power piston. Slide the gasket slit past the hanger to remove the gasket.

10. Use long-nose pliers to press the main (rear) piston down. Then let it snap upward to remove the piston and metering rods.

11. Lift the plastic filler block from the bowl. The block prevents sloshing and minimizes fuel quantity in the bowl.

12. Raise the float retaining pin and lift the float and inlet needle from the bowl. Remove the seat with a jet remover or wide-blade screwdriver.

QUADRAJET OVERHAUL (*Continued*)

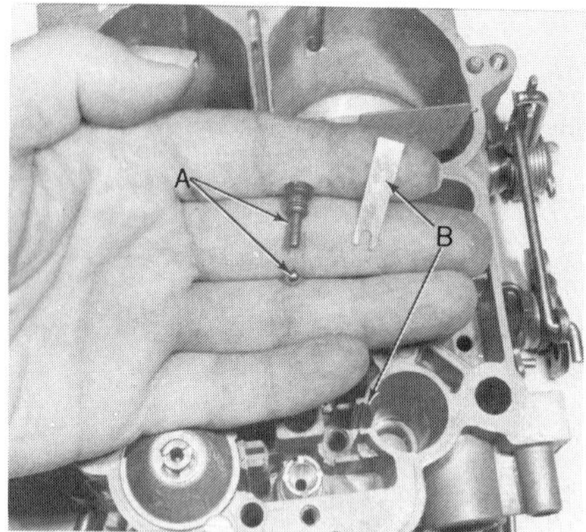

13. Remove the pump outlet retainer and check ball (A). Pull the baffle from the pump well (B) for cleaning. Remove primary jets with a removal tool.

14. Remove the adjustable-part-throttle (APT) or altitude-compensating aneroid from its well. Solvent will damage this part.

15. Remove screws or rivets and the retainer to remove the choke cap and spring of an integral choke.

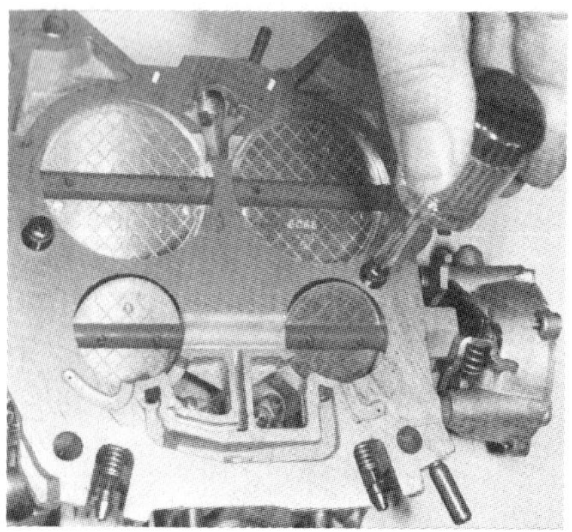

16. Turn the body over and remove the throttle body screws. Do not remove the throttles.

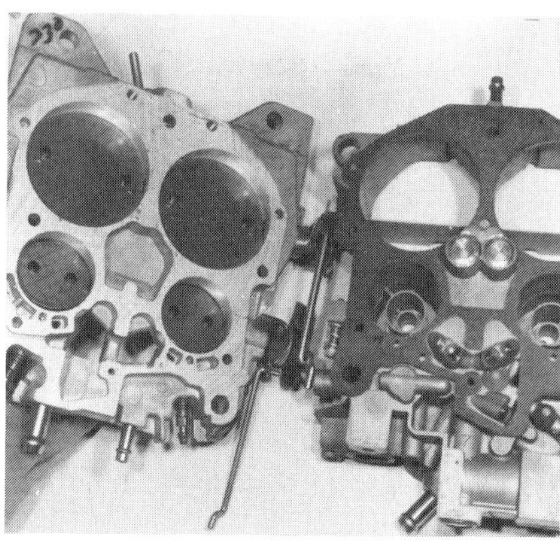

17. Separate the throttle body from the bowl (main body) and remove the gasket. Disconnect linkage as required.

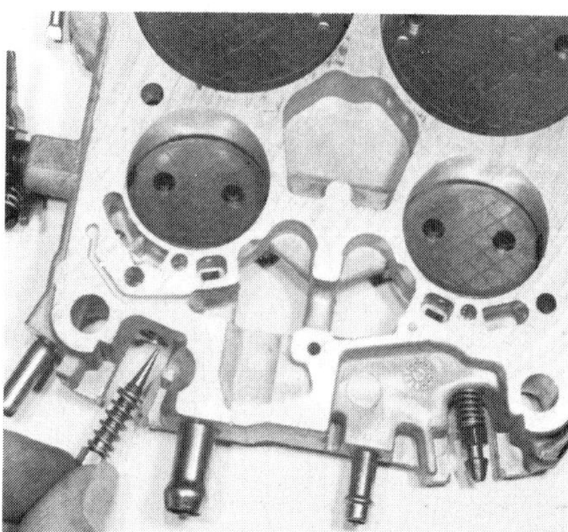

18. Remove and discard limiter caps if necessary. Gently turn the mixture screws inward and count the turns until lightly seated. Then remove them.

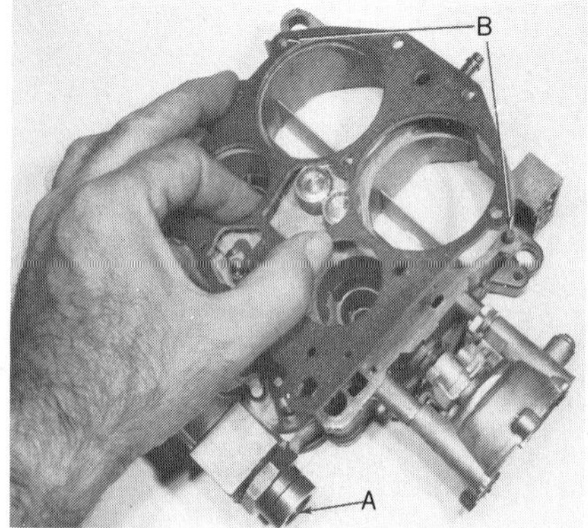

19. Remove the fuel inlet nut, gasket, filter, and spring (A). After cleaning, begin reassembly by placing the new gasket over dowels (B) on the bowl (main body).

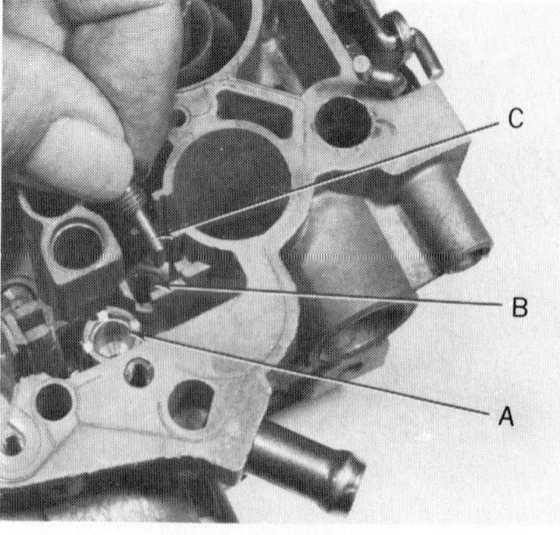

20. Install the needle seat and gasket (A) and put the ball into the passage (B). Install the retainer (C). Then assemble main and throttle bodies.

21. Install the APT aneroid and needle in the well with the tang in the slot. Do not change the factory adjustment.

QUADRAJET OVERHAUL (*Continued*)

22. Install the baffle in the pump well slot (A). Install the primary main jets and auxiliary power jet (B) with a jet tool. Secondary jets can't be removed.

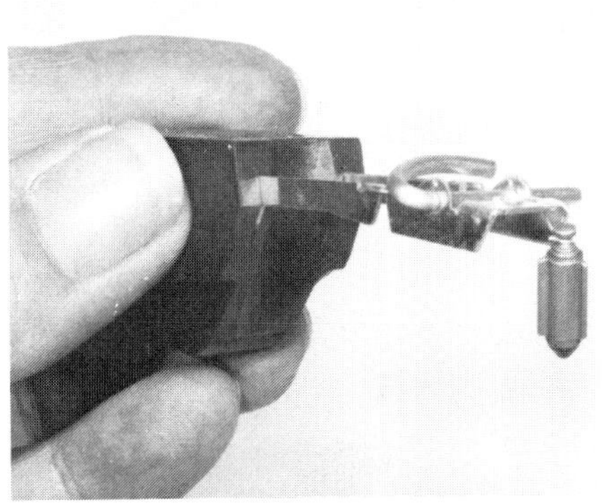

23. Clip the inlet needle to the float arm and insert the retaining pin through the float arm holes.

24. Install the float and needle in the bowl. Be sure the needle slides freely in the seat. Then install the retaining pin and adjust float level.

25. Press the filler block (A) into place until it seats. Install the primary power piston, spring, and rods. Be sure the rods enter the jets smoothly.

26. Install the auxiliary power piston and spring. Press down gently but firmly to seat the piston.

27. Using long-nosed pliers, press the piston down so that the raised plastic retainer is flush with the body casting.

28. Put a new gasket over the bowl dowel pins, pistons, and primary rods. Then install and connect the auxiliary metering rod with the piston arm behind the spring.

29. Raise the corner of the gasket and drop the pump spring into the well. Then install the pump plunger assembly.

30. Align the tubes on the airhorn with the openings in the main body and install the airhorn. Be sure the pump plunger fits freely through the airhorn.

QUADRAJET OVERHAUL (*Continued*)

31. Install the vent valve spring, gasket, and cover. Align the cover dimple with the spring and install the screw.

32. Use long-nosed pliers to install two screws next to venturis. Torque the 9 airhorn screws in the specified sequence. (See kit instructons.)

33. Carefully lower the secondary metering rods into place. Be sure they slide freely in the jets. Then install the bracket on the air valve cam.

34. Install the front vacuum diaphragm rod in the air valve lever (A) and attach the diaphragm bracket. Connect choke and pump linkage.

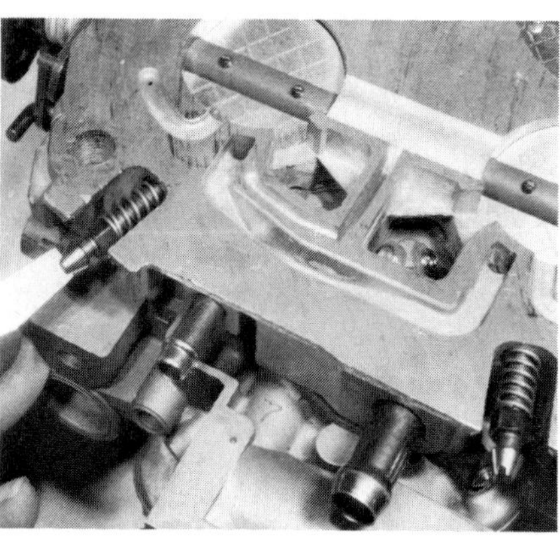

35. Lightly seat the mixture screws. Don't switch them from side to side. Back them out the required number of turns. Install limiter caps after adjustment on the engine.

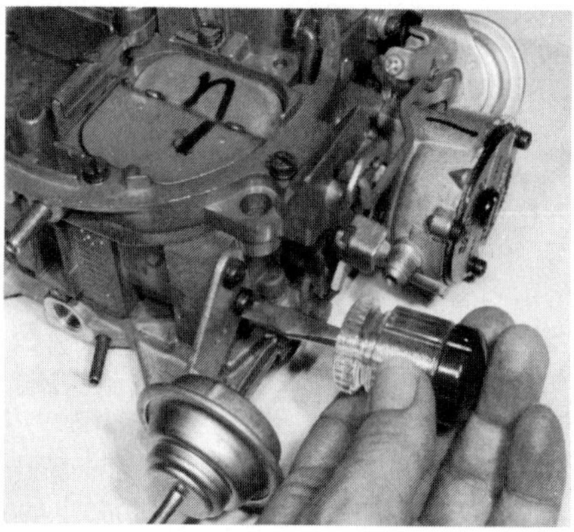

36. Install the rear vacuum diaphragm, choke cap, and other assist devices. Check linkages and other adjustments before carburetor installation.

REVIEW QUESTIONS

Multiple Choice

1. An engine is not running properly. The carburetor is the suspected problem. Mechanic A says the carburetor should be checked immediately. Mechanic B wants to do a general check of the other engine systems first. Who is right?

a. Mechanic A only
b. Mechanic B only
c. Both A and B
d. Neither A nor B

2. Which of the following jobs is not a part of routine, on-car carburetor adjustment:

a. float adjustment
b. choke test and adjustment
c. fast-idle adjustment
d. idle mixture adjustment

3. Nonsealed carburetor choke adjustments should be made:

a. never
b. when the engine is warm but not hot
c. when the thermostatic spring is cold
d. only if there is an index notch on the choke housing

4. An engine has a flooding problem when it is started cold. Mechanic A says that the choke unloader on the carburetor might need adjusting. Mechanic B says that the choke unloader is definitely the problem. Who is right?

a. Mechanic A only
b. Mechanic B only
c. Both A and B
d. Neither A nor B

5. Choke vacuum breaks are operated by:

a. ported vacuum
b. venturi vacuum
c. heat
d. manifold vacuum

6. An electric choke heater has been tested with an ohmmeter and a test lamp. The ohmmeter indicates that the continuity between the choke ground strap and carburetor ground is more than zero ohms. Mechanic A says that the choke heater is OK. Mechanic B says that the choke heater should be replaced. Who is right?

a. Mechanic A only
b. Mechanic B only
c. Both A and B
d. Neither A nor B

7. Mechanic A says that when you adjust the idle speed on a carburetor with an idle-air bypass screw, you turn the screw clockwise to decrease idle speed. Mechanic B says that the screw opens and closes the bypass passage to regulate airflow. Who is right?

a. Mechanic A only
b. Mechanic B only
c. Both A and B
d. Neither A nor B

8. A throttle stop solenoid is used to:

a. control fast idle
b. act as a governer
c. prevent dieseling
d. regulate throttle position at startup

9. A throttle stop solenoid has been checked by applying battery voltage to the solenoid lead. The solenoid was energized. Mechanic A says that the solenoid should be replaced. Mechanic B says that the solenoid is OK. Who is right?

a. Mechanic A only
b. Mechanic B only
c. Both A and B
d. Neither A nor B

10. Mechanic A says the lean-best-idle method of setting the idle mixture means adjusting the screws for best idle, then turning the screws counterclockwise till the speed decreases, then clockwise again until the lost speed is regained. Mechanic B says that Mechanic A is describing the lean-drop method of adjustment. Who is right?

a. Mechanic A only
b. Mechanic B only
c. Both A and B
d. Neither A nor B

11. A carburetor adjustment balance test has been performed. After adjusting the mixture screws the engine speed for one set of cylinders was within 10 rpm of the other set of cylinders. Mechanic A says that the carburetor is well balanced. Mechanic B says that the carburetor should be rebuilt. Who is right?

a. Mechanic A only
b. Mechanic B only
c. Both A and B
d. Neither A nor B

12. A carburetor has been adjusted by the propane injection method. Initially the speed increase was less than specified. Mechanic A says that the air-fuel mixture was too lean. Mechanic B says that the air-fuel mixture was too rich. Who is right?

a. Mechanic A only
b. Mechanic B only
c. Both A and B
d. Neither A nor B

13. Mechanic A says that you shouldn't use propane injection to test the idle mixture of cars with an EGO sensor and an engine computer because the computer will react and make the idle mixture too rich. Mechanic B says that in this case the computer will react to make the idle mixture too lean. Who is right?

a. Mechanic A only
b. Mechanic B only
c. Both A and B
d. Neither A nor B

14. Choke (cold enrichment) adjustment on a variable-venturi carburetor requires the use of a:

a. ohmmeter
b. dial indicator
c. voltmeter
d. CER and CVR

Fill in the Blank

15. Accelerator pump linkage has several adjustment positions to control pump stroke and __________.

16. Float __________ is the basic float adjustment.

17. If a carburetor float hangs from the air-horn it may require float __________ adjustment.

18. Electric choke heating elements help open the choke quickly to reduce __________.

19. The choke __________ opens the choke a specified amount after the engine starts.

20. Slow idle is also called __________ idle.

21. Three common methods of idle mixture adjustment for carburetors with exposed mixture screws are the __________ method, the __________ method, and the __________ method.

22. __________ is a carburetor adjustment method that adds a controlled amount of another fuel to the air-fuel mixture.

23. Carburetor jets should be removed with a __________.

24. When you install accelerator pump check balls the large one is usually for the __________.

18 DIESEL ENGINE PERFORMANCE SERVICE

INTRODUCTION

Previous chapters in this book have covered engine test equipment, troubleshooting principles, and basic services for cooling, exhaust, lubrication, and fuel systems. These chapters concentrated on gasoline engines but included some service procedures for diesel-powered vehicles. This chapter outlines performance diagnosis and basic fuel injection services exclusively for diesels. Without proper maintenance, a diesel engine can suffer the same problems as a gasoline engine: hard starting, poor performance, poor fuel economy, and generally poor driveability.

A light-duty diesel engine has a lot in common with a gasoline engine. Both are 4-stroke-cycle, internal-combustion, water-cooled engines. Both burn petroleum fuel mixed with air and rely on timed ignition for combustion. Gasoline and diesel engines can have the same problems of incorrect air-fuel ratios, incorrect timing, misfiring, and so on.

In this chapter, you will learn that diesel testing and service require the same skills as gasoline engine testing and service. The chapter begins with a summary of special diesel safety precautions and then goes through basic diesel troubleshooting, injection pump timing and adjustment, injector service, and glow plug testing.

GOALS

After studying this chapter, you should be able to do the following jobs in twice the flat-rate labor time or less:

1. Perform an organized troubleshooting sequence based on the driver's complaint and abnormal vehicle symptoms.

2. Using the manufacturer's procedures and specifications, as well as information in this chapter, test and adjust injection pump timing, idle speeds, and maximum speed.

3. Remove and replace injection nozzles.

4. Test nozzle operation and spray pattern.

5. Test, remove, and replace glow plugs using the carmaker's procedures and specifications.

DIESEL SERVICE SAFETY

Chapter 1 of this volume lists precautions that apply equally to gasoline and diesel vehicles. Additionally, diesel service requires the following special precautions:

1. Starting fluids or sprays (ether or similar materials) often are used on large industrial diesels with direct injection systems. *Do not* use starting fluids on light-duty automotive diesels. The glow plugs in these engines can ignite the fluid prematurely and damage the engine.

2. Do not wash or allow cold water to contact a diesel engine when it is warm. A rapid temperature change can cause the close-tolerance parts of the injection pump to seize.

3. Never run an injection pump when it does not contain fuel. Be sure the pump is primed before starting an engine.

4. Do not run an engine with loose injection pump mounting bolts. Never loosen the bolts when the engine is running.

5. Always wear safety glasses when working with high-pressure fuel injection components (pumps, lines, and injectors). Do not allow diesel fuel under high pressure to contact your skin. Fuel sprayed at pressures of 1,000 psi (7,000 kPa) or more can penetrate the skin and cause blood poisoning.

6. After servicing an injection system, be sure all injection line fittings are properly tightened at the pump and at the injectors, figure 18-1.

7. If you suspect a leak in an injection line or an injector, do not check for it with your hand. Place a piece of cardboard behind the line, figure 18-2. If there is a leak, fuel will spray onto the cardboard.

8. Do not run an engine with a bent or broken injection line. Do not try to remove an injection line or injector with the engine running.

9. Never run a diesel engine faster than its maximum governed speed.

10. A diesel intake manifold has no throttle or other restriction. Do not run

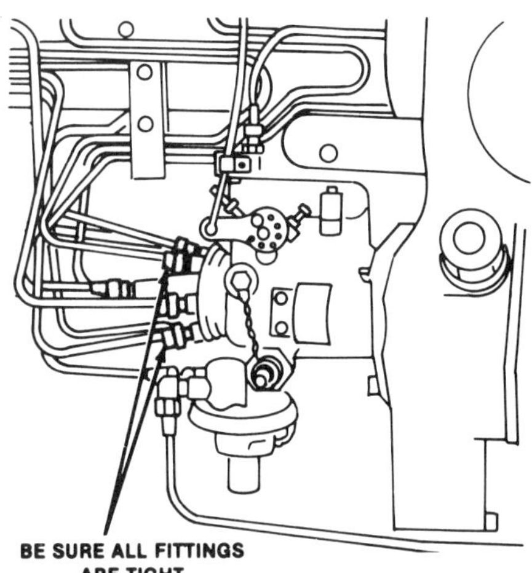

Figure 18-1. Be sure all fittings are tight after servicing any part of the injection system. (Ford)

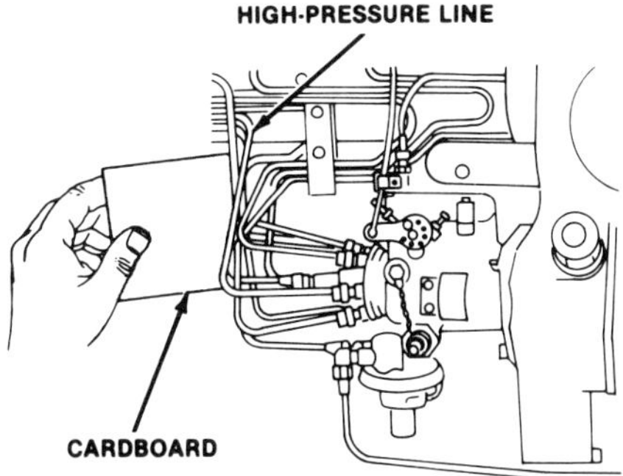

Figure 18-2. Hold a piece of cardboard, not your hand, near an injection line to check for a leak. (Ford)

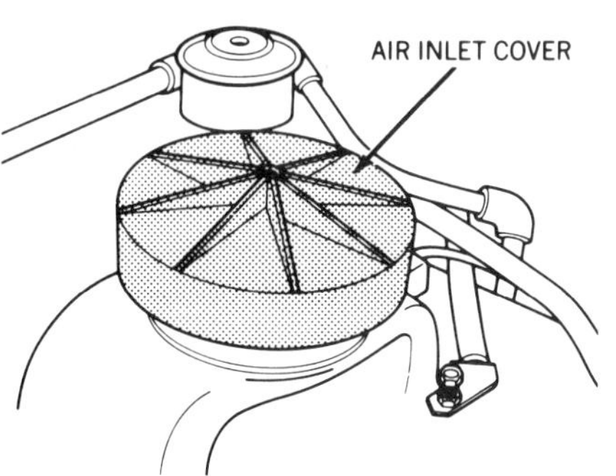

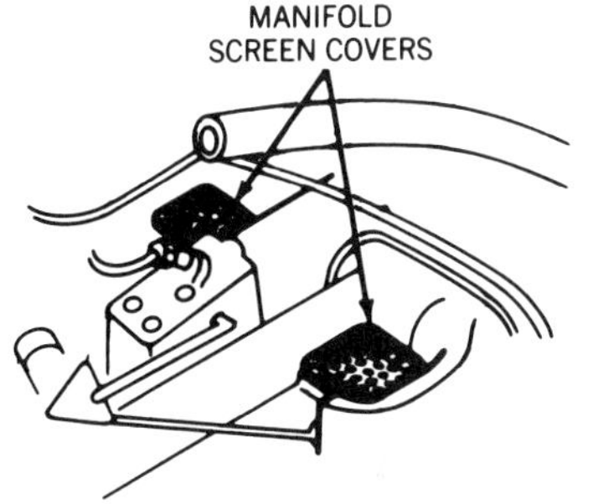

Figure 18-3. Cover the intake manifold opening whenever you remove the air cleaner. (Ford)

the engine with the air cleaner off. Dirt, cloths, small parts, and other objects can be drawn easily into a running engine. Always cover the intake air opening when the engine is off and the air cleaner is removed, figure 18-3.

11. Do not operate a diesel engine near a paint spray booth, a parts-cleaning tank, or other source of large amounts of airborne hydrocarbons. A large amount of airborne hydrocarbons can provide enough fuel for a diesel to cause the engine to run away, even when the fuel is shut off.

12. Do not try to stop a diesel by covering the air inlet with your hand. Remove the air filter and use a manifold inlet cover to block airflow.

DIESEL PERFORMANCE TROUBLESHOOTING

The following sections outline some diesel troubleshooting principles that you can combine with the 9-step diagnosis sequence that you learned in chapter 5 to isolate and correct performance problems.

Diesel Exhaust Smoke

Diesels naturally smoke, but manufacturers have made great efforts to reduce smoke from automotive diesels. Smoke is almost unnoticeable from a late-model diesel car or light truck. You may see some smoke from a diesel in normal operation, but excessive smoke can be a clue to engine problems.

Diesel smoke is classified in two categories: black smoke and blue or white smoke. Black smoke comes from unburned fuel. Blue, white, or blue-white smoke also can come from overfueling but more often from engine oil, a cold engine, or both.

A diesel can emit black smoke under normal heavy load (acceleration, hill climbing, or wide-open throttle operation). A diesel also normally emits more smoke at high altitudes because air density is less and there is a higher percentage of fuel in the air-fuel mixture.

You may see blue-white smoke when a diesel is started, either cold or warm. This smoke may continue at idle after a cold start, but it should disappear as soon as the engine warms. Below 50° F (10° C), blue smoke may reoccur during extended idling because the combustion chambers cool down. It should disappear, however, as soon as the engine is operated off idle.

Figure 18-4 lists common smoke symptoms that may indicate performance problems. Notice that black or gray smoke usually indicates too much fuel or not enough air, but it also can indicate low compression, which causes incomplete combustion. Blue smoke can indicate engine wear, among other problems.

Diesel Performance Diagnosis

Volkswagen and Audi divide performance diagnosis into three general categories of complaints, or symptoms, and their causes:

1. Hard-starting and cold-running problems

2. Idle problems

3. Warm-running problems.

Figure 18-5 summarizes these symptoms and the recommended diagnosis or repair. You can use these cate-

Diesel Exhaust Smoke Diagnosis

Black or Dark Gray Smoke

Symptom	*Probable Cause*	*Correction*	*Comments*
Smoke at full load, particularly at high and medium speeds; engine quieter than normal	Pump timing retarded	Check and adjust timing to specifications	Pump timing changes only if moved on purpose or if mountings are not properly tightened
Smoke at full load, particularly at low and medium speeds; engine noisier than normal	Pump timing advanced	Check and adjust timing to specifications	
Smoke at full load, particularly at high and medium speeds, probably with loss of power	Injection nozzle discharge hole fully or partly blocked	Clean or replace injection nozzles as required	
Smoke at full load, at higher speeds only	Air cleaner filter restricted	Replace air cleaner filter	Replace filter at scheduled intervals or more often under severe service conditions
Intermittent or puffy smoke sometimes with white or bluish tinge, usually along with engine knocking	Injector nozzle valve sticks open intermittently	Check injection nozzles for sticking valve, broken spring, or very low opening pressure; Also check for cross-threading in head	May be due to dirt or water in fuel; Injection nozzle should screw into head freely and must not be overtightened
Smoke at all speeds at high loads, mostly at low and medium speeds and probably along with hard starting	Loss of cylinder compression due to stuck rings, bore wear, burning, sticking valves, or incorrect valve setting	Engine requires overhaul if wear indications are present; Check valve setting Check compression	May be due to improper crankcase oil or incorrect valve clearance
Smoke at full load, mostly at medium and high speeds and probably along with low power	Injection fuel lines clogged or restricted by damage	Check injection fuel lines; Clean or replace as required	Fuel lines must be clear and unrestricted

Blue, Blue-White, or White Smoke

Symptom	*Probable Cause*	*Correction*	*Comments*
Blue or whitish smoke particularly when cold, and at high speeds and light load, but reducing or changing to black when hot and at full load, with loss of power at least at high speeds	Pump timing retarded	Check and adjust pump timing	Some engines show this symptom for less retard than causes black smoke, but usually substantial retard is required to produce blue smoke when running hot and under load
Blue or whitish smoke when cold, particularly at light loads but persisting when hot, probably with knocking	Injection nozzle stuck open or nozzle tip damaged and leaking	Examine nozzle for valve stuck open or nozzle tip damage	
Blue smoke at all speeds and loads, hot or cold	Engine oil passing by piston rings because of sticking rings or bore wear	Engine reconditioning as required	May be due to improper crankcase oil; Will be associated with high oil consumption
Blue smoke, particularly when accelerating from period of idling, tending to clear with running	Engine oil passing by worn inlet valve guides or valve stem umbrella seals worn or missing	Engine reconditioning as required	
Light blue smoke at high speed light loads, or running downhill, usually along with sharp odor	Engine running cold; Thermostat stuck open	Replace thermostat	Low temperatures also increase bore wear

Figure 18-4. Diesel exhaust smoke troubleshooting table.

Diesel Performance Diagnosis

Group 1—Cold-Running and Hard-Starting Problems

Symptom	Diagnosis Or Repair
Engine does not start—cranking speed is too slow	1. Verify correct starting procedure 2. Check the charging system 3. Check the starting system and starter connections 4. Check the starter motor 5. Check the engine oil viscosity 6. Install an engine or a battery heater (optional)
Engine does not start—cranking speed is normal	1. Verify correct starting procedure 2. Check the fuel supply system, including filters and tank strainer 3. Check the fuel shutoff solenoid on the injection pump 4. Check the fuel supply from the pump to the injectors 5. Verify proper fuel quality 6. Check the glow plug system
Engine starts, runs, dies— Engine, idles roughly— Engine smokes or misses— Engine starts hard, cranks too long	1. Verify correct starting procedure 2. Check the fuel supply system, including filters and tank strainer 3. Check fuel line union bolts or restrictor orifices 4. Verify fuel quality 5. Check the glow plug system 6. Check the air filter 7. Check and adjust valve clearance or valve timing 8. Check compression 9. Check and adjust injection timing 10. Check and adjust the cold-start advance and throttle linkage 11. Test the injection nozzles

Group 2—Idle Problems

Symptom	Diagnosis Or Repair
Idle speed varies up and down—Idle speed is too low or too high	1. Adjust idle speed 2. Check and adjust accelerator linkage
Warm engine has rough idle but runs okay above idle	1. Adjust idle speed 2. Check for fuel system air leaks 3. Check for fuel leaks 4. Check injection pump mounting brackets 5. Perform idle speed-drop test 6. Check compression 7. Check and adjust valve clearance or valve timing 8. Check and adjust injection timing 9. Check and adjust the cold-start advance and throttle linkage 10. Test the injection nozzles

Group 3—Warm-Running Problems

Symptom	Diagnosis Or Repair
Warm engine misfires or knocks at any speed—Warm engine surges at any speed	1. Verify proper fuel quality 2. Check for fuel system air leaks 3. Check for fuel leaks 4. Check injection pump mounting brackets 5. Perform idle speed-drop test 6. Check compression 7. Check and adjust injection timing 8. Test the injection nozzles 9. Check the air filter 10. Check and adjust the cold-start advance and throttle linkage

Figure 18-5. Diesel performance troubleshooting table.

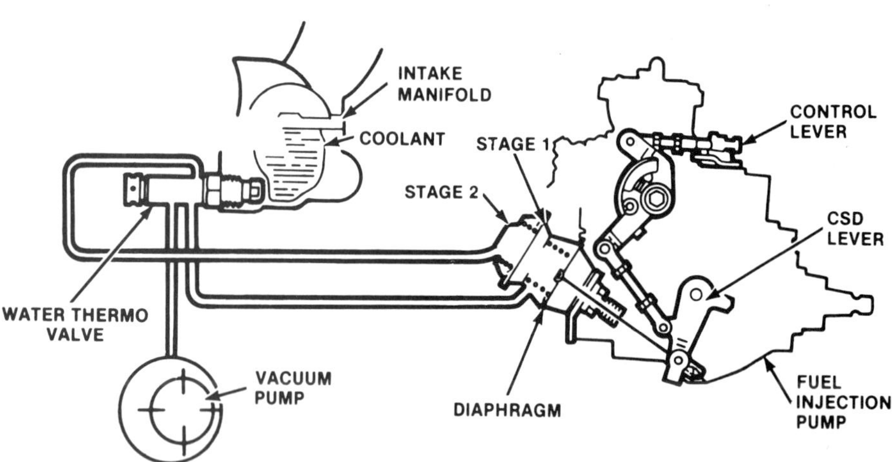

Figure 18-6. This cold-start advance system is operated by an engine-driven vacuum pump and controlled by a thermostatic valve. (Ford)

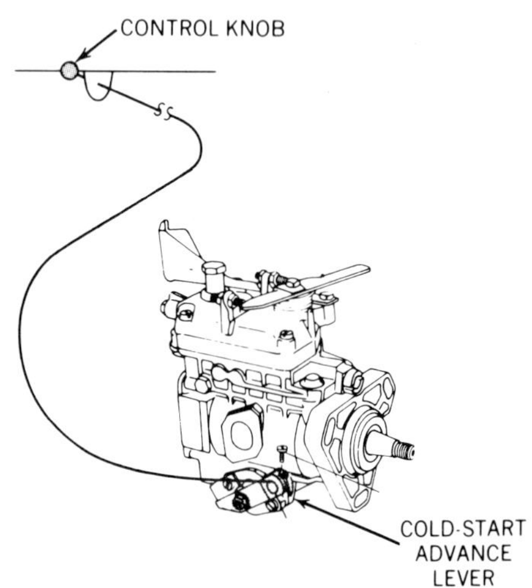

Figure 18-7. Engage the cold-start advance mechanism to start a cold diesel engine. (Volkswagen)

gories, along with a specific carmaker's test procedures, to organize your troubleshooting for any diesel engine.

Notice that many of the problems are the same as you might find with a gasoline engine. For example, a no-start or hard-start problem, along with slow cranking speed on any engine, can be caused by a discharged battery, a bad starter motor or starter connections, or a charging system defect. Also, several of the recommended tests or repairs are the same for more than one symptom. You have already learned how to do a compression test or a charging system output test as part of general engine diagnosis. The following sections summarize other unique diesel diagnosis and repair procedures.

Verify the Starting Procedure. You can't start a diesel engine just like a gasoline engine, particularly in very cold weather. Some car owners don't realize this, and failure to follow the right starting procedure often causes problems. Most diesel automobiles have a cold-start advance system and a glow plug system. The cold-start device advances injection timing a few degrees to aid starting. Some diesel vehicles have an automatic electronic or vacuum-operated cold-start advance system, figure 18-6. Many others have a manual cold-start system, operated by an instrument panel knob.

The glow plugs warm the combustion chambers to help heat the compressed air to ignition temperature. They do not ignite the air-fuel mixture. Follow this general procedure to start most diesel automobiles when cold.

1. If the vehicle has a manual cold-start system, pull out or otherwise engage the cold-start advance knob, figure 18-7.

2. Turn the ignition switch ON and verify that the glow plug lamp lights, figure 18-8.

3. Wait until the glow plug lamp goes out or flashes on and off. The time will vary depending on the vehicle model and year.

4. Put the transmission in neutral. Depress the clutch if the vehicle has a manual transmission.

5. Depress the accelerator pedal if temperature is below freezing.

6. Turn the ignition switch to START and crank the engine. Do not crank for more than 10 seconds.

7. If the engine does not start within 10 seconds, wait about 30 seconds and operate the glow plugs again. (Repeat steps 2 through 6.)

8. Push in, or otherwise disengage, the cold-start advance control after the engine has run about two minutes.

Check the Fuel Quality. Contaminated fuel or fuel of the wrong grade can cause cold-starting prob-

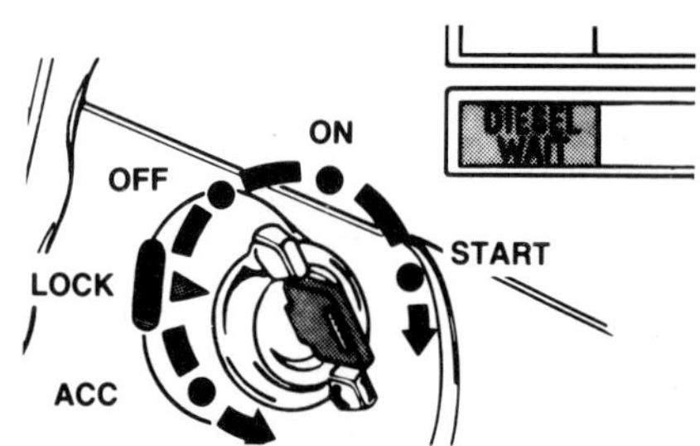

Figure 18-8. Wait for the glow plug lamp to go out or flash before starting a cold diesel. (AMC-Jeep)

lems and poor performance or smoke problems in warm weather. Carmakers recommend no. 2 diesel for most conditions except cold weather. Most manufacturers call for no. 1 diesel or no. 2 diesel mixed with kerosene in cold weather.

You can't tell no. 1 diesel from no. 2 diesel by looking at or smelling the fuel. But, you can check any diesel fuel for signs of major contamination.

Water can be a problem in either grade of diesel fuel in any weather. That is why carmakers put water separators and large filters in fuel systems. Cold weather, however, causes water to condense and collect in the fuel tank faster than in warm weather. If you suspect water contamination in the fuel, open the water drain to draw a fuel sample from the system. You can usually find a water drain on the fuel filter, figure 18-9, or on a fuel

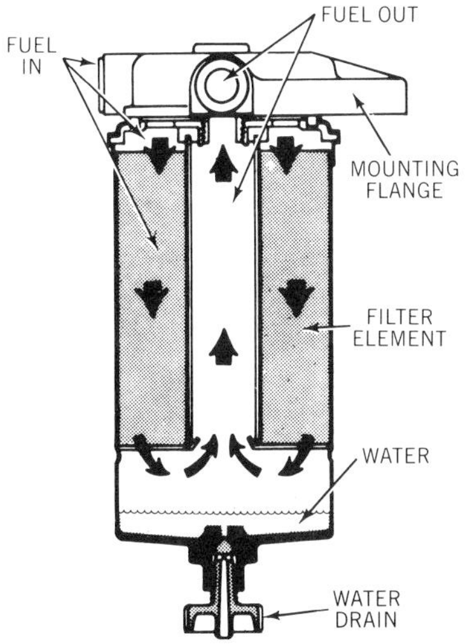

Figure 18-9. On some vehicles, the water drain is on the fuel filter. (Volkswagen)

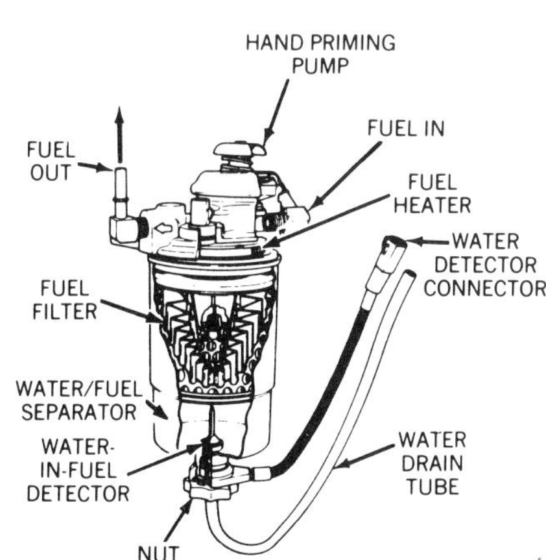

Figure 18-10. On some Ford and Japanese diesel vehicles, the water drain is on the fuel conditioner. (Ford)

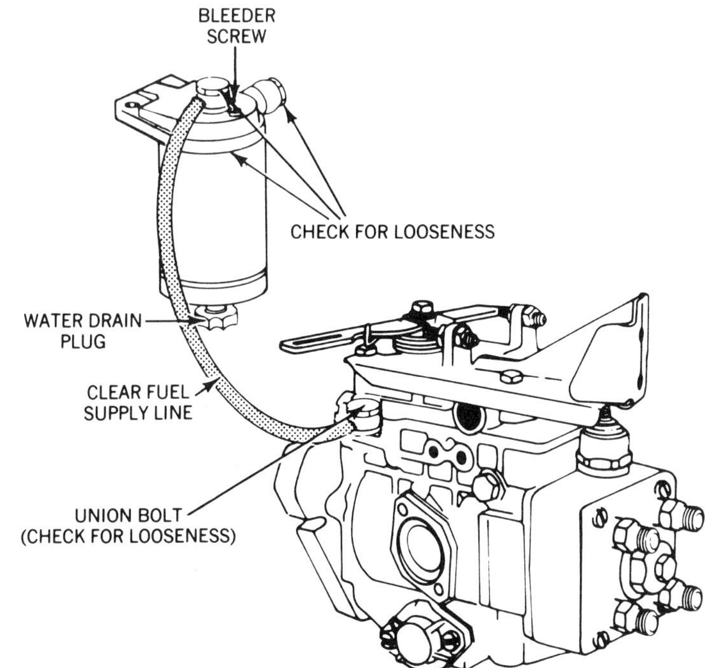

Figure 18-11. Many diesel vehicles have a clear section of fuel line that works as a sight glass to let you check fuel condition. (Volkswagen)

conditioner, figure 18-10. If the vehicle has an electric lift pump, you can disconnect the pump outlet line and operate the pump with battery voltage to draw fuel from the tank.

If you find water in the fuel, leave the drain open or operate the pump until clear fuel runs from the outlet. If the fuel is badly contaminated, you may have to drain the entire system and refill it. It also is a good idea to change the fuel filter if the system contains more water than you can remove simply by opening the drain for a few seconds.

Paraffin (wax) crystals can form in diesel fuel in very cold weather and clog fuel lines and filters. This happens at higher temperatures in no. 2 diesel than in no. 1 diesel. Examine a sample of fuel. If you see cloudiness or white particles, you are looking at paraffin crystals. If you find paraffin formation, you may have to drain the entire fuel system and refill it with diesel no. 1 fuel.

Some diesel vehicles, particularly Ford, Audi, and Volkswagen products, have a section of clear fuel line in the fuel system, figure 18-11. Often, you can check this clear section of line for signs of contamination, such as paraffin formation, water, and air bubbles.

Engine Temperature and Oil Viscosity. Because of its high compression, a diesel requires high cranking power. Anything that slows down cranking speed can keep the compressed air from reaching ignition temperature. High-viscosity oil, or oil thickened by low temperature, can reduce cranking speed. When you are troubleshooting a hard-starting, slow-cranking problem on a diesel, inspect the oil on the dipstick. Find out from the owner or car service records, if you can, what the oil viscosity is. Often a simple oil change will cure the problem.

Engine temperature also can keep a diesel from starting in cold weather. If the combustion chambers are too cold, they will cool the compressed air charge and keep it below ignition temperature. Glow plugs help overcome this problem, but at temperatures below freezing, glow plug heat may not be enough. Many carmakers sell engine block heaters as options for cold-weather use. You may want to consider installing such a heater to cure a cold-starting problem.

Some carmakers also install fuel heaters on diesel engines. Some are part of a fuel conditioner assembly, figure 18-10. Others are installed in the fuel line, figure 18-12. Fuel heaters are powered by the vehicle battery and controlled by a temperature-sensing thermostat. The heater keeps fuel temperature high enough to prevent paraffin formation. Again, when troubleshooting hard-starting or cold-running problems, check the fuel heater if the vehicle has one.

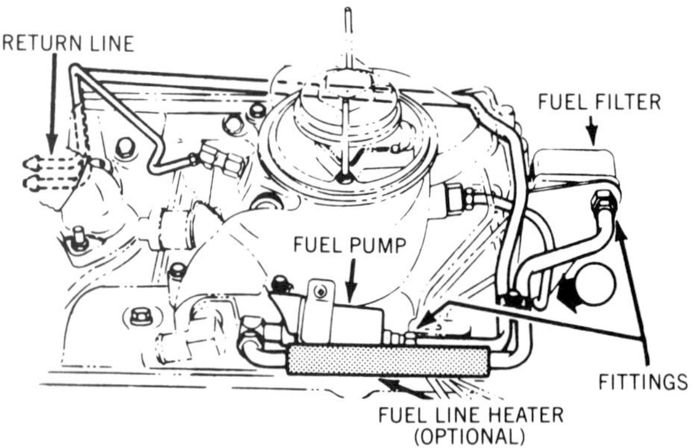

Figure 18-12. Some fuel system heaters are installed in a fuel line. (Cadillac)

Fuel System Inspection (Checking for Air Leaks). From your study of fuel systems, you know that all fuel delivery parts must work properly for any injection system or carburetor to do its job. Diesel fuel systems have the same basic requirements as gasoline systems and can suffer the same problems:

- Fuel leakage
- Dirt or contamination
- Restricted filters, lines, or fuel pickups.

Chapter 16 has procedures for fuel system inspection that apply to both gasoline and diesel vehicles.

One unique problem can occur on a diesel vehicle: air leakage. A diesel fuel system is a closed hydraulic system from the tank, to the injection system, and back to the tank. The return lines for the pump and injectors ensure constant fuel circulation for lubrication and proper injection pressure. Air will compress in injection lines and upset fuel metering to the injectors. The close-tolerance parts of the injection pump need constant lubrication by the fuel. Air in the pump prevents this lubrication and can cause serious damage. This is why you must never run an injection pump without priming it first to remove all air.

Air can enter the fuel system several ways:

- If the vehicle runs out of fuel
- Through a loose fitting or component connection
- If the system is opened for service.

You can check for air in the system and remove it in several ways. If the vehicle has a section of clear fuel line, figure 18-11, check it for air bubbles with the engine running. On some systems with water drains, you can attach a length of hose to the drain outlet and place the other end of the hose in a large container partly filled with fuel, figure 18-13. Air bubbles escaping from the hose with the engine running indicate an air leak between the drain and the tank. Some of the pressure tests that you will learn in this chapter also will help you find air leaks and check fuel delivery and return pressure.

Fuel System Priming. In Chapter 16, you learned something about priming a diesel fuel system when you change a filter. Different vehicles require different methods, and they may vary from year to year for the same model. When you open a system for service, check the carmaker's directions for priming the system before returning it to operation.

Some Bosch systems have priming pumps on the fuel filter housing. To prime these systems, operate the priming pump with the bleeder (vent) valve open, figure 18-11, until clear fuel flows from the vent. Fuel conditioners on Ford diesels have similar priming pumps, figure 18-13. If the vehicle has an electric lift pump, the directions may specify that you operate the electric pump with a specific fitting loosened to remove air. Systems with mechanical or electric lift pumps may be self-priming. However, always check and follow the manufacturer's priming procedures before starting a diesel engine after the fuel system has been opened. This is particularly important after injection pump replacement.

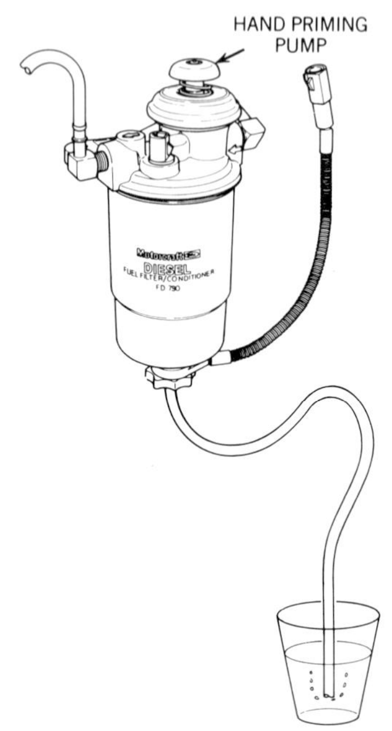

Figure 18-13. Connect a hose to a water drain and place the other end in a container of fuel to check for air in the system. (Ford)

Diesel Fuel System and Engine Pressure Tests

In chapter 16, you learned that a fuel pump pressure test is a valuable diagnostic check on a gasoline engine. Pressure testing is even more valuable on a diesel fuel system because it is a closed hydraulic system. Remember from your study of basic hydraulics that a hydraulic system must operate against resistance to do work. That work can be delivering fluid (fuel) in measured amounts to engine cylinders. The injectors provide the system resistance. To work properly, the fuel system must maintain different pressure levels in different parts of the system. Four basic fuel system pressure tests are:

1. Fuel pickup or pump inlet pressure test
2. Lift pump outlet pressure test
3. Injection pump internal pressure test
4. Fuel return pressure test.

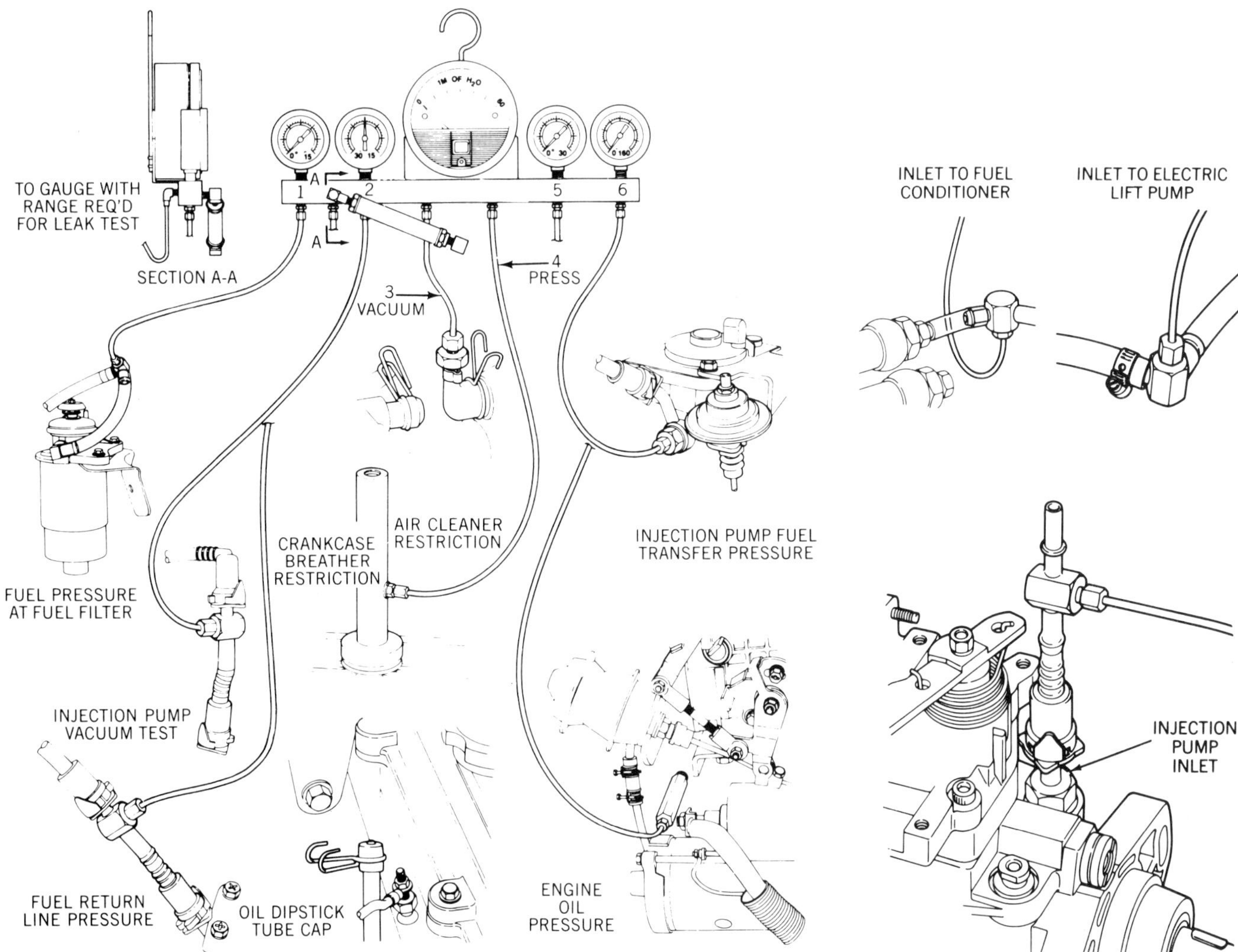

Figure 18-14. Ford recommends this special gauge bar equipment for diesel engine pressure tests. (Ford)

Figure 18-15. Typical fuel inlet pressure test points. (Ford)

You can't test injection pump outlet pressure on the engine. High pressures of more than 1,000 psi (7,000 kPa) require special injection pump test equipment, found only in injection pump rebuilding shops.

You can use pressure gauges, however, to test engine oil pressure and compression, which can cause performance problems if not correct. You learned about compression tests and oil pressure tests in chapters 5 and 6. The following sections outline the principles of other tests.

Ford Motor company recommends specific test sequences and a special gauge bar, or manifold, with five gauges, figure 18-14. If you do not have such a gauge bar, you can use separate gauges of comparable ranges for similar tests. Ford also recommends an air cleaner restriction, or pressure drop, test and a crankcase pressure test. The Ford gauge bar contains a special low-pressure gauge for these tests. Consult Ford test procedures for these engine checks.

Fuel Pump Pickup or Inlet Pressure Test. Fuel pickup pressure, or pressure at a lift pump inlet, should be a negative pressure, that is, below atmospheric pressure. You will need a vacuum or a compound gauge for this test. Test the inlet pressure at one of the following points, figures 18-14 and 18-15, according to system design and manufacturer's directions:

- At the inlet to a mechanical or electric lift pump
- At the inlet to the injection pump, if the vehicle does not have a separate lift pump
- At the fuel filter inlet or outlet, as directed, on the tank side of the lift or injection pump.

Connect the gauge into the fuel line with a T-fitting so that you can measure inlet pressure (pump suction) with the system operating. Negative pressure specifications range from 3 to 6 inches of mercury (in. Hg). If the gauge reads lower than specifications (closer to atmospheric pressure), you may have an air leak in the fuel delivery system. If the gauge reads higher than specifications, the fuel delivery system is restricted. Check for bent fuel lines or a clogged pickup in the tank. If a filter is in the line between the pump and

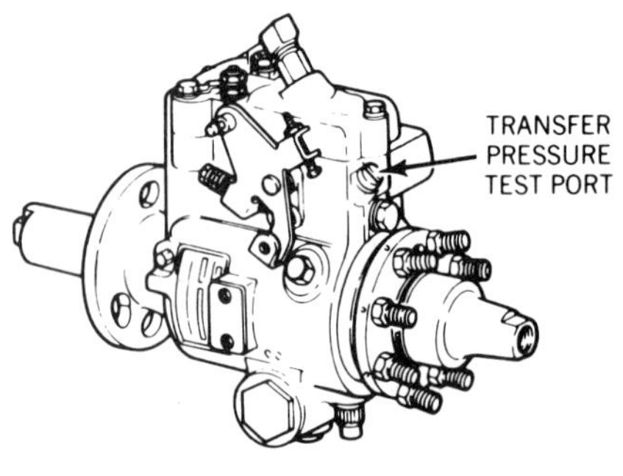

Figure 18-16. All injection pumps have a plug to remove to check internal pressure. (Ford)

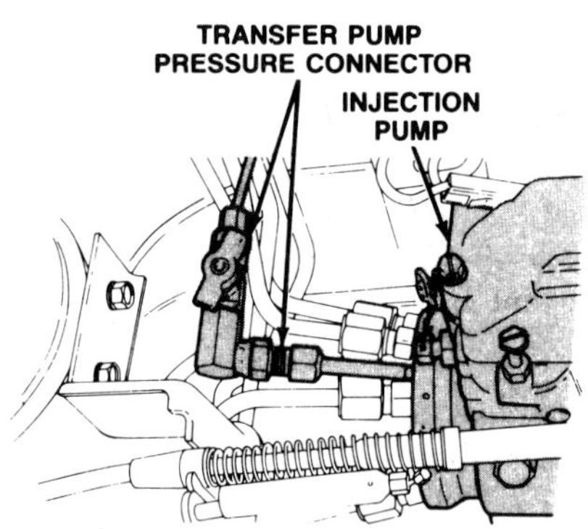

Figure 18-17. Connect the gauge and check pump internal pressure with the engine running. (Ford)

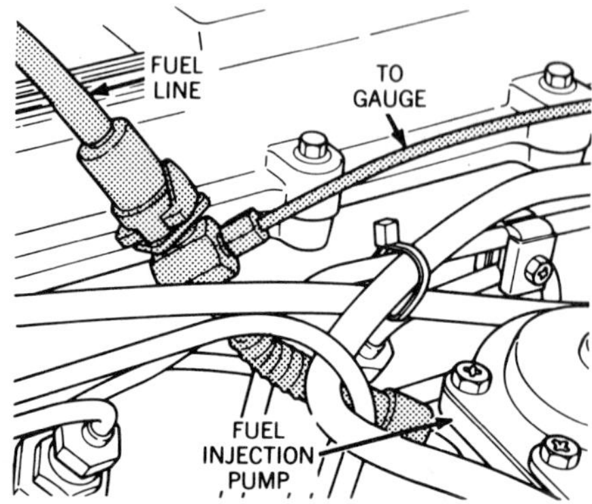

Figure 18-18. Test return pressure between the pump outlet and the fuel tank. (Ford)

the tank, move the test gauge to the filter inlet and outlet connections to check pressure drop across the filter. If you find high negative pressure (suction) at the filter outlet and normal pressure at the inlet, the filter is restricted.

Lift Pump Outlet Pressure Test. Testing the outlet pressure of a mechanical or electric lift pump is the same as testing the pressure of a gasoline engine fuel pump. Refer to the carmaker's specifications and the test procedures in chapter 16 for these tests.

Injection Pump Internal Pressure Tests. The transfer pump, built into the injection pump, draws fuel from the tank or receives it from the lift pump. The transfer pump fills the injection pump and develops low pressure to charge the high-pressure injection plunger. It also maintains fuel pressure and circulation for injection pump lubrication. Low internal pressure in the injection pump is one possible cause of poor-performance problems. All injection pumps have a pressure tap plug for gauge connection to check internal pressure, figure 18-16.

Internal pressure specifications vary for different pumps and engine speeds. They may be as low as 8 to 12 psi (55 to 85 kPa) at idle or 1,000 rpm, or as high as 60 to 100 psi (400 to 700 kPa) at higher speeds. Pressure specifications also vary depending on whether or not the cold-start advance mechanism is engaged. Using the carmaker's specifications and instructions, test internal pressure as follows:

1. Connect a suitable pressure gauge to the injection pump test port, figure 18-17.
2. Connect a diesel tachometer to the engine according to the equipment maker's directions.
3. Follow the carmaker's directions to engage or disengage the cold-start advance mechanism for pressure tests under different conditions.
4. Start the engine and run it at the specified test speed, or speeds. Note the gauge reading at each speed.

Pressures higher than normal usually indicate a clogged fuel return line. Pressures lower than normal can indicate a restricted fuel delivery line, filter, or lift pump. If the system passed previous fuel pickup pressure, filter restriction, or lift pump pressure tests, low pressure in the injection pump probably indicates an internal pump problem.

Fuel Return Line Pressure Test. The fuel return outlet from the pump contains a restriction to maintain the internal pump pressure. Return pressure on the tank side of the restriction should be low, usually about 2 psi (14 kPa). If return pressure is higher than normal, the return line is restricted between the pump and the tank. This increases fuel backpressure, which can raise pump internal pressure above normal. Use a low-pressure gauge to test return pressure. Connect it to the specified point in the return line, which will vary from vehicle to vehicle. Figure 18-18 shows one example.

The fuel inlet and outlet lines on many Bosch injection pumps (and those of similar design) have banjo fittings with hollow union bolts. The inlet and outlet bolts are the same size, but have different fuel flow orifices, figure 18-19. The inlet bolt has several large holes to allow unrestricted fuel inlet. The outlet bolt has a single small hole, calibrated to maintain the required pump internal pressure. Do not interchange these bolts when servicing the fuel lines. If pump pressure tests indicate a restriction on the pump inlet side and low internal pressure, remove the bolts to see if they have been interchanged previously.

Diesel Power Balance Testing

In chapter 5, you learned to do a power balance test on a gasoline engine by short circuiting the spark plugs in firing-order sequence and using a tachometer to measure the speed drop as each cylinder is killed. Obviously, you can't short circuit spark plugs on a diesel, but you can measure the relative power contribution of individual cylinders in two ways: an idle speed-drop test, or a glow plug resistance test. Before doing either test, check the carmaker's procedures for specific instructions.

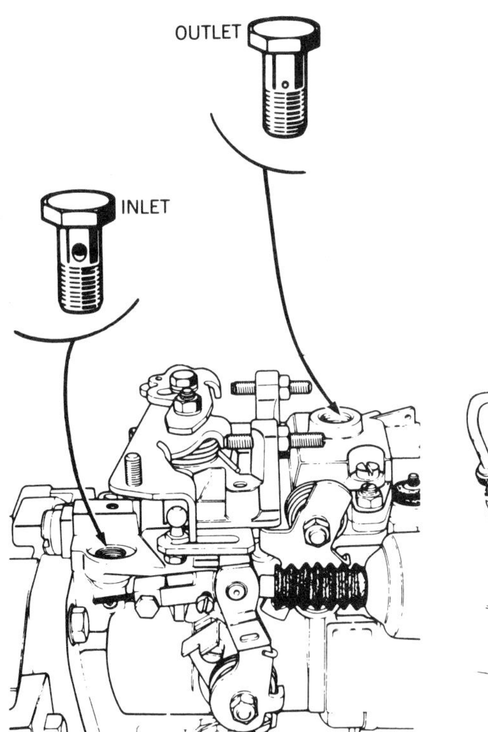

Figure 18-19. Don't interchange the inlet and outlet union bolts on a Bosch pump. (AMC-Jeep)

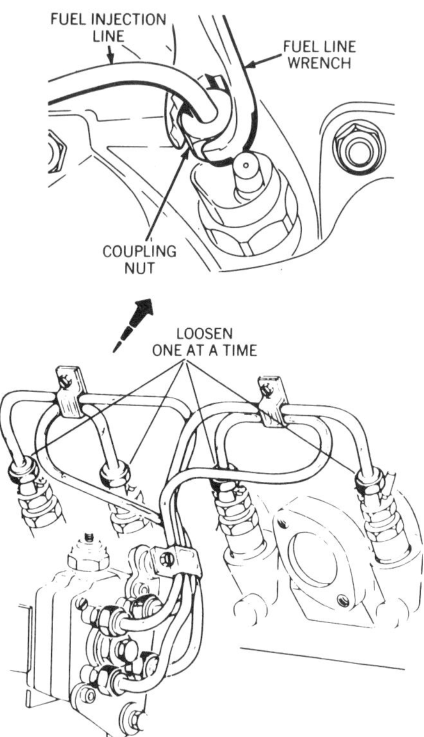

Figure 18-20. Loosen each injector fitting in firing-order sequence for a speed-drop test. (Ford) (Volkswagen)

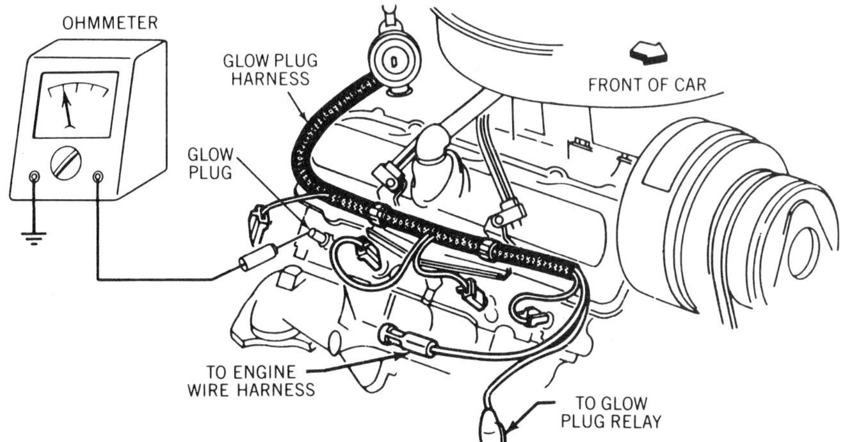

Figure 18-21. With the wiring harness disconnected, use an ohmmeter like this for the glow plug resistance power balance test. (Cadillac)

Idle Speed-Drop Test.

Warning. Wear safety glasses during this test. Do not let high-pressure diesel fuel contact your skin. Serious injury can result.

On many diesels, you can loosen a fitting at an injection nozzle, figure 18-20, or pump outlet line to relieve fuel pressure with the engine running at idle. This prevents injection at one cylinder and causes an rpm drop if the cylinder is delivering its share of engine power. Before doing the speed-drop test, hold a piece of cardboard near the injection line fittings to check for leaks.

If you find a cylinder that doesn't produce a speed drop when you relieve injection pressure, you may have one or more of the following problems:

- Low compression—worn cylinders, rings, or valves
- Incorrect valve adjustment or timing
- An injector that is leaking, sticking, or otherwise malfunctioning.

Use a compression test, as you learned in chapter 8, or the injector tests that you will learn later in this chapter to isolate the problem.

Glow Plug Resistance Test. An idle speed-drop test won't work on a 1980 or later GM diesel because the pump has a pressure-relieving rotor. When you reduce pressure to one cylinder, all pump pressure will drop, and the engine will stop.

You can use an ohmmeter, however, to do a power balance test on 1980 and later GM diesels and other diesel engines. The general test method is as follows, but check the carmaker's procedures before doing this test.

1. Run the engine to normal operating temperature; then shut it off.

2. Disconnect the alternator so the ohmmeter won't be damaged by voltage through the engine ground of the electrical system.

3. Disconnect the electrical leads to all glow plugs.

4. Start and run the engine at idle. Turn on the heater for complete coolant circulation.

5. Set the ohmmeter on its lowest range and connect one lead to a good engine ground.

6. Touch the other ohmmeter lead to the terminal of each glow plug, figure 18-21, and note the meter reading.

Engine speed will not change during this test, but resistance should be about 1.8 to 3.4 ohms on 1980 and later GM diesels. Resistance higher or lower than this range indicates a bad glow plug and invalidates the test. High resistance in the 1.8- to 3.4-ohm range shows high temperature in a cylinder. Low resistance within the range shows low cylinder temperature. Resistance specifications for glow plugs from other manufacturers may be slightly different.

Glow plug resistance, cylinder temperature, and injector opening pressure are directly related. If one injector opens at slightly lower pressure, it admits more fuel and causes higher temperature and resistance. Each 0.1-ohm resistance difference between cylinders indicates about a 30-psi difference in injector opening pressure. Unequal injector opening pressures can cause uneven engine performance. High resistance and high temperatures, along with smoke, can indicate overfueling. Low resistance and low temperatures can indicate insufficient injection. If resistance is about 1.2 to 1.3 ohms, check the compression in that cylinder.

Often, you can cure a rough-idle problem by changing injectors in selected cylinders and retesting the glow plug resistance until you get a smooth idle. Before changing injectors, test the opening pressure of the existing and the replacement injectors. Install an injector with higher pressure to lower the ohm reading, or an injector with a lower opening pressure to raise the ohm reading.

Diesel Air Cleaner and Crankcase Tests

A restricted air filter or excessive crankcase blowby can cause unique problems with a diesel engine. The following sections outline basic tests for these areas of a diesel engine.

Air Cleaner Restriction Test. A clogged air filter or restricted air inlet can cause performance problems on any engine, gasoline or diesel. Diesels, however, are particularly susceptible to restricted air intake problems because they operate with a high volume of air, drawn in with a low-pressure differential. A restricted air filter can cause low power, smoking, and poor fuel economy.

The pressure drop across a diesel air filter is very slight. You can't measure it with a vacuum gauge graduated in inches of mercury (in. Hg). You will need a gauge graduated in inches of water or millimeters of water. One in. Hg equals 13.6 in. H_2O. Ford's gauge bar test equipment contains such a gauge, called a magnehelic gauge. This is the large gauge at the top of figure 18-14. It has both vacuum and pressure connections to measure low negative or positive pressure.

To make this test, you must know the specifications for the air filter pressure drop. Some carmakers give the specifications; others do not. Connect the gauge to a specified point in the air intake or air cleaner between the filter and the intake manifold, figure 18-22, and run the engine at idle. Allowable pressure drop specifications range from 5 to 10 in. H_2O for a new filter to 20 to 25 in. H_2O for a used filter.

Crankcase Pressure Test. Pressure develops in the crankcase of any engine. The crankcase ventilation system circulates fresh air through the crankcase and meters vapors back to the intake airflow. Normal crankcase pressure is low and usually is a slight vacuum when the engine is running and the ventilation system is working. Excessive crankcase pressure can cause a unique problem with a diesel engine. If excessive oil vapors enter the intake airflow, a diesel can run on its own lubricating oil, even with the fuel supply shut off. While this problem is not common, it can destroy an engine and cause injury if it occurs. Less seriously, excessive oil vapors drawn into the air intake system can cause smoking and poor performance. The general cause of most such problems is excessive combustion blowby in the crankcase due to a worn engine.

You can check crankcase pressure with a low-pressure gauge graduated in inches of water. Remove the oil filler cap from the engine and connect the gauge to the opening with an adapter, figure 18-23. Disconnect and block the crankcase ventilation line to the intake manifold or air inlet. Also, remove the dipstick and plug the dipstick tube. These steps ensure that pressure doesn't escape from the crankcase during the test. Start the engine and run it at idle or a specified speed until the gauge reading stabilizes. Pressure higher than specifications indicates excessive blowby past the piston rings.

DIESEL INJECTION PUMP TIMING

Fuel must leave the injection pump at a precise time for the power stroke of each cylinder. This is the basic ignition timing of a diesel engine. Fuel must leave the pump before tdc on the compression stroke to allow for injection time lag and ignition lag. Pump timing should not change during normal engine use unless there is a mechanical problem in the pump or the engine. You will have to time the pump, however, anytime you remove and reinstall it. You can time a pump by either the static method (engine off)

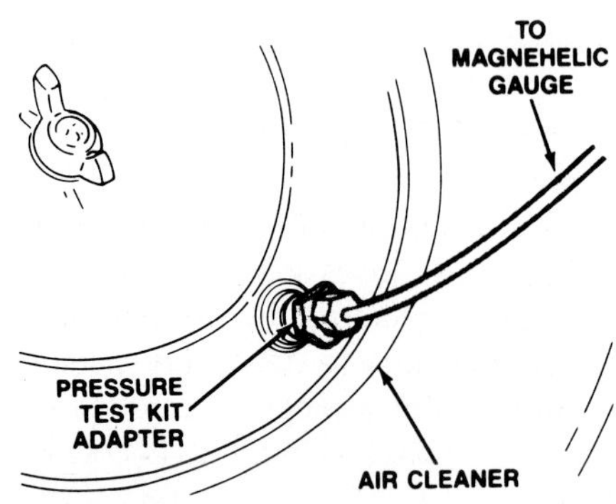

Figure 18-22. Connect the low-pressure magnehelic gauge to the air cleaner to measure air filter restriction. (Ford)

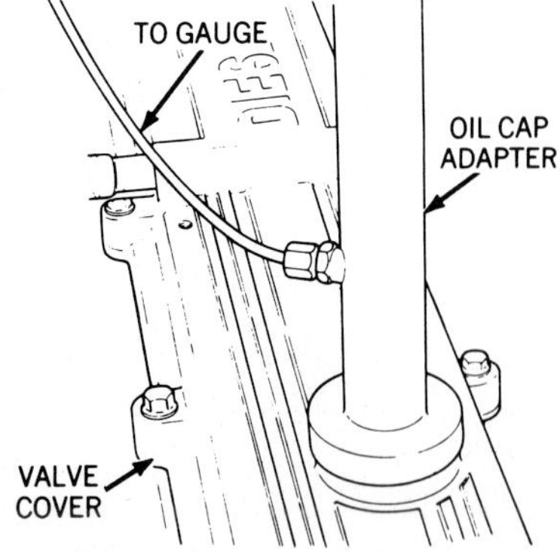

Figure 18-23. Connect the low-pressure gauge, or a compound low-pressure gauge, to the oil filler opening to measure crankcase pressure. (Ford)

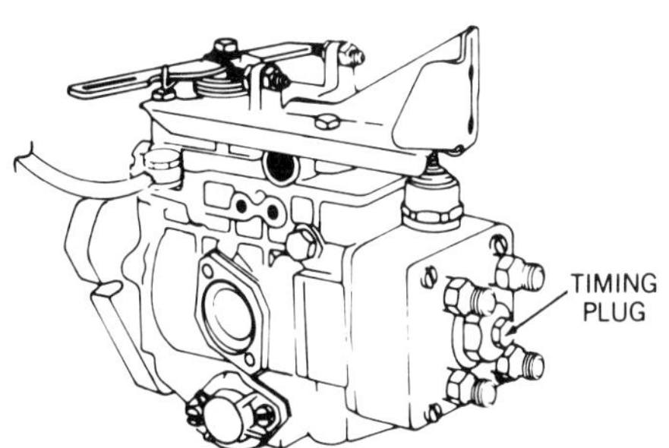

Figure 18-24. Remove this plug to install a dial indicator to time a Bosch distributor pump. (Volkswagen)

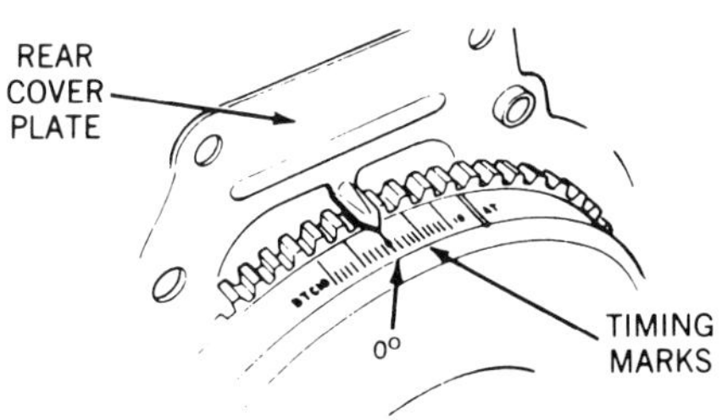

Figure 18-25. Some diesel timing marks are on the flywheel and bellhousing. (Ford)

Figure 18-26. Some diesel timing marks are on the camshaft and pump drive sprockets. (AMC-Jeep)

or the dynamic method (engine running). You will use the static timing method whenever you install a pump, and most carmakers recommend static timing for normal service adjustments. For dynamic timing, you need a special meter to check timing with the engine running. You adjust timing with the engine off.

Caution. Never run a diesel with the injection pump mounting bolts loose. Do not loosen the bolts with the engine running. Serious engine damage will result.

Manufacturers' timing instructions vary slightly, but most include the principles in the following procedures.

Timing a Bosch Distributor Pump

To time a Bosch distributor pump, you must position the engine at tdc of a specified cylinder (usually number 1) on the compression stroke. Then you must locate the bottom of the pump plunger stroke and rotate the pump to advance the plunger a specified distance from the bottom of its stroke. To do this you will align timing marks on the engine and on the pump. You must know the carmaker's specification for initial advance (pump stroke), and you must use a metric dial indicator to measure it. Bosch pumps have a plug in the center of the distributor head that you remove to insert the dial indicator, figure 18-24. Follow these basic steps, along with the manufacturer's instructions.

1. Disconnect the battery ground cable.

2. Disconnect the glow plug connectors and remove the glow plugs.

3. Disconnect the cold-start advance linkage and any other linkage specified by the carmaker.

4. Using a wrench on the crankshaft damper bolt or other suitable method, rotate the engine 2 turns clockwise to align engine tdc timing marks on the damper or on the flywheel, figure 18-25, or on the pump and camshaft sprockets, figure 18-26. AMC-Jeep also specifies that you place a locking pin through the block, into a hole on a crankshaft counterweight, figure 18-27. Remove the pin before rotating the engine again.

5. Remove the plug from the center of the pump distributor head, figure 18-24. Some carmakers also specify disconnecting the injection lines from the pump head or the injectors.

6. Install a metric dial indicator through the hole so that the indicator plunger contacts the pump plunger, figure 18-28. Some engines require adapters to mount the dial indicator.

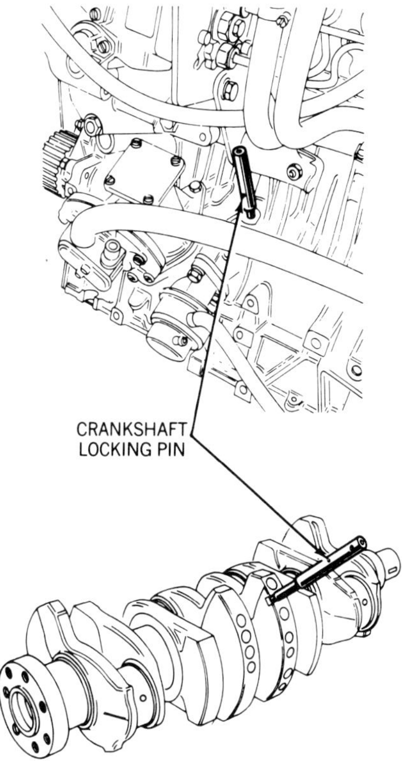

Figure 18-27. Install this locking pin to locate tdc on a Jeep 2-liter diesel. Remove it before rotating the engine. (AMC-Jeep)

7. Preload the dial indicator about 3 mm.

8. Rotate the engine counterclockwise (opposite normal rotation) until the dial indicator needle stops moving. Adjust the indicator to zero.

9. Slowly rotate the engine clockwise until the timing marks are realigned at tdc or until you can install the crankshaft locating pin.

10. Note the dial indicator reading and compare it to specifications. These range from about 0.65 to 1.15 mm, and adjustments must be exact to one-hundredth of a millimeter (0.01 mm).

11. If adjustment is not correct, loosen the pump mounting bolts, figure 18-29, and rotate the pump one direction or the other, as specified, to increase or decrease the dial indicator reading.

12. Tighten the pump bolts, remove the dial indicator, reinstall the pump plug (using a new washer), reconnect injection lines, and remove the crankshaft pin (if used).

13. Reinstall and reconnect the glow plugs, connect the battery, and connect any pump linkage. Start the engine and check operation.

Timing a Roosa-Master or CAV-Lucas Pump

These pumps are used on GM automotive diesels and some Ford and IHC V-8 diesel truck engines. Timing them requires aligning marks on the pump body and the pump mounting adapter, figure 18-30. GM and Ford recommend a dynamic timing meter to check timing with the engine running, figure 18-31. The meter includes a luminosity probe, which you install in place of one glow plug, and a magnetic tachometer with a probe that you install in a timing mark socket on the engine. The luminosity probe senses combustion to measure ignition timing.

Specific test and adjustment procedures vary from engine to engine. If timing is not correct when measured with the dynamic timing meter, stop the engine and loosen the pump mounting bolts. Rotate the pump in one direction or the other, as specified, to advance or retard the timing. Tighten the mounting bolts, restart the engine, and recheck the timing.

ENGINE SPEED ADJUSTMENTS

You may have to adjust slow- and fast-idle speeds as part of routine diesel engine service. You will have to adjust idle speeds after you install or time a pump, and you may have to adjust throttle linkage as well. Some pumps allow you to adjust maximum governed speed, but many carmakers seal the maximum speed adjustment screws and do not permit adjustments in the field.

Most diesel automobiles have cable throttle (or accelerator) linkage, figure 18-32. You may have to adjust the linkage before or after adjusting idle speed as specified by the carmaker. You will need the manufacturer's procedures for these adjustments. Most

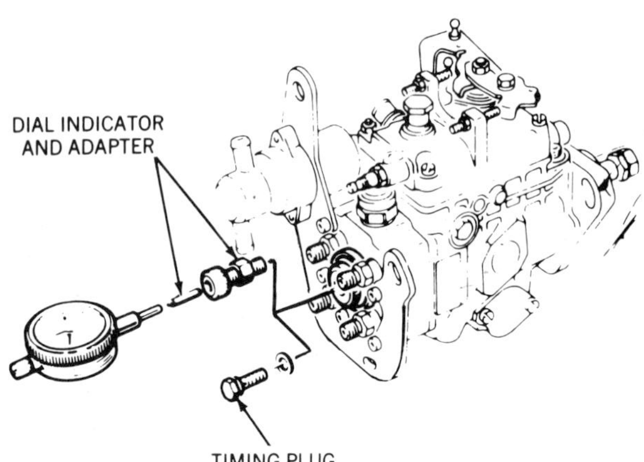

Figure 18-28. Remove the plug from the center of the pump distributor head and install a dial indicator. (AMC-Jeep)

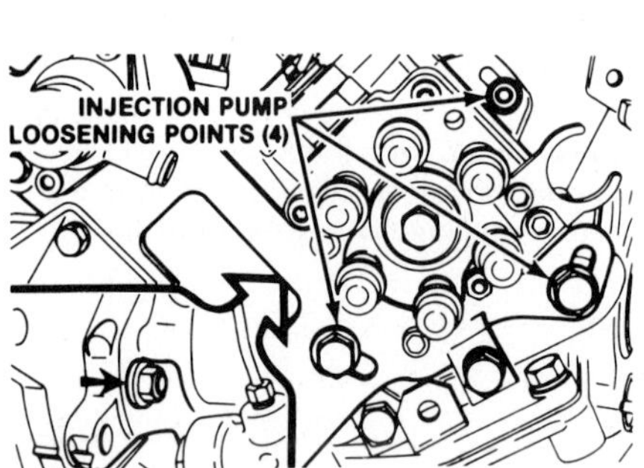

Figure 18-29. With the engine stopped, loosen the pump mounting bolts to adjust timing. (Ford)

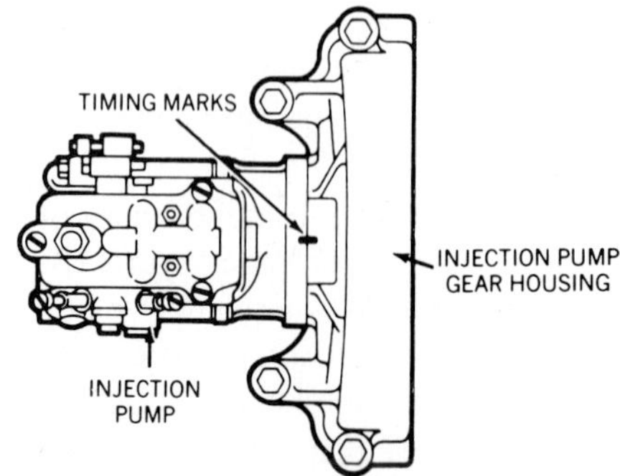

Figure 18-30. Timing marks for Roosa-Master and CAV-Lucas pumps are on the pump and pump adapter. (Ford)

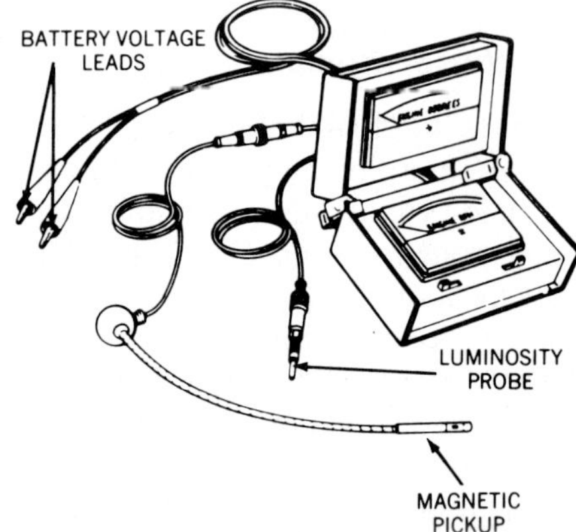

Figure 18-31. Use this dynamic timing meter to check injection timing on GM and some Ford diesels. (Cadillac)

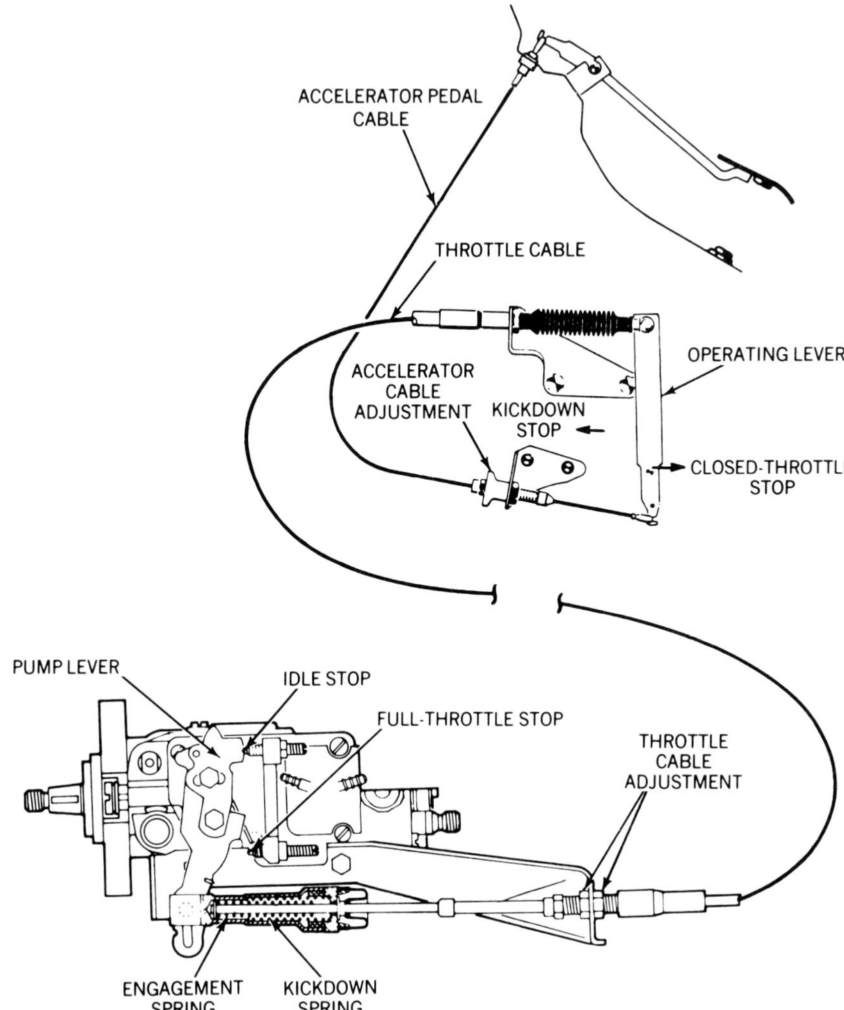

Figure 18-32. Typical diesel throttle (accelerator) linkage and adjustment points. (Volkswagen)

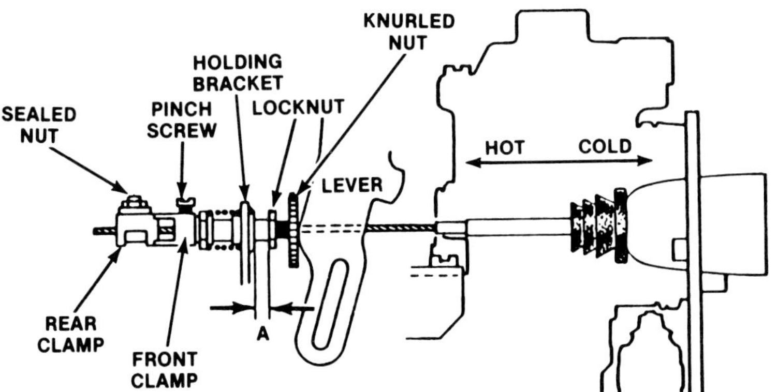

Figure 18-33. Many linkage adjustments require measuring and adjusting the distance between brackets and linkage pivot points. (Ford)

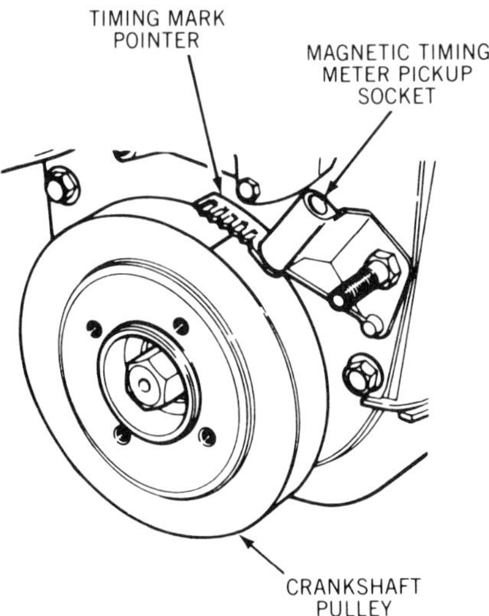

Figure 18-34. Install the probe for a magnetic tachometer in the timing mark socket. (Oldsmobile)

require that you measure and adjust the distance between a bracket and the cable connection on the pump throttle lever, figure 18-33.

To measure diesel engine speed, you will need a diesel tachometer. Two kinds are generally available:

1. A magnetic pickup tachometer
2. A photoelectric tachometer.

If the engine has a magnetic probe socket near the crankshaft damper or flywheel, figure 18-34, connect a magnetic tachometer according to the equipment maker's directions. To use a photoelectric tachometer, put a small piece of reflective tape on a specified engine timing mark and connect the equipment according to the maker's instructions.

Slow-Idle Speed Adjustment

Slow-idle speed is the normal idle speed of a fully warm engine. Adjust slow idle on most engines as follows:

1. Run the engine to normal operating temperature with all electrical accessories off.
2. With the parking brake applied and drive wheels blocked, put the transmission in neutral, park, or drive, as specified.
3. Be sure that the cold-start advance is disengaged and the throttle lever is

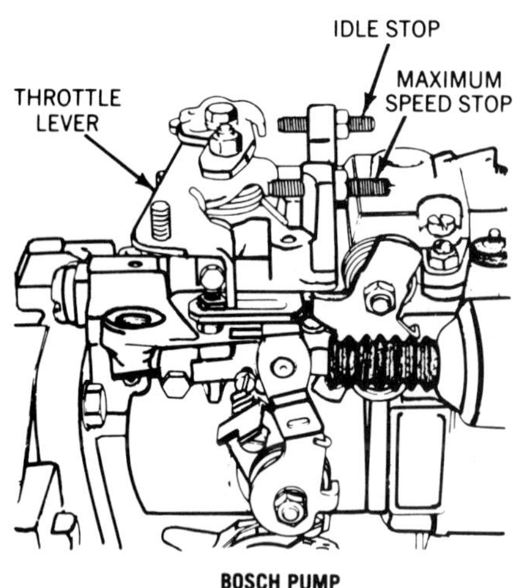

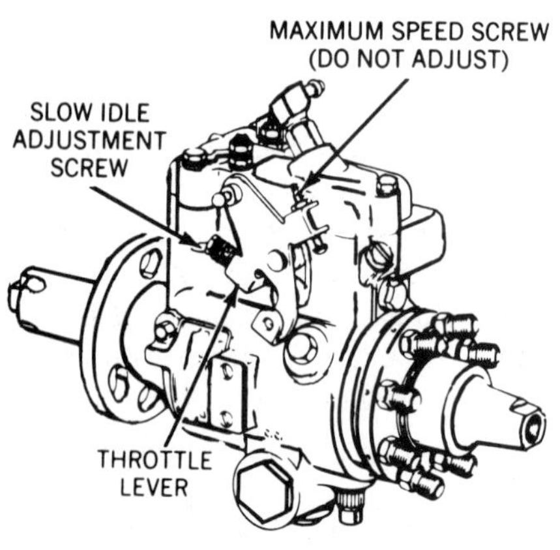

Figure 18-35. Typical slow-idle adjusting screws and maximum-speed screws for Bosch and Roosa-Master pumps. (AMC-Jeep) (Cadillac)

in the idle position, contacting the adjusting screw.

4. Note the idle speed reading on the tachometer and compare it to specifications. You can find these on the engine tuneup decal.

5. If idle speed is not correct, loosen the locknut on the slow-idle adjusting screw and turn the screw clockwise to increase speed or counterclockwise to decrease speed, figure 18-35.

6. Stop the engine and disconnect the tachometer.

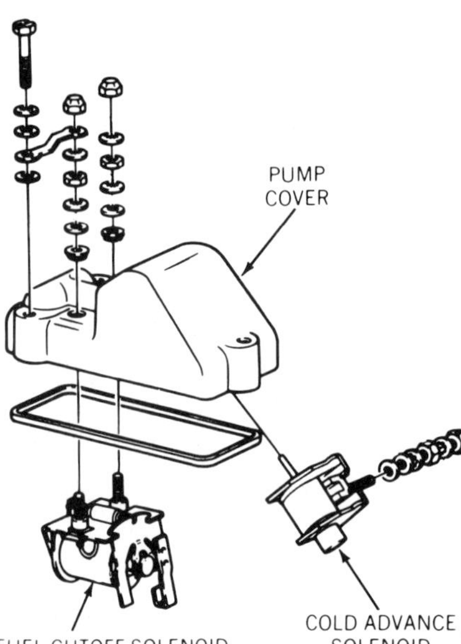

Figure 18-36. The cold-advance solenoid for a Roosa-Master pump in the pump cover. (Cadillac)

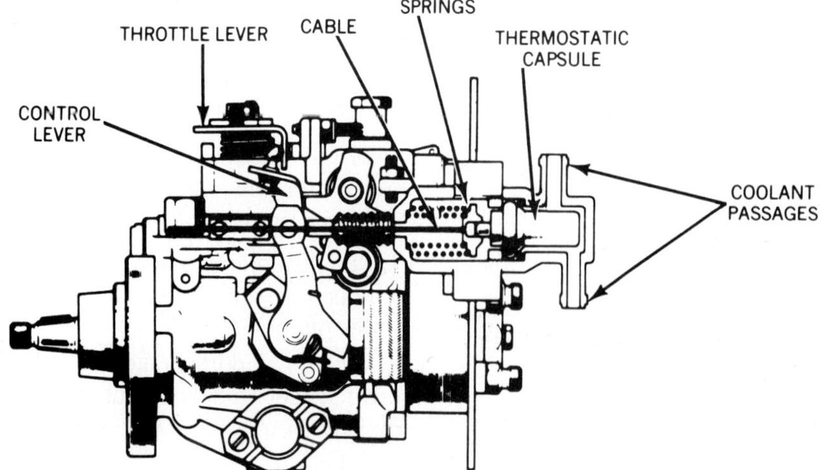

Figure 18-37. This pump has a fast-idle system controlled by springs and a thermostatic capsule. (AMC-Jeep)

Fast-Idle Speed Adjustment

Just as carburetors have fast-idle linkage to increase idle speed when the choke is closed, diesel injection pumps have fast-idle linkage for cold-engine operation. The fast-idle linkage is usually part of, or works with, the cold-start advance mechanism. By advancing injection timing for cold starting, the mechanism also increases idle speed.

Roosa-Master and CAV pumps have a cold-start advance solenoid in the top of the pump, figure 18-36. Jeep 4-cylinder diesels have a spring and a thermostatic element in a coolant chamber to control fast idle in relation to engine temperature, figure 18-37. Some Ford and Japanese engines use a vacuum diaphragm (operated by a vacuum pump) to move the throttle linkage, figure 18-38.

Many diesels also have solenoids or vacuum diaphragms to advance the throttle lever and control idle speed when the air conditioning is on. Figure 18-38 shows an example of such a "vacuum kicker." These idle-assist devices are usually separate from cold-start, fast-idle devices and require separate adjustments.

Linkage designs and adjustment methods vary. You will need the carmaker's instructions and specifications for specific engines. Fast-idle rpm specifications are on the engine tuneup decal. Always adjust slow-idle speed before adjusting fast idle.

Maximum Speed Check and Adjustment

A diesel engine must never be run faster than its maximum no-load governed speed. All manufacturers have procedures to check maximum speed; some allow service adjustment. Maximum speed is limited by an adjusting screw that contacts the pump throttle lever at its wide-open position, figure 18-35. Before checking maximum no-

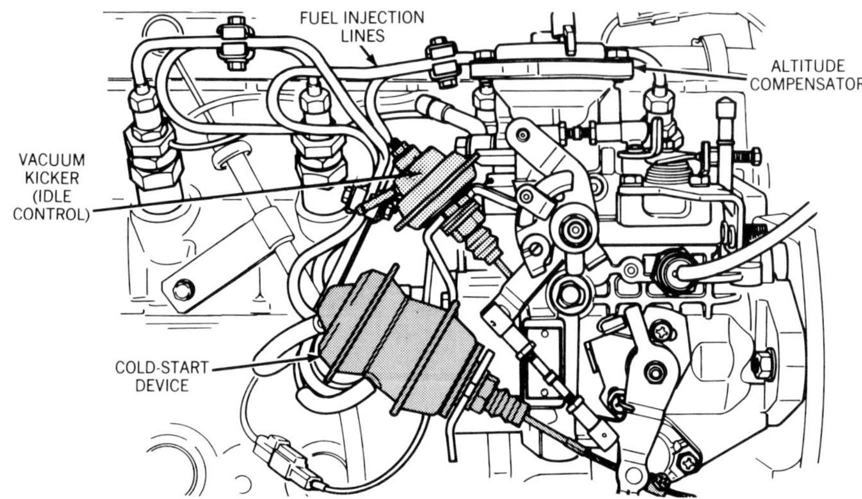

Figure 18-38. This engine has a vacuum-operated cold-start device and a vacuum kicker to control idle speed when the air conditioning is on. (Ford)

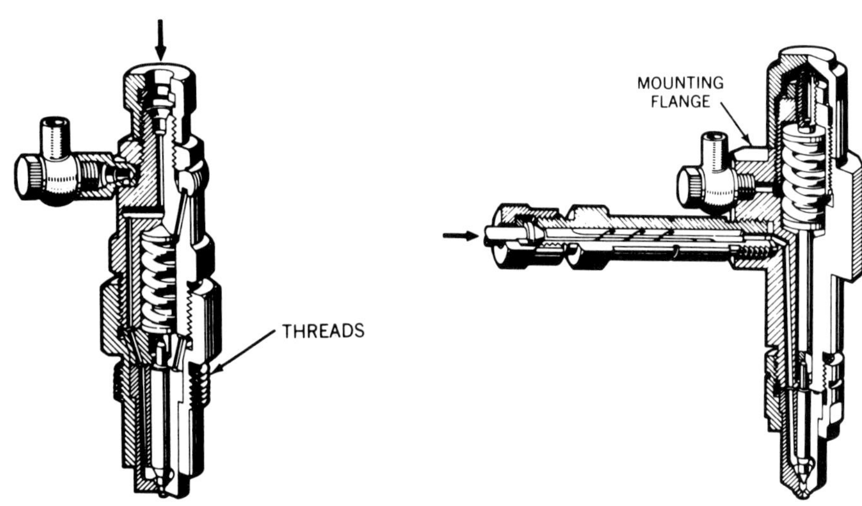

Figure 18-39. Most injectors screw into the cylinder head. (Bosch)

Figure 18-40. Some injectors are held by clamps on the injector body flange. (Bosch)

load speed, check and adjust slow-idle speed and the throttle linkage. Then proceed as follows:

1. Put the transmission in neutral, set the parking brake, and block the drive wheels. Turn off all accessories.

2. Slowly accelerate the engine until the pump throttle lever contacts the maximum speed screw.

3. Check engine speed on your tachometer and compare it to specifications.

4. If the speed is not correct and the carmaker permits adjustment, loosen the locknut and turn the screw *clockwise to decrease speed, counterclockwise to increase speed.*

5. Tighten the locknut and return the engine to slow idle. Check and readjust slow-idle speed if necessary.

INJECTOR AND INJECTION LINE SERVICE

You may have to replace an injection line if it is damaged or leaking or an injector if it is not delivering fuel properly. Also, you can remove an injector and test it for correct opening pressure, leakage, and spray pattern.

Injection Line Removal and Replacement

Injection line replacement looks simple enough, and it does require basically disconnecting fittings and brackets and removing the line. However, there are some guidelines and precautions that you must follow:

1. Always cap or plug the fittings on pumps, lines, and nozzles when you disconnect a line. Cleanliness is extremely important in diesel service. The smallest dust particle can upset proper fuel injection.

2. Injection lines are made of welded steel tubing to withstand high pressures. Never substitute copper tubing or any other tubing not made for fuel injection use.

3. Injection lines must be a precise length to maintain correct injection timing, and all lines for any engine are *exactly* the same length. Moreover, the line for each cylinder is bent for the most efficient fuel delivery. Most manufacturers recommend that you install only factory-made lines as replacements.

4. Always tighten fittings to the specified torque. Use a torque wrench with a flare-nut socket to install fuel injection fittings.

Injection Nozzle Removal and Replacement

Injection nozzles, or injectors, are mounted in the engine in two ways:

1. Screwed in, similar to spark plugs, figure 18-39.

2. Held by two bolts and clamps against a flange on the nozzle body, figure 18-40.

To remove and install an injector, follow the manufacturer's directions and these points:

1. Using compressed air, blow dirt away from the injector location in the head.

2. Disconnect and cap the fuel injection line and the fuel return line at the injector.

3. If the injector is mounted with flange clamps, remove the two bolts and clamps.

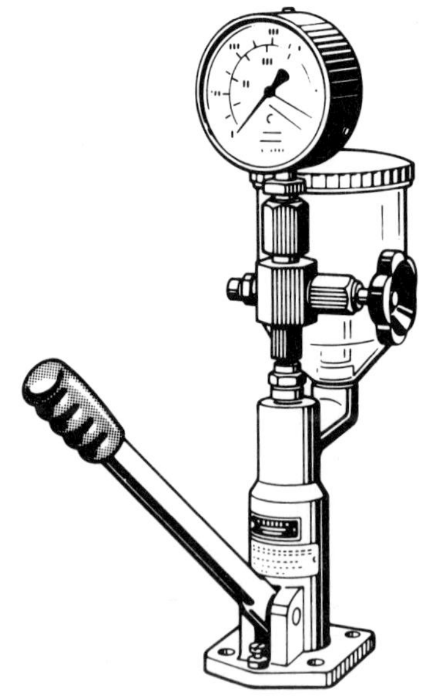

Figure 18-41. Use this special tester to check injector operation. (Bosch)

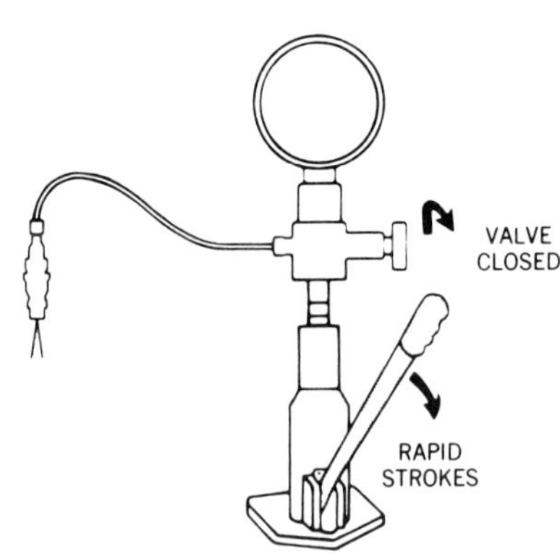

Figure 18-42. Close the gauge valve and operate the pump to remove air from the tester and injector. (Volkswagen)

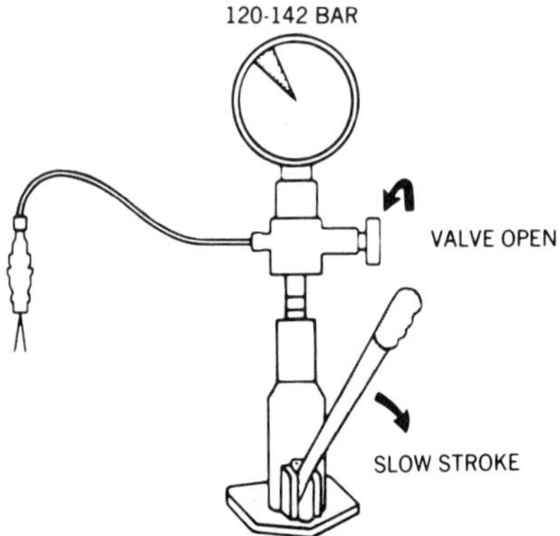

Figure 18-43. With the gauge open, operate the pump slowly to check opening pressure. (Volkswagen)

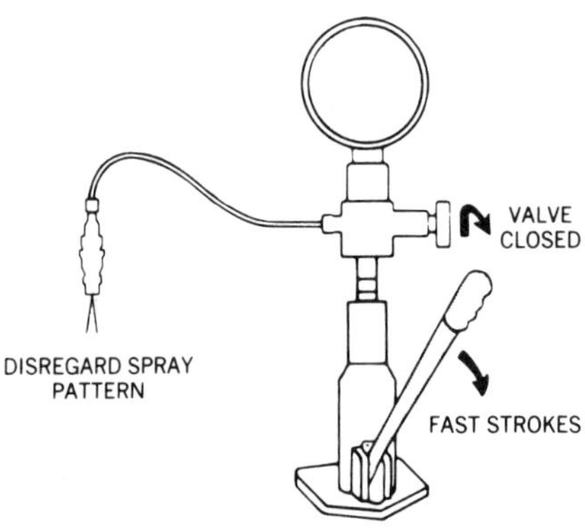

Figure 18-44. With the gauge closed, operate the pump quickly and listen to the injector sound. (Volkswagen)

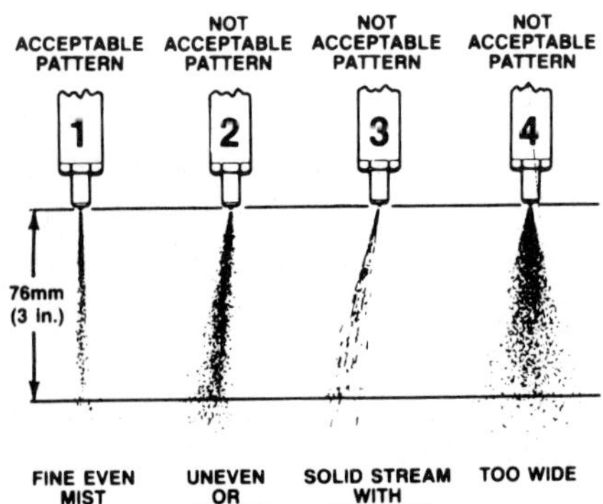

Figure 18-45. A good spray pattern is a narrow, evenly atomized cone. (Ford)

4. If the injector screws into the head, remove it with a wrench on the largest hex section of the nozzle holder.
5. Remove the injector and its heat shield or gasket.
6. Install the injector, using a new heat shield or gasket if required, and torque to specifications.
7. Reconnect the injection line and fuel return lines.

Injector Testing

Testing injector operation requires an injector test stand. This is a high-pressure hydraulic pump with a gauge, fluid reservoir, and mounting adapters for the injectors, figure 18-41.

Warning. Wear safety glasses when testing injectors, be sure injectors are shielded or placed in protective enclosures. Do not allow high-pressure fluid to contact your skin. Serious injury could result.

Caution. For accurate testing, use calibration oil (SAE J967D or ISO 4113), not diesel fuel. Never use gasoline or other fluids in an injector tester.

Test Preparation. You can clean the injector before or after testing, depending on the purpose of your test. To clean it, remove carbon deposits from the tip with a small brass brush. Mount the injector in the tester.

Fill the tester with calibration oil. Turn the gauge valve clockwise to close it and operate the pump handle rapidly to bleed air from the tester and prime the injector, figure 18-42.

Opening Pressure. Turn the gauge valve counterclockwise to open the gauge. Depress the pump handle slowly until the injector just begins to spray fuel. Note the gauge reading and compare it to specifications, figure 18-43. Typical specifications range from 1,200 to 2,000 psi (80 to 140 bars or 7,000 to 14,000 kPa) for a new injector, slightly less for a used injector.

Leakage. Install hose in the injector return fittings to keep oil from running down the injector body. Slowly depress the pump handle until the gauge reading is 150 to 300 psi (10 to 20 bars or 1,000 to 2,000 kPa) below the opening pressure. Maintain this pressure for 10 seconds and examine the injector for leakage from the tip or around the body. Slight dampness on the tip is normal. Replace the injector if you find any other leakage.

Chatter. Turn the gauge valve clockwise to close it, figure 18-44. Operate the pump handle at 1 to 2 strokes per second and listen to the injector. It should make a chattering or creaking sound if it is in good condition. Disregard the spray pattern.

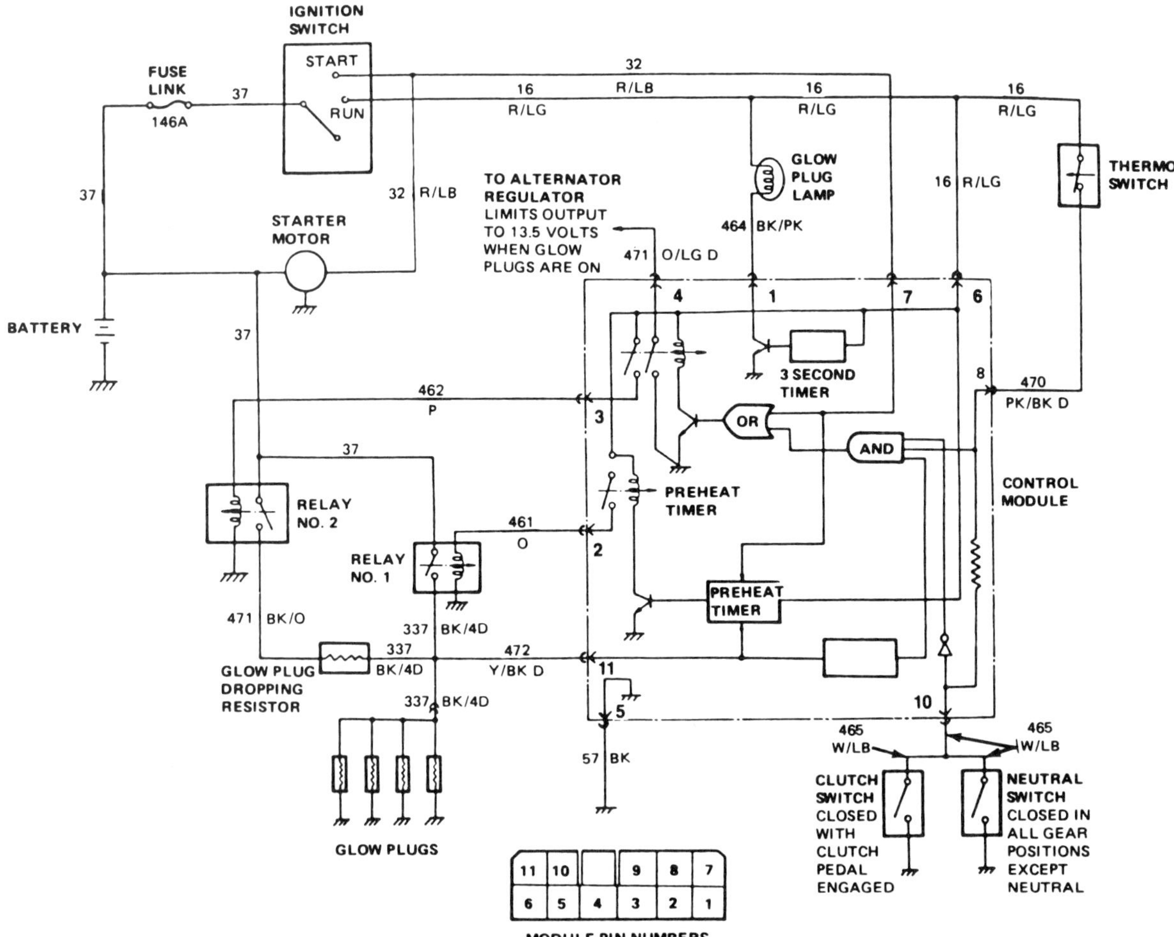

Figure 18-46. Check the manufacturer's test procedures and diagrams to troubleshoot a glow plug system. (Ford)

Spray Pattern. Proper injector spray pattern is critical for good engine performance. Close the gauge valve and operate the pump handle at 5 to 6 strokes per second. Observe the spray pattern, figure 18-45. It should be evenly and finely atomized in a narrow cone and centered on the injector tip. If it is a solid stream, too wide, unevenly atomized, or off center, clean or replace the injector.

GLOW PLUG SERVICE

Proper glow plug operation is essential for starting a cold diesel engine. A glow plug is a simple resistance heater, powered by the vehicle battery. The glow plug circuit is activated by the vehicle ignition switch and controlled by timers and thermostats. Some glow plugs are 12-volt items. Others are rated at 6 volts but may be powered by 12 volts. These are used in some fast-glow systems because they heat faster, but they require circuit timers to avoid overheating. Observe manufacturers' precautions about applying 12 volts to a 6-volt glow plug during testing.

We can't give you a general test procedure for the glow plug systems on all vehicles because they vary widely. You will need the carmaker's procedures and a circuit diagram, figure 18-46, for a complete test. Generally, you will use a voltmeter to test voltage at specified test points.

You can use an ohmmeter to measure resistance of a glow plug, either in or out of an engine. To check resistance with the plug installed, disconnect the connector from the plug and measure resistance between the plug terminal and an engine ground. If glow plug resistance is not within specifications, replace it. A glow plug cannot be repaired.

Proper fuel injection operation is necessary for correct glow plug service life. If a cylinder receives too much fuel, cylinder temperature may be high enough to burn the glow plug. An incorrect injector spray pattern also can overheat and burn out a glow plug.

SUMMARY

Diesel engine troubleshooting and adjustment are no more difficult than gasoline engine service. A diesel can suffer the same problems of incorrect air-fuel ratios, fuel system contamination, and improper ignition (injection pump) timing. Many diagnostic routines are similar for gasoline and diesel engines.

A diesel does not require a tuneup in the traditional sense that a gasoline engine does. However, regular replacement of fuel, oil, and air filters is critical for diesel engine efficiency. Diesels are particularly susceptible to fuel system contamination from water, paraffin formation, and dirt. Many troubleshooting and repair jobs involve removing fuel contamination.

The procedures in this chapter cover most basic diesel troubleshooting and adjustments and can be done in most automobile service shops. Repair and rebuilding of fuel injection pumps and injectors require special training and equipment.

REVIEW QUESTIONS

Multiple Choice

1. Mechanic A says that you can check for leaking diesel injector lines by feeling them with your hand. Mechanic B says that you can cut off airflow to a diesel in a run-on condition by blocking the air intake with your hand. Who is right?

a. A only
b. B only
c. both A and B
d. neither A nor B

2. Black smoke from a diesel engine is usually caused by:

a. a cold engine
b. too much fuel
c. engine oil in the combustion chambers
d. a lean air-fuel mixture

3. A diesel engine emits black smoke at full load, particularly at medium and high speeds. The engine is quieter than normal. Of the following, the most probable cause is:

a. the air filter is restricted
b. pump timing is retarded
c. an injection nozzle, or nozzles, is partially restricted
d. engine oil is leaking past worn valve guides

4. A diesel engine misfires or surges at any speed. All of the following could cause this problem EXCEPT:

a. poor fuel quality
b. water in the fuel
c. inoperative glow plugs
d. an air leak in the fuel system

5. A warm diesel engine idles roughly but runs okay above idle. All of the following could cause this EXCEPT:

a. cold-start advance and accelerator linkage
b. a restricted air filter
c. water in the fuel
d. incorrect injection pump timing

6. Mechanic A says that you should not use starting fluids on a diesel engine that has glow plugs. Mechanic B says that the glow plugs ignite the mixture of diesel fuel and air in a cold engine. Who is right?

a. A only
b. B only
c. both A and B
d. neither A nor B

7. Mechanic A says you can cure cold starting problems on some diesels in the winter by using number 1 diesel fuel. Mechanic B says that you can cure winter cold-starting problems for some engines by adding a small amount of kerosene to number 2 diesel fuel. Who is right?

a. A only
b. B only
c. both A and B
d. neither A nor B

8. You are inspecting a diesel fuel system at a section of clear fuel line and find that the fuel looks cloudy. This probably indicates:

a. water in the fuel
b. a fuel system air leak
c. wax formation in the fuel
d. kerosene in the fuel

9. You can test pressure at all of the following points with the injection pump running on the engine EXCEPT:

a. injection pump internal pressure
b. lift pump outlet pressure
c. fuel return pressure
d. injection pump outlet pressure

10. Specifications for fuel pump inlet pressure (suction) on a diesel engine are 6 in. Hg. Your test gauge reads 11 in. Hg. This could be caused by:

a. low injection pressures
b. air in the fuel
c. a clogged fuel pickup in the tank
d. water in the fuel

11. Your test gauge shows that internal fuel pressure in a diesel injection pump is much higher than specifications. This may be caused by:

a. a clogged fuel return line
b. air in the fuel
c. a clogged fuel filter
d. improper governor adjustment

12. During an idle speed-drop test, engine speed does not drop when the fuel injection fitting is loosened for one cylinder. All of the following could cause this EXCEPT:

a. a leaking injector
b. low compression
c. high injection pressure
d. incorrect valve adjustment

13. With the engine warmed up and running, glow plug resistance on 1980 and later GM diesels should be:

a. 1.2 to 1.3 ohms
b. 1.8 to 3.4 ohms
c. 120 to 150 ohms
d. 1.0 to 2.5 kilohms

14. Mechanic A says that high crankcase pressure in a diesel engine can cause the engine to smoke. Mechanic B says that high crankcase pressures can cause the engine to run on its own lubricating oil when fuel is shut off. Who is right?

a. A only
b. B only
c. both A and B
d. neither A nor B

15. Among the tools needed to time a Bosch distributor injection pump is:

a. a magnetic tachometer
b. a magnetic timing meter

c. a dial indicator

d. a luminosity probe

16. Mechanic A says that all fuel injection lines on an engine must be equal and precise lengths to maintain correct injection timing. Mechanic B says that when you cut copper tubing for a replacement injection line, you must allow exactly 3/16 inch (0.47 mm) extra length for flaring. Who is right?

a. A only

b. B only

c. both A and B

d. neither A nor B

17. Mechanic A says that blue or blue-white smoke from a diesel may be due to engine oil in the combustion chambers. Mechanic B says that blue or blue-white smoke may occur during cold-engine operation. Who is right?

a. A only

b. B only

c. both A and B

d. neither A nor B

18. A diesel engine emits black smoke at full load, particularly at low and medium speeds. The engine is noisier than normal. Of the following, the most probable cause is that:

a. the air filter is restricted

b. an injection nozzle, or nozzles, is partially restricted

c. pump timing is retarded

d. engine oil is leaking past worn valve guides

19. A diesel engine emits blue smoke at all speeds and load, hot or cold. Of the following, the most probable cause is that:

a. pump timing is advanced

b. an injection nozzle, or nozzles, is partially restricted

c. the air filter is restricted

d. engine oil is leaking past worn piston rings

20. Mechanic A says that air in diesel fuel injection lines will compress and upset fuel metering. Mechanic B says that air in the fuel lines can prevent proper lubrication for the injection pump. Who is right?

a. A only

b. B only

c. both A and B

d. neither A nor B

19 AIR INJECTION, EXHAUST GAS RECIRCULATION, AND CATALYTIC CONVERTER SERVICE

INTRODUCTION

Exhaust gas recirculation (EGR) and catalytic converters have been important emission controls for more than a decade. Air injection was one of the first methods used to control HC and CO emissions 25 years ago. Although these systems were introduced at different periods in the history of pollution control, today they are integrated parts of the total combustion control system.

In the mid-1960's air injection was introduced as a principal HC and CO control system. EGR appeared in the early 1970's to control NO_x emissions. The first catalytic converters of 1975 had an oxidation catalyst to control HC and CO. In the late 1970's carmakers introduced 3-way or 2-stage converters with a reduction catalyst to control NO_x. Today, air injection begins oxidizing HC and CO on a cold engine and helps to warm up the converter. The system then switches to supply air to the oxidation converter for maximum HC and CO control when the engine is warm. Air injection, thus, supports the converter system on a modern vehicle. Both must work together for proper emission control.

Similarly, EGR limits NO_x production in the combustion chambers. The reduction catalyst further reduces NO_x emissions at the tailpipe. All three systems on a late-model car depend on the electronic control system to maintain a stoichiometric 14.7:1 air-fuel ratio and optimum ignition timing. Overall engine performance requires electronic control of air injection and EGR operating modes.

From a service standpoint, you can repair and replace parts of these systems independently. On older vehicles, you can test their operation separately. On late-model automobiles, however, the air injection and EGR systems are controlled by the central engine computer, and accurate troubleshooting is part of total system diagnosis.

This chapter outlines the principles of testing and component repair and replacement for these three emission systems. The information in this chapter will help you use the carmakers' specific procedures quickly and accurately.

GOALS

After studying this chapter, you should be able to do the following jobs in twice the flat-rate labor time or less, using the manufacturer's procedures.

1. Inspect air injection components and test the system for proper operation; interpret test results and repair or replace parts as required.

2. Remove an air pump and replace a filter fan. Replace a relief valve and an outlet tube, if required. Reinstall and adjust the pump.

3. Inspect EGR components and test the system for proper operation; interpret test results and repair or replace parts as required.

4. Using an exhaust gas analyzer, test exhaust emissions on a converter-equipped car; evaluate test results and replace a converter if required.

AIR INJECTION TESTING AND SERVICE

Air injection systems have no parts that are scheduled for regular replacement or adjustment. Carmakers include regular inspection of the system in their maintenance schedules, however. Air injection service consists of:

1. Inspecting system parts
2. Testing system operation
3. Replacing valves, hoses, and belts and replacing or repairing an air pump.

Figure 19-1. Inspect the air pump drive belt for damage and wear.

Figure 19-2. Inspect air system hoses for wear, loose connections, and signs of burning at the check valves.

Air Injection System Testing

Before testing individual parts, inspect the complete system as follows:

1. Inspect the air pump drive belt for wear, damage, and looseness, figure 19-1. Check belt tension and adjust it to specifications. Do not pry on the pump housing when adjusting belt tension.
2. Inspect all air hoses, figure 19-2, for looseness, wear, and burning or heat damage. Tighten loose connections; replace damaged hoses.
3. Confirm a suspected air or exhaust leak by applying soapy water to the hose or tubing connection and looking for air bubbles.
4. Inspect check valves for signs of exhaust leakage, such as a burned hose. Replace a check valve with any sign of leakage.
5. Inspect all vacuum control hoses for air-switching valves and antibackfire valves for looseness and damage. Replace defective hoses.
6. Inspect air-switching tubing and hoses to the catalytic converter, figure 19-3, for damage and looseness. Repair or replace as required.
7. Disconnect the air pump outlet hose from the check valve or the inlet to the antibackfire or switching valve. Start the engine and check for airflow from the hose. If there is little or none, you will have to test the pump and system valves.
8. With the hose still disconnected, increase engine speed to about 1,500 rpm. If airflow does not increase, test the pump and system valves.
9. Recheck the drive belt for slippage if airflow is low in step 6 or 7.
10. With the engine at fast idle, briefly pinch or block the pump outlet hose. Listen for a hissing or light popping as the relief valve opens. The relief valve may be on the pump, figure 19-4, or the antibackfire valve, figure 19-5.

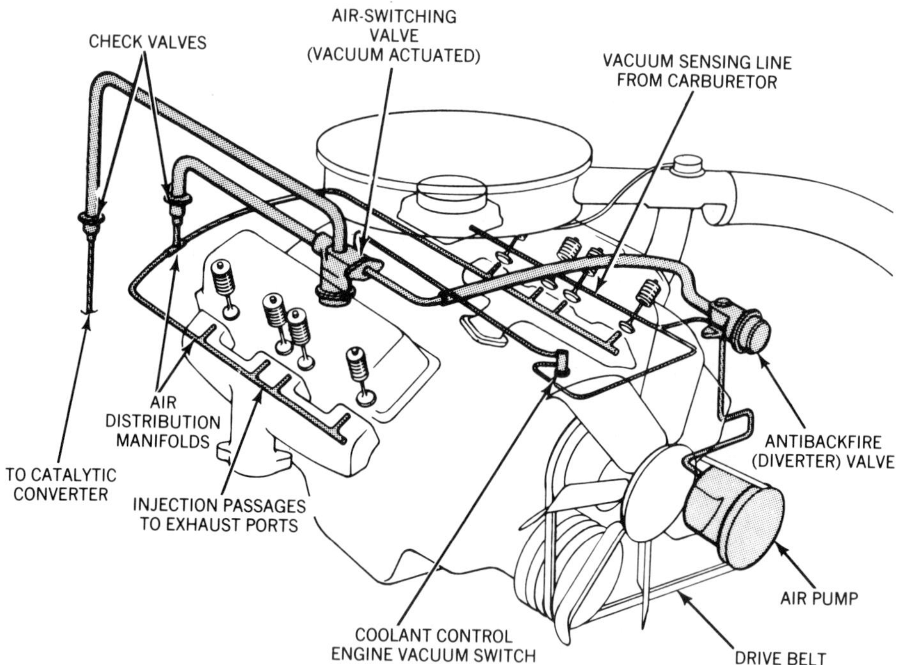

Figure 19-3. Inspect all tubing and hose connections, including those at the converter, for damage and looseness. (Chrysler)

Figure 19-6 summarizes some of the symptoms and possible causes of air injection system problems.

Pulse Air Injection System Inspection. Make the following basic checks of a pulse air injection system:

1. Inspect all hoses as for a pump-supplied system and inspect the pulse air valves for heat damage and signs of exhaust leakage.
2. Disconnect the air valve inlet hose,

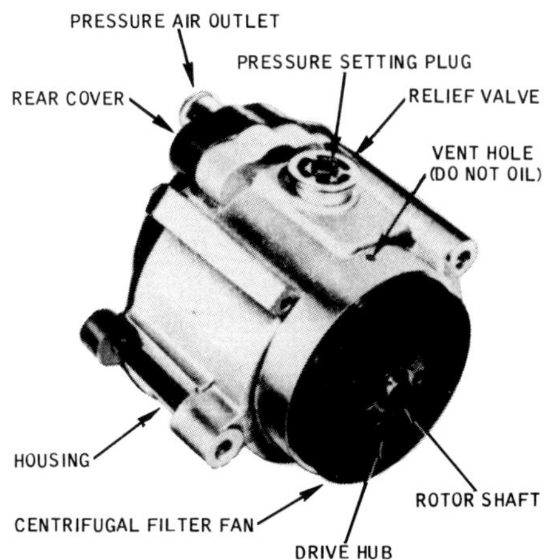

Figure 19-4. Older air pumps have built-in relief valves. (Ford)

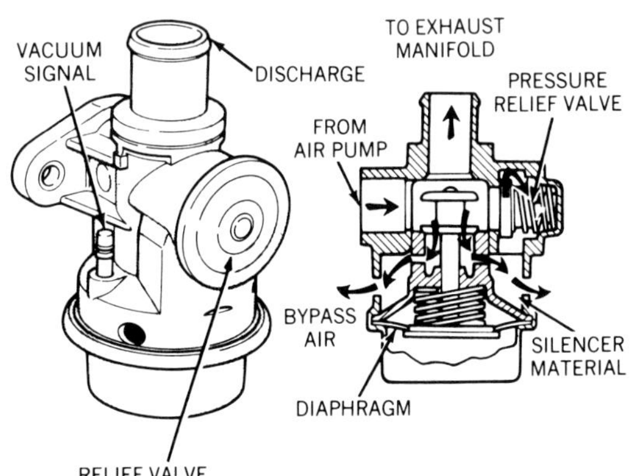

Figure 19-5. Most relief valves in late-model air injection systems are in the anti-backfire valve. (Chrysler)

SYMPTOM—Backfire during deceleration	
Possible Cause	*Correction*
Defective antibackfire valve (gulp or diverter)	Test and replace the valve.
SYMPTOM—No air supply (check at 1,500 rpm)	
Possible Cause	*Correction*
1. Loose or damaged drive belt	1. Tighten or replace the belt.
2. Leaking hoses or hose connections	2. Tighten or replace hoses.
3. Leaking diverter valve or relief valve	3. Test and replace the diverter valve or relief valve.
4. Check valve stuck closed	4. Test and replace the check valve.
5. Leaking pressure relief valve	5. Test and replace the relief valve.
SYMPTOM—Drive belt noise	
Possible Cause	*Correction*
1. Loose or glazed belt	1. Tighten or replace the belt.
2. Seized pump	2. Replace the pump.
SYMPTOM—Pump or system noise (knocking, rumbling, chirping)	
Possible Cause	*Correction*
1. New pump, not broken in	1. Run the car 20 miles and reinspect.
2. Leaking hose	2. Locate the leak; tighten or replace the hose.
3. Defective check or antibackfire valve	3. Locate and replace the bad valve.
4. Loose pump or hose bracket or mount	4. Locate and tighten the mounting.
5. Defective pump	5. Replace the pump.

Figure 19-6. Basic air injection troubleshooting table.

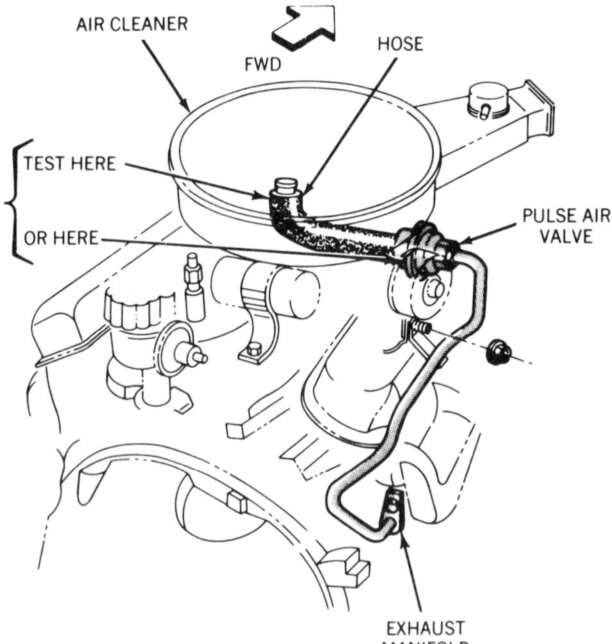

Figure 19-7. Check a pulse air valve at the points indicated for intake suction and exhaust leakage. (Chrysler)

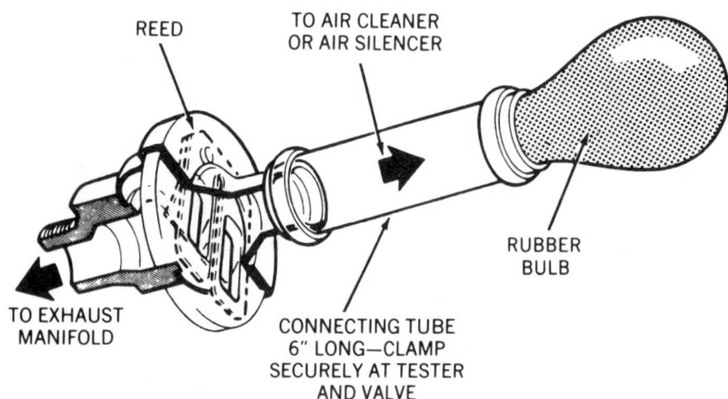

Figure 19-8. Use a large rubber bulb to check airflow through a pulse air valve. (Ford)

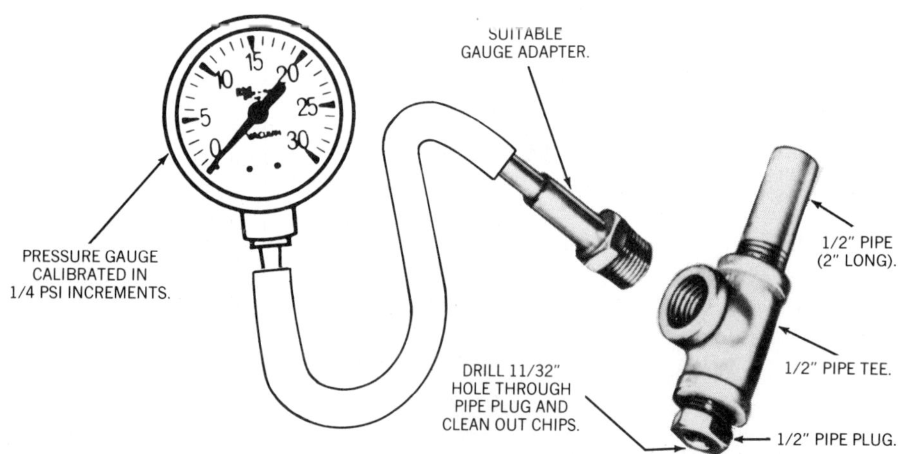

Figure 19-9. You can make this adapter to measure air pump output. (Ford)

or hoses; start the engine and run it at idle.

3. Feel for intake air pulses (intermittent suction) at the valve inlet, figure 19-7. A good valve should make a regular, light popping sound as it operates. Replace the valve if you do not feel intake pulses or if you feel hot exhaust gases.

4. If available, place a large rubber syringe bulb over the valve inlet and squeeze it, figure 19-8. Air should flow freely through the valve, and the bulb should stay collapsed after you release it. If airflow is restricted or the bulb does not stay collapsed, replace the air valve.

Air Pump Output Testing

Usually, checking pump belt tension, air delivery, and relief valve operation as outlined above will detect an air pump problem if one exists. If necessary, you can measure output pressure directly at the pump with a pressure gauge and test adapter, as follows:

1. Make an adapter with pipe fittings as shown in figure 19-9. The hole in the pipe plug regulates pump output pressure at normal operating levels.

2. Disconnect the pump outlet hose at the antibackfire valve inlet and connect it to the adapter and gauge with a hose clamp.

Caution. Keep the hose and gauge away from the moving fan and belts. Direct pump output air away from yourself.

3. Connect a tachometer to the engine. Start the engine and run it at a steady 1,000 rpm or at fast idle.

4. Note the gauge reading. If it is 1 psi or more and steady, the pump is okay. If pressure is below 1 psi or unsteady, replace the pump.

Air Pump Service

Air pumps require no periodic service and cannot be rebuilt. Observe the following precautions, however, to protect an air pump and ensure proper operation:

1. Never try to oil an air pump. Most pumps have a small vent hole in the body, figure 19-10. This is *not* an oil hole.

2. Do not pry against an air pump to adjust belt tension. The pump body is an aluminum casting and easily cracked.

3. When washing or steam cleaning an engine, cover the front of the pump and the vent hole to keep water out.

Pumps are usually replaced if defective, but the relief valve, the outlet tube, and the centrifugal filter fan can be changed on many pumps, if necessary.

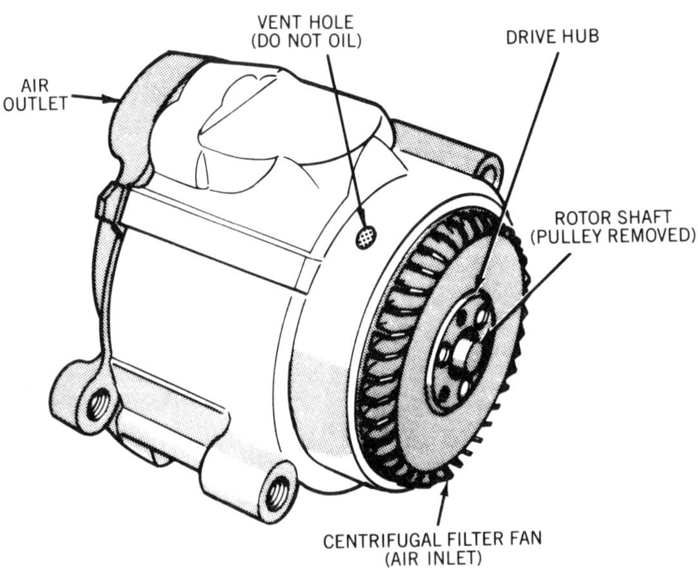

Figure 19-10. The principal parts of a typical air pump. Do not add oil to the vent hole. (Chrysler)

Air Pump Removal and Installation. Replacing an air pump is similar to removing and installing any engine-driven accessory. You probably will have to remove other accessories and drive belts for access to the air pump. If the pump is mounted low on the engine, you may have to remove it from underneath. Disconnect the output hose and the antibackfire valve and any vacuum lines if they are attached to the pump. Loosen the pump adjustment bolt and mounting bolts to relieve belt tension, figure 19-11. Disengage the belt from the pump. Then remove the bolts and remove the pump from the engine. To get clearance for pump removal, you may have to remove the drive pulley from a pump before removing the pump from the engine compartment.

To install an air pump, connect the antibackfire valve if mounted on the pump. Use a new gasket. Place the pump on its engine-mounting bracket and loosely install the bolts. Mount the pump pulley before or after placing the pump on its brackets, depending on engine clearance. Torque the pulley mounting bolts to specifications. Install the drive belt and adjust its tension to specifications. Use an

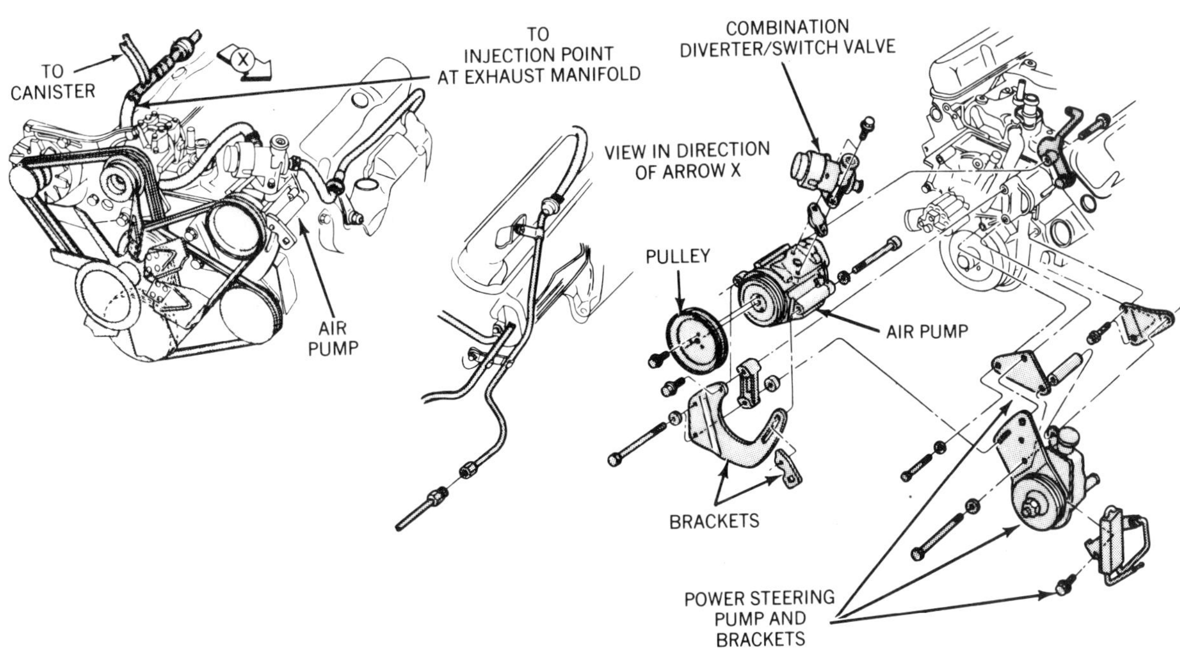

Figure 19-11. You may have to remove the pulley and other engine accessories before removing the pump to get enough clearance. (Chrysler)

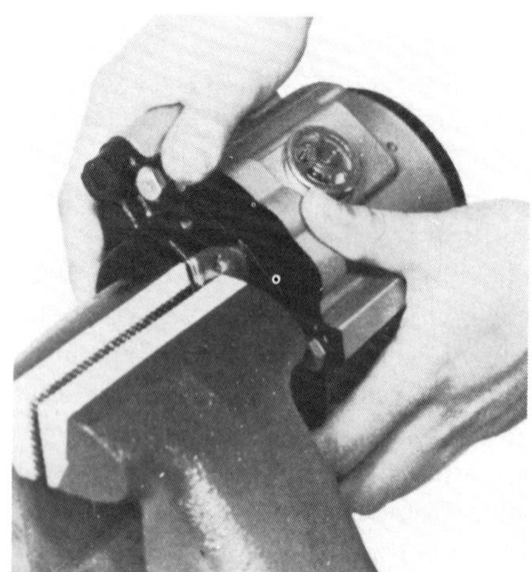

Figure 19-12. Removing the air pump outlet tube. (Saginaw)

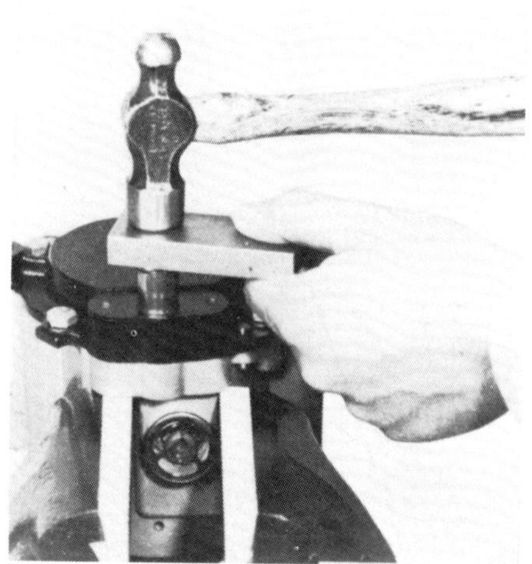

Figure 19-13. Installing the air pump outlet tube. (Saginaw)

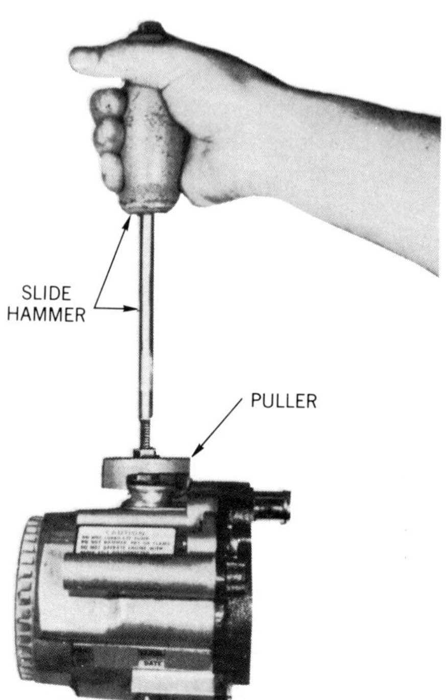

Figure 19-14. Removing the air pump relief valve. (Saginaw)

adjustment tool or otherwise follow the carmaker's instructions to adjust tension. If the air pump is driven by the same belt as another accessory, you may move that accessory to adjust belt tension. Torque the pump mounting bolts to specifications and connect the air output hose and any vacuum lines. Reinstall any other accessories removed for pump replacement and adjust belt tension.

Air Outlet Tube and Relief Valve Replacement. Some pumps have outlet tubes pressed into the rear of the pump body. You can replace such a tube by grasping it in a vise or with pliers and pulling it out with a twisting motion, figure 19-12. Support the pump body on the edge of a workbench or vise and place a new tube in the outlet opening. Tap the tube into place with a block of wood and a hammer until it extends about 7/8 inch from the pump body, figure 19-13.

If a pump has a relief valve pressed into its body, you can remove it with a small puller, figure 19-14. Install a new relief valve in the opening and tap it into place with a block of wood, figure 19-15, or a wrench socket with a diameter that matches the outside diameter of the valve. Relief valves on most late-model systems are part of the system antibackfire or air-switching valve. If such a relief valve malfunctions, you must replace the complete valve assembly.

Filter Fan Replacement. Saginaw air pumps used by domestic carmakers have centrifugal filter fans that remove dirt from incoming air by centrifugal force. The fan is plastic and can be replaced if damaged.

1. Remove the pulley from the pump and the outer disc from the fan if one is installed.

2. Use long-nosed pliers to pull the fan from the pump shaft, figure 19-16. It will break in the process. Do not let fragments fall into the pump.

3. Place a new fan on the pump shaft and install the pulley and its mounting bolt, or bolts.

4. Draw the new fan into place by tightening the pulley bolts, figure 19-17. The outer edge of the fan must fit into the pump housing bore. Do not hammer on the fan, or it will break.

5. Torque the pulley mounting bolts to specifications. The new fan may squeal for a few miles of use until the outer diameter seats in the pump housing bore.

Figure 19-15. Installing the air pump relief valve. (Saginaw)

Air System Valve Testing

Correct operation of antibackfire, air-switching, and check valves is essential for proper system operation. All air injection systems have one or more check valves to keep exhaust from

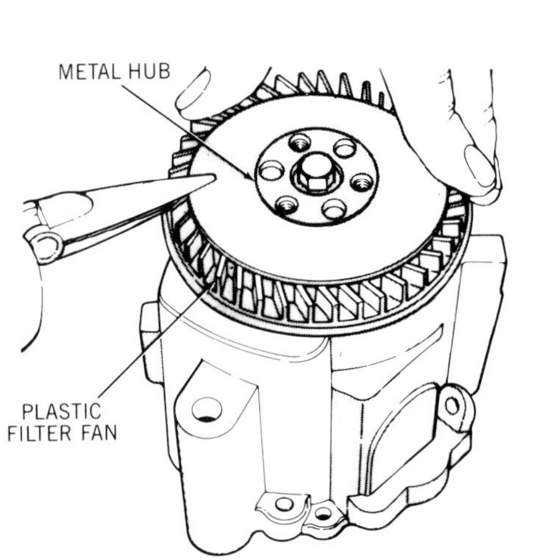

Figure 19-16. Removing the air pump filter fan. (Chrysler)

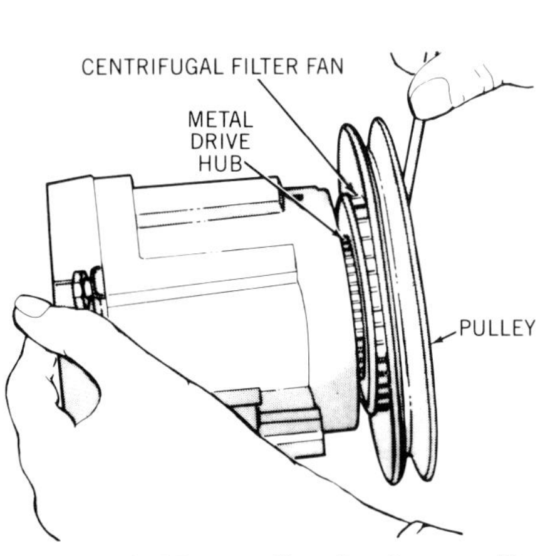

Figure 19-17. Installing the air pump filter fan and pulley. (Chrysler)

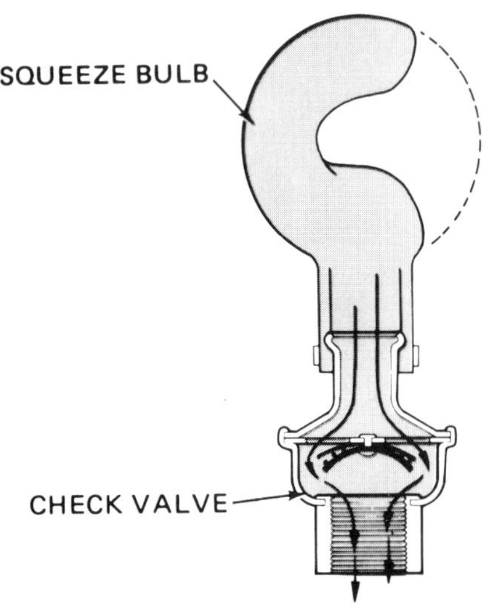

Figure 19-18. Use a large rubber bulb to test a check valve for airflow and leakage. (Ford)

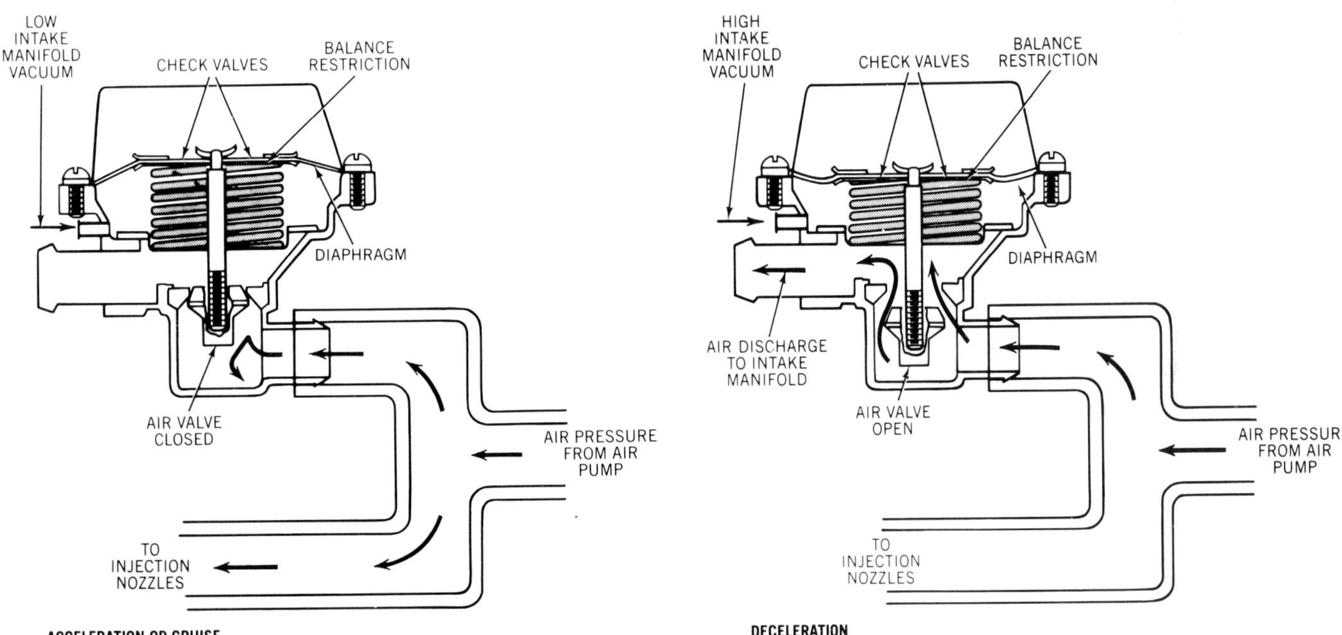

Figure 19-19. Basic gulp valve air injection system.

flowing back into the air lines and hoses. The check valve is opened by air pressure from the pump and closed when exhaust backpressure exceeds air pressure. In a pulse air injection system, the pulse air valve usually does the double job of air delivery and exhaust check valve. A pulse valve opens when exhaust pressure is lower than atmospheric pressure.

Check Valve Test. If a check valve leaks, it will allow hot exhaust back into the air injection system to damage hoses, other valves, and the pump. Your visual inspection of the system will usually detect a defective check valve, but if you are in doubt, you can test a valve by blowing air through the inlet and then sucking back through the valve. Use a large rubber bulb to do this, figure 19-18, or attach a hand-operated vacuum pump to the valve inlet with an adapter. A check valve should allow free air movement through the inlet, but no reverse flow. If it allows reverse airflow or if it does not hold 15 to 20 in. Hg of vacuum at the inlet for about 30 seconds, it is leaking and must be replaced.

Antibackfire Valve Tests. Early air injection systems, used without catalytic converters, simply had an antibackfire valve that diverted air away from the exhaust during deceleration.

- A gulp valve diverts air pump output to the intake manifold, figure 19-19.
- A diverter, or bypass, valve diverts air pump output to the atmosphere, either through an air muffler or to the engine air cleaner, figure 19-20.

Early systems of 20 years ago used gulp valves and diverter valves about equally. By the mid-1970's, diverter

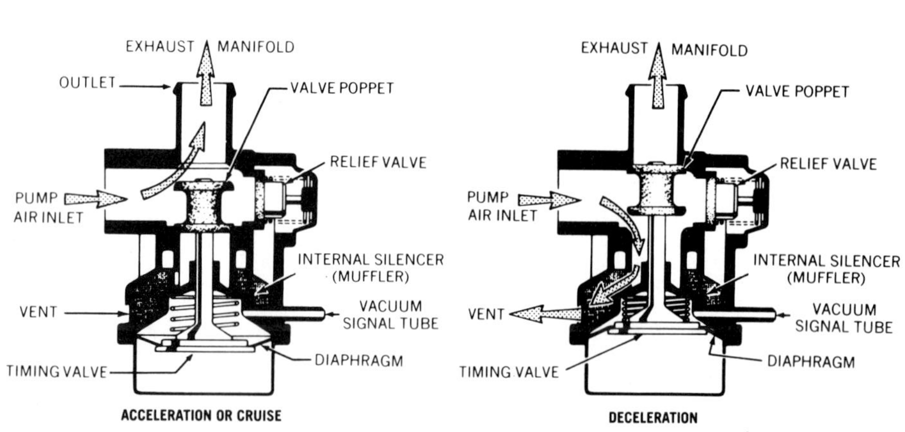

Figure 19-20. Basic diverter valve air injection operation. (GM)

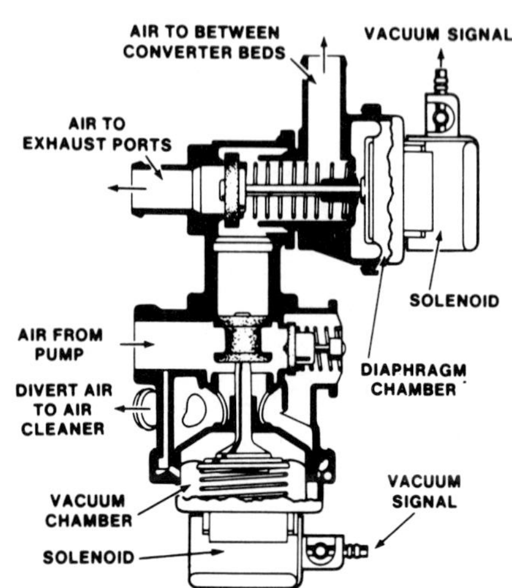

Figure 19-21. This air management valve uses vacuum and electrical control to provide backfire protection and air switching for open- and closed-loop operation. (GM)

valves were more common. With the development of air-switching and 3-way converters, gulp-type antibackfire valves have reappeared on many systems.

Simple gulp and diverter valves use a diaphragm operated by manifold vacuum to switch the air injection airflow, figures 19-19 and 19-20. When manifold vacuum rises during deceleration, the diaphragm actuates the valve to direct airflow to the intake manifold or to the atmosphere for 1 to 5 seconds. Vacuum is applied to the valve at all times, but it is only strong enough to actuate the valve during deceleration. When vacuum decreases, the valve switches air back to the exhaust manifold. Use the following guidelines and the carmaker's procedures to check either a gulp or a diverter valve:

Valve Name	*Abbreviation*
Nylon divert valve	NDV
Zinc divert valve	ZDV
Standardized diverter valve	SDV
Low-vacuum air control valve	LAC
High-vacuum air control valve	HAC
Electric air control valve	EAC
Electric air control valve plus low-vacuum air control valve	EAC & LACN
Air intake control valve	AIC
Low-vacuum air control valve plus electric air control valve	LAC & EACN
Low-vacuum air control valve plus low-vacuum air control valve	LAC & LACN
Electric air control valve plus electric air-switching valve	EAC & EAS
Electric air control—electric switching valve	ECES
Electric divert valve—electric air-switching valve	EDES
Remote divert—electric switching valve	RDES
Electric air control—electric air-switching valve	CSTSE
Pressure-operated electric divert and electric switching	PEDES

Figure 19-22. These are the most common air control valves used with GM CCC engine systems.

1. Inspect vacuum lines for damage or loose connections. If the amount of vacuum at the valve is not correct, it will not work right even if it is in good condition.

2. As with any vacuum-operated device, apply 15 to 20 in. Hg of vacuum to the diaphragm connection with a hand-operated pump and hold the vacuum for about 30 seconds. If vacuum leaks down, the diaphragm is leaking.

3. With the engine running at idle, disconnect the manifold vacuum line at the valve diaphragm and feel for vacuum. If there is no suction, the line is plugged or leaking. Repair the vacuum line before going further.

4. Reconnect the vacuum line to the valve and run the engine at idle:

a. For a gulp valve, if idle speed changes or the engine idles roughly, the valve is bypassing air to the intake manifold. Replace it.

b. For a diverter valve, listen and feel for airflow around the valve muffler or vent outlet. If you detect airflow, the valve is bypassing air to the atmosphere. Replace it.

5. Accelerate the engine to about 2,000 rpm and quickly close the throttle:

a. If the engine backfires with a gulp valve, replace the valve. To check a gulp valve further, disconnect the air outlet hose and repeat the test. You should feel and hear a short burst of air as the valve opens.

b. For a diverter valve, listen and feel for airflow around the valve muffler or vent outlet. If you detect a burst of air for about 1 to 5 seconds, the valve is diverting air correctly. If not, replace the valve.

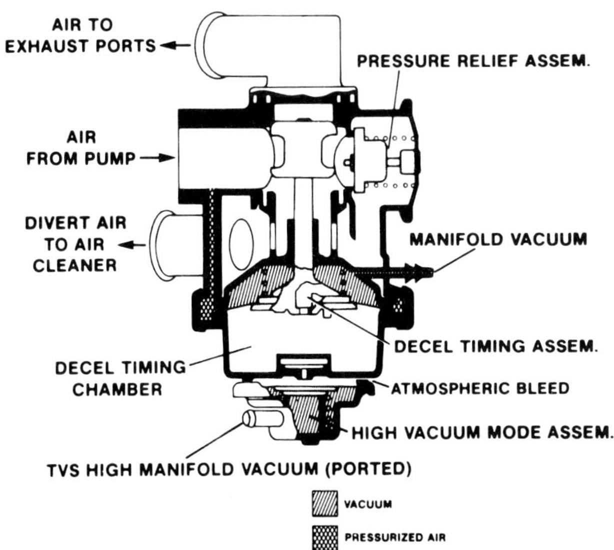

Figure 19-23. This valve directs airflow with vacuum signals on two different diaphragms. (GM)

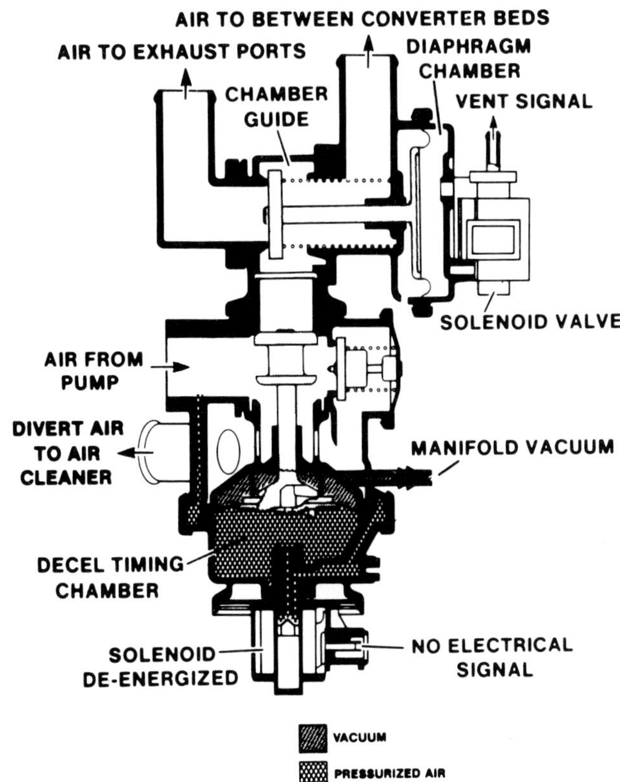

Figure 19-24. This valve directs airflow with vacuum and air pressure signals on opposite sides of a diaphragm. (GM)

Combination and Air-Switching Valve Tests. To test the air control valves of a late-model air injection system, you will need the manufacturer's procedures and system and valve descriptions for a particular vehicle. All late-model systems have the anti-backfire protection of a simple gulp or diverter valve, but the system valves do other jobs as well, figure 19-21. All functions may be performed by a single valve, or the system may have several valves for one or more of the following jobs:

- Direct air to the exhaust manifolds or ports during cold-engine operation to warm up the EGO sensor and the converter, as well as aid HC and CO control.
- Switch air downstream to the oxidation converter to aid warm-engine control of HC and CO.
- Divert air out of the exhaust during deceleration to prevent backfiring and protect the converter from overheating and other damage.
- Divert air out of the exhaust during wide-open throttle or heavy load and in cases of overheating or engine system malfunction that could damage a converter if overheated.

General Motors has 16 different air control valves available for use on current engines, figure 19-22. Ford, Chrysler, and other manufacturers have similar variations. Valve switching and bypass, or diverter, functions are controlled by:

1. Simple vacuum diaphragms similar to basic gulp and diverter valves.
2. Vacuum signals of different levels (ported and manifold vacuum, for example) on opposite sides of a diaphragm to modulate or time valve action, figure 19-23.
3. Vacuum and air pump pressure or atmospheric pressure on opposite sides of a diaphragm, figure 19-24.
4. Coolant-temperature-controlled vacuum valves in the vacuum line to a diaphragm.
5. Computer-operated solenoids in the vacuum lines to various diaphragms, figure 19-25.

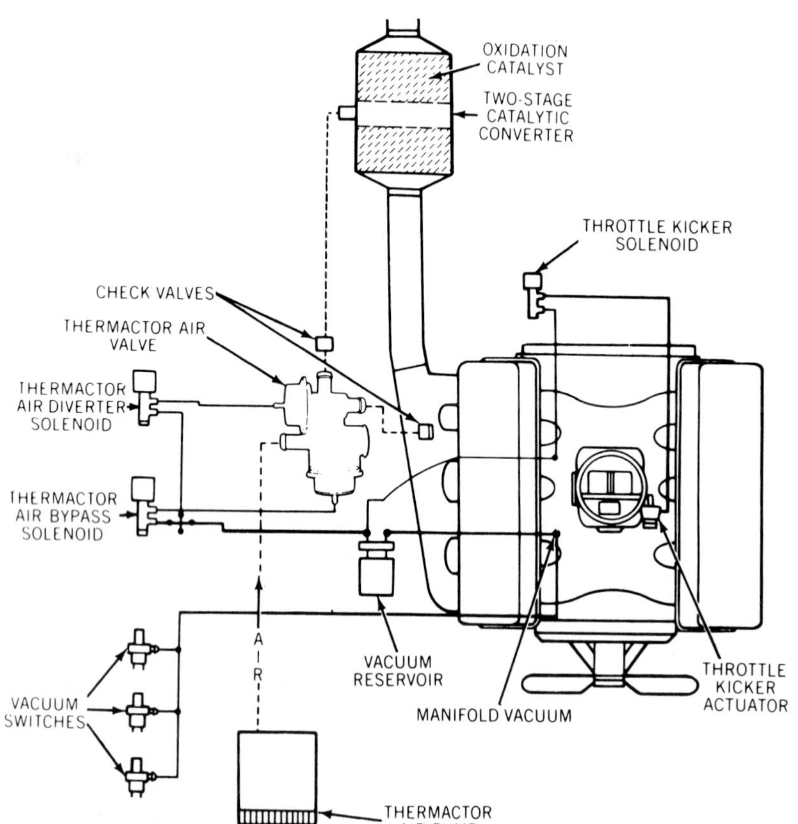

Figure 19-25. This system has two solenoid-operated vacuum valves to control the air valve. (Ford)

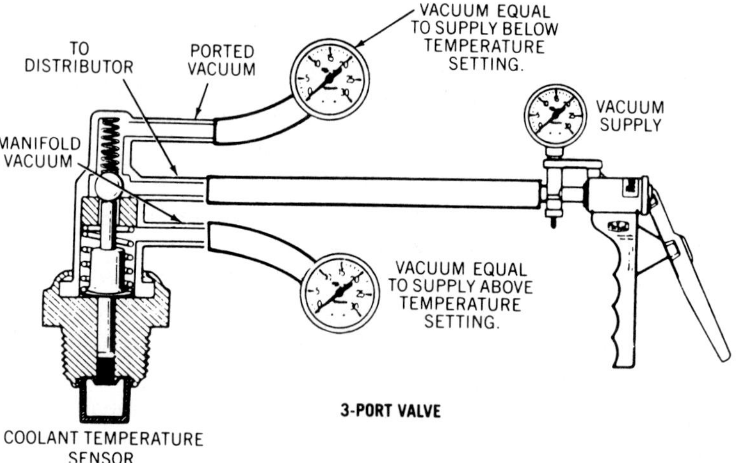

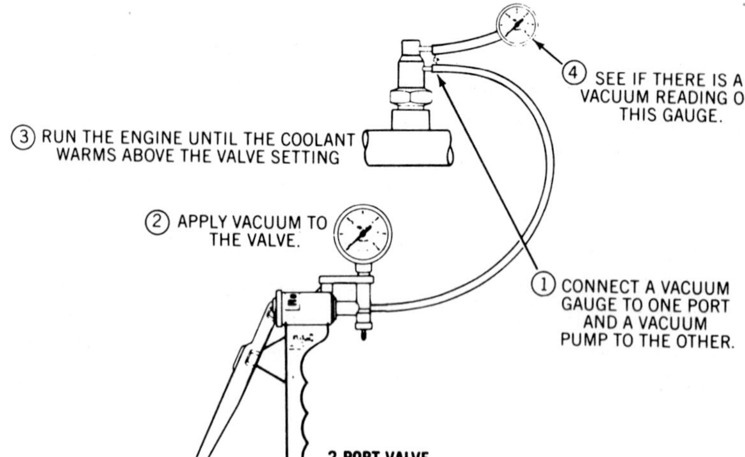

Figure 19-26. Test a thermostatic vacuum valve for an air injection system as you would the same valve used for vacuum advance control. (Ford)

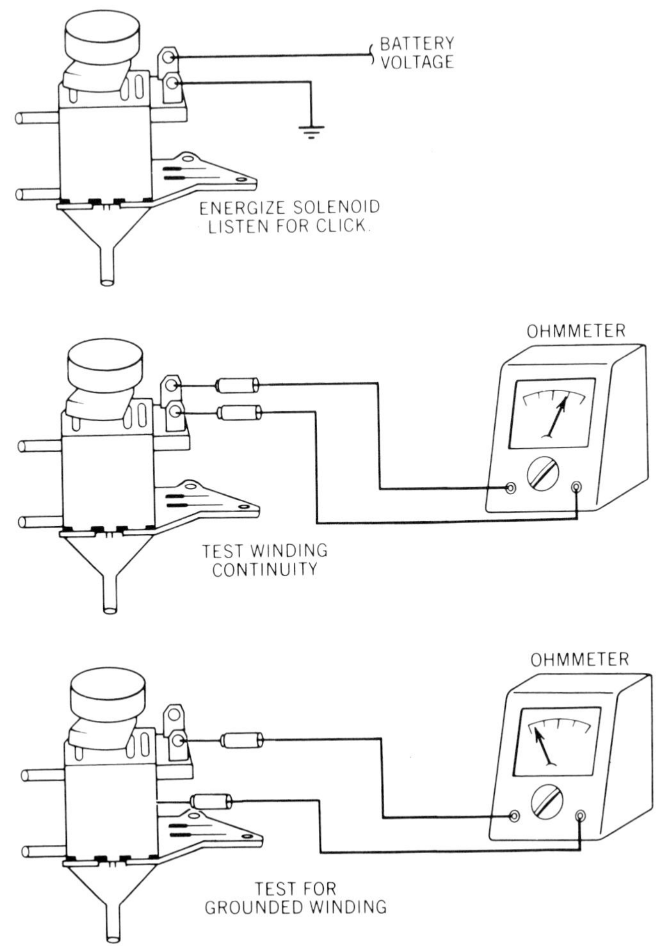

Figure 19-27. Solenoids used for air injection and EGR control work the same as solenoids in any other vacuum system.

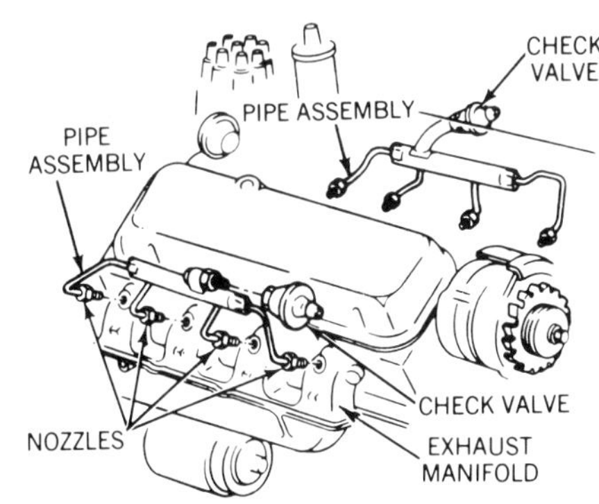

Figure 19-28. Typical air injection manifold and nozzles. (Chevrolet)

Consult the manufacturer's system and valve descriptions and follow specific test procedures to check system operation. Most procedures include several of the following principles:

1. Test vacuum diaphragms as explained for gulp and diverter valves. If the diaphragm has a timed air bleed or operates on vacuum differential, test its operation over a specific number of seconds.

2. Temperature-controlled vacuum valves work the same way for an air injection system as they do for vacuum spark advance control. Test them both warm and cold, figure 19-26. Use a hand-operated vacuum pump to apply vacuum to the valve and check the manufacturer's specifications for valve operation when hot and cold.

3. Use an ohmmeter to test solenoid coil resistance, figure 19-27, and compare to manufacturer's specifications. Computer-operated solenoids have a minimum, as well as a maximum, coil resistance.

4. Use an ohmmeter to check for a solenoid winding that is shorted to the solenoid housing.

5. Use a 12-volt test lamp or a voltmeter to test for output voltage signals from the computer and at air control solenoids.

6. Test the input signal voltages from coolant temperature, throttle position, and EGO sensors that affect air injection control.

7. Run the engine through various operating modes and check for air output from hoses and valves under different conditions. Compare your findings to the carmaker's specific requirements.

Air Injection Component Replacement

Other than minor repairs to an air pump, no components are repairable. All system parts must be replaced if defective. We have outlined pump replacement already. To replace hoses and valves, remove hose clamps or otherwise disconnect hoses and lines and remove the parts to be replaced. Note the routing and connections of all lines so that you reinstall them correctly. When you replace a hose, be sure the hose material is suitable for air injection use and will withstand high temperatures.

All air injection tubing and air nozzle manifolds are steel. It is best to replace these parts with preformed OEM components. Air tubing that runs to a converter is usually supported by brackets on the engine or exhaust system. Be sure that bracket clamps and tubing connections are tight so the tubing does not loosen from vibration. Remember that catalytic converters stay hot after an engine is shut off. Be careful when replacing air tubing at the converter.

You can remove individual air nozzles from cylinder heads or exhaust manifolds for cleaning or replacement. The air manifolds for many nozzles are 1-piece assemblies and must be removed along with all nozzles, figure 19-28. Heat and exhaust corrosion can cause air nozzles to seize in heads or manifolds. Before removing a nozzle, apply penetrating oil to

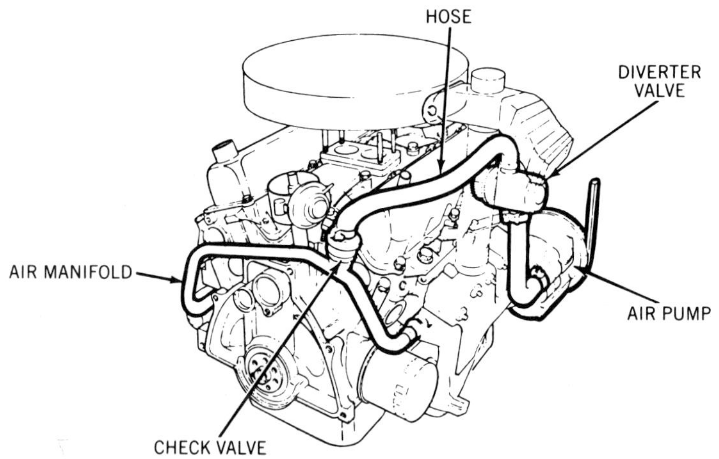

Figure 19-29. These air injection passages are cast into the cylinder heads. Cleaning requires head removal. (Ford)

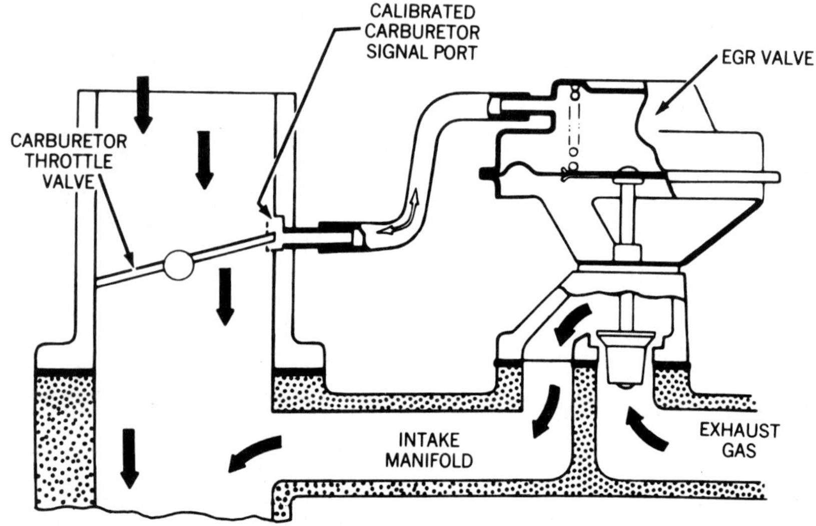

Figure 19-30. Basic EGR valve operation. (Chevrolet)

loosen it. Use tubing (flare) wrenches and sockets to remove air manifolds and nozzles. You can remove carbon deposits from air manifolds and nozzles by soaking them in carburetor cleaner and cleaning them with wire brushes. Remove the check valve from a manifold before cleaning it. Clean openings and passages with small wire brushes or drill bits. Apply antiseize compound to the threads of nozzles and air manifolds before installation. Tighten nozzles and fittings securely, but do not overtighten.

Air distribution passages cast into cylinder heads or exhaust manifolds can only be cleaned if the head or manifold is removed from the engine, figure 19-29. When you remove a head or manifold, inspect air passages for carbon deposits and remove any with a small-diameter wire brush.

EXHAUST GAS RECIRCULATION (EGR) TESTING AND SERVICE

Like air injection, most EGR systems have no parts scheduled for regular replacement or adjustment. Carmakers include regular inspection of the system in their maintenance schedules, however. EGR service consists of:

1. Inspecting the system
2. Testing system operation
3. Testing individual parts
4. Replacing defective parts.

Gasoline engine EGR systems are like vacuum advance controls and air injection systems. Carmakers have used many variations over the years, and you will need specific system descriptions and test procedures for thorough troubleshooting. Also like other systems, however, all EGR systems operate on common principles and have common goals. If you understand the fundamentals of system operation and service, you can use the manufacturers' instructions more effectively.

EGR dilutes the air-fuel mixture and limits NO_x formation when combustion temperatures are high and air-fuel ratios are lean. On a gasoline engine, EGR should operate during moderate acceleration and at cruising speeds from 30 to 70 mph. EGR should *not* operate on a gasoline engine at:

- Low temperature (cold-engine warm-up)
- Idle (rich mixture, low temperature)
- Low speed (low combustion temperature)
- High speed, wide-open throttle (rich mixture)
- Deceleration (rich mixture, low combustion temperature).

Because a diesel engine has its leanest air-fuel ratios and high combustion temperatures at idle, diesel EGR should operate at idle and decrease as speed increases.

Remembering these distinctions between when EGR *should* and *should not* operate is the first step toward troubleshooting any system. The second step is to remember that drivers seldom complain about *lack of EGR*. Most EGR-related driveability complaints are due to exhaust recirculation when the system should *not* be operating. Most often, you will be troubleshooting a system that causes the valve to open when it should not or to open too much.

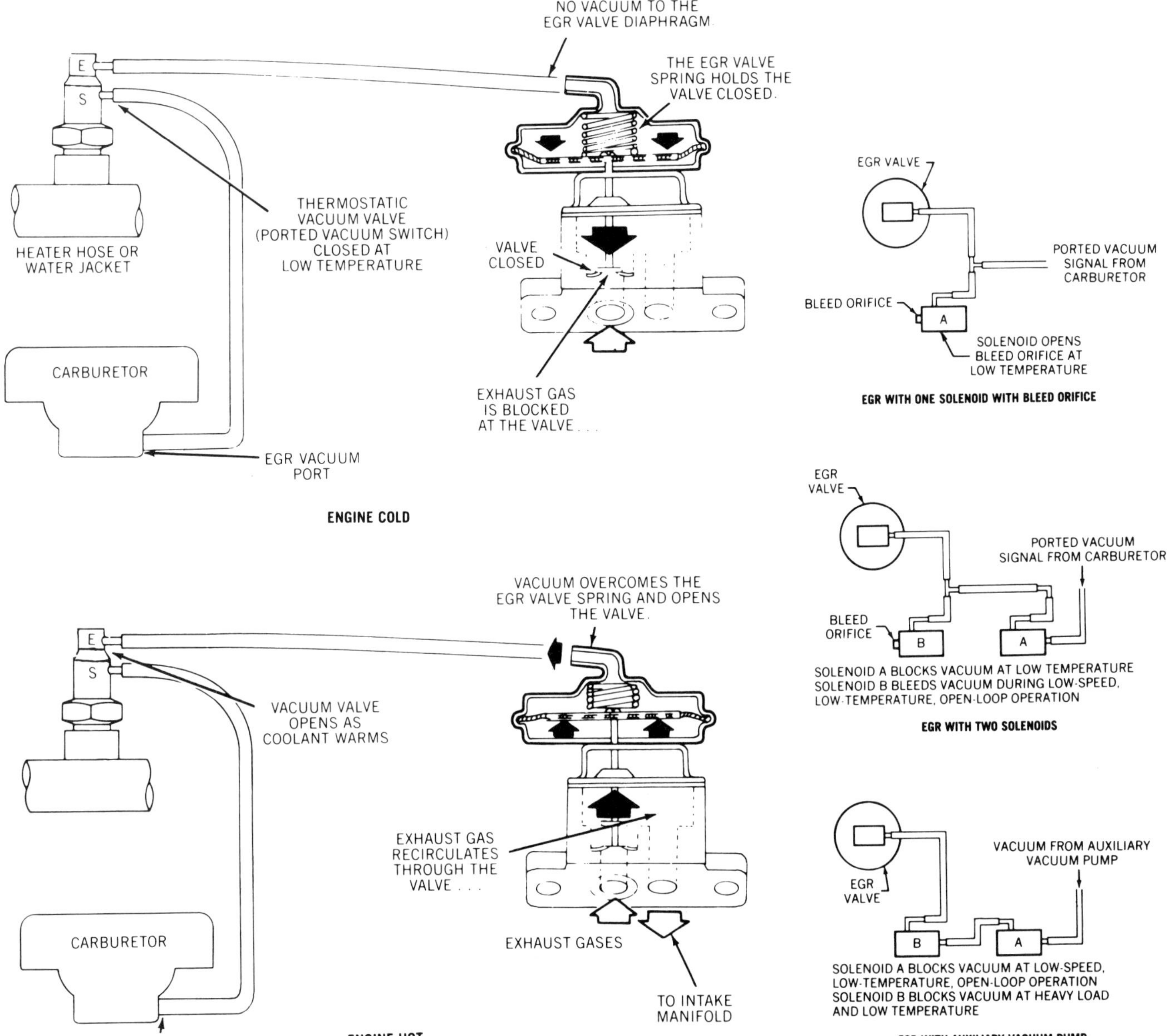

Figure 19-31. Some engines have temperature-controlled (thermostatic) vacuum valves to block EGR during warmup. (Ford)

Figure 19-32. Other engines have solenoid-controlled vacuum valves in the EGR system. (GM)

Remember, however, that EGR helps to control detonation on late-model engines. If a driver complains about detonation, or pinging, the EGR system is one of the first things to check. Also, on some late-model cars lack of EGR can reduce fuel economy slightly by causing the fuel feedback system to remain in open loop when it should not. The remaining steps of EGR troubleshooting are to understand and check the operation of the system controls.

All EGR systems except Chrysler's earliest floor jet systems of 1972–73 have a valve to meter exhaust into the air-fuel mixture, figure 19-30. All EGR valves except those on Ford's EEC-I systems are operated by vacuum. All systems have one or more of the following kinds of control devices to regulate EGR valve operation:

- Vacuum control valves, figure 19-31, operated directly by air or coolant temperature to shut off vacuum to the EGR valve when the engine is cold and apply vacuum when the engine is warm.
- Vacuum control solenoids, figure 19-32, that control vacuum for cold or warm engine operation and also may turn off vacuum at idle, wide-open throttle, or deceleration. Solenoids can be operated by an engine computer or by temperature or throttle-position switches. A solenoid can block vacuum, or it can open an air bleed to relieve vacuum.
- Dual-area vacuum diaphragms that modulate EGR flow by applying different amounts of vacuum to diaphragms of different areas, figure 19-33.

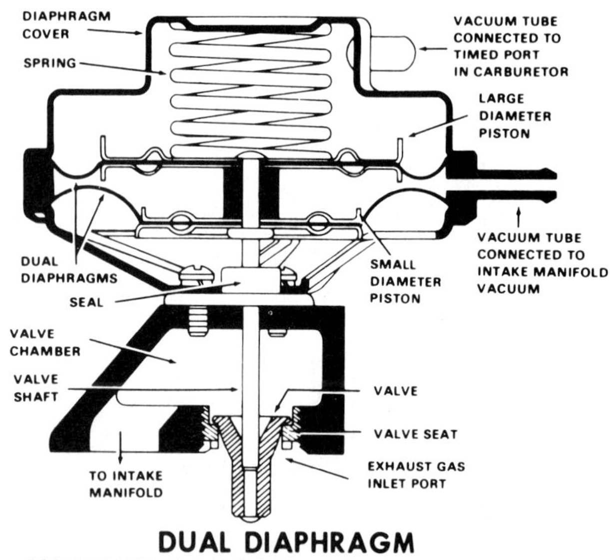

Figure 19-33. This EGR valve uses different vacuum signals on different-sized diaphragms to modulate EGR flow. (Buick)

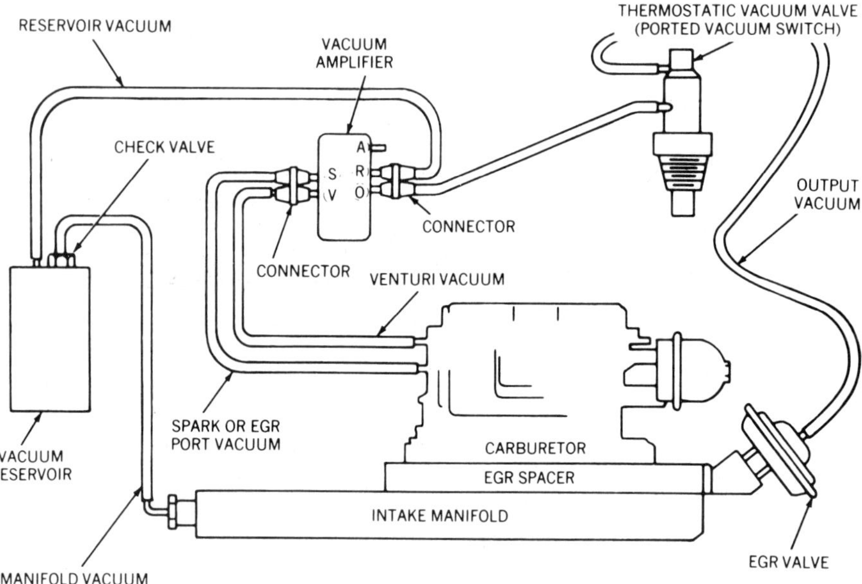

Figure 19-34. Typical EGR vacuum amplifier installation. (Ford)

• Vacuum amplifiers that use different combinations of manifold, ported, and venturi vacuum to modulate EGR valve action, figure 19-34.

• Exhaust backpressure transducers that modulate EGR valve action in relation to engine load and the amount and pressure of exhaust flow, figure 19-35.

• A vacuum aspirator that develops a vacuum signal from air injection pump airflow passing through a venturi, figure 19-36. This provides vacuum to the EGR valve when there is little engine vacuum and helps to control detonation.

Regardless of the number and kinds of control devices, you can begin troubleshooting an EGR system with basic inspection.

EGR System Inspection and Basic System Tests

Because EGR systems are basically vacuum-operated systems with many electrical controls, you can begin with the two basic vacuum and electrical inspection steps:

1. Inspect all vacuum hoses for damage and loose connections. Replace defective hoses.

2. Inspect electrical connections at solenoids and timer relays for looseness and damage. Repair as required.

After assuring yourself that vacuum and electrical connections are in good order, proceed with a general system test.

General EGR System Test. Use the following guidelines to test most EGR systems that have simple vacuum-operated valves *without backpressure transducers*. This check will tell you if the valve is opening and closing when it should and, in many cases, will indicate a blocked or leaking valve.

1. Connect a tachometer to the engine. You may have to remove the air cleaner for access to the EGR valve. If so, plug or cap any vacuum hoses that you disconnect.

2. Start the engine and warm it to normal operating temperature. Put the transmission in neutral or park, set the parking brake, and block the drive wheels.

3. Run the engine at a steady 1,800 to 2,000 rpm. Block the throttle or place the carburetor linkage on the high step of the fast-idle cam to hold this speed.

4. Perform either or both of the following steps to determine if the EGR valve is opening at cruising speed, as it should:

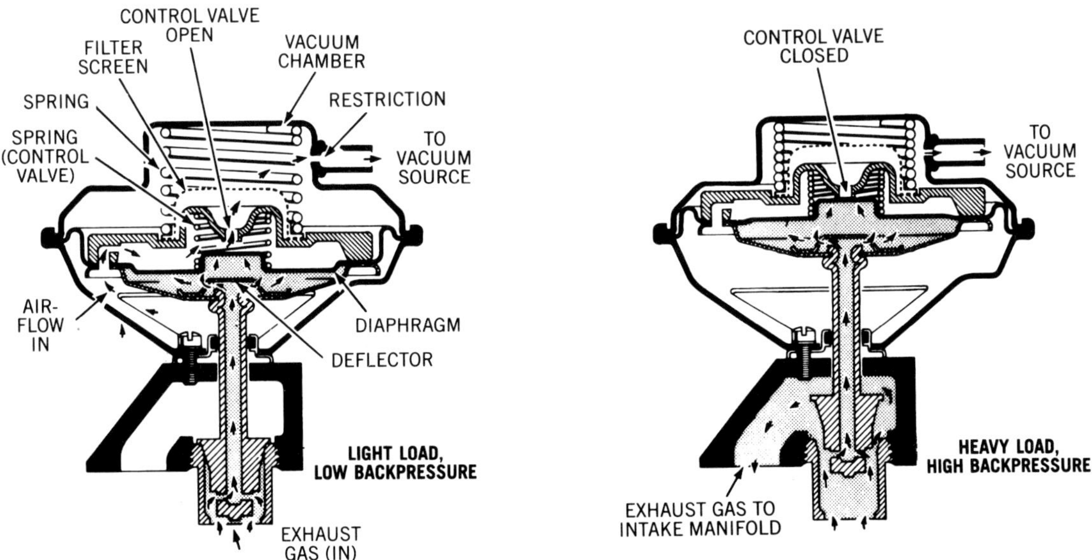

Figure 19-35. Most late-model EGR valves are regulated by exhaust backpressure transducers. (GM)

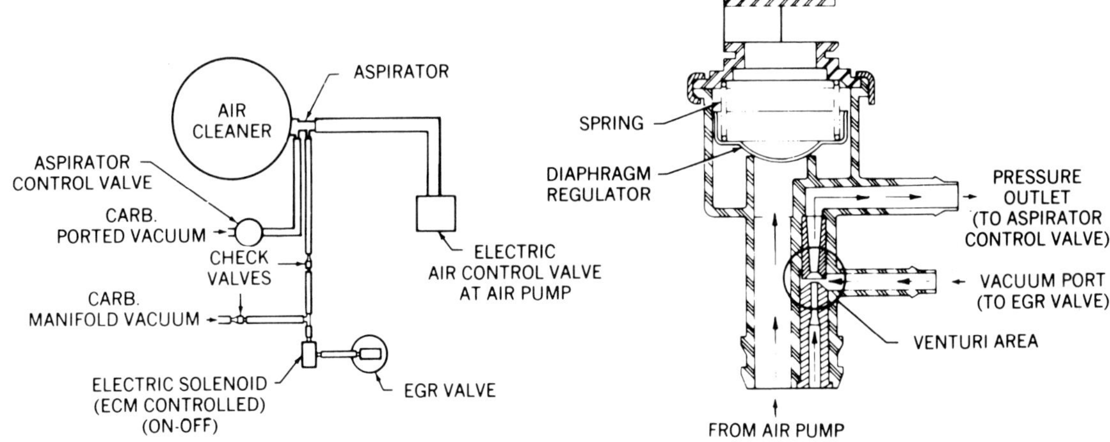

Figure 19-36. The aspirator uses air pump airflow to develop EGR vacuum. This system improves engine detonation control. (GM)

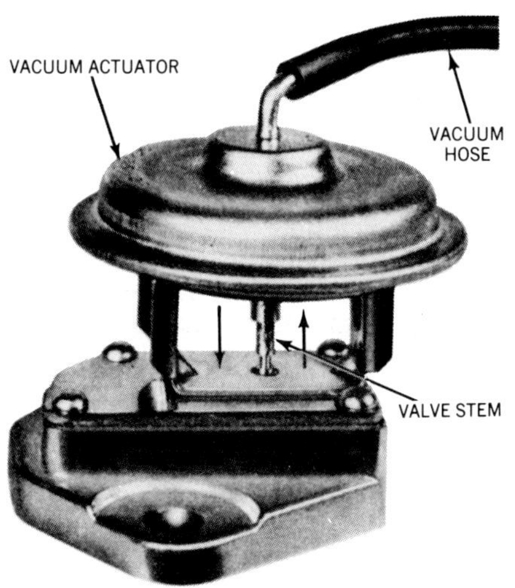

Figure 19-37. If the valve stem is visible, watch its movement as you apply and release vacuum. (Chrysler)

a. If the EGR valve stem is visible, figure 19-37, watch its movement as you accelerate to about 2,000 rpm. The stem should move upward slowly and steadily as the valve opens.

b. With the engine at a steady cruising rpm, disconnect and plug the vacuum hose at the EGR valve. Watch the tachometer. If the valve is working right, engine speed should increase 100 to 150 rpm as EGR is cut off.

5. Return the engine to idle speed. If the EGR valve stem is visible, watch it as you reduce engine speed. It should move downward as the valve closes.

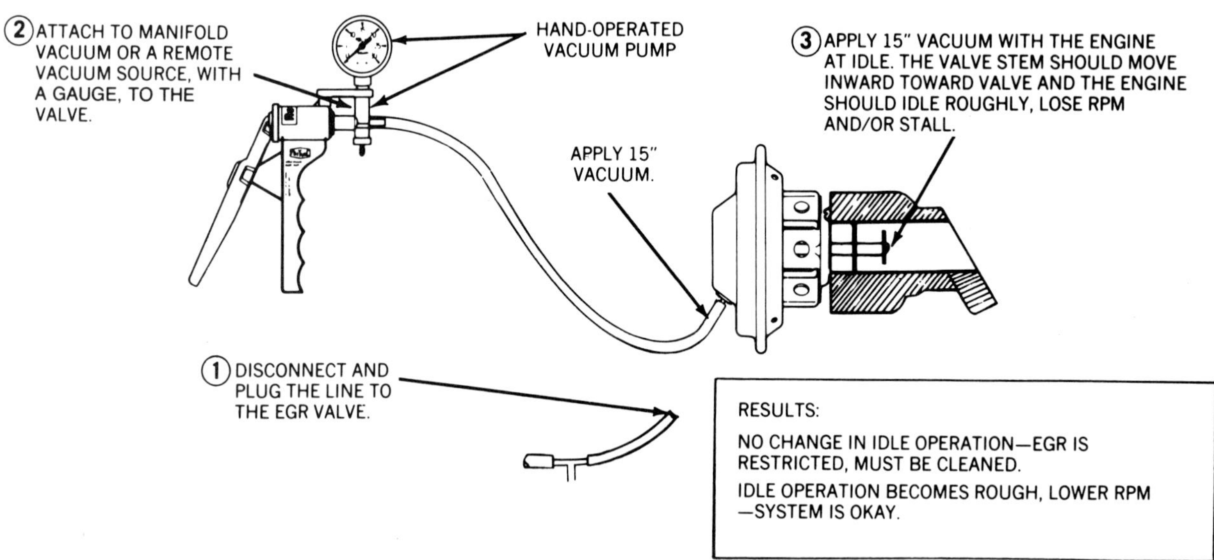

Figure 19-38. Engine rpm should drop when you apply vacuum to open the EGR valve at idle. (Ford)

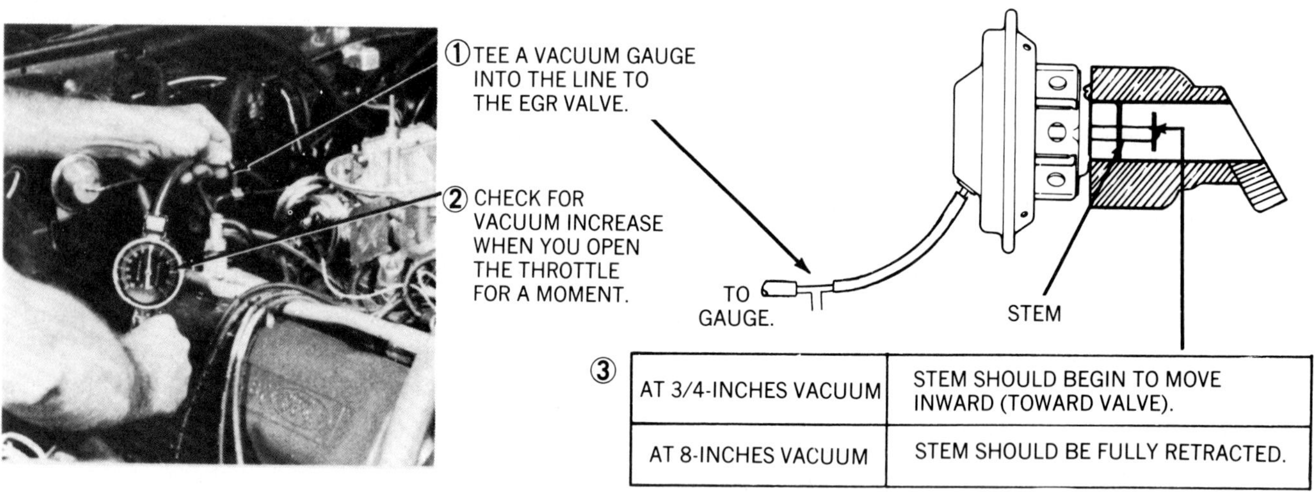

AT 3/4-INCHES VACUUM	STEM SHOULD BEGIN TO MOVE INWARD (TOWARD VALVE).
AT 8-INCHES VACUUM	STEM SHOULD BE FULLY RETRACTED.

Figure 19-39. Use a T-fitting to connect a gauge into the line to check actual EGR valve vacuum. (Ford)

6. With the engine at normal slow (curb) idle, disconnect and plug the EGR valve vacuum hose.

7. Connect a source of 8 to 10 in. Hg of vacuum to the valve. Use either a hose connected to a manifold vacuum port or a hand-operated vacuum pump, figure 19-38.

8. Idle speed should drop 100 to 150 rpm and become rough as the EGR valve opens. The engine may even stall.

9. If the EGR valve does not appear to be working right in steps 4 and 8, connect a vacuum gauge to the EGR valve vacuum hose with a T-fitting, figure 19-39, and check the vacuum at idle and cruising speed. You should have no vacuum at idle and 8 to 10 in. Hg (or the carmaker's specified amount) at cruising speed.

a. If vacuum is correct but the valve does not work right, clean or replace the EGR valve and be sure that exhaust passages to the valve are not restricted or leaking.

b. If the general vacuum levels are not correct in step 9, or if EGR problems occur at high speed or with a cold engine, continue with detailed system vacuum tests.

EGR System Test with Backpressure Transducer. The preceding tests may not work for a system with an exhaust backpressure transducer. Even with high vacuum applied to the valve at idle, exhaust backpressure may be too low to close the valve air bleed and allow the diaphragm to operate. Therefore, engine speed may not change at idle, even with a good valve. EGR valve and transducer calibrations vary from engine to engine, and carmakers have specific test procedures for different vehicles.

A transducer-controlled EGR valve should, however, open at cruising speed regardless of its calibration. Therefore, step 4 of the previous test should give you a general indication of valve operation for such a system.

EGR System Vacuum Tests. If the general system test indicates incorrect EGR vacuum, or if other driver complaints suggest an EGR problem, check the system vacuum with a cold engine and at wide-open throttle, as well as at idle and cruising speed. Some driveability complaints that may indicate an EGR problem are:

- Poor engine warmup or a rough-running cold engine
- Acceleration stumble
- Surging at medium and high speeds
- Lack of high-speed performance
- Detonation on acceleration (due to lack of EGR).

To test system vacuum, you must know the kinds of control devices (valves and solenoids) in the system and their operating modes. If possible, also look up the manufacturer's specified vacuum at the EGR valve under different conditions. Generally, look for the following vacuum conditions at the EGR valve in various engine operating modes. You can make most of these checks with a vacuum gauge connected to the EGR valve vacuum line with a T-fitting, figure 19-39.

1. With the engine at idle, hot or cold, there should be no vacuum at the EGR valve. If you find any vacuum, look for:

a. A defective vacuum amplifier if the system has one.

b. Incorrect idle speed or otherwise misadjusted throttle that exposes an EGR vacuum port at low speed.

2. If vacuum is absent or below specifications at the valve at cruising speed with a warm engine, look for:

a. Overall low engine vacuum (worn engine).

b. A vacuum system leak.

c. A defective vacuum amplifier.

d. A problem in a solenoid or vacuum control valve.

3. Repeat the general system test with a cold engine. With coolant temperature below 100° to 125° F (38° to 52° C) for most engines, there should be *no* vacuum at the valve at any speed. If any vacuum is present, look for a problem with a temperature vacuum control valve or solenoid.

4. Run the engine at high cruising speed (2,500 to 3,000 rpm) and check vacuum at the EGR valve. If vacuum is above specifications, too much EGR can cause surging. If you find too much vacuum, look for:

a. A defective vacuum amplifier.

b. A defective vacuum bleed solenoid.

c. A defective backpressure transducer.

5. With the engine at normal idle, open the throttle wide and accelerate to about 3,000 rpm. Close the throttle. Read the vacuum gauge during the snap acceleration. If vacuum is above specifications, too much EGR can cause acceleration stumble. If you find too much vacuum, look for the same possible causes as in step 4. A weak or broken diaphragm spring in the EGR valve also can cause a stumble on acceleration because EGR will come on too soon.

Caution. Do a wide-open throttle test during a road test or on a chassis dynamometer for vehicle safety and accurate results.

6. Operate the engine at wide-open throttle with a load and read the vacuum gauge. There should be 4 in. Hg or less vacuum. If you find vacuum higher than specifications at wide-open throttle, look for the same possible causes as in step 4.

If EGR vacuum is out of limits for any operating condition, you must test individual valves, solenoids, and other control devices to isolate the problem. If EGR vacuum is within manufacturer's specifications for different operating conditions, problems can be caused by a stuck or leaking EGR valve.

The next sections outline basic tests for EGR system parts.

EGR System Component Tests

Use the following guidelines and the carmaker's procedures to test vacuum control valves, solenoids, and EGR valves.

Vacuum Valve and Solenoid Tests

1. Using a 12-volt test lamp or a voltmeter, along with the carmaker's specifications and diagrams, check for available voltage at solenoids and relays under specific conditions. A solenoid in an EGR system controls vacuum much as it does for air injection control or vacuum spark advance.

2. Use an ohmmeter to test solenoid coil resistance and compare to manufacturer's specifications, figure 19-27. Computer-operated solenoids have a minimum, as well as a maximum, coil resistance.

3. Use an ohmmeter to check for a solenoid winding that is shorted to the solenoid housing.

4. Apply specified voltage from the battery or from the computer to a solenoid and listen for a click to determine when it energizes. Use a vacuum gauge to check vacuum through a solenoid, both energized and deenergized, and compare it to the carmaker's specifications.

5. Test temperature-controlled vacuum valves both warm and cold, figure 19-26. Use a hand-operated vacuum pump to apply vacuum to the valve and check the manufacturer's specifications for valve operation when hot and cold.

6. Test the input signal voltages from coolant temperature, throttle position, and EGO sensors that affect EGR control.

Vacuum Amplifier Tests. Many EGR systems apply vacuum to the EGR valve that is proportional to a vacuum signal at the carburetor venturi. This allows EGR flow to be closely proportional to intake airflow. However,

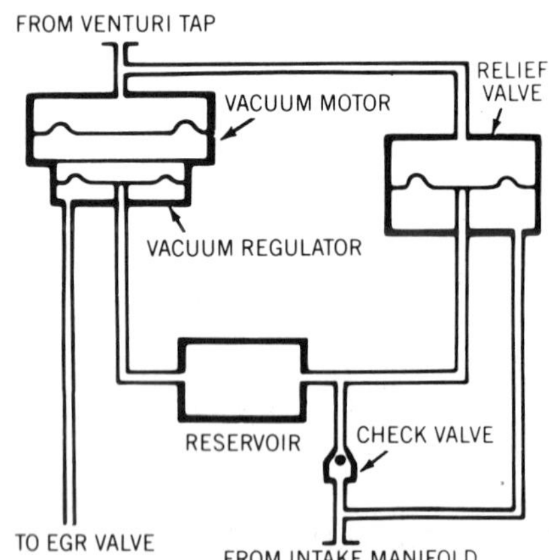

Figure 19-40. A vacuum amplifier uses a weak venturi vacuum signal to open a vacuum reservoir to the EGR valve. (Ford)

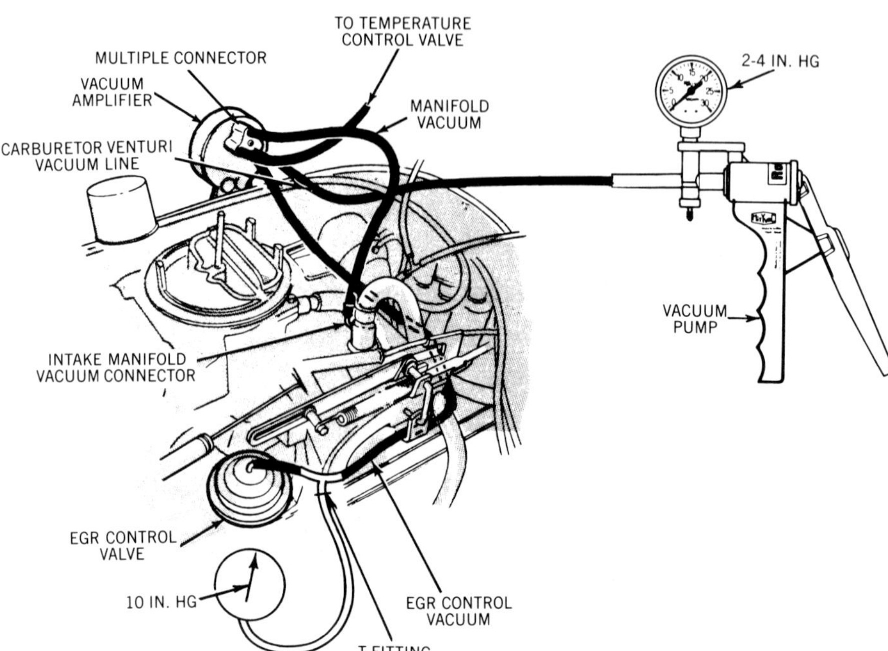

Figure 19-41. Use a vacuum pump to simulate signal vacuum and check amplifier operation. (Chrysler)

venturi vacuum is not strong enough to move an EGR valve diaphragm. These systems use a vacuum reservoir and amplifier to control the EGR valve. The reservoir holds manifold vacuum, which is used to operate the valve. Vacuum does not reach the EGR valve, however, until a vacuum signal reaches an amplifying valve at the reservoir. A low vacuum signal opens a large diaphragm valve with a light spring that allows stronger manifold vacuum to reach the EGR valve, figure 19-40. The signal vacuum can come from a carburetor venturi port or it can be a combination of venturi vacuum and ported vacuum.

Test procedures for vacuum amplifier systems are based on these principles:

1. Connect a vacuum gauge to the EGR vacuum hose with the engine at idle. There should be no vacuum.

2. Using a hand-operated vacuum pump, apply a specified vacuum (usually 2 to 4 in. Hg) to the venturi signal port of the vacuum amplifier, figure 19-41.

3. Read the gauge at the EGR valve. Now there should be 10 in. Hg or more of vacuum at the EGR valve hose.

4. If you find vacuum at the EGR valve with no signal vacuum at the amplifier, or no vacuum at the valve with signal vacuum applied, you must trace and test the manifold vacuum supply to the reservoir and the signal vacuum to the amplifier.

EGR Valve Tests. If the vacuum system and the vacuum controls for the EGR valve are working properly, a vehicle may still have problems with the EGR valve itself. The vacuum diaphragm may leak and not open the valve correctly, or the valve may be stuck or leaking and allow exhaust recirculation when it should not.

You can test a simple EGR valve *without a backpressure transducer* by applying about 10 in. Hg of vacuum with a hand-operated pump, figure 19-38. If the valve stem is visible, watch it as you apply vacuum. It should move as the valve opens. Hold vacuum on the valve for about 30 seconds. The valve should remain open with no vacuum decrease. If the valve does not pass these tests, the diaphragm is leaking. Replace the valve.

You *cannot* test an EGR valve with a backpressure transducer by applying vacuum in this way. A good valve will not open and hold vacuum because without exhaust backpressure, the transducer will not close the diaphragm vacuum bleeds. Test a transducer-controlled valve according to the carmaker's instructions or remove it from the engine and apply vacuum to the diaphragm while applying about 20-psi air pressure to the transducer to close the vacuum bleeds.

Testing Computer-Controlled EGR Systems and Parts

Previous references to testing backpressure-modulated EGR valves referred to positive-backpressure valves because these are the more common of the two kinds. The following paragraphs outline some quick tests for both positive- and negative-backpressure EGR valves used on computer-controlled engines. This section also provides some quick tests for computer-controlled solenoids and solenoid-type EGR valves. Most examples relate to GM and Ford systems, but the principles apply to all EGR systems that are part of an electronic engine control system.

Testing Positive- and Negative-Backpressure EGR Valves. You can identify GM positive- and negative-backpressure EGR valves made since 1984 by the letter N or P as part of the identification number stamped in the top of the EGR valve. If no letter is present, the valve is a simple ported vacuum valve with no transducer. You can identify Ford valves by the diameter of the vacuum diaphragm. Figure 19-42 shows typical GM and Ford EGR valve identification.

You can test either a positive- or a negative-backpressure EGR valve off the engine by using a vacuum pump that can provide at least 7 in. Hg of vacuum. To test a positive-backpressure valve, attach a hose over the exhaust gas inlet and apply air pressure by blowing into the hose, figure 19-43. Then apply vacuum to the vacuum nipple and watch for valve stem movement.

To test a GM negative-backpressure valve, apply the vacuum to the vacuum nipple on top of the valve. Check to see if the valve holds vacuum and that the valve pintle moves when vacuum is applied, figure 19-43.

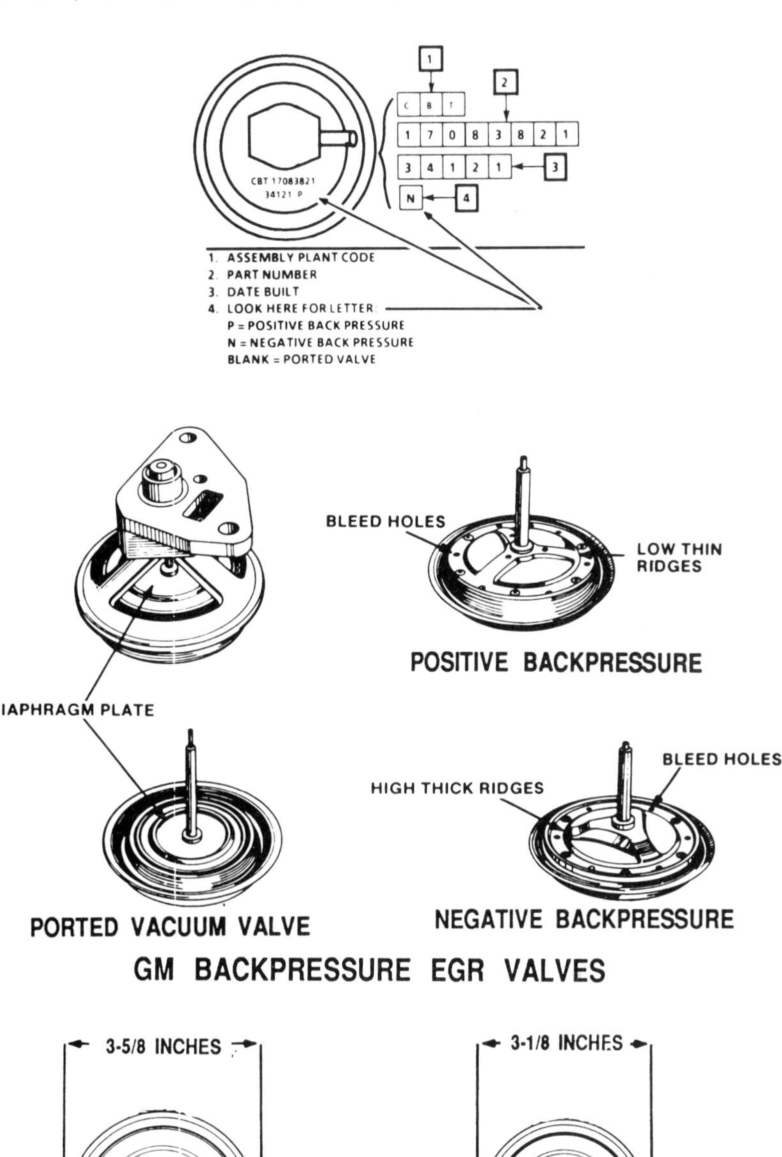

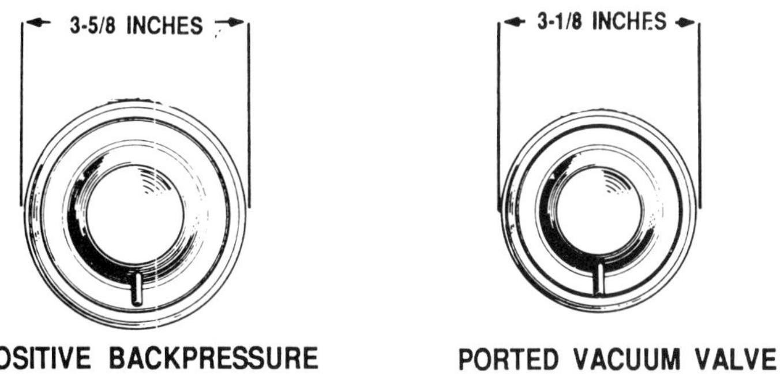

Figure 19-42. General Motors and Ford EGR valve identification.

Testing GM Computer-Controlled EGR Systems. The most common type of system uses a normally deenergized (open) solenoid to control the vacuum source. The computer energizes the solenoid when the engine is cold or in open loop. This *blocks* the vacuum to the EGR valve to prevent poor driveability when cold. If the circuit to the solenoid fails (goes open), this system allows vacuum to the EGR valve as a default condition.

If the EGR valve is opening when the engine is cold, check for an open ground wire to the computer or for lack of battery voltage at the solenoid. If the valve never opens, check for a shorted ground between the solenoid and the computer.

The second common type of GM EGR vacuum control uses a pulse-width-modulated (pwm) solenoid. The computer controls the duty cycle of this vacuum solenoid. The vacuum source may be ported or manifold vacuum. This system also may use a thermal vacuum switch (TVS) to control vacuum to the solenoid. The computer pulses the ground of the solenoid when the engine is running in closed loop.

You can easily check this computer output with a test light across the solenoid connector. The light should pulse as the computer cycles the ground circuit.

Figure 19-44 shows a simple circuit diagram for a GM computer-controlled EGR solenoid. The diagram by itself does not tell you whether it is a simple on-off solenoid or a pwm solenoid. This is another example of why you must know the system operating principles as well as the components used in the system.

The EGR, canister purge, and early fuel evaporation (EFE) solenoids may

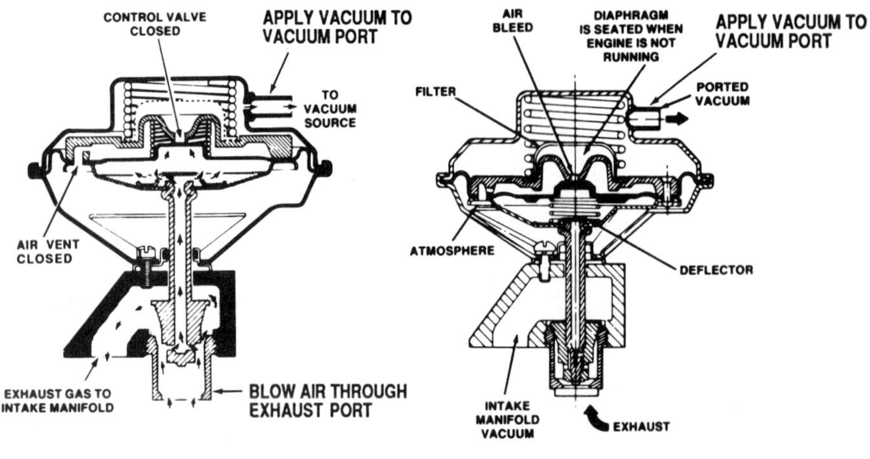

Figure 19-43. Test a GM backpressure EGR valve with vacuum or a combination of vacuum and air pressure as shown here.

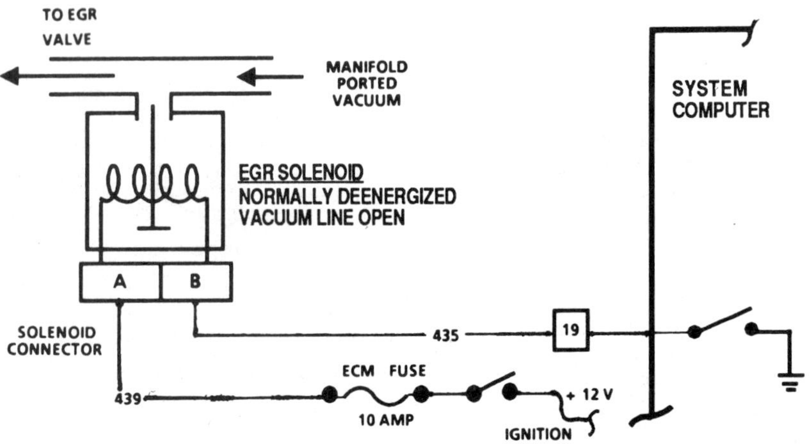

Figure 19-44. Typical circuit diagram for a computer-controlled EGR vacuum solenoid.

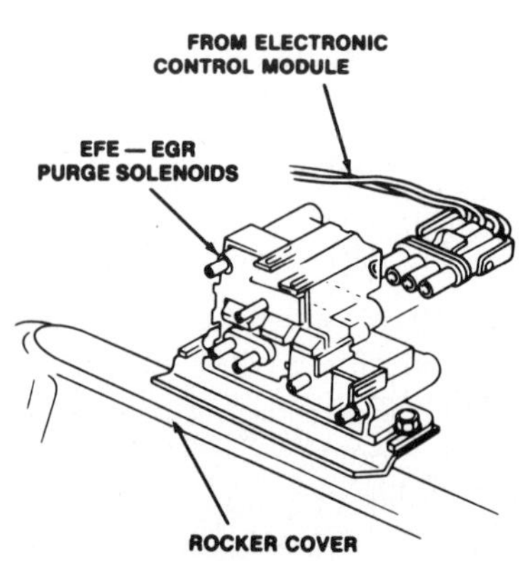

Figure 19-45. These EGR, EFE, and canister purge solenoids are mounted on a common bracket and share a common electrical connector.

be mounted in an assembly, figure 19-45. All of the solenoids receive battery positive voltage on one wire. The computer controls each solenoid ground and closes the circuit to energize each solenoid when required.

Feedback Diagnostic Controls. A feedback diagnostic switch is used on some GM engines with a pwm solenoid, figure 19-46. When the computer pulses the ground on the solenoid, it allows vacuum to the EGR valve and to the diagnostic switch. The computer monitors the signal from the switch and looks for a low voltage when vacuum is applied. If the computer receives 12 volts when EGR vacuum is commanded, it will set a trouble code.

To check the diagnostic switch, connect an ohmmeter to pins C and D at the 4-wire connector on the switch and apply vacuum to the diagnostic switch port, figure 19-46. The ohmmeter should read no continuity without vacuum and continuity with vacuum. You also can connect a test light or a digital volt-ohmmeter (DVOM) across pins C and D. The light should be on or the meter should read 12 volts with the ignition on and no vacuum applied to the switch. When vacuum is applied, the light should go out or the meter should go low.

Testing the GM Temperature Diagnostic Switch. Some Chevrolet-built 5.0-liter and 5.7-liter V-8 engines and 4.3-liter V-6 engines use a pwm solenoid and a temperature diagnostic switch, figure 19-47. The computer sends a 12-volt signal to the temperature switch. When the computer allows vacuum to the EGR valve by pulsing the vacuum control solenoid, the valve opens and the temperature switch closes. This pulls the 12-volt reference signal down near 0 volt.

The computer looks for low voltage from the temperature switch when EGR is commanded on. If the computer receives 12 volts when the EGR should be flowing, it will set a code. If circuit 935 in figure 19-47 is disconnected or open, the computer will never see the 12 volts go low and it will set a code. If the same circuit is grounded, the circuit would always be low and this also would set a code.

To check this switch, disconnect its connector and connect an ohmmeter between the switch terminal and ground. Then operate the EGR valve with a hand-operated vacuum pump while the engine is running. When the valve opens and EGR is flowing, the switch should show continuity on the ohmmeter. When EGR is off, the switch should be open.

Testing the Integrated Electronic EGR Valve. The integrated electronic EGR valve has a pwm vacuum solenoid built into the valve assembly, which controls a vacuum

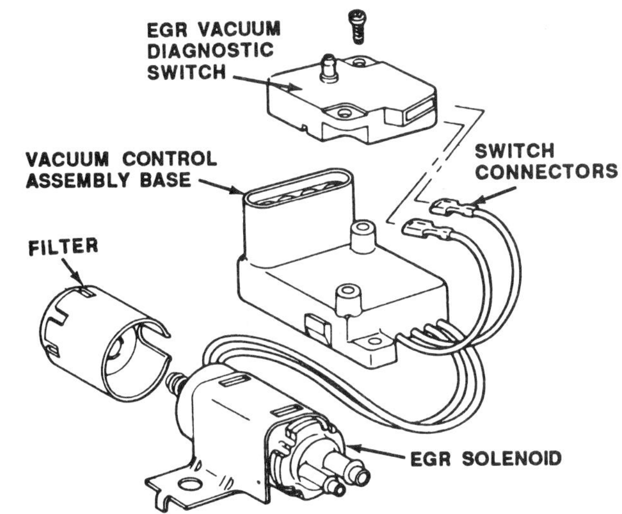

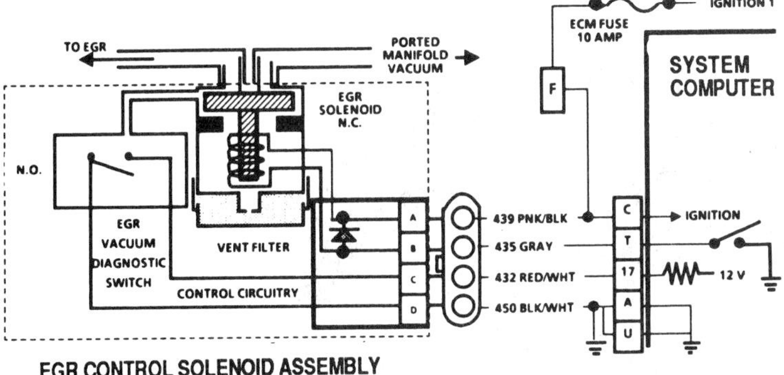

Figure 19-46. This GM EGR control assembly contains a pulse-width-modulated vent solenoid and a vacuum diagnostic switch to send a feedback signal to the computer.

bleed. A 0-percent duty cycle for this solenoid equals complete vacuum bleed. The solenoid valve is always open. A 100-percent duty cycle equals full vacuum applied to the EGR valve. The solenoid valve is always closed.

The feedback device is a potentiometer mounted on top of and inside this valve assembly, figure 19-48. The potentiometer operates on a 5-volt reference that is modified as the EGR pintle moves up and down in response to the vacuum control signal. The computer tracks the operation of the EGR valve by monitoring the pintle position signal. When the computer sends a vacuum control signal to the EGR valve, it expects to see a response on circuit 911 in figure 19-48. If no response is received, the computer will set a trouble code.

Testing the Digital EGR Valve. The digital EGR valve uses two or three separate solenoid-operated valves to control EGR, figure 19-49. The solenoids have a common voltage supply, and the grounds are controlled by the computer. Exhaust flows through two or three orifices of different sizes. The computer operates each solenoid independently or in combinations to produce different flow combinations.

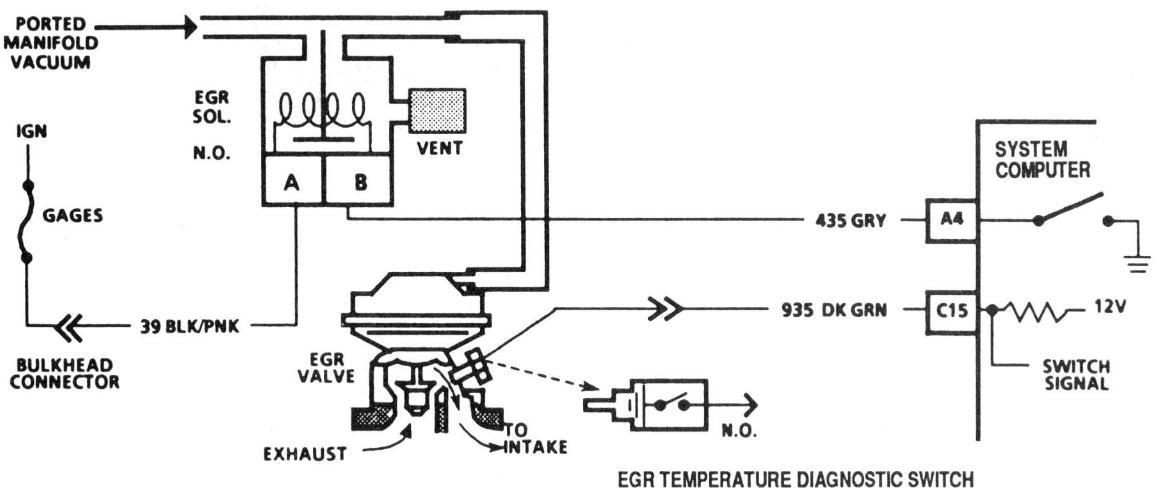

Figure 19-47. Some GM EGR valves on V-8 engines have this temperature diagnostic switch to sense exhaust flow.

The computer monitors the operation of the digital EGR valve while the vehicle is decelerating. As the computer operates the solenoids, it looks for a calibrated increase in manifold pressure. This indicates that the valve is working. If the computer sees no change in the MAP sensor voltage during this test, it will set a trouble code.

You can test the digital EGR valve by grounding the solenoids and monitoring MAP voltage as the computer does during self-test. You also can watch the rpm drop as the solenoids are operated. Be sure the engine is fully warmed up and the idle air control (IAC) motor is disconnected to prevent the computer from altering the idle during the test.

Identify the solenoid feed wire first to avoid blowing the 10-ampere fuse during testing. Ground solenoid number 1 first; the engine rpm should drop and then return to normal when the ground is removed. Repeat this step on solenoid number 2; the engine rpm should drop again, and the engine should run a little rougher than when testing solenoid 1. Ground solenoid number 3. This opens the largest orifice and may stall the engine. It also may set a trouble code for the quad driver circuit or for the EGR system.

While performing these tests, note any solenoid that does not alter the way the engine runs when it is energized. Check the continuity of that solenoid with an ohmmeter. If solenoid resistance is within specifications, the orifice may be leaking, which explains why there was little or no change in rpm during the test.

Other EGR Controls. Other EGR controls include the park-neutral switch and the vehicle speed sensor.

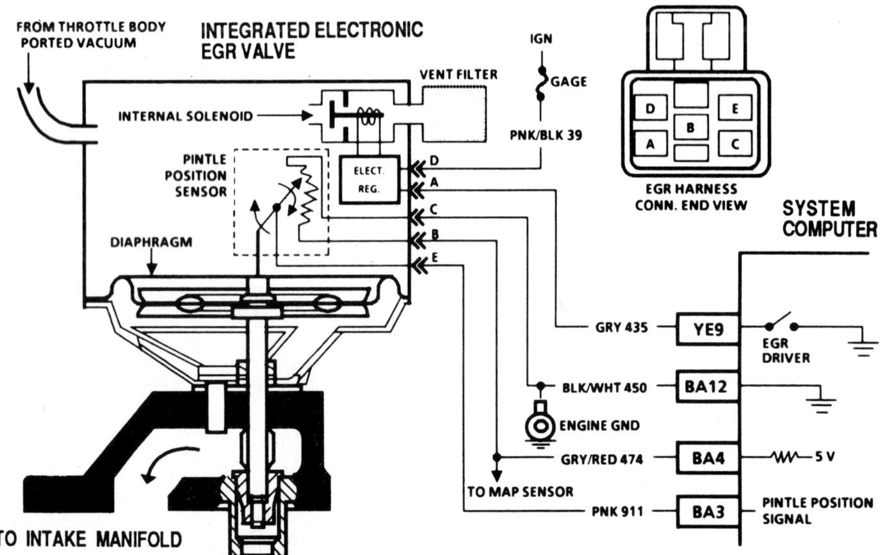

Figure 19-48. The General Motors integrated electronic EGR valve contains a pulse-width-modulated vent solenoid and a potentiometer used as a pintle position sensor.

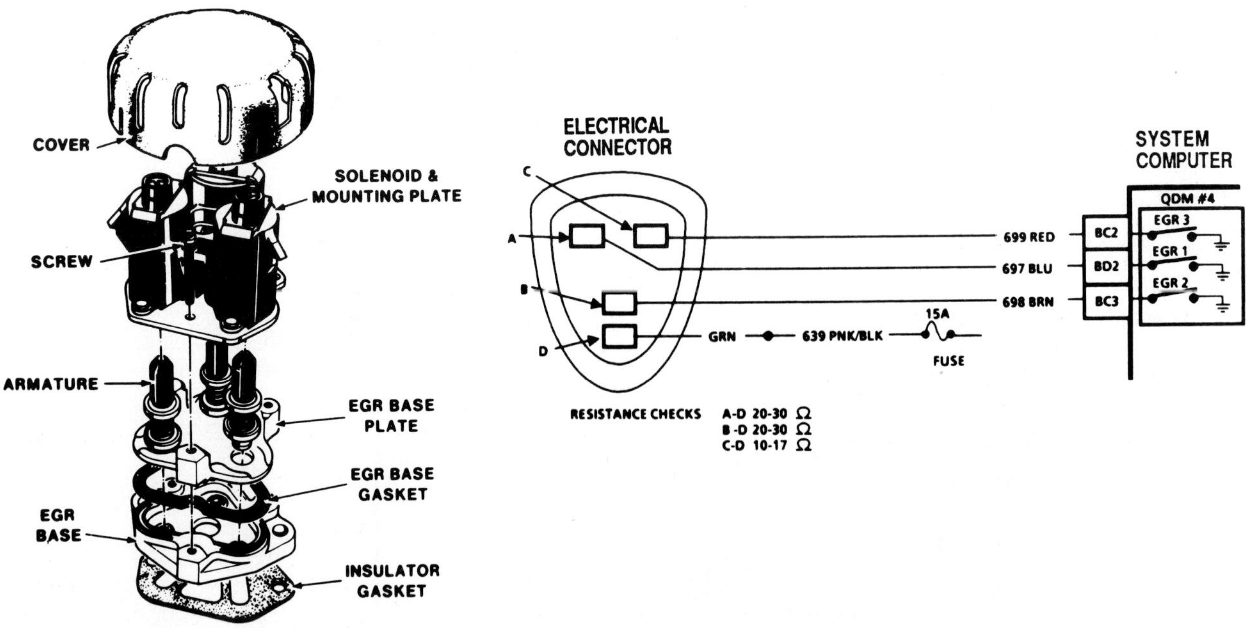

Figure 19-49. The General Motors digital EGR valve has two or three solenoids that open different sized orifices for EGR flow. The solenoids receive battery voltage on a common wire but are switched individually by the system computer.

Some computers will not apply an EGR signal to a vacuum solenoid unless the car is in gear or moving above a minimum speed.

EGR also may be blocked at wide-open throttle for maximum power. Some vehicles turn off EGR when the torque converter clutch is engaged. A brake switch cutout relay is another common control. You must identify a system visually and electrically using a wiring diagram before troubleshooting EGR problems.

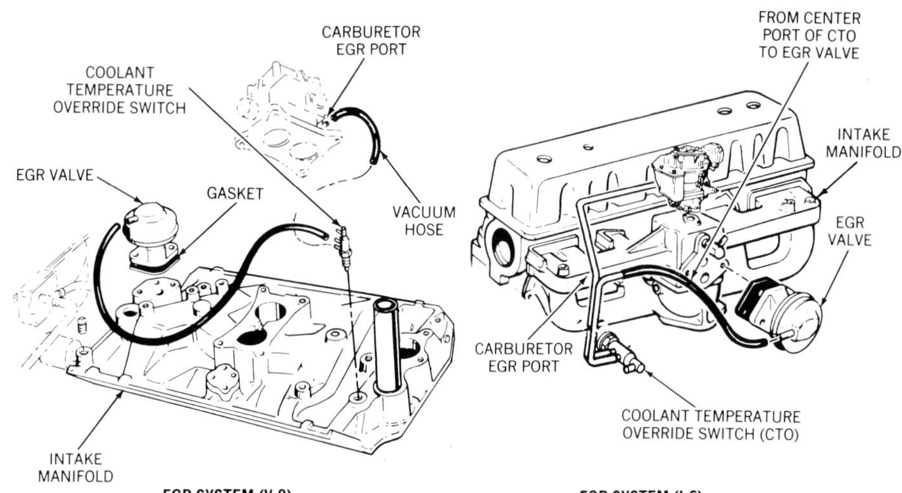

Figure 19-50. Most EGR valves are mounted on, or close to, the intake manifold. (AMC)

Valve Service

An EGR valve and its vacuum system can pass all operating tests, but a vehicle can still have driveability problems if the valve is stuck or leaking. Often a test on the engine will confirm that a valve is bad. Sometimes, however, you must remove a valve and inspect it to confirm that it is bad.

EGR Valve Removal and Replacement. Most EGR valves are mounted on or near the intake manifold, close to an exhaust crossover passage, figure 19-50. Follow these general steps to remove and replace a valve:

1. On many engines, remove the air cleaner for access to the EGR valve. Tag any vacuum lines that you disconnect for correct reinstallation.

2. Disconnect and plug the EGR valve vacuum line.

3. Remove the capscrews (usually two) that secure the valve.

4. If the valve has a remote transducer or exhaust tubing leading to or from the manifolds, remove it carefully with the valve.

5. Remove the valve from the engine.

6. When installing a valve, apply antiseize compound to the mounting capscrew threads. Use a new valve gasket.

7. If the valve has connecting exhaust tubing, be sure that it is fully seated and correctly installed when you install the valve.

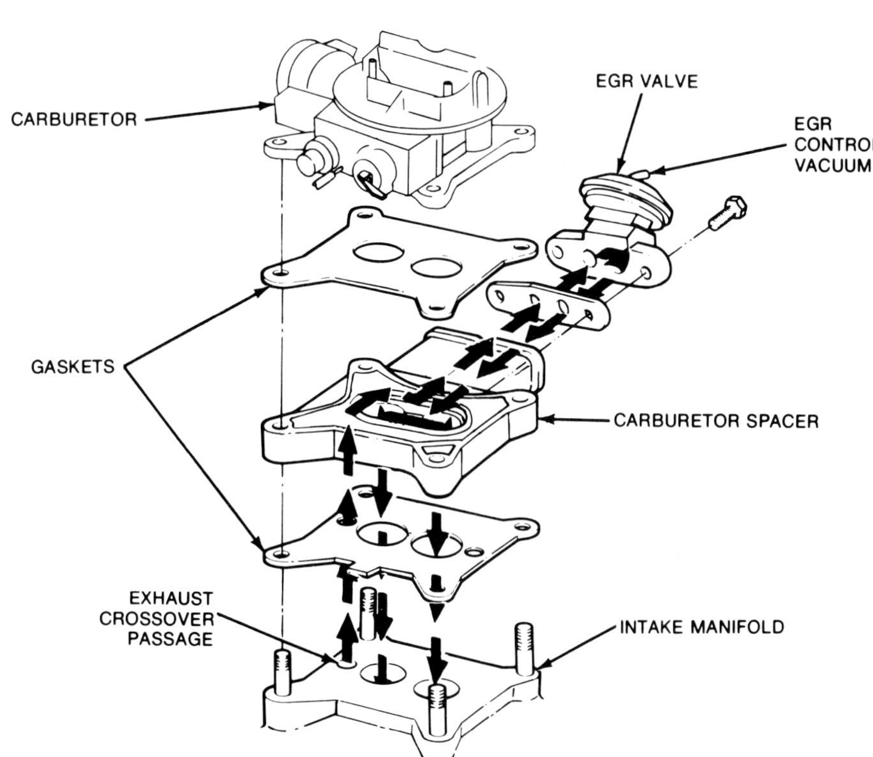

Figure 19-51. Most Ford V-8 engines have EGR valves mounted on a spacer between the carburetor and the manifold. (Ford)

Many Ford engines recirculate exhaust through a spacer plate under the carburetor or injection throttle body, figure 19-51. Leaks and restrictions can develop in the spacer while the valve is in good condition. If such an engine has symptoms of EGR leakage, remove the carburetor and spacer plate for inspection and possible replacement, figure 19-52.

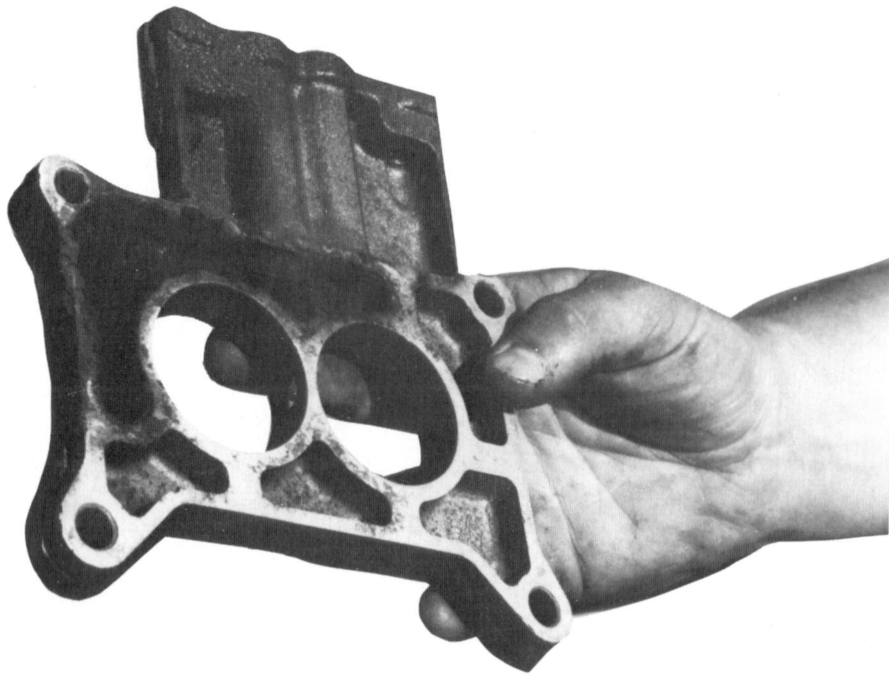

Figure 19-52. Inspect the EGR spacer for corrosion damage like this.

$CO + CO_2$ Percentage		Approximate Air-Fuel Ratio
13.5	↑	16.0:1
14.0	LEAN	15.5:1
14.5	↑	15.0:1
14.7		**14.7:1**
15.0		14.2:1
15.5		13.7:1
16.0	↓	13.0:1
16.5	RICH	12.5:1
17.0	↓	11.7:1

Figure 19-53. CO plus CO_2 percentages indicate approximate air-fuel ratios for testing converter-equipped cars.

	Emission Levels			
Condition	HC	CO	CO_2	O_2
Normal	Within Specifications		13-15%	1.0-2.0%
HC Problem	High	High or Low	Below 10%	Above 2%
CO Problem	High	High	Below 10%	Below 1.0%
HC & CO Problem	High	High or Low	Below 10%	Above 2.0% or Below 1.0%

Figure 19-54. Four-gas relationships for exhaust analysis of converter-equipped cars.

EGR Valve Leak Test (off the Engine). With the EGR valve removed from the engine, you can check it for leakage as follows:

1. Inspect for any obvious damage to the pintle.

2. Turn the valve upside down and pour *nonflammable* solvent, or a similar liquid, into the pintle cavity. (Do not use water.) Be careful not to overfill the pintle cavity.

3. Watch for rapid fluid loss past the pintle into the lower cavity. Some leakage may occur after 15 to 20 seconds on a valve that has been in regular service for some time. This does not necessarily indicate a leaky pintle. Rapid fluid loss is a good indication that enough exhaust can leak past the pintle to cause a rough idle or other driveability problems.

4. As an alternate test, place a large vacuum hose over the pintle housing. Apply approximately 15 in. Hg of vacuum to check the pintle seal. A slight, steady vacuum loss is not abnormal for a valve that has been in service. Immediate vacuum loss usually indicates a valve that will leak enough exhaust to cause driveability problems.

EGR Valve Cleaning. It is usually more economical and practical to replace an EGR valve if it is defective in any way. However, exhaust deposits that cause a valve to stick open and leak can be cleaned from many valves. You can clean an EGR valve in one of three ways:

- By scraping deposits from the valve with a small wire brush or scraping tool
- By sandblasting the valve in a spark plug cleaner
- By soaking the exhaust valve portion of the EGR valve in manifold heat control valve solvent or carburetor cleaner to loosen deposits and then removing them with a wire brush.

If you try to clean any EGR valve, observe these two precautions:

1. Do not let any solvent contact the diaphragm part of any valve. Solvent can damage the diaphragm and cause it to leak.

2. Do not sandblast any valve with a built-in backpressure transducer. Abrasive material in the transducer passage will block the transducer and vacuum bleeds, upsetting valve operation.

CATALYTIC CONVERTER SERVICE

Catalytic converters have no moving parts and require no periodic service.

Federal law requires that converters work effectively for at least 50,000 miles (80,000 km) on a properly maintained and operating engine. A converter on even the best maintained engine can wear out, however, at some point beyond 50,000 miles when the catalysts eventually lose their chemical effectiveness. Several vehicle problems can cause a converter to lose effectiveness or fail in fewer than 50,000 miles. Most problems fall into one of these three general categories:

1. Use of leaded gasoline
2. Converter overheating, caused by continually rich air-fuel ratios or engine misfire
3. Physical damage that breaks up or restricts the converter substrate. This can be caused by driving over a curb and hitting the converter or by a collision that damages the exhaust system.

Engine vacuum tests outlined in chapter 8 will indicate an exhaust system restriction. A restriction can be in a pipe, a muffler, or *a damaged converter.* You will have to disassemble the exhaust system to isolate the restriction and replace the defective part.

Using an Infrared Exhaust Analyzer to Test Catalytic Converter Condition

Chapter 8 outlines basic procedures for using an infrared exhaust analyzer. That chapter also has guidelines for interpreting test results at idle and cruising speeds, using a 2-, 3-, or 4-gas analyzer. Before using an infrared analyzer to check converter condition, review the information in chapter 8 on analyzer use.

If you use a 2-gas analyzer, you can check only HC and CO emissions. Converter-equipped cars, however, can reduce HC and CO to almost immeasurable levels at the tailpipe. If a 2-gas analyzer indicates HC and CO levels above the limits allowed for a particular year and kind of vehicle, you may have a converter problem, or you may have a problem elsewhere in the fuel, ignition, or emission control systems. You must then test and service the other combustion control systems before you can isolate the cause of high emissions to a faulty converter.

If you use a 3- or 4-gas analyzer to test a converter-equipped vehicle, CO_2 and O_2 readings can indicate converter condition. On a noncatalyst car, HC indicates an ignition problem or a lean condition. Although ignition problems are the more common cause of high HC emissions, either of these conditions can cause a misfire that results in unburned fuel passing through to the exhaust. On a converter-equipped car, O_2 indicates a lean condition. If O_2 is above 2.0 percent, the air-fuel mixture is probably lean. If O_2 is above 4 percent, the mixture definitely is too lean.

At the stoichiometric 14.7:1 air-fuel ratio, O_2 should be low, but CO_2 should be high. At lean air-fuel ratios, O_2 increases, and CO_2 decreases. Also at a 14.7:1 air-fuel ratio, the total percentage of CO and CO_2 is about 14.7 percent. Figure 19-53 shows the relationship of combined CO and CO_2 percentage to air-fuel ratio. Acceptable combustion in a converter-equipped car will produce the following 4-gas analyzer readings:

- HC and CO—within specifications (low or immeasurable)
- O_2—1.0 to 2.0 percent
- CO_2—above 10 percent (ideally 13 to 15 percent)

If O_2 exceeds CO and CO is above 0.5 percent, the catalytic converter may be defective. If CO is above 0.5 percent and also higher than O_2, the converter is probably okay, but the air-fuel mixture is rich. Figure 19-54 summarizes other relationships between HC, CO, CO_2, and O_2 readings on a 4-gas analyzer.

If exhaust system test and inspection or infrared exhaust analysis indicates a defective converter, you must usually replace it. The pellets in pellet-type AC-Delco converters used on many GM and AMC cars can be removed from the converter housing and replaced. It is more common, however, to replace the entire converter.

Catalytic Converter Replacement

Replacing a converter is about the same as replacing a muffler, a pipe, or any other exhaust system part. Most converters have flange-type or U-bolt connections to the exhaust system. Figure 19-55 shows typical examples. Use these general steps to remove and replace a converter.

Warning. To avoid burns, let the exhaust system and the converter cool before handling.

1. Raise the vehicle on a hoist to convenient height.
2. Remove heat shields above and around the converter.
3. Separate the air injection tube from the converter if so equipped. Some are pressed into converter housings; others are held by bolts or clamps.
4. Unbolt the converter front and rear flange bolts or U-bolts and separate the converter from the inlet and outlet pipes.
5. If necessary for easy installation, position upper heat shields above the converter location before installing the new converter.
6. Locate the new converter on the inlet and outlet pipes and install the flange bolts or U-bolts. Use new gaskets where required.
7. Install the air injection tube, if used, and align the converter with the rest of the exhaust system.
8. Tighten the converter mounting bolts and air injection tube. Install the rest of the heat shields.
9. Lower the vehicle, start the engine, and check for exhaust leaks.

Converter Pellet Replacement

It is possible to replace the catalyst pellets in the first generation of AC pellet-type converters. This requires the following special tools:

- An aspirator to create a vacuum in the converter
- A vibrator to shake the old pellets out and new pellets in.

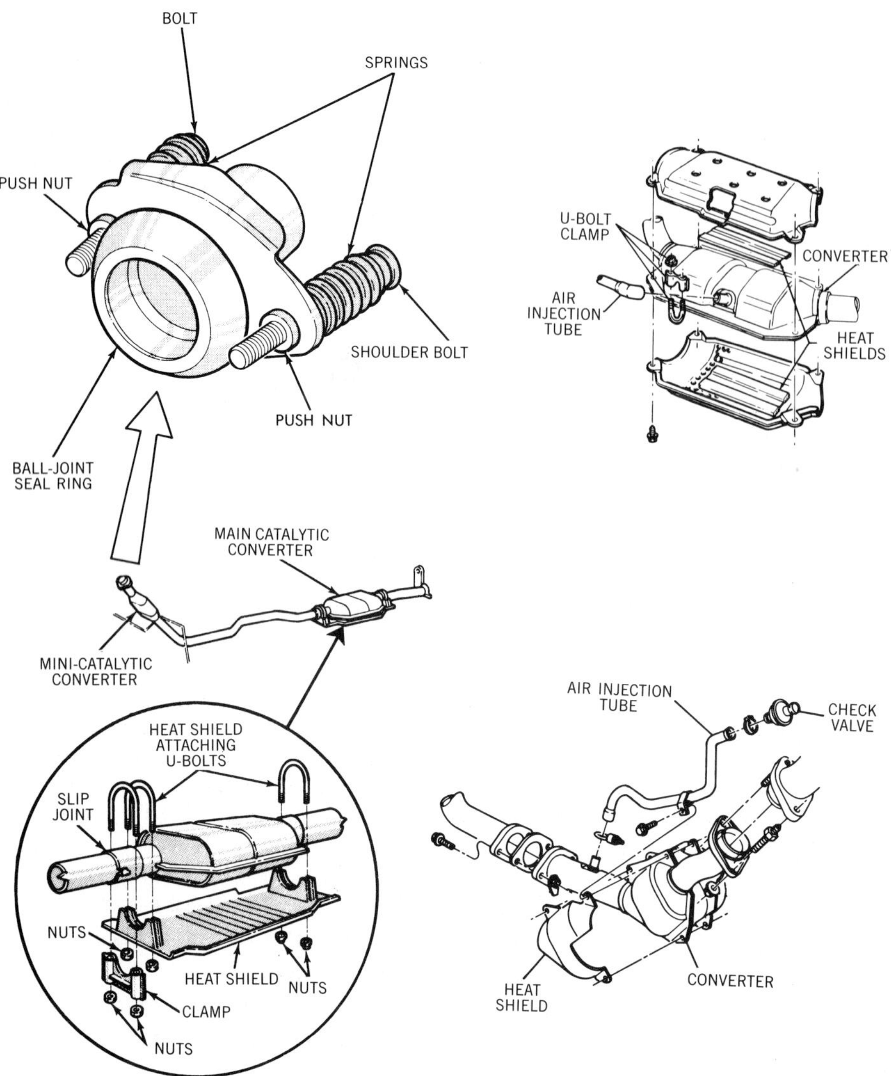

Figure 19-55. Typical catalytic converter installations.

These tools usually are available only in some General Motors dealerships. They are not commonly found in general repair shops. Today, catalyst pellet replacement is not a common service operation. To remove and replace catalyst pellets, refer to the General Motors procedures that accompany the special equipment for the job.

SUMMARY

Air injection, exhaust gas recirculation (EGR), and catalytic converters are interrelated emission controls on late-model cars. You can repair and replace parts of these systems independently, but, accurate troubleshooting requires total system diagnosis.

Air injection was introduced to control HC and CO. Today it supplements the action of the oxidation converter. EGR controls NO_x emissions and works with reduction converters on late-model cars to keep NO_x emissions at low levels. All three systems on a late-model car depend on the electronic control system to maintain a stoichiometric 14.7:1 air-fuel ratio and optimum ignition timing.

As with vacuum-advance control systems, fuel injection, and electronic controls, carmakers have modified and adapted air injection, EGR, and converter systems for different vehicles over the years. Regardless of system variations, however, all work on common principles. Service of any systems consists of:

- System inspection
- System operational (area) test
- Pinpoint tests of individual parts.

Most pinpoint tests are common electrical and vacuum checks of solenoids, relays, valves, and diaphragms.

REVIEW QUESTIONS

Multiple Choice

1. A burned air injection hose upstream from a check valve usually indicates:

a. excessive air pump pressure
b. a leaking check valve
c. a stuck relief valve
d. a broken diverter valve

2. With an engine running at fast idle, you briefly pinch the air delivery hose closed. You hear a light popping sound. This indicates that:

a. the relief valve is opening properly
b. the diverter valve is blocked
c. a check valve is leaking
d. the air pump is leaking internally

3. You hear a knocking sound from an air pump during an air injection system test. Any of the following could cause this EXCEPT:

a. a loose pump-mounting bracket
b. a leaking hose
c. a glazed drive belt
d. a defective antibackfire valve

4. Mechanic A says that you can test a pulse air valve with a large rubber bulb, or syringe. Mechanic B says that you can check for intake air pulses at a pulse air valve with the engine running. Who is right?

a. A only
b. B only
c. both A and B
d. neither A nor B

5. Mechanic A says that an air pump can be rebuilt in the shop. Mechanic B says that you can easily replace an air pump filter fan, relief valve, or outlet tube. Who is right?

a. A only
b. B only
c. both A and B
d. neither A nor B

6. Air valve switching and bypass (diverter) functions can be controlled by:

a. air pressure and atmospheric pressure on opposite sides of a diaphragm
b. computer-controlled vacuum solenoids
c. a simple vacuum diaphragm
d. all of the above

7. Exhaust gas recirculation (EGR) on a gasoline engine should *not* operate under all of the following conditions EXCEPT:

a. cold-engine warmup
b. deceleration
c. cruising speed between 30 and 50 mph
d. wide-open throttle

8. While testing a diesel engine, you find the EGR system operating at idle. This probably means that:

a. the system is operating normally
b. the EGR valve may be leaking
c. a vacuum control solenoid has failed
d. the EGR backpressure transducer has failed

9. An engine idles roughly. Carburetor and ignition adjustments do not cure the problem. You disconnect the EGR valve vacuum line, and the idle smooths out. This probably means that:

a. the EGR valve is leaking
b. a vacuum control solenoid has failed
c. the EGR valve diaphragm has failed
d. the engine has an intake manifold air leak

10. Mechanic A says that you can test an EGR valve with a built-in backpressure transducer by applying manifold vacuum to the valve at idle and watching for valve stem movement. Mechanic B says that you can test an EGR system with a backpressure transducer by disconnecting the vacuum-control solenoid at idle. Who is right?

a. A only
b. B only
c. both A and B
d. neither A nor B

11. An engine with EGR surges at cruising speed. This can be caused by:

a. a defective vacuum amplifier
b. a defective vacuum-bleed solenoid
c. a defective backpressure transducer
d. any of the above

12. Mechanic A says that an EGR vacuum amplifier should apply about 10 in. Hg of vacuum to the EGR valve with 2 to 4 in. Hg of vacuum at the amplifier signal port. Mechanic B says that if the signal vacuum is removed (a broken hose, for example), the amplifier applies manifold vacuum to the EGR valve continuously. Who is right?

a. A only
b. B only
c. both A and B
d. neither A nor B

13. When testing a converter-equipped car with a 4-gas exhaust analyzer, the O_2 reading is above 4%. This means that:

a. ignition timing is retarded
b. EGR is not operating
c. the air-fuel mixture is too lean
d. none of the above

20 TURBOCHARGER SERVICE

INTRODUCTION

Turbochargers have become commonplace since the early 1980's. Almost every major carmaker now builds one or more gasoline or diesel engines with turbochargers. As a professional mechanic, you must be able to troubleshoot turbocharged engines and perform routine maintenance. This chapter covers turbocharger troubleshooting and basic services.

GOALS

After studying this chapter, you should be able to do the following jobs in twice the flat-rate labor time or less:

1. Inspect and test a turbocharger and turbo system operation and interpret results.

2. Using a dial indicator, measure turbo radial and axial bearing clearances.

3. Using a pressure gauge and pump and a dial indicator, test and measure wastegate and actuator operation. Replace a wastegate actuator.

TURBOCHARGER SAFETY

All of the general fuel system safety practices outlined in chapter 1 apply to turbocharged engines. Additionally, observe these special precautions:

1. Turbochargers operate at high speeds (often more than 100,000 rpm) and high temperatures. Proper lubrication is critical. Turbos receive lubrication from the engine oil system, figure 20-1, and turbocharged engines require oil changes more often than naturally aspirated engines. Fix any lubrication problems immediately, and be sure that the oil supply to the turbo is working correctly.

2. After changing the oil on a turbocharged engine, or replacing a turbo, disable the ignition or disconnect the shutoff solenoid on a diesel injection pump. Then crank the engine until the oil pressure lamp goes out or the oil gauge shows steady pressure. Don't crank the engine for more than 15 seconds at a time. Fill the new oil filter with fresh oil before installing it to speed this priming time. You also can disconnect the oil drain line from the turbo to bleed air from the center housing. Use a small container to catch the oil overflow. On some systems, you can disconnect the turbo oil inlet and add a few ounces of oil directly to the center housing with a small funnel. Figure 20-2 shows a typical oil supply and return system for a turbocharger. After the turbo is primed, reconnect the drain line.

3. When starting any turbocharged engine, allow it to idle for about 20 seconds before accelerating it to be sure the turbo is getting a steady oil supply.

4. Never run a turbocharged engine with the air filter removed or the air intake duct disconnected from the turbo. The smallest dirt particles drawn into a turbo can damage close-tolerance compressor parts.

5. When servicing a turbo, be careful not to nick or bend the compressor or turbine blades. Slight damage to the blades can cause rotating imbalance and damage the bearings or compressor and turbine wheels. Always spin the compressor and turbine by hand before reconnecting an inlet or outlet duct to ensure that the rotating assembly does not bind and that there are no objects in the housings.

6. An engine decelerates rapidly when the throttle closes (or the pump returns to idle on a diesel). A turbocharger does not. The turbo continues to turn at high speed for up to 1 or 2 minutes after the engine returns to idle. Therefore, do not turn off a turbocharged engine immediately after it returns to idle. If you do, you shut off the oil supply to the turbo. Let the engine idle for 1 or 2 minutes to allow the turbo to slow down before stopping the engine.

CAUSES OF TURBOCHARGER FAILURE

Most causes of turbocharger failure can be summarized in one word—*dirt*—dirt in the intake or exhaust system or dirty oil. Of course you must look for specific kinds and causes of contamination to identify and fix a specific problem. We can group these into four major categories:

1. *Lack of lubrication or oil lag* due to—

- **a.** Restricted oil supply to the turbo
- **b.** High-viscosity oil in cold weather
- **c.** Sludge formation in the oil
- **d.** Low engine oil pressure due to wear

2. *Foreign material (dirt or contamination) in the oil system,* due to—

- **a.** Sludge formation caused by using oil of the wrong service classi-

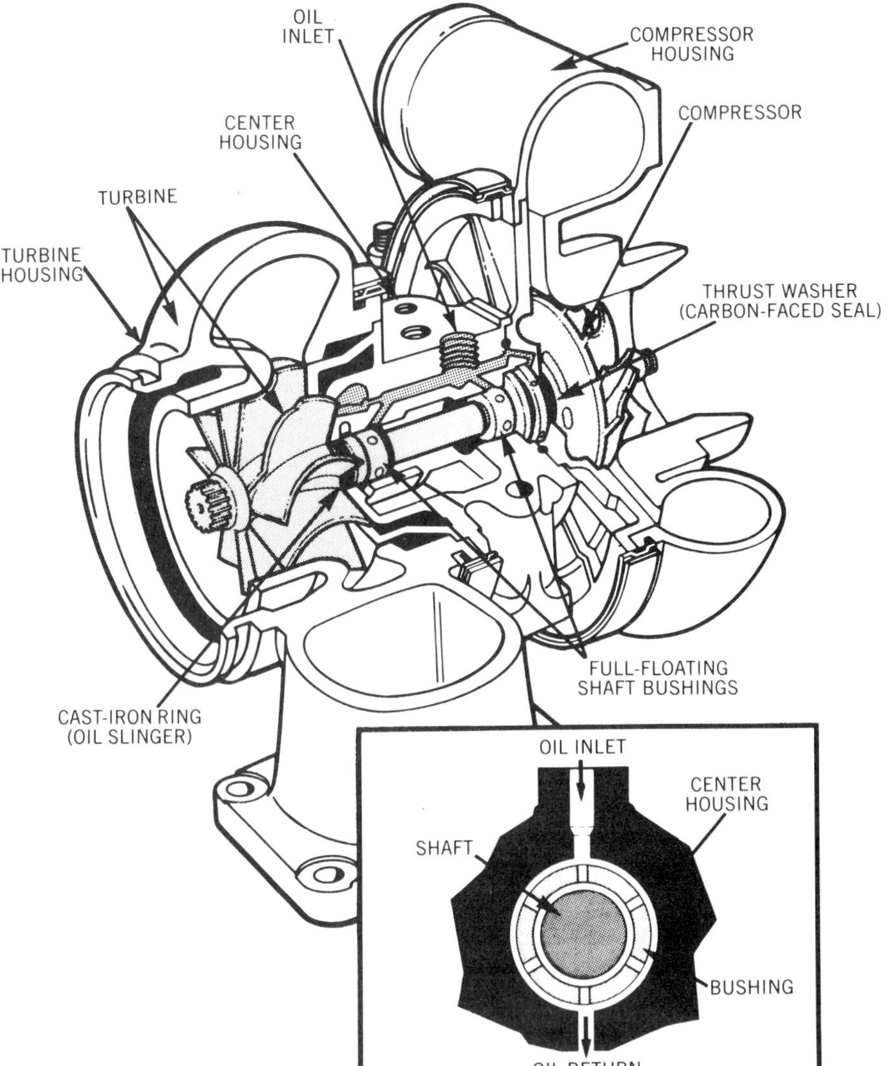

Figure 20-1. The engine lubrication system circulates oil to the turbo center housing for the bearings.

fication or viscosity or infrequent oil changes

b. Dirt and acid accumulation caused by not changing the oil often enough

c. Foreign material caused by engine bearing failure or other mechanical problems

d. An oil filter bypass valve stuck open or a ruptured filter

3. *Oil oxidation, or breakdown.* This is indicated by sludge in the oil and is caused by using oil of the wrong viscosity and classification or by not changing the oil often enough. Oil oxidation is very critical inside the turbocharger because the oil volume in the center housing is small and the turbine runs at very high temperatures. The spinning turbo shaft throws oil against the hot walls of the center housing where sludge can stick. Sludge can build up until it blocks lubrication from the bearings and blocks the oil drain line. Oil can then leak past the turbine and compressor seals. Oxidized oil also can form hard coke deposits that cause bearing and shaft wear. Besides the oil problems listed above, oxidation and sludge formation can be caused by:

a. Engine overheating

b. Coolant leaking into the oil

c. Excessive blowby in the crankcase.

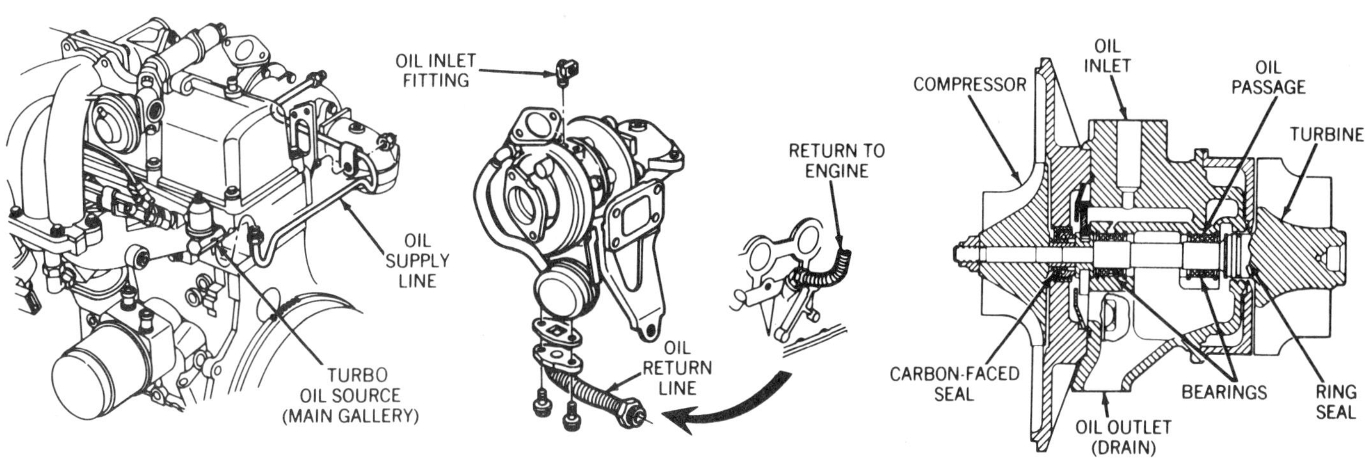

Figure 20-2. Typical turbocharger oiling system. You can prime the turbo by filling it through the inlet. Disconnect the outlet line to release trapped air. (Ford)

LACK OF POWER	
Possible Cause	***Correction***
Carburetor or fuel injection problem	Test and service—chapters 17, 18, and 21
Fuel delivery problem	Test and service—chapter 16
EGR problem	Test and service—chapter 19
Wrong ignition timing or advance or other ignition problem	Test and service—chapters 12 through 15
Restricted air filter	Replace air filter—chapter 16
Intake air leak or exhaust leak	Check turbocharger installation for air or exhaust leaks
Lack of compression or other engine wear	Test and service—chapter 5
Wrong valve timing or adjustment	Test and service—chapter 5
Exhaust system restriction	Test and service—chapter 7; Check manifold heat control valve for restriction
Turbocharger damage	Inspect the turbocharger for worn or bent turbine and compressor blades and lack of lubrication
Low boost pressure—wastegate leaking or opening at low pressure or adjusted wrong	Test boost pressure and wastegate operation; adjust or repair as necessary

Figure 20-3. If a turbocharged engine lacks power, check other engine systems, as well as the turbo.

EXHAUST SMOKE	
Possible Cause	***Correction***
Black Smoke (rich mixture)	
Restricted air filter	Replace air filter—chapter 16
Carburetor or fuel injection problem	Test and service—chapters 16, 18, and 21
Blue Smoke (oil)	
Excessive blowby or PCV problem	Test for engine wear—chapter 5; service PCV system—chapter 6
Worn engine (rings, cylinders, valves, and valve guides)	Test for engine wear and low compression—chapter 5
Leaking oil seals in the turbocharger	Check for oil leakage at the compressor and turbine ends of the turbo shaft
Restricted turbo oil return, overfull center housing	Check the oil return line for restrictions; blow out with the compressed air; check for sludge in turbo center housing; clean as required
White Smoke (coolant leakage into combustion chambers)	Check the engine for coolant leakage into combustion chambers—chapter 7

Figure 20-4. Check these possible causes for smoking problems on a turbocharged engine.

4. *Foreign material (dirt) in the air intake or exhaust systems,* due to—

a. Sand, debris, or dirt drawn into the compressor inlet because of a leaking air duct or missing air filter

b. Hard carbon particles or debris in the exhaust system that enters the turbine. A few engines have catalytic converters between the exhaust manifold and the turbine. If the converter is damaged, catalyst particles can destroy the turbine.

TURBOCHARGER TROUBLESHOOTING

A turbocharger does not basically change the characteristics of an engine. It only supplies an increased air volume for improved combustion. Before condemning a turbo, be sure the fuel, ignition, lubrication, cooling, and emission systems are working properly. Replacing a turbo will not fix problems in other engine systems.

Turbocharger Performance Complaints

Like any other engine diagnosis, turbocharger troubleshooting begins with evaluating the driver's complaint or the vehicle symptoms. Figures 20-3 through 20-6 list common performance complaints for turbocharged engines and their possible causes. Notice that many of the problems are the same as you might find on a naturally aspirated engine and are caused by systems other than the turbo.

The following sections contain procedures for troubleshooting a turbocharger when problems appear to be caused by the turbo system.

Listen to Turbocharger Operation and Check for Leakage

Before starting the engine, check the air cleaner, the filter, and all intake air ducts for loose connections and obvious leakage. Also inspect these areas for restrictions and contamination.

Check hoses for cracks, looseness, and leakage.

Inspect the exhaust system for leakage between the engine and the turbo. Look at gasket areas for burned spots that indicate exhaust leakage. Check fasteners and other parts for looseness. Inspect manifolds, exhaust pipes, and tubing for cracks.

Start the engine and listen to the turbo. As you work on turbocharged engines, you will learn to recognize the sounds of normal and abnormal turbo operation. A normal turbo makes a light, even whistling sound that increases steadily in frequency as the engine accelerates. The following sounds indicate problems:

- *Uneven noise and vibration* indicate possible shaft damage or damage to the compressor or turbine blades.
- *Grinding or rubbing sounds* indicate shaft or bearing damage or that the compressor or turbine is rubbing the housing.
- *Rattling from the wastegate area* indicates a loose exhaust pipe or outlet elbow or a damaged wastegate.
- *An uneven sound that goes up and down in pitch* indicates a restricted air intake—a clogged filter, a bent air duct, or dirt on the compressor blades or housing.
- *A higher than normal pitch* indicates an intake air leak.
- *A louder than normal sound, accompanied by hissing,* indicates an exhaust leak.
- *Sudden reduction in noise, accompanied by smoke and oil leakage,* indicates turbo failure.

Check for intake air leaks with the engine running by spraying a soap-and-water solution or a nonflammable solvent on suspected leakage areas. Engine speed will change if there is a leak because airflow will draw the liquid into the leak.

Check for exhaust leaks by applying lightweight motor oil or soapsuds to manifolds, pipes, and gasket areas. Look for bubbles that indicate leakage. You also can use a mechanic's stethoscope to listen for exhaust leakage and pinpoint the source.

SPARK KNOCK OR PINGING

Possible Cause	*Correction*
EGR problem	Test and service—chapter 19
Wrong ignition timing or advance or other ignition problem	Test and service—chapters 12 through 15
Restricted air filter	Replace air filter—chapter 16
Engine overheating	Test and service cooling system—chapter 7; check manifold heat control valve and test for exhaust restrictions
Cylinder deposits causing higher compression or hot spots	Test compression—chapter 5
Low-octane or high-cetane fuel	Use higher octane gasoline or lower cetane diesel fuel
Engine detonation sensor not working properly	Test and service detonation sensor according to the carmaker's instructions
Wastegate not opening properly (overboost)	Test and adjust or replace the turbocharger wastegate; check the wastegate hoses

Figure 20-5. Spark knock or pinging (detonation) can be caused by turbocharger problems and problems in other engine systems, as well.

HIGH OIL CONSUMPTION

Possible Cause	*Correction*
Engine oil leakage	Inspect the engine for oil leaks and repair as necessary
Excessive blowby or PCV problem	Test for engine wear—chapter 5; service PCV system—chapter 6
Worn engine (rings, cylinders, valves, and valve guides)	Test for engine wear and low compression—chapter 5
Leaking oil seals in the turbocharger	Check for oil leakage at the compressor and turbine ends of the turbo shaft
Restricted turbo oil return, overfull center housing causing seal leakage	Check the oil return line for restrictions; blow out with compressed air; check for sludge in turbo center housing; clean as required

Figure 20-6. Check these possible causes for oil consumption complaints on a turbocharged engine.

Figure 20-7. These turbine and compressor blades were broken by foreign objects in the exhaust and intake airflow. (Garrett)

Figure 20-8. The worn blade tips show that the turbine and compressor have been rubbing the housing. (Garrett)

Figure 20-9. Rotate the shaft and wheel assembly. Listen and feel for binding and rubbing. (Garrett)

Turbo Assembly Inspection

After inspecting intake and exhaust connections and listening to the engine and turbo run, disconnect the air inlet and exhaust outlet from the turbo housing. Using a flashlight, look inside the turbine and compressor housing for damage to the blades. Broken or bent blades, figure 20-7, indicate damage from foreign objects. Worn spots on blade tips and wear marks on the housings indicate that the blades have been rubbing the housing, figure 20-8. This means excessive bearing or shaft wear.

Rotate the shaft and wheel assembly by hand and listen and feel for rubbing or binding, figure 20-9. Push the shaft and wheels from side to side and continue rotating the assembly. It should not bind or rub in any position when you push on the shaft. Now, move both ends of the shaft up and down at the same time and feel the bearing journal radial clearance. You should feel very little movement, and the wheels should not touch the housing as you move the rotating assembly.

Inspect the shaft seal areas in the housings for oil leakage, figure 20-10. Leakage occurs more often at the exhaust turbine seal than at the compressor seal. Compressor pressure tends to force oil back into the center housing.

If the shaft and wheels rotate freely and you find no signs of the wheels rubbing the housing or oil leakage, the turbocharger is probably all right. You can check actual bearing clearances with a dial indicator as explained in the next section.

Checking Turbocharger Bearing Clearance

Turbocharger bearings are full-floating bushings. They operate with lubrication between the shaft and the bearings and between the bearings and their bores. Every turbo also has a thrust bearing, or thrust washer, at one end of the shaft to control endplay. Bearing clearances are very close to maintain the necessary close clearances of the turbine and compressor wheels in their housings. You can check the radial (up and down) and axial (end to end) clearances with a dial indicator and various adapters. Several tool companies sell dial indicator kits specifically for checking turbochargers.

To check bearing clearances, you must first look up the manufacturer's specifications. They vary slightly, but most turbos have 0.003 to 0.006 inch (0.076 to 0.152 mm) radial clearance and 0.001 to 0.003 inch (0.025 to 0.076 mm) axial clearance.

You can check bearing clearances on some turbos installed on the engine if you have access to the end housings and the oil drain line. Most carmakers, however, require that you remove the turbo from the engine. Follow the manufacturer's procedures to do this. Whether you check bearing clearances on or off the engine, prepare for the measurements as follows:

1. Remove the intake and exhaust tubes and ducts from the turbocharger housing.

2. Disconnect the pin or snapring that holds the wastegate actuator rod to the wastegate link. On some units, unbolt and remove the wastegate from the compressor housing.

3. Remove the exhaust outlet elbow, or housing, from the turbine housing for access to the shaft end to check axial clearance. The elbow is attached with 5 or 6 capscrews.

4. Remove the oil drain line and its fitting from the center housing for access to the shaft to check radial clearance.

CHECK FOR OIL AT COMPRESSOR OUTLET

CHECK FOR OIL AT TURBINE OUTLET

CHECK FOR OIL AT COMPRESSOR INLET

CHECK FOR OIL AT TURBINE INLET

Figure 20-10. Inspect for oil at these locations on the turbine and compressor housings. (Ford)

Radial Clearance Check

1. Attach an adapter on the dial indicator plunger and mount the dial indicator on the center housing with the plunger through the oil drain port, figure 20-11.

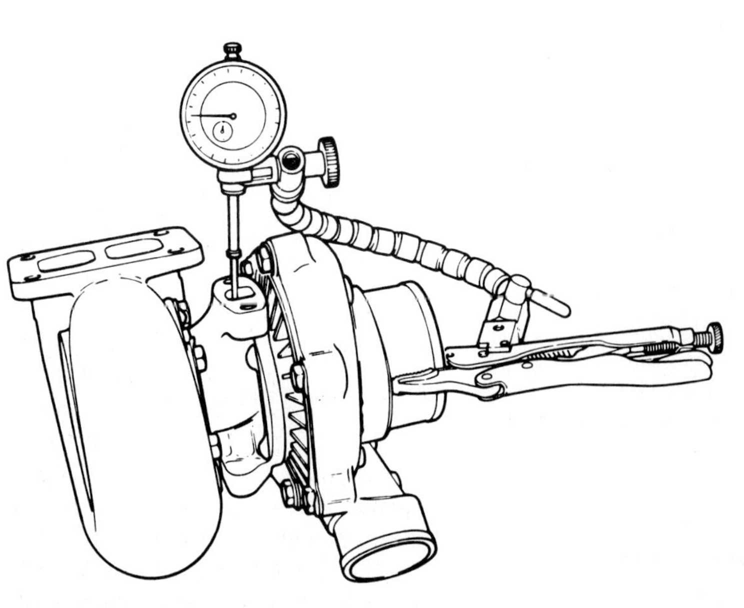

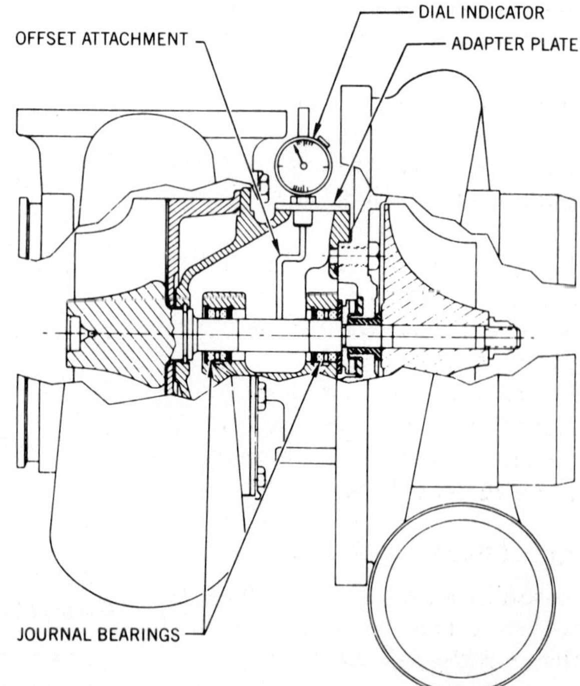

Figure 20-11. Mount a dial indicator like this to measure radial bearing clearance. (Garrett)

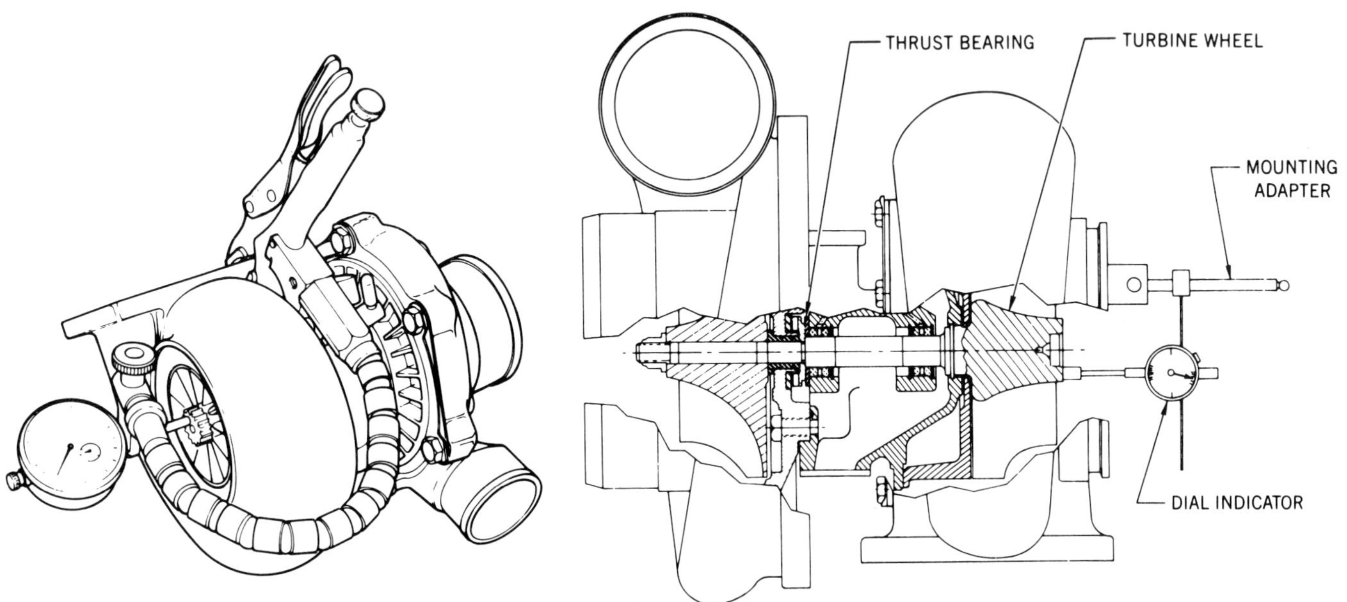

Figure 20-12. Mount a dial indicator like this to measure axial bearing clearance. (Garrett)

2. Be sure the plunger adapter contacts the shaft securely.
3. Push the compressor and turbine wheels equally to move the shaft away from the dial indicator as far as it will go. Movement will be very slight.
4. With pressure still applied to the shaft, adjust the dial indicator to zero.
5. Push the wheels equally in the opposite direction (toward the dial indicator) and note the gauge reading.
6. Move the shaft back and forth several times to verify the measurement. Be sure the indicator returns to zero each time you move the shaft away from it and has the same maximum reading each time you move the shaft toward it.
7. To check for a bent shaft or uneven bearing wear, rotate the wheels and take several readings at 90-degree intervals around the shaft.
8. If bearing radial clearance is less than the minimum or more than the maximum specification, replace or rebuild the turbocharger.

Axial Clearance Check

1. Mount the dial indicator on the turbo with the plunger contacting the end of the shaft through the turbine or the compressor housing. Most manufacturers specify measuring axial clearance at the turbine end of the shaft, figure 20-12.
2. Be sure the plunger adapter contacts the shaft squarely.
3. Push the compressor or turbine wheel evenly to move the shaft away from the dial indicator as far as it will go. Movement will be very slight.
4. With pressure still applied to the shaft, adjust the dial indicator to zero.
5. Rotate the shaft slowly with pressure applied to be sure the dial indicator setting stays at zero.
6. Push the shaft and wheels evenly in the opposite direction (toward the dial indicator) and note the gauge reading.
7. Move the shaft back and forth several times to verify the measurement. Be sure the indicator returns to zero each time you move the shaft away from it and has the same maximum reading each time you move the shaft toward it.
8. If thrust bearing axial clearance is less than the minimum or more than the maximum specification, replace or rebuild the turbo.

Boost Pressure and Wastegate Tests

You can check turbo boost pressure and wastegate operation with a pressure gauge and a dial indicator. You often check boost pressure and wastegate operation simultaneously, and these two factors are related to each other. Boost pressure below specifications often means that the wastegate is leaking or opening too far or too soon. Boost pressure higher than specifications usually means that the wastegate actuator or linkage is broken or stuck.

Boost Pressure Test. You can check maximum boost pressure most accurately during a road test or on a chassis dynamometer. Some carmakers, in fact, specify that boost pressure must be checked *only* under these conditions that apply a driving load to the engine. To check boost pressure, you need a pressure gauge graduated in psi, kPa, or bars, depending on the carmaker's specifications. Connect the gauge to a pressure port on the compressor side of the turbo. On some engines, you can put a T-fitting in the pressure line to the wastegate actuator or to the diesel fuel injection pump, figure 20-13, and connect the gauge at this point. On other cars, you can connect the gauge with a T-fitting in the line to an instrument panel boost gauge or warning lamp pressure switch, figure 20-14. In any case, connect your test gauge with a hose long enough to reach from the engine compartment to the driver's seat.

Start and warm the engine to normal operating temperature and check boost

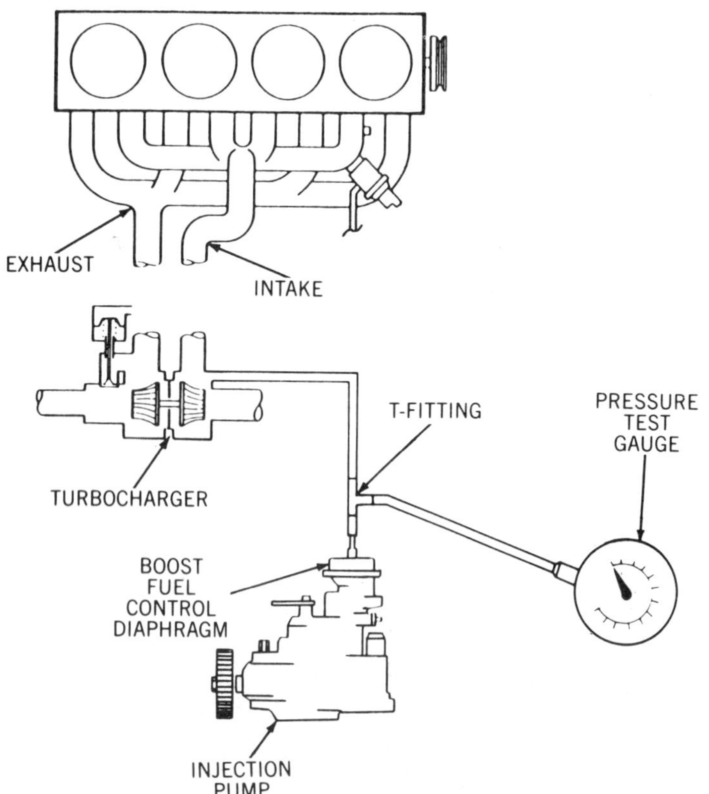

Figure 20-13. Install a gauge in the line to the injection pump boost control unit to check boost pressure on a turbo diesel. (Volkswagen)

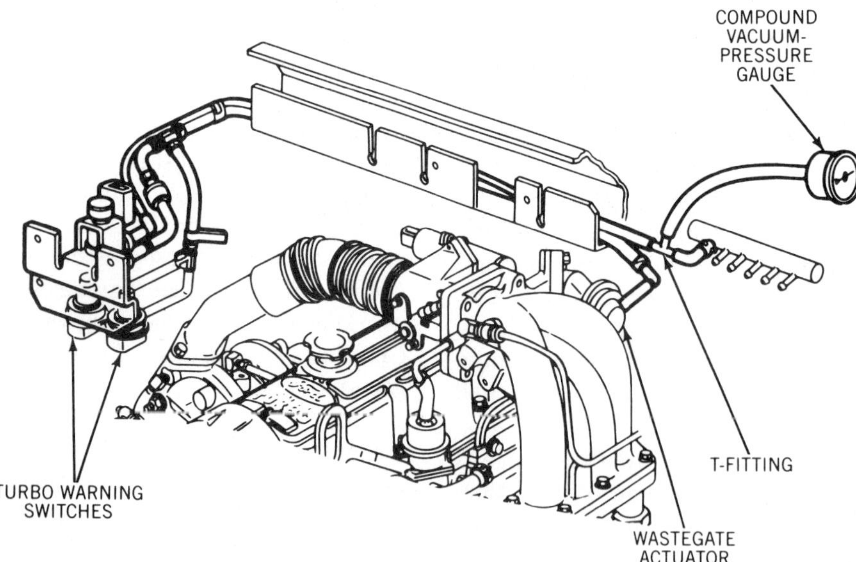

Figure 20-14. On some cars, connect your test gauge to the lines for vehicle boost gauge pressure switches. (Ford)

pressure on the road or on a dynamometer. Accelerate the vehicle at full throttle from zero to 40 or 50 mph and note the maximum gauge reading under load. Repeat the test two or three times to verify the reading. If maximum pressure is above or below specifications, test and replace or adjust the wastegate and actuator.

Wastegate Actuator Test. Most wastegate actuators have an external rod that connects to the wastegate valve link, or arm, figure 20-15. Rod length may be adjustable, but many carmakers seal the adjustment and specify actuator replacement if boost pressure or wastegate operation is out of limits. To check wastegate and actuator operation, you need a source of air pressure, a pressure gauge, and a dial indicator. A radiator pressure tester often works well for checking the actuator. Check operation with the engine off. Test procedures vary, but most include the following points:

1. Connect a hand-operated pressure pump and gauge, figure 20-15, or a compressed air line with a regulator and gauge, to the actuator.

2. Apply about 5 psi (35 kPa) of pressure to the actuator and hold the pressure for 1 minute. If pressure drops below 2 psi (14 kPa) in 1 minute or less, the actuator diaphragm is leaking.

3. Mount a dial indicator on the turbo housing with the plunger contacting the actuator rod, figure 20-16. With pressure removed from the actuator, adjust the indicator to zero.

4. Apply the specified test pressure or maximum boost pressure to the actuator and note the rod movement shown on the dial indicator. The rod travel is quite short, usually about 0.015 inch (0.385 mm).

5. If rod movement is out of limits, adjust or replace the actuator.

6. Disconnect the actuator rod from the wastegate arm, or link, and move the arm to verify that it moves freely through an arc of about 45 degrees, figure 20-17. Exhaust deposits on the wastegate can cause it to stick or

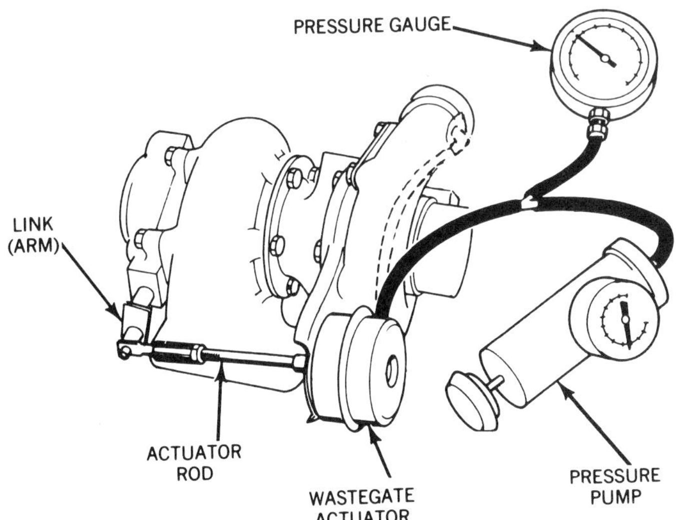

Figure 20-15. Connect a pressure pump to the wastegate actuator. (Volvo)

Figure 20-16. Mount a dial indicator like this to measure wastegate actuator rod travel.

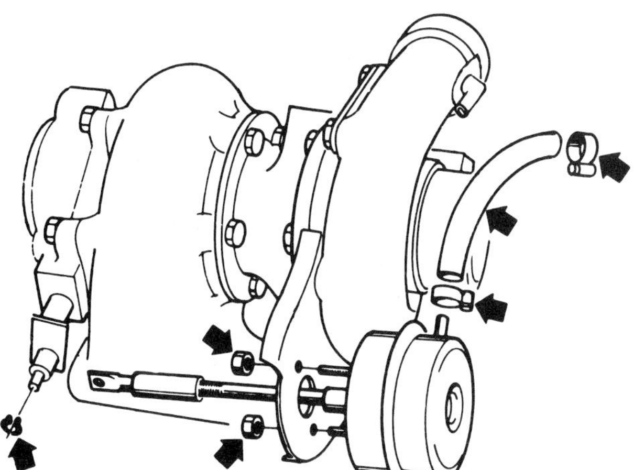

Figure 20-17. Disconnect these points to remove an actuator. Be sure the wastegate arm swings through about a 45-degree arc when loose. (Volvo)

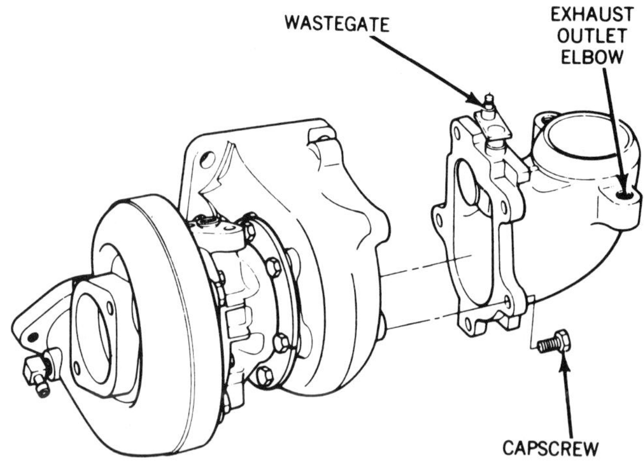

Figure 20-18. Replacing an exhaust outlet elbow. (Ford)

bind. If the arm does not move freely, replace the turbo outlet elbow and wastegate.

TURBOCHARGER REPAIR AND REPLACEMENT

A damaged or worn turbocharger can be overhauled or rebuilt. Many carmakers, however, specify replacing a damaged turbo and do not sell repair parts. Most manufacturers do provide replacement exhaust outlet and wastegate assemblies and wastegate actuators.

To replace an exhaust outlet, you must remove the turbo from the engine according to the carmaker's pro-

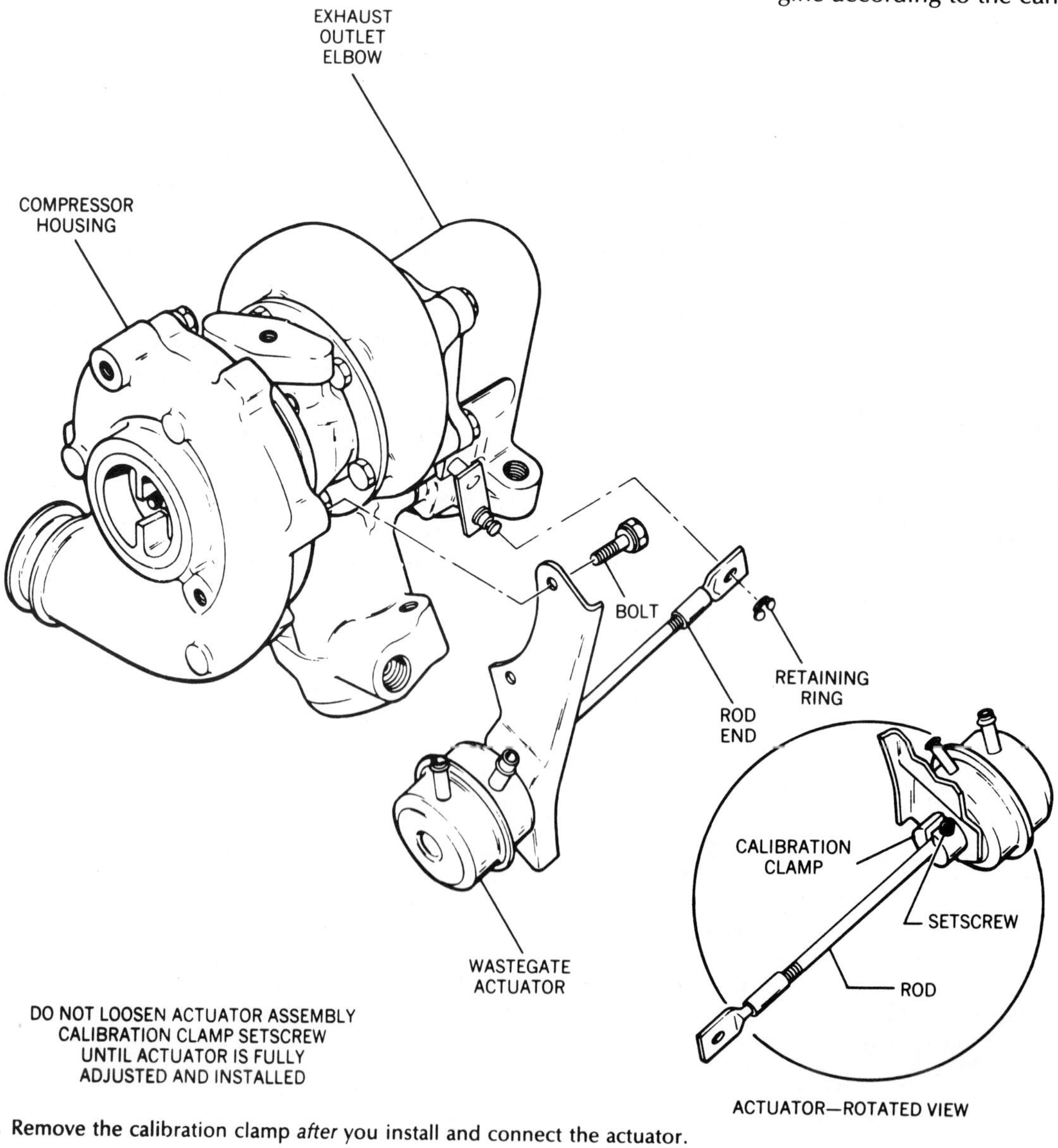

DO NOT LOOSEN ACTUATOR ASSEMBLY CALIBRATION CLAMP SETSCREW UNTIL ACTUATOR IS FULLY ADJUSTED AND INSTALLED

Figure 20-19. Remove the calibration clamp *after* you install and connect the actuator. (Garrett)

cedures; then replace the outlet assembly, figure 20-18. You can usually replace the actuator with the turbo either on or off the engine. Refer to figures 20-17 and 20-19:

1. Disconnect the hose, or hoses, from the actuator.
2. Remove the pin or snapring that connects the rod to the wastegate arm.
3. Remove 2 or 3 capscrews that attach the actuator to the turbo housing and remove the actuator.
4. Attach the new actuator to the turbo. Replacement actuators usually have a calibration clamp, figure 20-19, on the rod. Do not remove it.
5. Loosen the locknut on the actuator rod end and turn the rod end in or out until it aligns with the hole or pin on the wastegate arm.
6. Install the pin or snapring that holds the rod end to the arm and tighten the locknut.
7. Now, remove the calibration clamp from the actuator rod.
8. Connect the actuator hose, or hoses.

SUPERCHARGER SERVICE AND TROUBLESHOOTING

The Roots-type superchargers used by Ford and GM are not serviced in the field. If a seal or bearing fails, you must replace the entire unit. Ford recommends that the blower drive belts should be checked every 30,000 miles. Also, check the oil level in the blower sump every 30,000 miles. The check-and-fill plug is at the lower front of the blower housing. The Ford blower requires a special 90-weight synthetic lubricant (Ford part number E95Z19577A).

A major troubleshooting principle for supercharger installations, particularly Ford's, is to be sure that there are no leaks on either the intake or the outlet side. The plumbing maze on Ford's V-6, figure 20-20, requires secure connections at about a dozen clamps and flanges.

A leak on the intake side can let dust and dirt enter the blower and damage it. A leak anywhere between the mass airflow (MAF) sensor and the blower inlet will cause the engine to run lean. The engine computer will regulate fuel metering according to the MAF sensor signal, but an intake leak will let the blower draw in unmeasured air.

Conversely, a leak on the pressurized outlet side of the blower will cause a rich mixture because the engine will lose the extra air for which the computer is providing fuel. If a system self-test produces a lean-exhaust fault code, check for blower intake leaks. If the test produces a rich-exhaust code, check for outlet leaks.

If engine performance is sluggish on a Ford supercharged engine, check the vacuum-operated bypass valve. If it sticks open or opens too soon, boost is reduced. With 0 to 3 inches of vacuum at the actuator, the bypass valve should be closed. With 3 to 7 inches of vacuum, the valve is partly open. With more than 7 inches of vacuum, the valve should be fully open.

SUMMARY

Turbochargers are simple but precise devices that increase the intake air volume of an engine. A turbo can operate for tens of thousands of miles with no problem *if* it gets proper lubrication and is free from dirt. Dirt and poor lubrication are the chief causes of turbo failure.

Turbocharger troubleshooting consists of listening to the turbo as it runs, inspecting for air and exhaust leaks, and checking the compressor and turbine for rubbing or other damage. You can check bearing clearances with a dial indicator. You can check wastegate operation with a pressure source, a gauge, and a dial indicator. Some wastegate actuators are adjustable; others must be replaced.

Some carmakers specify that a damaged turbocharger must be replaced as a complete assembly. Other carmakers allow replacement of damaged parts, such as inlet and outlet elbows or housings, center housings, and wastegates. Although all turbochargers are similar in construction, specific models are tailored to specific engines. Always follow the carmaker's procedures for turbocharger service.

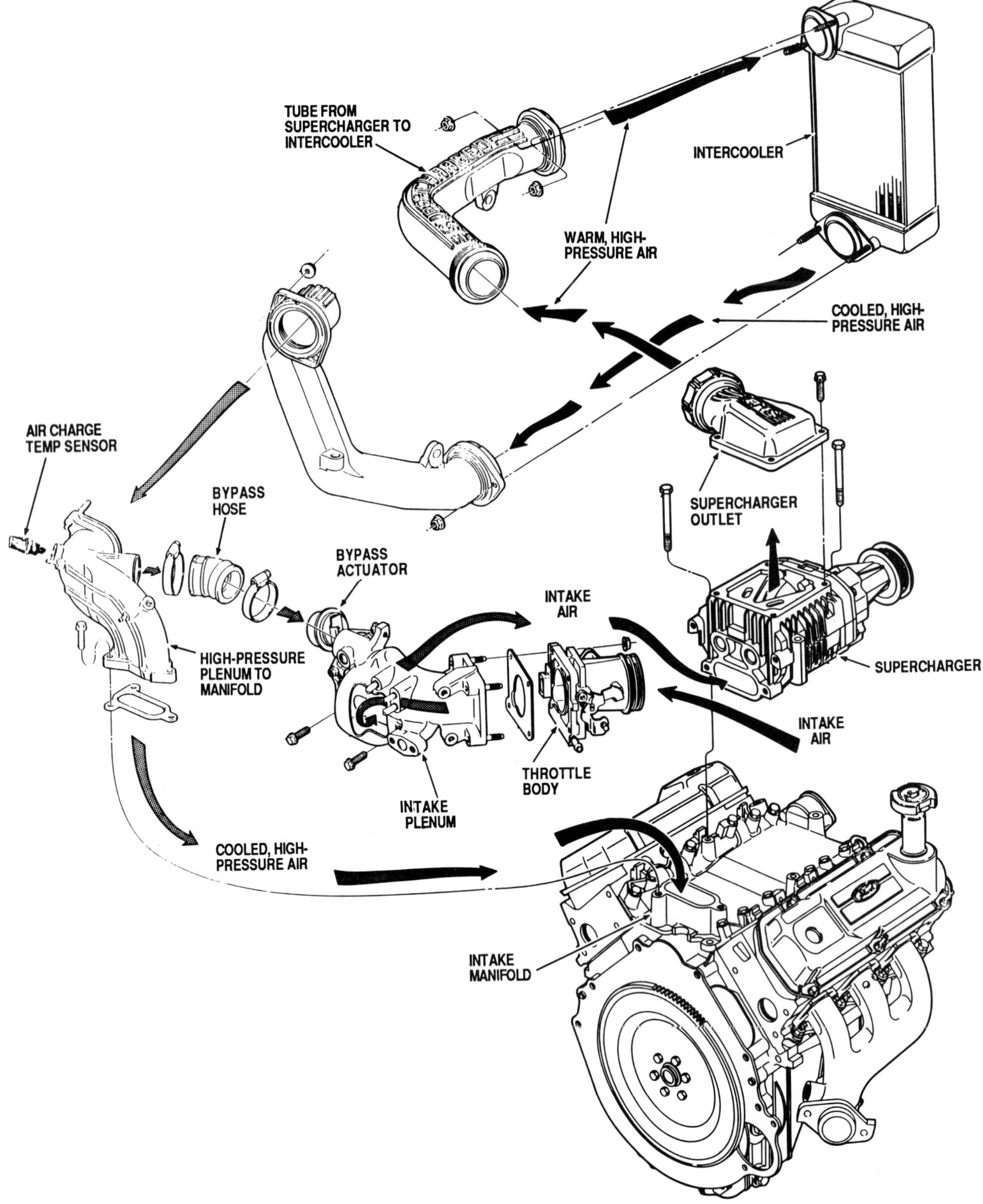

Figure 20-20. Supercharger installation and airflow through the intake system on Ford's V-6 engine. (Ford)

REVIEW QUESTIONS

Multiple Choice

1. Most turbocharger failures result from:

a. excessive high-speed operation
b. excessive exhaust temperatures
c. dirt
d. operation with leaded gasoline

2. Mechanic A says never run a turbocharged engine with the air filter removed. Mechanic B says never shut off a turbocharged engine immediately after returning to idle from high-speed operation. Who is right?

a. A only
b. B only
c. both A and B
d. neither A nor B

3. After changing oil in a turbocharged engine, Mechanic A cranks the engine without starting it for about 15 seconds to prime the turbo. Mechanic B fills the new oil filter with fresh oil to help prime the turbo. Who is doing the job correctly?

a. A only
b. B only
c. both A and B
d. neither A nor B

4. A driver complains of "lack of power" from a turbocharged engine. This could be caused by:

a. an EGR problem
b. wrong ignition timing or advance
c. a restricted exhaust system
d. any of the above

5. Blue exhaust smoke from a turbocharged engine could be caused by any of the following EXCEPT:

a. a restricted turbocharger oil return line
b. worn rings or valve guides in the engine
c. coolant leakage into the combustion chambers
d. a PCV system problem

6. A turbocharged engine suffers from spark knock, or pinging. This could be caused by:

a. an EGR problem
b. the turbo wastegate opening too soon or stuck open
c. a restricted turbocharger oil return line
d. an exhaust leak at the turbine inlet

7. A turbocharged engine suffers from high oil consumption. This could be caused by all of the following EXCEPT:

a. excessive blowby
b. worn rings or valve guides
c. a restricted turbocharger oil return line
d. the wastegate not opening

8. A turbocharger is making a light, steady whistling sound that increases in frequency as the engine accelerates. This probably indicates that:

a. a loose exhaust pipe or exhaust outlet elbow
b. an intake air leak
c. the turbo is operating normally
d. the bearings are damaged

9. A turbocharger is making a grinding or rubbing sound. This probably indicates:

a. a loose wastegate actuator
b. bearing damage
c. a loose exhaust outlet elbow
d. a restricted oil return line

10. Mechanic A says that worn spots on the turbine or compressor blade tips of a turbocharger indicate that the blades have been rubbing the housing. Mechanic B says that worn spots on the blade tips indicate bearing or shaft damage. Who is right?

a. A only
b. B only
c. both A and B
d. neither A nor B

11. Mechanic A says that most turbocharger shafts have 0.003 to 0.006 inch (0.076 to 0.152 mm) radial clearance. Mechanic B says that axial clearance is always slightly greater than radial clearance. Who is right?

a. A only
b. B only
c. both A and B
d. neither A nor B

12. Mechanic A says that you check radial bearing clearance on a turbo shaft by mounting a dial indicator to contact the end of the shaft. Mechanic B says that you check axial clearance by mounting a dial indicator to contact the side of the shaft through the center housing. Who is right?

a. A only
b. B only
c. both A and B
d. neither A nor B

13. Mechanic A says that you need a vacuum gauge to test boost pressure. Mechanic B says that you can use a radiator pressure tester to check wastegate actuator operation. Who is right?

a. A only
b. B only
c. both A and B
d. neither A nor B

14. Mechanic A says that exhaust deposits on the turbo wastegate can cause it to stick. Mechanic B says that exhaust deposits on the wastegate can cause high boost pressure. Who is right?

a. A only
b. B only
c. both A and B
d. neither A nor B

15. Black exhaust smoke from a turbocharged engine can indicate:

a. coolant leakage into the cylinders
b. leaking oil seals in the turbocharger
c. a clogged intake air filter
d. low boost pressure

16. At cruising speed, turbocharger noise suddenly decreases and the engine emits smoke. This probably indicates that:

a. the wastegate is stuck open
b. an intake air leak has developed
c. the air filter is restricted
d. the turbo has gone south

PART SIX
LATE-MODEL ELECTRONIC ENGINE TESTING AND SERVICE

INTRODUCTION

This final part of this book contains a single chapter that outlines the principles of electronic engine control system and gasoline fuel injection service. Do not get the idea that engine control system and fuel injection service is a simple operation. It is not.

This last chapter summarizes the general methods that you will use for engine performance service on late-model vehicles. All the service skills that you learned in the preceding 20 chapters apply equally to the fully integrated electronic control systems on current automobiles.

Chapter 21 does not attempt to provide specific troubleshooting instructions for all electronic engine controls on late-model cars. The Mitchell, Motor, and Chilton companies provide volumes of instructions that fill bookshelves for that purpose. Rather, this chapter contains examples of the diagnostic methods—and philosophies—used by General Motors, Ford, and Chrysler for their current engine control systems. After studying this material, you will recognize similarities in the specific test and service procedures provided by all carmakers. Moreover, you will learn that testing and servicing an electronic engine control system is based on the same principles of thorough troubleshooting that you learned in chapter 5 and applied to servicing all combustion control systems.

Chapter 21 ends with a summary of some of the special service requirements for electronic engine control systems.

21 ELECTRONIC ENGINE CONTROL SYSTEM AND FUEL INJECTION TESTING AND SERVICE

INTRODUCTION

Most late-model vehicles have all combustion control functions regulated by a central engine computer. When you service such a vehicle, you must practice "service integration." You must evaluate the entire system as you service its subsystems and parts. You must understand, for example, how:

- Air-fuel ratio affects ignition timing requirements
- Engine temperature affects EGR and air injection requirements
- Engine speed and load affect fuel metering, ignition timing, and emission control

You need the carmaker's test procedures and specifications to service a specific car. Moreover, you must know the kinds of sensors and actuators in a specific system, how each operates, and the kinds of signals it sends or receives. You have learned that sensors and actuators can be grouped in common families and that they measure or control similar engine functions. But manufacturers combine system components in different ways to reach similar goals. One system, for example, may use a variable resistor, while another may use an on-off switch to sense the same engine condition. If you understand the engine conditions that any system must monitor and control, however, you will be able to look for a sensor or actuator and recognize its operation.

The fundamentals of system operation that you learned in Volume 1 and the service principles that you have learned in this volume will help you sort out each carmaker's system. After studying this material, you will recognize similarities in the specific test and service procedures provided by all carmakers. While this chapter does not contain all service procedures for all systems, it will help you use any carmaker's information faster and more easily.

GOALS

After studying this chapter, you should be able to do the following jobs in twice the flat-rate labor time or less, using the carmaker's procedures:

1. Display and identify system trouble codes on systems with self-diagnostic capability.
2. Perform system functional diagnostic and performance tests (area tests).
3. Perform specific pinpoint tests to isolate and identify component problems.
4. Using manufacturers' test and repair procedures, test and adjust or replace fuel injection system sensors and actuators (fuel injectors).
5. Replace or adjust specific electronic engine control parts.

ENGINE TESTING SAFETY

When working around any automotive engine, safety should be your primary concern. Review the safety guidelines in chapter 1 and remember these key points:

1. Keep your hands, clothing, face, hair, and tools away from the fan and other moving engine parts whenever you work under the hood with the engine running.

2. Keep your body and clothes away from hot exhaust manifolds and other exhaust parts. Keep gasoline, other flammable liquids, and oily cloths away from running engines and hot exhaust parts. Catalytic converters remain very hot long after an engine is shut off.

3. You will be working around the fuel system and flammable fuel. Keep sparks and open flame away from fuel system components.

4. Keep the engine air cleaner installed whenever possible to avoid a backfire or the possibility of dropping

something into the engine. Keep small loose parts and tools away from the engine air intake.

5. Keep sparks and open flame away from batteries. Do not smoke.

6. Disconnect cables before charging the battery. Disconnect the ground cable before removing or replacing electrical components.

7. Operate the engine only in a well-ventilated area or use an exhaust collection system to remove fumes from the shop.

Electronic Safety

When you service an electronically controlled engine, you are working with all engine systems. Review chapter 1 and other preceding chapters for safety information and note the following precautions:

1. Disconnecting the battery on a car with electronic engine controls, or with any computer that has a memory, will erase the memory. During testing, retrieve any stored trouble codes from computer memory before disconnecting the battery. You must reset electronic clocks, radios, air conditioning controls, and other programmed electronic devices after disconnecting and reconnecting a battery.

2. Do not disconnect battery terminals with the ignition switch on. High-voltage surges can damage electronic parts.

3. Do not disconnect or connect electrical connectors with the ignition on unless specifically directed to do so. Breaking or making a connection can cause a high-voltage surge that may damage electronic parts.

4. Do not short circuit or ground any solid-state electronic parts or electrical terminals, or apply battery voltage directly to electronic parts unless specifically instructed to do so in a service procedure.

5. Always use a high-impedance digital volt-ohmmeter (DVOM) with an input impedance of at least 10 megohms per volt when testing electronic components or circuitry. Even a moderate current draw can short circuit and destroy such components or circuitry.

6. Take the necessary precautions against electrostatic discharges by grounding yourself when working on solid-state components such as computers.

7. Electronic systems operate with low voltage and very low current. They are very susceptible to high resistance in connectors, wiring, and components. After servicing a system, be sure that all connectors are clean and secure. You can often fix a problem simply by checking for and correcting loose, corroded, or broken wiring or connectors.

Fuel System and Fuel Injection Safety

Before servicing any part of a fuel system, review the shop and vehicle safety information in chapter 1. Particularly, observe these precautions:

1. Do not smoke when working on or near any part of a fuel system. Keep sparks, flame, and hot metal away from fuel system parts.

2. Work on fuel systems in a well-ventilated area. Clean up any spilled fluids promptly.

3. Cap or plug all fuel lines when disconnected to prevent fuel leakage and to keep dirt out of the system, figure 21-1.

4. Keep the air cleaner installed whenever possible to avoid a backfire or the possibility of dropping something into the engine.

5. Do not use gasoline as a cleaning solvent. Avoid prolonged skin contact with fuels and solvents.

6. Disconnect the battery ground cable before removing and replacing fuel system components on vehicles without electronic engine controls. If it is impractical to disconnect the battery, be sure the ignition is off and use extreme caution when servicing the fuel system.

7. Diesel fuel injection systems retain high pressure between the injection pump and the injectors when the engine is off. Most gasoline injection systems also hold pressure between the fuel pump and the injector body or fuel rail when the engine is off. Follow the carmaker's specific procedures for relieving fuel pressure before loosening any fuel lines or replacing any parts of these systems.

Figure 21-1. Cap or plug disconnected fuel lines to prevent spillage.

8. Most gasoline fuel injection systems retain high pressure between the pump and the injectors when the engine is off. Follow the carmaker's instructions to relieve system pressure before opening an injection system for service.

TROUBLESHOOTING PRINCIPLES

The first things you must know to service an electronic engine system are what functions the system monitors and what functions it controls. Most late-model systems are fully integrated. They monitor and control ignition, air-fuel ratio, EGR, air injection, and other functions. Some also control automatic transmission shifting, torque converter lockup, and cruise control. Many systems, particularly those of the late 1970's, control only selected functions. Some control *just* ignition timing; others control only fuel metering. Regardless of the number of system functions, operation and service principles are similar. The manufacturer's descriptions and the

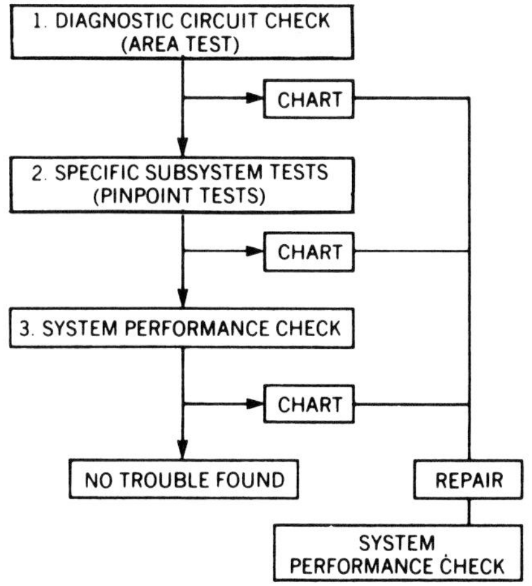

Figure 21-2. These three major steps of troubleshooting apply to any engine system. (GM)

information in Volume 1 of this text will help you recognize the functions of any system.

Area Tests and Pinpoint Tests

Earlier in this volume you learned how to test from general system operation (area tests) to specific subsystems, circuits, or parts (pinpoint tests). In chapter 9, you learned some basic engine area tests. All of these principles apply to troubleshooting electronic engine controls. All carmakers organize their system test procedures this way. General Motors, in fact, divides testing of the C-4 and CCC engine control systems into three steps, figure 21-2, which you can use to test any electronic engine control system:

1. Evaluate the driver's complaint and do an overall system diagnostic check (area test).

2. Use diagnostic charts or procedures to do specific subsystem or circuit tests and repairs (pinpoint tests).

3. After system repairs or adjustments, do a system performance check to verify proper system operation.

FUEL INJECTION TROUBLESHOOTING

To test and service a gasoline injection system, you will need the carmaker's test procedures and system specifications for the specific model and year of car. Injection systems don't vary radically from one vehicle to another, but manufacturers do make detail changes that affect troubleshooting and service. Carmakers publish extensive repair and adjustment instructions and test sequences each year, and you can find these in factory shop manuals or independent service guides.

Many gasoline injection systems are part of fully integrated electronic engine control systems that include electronic spark control and feedback fuel control with EGO sensors. When you service such vehicles, you must consider the *complete* engine control system. You know from your study of ignition systems and other engine systems that one general symptom or complaint can be caused by problems in several systems or multiple problems in one system. When a driver complains of "poor gas mileage," don't rush to condemn the fuel injection system and ignore other areas.

The electronic parts of injection systems operate on low voltages, and any high resistance in a circuit causes problems. Always check wiring and connectors closely for corrosion, damage, loose connections, and broken connections.

Basic Complaints and Their Causes

As with other service jobs, injection troubleshooting usually begins with a driver complaint. Most carmakers organize their procedures with the basic complaint as a starting point. We can't give you all of these test routines here, but figures 21-3 and 21-4 summarize the most common complaints and the injection system components that *may* be involved. Figure 21-3 is for the Bosch K-Jetronic mechanical injection system. This is the most common mechanical continuous injection system and is used by many European carmakers. Figure 21-4 is a basic list of complaints and related components for most electronic injection systems, both throttle body injection and port type. You can use these summaries to help yourself follow and understand carmakers' specific procedures.

Notice that several components can be involved in a single complaint and that a problem with one part can cause several symptoms. To make your testing faster and most accurate, you must know the function of each sensor and actuator in the system. Review the explanations in Volume 1 of this text and the manufacturer's descriptions before you begin servicing the car. Also, fuel-injected cars often require special starting procedures. They aren't difficult, but if the driver doesn't follow them, he or she may have a starting problem. Always verify that the driver is starting the car correctly.

SELF-DIAGNOSTIC SYSTEMS

Engine control computers can be built and programmed to test their own operation and the operation of each sensor and actuator circuit. Most late-model systems have this capability to one extent or another. The computer does one or more of three things:

1. It recognizes the absence of an input or output voltage signal for a sensor or actuator circuit—no engine speed or tachometer signal, for example.

2. It recognizes a signal that is improbable or that stays out of limits for a period of time, such as:

- **a.** A barometric pressure signal that indicates the car is operating at 30,000 feet
- **b.** Input signals from an idle switch and a wide-open-throttle switch that are closed simultaneously
- **c.** A constantly rich or lean EGO sensor signal that isn't corrected by fuel-metering adjustment.

Several test procedures call for you to create signals such as the first two examples (a and b) to put the computer into a self-test mode. You may have to apply vacuum to a barometric pressure sensor or close idle and throt-

	Cold Engine Doesn't Start or Hard to Start	*Hot Engine Doesn't Start or Hard to Start*	*Cold Engine Starts and Stalls or Runs Poorly*	*Warm Engine Runs Poorly*
Cold-Start Valve (Injector)	X			
Thermo-Time Switch	X			
Airflow Sensor Adjustment (off center or binding)	X	X	X	X
EGR Problem		X		
EEC Vapor Canister Purge Problem		X		
Idle Speed and CO Adjustment		X	X	X
Injector Operation and Fuel Quantity		X	X	X
Fuel Delivery Volume and Pressure	X	X	X	X
System Pressure				X
Control Pressure		X	X	X
Residual Pressure		X		
Auxiliary Air Regulator			X	
Vacuum Leaks	X	X	X	X
Vapor Lock		X		

Figure 21-3. Troubleshooting table for Bosch K-Jetronic continuous injection system. (CIS)

tle switches by hand to trigger a self-test sequence.

3. The computer can send a test voltage signal to a sensor or actuator to test circuit continuity or check a return voltage signal.

If the computer recognizes something that is out of limits ("not right") it records a fault code, or trouble code. Many systems will light an instrument panel warning lamp, or CHECK ENGINE lamp, figure 21-5. The system also stores the trouble code in the computer memory. Troubleshooting these systems begins with reading the trouble codes. They are usually accompanied by a driver complaint, or a driver complaint will cause you to look for the codes. Checking the trouble codes or driver complaints is your *area test* for a self-diagnostic electronic engine system.

General Motors provides self-diagnostic capabilities for its computer-controlled catalytic converter (C-4) and computer command control (CCC) engine control systems. Ford has similar capabilities in its family of electronic engine control (EEC) systems, as does Chrysler. Other carmakers have similar self-diagnostic systems, but we will use the GM, Ford, and Chrysler systems as examples in this chapter. You will learn how to read these codes in a later section. First, however, you should know some precautions about using and interpreting the codes and about different kinds of codes.

Trouble Code Precautions

Precautions for self-diagnostic trouble codes deal with the success of your service work, rather than with safety. Trouble codes are valuable diagnostic aids that help you organize your work effectively. They do not, however, do your testing and *thinking* for you.

A trouble code can indicate a fault in a particular circuit or engine function, or an abnormal condition that affects system operation. The code does not pinpoint the problem for you. Here are two examples.

	Open Circuit (Blown Fuse or Open Relay or Circuit Breaker)	*Airflow or Manifold Pressure Sensor*	*Speed Sensor or Distributor Signal*	*Coolant Temp Sensor*	*Air Temp Sensor*	*Throttle Position Switch or Sensor*	*Fast-Idle Device*	*Idle Speed Adjustment or Motor*	*Idle Air Compensator*	*Fuel Pressure Regulator or Delivery Pressure*	*EGR System*	*EEC Vapor Canister Purge*	*Injectors*
Engine Cranks But Won't Start	X	X	X			X				X			X
Hard Cold Starting		X	X	X	X	X	X	X		X			X
Hard Hot Starting		X	X	X		X				X	X	X	X
Starts Cold, Then Stalls		X		X	X		X	X	X	X	X	X	X
Starts Hot, Then Stalls		X	X					X		X	X	X	X
Rough or High Idle		X	X	X	X	X		X		X	X	X	X
No Fast Idle						X	X						
Prolonged Fast Idle						X							
Poor Acceleration (Hesitation)		X	X			X					X		X
Surging		X				X				X	X		X
Poor High-Speed Performance		X	X			X				X			X
Poor Fuel Economy		X		X	X					X			X

Figure 21-4. Basic troubleshooting table for electronic fuel injection (EFI) systems.

Figure 21-5. The CHECK ENGINE lamp lights to warn of a problem in the GM CCC system.

Rich or Lean Exhaust Indication. Most systems have a code, or two separate codes, to indicate exhaust oxygen content that varies from the desired 14.7:1 air-fuel ratio in closed-loop operation. These codes tell you that general engine operation is not right. They may be accompanied by driver complaints of "poor gas mileage," "surging," "poor high-speed performance," or "poor acceleration." You can begin to see that a trouble code doesn't isolate a fault, but points you in a general direction.

A "rich exhaust" code can indicate one or more of the following problems:

- Bad EGO sensor
- Bad connectors or wiring in the EGO sensor circuit
- Dirty air filter
- Stuck choke
- Too much blowby, overloaded PCV system, or a worn engine
- Excessive vapor canister purging
- Faulty carburetor or injection fuel metering control
- Faulty system computer

A "lean exhaust" code can indicate one or more of the following problems:

- Bad EGO sensor
- Bad connectors or wiring in the EGO sensor circuit

- Intake air (vacuum) leaks
- EGR fault
- Engine misfire
- Faulty carburetor or injection fuel-metering control
- Faulty system computer

After you identify a trouble code, or codes, in a general diagnostic (area) test, you must go through one or more pinpoint tests to find the faulty part, or parts. Although a single code can indicate several possible faults, it also can eliminate several others. Carmakers organize their test sequences to start at the most probable cause.

High or Low Coolant Temperature Indication. Most self-diagnostic systems also have trouble codes to indicate a fault, or faults, with the coolant temperature or temperature-sensing circuit. Some systems show only that there is a fault in this area. Others show either a high-temperature or a low-temperature problem. If the system indicates an open-circuit signal from the sensor, for example, you must know if high resistance (an open circuit) is equivalent to high or low temperature. This depends on the kind of sensor used.

All of this may start to sound confusing, but it won't be if you understand the fundamentals of the system. You must know the kind of sensor the carmaker uses. (You can look that up easily enough.) And, you must know the engine conditions that can cause that kind of sensor to send an improper signal. (You can find that in the carmaker's specifications and procedures or figure it out from your knowledge of sensor operation.)

High-temperature faults can be caused by:

- Engine overheating (this leads to several more test and repair possibilities)
  - —Thermostat stuck closed or opening at high temperature
  - —Low coolant level
  - —Blocked radiator airflow
  - —Clogged hoses
  - —Restricted water jackets
- Bad coolant temperature sensor
- Bad connectors or wiring in the collant temperature sensor circuit
- Faulty system computer

Low-temperature faults can be caused by:

- A missing thermostat or a thermostat that opens at low temperature
- Very cold ambient air temperature
- Bad coolant temperature sensor
- Bad connectors or wiring in the coolant temperature sensor circuit
- Faulty system computer

If you begin to see the logic of a testing method similar to a troubleshooting chart, you are right. Most carmakers will lead you from a system diagnostic area test to pinpoint tests in the form of troubleshooting charts or procedures, figures 21-6 and 21-7. You will find that parts of many pinpoint tests dupli-

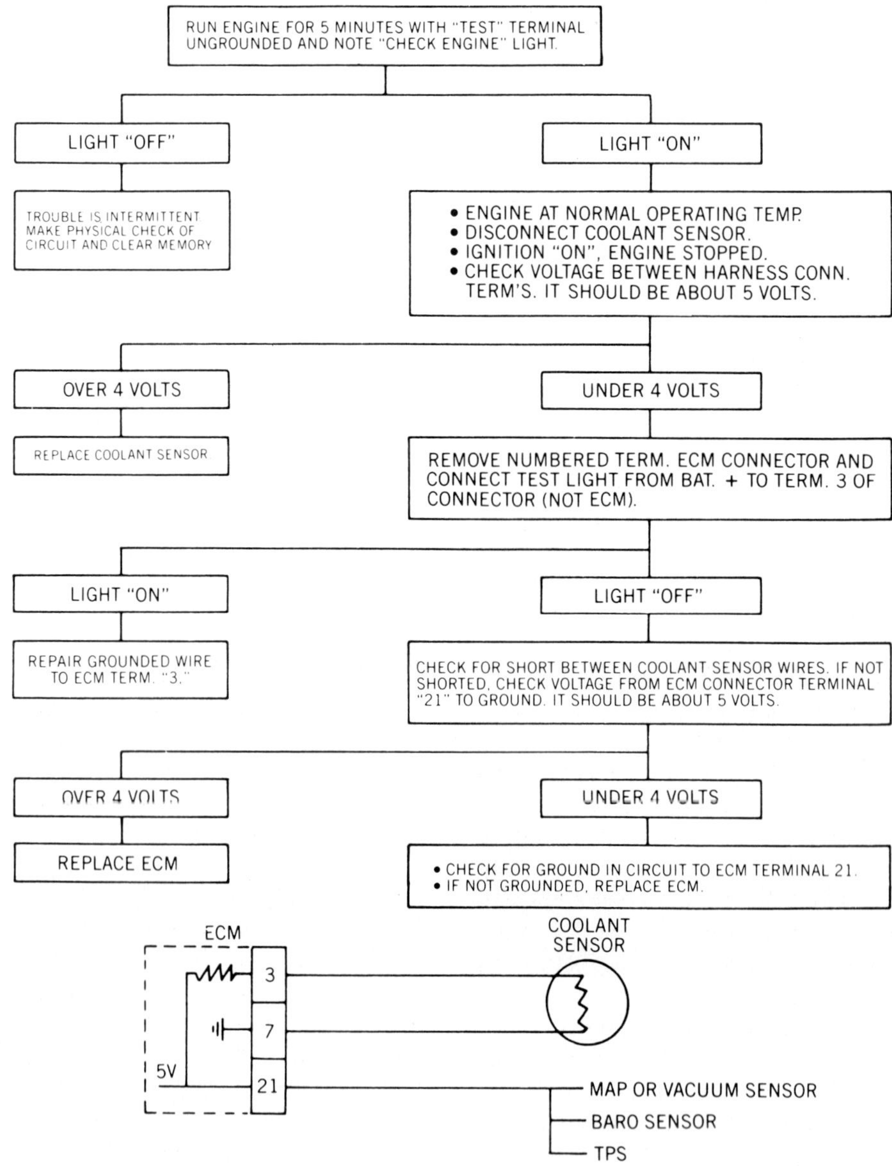

Figure 21-6. Typical GM troubleshooting chart for CCC system trouble code 14, a shorted coolant sensor. (Chevrolet)

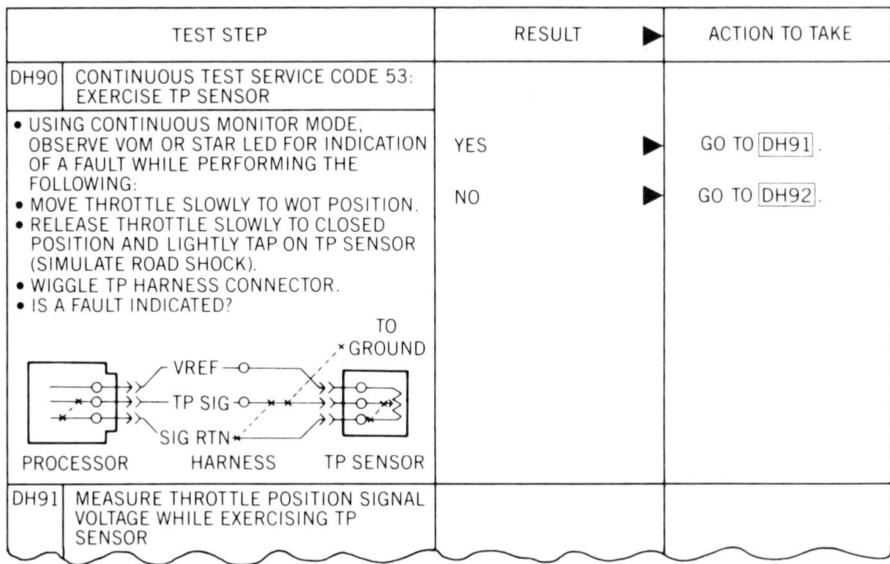

TEST STEP	RESULT ▶	ACTION TO TAKE
DH90 CONTINUOUS TEST SERVICE CODE 53: EXERCISE TP SENSOR		
• USING CONTINUOUS MONITOR MODE, OBSERVE VOM OR STAR LED FOR INDICATION OF A FAULT WHILE PERFORMING THE FOLLOWING: • MOVE THROTTLE SLOWLY TO WOT POSITION. • RELEASE THROTTLE SLOWLY TO CLOSED POSITION AND LIGHTLY TAP ON TP SENSOR (SIMULATE ROAD SHOCK). • WIGGLE TP HARNESS CONNECTOR. • IS A FAULT INDICATED?	YES ▶ NO ▶	GO TO DH91. GO TO DH92.
DH91 MEASURE THROTTLE POSITION SIGNAL VOLTAGE WHILE EXERCISING TP SENSOR		

Figure 21-7. This is the first step of Ford's pinpoint test for an EEC-IV throttle position sensor fault. (Ford)

cate each other because a particular problem, or combination of problems, can cause several trouble codes. Some test procedures may seem complicated; but if you follow them carefully, do not skip steps, and do not mix steps, they will lead you through your testing in the quickest, most accurate way possible.

Open-Loop and Closed-Loop Faults. We should point out one other example of something that can happen with a trouble code such as a "coolant temperature sensor fault." This, and several other kinds of trouble codes, indicates a fault that will keep an electronic engine system in open-loop operation. With high or low engine temperatures, the computer does not respond to EGO sensor signals and does not provide fuel-metering feedback control for the carburetor or injection system. The injectors or the carburetor solenoid operates at a fixed duty cycle. Some throttle position sensor faults or speed sensor faults can cause the same condition.

One important thing to remember about such open-loop faults is that the system can have other closed-loop faults, which may not be shown by trouble codes until you fix the open-loop problem. That is why the final step of any troubleshooting is an overall system performance check. This lets you find *all* faults in a system and fix the owner's general complaint.

Remember these precautions for interpreting and using trouble codes as you work through your system troubleshooting and repair.

Kinds of Trouble Codes

Carmakers categorize system trouble codes several ways. The most common distinction is between hard failures (codes) and soft failures (codes). A hard failure, or code, is one that is present at the time of testing. It may cause the engine to run poorly, or it may stop the engine from running. Ford refers to hard codes as "on-demand" codes because they are available on demand when you run a system self-test. The problem is not intermittent. It is present until you fix it.

A soft failure, or code, is one that occurred sometime before testing. It is recorded in computer memory but is not present at the time of testing. Ford refers to soft codes as "continuous memory" codes because they are stored continuously in computer memory *after* the fault has occurred. Other carmakers refer to soft codes as intermittent codes. A soft failure may cause the engine to run poorly, but it usually will not stop the engine from running.

A soft failure may be as serious as a hard failure because it may reoccur when driving conditions are the same as they were when it first occurred. "Soft," or "intermittent," simply means that it is not immediately present at the time of testing. A soft failure may decrease gas mileage or top-speed power. Some are more noticeable than others. If the system has self-diagnostic capabilities, it will set a trouble code even if the cause of the problem goes away.

If the computer get a continuously rich or lean EGO sensor signal, it will go into open-loop operation. The engine will still run, but gas mileage or performance may suffer. If the EGO sensor has failed, this will be a hard failure. If some other combination of causes is the reason for the rich or lean condition, the trouble code may be a soft failure. The conditions may not be present when you test the system.

A few hard failures will cause the engine to stop but allow it to restart. These produce an intermittent trouble code. Many intermittent faults occur only when the engine is in closed-loop operation. They do not produce trouble codes when the engine returns to open-loop operation.

For a system to record an intermittent trouble code, the computer must have a long-term memory. This is provided by wiring the system so that battery voltage stays applied to the computer memory when the ignition is off. Most late-model systems have long-term memory; many early systems do not. GM's first C-4 systems on 1980 X-cars do not have long-term memory, but you can connect a jumper wire to the computer memory, figure 12-8, to keep voltage applied when the ignition is off. All GM CCC systems, except "minimum function" systems on T-cars (Chevettes), have

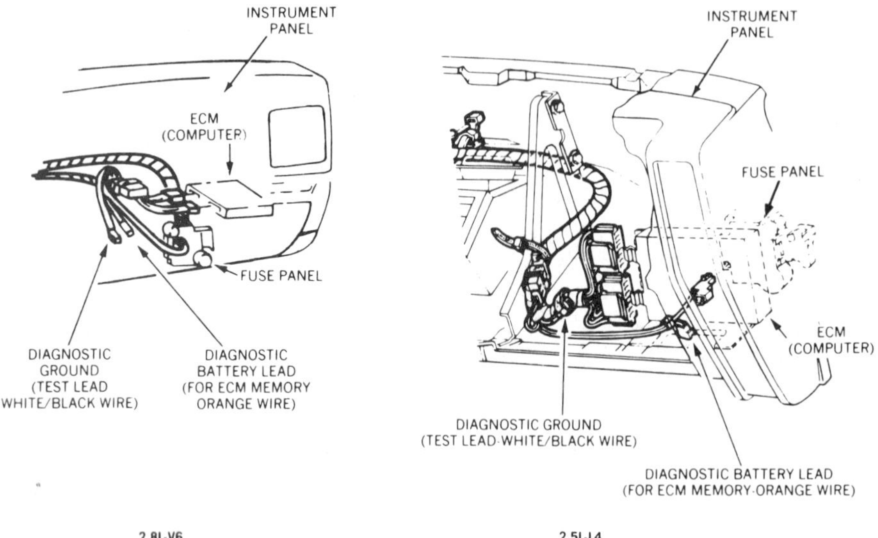

Figure 21-8. On early GM C-4 systems, connect a jumper wire to the diagnostic connector to turn on the computer's long-term memory. (Oldsmobile).

Figure 21-9. Typical diagnostic connector location for GM CCC systems.

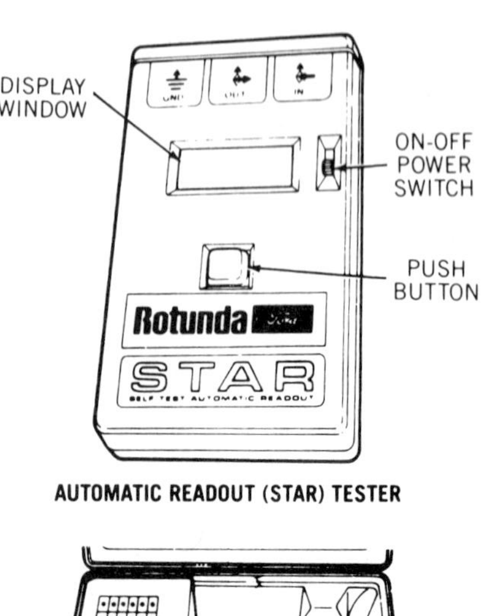

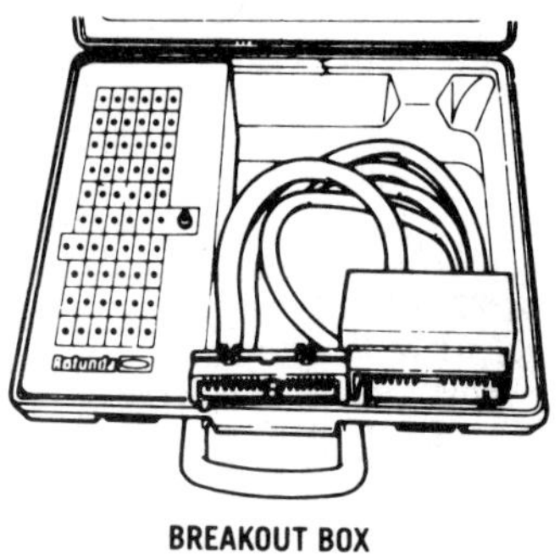

Figure 21-10. Ford's self-test and automatic readout (STAR) tester for EEC systems. The breakout box connects a digital volt-ohmmeter in parallel with the EEC system. (Ford)

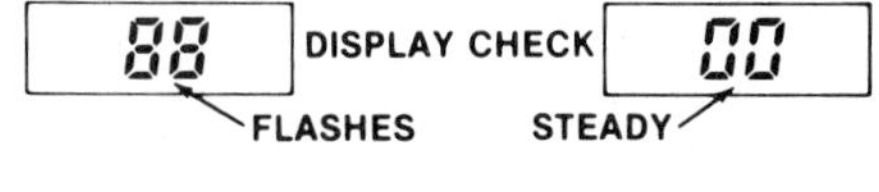

COLON DISPLAY

: 00 — COLON MUST BE DISPLAYED TO RECEIVE SERVICE CODES.

LO BAT INDICATOR

LO BAT : 23 — IF LO BAT SHOWS STEADILY WITH SERVICE CODE, REPLACE TESTER'S 9V BATTERY

Figure 21-11. Ford's STAR tester displays digital trouble codes. (Ford)

long-term memory and will store intermittent trouble codes.

Reading Trouble Codes

Each carmaker has a specific list of trouble codes for each system and specific procedures to read them. Most cars have diagnostic, or test, connectors to which you connect your equipment, figure 21-9. Carmakers and equipment companies sell special testers, figure 21-10, that you can plug into test connectors of specific vehicles to read the codes. These testers make troubleshooting faster and easier, but you can use an analog (dial-indicating) voltmeter and other general-purpose equipment to test most systems. Depending on the system test procedures and the equipment you use, you will usually see a trouble code in one of four basic ways:

1. As a digital, numerical display on a special tester, figure 21-11
2. As a pulsating signal from a voltmeter needle, figure 21-12
3. As a pulsating signal from an instrument panel CHECK ENGINE lamp as it flashes on and off
4. As a digital, numerical display on

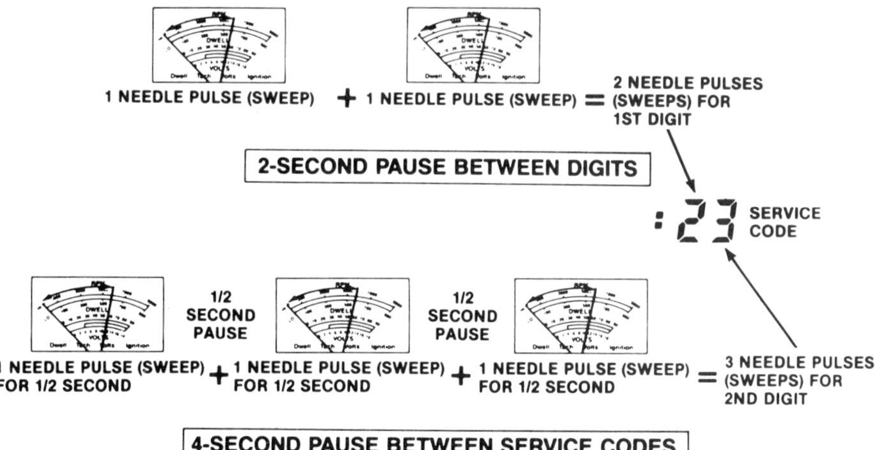

Figure 21-12. A pulsating voltmeter needle will signal numerical trouble codes. (Ford)

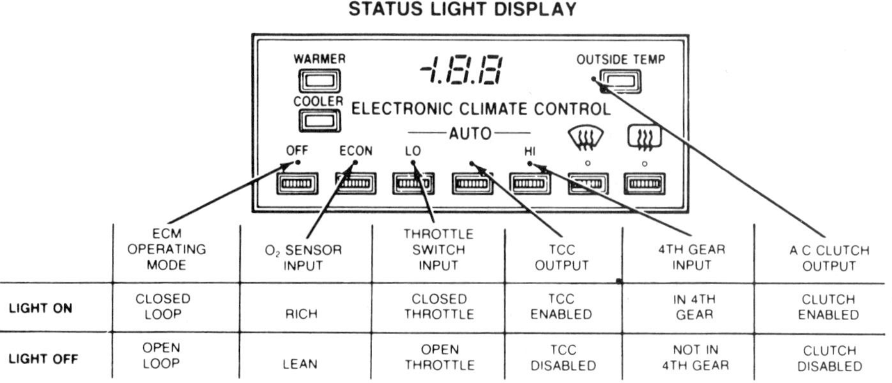

	ECM OPERATING MODE	O_2 SENSOR INPUT	THROTTLE SWITCH INPUT	TCC OUTPUT	4TH GEAR INPUT	A C CLUTCH OUTPUT
LIGHT ON	CLOSED LOOP	RICH	CLOSED THROTTLE	TCC ENABLED	IN 4TH GEAR	CLUTCH ENABLED
LIGHT OFF	OPEN LOOP	LEAN	OPEN THROTTLE	TCC DISABLED	NOT IN 4TH GEAR	CLUTCH DISABLED

Figure 21-13. Cadillac trouble codes are shown on the digital display of the air conditioning control panel. (Cadillac)

an instrument panel display area, figure 21-13.

TESTING THE GM CCC AND C-4 SYSTEMS

Late-model GM "full-function" CCC systems monitor several dozen engine and vehicle conditions and control as many output functions. Figure 21-14 is a block diagram of a full-function carbureted CCC system. Not all vehicles have all functions monitored or controlled. "Minimum-function" CCC systems on T-cars and earlier C-4 systems are examples.

Reading GM Trouble Codes—Diagnostic (Area) Tests

All CCC and C-4 systems have a CHECK ENGINE lamp on the instrument panel that lights to warn the driver of a malfunction. You can use the CHECK ENGINE lamp to read the codes by grounding a test terminal in a connector under the instrument panel. The various GM divisions call this connector the "assembly line communications link" (ALCL) or "assembly line diagnostic link" (ALDL) connector. The connector may have 4, 5, or 12 terminals, figure 21-15. The system test terminal and a ground terminal are usually next to each other on the right side or bottom of the connector. The connector is usually under the left side of the instrument panel, figure 21-9. Some C-4 systems have a single-wire connector near the control module, or computer, figure 21-8.

When you ground the test terminal with the ignition key on, the CHECK ENGINE lamp flashes to signal trouble codes. Full-function CCC systems will display both continuous (hard) and intermittent (soft) codes. C-4 systems and minimum-function CCC systems display codes only for faults present during testing. Check the operation of the diagnostic circuit and read trouble codes as follows:

1. Turn the ignition key on. Do not start the engine. The CHECK ENGINE lamp should light, along with other warning lamps. This is a bulb check to ensure that the lamp works.
2. Turn the ignition key off.
3. Connect the test terminal to the ground terminal in the connector, figure 21-16.
4. Turn the key on. Do not start the engine.
5. Observe that the CHECK ENGINE lamp flashes one short flash, followed by a pause, followed by two short, rapid flashes, figure 21-17. This is code 12 and indicates no distributor pulses (tach signal) to the computer.
6. Allow code 12 to flash three times.
   a. If other trouble codes are present in the system, the lamp will pause and then flash the other codes three times each. It will then pause and flash code 12 four times.
   b. If the lamp flashes code 12 four times, pauses, and then flashes code 12 four times again, there are no trouble codes in the system.

Because the engine is off, there should be no tach signal to the computer during this test. Code 12 is a system self-check, which indicates that the diagnostic function is working properly. Figures 21-18 and 21-19 list typical trouble codes for full-function and minimum-function CCC systems. Figure 21-20 lists trouble codes for most C-4 systems. Notice that many codes are the same for all systems. All use code 12 as a self-test.

These lists are only examples of typical trouble codes. As GM engine control systems have become more sophisticated throughout the 1980's

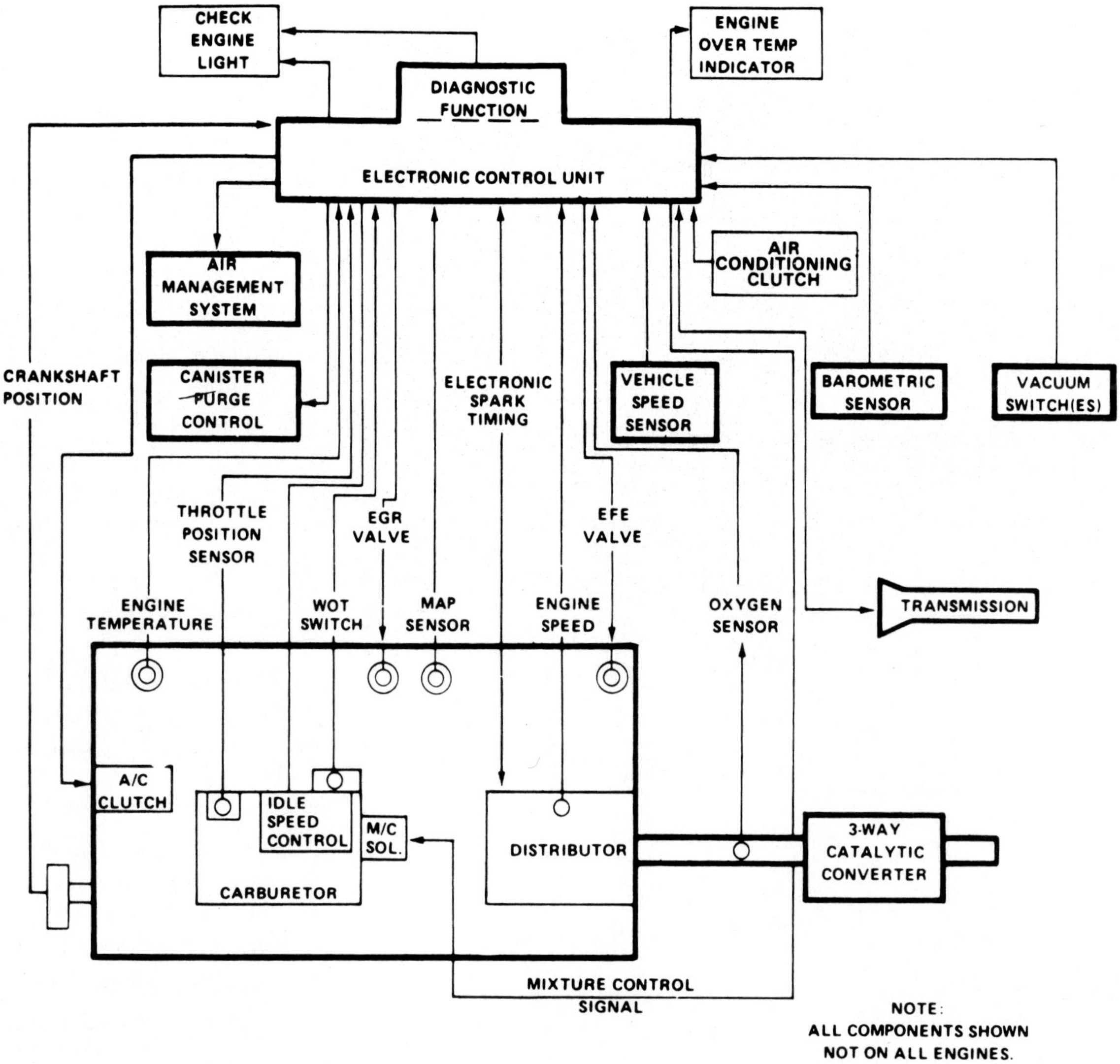

Figure 21-14. This block diagram shows the input and output functions of a CCC system. (Chevrolet)

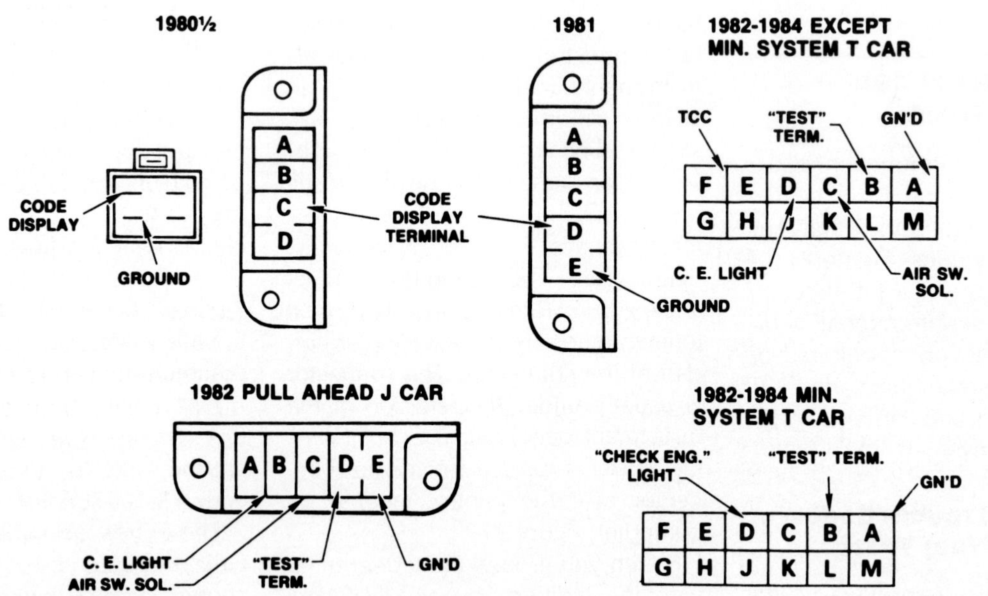

Figure 21-15. Typical GM CCC diagnostic (test) connector terminal locations. (GM)

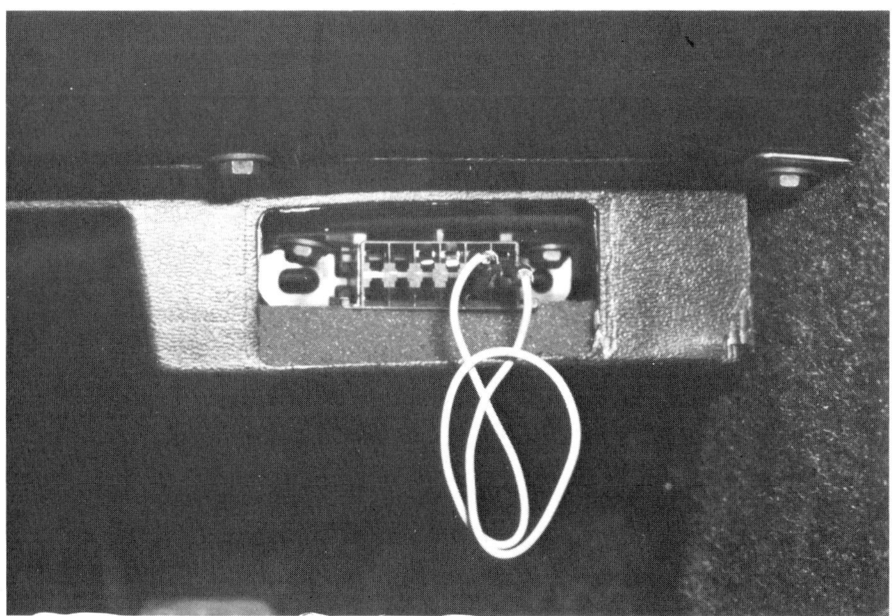

Figure 21-16. Ground the test terminal in the connector.

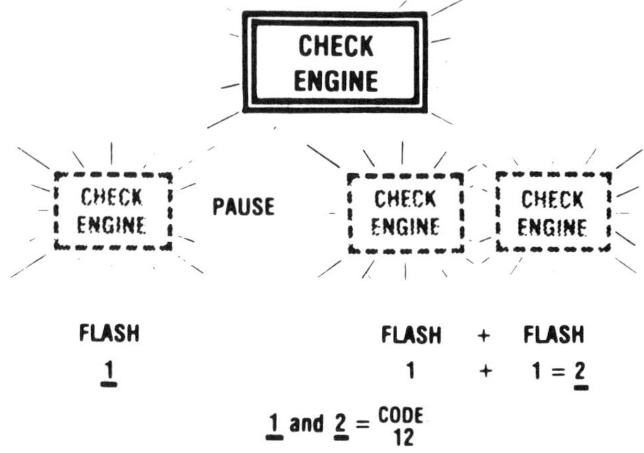

Figure 21-17. The CHECK ENGINE lamp flashes like this to indicate codes. (GM)

and into the 1990's, their abilities to monitor their own functions have increased yearly. GM systems of the early 1990's are able to display more than 100 different engine trouble codes. Also, those models with body computer modules can display 50 or more additional codes for body electrical problems. All of these facts are another reminder that you must work with the system specifications and test procedures for a specific vehicle to diagnose a problem accurately.

All GM systems flash codes in numerical sequence, starting with the lowest, except for 50-series codes. A 50-series code will flash before any other and indicates a problem in the computer, or electronic control module (ECM). You must diagnose 50-series codes and correct problems before proceeding with any other code troubleshooting. In any case, use the diagnostic charts or procedures to check out codes in the sequence in which they flash.

Write down the trouble codes as they flash because you must clear the computer memory to determine whether a code is continuous or intermittent. The long-term memory stores intermittent codes for 50 or 60 engine-start cycles (key off, on, and off) after they are recorded. Continue with the general diagnostic test as follows:

1. After writing down any trouble codes and seeing code 12 repeat, turn the ignition key off and unground the test terminal.

2. With the ignition off, clear the computer memory in one of the following ways, depending on the vehicle:

- **a.** Remove a specified fuse from the fuse block (courtesy lamps, computer, turn signals, etc.), figure 21-21.
- **b.** Disconnect a specified connector at the computer.
- **c.** Disconnect the computer memory connector at the battery positive terminal, figure 21-22.
- **d.** Disconnect the battery ground cable. This is the least preferred method because it erases other electronic memory devices on the vehicle (clocks, radios, etc.)

3. Reconnect the fuse or connector and start the engine.

4. Watch the CHECK ENGINE lamp. If the lamp lights, a continuous fault is present. If the lamp does not light, any codes displayed earlier were for previous intermittent faults.

Before we go to pinpoint tests, we should mention a few other things that can happen during a GM system diagnostic check. If the CHECK ENGINE lamp doesn't light during the bulb check (engine off, key turned to Run), GM provides a specific diagnostic chart that you should use before continuing with any other tests. If the light glows dimly but displays no codes, look for a poor ground at the computer or for

TROUBLE CODE IDENTIFICATION

The "CHECK ENGINE" light will only be "ON" if the malfunction exists under the conditions listed below. It takes up to five seconds minimum for the light to come on when a problem occurs. If the malfunction clears, the light will go out and a trouble code will be set in the ECM. Code 12 does not store in memory. Any codes stored will be erased when the ignition is turned "OFF."

The trouble codes indicate problems as follows:

TROUBLE CODE 12 No distributor reference pulses to the ECM. This code is not stored in memory and will only flash while the fault is present.

TROUBLE CODE 13 Oxygen sensor circuit—The engine must run up to five minutes at part throttle, under road load, before this code will set.

TROUBLE CODE 14 Shorted coolant sensor circuit—The engine must run up to five minutes before this code will set.

TROUBLE CODE 15 Open coolant sensor circuit—The engine must run up to five minutes before this code will set.

TROUBLE CODE 21 Throttle position sensor circuit—The engine must run up to 25 seconds, at specified curb idle speed, before this code will set.

TROUBLE CODE 23 Open or grounded M/C solenoid circuit.

TROUBLE CODE 24 Vehicle speed sensor (VSS) circuit—The car must operate up to five minutes at road speed before this code will set.

TROUBLE CODE 32 Barometric pressure sensor (BARO) circuit low, or altitude compensator low on J-Car.

TROUBLE CODE 34 Manifold absolute pressure (MAP) or vacuum sensor circuit—The engine must run up to five minutes, at specified curb idle speed, before this code will set.

TROUBLE CODE 35 Idle speed control (ISC) switch circuit shorted. (Over 50% throttle for over 2 sec.)

TROUBLE CODE 41 No distributor reference pulses to the ECM at specified engine vacuum. This code will store in memory.

TROUBLE CODE 42 Electronic spark timing (EST) bypass circuit or EST circuit grounded or open.

TROUBLE CODE 43 ESC retard signal for too long; causes a retard in EST signal.

TROUBLE CODE 44 Lean exhaust indication—The engine must run up to five minutes, in closed loop, at part throttle and road load before this code will set.

TROUBLE CODES 44 & 55 (At same time)—Faulty oxygen sensor circuit.

TROUBLE CODE 45 Rich exhaust indication—The engine must run up to five minutes, in closed loop, at part throttle and road load before this code will set.

TROUBLE CODE 51 Faulty calibration unit (PROM) or installation. It takes up to 30 seconds before this code will set.

TROUBLE CODE 54 Shorted M/C solenoid circuit or faulty ECM.

TROUBLE CODE 55 Grounded V ref (terminal "21"), faulty oxygen sensor or ECM.

Figure 21-18. Basic trouble codes for full-function CCC systems. (GM)

computer supply voltage lower than 9 volts. If the lamp flashes erratically and displays unlisted codes, look for voltage interference from one of these sources:

• The CCC wire harness is too close to the other engine wiring

• A diode is open in the air conditioning compressor clutch circuit, causing high-voltage interference whenever the clutch engages

• A citizens' band (CB) radio transmitter is causing high-voltage interference

• The ignition system is causing high-voltage interference.

If the CHECK ENGINE lamp does not light, but the driver has a performance complaint, GM provides a "driver comment" diagnostic chart that you can use to test for the most probable causes of certain symptoms, figure 21-23.

On carbureted engines, the serial data line from the ECM and the CHECK

TROUBLE CODE IDENTIFICATION

The "CHECK ENGINE" light will only be "ON" if the malfunction exists under the conditions listed below. It takes up to five seconds minimum for the light to come on when a problem occurs. If the malfunction clears, the light will go out and a trouble code will be set in the ECM. Code 12 does not store in memory. Any codes stored will be erased if no problem reoccurs within 50 engine starts.

The trouble codes indicate problems as follows:

Trouble Code 12 No distributor reference pulses to the ECM. This code is not stored in memory and will only flash while the fault is present.

Trouble Code 15 Open coolant switch circuit—The engine must run for five minutes before this code will set.

Trouble Code 21 Throttle position sensor circuit at WOT—The engine must run for 10 seconds, below 1000 RPM, before this code will set.

Trouble Code 23 M/C solenoid circuit. (Must be in closed loop mode to set code.)

Trouble Code 44 Lean exhaust indication—The engine must run up to one minute, in closed loop, at part throttle above 2000 RPM before this code will set.

Trouble Code 45 Rich exhaust indication—The engine must run up to one minute, in closed loop, at part throttle above 2000 RPM before this code will set.

Trouble Code 51 Faulty calibration unit (PROM) or installation. Turns ECM off.

Figure 21-19. Basic trouble codes for minimum-function CCC systems. (GM)

GM C-4 System Trouble Codes

Trouble Code	*Problem—See Specific Diagnostic Chart*
12	No tachometer signal to ECM.
13	Oxygen sensor circuit. The engine must run for about 5 minutes at part throttle for this code to show.
14	Shorted coolant sensor circuit. The engine must run for about 2 minutes for this code to show.
15	Open coolant sensor circuit. The engine must run for about 5 minutes at part throttle for this code to show.
21	Throttle position sensor circuit (V-6).
22	Grounded closed-throttle or wide-open-throttle switch circuit (4-cylinder).
21 & 22 (same time)	Grounded wide-open-throttle switch circuit (4-cylinder).
23	Carburetor solenoid circuit.
44	Lean oxygen sensor signal
45	Rich oxygen sensor signal
51	On service-replacement ECM,.check calibration unit. On original-equipment ECM, replace ECM.
54	Faulty carburetor solenoid or ECM.
52, 53	Replace ECM.
55	Faulty throttle position sensor (V-6) or ECM.

Figure 21-20. Basic trouble codes for C-4 systems on GM X-cars.

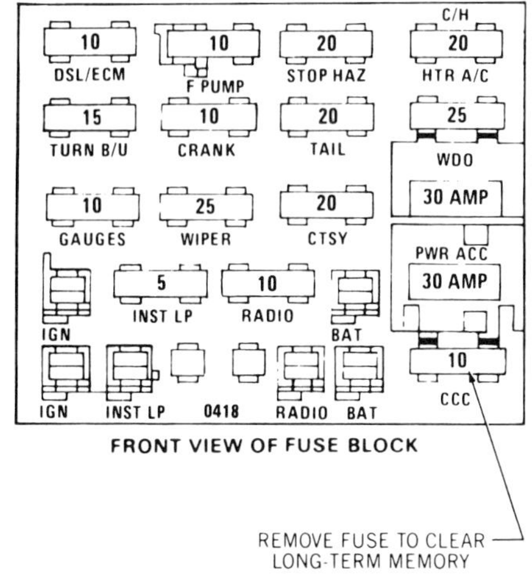

Figure 21-21. Remove a specified fuse on some cars to clear the computer memory. (Chevrolet)

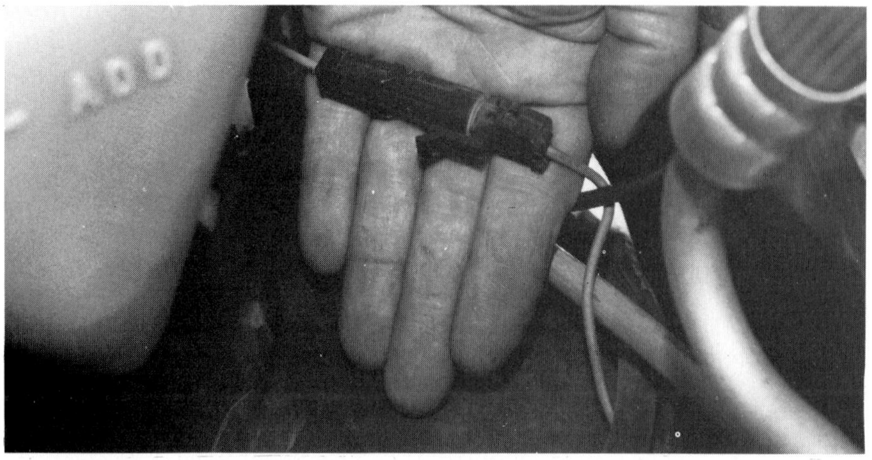

Figure 21-22. On most GM cars, disconnect the computer memory connector near the battery to clear the long-term memory.

DRIVER COMMENTS

ENGINE PERFORMANCE PROBLEM (ODOR, SURGE, FUEL ECONOMY ...)
EMISSION PROBLEM

IF THE "CHECK ENGINE" LIGHT IS NOT ON, NORMAL CHECKS THAT WOULD BE PERFORMED ON VEHICLE WITHOUT THE SYSTEM SHOULD BE DONE FIRST.

IF GENERATOR OR COOLANT LIGHT IS ON WITH THE "CHECK ENGINE" LIGHT, THEY SHOULD BE DIAGNOSED FIRST.

INSPECT FOR POOR CONNECTIONS AT COOLANT SENSOR, M/C SOLENOID, ETC., AND POOR OR LOOSE VACUUM HOSES AND CONNECTIONS. REPAIR AS NECESSARY.

- INTERMITTENT "CHECK ENGINE" LIGHT BUT NO TROUBLE CODE STORED.
  - CHECK FOR INTERMITTENT CONNECTION IN CIRCUIT FROM:
    - IGNITION COIL TO GROUND AND ARCING AT SPARK PLUG WIRES OR PLUGS.
    - BAT. TO ECM TERMS. 'C' AND 'R'.
    - ECM TERMS. 'A' AND 'U' TO ENGINE GROUND.
  - LOSS OF TROUBLE CODE (LONG-TERM MEMORY).
    GROUNDING DWELL LEAD FOR 10 SECONDS WITH "TEST" LEAD UNGROUNDED SHOULD GIVE CODE 23 WHICH SHOULD BE RETAINED AFTER ENGINE IS STOPPED AND IGNITION TURNED TO "RUN" POSITION.
    IF IT IS NOT, ECM IS DEFECTIVE.
  - EST WIRES SHOULD BE KEPT AWAY FROM SPARK PLUG WIRES, DISTRIBUTOR HOUSING, COIL AND GENERATOR. WIRES FROM EDM TERM. 13 TO DIST. AND THE SHIELD (IF USED) AROUND EST WIRES SHOULD BE A GOOD GROUND.
  - OPEN DIODE ACROSS A/C COMPRESSOR CLUTCH.

- STALLING, ROUGH IDLE, DIESELING OR IMPROPER IDLE SPEED.
  SEE IDLE SPEED CONTROL (ISC) CHECK.

- DETONATION (SPARK KNOCK)
  CHECK: ESC SYSTEM CHECK, IF APPLICABLE.
  MAP OR VACUUM SENSOR OUTPUT—CHART 7 FOR MAP SENSOR OR CHART 8 FOR VACUUM SENSOR.
  EGR CHECK.
  TPS ENRICHMENT OPERATION—SEE CHART 4.
  HEI OPERATION.

- POOR PERFORMANCE AND/OR FUEL ECONOMY.
  SEE EST PERFORMANCE CHECK.
  SEE ESC SYSTEM CHECK IF APPLICABLE.

- POOR FULL THROTTLE PERFORMANCE.
  SEE CHART 4 IF EQUIPPED WITH TPS.

- INTERMITTENT NO-START.
  - INCORRECT PICK-UP COIL OR IGNITION COIL. SEE "CRANKS, BUT WON'T RUN" CHART.
  - INTERMITTENT GROUND CONNECTIONS IN ECM.

- ALL OTHER COMPLAINTS
  MAKE SYSTEM PERFORMANCE CHECK ON WARM ENGINE
  (UPPER RADIATOR HOSE HOT).

THE SYSTEM PERFORMANCE CHECK SHOULD BE PERFORMED AFTER ANY REPAIRS TO THE SYSTEM HAVE BEEN MADE.

Figure 21-23. This driver complaint chart will help you begin troubleshooting a CCC system with no trouble codes. (GM)

ENGINE lamp are connected to the same terminal in the ALDL connector and receive voltage from a driver circuit in the ECM. If the lamp driver circuit becomes open, the lamp will not light and codes cannot be displayed. GM provides a diagnostic chart to pinpoint this problem, and the final fix is to replace the ECM.

Cadillac and GM 30 Diagnostic Displays

An easier and clearer way to display trouble codes is used on late-model Cadillacs with CCC systems and throttle body, digital fuel injection (DFI), as well as 1986 and later Eldorado, Seville, and Allanté models. The method can be used while driving the vehicle. You will need the Cadillac test procedures to use the extensive list of trouble codes and test procedures the system provides. System operation differs slightly depending upon vehicle model and year, but its basic operation is essentially the same. After turning the ignition on, you simultaneously depress and hold the OFF and WARMER buttons on the air conditioning electronic climate control (ECC) panel. The digital display, which normally shows time or temperature, then displays CCC system trouble codes, figure 21-13.

In addition to the CCC system control functions, the Cadillac system lets you check cruise control operation and transmission switch functions. Instead of a single CHECK ENGINE lamp, most Cadillacs have separate SERVICE NOW and SERVICE SOON lamps to distinguish between hard, or serious, failures and soft failures. Figure 21-24 is an example of trouble codes and other test displays for Cadillac models with DFI. Eldorado, Seville, and Allanté models have additional features also incorporated in the Buick and Oldsmobile versions of the GM 30 models.

The 1986 and later Oldsmobile Toronado system uses the air conditioning ECC panel for its display and is accessed in the same way as the

1983 DIAGNOSTIC CODES	
CODE	CIRCUIT AFFECTED
■■ 12	NO DISTRIBUTOR (TACH) SIGNAL
13	O_2 SENSOR NOT READY
14	SHORTED COOLANT SENSOR CIRCUIT
15	OPEN COOLANT SENSOR CIRCUIT
■■ 16	GENERATOR VOLTAGE OUT OF RANGE
18	OPEN CRANK SIGNAL CIRCUIT
19	SHORTED FUEL PUMP CIRCUIT
■■ 20	OPEN FUEL PUMP CIRCUIT
21	SHORTED THROTTLE POSITION SENSOR CIRCUIT
22	OPEN THROTTLE POSITION SENSOR CIRCUIT
23	EST/BYPASS CIRCUIT PROBLEM
24	SPEED SENSOR CIRCUIT PROBLEM
26	SHORTED THROTTLE SWITCH CIRCUIT
27	OPEN THROTTLE SWITCH CIRCUIT
28	OPEN FOURTH GEAR CIRCUIT
29	SHORTED FOURTH GEAR CIRCUIT
30	ISC CIRCUIT PROBLEM
■■ 31	SHORTED MAP SENSOR CIRCUIT
■■ 32	OPEN MAP SENSOR CIRCUIT
■■ 33	MAP/BARO SENSOR CORRELATION
■■ 34	MAP SIGNAL TOO HIGH
35	SHORTED BARO SENSOR CIRCUIT
36	OPEN BARO SENSOR CIRCUIT
37	SHORTED MAT SENSOR CIRCUIT
38	OPEN MAT SENSOR CIRCUIT
39	TCC ENGAGEMENT PROBLEM
■■ 44	LEAN EXHAUST SIGNAL
■■ 45	RICH EXHAUST SIGNAL
■■ 51	PROM ERROR INDICATOR
▼ 52	ECM MEMORY RESET INDICATOR
▼ 53	DISTRIBUTOR SIGNAL INTERRUPT
▼ 60	TRANSMISSION NOT IN DRIVE
▼ 63	CAR AND SET SPEED TOLERANCE EXCEEDED
▼ 64	CAR ACCELERATION EXCEEDS MAX. LIMIT
▼ 65	COOLANT TEMPERATURE EXCEEDS MAX. LIMIT
▼ 66	ENGINE RPM EXCEEDS MAXIMUM LIMIT
▼ 67	SHORTED SET OR RESUME CIRCUIT
.7.0	SYSTEM READY FOR FURTHER TESTS
.7.1	CRUISE CONTROL BRAKE CIRCUIT TEST
.7.2	THROTTLE SWITCH CIRCUIT TEST
.7.3	DRIVE (ADL) CIRCUIT TEST
.7.4	REVERSE CIRCUIT TEST
.7.5	CRUISE ON/OFF CIRCUIT TEST
.7.6	"SET/COAST" CIRCUIT TEST
.7.7	"RESUME/ACCELERATION" CIRCUIT TEST
.7.8	"INSTANT/AVERAGE" CIRCUIT TEST
.7.9	"RESET" CIRCUIT TEST
.8.0	A/C CLUTCH CIRCUIT TEST
-1.8.8	DISPLAY CHECK
.9.0	SYSTEM READY TO DISPLAY ENGINE DATA
.9.5	SYSTEM READY FOR OUTPUT CYCLING OR IN FIXED SPARK MODE
.9.6	OUTPUT CYCLING
.0.0	ALL DIANOSTICS COMPLETE
■■	TURNS ON "SERVICE NOW" LIGHT
	TURNS ON "SERVICE SOON" LIGHT
▼	DOES NOT TURN ON ANY TELLTALE LIGHT
NOTE: CRUISE IS DISENGAGED WITH ANY "SERVICE NOW" LIGHT OR WITH CODES 60-67.	

Figure 21-24. Cadillac CCC system trouble codes include checks for air conditioning and cruise control systems, as well as engine trouble codes. (Cadillac)

SERVICE DIAGNOSTICS
SYSTEM OVERVIEW

SERVICE DIAGNOSTICS
BASIC OPERATION

- ENTER DIAGNOSTICS BY SIMULTANEOUSLY PRESSING THE "OFF" AND "WARM" BUTTONS ON THE ECC FOR 3 SECONDS.
- DIAGNOSTIC CODES BEGIN WITH ANY ECM CODES FOLLOWED BY BCM CODES.
- TO PROCEED WITH DIAGNOSTICS PRESS THE INDICATED ECC BUTTONS.
- HI AND LO REFER TO THE FAN UP AND FAN DOWN BUTTONS.
- EXIT DIAGNOSTICS BY PRESSING "BI-LEV"

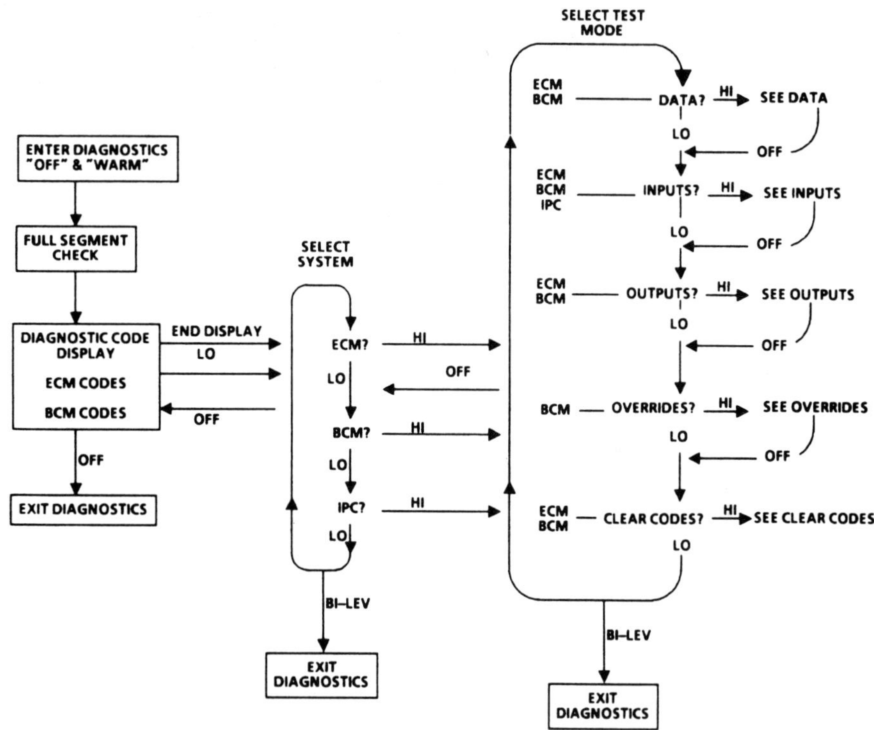

Figure 21-25. The self-diagnostic system flow chart for Cadillac Allanté, Eldorado, Seville, and Oldsmobile Toronado models with onboard diagnostics. (General Motors)

Cadillac models. The 1986 and later Buick Riviera has a menu-driven cathode ray tube controller (CRTC). This device requires a slightly different procedure to access the diagnostic mode. After turning the ignition on, you simultaneously depress and hold the OFF and WARM pads on the CRT's climate control page until two beeps are heard or until the "Service Mode" page appears on the display.

The Cadillac Allanté and GM 30 models (Eldorado, Seville, Toronado, and Riviera) all have a body computer module (BCM), which expands the diagnostic capability over the system used in DFI Cadillacs. Trouble codes can be set and stored by the electronic control module (ECM) and the BCM. When the diagnostic mode is entered, ECM codes are displayed first. If none are stored, a "No ECM Codes" message is displayed. BCM codes are displayed after all ECM codes have been displayed. If a problem still exists, the display will indicate "Current" in addition to the code, a capability the ECM does not have.

The GM 30 system also allows you to access and display a variety of ECM and BCM parameters, discrete inputs and outputs, and BCM output override messages. After all trouble codes have been displayed, you can access the ECM, the BCM, or the instrument panel cluster (IPC) for a series of tests to check data input, switch input and output cycling or override information, figures 21-25 and 21-26.

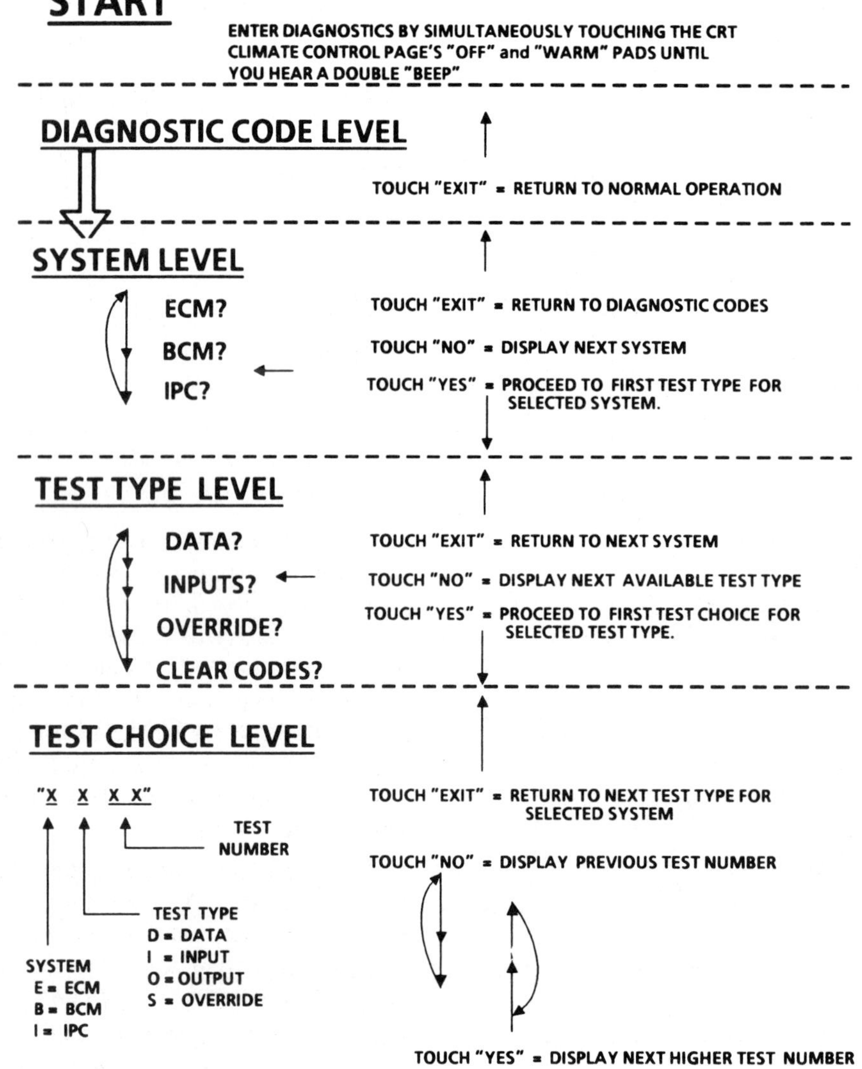

Figure 21-26. The self-diagnostic system flow chart for Buick Riviera models with onboard diagnostics. (General Motors)

GM CCC System Pinpoint Tests

After recording the system trouble codes and determining whether they are continuous or intermittent, you must go to a specific diagnostic chart for each code. Notice in figures 21-18 and 21-24 that some circuits have two codes for different faults. Codes 14 (shorted coolant sensor) and 15 (open coolant sensor) are examples. Each code has a separate chart to help you find the problem quickly. Some early CCC circuits or subsystems may not have trouble codes, such as air injection switching and vapor canister purging. The "driver comment" charts for common symptoms will lead you to charts for these circuits or subsystems.

You don't need special testers to troubleshoot CCC system problems. You can use basic equipment that you use to test electrical and vacuum systems, such as:

- A hand-operated vacuum pump and vacuum gauge
- A dwellmeter and tachometer
- An ohmmeter
- A 12-volt test lamp
- Assorted jumper wires
- A digital volt-ohmmeter (DVOM) with at least 10 megohms-per-volt input impedance
- Special hand tools to remove the EGO sensor and to service Weather-Pack electrical connectors.

Figure 21-27 shows the tools needed for CCC system service. Notice that the illustration shows the GM special tool numbers for many of the items. However, you can get equivalent tools from several different manufacturers. You can make many of the items yourself, such as jumper wires. Several tool companies sell special jumper wires with terminals to match the special connectors used on CCC system parts, figure 21-28.

You will use the dwellmeter to test carburetor mixture control solenoid operation. We will explain dwellmeter use and operation in the system performance check procedure.

Three conditions *must* exist before you can use a diagnostic chart:

1. All vacuum hoses must be connected correctly and free from leaks.
2. All electrical connectors must be connected correctly and free from high or low resistance.
3. Most importanty, *the fault must exist at the time of testing.*

Diagnostic charts usually start with steps to check vacuum and electrical connections. The charts *will not work,* however, for an intermittent fault that has been recorded earlier but is not present during testing. Notice in figure 21-18 that some codes require engine operation under specific conditions for a certain time before they will record in memory. Code 45, rich exhaust, is an example. An intermittent code 45 may record in memory,

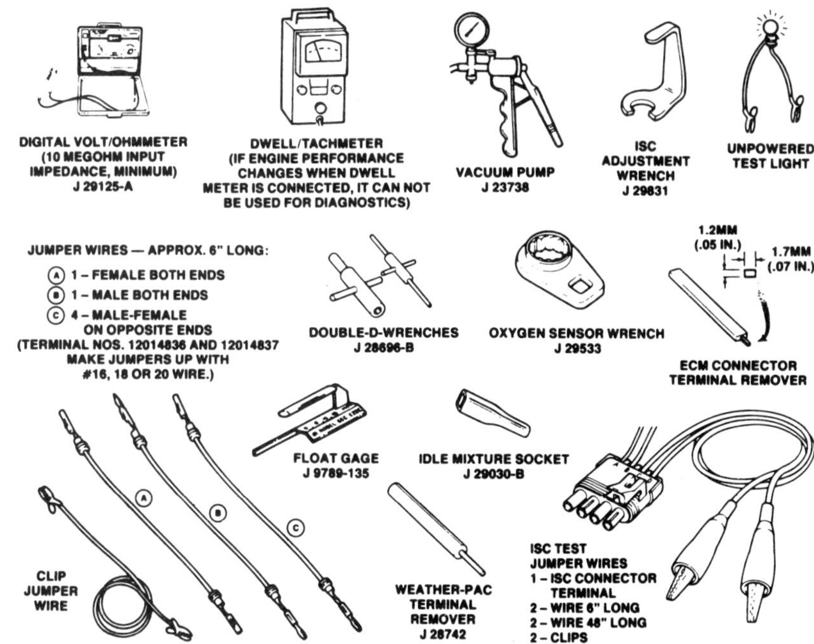

Figure 21-27. These are the tools you need for CCC system testing and repair. (GM)

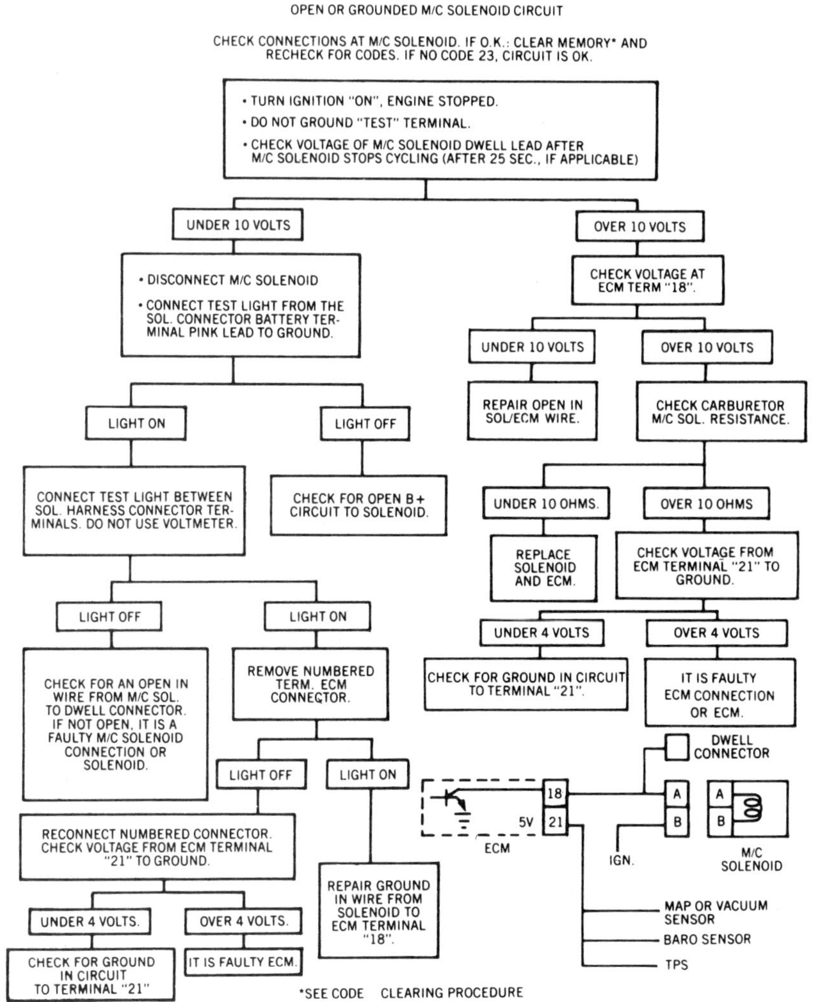

Figure 21-29. This is the CCC troubleshooting chart for code 23, an open or grounded mixture control solenoid circuit. (GM)

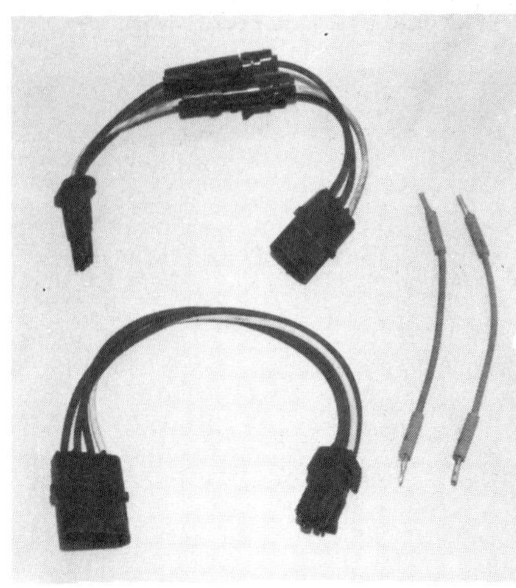

Figure 21-28. These jumper wires have connectors that match the Weather-Pack connectors of GM CCC systems.

but to find the cause, you must get it to occur during testing.

To explain the use of a diagnostic chart, let's take trouble code 23 as an example. Code 23 shows an open or shorted mixture control (MC) solenoid circuit for a CCC feedback carburetor. Figure 21-29 is the chart for troubleshooting this code. Your basic steps are as follows:

1. Check the connections at the MC solenoid for looseness, corrosion, or broken wiring. Repair if necessary.

2. With the CHECK ENGINE lamp test connector ungrounded, turn the ignition key on. Do not start the engine.

3. Connect a voltmeter to the MC solenoid dwell test connector. This is the green, single-terminal open connector at the rear of the engine compartment, figure 21-30. (You will use this connector to measure MC solenoid dwell, or duty cycle, during a system performance test.)

4. Note the voltmeter reading. It will be under or over 10 volts:

  a. If over 10 volts, check the voltage at computer (ECM) terminal 18.

  b. If under 10 volts, test for and repair an open circuit between the ECM and the MC solenoid.

5. If voltage at ECM terminal 18 is over 10 volts, use your ohmmeter to test resistance in the MC solenoid:

  a. If resistance is less than 10 ohms, replace both the ECM and the MC solenoid.

Figure 21-30. This is the mixture control solenoid test connector on a late-model GM car.

b. If resistance is more than 10 ohms, use your voltmeter to measure voltage between ECM terminal 21 and ground.

6. If voltage between terminal 21 and ground is less than 4 volts, check for a ground in the circuit to terminal 21. If voltage is more than 4 volts, repair the ECM connection or replace the ECM.

This diagnostic chart and the preceding steps are just one short example of the pinpoint tests that you will use to isolate problems. Some cars have 50 or more specific procedures. Notice that the diagnostic chart for this trouble code is arranged in "Christmas Tree" fashion. Each decision point leads you to a new checking sequence. You may go slowly the first few times you use one, but as you get familiar with the tests, you will go through them quickly and accurately. You must go through each diagnostic routine for each trouble code that you find in the system.

GM CCC System Performance Test

You can use the CCC system performance test to check overall system operation in four basic situations:

1. When a driver has a performance complaint, but the system displays no trouble codes

2. After each pinpoint test and service to verify a specific repair

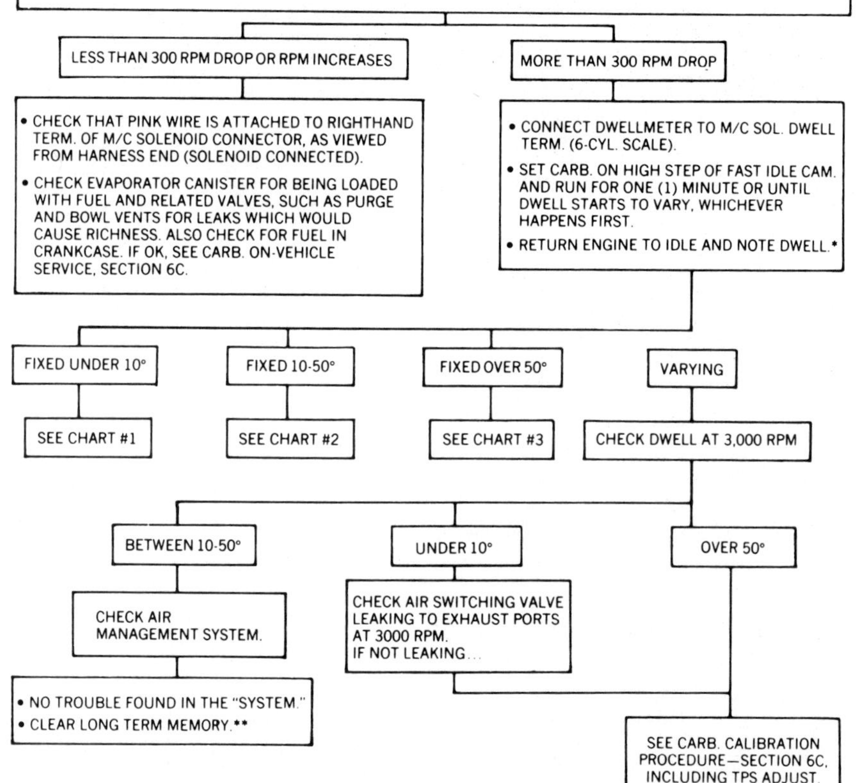

Figure 21-31. This is the chart for a CCC system performance test. (GM)

3. After all repairs to verify complete system operation

4. As a routine performance check (preventive maintenance)

This test checks the fuel control at the feedback carburetor. It also verifies the operation of system sensors, because they must work properly to have correct fuel metering. Figure 21-31 is the chart for the CCC performance check. It has three basic parts:

1. A carburetor lean-drop check at part throttle (3,000 rpm)

2. An MC solenoid dwell test at idle

3. An MC solenoid dwell test at 3,000 rpm

Caution. Do not ground the test terminal in the system diagnostic connector before starting the engine. Start the engine first. The computer may be damaged if the engine is started with the test terminal grounded.

For part 1 of the test, follow the 7 steps at the top of the chart. Disconnecting the MC solenoid forces the carburetor to a fully rich condition. When you reconnect the solenoid with the dwell test lead grounded, the carburetor should go to a fully lean position and engine speed should drop. If the mixture goes lean correctly at 3,000 rpm, speed should drop 400 to 1,000 rpm, depending on the engine. In any case, speed will change less than 300 rpm or more than 300 rpm.

- If speed drops less than 300 rpm, the carburetor or the solenoid circuit has a problem that prevents lean mixture control.
- If speed drops more than 300 rpm, lean mixture control is okay at part throttle.

Part 2 of the test leads you to fixing problems with the MC solenoid, the vapor canister purging, or the PCV system; *or* it leads you to another series of solenoid dwell tests at idle. If the dwell reading is fixed (steady) in one of three areas, go to another GM diagnostic chart. If the solenoid dwell varies between 10° and 50° at idle, go to part 3 of the performance test.

In part 3, you check MC solenoid dwell at 3,000 rpm. If it varies from 10° to 50°, the feedback carburetor is controlling fuel metering properly. However, check the air injection switching to be sure it works properly. If MC solenoid dwell is fixed over or under 10°, the performance test chart leads you to test the air injection system or service the carburetor.

This preceding example is for a carbureted engine. System performance tests for later-model fuel-injected engines differ in their details, but the principles are similar.

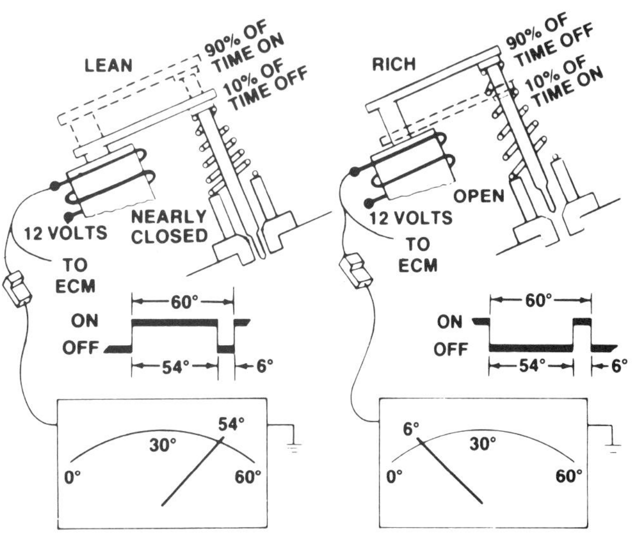

Figure 21-32. Dwell measurement indicates solenoid duty cycle. (GM)

Testing a Mixture Control Solenoid with a Dwellmeter

General Motors and some other carmakers specify the use of a dwellmeter to test the operation of mixture control solenoids and other solenoids that operate continually with varying duty cycles. To do this effectively, you must know how "dwell" readings relate to solenoid operation.

In chapter 12, you learned to use a dwellmeter to measure ignition dwell—the time when the primary circuit is closed and conducting current to the coil. This is the period when the breaker points are closed or the power transistor is on. Because ignition dwell relates to engine and distributor rotation, a dwellmeter measures it in degrees. Dwell also is ignition primary current pulse width and duty cycle. At a steady engine speed, the breaker points or electronic ignition module switches primary current on and off a specific number of times per second. The amount of on-time is the pulse width. The percentage of on-time to total cycle time is the duty cycle. Breaker-point ignitions have a fixed dwell (a fixed duty cycle). Some electronic ignitions have a variable dwell (a variable duty cycle). A dwellmeter can measure either condition.

Because the switching action of an ignition primary circuit and the switching of a solenoid circuit have measurable duty cycles, you can use a dwellmeter to test either. A mixture control solenoid operates at a fixed rate of 10 cycles per second, but the duty cycle (percentage of on-time) varies, figure 21-32. The duty cycle is the dwell reading that you get on the meter. It is the time when the solenoid is energized to move the carburetor metering rods to the lean-mixture position.

To test an MC solenoid with a dwellmeter, connect the meter to the solenoid supply voltage lead. GM provides a test connector in the engine compartment, figure 21-30, to do this. Then set the meter on the 6-cylinder scale. This provides a dwell range from 0° to 60°. Even though you read the solenoid "dwell" in degrees, you are really looking at the percent of on-time, the duty cycle. You can convert the degree readings to percentage readings easily, as follows:

0° = 0%
6° = 10%
15° = 25%
30° = 50%
45° = 75%
54° = 90%
60° = 100%

Figure 21-33 shows this relationship. Each 1-degree dwell change equals a 1.67-percent duty cycle change.

The MC solenoid moves the carburetor metering rods up and down 10 times per second. When the solenoid is on, the rods are down in the lean position. When the solenoid is off, the rods are up in the rich position. The system computer varies the duty cycle in response to signals from the EGO sensor in closed loop and other engine sensors in open loop. Figure 21-34 lists the important relationships of exhaust oxygen content, EGO sensor signal, MC solenoid operation, and the change in air-fuel ratio.

Figure 21-35 shows how solenoid operation changes for different open-loop (dwell fixed) and closed-loop (dwell varying) conditions. It also shows the relationships of solenoid

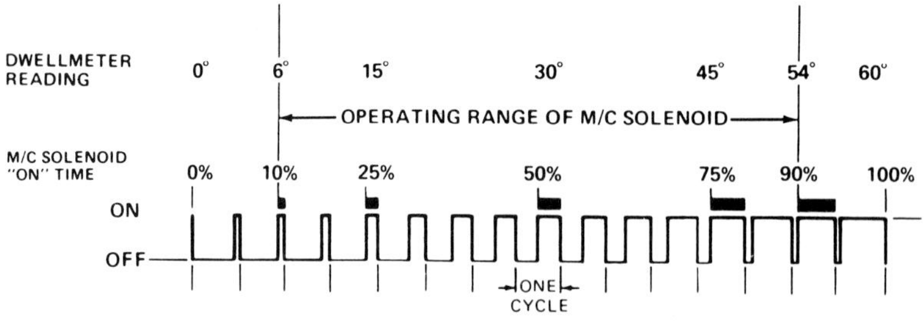

Figure 21-33. Dwellmeter readings indicate the duty cycle percentage. (GM)

Air-Fuel Mixture	*Exhaust Oxygen Content*	*Oxygen Sensor Voltage*	*Computer Output Signal*	*Mixture Control Solenoid*	*Fuel Metering Change To*	*Idle Airflow*
Rich	Low	High	Increase Pulse Time	On (High Dwell)	Lean	Increase
Lean	High	Low	Decrease Pulse Time	Off (Low Dwell)	Rich	Decrease

Figure 21-34. This table shows the relationships of air-fuel mixture, exhaust O_2, sensor voltage, computer response, and system response.

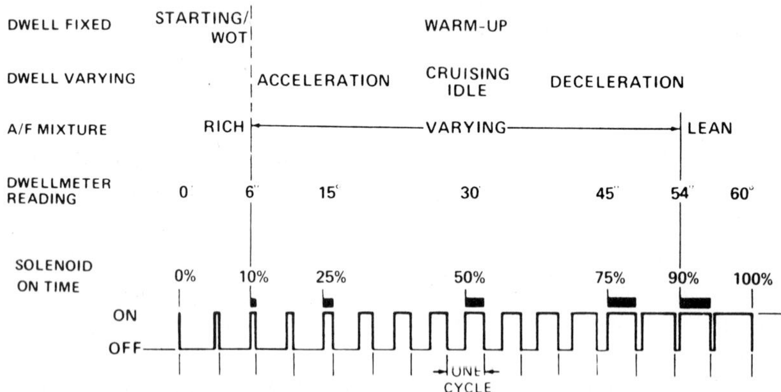

Figure 21-35. Solenoid duty cycles and dwellmeter readings change for different open- and closed-loop operating conditions. (GM)

on-time to dwellmeter readings. On CCC systems, the MC solenoid dwell should fluctuate between 5° and 6° (full rich) and 54° or 55° (full lean) during closed-loop operation. Open-loop fixed dwell may be anywhere between 5° and 55°, depending on the system computer program.

Testing GM Fuel Injection Block Learn and Integrator

Fuel injection systems do not have MC solenoids to control fuel metering, but GM has included system operating indicators in its fuel injection systems that are similar to MC dwell readings for carbureted engines. These are called **integrator** and **block learn**.

Integrator and block learn readings are available only with a scan tool, which you will learn about later in this chapter. You can't read these indications with a dwellmeter, but we will introduce them here because they are similar to the MC solenoid dwell readings.

Integrator and block learn are index numbers that indicate whether the fuel injection system is providing a lean or a rich mixture. Integrator numbers indicate a short-term correction; block learn numbers indicate a long-term, "learned" correction. On a scan tool, both integrator and block learn are displayed as numbers from 1 to 256. If the injection system is controlling fuel metering at the midpoint of the range, both numbers will be 128 (half of 256).

If the injection system is running lean, either the integrator or the block learn number will be lower than 128. If the system is rich, the number, or numbers, will be higher than 128. If the rich or lean condition is a long-term correction, the block learn number will be high or low, but the integrator number may be close to 128. If the rich or lean condition is a short-term correction, the block learn number will be close to 128, but the integrator number may be high or low.

Remember these guidelines for interpreting block learn and integrator numbers for GM fuel injections systems:

- High integrator (above 128) plus normal block learn (close to 128) means a short-term rich fuel-metering condition.
- Low integrator (below 128) plus normal block learn (close to 128) means a short-term lean fuel-metering condition.
- Normal integrator (close to 128) plus high block learn (above 128) means a long-term rich fuel-metering condition.
- Normal integrator (close to 128) plus low block learn (below 128)

means a long-term lean fuel-metering condition.

Generally, integrator can go as low as 50 or as high as 200 before an engine will stall from an overly lean or rich mixture.

A variety of conditions can cause block learn and integrator numbers to deviate from 128. GM provides troubleshooting procedures for the driver complaints (symptoms) and trouble codes that can indicate these problems. Both MC solenoid dwell (on carbureted engines) and block learn and integrator (on fuel-injected engines) are measurements that you take of an engine's operating condition. They are like measuring blood pressure or pulse rate of a doctor's patient.

No patient ever went to a doctor and said, "My blood pressure is irregular." No car owner ever went to a mechanic and said, "My integrator number is high." Patients and car owners complain of symptoms like, "I feel worn down and tired," or "I get poor gas mileage." Doctors test blood pressure (and other basic functions) to locate the cause of a symptom. Mechanics test MC dwell or block learn and integrator for the same reasons.

FORD ELECTRONIC ENGINE CONTROL (EEC) SYSTEM TESTING

Since its first feedback carburetor (FBC) system in 1978, Ford has used several electronic engine control systems. Among these are the microprocessor control unit (MCU), introduced in 1980 for feedback fuel control, and four generations of electronic engine control (EEC) systems. Ford's EEC-I, II, III, and IV systems are used with carburetors and fuel injection and various sensors and actuators to control ignition timing or feedback fuel metering, or both, along with other engine functions. EEC-III and IV have self-diagnostic capabilities, similar to GM's C-4 and CCC systems. EEC-IV is most common on late-model Ford products, and we will concentrate on this system in the following sections. Testing principles are similar, however, for EEC-III and other comparable systems.

Ford EEC-IV System Diagnostic Test

Ford calls the EEC-IV diagnostic area tests the "quick test" sequence, which includes self-diagnostic tests and the display of service codes. Not all Ford codes are trouble codes. Some are instruction codes, separator codes, or instructions to the operator. Before doing these tests, you must test battery voltage and condition and check all vacuum and electrical connections for looseness and damage. Ford divides the self-test sequence into three parts:

1. Key on, engine off
2. Engine running
3. Continuous (wiggle) testing for intermittent faults

To read codes from the EEC computer memory, you can use Ford's self-test automatic readout (STAR) tester, figures 21-36 and 21-37, or any analog (dial-indicating) voltmeter. Ford products before 1986 do not have a CHECK ENGINE lamp on the instrument panel that flashes the codes, but they all have a self-test connector in the engine compartment. If you have a Ford STAR tester, connect it as shown in figure 21-36. The tester will display codes in its digital display window, figure 21-37. Follow the operating instructions included with the STAR tester.

To use an analog voltmeter, connect it as shown in figure 21-38. The voltmeter indicates service codes as pulsing movements of the meter needle, figure 21-39. A code 23, for example, registers as two fast needle pulses, a 2-second pause, and three fast needle pulses. A 4-second pause separates codes when more than one code is present.

Key-On, Engine-Off Test. With the ignition and air conditioning off and the transmission in Park or Neutral, connect your voltmeter as described above. Then turn the key on. The volt-

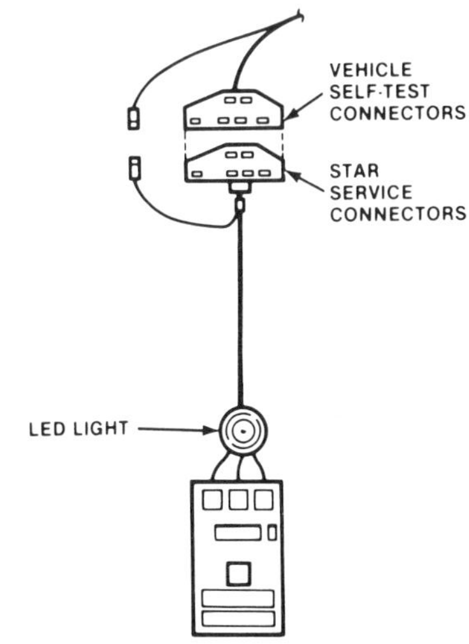

Figure 21-36. Typical Ford STAR tester connection to vehicle self-test connector. (Ford)

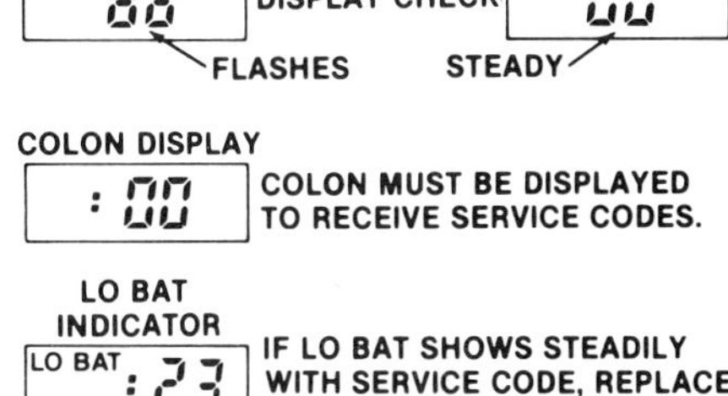

Figure 21-37. Ford STAR tester displays (readouts). (Ford)

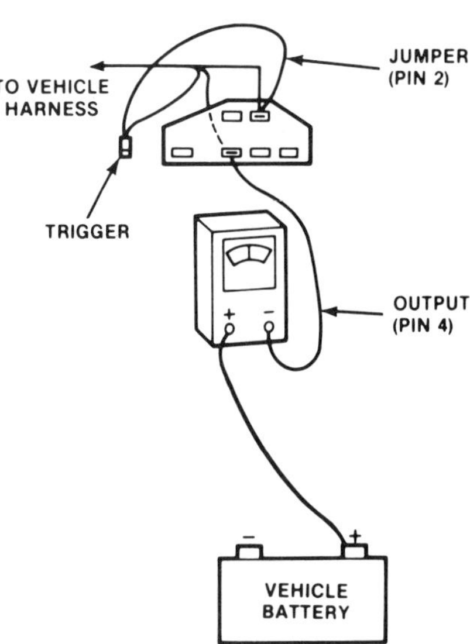

Figure 21-38. Voltmeter connections for Ford EEC self-test. (Ford)

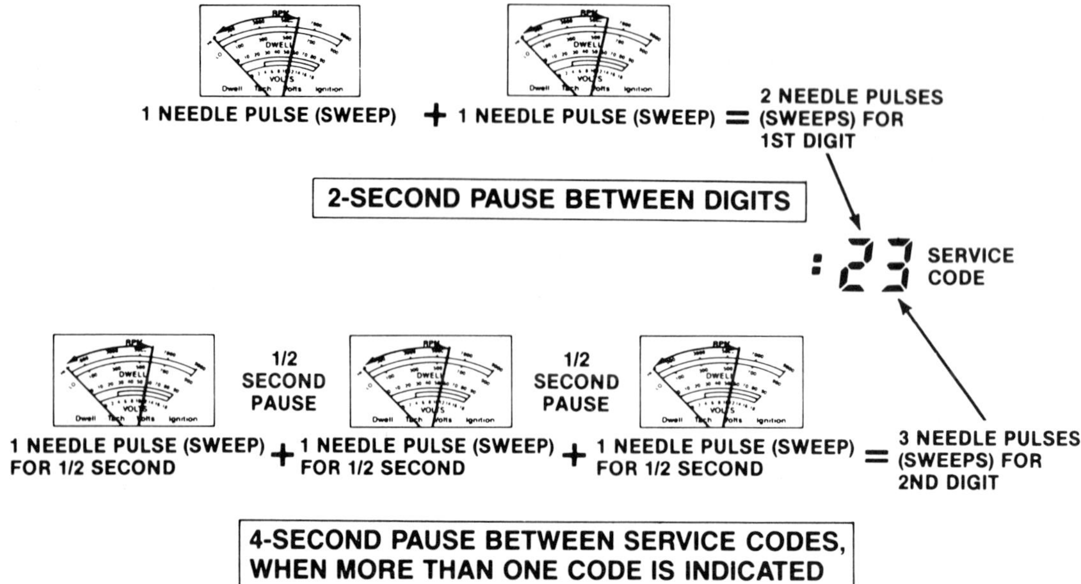

Figure 21-39. Voltmeter indications of Ford EEC service codes. (Ford)

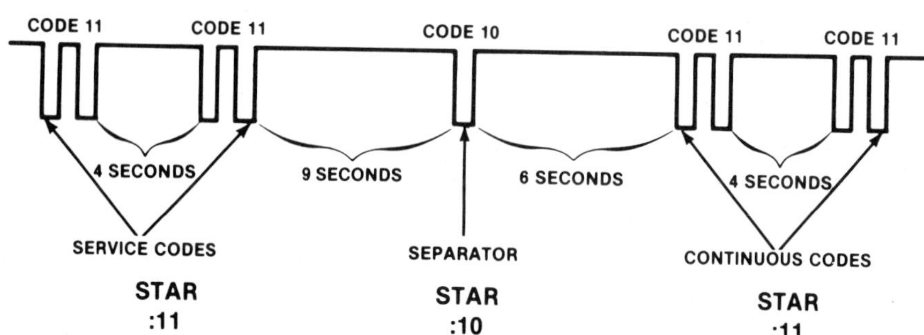

Figure 21-40. Ford EEC key-on, engine-off test code format. (Ford)

meter may produce one quick pulse from a "fast code," used for factory testing. Then it will indicate service codes. If it shows code 11, followed by a separator pulse, and another code 11, figure 21-40, the system is okay. You can go to the engine-running test. Ford's code 11 is a "system-pass" code, similar to GM's CCC code 12.

Codes before the separator pulse are "on-demand" codes that show a fault present during this test. These are mostly electronic faults, and Ford has pinpoint test procedures for each code. Codes after the separator pulse indicate intermittent faults that may or may not be present at this time. The codes will lead you to the:

- Engine-running test
- Continuous testing
- Specific pinpoint tests

Engine-Running Test. Do this test after the key-on, engine-off test unless a specific code or problem directs you to a pinpoint test.

1. Turn the ignition off.

2. Remove the jumper from the self-test connector to deactivate the test function, figure 21-36.

3. Start and run the engine at 2,000 rpm for 2 minutes to warm the EGO sensor.

4. Stop the engine and wait 10 seconds.

5. Reconnect the self-test jumper wire removed in step 2.

6. Start the engine and watch your voltmeter, figure 21-41. You will see an engine identification code (one-half the number of engine cylinders), followed in 6 to 20 seconds by a response code of one pulse. This means the engine is ready to test.

7. Open the throttle wide open and release it (snap acceleration).

8. Watch the voltmeter for additional service codes. If you see code 11, the system has found no problem. Other codes will lead you to pinpoint tests. These indicate a problem present during the test.

EEC-IV installations on some cars eliminate the snap acceleration step by activating the throttle position solenoid, or kicker, to start the self-test for you.

Continuous Monitor (Wiggle) Tests. Use this test to find intermittent open and short circuits in the EEC

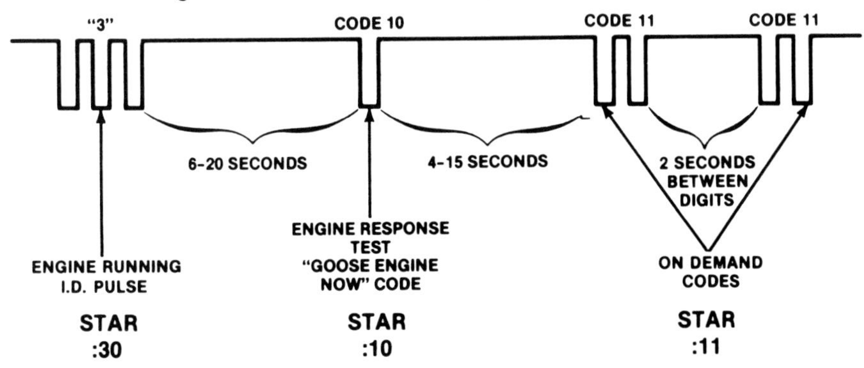

Figure 21-41. Ford EEC key-on, engine-running test code format. (Ford)

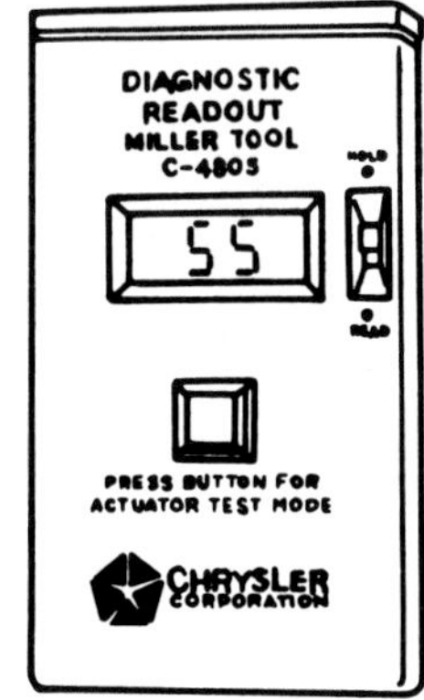

Figure 21-43. The Chrysler DRB I has a digital display similar to the Ford STAR tester. (Chrysler)

TEST STEP	RESULT ▶	ACTION TO TAKE
KC1 SERVICE CODES 44, 45 AND 46: VERIFY VACUUM LINE ROUTING		
• VERIFY PROPER VACUUM LINE ROUTING TO THE TAB/TAD SOLENOIDS AND TO THE BYPASS DIVERTER VALVE. REFER TO VECI DECAL.	NO ▶	SERVICE ROUTING OR FAULTS. REPEAT QUICK TEST.
• CHECK FOR KINKED OR BLOCKED VACUUM LINES. • CHECK FOR KINKED OR BLOCKED AIR HOSES. • CHECK FOR DISCONNECTED VACUUM LINES. • ARE VISUAL CHECKS SATISFACTORY?	YES ▶	SERVICE CODE 44, GO TO KC4. SERVICE CODE 45, GO TO KC2. SERVICE CODE 46, GO TO KC3.

Figure 21-42. This is an example of Ford pinpoint test procedures. This is the first step for codes 44, 45, and 46. (Ford)

system. You also can use it in many pinpoint test sequences.

1. Turn the ignition off.

2. Remove the jumper wire from the self-test connector.

3. Turn the ignition on but do not start the engine.

4. Move, wiggle, or tap the system connectors and components. The computer will record a service code and the voltmeter will display one pulse if an intermittent fault occurs. Repeat the key-on, engine-off test to display and record the codes.

5. Reconnect the self-test jumper wire and start the engine. Let the system go through the engine-running self-test and wait 2 minutes.

6. Repeat step 4 to check for intermittent faults with the engine running.

Ford EEC-IV Pinpoint Tests

As with General Motors' CCC systems, the service codes for Ford's EEC-IV will lead you from area diagnostic tests to specific pinpoint tests. Instead of being in "Christmas Tree" chart form, Ford's test procedures have three columns that explain each test step, the possible results, and the next step to take, figure 21-42. When you finish each pinpoint test and repair sequence, repeat the self-diagnostic tests to be sure the problem has been fixed and no more trouble codes are recorded in the system.

CHRYSLER ONBOARD DIAGNOSTIC SYSTEM TESTING

Chrysler introduced its first onboard diagnostic (OBD) system on late 1983 fuel-injected, front-wheel-drive models. The system parameters were expanded on 1985 engines with additional fault codes and increased test capabilities. All systems through 1987 were activated with a diagnostic readout box (DRB), figure 21-43, or a scan tool connected to a test connector in the engine compartment. Fault codes also can be retrieved through the POWER LOSS or POWER LIMITED lamp on the instrument panel.

The OBD system was greatly expanded on 1988 Dynasty, New Yorker, and New Yorker Landau models with a body computer module (BCM). A revised diagnostic readout box (DRB II), figure 21-44, replaced the DRB I and can be used to test all systems from 1984 on. The DRB II connects to the same engine test connector in the engine compartment to test engine control system operation. On BCM-equipped cars, the DRB II can be connected to a similar connector under the instrument panel to check body computer functions. Fault codes can also be retrieved through the CHECK ENGINE lamp on the instrument panel.

The computer used on all 1984-on and some 1987 4-cylinder models is a

Figure 21-44. The Chrysler DRB II uses interchangeable software cartridges and the same microprocessor as the SMEC.

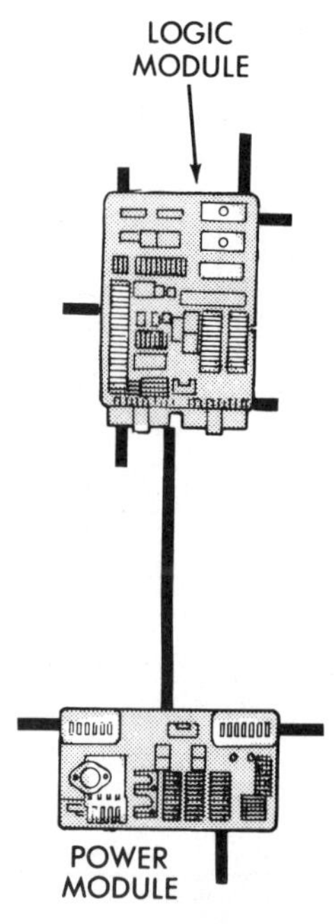

Figure 21-45. The Chrysler modular engine control system computer is a two-piece unit. (Chrysler)

2-piece unit called the logic module and the power module, figure 21-45. The logic module controls the fuel and ignition systems, provides memory for the onboard diagnostics (OBD), supplies the 5-volt sensor reference voltage, and activates the "limp-in" mode when a system fails. The power module is under the hood in the left front fenderwell. It supplies ground for the fuel injection and ignition coil circuits and provides 8 volts for logic module and Hall-effect distributor pickup operation.

The logic module was given a new and faster microprocessor chip and incorporated inside the power module housing on some 1987 models. The 1-piece unit was renamed the single-module engine controller (SMEC). On most 1989 models, the power module and the logic module functions are combined on a single circuit board assembly, called the single-board engine controller (SBEC). The SMEC or SBEC is used on all 1988 and later models. Starting with 1989 models, it identifies itself to the DRB II tester by model year, part number, and vehicle powertrain.

Fault Codes

Critical input and output circuits of the engine control system are monitored by the logic module. Some circuits are checked continuously; others are monitored only under specific conditions. If a circuit falls out of system parameters while being monitored and remains out of bounds for a predetermined length of time, a fault code is set into computer memory. This fault code will remain in memory as long as the system is out of parameters, or until the battery is disconnected. However, if the problem does not reoccur, the logic module will remove the fault code after 20 to 40 engine starts.

The Chrysler OBD system uses four types of 2-digit codes:

1. Fault codes identify the faulty circuit (but not the component in the circuit). A faulty code is the result, not the cause of the problem.

2. Indicator codes identify certain sequences or conditions that have taken place. An example is code 88, which indicates the beginning of the diagnostic mode.

3. Actuator test mode (ATM) test codes identify various circuits to be used during diagnostics.

4. Sensor test codes access a sensor readout.

Figure 21-46 is part of a Chrysler fault code chart showing codes, circuits, when they are monitored, and when a code is set in memory.

Power Loss or Power Limited Lamp

Fault codes can be retrieved through the POWER LOSS or POWER LIMITED lamp by turning the ignition key on-off-on-off-on within five seconds. The lamp will light for two to three seconds as a bulb check and then turn off. If no codes are stored in memory, the lamp will remain off. If codes are stored, the lamp will start flashing. Short pauses are inserted between flashes; a longer pause between digits and a four second pause between codes. When all codes are displayed, the lamp will flash a code 55 to indicate that the code display is over. Once the lamp starts flashing, it cannot be stopped. If you lose count of the flashes, you must wait until the lamp finishes its message, then repeat the procedure. While the lamp will retrieve the stored codes, it cannot be used as a substitute for the DRB when performing a driveability test.

OBD Pinpoint Tests

Chrysler provides a step-by-step driveability test procedure booklet for each different OBD application every model year. The procedure is a systematic approach, figure 21-47, to checking the function of each engine control subsystem. It allows you to see if the subsystem is working properly and determine where the problem is. The booklet contains the adjustment

Code	Type	Power Loss/ Limit Lamp	Circuit	When Monitored By The Logic Module	When Put Into Memory	ATM Test Code	Sensor Access Code
36	Fault	Yes	Wastegate Control Solenoid	All the time when the ignition switch is on.	If the solenoid does not turn on and off when it should.	09	None
37	Fault	No	Baro Read Solenoid	All the time when the ignition switch is on.	If the solenoid does not turn on and off when it should.	10	None
41	Fault	No	Alternator Field Control (Charging System)	All the time when the ignition switch is on.	If the field control fails to switch properly.	None	None
42	Fault	No	Auto Shutdown	All the time when the ignition switch is on.	If the control voltage of the relay pull in coil in the power module is not correct.	06	None
43	Fault	No	Spark Control	All the time when the ignition switch is on.	If the spark control interface fails to switch properly.	01	None
44	Fault	No	Battery Temperature Sensor (Charging System)	All the time when the ignition switch is on.	If the battery temperature sensor signal is below .04 or above 4.9 volts.	None	01
45	Fault	Yes	Overboost Monitor	All the time when the engine is running.	When M.A.P. sensor signal exceed a predetermine amount of boost indication.	None	None
46	Fault	Yes	Battery Voltage Sensing (Charging System)	All the time when the engine is running.	If the battery sense voltage is more than 1 volt above the desired control voltage for more than 20 seconds.	None	None
47	Fault	No	Battery Voltage Sensing (Charging System)	When the engine has been running for more than 6 minutes, engine temperature above 160°F and engine rpm above 1,500 rpm.	If the battery sense voltage is less than 1 volt below the desired control voltage for more than 20 seconds.	None	None
51	Fault	No	Oxygen Feedback System	During all closed loop conditions.	If the system stays lean for more than 12 minutes.	None	None
52	Fault	No	Oxygen Feedback System	During all closed loop conditions.	If the system stays rich for more than 12 minutes.	None	None
53	Fault	No	Logic Module	All the time in the diagnostic mode.	If the logic module fails.	None	None
54	Fault	No	Distributor Sync. Pickup	All the time when the engine is running.	If there is no distributor sync. pickup signal.	None	None
55	Indication	No			Indicates end of diagnostic mode.	None	None
88	Indication	No			Indicates start of diagnostic mode. **NOTE:** This code must appear first in the diagnostic mode or fault codes will be inaccurate.		
0	Indication	No			Indicates oxygen feedback system is lean with the engine running.	None	None
1	Indication	No			Indicates oxygen feedback system is rich with the engine running.		
8	Indication	No	Knock Circuit		Indicates knock sensor system is detecting knock.	None	06

Figure 21-46. A portion of a typical Chrysler code table providing fault codes, indicator codes, ATM test, and sensor access codes. (Chrysler)

procedures, specifications, and wiring diagrams required to diagnose the particular system. Diagnosis of the OBD system without the correct driveability test procedure booklet is not recommended.

Before starting the driveability test procedure, you should make sure that the battery is fully charged. If a cold test is specified, the engine must not be started for seven hours before the test. The warm test is performed at normal operating temperature. Some tests will specify the use of either a digital or an analog voltmeter. Either can be used if the test does not specify a particular type, but failure to use the type specified will result in an incorrect reading.

The DRB I or II is used to put the system into five different modes:

1. Diagnostic test
2. Actuator test mode (ATM) test
3. Switch test
4. Sensor test
5. Engine running test

Each test mode is required at certain points of the driveability test procedure.

The test modes available on Chrysler's DRB I and DRB II also are available on aftermarket scan tools. The DRB II and most scan tools also can display the logic module data list, similar to the serial data display from GM systems. Refer to the later section of this chapter, "Chrysler Engine Control System Scan Tool Tests," for more information on testing Chrysler computer systems.

AUTOMATIC SYSTEM TESTERS (SCAN TOOLS)

The preceding sections outlined the diagnostic principles of area and pinpoint testing for GM, Ford, and Chrysler electronic engine control systems. The test methods are based on using common test equipment, such as a voltmeter and a dwellmeter. Several automatic system testers are available, however, that can make troubleshooting faster and easier.

Ford's self-test and automatic readout (STAR) tester is an example of such a tester. So are Chrysler's diagnostic readout boxes (DRB I and DRB II) and GM's Tech 1 scan tool. All of these test instruments, and several similar ones built by aftermarket equipment companies, have come to be called "scan tools," or "scanners." They were given this name because they scan the overall operation of the computer system and read the input and output signals to and from the computer. The following sections explain the specific uses of these tools for checking engine control systems.

Scan Tool Features

A scan tool plugs into the vehicle diagnostic connector. For GM cars and trucks, this is the ALDL connector under the instrument panel (or else-

TEST 11	STEP A	CHECKING INJECTOR 1 AND 2 CONTROL CIRCUIT	
PROCEDURE		TEST INDICATION	ACTION REQUIRED
REMINDERS • Diagnostic readout box connected to the engine harness connector.	• Press the ATM button on the readout box. • Touch the other end of the jumper wire to a good ground. Make and break this connection several times.	• Voltmeter should pulsate as you make and break the connection.	• Voltmeter pulsates, replace the logic module. **CAUTION:** Before replacing the logic module check the terminal in cavity No. 2 of the red connector to make sure it is not crushed causing a poor connection.
• Turn the ignition switch off. • Connect a voltmeter to the white wire of the injector 6-way connector and ground. • Disconnect the electrical leads from injectors 3 and 4. • Disconnect the red connector from the logic module. • Connect one end of a jumper wire to cavity No. 2 of the red connector. • Turn the ignition switch to the run position.	DIAGNOSTIC READOUT BOX VOLTMETER WT INJECTOR 6-WAY CONNECTOR		• Voltmeter does not pulsate, **Perform STEP B.**

Figure 21-47. This sample Chrysler pinpoint test procedure is step A in checking the injector control circuit. (Chrysler)

where in the passenger compartment) on CCC-equipped vehicles, figure 21-9. On Ford products, the connection point is a "doghouse-shaped" connector in the engine compartment, figure 21-36. Chrysler diagnostic connectors also are in the engine compartment, usually near the left front strut tower or the firewall, figure 21-48.

Whether a scan tool is a carmaker's own special test instrument or an aftermarket tool, it is limited to displaying what the vehicle computer makes available. That is, the output signals are controlled by the car, not the scanner.

GM and Chrysler provide a long list of sensor signals and output commands to actuators. You can run the engine in the shop or drive the car to check system operation. In either case, you must be able to interpret system operating conditions to locate problems. You must know what the voltage readings or other data from various sensors mean and what they should be in different operating conditions. You also must know whether certain transmission or cruise control switches should be open or closed at certain times. The vehicle computer provides individual signals on a data stream that the scan tool reads. You must understand the readings and evaluate them as part of the troubleshooting.

The GM and Chrysler computer datastreams are often called "serial datastreams" because the computer sends out a series of signals that it receives from its sensors or sends to its actuators. These input and output signals are the same ones that the computer uses to control engine operation. The serial datastream is simply a parallel path that makes the same signals available to check system operation. The manufacturers use the serial datastream as part of final testing when the vehicles are built. By making the serial datastream available on a test connector, GM and Chrysler provide the same system operating data for service and repair.

Ford's approach to self-diagnostic systems is to program a series of tests into the vehicle computer and let the scan tool trigger the tests and read any fault codes that may result. This method assures that all vehicles are tested in the same way. The onboard computer cycles the engine through a series of operating modes and looks for normal or abnormal conditions. If fault codes are recorded, you must follow Ford's pinpoint tests to check specific circuits or components. Ford does not provide a serial datastream on any model through the 1980's to monitor the control system operation.

Several independent equipment companies make special testers for electronic engine control systems. Major manufacturers of scan tools are the OTC Division of Sealed Power Industries, Micro Processor Systems Incorporated (MPSI), General Motors Expertec Division (the Tech 1 scanner), and Snap-on Tools.

Figures 21-49 and 21-50 show several scan tools for late-model engine control systems. These testers have varying capabilities, depending upon design and price. The tester manufacturers provide operating instructions that follow the carmaker's test procedures.

Figure 21-48. The Chrysler scanner test connector is in the engine compartment, near the left strut tower.

Figure 21-49. These three instruments are typical of scan tools available to test late-model engine control systems.

Most modern scanners have interchangeable cartridges and cable adapters that allow them to be used on different vehicles. To test a GM car or truck, for example, you plug the GM test cartridge into the scanner. The scanner cartridge contains test programs for that car. Next, you attach the GM connector adapter to the scanner test lead and plug the scanner power cable into the cigarette lighter or into an adapter that attaches to the battery terminals. You then enter the vehicle identification into the scanner through a keypad. The scanner then displays instructions for you to choose a specific test sequence or to read sensor and actuator signals from the computer datastream.

To test a Ford or a Chrysler product, you remove the GM cartridge from the scanner and insert a Ford or Chrysler cartridge. Then remove the GM connector adapter from the scanner connector cable and attach a Ford or Chrysler adapter. When you plug into the Ford or Chrysler, you again enter the vehicle identification into the scanner. The cartridge then causes the scanner to display instructions for you to choose a specific test sequence for that car or truck.

The Snap-on scanner, figure 21-50, is the most advanced unit currently on the market. It operates with "menu-driven" displays of vehicle data and test programs that follow the carmakers' test sequences or diagnostic routines. It works much like the easiest personal computers.

Although all modern scanners are quite versatile in their capabilities, all are limited to displaying the information or running the tests made available by the onboard computer of the vehicle you are testing. No scanner, for example, can read serial computer data from a car that does not make those signals available through a test connector. The following sections outline the general uses of a scan tool on Ford, Chrysler, and GM engine control systems.

Data Recording

Most modern scanners have a feature that lets you record serial data from the vehicle computer for GM and Chrysler products. Data recording often is called a "snapshot" or a "movie" operation, depending on the amount of data recorded. The Snap-on scanner, for example, lets you record a 101-frame movie of engine operating data. Each data "frame" is a complete transmission cycle from the vehicle computer. Each frame contains one sample of every sensor or actuator signal on the serial datastream.

Figure 21-50. The Snap-on scanner displays four lines of trouble codes, sensor data, and test instructions. Each line is up to 40 characters long. You scroll the thumbwheel to move through code and data lists or select tests from menus. Press the buttons to answer questions and move through test programs.

Data recording is a valuable feature for pinpointing intermittent faults. To record a movie, you connect the scanner to the car and operate it in the shop or drive it on the road. When a noticeable problem occurs—a stumble, a surge, or a stall, for example—you press a button to start the recording. The scanner actually retains previous serial data in its memory and records these as part of the movie. It continues to record data frames after the trigger point until it reaches its memory capacity. The Snap-on scanner records 75 data frames before the trigger point and 25 frames after the trigger.

When you play the movie back from scanner memory, the scanner displays the recorded data just as if it were a "live" test. This lets you analyze the engine sensor and actuator data, along with any recorded trouble codes, to pinpoint the cause of an intermittent problem.

A scanner can make your troubleshooting faster and easier, but you must remember some basic precautions. The scanner will not do your thinking for you. You must understand the carmaker's test procedures and system operation to interpret the test results correctly. Although many testers will check individual sensor and actuator circuits, they simply tell you that the circuit is okay or that there is a problem somewhere in the circuit. The tester will not pinpoint the cause. You must still use traditional electrical and mechanical troubleshooting methods to identify and fix the cause of a problem.

A scanner won't identify traditional mechanical or electrical faults in areas not controlled by the computer. For example, a scanner will not pinpoint fouled spark plugs, bad plug wires, vacuum leaks, plugged fuel filters, or many other common problems. A scanner will, however, point out abnormalities in computer-controlled circuits that result from problems elsewhere in the engine. This information, along with your own troubleshooting knowledge, can lead you back to the basic cause of the problem.

GENERAL MOTORS SCAN TOOL TESTS

To test GM vehicles, a scan tool must communicate with the system computer, or electronic control module (ECM). The scan tool receives ECM data over a serial data link, through the assembly line diagnostic link (ALDL) connector. "Serial" means that data items, or **parameters,** are transmitted one after the other, in series.

The speed at which the scan tool operates and displays data depends on the length of the serial datastream and on the **baud rate** of the vehicle ECM. The baud rate is the data transmission speed in digital bits per second. Typical GM baud rates are:

- 80 baud for minimum-function systems on Chevettes and T-1000 models
- 160 baud for early carbureted and fuel-injected systems, including late-model throttle body-injected (TBI) systems
- 8192 baud for many 1986 and later fuel-injected engines.

The low-baud (slow-speed) systems are the basic C-3 systems. The high-baud (high-speed) systems are often called GM-P4 systems.

The baud rate determines how fast the scan tool responds to the ECM and how fast the data readings change on the scan tool display. It also affects the length of time that it takes to record data transmitted from the vehicle. Data readings from a high-baud ECM may appear to change almost instantly. Readings from a low-baud ECM will appear to change much more slowly. This display speed, or "data update rate," depends on the ECM; it is not controlled by the scan tool.

Some GM vehicles transmit slightly different data lists for different test functions. Whether or not a vehicle transmits an identical data list or variable data lists for different test conditions will affect some of the readings on a scan tool.

Bidirectional ECM's

Many late-model GM vehicles have bidirectional, 8192-baud P4 ECM's. This means that the ECM not only communicates with a scan tool, but it accepts commands from some test equipment. Bidirectional ECM's transmit complete datastreams to a scan tool and provide many special test capabilities. A few special test commands that override normal ECM operation, however, are restricted to GM test equipment.

Scan Tool Capabilities

Most scan tools offer these capabilities on GM domestic models with C-3 and later GM-P4 control systems from mid-1980 to the current model year:

- Read trouble codes and ECM data in all diagnostic modes available from a given vehicle.
- Read data from bidirectional computers.
- Data recording.
- Perform special (functional) tests, including ECM backup mode tests, air injection and EGR tests, and timing tests and adjustments.

Many scan tools also can perform these other electronic system test functions:

- Test antilock brakes on many 1988 and later cars and light trucks.
- Read body computer module (BCM) codes and data on many vehicles.
- Test supplemental inflatable restraint (SIR), or airbag, systems on late-model vehicles.
- Read transmission codes and data on 1991 and later cars and light trucks.

Scan Tool Communication with the ECM. You must do these three things to let a scan tool communicate with the ECM on a GM vehicle:

- Connect the scan tool to vehicle power (the cigarette lighter or the battery).
- Connect the scan tool to the ALDL connector.
- Turn the ignition on.

If the scan tool does not establish communication with the ECM, do not rush to condemn the ECM as faulty. More likely, vehicle identification may be entered incorrectly into the scan tool. Check the vehicle identification and correct it if necessary. If the identification is correct, disconnect the scan tool and check the vehicle connector for damaged terminals and open wiring. Lack of communication simply means that the scan tool is not receiving data from the ECM. The cause also may be as simple as a blown fuse or wiring fault on the car. In some cases, lack of communication may indicate an ECM problem. Other causes are more common, however.

Diagnostic Modes

A scan tool communicates with the ECM's on various GM vehicles in one or two of three basic ways. On earlier systems, the scan tool puts a resistive load across pins A and B in the ALDL connector. Late-model fuel-injected systems, on the other hand, transmit data continuously with no resistive load. Other systems require a pulse-width-modulated (PWM) communication signal from the scan tool. In all cases, the scan tool applies the required signal or load to communicate with the ECM.

Diagnostic Mode with 10-Kilohm Load. In the most common diagnostic mode for earlier GM C-3 systems, the scan tool puts a 10-kilohm resistive load across pins A and B in the ALDL connector. This places the ECM in the diagnostic mode so that it will transmit data and codes to the scan tool. *All* scan tools must apply this 10-kilohm load to gain access to the datastream. This diagnostic mode is used on most 1980–85 vehicles and many fuel-injected engines through the present models. *All* carbureted engines from 1980 to the present require the 10-kilohm load to transmit data. The 10-kilohm diagnostic mode was designed originally by GM for assembly line testing, and so it alters ECM operation in various ways. On *carbureted* engines, the 10-kilohm load may produce the following effects:

- The CHECK ENGINE light flickers.
- The closed-loop time delay is bypassed.
- New trouble codes may not be set on some vehicles.
- Cold-start enrichment may be disabled on some vehicles.
- Additional spark advance may be added on some engines.
- Canister purge operation may be reduced or disabled on some engines.

On *fuel-injected* engines, the 10-kilohm load may produce these effects:

- Closed-loop time delay is bypassed on all engines.
- New trouble codes may not be set on some engines.
- Idle speed may be increased and held at a fixed rpm on some engines.
- Additional spark advance may be added on some engines.
- Cold-start enrichment may be disabled on some engines.
- Cold fast idle may be bypassed, and cold-start enrichment may be disabled, on some engines.
- Closed-loop idle may be forced on some engines that normally run in open loop at idle.
- Electronic spark control (ESC) may operate at all temperatures on some engines.
- The ECM internal ESC functional test may be bypassed on some engines.
- The torque converter clutch (TCC) timer may be bypassed on some vehicles.

This operating mode may be called the ALDL (or ALCL) mode or the "special" or "10k" mode in some service manuals. It is most often called the "diagnostic" mode. Because the ECM affects vehicle operation in these ways in the 10-kilohm diagnostic mode, GM advises that these vehicles should not be driven during scan tool testing.

Unloaded Diagnostic Modes. Many 1986 and later fuel-injected systems transmit data continuously with no resistive load. Others require a pulse-width-modulated (PWM) communication signal from the scan tool. Some fuel-injected engines have both the loaded, 10-kilohm diagnostic mode and an open, unloaded, mode. Datastream length may differ between the two modes.

Some service manuals call this mode the "road test," the "open," or the "normal" mode because the scan tool does not cause the ECM to change engine operation. Whether or not a vehicle will transmit data in an unloaded mode depends on the specific ECM, not on the scan tool.

Trouble Codes

Trouble code information is part of the ECM data list. If codes are present, a scan tool will display them in numerical order. Most scan tools also provide a description. On some 1986 and later GM vehicles, you can clear codes through a scan tool.

Hard Codes, Soft Codes, and Historical Codes. Trouble codes may be either "hard" codes, indicating a problem that exists at the time of testing, or "soft" codes, indicating a problem that occurred in the past but is not present now. To distinguish between hard and soft codes on most vehicles, clear the ECM memory and reenter the diagnostic mode. Watch for codes to reappear. A hard code will reappear quickly. A soft code will not reappear until the problem that caused it reoccurs. Some 1988 and later GM systems separate hard codes from soft codes in ECM memory and distinguish them as "current codes" or "historical codes."

Cadillac soft codes are available only through the onboard diagnostic system, except for soft code 52 on 1989 and earlier cars and soft codes 52 and 72 on 1990 and later cars.

Special Tests

Most GM cars and light trucks have one or more special test functions available in the ECM. All have at least the "field service" mode, in which the scan tool shorts pin B in the ALDL connector to pin A. A scan tool can provide only those special tests available from the ECM of a given vehicle. Most current scan tools take full advantage of the test capabilities of all ECM's, however. The following paragraphs summarize GM special tests.

> **Warning.** Do not select any special test while driving a vehicle on a road test. ECM changes to ignition timing, fuel delivery, and other engine functions may affect engine operation and vehicle control.

Field Service. In the field service test, the scan tool grounds the diagnostic pin B in the ALDL connector to pin A. The ECM does not transmit data in this mode and will not set new trouble codes. On some vehicles, you can use field service to check or adjust ignition timing or the minimum idle speed.

With the key on and the engine off for all engines, the CHECK ENGINE lamp will flash stored trouble codes if any are present or code 12 if no codes are present. The ECM also energizes all solenoids, such as the MC solenoid, with the key on and the engine off. Therefore, you can use field service to test solenoid operation.

With a *carbureted* engine running in field service, the CHECK ENGINE lamp stops flashing code 12; and new trouble codes cannot be set. The ECM also fixes ignition timing at a specific degree of advance. This lets you check and adjust timing in field service for some carbureted engines. You also can use field service for a system performance check on carbureted engines.

For a *fuel-injected* engine, the CHECK ENGINE lamp flashes rapidly when the engine is running in open loop and slowly when in closed loop. Also in closed loop, the length of the CHECK ENGINE lamp flash indicates whether the exhaust is rich or lean. The lamp flash is longer if the exhaust is rich.

Air Injection Switching. The air injection switching test lets you energize the air switching solenoid to direct air into the exhaust manifold. This lets you check the operation of the exhaust oxygen sensor and the response of the MC solenoid or the fuel integrator, block learn, and injector pulse width. The engine must be warmed up and in closed loop before this test is valid. The ECM must be directing air downstream to the catalytic converter. If the ECM is directing the air-divert solenoid to route air to the atmosphere, this test cannot be performed reliably. When you press a specified button on the scan tool, it activates the air-switching solenoid to direct air to the exhaust manifold.

Because of vagaries in ECM functions and air injection system installations, some scan tools may list this test for engines without air switching. Always verify the type of air injection control solenoids and air system operation before relying on this test.

Fixed 10-Degree Spark. The fixed 10-degree spark test is available only on 1985 and later Cadillac C-cars with 4.1-liter engines. In this test mode, the ECM sets a fixed 10 degrees of spark advance (which is the base timing setting) and disables EGR. The following conditions must exist before the ECM will maintain 10 degrees of spark advance in this test:

1. Engine temperature must be above 185° F (85° C).

2. Engine speed must be under 900 rpm.

3. The transmission must be in park.

You can use this test to set basic timing on these Cadillac engines. Follow the instructions on the vehicle emission control information (VECI) decal.

Full-Lean and Full-Rich Mixture Tests. The full-lean and full-rich mixture tests are available only for minimum-function, carbureted T-cars (Chevettes, Acadians, and T-1000 models). In the full-lean test, the ECM commands the MC solenoid to a fixed 54° dwell (90% duty cycle). This test lets you check oxygen sensor operation and other engine operating conditions while the fuel system is held in a full-lean condition.

In the full-rich test, the ECM commands the MC solenoid to a fixed 6° dwell (10% duty cycle) condition. This lets you check O_2 sensor operation and other engine operating conditions while the fuel system is held in a full-rich condition.

In both tests, the scan tool displays the standard diagnostic data list for these cars.

Backup-Fuel and Backup-Spark-and-Fuel Tests. These tests let you check the operation of the backup fuel, or spark and fuel, programs in the ECM of some fuel-injected vehicles. These backup programs are failsafe, or limp-in, programs that set a fixed injector pulse width or ignition timing, or both. The failsafe program lets the vehicle be driven to a shop for repair in case of a major sensor failure.

These tests principally verify that the backup program is operational in the ECM. The backup test also can be used to doublecheck the operation of the fuel injection system, however. If a car with a driveability problem seems to run better in the backup condition than in normal operation, you should check fuel metering and air intake parameters closely. The scan tool does not display data list during this test.

EGR Control Tests. Some scan tools offer special EGR bidirectional tests for some 1988 and later engines. In these tests, the scan tool commands the ECM to cycle the EGR valve open and closed at fixed intervals. This provides a special test of EGR system operation. Tests are available for the integrated electronic EGR valve and the digital EGR valve.

FORD EEC-IV AND MCU SCAN TOOL TESTS

Unlike GM and Chrysler control systems, Ford electronic engine controls IV (EEC-IV) and microprocessor control unit (MCU) systems do not transmit operating data to a scan tool. Instead, Ford includes special system test programs in the electronic control assembly (ECA) of the vehicle. A scan tool acts as a switch to turn on the special test program. The ECA runs the self-test and then transmits service codes to the scan tool.

Ford electronic EEC-IV and MCU systems control many engine functions. But you should make basic fuel, ignition, and electrical tests before, or along with, control system testing. Be sure that these systems and parts are in good working order:

- Fuel delivery
- Battery condition, electrical connectors, and wiring harnesses
- Ignition primary and secondary circuits
- Vacuum lines and connectors
- Cooling system
- General engine mechanical condition.

Ford Service Codes

Ford refers to service codes as "on-demand" codes and "continuous memory" codes, and the system ECA transmits them in these groups during self-tests. On-demand codes are "hard" codes that indicate faults that are present at the time of testing. Continuous memory codes are "soft" codes from the ECA memory of EEC-IV systems. These indicate intermittent problems that have occurred in the past but which are not present at the time of testing. MCU systems do not have long-term memory and do not store soft (continuous) codes.

Troubleshoot Ford codes in the order in which they are transmitted by the vehicle in these three groups:

First—engine-off hard codes
Second—engine-running hard codes
Third—Soft codes from ECA memory

After fixing a problem, repeat the self-tests to be sure the code does not reappear. Some codes may be present as both hard and soft codes. Fixing the hard codes first may also correct problems that caused soft codes.

For the key-on, engine-off self-test of an EEC-IV system, the ECA transmits hard (on-demand) codes first, followed by soft (continuous) codes. The ECA does not transmit continuous codes for the engine-running test except on some 1983 1.6-liter Escort engines.

Most EEC-IV and MCU systems transmit codes at both slow and fast speeds to a scan tool. On most vehicles, the only difference between the slow and fast codes is the transmission speed from the ECA. A few cars transmit only slow codes, however, and fast code transmission on a few others may be unreliable for scan tool testing. Slow code transmission is intended primarily for code reading with an analog voltmeter. Fast code transmission is primarily for assembly line testing. Most current scan tools can read either slow codes or fast codes.

Three-Digit Service Codes. Many 1991 and later Ford cars and trucks transmit 3-digit service codes. Most scan tools can receive and display these 3-digit codes similarly to comparable 2-digit codes. Self-test procedures are basically the same for vehicles that transmit either 2-digit or 3-digit codes.

Ford Troubleshooting Sequence

Ford test procedures are very specific about the order in which self-tests should be performed and codes should be diagnosed and serviced.

A basic guideline for Ford testing is to turn the ignition key off, wait 10 seconds, and then turn it back on before repeating any particular test. This is not absolutely required on all vehicles, but it is good practice in order to avoid erroneous self-test actions by the ECA.

> **Warning.** Keep hands and test leads away from the electric cooling fan and other engine parts during all engine tests. The electric cooling fan and other actuators may operate without warning.

Key-On, Engine-Off (KOEO) Test. This test displays on-demand hard codes present with the ignition on, but the engine not running. These are usually electrical open and short circuits and must be serviced *first,* before any other codes. For EEC-IV systems, the key-on, engine-off test also displays continuous memory codes of intermittent faults from ECA memory. These memory codes should be serviced *last,* after any other hard codes.

The entire engine-off test should take 45 to 60 seconds if you are reading fast codes. It may take more than a minute if you are reading slow codes, depending on the number of codes present. When you start the test, you should hear solenoids and relays click on most engines. You also may hear the idle speed control device cycle through its range. If you do not hear actuators operate and do not receive a code transmission in about one minute, the self-test probably did not trigger. Refer to Ford service procedures for information about troubleshooting a self-test that does not run.

The scan tool displays hard, on-demand codes first, followed by continuous memory codes. This is the order in which they must be fixed. If faults are not present in either group, the scan tool displays code 11, which means no faults are present. If codes are present, the scan tool displays them by number in the order in which they should be diagnosed. Most EEC-IV systems built since 1988 require that hard, on-demand codes found during the key-on, engine-off test be corrected before you can do the engine-running self-test.

If you choose to read slow codes during the KOEO test, it will take slightly longer to receive all codes. All code digit pulses are 1/2 second on and 1/2 second off. A 2-second pause separates the digits of each code, and a 4-second pause separates one code from another. The last hard code is followed by a 6- to 9-second pause, then a single pulse as a separator code, and another 6- to 9-second pause. The ECA then transmits soft codes from its memory.

Key-On, Engine-Running (KOER) Test. This test displays on-demand hard codes present with the engine running. These should be serviced *second,* after any KOEO hard codes and before any memory codes. Most 1988 and later Ford vehicles will not perform an engine-running test if any engine-off hard codes are present and uncorrected.

The engine must be fully warm for the engine-running test to be valid. You must run the engine for two minutes to bring the coolant and the exhaust gas oxygen (EGO) sensor to operating temperature. This also ensures that the control system is in closed-loop operation.

During the engine-running test, the ECA puts the engine through a series of tests to check various sensors and actuators in the system. Depending on the particular engine being tested, these operations may include changing the ignition timing, changing the idle speed, and operating the EGR. These are dynamic tests that cannot be made just with electrical power applied to the system.

The entire engine-running test usually takes approximately 1 to 4 minutes if you are reading fast codes. The time varies depending on the particular engine system and the number of codes present. The test will take longer if you are reading slow codes. You should hear engine speed change with most EEC-IV systems during the test if the test is progressing normally.

On many vehicles, the ECA checks power steering switch and brake switch circuits during the engine-running test. If you do not operate the brake and power steering switches during the test, the ECA will transmit codes. If the scan tool displays a code indicating a fault in either of these areas, repeat the test and apply the brake and turn the steering wheel at any time after the first 10 seconds of the test. If a power steering or brake switch code reappears after doing this, it is a valid code.

After a few more seconds, the scan tool will display a code 10 for many vehicles. The code 10 test checks the throttle position sensor (TPS), the manifold absolute pressure (MAP) sensor, and other dynamic sensors. If you do not snap the throttle open quickly and sharply within a few seconds after the instruction, the test results will report an operator error code. Do not press and hold the accelerator to the floor. Engine damage may result. Snap the throttle open sharply and release it quickly, before the engine reaches high rpm.

The snap acceleration test is not required for most sequential electronic fuel injection (SEFI) engines (except the 3.8-liter SHO V-6) and a few other late-model Ford vehicles. Do not perform the snap acceleration test unless the scan tool directs you to do so.

Other engine systems may require other operator actions during the engine-running test. Some MCU systems, for example, require you to hold engine speed at a specified rpm during the test. The scan tool will display any special instructions on the screen when required.

The scan tool displays only hard, on-demand codes for faults present during the engine-running test. Diagnose and repair problems in the order in which codes are displayed. Repair any hard, on-demand codes from the key-on, engine-off test before diagnosing any codes from the engine-running test. If codes are not present, the scan tool displays code 11, which means no faults present.

If you choose to read slow codes during the KOER test, all code digit pulses are 1/2 second on and 1/2 second off. A 2-second pause separates the digits of each code, and a 4-second pause separates one code from another. The entire test may take several minutes, depending on the number of codes present.

Wiggle Tests. The engine-off and engine-running wiggle tests place the scan tool and the ECA in a standby mode to indicate an intermittent problem caused by wiggling electrical harnesses. If a fault occurs during a wiggle test, it is recorded in ECA memory as a soft intermittent code. You must repeat the engine-off test to read the code.

The engine-running wiggle test is not valid on 1984–85 1.6-liter engines, 1984–85 2.3-liter SVO engines, 1985–86 2.3-liter turbo engines, and 1986 3-liter engines. The self-test output line continuously switches from high to low so fast that it masks any faults that may be detected by the wiggle test.

Clearing Codes. To clear codes from the ECA memory, you simply do the engine-running self-test and interrupt the self-test input (STI) line while the ECA is transmitting slow codes. Ford service procedures state that you should clear all codes after making repairs and then repeat the self-test to verify the repair. Be sure to note any continuous codes displayed during the self-test, however. If you clear codes and a problem does not reoccur as an on-demand code when a self-test is repeated, the ECA will not transmit the code.

Remember that only soft (continuous memory) codes can be cleared. If a code reappears when you clear codes and repeat a test, it is a hard (on-demand) code that must be serviced.

Other EEC-IV System Tests

The ECA of an EEC-IV system contains other special test functions, and most current scan tools let you take full advantage of these capabilities. The following paragraphs outline these functions.

Output State Test. The output state test is a continuation of the engine-off test. It lets you switch the ECA signals to the engine actuators on and off so that you can test them with a voltmeter. If the engine is running, turn it off before selecting this test. When you begin the test, all actuators should be off and the control circuits from the ECA should be high (above 10 volts). Depress the throttle to wide-open throttle (WOT) to switch all engine actuators from off to on (high to low) or from on to off. The actuators will stay on or off until you depress the throttle again. Use a digital volt-ohmmeter (DVOM) to test the actuators in one state or the other as required.

Computed Timing. The computed timing test actually is a continuation of the engine-running self-test. It lets you check ignition timing with the engine running at a controlled idle speed. It also verifies the ability of the ECA to advance and retard timing. The ECA provides a fixed advance of 20 degrees above the base timing specification in this test mode.

Before entering this test, connect a timing light or a magnetic timing meter to the engine. Be sure the engine is fully warm and allow it to run at normal idle speed. Check the timing with a timing light or timing meter within 2 minutes after you start the test.

On most EEC-IV engines, the ECA advances timing 20 degrees above the base timing setting. For example, if the base timing specification is 10° btdc, you should read 30° btdc with the timing light or meter. Refer to Ford service manuals for timing specifications and test procedures.

Cylinder Balance for SEFI Engines. The cylinder balance test is a continuation of the engine-running test. It lets you check the operation of individual fuel injectors for sequential electronic fuel injection (SEFI) systems. The ECA disables each injector individually and compares the engine speed drop from cylinder to cylinder. At the end of the test stages, the scan tool will display the results recorded by the vehicle ECA. The engine must be fully warm and in closed loop for the test results to be valid.

The ECA actually repeats the engine-running self-test and displays the cylinder identification. All service codes must be transmitted. After a minute or two, the scan tool will give you instructions to increase the engine speed for several seconds. The ECA starts the test and disables all injectors in firing order sequence. The scan tool displays code 90 if all cylinders pass. If any cylinder or cylinders fail the test, the scan tool displays the cylinder numbers. These are the actual cylinder numbers, not the firing order. Cylinder "failure" is based on test limits programmed into the EEC-IV system, *not* on limits provided by the scan tool.

For 1986 vehicles, the cylinder balance test ends after one stage. For 1987 and later vehicles, the cylinder balance tests continue with stages two and three and resulting pass or fail messages. If any cylinder fails at stage two or three, the display at the end of that stage identifies the cylinder. A weak cylinder may fail stage one or two and pass stage two or three. A dead cylinder usually fails all three stages.

Idle Speed Adjustment. Some 1990 and later engines have an idle speed adjustment test as a continuation of the engine-running test. Start and run the engine at 2,000 rpm for 2 minutes and then enter the test. After the ECA completes the engine-running test, the scan tool sends a signal to start the idle speed adjustment test. During the test, the scan tool will flash an indicator (usually an LED) or sound a beeper. A continuously lit LED or continuous tone indicates that idle speed is correct. Fast flashing or beeping indicates high idle speed. Slow flashing or beeping indicates low idle speed. Turn the throttle-stop screw to adjust idle speed until the scan tool sounds a continuous tone or lights the LED continuously. If the throttle position sensor is out of adjustment at idle, the ECA will transmit a code to indicate this fault. You must service the TPS before you adjust idle speed.

CHRYSLER ENGINE CONTROL SYSTEM SCAN TOOL TESTS

This section explains how to use a scan tool on the engine control system variations of Chrysler carbureted and fuel-injected engines from 1983 to the present. Testing principles are similar, but carbureted engines have limited diagnostic capabilities.

Carbureted Engines

Domestically built Chrysler carbureted 4-cylinder engines in cars from 1985 through 1987 and V-8, V-6, and domestic 4-cylinder engines in 1985–88 trucks have minimal scan tool diagnostic capabilities. V-8 and 6-cylinder engines in rear-wheel-drive cars have no scan tool diagnostic capabilities.

Carbureted engines use a spark control computer (SCC) that contains computer logic and power functions in a single housing. Diagnostic capabilities are limited to sensor, switch, and actuator tests. These engine systems do not transmit computer operating data to a scan tool. You place a carbureted engine in the diagnostic mode by blocking the carburetor idle switch open. Vehicles with carbureted engines do not have a POWER LOSS or POWER LIMITED lamp on the instrument panel.

Fuel-Injected Engines

The engine control systems on most Chrysler fuel-injected engines built from mid-1983 through late 1987 have a computer divided into two separate modules: the logic module (in the passenger compartment) and the power module (under the hood). The logic module contains the system programs. It receives all sensor signals and processes all information to control the system actuators. The power module controls the high-current actuators, such as the ignition coil, the fuel injectors, and the auto shutdown relay.

In late 1987, Chrysler introduced a new computer with the logic module and the power module functions combined into a single assembly. Chrysler calls this the single-module engine controller (SMEC). In 1989, Chrysler introduced another revised computer that combines all logic and power functions onto a single circuit board. This is called the single-board engine controller (SBEC).

All fuel-injected engines provide access to computer data in a data-display mode. All vehicles with fuel-injected engines have a POWER LOSS, POWER LIMITED, or CHECK ENGINE lamp on the instrument panel.

Starting with the 1991 model year, most fuel injected Jeep and Eagle vehicles used the Chrysler SBEC control system. Earlier Jeep and Eagle vehicles used a Renix control system, which most scan tools cannot test. Only Chrysler's DRB tester provides test programs for these systems.

Although Chrysler has changed the design and appearance of the engine computer and added more capabilities since it was introduced, the operating principles and the logic and power control functions have stayed basically the same. We will continue to refer to the "logic module" and the "power module" functions in this book. All domestic Chrysler vehicles have the same engine diagnostic connector, which is under the hood near the right strut tower on cars or near the firewall on trucks.

Data Display

In the data-display mode, a scan tool reads all data available on the logic module datastream, including any fault codes that may be in memory. Because the scan tool does not affect logic module or engine operation, the vehicle can be driven in this mode to test for intermittent problems.

Before the scan tool can display data, it must communicate with the logic module in a certain sequence. The ignition must be on. The logic module will transmit data both with the engine off and the engine running.

In the data-display mode for 1988 and earlier fuel-injected systems, the scan tool displays the parameter values on which the logic module is operating. If certain sensors fail, the logic module will substitute a value from its own program for the faulty sensor signal. This is called a "default" value, and this is what the scan tool will display in the data-display mode. If any parameter value in this mode appears to be quite different from what you expect, switch to the engine-off sensor tests and compare the reading for the same sensor. In the engine-off sensor tests, the scan tool displays the actual signal sent by the sensor, not a default value. If the values for the same sensor are different in the two modes, the logic module is operating on a default value.

Fault Codes. Fault codes may be either "hard" codes, indicating a problem that exists at the time of testing, or "soft" codes, indicating a problem that occurred in the past but is not present now. To distinguish between hard and soft codes, clear the logic module memory and reenter the data-display mode. Watch for codes to reappear. A hard code will reappear quickly; a soft code will not reappear until the problem that caused it reoccurs.

Some Chrysler vehicles may display false fault codes in the data-display mode. These usually are the result of accessory circuits in the control system for options that are not installed on the vehicle. In these cases, the circuit is continuously open. If that circuit is included in the fault code library for the vehicle, the logic module will transmit a code to the scan tool. For example, if the logic module monitors the cruise control circuit but the vehicle does not have this option, it may transmit a false code for a problem with the cruise control servo. To avoid confusion by false codes, always verify that the vehicle has the component for which the logic module transmits a code.

Clearing Codes. For most 1987 and later vehicles, you can clear codes through a scan tool. For 1987 and earlier vehicles, you must disconnect the battery ground cable for 10 seconds or remove a specified fuse or disconnect a computer power connector. If possible, remove a fuse or unplug a connector to clear codes. Disconnecting the battery will erase other computer memory circuits.

Diagnostic Tests for 1988 and Earlier Systems

Although the data transmission capability has always been included in the logic modules for fuel-injected engines, Chrysler did not include the display capability in the first system testers used by dealer technicians. Instead, the Chrysler diagnostic readout (CDR) tester provided a series of diagnostic mode tests for fault codes, switches, sensors, and actuators. Only with the introduction of the second diagnostic readout box (DRB-II) in 1987 did Chrysler provide a data list display similar to GM's on the factory test equipment.

The diagnostic tests begin with a sequence that reads fault codes from the logic module and then tests individual switches and sensors (both with the engine off and running) and tests the operation of computer-controlled actuators. Most 1988 and earlier systems are placed in the diagnostic mode by cycling the ignition on-off-on-off-on. This puts the logic module in a self-test state and provides the following capabilities:

- A review of fault codes from the logic module
- A group of switch tests to check the operation of driver-controlled switches
- An actuator test mode (ATM) to check the operation of actuators, such as fuel injectors, the coolant fan and relay, and others
- A group of tests to check the operation of sensors, such as the exhaust oxygen sensor, the throttle position sensor, and others.

Reading Fault Codes. After cycling the ignition switch three times, leave it on. The scan tool will then read the fault codes from the logic module and display them on its screen. Chrysler systems begin the code display with 88, which means, "start of code report." The display ends with 55, which means, "end of code report." If codes for faults are present in the module, the scan tool will display them between the 88 and 55.

Switch Tests. After the code display, the logic module gives you the opportunity to test driver-controlled switches. On later-model vehicles, the list of switches that can be tested is more extensive than on earlier models.

> **Warning.** Keep your hands and equipment test leads away from the electric cooling fan and other engine components during the switch tests. The cooling fan and other actuators may operate without warning.

Actuate any of the switches listed on the scan tool display to check its operation. As you change the switch position, the scan tool will indicate whether the switch is on or off.

ATM Tests. The actuator test mode (ATM) tests let you check the operation of actuators in the engine control system. The ignition key must be on, and the engine must be off to run the ATM tests.

> **Warning.** Keep your hands and equipment test leads away from the electric cooling fan and other engine components during the ATM tests. The cooling fan and other actuators may operate without warning.

The logic module switches the selected actuator at a programmed cycle. Switching time varies for different actuators: 2 seconds for some, 4 seconds for others, and so on. With the logic module commanding the actuator in a known state, you can test the actuator operation with a voltmeter.

For all 1983–88 vehicles (except 1983–84 EFI and turbo models), the logic module will stay in the last ATM selection that you make until you turn off the ignition or choose another test. You can stop the actuation by turning off the ignition or selecting another test function. For 1989 and later vehicles, the ATM test operation will stop when you exit the individual test.

For these 1983–84 EFI and turbo models, the logic module energizes the ignition coil, the automatic idle speed (AIS) motor, and the fuel injectors simultaneously. For all other vehicles, the logic module provides a list of available ATM tests. Tests will vary for different vehicles; but they usually include the ignition coil, the fuel injectors, the auto shutdown relay, and other relays and solenoids.

Engine-Off Sensor Tests. These sensor tests let you check the individual operation of sensors with the engine off. Sensor tests are not available on 1983–84 models with limited ATM tests. The values displayed in these tests are the *actual* sensor signals. They are not *default* values used by the logic module in case of sensor failure.

Turn the ignition key on but leave the engine off to begin the tests. The scan tool will display a list of individual sensors. Selections will vary for different vehicles, but they usually include the exhaust oxygen sensor, the coolant sensor, the MAP sensor, and the throttle sensor.

Sensor tests are available on 1986 and later carbureted engines, and they vary for different models. These tests indicate general conditions such as low-medium-high or open-closed.

Engine-Running Tests. The engine-running tests on 1985 and later systems let you check the operation of the automatic idle speed (AIS) motor and the sensors. The AIS motor test lets the scan tool command the logic and power modules to cycle the AIS motor for increasing and decreasing engine rpm. The sensor tests provide access to readings from the same list of individual sensors available with the engine off. Again, these are actual sensor readings.

Diagnostic Tests for 1989 and Later Systems

Diagnostic mode tests for 1989 and later vehicles with SBEC computers are different from the tests for 1988 and earlier models. Fault codes are not displayed through this test mode. All codes are displayed in the data display. Codes can be cleared through the scan tool.

The 1989 and later diagnostic tests do not distinguish between engine-off and engine-running conditions. If a particular test cannot be performed with the engine either off or running, the SBEC will not accept the test command.

For the AIS motor tests, the SBEC commands the AIS motor to extend and retract to obtain a desired rpm. The scan tool can run the motor through its full range.

ATM tests must be performed with the engine off. The vehicle will not accept ATM test commands with the engine running. ATM tests that fire each ignition coil are available for the direct (distributorless) ignition systems on some 1990 and later engines. Using these tests, it is possible to fire a spark plug in a cylinder that contains an air-fuel charge. This could cause a backfire through the throttle body, or it could cause a car with a manual transmission to lunge slightly if the transmission is in gear and the parking brake is not set. Do not operate the DIS ATM tests after operating the fuel injector ATM tests. Start and run the engine to remove any residual air-fuel charge before operating DIS ATM tests.

The minimum idle speed test is a special test for most 1989 and later models. When you select this test, the AIS motor retracts to close the throttle and obtain the minimum idle rpm. The air-fuel mixture also is enriched. Most scan tools display engine rpm during the test. Refer to Chrysler service procedures and specifications for test and adjustment information.

The emission maintenance reminder (EMR) lamp on some 1989 and later vehicles will light at approximately 60,000 miles to alert the driver that the vehicle should have certain emission-related services performed. The EMR lamp *cannot* be turned off manually; it must be reset through the engine computer. Chrysler's DRB-II and some other scan tools have a special test program to reset the EMR lamp.

ELECTRONIC CONTROL SYSTEM SERVICE

The principles of testing, adjusting, and repairing engine system parts that are explained throughout this book also apply to engines with electronic controls. The basic methods that you have learned for fuel, ignition, and emission control service will guide you through maintenance jobs on late-model vehicles. However, some features of engine electronic systems have brought new requirements to the service profession. The following sections outline the key points of these jobs.

General Service Guidelines

Most sensors in fuel injection systems (and all electronic systems) are resistors. The control modules, or computers, receive changing voltage signals from the sensors that change as the resistance changes. Resistance can be changed by:

- Physical motion (throttle movement, for example)
- Temperature (air or coolant, for example)
- Fuel or air pressure.

You can test sensor condition and operation by using an ohmmeter to measure resistance of the sensor or resistance between the sensor and ground, figure 21-51. You also can use a voltmeter or oscilloscope to measure supply and return voltage at a sensor and voltage drop across the sensor. Most manufacturers advise using a high-impedance digital voltmeter for voltage tests. Carmakers include all of these kinds of tests in their procedures.

Temperature sensors usually have specific resistance values at specific temperatures. For example:

2.1 to 2.9 kilohms at 68° F (20° C)
250 to 380 ohms at 176° F (80° C)

You can heat and cool most temperature sensors in water to check resistance at different temperatures.

Most actuators in electronic systems are solenoids. Again, you can use an ohmmeter to test solenoid winding resistance for open and short-circuit problems. You also can use a voltmeter to check supply voltage to a solenoid.

Mechanical injection systems use injectors controlled by fuel pressure. You can make several basic tests on

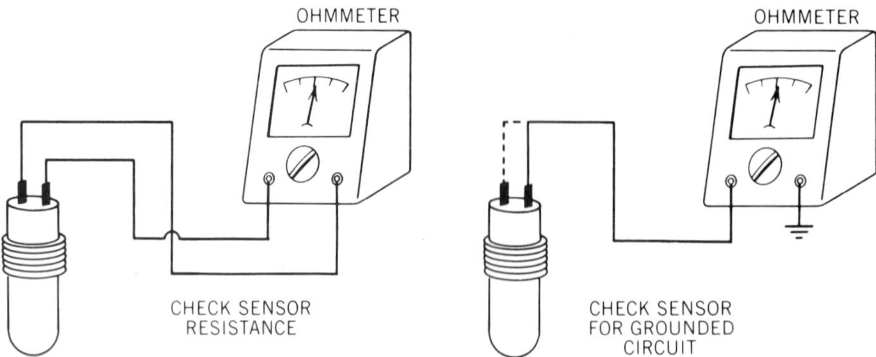

Figure 21-51. Use an ohmmeter to measure sensor resistance and to check for a grounded sensor.

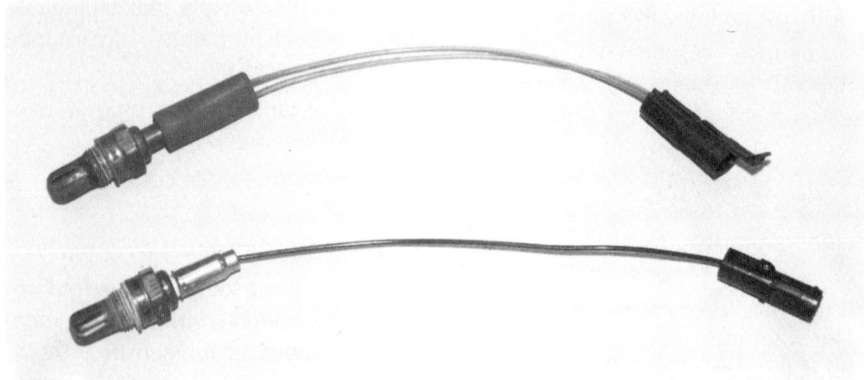

Figure 21-52. Test procedures may be different for 1-wire and 2-wire EGO sensors.

these systems with your pressure gauges. No fuel system for a carburetor, diesel injection, or gasoline injection can work right if fuel delivery pressure is out of limits. Sometimes fuel pressure is too high, but more often, it is too low. Low pressure is usually caused by:

- A bent or clogged fuel line
- A clogged fuel pickup in the tank
- A restricted fuel filter
- A bad fuel pump
- A leak in the delivery system
- A bad pressure regulator.

The fuel pump pressure tests that you learned in chapter 16 are important parts of injection system testing, as is fuel filter service. You can use your gauges to measure pressure at different points of the system and check for pressure drops or restrictions in filters and lines.

All gasoline injection systems have pressure regulators and operate with controlled pressure at the injectors, as diesel systems do. However, the pressures are much lower. Later in this chapter you will learn the principles for checking regulator and return pressure.

EGO Sensor Service

Some carmakers schedule EGO sensor replacement at specific time or mileage intervals on early engine control systems. Most EGO sensors on late-model systems are warranted for 50,000 miles, and replacement is not a scheduled service. You may have to test an EGO sensor, however, to isolate a system problem and replace it if it fails.

Caution. Do not short circuit or ground an EGO sensor output lead or test sensor output with an analog voltmeter. Many analog voltmeters have low impedance and will draw too much current. This can burn out an EGO sensor. Do not test an EGO sensor with an ohmmeter. It is *not* a resistor. Any input current may damage the sensor. Use a digital voltmeter with at least 10-megohms-per-volt impedance, following the carmaker's procedures to check sensor signal voltage.

Manufacturers' test procedures vary slightly for EGO sensors in different systems. Test procedures may be different for 1-wire and 2-wire sensors, figure 21-52. All carmakers, however, provide some method to check sensor output signal voltage and computer response to sensor signals. Heated EGO sensors may require a test for heating current or supply voltage.

EGO Sensor Removal and Replacement. An EGO sensor is about the size of a spark plug, and most have 18-mm threads similar to some spark plugs. Some sensors are installed in exhaust manifold outlets, low in the engine compartment. Access can be difficult, and you may have to reach the sensor from under the vehicle. Some sensors, on the other hand, are easily accessible in the center of a manifold, particularly on front-wheel-drive cars, figure 21-53. Several tool companies make special EGO sensor wrench sockets, figure 21-54, that slip over the connector leads and have an offset drive lug for a 1/2-inch ratchet or breaker bar.

Figure 12-53. This EGO sensor is easily accessible at the front of the engine.

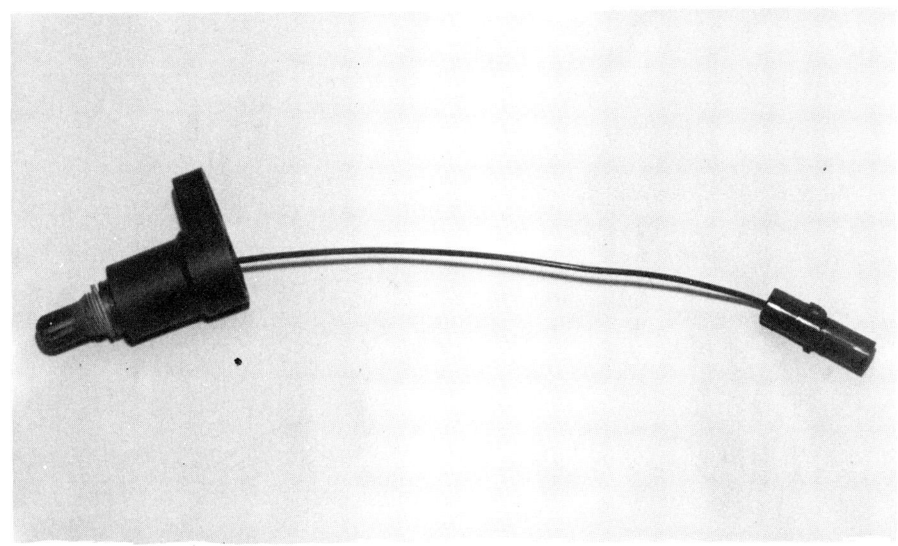

Figure 21-54. Special sockets are available for EGO sensor removal and installation.

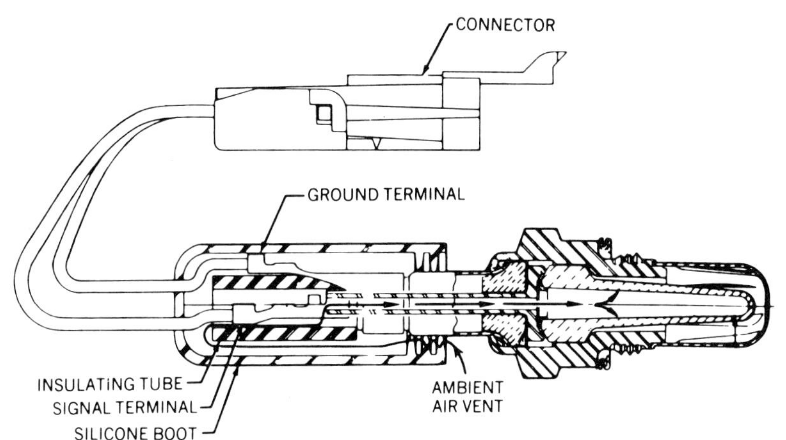

Figure 21-55. Be sure the boot is located properly to protect the sensor but not blocking the air vent. (GM)

When you install an EGO sensor, be sure the threads in the manifold or exhaust pipe are clean and free from burrs. You may have to clean them with an 18-mm thread chaser. All EGO sensors require antiseize compound on the threads when you install them. Most antiseize compounds for sensors contain small glass beads suspended in graphite. The graphite burns away after installation, but the glass beads remain on the threads to make removal easier. New replacement sensors usually come with antiseize compound already on the threads. When you reinstall a used sensor, clean the threads and apply the specified compound. Without the antiseize compound, sensor removal is difficult—in some cases, impossible.

Always follow the carmaker's torque specifications for sensor installation. These usually are in the 20 to 30 foot-pound (27 to 41 Nm) range. *Do not overtighten the sensor.* Because of the action of hot exhaust, a sensor usually requires two to three times more torque to remove than to install. If the sensor has a rubber boot with an air vent, figure 12-55, position it correctly to protect the sensor but not to block the air vent.

Mixture Control Solenoid Service

On an electronically controlled engine with a feedback carburetor, the mixture control (MC) solenoid regulates air-fuel metering in one of two ways:

1. By regulating a vacuum signal to carburetor diaphragms that control fuel-metering rods and air bleeds
2. By direct control of fuel-metering rods and air bleeds in response to computer signals

Rochester carburetors used with GM CCC systems have MC solenoids that control fuel metering and idle air bleeds within limits established by factory-adjusted limit screws. The most common CCC carburetors are:

- E2SE Varajet
- E2ME Dualjet
- E4ME Quadrajet

Figures 21-56 and 21-57 show the MC solenoid and the limit screws for the idle and main-metering systems of an E2SE Varajet. These limit screws should not require adjustment except for a major carburetor overhaul or to correct a specific performance problem. Do not try to adjust the E2SE limit screws unless you have the necessary service equipment and follow the factory procedures exactly.

The E2ME and E4ME Dualjet and Quadrajet carburetors have MC solenoids with adjustable rich-stop and lean-stop screws, figure 21-58. Again, these screws should be adjusted only to cure a specific performance problem or if measurements show that solenoid travel adjustments are out of limits.

To check MC solenoid travel, you will need a special float gauge. Insert the gauge through the D-shaped vent hole in the airhorn and be sure it moves freely, figure 21-59. Read the gauge with the solenoid released (up). Then press down gently until the solenoid bottoms at the lower limit and read the gauge again. If the solenoid travel is not within limits (the difference between the gauge readings up

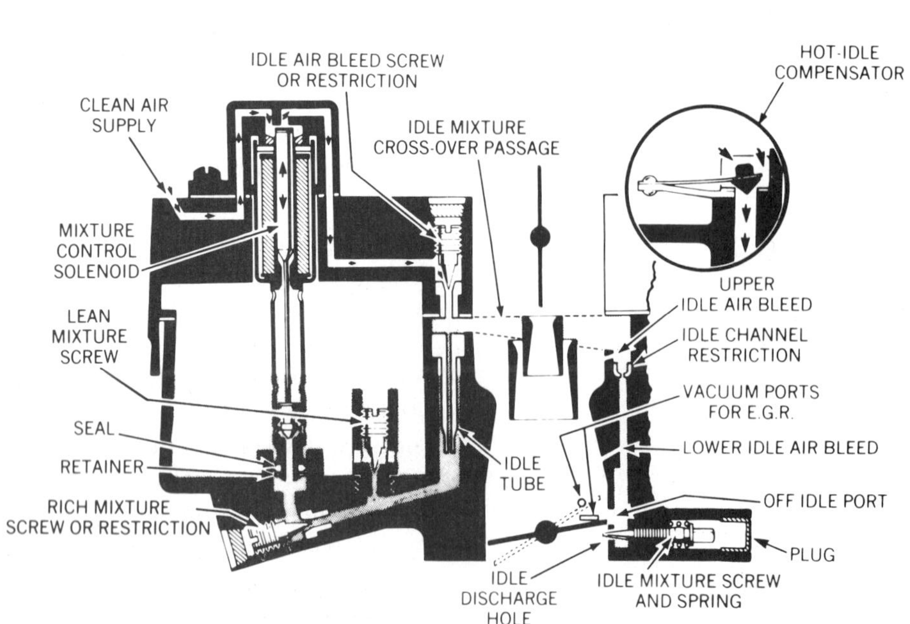

Figure 21-56. Rochester Varajet E2SE mixture control solenoid and idle system. (Rochester)

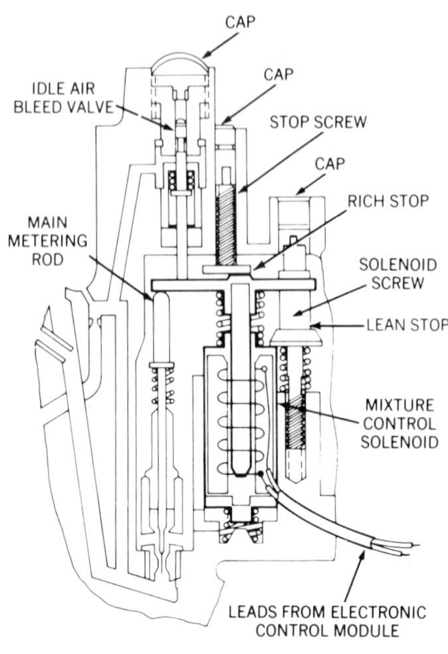

Figure 21-58. Rochester E4ME Quadrajet and Dualjet E2ME mixture control solenoid installation. (Rochester)

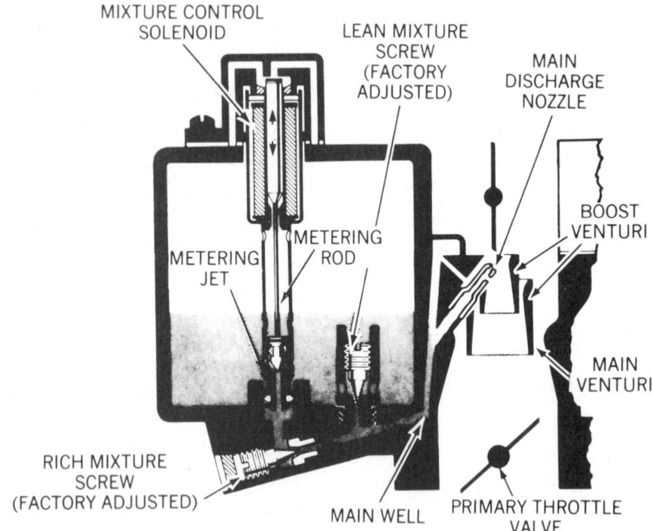

Figure 21-57. Rochester Varajet E2SE mixture control solenoid and main-metering system. (Rochester)

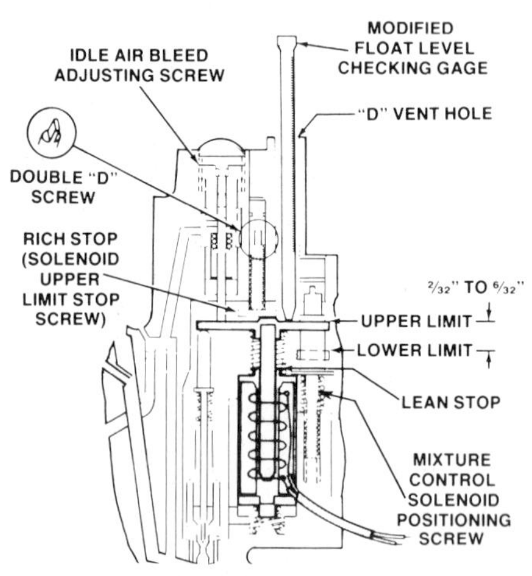

Figure 21-59. Rochester E4ME Quadrajet and Dualjet E2ME mixture control solenoid check and adjustment. (Rochester)

and down), you may have to adjust the lean stop or the rich stop. If the gauge readings are within limits, but performance problems continue, you may have to adjust the idle air bleed valve. Figures 21-60, 21-61, and 21-62 outline these adjustments, but you must check the carmaker's procedures for specific procedures and specifications.

Sensor (Resistor) Tests

Many sensors are potentiometers, which have 3-wire connectors. One wire supplies the reference voltage; another is the return voltage, or ground side, of the potentiometer. The third wire carries the sensor signal voltage to the computer. Potentiometers are built like this to maintain uniform current and temperature across the resistor. The return voltage signal is proportional to sensor movement or

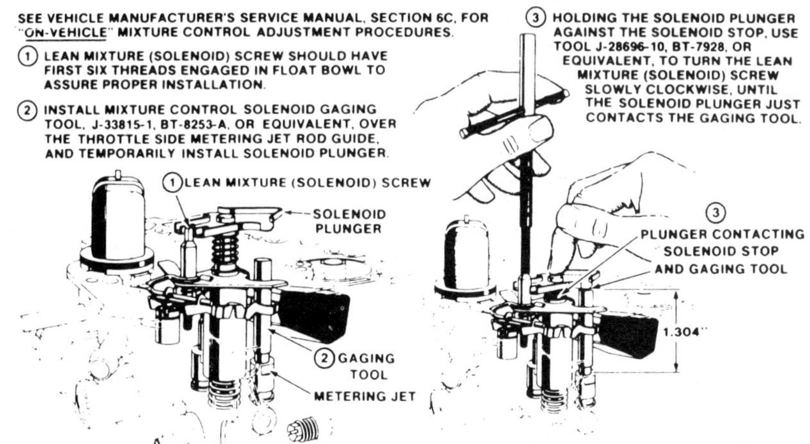

LEAN MIXTURE SCREW (BENCH) ADJUSTMENT
E2M & E4M MODELS ONLY

Figure 21-60. Typical lean-stop adjustment for Rochester carburetor mixture control solenoids. (Rochester)

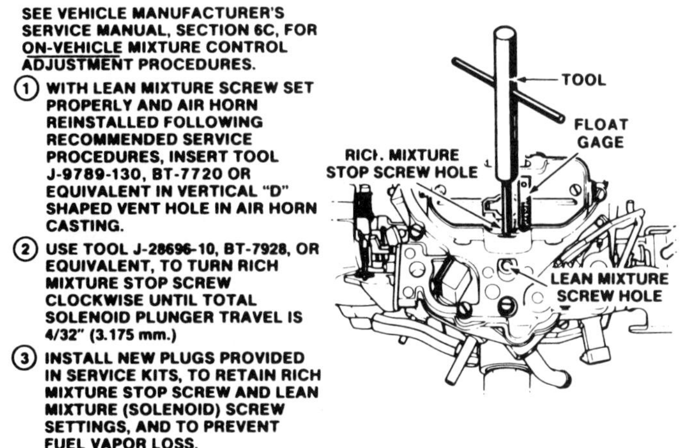

RICH MIXTURE STOP SCREW (BENCH) ADJUSTMENT
E2M & E4M MODELS ONLY

Figure 21-61. Typical rich-stop adjustment for Rochester carburetors with mixture control solenoids. (Rochester)

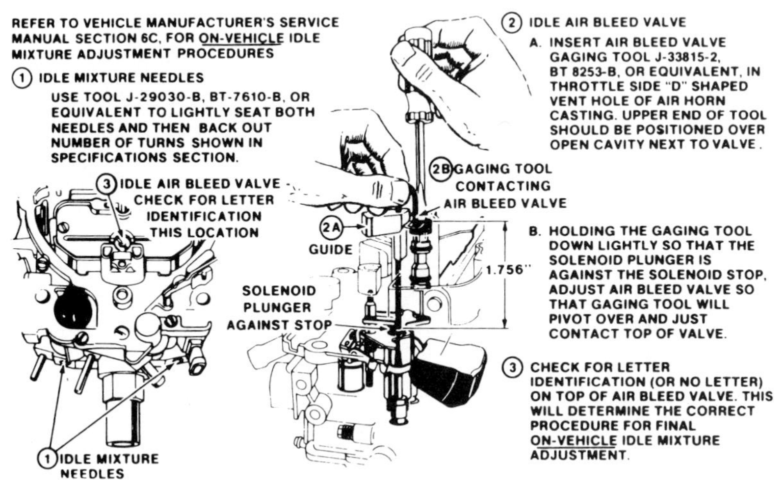

IDLE MIXTURE (BENCH) ADJUSTMENT
E2M & E4M MODELS ONLY

LEAN MIXTURE SCREW (BENCH) ADJUSTMENT—E2M & E4M Models Only

Figure 21-62. Typical idle air bleed screw adjustment for Rochester feedback carburetors. (Rochester)

pressure and is not affected by temperature changes. However, an open circuit or short circuit anywhere in the potentiometer or its wiring will upset the sensor signal. Figure 21-63 shows where to use your ohmmeter to check for opens and shorts in a potentiometer, or 3-wire sensor. Manufacturers give potentiometer resistance specifications through an operating range. If the sensor measures motion, you can operate it by hand while checking resistance. If the sensor measures pressure, operate it with a vacuum or pressure pump to test resistance.

Two-wire sensors are:

- Simple switches (high or low resistance, open- or closed-circuit signal)
- Thermistors (resistance changes proportionally to temperature)

Figure 21-64 shows where to use your ohmmeter to check for opens and shorts in any 2-wire sensor.

Many carmakers also give you procedures to use a voltmeter on a sensor to measure supply voltage and voltage drop during various operating conditions. Most of these procedures specify using a high-impedance digital volt-ohmmeter (DVOM). Analog (dial-type) meters may draw too much current and give inaccurate readings or damage electronic parts. You also can use an oscilloscope to measure voltage changes, and this often can indicate problems not shown by a DVOM. A meter indicates average voltage over a few milliseconds. An oscilloscope indicates dynamic voltage changes and possible signal interference, not shown by a DVOM.

Bosch K-Jetronic Air Sensor Adjustment

The air sensor in a Bosch K-Jetronic system measures airflow into the engine and controls fuel pressure proportionally to it. The sensor must be centered in its cone and at a precise height in relation to any airflow. It must not bind in any position. Measure and adjust these positions with the engine off.

1. Run the engine to normal operating temperature. Then shut it off.

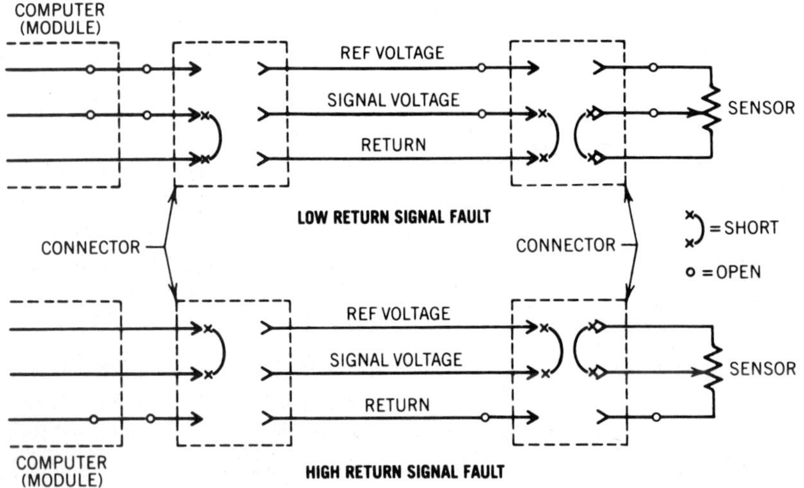

Figure 21-63. Use your ohmmeter to test a 3-wire sensor for open and shorted circuits as shown here.

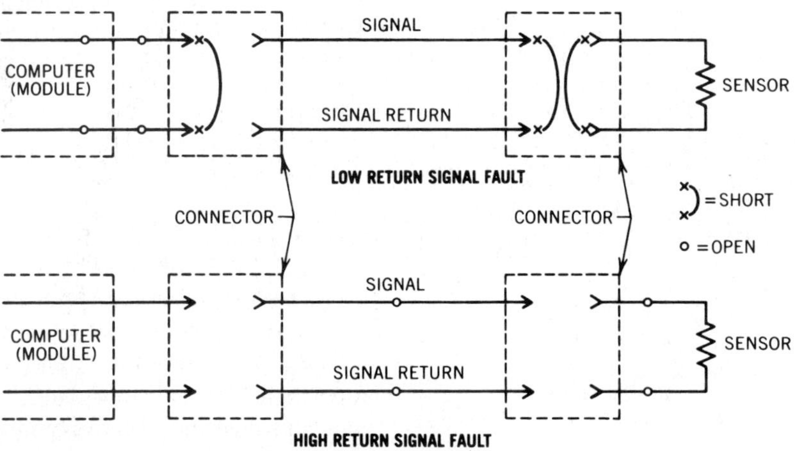

Figure 21-64. Use your ohmmeter to test a 2-wire sensor for open and shorted circuits as shown here.

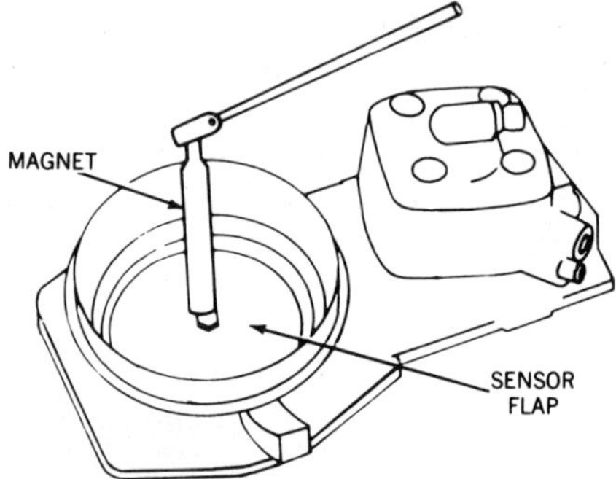

Figure 21-65. Use a magnet to raise and lower the K-Jetronic airflow sensor flap. (Volkswagen)

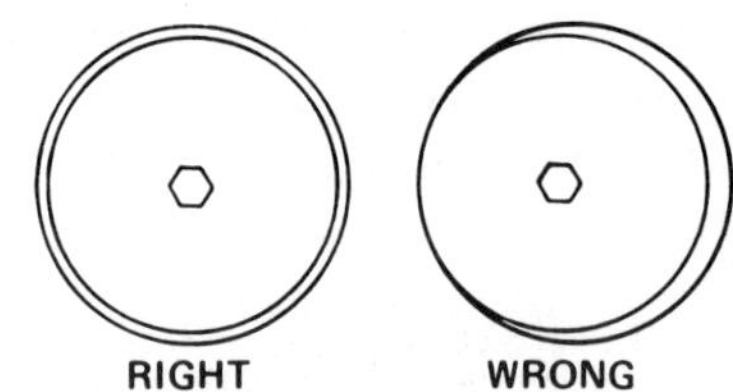

Figure 21-66. Be sure the airflow sensor is centered in the cone. (Volkswagen)

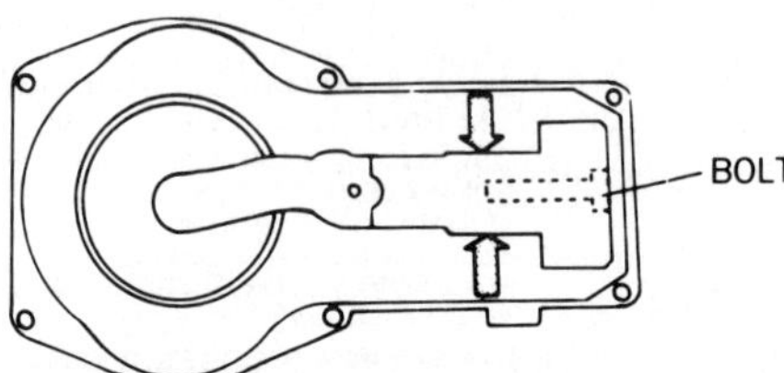

Figure 21-67. Loosen the airflow sensor lever bolt and slide the lever to a midposition on its shaft. (Volkswagen)

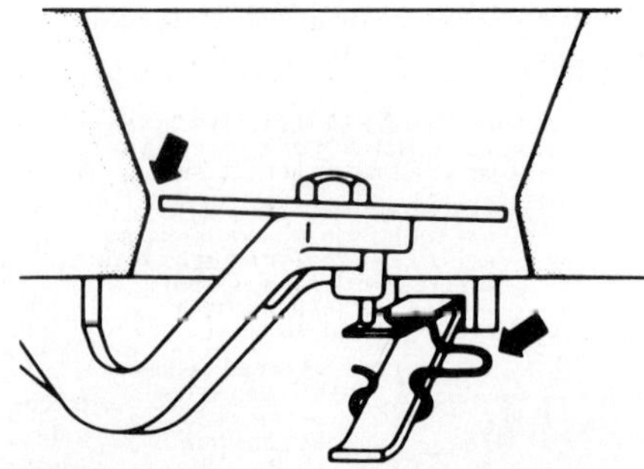

Figure 21-68. With the engine off, the plate must be at the narrowest part of the cone on the lever side. Bend the spring-loaded clip to adjust the height. (Volkswagen)

2. Remove the rubber boot from the sensor.

3. Using a strong magnet on the sensor bolt (first choice) or small pliers (second choice), gently move the sensor up and down to check for binding, figure 21-65.

4. If the sensor binds in both directions, check its centering and height adjustment. If it binds only upward, remove the fuel distributor and check the control plunger according to the carmaker's directions.

5. Look downward into the sensor cone at the sensor plate. It must be centered exactly, figure 21-66. Check the clearance with the specified feeler gauge (usually 0.01 mm or 0.004 inch) around the entire perimeter.

6. If the sensor is not centered, loosen the 10-mm bolt in its center and use a Bosch centering tool or the specified feeler gauge to position it properly. Then retorque the bolt to specifications.

7. If you can't center the plate and the lever appears off center in the cone, remove the sensor housing and the clamping bolt on the counterweight.

8. Slide the lever to a center position (side to side in its housing), figure 21-67, and retorque the clamping bolt to specifications.

9. After centering the sensor plate, check its height in the cone. With the engine off, the plate edge must be at the narrowest part of the cone on the counterweight side, figure 21-68.

10. If sensor height is incorrect, bend the spring-loaded stop clip up or down with small pliers to reposition it, figure 21-68. Do not scratch the plate or the cone. Scratches will upset airflow measurement.

Electronic Airflow Sensor Test and Adjustment

Bosch L-Jetronic systems and port injection systems from Ford, Chrysler, AMC, and several Japanese manufacturers have an airflow sensor. This is a movable flap in the air intake between the air filter and the throttle. The sensor flap moves a potentiometer that sends a voltage signal to the computer that is proportional to airflow volume, figure 21-69. Incorrect sensor operation can cause hard starting, erratic performance, and poor fuel economy. The following steps will help you use manufacturers' specific instructions to check and adjust this kind of sensor:

1. Turn the ignition on but do not start the engine.

2. Remove the air filter or sensor air intake duct.

3. Gently and slowly, push the sensor flap fully open. The electric fuel pump should run when the flap opens slightly. The flap must move freely in both directions.

4. Turn off the ignition and disconnect the sensor harness connector.

5. Use an ohmmeter to check resistance across specified terminals and compare to specifications. Some procedures specify that you move the flap by hand and check the resistance values as the potentiometer moves through its range.

Regulated Fuel Pressure Tests

All gasoline injection systems have pressure regulators to maintain uniform fuel pressure at the injectors. The regulator is on the downstream (return) side of the injectors. It maintains the required pressure at the injectors and releases excess fuel through the return line to the tank. Incorrect fuel pressure can cause hard starting (hot or cold) and a range of poor performance problems. Before testing the regulated fuel pressure, check the fuel pump delivery pressure and check for fuel line restrictions as explained in chapter 16.

Bosch K-Jetronic Pressure Tests. Fuel pressure in a Bosch K-Jetronic system is controlled by a primary pressure regulator, or relief valve, built into the mixture control unit and by a warmup regulator, figure 21-70. Follow the specific carmaker's procedures to check the following:

- System delivery pressure (primary pressure)
- Control pressure
- Injection pressure
- Return pressure
- Residual pressure (system pressure with the engine off).

You can adjust system (delivery) pressure by changing shims in the primary pressure regulator, figure 21-71, or by replacing the regulator valve. If control pressure is out of limits, replace the control pressure regulator.

Electronic Injection Pressure Tests. After testing fuel pump delivery pressure and checking for fuel line restrictions, check the regulator according to the manufacturer's instructions and these guidelines. Most regulators are controlled by a manifold vacuum diaphragm and regulate fuel pressure in relation to manifold pressure. The test procedures may specify that you apply a certain amount of vacuum to the regulator during testing.

1. Connect a pressure gauge to the system in one of these two ways:

a. If the fuel rail or throttle body has a test fitting with a schrader valve, figure 21-72, use a hose with a valve depressor to connect a gauge.

b. If the system does not have a schrader valve, use a T-fitting to connect a gauge on the inlet line to

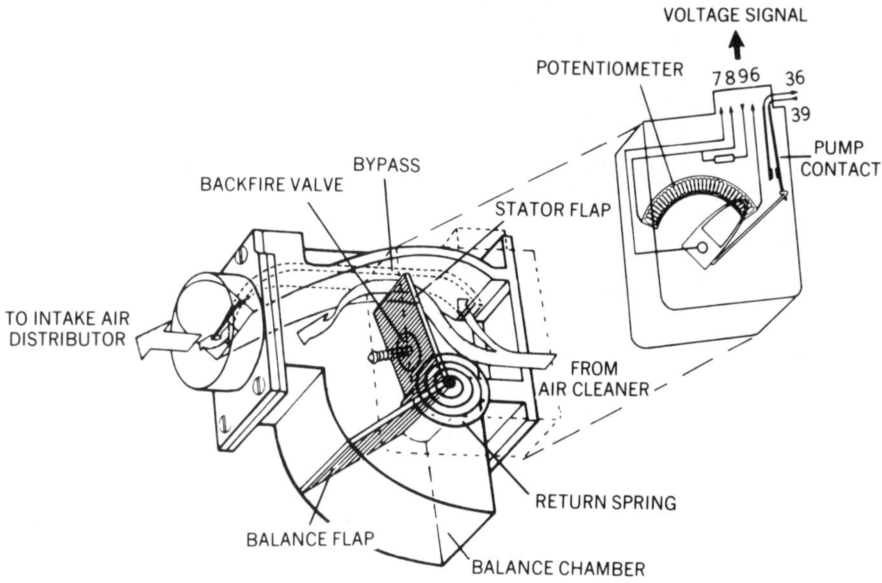

Figure 21-69. Use an ohmmeter to check airflow sensor resistance at the connector terminals. (Volkswagen)

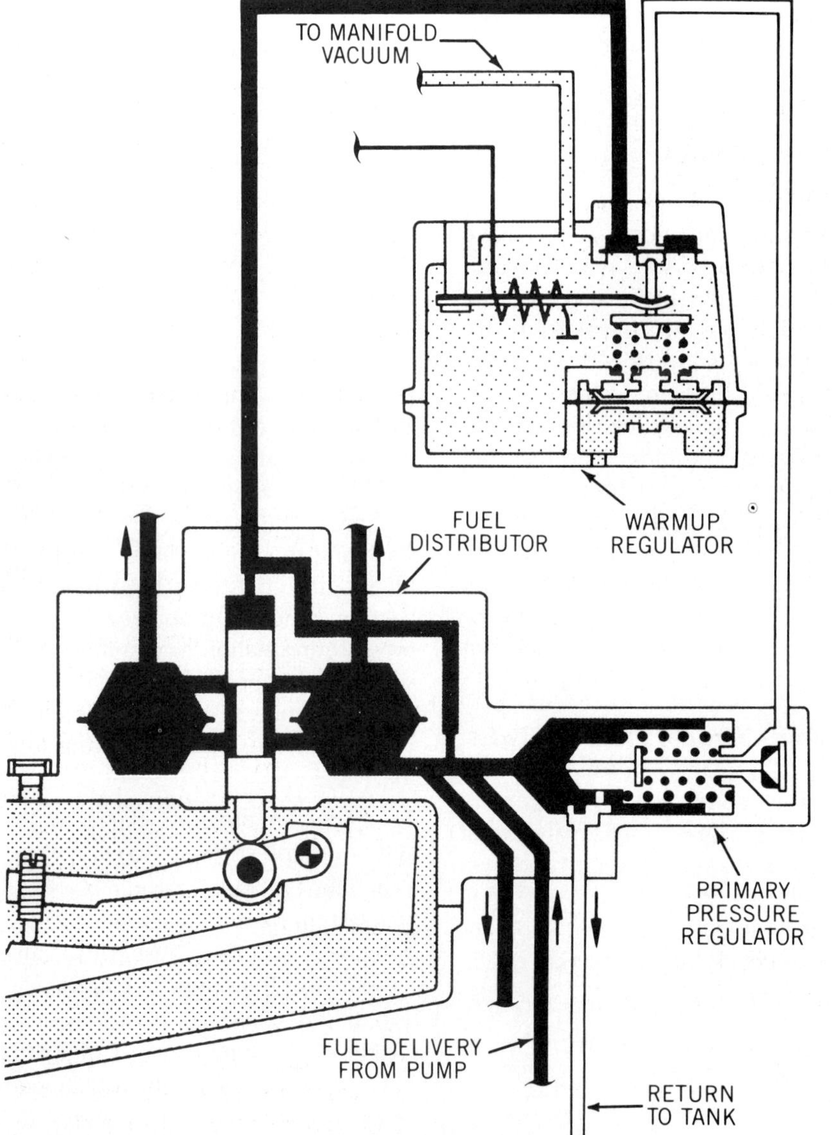

Figure 21-70. The primary pressure regulator and the control pressure regulator control pressure in a K-Jetronic system. (Bosch)

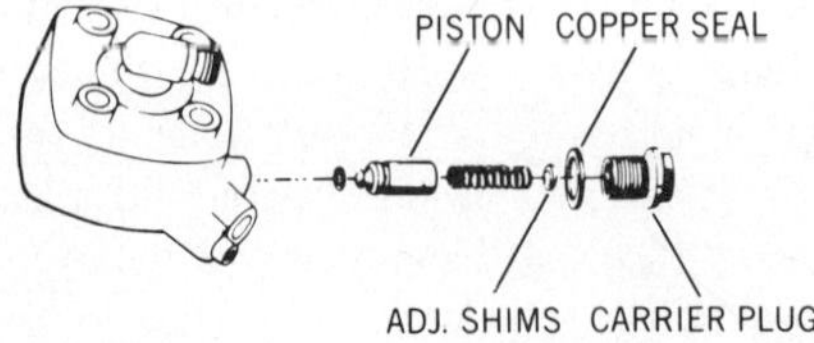

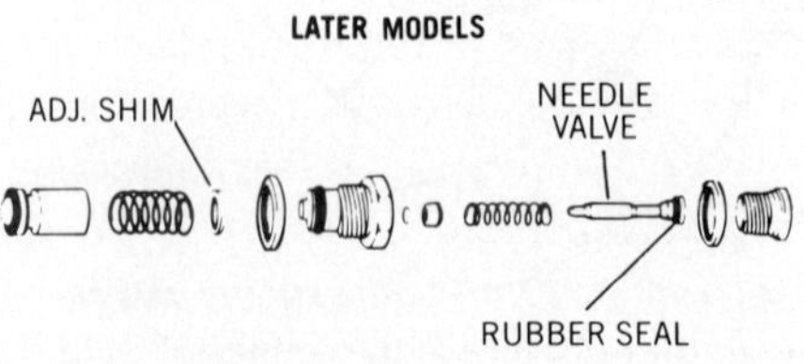

Figure 21-71. Add or subtract shims in the primary regulator to adjust system pressure. (Volkswagen)

the fuel rail or throttle body, figure 21-73.

2. Check regulated fuel pressure in one of these ways:

a. Start and run the engine at idle or a specified speed and note the gauge reading.

b. Turn the ignition key on and off several times to activate the fuel pump and build up system pressure. Then apply vacuum to the regulator with a hand-operated pump, figure 21-73, and note the fuel pressure at several vacuum readings.

3. Apply 15 to 20 in. Hg of vacuum to the regulator and close the vacuum pump valve. If the vacuum pump gauge reading drops, the regulator diaphragm is leaking.

If fuel pressure is *below* specifications, you may have a bad regulator, a bad fuel pump, or a restriction in the fuel delivery system. Test pump pressure and suction before replacing the regulator.

If fuel pressure is *above* specifications, you may have a bad regulator or a restricted fuel return line. Check the return line pressure and inspect for restrictions before replacing the regulator.

Fuel Pressure Relief

Remember that gasoline injection systems retain fuel pressure when the engine is off. Before you open a fuel line for testing or service, you must relieve this pressure. Carmakers have specific procedures for this, which you *must* follow. Most work on one or more of the following methods:

1. If the system has a schrader valve (similar to a tire valve) on the injector throttle body, fuel rail, or fuel manifold, figure 21-72, connect a hose with a valve depressor to the valve and release fuel into a suitable container.

2. Remove the electric fuel pump fuse or disconnect the pump wiring connector. Start and run the engine until it stalls to use the fuel held in the lines. Then crank the engine for 3 to 5 seconds to be sure that all pressure is relieved.

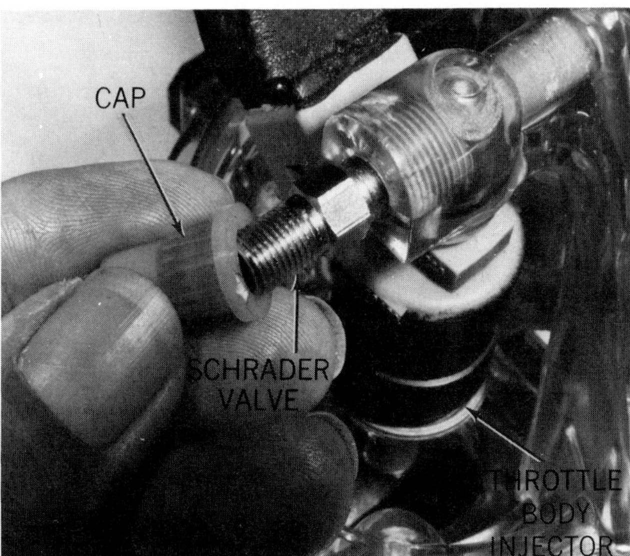

Figure 21-72. Use the schrader valve to relieve fuel pressure and to connect a gauge to test fuel pressure.

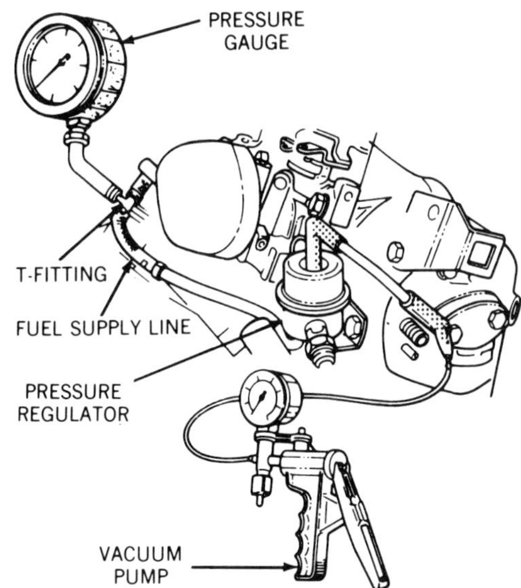

Figure 21-73. Use a T-fitting to connect a test gauge to the fuel delivery line. Check regulator operation with a vacuum pump. (Renault)

3. Disable the ignition and crank the engine to operate the injectors and release fuel into the intake manifold. This method is not recommended for cars with catalytic converters.

Fuel Injector Testing

You can test the hydraulically operated injectors of Bosch K-Jetronic systems for opening pressure, leakage, and delivery volume. These tests are similar to tests for diesel injectors. They require special test equipment, figure 21-74, and procedures.

Electronic injectors are solenoid-operated actuators. Therefore, you can apply the same test principles that you would use for any solenoid-operated device. Before checking the injectors, be sure that fuel delivery pressure and regulated pressure are correct. Also be sure that the battery is fully charged, the charging system is working correctly, and the computer or electronic module is getting the required supply voltage. Supply voltage to the injectors is regulated to a constant level and applied to the injectors through a relay or through one circuit in the computer or module. Most Bosch systems have resistors in series with the injectors to lower supply voltage to 3 to 5 volts. The computer or module switches the ground side of the injectors to turn them on and off, figure 21-75. Manufacturers' test procedures include one or more of the following kinds of checks:

1. With the engine running at idle, use a stethoscope to listen to each injector. You should hear a steady clicking sound. If you don't, the solenoid is not operating.

2. Throttle body injectors are usually visible with the air intake duct or air cleaner cover removed. Observe the injector spray pattern with the engine at idle. It should be steady and uniform. Connect a timing light to the engine and aim the beam at the injector spray. This will visually "freeze" injector operation for better examination.

3. Each injector has a 2-wire connector. Disconnect it and use an ohmmeter to measure resistance of the solenoid coil, figure 21-76. Use the ohmmeter to measure between each injector terminal and the injector shell or the engine to check for a shorted injector, figure 21-76. If coil resistance is out of limits or if the injector is shorted, replace it.

4. You also can use an ohmmeter to check the injector harness for open or short circuits, with the harness disconnected from the computer. With the harness connected to the injectors, connect your ohmmeter to the supply and return terminals for each injector or group of injectors, figure 21-77. If the injectors are energized in groups of 2, 3, or 4, the resistance through the harness and the injector group will be lower than for a single injector. Resistance higher than specifications indicates one or more open injectors or circuits. Low resistance indicates one or more shorted injectors or circuits.

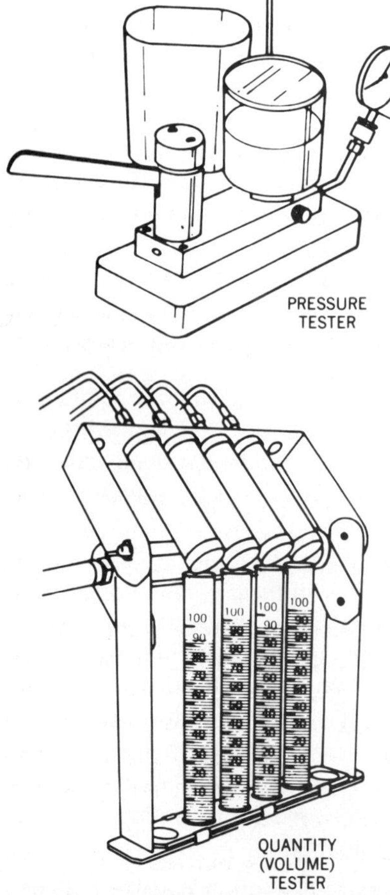

Figure 21-74. You need special equipment to check K-Jetronic injector pressure and volume. (Volkswagen)

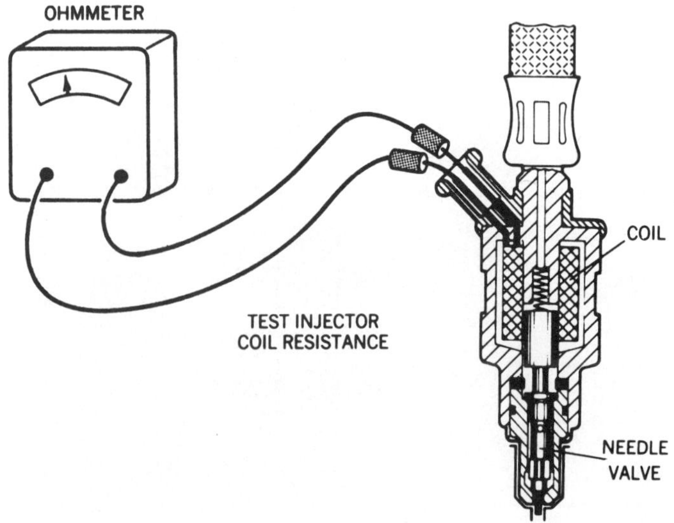

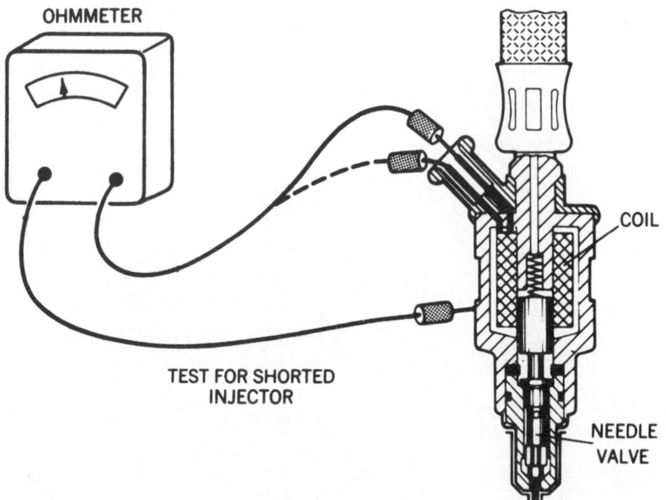

Figure 21-76. Use an ohmmeter to check an injector for shorted and open circuits as you would any solenoid. (Volkswagen)

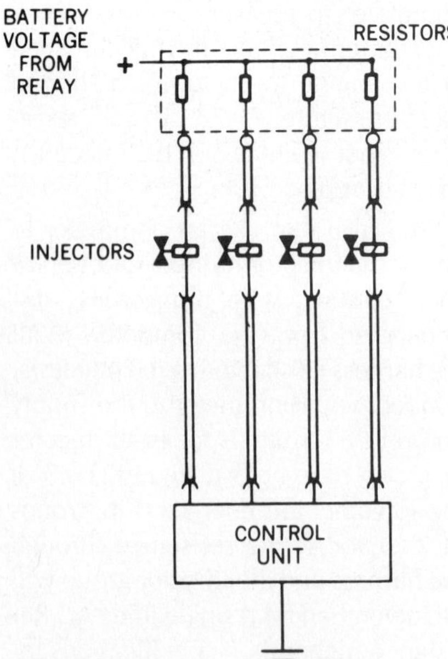

Figure 21-75. The control module switches the ground side of the injectors. (Volkswagen)

Caution Unless specifically directed, do not apply battery voltage directly to an injector. Many injectors operate on only 3 to 5 volts. Full 12-volt battery voltage may burn the coil windings.

5. On some systems, you can use a 12-volt test lamp to check for supply voltage at the injector. Connect the lamp between the supply wire connector and the injector connector or between the injector connector and ground, figure 21-78. Crank the engine. The lamp should light each time the injector, or group of injectors is energized.

6. You can use an oscilloscope to examine injector pulse width and duty cycle. Pulse width is the time in milliseconds that the injector is energized. Duty cycle is the percentage of on-time to total cycle time. You will need the equipment maker's instructions and specifications for this test. Basically, connect the scope positive (+) lead to the injector supply wire and the negative (−) lead to a ground on the engine. Set the scope on the low-voltage scale and adjust the pattern to fill the screen. If you see a steady unbroken trace with the engine running or cranking, figure 21-79, that

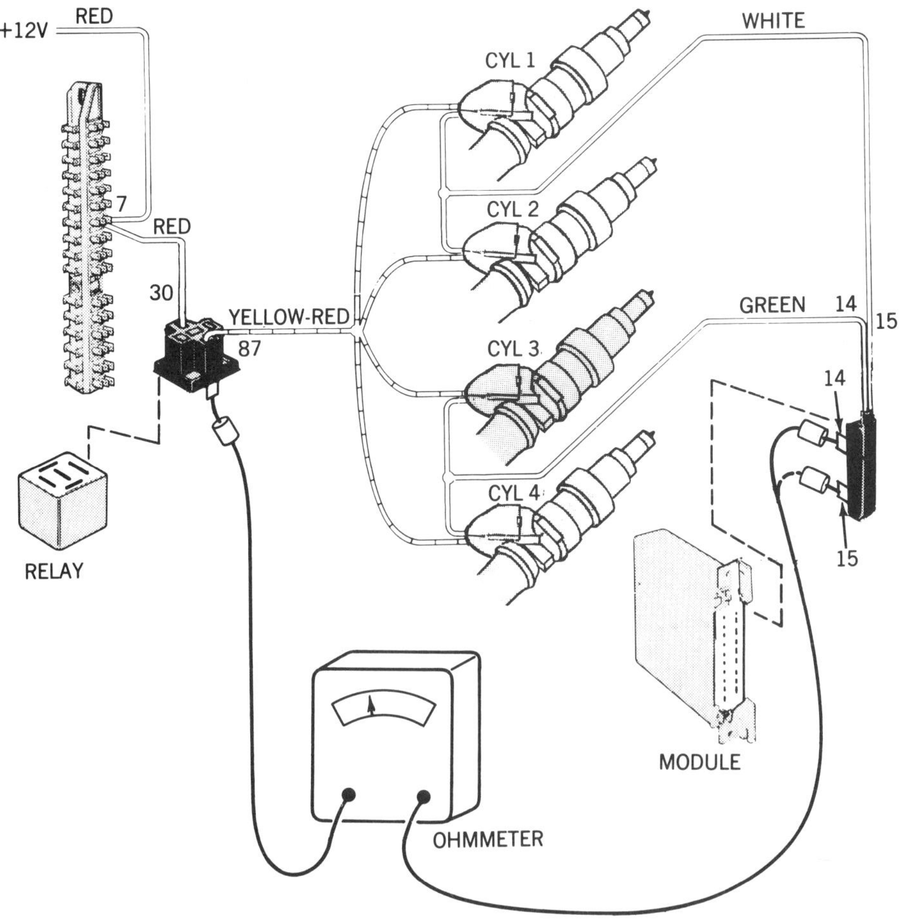

Figure 21-77. Connecting your ohmmeter like this allows you to check complete circuit resistance. (Volvo)

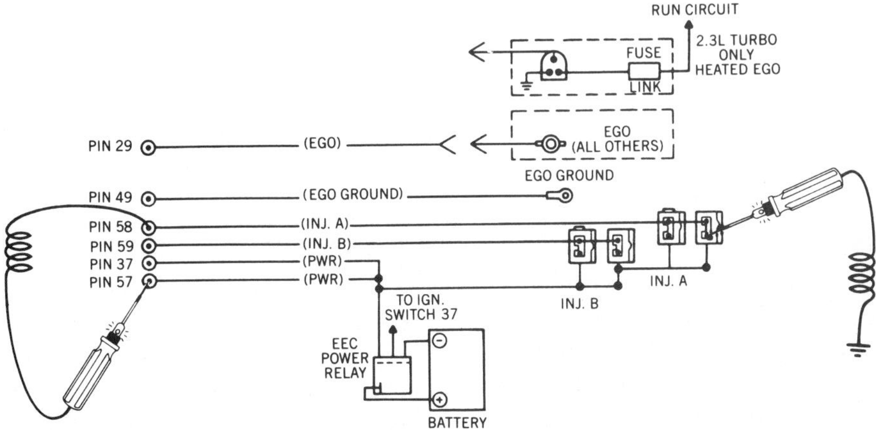

Figure 21-78. Using a test connector, you can check applied voltage at the injectors with a test lamp. (Ford)

injector has an open circuit. The solenoid is not switching on and off. If the voltage trace is too high, the injector circuit has high resistance; if it is too low, the circuit has low resistance, figure 21-80.

Injector Power Balance (Speed-Drop) Test. In chapter 5, you learned to do a power balance test by killing the ignition to each cylinder, in turn, and measuring the speed drop with a tachometer. In chapter 18, you learned to do a similar speed-drop test on some diesels by loosening each injector fitting, in turn, and measuring speed drop. If engine speed drops the same rpm for each cylinder, that cylinder is producing its share of power. If speed doesn't drop when you kill a cylinder, that cylinder is not producing full power. You can apply these same principles to some electronic port injection systems, if manufacturer's instructions permit. Some carmakers advise against this test for vehicles with catalytic converters. If the test is recommended by the carmaker, proceed as follows:

1. Connect a tachometer and run the engine to normal operating temperature.

2. Run the engine at a steady idle speed or other specified rpm.

3. Disconnect and reconnect the injector electrical connectors one at a time to shut off fuel to each cylinder.

4. If speed does not drop uniformly for all cylinders, the injectors may not be operating properly.

You must remember that this test basically indicates that a cylinder is not producing full power if speed does not decrease. The absence of a speed drop may be caused by ignition or me-

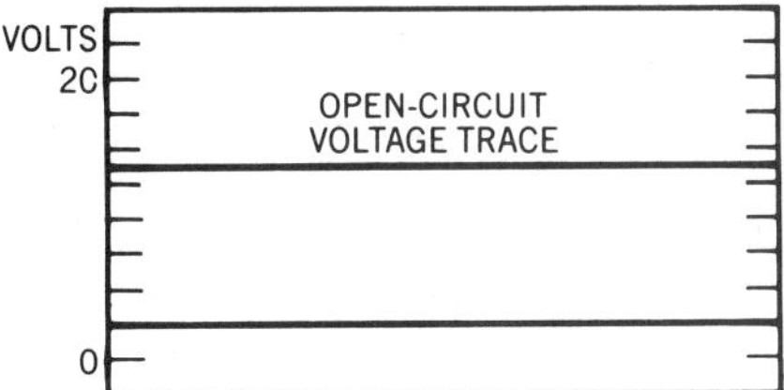

Figure 21-79. A straight, unbroken scope trace indicates an open circuit.

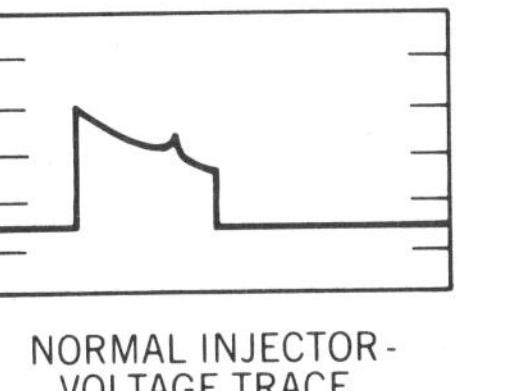

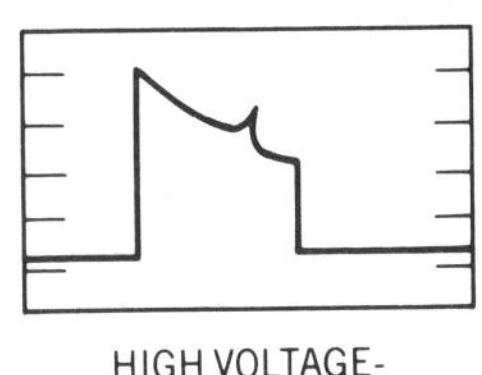

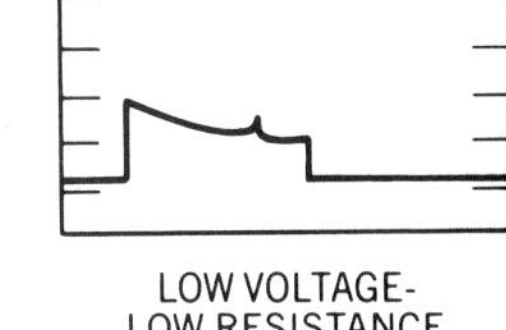

Figure 21-80. Injector voltage oscilloscope patterns.

chanical problems. Before condemning an injector, verify that the ignition is operating properly in that cylinder. Also do a compression or cylinder leakage test to check for mechanical problems.

EFI Pressure Drop Test

Electronic injectors used in multiport systems also can be checked on the vehicle with a pressure drop test. Carmakers may specify different testers and slightly varied procedures, but all essentially measure the pressure drop that occurs after an injector is fired.

The General Motors test is shown in figure 21-81. A pressure gauge connected to the fuel rail provides an initial system pressure as a basis for comparison. Connect a balance tester to one injector according to the test equipment instructions. Turn the ignition key on and note the gauge pressure recorded as the initial pressure (first reading in figure 21-81). Energize the tester to turn the injector on for a predetermined length of time. This causes the injector to spray a measured amount of fuel into the manifold. Record the fuel rail pressure measured after the fuel is sprayed (second reading in figure 21-81) for comparison with the base pressure and the other injectors. Test each injector in the same way and record the results. Subtract the second reading from the first reading to determine the pressure drop of each injector.

All good injectors should show a similar pressure drop specified by the manufacturer. This is generally 10 kPa. An injector that shows a pressure drop of 10 kPa or more higher or lower than the average drop of the other injectors is considered defective and should be replaced.

For example, suppose that the basic pressure (first reading) on a V-6 engine is 200 kPa. After the test, four injectors give a second reading of 100 kPa. However, the second reading is 85 kPa on the fifth injector, and 110 kPa on the sixth injector. Thus, the average pressure drop is 100 kPa.

The fifth and sixth injectors should both be replaced, but for different reasons. The fifth injector's pressure drop is too great, which means that it is spraying too much fuel (excessively rich). The sixth injector's pressure drop is too little. It is not spraying enough fuel (excessively lean).

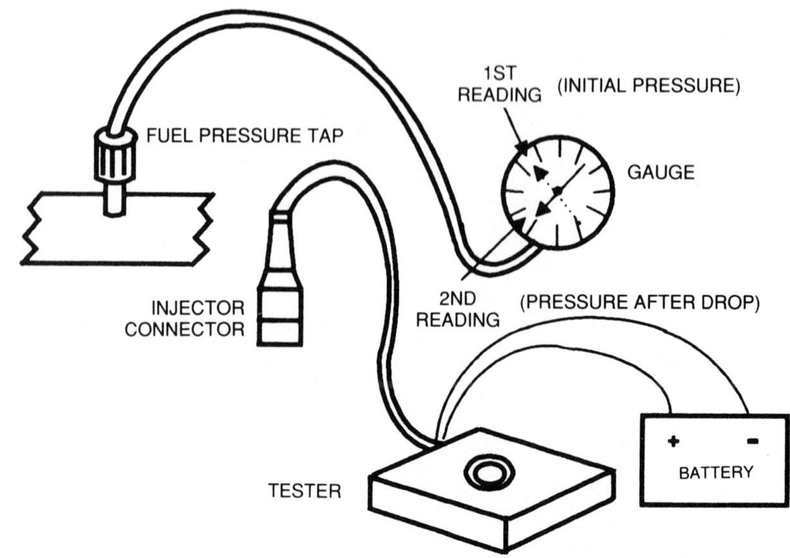

Figure 21-81. The GM fuel injector pressure drop test. (General Motors)

Carmakers generally recommend that you retest any injector that fails a balance test before replacing it. However, remember that the balance test has sprayed a quantity of unburned fuel into the engine. Before retesting, you should start and run the engine to eliminate the unburned fuel and prevent the possibility of flooding.

Throttle Position Sensor Adjustment

Every electronic engine system has a throttle position sensor. It may be a simple closed-throttle (idle-stop) switch, a wide-open throttle switch, or a variable potentiometer sensor. Some systems have a combination of two or all three of these. On most systems, the throttle position sensor is the only adjustable sensor. If the sensor or switch is out of adjustment, it can cause incorrect idle speeds or fuel metering, and it can keep the system from going into open- or closed-loop modes when it should.

Although procedures vary for different switches and sensors, the throttle position sensor in Rochester feedback carburetors provides a good example of adjustment principles. The sensor is mounted in the carburetor fuel bowl, and its shaft contacts the accelerator pump lever, figure 21-82. The sensor receives a reference voltage input from the computer and returns a signal voltage that varies with throttle position. At closed throttle, the return voltage is 1 volt or less. At wide-open throttle, the return voltage approximately equals the 5-volt reference voltage. To test and adjust the sensor:

1. Drill and remove the plug above the adjustment screw.

2. Connect a digital voltmeter across terminals B and C of the sensor connector. Use jumper wires or connector adapters, if necessary.

3. With the ignition on and the engine and air conditioning off, move the throttle to the position specified in the carmaker's instructions and read the signal voltage.

4. If voltage is out of limits, turn the adjustment screw to get the correct signal voltage.

5. After adjustment, install a new plug over the adjustment screw.

Many throttle position sensors used with fuel injection systems are not adjustable. They are simply replaced if out of specification as determined by your testing. Adjustable sensors have slotted mounting screw holes that

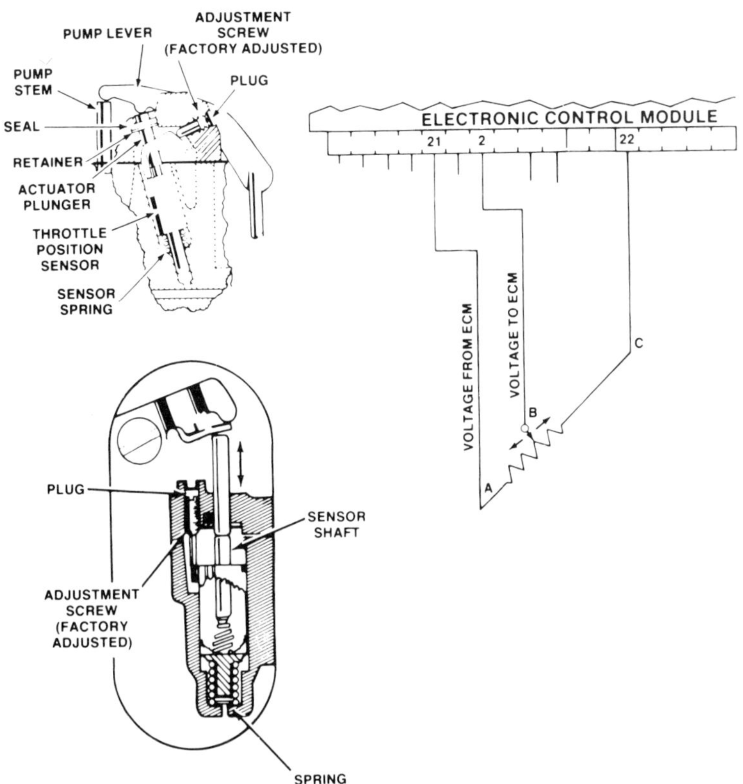

Figure 21-82. Typical Rochester carburetor throttle position sensor. (Rochester)

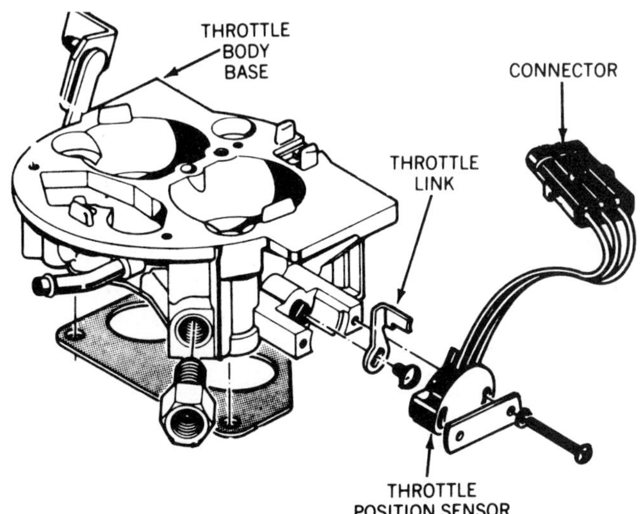

Figure 21-83. Every gasoline injection system has a throttle position sensor or switch. Test it as you would any potentiometer or on-off switch. (Rochester)

allow them to be repositioned to provide a specified voltage reading (usually 1 volt or less) with the throttle in its normally closed idle position. The voltage reading is then checked with the throttle in its wide-open position and compared to specifications (usually just slightly less than the reference voltage). If the reading is within specifications at each end of the range, the sensor is good. If the reading is within specifications at the closed idle position, but out of specifications at the wide-open position, the sensor is defective and cannot be satisfactorily adjusted. Figure 21-83 shows a typical fuel injection TPS.

Injection System Idle Adjustments

As with a carburetor or a diesel injection system, a gasoline injection system must maintain a steady idle speed and air-fuel ratio. These are routine adjustments on some systems. On many late-model cars, however, the electronic engine control system governs idle speed and mixture, and these functions are not adjustable. We can't give you all of the procedures for specific injection systems, but the following sections outline the idle adjustment points for some common systems. Most systems that have idle adjustments require these preliminary steps:

1. Connect a tachometer to the engine.
2. Set the parking brake and block the drive wheels. If the car has a vacuum-release parking brake, disconnect and plug the hose.
3. Run the engine to normal temperature. Be sure that cold-start enrichment and fast-idle devices are off.
4. Check the engine decal for basic specifications and instructions.
5. Follow the carmaker's directions for the following points:
    a. Disconnecting and plugging EEC vapor canister purge hoses to the engine
    b. Air cleaner position (on or off)
    c. Transmission position (drive, neutral, or park)
    d. Air conditioning and headlamps (on or off).
6. If you are using an infrared exhaust analyzer, place the probe in the tailpipe.
7. Check or adjust ignition timing before adjusting the idle.

Bosch K-Jetronic Idle Adjustment. K-Jetronic systems have adjustment points for idle speed and exhaust CO percentage. Adjust idle speed by turning a restrictor screw in an idle air bypass passage around the throttle, figure 21-24. Turn the screw inward (clockwise) to decrease the idle speed

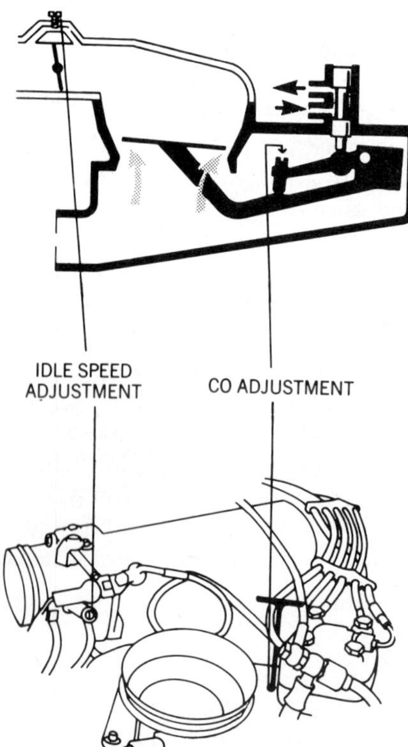

Figure 21-84. Typical idle speed and CO adjustment points for a Bosch K-Jetronic system. (Volkswagen)

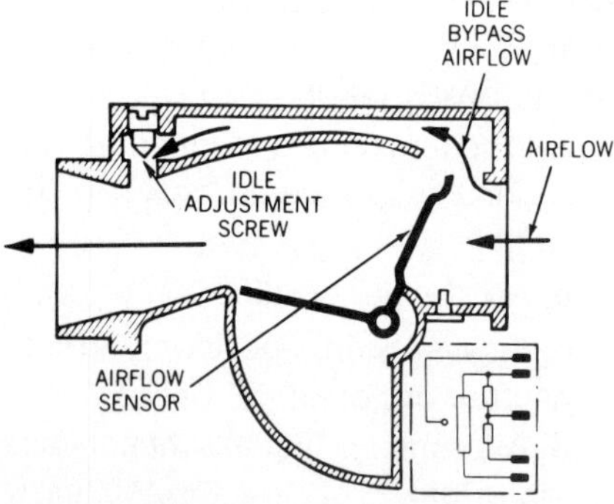

Figure 21-85. The idle mixture screw controls both speed and mixture on an L-Jetronic system. (Renault)

and outward (counterclockwise) to increase idle speed.

The CO adjustment changes fuel pressure and thus affects idle mixture. The adjusting screw is inside the air sensor housing of the mixture control unit and accessible by removing a plug, figure 21-84. Turn the screw clockwise to richen the mixture (increase CO) and counterclockwise to lean the mixture (decrease CO).

Adjustment instructions vary from car to car, particularly for those with EGO sensors and feedback control as part of the K-Jetronic system. Follow instructions for disconnecting idle stabilizers and PCV and EEC hoses. The EGO sensor must be connected and operating in closed loop to adjust CO.

Bosch L-Jetronic Idle Adjustment. Idle speed and mixture on early L-Jetronic systems and similar airflow-controlled, port injection systems are controlled by an air bypass screw, figure 21-85. Since the idle air bypasses the sensor, it doesn't affect the signal to the fuel control module. Adjusting the screw, therefore, changes the air-fuel ratio as well as speed. The adjustment changes the idle airflow as fuel flow stays constant. Many service manuals refer to this as a "CO adjustment." Turn the screw inward to richen the mixture (increase CO) and outward to lean the mixture (decrease CO). Late-model L-Jetronic systems have constant-speed idle air control valves, operated by the engine computer.

Minimum Idle Speed Adjustment for Fuel Injected Engines. All carmakers have procedures for adjusting the minimum idle speed on fuel injected engines. This often is called the minimum air rate. This is the minimum speed at which the engine will idle on airflow around the throttle plate, with the idle air control (IAC) valve closed or the idle speed control (ISC) motor fully retracted.

Minimum idle speed should be checked and adjusted if necessary when an engine suffers from a rough or rolling idle or stalling. It also should be checked if the throttle body is serviced in any way, especially after TPS adjustment or replacement. The engine should be at normal operating temperature when minimum idle speed is adjusted.

Before minimum idle speed adjustment on port injection systems, inspect the throttle body and idle speed control (ISC) air bypass valve for sludge or varnish buildup. If sludge is present, clean it as outlined in the next section.

Because most late-model EFI systems have adaptive memory that modifies idle speed control as the engine ages, the computer idle speed control program must be reset after minimum idle speed adjustment. The manufacturer's instructions must be followed in all cases, or idle speed control will not be correct after adjustment. In some cases, it may be necessary to drive the vehicle for the processor to learn the idle speed control program, or strategy.

Minimum idle speed is adjusted by the throttle stop screw in the throttle body, similar to an idle speed screw on a carburetor. It is *not* adjusted by repositioning the IAC valve or ISC motor. Throttle stop screws on most Ford engines are on the outside of the throttle body and bear against the throttle linkage. They may be sealed with tamper-resistant paint. GM throttle bodies for TBI systems and for PFI systems have the throttle stop screw located behind a concealment plug in the base of the throttle body. You must remove the plug for access to the screw, which usually can be done without removing the throttle body from the engine. GM throttle bodies built before mid-1984 had a hardened steel plug over the screw, which must be removed with a punch and a drill. Throttle bodies built after mid-1984 have a soft plug that can be removed by piercing it with an awl or small punch and prying it out.

When checking and adjusting minimum idle speed, you must follow the manufacturer's procedure and specification exactly for a specific engine. We can't give you the instructions here because of the variety. For example, Ford has more than 25 different procedures for engines from 1981 to the present; GM has 9 adjustment procedures and 11 different ways to reset the idle control program.

Throttle Body Cleaning

Throttle body cleaning should be on the preventive maintenance schedule for any port fuel injected (PFI) engine. Unfortunately, it usually is not.

PFI engines tend to accumulate a thin varnish, or sludge, deposit around the throttle plate in the throttle body. Some carmakers refer to this as "throttle body coking." This residue may accumulate in the throttle body bore, on the throttle plates, or in the idle air bypass valve.

Throttle body sludge deposits are caused by combustion byproducts drawn in through the PCV system. These deposits can reduce airflow, which is particularly critical during cold-engine warmup, idle, and deceleration. The deposits also can cause a rough idle and stalling after a cold start, on deceleration, or during idle. They also may cause the engine to stumble on light acceleration. Sludge deposits often are hard to recognize because they can appear as a thin transparent film of oil.

Throttle body cleaning is an easy service and should be performed every 20,000 to 30,000 miles. The exact steps will vary for different engines, but you can apply the following guidelines.

First, use a high-quality carburetor choke spray cleaner that is safe for use with catalytic converters and oxygen sensors and does not contain methylethylketone (MEK). Check carmakers' recommendations for specific cleaners.

Second, save your old toothbrushes or buy some new ones for throttle body cleaning. Small bottle brushes, cotton swabs, Scotch Brite cleaning pads, and clean lint-free clothes are the other simple cleaning tools you will need.

Third, follow the specific carmaker's procedure to test minimum idle speed (or minimum air rate) before cleaning the throttle body. If the throttle body is dirty, you will probably find idle speed out of specification. In fact, the engine may stall during testing. Don't adjust the minimum idle speed before cleaning the throttle body. That's the last step of the service.

To clean the throttle body, you will have to remove the air intake duct. After that first step, manufacturers' variations will affect how you proceed. You can clean some throttle bodies while they are mounted on the manifold air intake plenum. You must remove others to clean both sides of the throttle plate satisfactorily, figure 21-86. Many Toyota throttle bodies must be removed, for example, while most GM, Ford, and Chrysler units can be cleaned while mounted on the manifold.

If you have to remove the throttle body from the manifold, disconnect the throttle linkage and any wiring harness connectors for the throttle position sensor (TPS) and other electrical devices. You can usually leave vacuum lines attached.

Do not spray the cleaner onto or into electrical components such as the TPS and idle air control (IAC) valves. General Motors says to cover the TPS and the IAC valve. Ford cautions against spraying cleaner on black plastic IAC valves.

Starting with some 1989 models, Ford introduced a "sludge-tolerant" throttle body and IAC valve design that has a special coating on the throttle plates and bore, as well as on some IAC valve parts. This coating must remain in place for correct closed-throttle idle airflow. Any attempt to clean these throttle bodies or bypass valves will damage the coating and alter the idle airflow calibration. These components are identified with a yellow-and-black ATTENTION decal. Do not try to clean throttle body parts so marked.

Do not run an engine with a vane airflow meter or mass airflow sensor during the cleaning procedure. With these precautions in mind, proceed with throttle body cleaning along these guidelines:

1. Start and run the engine to normal operating temperature.

2. Remove the air intake duct. Remove the throttle body if necessary.

3. Spray enough cleaner into the throttle body to wet the bore. Do not spray plastic and electrical parts. Let the cleaner soak for about 15 minutes.

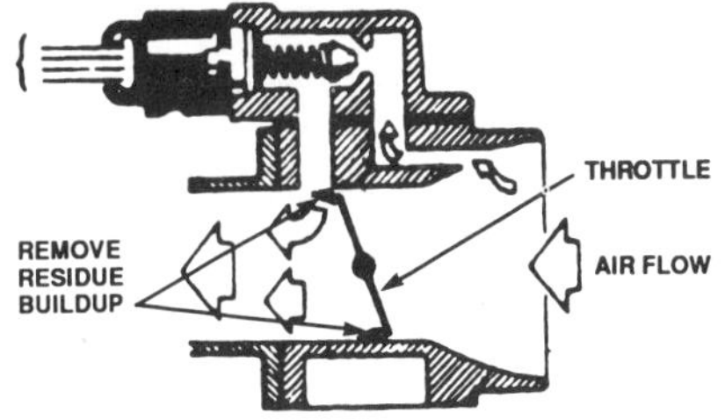

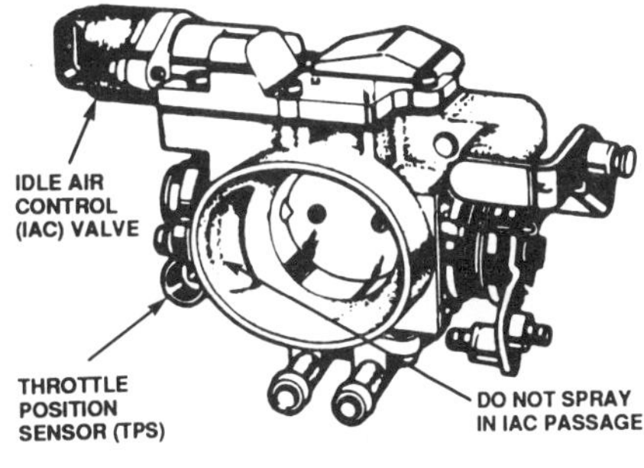

Figure 21-86. Typical throttle body cleaning areas.

4. Saturate a brush or a Scotch Brite pad with cleaner and scrub the bore and throttle plate firmly to remove all sludge. Be sure to clean the following areas:

- Where the throttle shaft meets the bore wall
- Both sides of the throttle plate and the edges that contact the bore wall
- The area of the bore immediately behind the throttle plate.

5. Rinse the throttle body by spraying more cleaner through it. Be careful to keep the cleaner off the TPS, the IAC valve motor and other electrical and plastic parts. Dry the throttle body with compressed air.

6. To check the effectiveness of the cleaning, open the throttle completely and let it snap closed. Then open the throttle lightly and check for a sticking throttle plate. If the throttle plate still sticks, repeat the cleaning operation to remove any remaining deposits.

7. Reconnect any hoses or connectors removed for cleaning and reinstall the throttle body if removed. Reinstall the intake air duct.

8. Check and adjust minimum idle rpm or airflow according to the carmaker's procedures.

GM CCC Solenoids and Relays

Electronic system actuators must have resistance designed to match system voltage and current capacity. Again, the GM CCC system provides a good example. All CCC solenoids and relays have a minimum resistance of 20 ohms. This is necessary because the computer transistors can handle a maximum current of 0.8 ampere (800 milliamperes) at a maximum system voltage of 16 volts. If actuator or relay resistance were lower than 20 ohms, high current would burn out the computer transistors. Ohm's law proves this, because:

$$16 \text{ volts} \div 0.8 \text{ ampere} = 20 \text{ ohms.}$$

A shorted or grounded solenoid or relay can damage the computer. Check solenoid and relay with an analog ohmmeter on the low-resistance scale. Some solenoids have diodes as reverse current protection for the computer. Therefore, check solenoid resistance in both directions and use the higher reading to determine solenoid condition.

PROM, CALPAK, MEMCAL, and Computer Replacement

The programmable read-only memory (PROM) of an engine computer contains the calibration program for a particular vehicle, engine, transmission, and accessory combination. Depending upon the carmaker and the purpose of the PROM, it may be called a PROM, a calibration assembly, CALPAK, or MEMCAL. Ford calls its PROM a calibration assembly. A CALPAK is a separate PROM used with GM TBI fuel injection systems. It provides backup logic to control fuel delivery in case the ECM is not working properly. GM also uses a MEMCAL that combines the PROM, CALPAK, and electronic spark control (ESC) module logic all in one plug-in unit.

The computer itself contains the basic system operating program and performs the system switching functions. By using an interchangeable PROM, MEMCAL, or calibration assembly designed for specific applications, carmakers can adapt a single computer for a wide range of car models.

When a system computer problem occurs, it can be in the computer or in the PROM, or both. Most carmakers provide replacement computers and PROM's as separate parts. For example, if you replace a computer, you probably will have to remove the original PROM and install it in the replacement computer. If the computer uses a CALPAC or MEMCAL, it also will have to be transferred to the replacement computer. Because this type of memory is tailored to a specific vehicle model, you must always be sure that the right part number PROM is installed in a particular car. Manufacturers' parts books and service manuals provide this information.

The PROM's or calibration devices for some systems are built into separate modules that plug into the computer and are secured by screws. The PROM's for GM C-4 and CCC systems are mounted in small, dual-inline-package (DIP) devices with two rows of pins. The pins plug directly into socket openings in early computers. With late-model computers, the PROM plugs into a small plastic PROM carrier, and the carrier plugs into the computer. A special tool is available to aid in PROM removal and replacement. The tool is insurance against damaging the PROM pins during replacement. Starting with some 1986 models, the PROM pins are lightly soldered into the carrier, which contains circuitry that connects the PROM and CALPAC or MEMCAL. This type of carrier and PROM is replaced as an entire assembly.

To remove and replace a CCC PROM, CALPAC, or MEMCAL, you must make sure the ignition is off, then disconnect the battery negative cable. If the ignition is on when you disconnect or reconnect the battery cable, the voltage surge may destroy the computer. Computer locations vary from car to car. Disconnect the wiring harness connectors from the computer and remove the fasteners holding it under the instrument panel, behind a kick panel or from the engine compartment, figure 21-87. Be sure to take the necessary precautions to prevent electrostatic discharge while removing or installing the computer. Figures 21-88 and 21-89 summarize the steps for PROM, CALPAC, or MEMCAL replacement.

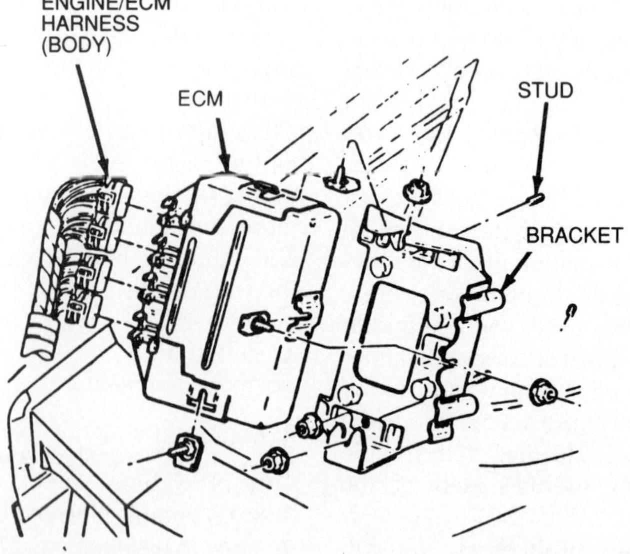

Figure 21-87. GM ECM locations vary, but computer removal is the same on all models. (Oldsmobile)

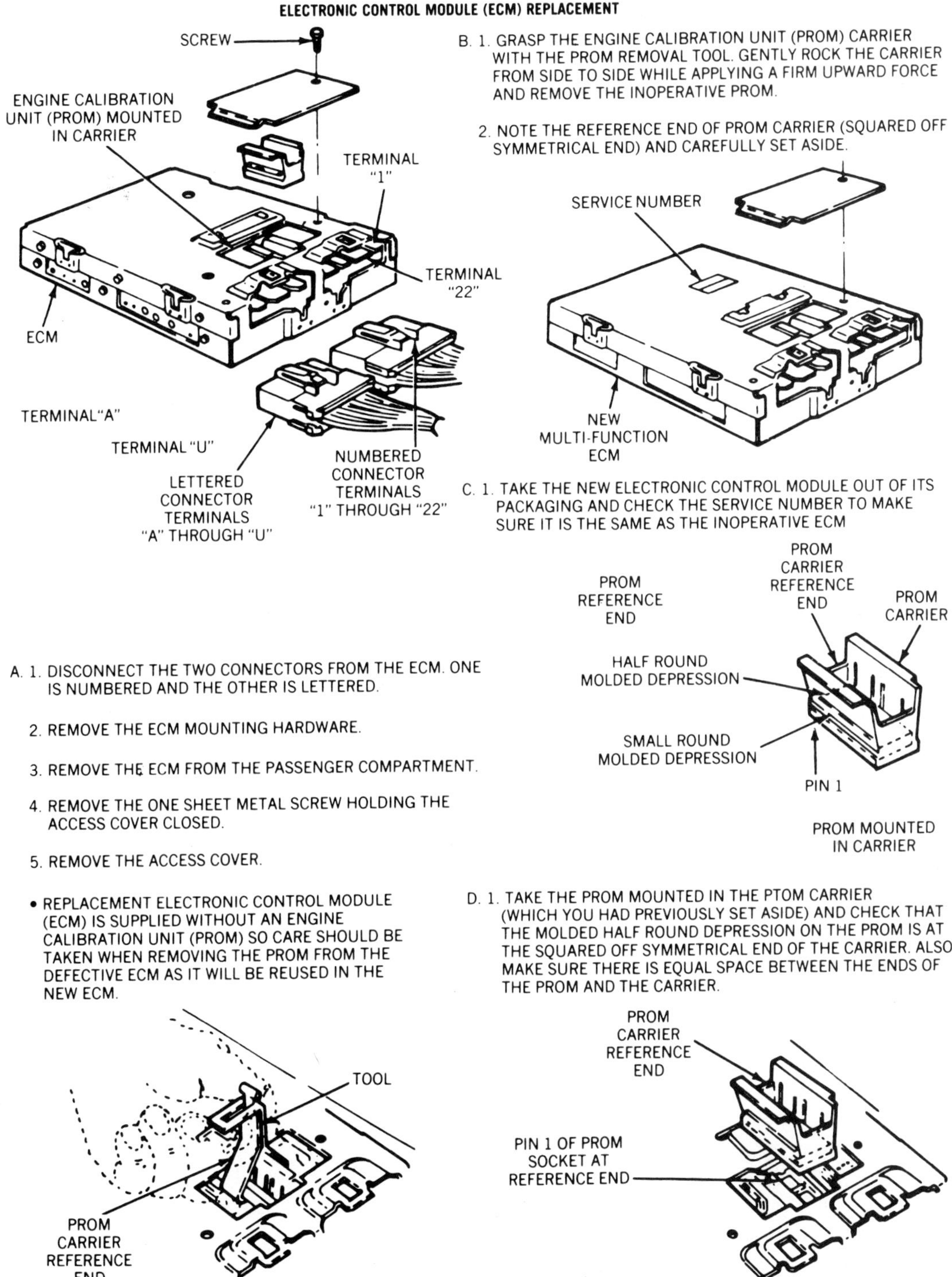

Figure 21-88. These instructions summarize PROM replacement in one GM CCC system. (Chevrolet)

The PROM will fit into its socket in either direction. When installing a new PROM, however, you must be very careful to install it in the right direction. If it is installed backwards, it will be destroyed the first time the ignition is turned on. This also may damage the computer.

GM provides reference notches on its PROM's, CALPAC's and MEM-CAL's. These notches must be aligned with a specified end of the carrier. The carrier must then be aligned in a specified position in the computer. Figures 21-88 and 21-89 show one example. PROM's are installed in opposite directions for systems with carbureted or with fuel-injected engines, figure 21-90. PROM package designs and computer socket designs may vary from year to year. For these reasons, *you must always check the carmaker's instructions for a specific vehicle before removing or replacing a PROM.*

Late-Model MEMCAL Installation. When installing a MEMCAL in a 1986 or later General Motors ECM, do not exert excessive vertical force while engaging the MEMCAL with its socket on the ECM circuit board. Too much force can flex the circuit board and damage other parts in the ECM. Install the MEMCAL as follows, using only enough force to latch the MEMCAL into its socket.

1. Remove the MEMCAL access cover from the ECM, figure 21-91.

2. Rotate the latches outboard to unlock the old MEMCAL. Lift it out gently.

3. If a cork spacer is glued to the top of the MEMCAL, remove it before installation.

4. Align the MEMCAL notches with matching notches in the ECM socket, figure 21-91.

5. Gently press down on the ends of the MEMCAL until the latches rotate toward the sides of the MEMCAL.

E. 1. POSITION THE CARRIER SQUARELY OVER THE PROM SOCKET WITH THE SQUARED OFF SYMMETRICAL END OF THE CARRIER ALIGNED WITH THE SMALL NOTCH IN THE SOCKET AT THE PIN 1 END.

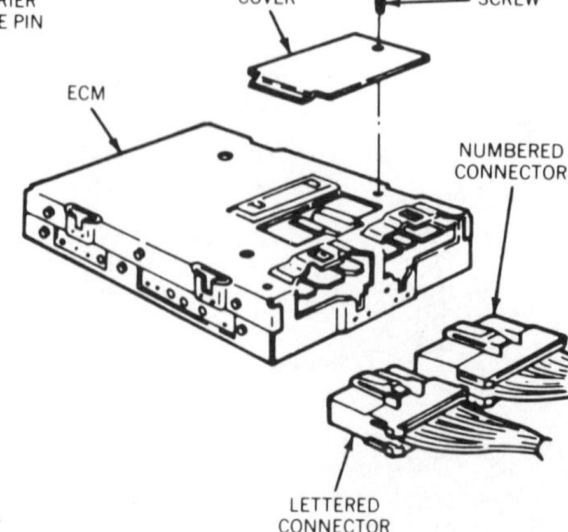

2. PRESS DOWN FIRMLY ON THE TOP OF THE CARRIER.

3. WHILE FIRMLY PRESSING DOWN ON THE CARRIER, TAKE A NARROW BLUNT TOOL AND PRESS DOWN ON THE BODY OF THE PROM. TRY TO SEAT THE PROM IN THE SOCKET SQUARELY BY ALTERNATELY PRESSING ON EITHER END OF IT.

F. 1. REPLACE ACCESS COVER ON THE NEW ECM.

2. REINSTALL ACCESS COVER FASTENING SCREW.

3. REINSTALL NEW ECM IN PASSENGER COMPARTMENT.

4. CONNECT THE TWO CONNECTORS TO THE NEW ECM.

5. TURN IGNITION ON AND START ENGINE.

6. GROUND DIAGNOSTIC TEST LEAD.

7. LOOK FOR TROUBLE CODE 51. IF THIS OCCURS, THE PROM IS NOT FULLY SEATED, INSTALLED BACKWARDS, HAS BENT PIN, OR IS DEFECTIVE.

- IF A TROUBLE CODE DOES NOT OCCUR, THE PROM IS INSTALLED PROPERLY.
- IF IT IS NECESSARY TO REMOVE THE PROM, FOLLOW INSTRUCTION IN STEP "A" & "B".
- IF NOT FULLY SEATED, PRESS FIRMLY ON THE PROM.
- IF PINS BEND, REMOVE PROM, STRAIGHTEN PINS AND REINSTALL. IF BENT PINS BREAK OR CRACK DURING STRAIGHTENING, DISCARD PROM AND REPLACE IT.
- IF FOUND INOPERATIVE, REPLACE PROM.
- IF INSTALLED BACKWARDS, REPLACE THE PROM.

ANYTIME THE PROM IS INSTALLED BACKWARDS AND THE IGNITION SWITCH IS TURNED ON, THE PROM IS DESTROYED

Figure 21-89. GM PROM replacement, continued. (Chevrolet)

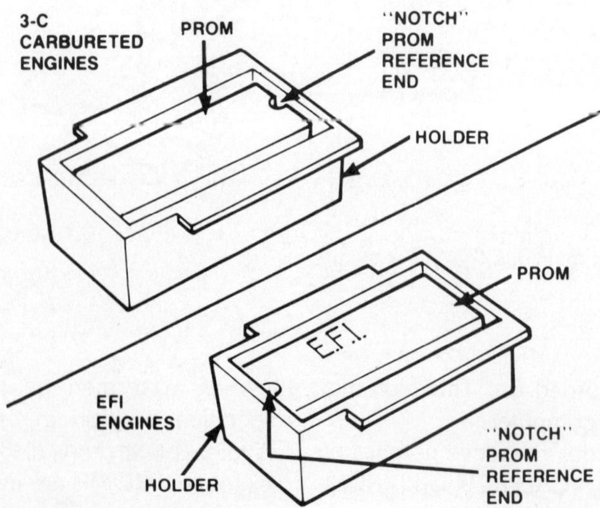

Figure 21-90. The identifying notch on the PROM is located differently for carbureted and fuel-injected engines. (Champion Spark Plug)

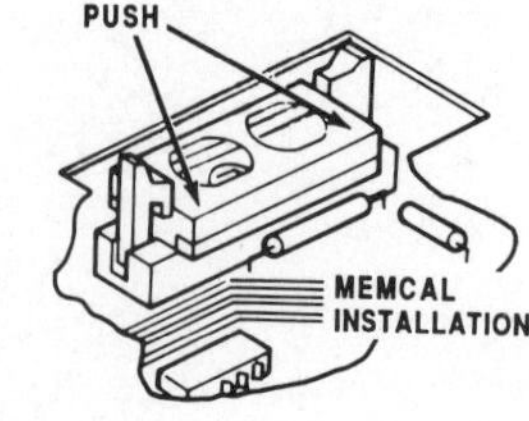

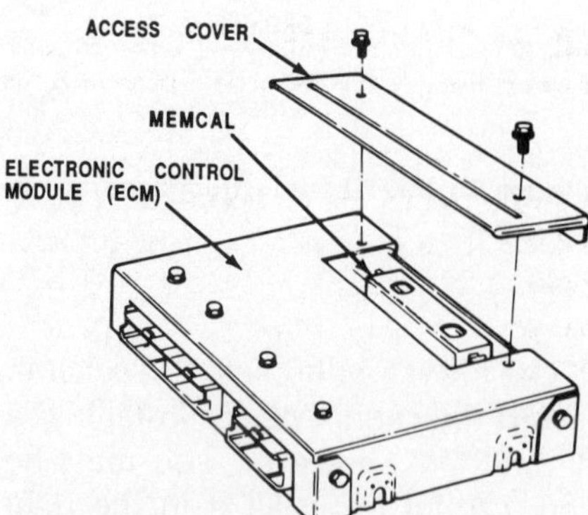

Figure 21-91. Replace a General Motors MEMCAL carefully, as shown here.

6. Continue pressing lightly on the ends of the MEMCAL and press the latches inward until they snap into place. Listen for a click. *Do not press hard.*

7. Reinstall the access cover.

SECOND GENERATION ONBOARD DIAGNOSTICS

The explanations of the self-test capabilities and troubleshooting methods for engine control systems of the 1980's and early 1990's show that there has been a lot of variety among the diagnostic philosophies of different carmakers. Several industry groups have worked to establish a basic set of uniform test capabilities for current vehicles. Among these are the Society of Automotive Engineers (SAE), the California Air Resources Board (CARB), the Environmental Protection Agency (EPA), and the International Standards Organization (ISO). The OBD-II requirements apply to all domestic, European, and Asian cars and light trucks sold in the United States. European and Asian countries have similar standards that make this a truly international program.

Starting with the 1994 model year, the second generation of onboard diagnostic systems will appear at least on vehicles sold in California. Known as onboard diagnostics II (OBD-II), these systems have a standard test connector for the engine computer, figure 21-92. The 16-pin connector has standard pin assignments for power (B+), power ground, signal ground, data transmission, and data reception.

Figure 21-92. This is the standard 16-pin test connector for OBD-II systems. The female half is the vehicle connector. The male half is the scan tool portion.

All vehicles must transmit a minimum standard list of trouble codes and computer system data, such as oxygen sensor voltage, engine temperature, and fuel and air control parameters to a scan tool. Beyond the minimum standard requirements, carmakers are free to include other codes, data, and special tests for their own systems.

SUMMARY

Because a central engine computer controls fuel metering, ignition timing, and emission devices on most late-model cars, you must practice "service integration" for testing and repair. During troubleshooting, you must consider the entire system and the effect of each part on total engine operation.

General Motors CCC, Ford EEC, and Chrysler OBD systems are typical of electronic engine control systems with self-diagnostic capabilities. The carmakers provide detailed system test procedures that are based on the principles of:

- Area testing to identify a faulty circuit, subsystem, or general condition
- Pinpoint testing to isolate a faulty part

Reading the system self-diagnostic trouble codes or evaluating the driver's complaints is area testing. Pinpoint testing is following the detailed troubleshooting procedures. Electronic engine control service ends with a general performance test to ensure that problems are corrected and that the system is operating properly.

The principles of testing, adjusting, and repairing engine systems that are explained throughout this book apply to engines with electronic controls. The basic methods for fuel, ignition, and emission control service will guide you through maintenance jobs on late-model vehicles. Some features of engine electronic systems, however, have brought new requirements to the service profession. The carmakers' procedures and the principles outlined in this chapter will guide you through these jobs successfully.

REVIEW QUESTIONS

1. Mechanic A says that you should do an overall system performance test as the first step of troubleshooting an electronic engine control problem. Mechanic B says that you should do an overall system performance test as the final step of electronic engine control service to check the repair. Who is right?

a. A only
b. B only
c. both A and B
d. neither A nor B

2. Checking the trouble codes registered by an engine computer is equivalent to a basic:

a. timing test
b. pinpoint test
c. area test
d. emission test

3. A "lean exhaust" trouble code may indicate any of the following EXCEPT:

a. an engine misfire
b. an EGR problem
c. excessive vapor canister purging
d. intake air (vacuum) leaks

4. Mechanic A says that a coolant-temperature trouble code can indicate either a high-temperature or a low-temperature fault. Mechanic B says that a coolant-temperature trouble code means a bad coolant-temperature sensor. Who is right?

a. A only
b. B only
c. both A and B
d. neither A nor B

5. Mechanic A says that an engine computer will record both open-loop faults and closed-loop faults. Mechanic B says that you may have to fix an open-loop fault before identifying a closed-loop fault. Who is right?

a. A only
b. B only
c. both A and B
d. neither A nor B

6. An engine control system fault that keeps an engine from running is:

a. a soft failure
b. an intermittent failure
c. a hard failure
d. a RAM failure

7. To record and store an intermittent trouble code, an engine computer must have:

a. PROM memory
b. RAM memory
c. hard memory
d. long-term memory

8. Mechanic A says that trouble code 12 for a GM CCC or C-4 system indicates no distributor pulses or tach signal. Mechanic B says that GM's trouble code 12 indicates that the system is in the self-test mode. Who is right?

a. A only
b. B only
c. both A and B
d. neither A nor B

9. A 50-series trouble code for a GM CCC or C-4 system indicates:

a. completion of the self-test sequence
b. electronic spark timing problems
c. a computer problem
d. an engine temperature problem

10. The *least preferred* way to clear the computer memory for a GM CCC system is to:

a. disconnect the computer memory connector near the battery + terminal
b. disconnect the battery ground cable
c. remove the control system fuse from the fuse block
d. disconnect a particular connector at the computer

11. All of the following conditions must exist before you can use a GM CCC trouble-code diagnostic chart EXCEPT:

a. all vacuum hoses must be connected correctly
b. all electrical connectors must be connected correctly
c. computer long-term memory must be cleared of residual trouble codes
d. the fault must exist at the time of testing

12. When you ground the dwell test lead during a GM CCC system performance test, the carburetor:

a. goes fully rich
b. goes fully lean
c. operates in closed loop in response to EGO sensor signals
d. disregards the idle speed control motor

13. A dwellmeter is used during GM CCC testing to measure:

a. mixture control solenoid frequency
b. computer reference voltage pulse width
c. mixture control solenoid duty cycle
d. EGO sensor duty cycle

14. You can read trouble codes for a Ford EEC IV system through:

a. the CHECK ENGINE lamp on the instrument panel
b. a digital display module on the instrument panel
c. a self-test connector in the engine compartment
d. any of the above

15. Mechanic A says that you can test an EGO sensor with an ohmmeter. Mechanic B says that you can test EGO sensor current draw with an inductive ammeter. Who is right?

a. A only
b. B only
c. both A and B
d. neither A nor B

16. The only adjustable sensor on most electronic engine control systems is the:

a. throttle position sensor
b. EGO sensor
c. coolant temperature sensor
d. manifold absolute pressure (MAP) sensor

17. A "rich exhaust" trouble code may indicate any of the following problems EXCEPT:

a. intake air (vacuum) leaks
b. a dirty intake air filter
c. a stuck choke
d. an overloaded PCV system

18. Mechanic A says that you can do a quick test of exhaust gas oxygen (EGO) sensor operation by grounding the sensor output lead and noting a change in engine speed. Mechanic B says that you can test EGO sensor voltage with a high-impedance, digital voltmeter. Who is right?

a. A only
b. B only
c. both A and B
d. neither A nor B

19. Mechanic A says that leaded gasoline used in an engine with an EGO sensor will leave lead deposits on the sensor and cause a false rich-mixture signal. Mechanic B says that silicone lubricant sprays can be used on EGO sensor threads to ease installation. Who is right?

a. A only
b. B only
c. both A and B
d. neither A nor B

20. Engine control systems may display trouble codes by:

a. flashing a CHECK ENGINE lamp on the instrument panel
b. sending voltage pulses to an analog voltmeter
c. displaying a digital readout on an instrument panel console
d. any of the above

21. A trouble code that appears and then disappears at different times and under different conditions is:

a. a PROM failure
b. a soft failure
c. an intermittent failure
d. a RAM failure

22. Any of the following can cause a GM CHECK ENGINE lamp to flash erratically or display unlisted codes EXCEPT:

a. computer supply voltage lower than 9 volts
b. an open diode in the air conditioning compressor clutch circuit
c. radiofrequency interference from the ignition system
d. routing the CCC system wiring harness too close to other engine wiring

23. Mechanic A says that a system performance test for the GM CCC system verifies the operation of system sensors. Mechanic B says that this test checks fuel control at the feedback carburetor. Who is right?

a. A only
b. B only
c. both A and B
d. neither A nor B

24. A 30° dwell reading for a GM mixture control solenoid equals a duty cycle of:

a. 25%
b. 50%
c. 75%
d. 100%

25. Code 11 for Ford's EEC IV system is similar to GM's CCC system code:

a. 12
b. 42
c. 14
d. 51

26. Mechanic A says that you can clean an EGO sensor with carburetor choke-cleaning spray and low-pressure compressed air. Mechanic B says that you should install an EGO sensor with antiseize compound on the threads. Who is right?

a. A only
b. B only
c. both A and B
d. neither A nor B

27. Mechanic A says that a computer PROM must be installed in one specified direction only. Mechanic B says that a computer PROM should be removed and installed with the ignition switch on. Who is right?

a. A only
b. B only
c. both A and B
d. neither A nor B

28. Mechanic A says that you should not crank the engine on a converter-equipped car for more than 15 seconds before starting it. Mechanic B says that after killing a cylinder during a power balance test, you should run the engine at fast idle for 30 seconds to clear the converter. Who is right?

a. A only
b. B only
c. both A and B
d. neither A nor B

29. A potentiometer sensor almost always has a connector with:

a. 1 wire
b. 2 wires
c. 3 wires
d. 4 wires

30. To check a TBI solenoid injector, you can use:

a. a stethoscope
b. a timing light
c. an ohmmeter
d. all of the above

31. An oscilloscope is connected to an injector solenoid. The scope pattern shows a steady, unbroken trace with the engine cranking. This indicates:

a. the solenoid wiring harness has high resistance
b. the solenoid has an open circuit
c. supply voltage is too high
d. the solenoid has low resistance

32. Mechanic A says that Chrysler fault codes can be retrieved either with the DRB I or the power loss lamp. Mechanic B says that the Chrysler Driveability Test Booklet is not necessary if you have a DRB II. Who is right?

a. A only
b. B only
c. both A and B
d. neither A nor B

33. Mechanic A says that a GM CALPAK is a separate PROM used with GM TBI fuel injection systems that provides back-up logic to control fuel delivery in case the ECM is not functioning properly. Mechanic B says that a GM MEMCAL combines the PROM, CALPAK, and electronic spark control (ESC) module logic all in a single plug-in unit. Who is correct?

a. A only
b. B only
c. both A and B
d. neither A nor B

34. Mechanic A says that the Tech 1 will run all GM CCC diagnostic systems from under the hood. Mechanic B says that Chrysler's DRB II tester will put the diagnostic system into four different test modes. Who is right?

a. A only
b. B only
c. both A and B
d. neither A nor B

APPENDIX METRICS*

*Reprinted through the courtesy of Ford Motor Company

INTRODUCTION

The specifications for threaded fasteners define mechanical properties such as tensile strength, yield strength, proof load, and hardness. For vehicle safety, replacement fasteners must have the same specifications as original-equipment fasteners.

Most customary inch-size and metric fasteners have markings that indicate the strength of the fastener. The illustrations on the following pages describe and explain these markings.

The following pages also contain tables of English and metric equivalencies and conversion factors that are useful in the automotive service profession.

NOMENCLATURE FOR BOLTS

(ENGLISH) INCH SYSTEM
Bolt, 1/2-13×1

METRIC SYSTEM
Bolt M12-1.75×25

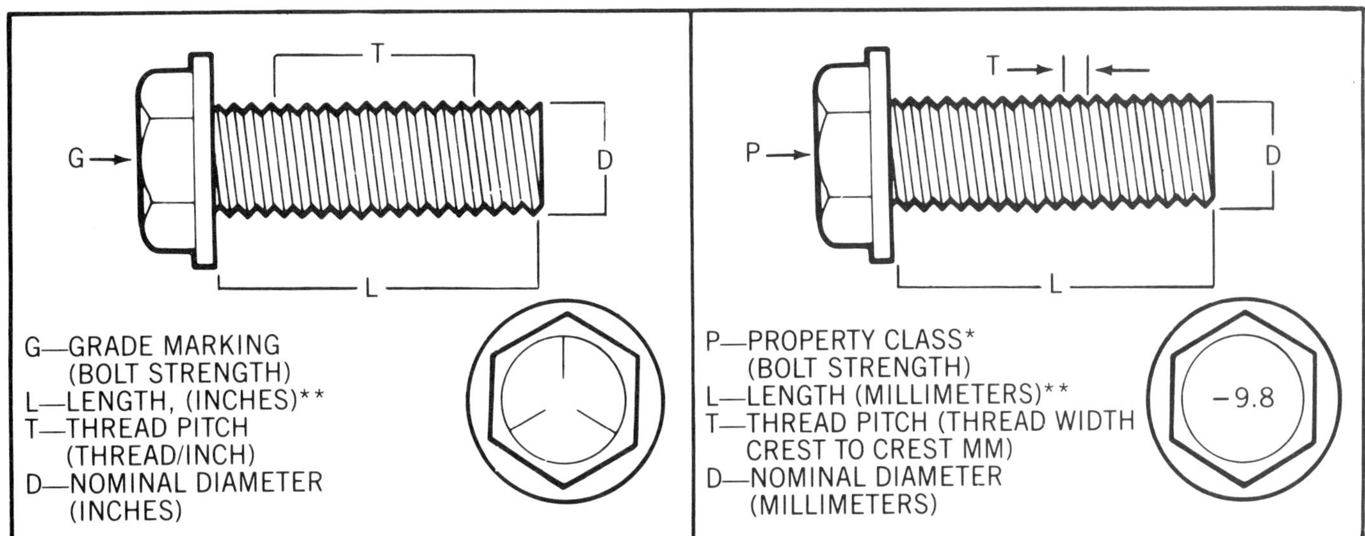

*THE PROPERTY CLASS IS AN ARABIC NUMERAL DISTINGUISHABLE FROM THE SLASH SAE ENGLISH GRADE SYSTEM.
**THE LENGTH OF ALL BOLTS IS MEASURED FROM THE UNDERSIDE OF THE HEAD TO THE END.

BOLT STRENGTH IDENTIFICATION

(ENGLISH) INCH SYSTEM

English (Inch) bolts—Identification marks correspond to bolt strength—increasing number of slashes represent increasing strength.

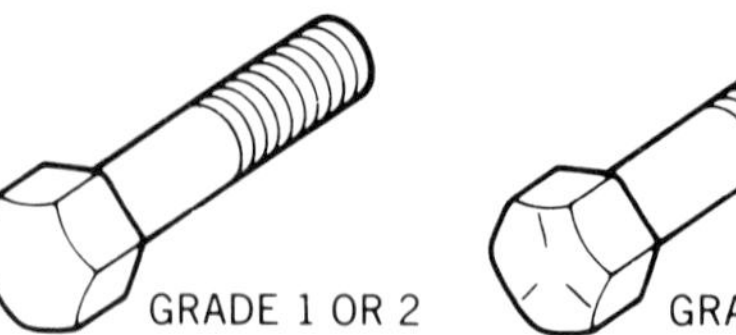

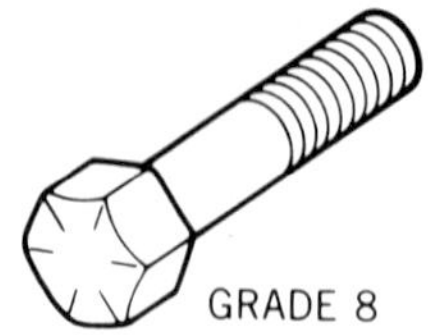

METRIC SYSTEM

Metric bolts—Identification class numbers correspond to bolt strength—increasing numbers represent increasing strength. Common metric fastener bolt strength properties are 9.8 and 10.9 with the class identification embossed on the bolt head.

HEX NUT STRENGTH IDENTIFICATION

(ENGLISH) INCH SYSTEM

GRADE	HEX NUT GRADE 5	HEX NUT GRADE 8
IDENTIFICATION	3 DOTS	6 DOTS

INCREASING DOTS REPRESENT INCREASING STRENGTH.

METRIC SYSTEM

CLASS	HEX NUT PROPERTY CLASS 9	HEX NUT PROPERTY CLASS 10
IDENTIFICATION	9 ARABIC 9	10 ARABIC 10

MAY ALSO HAVE BLUE FINISH OR PAINT DAUB ON HEX FLAT. INCREASING NUMBERS REPRESENT INCREASING STRENGTH.

OTHER TYPES OF PARTS

Metric identification schemes vary by type of part, most often a variation of that used of bolts and nuts. Note that many types of English and metric fasteners carry no special identification if they are otherwise unique.

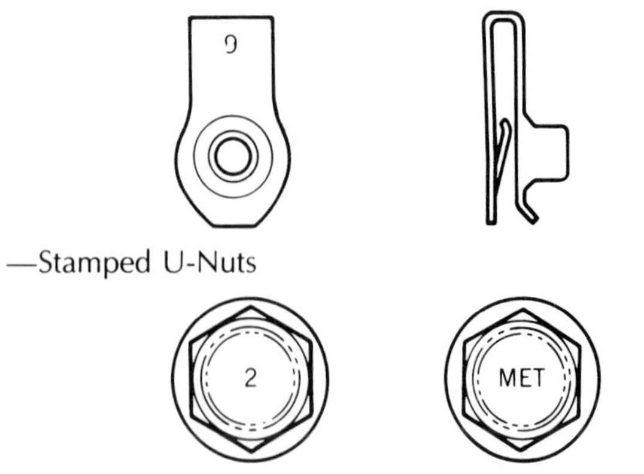

—Stamped U-Nuts

—Tapping, thread forming and certain other case hardened screws

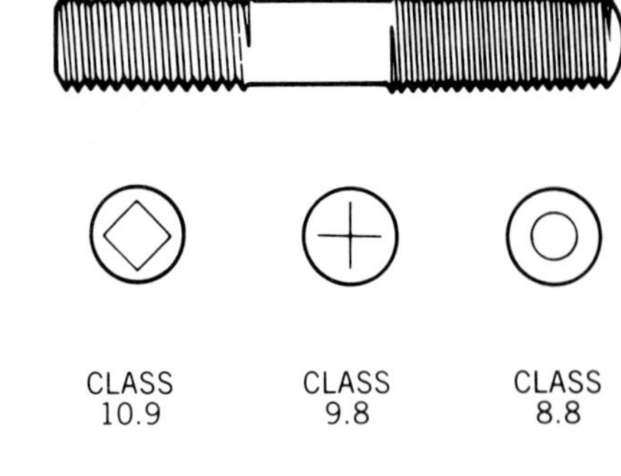

—Studs, Large studs may carry the property class number. Smaller studs use a geometric code on the end.

ENGLISH METRIC CONVERSION

DESCRIPTION	MULTIPLY	BY	FOR METRIC EQUIVALENT
ACCELERATION	FOOT/SEC^2	0.304 8	METER SEC^2 (M/S^2)
	INCH/SEC^2	0.025 4	METER SEC^2
TORQUE	INCH-POUND	0.112 98	NEWTON-METERS (N·M)
	FOOT-POUND	1.355 8	NEWTON-METERS
POWER	HORSEPOWER	0.746	KILOWATTS (KW)
PRESSURE OR STRESS	INCHES OF WATER	0.2488	KILOPASCALS (KPA)
	POUNDS/SQ.IN. (PSI)	6.895	KILOPASCALS (KPA)
ENERGY OR WORK	BTU	1.055.	JOULES (J)
	FOOT-POUND	1.355 8	JOULES (J)
	KILOWATT-HOUR	3 600 000. OR 3.6 × 10^6	JOULES (J = ONE W'S)
LIGHT	FOOT CANDLE	10.76	LUMENS/METER^2 (LM/M^2)
FUEL PERFORMANCE	MILES/GAL	0.425 1	KILOMETERS/LITER (KM/L)
	GAL/MILE	2.352 7	LITERS/KILOMETER (L/KM)
VELOCITY	MILES/HOUR	1.609 3	KILOMETERS/HR. (KM/H)
LENGTH	INCH	25.4	MILLIMETERS (MM)
	FOOT	0.304 8	METERS (M)
	YARD	0.914 4	METER (M)
	MILE	1.609	KILOMETERS (KM)
AREA	INCH^2	645.2	MILLIMETERS^2 (MM^2)
		6.45	CENTIMETERS (CM^2)
	FOOT^2	0.092 9	METERS^2 (M^2)
	YARD^2	0.836 1	METERS^2
VOLUME	INCH^3	16 387.	MM^3
	INCH^3	16.387	CM^3
	QUART	0.016 4	LITERS (1)
	QUART	0.946 4	LITERS
	GALLON	3.785 4	LITERS
	YARD^3	0.764 6	METERS^3 (M^3)
MASS	POUND	0.453 6	KILOGRAMS (KG)
	TON	907.18	KILOGRAMS (KG)
	TON	0.90718	TONNE
FORCE	KILOGRAM	9.807	NEWTONS (N)
	OUNCE	0.278 0	NEWTONS
	POUND	4.448	NEWTONS
TEMPERATURE	DEGREE FAHRENHEIT	0.556 (°F −32)	DEGREE CELSIUS (°C)

DECIMAL AND METRIC EQUIVALENTS

FRACTIONS	DECIMAL INCH	METRIC MM
1/64	.015625	.397
1/32	.03125	.794
3/64	.046875	1.191
1/16	.0625	1.588
5/64	.078125	1.984
3/32	.09375	2.381
7/64	.109375	2.778
1/8	.125	3.175
9/64	.140625	3.572
5/32	.15625	3.969
11/64	.171875	4.366
3/16	.1875	4.763
13/64	.203125	5.159
7/32	.21875	5.556
15/64	.234375	5.953
1/4	.250	6.35
17/64	.265625	6.747
9/32	.28125	7.144
19/64	.296875	7.54
5/16	.3125	7.938
21/64	.328125	8.334
11/32	.34375	8.731
23/64	.359375	9.128
3/8	.375	9.525
25/64	.390625	9.922
13/32	.40625	10.319
27/64	.421875	10.716
7/16	.4375	11.113
29/64	.453125	11.509
15/32	.46875	11.906
31/64	.484375	12.303
1/2	.500	12.7

FRACTIONS	DECIMAL INCH	METRIC MM
33/64	.515625	13.097
17/32	.53125	13.494
35/64	.546875	13.891
9/16	.5625	14.288
37/64	.578125	14.684
19/32	.59375	15.081
39/64	.609375	15.478
5/8	.625	15.875
41/64	.640625	16.272
21/32	.65625	16.669
43/64	.671875	17.066
11/16	.6875	17.463
45/64	.703125	17.859
23/32	.71875	18.256
47/64	.734375	18.653
3/4	.750	19.05
49/64	.765625	19.447
25/32	.78125	19.844
51/64	.796875	20.241
13/16	.8125	20.638
53/64	.828125	21.034
27/32	.84375	21.431
55/64	.859375	21.828
7/8	.875	22.225
57/64	.890625	22.622
29/32	.90625	23.019
59/64	.921875	23.416
15/16	.9375	23.813
61/64	.953125	24.209
31/32	.96875	24.606
63/64	.984375	25.003
1	1.00	25.4

TORQUE CONVERSION

NEWTON METERS (N·M)	FOOT-POUNDS
1	0.7376
2	1.5
3	2.2
4	3.0
5	3.7
6	4.4
7	5.2
8	5.9
9	6.6
10	7.4
15	11.1
20	14.8
25	18.4
30	22.1
35	25.8
40	29.5
50	36.9
60	44.3
70	51.6
80	59.0
90	66.4
100	73.8
110	81.1
120	88.5
130	95.9
140	103.3
150	110.6
160	118.0
170	125.4
180	132.8
190	140.1
200	147.5
225	166.0
250	184.4

FOOT-POUNDS	NEWTON METERS (N·M)
1	1.356
2	2.7
3	4.0
4	5.4
5	6.8
6	8.1
7	9.5
8	10.8
9	12.2
10	13.6
15	20.3
20	27.1
25	33.9
30	40.7
35	47.5
40	54.2
45	61.0
50	67.8
55	74.6
60	81.4
65	88.1
70	94.9
75	101.7
80	108.5
90	122.0
100	135.6
110	149.1
120	162.7
130	176.3
140	189.8
150	203.4
160	216.9
170	230.5
180	244.0

GLOSSARY

absolute zero. The temperature at which molecular motion stops and all substances possess minimal electron energy. Equal to −273° C or −459.67° F.

absorption muffler. A kind of muffler that reduces noise by changing sound energy to friction energy through sound-absorbing material.

ac ripple. The small waves at the top of a rectified dc voltage oscilloscope pattern. They are the peaks of the ac voltage sine waves. Any dc voltage from a rectified ac source always has a slight ac ripple.

actuator. Any device that receives an output signal from a computer and does something in response to the signal.

adaptive learning strategy. A strategy that modifies the original computer program to compensate for changes in system parameters caused by various factors such as engine wear, driving habits, poor fuel quality, or component malfunctioning

adsorption. The collection of gas, vapor, or dissolved solid on the *surface* of a solid material.

aeration. The process of exposing liquid to air circulation and trapping air bubbles in the liquid. Mixing air or another gas with liquid.

aerobic. Curing in the presence of oxygen.

after top dead center (atdc). The position of a piston after it has passed top dead center of a stroke, measured in degrees of crankshaft rotation.

air bleed. A small opening that allows air to enter (bleed into) a fuel or vacuum passage in a carburetor.

alternating current (ac). Current flow that continually changes direction between two points through a conductor in a given amount of time.

ambient air temperature. Temperature of the air surrounding an area or device.

American Wire Gauge (AWG). A system for specifying wire size (conductor cross section area) by a series of gauge numbers.

ampere (A). The unit for measuring current flow. One ampere equals a current flow of 6.28×10^{18} electrons per second.

ampere-hour rating. A battery current-capacity rating that indicates the amperes of current a battery can supply steadily for 20 hours without voltage dropping below 1.75 volts per cell. It also is called the "20-hour discharge rating."

anaerobic. Curing in the absence of oxygen.

analog. A signal that varies proportionally with the information that it measures. In a computer, an analog signal is voltage that fluctuates over a range from high to low.

analogy. A likeness. A similarity in some respects between two things that are otherwise dissimilar.

aneroid. An evacuated metal bellows containing a vacuum that changes in length as a response to changes in atmospheric pressure.

anode. The positively charged electrode in a voltage cell. The positive-polarity terminal of a component.

antioxidant. A kind of oil additive that reduces oil oxidation and the formation of varnish and carbon deposits.

API service classifications. The system for classifying motor oil resistance to engine wear, oxidation, corrosion, and dirt and carbon formation.

area test. A general test made to a complete vehicle system, or systems, to determine overall system operation.

armature. The movable part in a relay or the rotating part of a generator or motor consisting of a conductor wound around a laminated iron core.

asynchronous. Data transmission that incorporates a constant tone between sending and receiving. A start pulse precedes the transmission of new data, and a stop pulse tells the computer that the transmission has been completed.

atmospheric pressure. Pressure on the surface of the earth created by the weight of air in the atmosphere. Atmospheric pressure changes with altitude and temperature. At sea level it is 14.7 psi at 68° F (760 mm Hg at 20° C).

atom. The smallest part of a chemical element that has all the properties and characteristics of the element.

atomize. The action of breaking a liquid into a fine mist or small droplets.

autoignition. The spontaneous ignition of a fuel due to high temperature. The autoignition temperature of a fuel is the temperature at which the fuel will ignite by itself.

available voltage. The maximum high-voltage capability of an ignition coil. Also called open-circuit voltage, peak voltage, or no-load voltage.

axis. The rotation centerline of a wheel, a gear, or a shaft.

backpressure. Pressure that slows the flow of exhaust gas from an engine. Backpressure is caused by restrictions in the exhaust system. Also, any similar pressure caused by restrictions that slow gas or liquid flow in a system.

balance tube. The vent that opens the carburetor fuel bowl to atmospheric pressure at the airhorn and thus provides balanced pressure for fuel metering.

base. The center layer of a bipolar transistor, made of doped material opposite from the collector and the emitter.

base unit. One of the seven basic units that are the foundation of SI metric measurements.

basic (initial) timing. The ignition timing setting of an engine at idle speed.

battery cell. An assembly of a positive plate group, a negative plate group, separators, and connectors in a case with electrolyte. A complete chemically and electrically active unit of a battery.

baud rate. Computer data transmission speed in digital bits per second or characters per second.

BCI group number. A battery size identification number that indicates the length, width, height, terminal design and location, and other physical features of a battery.

bearing web (saddle). The area of an engine crankcase that holds a main bearing for the crankshaft.

before top dead center (btdc). The position of a piston before it reaches top dead center of a stroke, measured in degrees of crankshaft rotation.

bias voltage. Voltage applied across a semiconductor PN junction, such as the PN junction of a diode.

bimetal. Two strips of metal, each of which expands and contracts at a different rate when exposed to changes in temperature.

bimetal element. A strip made of different metals that expand at different rates when heated. Commonly used in circuit breakers and temperature switches to open the contacts when high current flows through the circuit.

bimetal temperature sensor. A temperature sensor (thermostat) made by joining two metals with different heat expansion rates. As the metals heat, they expand at different rates and bend to open a valve or move clutches, air cleaner vacuum valves, and choke coils.

binary system. The mathematical system that uses only the digits 0 and 1 to present information. The mathematical system used by digital computers.

bipolar. The general name for NPN and PNP transistors because current flows through semiconductor materials of both polarities. Holes and electrons are both used as current carriers.

bit. A binary digit (0 or 1). Bit combinations are used to represent letters and numbers in digital computers. Eight bits equal one byte.

block learn. The factor that represents long-term operation and long-term correction of fuel metering on a General Motors fuel injection system.

blowby. Combustion gases that leak past piston rings into the crankcase. Blowby can contain unburned fuel, water vapor, acids, and exhaust.

blowoff valve. A large pressure relief valve on the outlet side of a turbocharger compressor. It opens to relieve pressure when boost reaches a desired maximum pressure.

bonding straps. Additional ground cables installed between the engine and chassis to provide a low-resistance ground path for various electrical circuits.

boost pressure. The amount of air pressure increase above atmospheric pressure provided by a supercharger or turbocharger.

bore. The diameter of a cylinder. Also the machining process of enlarging or finishing a hole.

boss. A machined circular hole in a part, into which another part is installed.

bottom dead center (bdc). The exact bottom of a piston stroke.

bound electrons. Five or more tightly held electrons in the valence shell of an atom.

bourdon tube. A small coiled tube attached to a gear mechanism that turns a gauge needle. When pressure is applied to the tube, it flexes and moves the needle.

brake horsepower (bhp). The horsepower developed at an engine's crankshaft.

brake mean effective pressure. Mean effective pressure measured as force at the engine crankshaft.

breakdown voltage. The prescribed, or designed, voltage at which a zener diode allows reverse current flow.

British thermal unit (Btu). A standard unit for measuring heat energy. One Btu is the amount of heat needed to raise the temperature of one pound of water one degree Fahrenheit at atmospheric pressure.

burn time. The time in milliseconds during which the fuel ignites and burns to develop cylinder pressure.

bus. A common conductor, or transmission path, shared by several circuits, or components, in a computer.

bus bar. A solid metal strap or bar in a fuse block that acts as a common conductor for several fuses.

bypass oil filter. An oil filter that receives only part of the oil supply on a bypass line from the main oil gallery.

bypass or module timing. Ignition timing provided by the module of a direct ignition system during cranking until the synchronization signal is received and ignition coil setup is achieved.

byte. A unit of computer data consisting of eight digital bits.

cam. An irregularly shaped wheel, or other object, on a rotating shaft or other movable part; used to produce reciprocating motion in a contacting part. In an engine valve train, a cam moves a valve lifter to open a valve.

camshaft. A rotating shaft or other movable part with irregularly shaped protrusions or lobes used to produce reciprocating motion in a contacting part. In an engine valve train, camshaft motion moves a valve lifter to open a valve.

capacitance. The ability of two conducting surfaces to accumulate a voltage charge and store that voltage when separated by an insulator.

capacitive-discharge (CD) ignition. An ignition system that stores its primary energy in a capacitor and then discharges the capacitor to create an expanding magnetic field in the coil.

case ground. A ground connection directly from the case of a part to the vehicle engine or chassis.

catalytic cracking. The process of cracking hydrocarbon molecules by using a chemical catalyst.

catalytic reaction. A chemical reaction caused by a material (a catalyst) that is not, itself, changed by the reaction.

cathode. The negatively charged electrode in a voltage cell.

cathode ray tube (CRT). An electron-beam tube in which the beam is focused on a small area of a luminescent screen and varied in position and intensity to produce a visible pattern. Used as the display screen in a TV set, an oscilloscope, or a computer monitor.

central processing unit (CPU). The calculating part of a computer that makes calculations and logical decisions by comparing input data with data in memory.

centrifugal (mechanical) advance. A distributor mechanism that uses centrifugal weights to advance timing in relation to engine speed.

centrifugal force. A force that tries to keep moving objects traveling in a straight line. When combined with rotating force, centrifugal force causes an object to move away from the center of rotation and circle the center.

centripetal force. Force acting on a rotating body that is applied at the outer edge and directed inward. The opposite of centrifugal force, which moves outward.

cetane number. A measurement of diesel fuel's ability to ignite and burn quickly.

check valve. A pressure-regulating valve

that allows gas or fluid flow in one direction but prevents flow in the opposite direction.

choke coil. A coil wound with fine wire, used to absorb oscillations that occur in a circuit when the circuit is opened or closed.

circuit. An unbroken circular path through which electric current can flow. Any path through which force can be applied or matter can move. A hydraulic system also is a circuit.

claw pole. The finger, or claw, of an alternator rotor that concentrates the magnetic flux and forms the magnetic field. A claw-pole rotor has interlaced sets of claw poles to form alternating north-south poles.

clearance volume. The volume of the combustion chamber and area around the top of the piston with the piston at top dead center.

closed loop. The system operating condition when an output error signal is fed back to the system controller, which then readjusts the system to eliminate the error.

closed-loop dwell control. A feature of a distributorless ignition at which the ignition control module adjusts dwell time to provide full coil saturation based on the previous coil buildup.

cloud point. The low temperature at which wax crystals form in diesel fuel or motor oil.

code. A numerical code, generated by a vehicle control system to indicate that a fault has occurred in a particular subsystem, circuit, or part. Carmakers refer to codes as follows: GM, *trouble* codes; Ford, *service* codes; Chrysler, *fault* codes.

cold-cranking rating. A battery current-capacity rating that indicates the amperes of current a battery can supply for 30 seconds at 0° F (−18° C) with no cell voltage dropping below 1.2 volts.

collector. The outer layer of semiconductor material that receives the majority current carriers in a bipolar transistor.

commutator. A segmented ring attached to one end of an armature shaft in a motor (or a dc generator). A commutator acts as a rotary switch to provide a continuous current path from an external circuit through carbon brushes and through the armature windings.

compound. A substance, or form of matter, formed by molecules of two or more different elements. Molecules of compounds are chemically bound together by electron sharing.

compression pressure. Pressure created by the piston as it moves from bdc to tdc on the compression stroke without having the air-fuel mixture ignited.

compression ratio. The ratio of the total volume of a cylinder and combustion chamber to the clearance volume of the chamber alone.

computer hardware. The mechanical, magnetic, electrical, and electronic devices that make up the physical structure of a computer.

condenser. A capacitor that usually is made of two sheets of metal foil that are separated by an insulator and housed in a small metal can.

conductor. A material that has many free electrons and provides easy current flow. Metals are good electrical conductors.

constant depression. The constant pressure drop, or vacuum, developed in a variable-venturi carburetor.

continuity. Unbroken or continuous. A circuit or electrical device has continuity if it has an unbroken path for current flow.

conventional theory. The theory that electric current flows from positive to negative. Also called the positive current flow theory.

conversion factor. Any number used to multiply or divide a measurement in one set of units to change it to an equal measurement in another set of units.

counterelectromotive force (CEMF). An induced voltage that opposes the source voltage and any increase or decrease in source current flow.

cracking. The oil refining process of breaking large hydrocarbon molecules into smaller ones.

crankshaft throw. The offset portion of a crankshaft to which a connecting rod and rod bearing are attached. Each throw creates a lever with the crankshaft centerline as the pivot point (fulcrum).

crossflow head. A cylinder head with intake ports on one side and exhaust ports on the other side.

curb idle. The normal slow-idle speed of a fully warmed up engine.

cycling. The action of changing from one condition to an opposite condition and back to the original condition. Battery cycling is the electrochemical action from fully charged to discharged and back to fully charged.

darlington pair (circuit). A current-amplifying arrangement whereby one transistor acts as a preamplifier and creates a larger base current for a second transistor. Often described as a "piggyback" power transistor, which requires only three external circuit connections.

d'Arsonval movement. A small current-carrying coil in the field of a permanent magnet. Used as the measuring movement in electrical meters.

data packet. One complete package, or serial data transmission cycle, from a vehicle that provides control system operating parameters. Also called a data *frame*.

datum line. The primary reference dimension for an object, from which all other dimensions are determined.

delta stator. A 3-winding alternator stator with the ends of each winding connected to each other; there is no neutral junction. Schematically, the arrangement looks like the Greek letter delta, or a triangle. A delta stator produces high current at low speed.

demultiplexer (DEMUX). The solid-state switching device that handles the output data from a body computer module (BCM) to an actuator when the BCM is engaged in the process of sequential sampling.

derived unit. Any compound metric unit formed by multiplying two or more base, supplementary, or other derived units.

detent. A position of a multiple-pole switch that can be felt or in which the switch will stay until moved to another position.

detented. A switch with multiple detent positions.

detergent-dispersant. A group of oil and fuel additives that reduce sludge and dirt formation by suspending dirt particles in the oil or fuel.

detonation. An unwanted, uncontrolled explosion of the air-gasoline mixture due to high temperature and pressure after spark ignition.

dielectric. An insulating material between the plates of a capacitor.

dieseling. The tendency of a gasoline engine to keep running after the ignition is off. Also called afterrun. Caused by high combustion chamber temperatures and a throttle that is slightly open.

digital. A signal that is either on or off and that is translated into the binary digits 0 and 1. In a computer, a digital signal is voltage that is either low or high, or current flow that is on or off.

digital microprocessor. A miniature digital computer on a single silicon chip integrated circuit (IC).

diode. An electronic device made by joining P-material to N-material at a junction. It allows current flow in one direction and blocks it in the other.

direct current. Current flow that is always in one direction through a conductor.

direct drive. A 1 : 1 gear ratio where both drive and driven gears turn at the same speed.

direct injection. A diesel fuel injection system that injects fuel directly into a combustion chamber in the cylinder. The chamber often is in the piston crown.

discrete device. An individual diode, transistor, or other electronic part that performs only one function and is connected to a circuit by wire leads.

displacement. The measurement of the volume of air moved by a piston as it travels from bottom to top of its stroke. Engine displacement is the displacement of one cylinder multiplied by the number of cylinders in the engine.

distillation. The process of using heat to vaporize different parts (fractions) of a liquid mixture at different temperatures, then the condensing and collecting of the distilled liquids.

diverter valve. An air injection antibackfire valve that diverts airflow to the atmosphere during deceleration.

dopant. The other elements added to pure silicon or germanium in the doping process to change the electrical characteristics of the semiconductor material.

doping. The addition of a second element to a pure semiconductor. The second element has either three or five valence electrons.

double-overhead-cam (dohc) engine. An overhead-cam engine with separate camshafts for intake and exhaust valves.

drain. The part of a field-effect transistor that receives the current-carrying holes or electrons. Similar to the collector in a bipolar transistor.

driver transistor. The smaller transistor in an ignition control module that receives a timing signal from the distributor and switches the power transistor off.

dry-sump oil system. An engine lubrication system in which the oil supply is stored in a tank separate from the engine.

duplex serial data line. A data link between two computers that allows each computer to send and receive data from the other. An external clock line between the two computers controls which one sends and which one receives.

duration spring. The spring that operates the carburetor accelerator pump plunger or diaphragm.

duty cycle. The percentage of time that a solenoid is energized during one complete on-off cycle. Duty cycle is the ratio of on-time to total cycle time.

dwell angle. The number of degrees of distributor cam rotation during which the points and the primary circuit are closed and current flows through the coil primary winding. Also called cam angle or simply, dwell.

dynamic inertia. The tendency of an object in motion to keep moving.

dynamic range. The operating range of a sensor.

dynamometer. An instrument used to measure the torque and power output of an engine.

eccentric. Off center. A circular shaft lobe that has a center different from the centerline of the shaft.

eddy current. Small induced current in the armature of a motor (or a dc generator), created by CEMF. Eddy currents oppose circuit current flow.

electric current. The controlled, directed movement of electrons from atom to atom in a conductor. It is caused by electromotive force (voltage).

electrical load. The working devices of an electrical circuit, such as lamps and motors, that change electrical energy to heat, light, or motion.

electrically erasable programmable read-only memory (EEPROM). Computer memory program circuits that can be erased electrically with a voltage pulse and then reprogrammed.

electrochemical. A reaction or condition in which chemical energy is changed to electrical energy, or vice versa. Electrochemical action in a battery produces voltage through the reaction of two dissimilar electrodes in a conductive electrolyte.

electrochemistry. Voltage developed by the chemical action of two dissimilar materials in a conductive chemical solution.

electrode. A conductor through which electric current enters or leaves a device. A positive or negative electrically charged terminal. A conductive, electrochemically active part of a battery.

electrolysis. Chemical decomposition of a material in the presence of a liquid (electrolyte) and electric current.

electrolyte. The chemical solution in a battery that provides a medium for plate materials to create voltage and conduct current. Automobile battery electrolyte is sulfuric acid (H_2SO_4) plus water (H_2O).

electromagnet. An iron bar or rod surrounded by a current-carrying coil. Current flow creates a magnetic field that concentrates in the iron and magnetizes it.

electromagnetic. The relationship of electrical energy and magnetic energy.

electromagnetic induction. The generation of voltage in a conductor by relative motion between the conductor and a magnetic field.

electromagnetic interference (EMI). Electrical impulses emitted as high-frequency transmission or radiation from electromagnetic devices. EMI can interfere with computer data transmission, as well as with radio and TV reception.

electromagnetism. Magnetism caused by electric current flow.

electomotive force (emf). The electrical force that causes current flow (voltage).

electron. The smallest individual part of an atom. Electrons have a negative (−) charge and orbit the nucleus of an atom.

electron theory. The theory that electric current flows from negative to positive.

electrostatic field. The electrically charged area around two points or surfaces that have a voltage difference between them.

element. A chemical element is a substance that contains all identical atoms. A fundamental unit of matter. There are 103 known elements, of which 92 exist in nature and 11 are manmade.

emitter. The outer layer of semiconductor material that supplies the majority current carriers in a bipolar transistor.

energy. The ability, or capacity, to do work. Energy is the work that a system or object *can* do by changing from a given state or position into another state or position.

engine mapping. The process of measuring the best combinations of timing, air-fuel ratio EGR, and other controllable variables for various speeds, loads, and temperatures.

equivalent resistance. Total series resistance of all branches of a parallel circuit.

erasable programmable read-only memory (EPROM). Computer memory program circuits that can be erased and reprogrammed. Erasure is done by exposing the IC chip to ultraviolet light.

ethylene glycol. The compound that resists boiling and freezing and that is mixed with water to form engine coolant.

excitation current. Field current of an alternator. It magnetically excites the field. During starting, it comes from the battery; when running, it comes from the alternator output.

excitation voltage. Battery or charging system voltage that delivers field current to the alternator.

exponent. Any number written to the right and above another number to indicate how many times the other number is multiplied or divided by itself. Also called a "power."

extended-core spark plug. A spark plug with the insulator and both electrodes made to extend farther into the combustion chamber than they do on a standard spark plug.

farad (F). The unit for measuring capacitance.

feedback. The process of sending an output error signal back to the system controller, or computer, so that the computer can readjust system output control.

field-effect transistor (FET). A transistor through which current flow is controlled by voltage in a capacitive field. A unipolar transistor.

firing order. The sequence in which ignition occurs in cylinders of an engine.

firing (ionization) voltage. The high voltage needed to ionize the air and fuel molecules at the spark plug gap and create an ignition spark across the gap.

fixed-dwell electronic ignition. An electronic ignition with a ballast resistor to control primary voltage and current to the coil. The dwell angle in degrees is fixed and does not vary with engine speed.

flame front propagation. The controlled burning process, or spread of the combustion flame front, from the spark plug throughout the combustion chamber.

flux density. The number of magnetic flux lines that pass through one square centimeter of any area.

flux lines. Magnetic lines of force.

forward bias. Voltage applied across a diode that causes current to flow across the junction.

four-stroke cycle. The engine operating cycle in which a piston makes four full strokes (intake, compression, power, and exhaust) and 720 degrees of crankshaft revolution to develop power from the combustion process. Also called the Otto cycle.

frame. One complete serial data transmission cycle, or package from a vehicle that provides control system operating parameters. Also called a *data packet*.

free electrons. Three or fewer loosely held electrons in the valence shell of an atom.

friction. The force that resists motion between the surfaces of two objects, or forms of matter.

friction horsepower. Engine power lost to internal friction of the engine. Power consumed to overcome internal engine friction.

full-flow oil filter. An oil filter through which the entire oil supply passes from the pump before entering the engine oil galleries and lines.

gain. The ratio of amplification in an electronic device.

galvanic battery. A source of dc voltage. In an exhaust gas oxygen sensor, voltage is generated by a difference in oxygen content near two electrodes.

gasket creep. The compression of a gasket that occurs after bolts are tightened.

gate. The part of a field-effect transistor that controls the capacitive field and current flow. Similar to the base in a bipolar transistor.

gear ratio. The number of revolutions made by a drive gear compared to the number of revolutions made by a driven gear. It is usually expressed as a ratio of drive revolutions to one driven revolution, such as 3 : 1 or 0.75 : 1.

gear reduction. A gear ratio where the drive gear rotates faster than the driven gear. The driven gear turns at reduced speed with increased torque. A ratio of 2 : 1 is a gear reduction ratio.

gross brake horsepower. Brake horsepower of an engine with only the fuel, oil, and water pumps and built-in emission controls as accessory loads.

grounded circuit (ground). An unwanted connection between part or all of a circuit that bypasses current directly to ground. A short circuit to ground.

gulp valve. An air injection antibackfire valve that redirects airflow to the intake manifold during deceleration.

half-wave rectification. The process of rectifying only one-half of an ac voltage (either positive or negative) and producing pulsating dc.

Hall-effect switch (sensor). A semiconductor device that produces a voltage pulse dependent on the presence of a magnetic field.

hard code. A trouble code from a vehicle computer system that indicates a problem that is permanently present at the time of testing. A hard code may or may not keep the system from operating.

hard conversion. Redesigning a part to convert its design dimensions from one measurement system to another.

header. A low-restriction exhaust manifold, usually made of welded steel tubing.

heat sink. Any object that absorbs and dissipates heat from another object.

heat range. The indication of a spark plug's ability to dissipate, or transfer, heat from its center electrode to the engine.

helical gear. An external or internal gear on which the teeth are at an angle to the gear axis, or centerline. All teeth are at the same angle.

helix. A spiral or three-dimensional curve. In a fuel injection pump, the spiral groove that opens the spill port and controls fuel metering.

hertz (Hz). A unit of frequency equal to 1 cycle per second.

hole. The place in a semiconductor valence shell from which an electron is missing. A positive charge carrier.

horsepower (hp). One horsepower equals 33,000 foot-pounds of work done in one minute. The amount of power to do an equivalent amount of work in an equivalent amount of time.

hybrid. Something of mixed composition. A catalytic converter composed of both oxidation and reduction catalysts.

hybrid circuit. An electronic circuit built from discrete devices and IC chips.

hydraulics. The science, or study, of using liquids to do work.

hydrogenation. The chemical process of adding hydrogen to a compound.

hydrometer. An instrument used to measure specific gravity of liquids, such as battery electrolyte or engine coolant.

idler. A gear, pulley, or sprocket that turns with drive and driven members but does not change the overall drive-to-driven ratio. An idler gear changes direction of rotation. Idler pulleys or sprockets are used as tensioners in belt and chain drives.

ignition control module. The electronic assembly of transistors and other solid-state devices that controls ignition primary current.

ignition interval. The degrees of crankshaft rotation between ignition in cylinders in firing order sequence.

ignition lag. In a diesel engine, the time from the injection of fuel until the fuel vaporizes and ignites; usually about 2 milliseconds.

impedance. The combined opposition to current flow created by resistance, capacitance, and inductance in a circuit. In dc electrical systems, impedance is largely resistance.

indicated horsepower. The theoretical maximum power produced inside an engine's cylinders, calculated by measuring combustion pressure.

indirect injection. A diesel fuel injection system that injects fuel into a small prechamber separate from the cylinder.

induced current. Current in a conductor generated by magnetic flux lines due to relative movement of either the magnetic field or the conductor.

induced voltage. Voltage in a conductor generated by magnetic flux lines due to relative movement of either the magnetic field or the conductor.

induction coil. Two conductors wound into coils, with one placed inside the other. Transformers and ignition coils are examples.

inductive-discharge ignition system. An ignition system based on the induction of high voltage in a coil.

inertia. The tendency of an object in motion to keep moving and the tendency of an object at rest to remain at rest.

inertia engagement. A general type of starter motor that uses the rotating inertia of the motor armature shaft to engage the drive pinion with the engine flywheel. A Bendix drive is the most common inertia-engagement starter drive.

insulator. A material that has many bound electrons and therefore opposes current flow. Many nonmetallic compounds such as glass, rubber, and plastics are good insulators.

integrated circuit (IC). A complete electronic circuit of many transistors and other devices, all formed on a single silicon chip.

integrator. The factor that represents short-term operation and short-term correction of fuel metering on a General Motors fuel-injection system.

intercooler. An air-to-air or air-to-water heat exchanger. In a turbocharger installation, an intercooler lowers the temperature of the intake air charge.

intermittent code. An electronic engine control system trouble code that indicates a problem that may or may not exist at the time of testing. An intermittent code occurs on and off, but will be recorded in long-term memory, if the computer has such memory.

internal combustion engine. An engine in which the air-fuel mixture is burned inside the engine power chamber.

internal ring gear. A gear wheel with an open center and teeth cut on its inner circumference. Also called an *annulus gear* or simply a ring gear.

ion. An atom that has become electrically unbalanced by losing or gaining electrons. An ion can be positive (fewer electrons than protons) or negative (more electrons than protons).

keep-alive memory (KAM). Random-access memory that is retained by keeping a voltage applied to the circuits when the engine is off.

kinetic energy. The specific energy of motion.

lambda. The ratio of one number to another. For exhaust gas measurement, lambda is the ratio of excess air to ideal air for complete combustion.

left-hand rule. The current flow and flux direction rule based on the electron theory of current flow. Grasping a conductor with the left-hand thumb pointing in the direction of current flow (toward +) causes the fingers to wrap around the conductor in the direction of the flux lines.

light-emitting diode (LED). A gallium-arsenide diode that releases energy as light when holes and electrons collide.

lightoff temperature. The temperature at which a catalytic converter becomes 50 percent effective, approximately 400° to 500° F (204° to 260° C).

linear. In a straight line.

linearity. The expression of sensor accuracy throughout its dynamic range.

liquid crystal display (LCD). A display that sandwiches electrodes and polarized fluid between two pieces of glass. The application of voltage to the electrodes rearranges the light slots in the fluid and causes light to pass through it.

logic gate. Logic circuits used by a computer to manipulate data bits and make variable decisions. The three basic gates are the NOT, AND, and OR gates.

magnetic field. The area around a magnet formed by magnetic lines of force. The area that is influenced by the magnet's energy.

magnetic flux. A magnetic field.

magnetic saturation. The full strength (maximum flux density) of a magnetic field with a steady amount of current.

magnetism. The form of energy caused by the alignment of atoms in some materials. The ability of a material to attract iron.

manifold absolute pressure (MAP). A combination of atmospheric (barometric) pressure and vacuum. MAP equals barometric pressure minus vacuum.

mass. The measure of the inertia of an object or form of matter or its resistance to acceleration.

mass airflow (MAF). The measurement of the molecular mass, or weight, of air entering an engine

matter. All things in the universe that occupy space, have mass and weight, and are subject to inertia and momentum. Matter may be in gas, liquid, or solid form.

mean effective pressure. The average cylinder pressure developed throughout one complete 4-stroke cycle.

menu. A list of vehicle tests or programs available from a scan tool or other computerized test equipment. A selection is made from the menu to perform a specific test or to read computer data.

menu driven. Referring to a computer program that provides the user with a "menu" or list of choices. Making a choice brings up another page of the menu and requires that another choice be made. In this way, the user tells the computer what he wishes to know and the program displays the desired information.

micron. A unit of length equal to one-millionth (10^{-6}) of a meter, or 0.00254 inch.

microprocessor. A miniature digital computer built as an integrated circuit on a single silicon chip. The microprocessor is the central processing unit (CPU) of an automotive computer.

minus rule. The rule that indicates a low-capacity ignition condenser. "Minus metal (pitting) on the minus (−) point indicates a minus (low) capacity condenser."

mixture. A substance, or form of matter, consisting of two or more elements or compounds that are not bound together chemically by shared electrons.

molecule. Two or more atoms of the same or different elements chemically joined by sharing valence electrons.

momentary contact. A kind of switch that operates only when held in position. When released, it returns to its normally open or normally closed position.

momentum. The force of continuing motion. The momentum of a moving object equals its mass times its speed.

monolithic converter. A catalytic converter that has the catalyst deposited on a single large honeycomb block (a monolith), through which the exhaust flows.

MOSFET. A metal oxide semiconductor, field-effect transistor. A type of integrated-circuit (IC) device used in a microprocessor

motor octane. Gasoline antiknock quality tested under conditions of heavy load, high speed, high temperature, and full throttle.

movie. A recording of data transmitted from a vehicle to a scan tool. Also called a *snapshot.*

multigrade (multiviscosity) oil. An oil whose viscosity is tested and rated at both low temperature and 210° F (−18° and 99° C).

multimeter. An electrical test instrument that contains two or more different meters or performs two or more test functions. Often a combined ammeter, voltmeter, and ohmmeter.

multiple. A measurement unit that is larger than the stem unit through multiplying by a power of 10.

multiplexer (MUX). A solid-state switching device that handles the input data from a given sensor for a body computer module (BCM) when it is engaged in the process of sequential sampling. Multiplexing is the process of transmitting several signals simultaneously over the same data line.

multiplex wiring system. An electrical circuit in which signals are transmitted simultaneously by a peripheral serial bus or over optical fiber cables. Several devices share signals on a common conductor.

mutual induction. The creation of voltage in one coil by the increase and decrease in the magnetic field around another coil.

N-material. Silicon or germanium doped with phosphorus or arsenic so that it has excess free electrons and a negative charge.

naturally (normally) aspirated. The natural process of breathing, or taking in air, from the force of atmospheric pressure. A naturally aspirated engine is one without a supercharger to increase intake air pressure.

negative-ground electrical system. A dc electrical system in which the battery negative terminal is connected to ground. Most modern automobiles have negative-ground systems.

negative temperature coefficient (NTC) resistor (thermistor). A thermistor whose resistance decreases as temperature increases—high resistance at low temperature and low resistance at high temperature.

net brake horsepower. Brake horsepower of a fully equipped engine with all accessory loads.

neutral junction. The center tap to which the common ends of Y-type stator windings are connected.

neutron. A neutral part of an atom with neither positive nor negative charge. Neutrons are located in the nucleus, along with protons.

noble metals. The group of metallic elements that includes platinum, paladium, and rhodium. These elements resist corrosion and do not combine easily in chemical reactions, but they serve as chemical catalysts.

normally closed (NC) switch. A switch whose contacts are closed until an outside force opens them to open the circuit.

normally deenergized. The condition of a solenoid or other electrical load that receives electrical power only after a switch is activated.

normally energized. The condition of a solenoid or other electrical load that receives power directly from an electrical system without having a switch activated.

normally open (NO) switch. A switch whose contacts are open until an outside force closes them to complete the circuit.

nucleus. The center of an atom, made up of protons and neutrons. The center, or core, of any body of matter.

octane rating. The measurement of a gasoline's antiknock qualities, compared to 100% isooctane.

ohm. The unit used to measure the amount of electrical resistance in a circuit. One ohm is the amount of resistance present when one volt forces one ampere of current through a circuit.

open circuit. An incomplete, or broken, circuit. An open circuit has infinite resistance and allows no current flow.

open-circuit (no-load) voltage. Applied voltage measured when a circuit is open and no current is flowing. Usually battery voltage in automobile electrical systems. Also the voltage discharge from a coil secondary winding when not connected to a complete circuit. The same as available voltage for given primary voltage and current.

open loop. The system operating condition when an output error signal is not fed back to the system controller.

orifice. A restriction in a line (gas or liquid) that regulates pressure by creating a pressure differential across the restriction.

oscillate. To move back and forth between two points in regular cycles.

oscilloscope. A test instrument that produces a trace of electron motion on a cathode ray tube. The trace is an analog voltage representation that is proportional to amplitude and frequency.

overdrive. A gear ratio where the drive gear rotates slower than the driven gear.

The driven gear is overdriven at increased speed with decreased torque. A ratio of 0.75 : 1 is an overdrive ratio.

overhead-cam (ohc) engine. An engine with the camshaft mounted in the cylinder head, rather than in the block.

overhead valve (ohv) engine. An engine with the intake and exhaust valves in the cylinder head, above the combustion chambers.

oxidation reaction. Any chemical reaction in which a compound is formed by adding oxygen to another element or compound.

P-material. Silicon or germanium doped with boron or gallium so that it has a shortage of free electrons and a positive charge.

paladium. An oxidation catalyst used in catalytic converters.

parallel circuit. A circuit with two or more branches or paths through which current can flow.

parallel data. Data transmission in which each component of the data is sent at the same time on its own line.

parameter. A measured value of control system input or output operation. Parameters include voltage signals, as well as temperature, pressure, speed, and other data.

particulate. Microscopic solid particles emitted to the air. A form of pollution.

pellet-type converter. A catalytic converter that has the catalyst deposited on small ceramic pellets, through which the exhaust flows.

percolation. Boiling and vaporization of a liquid that causes the liquid to move from one area to another.

peripheral serial bus. A data link between two computers in which one link sends and the other link receives data transmissions.

permeability. The ease with which a material can be penetrated by magnetic flux lines. Iron is more permeable than gases such as air.

permanent magnet (PM) generator sensor. A sensor using a pickup coil wound around a permanent magnet polepiece to generate an ac voltage signal as a toothed wheel (reluctor or armature) rotates past it. Also called a reluctance sensor or a pickup coil sensor.

phase angle. The angular location of alternator conductors in relation to the rotating field. A phase is a fraction of a complete ac cycle, usually expressed as an angle between sine waves. A 3-phase alternator has three windings 120 degrees apart and produces overlapping sine wave voltage.

photochemical reaction. A chemical reaction in the presence of sunlight. HC and NO_X react in the presence of sunlight to form smog.

photoelectricity. Voltage developed by light striking the surface of some metals.

piezoelectric crystal. A substance that can generate a voltage when subjected to mechanical pressure.

piezoelectricity. Voltage developed by pressure applied to certain crystal materials.

piezoresistive sensor. A substance whose electrical resistance changes when subjected to mechanical pressure.

pinpoint test. A specific test made to a single component or a single function to isolate a malfunction.

pintle valve. Any valve formed by a pin that passes through a round opening. A diesel fuel injection valve of this design.

platinum. An oxidation catalyst used in catalytic converters. Also can be a reduction catalyst.

pneumatic. Systems using air or other gases for mechanical work. The study of the mechanical properties of gases, especially compressed air.

PN junction. The point where two opposite kinds of semiconductor material (P and N) join together.

point gap. The distance between the breaker points when they are completely open. A wide gap creates a short dwell angle; a narrow gap creates a longer dwell angle.

polarity. The possession or existence of two opposing forces or characteristics, such as the north and south poles of a magnet or the positive and negative terminals of an electrical circuit.

pole. The opposing ends of a magnet, identified as north (N) and south (S). The areas of a magnetic field where the lines of force are concentrated.

poppet valve. A valve that opens and closes by back-and-forth linear motion.

ported vacuum. Vacuum taken from a port above the throttle. The amount and timing of vacuum application depends on throttle position. Ported vacuum is shut off at idle.

positive displacement pump. A pump that displaces, or delivers, the same amount (volume) of liquid with each revolution regardless of speed.

positive engagement. A general type of starter motor that uses the positive mechanical action of a solenoid or a movable pole shoe to engage the drive pinion with the engine flywheel.

positive-ground electrical system. A dc electrical system in which the battery positive terminal is connected to ground.

positive temperature coefficient (PTC) resistor (thermistor). A thermistor whose resistance increases as temperature increases—low resistance at low temperature and high resistance at high temperature.

potential energy. Stored, unreleased energy.

potentiometer. A variable resistor that acts as a voltage divider to produce a continuously variable output signal proportional to mechanical position.

pour point. The low temperature (lower than the cloud point) at which diesel fuel or motor oil will barely pour or flow.

power. The rate, or speed, at which work is done. Power equals the amount of work times the amount of time needed to do it.

powers of 10. The number of times that 10 is multiplied or divided by itself to get a larger or smaller number. Powers of 10 are indicated by exponents.

power transistor. The high-current transistor, or transistors, that carry primary current and switch it on and off in an ignition control module. Any transistor that carries high current.

preignition. An unwanted, uncontrolled ignition of the air-gasoline mixture from a combustion chamber hot spot before spark ignition.

pressure. Force exerted on a given unit of surface area. Measured in pounds per square inch (psi) or kilopascals (kPa). Pressure = force/area.

pressure differential. A difference in pressure between two points. The pressure differential between 50 psi and 10 psi is 40 psi.

pressure-regulating valve. A hydraulic valve that controls pressure developed in a system and the release of that pressure.

primary battery. A battery whose electrochemical reaction cannot be reversed. The battery dies when one of the electrodes is destroyed by the electrochemical action; it is not rechargeable.

primary circuit. The low-voltage ignition circuit, originating at the battery, that induces high voltage in the coil.

primary (ballast) resistor. The resistor in an ignition primary circuit that stabilizes voltage and current to the coil primary winding.

primary winding. A coil winding that uses voltage and current flow to create a magnetic field and induce a voltage in a second winding.

primary wiring. Low-voltage automobile wiring as distinguished from high-voltage ignition secondary wiring.

program. The job instructions for a computer. The program is usually stored in programmable read-only memory (PROM) devices for a vehicle computer.

programmable read-only memory (PROM). A computer memory IC chip that can be programmed once to store the computer program.

proton. The part of an atom with a positive (+) charge. Protons are located in the center of an atom, along with neutrons.

pulley. A grooved, or toothed, wheel that engages a belt in a belt-drive system. Also called a *sheave*.

pulse air injection. An air injection system that uses negative pressure (vacuum) pulses in the exhaust to draw air into the exhaust system.

pulse width. The operating time, usually measured in milliseconds, of an intermittent device, such as a solenoid.

pulse width modulation (pwm). A continuous on-off cycling of a solenoid a specified number of times a second. The pulse width (on-time) can be changed, or modulated, to produce a variable duty cycle. (See *duty cycle*.)

pumping losses. Power used and wasted by a gasoline engine to take in air past a restricted throttle.

radiofrequency interference (RFI). High-voltage discharges from ignition systems (and other sources) that interfere with radio and television transmission.

random-access memory (RAM). Computer read-write memory on which information can be written and from which it can be read.

read-only memory (ROM). The permanent program memory of a computer. Instructions can be read from ROM, but nothing can be written into it and it cannot be changed.

reciprocating engine. An engine in which pistons move up and down in cylinders. Combustion of an air-fuel mixture at one end of the cylinders develops power and causes a reciprocating motion. Also called a piston engine.

rectifier bridge. A single assembly that contains three or all six rectifying diodes for an alternator.

rectify. To change alternating current to direct current.

reduction reaction. Any chemical reaction in which oxygen is removed from a compound.

reference voltage. The constant voltage applied by a computer to a resistive sensor. The sensor sends a varying return voltage related to the quantity being measured.

relative motion. A difference in movement between two objects or forces.

relay. An electromagnetic switch that uses a small amount of current in one circuit to open or close a circuit with greater current flow. Often used for remote control of a circuit.

relief valve. A valve that releases system pressure (gas or liquid) when it exceeds a predetermined value.

reluctance. The opposition, or resistance, to magnetic flux lines or a magnetic field.

reluctance sensor. A magnetic pulse generator that sends a voltage signal in response to varying reluctance of a magnetic field.

remote ground. A ground connection through a wire between an electrical part and the vehicle engine or chassis.

required voltage. The specific high voltage required to ionize and fire a spark plug under any given combination of operating conditions.

research octane. Gasoline antiknock quality tested under conditions of low load, low temperature, full throttle, and low to medium speed.

reserve-capacity rating. A battery current-capacity rating that indicates the number of minutes a battery can supply 25 amperes of current with no cell voltage dropping below 1.75 volts.

resistor spark plug. A spark plug with a resistor of about 10,000 ohms built into the center electrode.

resonator. A kind of muffler that reduces noise by reflecting sound waves back toward the noise source.

retraction valve. A diesel fuel injection pump delivery valve, so called because it causes fuel to retract in an injection when it closes and reduces injection pressure.

reverse bias. Voltage applied across a diode that prevents current flow across the junction.

reverse polarity. The condition in which the ignition primary circuit has a reversed connection (usually at the coil). This causes the secondary voltage polarity at the spark plug to be reversed and force current from a positive (+) center electrode to a negative (−) ground electrode.

reversing relay. A relay that reverses the action of a switch. Closing a switch applies current to the control contacts of the relay to open the normally closed power contacts of the relay.

revolutions per minute (rpm). A measurement of engine rotating speed. The number of complete revolutions made by the crankshaft in one minute.

rheostat. A variable resistor used to regulate the strength of an electrical current.

rhodium. A reduction catalyst used in catalytic converters.

right-hand rule. The current flow and flux direction rule based on conventional current flow theory. Grasping a conductor with the right-hand thumb pointing in the direction of current flow (toward −) causes the fingers to wrap around the conductor in the direction of the flux lines.

rocker arm ratio. The ratio of the distance from the rocker arm pivot point to the pushrod center on one end and the valve stem center on the other. Rocker arm ratio multiplies cam lift to determine actual valve lift.

roller chain. A drive chain in which the cross piece of each link is a small roller that rotates in a pin to reduce friction as it engages the teeth of a sprocket.

Roots blower. A positive-displacement blower or supercharger that uses two lobe-shaped rotors to draw in and compress air.

running change. A change made by the

manufacturer to a particular car model during the course of a model year. Running changes often cause revised specifications and service procedures.

running mates. Cylinders in which ignition occurs 360° from each other.

scan tool (scanner). A diagnostic computer that is used to communicate with and test a vehicle onboard computer. Because it "scans the serial datastream" on a GM car, it came to be called a "scan tool, or scanner."

secondary battery. A battery whose electrochemical reaction can be reversed. The battery electrodes and electrolyte can be restored by reversing the current flow.

secondary circuit. The high-voltage ignition circuit, originating at the coil secondary winding, that delivers high voltage to the spark plugs.

secondary winding. A coil winding that receives an induced voltage from the magnetic field created by current flow in a primary winding.

semiconductor. An element with four valence electrons. Semiconductors are neither good conductors nor good insulators but will conduct current under controlled conditions.

sensor. Any device that sends an input signal to a computer.

sequential sampling. A complex switching method involving a multiplexer and demultiplexer, which allows a body computer module (BCM) to deal with sensor input and actuator output in sequence and in a logical manner.

serial data line. A data link between two computers that transmits the digital words one after another, or serially, from one computer to the other. The computers are clocked together and synchronized to the data being transmitted.

serial datastream. The stream of computer data transmitted serially from a General Motors car to a scan tool.

series circuit. A circuit with only one path for all current flow.

series-parallel circuit. A circuit with some loads in series with the power source and other loads, and some loads in parallel with each other.

serpentine drive belt. A V-ribbed drive belt installation that uses a single belt to drive several engine accessories.

short circuit (short). An unwanted connection between two conductors that bypasses part of a circuit.

shunt. A low-resistance electrical connection or branch circuit in parallel with another connection or branch circuit.

silent chain. A drive chain made from a series of flat metal links pinned together so that the teeth of each link align to form teeth similar to those on an internal ring gear.

silicon-controlled rectifier (SCR). A common type of thyristor used in automotive circuits.

SI metric. The modern international metric system, used by the automobile industry and other industries.

sine wave voltage. The changing voltage polarity between positive and negative of alternating current. The voltage pattern is based on the trigonometry sine function.

single-grade (straight-grade) oil. An oil whose viscosity is tested and rated at either, but not both, 0° or 210° F (−18° or 99° C).

single-overhead-cam (sohc) engine. An overhead cam engine with a single camshaft for intake and exhaust valves.

single-phase voltage. The complete sine wave voltage induced in one conductor during one revolution of an alternator rotor.

siphon. The use of a tube or hose to allow atmospheric pressure to force a liquid upward, over a barrier, and then downward to a point lower than its original level.

sludge. Heavy, black engine deposits formed when oil mixes with blowby and water vapor.

snapshot. A recording of data transmitted from a vehicle to a scan tool. Also called a *movie*.

soft code. A trouble code from a vehicle computer system that indicates a problem that is not present at the time of testing. A soft code indicates an intermittent problem that occurred sometime before testing, but which is not present now. Soft codes are stored in long-term computer memory. Usually they are automatically erased after 50 or 100 ignition on-off cycles.

soft conversion. Restating the dimensions of a part or some measured value in units of a measurement system other than the system originally used. The part is not physically changed.

software. The various programs and data in ROM and RAM that provide a computer with memory and operating instructions

solenoid. A device similar in operation to a relay, but movement of the armature or movable iron core changes electrical energy into mechanical motion.

solid-state electronics. Electronic systems based on solid semiconductor devices such as diodes and transistors.

spark plug reach. The dimension from the spark plug seat to the tip of the shell (the end of the threaded portion). Spark plug reach must match the depth of the spark plug hole in the cylinder head.

spark voltage. The voltage, lower than firing voltage, needed to sustain a spark across a plug gap after initial ionization and firing.

specific energy. Energy that has been released from its potential form and is doing specific work as heat, light, or mechanical energy.

specific gravity. The weight of a volume of liquid divided by the weight of an equal volume of water at equal temperature and pressure. The ratio of the weight of any liquid to the weight of water, which has a specific gravity of 1.000.

speed density. More correctly, speed *and* density. A fuel injection control method that uses engine speed (rpm) and air density measured by a MAP sensor to regulate fuel metering in relation to engine air intake.

sprocket. A toothed wheel that engages a chain in a chain-drive system.

spur gear. An external or internal gear on which the teeth are parallel with the gear axis, or centerline.

starting bypass. The parallel circuit path that bypasses the ballast resistor when an engine is started.

state of charge. The percentage of charge present in a battery at any time. It is indicated by electrolyte specific gravity and battery voltage.

static electricity. Opposite voltage charges present on two opposing surfaces. The charges do not move; they are static. Usually caused by the friction of rubbing two materials together. Also the kind of charge present in a capacitive field.

static inertia. The tendency of an object at rest to remain at rest.

stem unit. Any metric unit to which a prefix can be added to indicate larger or smaller measurements to some power of 10. Kilopascal is formed by adding the prefix, "kilo," to the stem unit, "pascal."

stepper motor. A dc motor that moves in incremental steps between a no-voltage (deenergized) position and a full-voltage (energized) position.

stoichiometric ratio. An air-fuel ratio of approximately 14.7 : 1, which provides the most complete combustion and recombination of air and fuel.

stroboscopic. The effect of a rapidly flashing light that makes moving objects appear to stand still.

stroke. The distance that a piston travels from bottom to top in a cylinder. Also, one complete bottom-to-top or top-to-bottom movement of a piston.

submultiple. A measurement unit that is smaller than the stem unit through dividing by a power of 10.

substrate. A foundation material upon which other materials are deposited. The ceramic core (pellets or monolith) of a catalytic converter, or the basic silicon wafer of a semiconductor device.

sulfur oxides. Compounds formed by the combination of sulfur and oxygen. A form of air pollution.

supercharging. The process of using a mechanical pump to increase intake air pressure for an internal combustion engine.

supplementary unit. The two SI metric units for angular measurement that are based on abstract geometrical concepts rather than on physical measurement of a quantity.

switching transistor. The power transistor in an ignition control module. Any transistor used as a switch or relay to turn current on and off. (See *power transistor*.)

symmetrical. The same pattern on each side of a centerline. The light beam of a symmetrical high-beam headlamp is positioned the same distance on each side of the center.

synchronous data. A constant data flow with regular clock pulses that incorporates a sync pulse at the beginning of each new piece of data to tell the computer that this is a new word.

synthetic motor oil. Oil compounded by breaking down and recombining petroleum molecules or by formulating lubricant molecules from nonpetroleum materials.

television-radio-suppression (TVRS) cables. High-resistance spark plug cables that have nonmetallic conductors to reduce radiofrequency interference.

temperature inversion. A layer of warm air at high altitude that traps cooler air near the ground.

tensioner. An idler pulley or sprocket that maintains necessary tension on a drive belt or chain.

tetraethyl lead. A gasoline antiknock additive that raises the gasoline octane rating.

thermal cracking. The process of cracking hydrocarbon molecules by applying heat.

thermal energy. The specific energy of heat.

thermistor. A variable resistor whose resistance changes when temperature changes.

thermocouple. A temperature-measuring device made of two dissimilar metals joined so that voltage is generated between two points that is directly proportional to the temperature difference between the two points.

thermoelectricity. Voltage developed by the flow of heat through two dissimilar metals.

three-phase alternating current. Three overlapping, evenly spaced, single-phase ac voltage sine waves that make up the complete ac output of an alternator.

three-phase voltage. Three overlapping ac voltage sine waves generated by an alternator with a 3-winding stator. The sine waves are at equally spaced, varying phases.

threshold voltage. The minimum conduction voltage for junction diode operation.

throttling. The process of regulating the amount of air-fuel mixture entering an engine to control speed and power.

timing (spark) advance. Ignition occurring sooner before top dead center than initial timing. Any advance in timing is measured in degrees of crankshaft rotation.

titania. Titanium dioxide. A compound used as a temperature-sensing device in one kind of exhaust oxygen sensor.

top dead center (tdc). The exact top of a piston stroke.

torque. Twisting or turning force. The action of a force that produces rotating mechanical work around an axis. Torque equals the distance to the axis times the force applied.

transducer. A device that changes one form of energy to another. For example, a transducer can change motion to voltage, or air pressure to motion. An automotive sensor, actuator, or display device is a transducer.

transfer port. The slot-shaped metering port, or ports, just above the throttle in a carburetor barrel that transfers fuel flow for low-speed operation.

transistor. A 3-element semiconductor device of NPN or PNP materials that *trans*fers electrical signals across a res*is*tance.

turbocharger. An exhaust-turbine-driven centrifugal (variable-displacement) supercharger.

turbo lag. The brief hesitation from the time the throttle opens until a turbocharger starts delivering boost.

turns ratio. The ratio of the number of turns in a coil secondary winding to the number of turns in the primary winding. Turns ratio indicates voltage multiplication in a coil.

two-stage (three-way) catalytic converter system. A system that has two catalysts, oxidation and reduction, to work against three pollutants, HC, CO, and NO_X.

two-stroke cycle. The engine operating cycle in which a piston makes two strokes (intake-compression and power-exhaust) and one 360-degree revolution to develop power from the combustion process.

vacuum. Air or gas pressure lower than atmospheric pressure. Complete vacuum is the complete absence of gaseous matter.

vacuum advance. A distributor advance mechanism that uses a vacuum diaphragm and linkage to advance timing in relation to engine load.

vacuum lock. A low-pressure condition in a fuel tank or other reservoir that keeps fluid from flowing, or being pumped, from the tank.

valence shell. The outer electron shell in any atom that contains the electrons that join with electrons of other atoms to create molecules. The valence shell contains 1 to 8 electrons.

valve duration. The number of crankshaft degrees during which a valve is open.

valve lift. The maximum distance that a valve opens, measured in a decimal-inch fraction or millimeters.

valve overlap. The period (in crankshaft degrees) when the intake and exhaust valves for one cylinder are open simultaneously. Overlap occurs between the exhaust and the intake strokes.

valve timing. The relationship between crankshaft position and the opening and closing positions of the intake and exhaust valves.

valve train. All of the parts from camshaft to valves that open and close the valves in an engine.

vaporize. To change solid or liquid matter into a gaseous form. Vapor is gaseous matter.

vapor lock. A condition in which gasoline vaporizes in the fuel pump, the lines, or the carburetor fuel bowl due to high temperature. Fuel flow stops because of vapor lock.

variable displacement pump. A pump such as a centrifugal pump, whose output volume changes with speed.

variable-dwell electronic ignition. An ignition that uses full battery voltage on the coil with no ballast resistor. Dwell time and coil saturation are controlled by circuits in the ignition module. Dwell angle varies with engine speed.

varnish. Engine deposits formed by high-temperature oxidation of motor oil.

venturi. A restriction in an airflow passage, such as a carburetor barrel, that increases airflow velocity and reduces pressure.

venturi effect. The relationship of increased airflow velocity and reduced pressure created by a venturi.

vernier scale. A fine, auxiliary scale that indicates fractional parts of a larger scale.

viscosity. The measurement of a liquid's tendency to resist flowing.

viscous drive. A drive mechanism, such as a fan clutch, that connects drive and driven members through the viscosity of a fluid.

volatility. A measure of a liquid's ability to change into a vapor.

volt. The unit used to measure the amount of electrical force or energy.

voltage. The electromotive force that causes current to flow. The potential force that exists between two points when one is positively charged and the other is negatively charged.

voltage cell. Two dissimilar materials suspended in a conductive and reactive substance.

voltage reserve. The difference between the required voltage needed to fire a spark plug and the maximum available voltage that a coil can produce.

volume. The physical size of an object, or form of matter, for a given mass. An object's length, width, and height.

volumetric efficiency. Expressed as a percentage, the comparison of the actual volume of air taken into an engine to the theoretical maximum amount that the engine could draw in (its displacement).

wastegate. A pressure-operated valve that diverts part of the exhaust around a turbocharger turbine. It opens to limit turbine speed when boost reaches a desired maximum pressure.

waste spark. An ignition system in which one coil fires two spark plugs at one time. The spark plug in the cylinder on compression ignites the mixture; the spark fired in the cylinder on its exhaust stroke is wasted.

watt (W). The unit of electric power. The amount of electrical work done in a given time. Watts = amperes × voltage.

weight. The measure of the earth's gravitational force, or pull, on an object.

wet-sump oil system. An engine lubrication system in which the oil supply is stored in the engine oil pan.

Wheatstone bridge. An arrangement of resistors that allows resistance to be measured by measuring current and voltage.

wiring harness. A collection of wires encased in a plastic covering or conduit and routed to specific areas of the vehicle. Most harnesses terminate in plug-in connectors.

work. The transfer of energy from one object or system to another as potential energy becomes specific energy. Mechanical work is the application of energy through force to produce motion.

work harden. The effects of vibration and bending that cause a piece of metal to become hard and brittle.

Y-type stator. A 3-winding alternator stator in which one end of each winding is connected to a neutral junction. The other end of each winding is connected between a positive and a negative diode. Schematically, the windings are arranged like a letter "Y." A Y-type stator produces high voltage at low speed.

zener diode. A diode that allows reverse current flow above a prescribed voltage without being damaged.

zirconia. Zirconium dioxide. A compound used as an oxygen-sensing device in one kind of exhaust oxygen sensor. When exposed to oxygen, zirconia acts as a galvanic battery and generates a voltage.

INDEX